Handbook of Optoelectronic Device Modeling and Simulation

Series in Optics and Optoelectronics

Series Editors: **Robert G W Brown**
University of California, Irvine, USA

E Roy Pike
Kings College, London, UK

Handbook of Optoelectronic Device Modeling and Simulation

Lasers, Modulators, Photodetectors, Solar Cells, and Numerical Methods

VOLUME TWO

Edited by
Joachim Piprek

CRC Press
Taylor & Francis Group
Boca Raton London New York

CRC Press is an imprint of the
Taylor & Francis Group, an **informa** business

CRC Press
Taylor & Francis Group
6000 Broken Sound Parkway NW, Suite 300
Boca Raton, FL 33487-2742

First issued in paperback 2020

© 2018 by Taylor & Francis Group, LLC
CRC Press is an imprint of Taylor & Francis Group, an Informa business

No claim to original U.S. Government works

ISBN-13: 978-1-4987-4956-5 (hbk)
ISBN-13: 978-0-367-78180-4 (pbk)

Library of Congress Cataloging-in-Publication Data

Names: Piprek, Joachim, editor.
Title: Handbook of optoelectronic device modeling and simulation / edited by Joachim Piprek.
Other titles: Series in optics and optoelectronics (CRC Press) ; 27.
Description: Boca Raton, FL : CRC Press, Taylor & Francis Group, [2017] |
Series: Series in optics and optoelectronics ; 27 | Includes bibliographical references and index.
Contents: volume 1. Fundamentals, materials, nanostructures, LEDS, and amplifiers – volume 2. Lasers, modulators, photodetectors, solar cells, and numerical methods.
Identifiers: LCCN 2016058063 | ISBN 9781498749466 (v. 1; hardback ; alk. paper) |
ISBN 1498749461 (v. 1; hardback ; alk. paper) | ISBN 9781498749565 (v. 2 ; hardback ; alk. paper) |
ISBN 1498749569 (v. 2 ; hardback ; alk. paper)
Subjects: LCSH: Optoelectronic devices–Mathematical models–Handbooks, manuals, etc. |
Optoelectronic devices–Simulation methods–Handbooks, manuals, etc. |
Semiconductors–Handbooks, manuals, etc. | Nanostructures–Handbooks, manuals, etc.
Classification: LCC TK8304 .H343 2017 | DDC 621.381/045011–dc23
LC record available at https://lccn.loc.gov/2016058063

Visit the Taylor & Francis Web site at
http://www.taylorandfrancis.com

and the CRC Press Web site at
http://www.crcpress.com

Contents

PART VI   Laser Diodes

PART VII   Photodetectors and Modulators

PART VIII   Solar Cells

PART IX   Novel Applications

PART X   Mathematical Methods

Series Preface

This international series covers all aspects of theoretical and applied optics and optoelectronics. Active since 1986, eminent authors have long been choosing to publish with this series, and it is now established as a premier forum for high-impact monographs and textbooks. The editors are proud of the breadth and depth showcased by the published works, with levels ranging from advanced undergraduate and graduate student texts to professional references. Topics addressed are both cutting edge and fundamental, basic science and applications-oriented, on subject matter that includes lasers, photonic devices, nonlinear optics, interferometry, waves, crystals, optical materials, biomedical optics, optical tweezers, optical metrology, solid-state lighting, nanophotonics, and silicon photonics. Readers of the series are students, scientists, and engineers working in optics, optoelectronics, and related fields in the industry.

Proposals for new volumes in the series may be directed to Lu Han, senior publishing editor at CRC Press/Taylor & Francis Group (lu.han@taylorandfrancis.com).

Preface

Optoelectronic devices have become ubiquitous in our daily lives. For example, light-emitting diodes (LEDs) are used in almost all household appliances, in traffic and streetlights, and in full-color displays. Laser diodes, optical modulators, and photodetectors are key components of the Internet. Solar cells are core elements of energy supply systems. Optoelectronic devices are typically based on nanoscale semiconductor structures that utilize the interaction of electrons and photons. The underlying and highly complex physical processes require mathematical models and numerical simulation for device design, analysis, and performance optimization. This handbook gives an introduction to modern optoelectronic devices, models, and simulation methods.

Driven by the expanding diversity of available and envisioned practical applications, mathematical models and numerical simulation software for optoelectronic devices have experienced a rapid development in recent years. In the past, advanced modeling and simulation was the domain of a few specialists using proprietary software in computational research groups. The increasing user-friendliness of commercial software now also opens the door for nontheoreticians and experimentalists to perform sophisticated modeling and simulation tasks. However, the ever-growing variety and complexity of devices, materials, physical mechanisms, theoretical models, and numerical techniques make it often difficult to identify the best approach to a given project or problem. This book presents an up-to-date review of optoelectronic device models and numerical techniques. The handbook format is ideal for beginners but also gives experienced researchers an opportunity to renew and broaden their knowledge in this expanding field.

Semiconductors are the key material of optoelectronic devices, as they enable propagation and interaction of electrons and photons. The handbook starts with an overview of fundamental semiconductor device models, which apply to almost all device types, followed by sections on novel materials and nanostructures. The main part of the handbook is ordered by device type (LED, amplifier, laser diode, photodetector, and solar cell). For each device type, an introductory chapter is followed by chapters on specialized device designs and applications, describing characteristic effects and models. Finally, novel device concepts and applications are reviewed. At the end of the handbook, an overview of numerical techniques is provided, both for electronic and photonic simulations.

I would like to thank the publisher for initiating this important handbook project and for giving me the opportunity to serve as editor. Many years of organizing the annual international conference on *Numerical Simulation of Optoelectronic Devices (NUSOD)* enabled me to attract a large number of experts from all over the world to write handbook chapters on their research area. I sincerely thank all authors for their valuable contributions.

Joachim Piprek
Newark, Delaware, USA

MATLAB® is a registered trademark of The MathWorks, Inc. For product information, please contact:

The MathWorks, Inc.
3 Apple Hill Drive
Natick, MA 01760-2098 USA
Tel: 508-647-7000
Fax: 508-647-7001
Email: info@mathworks.com
Web: www.mathworks.com

Editor

Joachim Piprek received his diploma and doctoral degrees in physics from the Humboldt University in Berlin, Germany. For more than two decades, he worked in industry and academia on modeling, simulation, and analysis of various semiconductor devices used in optoelectronics. Currently, he serves as president of the NUSOD Institute, Newark, Delaware (see http://www.nusod.org). During his previous career in higher education, Dr. Piprek taught various graduate courses at universities in Germany, Sweden, and the United States. Since 2001, he has been organizing the annual international conference on *Numerical Simulation of Optoelectronic Devices (NUSOD)*. Thus far, Dr. Piprek has published three books, six book chapters, and about 250 research papers, which have received more than 6000 citations. He was an invited guest editor for several journal issues on optoelectronic device simulation and currently serves as an executive/associate editor of two research journals in this field.

Contributors

Urs Aeberhard
IEK-5 Photovoltaik
Forschungszentrum Jülich
Jülich, Germany

Tim Albes
Department of Electrical Engineering
Technical University of Munich
Munich, Germany

Matthias Auf der Maur
Department of Electronic Engineering
University of Rome Tor Vergata
Rome, Italy

Eugene Avrutin
University of York
York, United Kingdom

Prasanta Basu
Institute of Radio Physics and Electronics
University of Calcutta
Kolkata, India

Steve Bull
Department of Electrical and Electronic
    Engineering
University of Nottingham
Nottingham, United Kingdom

Kwong-Kit Choi
Sensors and Electron Devices Directorate
U.S. Army Research Laboratory
Adelphi, Maryland

Weng W. Chow
Sandia National Laboratories,
Albuquerque, USA

Tomasz Czyszanowski
Institute of Physics
Lodz University of Technology
Lodz, Poland

Robin Daugherty
Arizona State University
Tempe, Arizona

Maciej Dems
Institute of Physics
Lodz University of Technology
Lodz, Poland

Duy Hai Doan
Research Group Partial Differential Equations
Weierstrass Institute
Berlin, Germany

Ignacio Esquivias
Center of Advanced Materials and
    Devices for ICT Applications
Universidad Politécnica de Madrid
Madrid, Spain

Patricio Farrell
Research Group Numerical Mathematics and
    Scientific Computing
Weierstrass Institute
Berlin, Germany

Leszek Frasunkiewicz
Institute of Physics
Lodz University of Technology
Lodz, Poland

Jürgen Fuhrmann
Research Group Numerical Mathematics and
    Scientific Computing
Weierstrass Institute
Berlin, Germany

Alessio Gagliardi
Department of Electrical Engineering
Technical University of Munich
Munich, Germany

Dominic F.G. Gallagher
Photon Design Ltd.
Oxford, United Kingdom

Christopher Gies
Institute for Theoretical Physics
University of Bremen
Bremen, Germany

Stephen M. Goodnick
Arizona State University
Tempe, Arizona

Niels Gregersen
Technical University of Denmark
Lyngby, Denmark

Raghuraj Hathwar
Arizona State University
Tempe, Arizona

Mohamad Anas Helal
Department of Electrical and
    Electronic Engineering
University of Nottingham
Nottingham, United Kingdom

Ortwin Hess
Imperial College
London, United Kingdom

Karin Hinzer
School of Electrical Engineering and
    Computer Science
University of Ottawa
Ottawa, Canada

Weida Hu
State Key Laboratory of Infrared Physics
Shanghai Institute of Technical Physics
Shanghai, China

Frank Jahnke
Institute for Theoretical Physics,
University of Bremen
Bremen, Germany

Julien Javaloyes
Universitat de les Illes Balears,
Palma de Mallorca, Spain

Olafur Jonasson
Department of Electrical and Computer
    Engineering
University of Wisconsin
Madison, Wisconsin

Markus Kantner
Research Group Laser Dynamics
Weierstrass Institute
Berlin, Germany

Simeon N. Kaunga-Nyirenda
Department of Electrical and Electronic
    Engineering
University of Nottingham
Nottingham, United Kingdom

Irena Knezevic
Department of Electrical and Computer
    Engineering
University of Wisconsin
Madison, Wisconsin

Mirella Koleva
Department of Physics
University of Oxford
Oxford, United Kingdom

Thomas Koprucki
Research Group Partial Differential Equations
Weierstrass Institute
Berlin, Germany

Dmitry Labukhin
Photon Design Ltd.
Oxford, United Kingdom

Eric Larkins
Department of Electrical and
  Electronic Engineering
University of Nottingham
Nottingham, United Kingdom

Akash Laturia
Arizona State University
Tempe, Arizona

Kuan-Chen Lee
National Chiao Tung University
Hsinchu, Taiwan

Xun Li
Department of Electrical and Computer
  Engineering
McMaster University
Hamilton, Canada

Michael Lorke
Institute for Theoretical Physics
University of Bremen
Bremen, Germany

Dara P.S. McCutcheon
Quantum Engineering Technologies Labs
University of Bristol
Bristol, United Kingdom

Song Mei
Department of Electrical and Computer
  Engineering
University of Wisconsin
Madison, Wisconsin

Jesper Mørk
Technical University of Denmark
Lyngby, Denmark

Matthias Müller
Institute of Applied Physics
Technical University Bergakademie Freiberg
Freiberg, Germany

A. Freddie Page
Imperial College London
London, United Kingdom

Antonio Pérez-Serrano
Center of Advanced Materials and Devices for
  ICT Applications
Universidad Politécnica de Madrid
Madrid, Spain

Joachim Piprek
NUSOD Institute LLC
Newark, Delaware

Suleman S. Qazi
Arizona State University
Tempe, Arizona

Mindaugas Radziunas
Research Group *Laser Dynamics*
Weierstrass Institute
Berlin, Germany

Katerina Raleva
University Sts. Cyril and Methodius
Skopje, Macedonia

Nella Rotundo
Research Group Partial Differential
  Equations
Weierstrass Institute
Berlin, Germany

Frank Schmidt
Zuse Institute
Berlin, Germany

Abdul R. Shaik
Arizona State University
Tempe, Arizona

Yanbing Shi
Department of Electrical and Computer
    Engineering
University of Wisconsin
Madison, Wisconsin

Gabriela Slavcheva
Centre for Photonics and Photonic
    Materials
Department of Physics
University of Bath
Bath, United Kingdom

José-Manuel G. Tijero
Center of Advanced Materials and Devices
    for ICT Applications
Universidad Politécnica de Madrid
Madrid, Spain

Dragica Vasileska
Arizona State University
Tempe, Arizona

Hans Wenzel
Ferdinand Braun Institute
Berlin, Germany

Matthew Wilkins
School of Electrical Engineering and Computer
    Science
University of Ottawa
Ottawa, Canada

Bernd Witzigmann
Department of Electrical Engineering and
    Computer Science
University of Kassel
Kassel, Germany

Shun-Tung Yen
National Chiao Tung University
Hsinchu, Taiwan

Anissa Zeghuzi
Ferdinand Braun Institute
Berlin, Germany

VI

Laser Diodes

26

Laser Diode Fundamentals

Joachim Piprek

26.1 Introduction

Semiconductor laser diodes are key components in optical fiber communication, data storage, sensing, material processing, and other applications. They are based on sophisticated interactions of electrons and photons in semiconductor nanostructures (see Chapter 3). Advanced theoretical models and simulation tools are required for the development and analysis of future generations of laser diodes. The following chapters describe some of these complex models in detail. This introductory chapter is aimed at readers who are not yet familiar with basic models and parameters of semiconductor lasers. By exploring simple analytical models, this chapter tries to develop an intuitive understanding of internal laser physics that will help to digest the more complicated theory outlined in subsequent chapters. While these analytical models are quite popular, they often have limits beyond which numerical simulations are required, as shown below.

Section 26.2 examines basic formulas for optical gain and optical losses in semiconductor lasers, followed by the introduction of key performance parameters, threshold current, and slope efficiency. As practical applications often suffer from undesired self-heating of the laser, temperature effects are discussed in Section 26.4 and the resulting changes in laser efficiency in Section 26.5. The rate equation model for dynamic lasing processes is introduced in Section 26.6, including formulas for small signal analysis. Section 26.7 briefly reviews basic laser cavity designs.

26.2 Optical Gain and Optical Loss

Traveling through a semiconductor, a single photon with an energy close to the band gap is able to generate an identical second photon by stimulating the recombination of an electron–hole pair. This is the basic physical mechanism of lasing. The second photon exhibits the same wavelength and the same phase as of the first photon, doubling the amplitude of their monochromatic wave. Subsequent repetition of this process leads to strong light amplification. However, the competing process is the absorption of photons by the generation of new electron–hole pairs (see Figure 26.1). Stimulated emission prevails when more electrons are present at the higher energy level (conduction band) than at the lower energy level (valence band).

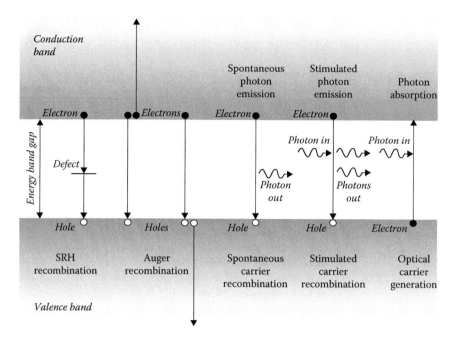

FIGURE 26.1   Electron–hole recombination and generation mechanisms in semiconductors. (SRH, Shockley–Read–Hall.)

This population inversion is one of the key requirements for lasing. Semiconductor lasers typically employ *pin* junctions with a thin active layer of lower band gap (Figure 26.2). At forward bias, electrons and holes are collected in the active layer to achieve inversion. Continuous current injection into the device leads to a continuous stimulated emission of photons, but only if enough photons are constantly present in the device to trigger this process. Thus, only part of all photons can be allowed to leave the laser diode as a lasing beam, the rest must be reflected to remain inside the diode and to generate new photons (Figure 26.3). This optical feedback and confinement of photons in an optical resonator is the second basic requirement of lasing.

The light amplification in the active layer is described by the optical gain $g(n, p, T, \lambda, S)$ as a function of the density of electrons n and holes p inside the active layer, the optical wavelength λ (or photon energy), the photon density S, and the temperature T. This gain function is the heart of laser physics, and a realistic calculation may require sophisticated models (see Chapter 3 for details). We here briefly discuss some popular analytical approximations. The linear gain approximation $g(N) = a(N - N_{tr})$ is often used, employing the transparency density N_{tr} and assuming a fixed differential gain $a = dg/dN$ as well as identical densities of electrons and holes ($N = n = p$). For $N = N_{tr}$, the absorption and gain are the same and the material is transparent. The differential gain dg/dN is a key parameter for laser light modulation (which is discussed Section 26.6). It can only be considered constant for small variations of the carrier density since dg/dN is known to decline with increasing carrier densities (Figure 26.4). This is described by the more general logarithmic function (Coldren and Corzine 1995)

$$g(N, S) = \frac{g_o}{1 + \varepsilon S} \ln\left(\frac{N + N_s}{N_{tr} + N_s}\right) \tag{26.1}$$

The gain compression factor ε describes the gain saturation at high photon densities, e.g., due to carrier depletion (g_o and N_s are fit parameters). All parameters in Equation 26.1 need to be extracted from more fundamental gain models, as measurements are difficult (Shtengel et al. 1998). However, the lack of

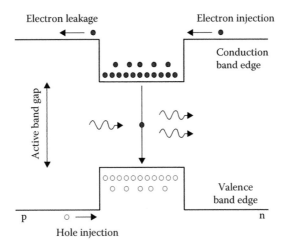

FIGURE 26.2   Electron and hole injection into the light-emitting active layer that is sandwiched between a p-doped and an n-doped material of higher energy bandgap. (Adapted from Piprek, J., In S. S. Sun and L.R. Dalton [eds.], *Introduction to Organic Electronic and Optoelectronic Materials and Devices*, Boca Raton, FL: CRC Press, 2008.)

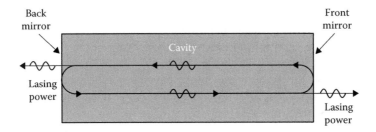

FIGURE 26.3   Optical wave propagation in an Fabry–Pérot (FP) laser cavity (resonator) formed between two reflecting facets. (Adapted from Piprek, J., In S. S. Sun and L. R. Dalton [eds.], *Introduction to Organic Electronic and Optoelectronic Materials and Devices*, Boca Raton, FL: CRC Press, 2008.)

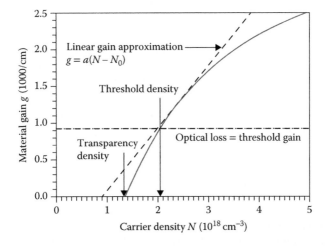

FIGURE 26.4   Optical gain provided by the active layer material as function of the carrier density.

experimental validation of the gain model creates a major uncertainty in any laser simulation. The analysis of laser measurements typically only delivers the relationship $g_m(j) = G_o \ln(j/j_{tr})$ between the modal gain $g_m = \Gamma_a g$ and the current density j, with G_o and the transparency current density j_{tr} as fit parameters (see Chapter 27). The optical confinement factor Γ_a gives the overlap of the active layer and the lasing mode, which is usually extracted from waveguide simulations (see Chapter 4).

Figure 26.4 indicates optical losses that are caused by photon emission from the laser as well as by the internal absorption and photon scattering. For the simple Fabry–Pérot (FP) laser structure shown in Figure 26.3, the optical loss at the two cavity mirrors is given by

$$\alpha_m = \frac{1}{2L} \ln \left(\frac{1}{R_f R_b} \right) \tag{26.2}$$

with the cavity length L and the reflectances R_f and R_b of front and back mirrors, respectively. The internal optical loss α_i can be extracted from laser measurements (see Chapter 27) but the microscopic mechanisms causing this loss are often hard to identify. They are typically approximated as

$$\alpha_i = \alpha_b + \sum_i \Gamma_i \left(k_{n,i} n + k_{p,i} p \right) \tag{26.3}$$

with the background loss α_b (e.g., due to photon scattering) and the sum over all free-carrier-related loss in any individual layer i of the laser structure (where k is a free-carrier absorption parameter). The confinement factor Γ_i is the ratio of the layer volume to the volume of the optical lasing mode. Photon absorption by free carriers depends on the energy band structure of the conduction and valence bands. It is typically stronger for holes due to intervalence band absorption (Piprek et al. 2000).

To reach the lasing threshold, the optical gain must compensate for the internal optical loss (α_i) and for photon emission from the device (α_m). Both loss parameters apply to the whole lasing mode so that the threshold gain g_{th} is defined by

$$\Gamma_a g_{th} \left(N_{th} \right) = \alpha_m + \alpha_i \tag{26.4}$$

with the active layer confinement factor Γ_a and the threshold carrier density N_{th} (see Figure 26.4).

26.3 Threshold Current and Slope Efficiency

The threshold current I_{th} provides the threshold carrier density N_{th} and compensates for various carrier loss mechanisms, some of which are illustrated in Figure 26.1. A spontaneous electron–hole recombination is needed to provide initial photons for stimulated recombination (lasing), but most spontaneously emitted photons are lost. In a common analytical approach, the spontaneous emission rate $R_{spon} = BN^2$ is proportional to the square of the carrier density. Nonradiative recombination mechanisms are either defect-related Shockley–Read–Hall (SRH) recombinations ($R_{SRH} = AN$) or Auger recombinations ($R_{Aug} = CN^3$). The former transfers the recombination energy to lattice vibrations (phonons) and the latter to other free carriers. This simple but very popular ABC recombination model leads to the threshold current

$$I_{th} = \frac{eV_a}{\eta_a} R_{ABC}(N_{th}) = \frac{eV_a}{\eta_a} \left(AN_{th} + BN_{th}^2 + CN_{th}^3 \right) = I_{SRH} + I_{spon} + I_{Aug} + I_{leak} \tag{26.5}$$

with the injection efficiency η_a giving the fraction of electrons that recombines within the active layer of volume V_a, thereby accounting for carriers that recombine outside the active layer, e.g., due to electron leakage (see Figure 26.2). Leakage can occur by various mechanisms (Piprek et al. 2000). It is hard to asses

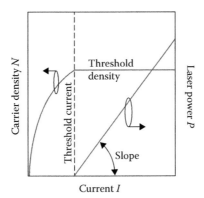

FIGURE 26.5 Illustration of carrier density and lasing power as function of the injected current without self-heating.

with analytical models and typically requires full solutions of the semiconductor transport equations (see Chapter 2). Simpler models often neglect leakage or use a fictitious parameter η_a.

With a stronger current injection $I > I_{th}$, the carrier density remains constant at N_{th} as additional electron–hole pairs are consumed by stimulated recombination (Figure 26.5). The stimulated recombination rate is $R_{stim} = v_g g S$ (where v_g is photon group velocity). Under ideal conditions without self-heating, the laser power rises proportional to $I_{stim} = I - I_{th}$ as

$$P = \eta_d \frac{h\nu}{e} \left(I - I_{th}\right) = \eta_i \frac{\alpha_m}{\alpha_m + \alpha_i} \frac{h\nu}{e} \left(I - I_{th}\right) \tag{26.6}$$

with the photon energy $h\nu$. Note that P gives the total emission from both facets. The differential quantum efficiency η_d is the fraction of carriers injected above threshold that contributes photons to the laser beams. It can be separated into internal differential efficiency η_i and optical efficiency η_{opt}. The latter is equal to $\alpha_m(\alpha_m + \alpha_i)^{-1}$ and gives the fraction of stimulated photons that leaves the laser. The internal differential efficiency η_i is often close to unity above threshold as there are no further recombination losses with constant carrier density N_{th} in the active layer. However, the leakage current may rise above the threshold, especially at higher temperatures (Piprek et al. 2000).

26.4 Temperature Effects

The current flow through the laser diode as well as nonradiative recombination processes generate heat inside the laser and elevate the internal temperature distribution $T(x, y, z)$. This temperature rise more or less affects all material parameters. One of the most fundamental changes occurs with the semiconductor bandgap $E_g(T)$, which is commonly modeled by the Varshni formula

$$E_g(T) = E_g(0) - \frac{AT^2}{B + T} \tag{26.7}$$

using the phenomenological parameters A and B (Piprek 2003). As the band gap shrinks, the lasing wavelength increases (redshifts). In addition, the Fermi distribution of carriers inside the energy bands broadens with higher temperature. Both these effects change the gain spectrum. Figure 26.6 illustrates this change for a fixed carrier density. The gain peak determines the emission wavelength of FP lasers, and it redshifts with higher temperature. But the gain peak also declines, so that more carriers are needed to maintain the lasing threshold (cf. Figure 26.4). Consequently, the threshold current rises with higher temperature, which is often described by the characteristic temperature $T_0 = (T_2 - T_1)/\ln(I_2/I_1)$ that can be extracted

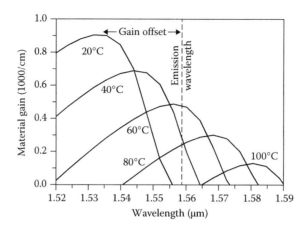

FIGURE 26.6 Optical gain spectrum shift with increasing temperature and constant carrier density.

from measurements of $I_{th}(T)$. The decline of the slope efficiency is calculated in a similar way. These phenomenological parameters are useful in characterizing the temperature sensitivity of a given laser diode (see Chapter 27). However, they don't reveal the physical mechanism behind temperature effects and are not suitable for predictive simulations. For instance, if the lasing wavelength is fixed by the optical cavity design, the impact of the gain shift very much depends on the initial difference between emission wavelength and gain peak wavelength (gain offset). An example is illustrated in Figure 26.6 (Piprek et al. 1998). Here, the rising temperature first increases the gain available at the emission wavelength. The threshold current initially declines and reaches a minimum near 60°C when the gain offset is zero. With further heating, the threshold current rises rapidly. Thus, T_0 depends on the temperature in this case and fails even as a descriptive parameter.

More advanced laser models don't employ fit parameters such as T_0 and describe the underlying physical mechanisms instead, starting with the shift of the gain spectrum (Piprek et al. 2000). An increasing carrier density in the active layer goes hand in hand with increasing carrier losses. For instance, the Auger recombination rate is not affected only by the increasing carrier density, but also by the temperature sensitivity of the Auger process, which depends on its activation energy E_a: $C(T) \propto \exp(-E_a/kT)$. Carrier leakage is also sensitive to temperature changes, and it may raise the threshold current and reduce the slope efficiency. The slope efficiency also depends on the free-carrier absorption, which increases with the carrier density. All these interdependencies require advanced numerical laser models for a more reliable analysis of the temperature sensitivity (see Chapter 27).

26.5 Efficiency Analysis

As an illustrative example for this section, Figure 26.7 shows the simulated power–current and bias–current characteristics of a GaN-based laser diode in continuous-wave (CW) operation (Piprek 2016). Self-heating apparently causes a decline of the slope efficiency, which limits the maximum power achievable. Another key performance parameter is the power conversion efficiency η_{PCE}. It is defined as the ratio of light output power P to electrical input power IV (where V is bias). It can be separated into electrical efficiency $\eta_{ele} = h\nu/eV$ and external quantum efficiency η_{EQE}. The latter is the ratio of emitted photon number to injected number of electron–hole pairs. Equation 26.6 leads to the following power conversion efficiency formula:

$$\eta_{PCE}(I) = \frac{h\nu}{eV}\eta_i\frac{\alpha_m}{\alpha_m + \alpha_i}\frac{I - I_{th}}{I} = \eta_{ele}\eta_{EQE} = \eta_{ele}\eta_s\eta_{th} \qquad (26.8)$$

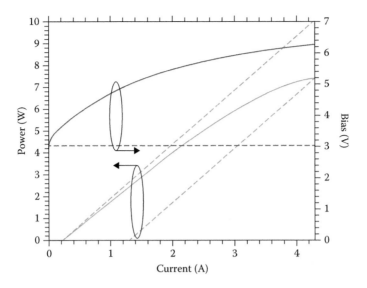

FIGURE 26.7 Lasing power and bias versus current. The dashed bias line indicates the photon energy. The dashed power lines indicate the change in threshold current for constant slope efficiency.

including the unitless slope efficiency η_s and the threshold current efficiency $\eta_{th} = (I - I_{th})/I$. Note that η_s is different from the differential slope efficiency dP/dI and from the averaged slope efficiency $P/(I - I_{th})$, which are both given in W/A and not used here. However, this popular analytical model is somewhat ambiguous when the laser experiences relevant self-heating, which causes a sublinear $P(I)$ characteristic as shown in Figure 26.7. Most parameters in Equation 26.8 change as the internal laser temperature rises with increasing current. The threshold current rises together with the threshold carrier density due to declining material gain. The slope efficiency declines due to increasing carrier leakage and/or rising internal absorption.

Figure 26.8 illustrates these efficiencies as simulated for an InGaN/GaN laser diode (Piprek 2016). At low current, η_{PCE} is mainly limited by the threshold current efficiency (which is zero for $I < I_{th}$). At high current, the strongest efficiency limitation is caused by the decline of the electrical efficiency due to the increasing excess bias above the minimum required bias $h\nu/e$ (dashed bias line shown in Figure 26.7). This excess bias is relatively high in GaN-based lasers, partially due to the low hole conductivity. However, the influence of slope efficiency and threshold efficiency depends on the assumption made in the analysis. The assumption of a constant, temperature-insensitive threshold current leads to the solid curves shown in Figure 26.8, so that temperature effects are mainly reflected by the slope efficiency. This is a convenient and common approach to extract efficiency plots directly from the measured *PI* and *VI* characteristics, without any simulation (Crump et al. 2013). However, with strong self-heating, the dependence on $I_{th}(T)$ needs to be considered, which is hard to extract directly from experimental results. Numerical simulations are indicated by the dashed lines in Figure 26.8, which reveal that the threshold efficiency is more temperature sensitive than the slope efficiency (Piprek 2016). For constant, temperature-insensitive slope efficiency, an approximate method of extracting $I_{th}(I)$ directly from measurements is illustrated by the dashed power lines shown in Figure 26.7. Starting with the measured slope dP/dI at threshold, the dashed line is shifted parallel to the current axis so that each power $P(I)$ is connected to a threshold current $I_{th}(I)$.

26.6 Rate Equation Analysis

Time-dependent effects are often analyzed in terms of rate equations considering all physical processes that change the densities of photons and carriers. We here discuss a set of two rate equations for the

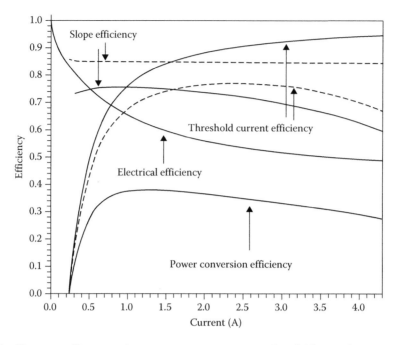

FIGURE 26.8 Changes in efficiency with increasing injection current. The solid lines indicate constant threshold current and the dashed lines indicate constant slope efficiency.

single-mode photon density S averaged over the modal volume V_m and the carrier density N averaged over the active volume V_a:

$$\frac{dN}{dt} = \frac{\eta_a I}{eV_a} - \left[AN + BN^2 + CN^3\right] - v_g g\,(N, S)\,S \tag{26.9}$$

$$\frac{dS}{dt} = \Gamma_a v_g g\,(N, S)\,S + \beta\Gamma_a BN^2 - v_g\left[\alpha_i + \alpha_m\right]S \tag{26.10}$$

The active layer carrier density $N(t)$ in Equation 26.9 is increased by current injection (first term, including the injection efficiency η_a) and is reduced by all four recombination processes (see Figure 26.1) that limit the carrier lifetime. The photon density $S(t)$ in Equation 26.10 is increased by stimulated emission (first term) and by the small fraction β of spontaneously emitted photons that enters the lasing mode (second term). $S(t)$ is reduced by photon emission and internal photon losses (third term in Equation 26.10), which limit the photon lifetime τ_p defined by $\tau_p^{-1} = v_g(\alpha_i + \alpha_m)$. The dynamic response of both densities can be understood from these rate equations. For instance, when $N(t)$ increases, $S(t)$ increases due to the rising gain $g(N)$. But that decreases the carrier density according to the last term in Equation 26.9. $N(t)$ is also reduced by ABC recombination. Consequently, the photon density $S(t)$ drops again, also due to photon losses. Thus, the dynamic behavior of both densities is strongly influenced by loss mechanisms.

However, this approach neglects the nonuniform distribution of carriers and photons (Carroll et al. 1998). A nonuniform distribution of photons can result in spatial hole burning into the carrier distribution in regions with high photon density, reduce the gain, and increase the refractive index (see Chapter 27). Multiple optical modes would require multiple rate equations for calculating each photon density (Petermann 1988). Lateral diffusion of carriers out of the active layer is partially considered by the injection efficiency η_a and is often minimized by lateral carrier confinement.

Under steady-state conditions ($dS/dt = 0$) with vanishing β, Equation 26.10 gives the relation $\Gamma_a g(N, S) = \alpha_i(N) + \alpha_m$ describing the balance of gain and losses required for lasing (see Figure 26.4).

The first equation $dN/dt = 0$ then yields the steady-state photon density $S_o = \eta_i \tau_p (I - I_{th})(eV_m)^{-1}$. The steady-state optical power emitted through both mirrors is given by $P_o = v_g \alpha_m h\nu V_m S_o$ in agreement with Equation 26.6.

With analog modulation, sinusoidal variations are added to the steady-state input current I_o. In the simple case of just one angular frequency $\omega = 2\pi f$ and constant amplitude ΔI, the injection current in Equation 26.9 becomes $I(t) = I_o + \Delta I \times \sin(\omega t)$. If the period of the modulation is much larger than any time constant, the output power still follows the steady-state solution (Equation 26.6). But in the general dynamic case, analytical solutions of the rate equations cannot be found and numerical methods need to be applied.

Current modulations ΔI well below $(I_o - I_{th})$ lead to variations ΔN, ΔS, and ΔP, which are much smaller than the steady-state values N_{th}, S_o, and P_o, respectively. This small signal case allows the rate equations to be solved analytically (Coldren and Corzine 1995) using the linear gain approximation illustrated in Figure 26.4. Assuming $\beta = 0$, the small signal solution to the rate equations is

$$\Delta P(\omega) = M(\omega) \times \Delta P = \frac{\omega_r^2}{\omega_r^2 - \omega^2 + i\omega\gamma} \times \eta_i \frac{\alpha_m}{\alpha_i + \alpha_m} \frac{h\nu}{e} \Delta I \qquad (26.11)$$

with the angular electron–photon resonance frequency $\omega_r = 2\pi f_r$ given by

$$\omega_r^2 = \frac{a v_g S_o}{\tau_p (1 + \varepsilon S_o)} \left(1 + \frac{\varepsilon}{v_g a \tau_c}\right) \qquad (26.12)$$

including the photon lifetime τ_p and the gain compression factor ε from Equation 26.1. At low photon densities ($\varepsilon = 0$), Equation 26.12 is reduced to $\omega_r^2 = (\tau_{st}\tau_p)^{-1}$ with the differential stimulated emission time $\tau_{st} = dR_{stim}/dN = (a v_g S_o)^{-1}$. The damping constant is given by $\gamma = \tau_{st}^{-1} + \tau_c^{-1}$ including the differential carrier lifetime τ_c, with $\tau_c^{-1} = dR_{ABC}/dN = A + 2BN_{th} + 3CN_{th}^2$.

Figure 26.9 illustrates the normalized modulation response $|M(\omega)|$ as a function of modulation frequency for different gain compression factors ε (Piprek and Bowers 2002). At low frequencies, the photon

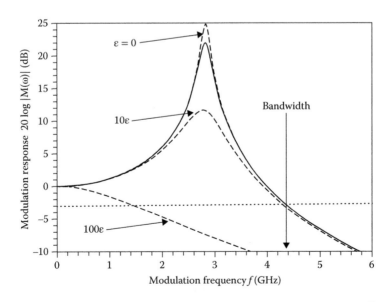

FIGURE 26.9 Modulation response versus modulation frequency. The dashed lines show the effect of changing the gain suppression factor ε.

density can easily follow the current modulation and the response function is quite flat. The response is most intense near the resonance frequency ω_r. The peak frequency is given by $\omega_p^2 = \omega_r^2 - \gamma^2/2$. At even faster current modulation, the photons cannot follow any more and the response function declines. The modulation bandwidth f_b is the frequency at which the response drops to $|M(2\pi f_b)| = 2^{-1/2} = -3dB$ (see Figure 26.9). The bandwidth rises with the steady-state current but saturates at high currents due to increased damping, device heating, gain compression, transport effects, or parasitics (Kjebon et al. 1996).

26.7 Basic Laser Cavity Designs

Chapters 27 through 34 feature specific laser designs and applications in much more detail, we here only give a brief overview of the main types of optical resonators. The most simple cavity design uses the reflection at the two laser facets for optical feedback (FP laser, see Figures 26.3 and 26.10, and Chapter 27). Constructive interference of forward and backward traveling optical waves is restricted to specific wavelengths. Those wavelengths constitute the longitudinal mode spectrum of the laser. The cavity length is typically on the order of several hundred microns, much larger than the lasing wavelength, so that many longitudinal modes may exist. The actual lasing modes are those receiving strong optical gain. Single-mode lasing is hard to achieve in simple FP structures, especially under modulation. Dynamic single-mode operation is required in many applications and is achieved using optical cavities with selective reflection. The distributed feedback (DFB) laser is widely used in single-mode fiber optic applications (Figure 26.10). Typical DFB lasers exhibit a periodic longitudinal variation of the refractive index within one layer of the edge-emitting waveguide structure. This index variation provides continuous (distributed) reflection at a wavelength given by the variation period. Facet reflection is not needed in DFB lasers; however, facet coating may be used to increase the light emission from one end of the cavity. Other laser resonators terminate the optical cavity by two distributed Bragg reflectors (DBRs) with a stepwise alternating index. More details on DFB and DBR lasers are given in Chapters 30 and 31.

A special type of DBR laser is the vertical-cavity surface-emitting laser (VCSEL) which emits light through the bottom and/or top surface of the layered structure (see Figure 26.10 and Chapters 34 and 45). The light travels perpendicular to the active layer and receives optical gain only over a very short travel distance. Thus, many more photon roundtrips and highly reflective VCSEL mirrors with more than 99% reflectivity are needed for lasing.

In transversal directions, the optical wave is typically confined by the refractive index profile, using a ridge (DFB laser) or a pillar (VCSEL) to form the waveguide. Even with restriction to one longitudinal mode, multiple transversal optical modes may occur in all three types of lasers.

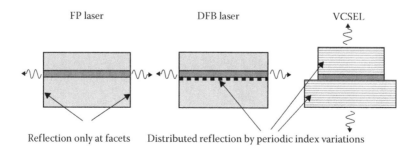

FIGURE 26.10   Illustration of basic laser cavity designs. [DFB, distributed feedback; FP, Fabry–Pérot; VCSEL, vertical-cavity surface-emitting laser. (Adapted from Piprek, J., In S. S. Sun and L. R. Dalton [eds.], *Introduction to Organic Electronic and Optoelectronic Materials and Devices*, Boca Raton, FL: CRC Press, 2008.)]

References

Carroll, J., Whiteaway, J., and Plumb, D. *Distributed Feedback Semiconductor Lasers*. London/Washington: IEE/SPIE Press, 1998.

Coldren, L. A., and Corzine, S. W. *Diode Lasers and Photonic Integrated Circuits*. New York, NY: Wiley, 1995.

Crump, P., Erbert, G., Wenzel, H. et al. Efficient high-power laser diodes. *IEEE Journal of Selected Topics in Quantum Electronics*, vol. 19, p. 1501211, 2013.

Kjebon, O., Schatz, R., Lourdudoss, S., Nilsson, S., and Stalnacke, B. Modulation response measurement and evaluation of MQW InGaAsP lasers of various design. *SPIE Proceedings*, vol. 2687, pp. 138–152, 1996.

Petermann, K. *Laser Diode Modulation and Noise*. Dordrecht, the Netherlands: Kluwer Academic Publishers, 1988.

Piprek, J., Akulova, Y. A., Babic, D. I., Coldren, L. A., and Bowers, J. E. Minimum temperature sensitivity of 1.55-micron vertical-cavity lasers at −30 nm gain offset. *Applied Physics Letters*, vol. 72, no. 15, pp. 1814–1816, 1998.

Piprek, J., Abraham, P., and Bowers, J. E. Self-consistent analysis of high-temperature effects on strained-layer multi-quantum well InGaAsP/InP lasers. *IEEE Journal of Quantum Electronics*, vol. 36, no. 3, pp. 366–374, 2000.

Piprek, J. and Bowers, J. E. Analog modulation of semiconductor lasers, Chapter 3. In *RF Photonic Technology in Optical Fiber Links*. Chang, W., ed. Cambridge, UK: Cambridge University Press, 2002.

Piprek, J. *Semiconductor Optoelectronic Devices: Introduction to Physics and Simulation*. San Diego, CA: Academic Press, 2003.

Piprek, J. Introduction to optoelectronic device principles, Chapter 2. In *Introduction to Organic Electronic and Optoelectronic Materials and Devices*. Sun, S. S. and Dalton, L. R, eds. Boca Raton, FL: CRC Press, 2008, pp. 25–46.

Piprek, J. Analysis of efficiency limitations in high-power InGaN/GaN laser diodes, *Optical Quantum Electron*, vol. 48, p. 471, 2016.

Shtengel, G. E., Kazarinov, R. F., Belenky, G. L., Hybertsen, M. S., and Ackerman, D. A. Advances in measurements of physical parameters of semiconductor lasers. *International Journal of High Speed Electronics and Systems*, vol. 9, pp. 901–940, 1998.

27

High-Power Lasers

Hans Wenzel
and
Anissa Zeghuzi

27.1 Introduction

High-power diode lasers deliver the energy to all high-performance laser systems, either as a pump source or as a source for direct material processing. Comprehensive descriptions of their manufacturing and applications can be found in, for example, References [1–4].

The lasers are constructed like all edge-emitting diode lasers as shown in Figure 27.1. The layer structure grown by metal-organic vapor phase epitaxy (MOVPE) or molecular beam epitaxy (MBE) on a crystalline substrate (e.g., GaAs, InP, GaN) consists basically of n-doped cladding and optical confinement layers, an active region (typically a single quantum well [QW]) and p-doped optical confinement and cladding layers and is completed by a highly p-doped cap (contact) layer. However, there are three peculiarities compared to other lasers: first, the vertical waveguide is weak (large total thickness of confinement layers or small index step between confinement and cladding lasers). Second, the emitting aperture is very broad, ranging from tens to hundreds of micrometers, which results in a nonstationary behavior [5,6]. Third, the cavity between the cleaved facets is very long, reaching values of several millimeters and the two facets are extremely differently coated.

The lateral optical and current confinement can be achieved by several means. In the most simple case, there is no built-in waveguide at all so that the optical field is confined to the region below the contact

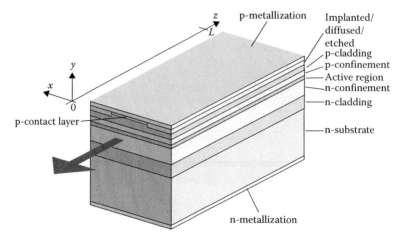

FIGURE 27.1　Schematic view of a high-power broad-area semiconductor laser.

stripe by gain guiding at low power and under pulsed operation, but is strongly influenced by the thermally induced waveguide created under continuous-wave (CW) operation at high power. In order to restrict the current spreading, the electrical conductivity of parts of the p-doped region besides the contact stripe can be reduced by ion implantation, impurity diffusion, or by implementation of a reverse-biased p-n junction. If the semiconductor besides the p-contact is etched away up to a defined depth and subsequently filled with an insulator, the index contrast between the semiconductor below the contact stripe and the insulator results in a lateral built-in index guide, which stabilizes the optical field at low power and under pulsed operation, but increases the far-field divergence.

The cleaved facets located at $z = 0$ and $z = L$ in Figure 27.1 are low reflection coated on the front side ($R_0 \propto 0.01$) and high reflection coated on the rear side ($R_L > 0.9$). If a small optical spectrum and a reduced drift of the wavelength with temperature and injection current are required, Bragg gratings could be implemented into the cavity, resulting in distributed feedback (DFB) or distributed Bragg-reflection (DBR) lasers [7]. However, here we focus on the Fabry–Pérot (FP) type of high-power lasers where the coated facets provide the feedback.

The simulation of broad-area lasers is challenging because of the different temporal and spatial scales involved. The timescales for the variations of the optical field, the carrier densities, and the temperature are ps, ns, and μs, respectively. The spatial scales range from nanometers (active QW) and micrometers (epitaxial layers and lateral waveguides) up to millimeters (cavity). Another difficulty arises from the highly nonlinear behavior because of the coupling of the optical, electronic, and thermal phenomena. Therefore, until now, no simulation tool covering all spatiotemporal scales and physical phenomena has been available. In this chapter, we survey models with different complexity for the simulation of high-power lasers [6].

This chapter is organized as follows. In Section 27.2, we present a model for high-power lasers based on measurable parameters, which can be used to compare different laser structures and to predict the electro-optical characteristics as a function of cavity length and facet reflectivities, for example. In Section 27.3, we investigate the profile of the optical power in the cavity in more detail and derive some of the equations used in Section 27.2. Furthermore, we discuss several effects responsible for the saturation of the output power with increasing injection current. The parabolic paraxial wave equation based on the slowly varying amplitude and rotating wave approximations, taking into account gain dispersion, spontaneous emission, and a third-order nonlinear susceptibility, is derived in Section 27.4. A balance equation for the electromagnetic energy density will be given. In Section 27.5, we present the equations that can be used to calculate the nonlinear lateral optical modes of broad-area lasers and discuss several root causes for the multipeaked and not diffraction-limited lateral field profile of broad-area lasers. Finally, in Section 27.6, we summarize a thermodynamic-based model for the transport of the charged carriers and the temperature flow. Particular attention is paid to a consistent formulation with the model for the optical field presented in Section 27.4.

27.2 Phenomenological Model

In this section, a model for high-power lasers based on measurable parameters is presented. The model can be used to compare different laser structures and to predict the electro-optical characteristics in dependence on cavity length and facet reflectivities, for example.

27.2.1 Summary of the Governing Equations

The total output power P_{out} of a semiconductor laser in dependence on the injection current I is given by

$$P_{out} = \frac{\hbar\omega}{e}\eta_{ext}\left(I - I_{thr}\right) \tag{27.1}$$

where η_{ext} is the external differential efficiency, I_{thr} the threshold current, ω the angular lasing frequency, $\hbar$ the reduced Planck constant, and e the elementary charge. Here and in what follows $I \geq I_{thr}$ is assumed. The photon energy $\hbar\omega$ can be written as (c vacuum speed of light, h Planck constant)

$$\hbar\omega = \frac{hc}{\lambda} \tag{27.2}$$

where the vacuum lasing wavelength λ is assumed to vary linearly with the temperature T in the cavity as

$$\lambda = \lambda_{ref} + \frac{d\lambda}{dT}\Delta T \qquad \text{with} \qquad \Delta T = T - T_{ref} \tag{27.3}$$

with T_{ref} being the reference (heat sink or ambient) temperature. The output powers P_0 at the left ($z = 0$) facet and P_L at the right ($z = L$) facet with intensity reflection coefficients R_0 and R_L, respectively, are obtained from the total power as

$$P_0 = \frac{P_{out}}{1 + (1 - R_L)(1 - R_0)^{-1}\sqrt{R_0 R_L^{-1}}} \quad \text{and} \quad P_L = \frac{P_{out}}{1 + (1 - R_0)(1 - R_L)^{-1}\sqrt{R_L R_0^{-1}}} \tag{27.4}$$

(cf. Equation 27.40).

To calculate the threshold current, a model for the modal gain is required. First one should be aware that the gain depends on both the wavelength and the current density. For a fixed wavelength λ, the dependence of the modal gain of QW lasers on the current density can be well approximated by a logarithmic relation,

$$g_m = G_0 \ln\frac{j}{j_{tr}} \tag{27.5}$$

where G_0 is the gain prefactor (proportional to the differential gain) and j_{tr} the transparency current density. At $j = j_{tr}$ the gain vanishes, i.e., a wave with the wavelength λ propagating along the QW embedded in a nonabsorbing medium is neither absorbed nor amplified (transparency). Due to the fact that FP lasers lase at the maximum of the gain spectrum, we need the dependence of the peak gain on the current density for which the relation Equation 27.5 can be used, too. However, one must keep in mind that the peak gain never becomes negative, but approaches zero for vanishing current. Therefore, $j \gg j_{tr}$ must be observed and the transparency current density obtained by the analysis of the length dependencies of external differential efficiency and threshold current as described in Section 27.2.2 has no direct physical interpretation. One should mention that of course any other functional dependence $g_m(j)$ that can be solved analytically for j, such as a linear one, could be used. However, then the procedure presented in Section 27.2.2 would yield differing parameters G_0 and j_{tr}.

Assuming the logarithmic dependence of the modal gain on the current density and an exponential dependence of the threshold current on the temperature, the threshold current can thus be written as

$$I_{thr} = WL j_{tr} e^{\frac{g_{thr}}{G_0}} e^{\frac{\Delta T}{T_0}} \tag{27.6}$$

where W is the contact width, L the contact length (assumed equal to the cavity length), and T_0 the characteristic temperature of the threshold current. Similarly, the external differential efficiency can be written as (cf. Equation 27.44)

$$\eta_{ext} = \frac{\alpha_{out}}{g_{thr}} \eta_i e^{-\frac{\Delta T}{T_1}} \tag{27.7}$$

where T_1 is the characteristic temperature of the external differential efficiency. The internal efficiency η_i gives the fraction of the total current increment that results in stimulated emission of photons. It describes the effect that not all electron–hole pairs *additionally* injected above threshold are converted into photons by stimulated emission. The threshold gain g_{thr} is

$$g_{thr} = \alpha_{out} + \alpha_i \tag{27.8}$$

with the internal losses α_i and the outcoupling (mirror) losses

$$\alpha_{out} = -\frac{1}{2L} \ln(R_0 R_L) \tag{27.9}$$

Finally, the temperature rise can be calculated from the dissipated power Q and the thermal resistance r_{th} related to the contact area,

$$\Delta T = \frac{r_{th}}{WL} Q \tag{27.10}$$

The dissipated power is given by the difference between electrical input power and optical output power,

$$Q = UI - P_{out} \tag{27.11}$$

with the voltage–current characteristics

$$U = \Delta U + \frac{\hbar\omega}{e} + \frac{r_s}{WL} I \tag{27.12}$$

where ΔU is the so-called defect voltage and r_s the series resistance related to the contact area. The main contribution to the defect voltage arises from the spacing of the quasi-Fermi potentials of electrons and holes in the active region (Fermi voltage), which is assumed to be clamped above threshold and is always larger than the photon energy divided by the elementary charge. The last term in Equation 27.12 is due to the Ohmic voltage drop in the bulk semiconductor layers, at the heteroboundaries, and at the semiconductor–metal junctions.

Instead of Equation 27.11 sometimes other expressions are reported. For example, by inserting Equations 27.1 and 27.12 into Equation 27.11 and rearranging the terms,

$$Q = \frac{\hbar\omega}{e} I_{thr} + \frac{\hbar\omega}{e}(1 - \eta_i e^{-\frac{\Delta T}{T_1}})(I - I_{thr}) + \frac{\hbar\omega}{e}(\eta_i e^{-\frac{\Delta T}{T_1}} - \eta_{ext})(I - I_{thr}) + \Delta U I + \frac{r_s}{WL} I^2 \tag{27.13}$$

is obtained. The third term can be also written as

$$\frac{\hbar\omega}{e}(\eta_i e^{-\frac{\Delta T}{T_1}} - \eta_{\text{ext}})(I - I_{\text{thr}}) = \alpha_i L \bar{P} \tag{27.14}$$

where we have used Equations 27.7, 27.8, and 27.43. The averaged internal power $\bar{P}$ is defined in Equation 27.32. The dissipated power is thus the sum of the heat generated at threshold by nonradiative and spontaneous recombination (first term in Equation 27.13), the heat generated above the threshold due to additional nonstimulated recombination and carrier leakage (second term), the heat generated by absorption (third term), the heat caused by the defect voltage (fourth term), and the Joule heat (last term). It is worth to mention that in Equations 27.11 and 27.13, it is assumed that the spontaneously emitted radiation is completely absorbed in the cavity so that P_{out} contains only the stimulated emission.

All equations have been written in such a manner that the geometrical scaling given by W and L is explicitly separated. Note, however, that the dependence of r_{th} on W cannot be neglected because the thermal resistance $R_{\text{th}} = r_{\text{th}}/(WL)$ is not inverse proportional to W. For example, for a rectangular-shaped infinitely thin heat source,

$$r_{\text{th, 2D}} \approx \frac{W}{\kappa\pi}\ln\left(\frac{4h}{W}\right) \tag{27.15}$$

holds under the condition $W \ll h < W_s$, where κ is the thermal conductivity of the material with thickness h and width $2W_s$ separating the heat source of width W from the heat sink [8]. Only for a purely one-dimensional heat flow

$$r_{\text{th, 1D}} = \frac{h}{\kappa} \tag{27.16}$$

is independent of W.

For shallow-edged samples with substantial lateral current spreading, the transparency current density obtained by the procedure presented in Section 27.2.2 depends on the p-contact width, too [9]. In order to determine the transparency current density for an infinite p-contact width $j_{\text{tr},W\to\infty}$ without current spreading, the transparency *current* has to be plotted and linearly fitted in dependence on W. From the slope dI_{tr}/dW one obtains $j_{\text{tr},W\to\infty}$ and from the extrapolation $W \to 0$ the spreading current $I_{\text{tr, spread}}$ according to the relation [9]

$$I_{\text{tr}}(W) = j_{\text{tr},W\to\infty}LW + I_{\text{tr, spread}} \tag{27.17}$$

The basic assumption underlying Equation 27.17 is that $I_{\text{tr, spread}}$ is independent of W. If different cavity lengths are considered, Equation 27.17 has to be divided by L.

For a numerical calculation of the electro-optical characteristics, Equation 27.10 is best suited because $\Delta T(I)$ is a monotonous function (in contrast to $P(I)$). Knowing ΔT, the threshold current and external efficiency can be determined from Equations 27.6 and 27.7, respectively, and finally the output power from Equation 27.1.

27.2.2 Determination of the Parameters Entering the Model

In what follows we consider an asymmetric super-large optical-cavity structure published in Reference [10]. The parameters η_i, α_i, G_0, and j_{tr} are determined by measuring the power–current characteristics of as-cleaved lasers having different cavity lengths L, operated under pulsed conditions to avoid self-heating ($\Delta T = 0$). Thus pulse lengths below 1 μs and duty cycles below 1% are required. Note that the lasing wavelength could substantially vary with L, which must be taken into account in the calculation of the optical power from the measured photovoltage of the detector by adjusting the calibration factor. The measured *P–I* characteristics have to be linearly fitted to extract the slope efficiencies and threshold currents for the different cavity lengths.

If the inverse external differential efficiency is plotted versus cavity length (cf. Figure 27.2a),

$$\eta_{\text{ext}}^{-1}(L) = \eta_{\text{i}}^{-1} \left(1 - \frac{\alpha_{\text{i}}}{\ln(R)} L \right) \tag{27.18}$$

a linear fit delivers $-\eta_{\text{i}}^{-1}\alpha_{\text{i}}/\ln(R)$ from the slope $\eta_{\text{ext}}^{-1}(L)$ and η_{i}^{-1} from the extrapolation of η_{ext}^{-1} to $L \to 0$. A linear fit of the logarithm of the threshold current density versus the inverse cavity length

$$\ln(j_{\text{thr}})(L^{-1}) = \ln(j_{\text{tr}}) + G_0^{-1} \left(\alpha_{\text{i}} - \ln(R)L^{-1} \right) \tag{27.19}$$

as shown in Figure 27.2b yields $-G_0^{-1} \ln(R)$ from the slope $\ln(j_{\text{thr}})(L^{-1})$ and the threshold current density for an infinite cavity length $\ln(j_{\text{thr},\infty}) = \ln(j_{\text{tr}}) + G_0^{-1}\alpha_{\text{i}}$ from the extrapolation of $\ln(j_{\text{thr}})$ to $L^{-1} \to 0$. The results of this procedure are contained in Table 27.1, where facet reflectivities $R_0 = R_L = 0.3$ have been assumed in the evaluation. For other cases they must be correspondingly chosen.

The cavity lengths L have to be carefully chosen to ensure linear dependencies $\eta_{\text{ext}}^{-1}(L)$ and $\ln(j_{\text{thr}})(L^{-1})$. If L is too small (large threshold gain), η_{ext}^{-1} could increase with decreasing L due to an increase of α_{i} (enhanced free-carrier absorption) and a decrease of η_{i} (enhanced carrier leakage). If L is too large (small threshold gain), the threshold gain does not depend logarithmically on the threshold current density because $j_{\text{th}} \to j_{\text{tr}}$ so that the model fails, too.

The characteristic temperatures should be determined by measuring the power–current characteristics at different chip temperatures under pulsed conditions. The outcoupling losses (determined by cavity length and facet reflectivities) of the laser under investigation should coincide, nearly, with the intended values for CW operation, because T_0 and T_1 decrease with increasing threshold gain. The temperature range must be correspondingly chosen, too, because T_0 and T_1 decrease with increasing temperature. The results of the measurements and the linear fits of the logarithms of the threshold current density and external

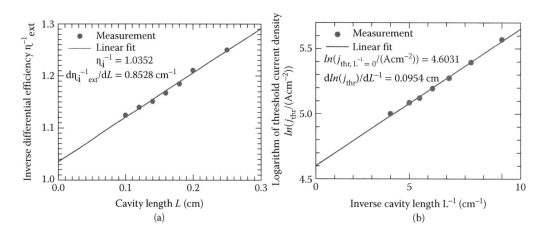

FIGURE 27.2 Determination of internal efficiency η_{i}, internal optical losses α_{i}, gain prefactor G_0, and transparency current density j_{tr}. (a) Inverse external differential efficiency versus cavity length. Bullets indicate measurements and solid line indicates linear fit. (b) Logarithm of threshold current density versus inverse cavity length. Bullets indicate measurements and solid line indicate linear fit.

TABLE 27.1 Phenomenological Laser Parameters

Parameter	Symbol	Value	Unit
Front facet reflectivity	R_0	0.012	
Rear facet reflectivity	R_L	0.95	
Width of active region	W	90	μm
Cavity length	L	4000	μm
Internal efficiency	η_i	0.97	
Internal optical losses	α_i	0.99	cm^{-1}
Gain prefactor	G_0	12.6	cm^{-1}
Transparency current density	j_{tr}	92.2	A/cm^2
Characteristic temperature of threshold current	T_0	98	K
Characteristic temperature of external efficiency	T_1	341	K
Reference wavelength	λ_{ref}	899	nm
Change of wavelength with temperature	$d\lambda/dT$	0.331	nm/K
Series resistance	r_s	$0.767 \cdot 10^{-4}$	$\Omega \cdot cm^2$
Thermal resistance	r_{th}	$0.936 \cdot 10^{-2}$	$K \cdot cm^2/W$
Defect voltage	ΔU	0.033	V
Differential threshold current	j'_{thr}	5	cm^{-2}
Saturation current density	j_{sat}	$5 \cdot 10^{-6}$	cm^2/A
Differential series resistance	r'_s	$-5 \cdot 10^{-3}$	$\Omega \cdot cm^4/A$

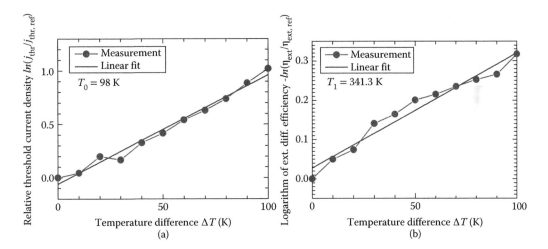

FIGURE 27.3 Determination of characteristic temperatures. Cavity length and facet reflectivities are given in Table 27.1. (a) Logarithm of threshold current density versus temperature rise. Bullets indicate measurements and solid line indicate linear fit. (b) Logarithm of external differential efficiency versus temperature rise. Bullets indicate measurements and solid line indicate linear fit.

differential efficiency versus temperature rise,

$$\ln\left(\frac{j_{thr}(T)}{j_{thr}(T_{ref})}\right) = \frac{\Delta T}{T_0} \quad \text{and} \quad -\ln\left(\frac{\eta_{ext}(T)}{\eta_{ext}(T_{ref})}\right) = \frac{\Delta T}{T_1} \qquad (27.20)$$

are shown in Figure 27.3a and 27.3b.

The series resistance and defect voltage are obtained from a fit of the linear part of the voltage–current characteristics above threshold:

$$U = U_0 + \frac{r_s}{WL}I \quad \text{and} \quad \Delta U = U_0 - \frac{hc}{e\lambda_{ref}} \quad (27.21)$$

The result is shown in Figure 27.4a. The voltage must be measured with care to ensure that the true series resistance of the chip is obtained. Typically a so-called four-terminal or 4-wire sensing separating the current and voltage electrodes has to be applied. Finally, the thermal resistance can be determined from a linear fit of the lasing wavelength versus the dissipated power,

$$\lambda(Q) = \lambda_{ref} + \frac{d\lambda}{dT}\frac{r_{th}}{WL}Q, \quad (27.22)$$

where Q is calculated according to Equation 27.11

There are two possibilities to employ Equation 27.22. The first possibility is based on the measurement of the center wavelength λ_{center} of the emission spectrum above threshold versus injection current. This yields a correct result, if a sufficiently large number of modes is lasing and the envelope of the spectrum exhibits a Gaussian-like shape. The shift of λ_{center} is mainly determined by the temperature dependence of the energy gap of the active region, but sometimes also by the temperature dependence of the confinement factor or screening of polarization fields in GaN-based lasers, for example. The method has the advantage, that the requirement on the spectral resolution of the optical spectrometer is modest. A disadvantage is that a unique center wavelength cannot always be determined.

The second possibility is to trace the wavelength of a single longitudinal mode while increasing the current, using a high-resolution optical spectrometer. The shift of the wavelength λ_{mode} of a mode is determined by the temperature dependence of its modal phase index, which results in a much weaker coefficient $d\lambda_{mode}/dT$ than for the center wavelength $d\lambda_{center}/dT$ of the emission spectrum. This method yields more reliable results for the thermal resistance. In any case, $d\lambda_{center}/dT$ or $d\lambda_{mode}/dT$ has to be determined in advance by measuring the functions $\lambda_{center}(T)$ or $\lambda_{mode}(T)$, respectively, where T is the chip temperature.

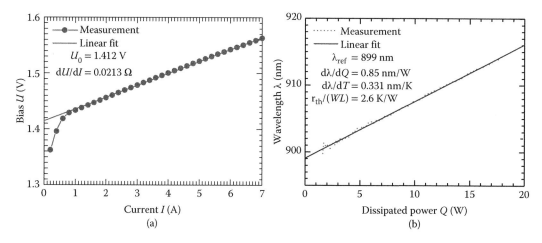

FIGURE 27.4 Determination of series and thermal resistances. (a) Voltage–current characteristic. Bullets indicate measurements. Solid line indicates linear fit. (b) Center wavelength versus dissipated power. Light grey line indicates measurements. Straight line indicates linear fit. Cavity length and facet reflectivities are given in Table 27.1.

27.2.3 Comparison of Simulated and Measured CW Characteristics

In Figure 27.5 the measured $P - I$, $U - I$ and $\eta_c - I$ characteristics (solid lines) are compared with the results obtained with the phenomenological model (dashed lines) presented above in Section 27.2.1. The conversion efficiency η_c is defined as

$$\eta_c = \frac{P_{out}}{UI} \tag{27.23}$$

The bending of the power–current characteristics at high bias is caused by the temperature-induced increase of the threshold current and decrease of the slope efficiency (determined by the parameters T_0 and T_1 here) due to the power dissipation. It leads to what is commonly referred to as "thermal roll-over." However, the measured bending is stronger than the simulated one.

This discrepancy is caused by the fact that there are nonthermal reasons for the bending of the P–I characteristic at high bias (also referred as "power saturation") which cannot be addressed by the phenomenological laser models based on parameters measured at low bias. For example, bending of the conduction and valence bands and carrier accumulation effects occurring at large bias leading to enhanced nonstimulated recombination and internal absorption [11] result in an increased threshold current and a decreased external differential efficiency. Other effects are longitudinal spatial holeburning (LSH), nonlinear gain compression, and two-photon absorption as discussed in Sections 27.3.2 and 27.3.5, which result in a saturation of the output power.

These effects could be modeled phenomenologically by additional explicit current dependencies $I_{thr}(I)$ and $\eta_{ext}(I)$ with parameters j'_{thr} and j_{sat},

$$I_{thr} = I_{thr,0} + WLj'_{thr}\left(I - I_{thr,0}\right) \tag{27.24}$$

and

$$\eta_{ext} = \frac{\eta_{ext,0}}{1 + \frac{I - I_{thr,0}}{WLj_{sat}}}, \tag{27.25}$$

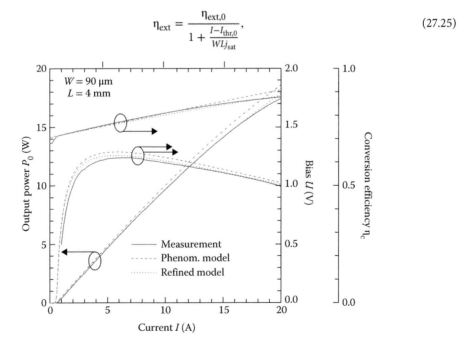

FIGURE 27.5 Output power at $z = 0$ (left axis), applied voltage (right axis), and conversion efficiency (right-most axis) versus injection current. Solid line indicates measurement and dashed line indicates phenomenological laser model. Dotted line indicates refined model.

respectively. The proportionality factor $j'_{\text{thr}} > 0$ models the increase of the threshold current due to non-thermal effects with $I_{\text{thr},0}$ given by Equation 27.6. Similarly, the parameter $j_{\text{sat}} > 0$ causes a reduction of the external differential efficiency where $\eta_{\text{ext},0}$ is given by Equation 27.7.

Furthermore, the series resistance varies also far above threshold, either caused by carrier accumulation (differential resistance $r'_s < 0$) or temperature effects ($r'_s > 0$),

$$r_s = r_{s,0} + \frac{r'_s}{WL}\left(I - I_{\text{thr},0}\right) \tag{27.26}$$

With the refined model (dotted lines), a better match can be achieved as Figure 27.5 reveals. Note that the used parameters j'_{thr}, j_{sat}, and r'_s given in Table 27.1 are guess values. An even better agreement between theory and experiment could be achieved by employing a fitting procedure, probably.

27.3 Models for the Optical Power

In this section, we investigate the profile of the optical power in the cavity in more detail and derive some of the equations used in Section 27.2. Furthermore, we discuss the nonthermal effects responsible for the saturation of the output power with increasing injection current.

27.3.1 Basic Relations

The total power P can be always separated into forward and backward propagating parts,

$$P(z,t) = P^+(z,t) + P^-(z,t) \tag{27.27}$$

In an FP cavity, they fulfil the partial differential equations ($\partial_t = \partial/\partial t$, $\partial_z = \partial/\partial z$)

$$\partial_t P^{\pm} \pm v_g \partial_z P^{\pm} = v_g(g_m - \alpha_m)P^{\pm} + \frac{\dot{P}_{\text{sp}}}{2} \tag{27.28}$$

subject to the boundary conditions

$$P^+(0,t) = R_0 P^-(0,t)$$
$$P^-(L,t) = R_L P^+(L,t) \tag{27.29}$$

Here $g_m - \alpha_m$ is the modal net gain in the cavity with g_m being the modal gain and α_m the modal losses, L the cavity length, $v_g = c/n_g$ the group velocity with n_g being the modal group index, and $\dot{P}_{\text{sp}}$ the rate of spontaneous emission coupled into the lasing modes. The total output power at the facets is $P_{\text{out}} = P_0 + P_L$ with

$$P_0 = (1 - R_0)P^-(0), \quad P_L = (1 - R_L)P^+(L) \tag{27.30}$$

Integrating Equation 27.28 along z, adding both equations, and taking into account Equation 27.29 we obtain the power balance

$$\frac{1}{v_g}\frac{d\bar{P}}{dt} = -\frac{P_{\text{out}}}{L} + \frac{1}{L}\int_0^L (g_m - \alpha_m)P\,dz + \frac{1}{v_g L}\int_0^L \dot{P}_{\text{sp}}\,dz \tag{27.31}$$

with the average internal power

$$\bar{P} = \frac{1}{L} \int_0^L (P^+ + P^-)\, dz. \tag{27.32}$$

In what follows, we consider the steady state ($\partial_t = 0$) and two special cases, namely the case around threshold and the case above threshold.

27.3.1.1 Steady State around Threshold

Around threshold both $g_m - \alpha_m$ and $\dot{P}_{sp}$ can be assumed to be constant (z-independent). Therefore, the solution of Equation 27.28 taking into account Equation 27.29 is

$$P^\pm(z) = \frac{\dot{P}_{sp}}{2v_g} \cdot \frac{C^\pm e^{\pm(g_m - \alpha_m)z} - 1}{g_m - \alpha_m} \tag{27.33}$$

with

$$C^+ = \frac{1 - R_0 + (1 - R_L)R_0 e^{(g_m - \alpha_m)L}}{1 - R_0 R_L e^{2(g_m - \alpha_m)L}} \tag{27.34}$$

$$C^- = \frac{1 - R_L + (1 - R_0)R_L e^{(g_m - \alpha_m)L}}{1 - R_0 R_L e^{2(g_m - \alpha_m)L}} e^{(g_m - \alpha_m)L}. \tag{27.35}$$

The ratio between the outcoupled powers at the facets is given by

$$\frac{P_L}{P_0} = \frac{1 - R_L}{1 - R_0} \cdot \frac{1 + R_0 e^{(g_m - \alpha_m)L}}{1 + R_L e^{(g_m - \alpha_m)L}} \tag{27.36}$$

If the spontaneous emission approaches zero, $C^\pm$ must go to infinity to obtain a nonzero output power, which yields the so-called threshold condition $g_m - \alpha_m = \alpha_{out}$.

27.3.1.2 Steady State above Threshold

Above threshold $g_m - \alpha_m$ cannot longer be assumed to be constant due to LSH, but the spontaneous emission can be neglected, $\dot{P}_{sp} = 0$. Hence, the solution of Equation 27.33 is

$$P^\pm(z) = P^\pm(0)e^{\pm \int_0^z [g_m(z') - \alpha_m(z')]\, dz'}. \tag{27.37}$$

In order to obey the boundary conditions Equation 27.29, the threshold condition

$$g_{thr} = \frac{1}{L} \int_0^L \alpha_m\, dz + \alpha_{out} \tag{27.38}$$

must hold, where α_{out} are the outcoupling losses Equation 27.9 and

$$g_{thr} = \frac{1}{L} \int_0^L g_m\, dz \tag{27.39}$$

is the threshold gain. The ratio between the outcoupled powers

$$\frac{P_{\mathrm{L}}}{P_0} = \frac{1 - R_{\mathrm{L}}}{1 - R_0} \sqrt{\frac{R_0}{R_{\mathrm{L}}}} \tag{27.40}$$

follows from the boundary conditions Equation 27.29, the general relation

$$\partial_z(P^+ P^-) = 0 \tag{27.41}$$

and from the ratio between the internal and external powers given by Equation 27.30.

27.3.1.3 External Differential Efficiency

From the threshold condition Equation 27.38, the modal threshold gain g_{thr} can be calculated and from that, depending on the model used, the threshold carrier density N_{thr}, the threshold voltage U_{thr}, or the threshold current I_{thr}. Finally, the external differential efficiency is the ratio between the energy leaving the cavity and the energy generated by stimulated recombination (cf. Equation 27.47),

$$\eta_{\mathrm{ext}} = \frac{P_{\mathrm{out}}}{\hbar\omega \int R_{\mathrm{st}}\,\mathrm{d}V} = \frac{P_{\mathrm{out}}}{\int g_{\mathrm{m}} P\,\mathrm{d}z}. \tag{27.42}$$

Assuming z-independent $g_{\mathrm{m}} - \alpha_{\mathrm{m}}$, the relation between the total output power and the average internal power follows from Equations 27.31 and 27.38 to

$$P_{\mathrm{out}} = \alpha_{\mathrm{out}} L \bar{P} \tag{27.43}$$

and the external differential efficiency is

$$\eta_{\mathrm{ext}} = \frac{\alpha_{\mathrm{out}}}{g_{\mathrm{thr}}} \tag{27.44}$$

27.3.2 Rigrod Model

In order to determine the variation of the modal gain along z above threshold, a model for the carrier densities is required. The simplest model consists of a rate equation for the excess carrier density N in the active region

$$\frac{\mathrm{d}N}{\mathrm{d}t} = \frac{j}{ed} - R(N, P) \tag{27.45}$$

with a constant injection current density j. The recombination rate [12]

$$R = R_{\mathrm{non\text{-}rad}} + R_{\mathrm{sp}} + R_{\mathrm{st}} \tag{27.46}$$

consists of nonradiative (Shockley–Read–Hall, Auger) recombination $R_{\mathrm{non\text{-}rad}}$, radiative spontaneous recombination R_{sp}, and radiative stimulated recombination:

$$R_{\mathrm{st}} = \frac{g_{\mathrm{m}} P}{dW\hbar\omega} \tag{27.47}$$

where $P = P^+ + P^-$, g_m is the modal gain, $\hbar\omega$ the photon energy, d the thickness of the active region, and W the width of the active region.[†]

Even for the steady-state above threshold, there exists no analytical solution of Equations 27.28 and 27.45. However, if the nonstimulated recombination rate and the gain are linearized,

$$R_{\text{non-rad}} + R_{\text{sp}} = \frac{N}{\tau_N} \tag{27.48}$$

and

$$g_m = g'_m (N - N_{\text{tr}}), \tag{27.49}$$

respectively, and inserted into Equation 27.45, the excess carrier density N can be determined. Introducing the obtained expression for N again into Equation 27.49 yields

$$g_m = \frac{g_0}{1 + \dfrac{P}{P_{\text{sat}}}} \tag{27.50}$$

with the unsaturated gain

$$g_0 = \frac{g'_m \tau_N}{ed}(j - j_{\text{tr}}) \tag{27.51}$$

where g'_m is the differential gain, τ_N the effective carrier lifetime, $j_{\text{tr}} = edN_{\text{tr}}/\tau_N$ the transparency current density, N_{tr} the transparency carrier density, and

$$P_{\text{sat}} = \frac{dW\hbar\omega}{g'_m \tau_N} \tag{27.52}$$

is the saturation power. For the typical values of $d = 10\,\text{nm}$, $W = 100\,\mu\text{m}$, $g'_m = 10 \cdot 10^{-18}\,\text{cm}^2$, $\tau_N = 1\,\text{ns}$, and $\hbar\omega = 1.24\,\text{eV}$, we obtain $P_{\text{sat}} = 0.2\,\text{W}$, which is much smaller than the internal power of state-of-the-art broad-area lasers operated far above threshold.

The resulting equation for the steady-state

$$\pm \frac{dP^\pm}{dz} = \left(\frac{g_0}{1 + \dfrac{P^+ + P^-}{P_{\text{sat}}}} - \alpha_m \right) P^\pm + \frac{\dot{P}_{\text{sp}}}{2v_g} \tag{27.53}$$

has been analytically solved by Rigrod [13] neglecting spontaneous emission and modal losses ($\dot{P}_{\text{sp}} = \alpha_m = 0$). For the general case a numerical solution has to be employed.

For $P_{\text{sat}} \ll P$, $P^+ \ll P^-$, $g_m \approx g_0 P_{\text{sat}}/P^-$ and the resulting equation

$$-\frac{dP^-}{dz} = g_0 P_{\text{sat}} - \alpha_m P^- + \frac{\dot{P}_{\text{sp}}}{2v_g} \tag{27.54}$$

[†] Equation 27.47 follows from Equation 27.140 by averaging over the active region $1/(dW) \iint dx dy$, taking into account Equation 27.106 and $n_m = \bar{n}$.

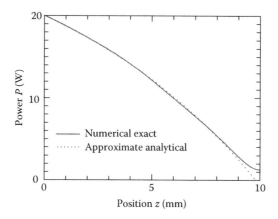

FIGURE 27.6 Longitudinal profiles of the total power for the parameters in Table 27.2 based on the numerical solution of Equation 27.53 (solid) and the analytical solution of Equation 27.55 (dashed).

has an analytical solution, too, namely

$$P(z) \approx P^-(z) = P^-(0)e^{\alpha_m z} + \left(g_0 P_{\text{sat}} + \frac{\dot{P}_{\text{sp}}}{2v_g}\right)\left(\frac{1 - e^{\alpha_m z}}{\alpha_m}\right) \qquad (27.55)$$

In Figure 27.6, the exact and approximate solutions are compared. Except near the right facet located at $z = 10$ mm where the condition $P^+ \ll P^-$ is not fulfilled, a good agreement can be noted. Note that according to Equation 27.55 $P(z)$ varies linearly for vanishing modal absorption $\alpha_m = 0$.

27.3.3 Treat-Power-as-a-Parameter Method

The equation for the steady-state longitudinal power profile above threshold

$$\pm\frac{dP^\pm}{dz} = \left[g_m(P) - \alpha_m\right]P^\pm \qquad (27.56)$$

subject to the boundary conditions Equation 27.29 can be conveniently solved numerically by the "treat-power-as-a-parameter" (TPP) method as introduced for the simulation of DFB lasers in Reference [14] and applied to high-power Fabry–Pérot lasers in Reference [15]. In a first step, g_m is calculated as a function of an external parameter (such as injection current I or bias U) and the power P, and stored in a look-up table, together with other quantities of interest (e.g., λ and I). In a second step, the boundary value problem (Equation 27.56) is solved by interpolating g_m in the look-up table. For given I or U, one chooses a guess value $P^-(0)$ and integrates Equation 27.56 from $z = 0$ to $z = L$ where typically the boundary condition is not fulfilled. Therefore, the function $P^-(L) - R_L P^+(L)$ has to be nullified by varying $P^-(0)$. It is also possible to give $P^-(0)$ and to vary I or U to fulfil the boundary condition.

The lasing wavelength can be determined approximately by searching the maximum of integral $\int_0^L (g_m - \alpha_m)\,dz$ with respect to λ. LSH is included automatically via the power dependence of g_m in Equation 27.56. If g_m is evaluated at the average power $\bar{P}$ in the cavity, the usual model neglecting LSH is recovered.

27.3.4 LSH and the Impact of Series Resistance and Internal Loss

The assumption of a constant injection current density in Equation 27.45 results in an overestimation of spatial hole burning. At the ohmic contacts, the quasi-Fermi potentials of electrons and holes, φ_n and

φ_p, respectively, are fixed and given by the applied bias, $\varphi_n = 0$ at the n-contact and $\varphi_p = U$ at the p-contact. The injection current density can be set equal to the hole current density and thus proportional to the gradient of the quasi-Fermi potential of the holes at the boundary between the active region and the p-doped region. Assuming isothermal conditions, one-dimensional current flow, and vanishing recombination outside the active region as well as infinitely high electron conductivity ($\sigma_n \to \infty$) in the n-doped region,

$$j(N) = \frac{U - \varphi_F(N)}{r_s} \tag{27.57}$$

can be derived from Equations 27.138 and 27.149 for the region beneath the p-contact, where $\varphi_F(N) = \varphi_p(N) - \varphi_n(N)$ is the Fermi voltage and

$$r_s = \sum_i \frac{d_i}{\sigma_{p,i}} \tag{27.58}$$

is the area-related total series resistance of the p-doped layers with thicknesses d_i and hole conductivities $\sigma_{p,i}$ between the active region and p-contact stripe.

For parabolic bands the relations between quasi-Fermi potentials φ_n and φ_p and the electron and hole densities n and p, respectively, are given by

$$e\varphi_n = -k_B T \mathcal{F}_i^{inv}\left(\frac{n}{N_c}\right) - E_c + e\varphi \quad \text{and} \quad e\varphi_p = k_B T \mathcal{F}_i^{inv}\left(\frac{p}{N_v}\right) - E_v + e\varphi, \tag{27.59}$$

where E_c, E_v are the (effective) conduction and valence band edges, N_c and N_v the conduction and valence band edge density of states, $\mathcal{F}_i^{inv}$ the inverse Fermi integrals with $i = 1/2$ for bulk and $i = 0$ for QW active regions and φ the electrostatic potential. Hence, the Fermi voltage is given by

$$\varphi_F = \frac{k_B T}{e}\left[\mathcal{F}_i^{inv}\left(\frac{p}{N_v}\right) + \mathcal{F}_i^{inv}\left(\frac{n}{N_c}\right)\right] + \frac{E_g}{e} \tag{27.60}$$

where $E_g = E_c - E_v$ is the energy gap. In Equation 27.57, local charge neutrality $p - n + p_D - n_A = 0$ is assumed so that the electron and hole densities are given by $n = n_0 + N$ and $p = p_0 + N$, respectively, with n_0, p_0 being the equilibrium densities and N the excess carrier density.

For $r_s \to \infty$, the usual model of a constant current density is recovered, whereas for $r_s \to 0$, the carrier density becomes constant. This can be more readily seen by expanding $\varphi_F(N)$ around the average carrier density $\bar{N}$ which yields

$$j(N) = \bar{j} - \frac{\varphi_F'}{r_s}(N - \bar{N}) \tag{27.61}$$

where the transition between both models (constant current density and constant carrier density, respectively) is governed by the parameter φ_F'/r_s.

Equations 27.56 and 27.45 with Equations 27.46, 27.47, and 27.57 assuming linear recombination and gain models, Equations 27.48 and 27.49, respectively, have been solved using the TPP method. The parameters used are given in Table 27.2. In Figure 27.7a, the influence of the series resistance on the power–current characteristics is clearly visible. For $r_s = 10^{-6}\ \Omega \cdot cm^2$ almost the same characteristics as without LSH is obtained.

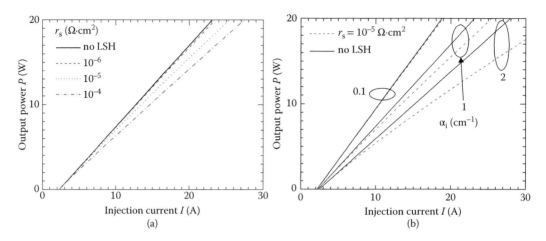

FIGURE 27.7 Power–current characteristics without (black solid) and with longitudinal spatial holeburning for different series resistances r_s and modal losses α_m. (a) Influence of the series resistance for a modal loss of $\alpha_m = 1$ cm^{-1} and (b) Influence of modal losses for a series resistance of $r_s = 10^{-5}\Omega \cdot$ cm^2.

TABLE 27.2 Laser Parameters Used to Investigate Power Saturation Effects

Parameter	Symbol	Value	Unit
Front facet reflectivity	R_0	0.001	
Rear facet reflectivity	R_L	0.95	
Width of active region	W	100	μm
Thickness of active region	d	10	nm
Cavity length	L	1	cm
Modal losses	α_m	1	cm^{-1}
Rate of spontaneous emission	$\dot{P}_{sp}$	0	
Differential modal gain	g'_m	$10 \cdot 10^{-18}$	cm^2
Transparency carrier density	N_{tr}	10^{18}	cm^{-3}
Carrier lifetime	τ_N	1	ns
Wavelength	λ	1000	nm
Series resistance	r_s	10^{-5}	$\Omega \cdot$ cm^2
Temperature	T	300	K
Conduction band density of states	N_c	$0.4 \cdot 10^{18}$	cm^{-3}
Valence band density of states	N_v	$13 \cdot 10^{18}$	cm^{-3}

If there are no internal losses, LSH has no impact on the slope efficiency at all, because from Equation 27.31

$$P_{out} = \int_0^L g_m P \, dz \tag{27.62}$$

follows and Equation 27.42 gives $\eta_{ext} = 1$ independent of the spatial profiles of g_m and P. As Figure 27.7b reveals, the difference between the power–current characteristics calculated with and without spatial hole burning increases with increasing modal losses.

The longitudinal profiles of forward, backward, and total power; modal gain; and injected current density are shown in Figure 27.8 for the extreme values of r_s. For a high value of r_s (Figure 27.8a), the injected current density is almost constant, so that the gain varies strongly. In contrast, for a low value of r_s (Figure 27.8b) the injected current density varies strongly so that the gain is almost constant. This results in quite different power profiles. In particular, the averaged internal power differs by almost a factor of 2 ($\bar{P} = 11$ W versus $\bar{P} = 6$ W). Besides by a reduction of the internal losses, the series resistance or the asymmetry in the facet coating ("unfolding the cavity" [16]) LSH could be mitigated by a tapered lateral waveguide design where the contact width W increases from the rear facet toward the front facet [17,18].

27.3.5 Discussion of Further Nonthermal Power Saturation Effects

27.3.5.1 Nonlinear Gain Compression

There are a couple of effects that can be described by an effective decrease of the optical gain with increasing photon density $|u|^2$ or power P, described by the functions

$$g = \frac{g_0}{1 + \epsilon_s |u|^2} \quad \text{or} \quad g = \frac{g_0}{1 + \frac{P}{P_s}} \qquad (27.63)$$

similar to Equation 27.50. The relation between the gain compression factor ϵ_s and the saturation power P_s is given by

$$P_s = \frac{d W v_g \hbar \omega}{\Gamma \epsilon_s} \qquad (27.64)$$

where Γ is the optical confinement factor of the active region. The gain compression factor is of the order $\epsilon_s \propto 10^{-17}$ cm^3, which results in a saturation power of $P_s \propto 150$ W for typical values $d = 10$ nm, $W = 100$ μm, $\Gamma = 0.01$, $n_g = 4$, $\lambda_0 = 1$ μm. Hence gain compression effects are of minor importance for CW lasers, but might result in power saturation of pulsed lasers where $P = 100$ W from a stripe width of $W = 100$ μm can be achieved [19].

The first effect contributing to nonlinear gain compression is spectral hole burning, which arises due to the interplay of the depletion of the carriers at the lasing wavelength due to stimulated emission and

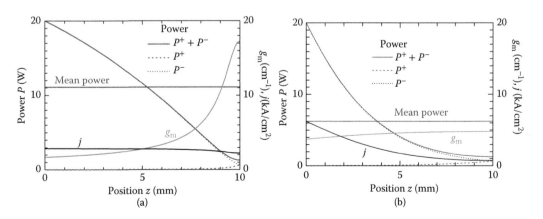

FIGURE 27.8 Longitudinal profiles of forward (P^+), backward (P^-), and total ($P = P^+ + P^-$), power, modal gain g_m, and injected current density j for two series resistances r_s at an output power of $P_{out} = 20$ W. (a) Series resistance $r_s = 10^{-4}$ cm^2 and (b) series resistance $r_s = 10^{-6}$ cm^2. The horizontal line is the averaged internal (mean) power $\bar{P}$.

the relaxation of injected carriers into the depleted spectral region. At high power densities, the relaxation processes due to intraband carrier–carrier scattering with a time scale of 50–100 fs are not sufficiently fast to fill the spectral hole formed by the optical transitions. This results in a reduction of the optical gain around the lasing wavelength.

A second effect is the increase of the temperatures of the carrier distributions beyond the lattice temperature by the removal of "cold" carriers near the band-edges due to stimulated emission or by the transfer of carriers to high energies within the bands by free-carrier, inter-valence band and two-photon absorption. The temperature increase results in a reduction of the optical gain, too.

Nonlinear gain compression also can be caused by lateral spatial hole burning [20] and the Bragg grating induced by the standing waves in Fabry–Pérot lasers [21]. The impact of both the gain compression and two-photon absorption on the power–current characteristics is discussed in the next section (cf. Figure 27.9).

27.3.5.2 Two-Photon Absorption

The two-photon absorption coefficient β is described in more detail in Section 27.5.2. The resulting modal absorption for a broad-area laser with a lateral top-hat intensity profile is given by (cf. Equation 27.121)

$$\alpha_{2,m} = \frac{\int n_r(x,y)\beta(x,y)|\Phi(x,y)|^4\,dxdy}{n_m(\int |\Phi(x,y)|^2\,dxdy)^2}P \approx \frac{\int n_r(y)\beta(y)|\phi(y)|^4\,dy}{n_m(\int |\phi(y)|^2\,dy)^2}\frac{P}{W} = \frac{\bar{\beta}_m}{A_m}P, \qquad (27.65)$$

where ϕ is the profile of the vertical mode, W the lateral width, n_m the modal index, and A_m the mode area given by

$$A_m = \frac{W(\int |\phi(y)|^2\,dy)^2}{\int |\phi(y)|^4\,dy} \qquad (27.66)$$

and

$$\bar{\beta}_m = \frac{\int n_r(y)\beta(y)|\phi(y)|^4\,dy}{n_m \int |\phi(y)|^4\,dy} \qquad (27.67)$$

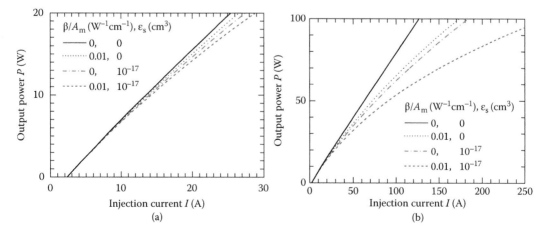

FIGURE 27.9 Impact of nonlinear-gain compression ϵ_s and two-photon absorption $\bar{\beta}_m/A_m$ on the power–current characteristics. (a) Power range corresponding to CW operation. (b) Power range corresponding to pulsed operation.

Approximating the intensity profile of the vertical mode by a Gaussian with a full width d_{1/e^2} at $1/e^2$ maximum,

$$\phi^2(y) = e^{-8y^2/d_{1/e^2}^2} \tag{27.68}$$

the mode area is $A_m = \frac{\sqrt{\pi}}{2} d_{1/e^2} W \approx d_{1/e^2} W$. Assuming an averaged two-photon absorption coefficient of $\bar{\beta}_m = 15\,\text{cm}\cdot\text{GW}^{-1}$ (cf. Figure 27.12a) and $d_{1/e^2} = 1.5\,\mu\text{m}$ as appropriate for a high-power laser we obtain $\bar{\beta}_m/d_{1/e^2} = 10^{-4}\,\text{W}^{-1}$. For $W = 100\,\mu\text{m}$ and $P = 10\,\text{W}$ this results in a modal absorption coefficient of $\alpha_{2,m} = 0.1\,\text{cm}^{-1}$. Note that the modal two-photon absorption is maximal at minimum d_{1/e^2}, i.e., for a strong optical confinement. The secondary two-photon loss mechanism due to intraband absorption caused by the free carriers generated by two-photon absorption is strongly dependent on the carrier lifetime and the effective drift-diffusion length. In most cases, the resulting absorption coefficient can be neglected, except in structures where the intensity peak of the vertical mode is not located at the position of the active layer [22].

Figure 27.9 shows the impact of nonlinear gain compression and two-photon absorption on the power–current characteristics using the model presented in Section 27.3.4 with the parameters given in Table 27.2. The following conclusions can be drawn: First, a gain compression factor $\epsilon_s = 10^{-17}\,\text{cm}^3$ (saturation power $P_s = 150\,\text{W}$) or a two-photon absorption coefficient $\bar{\beta}/A_m = 0.01\,(\text{Wcm})^{-1}$ act very similar. Second, at a current of $I = 15\,\text{A}$, which is typical for lasers operating in CW mode, the combination of both effects results in a reduction of the optical power by 10% (2 W; cf. Figure 27.9a). However, at a current of $I = 125\,\text{A}$, which can be reached under pulsed operation [19], the power is reduced from 100 to 60 W, which seems a little bit too large, considering that leakage currents are not accounted for at all. Hence, either the gain compression factor or the two-photon absorption coefficient or both are smaller in reality than assumed here.

27.3.5.3 Leakage Currents

Leakage currents caused by the transport of carriers into regions without stimulated recombination can be divided into vertical and lateral ones. Vertical leakage currents are minority currents caused by the accumulation and subsequent recombination of electrons and holes in the p-doped and n-doped, respectively, optical confinement layers. This effect results in an increased free-carrier absorption, too. An analytical investigation of the carrier accumulation in the confinement layers and its impact on the power–current characteristics can be found in Reference [23] and an exact treatment based on the numerical solution of the drift-diffusion equations has been given in Reference [11,24].

The electron current flowing into the p-doped region can be obtained by integrating the continuity equation for the electron current density (Equation 27.137) over the p-doped region between the upper boundary of the active region $y = y_p$ and the p-metallization for the steady state,

$$\int_{\partial(\text{p-region})} \boldsymbol{j}_n \cdot \boldsymbol{n} dS = e \int_{\text{p-region}} R\, dV \tag{27.69}$$

where $\partial(\text{p-region})$ denotes the surface of the p-doped region with $\boldsymbol{n}$ being the normal vector. Assuming that the normal components of the electron current density vanish at the outer boundaries including the p-contact, we obtain the electron leakage current

$$I_{n,\text{leak}} = \iint j_{n,y}|_{y=y_p}\, dxdz. \tag{27.70}$$

Hence, the electron leakage current can be calculated either from the electron current density at the boundary between active region and p-doped region or by integrating the recombination rate over the corresponding volume.

Figure 27.10a and b shows the vertical profiles of the current densities and the recombination rate of the asymmetric super-large optical-cavity structure published in [10] at a current of $I = 15$ A. Figure 27.10a reveals that for a properly designed structure the hole current flowing into the n-doped region can be neglected. The leakage current calculated from the electron current density at $y = 1.9\,\mu$m is $I_{n,leak} = 0.44$ A, which is also obtained from the integration of the recombination rate. The relative fraction of the leakage current is $(I - I_{n,leak})/I = 0.97$, which coincides with the internal efficiency $\eta_i = 0.97$, obtained from the length-dependent measurement of the external differential efficiency as described in 27.2.

With increasing bias the leakage current increases due to the bending of the band edges as shown in Reference [11], which contributes to the roll-over of the power–current characteristics. The electron leakage current can be reduced by an increase of the energy gap and doping level or a decrease of the thickness of the p-doped optical confinement layer as it is the case in extreme-asymmetric super large optical cavities. However, one should keep in mind that an increase of the energy gap results in a decrease of the electrical conductivity in the AlGaAs material system, an increase of the doping results in an increase of the free-carrier absorption and an increase of the waveguide asymmetry results in a decrease of the field intensity at the position of the active region.

The lateral leakage current $I_{p,spead}$ caused by current spreading in the p-doped region can be reduced by an insulation of the highly doped p-contact and p-cladding layers beyond the p-metallization by implantation, diffusion of impurities, or etching as indicated in Figure 27.1 or by implementing a reverse-biased p-n junction. The lateral leakage current due to the diffusion of the carriers along the active region is more difficult to mitigate.

An approximate expression for $I_{p,spead}$ has been derived in Reference [25],

$$\frac{I_{p,spead}}{L} \approx -2\frac{\partial \varphi_F}{\partial x}\bigg|_{x=W/2} \sum_i \sigma_{p,i} d_i \tag{27.71}$$

where φ_F is the Fermi voltage given in Equation 27.60, $\sigma_{p,i}$ the hole conductivity of layer i and d_i the corresponding thickness. The derivative has to be taken at the edge of the p-contact stripe, and the summation includes all p-doped continuous semiconductor layers above the active region.

27.3.5.4 Carrier Capture

As it is well known [26–30], the carriers belonging to bound states in the QW(s) of the active region are not necessarily in thermal equilibrium with the carriers resulting from the continuum states located

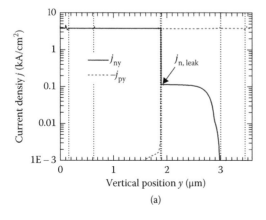

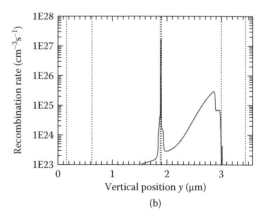

FIGURE 27.10 Vertical profiles of (a) current densities and (b) recombination rate for p-contact width $W = 100\,\mu$m, cavity length $L = 4000\,\mu$m, and output power $P = 15$ W. Heteroboundaries are indicated by dotted vertical lines. Note the logarithmic scales of the ordinates.

energetically above the QWs. It can be assumed that both types of carriers coexisting in the QW region are in quasi-equilibrium among themselves described by Fermi–Dirac statistics with independent quasi-Fermi potentials and transferred into each other by capture-escape processes. The transport of both types of carriers in the in-plane directions (x, z) can be described by classical drift diffusion. Most conveniently the bound carriers are described by sheet densities (unit $1/m^2$) and appear in the rhs of the Poisson equation (27.141) as interface states.

An expression for the capture–escape rate that fulfils universal conditions such as the balance between capture and escape in case of equal quasi-Fermi potentials and no capture but finite escape in case of a fully occupied bound states has been given in [30]. For electrons (similarly for holes) it reads

$$Q_n = \frac{dn_f}{\tau_c} \left[1 - \frac{n_b}{N_b} \right] \left[1 - e^{\frac{e(\varphi_f - \varphi_b)}{kT}} \right] \tag{27.72}$$

where n_f and n_b are the free and bound, electron densities, respectively; φ_f and φ_b are the corresponding quasi-Fermi potentials; d is the thickness of the QW(s); τ_c is the capture time; and N_b is the maximum bound electron density ($\varphi_b \to -\infty$). Whereas the quasi-Fermi potentials of the free carriers depend directly on the bias applied to the contacts, the quasi-chemical potentials of the bound carriers are determined by the capture-escape rates.

The impact of the nonequilibrium between confined and unconfined carriers in high-power lasers on the power–current characteristics as a function of the hole capture time has been investigated in Reference [28], using, however, a capture-escape rate that differs from the one given in Equation 27.72.

27.4 Model for the Optical Field

We derive the parabolic paraxial wave equation based on the slowly varying amplitude and rotating wave approximations, taking into account gain dispersion, spontaneous emission, and a third-order nonlinear susceptibility. A balance equation for the electromagnetic energy density will be given.

27.4.1 Basic Three-Dimensional Equations

We start with the homogeneous Maxwell equation for the electric field E and the magnetic field H:

$$\nabla \cdot D = 0 \tag{27.73}$$

$$\nabla \cdot H = 0 \tag{27.74}$$

$$\nabla \times E + \mu_0 \partial_t H = 0 \tag{27.75}$$

$$\nabla \times H - \partial_t D = 0 \tag{27.76}$$

where $D = \varepsilon_0 E + P$ denotes the electric displacement and P the macroscopic polarization density of the material. The fields E, H, D depend on three spatial variables $r = (x, y, z)$ and the time t. By applying $\nabla \times$ to Equation 27.75, ∂_t to Equation 27.76, and using Equation 27.73 we get the wave equation

$$\frac{1}{\varepsilon_0} \nabla(\nabla \cdot P) + \nabla^2 E = \frac{1}{c^2} \partial_{tt} E + \mu_0 \partial_{tt} P \tag{27.77}$$

The polarization density P contains a linear convolution of the electric field history $E(t - \tau)$, $\tau \geq 0$, with a susceptibility function $\chi(r, \tau)$ and a second-order part which models dispersion (P_{disp}), nonlinear effects

(P_{NL}), and spontaneous emission (P_{sp}):

$$P(r, t) = \varepsilon_0 \int_0^\infty \chi(r, \tau) E(r, t - \tau) \, d\tau + P_{disp}(r, t) + P_{NL}(r, t) + P_{sp}(r, t) \tag{27.78}$$

The spontaneous emission is modeled by introducing a spontaneous current density,

$$\partial_t P_{sp}(r, t) = J_{sp}(r, t) \tag{27.79}$$

We assume that P either is slowly varying in space, so that $\nabla \cdot P \approx 0$ holds, or has discontinuities at boundaries between different materials. At these boundaries E and H have to fulfill certain continuity rules that can be derived from Maxwell equations. Due to the nearly planar geometry of edge-emitting lasers defined by the epitaxial layer structure, the electromagnetic field is either mainly transverse electric (TE) or transverse magnetic (TM) polarized. By assuming TE polarization and choosing a reference frequency ω_0, we have

$$E(r, t) = e_x \frac{1}{2} E(r, t) e^{i\omega_0 t} + \text{c.c.}$$

$$P_{disp}(r, t) = e_x \frac{1}{2} P_{disp}(r, t) e^{i\omega_0 t} + \text{c.c.}$$

$$P_{NL}(r, t) = e_x \frac{1}{2} P_{NL}(r, t) e^{i\omega_0 t} + \text{c.c.} \tag{27.80}$$

$$J_{sp}(r, t) = e_x \frac{1}{2} j_{sp}(r, t) e^{i\omega_0 t} + \text{c.c.}$$

where e_x denotes the unit vector in x direction, c.c. stands for complex conjugation, and $E, P_{disp}, P_{NL}, j_{sp}$ are scalar complex-valued functions. In the slowly varying amplitude approximation the corresponding derivatives are given by

$$\partial_{tt} E \approx -e_x \frac{1}{2} \left(\omega_0^2 E - 2i\omega_0 \partial_t E \right) e^{i\omega_0 t} + \text{c.c.}$$

$$\partial_{tt} P_{disp} \approx -e_x \frac{1}{2} \omega_0^2 P_{disp} e^{i\omega_0 t} + \text{c.c.}$$

$$\partial_{tt} P_{NL} \approx -e_x \frac{1}{2} \omega_0^2 P_{NL} e^{i\omega_0 t} + \text{c.c.} \tag{27.81}$$

$$\partial_{tt} P_{sp} \approx e_x \frac{1}{2} i\omega_0 j_{sp} e^{i\omega_0 t} + \text{c.c.}$$

Let $\chi(r, \omega)$ denote the Fourier transform of $\chi(r, t)$ evaluated at frequency ω. First we insert Equation 27.80 into Equation 27.78 and replace $E(r, t - \tau)$ by its first-order approximation $E(r, t) - \partial_t E(r, t)\tau$. Then we insert the result and Equation 27.80 into Equation 27.77 taking into account Equation 27.81. Finally, we multiply by $e^{-i\omega_0 t}$ and neglect the rapidly varying terms containing $e^{-i2\omega_0 t}$ (rotating wave approximation) and get

$$\Delta E = -\frac{\omega_0^2}{c^2} E + \frac{2i\omega_0}{c^2} \partial_t E - \frac{\omega_0^2}{c^2} \chi(r, \omega_0) E + \frac{i\omega_0}{c^2} \left(2\chi(r, \omega_0) + \omega_0 \partial_\omega \chi(r, \omega)_{|\omega = \omega_0} \right) \partial_t E$$

$$- \mu_0 \omega_0^2 \left(P_{disp} + P_{NL} \right) + i\mu_0 \omega_0 j_{sp} \tag{27.82}$$

Consistent with the neglection of $\partial_{tt}E(r, t)$ we approximate in front of $\partial_t E(r, t)$:

$$2\chi(r, \omega_0) + \omega_0 \partial_\omega \chi(r, \omega)|_{\omega = \omega_0} \approx 2\bar{\chi}(\omega_0) + \omega_0 \partial_\omega \bar{\chi}(\omega)|_{\omega = \omega_0} \qquad (27.83)$$

where $\bar{\chi}$ is an averaged real-valued susceptibility of the medium. This means that the spatial variation and the imaginary part of χ is considered to be a first-order correction, which can be neglected in front of the time derivative.

Introducing the complex refractive index $n^2(r, \omega) = 1 + \chi(r, \omega)$, the real-valued reference index $\bar{n}^2 = 1 + \bar{\chi}$, the real-valued group index $n_g = \bar{n} + \omega_0 \partial_\omega \bar{n}(\omega)|_{\omega = \omega_0}$ and $k_0 = \omega_0/c$, we get

$$\Delta E = 2ik_0 \frac{\bar{n}n_g}{c} \partial_t E - k_0^2 n^2 E - \mu_0 \omega_0^2 \left(P_{\text{disp}} + P_{\text{NL}} \right) + i\mu_0 \omega_0 j_{\text{sp}} \qquad (27.84)$$

Next we remove the rapid oscillations along the longitudinal z direction with the Ansatz

$$E(r, t) = E^+(r, t)e^{-i\bar{n}k_0 z} + E^-(r, t)e^{i\bar{n}k_0 z} \qquad (27.85)$$

Inserting Equation 27.85 into Equation 27.84, neglecting $\partial_{zz}E^\pm$, multiplying the result with $e^{\pm i\bar{n}k_0 z}$, dropping rapidly varying terms containing $e^{\pm 2i\bar{n}k_0 z}$, and dividing by $2i\bar{n}k_0$, we get

$$\frac{1}{v_g}\partial_t E^\pm(r, t) \pm \partial_z E^\pm(r, t) = \frac{1}{2i\bar{n}k_0}\left[(\partial_x)^2 + (\partial_y)^2\right]E^\pm(r, t) - ik_0\frac{n^2(r, \omega_0) - \bar{n}^2}{2\bar{n}}E^\pm(r, t)$$

$$+ \frac{\mu_0 \omega_0^2}{2i\bar{n}k_0}\left[P_{\text{disp}}(r, t) + P_{\text{NL}}(r, t)\right]e^{\pm i\bar{n}k_0 z} - \frac{\mu_0 \omega_0}{2\bar{n}k_0}j_{\text{sp}}(r, t)e^{\pm i\bar{n}k_0 z} \qquad (27.86)$$

where $v_g = c/n_g$.

Equation 27.86 must be supplemented by appropriate boundary conditions. At the plane facets of the laser located at $z = 0$ and $z = L$

$$E^+(x, y, 0, t) - r_0 E^-(x, y, 0, t) = 0$$
$$E^-(x, y, L, t) - r_L e^{-i2\bar{n}k_0 L}E^+(x, y, L, t) = 0 \qquad (27.87)$$

hold. We should mention that in the paraxial approximation, the amplitude reflection coefficients r_0 and r_L are input parameters, which have to be calculated in advance.

At the transverse boundary denoted by Γ one can assume, for example, decaying fields or a perfect electric wall,

$$\lim_{|r_t| \to \infty} E^\pm = 0 \quad \text{or} \quad E^\pm|_{r_t \in \Gamma} = 0, \qquad (27.88)$$

respectively. If only a part of the cross section of the cavity is simulated, a nonreflecting or transparent boundary condition has to be used, which models the fact that only outgoing waves should be present. A very popular method to implement a boundary condition of this type is the introduction of a so-called perfectly matched layer (PML) [31]. However, in the frequency domain, it results in a large number of spurious modes [32]. Within the slowly varying approximation discontinuities of n can only be treated by employing corresponding transition conditions, except for resonant Bragg waveguide gratings, which can be described by extra terms coupling the forward and backward propagating waves.

27.4.2 Ansatz for Dispersion, Nonlinear Susceptibility, and Spontaneous Emission

27.4.2.1 Dispersion

We use the following Ansatz for $P_{\mathrm{disp}}(\boldsymbol{r}, t)$:

$$P_{\mathrm{disp}}(\boldsymbol{r}, t) = -i\frac{n_{\mathrm{r}}(\boldsymbol{r}, t)g_{\mathrm{r}}(\boldsymbol{r}, t)k_0}{\omega_0^2 \mu_0}[\mathcal{P}(\boldsymbol{r}, t) - E(\boldsymbol{r}, t)] \tag{27.89}$$

where the electric field E is subtracted to ensure that the contribution to the dispersion vanishes at the gain peak. We model dispersion via the following response function in the frequency domain [33–35]

$$\mathcal{P}(\boldsymbol{r}, \omega) = \mathfrak{L}(\boldsymbol{r}, \omega)E(\boldsymbol{r}, \omega) \tag{27.90}$$

with

$$\mathfrak{L}(\boldsymbol{r}, \omega) = \frac{\gamma(\boldsymbol{r})}{i[\omega - \omega_0 - (\omega_{\mathrm{p}}(\boldsymbol{r}) - \omega_0)] + \gamma(\boldsymbol{r})} \tag{27.91}$$

The Lorentzian $\mathfrak{L}(\boldsymbol{r}, \omega)$ achieves its maximum value 1 at the frequency ω_{p}. The imaginary part Im $\mathfrak{L}(\boldsymbol{r}, \omega)$ has a half width γ at half of the maximum. The approximation is valid only within a small frequency region around ω_0 corresponding to the frequency range of optical transitions in the active material. By multiplying Equation 27.90 with the denominator and taking the inverse Fourier transform with respect to $\omega - \omega_0$ we get the ordinary differential equation

$$\partial_t \mathcal{P}(\boldsymbol{r}, t) = \left[i(\omega_{\mathrm{p}}(\boldsymbol{r}) - \omega_0) - \gamma(\boldsymbol{r})\right]\mathcal{P}(\boldsymbol{r}, t) + \gamma(\boldsymbol{r})E(\boldsymbol{r}, t) \tag{27.92}$$

We use similar decompositions for the polarization $\mathcal{P}(\boldsymbol{r}, t)$ as for the electric field $E(\boldsymbol{r}, t)$,

$$\mathcal{P}(\boldsymbol{r}, t) = \mathcal{P}^+(\boldsymbol{r}, t)\mathrm{e}^{-i\bar{n}k_0 z} + \mathcal{P}^-(\boldsymbol{r}, t)\mathrm{e}^{i\bar{n}k_0 z} \tag{27.93}$$

Inserting Equations 27.93 and 27.85 into Equation 27.92, multiplying with $\mathrm{e}^{\pm i\bar{n}k_0 z}$, and again neglecting $\mathrm{e}^{\pm 2i\bar{n}k_0 z}$ we get

$$\partial_t \mathcal{P}^\pm(\boldsymbol{r}, t) = i(\omega_{\mathrm{p}}(\boldsymbol{r}) - \omega_0)\mathcal{P}^\pm(\boldsymbol{r}, t) + \gamma(\boldsymbol{r})\left[E^\pm(\boldsymbol{r}, t) - \mathcal{P}^\pm(\boldsymbol{r}, t)\right] \tag{27.94}$$

Dispersion can also be taken into account with higher order time derivatives [36], on a microscopic level [37–39], by means of a digital filter [40], or by a convolution integral [41].

27.4.2.2 Nonlinear Susceptibility

Considering a third-order nonlinearity and assuming an isotropic medium the nonlinear polarization density reads [42–44]

$$P_{\mathrm{NL}}(\boldsymbol{r}, t) = \varepsilon_0 \frac{3}{4}\chi_{xxxx}^{(3)}(\boldsymbol{r}, t)|E(\boldsymbol{r}, t)|^2 E(\boldsymbol{r}, t) \tag{27.95}$$

with the third-order susceptibility $\chi_{xxxx}^{(3)}(\omega_0; -\omega_0, -\omega_0, \omega_0)$. The optical Kerr coefficient $\tilde{n}_2$ and the two-photon absorption coefficient $\tilde{\beta}$ are defined as

$$\tilde{n}_2 - i\frac{\tilde{\beta}}{2k_0} = \frac{3}{8n_r}\chi_{xxxx}^{(3)} \equiv \Delta\tilde{n}. \tag{27.96}$$

The connection with the commonly used coefficients n_2 and β related to the time-averaged intensity $I = \varepsilon_0 \bar{n}c|E|^2/2$ and discussed in Section 27.5.2 is given by

$$n_2 - i\frac{\beta}{2k_0} = \frac{2\Delta\tilde{n}_2}{\varepsilon_0\bar{n}c} \equiv \Delta n_2 \tag{27.97}$$

where n_2 has the unit m^2/W and β the unit m/W.

27.4.2.3 Spontaneous Emission

The stochastic forces

$$F_{sp}^{\pm}(r, t) = -\frac{\mu_0\omega_0}{2\bar{n}k_0}j_{sp}(r, t)e^{\pm i\bar{n}k_0 z} \tag{27.98}$$

in Equation 27.86 have the properties

$$\langle F_{sp}^+(r, t)\rangle = \langle F_{sp}^-(r, t)\rangle = 0 \tag{27.99}$$

$$\langle F_{sp}^+(r, t)F_{sp}^{+*}(r, t')\rangle = \langle F_{sp}^-(r, t)F_{sp}^{-*}(r', t')\rangle = \frac{2\hbar\omega_0 n_r(r, t)g(r, t)n_{sp}(r, t)}{\bar{n}^2\varepsilon_0 c}\delta(r - r')\delta(t - t') \tag{27.100}$$

where $\langle\rangle$ denotes ensemble average and $\delta(x)$ is the Dirac delta function. Equation 27.100 is not suited for a numerical evaluation, because in dependence on the carrier densities the inversion (or spontaneous emission) factor n_{sp} has a singularity when the gain g changes its sign (transparency) so that the product $g \cdot n_{sp}$ is undefined. One possibility to circumvent this is to take the second moment (Equation 27.100) proportional to $\beta_{sp}R_{sp}$ where R_{sp} given in Equation 27.139 is the rate of spontaneous emission into all modes and the dimensionless factor β_{sp} is chosen such that at threshold the correct values of the second moments are obtained [8].

27.4.3 Final Field Equation and Balance of Radiative Energy

Summarizing the results of the previous subchapter, the paraxial wave equation can be written as

$$\frac{1}{v_g}\partial_t E^{\pm}(r, t) \pm \partial_z E^{\pm}(r, t) = -\frac{i}{2\bar{n}k_0}\left(\partial_{xx} + \partial_{yy}\right)E^{\pm}(r, t)$$

$$- \frac{ik_0}{2\bar{n}}\Delta n^2(r, t)E^{\pm}(r, t) - \frac{n_r(r, t)g_r(r, t)}{2\bar{n}}\left[E^{\pm}(r, t) - \mathcal{P}^{\pm}(r, t)\right]$$

$$- \frac{ik_0 n_r(r, t)}{\bar{n}}\Delta\tilde{n}_2(r, t)\left[|E^+|^2 + |E^-|^2\right]E^{\pm}(r, t) + F_{sp}^{\pm}(r, t) \tag{27.101}$$

where

$$\Delta n^2(\boldsymbol{r}, t) \equiv n^2(\boldsymbol{r}, t, \omega_0) - \bar{n}^2 = \Delta n_r^2 + i \frac{n_r (g - \alpha)}{k_0} \tag{27.102}$$

Here Δn_r^2 is the real part of Δn^2, n_r the real part of n, g the coefficient of optical gain originating from the emission or absorption of photons due to transitions between the conduction and valence bands, and α the coefficient of absorption of photons due to transitions within the conduction and valence bands (such as free-carrier absorption and intervalence band absorption). Optical losses due to scattering on waveguide imperfections could be included in α, too.

Multiplying Equation 27.101 with the complex conjugate $E^{\pm *}$, multiplying the complex conjugate of Equation 27.101 with $E^{\pm}$ and adding all equations results in

$$\frac{1}{v_g} \frac{d\|E\|^2}{dt} = \frac{1}{\bar{n} k_0} \mathrm{Im} \left[\partial_x (E^{+*} \partial_x E^+ + E^{-*} \partial_x E^-) + \partial_y (E^{+*} \partial_y E^+ + E^{-*} \partial_y E^-) \right]$$

$$+ \partial_z \left[|E^-|^2 - |E^+|^2 \right]$$

$$+ \frac{n_r(g - \alpha)}{\bar{n}} \|E\|^2 + \frac{n_r g_r}{\bar{n}} \left[\mathrm{Re}(E^{+*} \mathcal{P}^+ + E^{-*} \mathcal{P}^-) - \|E\|^2 \right] \tag{27.103}$$

$$- \frac{n_r \tilde{\beta}}{\bar{n}} \|E\|^4 + 2\mathrm{Re}(E^{+*} F_{sp}^+ + E^{-*} F_{sp}^-)$$

with $\|E\|^2 = |E^+|^2 + |E^-|^2$. This equation can be interpreted as a balance equation for the radiative energy density

$$u_{rad}(\boldsymbol{r}, t) = \frac{\varepsilon_0 \bar{n} n_g}{2} \|E(\boldsymbol{r}, t)\|^2 \tag{27.104}$$

Integrating Equation 27.103 over the device volume V and using the boundary conditions Equations 27.88 and 27.87 leads to

$$\frac{d \int_V u_{rad} \, dV}{dt} = -P_{out} + \frac{\varepsilon_0 c}{2} \int_V n_r (g - g_r - \alpha) \|E\|^2 \, dV$$

$$+ \frac{\varepsilon_0 c}{2} \int_V n_r g_r \left[\mathrm{Re}(E^{+*} \mathcal{P}^+ + E^{-*} \mathcal{P}^-) \right] dV - \frac{\varepsilon_0 c}{2} \int_V n_r \tilde{\beta} \|E\|^4 \, dV$$

$$+ \frac{\varepsilon_0 c}{\bar{n}} \int_V \mathrm{Re}(E^{+*} F_{sp}^+ + E^{-*} F_{sp}^-) \, dV \tag{27.105}$$

which is a generalization of Equation 27.31. The output power $P_{out} = P_0 + P_L$ is given in Equation 27.30 where $R_0 = |r_0|^2$ and $R_L = |r_L|^2$. The optical power is

$$P^{\pm}(z) = \frac{\varepsilon_0 \bar{n} c}{2} \iint |E^{\pm}(x, y, z)|^2 \, dxdy. \tag{27.106}$$

Thus the first term in Equation 27.103 is the divergence of the transverse energy flux density, which is assumed to vanish on the transverse surface of the device. The second term gives the radiation leaving the cavity. The third and fourth terms describe the increase or decrease of the energy due to stimulated emission or absorption, respectively, and the last two terms due to two-photon absorption and spontaneous emission.

27.4.4 Cavity Modes, Beam Propagation Method, and Roundtrip Operator

The cavity modes are time-periodic solutions of the form $E_m^\pm \exp(i\Omega_m t)$, $P^\pm \exp(i\Omega_m t)$ and obey the linear equations

$$\pm i\partial_z E_m^\pm(r) = \left[\frac{\Omega_m}{v_g} + H(r, z) \right] E_m^\pm(r) \tag{27.107}$$

subject to the boundary conditions (Equations 27.87 and 27.88). The operator H is given by

$$H(r, z) = \frac{1}{2\bar{n}k_0} \left(\partial_{xx} + \partial_{yy} \right) + \frac{k_0}{2\bar{n}} \Delta n^2(r) + i \frac{n_r(r)g_r(r)}{2\bar{n}} \left[\frac{\gamma(r)}{i\Omega_m - i(\omega_p(r) - \omega_0) + \gamma(r)} - 1 \right] \tag{27.108}$$

The nontrivial solutions of (Equation 27.107) may depend on time via the dependence of the complex-valued refractive index on temporally varying carrier densities and temperature. The complex-valued relative mode frequencies Ω_m are the eigenvalues and the mode profiles $E_m^\pm$ are the eigenfunctions of Equation 27.107. The real parts of Ω_m give the wavelengths relative to the reference wavelength λ_0,

$$\Delta\lambda_m = \frac{d\lambda}{d\omega}\bigg|_{\lambda_0} \text{Re}(\Omega_m) \tag{27.109}$$

and the imaginary parts describe the damping of the modes. For a passive cavity, $\text{Im}(\Omega_m) > 0$ must hold. Lasing modes of an active cavity are distinguished by vanishing damping, $\text{Im}(\Omega_m) = 0$, due to the balance of the outcoupling and internal losses and the gain.

It can be shown that the cavity modes fulfil an orthogonality relation which does not define a scalar product because of the non-Hermitian character of Equation 27.107. Due to the dispersion term in Equation 27.107 the orthogonality relation differs from that given in [6], but can be derived in the same manner.

If the operator H depends only on the transverse coordinates r_t, the solution of Equation 27.107 can be formally written as

$$E_m^\pm(r_t, z') = e^{\mp i\left(\frac{\Omega_m}{v_g} + H\right)(z' - z)} E_m^\pm(r_t, z) \tag{27.110}$$

The numerical evaluation of Equation 27.110 is the basis of what is known as the beam propagation method (BPM) [45–47]. For the case of a spatially and temporally constant index, $n = \text{const.}$, Equation 27.110 can be evaluated exactly to yield

$$E_m^\pm(r_t', z') = e^{\mp i\left(\frac{\Omega_m}{v_g} + \frac{k_0}{2\bar{n}}\Delta n^2\right)(z' - z)} \int G^\pm(r_t' - r_t, z' - z) E_m^\pm(r_t, z)\, dxdy \tag{27.111}$$

with

$$G^\pm(r_t' - r_t, z' - z) = \Theta\left(\pm (z' - z) \right) \left[\sqrt{\pm \frac{i\bar{n}k_0}{2\pi(z' - z)}} \right]^2 e^{\mp \frac{i\bar{n}k_0 |r_t' - r_t|^2}{2(z' - z)}} \tag{27.112}$$

where Θ denotes the Heaviside step function. The integral equation (Equation 27.111) together with the propagator Equation 27.112 is known as Huygen's integral in the Fresnel approximation.

Based on Equation 27.110 and the boundary conditions (Equation 27.87), it is possible to construct roundtrip operators $\mathbf{M}^{\pm}$. One starts at some position $z = z_0$ within the cavity and performs a full roundtrip. Depending on whether we start into forward $(+)$ or backward $(-)$ directions, the eigenvalue problems

$$\mathbf{M}^{\pm}(\Omega_m)E_m^{\pm}(r_t, z_0) = e^{i\frac{2\Omega_m}{v_g}L}E_m^{\pm}(r_t, z_0) \tag{27.113}$$

are obtained. Note that the roundtrip operators $\mathbf{M}^{\pm}$ depend on the eigenvalues Ω_m if dispersion is taken into account. The eigenfunctions of $\mathbf{M}^{\pm}$ are the mode distributions $E_m^{\pm}(r_t, z_0)$ at the position z_0. A very popular method for solving Equation 27.113 is based on the Fox–Li approach, cf. [48] and the references therein. The idea is to choose a normalized, more or less arbitrary start distribution $E_m^{\pm}(r_t, z_0)$ and to apply the roundtrip operator $\mathbf{M}^{\pm}$ recurrently until one arrives (hopefully) at a steady state. It is known that the algorithm fails if there are cavity modes having identical or nearly identical damping $\text{Im}(\Omega_m)$, as it is the case in broad-area lasers. The association of the nonconvergence of the Fox–Li iteration with an dynamically unstable laser behavior should be done with care. Instead, for multimode high-power lasers, a time-dependent approach based on Equation 27.101 should be preferred, although it is numerically more challenging.

27.5 Models for Nonlinear Modes and Filamentation

The multipeaked and not diffraction-limited lateral field profile of wide-aperture semiconductor lasers has been a long-standing problem and has been investigated in the past by numerous authors [49–55]. Although the broadening of the far field of CW operating lasers with increasing power (also called far-field blooming) can be at least partially attributed to the thermal lensing effect [56–58], a complete picture of the origin and mechanism has not revealed yet.

One mechanism is believed to be due to carrier-induced antiguiding, i.e., the reduction of the refractive index with increasing carrier density. This leads to a self-focusing mechanism because the index increases in regions of high intensity due to a depletion of the injected carrier density and can result in the formation of what is sometimes called lasing filaments.

Another mechanism that could explain the multipeaked structure and the broadening of the farfield is the simultaneous lasing of a large number of waveguide modes, originating from a built-in or thermally induced waveguide. Indeed, recent experiments reveal, that even at currents several times above threshold the lateral modes can be clearly identified by spectrally resolved near- and far-field measurements [58,59]. However, with increasing current, a broadening of the individual modes and the appearance of new modes with broad near and far fields can be observed [60].

Here, we derive the equation for the description of the lateral field profile and discuss two basic nonlinearities due to the virtual transitions and due to lateral spatial hole burning.

27.5.1 Longitudinal–Lateral Approximate Projected Equations

First we project the basic equations onto the dominant vertical mode using the Ansatz

$$E^{\pm}(r, t) = \sqrt{\frac{2d\hbar\omega_0}{\varepsilon_0 \bar{n}n_g}}\phi(y, x, z)u^{\pm}(x, z, t) \tag{27.114}$$

where the mode profile $\phi(y, x, z)$ is normalized according to $\int |\phi|^2 \, dy = 1$. The scaling in Equation 27.114 is chosen such that $|u^{\pm}|^2$ is a photon density (unit m^{-3}). The power is given by

$$P^{\pm} = \hbar \omega_0 v_g d \int |u^{\pm}|^2 \, dx \tag{27.115}$$

where d is the thickness of the active region.

For each (x, z), $\phi(y, x, z)$ is a solution of a vertical waveguide equation with a real-valued index profile n_0 not dependent on carrier densities and temperature,

$$\partial_{yy} \phi(x, y, z) + k_0^2 n_0^2(x, y, z) \phi(x, z, y) = k_0^2 n_{\text{eff},0}^2(x, z) \phi(x, y, z) \tag{27.116}$$

where the real-valued effective index $n_{\text{eff},0}$ is the eigenvalue of Equation 27.116. Inserting Equation 27.114 into Equation 27.101, multiplying with ϕ, and integrating along y yields the equations

$$\frac{1}{v_g} \partial_t u^{\pm}(x, z, t) = -\frac{i}{2 \bar{n} k_0} \partial_{xx} u^{\pm}(x, z, t) \mp \partial_z u^{\pm}(x, z, t) - \frac{i k_0}{2 \bar{n}} \left[n_{\text{eff}}^2(x, z, t) - \bar{n}^2 \right] u^{\pm}(x, z, t)$$

$$- i k_0 \Delta n_{2,\text{eff}}(x, z, t) \hbar \omega_0 v_g d \| u(x, z, t) \|^2 u^{\pm}(x, z, t) \tag{27.117}$$

where we omitted the dispersion and spontaneous emission terms for simplicity and

$$n_{\text{eff}}^2 = n_{\text{eff,r}}^2 + i \frac{\bar{n} \left(g_{\text{eff}} - \alpha_{\text{eff}} \right)}{k_0}, \tag{27.118}$$

$$n_{\text{eff,r}}^2 = n_{\text{eff},0}^2 + \int (n_r^2 - n_0^2) |\phi|^2 \, dy, \tag{27.119}$$

$$g_{\text{eff}} - \alpha_{\text{eff}} = \frac{\int n_r (g - \alpha) |\phi|^2 \, dy}{\bar{n}} \tag{27.120}$$

and

$$\Delta n_{2,\text{eff}} = \frac{\int n_r \Delta n_2 |\phi|^4 \, dy}{\bar{n}} \approx \frac{\Delta \bar{n}_2}{d_{1/e^2}} \tag{27.121}$$

The second term in Equation 27.119 treats effects, which are not included in n_0 and thus $n_{\text{eff},0}$, for example, the dependence of the real and imaginary parts of n_r on the carrier densities or the temperature. The vertical mode size d_{1/e^2} has been introduced in Equation 27.68. The approach sketched is called the "effective index method" in the semiconductor laser community. Inserting Equation 27.114 into the rate of stimulated recombination Equation 27.140 and averaging over the active region $1/d \int_0^d \, dy$ yields

$$R_{\text{st}} = v_g g_{\text{eff}} \| u \|^2 \tag{27.122}$$

with $\| u \|^2 = |u^+|^2 + |u^-|^2$.

Equation 27.117 together with correspondingly projected equations for the polarization (Equation 27.94) and a lateral diffusion equation for the excess carriers have been successfully applied to the simulation of a large variety of high-power laser structures [61–66]. A discussion on numerical issues can be found in [67].

27.5.2 Nonlinearities due to Virtual Transitions

The energy gaps of the confinement and cladding layers of a laser structure are larger than the energy of the photons generated by stimulated emission. In this situation, the photon energy is too small to allow absorption of single-photons due to transitions of electrons between the valence and conduction bands, as intended. However, virtual transitions involving two photons are still possible, which result in a third-order susceptibility defined in Equation 27.96. An approximate expression for the dispersion of the two-photon absorption coefficient β for direct transitions has been given in [68]:

$$\beta(\omega) = K \frac{\sqrt{E_p}}{n_r^2 E_g^3} F_2\left(\frac{\hbar\omega}{E_g}\right) \tag{27.123}$$

where E_p is the energy equivalent of the momentum matrix element for direct transitions between the valence and conduction bands, E_g is the (direct) energy gap and

$$F_2(x) = \frac{(2x-1)^{1.5}}{(2x)^5} \quad \text{for} \quad 2x > 1 \tag{27.124}$$

The factor K can be considered to be a free parameter. The two-photon absorption coefficient $\beta = 260$ m/TW experimentally determined for GaAs at a wavelength of $\lambda = 1064$ nm [44] is obtained with $K = 41,200$ m/TW $\times$ eV$^3/\sqrt{\text{eV}}$, using $E_g = 1.42$ eV, $E_p = 26.1$ eV, and $n_r = 3.48$. Note the E_g^{-3} dependence of β.

Two-photon absorption as well as Raman and Stark effects result also in an intensity-dependent contribution to the real part of the refractive index expressed as the optical Kerr coefficient n_2,

$$n_2(\omega) = \tilde{K} \frac{\hbar c \sqrt{E_p}}{2 n_r^2 E_g^4} G_2\left(\frac{\hbar\omega}{E_g}\right) \tag{27.125}$$

where the function $G_2(\hbar\omega/E_g)$ given in [68] is shown in Figure 27.11 together with the function $F_2(\hbar\omega/E_g)$. It can be seen, that G_2 is maximal around $\hbar\omega = 0.5 E_g$ and changes its sign at $\hbar\omega = 0.7 E_g$, where F_2 is

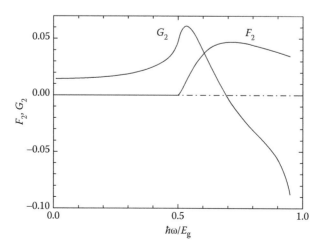

FIGURE 27.11   Functions F_2 and G_2 used in Equations 27.123 and 27.125.

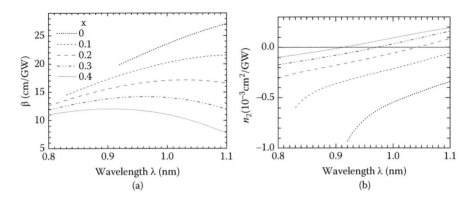

FIGURE 27.12 Nonlinear coefficients calculated by Equations 27.123 and 27.125 versus wavelength for different Al compositions x in $Al_xGa_{1-x}As$. (a) Two-photon absorption coefficient and (b) Optical Kerr coefficient.

maximal[†]. The factor $\tilde{K}$ can again be considered as a free parameter. The optical Kerr coefficient $n_2 = -4.1 \cdot 10^{-17}$ m/W experimentally determined for GaAs at a wavelength of $\lambda = 1064$ nm [44] is obtained with $\tilde{K} = 7.212 \cdot 10^{11}$ m/W $\times 1/(Ws\ m) \times eV^4/\sqrt{eV}$.

The dependencies of β and n_2 of $Al_xGa_{1-x}As$ on λ are shown in Figure 27.12a and 27.12b, respectively. The two-photon absorption increases from $\beta \propto 10$ cm/GW to $\beta \propto 25$ cm/GW if the Al composition is decreased from $x = 0.4$ to $x = 0$ due to the decrease of E_g. It shows a nonmonotonous dependence on the wavelength given by $F_2(\lambda)$. The optical Kerr coefficient decreases with increasing x and changes its sign in dependence on λ at higher Al compositions within the wavelength range investigated. For smaller Al compositions and wavelengths n_2 is negative, but for larger Al compositions and wavelengths n_2 is positive. For $\lambda = 980$ nm and relevant compositions $|n_2| < 2 \cdot 10^{-4}$ cm²/GW. Under CW operation, the maximum power density of broad-area lasers, which are state of the art during the writing of the book is of the order 10^{-2} GW/cm². Hence the resulting index change $|\Delta n| < 4 \cdot 10^{-6}$ should not have a big impact on the optical field.

It should be noted that in nonisotropic media the relation between the nonlinear polarization and the electric field is more complicated than given in Equation 27.95. Hence the two-photon absorption and the optical Kerr effect depend on the crystallographic orientation and on the polarization direction of the optical field.

27.5.3 Nonlinearity Induced by Lateral Spatial Hole Burning

We now derive equations for the right and left traveling fields alone, eliminating the carrier density. This is only possible for a (hypothetical) steady state. To this end, the real and imaginary parts of Δn_{eff}^2 are linearized around a reference carrier density N_{ref}, which is typically set to the transparency carrier density N_{tr} were $g_{eff}(N_{tr}) = 0$. Other possible choices could be the carrier density for waveguide transparency (solution of $g_{eff}(N_{ref}) = \alpha_{eff}$) or the threshold carrier density of a laterally infinite laser (solution of $g_{eff}(N_{thr}) = \alpha_{eff} + \alpha_{out}$).

For the steady state, linearizing the recombination rate as in Section 27.3.2, neglecting drift-diffusion, and assuming a constant injection current density the rate equation can be solved for the excess

---

[†] The functions F_2 and G_2 due to indirect transitions behave differently [69]: The maximum of F_2 occurs slightly above the indirect bandgap, and G_2 is positive throughout the transparent wavelength range and crosses zero only above the indirect gap. The two-photon absorption coefficient is dominated by direct transitions for $\hbar\omega > E_{g,direct}/2$ if $E_{g,indirect} > E_{g,direct}/2$.

carrier density,

$$N - N_{tr} = \frac{\frac{j\tau_N}{ed} - N_{tr}}{1 + v_g g'_{eff}\tau_N\|u_s\|^2}. \tag{27.126}$$

From Equation 27.126 it can be seen that $N \to N_{tr}$ for $\|u_s\|^2 \to \infty$, i.e., the carrier density cannot be depleted below the transparency carrier density. Inserting Equation 27.126 into $\Delta n^2_{eff,r}(N)$, $g_{eff}(N)$, and $\alpha_{eff}(N)$ yields

$$g_{eff}(j) = \frac{g'_{eff}\tau_N(j - j_{tr})}{ed(1 + \epsilon_{sat}\|u_s\|^2)}, \tag{27.127}$$

$$\alpha_{eff}(j) = \alpha_{eff}(N_{tr}) + \frac{\alpha'_{eff}\tau_N(j - j_{tr})}{ed(1 + \epsilon_{sat}\|u_s\|^2)}, \tag{27.128}$$

$$n^2_{eff,r}(j) = n^2_{eff,r}(N_{tr}) + \frac{\alpha_H \bar{n} g'_{eff}\tau_N(j - j_{tr})}{edk_0(1 + \epsilon_{sat}\|u_s\|^2)} \tag{27.129}$$

with the saturation parameter

$$\epsilon_{sat} = v_g g'_{eff}\tau_N \tag{27.130}$$

and Henry's α-factor

$$\alpha_H = 2k_0 n'_{eff}/g'_{eff}. \tag{27.131}$$

For typical values $n_g = 4$, $g'_{eff} = 10 \cdot 10^{-18}$ cm^2, and $\tau_N = 1$ ns we obtain $\epsilon_{sat} = 7.5 \cdot 10^{-17}$ cm^3. In general, $g'_{eff} > 0$ and $n'_{eff} < 0$ and hence $\alpha_H < 0$ for frequencies around the gain peak. Typical values of α_H range from -1 to -10.

The stationary states correspond to time-harmonic solutions

$$u^\pm(x, z, t) = u_s^\pm(x, z)e^{i\Omega t} \tag{27.132}$$

of Equation 27.117 with $\text{Im}(\Omega) = 0$. Inserting Equations 27.132, 27.127, and 27.129 into Equation 27.117, we obtain

$$\pm i\partial_z u_s^\pm = \frac{1}{2\bar{n}k_0}\partial_{xx}u_s^\pm(x, z, t) + \frac{\Omega}{v_g}u_s^\pm + \frac{k_0}{2\bar{n}}\left[n^2_{eff,r}(N_{tr}) - \bar{n}^2\right]u_s^\pm - \frac{i}{2}\alpha_{eff}(N_{tr})u_s^\pm$$

$$+ \frac{\tau_N\left[(i + \alpha_H)g'_{eff} - i\alpha'_{eff}\right](j - j_{tr})}{2ed(1 + \epsilon_{sat}\|u_s\|^2)}u_s^\pm + k_0\Delta n_{2,eff}\hbar\omega_0 v_g d\|u_s\|^2 u_s^\pm \tag{27.133}$$

For $\epsilon_{sat}\|u\|^2 \ll 1$ we can expand

$$(1 + \epsilon_{sat}\|u_s\|^2)^{-1} \approx 1 - \epsilon_{sat}\|u_s\|^2 \tag{27.134}$$

so that the second-last term in Equation 27.133 resembles the last term with

$$\Delta n_{2,SHB} = -\frac{\tau_N\left[(\alpha_H + i)g'_{eff} - i\alpha'_{eff}\right](j - j_{tr})\epsilon_{sat}}{2ek_0\hbar\omega_0 v_g d^2} \tag{27.135}$$

For typical values $\alpha_H = -2$, $g'_{eff} = 10 \cdot 10^{-18}$ cm^2, $j = 10$ kA/cm^2, $d = 10$ nm, $\tau_N = 1$ ns, $j_{tr} = 0.16$ kA/cm^2 (corresponding to $N_{tr} = 10^{18}$ cm^{-3}), $\lambda_0 = 1$ μm, $n_g = 4$, $\varepsilon_{sat} = 75 \cdot 10^{-18}$ cm^3 we obtain Real($\Delta n_{2,SHB}$) $= 5 \cdot 10^{-4}$ cm/W, which is several orders of magnitudes larger than the absolute value of the effective optical Kerr coefficient due to non-resonant virtual transitions Real($\Delta n_{2,eff}$) $\approx -7 \cdot 10^{-10}$ cm/W for $\bar{n}_2 = -10^{-13}$ cm^2/W (cf. Figure 27.12b) and $d_{1/e^2} = 1.5$ μm (cf. Equation 27.121).

Due to the fact that Real($\Delta n_{2,SHB}$) is positive, two basic instabilities sometimes termed "filamentation" arise [44]. First, a local intensity maximum induces a variation of the real part of the effective index with a larger value at the position of the intensity peak than outside. Thus a local index waveguide is created (self-focusing) [49]. However, at the position of the intensity peak the effective gain is decreased which results in an reduced amplification in the local-waveguide core and an enhanced amplification outside (self-defocusing).

The other instability, first described by Bespalov and Talanov [70], is the breakup of a laser beam with a homogeneous intensity distribution into a beam with a random intensity distribution as a consequence of the growth of irregularities initially present on the laser wavefront. For a mathematical description of the effect, the forward and backward propagating waves have to be expressed as a sum of three plane-wave components each [44,71].

In [51,72] the mean-field approximation, $\bar{u}_s^\pm = 1/L \int_0^L u_s^\pm \, dz$ has been applied to Equation 27.133 and the resulting nonlinear second-order ordinary differential equation has been solved numerically. In both references, the case of a purely gain-guided laser is considered, i.e., any impact of a built-in or thermally induced index guide has not been investigated. In [72], a zoo of solutions of different types, including asymmetric ones, has been found. The basic stationary states are the linear modes with the wavelengths

$$\Delta\lambda_{vk} = -\frac{\lambda_0}{Lk_0 n_g}\left[\frac{\varphi_0 + \varphi_L}{2} + \pi k - Lk_0\bar{n} - \frac{Lk_0}{2\bar{n}}\text{Re}(n_v^2 - \bar{n}^2)\right] \tag{27.136}$$

where k denotes the longitudinal mode, n_v is the modal index of the vth lateral mode, φ_0 and φ_L are the phases of the reflectivities. The other types of modes exist only due to the nonlinearity in the complex effective index and arise above the thresholds of the linear-guided modes.

There are a number of shortcomings of the theory presented so far. The assumption of a constant injection current density, i.e., an infinite series resistance as in Section 27.3.2 discussed, and the neglect of drift-diffusion result in an overestimation of spatial hole burning. A local variation of the complex effective index is not only caused by its dependence on the carrier density but also on the temperature ($\partial n/\partial T > 0$). Furthermore, in all analytical investigations, we are aware of no built-in or thermally induced index waveguides have been taken into account. These waveguides stabilize the linear guided modes, which are thus observable even far above threshold [60].

Based on an expansion of the optical field on the linear-guided modes and a numerical solution of the drift-diffusion equations in the mean-field approximation the far-field blooming of an index-guided broad-area laser has been investigated in [73]. The simulations revealed that a substantial part of the far-field blooming is not caused by self-heating but by increasing gain nonuniformity due to lateral spatial holeburning and laterally varying hole injection into the QWs. A discussion of filamentation effects based on a numerical solution of Equation 27.117 can be found in [53–55].

27.6 Thermodynamic-Based Energy-Transport Model

In high-power lasers operated in CW mode the transport of the charged carriers (electrons and holes) and the photons must be consistently formulated with the temperature flow in order to describe self-heating effects such as thermal roll-over and thermal lensing properly. A derivation of such a self-consistent energy-transport model by applying fundamental thermodynamic principles has been given in [74] using Boltzmann statistics. The model fulfils the first and second laws of thermodynamics as well as Onsager's reciprocity relations for the current densities. We note, that in this model electron, hole, and lattice temperatures are assumed to be equal, opposed to other energy-transport models [29,75]. A more general energy-transport model for semiconductor devices has been derived in [76] taking into account Fermi statistics and different temperatures for the charged carriers and the lattice, but disregarding optical fields. Previous formulations have been also given in [77,78], for example. In what follows we will summarize the energy-transport model paying particular attention to a consistent formulation with the model for the optical field presented in Section 27.4.

27.6.1 Basic Equations

The particle current flow is governed by the continuity equations for electrons and holes,

$$\nabla j_n = +e \left(R + \partial_t n \right) \tag{27.137}$$

$$\nabla j_p = -e \left(R + \partial_t p \right) \tag{27.138}$$

were j_n and j_p are the current densities for electrons and holes, respectively. The recombination rate R given in Equation 27.46 consists of nonradiative recombination $R_{\text{non-rad}}$, radiative spontaneous recombination R_{sp}, and radiative stimulated recombination R_{st}. The rate of radiative spontaneous recombination is often written as

$$R_{\text{sp}} = B(np - n_0 p_0) \tag{27.139}$$

with the equilibrium electron and hole densities n_0 and p_0, respectively. In the case of Boltzmann statistics, the coefficient B is constant whereas in the general case of Fermi statistics B decreases with increasing carrier densities.

The rate of stimulated recombination is the rate by which the energy density of the optical field changes by stimulated emission or absorption of a photon due to transitions between the conduction and valence bands, divided by the energy of the emitted or absorbed photon and follows from Equations 27.104 and 27.103 to

$$R_{\text{st}} = \frac{\varepsilon_0 c n_{\text{r}}}{2 \hbar \omega_0} \left[g \|E\|^2 + g_{\text{r}} \sum_{\nu=+,-} \text{Re} \left(E^{\nu*} \mathcal{P}^{\nu} - E^{\nu*} E^{\nu} \right) \right] \tag{27.140}$$

The electrostatic field itself is affected by the charge distribution of mobile (n and p) and fixed (n_A and p_D) carrier densities. The corresponding electrostatic potential φ solves the Poisson equation

$$-\nabla(\varepsilon_0 \varepsilon_s \nabla \varphi) = e(p - n + p_D - n_A) \tag{27.141}$$

with the relative static dielectric constant ε_s. For parabolic bands the relations between quasi-Fermi potentials φ_n and φ_p, electrostatic potential φ and carrier densities n and p are given by Equation 27.59.

In what follows we introduce the entropies per particle[†]

$$s_n = k_B \left[1 + R_n - \frac{\partial_T E_c}{k_B} + \frac{e\varphi_n + E_c - e\varphi}{k_B T} \right] \qquad (27.142)$$

$$s_p = k_B \left[1 + R_p + \frac{\partial_T E_v}{k_B} - \frac{e\varphi_p + E_v - e\varphi}{k_B T} \right] \qquad (27.143)$$

and the energies per particle

$$u_n = k_B T R_n - T\partial_T E_c + E_c = T(s_n - k_B) + e(\varphi - \varphi_n) \qquad (27.144)$$

$$u_p = k_B T R_p + T\partial_T E_v - E_v = T(s_p - k_B) + e(\varphi_p - \varphi) \qquad (27.145)$$

The functions R_n and R_p are given by the temperature derivatives of the inverse Fermi integrals,

$$R_n = -k_B T \partial_T \mathcal{F}_i^{\mathrm{inv}} \left(\frac{n}{N_c} \right) \qquad (27.146)$$

$$R_p = -k_B T \partial_T \mathcal{F}_i^{\mathrm{inv}} \left(\frac{p}{N_v} \right) \qquad (27.147)$$

For the bulk case ($i = 1/2$), Boltzmann statistics, parabolic bands, and temperature-independent electron and holes masses R_n and R_p are equal to $3/2$.

The electron and hole current densites are given by

$$j_n = -\sigma_n(\nabla\varphi_n - P_n\nabla T) \qquad (27.148)$$

$$j_p = -\sigma_p(\nabla\varphi_p + P_p\nabla T) \qquad (27.149)$$

where σ_n and σ_p are the electrical conductivities and P_n and P_p are the Seebeck coefficients or thermoelectric powers being the entropies per particle divided by the elementary charge e,

$$P_n = \frac{s_n}{e} \quad \text{and} \quad P_p = \frac{s_p}{e} \qquad (27.150)$$

If the coefficients in front of the temperature derivatives in Equations 27.148 and 27.149 are derived from the Boltzmann equation in relaxation time approximation, then the factorization $\sigma_n \cdot P_n$ and $\sigma_p \cdot P_p$ is only possible for parabolic bands and Boltzmann statistics. The same holds for the factorization of the electrical conductivities into products of carrier-density independent mobilities and carrier densities, $\sigma_n = e\mu_n \cdot n$ and $\sigma_p = e\mu_p \cdot p$. Furthermore, the magnitudes of P_n, P_p similarly as μ_n, μ_p depend on the scattering processes involved [79]. For example, if the dependence of the relaxation time on the energy is given by $\tau_0[E/(k_B T)]^r$ where r ranges typically between $-3/2$ and $+3/2$, then $R_n = R_p = 3/2+r$ in Equations 27.142 and 27.143. Thus $R_n = R_p = 3/2$ holds only for an energy-independent relaxation time. The temperature derivatives $\partial_T E_c$ and $\partial_T E_v$ of the conduction and valence band edges are often not included in the definition of the Seebeck coefficients.

According to Reference [74] the heat flow equation reads

$$c_h \partial_t T - \nabla \kappa_L \nabla T = h \qquad (27.151)$$

[†] The entropies can be also written as $s_n = e\partial_T(\varphi_n)_{(n,p)} + k_B$, $s_p = -e\partial_T(\varphi_p)_{(n,p)} + k_B$, where n, p are kept constant in the differentiation.

with the thermal conductivity of the crystal lattice κ_L, the heat capacity

$$c_h = c_L + n\partial_T u_n + p\partial_T u_p \tag{27.152}$$

with c_L being the heat capacity of the lattice and the heat source density

$$h = \frac{k_B T}{e}\boldsymbol{\nabla}(\boldsymbol{j}_n - \boldsymbol{j}_p) + T(\boldsymbol{j}_n\boldsymbol{\nabla}P_n - \boldsymbol{j}_p\boldsymbol{\nabla}P_p)$$
$$+ \frac{1}{\sigma_n}\boldsymbol{j}_n^2 + \frac{1}{\sigma_p}\boldsymbol{j}_p^2 + (u_n + u_p)R - \partial_t u_{\text{rad}} - \gamma_{\text{rad}} \tag{27.153}$$

The first term in Equation 27.153 is related to thermodiffusion and can be written as

$$\frac{k_B T}{e}\boldsymbol{\nabla}\cdot(\boldsymbol{j}_n - \boldsymbol{j}_p) = 2k_B TR + k_B T(\partial_t n + \partial_t p) \tag{27.154}$$

taking into account Equations 27.137 and 27.138. The second term describes Thomson–Peltier heat,

$$h_{\text{TP}} = T(\boldsymbol{j}_n\boldsymbol{\nabla}P_n - \boldsymbol{j}_p\boldsymbol{\nabla}P_p) = h_{\text{Thomson}} + h_{\text{Peltier}} \tag{27.155}$$

which is generated by a current flow along the gradients of the Seebeck coefficients ∇P_n and ∇P_p. By applying the chain rule to the gradients, the contributions due to Thompson heat and Peltier heat can be separated,

$$h_{\text{Thomson}} = T\boldsymbol{\nabla}T\left(\boldsymbol{j}_n\partial_T P_n|_{(n,p)} - \boldsymbol{j}_p\partial_T P_p|_{(n,p)}\right) \tag{27.156}$$

$$h_{\text{Peltier}} = T\left(\boldsymbol{j}_n\partial_n P_n|_T\boldsymbol{\nabla}n - \boldsymbol{j}_p\partial_p P_p|_T\boldsymbol{\nabla}p\right) \tag{27.157}$$

The third and forth terms correspond to Joule heat:

$$h_J = \frac{1}{\sigma_n}\boldsymbol{j}_n^2 + \frac{1}{\sigma_p}\boldsymbol{j}_p^2 \tag{27.158}$$

generated by scattering of the carriers on phonons resulting in a energy loss to the lattice. The last term is due to contributions of the recombination of electron–hole pairs which sets free the energy $u_n + u_p$ that is either transferred to the lattice as heat or transferred to the radiative field. The latter part is described by the term $\partial_t u_{\text{rad}}$, which has to be subtracted from the source term, like the term γ_{rad}, which denotes the energy loss from the cavity. Inserting Equations 27.154, 27.155, and 27.158 into Equation 27.153 gives

$$h = k_B T(\partial_t n + \partial_t p) + h_{\text{TP}} + h_J + h_{\text{rec+abs}} \tag{27.159}$$

with the recombination and absorption heat

$$h_{\text{rec+abs}} = e(TP_n + TP_p + \varphi_p - \varphi_n)R - \partial_t u_{\text{rad}} - \gamma_{\text{rad}} \tag{27.160}$$

where $R = R_{\text{non-rad}} + R_{\text{sp}} + R_{\text{st}}$. Likewise the rate of radiative recombination, the radiative energy density u_{rad} and the radiative energy loss γ_{rad} have contributions from spontaneous and stimulated emission, $u_{\text{rad}} = u_{\text{sp}} + u_{\text{st}}$ and $\gamma_{\text{rad}} = \gamma_{\text{sp}} + \gamma_{\text{st}}$. Here, u_{sp} is the energy density and γ_{sp} the corresponding cavity loss of the radiation generated by spontaneous emission into all modes (i.e., spatial directions, polarization directions of the field and frequencies) not included in $E^{\pm}$ given by Equation 27.101.

The balance equation for the energy density of the radiation generated by stimulated emission is given by Equations 27.103 and 27.104. As it has been stated before the first term on the rhs of Equation 27.103 gives the divergence of the transverse energy flux density and the second term the radiation leaving the cavity in propagation direction. Both terms combined give the total stimulated-energy loss from the cavity $-\gamma_{st}$,

$$-\gamma_{st} = \frac{\varepsilon_0 c}{2k_0} \mathrm{Im} \left[\partial_x (E^{+*} \partial_x E^+ + E^{-*} \partial_x E^-) + \partial_y (E^{+*} \partial_y E^+ + E^{-*} \partial_y E^-) \right]$$
$$+ \frac{\varepsilon_0 \bar{n} c}{2} \partial_z \left[|E^-(r,t)|^2 - |E^+(r,t)|^2 \right] \tag{27.161}$$

and hence

$$\partial_t u_{st} = \frac{\varepsilon_0 c n_r}{2} \left((g - g_r - \alpha)\|E\|^2 + g_r \mathrm{Re}(E^{+*} \mathcal{P}^+ + E^{-*} \mathcal{P}^-) \right) - \frac{\varepsilon_0 c n_r \tilde{\beta}}{2} \|E\|^4 - \gamma_{st}. \tag{27.162}$$

Inserting Equations 27.162 and 27.140 into Equation 27.160 gives

$$h_{rec+abs} = e(\mathrm{TP}_n + \mathrm{TP}_p + \varphi_p - \varphi_n)R_{non\text{-}rad} + \frac{\varepsilon_0 c n_r \alpha}{2} \|E\|^2 + \frac{\varepsilon_0 c n_r \tilde{\beta}}{2} \|E\|^4$$
$$+ \left[e(\mathrm{TP}_n + \mathrm{TP}_p + \varphi_p - \varphi_n) - \hbar\omega_0 \right] R_{st}$$
$$+ e(\mathrm{TP}_n + \mathrm{TP}_p + \varphi_p - \varphi_n)R_{sp} - \partial_t u_{sp} - \gamma_{sp} \tag{27.163}$$

with the rate of stimulated recombination Equation 27.140. Equation 27.163 is the net heat source caused by recombination and absorption. The first term is the heat generated by nonradiative recombination. The second and third terms describe the heat due to absorption of the stimulated radiation. The fourth term is caused by a possible incomplete energy transfer from the carrier ensemble to the radiation field during stimulated emission, also referred to as quantum defect energy. The last terms describe the heat due to the absorption of the spontaneous radiation, which could be treated similarly as the stimulated radiation but approximations have to be employed because the field generated by spontaneous emission is more challenging to calculate.

27.6.2 Spatial Distributions of the Heat Sources

The profiles of the heat sources of an asymmetric super-large optical-cavity structure similar to that published in [10] are shown in Figure 27.13a through 27.13d for the steady state for an output power of $P_{out} = 18$ W and an averaged internal power of $\bar{P} = 8$ W. Some of the parameters are given in Table 27.1. The one-dimensional simulation (along y) has been performed with the simulator WIAS-TeSCA [80]. The absorption heat

$$h_{abs} = (f_{c,n} n + f_{c,p} p) \bar{P} \frac{|\phi|^2}{W} \tag{27.164}$$

with $\phi(y)$ being the vertical normalized mode profile is shown in Figure 27.13. The cross sections for free-carrier absorption are $f_{c,n} = 4 \cdot 10^{-18}$ cm^2 and $f_{c,p} = 12 \cdot 10^{-18}$ cm^2. The main contributions arise in the active layer located at $y \approx 1.9$ μm due to the high nonequilibrium electron and hole densities and in the adjacent p- and n-doped confinement layers where the optical mode resides. The recombination heat

$$h_{rec} = (5k_B T + E_g)\left(R_{non\text{-}rad} + R_{sp}\right) \tag{27.165}$$

in Boltzmann approximation is maximal in the active layer as expected. Note that the temperature dependencies of E_c and E_v were not taken into account. The recombination of leaky electrons and holes in the p-confinement layer (cf. Figure 27.10b) generates heat, too. The quantum defect heat

$$h_{\text{defect}} = (5k_B T + E_g - \hbar\omega_0)\frac{g\bar{P}|\phi|^2}{\hbar\omega_0 W} \tag{27.166}$$

in a Boltzmann approximation, not shown here, is nonvanishing only in the active region and has the same order of magnitude like the recombination heat there ($h_{\text{defect}} \approx 8 \cdot 10^{14}$ W/m^3). The Joule heat (Equation 27.158) shown in Figure 27.13c is mainly generated in the p-confinement layer due to its low doping and the small mobility of the holes. Finally, Thomson heat

$$h_{\text{Thompson}} = \frac{3}{2}\frac{k_B}{e}\frac{dT}{dy}\left(j_{n,y} - j_{p,y}\right) \tag{27.167}$$

and Peltier heat

$$h_{\text{Peltier}} = -\frac{k_B T}{e}\left[\frac{d\ln(n)}{dy}j_{n,y} - \frac{d\ln(p)}{dy}j_{p,y}\right] \tag{27.168}$$

in Boltzmann approximations are shown in Figure 27.13d. The Thompson heat is negative due to the temperature gradient. The Peltier heat is positive or negative corresponding to the signs of the gradients of the carrier densities. Except in the p-cladding layer ($y > 3$ μm) it dominates over the Peltier heat. Note that the Peltier heat results in a cooling of the p-confinement layer. The integrated heat powers are $Q_{\text{Joule}} = 10.2$ W, $Q_{\text{abs}} = 2.7$ W, $Q_{\text{rec}} = 1.3$ W, and $Q_{\text{defect}} = 1.9$ W. Hence Joule heat amounts to more than 50% of the total heat.

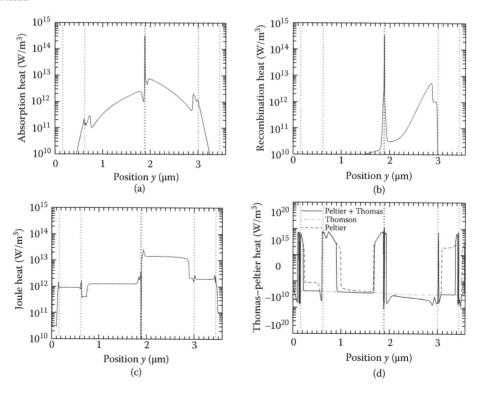

FIGURE 27.13 Vertical profiles of the heat sources of a high-power laser. (a) Absorption heat, (b) recombination heat, (c) Joule heat, and (d) Thomson–Peltier heat. Heteroboundaries are indicated by dotted vertical lines.

27.6.3 Energy Conservation

Inserting Equations 27.149 and 27.148 into the heat sources (Equations 27.155, 27.158 through 27.160) and integrating over the device volume V leads to

$$\int_V c_h \partial_t T \, dV - \int_V \mathbf{\nabla} \kappa_L \mathbf{\nabla} T \, dV = k_B \int_V T(\partial_t n + \partial_t p) \, dV$$

$$+ \int_V j_n \mathbf{\nabla}(TP_n - \varphi_n) \, dV - \int_V j_p \mathbf{\nabla}(TP_p + \varphi_p) \, dV + \int_V eR(TP_n - \varphi_n) \, dV$$

$$+ \int_V eR(TP_p + \varphi_p) dV - \partial_t \int_V u_{rad} \, dV - \int_V \gamma_{rad} \, dV \qquad (27.169)$$

Using a Green's identity we can convert

$$\int_V j_n \mathbf{\nabla}(TP_n - \varphi_n) \, dV = \int_{\partial V} (TP_n - \varphi_n) j_n \cdot n dS - \int_V (TP_n - \varphi_n) \underbrace{\mathbf{\nabla} j_n}_{e(R + \partial_t n)} dV \qquad (27.170)$$

and

$$- \int_V j_p \mathbf{\nabla}(TP_p + \varphi_p) \, dV = - \int_{\partial V} (TP_p + \varphi_p) j_p \cdot n dS + \int_V (TP_p + \varphi_p) \underbrace{\mathbf{\nabla} j_p}_{-e(R + \partial_t p)} dV \qquad (27.171)$$

where ∂V denotes the surface of the device with n being the normal vector. Hence the terms containing the recombination rate cancel, and after rearranging we obtain

$$\int_V c_h \partial_t T \, dV - \int_{\partial V} \kappa_L \mathbf{\nabla} T \cdot n dS$$

$$= \int_V \underbrace{(Tk_B - eTP_n + e\varphi_n)}_{-u_n + e\varphi} \partial_t n \, dV + \int_V \underbrace{(Tk_B - eTP_p - e\varphi_p)}_{-u_p - e\varphi} \partial_t p \, dV$$

$$+ \int_{\partial V} (TP_n - \varphi_n) j_n \cdot n dS - \int_{\partial V} (TP_p + \varphi_p) j_p \cdot n dS$$

$$- \partial_t \int_V u_{rad} \, dV - \int_V \gamma_{rad} \, dV \qquad (27.172)$$

From the Poisson equation (Equation 27.141) we can derive the relation

$$e \int_V \varphi(\partial_t n - \partial_t p) \, dV = \varepsilon_0 \int_{\partial V} \varphi \varepsilon_s \partial_t \mathbf{\nabla} \varphi \cdot n dS - \frac{\varepsilon_0}{2} \int_V \varepsilon_s \partial_t |\mathbf{\nabla}\varphi|^2 \, dV \qquad (27.173)$$

where we applied again a Green's identity. Using this relation and the expression for the heat capacity Equation 27.152 we obtain the energy balance equation

$$\frac{d \int_V u \, dV}{dt} = - \int_{\partial V} j_u \cdot n dS - \int_V \gamma_{rad} \, dV \qquad (27.174)$$

where u is the total energy density (sum of electro-static, internal, and radiative energy)

$$u = \frac{\varepsilon_0 \varepsilon_s}{2} |\nabla \varphi|^2 + c_L T + u_n n + u_p p + u_{rad}, \tag{27.175}$$

and j_u is the energy current density,

$$j_u = -\varepsilon_0 \varepsilon_s \varphi \partial_t \nabla \varphi - \kappa_L \nabla T - (TP_n - \varphi_n) j_n + (TP_p + \varphi_p) j_p \tag{27.176}$$

both already introduced in [74]. The first term in Equation 27.176 is related to the displacement current density. The other terms can be reduced to the entropy current density multiplied by T and the particle flux multiplied by the respective quasi-Fermi potentials.

In what follows, we consider the steady state ($d/dt = \partial_t = 0$) and evaluate the surface integral in Equation 27.174. We assume no flow of electrical current through the surface, except at the n-contact located at $y = 0$ and the p-contact at $y = H$:

$$\left.\begin{array}{r} j_n \cdot n = 0 \\[4pt] j_p \cdot n = 0 \end{array}\right\} \quad \text{for} \quad y \neq 0 \quad \text{and} \quad y \neq H \tag{27.177}$$

Between the contacts a forward bias U is applied, so that

$$\varphi_n|_{y=H} = \varphi_p|_{y=H} = U \qquad \text{and} \qquad \varphi_n|_{y=0} = \varphi_p|_{y=0} = 0 \tag{27.178}$$

hold. The normal components of the electron current density at the p-contact $y = H$ and the hole current density at the n-contact $y = 0$ are assumed to vanish,

$$j_{n,y}|_{y=H} = j_{p,y}|_{y=0} = 0 \tag{27.179}$$

Similarly, we assume no heat flow through the surface, except at the surface, located at $y = H$, attached to the heatsink where

$$\kappa_L \partial_y T|_{y=H} = \frac{T_{ref} - T|_{y=H}}{r_{th}} \tag{27.180}$$

with r_{th} being the thermal transmission resistance (unit Km^2/W). By inserting the boundary conditions into Equation 27.174 we get for the steady state

$$\iint \frac{T|_{y=H} - T_{ref}}{r_{th}} dxdz = UI - \int_V \gamma_{rad} dV + \iint [(TP_p)|_{y=H} - (TP_n)|_{y=0}] j \, dxdz \tag{27.181}$$

with $j = j_{n,y}|_{y=0} = -j_{p,y}|_{y=H}$ and $I = \iint j \, dxdz$. The lhs of Equation 27.181 is the heat flow to the heatsink. The first term on the rhs is the electric input power UI and the second term is the optical power that leaves the cavity. The last term can be considered as Peltier power, which is generated between the electric contacts in the presence of a current flow and results into heating or cooling. Assuming that the spontaneous emission is absorbed in the cavity so that $\gamma_{rad} = \gamma_{st}$, using Equation 27.161 and the boundary conditions Equations 27.88 and 27.87, the second term can be shown to be (cf. Equation 27.105)

$$\int_V \gamma_{rad} dV = P_{out} \tag{27.182}$$

Thus in the steady state the total heat generated in the cavity is given by the electrical input power UI minus the optical output power, which coincides with Equation 27.11, minus or plus the Peltier heat.

Acknowledgment

The authors are indebted to K. H. Hasler, M. Lichtner, M. Platz, R. Staske, and H. J. Wünsche for their contributions to the chapter.

References

1. D. F. Welch. A brief history of high-power semiconductor lasers. *IEEE J. Sel. Topics Quantum Electron.*, 6(6):1470–1477, 2000.
2. R. Diehl. *High-power Diode Lasers: Fundamentals, Technology, Applications*, volume 78. Berlin, Heidelberg, New York, NY: Springer, 2000.
3. M. Behringer. High-power diode laser technology and characteristics. In F. Bachmann, P. Loosen, and R. Poprawe, editors, *High Power Diode Lasers*, volume 128 of *Springer Series in Optical Sciences*, pp. 5–74. New York, NY: Springer, 2007.
4. X. Liu, W. Zhao, L. Xiong, and H. Liu. *Packaging of High Power Semiconductor Lasers*. Berlin, Heidelberg, New York, NY: Springer, 2014.
5. I. Fischer, O. Hess, W. Elsäßer, and E. Göebel. Complex spatio-temporal dynamics in the near-field of a broad-area semiconductor laser. *Europhys Lett.*, 35(8):579–584, 1996.
6. H. Wenzel. Basic aspects of high-power semiconductor laser simulation. *IEEE J. Sel. Topics Quantum Electron.*, 19(5):1–13, 2013.
7. P. Crump, O. Brox, F. Bugge, J. Fricke, C. Schultz, M. Spreemann, B. Sumpf, H. Wenzel, and G. Erbert. High-power, high-efficiency monolithic edge-emitting GaAs based lasers with narrow spectral widths. In J. J. Colemann, A. C. Bryce, and C. Jagadish, editors, *Advances in Semiconductor Lasers*, volume 86 of *Semiconductor and Semimetals*, chapter 2. Amsterdam: Elsevier, 2012.
8. L. A. Coldren, S. W. Corzine, and M. L. Mashanovitch. *Diode Lasers and Photonic Integrated Circuits*. Hoboken, NJ: John Wiley & Sons, 2012.
9. H. Wenzel, G. Erbert, A. Knauer, A. Oster, K. Vogel, and G. Tränkle. Influence of current spreading on the transparency current density of quantum-well lasers. *Semicond. Sci. Technol.*, 15(6):557, 2000.
10. K. H. Hasler, H. Wenzel, P. Crump, S. Knigge, A. Maaßdorf, R. Platz, R. Staske, and G. Erbert. Comparative theoretical and experimental studies of two designs of high-power diode lasers. *Semicond. Sci. Technol.*, 29(4):045010, 2014.
11. H. Wenzel, P. Crump, A. Pietrzak, X. Wang, G. Erbert, and G. Tränkle. Theoretical and experimental investigations of the limits to the maximum output power of laser diodes. *New J. Phys.*, 12(8):085007, 2010.
12. J. Piprek. *Semiconductor Optoelectronic Devices: Introduction to Physics and Simulation*. London, Oxford, Boston, New York and San Diego: Academic Press, 2003.
13. W. W. Rigrod. Saturation effects in high-gain lasers. *J. Appl. Phys.*, 36(8):2487–2490, 1965.
14. H.-J. Wünsche, U. Bandelow, and H. Wenzel. Calculation of combined lateral and longitudinal spatial hole burning in $\lambda/4$ shifted DFB lasers. *IEEE J. Quantum Electron.*, 17:1751–1760, 1993.
15. H. Wenzel and G. Erbert. Simulation of single-mode high-power semiconductor lasers. *Proc. SPIE.*, 2693:418–429, 1996.
16. A. Demir, M. Peters, R. Duesterberg, V. Rossin, and E. Zucker. Semiconductor laser power enhancement by control of gain and power profiles. *IEEE Photon Techn. Lett.*, 27(20):2178–2181, 2015.
17. A. Guermache, V. Voiriot, D. Locatelli, F. Legrand, R.-M. Capella, P. Gallion, and J. Jacquet. Experimental demonstration of spatial hole burning reduction leading to 1480-nm pump lasers output power improvement. *IEEE Photon. Techn. Lett.*, 17(10):2023, 2005.

18. Z. Chen, L. Bao, J. Bai, M. Grimshaw, R. Martinsen, M. DeVito, J. Haden, and P. Leisher. Performance limitation and mitigation of longitudinal spatial hole burning in high-power diode lasers. In *Proceedings SPIE*, pp. 82771J–82771J, 2012.

19. X. Wang, P. Crump, H. Wenzel, A. Liero, T. Hoffmann, A. Pietrzak, C. M. Schultz, A. Klehr, A. Ginolas, S. Einfeldt, et al. Root-cause analysis of peak power saturation in pulse-pumped 1100 nm broad area single emitter diode lasers. *IEEE J. Quantum Electron.*, 46(5):658–665, 2010.

20. S. F. Yu, R. G. S. Plumb, L. M. Zhang, M. C. Nowell, and J. E. Carroll. Large-signal dynamic behavior of distributed-feedback lasers including lateral effects. *IEEE J. Quantum Electron.*, 30(8):1740–1750, 1994.

21. H. E. Lassen, H. Olesen, and B. Tromborg. Gain compression and asymmetric gain due to the Bragg grating induced by the standing waves in Fabry-Perot lasers. *IEEE Photon Technol. Lett.*, 1(9):261–263, 1989.

22. E. A. Avrutin and B. S. Ryvkin. Theory of direct and indirect effect of two photon absorption on nonlinear optical losses in high power semiconductor laser. *Semicond. Sci. Technol.*, 32(1): 015004, 2016.

23. E. A. Avrutin and B. S. Ryvkin. Theory and modelling of the power conversion efficiency of large optical cavity laser diodes. In *2015 IEEE High Power Diode Lasers and Systems Conference (HPD)*, pp. 9–10, 2015.

24. H. Wenzel, P. Crump, A. Pietrzak, C. Roder, X. Wang, and G. Erbert. The analysis of factors limiting the maximum output power of broad-area laser diodes. *Opt. Quantum Electron.*, 41(9):645–652, 2009.

25. W. B. Joyce. Current-crowded carrier confinement in double-hetero-structure lasers. *J. Appl. Phys.*, 51(5):2394–2401, 1980.

26. M. Grupen and K. Hess. Simulation of carrier transport and nonlinearities in quantum-well laser diodes. *IEEE J. Quantum Electron.*, 34(1):120–140, 1998.

27. B. Witzigmann, A. Witzig, and W. Fichtner. A multidimensional laser simulator for edge-emitters including quantum carrier capture. *IEEE Trans. Electron Devices*, 47(10):1926–1934, 2000.

28. L. Borruel, J. Arias, B. Romero, and I. Esquivias. Incorporation of carrier capture and escape processes into a self-consistent cw model for quantum well lasers. *Microelectron J.*, 34(5):675–677, 2003.

29. Y. Liu, W. C. Ng, K. D. Choquette, and K. Hess. Numerical investigation of self-heating effects of oxide-confined vertical-cavity surface-emitting lasers. *IEEE J. Quantum Electron.*, 41(1):15–25, 2005.

30. S. Steiger, R. G. Veprek, and B. Witzigmann. Unified simulation of transport and luminescence in optoelectronic nanostructures. *J. Comp. Electron.*, 7(4):509–520, 2008.

31. A. F. Oskooi, L. Zhang, Y. Avniel, and S. G. Johnson. The failure of perfectly matched layers, and towards their redemption by adiabatic absorbers. *Optics Express.*, 16(15):11376–11392, 2008.

32. T. Tischler. *Die Perfectly-Matched-Layer-Randbedingung in der Finite-Differenzen-Methode im Frequenzbereich: Implementierung und Einsatzbereiche*. Innovationen mit Mikrowellen und Licht. Goettingen: Cuvillier, 2004.

33. C. Z. Ning, R. A. Indik, and J. V. Moloney. Effective Bloch equations for semiconductor lasers and amplifiers. *IEEE J. Quantum Electron.*, 33(9):1543–1550, Sep 1997.

34. U. Bandelow, M. Radziunas, J. Sieber, and M. Wolfrum. Impact of gain dispersion on the spatio-temporal dynamics of multisection lasers. *IEEE J. Quantum Electron.*, 37(2):183–188, 2001.

35. M. Lichtner, M. Radziunas, U. Bandelow, M. Spreemann, and H. Wenzel. Dynamic simulation of high brightness semiconductor lasers. *Proc NUSOD '08*, pp. 65–66, 2008.

36. S. Balsamo, F. Sartori, and I. Montrosset. Dynamic beam propagation method for flared semiconductor power amplifiers. *IEEE J. Quantum Electron.*, 2(2):378–384, Jun 1996.

37. C. Z. Ning, R. A. Indik, J. V. Moloney, W. W. Chow, A. Girndt, S. W. Koch, and R. H. Binder. Incorporating many-body effects into modeling of semiconductor lasers and amplifiers. *Proc. SPIE*, 2994:666–677, 1997.

38. E. Gehrig and O. Hess. Spatio-temporal dynamics of light amplification and amplified spontaneous emission in high-power tapered semiconductor laser amplifiers. *IEEE J. Quantum Electron.*, 37(10):1345–1355, Oct 2001.

39. W. W. Chow and H. Amano. Analysis of lateral-mode behavior in broad-area InGaN quantum-well lasers. *IEEE J. Quantum Electron.*, 37(2):265–273, 2001.

40. M. Kolesik and J. V. Moloney. A spatial digital filter method for broad-band simulation of semiconductor lasers. *IEEE J. Quantum Electron.*, 37(7):936–944, Jul 2001.

41. J. Javaloyes and S. Balle. Quasiequilibrium time-domain susceptibility of semiconductor quantum wells. *Phys. Rev. A*, 81(6):062505, 2010.

42. P. N. Butcher and D. Cotter. *The Elements of Nonlinear Optics*, volume 9. Melbourne, Australia: Cambridge University Press, 1991.

43. G. P. Agrawal. *Nonlinear Fiber Optics*. London, Oxford, Boston, New York and San Diego: Academic Press, 2007.

44. R. W. Boyd. *Nonlinear Optics*. London, Oxford, Boston, New York and San Diego: Academic Press, 2008.

45. R. März. *Integrated Optics: Design and Modeling*. Norwood, MA: Artech House, 1995.

46. R. Scarmozzino, A. Gopinath, R. Pregla, and S. Helfert. Numerical techniques for modeling guided-wave photonic devices. *IEEE J. Sel. Topics Quantum Electron.*, 6(1):150–162, Jan–Feb 2000.

47. T. M. Benson, B. B. Hu, A. Vukovic, and P. Sewell. What is the future for beam propagation methods? *Proc. SPIE.*, 5579:351–358, 2004.

48. A. E. Siegman. Laser beams and resonators: The 1960s. *IEEE J. Sel. Topics Quantum Electron.*, 6(6):1380–1388, Nov–Dec 2000.

49. G. H. B. Thompson. A theory for filamentation in semiconductor lasers including the dependence of dielectric constant on injected carrier density. *Opto-electronics.*, 4:257–310, 1972.

50. D. Mehuys, R. Lang, M. Mittelstein, J. Salzman, and A. Yariv. Self-stabilized nonlinear lateral modes of broad area lasers. *IEEE J. Quantum Electron.*, 23(11):1909–1920, 1987.

51. R. J. Lang, A. G. Larsson, and J. G. Cody. Lateral modes of broad area semiconductor lasers: Theory and experiment. *IEEE J. Quantum Electron.*, 27(3):312–320, 1991.

52. J. R. Marciante and G. P. Agrawal. Nonlinear mechanisms of filamentation in broad-area semiconductor lasers. *IEEE J. Quantum Electron.*, 32(4):590–596, 1996.

53. J. V. Moloney. Semiconductor laser device modeling. In B. Krauskopf and D. Lenstra, editors, *Fundamental Issues of Nonlinear Laser Dynamics*, pp. 149–172. College Park, MD: American Institute of Physics, 2000.

54. E. Gehrig and O. Hess. *Spatio-Temporal Dynamics and Quantum Fluctuations in Semiconductor Lasers*, volume 189. Springer Science & Business Media, 2003.

55. K. Böhringer. *Microscopic Spatio-Temporal Dynamics of Semiconductor Quantum Well Lasers and Amplifiers*. PhD thesis, Institut fuer Technische Physik, Deutsches Zentrum fuer Luft- und Raumfahrt, 2007.

56. J. V. Moloney, M. Kolesik, J. Hader, and S. W. Koch. Modeling high-power semiconductor lasers: From microscopic physics to device applications. *Proc. SPIE.*, 3889:120–127, 2000.

57. H. Wenzel, P. Crump, H. Ekhterei, C. Schultz, J. Pomplun, S. Burger, L. Zschiedrich, F. Schmidt, and G. Erbert. Theoretical and experimental analysis of the lateral modes of high-power broad-area lasers. *Proc NUSOD '11*, pp. 143–144, 2011.

58. P. Crump, S. Böldicke, C. M. Schultz, H. Ekhteraei, H. Wenzel, and G. Erbert. Experimental and theoretical analysis of the dominant lateral waveguiding mechanism in 975 nm high power broad area diode lasers. *Semicond. Sci. Technol.*, 27(4):045001, 2012.

59. N. Stelmakh and M. Flowers. Measurement of spatial modes of broad-area diode lasers with 1-GHz resolution grating spectrometer. *IEEE Photon Technol. Lett.*, 18(15):1618–1620, 2006.

60. P. Crump, M. Ekteraei, C. M. Schultz, G. Erbert, and G. Tränkle. Studies of limitations to lateral brightness in high power diode lasers using spectrally-resolved mode profiles. In *2014 IEEE International Semiconductor Laser Conference (ISLC)*, pp. 23–24, 2014.

61. M. Spreemann, M. Lichtner, M. Radziunas, U. Bandelow, and H. Wenzel. Measurement and simulation of distributed-feedback tapered master-oscillator power amplifiers. *IEEE J. Quantum Electron.*, 45(6):609–616, 2009.

62. C. Fiebig, V. Z. Tronciu, M. Lichtner, K. Paschke, and H. Wenzel. Experimental and numerical study of distributed-Bragg-reector tapered lasers. *Appl. Phys. B.*, 99(1):209–214, 2010.

63. M. Spreemann, H. Wenzel, B. Eppich, M. Lichtner, and G. Erbert. Novel approach to finite-aperture tapered unstable resonator lasers. *IEEE J. Quantum Electron.*, 47(1):117–125, Jan 2011.

64. S. Tronciu, V. Z Schwertfeger, M. Radziunas, A. Klehr, U. Bandelow, and H. Wenzel. Numerical simulation of the amplification of picosecond laser pulses in tapered semiconductor amplifiers and comparison with experimental results. *Optics Communications.*, 285(12):2897–2904, 2012.

65. M. Lichtner, V. Z. Tronciu, and A. G. Vladimirov. Theoretical investigation of striped and non-striped broad area lasers with off-axis feedback. *IEEE J. Quantum Electron.*, 48(3):353–360, 2012.

66. M. Radziunas, R. Herrero, M. Botey, and K. Staliunas. Far-field narrowing in spatially modulated broad-area edge-emitting semiconductor amplifiers. *J. Opt. Soc. Am. B.*, 32(5):993–1000, 2015.

67. M. Radziunas and R. Ciegis. Effective numerical algorithm for simulations of beam stabilization in broad area semiconductor lasers and amplifiers. *Math. Model. Anal.*, 19(5):627–646, 2014.

68. M. Sheik-Bahae and E. W. Van Stryland. Optical nonlinearities in the transparency region of bulk semiconductors. In R. K. Willardson, E. R. Weber, E. Garmire, and A. Kost, editors, *Nonlinear Optics in Semiconductors I*, volume 58 of *Semiconductor and Semimetals*, pp. 257–318. Amsterdam: Elsevier, 1998.

69. M. Dinu. Dispersion of phonon-assisted nonresonant third-order nonlinearities. *IEEE J. Quantum Electron.*, 39(11):1498–1503, 2003.

70. V. I. Bespalov and V. I. Talanov. Filamentary structure of light beams in nonlinear liquids. *ZhETF Pisma Redaktsiiu.*, 3:471, 1966.

71. A. P. Bogatov. Lateral field instability and six-wave mixing in a diode laser with broad active area. *J. Russian Laser Res.*, 15(5):417–453, 1994.

72. S. Blaaberg, P. M. Petersen, and B. Tromborg. Structure, stability, and spectra of lateral modes of a broad-area semiconductor laser. *IEEE J. Quantum Electron.*, 43(11):959–973, 2007.

73. J. Piprek and Z. M. S. Li. On the importance of non-thermal far-field blooming in broad-area high-power laser diodes. *Appl. Phys. Lett.*, 102(22):221110, 2013.

74. U. Bandelow, H. Gajewski, and R. Hünlich. Fabry-Perot lasers: Thermodynamics based modeling. In J. Piprek, editor, *Optoelectronic Devices—Advanced Simulation and Analysis*, pp. 63–85. New York, NY: Springer, 2005.

75. T. Grasser, T.-W. Tang, H. Kosina, and S. Selberherr. A review of hydrodynamic and energy-transport models for semiconductor device simulation. *Proc. IEEE.*, 91(2):251–274, 2003.

76. G. Albinus, H. Gajewski, and R. Hünlich. Thermodynamic design of energy models of semiconductor devices. *Nonlinearity.*, 15(2):367, 2002.

77. G. K. Wachutka. Rigorous thermodynamic treatment of heat generation and conduction in semiconductor device modeling. *IEEE Trans Comput-Aided Design Integr. Circuits Syst.*, 9(11):1141–1149, 1990.

78. J. E. Parrott. Thermodynamic theory of transport processes in semiconductors. *IEEE Trans. Electron Devices.*, 43(5):809–826, 1996.

79. K. Seeger. *Semiconductor Physics*. Berlin and Heidelberg: Springer Science & Business Media, 2013.

80. H. Gajewski, B. Heinemann, H. Langmach, R. Nürnberg, G. Telschow, K. Zacharias, H.-Chr. Kaiser, and U. Bandelow. WIAS-TeSCA Two-Dimensional Semi-Conductor Analysis Package, 2012.

28

High-Brightness Tapered Lasers

Ignacio Esquivias

Antonio
Pérez-Serrano

and

José-Manuel
G. Tijero

28.1 Introduction

The brightness of an optical source is commonly defined as the emitted power per unit of emitting area and per unit of the solid angle into which the power is emitted (Walpole 1996). Therefore, a high-brightness source requires not only a high value of the emitted power but also a high "beam quality" in terms of a low product of the beam size and the beam divergence. The product of the beam radius at waist and the beam divergence half angle is called "beam parameter product" and based on it, the most widely used figure of merit for beam quality, the beam propagation ratio M^2, is defined as the ratio of the beam parameter product of the beam of interest to the beam parameter product of a diffraction-limited, perfect Gaussian beam (TEM$_{00}$) of the same wavelength λ (ISO 2005; Siegman et al. 1998). Therefore, a value $M^2 = 1$ represents an ideal diffraction-limited source, while values higher than unity indicate a degradation of the beam quality.

Semiconductor lasers are optical sources with very well-known advantages over other types of optical sources: small size, high conversion efficiency, and low cost. There are many applications of semiconductor lasers demanding high brightness: material processing, optical pumping of solid state and fiber lasers, medical treatments, optical wireless communications, and in general all applications requiring high power launched into an optical fiber. However, the brightness of a semiconductor laser is usually limited due to two counteracting requirements: a large emitting area is required to produce high power with reduced bulk and surface heating, while reduced dimensions are required to maintain a single spatial mode and thus a high-quality beam. High-power semiconductor lasers are based on broad-area (BA) devices, with a poor beam quality along the lateral axis, while devices with reduced lateral dimensions and good beam quality, such as ridge waveguide (RW) lasers, suffer from a limited maximum output power.

As a consequence, an important research effort has been devoted to improve the brightness of semiconductor lasers during the last years and various new approaches have been proposed, including lasers with a tapered gain region (Walpole 1996; Wenzel et al. 2003; Sumpf et al. 2009), the master-oscillator power amplifier configuration (O'Brien et al. 1993; Spreemann et al. 2009), and the angled grating distributed feedback laser (Lang et al. 1998; Paschke et al. 2003). Tapered lasers, also called flared unstable cavity lasers, are possibly the best choice to achieve high brightness at moderate cost, due to the technological simplicity of their fabrication process.

The schematic of a typical tapered laser is shown in Figure 28.1a. It is similar to the tapered semiconductor optical amplifier (SOA) described in Chapter 22 of this book (Tijero et al. 2017). In brief, it is composed of a straight and narrow index-guided (IG) section, usually an RW structure, and a gain or IG tapered section where the beam is amplified. Ideally, the optical beam of a taper laser is a single lateral mode that diffracts at the exit of the RW section and is amplified in the tapered section while preserving its shape. The main difference with the tapered SOA is that the reflectivities of the facets are modified to provide laser oscillation at a reasonable value of the injection current. The output facet is usually antireflection (AR) coated while the back facet is coated to provide a high reflectance (HR) in order to decrease the threshold current and to maximize the output power. In many cases, the designs include beam spoiler elements in the form of trenches located in the neighborhood of the border between the RW and the tapered section (Figure 28.1a) (Kintzer et al. 1993). The main role of these elements is to filter the backward propagating field out of the RW. Most of the tapered lasers are based on a standard Fabry–Pérot cavity and therefore they present multiple longitudinal mode spectra. Some designs include a distributed Bragg reflector (DBR) at the end of the RW in order to provide narrow and stable emission spectra (Hasler et al. 2008).

In comparison with other semiconductor lasers, the specific design of tapered lasers leads to an optical beam with strong astigmatism. The virtual source for the vertical (y) axis (sometimes referred to as fast axis) is located at the output facet of the laser from where the beam diffracts in air. However, in the lateral (x) axis (sometimes referred to as slow axis), the beam diffracts from the exit of the RW section in the semiconductor medium and therefore, at low power, the separation between the two virtual sources (astigmatism) is approximately given by the taper section length L_{TAP} divided by the effective index n_{eff}. This is illustrated in Figure 28.1b. In the vertical axis, the beam can be considered as diffraction limited no matter the power. However, in the lateral axis the beam often degrades losing its diffraction-limited character when the power increases. As mentioned earlier, this degradation is usually characterized by the beam propagation ratio M^2. Referring to Figure 28.1c, the beam parameter product, bpp, and M^2 in the lateral axis are, respectively, given by

$$\text{bpp} = W_{0x}\theta_x \text{ and } M^2 = \frac{\pi}{\lambda}\text{bpp} \tag{28.1}$$

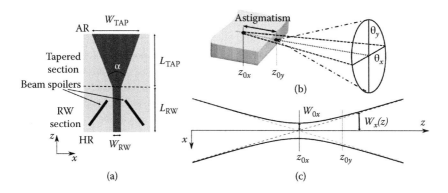

FIGURE 28.1 (a) Schematics of a gain-guided tapered laser. The shaded area is the contact region. Grooves acting as beam spoilers are also depicted. (b) Schematics showing the far field distribution of a tapered laser as a consequence of astigmatism. (c) Evolution of the beam size in the lateral direction along the propagation axis z.

(λ/π being the bpp of a Gaussian beam of wavelength λ). In Equation 28.1, W_{0x} is the beam half width at the virtual source position (beam waist half width), and θ_x is the beam divergence half angle (far-field half width) given by

$$\theta_x = \lim_{z \to \infty} \frac{W_x(z)}{(z - z_{0x})} \tag{28.2}$$

where $W_x(z)$ is the beam half width at the z position and z_{0x} the virtual source location. It is important to point out here that the beam sizes in Equations 28.1 and 28.2 (and therefore the beam divergence) are defined in terms of the second-order moment (variance) of the power density distribution by $W_x(z) = 2\sigma_x(z)$ where

$$\sigma_x^2(z) = \frac{\int\limits_{-\infty}^{\infty} \int\limits_{-\infty}^{\infty} P(x, y, z)(x - \bar{x})^2 \mathrm{d}x\,\mathrm{d}y}{\int\limits_{-\infty}^{\infty} \int\limits_{-\infty}^{\infty} P(x, y, z)\mathrm{d}x\,\mathrm{d}y} \tag{28.3}$$

and $\bar{x}$ is the first order moment of the power density distribution (x coordinate of the centroid) (ISO 2005). When $W_x(z)$ is defined in this way, and only in this case, the evolution of the beam lateral size along the propagation axis z for any simple astigmatic beam can be described by the hyperbola depicted in Figure 28.1c and given by

$$W_x(z) = W_{0x} \sqrt{1 + \left(\frac{M^2 \lambda (z - z_{0x})}{\pi (W_{0x})^2} \right)^2} \tag{28.4}$$

The fitting of the measured $W_x(z)$ to this expression is a method for the experimental determination of M^2, W_{0x}, and z_{0x} (ISO 2005).

Nevertheless, it is a common practice in the research groups having developed tapered lasers in the last years to characterize the beam quality by the parameter $M^2(1/e^2)$ (Krakowski et al. 2002; Sumpf et al. 2002), using the expression

$$M^2(1/e^2) = \frac{\pi}{\lambda} \theta_x(1/e^2) W_{0x}(1/e^2) \tag{28.5}$$

where $\theta_x(1/e^2)$ and $W_{0x}(1/e^2)$ are the half widths of the divergence and virtual source at $1/e^2$, respectively. For Gaussian beams $M^2(1/e^2) = M^2 = 1$ and in general $M^2(1/e^2)$ is a useful parameter to compare different lasers and to estimate the efficiency of the source to couple power into a reduced area device, such as a single mode fiber. However, $M^2(1/e^2)$ for non-Gaussian beams is usually much lower than M^2 and it is not actually a beam propagation ratio in the sense that it is not an invariant of the beam when it propagates in air or across passive, nonaberrating optical elements as required by a real beam propagation ratio. Therefore, the reader should be aware of this when interpreting the real relevance of this parameter. An example of the severe discrepancies between M^2 and $M^2(1/e^2)$ for a non-Gaussian beam will be commented in Section 28.3.5.

Two clearly different types of tapered lasers have been reported to date: gain-guided (GG) and IG lasers. The GG tapered lasers feature a relatively large full taper angle, α_{tap}, electrically defined in the p-contact layer. In these lasers, α_{tap} is designed to match the free diffraction angle (typically 4°–8°, depending on wavelength), as we will describe later. In the IG tapered lasers a small taper angle ($\alpha_{tap} < 1°$) is defined both, electrically in the p-contact layer and optically by an effective index step created by removing a fraction of the upper epitaxial layers. The beam properties of the two kind of tapered lasers are significantly different (Borruel et al. 2004a). GG tapered lasers with a taper angle close to the free diffraction angle have been fabricated at many different wavelengths from red (Blume et al. 2012) to around 2μm (Pfahler et al. 2006). A review of previous work on these devices can be found in Wenzel et al. (2003) and Sumpf et al. (2009). GG tapered lasers have demonstrated good beam quality and continuous wave (cw) output powers higher than 10 W at 980 nm (Fiebig et al. 2009), 1060 nm (Sumpf et al. 2009), and 1030 nm (Müller et al. 2016). Narrow IG tapered lasers have achieved more than 1 W at 980 nm (Krakowski et al. 2003) and 915 nm (Michel et al. 2005). Narrow IG lasers are usually combined in parallel arrays in high-power laser bars (Auzanneau et al. 2003; Wilson et al. 1999).

It is well known that the interaction between the optical field and the semiconductor gain media promotes a complex spatial-spectral dynamics in semiconductor lasers. In the case of tapered lasers, nonlinear effects, such as spatial hole burning (SHB) and thermal lensing add complexity to the physical phenomena involved and make the high power behavior to significantly deviate from the ideal low-power performance. Efficient and accurate modeling approaches are thus necessary to analyze and predict the beam properties of tapered lasers in order to design new geometries with improved performance. During last 20 years, different approaches have been applied to the modeling and simulation of tapered lasers and amplifiers. For a review see (Tijero et al. 2017) in this book.

Our group at the Universidad Politécnica of Madrid, in collaboration with the University of Nottingham, developed CONAN (Borruel et al. 2002, 2004a; Sujecki et al. 2003), a sophisticated simulator for tapered lasers that solves the electrical, optical, and thermal equations for these devices. Despite the assumptions needed to reduce the model complexity (steady state, single frequency, two-dimensional [2D] propagation of the optical mode), the simulations showed good qualitative and quantitative agreement with experimental results in tapered lasers with different geometries and based on different materials (Sujecki et al. 2003; Borruel et al. 2004a; Odriozola et al. 2009; Esquivias et al. 2010). Furthermore, the simulator demonstrated to be a useful tool to predict the behavior of novel designs prior to their fabrication (Borruel et al. 2005; Michel et al. 2009). Other models in literature (Williams et al. 1999; Mariojouls et al. 2000), based on similar approaches, have also reproduced the main trends observed experimentally.

In this chapter, we present a didactic overview of how the main beam characteristics of high-brightness tapered lasers can be accounted by simulation approaches with specific detail devoted to our simulation tool, and provide illustrative examples representative of some of the most typical behaviors of these devices. After this introduction, Section 28.2 presents a brief description of our simulation model and our procedure to calibrate the model in comparison with experimental results; in Section 28.3, we apply the model to three devices representative of some of the most common geometries and guiding mechanisms and analyze how geometry and guiding determine the beam characteristics. The chapter ends with a summary in Section 28.4.

28.2 Simulation Model

28.2.1 Model Overview

Our quasi-3D (three-dimensional) model (Borruel et al. 2002, 2004a; Sujecki et al. 2003) solves self-consistently the complete steady-state electrical, thermal, and optical equations for the tapered laser, assuming single-frequency operation. The laser simulator includes a 3D electrical solver of the Poisson and continuity equations coupled to a 3D thermal solver of the heat-flow equation with the local heat sources provided by the electrical solution. The optical fields in the tapered laser are solved using a 2D wide-angle finite-difference beam propagation method (WA-FDBPM) making use of the effective index approximation (Hadley 1992). Further details of the model can be found in Chapter 22 of this book, which is devoted to tapered SOAs (Tijero et al. 2017). Here, we briefly describe the model for completeness, in order to emphasize the differences when applying it to tapered lasers instead of tapered amplifiers. In fact, the main difference is that in the simulator for tapered amplifiers the WA-FDBPM is applied for the propagation of the optical solution only in the forward direction, while in the case of tapered lasers, the optical solution has to be propagated in both, the forward and then in the backward direction taking into account the corresponding reflectivities of the front and rear facets as corresponds to an optical resonator (Fox and Li 1961). Table 28.1 summarizes the main physical effects included in the model and how their dependence on temperature, wavelength, or carrier densities has been considered.

The simulator flow is shown in Figure 28.2. The solution procedure is initialized by a one-dimensional (1D) laser simulator (Harold, 3.0) which provides for each bias current the lasing wavelength, the bias voltage, and the average photon density in the laser cavity. The tapered laser is divided into 2D slices perpendicular to the z-axis at positions z_i where $i = 1, \ldots, N$. The 1D average photon density is used to define an initial guess optical field at the first slice, at the rear facet of the device. The photon density of this field is used as input for the electrothermal solver to calculate the lateral gain and refractive index perturbation profiles in the first slice. With these inputs, a 2D WA-FDBPM making use of the effective index approximation propagates the optical field through the first slice and provides the electrothermal solver with the photon density profile corresponding to the next slice. This procedure is repeated until arriving to the front facet of the device, $i = N$. After applying the electrothermal solver to the last slice, the output power P_{out} and the excess power P_{exc} are calculated (Tijero et al. 2017) and used as inputs for a 3D thermal solver that is applied to the entire cavity and provides a new temperature profile. The whole process is repeated backward, i.e., propagating the solution from the front to the rear facet. After a number of round trips or iterations, the steady state for all the electrical, thermal, and optical variables is found.

TABLE 28.1 Main Physical Effects Included in the Model

Physical Effect or Parameter	Included	Comments
Temperature dependence of energy gap	YES	Empirical Varshni form
Band-gap renormalization	NOT	
Contribution to current density from thermal gradients	YES	Defined by the electron and hole thermoelectric powers
Carrier capture/escape processes in the quantum well (QW)	YES	Defined by electron and hole capture times
Thermionic emission in heterojunctions	NOT	
Fermi–Dirac statistics in bulk materials	NOT	
Dependence of electron and hole mobilities on dopant concentration	YES	Electric field and temperature dependencies not included
Auger recombination	YES	Temperature-dependent Auger parameters
Shockley–Read–Hall (SRH) nonradiative recombination	YES	Complete SRH formula, dependent on both electron and hole carrier concentrations, and on trap properties (density, energy, and degeneration factor of the trap, considering a temperature-dependent capture cross section)
Wavelength dependence of refractive index	YES	Temperature dependence not included
Free-carrier absorption	YES	Linear with local carrier concentration, defined by electron and hole free-carrier absorption coefficients
Nonconstant linewidth enhancement factor	YES	Calculated in QW region from a carrier-dependent differential refractive index, and carrier and wavelength-dependent differential gain
Gain broadening	YES	Lorentzian function
Coulomb enhancement of the gain	NOT	
Spontaneous emission noise	NOT	
Detailed calculation of local heat sources	YES	Local heat sources: Joule effect, nonradiative recombination, free-carrier absorption
Temperature dependence of thermal conductivities	YES	Included in semiconductor layers

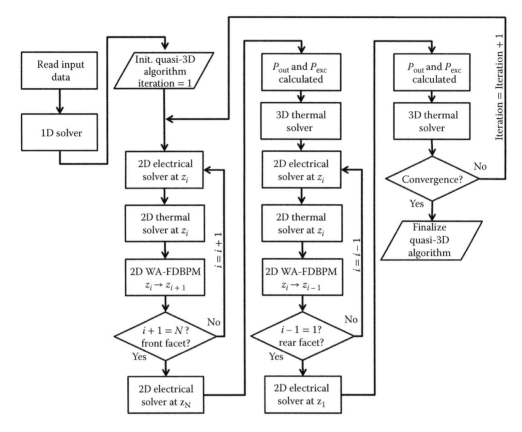

FIGURE 28.2   Main flow of the quasi-3D simulator for tapered lasers CONAN.

The convergence criterion is based on the stability of the optical field at the rear facet after consecutive iterations, both in shape and power. It can be expressed in terms of a parameter ε as

$$\varepsilon = \frac{\left\langle \left| E_n(x) - E_{n-1}(x) \right|, \left| E_n(x) - E_{n-1}(x) \right| \right\rangle}{\left\langle \left| E_n(x) + E_{n-1}(x) \right|, \left| E_n(x) + E_{n-1}(x) \right| \right\rangle} \tag{28.6}$$

where $E_n(x)$ is the optical field at the rear facet after n iterations and $\langle , \rangle$ denotes the scalar product. The iterative process finishes when ε is lower than a threshold value supplied by the user, typically 5×10^{-5}.

As commented previously, in order to filter the front facet back-reflected optical field reaching the ridge section, some tapered devices include beam spoilers (Walpole 1996). The beam spoilers are modeled by setting the optical field to zero at the beam spoiler locations (Mariojouls et al. 2000).

It should be noticed that our algorithm uses a fixed lasing wavelength provided by the 1D simulator. In a more rigorous approach, the lasing wavelength should be recalculated during the quasi-3D algorithm in order to take into account the spatial effects on it. Nevertheless, the comparison with experimental results has proven the approximate validity of our approach.

In some cases, the tapered lasers are fabricated with separate contacts for each section (Paschke et al. 2005; Odriozola et al. 2009; Michel et al. 2009). The model takes into account this possibility and allows the different sections to have different bias voltages. This is implemented as follows: after the initialization, the 1D simulator provides the laser wavelength, the initial photon density, and a bias voltage V_0. Then, for each section i, the applied bias voltage is recalculated as $V_i = V_0 + \Delta V_i$, where ΔV_i can be positive or negative. With these voltage inputs, the simulator proceeds normally and finally, the current at each section is calculated by integration of the current density.

The model can also be applied to the simulation of tapered lasers under patterned injection current (Borruel et al. 2004b). This is an interesting strategy aimed at counteracting the detrimental effects of the SHB. For this, a predefined function that laterally scales the epilayer material conductivities is introduced in the simulator. In this way, the local resistance is modified and therefore the current density profile is also modified according to the conductivity profile.

28.2.2 Model Options Regarding Symmetry

Our simulation procedure is based on launching a trial optical field and solving the forward and backward propagation until convergence (Fox and Li 1961). Depending on the symmetry of the initial trial optical field, the model has three versions:

- Half-cavity (HC) model: This is the basic version of the model. In this version, as in similar models (Williams et al. 1999; Mariojouls et al. 2000), the trial optical field is the fundamental mode of the passive RW section (an even function), although we have checked that the final optical field is independent of the shape of the initial field, provided it is an even function. Since the beam propagation method (BPM) preserves the parity of the field and the device is symmetric, only half of the cavity needs to be explicitly solved, thus reducing the computational effort. However, this version cannot take into consideration the effect of the odd components of the field and fails in reproducing some experimental features of the beam properties of IG tapered lasers, such as the excitation of secondary lateral modes. Therefore, the model was upgraded and two versions including optical fields with both odd and even components were created (Esquivias et al. 2010).
- Full cavity coherent coupling (FCCC) model: In this version, the equations are solved in the full cavity and the initial trial field is an asymmetric field $E_a(x)$, containing odd and even components, i.e., $E_a(x) = E_e(x) + E_o(x)$. The photon density is taken as proportional to $|E_a(x)|^2$ with $E_a(x)$ being the optical field after the propagation by the BPM through the previous slice. This photon density is in general asymmetric, although it can result in a final symmetric solution, depending on the particular device under study. This version often produces a snake-like intensity profile in the plane of the active layer. The corresponding near-field (NF) and far-field (FF) profiles are asymmetric and the intensity profile is not stable after subsequent roundtrips. This kind of behavior has been experimentally observed and attributed to the coherent coupling of frequency-locked lateral modes (Guthre et al. 1994).
- Full cavity incoherent coupling (FCIC) model: As in the FCCC version, the equations are solved in the full cavity and the initial trial field is an asymmetric field $E_a(x)$, but in this case, the photon density is calculated as proportional to $|E_s(x)|^2$, the addition of the field intensities of the even and odd components of $E_a(x)$: $|E_s(x)|^2 = |E_e(x)|^2 + |E_o(x)|^2$. In this way, the photon density is symmetric, and consequently also the carrier and temperature profiles. This approach is equivalent to the simultaneous propagation of an even and an odd field with slightly different frequencies and therefore only coupled through its interaction with the gain medium, i.e., incoherently coupled. This version is more appropriate than the HC model for IG lasers supporting higher order lateral modes besides the fundamental mode.

28.2.3 Model Calibration and Comparison with Experiments

A critical point in all sophisticated laser models is the high number of relatively unknown material parameters. Some of them (energy band parameters, refractive index, etc.) are specific of the different semiconductor materials used in the device. These are usually well known only for the most common binary materials (GaAs, InP), relatively well known for some ternary alloys and relatively unknown for many other alloys. Even for the best known materials, the temperature or wavelength dependence of some parameters has not been reported. In addition, other important parameters (scattering losses, trap characteristics, etc.) depend on the particular fabrication process, and can be different for nominally identical materials. As a consequence, it is always necessary to calibrate the model parameters with experimental results in order to reproduce the experiments and to predict the trends of the device performance when modifying the material composition or the device geometry.

The main idea is to find, if possible, a set of simulation parameters such that the model reproduces the main trends of relevant experimental results, such as power–voltage–current characteristics, FF and NF patterns, and evolution of M^2 and FF and NF patterns with current. If these simulation parameters conveniently account for the main physical effects causing the observed beam properties, then the model will predict qualitatively, and even quantitatively, the performance of new devices based on the same materials but with different design. In other words, it is not so important to use a complete set of correct material parameters, but rather to find out those parameters relevant for the model and related to the main internal process limiting the maximum power or degrading the beam quality.

Our calibration process includes three steps, the first one at theoretical level, the second one by comparing with experimental results in BA lasers, and the third one by comparing with results in tapered lasers. Prior to the calibration, it is important to consider the entire epilayer structure and find out in the literature as many material parameters as possible, especially in the most standard references (Adachi 1992; Vurgaftman et al. 2001). Unfortunately, as stated above, many important parameters, such as refractive index, Auger, and free-carrier absorption coefficients, have never been measured for new materials or alloy compositions, in which case we use judicious guess values based on the parameter values in other alloys with similar gap or composition.

The first step of the calibration procedure is the fitting of the parameters used to calculate the material gain spectra $g_{mat}(\lambda)$ and the spontaneous recombination rate R_{sp}, as a function of the electron and hole quasi-Fermi energies. As it is described in Tijero et al. (2017), the simulation tool operates with a parabolic band model for the calculation of these functions but takes into account band mixing effects by the following fitting procedure: first, we use a valence band (VB) mixing model (Coldren and Corzine 1995), to calculate the quantum well (QW) energy levels and the maximum gain and the spontaneous recombination rate versus carrier concentration, $g_{max}(n)$, and $R_{sp}(n)$. Then, these calculations are fitted by the results provided by the simulation tool using same formulation but considering parabolic valence subbands. The fitting parameters are the QW energy levels, the effective mass of each level, and two scaling parameters (multiplying factors) for $g_{mat}(\lambda)$ and for R_{sp}. Finally, the tool is fed with the parameters that best fit $g_{max}(n)$ and $R_{sp}(n)$ obtained by the VB mixing model. Very good agreement was achieved for a wide range of carrier densities.

The second step in the calibration procedure is to compare 1D simulations with experimental results in BA lasers fabricated with the same epitaxial material than the tapered lasers. It is important to include power–current (P–I) characteristics for lasers with different cavity lengths measured at different temperatures, as well as the FF patterns along the vertical axis. The goal of this step is to determine those simulation parameters related to the material quality (scattering losses and Shockley–Reed–Hall recombination parameters), and also to modify those relevant parameters, which are not well known, such as the Auger coefficient and its temperature dependence. The comparison is made between measured and simulated results in terms of the threshold current density (J_{th}) dependence on the cavity length L, internal quantum efficiency, internal losses, and characteristic temperature T_0. The comparison between measured and calculated FF patterns provides a method to modify the refractive indices, especially that of the QW, which is very important for the calculation of the optical confinement factor in the simulations. The comparison between experimental and simulated current–voltage characteristics provides information on the total resistance of the device, which is relevant for a correct estimation of the Joule heating in the simulations.

The final stage of our calibration procedure includes two parameters: one to account for the carrier-induced refractive index change, and the other one to account for the heat transfer efficiency. For the first one, we focus on its effect on the beam divergence and fit the experimental value of the beam divergence at a fixed output power to the simulated value. This can be achieved by using as fitting parameter the coefficient n_1 that relates the index change and the carrier density assuming a square root dependence (Borruel et al. 2004a). Regarding heat transfer, in order to avoid extending the thermal simulation region, we artificially consider a heat-sink area equal to the device area and use the heat-sink thickness as fitting parameter. We use it to fit the experimental average temperature increase when increasing the current to the simulated temperature increase (on average). The experimental average temperature increase is estimated from the shift of the lasing wavelength when increasing the current, assuming a standard dependence. We have checked that by changing the heat sink thickness, the average value of the temperature changes without important modifications in the lateral and longitudinal temperature profiles.

This procedure was applied to 975 nm (IG and GG with beam-spoilers) and to 735 nm (GG without beam spoilers) tapered lasers in Borruel et al. (2004a). We found a good agreement in the P–I characteristics, shape of the NF and FF patterns, and especially in the evolution of beam properties (M^2, astigmatism, widths of NF at waist and FF patterns) with the injection level. Furthermore, we also found a good agreement between the maximum measured power in the 975 nm GG devices and the maximum power with numerical convergence in the simulations, indicating that the physical mechanism limiting the power was correctly reproduced. This good agreement provided the basis for a new geometrical design, the clarinet laser (Borruel et al. 2005), which showed beam properties similar to those predicted by the simulations. Our model was also applied to simulate the 915-nm IG lasers described in Michel et al. (2005), but in this case the experiments showed a double peak in the NF and FF which was not reproduced by the simulations and gave rise to the upgraded FCIC version of the model previously described. Additional comparisons between experiments and simulations can be found in Odriozola et al. (2009), Michel et al. (2009), and Esquivias et al. (2010).

28.3 Simulation Examples

In order to illustrate the capabilities of the simulation tools for accounting for the behavior of typical tapered laser geometries and guiding mechanisms, we have selected three representative geometries and guiding mechanisms sharing the same epitaxial structure. In this section, we analyze with our simulation tool CONAN the effect of these design parameters on the device performance. We devote specific attention to the comparative analysis of the effects on the beam properties of the device geometry and injection conditions. Since this analysis is presented here mainly for illustrative purposes, some interesting effects will be just overviewed without a detailed study. The half-cavity version of the simulator was used in the three examples. The simulations were performed under isothermal conditions to concentrate the focus in carrier-related effects, since thermal effects are strongly dependent on the value of some relatively unknown material parameters. A detailed discussion on the role of thermal effects in tapered lasers can be found in Esquivias et al. (2010).

28.3.1 Device Geometries and Simulation Parameters

The epitaxial structure of the simulated devices corresponds to that of the 1060 nm GG tapered lasers reported in Ruiz et al. (2009). In brief, it consists of a strained InGaAs QW embedded in a large InGaAsP symmetric optical cavity with AlGaAs cladding regions. We will compare the beam properties of three devices: (1) a GG tapered laser with beam spoilers (GG-BS), (2) a GG tapered laser without beam spoilers (GG-NBS), and (3) a narrow IG laser without beam spoilers (IG).

The geometrical and material parameters used in the simulations are identical for the three devices except for the taper angle and guiding mechanism in the tapered section. The total cavity length is 3 mm and the taper angles are 6° and 1° for the GG and IG devices, respectively. Table 28.2 shows these parameters as well as a brief summary of the most influential material and device parameters used in the

TABLE 28.2 Geometrical Parameters of the Tapered Lasers and Summary of the Most Relevant Material and Device Parameters Used in the Simulation

Symbol	Parameter	Value	Units
L_{RW}	Length of the RW section	1	mm
W_{RW}	Width of the RW section	2.5	μm
L_{TAP}	Length of the taper section	2	mm
R_f	Front facet reflectivity	0.025	
R_b	Back facet reflectivity	0.95	
	Full aperture of beam spoilers	17	μm
	Distance from beam spoilers to back facet	1	mm
α_{tap}	Full taper angle	6 (GG); 1 (IG)	°
Δn_{eff}	Effective index step of the RW section	4.7×10^{-3}	
T_{HS}	Heat sink temperature	20	°C
Γ	Confinement factor	0.0084	
α_{scat}	Scattering losses coefficient	0.5	cm^{-1}
$C_n(C_p)$	Electron (hole) Auger recombination coefficient	$2(2) \times 10^{-30}$	cm^{6}s^{-1}
$k_e, (k_h)$	Electron (hole) free-carrier absorption coefficient	$3(7) \times 10^{-18}$	cm^2
n_I	Differential refractive index coefficient	4.5×10^{-11}	cm$^{3/2}$

simulation. These parameters were extracted from standard references or fitted after applying the calibration procedure described in Section 2.3 to the GG-BS device (Esquivias et al. 2010). The taper angle of the GG devices was selected so as to fit the calculated free diffraction angle assuming an index step $\Delta n_{eff} = 4.7 \times 10^{-3}$.

28.3.2 GG Tapered Laser with Beam Spoilers

The GG-BS device presented here is representative of GG tapered lasers with beam spoilers. At low power, these devices show single-lobed NF and FF patterns and low values of M^2. Figure 28.3a and b illustrates the evolution of the profile of the forward and backward optical field intensities along the cavity at low power (slightly above threshold). A more detailed view of the beam profiles at different longitudinal positions is provided in Figure 28.4a and b. The shape of the fundamental lateral mode of the RW section entering the gain section (see curve A in Figure 28.4a) can be approximated by a Gaussian function. In this example, the calculated full width of the mode at $1/e^2$ (W_{mode}) is 4 μm.

When entering the tapered section, the mode is subjected to two different effects: (1) amplification by the gain medium and (2) free diffraction if the full taper angle is larger than the free diffraction full angle θ_D (at $1/e^2$). The free diffraction angle of an ideal Gaussian beam is given by

$$tg\left(\frac{\theta_D}{2}\right) = \frac{2 \cdot \lambda}{\pi \cdot n_{eff} \cdot w_{mode}} \tag{28.7}$$

where n_{eff} is the effective index of the vertical waveguide. In the case of the GG-BS device under analysis, the full taper angle (6°) has been chosen to match the value of θ_D. Figure 28.4a shows the lateral intensity profile of the forward traveling light at cross sections taken at several positions along the cavity (curve A at $z = 1$ mm, curve B at $z = 2$ mm, and curve C at $z = 3$ mm, $z = 0$ and $z = 3$ mm being the back and the output facet, respectively). The beam expands smoothly as it propagates along the tapered section and reaches the output facet keeping its Gaussian-like profile (curve C), although the wave front has a convex shape, and therefore the phase at the facet is far from being uniform. The reflected (or backward) field continues diffracting in its way back but now the freely diffracting beam does not overlap any more with the gain region and therefore the beam becomes narrower as it propagates backward. This evolution is

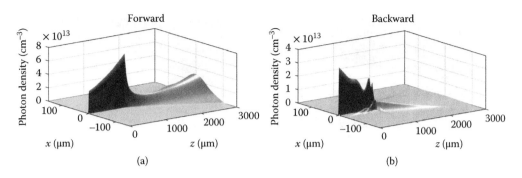

FIGURE 28.3 (a) Forward and (b) backward optical field intensity inside the cavity for the GG-BS tapered laser, when operated at low power (P_{out} = 38 mW).

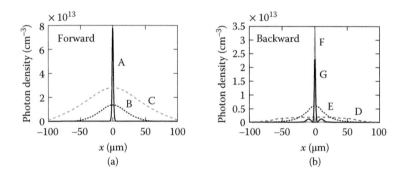

FIGURE 28.4 (a) Forward and (b) backward optical field intensities at different positions inside the cavity for the GG-BS device, when operated at low power (P_{out} = 38 mW). In panel (a): A at z = 1 mm, B at z = 2 mm, and C at z = 3 mm (output facet). In panel (b): D at z = 2 mm, E at z = 1 mm, F at z = 0.5 mm, and G at z = 0 mm (back facet).

illustrated in Figure 28.4b, where curves D and E show the intensity profiles at an intermediate position in the tapered region (z = 2 mm) and at the entrance of the RW section, respectively. As the beam entering the RW section is wider than the fundamental mode of this section (70 versus 4 µm at $1/e^2$), a substantial part of the power is not coupled producing the so-called coupling losses (Walpole 1996) or taper losses. The beam propagating along the RW section is filtered by the single-mode waveguide, with the help of the beam spoilers. Curves F and G are the profiles at the middle of the RW section and at the back facet of the device, respectively. The side lobes of curve F reveal that the filtering effect is still imperfect at the middle of the RW section (see also Figure 28.5b). However, at the back facet, the lobes have virtually disappeared and the beam is as narrow as the fundamental mode of the waveguide revealing that in this case the filtering effect is fully accomplished at the back facet. In other cases, the filtering could be still accomplished in the subsequent forward propagation along the RW and the second pass through the beam spoilers after the reflection at the back facet.

Even more insight into the evolution of the beam inside the cavity can be gained by the gray-scale plots in Figure 28.5. In these plots, the forward (Figure 28.5a) and backward (Figure 28.5b) photon densities in each slice perpendicular to the longitudinal axis have been normalized to their maximum value in the slice and white lines have been drawn at the border of the injected region and at the position of the beam spoilers. In comparison with the smooth and homogeneous expansion of the photon density profile of the forward field in the tapered section, the width of the backward field photon density profile increases from z = 3 mm to about z = 2.5 mm and decreases afterward due to the gain guiding in the narrower part of the tapered region. Nevertheless, at the entrance of the RW section, the photon density profile of the

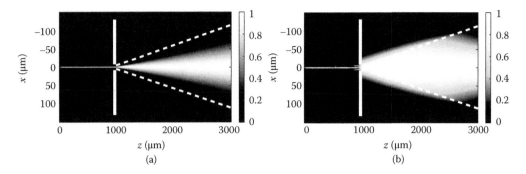

FIGURE 28.5 Normalized (a) forward and (b) backward optical field intensity inside the cavity at each z slice for the GG-BS device, when operated at low power (P_{out} = 38 mW). The white dashed lines indicate the pumped tapered region, while the white solid lines indicate the position of the beam spoilers.

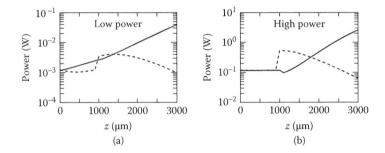

FIGURE 28.6 Optical power of the forward (solid) and backward (dashed) traveling optical fields along the cavity for the GG-BS laser, when operated at (a) low power (P_{out} = 38 mW) and (b) high power (P_{out} = 2.5 W).

backward field still expands far beyond the limits of the injected area. At this point, the filtering role of the beam spoilers is crucial and only tiny diffraction lobes survive after entering the RW section.

To complete the picture, the evolution along the cavity of the total powers carried by the forward and backward beams is shown in Figure 28.6. These powers were obtained by integration of the forward and backward photon densities in the lateral direction at each longitudinal position. In the logarithmic scale of Figure 28.6, an exponential growth of the power is represented by a straight line in which slope is proportional to the effective modal gain, defined as the difference between the modal gain and all the losses occurring in the beam propagation. At low power (Figure 28.6a), the forward field effective gain is constant in both sections. In this case, it is slightly lower in the RW section due to a lower material gain and a slightly worse overlapping with the optical mode in the RW section (not shown). The backward beam power shows initially an exponential growth up to about z = 2.5 mm due to the good overlapping with the gain region. The progressively worse overlapping makes the beam amplification to decreases down to negative values at the entrance of the RW section. At this point, the power drops down due to the filtering effect of the beam spoilers. The subsequent evolution of the backward propagating power is the result of the competing mechanisms of filtering and gain in the RW section, the balance being slightly positive at the back facet. The evolution of the forward and backward propagating powers at high power is illustrated in Figure 28.6b and is mentioned later in this section.

The nearly ideal behavior of the GG-BS device observed at low power changes dramatically when increasing the output power. The main reason for this is the mutual interaction in the semiconductor material between photons and carriers in a feedback loop leading eventually to power saturation and self-focusing at high output powers. Let us start with the optical mode profile. At high injection the mode narrows, thus concentrating a high photon density in the cavity axis. This high photon density depletes the carrier density along the cavity axis due to the higher stimulated recombination in this region, the so-called SHB effect. Figure 28.7a shows the simulated carrier density for the GG-BS device at the output facet for increasing output powers. The initially flat profile of the carrier density evolves when the injection increases to a "batman-ears" like profile, with maxima at the sides and minimum at the center. This profile has been experimentally observed through spontaneous emission measurements (Pagano et al. 2011). The carrier density minimum is limited by the transparency carrier density, 1.2×10^{18} cm^{-3} in our example. The initial simplified approach by Walpole (1996) suggested that the local gain saturation in the cavity axis caused by the SHB would induce an increase of the photon density in the side regions, leading to a top hat shape. But in semiconductor materials, a change of the carrier density produces simultaneously changes in the gain and in the refractive index, which are related by the linewidth enhancement factor. Figure 28.7b and c shows the corresponding gain and index profiles at the power levels of Figure 28.7a. The gain decreases at the cavity axis and the refractive index increases. The shape of the index profile produces a parasitic waveguide for the beam, with more important consequences on the beam shape than the gain profile. The carrier-induced waveguide produces a convergent lens effect during the beam propagation along the tapered region, which concentrates the power density at the center of the beam, thus closing the feedback loop (see Figure 28.4d).

At high power density, the strong feedback induces self-focusing of the beam, leading to saturation of the output power. This effect can be visualized with the help of Figure 28.8, where (as in Figure 28.5) we have plotted the forward (Figure 28.8a) and backward (Figure 28.8b) normalized photon densities at a high power level. In comparison with Figure 28.5a, the forward beam is much narrower, showing clearly the self-focusing. More difficult is the interpretation of the backward beam shape in the tapered section (Figure 28.8b). The expansion of the backward beam beyond the limits of the injected region adds to the gain guiding and the induced index guiding a new degree of complexity making the beam shape extremely difficult to interpret in simple terms. Again, the filtering role of the beam spoilers is apparent in the RW

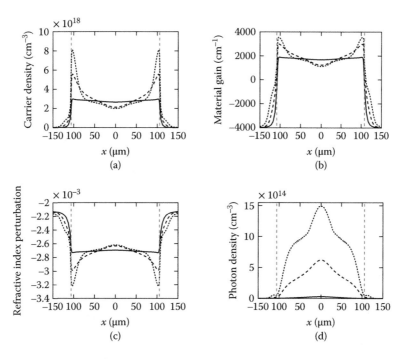

FIGURE 28.7 (a) Carrier density, (b) material gain, (c) refractive index perturbation, and (d) photon density profiles at the output facet for the GG-BS laser at different output powers: $P_{out} = 38$ mW (solid), $P_{out} = 955$ mW (dashed), and $P_{out} = 2.5$ W (dotted). Vertical dashed gray lines indicate the pumped region limits.

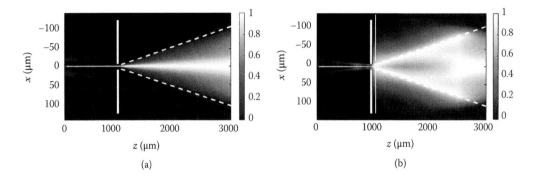

FIGURE 28.8 Normalized (a) forward and (b) backward optical field intensity inside the cavity at each z slice for the GG-BS device, when operated at high power (P_{out} = 2.5 W). The white dashed lines indicate the pumped tapered region, while the white solid lines indicate the position of the beam spoilers.

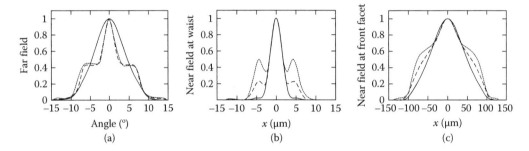

FIGURE 28.9 (a) Far field, (b) near field at waist, and (c) near field at the front facet for the GG-BS device at three power levels, P_{out} = 38 mW (solid), P_{out} = 955 mW (dashed), and P_{out} = 2.5 W (dotted).

section. The small diffraction lobes surviving at low power are now more noticeable and can even reach the back facet from where they are reflected as can be (hardly) seen in Figure 28.8a.

The evolution of the integrated forward propagating power at high injection (Figure 28.6b) shows also clear differences with respect to the low injection behavior. The nearly homogeneous effective gain for the forward field at low power (Figure 28.6a) becomes clearly different in the RW and tapered sections at high power. In the RW section, the effective gain vanishes revealing a strong gain saturation confirmed by a carrier density in this section slightly higher than the transparency value (not shown). As the total round-trip gain should be constant, the absence of gain in the RW section is compensated with the high values of the effective gain at the beginning of the tapered section. Further in the tapered section, the forward effective gain decreases slowly due once more to gain saturation. In contrast, the evolution of the backward propagating power at high injection is not significantly different from the low-injection behavior. In the RW section, after the sudden drop of the power due to the beam spoilers, the modal gain remains constant at a value close to zero as for the forward beam.

The SHB and self-focusing of the beam at high injection not only limit the maximum power, but also degrade the beam quality. Figure 28.9 shows the FF pattern and the NF patterns at waist and at the facet, at three power levels. Under simplifying ideal assumptions, the expected value of the FF width can be estimated by applying Snell's law to the beam at the output facet. In this case, this yields a full beam divergence angle θ_{out} (at $1/e^2$) $\sim n_{eff} \cdot \theta_D = 20°$ (n_{eff} = 3.34). The simulated value at low power is θ_{out} = 16.5°, not far from the previous estimation. It is clear how the shape of the beam is modified by the carrier-induced convergent lens when increasing the power: the NF patterns at waist and at the front facet develop shoulders and the FF patterns evolve into a narrower central lobe together with the apparition of side lobes. These effects produce also an increase of the astigmatism and the value of M^2 is shown in Section 3.5.

28.3.3 GG Tapered Laser without Beam Spoilers

The GG-NBS device presented here is representative of GG tapered lasers without beam spoilers. In spite of the advantages of the use of beam spoilers, they have also drawbacks due to the additional processing steps and also to the possible introduction of defects close to the active region. In fact, very high power levels have been reported for tapered lasers without beam spoilers (Sumpf et al. 2009; Müller et al. 2016). In this case, the filtering properties of the RW section had been improved by increasing its length, after a careful balance to optimize this design parameter (Wenzel et al. 2003). The main characteristic of the beam experimentally observed in tapered lasers without beam spoilers is a multilobed FF pattern, with an increasing number of lobes together with an increase of the peak to valley ratio when increasing the output power (Borruel et al. 2004a; Fiebig et al. 2009; Sumpf et al. 2009; Müller et al. 2016).

Figure 28.10 shows the simulated evolution of the FF and NF profiles of the GG-NBS device as a function of the output power. The relatively smooth profiles of the FF and NF at low power evolve when the injection increases to a more structured profile with an increasing number of more and more distinguishable lobes. The FF width at $1/e^2$ decreases slightly when the power increases, while the $1/e^2$ NF width at waist remains almost invariant, in agreement with what has been experimentally observed (Fiebig et al. 2009). These behaviors yield an almost invariant low value of $M^2(1/e^2)$. However, the value of M^2 (second moment) significantly increases with the power, mainly due to the tiny side lobes below the $1/e^2$ level in the NF at waist (see inset in Figure 28.10b). These differences will be commented later in Section 3.5. The agreement between the appearance and the evolution of the FF and NF lobes in simulation and experiments provides support for the use of the simulation tool to provide a physical understanding of the origin of these lobes in devices without beam spoilers.

Aiming at this, for an output power level of 0.26 W, Figure 28.11 shows gray-scale plots of the forward (Figure 28.11a) and backward (Figure 28.11b) propagating photon densities, normalized to their maximum value at each cavity position. The side lobes that are already apparent at the beginning of the tapered section, propagate toward the front facet, and finally result in the multilobed NF pattern shown in Figure 28.10c. During the backward propagation (Figure 28.11b), the highest intensity side lobes run away the tapered region where they extinguish without reaching the RW section. However, the backward field entering the RW section is not perfectly filtered and a fraction of it is diffracted by the RW section aperture producing side lobes in the field that reaches the back facet. This residual field at the sides of the RW section is reflected by the back facet and interferes in the tapered section with the forward field arising directly from the RW section, thus producing the multiple peaks observed in the beam characteristics.

The lack of a complete filtering in the RW section is more important at high power, as the intensity of the backward field increases and optically pumps the sides regions around the RW, thus reducing the absorption. This effect is further illustrated in Figure 28.12, where we have plotted the forward and backward photon densities (Figure 28.12a) and the carrier density (Figure 28.12b) at the interface between the

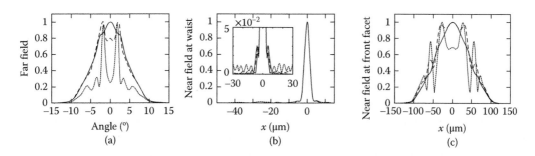

FIGURE 28.10 (a) Far field, (b) near field at waist, and (c) near field at the front facet for the GG-NBS device at three power levels, $P_{out} = 40$ mW (solid), $P_{out} = 260$ mW (dashed), and $P_{out} = 393$ mW (dotted). The inset in (b) is a magnification emphasizing the side lobes far from the axis appearing at $P_{out} = 260$ and 393 mW.

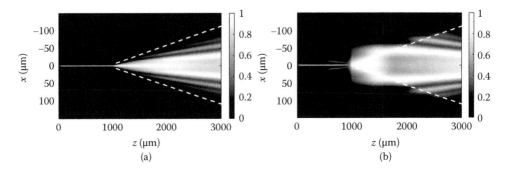

FIGURE 28.11 Normalized (a) forward and (b) backward optical field intensity inside the cavity at each z slice for the GG-NBS device, when operated at high power (P_{out} = 260 mW). The white dashed lines indicate the pumped tapered region.

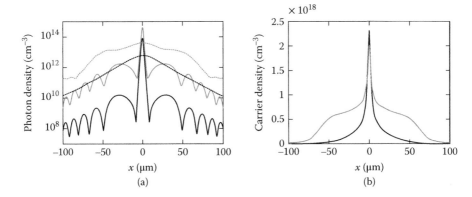

FIGURE 28.12 (a) Forward (solid) and backward (dotted) photon densities and (b) carrier density profiles at the interface between the RW (2.5μm wide) and tapered sections for the GG-NBS device at two output powers P_{out} = 40 mW (black), P_{out} = 260 mW (gray).

RW and tapered sections at two power levels. At low power, the side lobes of the forward field are around 40 dB below the main lobe in the RW (Figure 28.12a) and therefore, this results in a single-lobed NF at the facet (see Figure 28.10c). The backward field at the sides of the RW is low and therefore the induced carrier density (Figure 28.12b) is far below transparency; in consequence, this region is highly absorbent. However, at medium and high injection levels, the backward field intensity is high enough for pumping the sides of the RW to a carrier concentration close to transparency (Figure 28.12b), thus decreasing their absorption. As a result, the forward field intensity at the side lobes becomes only around 20 dB lower than the maximum (Figure 28.12a) giving rise to the multilobed profile obtained at medium and high power level.

Our previous simulations of GG tapered lasers without beam spoilers have shown the same trends as the experimental results (Borruel et al. 2004a; Odriozola et al. 2009), but the quantitative agreement is not as good as we have found in tapered laser with beam spoilers. Furthermore, the maximum output power with numerical convergence in the simulations is usually lower than the measured maximum power. We attribute these discrepancies to the limitations of our steady-state single-frequency model. We think that in the real device, there is a complex dynamics of the different lateral modes giving rise to rapidly varying NF and FF patterns. As the measured NF and FF patterns are temporal averages, it is expected that the narrow and pronounced lobes would average resulting in smoother profiles. In fact, the lack of convergence in the simulations is due to different shapes and positions of the lobes after subsequent round-trips, yielding a stable output power but a different field profile after each iteration. This is illustrated in Figure 28.13, where we have plotted the NF at the facet after two consecutive round-trips for simulation conditions in which a stable solution is not found. The averaging of the field intensity after different roundtrips would produce less pronounced lobes as they appear in the experimental results.

28.3.4 Narrow Index-Guided Tapered Laser

The IG device analyzed in this section is representative of narrow IG tapered lasers. Significant differences with respect to the behavior of large angle GG devices arise from the fact that the single lateral optical mode launched by the RW section does not just expand by free diffraction into an injected tapered region designed to match the free diffraction angle. Instead, the beam is guided in a narrow tapered section defined by the refractive index step, where injection takes place.

The gray-scale plots of Figure 28.14 are illustrative of the forward and backward beam propagation in the IG device at low power (notice the different lateral dimension with respect to the corresponding plots for GG devices). The forward beam expands preserving its shape, with most of the power (99.8% at $z = 2$ mm) inside the guiding region (Figure 28.14a). The propagation of the backward field is determined by a combination of competitive phenomena: diffraction, gain and index guiding, and reflections at the waveguide interface, resulting in a multilobed profile (Figure 28.14b). The index guiding, as well as the relatively small taper angle, produces a beam entering the RW which is narrower than that of the GG devices, hence reducing the taper losses. The RW section acts again as a spatial filter, and the beam recovers its original single-mode shape after a complete round trip.

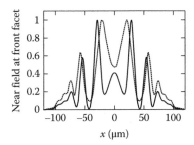

FIGURE 28.13 Example of the simulated NF profiles at the facet for two consecutive iterations of a simulation of the GG-NBS device under conditions in which a stable solution is not found.

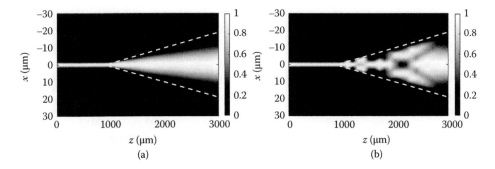

FIGURE 28.14 Normalized (a) forward and (b) backward optical field intensity inside the cavity at each z slice for the IG device, when operated at very low power ($P_{out} = 40$ mW). The white dashed lines indicate the pumped tapered region and the index step.

The main effect degrading the beam quality of IG tapered lasers is carrier-induced lensing and self-focusing, similar to the effect analyzed in Section 28.3.2 for GG devices, but with important differences. Figure 28.15 shows the simulated evolution of the FF and NF profiles of the IG device as a function of the output power. Both the profiles as well as the evolution with power are completely different than those of the GG devices shown in Figures 28.9 and 28.10. At very low power (40 mW), since the beam expands in the tapered section covering almost all the guiding region, the simulated FF width at $1/e^2$ is $\theta_{out} = 3.28°$, close to 3.34°, the value resulting from applying Snell's law to the taper angle. In consequence, the beam is clearly astigmatic (204 μm). But the picture changes dramatically when increasing the power: the carrier lensing reduces the width of the beam at the output facet down to sizes for which diffraction effects become relevant, and consequently, the angular width of the FF patterns increases with the output power while the NF at waist and the NF at the facet narrow. In addition, the carrier lensing also produces an almost collimated beam inside the cavity (see Figure 28.16). Therefore, the virtual source position shifts toward the output facet and even beyond, giving rise to a fast decrease of the astigmatism to zero or even to negative values. The evolution of the beam inside the tapered region, as a consequence of the carrier-induced graded index profile (Figure 28.16), resembles that of a graded-index lens or an optical fiber. The strong SHB at high power produces a self-focusing of the beam leading to saturation of the output power, as will be shown in the next section.

28.3.5 Comparison between Devices

In this section, the main performance parameters of the three simulated devices are comparatively analyzed. Figure 28.17a shows the P–I characteristics, and Table 28.3 summarizes the main parameters

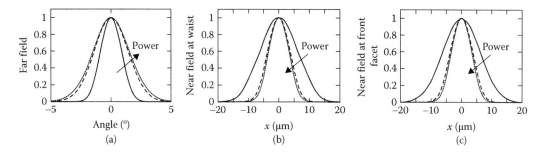

FIGURE 28.15 (a) Far field, (b) near field at waist, and (c) near field at the front facet for the IG device at three power levels, $P_{out} = 40$ mW (solid), $P_{out} = 623$ mW (dashed), and $P_{out} = 912$ mW (dotted).

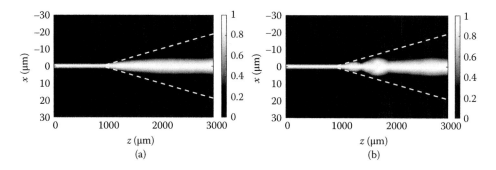

FIGURE 28.16 Normalized (a) forward and (b) backward optical field intensity inside the cavity at each z slice for the IG device, when operated at high power ($P_{out} = 912$ mW). The white dashed lines indicate the pumped tapered region and the index step.

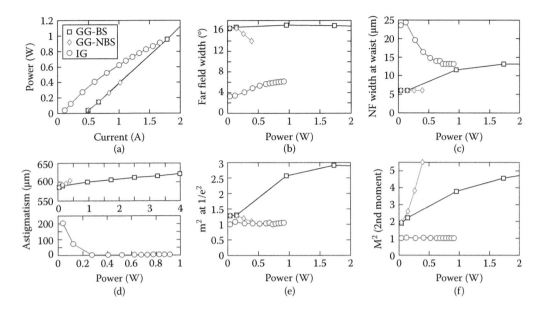

FIGURE 28.17 Comparison of the figures of merit for the different devices: GG-BS (squares), GG-NBS (diamonds), and IG (circles). (a) Power versus current characteristics, (b) far-field width versus power, taken at FWHM (open) and $1/e^2$ (closed), (c) near field at waist width versus power, taken at FWHM (open) and $1/e^2$ (closed), (d) astigmatism versus power, (e) M^2 $(1/e^2)$ versus power, and (f) M^2 (FWHM) versus power.

TABLE 28.3 Parameters Extracted from the Simulated P–I Characteristics of the Three Devices

	GG-BS	GG-NBS	IG
Threshold current (A)	0.43	0.43	0.066
Threshold current density (A/cm²)	198	197	156
Slope efficiency at threshold (W/A)	0.625	0.625	0.85
Taper losses (cm⁻¹)	8.6	8.6	5.4

extracted from these results: the threshold current and threshold current density, the slope efficiency, and the taper losses. The threshold currents of the GG-BS and GG-NBS devices are similar and much higher than that of the IG device, as expected from the comparison of the device area; however, the threshold current density of the IG device is lower, due to the lower taper losses. The taper losses at threshold can be estimated from the simulation results by considering the threshold current density and the modal gain versus current density characteristic, and taking into account the mirror and the internal losses. Values of 8.6 and 5.4 cm^{-1} are obtained for the GG and IG devices, respectively, the lower value for the IG device due to the better matching of the backward field. This difference in the taper losses is also the reason for the higher slope efficiency of the IG device. However, the strong narrowing of the beam in this device due to carrier lensing makes the slope efficiency to decrease when the injection increases (see Figure 28.17a).

Figure 28.17b and c shows the evolution of the widths of the FF and of the NF at waist (at $1/e^2$), respectively, for the three simulated devices. At low power, the FF widths of the GG and IG devices are 16.5° and 3.3°, respectively, not far from the expected value for a Gaussian beam applying Snell's law, as previously discussed. On the contrary, the NF at waist is much wider in the case of the IG device, as expected from its lower divergence. The astigmatism at low power (Figure 28.17d) is also quite different, with the virtual source (or beam waist) located at a distance behind the front facet $z_{vs} = 590$ μm and $z_{vs} = 204$ μm for the GG and IG devices, respectively. The former value is close to $L_{tap}/n_{eff} = 599$ μm, the expected value for the free diffraction of a beam at the output of the RW; the latter is lower, even at very low power, due to the index-guiding mechanism.

As previously mentioned, the devices show clear differences in the evolution of their beam characteristics when power increases. In summary: in the GG-BS device, the FF width at $1/e^2$ is approximately constant and the waist increases, in the GG-NBS the waist is constant and the FF width decreases, and in the IG device the FF broadens and the waist remains constant. These differences are also reflected in the behavior of the astigmatism (Figure 28.17d), which increases with the power for the GG devices while decreases for the IG laser. Finally, the evolution of $M^2(1/e^2)$ and M^2 (second moment) for the three devices is compared in Figure 28.17e and f, respectively. The best beam quality is obtained for the IG device with a M^2 (second moment) close to the unity, indicating a diffraction-limited beam. When power increases, the increase of the value of M^2 (second moment) is much faster for the GG-NBS device than for the GG-BS laser due to the fact that the multiple lobes in both the FF and the NF at waist of the first one strongly affect the second moment widths. On the contrary, the value of $M^2(1/e^2)$ is very low, close to the unity, for the GG-NBS device, even at the highest power in the simulations. Similar discrepancies between M^2 (second moment) and $M^2(1/e^2)$ have been reported experimentally for GG devices without beam spoilers (Fiebig et al. 2009).

28.4 Summary

We have presented an overview of the current state of the art in the modeling of high-power tapered lasers, with a detailed description of our steady-state single frequency quasi-3D simulation tool. We have explained the calibration procedure required for making meaningful comparisons with experiments and using the simulator as a predictive tool. The capabilities of the model have been illustrated by comparing the beam properties of three different types of tapered lasers emitting at 1060 nm: GG with beam spoilers, GG without beam spoilers, and narrow IG. The simulations reproduce the different behaviors experimentally observed in the three types of device and can be used for a better understanding of the interaction between carriers and photons that determines the operation of tapered lasers.

Acknowledgments

This work was supported by Ministerio de Economía y Competitividad of Spain under projects RANGER (TEC2012-38864-C03-02) and COMBINA (TEC2015-65212-C3-2-P); and by the Comunidad de Madrid under program SINFOTON-CM (S2013/MIT-2790). A. Pérez-Serrano acknowledges support from Ayudas a la Formación Posdoctoral 2013 program (FPDI-2013-15740). The authors acknowledge the contribution of L. Borruel, H. Odriozola, S. Sujecki, and E. C. Larkins to the development of the model and code for the simulation of tapered lasers.

References

Adachi, S. 1992. *Physical Properties of III-V Semiconductor Compounds: InP, InAs, GaAs, GaP, InGaAs, and InGaAsP*, New York: Wiley-Interscience.

Auzanneau, S., Krakowski, M., Berlie, F., et al. 2003. High-power and high-brightness laser diode structures at 980 nm using an Al-free active region. *Proc SPIE.* 4995:184–195.

Blume, G., Kaspari, C., Feise, D., et al. 2012. Tapered diode lasers and laser modules near 635 nm with efficient fiber coupling for flying-spot display applications. *Opt Rev.* 19(6):395–399.

Borruel, L., Esquivias, I., Moreno, P., et al. 2005. Clarinet laser: Semiconductor laser design for high-brightness applications. *Appl Phys Lett.* 87:101104.

Borruel, L., Sujecki, S., Esquivias, I., et al. 2002. A self-consistent electrical, thermal and optical model of high brightness tapered lasers. *Proc SPIE.* 4646:355–366.

Borruel, L., Sujecki, S., Moreno, P., et al. 2004a. Quasi-3-D simulation of high-brightness tapered lasers. *IEEE J Quantum Electron.* 40:463–472.

Borruel, L., Sujecki, S., Moreno, P., et al. 2004b. Modeling of patterned contacts in tapered lasers. *IEEE J Quantum Electron.* 40:1384–1388.

Coldren, L. A. and Corzine, S. W. 1995. *Diode Lasers and Photonic Integrated Circuits.* New York, NY: John Wiley & Sons.

Esquivias, I., Odriozola, H., Tijero, J. M. G., et al. 2010. Simulation of high brightness tapered lasers. *Proc SPIE.* 7616:76161E.

Fiebig, C., Blume, G., Uebernickel, M., et al. 2009. High-power DBR-tapered laser at 980 nm for single-path second harmonic generation. *IEEE J Sel Top Quantum Electron.* 15(3):978–983.

Fox, A. G. and Li, T. 1961. Resonant modes in a maser interferometer. *Bell Syst Tech J.* 40(2):453–488.

Guthre, J., Tan, G. L., Ohkubo, M., et al. 1994. Beam instability in 980 nm power laser: Experiment and analysis. *IEEE Photon Technol Lett.* 6(12):409–1411.

Hadley, G. R. 1992. Wide-angle beam propagation using Padé approximant operators. *Opt Lett.* 17(20):1426–1428.

Harold^{TM} 3.0 Reference Manual. 2001. Photon Design, Oxford (UK).

Hasler, K.-H., Sumpf, B., Adamiec, P., et al. 2008. 5-W DBR tapered lasers emitting at 1060 nm with a narrow spectral linewidth and a nearly diffraction-limited beam quality. *IEEE Photon Technol Lett.* 20:1648–1650.

ISO 11146. 2005. Lasers and laser-related equipment—Test methods for laser beam widths, divergence angles and beam propagation ratios.

Kintzer, E. S., Walpole, J. N., Chinn, S. R., et al. 1993. High-power, strained-layer amplifiers and lasers with tapered gain regions. *IEEE Photon Technol Lett.* 5:605–607.

Krakowski, M., Auzanneau, S. C., Berlie, F., et al. 2003. 1 W high brightness index guided tapered laser at 980 nm using Al-free active region materials. *Electronics Lett.* 39(15):1122–1123.

Krakowski, M., Auzanneau, S. C., Calligaro, M., et al. 2002. High power and high brightness laser diode structures at 980 nm using Al-free materials. *Proc SPIE.* 4651:80–91.

Lang, R. J., Dzurko, K., Hardy, A. A., et al. 1998. Theory of grating-confined broad-area lasers. *IEEE J Quantum Electron.* 34:2196–2210.

Mariojouls, S., Margott, S., Schmitt, A., et al. 2000. Modeling of the performance of high-brightness tapered lasers. *Proc SPIE.* 3944:395–406.

Michel, N., Hassiaoui, I., Calligaro, M., et al. 2005. High-power diode lasers with an Al-free active region at 915 nm. *Proc SPIE.* 5989:598909.

Michel, N., Odriozola, H., Kwok, C. H., et al. 2009. High modulation efficiency and high power 1060 nm tapered lasers with separate contacts. *Electron Lett.* 45(2):103–104.

Müller, A., Fricke, J., Bugge, F., et al. 2016. DBR tapered diode laser with 12.7 W output power and nearly diffraction-limited, narrowband emission at 1030 nm. *Appl Phys B.* 122:87.

O'Brien, S., Welch, D. F., Parke, R. A., et al. 1993. Operating characteristics of a high-power monolithically integrated flared amplifier master-oscillator power-amplifier. *IEEE J Quantum Electron.* 29:2052–2057.

Odriozola, H., Tijero, J. M. G., Borruel, L., et al. 2009. Beam properties of 980 nm tapered lasers with separate contacts: Experiments and simulations. *IEEE J Quantum Electron.* 45(1):42–50.

Pagano, R., Ziegler, M., Tomm, J. W., et al. 2011. Two-dimensional carrier density distribution inside a high power tapered laser diode. *Appl Phys Lett.* 98(22):221110.

Paschke, K., Bogatov, A., Drakin, A., et al. 2003. Modeling and measurements of the radiative characteristics of high-power α-DFB lasers. *IEEE J Select Topics Quantum Electron.* 9:835–843.

Paschke, K., Sumpf, B., Dittmar, F., et al. 2005. Nearly diffraction limited 980-nm tapered diode lasers with an output power of 7.7W. *IEEE J Select Topics Quantum Electron.* 11(5):1223–1227.

Pfahler, C., Eichhorn, M., Kelemen, M. T., et al. 2006. Gain saturation and high-power pulsed operation of GaSb-based tapered diode lasers with separately contacted ridge and tapered section. *Appl Phys Lett.* 89:021107.

Ruiz, M., Odriozola, H., Kwok, C. H., et al. 2009. High-brightness tapered lasers with an Al-free active region at 1060 nm. *Proc SPIE.* 7230:72301D.

Siegman, A., Nemes, G., and Serna, J. 1998. How to (maybe) measure laser beam quality. *OSA TOPS.* 17(2):184–199.

Spreemann, M., Lichtner, M., Radziunas, M., et al. 2009. Measurement and simulation of distributed-feedback tapered master-oscillator power-amplifiers. *IEEE J Quantum Electron.* 45:609–616.

Sujecki, S., Borruel, L., Wykes, J., et al. 2003. Nonlinear properties of tapered laser cavities. *IEEE J Sel Top Quantum Electron.* 9:823–834.

Sumpf, B., Hasler, K.-H., Adamiec, P., et al. 2009. High-brightness quantum well tapered lasers. *IEEE J Sel Top Quantum Electron.* 15(3):1009–1020.

Sumpf, B., Hülsewede, R., Erbert, G., et al. 2002. High-brightness 735 nm tapered diode lasers. *Electron Lett.* 38(4):183–184.

Tijero, J. M. G., Pérez-Serrano, A., del Pozo, G., and Esquivias, I. 2017. Tapered semiconductor optical amplifiers. In *Handbook of Optoelectronic Device Modeling and Simulation.* vol 1, (Piprek, J., ed.) pp.695–712. Boca Raton, FL: Taylor & Francis.

Vurgaftman, I., Meyer, J. R., and Ram-Mohan, L. R. 2001. Band parameters for III–V compound semiconductors and their alloys. *J Appl Phys.* 89(11):5815–5875.

Walpole, J. N. 1996. Semiconductor amplifiers and lasers with tapered gain regions. *Opt Quantum Electron.* 28:623–645.

Wenzel, H., Sumpf, B., and Erbert, G. 2003. High-brightness diode lasers. *C R Physique.* 4:649–661.

Wilson, F. J., Lewandowski, J. J., Nayar, B. K., et al. 1999. 9.5W cw output power from high brightness 980 nm InGaAs/AlGaAs tapered laser arrays. *Electron Lett.* 35:43–45.

Williams, K. A., Penty, R. V., White, I. H., et al. 1999. Design of high-brightness tapered laser arrays. *IEEE J Select Topics Quantum Electron.* 5:822–831.

29

High-Brightness Laser Diodes with External Feedback

Mohamad Anas Helal

Simeon N. Kaunga-Nyirenda

Steve Bull

and

Eric Larkins

High-power laser diodes first gained interest as pump sources for solid-state lasers, with cost/watt and reliability as the primary market drivers. In recent years, new markets have been opened by increases in the brightness of the laser diode sources and by advances in the optical systems for beam shaping and combining. The work in this chapter is motivated by the development of high-brightness laser diodes for direct-diode laser systems targeting industrial applications and a desire to illustrate the important role of laser simulation tools at both the device and system level. Traditionally, welding and sheet metal cutting are the most lucrative industrial laser markets, but they are also the most demanding in terms of brightness. Laser additive manufacturing processes, such as selective laser melting, are also quickly becoming a reality and allow the fabrication of structures that cannot be made by traditional means. Their automated nature is opening the door to new manufacturing paradigms.

Direct-diode laser systems for industrial applications combine the beams from many individual laser diodes (or diode arrays) to couple them into an optical fiber for delivery to the target. The role of the optical system is to combine the individual beams without losing their brightness. High-power direct-diode lasers rely on multiple beam-combining methods, including incoherent or "side-by-side" beam combining, polarization multiplexing, and spectral beam combining. The first commercial kW-class direct-diode laser system used an external cavity to stabilize the wavelengths of, and spectrally combine, the beams of a large number of broad-area (BA) lasers to couple a 1 kW beam into a 200 μm fiber (Huang et al., 2011). The brightness of the final system is ultimately limited by the brightness of the individual sources.

All beam-combining techniques couple light back into the laser diodes—deliberately, as in external wavelength stabilization and spectral beam combining, or accidentally due to reflections off the optics. This optical feedback is known to affect both the beam quality and the degradation of the laser diodes (Hempel et al., 2013), but little has been published on how this feedback affects the operation of the laser. As the power and performance requirements increase, the role of external optical feedback is becoming increasingly interesting.

This chapter focuses on the impact of external optical feedback on high-brightness laser diodes. We start by introducing the common beam quality metrics used for lasers and briefly review the diode laser technologies most commonly considered for high-brightness direct-diode laser systems, before focusing on the tapered laser and showing why they are strong contenders for future high-power direct-diode laser systems. We then describe the coupling of our continuous wave (CW) simulation tool, *Speclase*, to commercial optical design tools to self-consistently simulate high-brightness diode lasers with external optical feedback. We conclude with a case study exploring the impact of unintentional feedback on the excitation of higher order vertical modes and lateral beam quality in a large optical cavity (LOC) tapered laser.

29.1 Power Scaling and the Role of Beam Quality

The purpose of the beam-combining optics in a direct-diode laser is usually to couple the highest amount of power from the individual laser diode sources into the end of an optical fiber. The brightness of the combined beam (units = W cm^{-2} sr^{-1}) limits the power that can be coupled into the delivery fiber and depends on the brightness of the individual diode sources.

Metrics such as the beam parameter product (BPP) Q or beam propagation factor M^2 are common metrics used by manufacturers and industries to specify the beam quality of both the individual laser diode and direct-diode laser systems. The BPP is the product of beam radius (measured at the beam waist) and the half-angle beam divergence:

$$Q = \omega_0 \times \theta_{\text{div}}, \tag{29.1}$$

where ω_0 is the beam radius measured in millimeter and θ_{div} measured in milliradian and Q is the BPP measured in mm·mrad. The BPP allows optical designers to determine the number of individual beams that can be imaged onto the end of a fiber with a fixed diameter (physical aperture) and maximum acceptance angle (numerical aperture). The power coupled into the fiber also depends on the power of the combined beam (and hence of the individual emitters). The brightness (units: W cm^{-2} sr^{-1}) is defined as

$$B = \frac{P}{\pi^2 Q^2} = \frac{P}{\lambda^2 M^2} = \frac{P}{\lambda^2 M_x^2 M_y^2}, \tag{29.2}$$

where P is the laser output power, Q is the BPP, λ is the wavelength and M^2 is the beam propagation factor. M^2 is another measure of beam quality as defined by ISO 11146. The beam quality metrics, M^2 and Q are related by Equation 29.2, as shown in Table 29.1 for $\lambda = 975$ nm. (A detailed discussion of beam quality metrics is in Chapter 28.)

Figure 29.1 shows the power and beam quality needed for different industrial laser applications. The dashed lines show the improvement in laser diodes (and direct-diode laser systems based on them) between 2000 and 2014. A laser with a power of 20 W and $M^2 \sim 20$–30 (e.g., a good BA laser diode [Thestrup et al., 2003]) is suitable for printing and material processing, but not for additive manufacturing, welding, or cutting. Conversely, a laser with a power of 10 W and $M^2 < 1.5$ (e.g., a good tapered laser diode [Fiebig et al., 2008]) is suitable for sheet metal drilling, marking, additive manufacturing, welding, and cutting of metal.

TABLE 29.1 Correspondent Values of M^2 and Q

M^2	Q mm·mrad
1.0	0.31
1.5	0.47
2.0	0.62
3.0	0.93
5.0	1.6
10.0	3.1
20.0	6.2
30.0	9.3

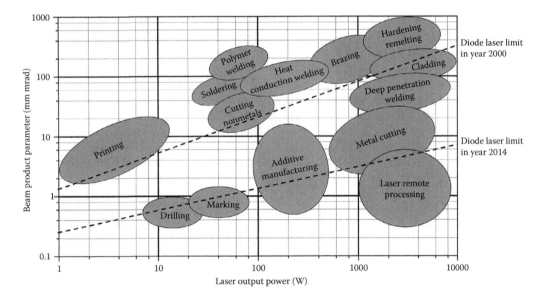

FIGURE 29.1 Beam parameter product versus laser power for different applications.

29.2 High-Brightness Laser Diode Sources

Laser diodes for the most demanding industrial applications should have high brightness (e.g., $P \geq 10$ W, $Q < 1$ mm·mrad, or $M^2 < 3$) and high-power conversion efficiency (PCE) (e.g., $PCE \geq 55\%$). The sources (and their performance) should be reliable and insensitive to reflections.

29.2.1 Vertical Cavity Design

The vertical cavity design of a laser diode for efficient, high-power operation in an external cavity is more complicated than that of an isolated high-power laser diode. First, a wide vertical mode profile (vertical near-field pattern) is needed for efficient external cavity coupling and reduced alignment tolerance. This also increases the tolerance to "smile" (a bend in the horizontal axis of a laser bar, which introduces pointing errors in the slow axis) across a laser array and allows the use of simpler, less costly optics. Second, the excitation and lasing of higher order vertical cavity modes by external optical feedback must be suppressed; photons coupled from external cavity will lead to stimulated emission. Third, a narrow vertical far-field divergence is needed to minimize the impact of optical aberrations, thereby improving the cost and performance of the optical system. Finally, the need for high-PCE means that Joule heating and free-carrier absorption (FCA) play critical roles in the vertical cavity design (Larkins et al., 2014).

High-brightness laser diodes typically use LOC waveguides to produce a wide vertical near-field pattern and narrow far-field divergence pattern. The LOC cavity reduces the power density at the facet (increasing the mirror damage threshold) and reduces the vertical confinement factor (reducing carrier-induced lensing). The confinement factor is the ratio of modal power contained in the active area to that contained in the structure. Below is the definition of the vertical confinement factor:

$$\Gamma(y) = \frac{\displaystyle\int_{-d/2}^{+d/2} E(y)^2 \mathrm{d}y}{\displaystyle\int_{-\infty}^{+\infty} E(y)^2 \mathrm{d}y}. \tag{29.3}$$

The LOC waveguide also facilitates external cavity feedback (e.g., for wavelength stabilization and/or lateral mode filtering), but also makes the device more sensitive to parasitic reflections.

As the LOC waveguide thickness increases beyond a certain point, the influence of the cladding layer diminishes and the waveguide is formed by the index contrast of the active region and the waveguide. Further increases in waveguide thickness raise the Joule heating and FCA losses, but do not significantly increase the near-field width or reduce the far-field divergence. Wider near-field patterns can be achieved (with thinner, more efficient waveguides) using other methods, including low-index quantum barriers to reduce the index contrast between the active region and the waveguide (Wang et al., 2013); or high-index optical traps to draw the field profile out into the waveguide (Buda et al., 1999).

LOC waveguides support multiple vertical modes, but, in the absence of external feedback, lasing is limited to the mode that reaches threshold first. When operated in a system with external feedback, however, back coupling to higher order modes can cause them to lase. Even if the higher order modes do not lase, the additional stimulated emission and FCA can reduce the laser's efficiency and increase self-heating. External feedback also affects the reliability of high-power laser diodes (Tomm et al., 2011). Thus, feedback is becoming important as diode laser systems grow in power and complexity.

The suppression of higher order vertical modes becomes more challenging when the laser is operated in an external cavity, where the optical feedback also excites the higher order vertical modes. First, as little power as possible must couple into the higher order vertical modes. Second, the higher order vertical modes must be prevented from reaching threshold. This can be achieved by engineering the laser cavity to make the confinement factors of higher order modes much lower than that of the fundamental mode, so that their modal gain is lower. Accordingly, we define a new figure of merit for modal discrimination (MD):

$$\mathrm{MD} = \Gamma(1)/\Gamma(n), \tag{29.4}$$

where Γ is the optical confinement factor and n refers to the higher order mode with the largest confinement factor. MD_n can also be used to describe the MD of a particular mode n. (*Note:* Propagation loss is neglected in MD, as it depends on the doping profile and operating bias—which are usually optimized *after* the initial cavity design.)

Power from the external cavity can be coupled into the higher order modes, even if they do not reach threshold. The modal gain of the higher order modes can also be reduced by increasing their propagation loss—either by FCA or by substrate leakage. Damping of the higher order vertical modes also reduces the total amplified spontaneous emission (ASE) from these parasitic modes. ASE and gain of light coupled into these modes do not contribute to the fundamental mode. Instead, they act as current leakage paths, contributing to self-heating and reducing the PCE.

LOC laser structures based on both the low-index quantum barrier and the high-index optical trap approaches were optimized (optically and electrically) for operation with external optical feedback (Larkins

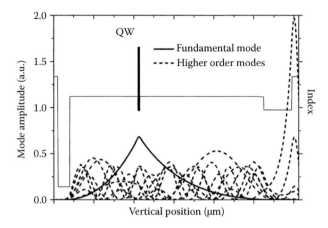

FIGURE 29.2 Vertical mode profiles in the ELoD2 LOC laser structure.

TABLE 29.2 Confinement Factor and MD Values for All Vertical Modes in ELoD2 Structure

Mode Number	Confinement Factor (Γ)	Modal Discrimination (MD)
1	0.0163	1
2	0.00376	4.33
3	0.000282	57.72
4	0.00856	1.90
5	0.00390	4.17
6	1.090E-05	1491.39
7	0.00130	12.58
8	0.00424	1.92

et al., 2014). The tapered lasers in this chapter are based on the ELoD2 vertical cavity with low-index quantum barriers (Crump et al., 2013c). This structure supports eight vertical modes, as shown in Figure 29.2. Their confinement factors and MD are given in Table 29.2.

29.2.2 Lateral Cavity Design

State-of-the-art high-power lasers need a large output power, high PCE, and excellent beam quality. Ridge waveguide (RW) lasers can produce a diffraction limited beam (single lateral mode), but have only achieved output powers of 1.6 W (Yang et al., 2004). Flared RW lasers have achieved a maximum power of 3 W and kink-free power of 2.2 W (Sverdlov et al., 2013). To achieve higher output powers, the laser diode needs a larger gain volume (energy reservoir). This can be most easily achieved by increasing the emitter width, which has the added benefit of lowering the power density at the facet—thereby increasing the catastrophic optical mirror damage (COMD) threshold.

Slab-coupled optical waveguide laser (SCOWL) diodes (Donnelly et al., 2003) use a shallower RW etch to allow the beam profile to expand laterally in the "slab" below. They have been used as high-brightness sources for power scaling by spectral beam combining (Huang et al., 2009). SCOWLs have good beam quality and low astigmatism, but their limitations are similar to those of the flared RW laser and comparatively low power (2.8 W) (Huang et al., 2007) limits their single-emitter brightness.

BA diodes have achieved powers of 29 W, reliable operation at 20 W, and record PCE (76%) at 0°C. BA lasers have simultaneously achieved high power *and* high PCE at 300 K (14.5 W, PCE >60%)

(Crump et al., 2009, 2013a). Their main limitation is their poor slow-axis beam quality ($M^2 \sim 20$–30) (Knigge et al., 2005).

Lim et al. used simulations to investigate the dependence of the beam quality of BA lasers on cavity length and width as a function of power (Lim et al., 2005). Lim's work showed that the brightness of gain-guided BA lasers improved with reduced cavity width and increased cavity length—establishing a new basis for the development of narrow broad-area (NBA) laser diodes. Since then, NBA laser diodes have received increasing attention (Crump et al., 2013b; Decker et al., 2014; Skidmore et al., 2016), with powers of 7.5 W with $Q_{\text{slow axis}} = 1.8$ mm·mrad and high efficiency (PCE = 57%) (Crump et al., 2013b). Despite the impressive performance of NBA laser diodes, they still fall short of the desired performance for the most demanding applications.

The beam quality of BA lasers can also be improved using asymmetric feedback to reduce the threshold of a particular higher order lateral mode (Pillai and Garmire, 1996; Thestrup et al., 2003; Wolff et al., 2003; Lang et al., 2008), exploiting a concept previously demonstrated for phase-locked laser arrays (). However, the usefulness of asymmetric feedback stabilization for high-brightness operation still needs to be demonstrated—in particular, whether they can maintain their high beam quality and PCE at high power

Tapered laser diodes have received great interest, since they combine the lateral mode confinement/filtering of the RW laser and the large gain volume of the BA laser. Thus, tapered lasers are able to produce a high output power (12 W) with high beam quality ($M^2 < 1.2$) and exceptional brightness ($B = 1.1$ GW·cm^{-2} sr^{-1}) (Walpole et al., 1992; Kelemen et al., 2005; Sumpf et al., 2010). This comes at the expense of lower PCE (43%–55%) and power-dependent "wandering" astigmatism (Dittmar et al., 2006; Fiebig et al., 2008). Thus, the tapered laser is promising for high-brightness direct-diode laser systems. The main challenges are to control their beam quality degradation at high power and to understand the role of optical feedback.

29.2.3 Evolution of Lateral Beam Quality of High-Brightness Tapered Laser Diode

The tapered laser diode comprises RW and tapered amplifier (TA) sections. The RW supports a single lateral mode, to inject a diffraction-limited beam into the amplifier. It also filters out higher order lateral modes from the backward traveling fields it receives from the tapered section. The TA provides a large gain volume with a large output aperture. The taper angle is chosen to match the diffraction angle from the RW (Pearson et al., 1969; Walpole et al., 1992), allowing the field to expand in the taper to produce a smooth output field profile, as illustrated in Figure 29.3.

A nearly diffraction-limited lateral beam is obtained from conventional tapered lasers at low to moderate power, but the beam quality degrades rapidly at high power. This beam quality degradation reveals degradation of the RW filter performance due to gain saturation inside the RW and absorption bleaching by the backward traveling fields outside of it (Sujecki et al., 2003; Kaunga-nyirenda et al., 2014; Larkins et al., 2014, 2016). The absorption bleaching (due to band filling by the generated carriers) renders the material transparent and allows the backward traveling fields outside the RW to reach the rear facet, where they are reflected and reenter the TA, as shown in Figure 29.4

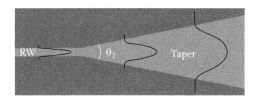

FIGURE 29.3 Schematic diagram of a tapered laser, showing the adiabatic expansion of the beam in the taper.

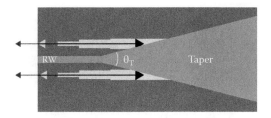

FIGURE 29.4 The beam outside the ridge waveguide (RW) bleaches the absorption in these regions, allowing the reflected beam from the back fact to travel back and couple into the tapered amplifier (TA)

These unfiltered fields create high-spatial-frequency features in the near-field pattern. These high-frequency features or filaments are caused by the propagation and diffraction of the backward-traveling fields in the absorption bleached regions outside the RW (Lim et al., 2012; Larkins et al., 2014). This diffraction is primarily due to the diffraction of the backward propagating fields that are *not* coupled into the RW. Figure 29.5a shows the back propagating field in a 975 nm-tapered laser at $I = 15$A. The backward propagating fields are amplified in the taper and gain guided at the edges of the taper to form a high-intensity spot where the RW joins the TA. The fields diffracting from this spot are clearly visible alongside of the RW. Figure 29.5b shows the backward propagating fields from this high-intensity spot in the absence of index and gain-guiding effects. The field pattern agrees with the corresponding region in Figure 29.5a (inside the white box)—except that the central lobe widens due to the absence of index guiding by the RW. In Figure 29.5c, only the central field lobe (the lobe that couples into the RW) was back propagated, while in Figure 29.5d only the outer lobes of the spot were propagated. Figure 29.5c and d shows that only the fields *not* coupled into the RW give rise to the diffraction pattern—and thus, for the absorption bleaching and beam quality degradation.

Finally, spatial hole burning at the center of the taper and electrical over-pumping at the edges creates high carrier densities and gain at the edges of the taper, as seen in quantitative intracavity spontaneous emission imaging measurements (Bull et al., 2004, 2006). Amplification and carrier-induced waveguiding in these regions cause the "batman" ears in the near-field pattern and amplify/guide the backward fields that degrade the RW filter performance (Williams et al., 1999; Sujecki et al., 2003) Different techniques have been suggested to overcome the degradation of beam quality, such as reducing the front facet reflectivity, using beam spoilers or using a longer RW section. Figure 29.6 shows the impact of reducing front facet reflectivity on the total photon density distribution (i.e. the sum of densities of both the forward and backward traveling waves). These techniques all reduce the backward traveling field and associated absorption bleaching. At the emitter powers sought for high-brightness direct-diode lasers (>10 W), however, these techniques are insufficient. (Beam quality degradation in tapered lasers is also discussed in Chapter 28.)

An integrated distributed Bragg reflector (DBR) mirror at the end of the RW provides a solution with better high-power performance. The rear facet of the laser must be antireflection (AR) coated for the DBR to work properly. The DBR (and AR-coated facet) act a spatial filter, allowing the backward traveling fields in the regions outside the RW to exit the rear facet, as illustrated in Figure 29.7. In order to study the spatial filtering performance of the RW section, and understand how the DBR RW section improves the spatial filtering performance of the conventional tapered laser, we define the RW filter response as the ratio of the forward and backward propagating field distributions at the interface between RW and TA sections:

$$H_{RW} = \log_{10}\left[\frac{P_{forward}}{P_{backward}}\right]. \tag{29.5}$$

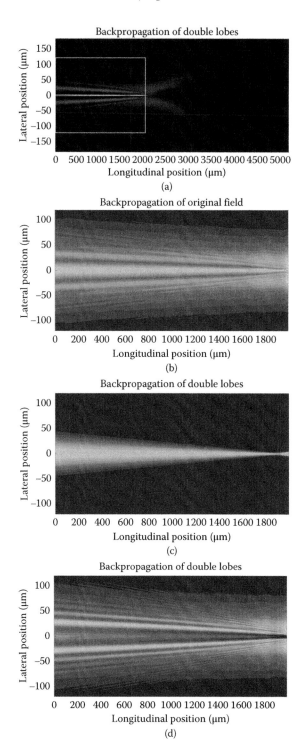

FIGURE 29.5 (a) Backward propagating field in 975 nm tapered laser ($I = 15$ A) for converged simulation. Backward-propagated photon distribution in the RW section for transparent, uniform index material for: (b) the total field; (c) the central lobe coupled into the RW; and (d) the fields *not* coupled into the RW.

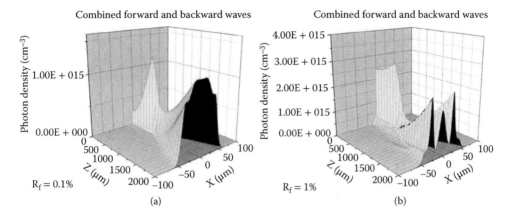

FIGURE 29.6 Impact of front facet reflectivity on the shape of forward and backward traveling waves. Both figures (a and b) show the total photon density distribution (forward + backward).

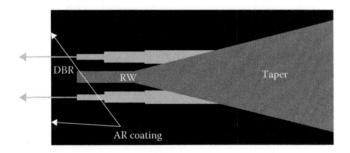

FIGURE 29.7 Schematic diagram of the DBR-tapered laser, showing how the back-propagating beams outside of the RW are allowed to leave the structure, instead of coupling back into the TA.

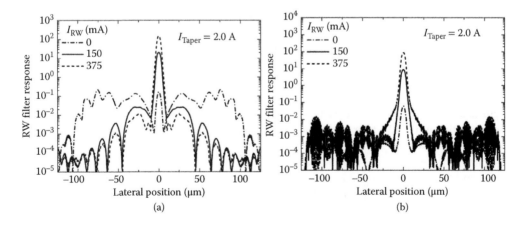

FIGURE 29.8 RW filter performance comparison of conventional (a) and DBR (b) tapered lasers.

Figure 29.8 compares the RW filter responses of DBR and conventional tapered lasers with dual contacts for a constant taper current as a function of RW current. In the conventional tapered laser, the fields in the RW are not amplified at the lowest RW current, but those outside the RW propagate and are reflected—reducing the performance of the RW filter. In the DBRtapered laser, these fields are not reflected and exit

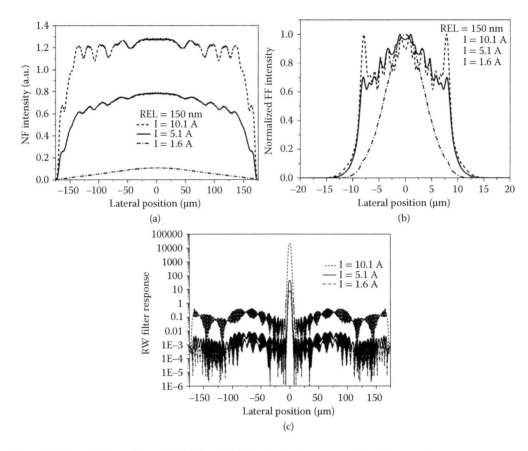

FIGURE 29.9   Evolution of near-field (a), far-field (b) and RW filter response (c) with current.

through the AR-coated rear facet. Thus, the filter performance of the DBRtapered laser remains good up to higher powers.

Despite the large performance improvement of the DBR-tapered laser at high powers, $M^2_{\text{2nd moment}}$ still increases with power. To understand this, we studied the evolution of the near- and far-field patterns and the RW response with current, as shown in Figure 29.9.

As the current increases, high spatial frequency features appear in the near-field pattern and grow in strength, causing degradation of the far-field pattern. The appearance of these features correlates with a sudden reduction in the RW filter performance. Furthermore, the RW width and etch depth have little impact on beam quality as seen in Figure 29.10. (Figure 29.10 also shows that $M^2_{e^{-2}}$ [beam radius w_0 in Equation 29.1 is measured at $1/e^2$] is independent of current, while $M^2_{\text{2nd moment}}$ [beam radius w_0 in Equation 29.1 is measured using second moment of area definition, as defined in Chapter 29] provides a more sensitive measure of the changes in the near and far-field patterns.)

Figure 29.11 shows the forward propagating field at the junction of the RW and TA sections. Although the fields outside the RW are small, they will be strongly amplified.

The fields outside the RW section explain the degradation of the filter function (and the changes in the near- and far-field patterns), but there is a problem: *If the backward propagating fields outside the RW all escape through the AR-coated rear facet, then where do these forward propagating fields come from?* The answer lies in the forward propagating photon distribution in Figure 29.12, which reveals a single aperture diffraction pattern—as confirmed by the positions of the nodes in the pattern. Although these simulations assume a simple patterned rear facet reflectivity, the DBR will also produce aperture diffraction.

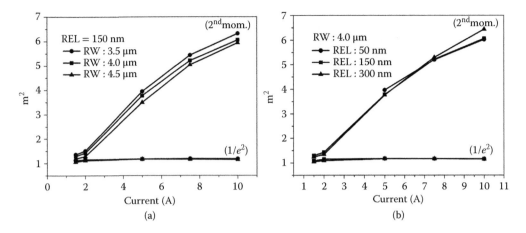

FIGURE 29.10 Impact of RW width (a) and RW etch depth (b) on beam quality.

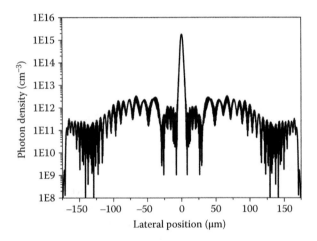

FIGURE 29.11 Forward propagating photon distribution at the RW/TA interface of a DBRtapered laser.

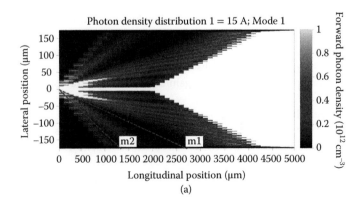

FIGURE 29.12 2D forward propagating photon density with a perfect AR-coating on the rear facet (a) and with an AR-coating with a power reflectivity of 0.1% (b). (*Continued*)

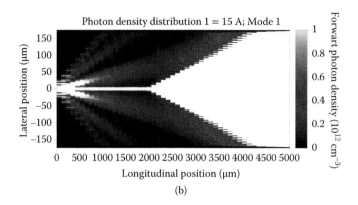

(b)

FIGURE 29.12 (Continued)   2D forward propagating photon density with a perfect AR-coating on the rear facet (a) and with an AR-coating with a power reflectivity of 0.1% (b).

29.3 External Cavity Laser Simulation

New simulation tools are needed to address the problems posed by high-brightness diode lasers operating in a system context with external optical feedback. The laser diode and the external cavity pose different modeling challenges and require the use of different modeling approaches and software tools. These approaches and tools must be brought together to model the laser diode and the external optical system self-consistently.

29.3.1 Laser Diode Simulation Tool *"Speclase"*

The simulation of high-brightness laser diodes is a challenging task, which requires self-consistent modeling of the electrical, optical, and thermal processes throughout the device (Williams et al., 1999; Lim, 2003; Sujecki et al., 2003; Lim et al., 2009). LOC devices with external feedback bring the additional challenge of simulating fields with different vertical mode profiles and their competition for the available gain.

The simulation of the laser diode is performed with an in-house laser simulation tool, *Speclase* (Lim et al., 2009), for the self-consistent quasi-3D optical, electrical, and thermal simulation of high-brightness laser diodes. *Speclase* uses the two-dimensional (2D) wide-angle finite-difference beam-propagation method (WA FD-BPM) for the optical field propagation in the longitudinal and lateral directions (x–z plane). Bipolar electrical simulations are performed in a series of planes orthogonal to the optical axis (x–y planes), neglecting heat and carrier flow in the z-direction. The electrothermal and optical solvers are coupled through the stimulated emission/absorption and ASE processes and refractive index perturbations (calculated from perturbations of the gain/absorption spectra using the Kramers–Kronig relations).

The optical solver models the spectral behavior of the device by propagating multiple wavelengths and their competition through the spectral-spatial gain distribution. Figure 29.13 shows the flow diagram of *Speclase*. The electrothermal and optical models are solved self-consistently, using an accelerated Fox–Li iterative approach (Agrawal, 1984).

In order to model the excitation and competition of fields with different vertical mode profiles, *Speclase* calculates each of the vertical field distributions and their vertical confinement factors. This allows *Speclase* to calculate the effective index and FCA distributions needed for the 2D propagation of the fields with each of the allowed vertical mode profiles. Figure 29.14 describes the flow diagram of *Speclase* for the simulation of a high-brightness LOC laser diode coupled to an external cavity.

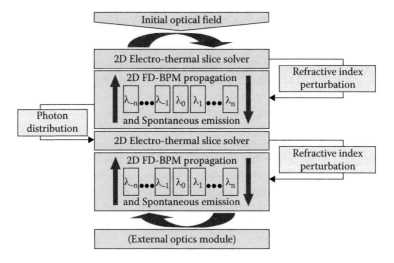

FIGURE 29.13 Flow diagram of stand-alone (SA) *Speclase* with multiple wavelengths.

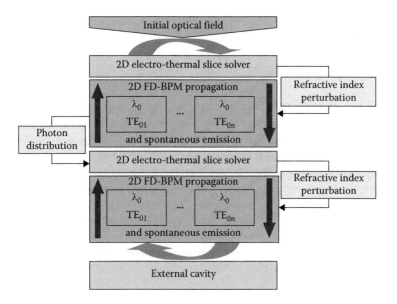

FIGURE 29.14 Flow diagram of *Speclase* with multiple vertical modes with external cavity feedback.

29.3.2 External Cavity Simulation Tool

The modeling of the external cavity requires modeling of free-space optical propagation, including the transmission, reflection, and scattering effects. The optical models used inside the laser diode are not well suited to free-space optical propagation, where Fourier optics (FO) and coherent ray tracing methods are more appropriate. FO models preserve the wave nature of the optical propagation, but the underlying paraxial approximation is best suited to low-divergence beams. Coherent ray tracing is better suited for external cavity laser diodes with large beam divergence. Coherent ray tracing methods are also fast and flexible, making them attractive for modeling complex optical systems. We describe an external cavity modeling tool (*OpticStudio*) based on coherent ray tracing. FO models are also used in this chapter, but do not need wave-ray transformation and are not discussed here.

29.3.3 Coupling of the Laser Diode Simulation Tool and Coherent Ray Tracing Tools

The coupling of the laser simulator and the coherent ray tracing tool for the external cavity is not straight forward, since the light is treated as a wave inside the laser diode, but as rays in the external cavity. Careful analysis of the output beam of the laser diode is needed to produce the input rays for the external cavity model, while maintaining all of the properties of the beam emitted from the laser diode (e.g., near- and far-field patterns, astigmatism). Bidirectional coupling is even more challenging, requiring conversion of the rays back into an optical field distribution, which contains both intensity and phase information.

First, the field data produced by *Speclase* (near-field, far-field and wavelength) must be converted into rays, which represent the laser source in *OpticStudio*. These rays each have a set of attributes: launching coordinates, direction cosines and intensity.

Speclase simulates the lateral field distribution at the facet using 2D WA FD-BPM, so the lateral near-field pattern is multiplied by the associated vertical field profile. The horizontal and vertical far-field patterns are calculated by Fourier transformation:

$$\psi\left(\theta_H\right) = \left(\cos\theta_H\right)^2 \left|\int_{-\infty}^{+\infty} E(x)e^{j(\sin\theta_H k_0 x)}\,dx\right|^2 \tag{29.6}$$

$$\psi(\theta_V) = (\cos\theta_V)^2 \left|\int_{-\infty}^{+\infty} E(y)e^{j(\sin\theta_V k_0 y)}\,dy\right|^2, \tag{29.7}$$

where ψ is far-field profile, θ is the divergence angle, E is the electric field, and k_0 is the wavenumber. Due to the astigmatic nature of the laser beam, the output rays appear to originate from two virtual line sources (nonastigmatic sources have a virtual point source). The output rays must intersect both of these line sources—one for the vertical beam divergence and the other for the horizontal beam divergence. The longitudinal positions of these line sources are obtained by FO back propagation (horizontal line source) and by back tracing the rays (vertical line source). Their transverse positions are the first moments of the near-field patterns

The second challenge is to convert the rays returning to the laser facet back into complex field distributions, taking the ray phases into account. A fraction of the optical energy returned to the laser facet at each lateral position (x) is coupled into the laser for each of the vertical modes supported by the LOC structure, as determined by the overlap integral:

$$\eta_n = \frac{\left|\int_{-\infty}^{+\infty}\left\{E_{\text{ext}} TE_n{}^*\right\}dy\right|^2}{\int_{-\infty}^{+\infty}\left|E_{\text{ext}}\right|^2 dy \int_{-\infty}^{+\infty}\left|TE_n\right|^2 dy}, \tag{29.8}$$

where E_{ext} is the distribution of the external field incident on the facet at this lateral position, TE_n is the nth vertical field profile, and η_n is the percentage of the reflected power coupled into cavity mode n. The phase of the coupled field at each position can be calculated by

$$\varphi = \tan^{-1}\left[\frac{Im\left\{\int_{-\infty}^{+\infty}\left\{E_{\text{ext}} TE_n{}^*\right\}dy\right\}}{Re\left\{\int_{-\infty}^{+\infty}\left\{E_{\text{ext}} TE_n{}^*\right\}dy\right\}}\right]. \tag{29.9}$$

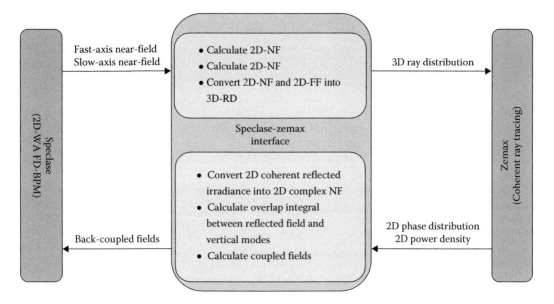

FIGURE 29.15 Schematic of coupling *Speclase* to *OpticStudio*.

This process is repeated for the vertical field distributions at each lateral position along the facet. The resulting one-dimensional (1D) field distributions are coherently added to the internally reflected fields and propagated backward within the laser diode using 2D WA FD-BPM. The entire external cavity simulation is repeated using Fox–Li iterations until it converges. Figure 29.15 shows a schematic diagram of the coupling of *Speclase* and *OpticStudio*

29.4 Case Study: The Impact of Unintentional Reflections on a DBR-Tapered Laser

High-brightness tapered lasers are usually characterized and simulated in isolation, without back reflections to degrade the slow-axis beam quality or couple into higher order vertical modes. In this section, we simulate the simple scenario of an LOC tapered laser with fast- and slow-axis beam collimation lenses with the light incident on the end of an uncoated optical fiber ($R = 5\%$), as illustrated in Figure 29.16. The purpose of this study is to:

1. Observe how the feedback affects the power, PCE, and beam quality
2. Observe the impact of the feedback on the excitation of other vertical modes
3. Reveal the impact of self-heating from the excitation of other vertical modes
4. Investigate approaches for suppressing the impact of higher order vertical modes

The laser diode in this study is a 975-nm LOC DBR-tapered laser based on the ELoD2 structure (Crump et al., 2013c), with a 2-mm RW and a 3-mm gain-guided amplifier with a 6° taper. The DBR is represented by a patterned reflectivity ($R = 31\%$ inside the RW, $R = 0.1\%$ adjacent to it). The front facet reflectivity is 1%. Stray external cavity reflections are included using ray splitting and scattering. Figure 29.17 shows the simulated and experimental power and PCE versus current behavior of an ELoD2 BA laser.

High-power laser diodes are often fabricated as linear bar arrays or laser "bars." Laser bars often bend slightly when they are soldered to a heatsink, so that the emitter near-field patterns deviate from a straight line—an effect referred to as "smile." Figure 29.18 shows how "smile" of packaged laser bars displaces the emitters ($\pm 0.5\,\mu m$) from the axis This displacement causes the beam to propagate through the external

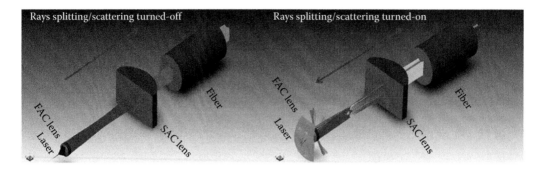

FIGURE 29.16 External cavity setup used in the simulation consists of a fast-axis collimation (FAC) lens, a slow-axis collimation (SAC) lens, and a fiber. Rays splitting and scattering turned off (left). Rays splitting and scattering turned on (right).

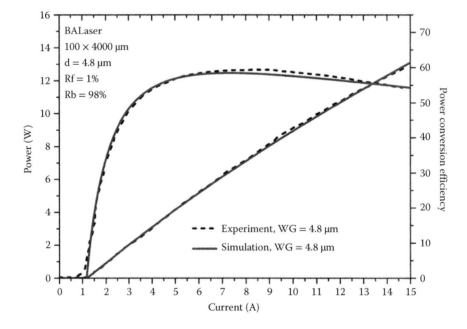

FIGURE 29.17 Simulation versus experiment for ELoD2 structure.

FIGURE 29.18 "Smile" of a linear laser array or bar, which is caused by bending as the bar is soldered to a heatsink.

optics at an angle to the optical axis, a phenomenon known as "pointing error." To emulate the impact of "smile" and "pointing error," the emitter position is shifted vertically with respect to the external optical system.

Two external cavity simulations were performed (on- and off-axis) and compared to a stand-alone (SA) simulation (no external feedback). The overlap integrals of the reflected field of the fundamental mode (from the SA simulation) and the cavity modes were calculated to estimate the coupling coefficients versus displacement, as shown in Figure 29.19. These curves show how displacement affects the feedback coupling to the vertical modes. The off-axis simulation was performed for a displacement of +0.5 µm,

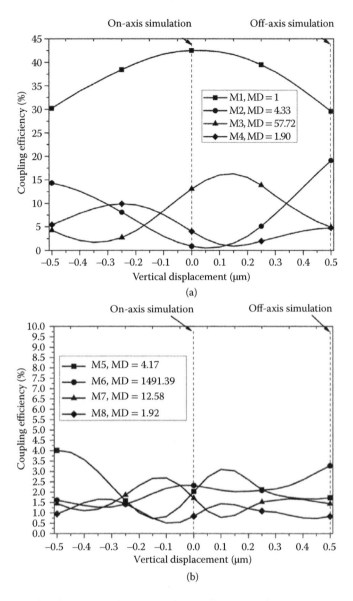

FIGURE 29.19 Impact of smile on external cavity coupling coefficients. Modes 1–4 (a) have coupling >10%, while modes 5–8 (b) have coupling <5%. An off-axis point of interest is displacement = +0.5μm.

where the fundamental mode coupling coefficient (29.5%) approaches that of a higher order mode (19%, mode 2).

29.4.1 Parasitic Reflections with On-Axis Alignment

Simulations were performed for two different external feedback conditions: no feedback; and feedback to the fields for all of the vertical mode profiles. Simulations were performed at bias currents of 2.5, 5.0, 10, 15, and 20 A. Figure 29.20 shows the power versus current and PCE versus current characteristics to compare the performance of the laser for these two feedback conditions. Our discussion will focus on the performance at a bias current of 15 A, since this produces an output power in the range of interest (10–12 W) for the targeted application.

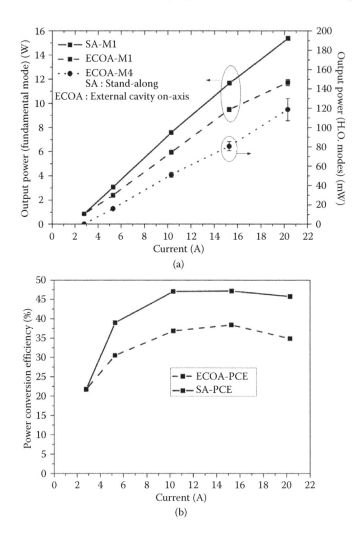

FIGURE 29.20 Simulated power (a) and power conversion efficiency (b) of the tapered laser with and without on-axis feedback.

Figure 29.20 shows that the optical feedback reduces the output power and the PCE. At $I = 15$ A, the on-axis feedback caused the fundamental mode power (mode 1) to drop from 11.67 to 9.49 W, while the PCE fell from 47.1% to 38.4%.

Figure 29.21 compares the near-field patterns (left) and far-field patterns (right) of the on-axis laser for the two feedback conditions. The feedback reduces the output power and increases the modulation depth of the high spatial frequency components in the near-field pattern, as observed experimentally (Hempel et al., 2013; Leonhäuser et al., 2014). The feedback may also lead to excitation and lasing of higher order modes—in this case mode 4.

Figure 29.22 shows the current dependence of the beam propagation parameter M^2 (both the e^{-2} and second moment definitions) for the on-axis external cavity feedback conditions. At $I = 15$ A, the feedback had little effect on $M^2_{e^{-2}}$, but $M^2_{2nd\ moment}$ increased from 6.32 to 12.03. This shows that aperture diffraction from the DBR dominates the beam quality degradation.

The total photon distributions (forward and backward propagating) with on-axis feedback are shown in Figure 29.23 for the first four vertical modes. Only modes 1 and 4 are lasing, but light is also coupled into the other modes. Although they do not lase or produce an output beam (and are difficult to characterize and

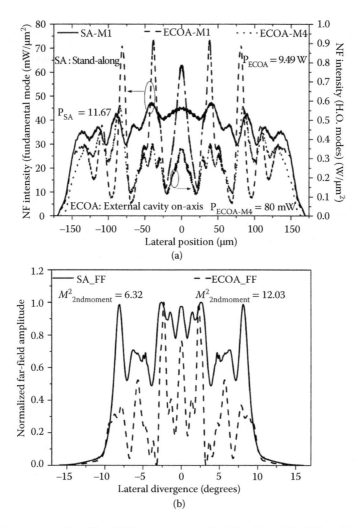

FIGURE 29.21 Lateral nearfield profiles of all lasing vertical modes (a) and lateral far-field profiles of the fundamental vertical mode (b) for different on-axis external cavity feedback conditions.

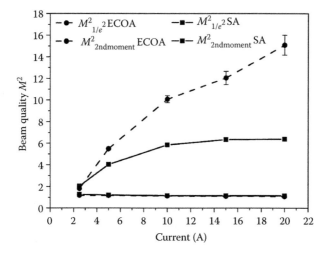

FIGURE 29.22 Dependence of M^2 on bias current for SA and external cavity on-axis simulations.

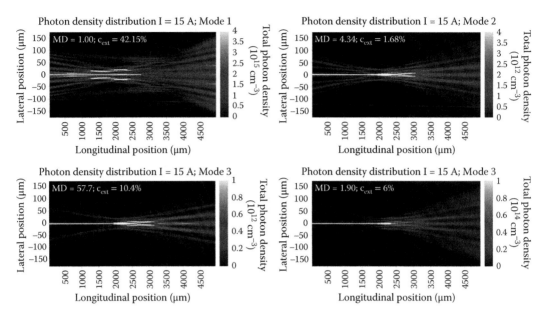

FIGURE 29.23　Total photon density. Mode 1 (top left). Mode 2 (top right). Mode 3 (bottom left). Mode 4 (bottom right). Modal discrimination (MD) and coupling coefficient (C_{ext}) are quoted for each mode.

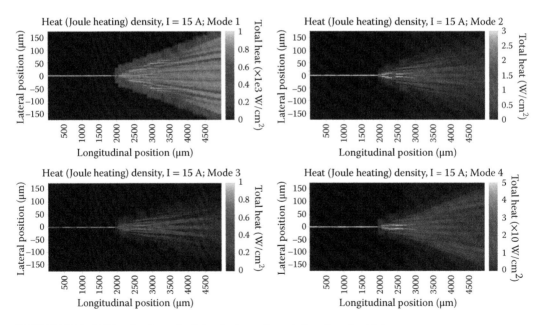

FIGURE 29.24　Self-heating due to Joule heating by each vertical mode at $I = 15$ A.

easy to overlook), they affect the laser through optical pumping (absorption bleaching), parasitic current leakage, and heat generation.

Figure 29.24 shows the Joule heating distributions caused by the currents supporting stimulated emission of the different modes at $I = 15$ A—as well as how the different vertical modes share/compete for the available gain. These distributions show where the stimulated emission occurs within the device for each of the vertical modes. For the higher order modes, this occurs solely because of the power fed back into

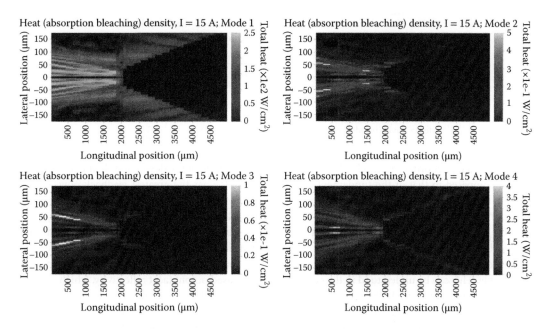

FIGURE 29.25 Self-heating due to optical pumping by each vertical mode at $I = 15$ A.

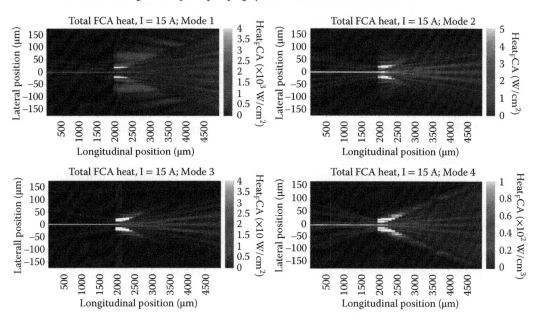

FIGURE 29.26 Self-heating due to free-carrier absorption by each vertical mode at $I = 15$ A.

them from the external cavity. Figure 29.25 shows the self-heating distributions due to optical pumping (absorption) at $I = 15$ A. This pumping causes the absorption bleaching, which plays a critical part in the degradation of the RW spatial filter performance. These distributions show the spatial distribution and relative contributions of the vertical modes to the absorption bleaching. The fundamental mode clearly plays a dominant role in the optical pumping, but the other modes also contribute. Figure 29.26 shows the spatial distributions of self-heating due to FCA. The drop in FCA to the left of the RW–TA interface is due to the RW etch. The higher order modes lose much more power due to FCA than to optical pumping.

TABLE 29.3 Impact of External Cavity Feedback on the Laser Performance (Power, Efficiency, and Beam Quality) for Coupling to All Modes and for Coupling Just to the Fundamental Mode

Simulation	Output Power (W)	PCE (%)	$M^2_{2\text{nd moment}}$
Stand-alone	11.67	47.13	6.32
External cavity (all modes)	9.49	38.39	12.04
External cavity (mode 1 only)	9.46	38.13	12.3

The excited higher order modes clearly play a role in the operation of the device, but their overall power is small compared to that of the fundamental mode (which also has the greatest feedback power). The question is, whether the degradation in PCE and beam quality is primarily controlled by feedback to the higher order vertical modes or to the fundamental mode. Table 29.3 compares the output power, PCE, and $M^2_{2\text{nd moment}}$ for the laser in isolation (SA), with on-axis feedback to all vertical modes and with on-axis feedback to just the fundamental mode. It is clear that the feedback to the fundamental mode has the greatest effect on both PCE and beam quality, while feedback to higher order vertical modes plays a smaller role. This can be understood in the context of an increased output facet reflectivity, which reduces the output power and PCE. At the same time, the optical feedback into the fundamental mode increases the absorption bleaching adjacent to the RW and also the power incident on the DBR reflector inside the RW, as discussed in Section 29.2.2. Both of these effects contribute to the beam quality degradation—the first by reducing the absorption filtering of the diffracted light outside the RW and the second by increasing the power in the side lobes of the diffraction from the DBR reflector.

29.4.2 Parasitic Reflections with Off-Axis Alignment

External feedback couples to the vertical modes differently, when its alignment to the waveguide changes. Here, we explore the impact of a vertical emitter shift of +0.5 μm due to "smile." Table 29.2 shows that modes 2 and 4 have moderate ($MD_2 = 4.33$) and low ($MD_4 = 1.90$) values of MD. Figure 29.19 shows they now have large (19%) and moderate (4.8%) coupling coefficients. The performances of these two modes should reveal the relative importance of the MD and the coupling coefficient. The coupling to fundamental mode also decreases from 42.15% (on-axis) to 29.5% (off-axis).

Figure 29.27 shows the power versus current and PCE versus current characteristics of the LOC tapered laser. With the off-axis feedback, the output power and PCE at $I = 15$ A dropped from 11.67 to 9.22 W, while the PCE fell from 47.1% to 37.3%. The coupling into mode 2 increased by a factor of ~11, allowing both modes 2 and 4 to lase.

Figure 29.28 shows the current dependence of the beam propagation parameter M^2 (both the e^{-2} and second moment definitions) for the off-axis external cavity feedback condition. At $I = 15$ A, the off-axis feedback did not change $M^2_{e^{-2}}$, but $M^2_{2\text{nd moment}}$ increased from 6.32 to 11.72.

29.4.3 Analysis and Discussion

In this section, we analyze and discuss what happens to the power fed back into the higher order modes and how it affects the performance of the laser. We are particularly interested in how the feedback effects depend on parameters that can be influenced during the design of the laser: MD, coupling coefficient, and FCA. We are also interested in how the external cavity optics can be designed to minimize feedback effects.

Tables 29.4 and 29.5 summarize a range of power-related performance metrics for each of the vertical modes at a bias current of 15 A. This bias was selected, as it produces an output power of 11.7 W in the absence of external feedback and is within our desired range of 10–12 W. Table 29.4 is for the on-axis simulation. Table 29.5 is for the off-axis simulation when the emitter is vertically displaced by 0.5μm. Modes in boldface font are lasing. H_{stm} is the Joule heating power of the current needed to support the stimulated

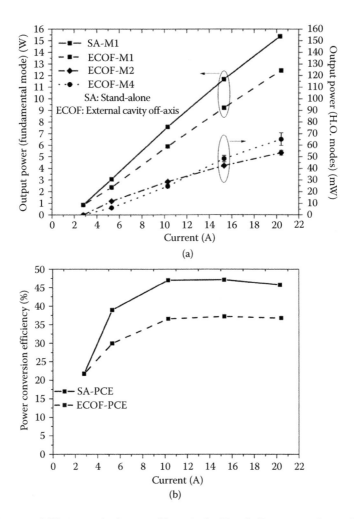

(a)

(b)

FIGURE 29.27 L–I curves of all lasing modes for tapered lasers in the SA and off-axis external cavity feedback (ECOF) configurations.

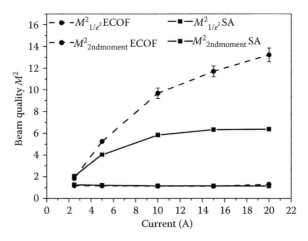

FIGURE 29.28 Dependence of M^2 on bias current for both SA and external cavity off-axis simulations.

TABLE 29.4　Summary of Impact of MD on Optical Metrics of All Modes for On-Axis Simulation for a Current of 15 A.

Mode	MD	Coupling Coefficient (%)	Coupled External Power (mW)	Output Power (W)	H_{stm} (W)	H_{abs} (W)	H_{FCA} (W)	I_{stm} (A)
1	**1.00**	**42.15**	**155.3**	**9.5**	**3.86**	**0.95**	**3.2**	**11.2**
2	4.34	1.68	6.2	3.1e-3	2.4e-3	3.14e-4	4.2e-3	7.1e-3
3	57.7	10.4	38.6	2.6e-3	5.8e-4	4.43e-5	27.7e-3	1.7e-3
4	**1.90**	**6**	**22.2**	**80.1e-3**	**47.7e-3**	**1.7e-3**	**124.6e-3**	**138.1e-3**
5	4.17	1.2	4.6	1e-3	1.5e-3	9.48e-5	10.9e-3	4.5e-3
6	1491.	2.2	8.4	4e-6	3.6e-6	4.83e-8	8e-3	7.6e-6
7	12.58	2.2	8.3	1.6e-4	4.6e-4	3.94e-6	11e-3	1.3e-3
8	1.92	0.65	2.4	5.9e-3	4.6e-3	2.43e-5	22.2e-3	13.4e-3
Total for higher order modes		90.7		92.9e-3	57.2e-3	3e-3	208.6e-3	166.1
Total for all modes		246.0		9.59	3.92	0.953	3.41	11.37

Note: Values in boldface font signify that the mode is lasing

TABLE 29.5　Summary of Impact of MD on Optical Metrics of All Modes for Off-Axis Simulation for a Current of 15 A.

Mode	MD	Coupling Coefficient (%)	Coupled External Power (mW)	Output Power (W)	H_{stm} (W)	H_{abs} (W)	H_{FCA} (W)	I_{stm} (A)
1	**1.00**	**29.5**	**142.6**	**9.22**	**3.71**	**0.96**	**3.03**	**10.96**
2	**4.34**	**19**	**92**	**42.3e-3**	**26.5e-3**	**2.7e-3**	**38.5e-3**	**76.6e-3**
3	57.7	5.23	25.28	1.5e-3	4.13e-4	3.81e-5	13.3e-3	1.2e-3
4	**1.90**	**4.8**	**23.2**	**48.2e-3**	**38.4e-3**	**4.7e-3**	**81.4e-3**	**111.2e-3**
5	4.17	1.73	8.4	1.4e-4	2.5e-3	2.62e-4	17.1e-3	7.1e-3
6	1491.	3.28	15.88	1.94e-6	4.12e-6	1.08e-7	13.3e-3	1.2e-5
7	12.58	1.44	6.97	5e-5	4.58e-4	1.08e-5	8.8e-3	1.3e-3
8	1.92	0.83	4	1.8e-3	4.6e-3	4.35e-4	31.8e-3	13.4e-3
Total for higher order modes		175.7		94.0e-3	72.9e-3	13e-3	204.2e-3	210.8
Total for all modes		318.3		9.31	3.78	0.973	3.23	11.17

Note: Values in boldface font signify that the mode is lasing

emission of the mode, I_{stm}. H_{abs} is the total heat power due to band-to-band absorption outside the gain regions and is responsible for absorption bleaching. H_{FCA} is the total heat power due to FCA. For the case of on-axis feedback, 91 mW of light is coupled into the higher order modes, which consume 166 mA of current and produce 93 mW of light and 269 mW of heat. For the case of off-axis feedback, 176 mW of light is coupled into the higher order modes, which consume 211 mA of current and produce 94 mW of light and 290 mW of heat.

The previous section showed that external feedback can cause higher order vertical modes to lase. For the laser diode and external cavity simulated here, only the fundamental mode lases without feedback, while mode 4 also lases in the case with on-axis feedback. For the case of off-axis feedback, two higher order vertical modes also lase (modes 2 and 4). The feedback reduces the power and PCE of the fundamental mode, which can be attributed partly to higher order mode lasing and partly to an increase in the effective facet reflectivity by feedback to the fundamental mode. For the device and cavity investigated here, Tables 29.4 and 29.5 suggest that modes with MD of MD < 2 are able to start lasing with comparatively small external cavity coupling coefficients of <5%, while modes with MD ~ 4 require larger coupling coefficients (e.g., 10%–20%). Additional simulations showed that the feedback required for higher order mode lasing also depends on the coupling of the reflected beam into the RW (i.e., into the RW aperture at the RW–TA interface). This is also apparent from the photon distributions in Figure 29.23.

As higher order modes began to lase more strongly, they began to compete with the fundamental mode for the spatial gain distribution and modal power oscillations were observed in the simulations. Although these are not true dynamic simulations, the oscillations have some relation to mode beating and mode partition noise. Despite that the higher order mode powers were small, they appear to affect spatial filamentation and contribute to the near- and far-field patterns of the fundamental mode—consistent with experimental observations (Hempel et al., 2013; Leonhäuser et al., 2014). For stable operation, M^2_{e-2} remains below 1.5 and is insensitive to feedback, while $M^2_{\text{2nd moment}} \sim 6$ for SA operation and nearly doubles with feedback (As discussed in Section 29.4.1, increased feedback to the fundamental mode has the greatest impact on beam quality, but excitation of higher order modes also affects $M^2_{\text{2nd moment}}$). For astable operation (modal power oscillations), there is significant degradation in both M^2_{e-2} and $M^2_{\text{2nd moment}}$.

The optical power back-coupled into the higher order vertical modes also creates heat. This occurs even if the modes are not lasing, but increases dramatically if they are. The heat power of the higher order modes is dominated FCA, followed by Joule heating associated with the current supporting stimulated emission into these modes. The heat generated per feedback photon decreases with MD. This shows the importance of both MD and FCA for suppressing feedback amplification and thus, heating by the higher order modes. At this bias level (15 A), the total heat generation in the higher order modes is 0.269–0.290 W (depending on the alignment of the laser diode), corresponding to a 3.2%–3.6% increase in heat power. The heat generation by the higher order modes alone is responsible for a ~2.3%–2.4% drop in PCE, which increases to ~2.5% if their optical output power is also considered. (This is ~20%–25% of the total drop in PCE.) The heat generated by the excitation of these higher order vertical modes is distributed along the entire cavity—irrespective of the MD. Thus, although the excitation of higher order modes affects the beam quality, it is probably not responsible for observed increases in device degradation. Instead, as shown by Kissel et al. (2016), the observed increase in degradation is probably due to feedback that is *not* coupled into the waveguide (i.e., into the substrate)—particularly when the substrate is strongly absorbing (i.e., $E_{\text{g\_substrate}} < E_{\text{photon}}$).

We have shown that external feedback can cause lasing of higher order vertical modes, resulting in self-heating, mode partition instabilities, and beam quality degradation. The question is *How can we design the laser diode and optical system to reduce the impact of optical feedback?* For the design of the laser diode, MD and FCA are key parameters for suppressing the impact of higher order modes. The external cavity coupling coefficients play a smaller role, but are still important. The stray light from aperture diffraction by the DBR at the back of the RW must be suppressed. For the design of the external cavity, the reflected power must be minimized—for example, by using good AR coatings and, perhaps, tilted facets to eliminate specular reflections. For high powers, however, other approaches may also be required, such as off-axis optical alignment to prevent reflected beams from coupling into the RW filter. Finally, care must also be taken to minimize the power coupled into the substrate—particularly if the substrate is absorbing.

29.5 Conclusions

This chapter discussed how, for many applications that employ high-brightness diode lasers, it is necessary to scale the output power through beam combination techniques, while maintaining an excellent beam quality. All beam combination techniques cause light to be fed back into the laser cavity. In some instances, this feedback can be exploited, such as in the wavelength stabilization of diode lasers and arrays. However, any laser placed in an external system will also suffer from unintentional feedback (parasitic reflections). This chapter describes a detailed method of modeling external cavity lasers with both intentional and unintentional feedback. This model, developed at the University of Nottingham, consists of an advanced laser simulation tool coupled to commercial optical design software.

In large vertical cavity lasers, as required to reach the highest output powers, the vertical cavity must be carefully considered. The structure must be designed to minimize the coupling of parasitic feedback to higher order modes. The effects of parasitic feedback on the performance (output power, PCE, and beam quality) of a tapered laser in a simple optical system are studied. The excitation and propagation of higher order modes increases the FCA causing self-heating, eventually causing degradation of the output power and efficiency. Moreover, the parasitic feedback causes more absorption bleaching, leading also to degradation of the lateral beam quality. Therefore, the inclusion of external parasitic feedback in the design process is highly significant. Coupling coefficients and MD values can be used to engineer a laser structure that is less affected by feedback. This can be accomplished by ensuring that modes with low-to-moderate MD values have the lowest coupling coefficients, thereby minimizing the light coupled into them.

The lateral cavity design also has a significant impact on brightness and the tapered laser is a strong contender for high-brightness laser systems. Tapered lasers offer, to a certain extent, the beam quality advantages of an RW laser and the high power advantages of a BA laser. However, despite this and even without external parasitic feedback, tapered lasers still suffer from beam quality degradation at higher powers. Investigations into the cause of this beam quality degradation reveal that diffraction of light at the back aperture leads to regions adjacent to the RW section becoming bleached from carriers—thereby, degrading the beam quality.

Acknowledgments

The authors thank J. Decker and P. Crump at the Ferdinand Braun Institute and N. Michel and M. Krakowski at Alcatel-Thales III–V Lab for experimental results and helpful discussions, and U. Witte and M. Traub at Fraunhofer Institute for Laser Technology for helpful discussions on optical systems for direct-diode lasers. We also thank G. Schimmel, G. Lucas-Leclin, and P. Georges at CNRS Institute d'Optique; and V. Vilokkinen and P. Uusimaa at Modulight Ltd. for helpful discussions. The authors acknowledge L. Borruel and I. Esquivias for their contributions to early collaborations on the simulation of tapered laser diodes during the EC ULTRABRIGHT project. The authors gratefully acknowledge funding from the European Commission projects: BRIDLE (IST–314719, 2012–15), WWW.BRIGHTer.EU (IST–2005–035266, 2006–10), FAST ACCESS (IST-004772, 2004–7 WWW.BRIGHT.EU (IST-511722; 2004–6), POWERPACK (IST-2000-29447, 2001–3), ULTRABRIGHT (IST-1999-10356, 2000–3).

References

Agrawal GP (1984) Fast-Fourier-transform based beam-propagation model for stripe-geometry semiconductor lasers: Inclusion of axial effects. *Journal of Applied Physics* 56:3100–3109.

Buda M, van der Vleuten WC, Iordache G, Acket GA, van de Roer TG, van Es CM, van Roy BH, Smalbrugge E (1999) Low-loss low-confinement GaAs-AlGaAs DQW laser diode with optical trap layer for high-power operation. *IEEE Photonics Technology Letters* 11:161–163.

Bull S, Andrianov A, Wykes JG, Lim JJ, Sujecki S, Auzanneau SC, Calligaro M, Lecomte M, Parillaud O, Krakowski M, Larkins EC (2006) Quantitative imaging of intracavity spontaneous emission distributions using tapered lasers fabricated with windowed n-contacts. *IEEE Proceedings—Optoelectronics* 153:2–7.

Bull S, Wykes JG, Andrianov AV, Lim JJ, Borruel L, Sujecki S, Auzanneau SC, Calligaro M, Krakowski M, Esquivias I, Larkins EC (2004) Imaging of spontaneous emission from 980 nm tapered lasers with windowed N-contacts. *European Physical Journal Applied Physics* 27:455–459.

Crump P, Blume G, Paschke K, Staske R, Pietrzak A, Zeimer U, Einfeldt S, Ginolas A, Bugge F, Häusler K, Ressel P, Wenzel H, Erbert G (2009) 20 W continuous wave reliable operation of 980 nm broad-area single emitter diode lasers with an aperture of 96 μm. *Proceedings of SPIE* 7198:719814–719819.

Crump P, Erbert G, Wenzel H, Frevert C, Schultz CM, Hasler KH, Staske R, Sumpf B, Maaßdorf A, Bugge F, Knigge S, Tränkle G (2013a) Efficient high-power laser diodes. *IEEE Journal of Selected Topics in Quantum Electronics* 19:1501211.

Crump P, Hasler KH, Wenzel H, Knigge S, Bugge F, Erbert G (2013b) High efficiency, 8 W narrow-stripe broad-area lasers with in-plane beam-parameter-product below 2 mm mrad. In: *2013 Conference on Lasers & Electro-Optics Europe & International Quantum Electronics Conference CLEO EUROPE/IQEC*, Munich, 2013, p. 1.

Crump P, Knigge S, Maaßdorf A, Bugge F, Hengesbach S, Witte U, Hoffmann H-D, Köhler B, Hubrich R, Kissel H, Biesenbach J, Erbert G, Traenkle G (2013c) Low-loss smile-insensitive external frequency-stabilization of high power diode lasers enabled by vertical designs with extremely low divergence angle and high efficiency. *Proceedings of SPIE* 8605:86050T–86013.

Decker J, Crump P, Fricke J, Maaßdorf A, Erbert G, Tränkle G (2014) Narrow stripe broad area lasers with high order distributed feedback surface gratings. *IEEE Photonics Technology Letters* 26:829–832.

Dittmar F, Sumpf B, Fricke J, Erbert G, Trankle G (2006) High-power 808-nm tapered diode lasers with nearly diffraction-limited beam quality of M/sup 2/=1.9 at P = 4.4 W. *IEEE Photonics Technology Letters* 18:601–603.

Donnelly JP, Huang RK, Walpole JN, Missaggia LJ, Harris CT, Plant JJ, Bailey RJ, Mull DE, Goodhue WD, Turner GW (2003) AlGaAs-InGaAs slab-coupled optical waveguide lasers. *IEEE Journal of Quantum Electronics* 39:289–298.

Fiebig C, Blume G, Kaspari C, Feise D, Fricke J, Matalla M, John W, Wenzel H, Paschke K, Erbert G (2008) 12 W high-brightness single-frequency DBR tapered diode laser. *Electronics Letters* 44:1253–1255.

Hempel M, Chi M, Petersen PM, Zeimer U, Tomm JW (2013) How does external feedback cause AlGaAs-based diode lasers to degrade? *Applied Physics Letters* 102:023502–023504.

Huang RK, Chann B, Glenn JD (2011) Extremely high-brightness kW-class fiber coupled diode lasers with wavelength stabilization. *Proceedings of SPIE* 8039:80390N–80310.

Huang RK, Chann B, Missaggia LJ, Augst SJ, Connors MK, Turner GW, Sanchez-Rubio A, Donnelly JP, Hostetler JL, Miester C, Dorsch F (2009) Coherent combination of slab-coupled optical waveguide lasers. *Proceedings of SPIE—the International Society for Optical Engineering* 7230:72301G–72312.

Huang RK, Donnelly JP, Missaggia LJ, Harris CT, Chann B, Goyal AK, Sanchez-Rubio A, Fan TY, Turner GW (2007) High-brightness slab-coupled optical waveguide lasers. *Proceedings of SPIE* 6485:64850F–64859.

Kaunga-nyirenda SN, Bull S, Lim JJ, Hasler KH, Fricke J, Larkins EC (2014) Factors influencing brightness and beam quality of conventional and distributed Bragg reflector tapered laser diodes in absence of self-heating. *IET Optoelectronics* 8:99–107.

Kelemen MT, Weber J, Kaufel G, Bihlmann G, Moritz R, Mikulla M, Weimann G (2005) Tapered diode lasers at 976 nm with 8 W nearly diffraction limited output power. *Electronics Letters* 41:1011–1013.

Kissel H, Leonhäuser B, Tomm JW, Hempel M, Biesenbach J (2016) Investigation of accelerated and catastrophic degradation of laser diodes caused by external optical feedback operation. European Semiconductor Laser Workshop, Darmstadt, Germany.

Knigge A, Erbert G, Jonsson J, Pittroff W, Staske R, Sumpf B, Weyers M, Trankle G (2005) Passively cooled 940 nm laser bars with 73% wall-plug efficiency at 70 W and 25/spl deg/C. *Electronics Letters* 41:250–251.

Lang L, Lim JJ, Sujecki S, Larkins EC (2008) Improvement of the beam quality of a broad-area diode laser using asymmetric feedback from an external cavity. *Optical Quantum Electronics* 40:1097–1102.

Larkins EC, Bull S, Kaunga-Nyirenda S, Helal MA, Vilokkinen V, Uusimaa P, Crump P, Erbert G (2014) Design optimisation of high-brightness laser diodes for external cavity operation in the BRIDLE Project. In: *2014 International Semiconductor Laser Conference*, Palma de Mallorca, 2014, pp. 21–22.

Larkins EC, Helal MA, Kaunga-Nyirenda SN, Bull S, Moss D (2016) Design and simulation of high-brightness diode lasers for operation in the presence of external feedback. In: Conference on Novel In-Plane Semiconductor Lasers XV, 2016. (Photonics West, San Francisco, USA) 9767

Leonhäuser B, Kissel H, Unger A, Köhler B, Biesenbach J (2014) Feedback-induced catastrophic optical mirror damage (COMD) on 976 nm broad area single emitters with different AR reflectivity. *Proceedings of SPIE* 8965:896506–896510.

Lim JJ (2003) Investigation of factors influencing the brightness of high-power laser diodes. PhD dissertation, University of Nottingham.

Lim JJ, Benson TM, Larkins EC (2005) Design of wide-emitter single-mode laser diodes. *IEEE Journal of Quantum Electronics* 41:506–516.

Lim JJ, Bull S, Kaunga-Nyirenda S, Sujecki S, Larkins EC, Hasler KH, Fricke J (2012) Factors influencing the brightness and beam quality of tapered laser diodes and bars. IEEE Summer Topical Meeting on High Power Semiconductor Lasers TuA4.1.

Lim JJ, Sujecki S, Lei L, Zhichao Z, Paboeuf D, Pauliat G, Lucas-Leclin G, Georges P, MacKenzie R, Bream P, Bull S, Hasler KH, Sumpf B, Wenzel H, Erbert G, Thestrup B, Petersen PM, Michel N, Krakowski M, Larkins EC (2009) Design and simulation of next-generation high-power, high-brightness laser diodes. *IEEE Journal of Selected Topics in Quantum Electronics* 15:993–1008.

Pearson JE, McGill TC, Kurtin S, Yariv A (1969) Diffraction of gaussian laser beams by a semi-infinite plane. *Journal of the Optical Society of America* 59:1440–1445.

Pillai RMR, Garmire EM (1996) Paraxial-misalignment insensitive external-cavity semiconductor-laser array emitting near-diffraction limited single-lobed beam. *IEEE Journal of Quantum Electronics* 32:996–1008.

Skidmore J, Peters M, Rossin V, Guo J, Xiao Y, Cheng J, Shieh A, Srinivasan R, Singh J, Wei C, Duesterberg R, Morehead JJ, Zucker E (2016) Advances in high-power 9XXnm laser diodes for pumping fiber lasers. *Proceedings of SPIE* 9733:97330B–97337.

Sujecki S, Borruel L, Wykes J, Moreno P, Sumpf B, Sewell P, Wenzel H, Benson TM, Erbert G, Esquivias I, Larkins EC (2003) Nonlinear properties of tapered laser cavities. *IEEE Journal of Selected Topics in Quantum Electronics* 9:823–834.

Sumpf B, Adamiec P, Fricke J, Ressel P, Wenzel H, Erbert G, Tränkle G (2010) Comparison of 650 nm tapered lasers with different lateral geometries at output powers up to 1 W. *Proceedings of SPIE* 7616:76161–76168.

Sverdlov B, Pfeiffer HU, Zibik E, Mohrdiek S, Pliska T, Agresti M, Lichtenstein N (2013) Optimization of fiber coupling in ultra-high power pump modules at $\lambda = 980$ nm. *Proceedings of SPIE* 8605:860508–860510.

Thestrup B, Chi M, Sass B, Petersen PM (2003) High brightness laser source based on polarization coupling of two diode lasers with asymmetric feedback. *Applied Physics Letters* 82:680–682.

Tomm JW, Ziegler M, Hempel M, Elsaesser T (2011) Mechanisms and fast kinetics of the catastrophic optical damage (COD) in GaAs-based diode lasers. *Laser & Photonics Reviews* 5:422–441.

Walpole JN, Kintzer ES, Chinn SR, Wang CA, Missaggia LJ (1992) High-power strained-layer InGaAs/AlGaAs tapered traveling wave amplifier. *Applied Physics Letters* 61:740–742.

Wang X, Erbert G, Wenzel H, Crump P, Eppich B, Knigge S, Ressel P, Ginolas A, Maaßdorf A, Tränkle G (2013) High power, high beam quality laser source with narrow, stable spectra based on truncated-tapered semiconductor amplifier. *Proceedings of SPIE* 8605:86050G–86011.

Williams KA, Penty RV, White IH, Robbins DJ, Wilson FJ, Lewandowski JJ, Nayar BK (1999) Design of high-brightness tapered laser arrays. *IEEE Journal of Selected Topics in Quantum Electronics* 5:822–831.

Wolff S, Rodionov A, Sherstobitov VE, Fouckhardt H (2003) Fourier-optical transverse mode selection in external-cavity broad-area lasers: Experimental and numerical results. *IEEE Journal of Quantum Electronics* 39:448–458.

Yaeli J, Streifer W, Scifres DR, Cross PS, Thornton RL, Burnham RD (1985) Array mode selection utilizing an external cavity configuration. *Applied Physics Letters* 47:89–91.

Yang G, Smith GM, Davis MK, Loeber DAS, Hu M, Chung-en Z, Bhat R (2004) Highly reliable high-power 980-nm pump laser. *IEEE Photonics Technology Letters* 16:2403–2405.

30

Single Longitudinal Mode Laser Diodes

Xun Li

30.1 Introduction

30.1.1 What Is a Single Longitudinal Mode Laser?

Laser is a light source that emits photons described by a "near coherent" optical field. By "coherent optical field," one means the monochromatic electromagnetic field at optical frequency (ω_0) with constant amplitude (A) and phase (θ). Therefore, a coherent optical field can be expressed as $A \cos(\omega_0 t + \theta)$, with phase θ determined by a reference starting time. Any laser built as an oscillator must take spontaneously emitted photons as its initial driven seed, since a laser doesn't have any coherent light as its input. Knowing the fact that the spontaneous emission is a random process, one therefore cannot expect that the laser will give an ideal coherent optical field output. Rather, the laser emits photons described by the ideal coherent optical field, driven by spontaneously emitted photons, for a certain amount of time τ_0 until the emerging of another group of spontaneously emitted photons. Consequently, its output field will experience a sudden change on its amplitude and phase at time τ_0 and this sequence keeps repeating indefinitely. As such, one can express the laser output as a "quasi-monochromatic" electromagnetic field at optical frequency ω_0 as $[A + \delta_a(t)] \cos[\omega_0 t + \theta(t)]$, with $\delta_a(t)$ as a random process with zero mean and $\theta(t)$ also a random process with uniform distribution between 0 and 2π. Once a laser is operated under a bias beyond its threshold, $A >> |\delta_a(t)|$ is satisfied. Hence the only nonideality of a laser output from a coherent optical field lies in its random phase. The difference becomes more apparent if one observes the output optical spectrum of a laser. Actually, as the Fourier transform of an ideal coherent optical field, its frequency-domain spectrum is a delta function that appears at ω_0 with an amplitude of $|A|$. The quasi-monochromatic field, however, has its averaged spectrum in the shape of a "broadened" delta function still peaked at ω_0, but with an amplitude in $\sqrt{1/2\pi}|A|\tau_0$ and a width of $\Delta\omega = 2\pi/\tau_0$. This is because the averaged duration between two consecutive abrupt phase changes is τ_0, which means the quasi-monochromatic field can be described as a series of truncated ideal coherent fields known as a wave train, with τ_0 as the truncation window or the averaged length of the wave train. The Fourier transform of a single truncated coherent field piece corresponds to the convolution between a delta function and a sampling function in the form of $\sin(\omega\tau_0/2)/(\omega\tau_0/2)$ as the Fourier transforms of the ideal coherent field and the flat window function (i.e., 1 for t inside τ_0 and 0 elsewhere), respectively, which gives the result as the aforementioned broadened delta function. Since distinct truncated coherent field pieces in the wave train differ by a time shift only, their Fourier transforms differ just by a phase. Hence, the frequency-domain amplitude spectra of the quasi-monochromatic field as a wave train composed of all these pieces are overlapped as a single peak as shown in Figure 30.1.

It is worth mentioning that the above description is in a "phenomenological" sense, as an accurate treatment of spontaneously emitted photons with a classical electromagnetic field theory is not possible. A physics-based description can be given by exploiting the full quantum mechanics theory, but it will have to involve the quantization of the electromagnetic field and is beyond the scope of this chapter.

The optical frequency ω_0 of the quasi-monochromatic field as the laser output is called the "lasing" frequency. Consequently, the aforementioned spectral peak is called the "lasing" mode or the "longitudinal" mode in order to make a distinction from the "transverse" mode that refers to a spatial field distribution. The term longitudinal comes from the fact that the lasing frequency is determined by the laser resonant cavity. In edge-emitting semiconductor laser diodes, the cavity is set along the longitudinal direction with its geometrical dimension significantly larger than any size in the cross-sectional area. The light is confined inside the cross-sectional area by a waveguide and is only allowed to propagate along the longitudinal direction, or along the cavity. Upon resonance, the lasting longitudinal optical field distribution gives the surviving pattern and determines the lasing frequency. Hence, we have the term longitudinal mode, as opposed to the transverse mode, indicating the confined optical field by the waveguide in the cross-sectional area. Because of the one-to-one correspondence between the optical spectral peak and the lasting longitudinal optical field distribution, conventionally the term longitudinal mode can be referred to either as the spectral peak (or simply as the lasing frequency) or as the spatial field pattern in the cavity

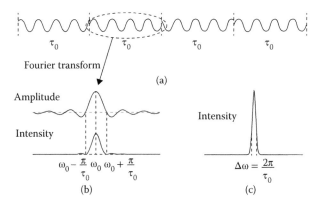

FIGURE 30.1 (a) The quasi-monochromatic field as a wave train in the time domain and a single piece of the truncated coherent field, (b) the frequency-domain amplitude and intensity spectra as the Fourier transform of the wave train (with different pieces all overlapped), and (c) the intensity spectrum with the peak indicating a single (spectral) longitudinal mode.

(longitudinal) direction, whereas the term transverse mode can only be referred to as the spatial field pattern in the cross-sectional area, without any spectral meaning.

Generally, a cavity can support multiple resonant patterns and consequently bear with multiple resonant frequencies. Without exception, a laser can also have multiple longitudinal modes, with multiple peaks shown in its output optical spectrum and multiple lasting optical field patterns inside its cavity. With special designs, however, one is able to leave only one surviving longitudinal mode by eliminating all other resonances inside a cavity. A laser with such a cavity is therefore called a single longitudinal mode (SLM) laser.

30.1.2 Why SLM Laser?

When used as the light source in fiber-optic communication systems, a semiconductor laser operated with multiple longitudinal modes suffers the mode partition noise (MPN) [1,2], which stands as the dominant limiting factor to the transmission span of the optical signal in fiber, as the MPN jeopardizes the signal by introducing the intersymbol interference (ISI) among the pulses in the stream—an effect that cannot be simply suppressed through increasing the laser output power. If a multiple longitudinal mode laser is directly modulated, the power allocated to each of its mode fluctuates in a random manner due to the mixed homogeneous and inhomogeneous gain broadening nature of direct bandgapped semiconductors, with the randomness originated from the spontaneous emission noise. Since the signal components carried by different longitudinal modes propagate at different speed due to fiber dispersion, these components won't arrive at the destination at the same time, which results in the pulse spreading over and spilling out of its allocated time slot, and consequently causes ISI. Such ISI bears a random nature due to the random power fluctuation of the multiple longitudinal modes as the signal carriers. Hence, it cannot be eliminated through linear equalization or phase delay compensation. Also, because the random fluctuation is in proportion to the total power, increasing the laser output power won't solve the problem, if doesn't make it worse. Normally, the power penalty soars even starting from a moderate MPN level. For example, at ~1550 nm (the center of the C-band), the maximum transmission capacity–distance product is only 5 Gbps-km, which means by using a typical multiple longitudinal mode Fabry–Pérot (FP) laser, the 2.5 and 10 Gbps optical signal can only be transmitted for 2 km and 500 m, respectively.

The only viable solution to this problem, therefore, lies in the replacement of the light source with the SLM laser where the MPN naturally disappears.

30.1.3 Current Application Status

Over the past three decades, many SLM laser structures have been proposed and demonstrated; a few dominant structures survived and became popular products on today's market. The super star of the SLM laser is no doubt the distributed feedback (DFB) laser as it takes over 99% shares of the SLM laser market—the throughput of DFB lasers have reached 10 million per year in the recent few years just by one supplier (Wuhan Telecommunication Device Co., Wuhan, China) among the few largest semiconductor laser manufactures.

30.1.4 Why Simple FP Laser Won't Work

It is well known that a piece of straight and smooth optical waveguide with its two mirror-like ends forms a simplest resonant cavity, and semiconductor laser diodes exploiting such a cavity are the FP lasers. A semiconductor optical waveguide, like the typical dielectric waveguide, has the high-pass filter characteristic that cuts off the guiding wave with wavelength longer than some critical value but imposes no constraint on the guiding waves with shorter wavelengths. Although such waveguide does only support discrete space distribution patterns in its cross section and each individual space distribution pattern is also called a transverse mode, one shouldn't mess up it (the transverse mode) with the concept of the longitudinal mode that corresponds a specific field distribution along the cavity and bears a discrete wavelength. Namely, a (dielectric) waveguide picks up the guided transverse mode, but doesn't select a single wavelength from a continuous spectrum of wavelengths that all takes the same transverse mode, except for cutting off the wavelengths below a lower bound. One can also understand it as a given waveguide structure only defines a continuous dispersion relation between the wave propagation constant (β) and the wavelength (λ), which means that, staying with a transverse mode represented by a continuous β–λ curve, one will always manage to find a β indicating a guiding (propagating along the waveguide) wave for a given wavelength λ. Nevertheless, an optical waveguide that only supports a single transverse mode stands as a necessary condition for achieving the SLM operation. As otherwise, each of the multiple nondegenerated transverse modes having its own nonidentical longitudinal mode will have to bear a different wavelength, which makes the SLM operation impossible. We are not going to dig out the waveguide concept further as the focus of this chapter is on the SLM laser. The interested readers can refer to, e.g., References [3,4].

With mirrors on the waveguide ends, the otherwise not interfered forward and backward propagating waves are coupled through the partial reflection. The waves retained inside the waveguide (i.e., the cavity) forms a standing wave pattern that corresponds to a possible longitudinal mode, as the standing wave pattern (other than a homogenous coefficient) doesn't change with time by satisfying the boundary condition at the waveguide ends and consequently bears with a static wavelength. However, in the case where the cavity length (L) is much longer than the wavelength (λ), the boundary condition can be satisfied simultaneously by multiple longitudinal modes. This is because the boundary condition of a simple FP cavity can be equivalent to a round-trip phase condition on aggregate, which reads as follows [5]:

$$2\beta L = 2\pi m \tag{30.1}$$

with m as any integer, $\beta = 2\pi n_{\text{eff}}/\lambda$ denoting the wave propagation constant and n_{eff} the effective index of the waveguide. In Equation 30.1, the facet (end mirror) phase delay has been set to zero as usually the cavity has an effective index higher than that of the surrounding medium, so that the wave experiences the internal (from high refractive index to low refractive index medium) rather than the external (from low refractive index to high refractive index medium) reflection in which the phase delay is indeed zero on reflection. From Equation 30.1, it is apparent that the number of allowed longitudinal modes in an FP cavity is directly proportional to the cavity length scaled by the wavelength, i.e., L/λ. To build an edge-emitting laser with acceptable coherence required by many applications, as well as for obtaining sufficient optical gain for achieving superior laser performance, one has to leave $L \gg \lambda$, hence multiple longitudinal modes exist in such FP cavities.

Other than the aforementioned phase condition, the amplitude condition still needs to be satisfied by the longitudinal mode that survived the phase condition to make it lase. One can therefore consider using the latter mechanism to eliminate the extra longitudinal modes. Unfortunately, the optical gain spectrum of the semiconductor material is much broader ($\sim$60 nm) than the FP longitudinal mode spacing that can readily be derived from Equation 30.1 as follows [5]:

$$\Delta\lambda = \frac{\lambda_0^2}{2n_g L} \tag{30.2}$$

with λ_0 as the center wavelength and $n_g = n_{\text{eff}} + (dn_{\text{eff}}/d\lambda)\lambda_0$ denoting the group index. For example, for typical InGaAsP/InP or AlGaInAs/InP semiconductor laser diodes, $n_g \sim 3.4$, their mode spacing vary from 0.6 to 1.2 nm in the O-band ($\lambda_0 \sim 1300$ nm), and from 0.9 to 1.8 nm in the C-band ($\lambda_0 \sim 1550$ nm), respectively, when the cavity length changes from 400 to 200 μm. As such, multiple longitudinal modes with their wavelengths closely packed adjacent to the material gain peak will share almost the same gain. Since the partially inhomogeneous gain in a semiconductor material cannot be fully clamped, small gain differences among these longitudinal modes won't be sufficient to effectively suppress the side modes, which eventually excites multiple longitudinal lasing modes. This is the main reason that the material gain dispersion in general cannot be used for longitudinal mode selection, not to mention that in the case of direct modulation, any dynamic gain change will disrupt the gain profile and consequently jeopardize any attempt of using the gain to discriminate longitudinal modes.

One therefore understands why conventional semiconductor FP laser diodes operate in multiple longitudinal modes. Above analysis also shed some light on the direction one needs to follow in the effort of developing SLM lasers, i.e., the simple FP cavity needs to be replaced by more complicated ones or modified with added structures with further built-in wavelength selectivity.

30.1.5 Classification of SLM Lasers

Canonically, there are three main approaches for semiconductor laser diodes to achieve SLM operation, all by introducing wavelength-selective cavities. These three categories of SLM lasers can be classified as the grating-assisted lasers that exploit the grating in their cavities, the coupled-cavity lasers that have extra optical (band-pass) filters in their cavities, and the external cavity lasers that use either one of the above-mentioned configurations, or the injection locking mechanism to purify the lasing spectrum but by placing the extra components outside of the main FP cavity through hybrid packaging. Theoretically, lasers in the latter category makes no difference from their counterparts in the first two categories, other than some quantitative difference on the coupling strength between the main FP cavity and the added wavelength selection components. Since they never become the mainstream product except in a few specific applications where the cost is not a concern, SLM lasers in this category are not discussed. The interested readers can refer to, e.g., References [6–8].

30.1.6 Organization of this Chapter

Sections 30.2 and 30.3 will be dedicated to discuss the grating-assisted lasers and the coupled-cavity lasers, respectively, covering their structures and working mechanisms, governing equations describing the device physics processes, and numerical simulation results on device performance.

The following section briefly describes the recent development on this topic, by showing a few advanced structures for emerging demands, their operating principles, the measured prototype device performance, in contrast to numerical simulation results.

30.2 Grating-Assisted SLM Laser Diodes

30.2.1 Grating Analysis

Noticing that a normal facet mirror formed by a dielectric (semiconductor)—dielectric (air) interface doesn't have wavelength dependence, which makes the FP cavity formed by a straight and smooth waveguide with such facet mirrors on both ends lacking the ability to select a single lasing wavelength, one would naturally think of exploiting a wavelength-selective reflector to replace the end facet mirror, which directly leads to the birth of distributed Bragg reflector (DBR) laser. This idea can logically be extended to turn the entire or part of the original straight and smooth waveguide inside the cavity into a corrugated waveguide (i.e., the waveguide grating) for lasing wavelength selection, which leads to the innovation of a whole family of DFB lasers, although the original idea of the DFB laser was independently proposed [9]. In all such grating-assisted SLM lasers, the grating obviously plays a center role. We therefore start this section by analyzing the grating itself.

A uniform (passive) grating with period Λ in any shape can be viewed as a linear superposition of many sinusoidal gratings with harmonic periods Λ/m, $m = 1, 2, 3, \ldots$, according to the Fourier expansion, simply because the refractive index change along the grating follows a periodic function. A phase-matching condition can therefore be found as follows [10]:

$$k_i \cos \theta_i = \pm \frac{2\pi}{(\Lambda/m)} + k_{om} \cos \theta_{om} \tag{30.3}$$

with k_i and k_{om} indicating the incident and mth order diffracted (by the mth order sinusoidal grating) wave numbers, respectively, θ_i and θ_{om} the incident and mth order diffracted wave angles as shown in Figure 30.2, respectively.

In this application, the incident light is propagating along the laser waveguide, $\theta_i = 0$ and $k_i = \beta = 2\pi n_{\text{eff}}/\lambda$. What one wants is to have the incident wave turn around by 180°, propagating along the waveguide in the opposite direction, which requires $\theta_{om} = \pi$ and $k_{om} = \beta = 2\pi n_{\text{eff}}/\lambda$, hence one has, from Equation 30.3,

$$2\beta = \frac{4\pi n_{\text{eff}}}{\lambda} = \frac{2\pi m}{\Lambda}, \text{ or } \Lambda = \frac{m\lambda}{2n_{\text{eff}}} \tag{30.4}$$

as the rule to select the grating period for coupling the forward and backward propagating waves in the waveguide at a given wavelength (λ), with n_{eff} still indicating the effective index of the waveguide. The wavelength that satisfies condition (Equation 30.4) is conventionally called Bragg wavelength. It is self-evident that for a given grating with fixed Λ, any wavelength deviated from λ wouldn't satisfy the phase-matching condition (Equation 30.4), consequently at such wavelength there won't be any coupling between the forward and backward propagating waves. As such, an ideal grating (with infinite length) would indeed select a single wavelength to reflect, which makes a perfect mirror as required by an SLM laser cavity. In reality, the grating cannot be made infinitely long. One would therefore expect some residue reflection at deviated wavelengths. Nevertheless, by making the grating sufficiently long, one can always expect a mirror with sufficient wavelength selection required by the SLM operation.

Yet another design consideration is how to choose the grating order. By choosing a sinusoidal grating component with order $m = M$ to be satisfied by Equation 30.4, one understands that those sinusoidal grating components with orders beyond M won't take any effect, simply because if $\lambda = 2n_{\text{eff}}\Lambda/M$, for any grating orders in $M + n$, $n = 1, 2, 3, \ldots$, their diffraction angles according to Equation 30.4,

$$\theta_{o(M+n)} = \arccos\left(1 + \frac{2n}{M}\right) \tag{30.5}$$

offer no real solution at all. Physically, the distance between the two consecutive reflections of the grating in order $M + n$ is $\Lambda/(M + n)$ shorter than half of the equivalent wavelength in waveguide $\lambda/(2n_{\text{eff}}) = \Lambda/M$, therefore, constructive addition of the waves on its reflection is impossible because there is no phase-matching condition (i.e., two identical phases) that can possibly be found within 2π. This can be seen more clearly if we show the round-trip phase delay between the two consecutive reflections explicitly by

$$2\frac{2\pi n_{\text{eff}}[\Lambda/(M+n)]}{\lambda} = 2\frac{2\pi n_{\text{eff}}[\Lambda/(M+n)]}{2n_{\text{eff}}\Lambda/M} = 2\pi\frac{M}{M+n} < 2\pi.$$

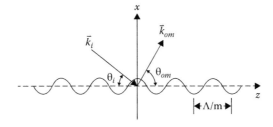

FIGURE 30.2 An illustrative diagram showing the light diffraction by a grating, with Λ/m indicating the mth order harmonic period, θ_i and θ_{om} the incident and mth order diffracted wave angles in respect to the grating plane, respectively, $\vec{k}_i$ and $\vec{k}_{om}$ the incident and mth order diffracted wave vectors, respectively; they must be in the same plane as required by the matching tangential field condition at the grating boundary.

Therefore, those sinusoidal grating components with orders higher than M virtually take no effect, because all the reflected waves by these higher order grating components cannot be added in a constructive way, rather, they all tend to be cancelled out.

For sinusoidal grating components with orders $n = 1, 2, \ldots, M - 1$, they will make the incident wave diffract into different directions, with their angles specified, according to Equation 30.4:

$$\theta_{on} = \arccos\left(1 - \frac{2n}{M}\right) \tag{30.6}$$

For example, if one chose the Bragg wavelength to match the second-order sinusoidal grating component $M = 2$ and $\lambda = n_{eff}\Lambda$. The second-order sinusoidal grating component therefore couples the forward going wave in the waveguide into the backward going wave, which offers the effect one needs, whereas the first-order sinusoidal grating component couples the forward going wave in the waveguide into a wave going along the vertical direction ($\theta_{o1} = 90°$). The latter appears to be a vertically radiating wave that leaves away from the waveguide. The rest sinusoidal components with their orders higher than 2 all take negligible effect.

In this specific application, one usually doesn't want to create any accompanying radiation that will likely bring in unwanted loss, hence one should align the Bragg wavelength with the first-order sinusoidal grating component $M = 1$ and $\lambda = 2n_{eff}\Lambda$. As such, the forward and backward waves are coupled at the lowest grating order and consequently it leaves no lower order grating components for the guided waves to be coupled to the radiating wave. Again, those higher order grating components ($M > 1$) virtually bring in no effect.

From the above qualitative analysis, one also knows that it is the period, rather than the shape, that dominates the grating characteristics. And the above analysis is valid under the assumption that the grating is sufficiently long. The required grating length to achieve certain performance, however, can only be evaluated in connection with the grating coupling strength in unit length through numerical approach. Also known as the grating coupling coefficient (κ), the unit length grating coupling strength is determined by the grating shape and the refractive index contrast, or the effective index contrast (Δn_{eff}) on aggregate. In the most general case, one needs to use the mode-matching method (MMM) [11,12] to calculate the reflection, transmission, and loss spectra of a given waveguide grating. A good approximation is to chop the waveguide grating into many short pieces along the guiding direction. By solving the mode in each piece as an eigenvalue problem defined in the local waveguide cross section, one will be able to extract the effective index in each piece as the eigenvalue solution. One can therefore use the one-dimensional (along the guiding direction) transfer matrix method (TMM) [13,14] to solve for the aforementioned grating spectra.

While numerical approaches would offer us an accurate result, one sees little physics and it is also hard to link the grating design parameters to its characteristics. For this reason, one usually uses the full numerical solver as a simulation tool only to validate or confirm a design. One therefore still needs an analytical or semi-analytical tool for the grating design purpose. For this reason, we will show following the expressions [15]. Actually, the refractive index distribution of a waveguide grating can be written in the Fourier series:

$$n^2(x, y, z) = n_0^2(x, y) + \sum_{m \neq 0} \Delta n_m^2(x, y)e^{jm\frac{2\pi z}{\Lambda}} \tag{30.7}$$

due to its periodicity along z (the guiding direction), where n_0^2 and Δn_m^2 indicate the DC and mth order coefficient of the square refractive index in the expansion, respectively, and can readily be found by taking the overlap integral (along z within one period Λ) between the square refractive index (n^2) and the respective sinusoidal grating order (in its complex form $e^{jm2\pi z/\Lambda}$, $m = 0, \pm1, \pm2, \ldots$) normalized by the period Λ. The cross-sectional (x, y) dependence of the refractive index reflects the waveguide structure. Consequently, the coupling coefficient of the mth order sinusoidal grating component can be found by

$$\kappa_m = \frac{\pi}{\lambda n_{\text{eff}}} \frac{\iint\limits_{\Sigma} \Delta n_m^2(x, y)|\varphi(x, y)|^2 dxdy}{\iint\limits_{\Sigma} |\varphi(x, y)|^2 dxdy} \tag{30.8}$$

where n_{eff} and $\varphi(x, y)$ denote the effective index and optical field distribution of the waveguide defined by the background refractive index distribution (i.e., the DC component in the Fourier expansion) $n_0^2(x, y)$, respectively, and can be found by solving the following eigenvalue problem (with n_{eff} and φ taken as the eigenvalue and eigenfunction, respectively):

$$\left(\frac{\partial^2}{\partial x^2} + \frac{\partial^2}{\partial y^2}\right)\varphi(x, y) + \left(\frac{2\pi}{\lambda}\right)^2 [n_0^2(x, y) - n_{\text{eff}}^2]\varphi(x, y) = 0 \tag{30.9}$$

In Equation 30.8, the integration area Σ extends to where φ can reach. In following discussions, we will stay with the first-order Bragg grating unless otherwise specified. Hence we will use $\kappa_{\pm}$ to indicate $\kappa_M = \kappa_{\pm 1}$. One also needs to note that Equation 30.7 is only valid for an infinitely long grating, consequently the grating coupling coefficient ($\kappa_{\pm}$) given in Equation 30.8 is subject to the same assumption. Practically for $n_{\text{eff}}L/\lambda > 500$, such extracted $\kappa_{\pm}$ in conjunction with the analytical spectrum expression (shown below) provides no appreciable difference from the accurate result obtained by the full numerical approaches (e.g., MMM or TMM).

For a fully confined guided mode (hence φ is real), both the grating symmetry and the nature of the complex material refractive index change dictate the coupling complexity. For a purely refractive index-coupled grating with real $n^2 - n_0^2$, one may easily find from Equation 30.7 that $\Delta n_{-1}^2 = (\Delta n_{+1}^2)^*$, hence $\kappa_- = \kappa_+^*$ according to Equation 30.8. One may conclude that, if the grating has a center symmetry, n^2 is an even function of z and $\Delta n_{\pm 1}^2$ are real, $\kappa_{\pm}$ are real and $\kappa_- = \kappa_+$, whereas if the grating has a center antisymmetry, n^2 is an odd function of z and $\Delta n_{\pm 1}^2$ are imaginary, $\kappa_{\pm}$ are imaginary and $\kappa_- = -\kappa_+$. For a purely gain- or loss-coupled grating with imaginary $n^2 - n_0^2$, one finds, however, $\Delta n_{-1}^2 = -(\Delta n_{+1}^2)^*$ from Equation 30.7, hence $\kappa_- = -\kappa_+^*$ following Equation 30.8. One therefore knows that, if the grating has a center symmetry, n^2 is an even function of z and $\Delta n_{\pm 1}^2$ are imaginary, $\kappa_{\pm}$ are imaginary and $\kappa_- = \kappa_+$, whereas if the grating has a center antisymmetry, n^2 is an odd function of z and $\Delta n_{\pm 1}^2$ are real, $\kappa_{\pm}$ are real and $\kappa_- = -\kappa_+$. Generally, for a complex-coupled grating with complex $n^2 - n_0^2$, or for a grating structure that is neither symmetric nor antisymmetric, $\Delta n_{\pm 1}^2$ are complex so are $\kappa_{\pm}$.

For a waveguide grating with a length of L without any extra end facet reflection [which can practically be realized by, e.g., antireflection (AR) coating on the end facet], one finds the amplitude reflection spectrum (defined as the ratio of the backward and forward going guided wave at the input port) and the amplitude transmission spectrum (defined as the ratio of the forward going guided wave at the output and input port) as [16]:

$$R_{DBR}^{\pm}(\lambda) = \frac{j\kappa_{\mp}\sinh(\gamma L)}{\gamma\cosh(\gamma L) - (\alpha + j\delta)\sinh(\gamma L)} \tag{30.10}$$

$$T_{DBR}^{\pm}(\lambda) = \frac{\gamma}{\gamma\cosh(\gamma L) - (\alpha + j\delta)\sinh(\gamma L)} \tag{30.11}$$

where $\gamma^2 = (\alpha + j\delta)^2 + \kappa_+\kappa_-$ and $\delta = 2\pi[n_{\text{eff}}/\lambda - 1/(2\Lambda)]$, with λ now indicating the wavelength variable and α the amplitude modal loss (< 0) or gain (> 0) variable of the waveguide grating. Figure 30.3

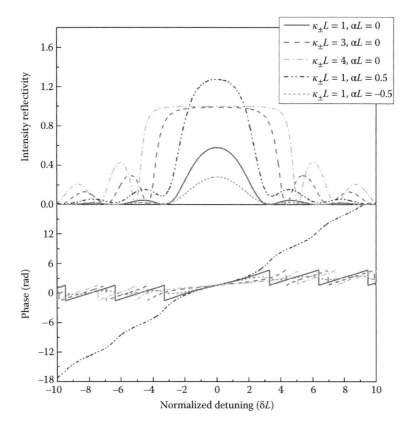

FIGURE 30.3 Reflection spectra of the Bragg grating under different normalized coupling coefficients $\kappa_\pm L = 1, 3, 4$ without gain or loss $\alpha L = 0$, and under $\kappa_\pm L = 1$ with gain $\alpha L = 0.5$ and loss $\alpha L = -0.5$.

shows respectively the amplitude and phase of the reflection spectrum of a purely refractive index-coupled grating, under different normalized (real) grating coupling coefficient ($\kappa_\pm L$) and normalized modal loss/gain (αL).

As can be seen from the spectrum, the maximum reflection happens at the Bragg wavelength. A Bragg stop-band can be identified by the main lobe with its width defined as the difference between the two zero-reflection wavelengths. Zero-reflection happens to the wave at the wavelength with its reflections from two adjacent grating teeth completely canceled out due to a phase delay of π, which makes the further cancel out of all reflections through the entire grating due to the equal tooth distance (i.e., the grating period Λ). Obviously, the wave at the zero-reflection wavelength will pass through the grating with 100% transmissivity as if the waveguide grating offers nothing more than a phase delay unit. For a purely passive, refractive index-coupled grating ($\kappa_+\kappa_- = \kappa_+\kappa_+^*$, hence $\kappa_+\kappa_-$ is real and can be written as $|\kappa|^2$) without any modal gain or loss ($\alpha = 0$), zero-reflections (maximum transmission) happen at $\gamma L = jk\pi$, $k = \pm 1, \pm 2, \pm 3, \ldots$, or $\delta L = \pm\sqrt{(|\kappa_+|L)^2 + (k\pi)^2}$. The Bragg stop-band width is therefore given by [16]

$$2\delta L = 2\sqrt{(|\kappa|L)^2 + \pi^2} \text{ or } \Delta\lambda_B = \frac{\lambda_B^2}{n_{\text{eff}}L}\sqrt{(|\kappa|L/\pi)^2 + 1} \qquad (30.12)$$

with $\lambda_B = 2n_{\text{eff}}\Lambda$ indicating the Bragg wavelength (center of the Bragg stop-band) given by the phase-matching condition (Equation 30.4) for the first-order Bragg grating. It is apparent that the Bragg stop-band width ($\Delta\lambda_B$) increases with the normalized coupling coefficient almost linearly if $|\kappa|L >> \pi$, or quadratically if $|\kappa|L << \pi$. It is worth mentioning that the measured stop-band width of the lasing spectrum in the DFB lasers is always smaller than the Bragg stop-band width, and the former approaches $\Delta\lambda_B$ as $|\kappa|L$ increases. We explain the underlying physics later in discussion of DFB lasers.

In the near infrared fiber-optic communication wavelength range, the first-order Bragg grating has its period in the submicrometer range (e.g., ~200 nm in the O-band centralized at 1300 nm and ~240 nm in the C-band centralized at 1550 nm). It is impossible to fabricate the waveguide surface-relief grating by the standard photolithography technique. Rather, one has to exploit more complicated technologies such as the holographic lithography [17] or the electron-beam lithography (EBL) [18]. While the former doesn't need expensive facility and fits the mass production mode, it can hardly create any nonuniform grating pattern. The latter, conversely, can create arbitrary user-defined grating pattern but needs expensive machine and doesn't fit the production mode for the long hours it needs to scan-write the whole wafer. A newly emerged technique, nano-imprinting [19], seems to be very promising as the ultimate solution to the fabrication of the waveguide grating. With the imprinting mask written by EBL with complicated patterns, one can use the mask repeatedly to transfer its pattern onto wafers by a pressing-and-developing process through the spin-coated pressure-sensitive deforming-resister on the wafer surface. As such, the expensive EBL writing just needs to be done once on the mask, the later pattern transfer process from the mask to wafers (imprinting) is fast and reliable, with an even better quality compared to the widely used holographic lithography technique [20]. It is also worth mentioning that the invention of the "floating" grating idea [21] solves the problem in the precise control of the grating coupling strength and guarantees its repeatability in the mass production mode, as the depth of the grating teeth is now determined by the grating layer thickness which can be made accurate (to sub-10 nm range) in material growth, as opposed to be given by the etching depth which is not only hard to control down to 10~20 nm but also hardly repeatable. This invention sets a milestone on the mass production as well as the cost reduction of DFB lasers.

30.2.2 Distributed Bragg Reflector Laser

A typical DBR laser has a structure shown in Figure 30.4. With the first-order Bragg grating placed on one or both ends of an FP laser as the reflector, the coupling between the forward and backward going waves inside the cavity becomes wavelength dependent.

Similar to the lasing condition derived for the FP laser [5], one can readily find the resonance condition for the DBR laser by following an approach of matching the traveling wave with itself after a

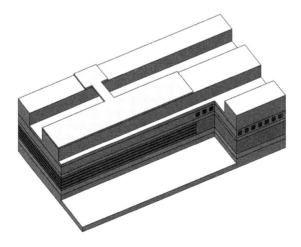

FIGURE 30.4 The DBR laser structure.

round trip inside the cavity:

$$R_l(\lambda)R_r(\lambda)e^{2[g-\alpha_i+j\frac{2\pi n_{\text{eff}}}{\lambda}]L_{ar}} = 1 \tag{30.13}$$

with R_l and R_r denoting the wavelength-dependent amplitude reflectivity of the Bragg grating reflector on the left and right end of the active region, respectively, given in the form of (Equation 30.10), g the modal gain, α_i the modal internal loss, and L_{ar} the active region length. One can further split the lasing condition (Equation 30.13) into the amplitude and phase condition as

$$g = g_{th} \equiv \alpha_i + \alpha_c \equiv \alpha_i + \frac{1}{2L_{ar}} \ln \frac{1}{|R_l(\lambda)||R_r(\lambda)|} \tag{30.14}$$

and

$$\frac{4\pi n_{\text{eff}}L_{ar}}{\lambda} + \phi_l(\lambda) + \phi_r(\lambda) = 2\pi m, \quad m = 1, 2, 3, ... \tag{30.15}$$

where ϕ_l and ϕ_r indicate the phase of the amplitude reflectivity of the Bragg grating on the left and right end, respectively. It is clear that, following Equation 30.14, only the wave at the Bragg wavelength sees the highest reflection and consequently has the smallest cavity loss defined by $\alpha_c \equiv -(0.5/L_{ar})\ln(|R_l||R_r|)$, since at least one of the end reflectors will offer the highest reflection (i.e., either R_l, or R_r, or both will take the largest value) at the Bragg wavelength following the grating reflection spectrum shown in Figure 30.3. It then requires the smallest modal gain to reach the threshold in the neighborhood of the Bragg wavelength. It is not as clear though, following Equation 30.15, that one can find a solution to λ in the neighborhood of the Bragg wavelength, not going beyond $\lambda_B^2/(2n_{\text{eff}}L_{ar})$, with λ_B indicating the Bragg wavelength ($2n_{\text{eff}}\Lambda$). Actually, if one takes $\lambda = \lambda_B + \Delta\lambda$ ($\Delta\lambda \sim \lambda_B^2/(2n_{\text{eff}}L_{ar}) << \lambda_B$) in Equation 30.15 to obtain

$$2\pi\Delta\lambda\left(\frac{2n_{\text{eff}}L_{ar}}{\lambda_B^2}\right) = \frac{4\pi n_{\text{eff}}L_{ar}}{\lambda_B} + \phi_l(\lambda) + \phi_r(\lambda) - 2\pi m \tag{30.16}$$

and also to notice that the total phase on the right hand side of Equation 30.16 is a slow-varying function of λ in the neighborhood of λ_B following Figure 30.3, one immediately finds that as $\Delta\lambda$ changes from 0 to $\lambda_B^2/(2n_{\text{eff}}L_{ar})$, the total phase on the left-hand side of Equation 30.16 sweeps over an entire range of 2π, which means that the phase on the two sides of Equation 30.16 will have a matching point within the interval between λ_B and $\lambda_B \pm \lambda_B^2/(2n_{\text{eff}}L_{ar})$, given the fact that the phase takes 2π as its modulo. One therefore reaches the conclusion that the phase condition (Equation 30.15) always has a solution in a close neighborhood of the Bragg wavelength bound by $\lambda_B^2/(2n_{\text{eff}}L_{ar})$. Consequently, one knows that the amplitude condition (Equation 30.14) and the phase condition (Equation 30.15) jointly select a single lasing wavelength near the Bragg stop-band center of the end grating reflector. This explains why the DBR laser operates under the SLM.

The main advantage of the DBR laser lies in the separation of its gain and wavelength selection (i.e., the passive Bragg grating) region. Since the grating region is not biased, the wavelength selection mechanism suffers little change with the injection current in the gain region. One then expects a stable SLM operation with high side-mode-suppression-ratio (SMSR) and small wavelength chirp, which is evidenced by many publications [22,23]. It is also quite convenient to introduce a lasing wavelength tuning in DBR laser, as one can readily bias the grating region by a separate electrode from the gain region. As such, the current injection in the grating region introduces an effective index change, since the carrier-induced gain/loss change will have an accompany refractive index change following the Kramers–Kronig relation. Therefore, the peak reflection wavelength (the Bragg wavelength) given by Equation 30.4 will change with the current injected in the grating region, which causes the lasing wavelength change accordingly. Usually, a phase adjustment section needs to be inserted between the gain and the grating region to separately tune the phase for matching the lasing condition, as otherwise the DBR laser may cease lasing or experience a lasing mode hopping due to the mismatch of the phase condition. However, since the grating region has to be made transparent to the lasing wavelength, as otherwise the light will be absorbed by the unbiased passive grating region if it has the same band-gap with the active region, the fabrication of DBR lasers inevitably involves the monolithic integration technique that still stands as an unsolved problem up to today, for its low yield. With the birth of silicon photonics in the recent decade, it is quite promising to make the Si photonics DBR laser by bonding a direct band-gaped III-V compound semiconductor gain block on top of the silicon-on-insulator (SOI) waveguide, with the Bragg grating engraved on SOI. There is a hope that the DBR laser will be reborn with the booming Si photonics.

30.2.3 Distributed Feedback Lasers

A typical pure refractive index-coupled uniform-grating DFB laser has a structure shown in Figure 30.5. Unlike the DBR laser with a gain region made of smooth waveguide separated from the passive grating region, the DFB laser has these two regions merged into one, which makes the laser cavity substantially different from the aforementioned FP or DBR structures in the sense that the forward and backward going waves are constantly coupled in a distributive manner inside the cavity rather than in a lumped sum manner at the two ends of the cavity only. A wavelength discrimination mechanism is therefore brought in by such distributed coupling, since only the waves at the phase matched wavelength(s) will possibly add constructively to establish a standing wave pattern (i.e., a longitudinal mode) inside the cavity, which imposes a stringent condition that may likely purifies the lasing spectrum by cutting off most of the cavity modes otherwise allowed in the FP cavity.

An immediate finding is that this structure won't have its lasing wavelength at the Bragg wavelength anymore. This is because the wave near the Bragg wavelength sees the highest reflection. It therefore cannot travel far along the waveguide grating that also provides the gain. Consequently, it cannot obtain sufficient gain as required for the lasing to happen. On the contrary, at the Bragg stop band edge where the grating offers zero-reflection as shown in Figure 30.3, a maximum transmission through the grating region is obtained. Although the wave at this particular wavelength experiences the highest single-pass gain as it sees the entire length of the grating (with gain), it cannot be the lasing mode either since no reflection happens to it so that no resonance can be established. In this sense, the lasing should happen somewhere in between, i.e., the traveling waves at the lasing wavelength should be partially reflected by the grating to

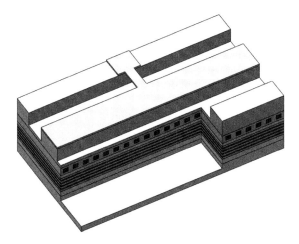

FIGURE 30.5   The DFB laser structure.

establish the resonance. It should be partially transmitted in order to experience the gain in the grating region as well. As such, the phase matching and amplitude sustenance (gain) conditions must be mixed.

Following the same thought, i.e., searching for the consistent condition between the wave and itself after a round trip inside the cavity, one can still obtain the resonance condition for the DFB cavity [24], which will stand as the lasing condition of the DFB laser. In the general case, however, one cannot expect to have analytical expressions for such lasing condition, so that we will cover the general case by the full numerical model introduced later. To gain an insight into the DFB laser, we will analyze a simple case with both of the end facet reflections and longitudinal spatial horn burning (LSHB) effect neglected. Actually, the lasing condition of such a simple DFB laser can be found by setting the denominator of Equation 30.10 and/or Equation 30.11 equal to zero. The associated infinite reflection and transmission indicate that the device can offer output without input—a feature must be carried by an oscillator. Hence one has

$$\gamma \coth(\gamma L) = \alpha + j\delta, \ \gamma^2 = (\alpha + j\delta)^2 + \kappa_+\kappa_- \qquad (30.17)$$

Usually, for a given $\kappa_\pm L$ as the grating design parameter, one solves for γL from

$$(\gamma L)^2 \coth^2(\gamma L) = (\gamma L)^2 - \kappa_+\kappa_- L^2 \qquad (30.18)$$

which has the same solution set as Equation 30.17. Hence the normalized gain (αL) and detuning (δL) can be found by

$$\alpha L + j\delta L = \pm\sqrt{(\gamma L)^2 - \kappa_+\kappa_- L^2} \qquad (30.19)$$

from which one will be able to obtain the lasing modal threshold gain (from $g_{th} \equiv \alpha_i + \alpha$, α_i the modal internal loss) and the lasing wavelength (by solving $\delta = 2\pi[n_{eff}/\lambda - 1/(2\Lambda)]$ for λ). The normalized gain (αL) and detuning (δL) for different normalized grating coupling coefficient ($\kappa_\pm L$) are plotted in Figure 30.6, as the solution to Equations 30.18 and 30.19 obtained by a numerical root searching program built on Muller's algorithm [25].

The threshold gain solution indicates that, in a similar role as $\alpha_c = -(0.5/L)\ln(|R_l||R_r|)$ to the FP or DBR cavity, α, as the cavity loss to the DFB cavity, is in proportion to $1/(\kappa L)$. For a DFB grating design

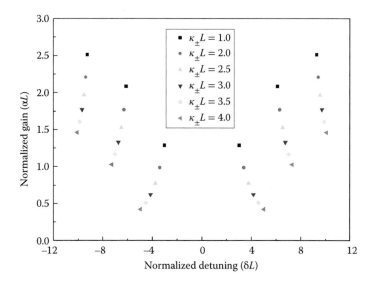

FIGURE 30.6 $\alpha L \sim \delta L$ for different $\kappa_\pm L$ as solutions to Equations 30.18 and 30.19.

with a higher κ_+, the cavity is more closed as an FP or DBR cavity with a higher end reflection R, and with the cavity length scaled in the same way. This conclusion, together with the above discussion on its lasing wavelength position, stands as the most basic design guidance for the DFB laser. Also for the DFB laser with high $\kappa_+ L$ design, its cavity loss reduces so as the required modal gain to reach the lasing threshold, the lasing wavelength will therefore approach more closely to the zero-reflection wavelengths, as the wave may have sufficient gain to reach its threshold through fewer passes, and consequently less reflection is necessary. Namely, as the threshold gain approaches to zero, the lasing will happen near the Bragg stop-band edge. Hence one finds that the lasing wavelength of the DFB laser with higher $\kappa_+ L$ should be closer to the zero-reflection wavelength of the unbiased grating, i.e., the Bragg stop-band edge.

A serious issue with the purely refractive index-coupled uniform-grating DFB laser is, however, its dual-mode operation nature originated from the double degeneracy caused by the center mirror-symmetry (or antisymmetry) of the cavity. It is quite obvious that, from Equation 30.18, for any real $\kappa_+ \kappa_-$, once γ is a solution, $-\gamma$ and $\pm\gamma^*$ are all solutions. Hence for any solution set (α, δ), $(\pm\alpha, \pm\delta)$ are all possible solutions according to Equation 30.19, which means that for any modal gain required to reach the lasing threshold, also known as the DFB laser cavity loss (α as a solution to the lasing condition), there is a double degeneracy of the lasing wavelength corresponding to $\pm\delta$, respectively. These two lasing wavelengths sit on each side of the Bragg wavelength, and approach to the Bragg stop-band edges as $\kappa_+ L$ increases, in consistency with previous analysis. Physically, this effect can be explained by the two equally possible standing wave patterns shown in Figure 30.7a. In a DFB laser cavity with the center mirror-symmetry ($L/\Lambda = n + 1/2$, n is an integer) or center mirror-antisymmetry ($L/\Lambda = n$ is an integer) and with zero end facet reflections (the latter condition can actually be further relaxed to identical end facet reflections), there exist two possible standing wave patterns as a result of the constructive addition of the distributively coupled forward and backward going waves. They share the same gain inside the cavity and have the identical cavity loss. The only difference between the two patterns is that one of them has its intensity peak aligned with the high refractive index region in every period, whereas the other one has its intensity peak aligned with the low refractive index region in every period. As such, the former and latter pattern will see the cavity mainly in high and low refractive index, respectively. Consequently, the former and latter will take the longer (red) and shorter (blue) lasing wavelength, respectively, on each side of the center Bragg wavelength. The double degenerated longitudinal mode with different standing wave pattern at different lasing wavelength but with

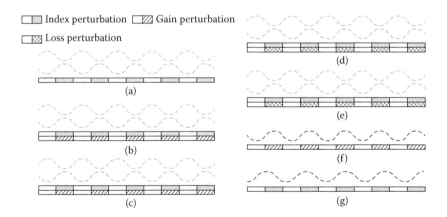

FIGURE 30.7 Schematic diagrams for (a) purely refractive index-coupled uniform-grating DFB, (b) partially gain-coupled in-phase DFB, (c) partially gain-coupled antiphase DFB, (d) partially loss-coupled in-phase DFB, (e) partially loss-coupled antiphase-grating DFB, (f) purely gain-coupled DFB, (g) quarter-wavelength phase-shifted (QWPS) DFB; the blue and red dashed lines in (a)–(e) show the intensity standing wave patterns of the longitudinal modes bearing the short and long wavelength (on the blue and red side of the Bragg wavelength), respectively; the black dashed line in (f) and (g) shows the intensity standing wave pattern of the longitudinal mode bearing the Bragg wavelength.

the same threshold gain breaches the SLM operation condition, which makes the purely refractive index-coupled uniform-grating DFB with a perfect center mirror-symmetric (or antisymmetric) cavity not a SLM laser.

This conclusion can also be generalized to any cavity with the perfect center mirror-symmetry or center mirror-anti-symmetry, once its reflection (or transmission) spectrum is symmetric, a necessary condition to reach the SLM operation is to have its lasing wavelength right in the center of the spectrum. Otherwise, dual-mode operation is inevitable. By recalling the lasing condition for the DBR laser, one may conclude that, in order to ensure its SLM operation with high SMSR, one has to either somehow break the symmetry (or anti-symmetry) of the cavity by, e.g., introducing two slightly different end reflection gratings (usually with slightly misaligned Bragg wavelengths) or use the grating reflector on one end only, or to ensure that the solution to the phase condition (Equation 30.15) is at the Bragg wavelength so that the lasing will happen exactly in the reflection spectrum center.

Since it is impossible to make the purely refractive index-coupled uniform-grating DFB lase at the Bragg wavelength, a conclusion not only drawn from the analysis at the beginning of the discussion on DFB lasers, but also known by the lasing condition (Equation 30.18). This is because for real $\kappa_+\kappa_-L^2$, real γL cannot be a solution as otherwise, the right-hand side is smaller than γL whereas the left-hand side is bigger than γL, hence γL must be complex, which leaves $\delta \neq 0$. The only viable way to reach the SLM operation for such DFB laser is then to break the cavity symmetry (or antisymmetry) by introducing asymmetric facet coatings on the two ends, usually a combination of the AR coating for the front facet and the high-reflection (HR) coating for the rear facet, respectively. However, the SLM yield with high SMSR for such DFB laser is still low since the grating phase at the two ends is usually random, given the condition that the grating period is in 1/5~1/4 of a micrometer whereas the uncertainty of the facet cleaving position is in a range of $\pm(2 \sim 3)\mu m$. Particularly for those DFB lasers with high κ_+L designs, they usually suffer from very poor SLM yield (only ~15% for *SMSR* > 30 dB with κ_+L > 3) [26]. This is because the DFB laser with high κ_+L tends to have the envelope of the standing wave pattern piled up in the middle, rather than at the two ends, of the cavity, due to the strong coupling, which makes the cavity less affected by the end facet condition as the waves inside the cavity simply don't "feel" much about the existence of the cavity ends. As a result, the end facet asymmetry applied on a strongly coupled DFB cavity is not sufficient to break the degenerated threshold gain between the two modes. Hence one finds that high κ_+L often triggers dual mode lasing in DFB lasers. Figure 30.8 shows the standing wave envelope distributions inside the DFB cavity under different coupling strengths, from which one finds that the field stays at the two ends, or piles up in the middle of the cavity, corresponding to the weak (with $\kappa_+L < \pi/2$), or strong (with $\kappa_+L > \pi/2$) coupling condition, respectively. And the field takes almost an even distribution along the cavity under the critical condition $\kappa_+L = \pi/2$, as it has to be.

To thoroughly solve the dual-mode operation problem associated with the purely refractive index-coupled uniform-grating DFB laser, one can have different approaches along with the thought of breaking the cavity symmetry, or forcing the lase at the Bragg wavelength. While the former led to the invention of various complex-coupled DFB lasers [27–30], the latter resulted in the popular product of the QWPS DFB laser [31].

Figure 30.7b shows the working principle of the in-phase partially gain-coupled DFB laser. As an extra gain is introduced periodically inside the high refractive index region, the longitudinal mode with its intensity peak in its standing wave pattern located in the high refractive index region in each period obtains more gain than the other mode, hence the degeneracy between the two longitudinal modes breaks and only the mode with the longer wavelength (on the right/red side of the Bragg wavelength) will lase, which makes such DFB laser operate under SLM. Figure 30.7c shows the working principle by taking the other option to boost the longitudinal mode bearing the shorter wavelength (on the left/blue side of the Bragg wavelength). Known as the antiphase partially gain-coupled DFB laser, the structure has an extra gain introduced periodically inside the low refractive index region, which again breaks the degeneracy between the two longitudinal modes since the longitudinal mode with its intensity peak in its standing wave pattern located in the low refractive index region in each period has more gain than the other

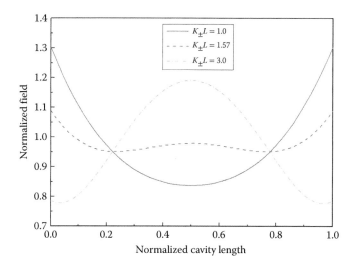

FIGURE 30.8   Field envelope distributions inside the DFB cavity for $\kappa_{\pm}L = 1.0(< 1.57)$, $\kappa_{\pm}L \sim 1.57$, and $\kappa_{\pm}L = 3.0(> 1.57)$.

mode. As a result, only the shorter wavelength mode will lase, which again makes such DFB laser operate under SLM.

Figure 30.7d and e shows the working mechanisms of the partially loss-coupled in-phase and antiphase DFB lasers, respectively. If an extra loss is introduced periodically inside the high (or low) refractive index region, the longitudinal mode with its intensity peak in its standing wave pattern located in the high (or low) refractive index region in each period suffers from more loss than the other mode, hence the degeneracy between the two longitudinal modes breaks and only the mode with the shorter (or longer) wavelength on the left/blue (or right/red) side of the Bragg wavelength will lase, due to the suppression of its counterpart on the other side of the Bragg wavelength, which makes such DFB laser operate under SLM.

One can also understand why the degeneracy breaks in such partially gain- or loss-coupled DFB lasers by the lasing condition (Equation 30.18). For complex-coupled DFBs, $\kappa_{+}\kappa_{-}L^2$ is also complex or $\kappa_{+}\kappa_{-}L^2 \neq (\kappa_{+}\kappa_{-}L^2)^*$. As such, if γ is a solution, γ^* is not necessarily a solution. Following Equation 30.19, one then finds that $\pm\delta$ cannot share the same α as the solutions.

Figure 30.7f shows the working principle of the purely gain-coupled DFB laser. Since the purely gain-coupled laser has a purely negative $\kappa_{+}\kappa_{-}$ ($\kappa_{-} = -\kappa_{+}^*, \kappa_{+}\kappa_{-} = -\kappa_{+}\kappa_{+}^* \equiv -|\kappa|^2$), which makes the lasing condition (Equation 30.18) turn into

$$(\gamma L)^2 \coth^2(\gamma L) = (\gamma L)^2 + (|\kappa|L)^2 \tag{30.20}$$

a real solution of γ is then possible which forces the lasing happen at the Bragg wavelength according to Equation 30.19, or

$$\alpha L + j\delta L = \sqrt{(\gamma L)^2 + (|\kappa|L)^2} \tag{30.21}$$

from which one readily achieve $\delta = 0$ for the right-hand side is real. The dual mode lasing in purely gain-coupled DFB laser is not possible as one cannot find two different standing wave patterns inside the cavity that share the same gain. It is obvious that only a single standing wave pattern is allowed inside the cavity with itself completely aligned with the gain grating. Any standing wave pattern in different shape with the gain grating will naturally be suppressed. Therefore, one knows that Equation 30.20 cannot have a complex γ as its solution associated with the longitudinal mode with the lowest cavity loss (α). As otherwise, γ^* will be an allowed solution as well, according to Equation 30.21, one will then be able to have a pair of lasing wavelengths corresponding to the doubly degenerated solutions ($\alpha, \pm\delta$), which is in conflict against our previous physics-based analysis on the purely gain-coupled DFB laser cavity. Namely, solution ($\alpha, 0$) must take the smallest cavity loss $\alpha = \alpha_{min}$, while any other possible doubly degenerated solutions in the form of ($\alpha, \pm\delta$) with $\delta \neq 0$ must have a higher cavity loss $\alpha > \alpha_{min}$ hence cannot be the lasing modes. This conclusion has been evidenced by a numerical searching for all possible solutions of Equation 30.20 [32]. Finally, one may conclude that the purely gain-coupled DFB laser will lase at the Bragg wavelength and naturally stands as the SLM laser.

Figure 30.7g shows the working mechanism of the QWPS DFB laser. The only difference from the uniform-grating purely refractive index-coupled DFB is that there is an extra half period of the grating inserted right at the cavity center. An extra half period of the grating is equivalent to a quarter-wavelength phase shift in the first-order grating following the relationship between the grating period and the Bragg wavelength $\Lambda = \lambda_B/(2n_{eff})$, which therefore introduces a π phase shift to the reflected wave or a π phase shift to the round trip wave. As such, the wave will have to make two round trips to recover its status, which leads to its resonance. This feature suggests us to fold back the cavity from the center to equivalent the original structure to a uniform-grating DFB with half of its original length and with a perfect facet with 100% reflectivity at one end (the original cavity center). In such a half cavity equivalent structure, if the standing wave pattern has its intensity peak originally aligned with the high refractive index region in every period, upon the 100% reflection at one end, the peak will have to be aligned with the low refractive index region in every period, and vice versa. As such, none of the two original standing wave patterns inside the uniform-grating DFB can exist inside the modified cavity with a quarter-wavelength phase shift introduced at the cavity center. Rather, only a standing wave pattern with its peak aligned at the edge of the grating tooth (i.e., the interface of the high and low refractive index region) in every period can stay. This standing wave pattern is unique inside the cavity and corresponds to the longitudinal mode at the Bragg wavelength, for it sees an averaged refractive index between the high and low index region in every period, which happens to be the effective index of the unperturbed waveguide by the grating. Also physically, a longitudinal mode with its envelope piled up in the middle of the cavity will see the highest reflection from both sides if it bears the Bragg wavelength. Hence this mode will be most tightly confined inside the cavity and sees the most gain. Consequently, it should have the lowest cavity loss and becomes the lasing mode. The numerically calculated longitudinal mode distribution along the cavity of such structure confirms this conclusion [33] as one finds that, with its lasing mode always at the Bragg wavelength, the QWPS DFB laser has a sharp longitudinal mode distribution peak at the center of the cavity. In a uniform-grating DFB without a phase-shifted center inside the cavity, this scenario can never happen, as the maximum reflection at the Bragg wavelength can only be possibly seen by the wave from one side rather than from both sides. Hence the wave at the Bragg wavelength cannot stay long inside the cavity to obtain sufficient gain for lasing, a same conclusion drawn earlier for the uniform-grating purely refractive index-coupled DFB laser. It is worth mentioning that a more detailed analysis would have to have a new lasing condition involved as Equation 30.17 is no longer valid for either a nonuniform grating (the grating with a phase shift) or a uniform grating with nonzero end facet reflection (the equivalently folded cavity with a uniform grating plus 100% reflection at one end). From the newly derived lasing condition for this structure, one will be able to find that $\delta = 0$ is indeed an allowed solution that has the minimum cavity loss ($\alpha = \alpha_{min}$) [34]. Therefore, the QWPS DFB laser always operates under the SLM and has its lasing mode at the Bragg wavelength.

30.2.4 Governing Equations for Design, Modeling, and Simulation

The following one-dimensional (along the propagation direction, z) traveling wave model is often used as the optical governing equation for the edge-emitting semiconductor lasers with various waveguide-grating structures [16]:

$$
\left(\frac{1}{v_g}\frac{\partial}{\partial t}+\frac{\partial}{\partial z}\right)e^f(z,t)=[j\left(\frac{2\pi n_{\mathrm{eff}}}{\lambda_0}-\frac{\pi}{\Lambda}\right)+j\frac{2\pi}{\lambda_0}\Gamma\Delta n(z,t,\omega_0)
$$

$$
+\frac{1}{2}\Gamma g(z,t,\omega_0)-\frac{1}{2}\alpha_L]e^f(z,t)+j\kappa_+ e^b(z,t)+s^f(z,t)
$$

$$(30.22)$$

$$
\left(\frac{1}{v_g}\frac{\partial}{\partial t}-\frac{\partial}{\partial z}\right)e^b(z,t)=[j\left(\frac{2\pi n_{\mathrm{eff}}}{\lambda_0}-\frac{\pi}{\Lambda}\right)+j\frac{2\pi}{\lambda_0}\Gamma\Delta n(z,t,\omega_0)
$$

$$
+\frac{1}{2}\Gamma g(z,t,\omega_0)-\frac{1}{2}\alpha_L]e^b(z,t)+j\kappa_- e^f(z,t)+s^b(z,t)
$$

where $e^{f,b}$ denote the slow-varying envelopes of the forward and backward going traveling wave field, respectively, g and Δn the material gain and refractive index change, respectively, α_L the non-interband optical modal loss, ω_0 the reference optical frequency, $\lambda_0 = 2\pi c/\omega_0$ the reference optical wavelength (i.e., the vacuum wavelength of the reference optical frequency), $v_g = c/n_g$ the group velocity, n_{eff} and Γ the effective index (of the waveguide) and confinement factor (of the active region), respectively, Λ and $\kappa_\pm$ the period, the backward-to-forward (+), and the forward-to-backward (−) coupling coefficient of the first-order waveguide Bragg grating, respectively, $s^f = s^b \equiv \tilde{s}$ the spontaneous emission contributions.

The associated optical field can be expressed as

$$
\vec{E}(x,y,z,t)=\frac{1}{2}\vec{s}\varphi(x,y)[e^f(z,t)e^{j\frac{\pi}{\Lambda}z}+e^b(z,t)e^{-j\frac{\pi}{\Lambda}z}]e^{-j\omega_0 t}+c.c.
$$

$$(30.23)$$

with $\vec{s}$ indicating the unit vector of along the (linear) field polarization direction, φ the field distribution of the waveguide mode as the solution of Equation 30.9. For a given waveguide-grating structure, one needs to solve Equation 30.9 first to find the effective index (n_{eff}) and the guided mode field distribution (φ). The coupling coefficients ($\kappa_\pm = \kappa_{\pm 1}$) can then be obtained from Equation 30.8 and the confinement factor is given as

$$
\Gamma=\int_{A.R.}|\varphi(x,y)|^2 dxdy/\int_{\Sigma}|\varphi(x,y)|^2 dxdy
$$

$$(30.24)$$

with the integration areas *A.R.* and Σ indicating the active region (quantum wells) only and the whole area where the optical field extends.

Equation 30.22 is obtained from the optical wave equation under the slow-varying envelope approximation with the second-order derivatives of the envelope ($\partial^2 e^{f,b}/\partial z^2$ and $\partial^2 e^{f,b}/\partial t^2$) all ignored, whereas the optical wave equation is directly obtained from the Maxwell equations with the coupling among the field polarization components $\nabla(\nabla \cdot \vec{E})$ ignored. The former is true for the edge-emitting device with its cavity length L much longer than the operating wavelength λ (usually L is in a few hundreds to a thousand of λ/n_{eff}), in which the optical wave is propagating along the cavity with its traveling wave factors $\phi_\pm = e^{j(\pm \pi z/\Lambda - \omega_0 t)}$ varying much faster as z and t, hence:

$$
|\partial^2 e^{f,b}/\partial z^2/e^{f,b}| << |\partial^2\phi_\pm/\partial z^2/\phi_\pm| = (\pi/\Lambda)^2
$$

$$
|\partial^2 e^{f,b}/\partial t^2/e^{f,b}| << |\partial^2\phi_\pm/\partial t^2/\phi_\pm| = \omega_0^2
$$

$$(30.25)$$

The latter assumption holds for the weakly confined waveguide, which is generally true of most of the III–V compound semiconductor laser diodes with either ridge waveguide or buried heterojunction structures.

In the field expansion (Equation 30.23), the single-guided-mode assumption has been invoked as a necessary condition for the SLM operation. As different guided modes have different effective indices, for the same grating period Λ, different wavelengths will be chosen to satisfy the phase-matching condition (Equation 30.4). As a result, multiple longitudinal modes in different lasing wavelengths, with each of them corresponding to a guided mode in a different effective index, will be excited, which breaches the SLM condition. A single polarization component is assumed for the same reason, but this is not necessary as the difference between the two effective indices corresponding to the two orthogonally polarized components in a guided mode is usually very small in weakly confined waveguides. As a result, the difference between their corresponding lasing wavelengths can hardly be appreciable. Actually, for most of the III–V compound semiconductor laser diodes, the active region is made of the compressively strained multiple quantum wells (CS-MQW). The compressively strained quantum well can only provide the optical gain in the (100) plane [35], which means only the TE mode with its optical E-field polarized in parallel to the slab waveguide interfaces, i.e., the interfaces of the layer stack grown in the usual <100>direction, can see the gain whereas the other polarization component, i.e., the TM mode with its E-field polarized in the perpendicular direction, will be completely suppressed. Therefore, it is usually sufficient to study the TE mode with its E-field horizontally polarized in parallel with the device top and bottom surfaces.

The material gain of semiconductors can generally be calculated by the physics-based first-principles model [16]. For performance simulation of the SLM laser, however, the following model that phenomenologically links the gain to the (minority) carrier density inside the active region is sufficient:

$$g(z, t, \omega_0) = a(\omega_0) \ln \frac{N(z, t)}{N_{\mathrm{tr}}(\omega_0)} / [1 + \varepsilon(|e^f(z, t)|^2 + |e^b(z, t)|^2)] \tag{30.26}$$

with $a(\omega_0)$ and $N_{\mathrm{tr}}(\omega_0)$ indicating the gain coefficient and the transparent (minority) carrier density for the active region comprising (strained-layer) multiple quantum wells, ε the nonlinear gain saturation factor. These parameters are usually extracted in the neighborhood of the reference optical frequency (ω_0) from the physics-based material gain model or measured experimentally.

The refractive index change and the material gain, corresponding to the real and imaginary part of the material susceptibility, are connected by the Kramers–Kronig relation. Actually, if one views the susceptibility as the frequency-domain transfer function of the material system in responding to the optical E-field as the input signal, the polarization excited by the optical E-field then becomes the output signal. Following the linear system theory, the time-domain polarization is given as the convolution of the time-domain optical E-field and the inverse Fourier transform of the susceptibility. Since a real physical system must be causal, which means the polarization at any time instant t can only be dependent on the optical E-field given before t, it imposes a strong constraint between the real and imaginary part of the susceptibility and is mathematically given in the form of the Kramers–Kronig transform. By noticing that the SLM laser is a typical narrow-band device with its lasing frequency in the neighborhood of the reference optical frequency (ω_0) and the Kramers–Kronig relation is linear, one can readily express the refractive index change as a linear function of the material gain:

$$\Delta n(z, t, \omega_0) = \alpha_{\mathrm{LEF}} g(z, t, \omega_0) \tag{30.27}$$

with α_{LEF} denoting the linewidth enhancement factor.

Finally, the carrier density inside the active region can be described by the following rate equation [16,36]:

$$\frac{\partial N(z, t)}{\partial t} = \frac{I(z, t)}{qV} - [AN(z, t) + BN^2(z, t) + CN^3(z, t)]$$

$$- \frac{n_{\mathrm{eff}}}{2\hbar\omega_0} \sqrt{\frac{\varepsilon_0}{\mu_0}} \frac{L\Gamma}{V} g(z, t, \omega_0)[|e^f(z, t)|^2 + |e^b(z, t)|^2] \tag{30.28}$$

where I represents the bias current; q the unit electron charge; V the active region volume; A, B, and C the minority carrier Shockley–Read–Hall (SRH); bimolecular and spontaneous emission, and Auger recombination coefficients, respectively; $\hbar$ Plank constant; $\sqrt{\varepsilon_0/\mu_0} \approx 1/377[\text{S}]$ the vacuum admittance; and L the active region length (along the cavity direction). Given the fact that the active region of the laser diode has a low (unintentional) doping concentration and the potential drop across it is usually negligible, a quasi-neutrality condition holds and consequently the electrons and holes have the same density inside the active region—we therefore don't have to distinguish the electron and hole density and simply use the term "carrier density" to indicate both.

As the seed of the lasing process, the spontaneous emission contribution cannot be ignored at the beginning as otherwise Equation 30.22 becomes homogenous and only the trivial zero solution exists. Once the lasing starts, i.e., for the device operated under a bias beyond the threshold, however, the contribution from the spontaneous emission becomes negligible, which is reflected as a fact that the self-consistent solution of the above set of equations always converges to the same value regardless of the excitation method of the spontaneous emission, as long as the spontaneously emitted noise power is self-consistently described. The spontaneous emission contribution is usually assigned as a Gaussian distributed zero-mean random variable with its autocorrelation function normalized by the spontaneously emitted noise power [16]:

$$< |\tilde{s}(z,t)||\tilde{s}(z',t')| >= \frac{2\hbar\omega_0}{n_{\text{eff}}}\sqrt{\frac{\mu_0}{\varepsilon_0}}\gamma\Gamma n_{sp}g(z,t,\omega_0)\delta(z-z')\delta(t-t') \qquad (30.29)$$

where γ denotes the dimensionless coefficient of the coupling from the spontaneous emission to the entire spatial sphere and over the whole frequency spectrum to the waveguide mode at the reference optical frequency, n_{sp} the dimensionless ratio of the spontaneous emission to stimulated emission gain, and δ the Dirac function.

Equations 30.22, and 30.26 through 30.29 form a closed loop for one to find a set of self-consistent solution on the (minority) carrier density inside the active region (N), the material gain (g), and the slow-varying envelopes of the forward and backward going traveling wave field ($e^{f,b}$), for any given bias (I), with Equations 30.9, 30.8 and 30.24 presolved for a given SLM laser structure. Consequently, the optical field inside the laser can be found through Equation 30.23 and other physics quantities for characterizing the laser, such as the output optical power and the lasing wavelength, can readily be found [16,37]. An efficient solution technique for solving Equation 30.22 is to use the time-domain split-step method [38], with Equation 30.28 solved by the well-known Runge–Kutta method [39].

To describe the thermal effect, one still needs to add on the thermal diffusion equation [40] and to modify the empirical formulas for the material gain and refractive index change by considering their temperature dependence [16]. However, we will exclude the thermal description in this model to focus our study on the SLM operation aspect of semiconductor laser diodes.

30.2.5 Examples of SLM Laser Characteristics

By exploiting the aforementioned model, one can calculate the device performance for SLM DFB lasers with a few different types of grating structures as discussed in Section 30.2.3, shown in Figures 30.5 and 30.7. Figures 30.9 through 30.14 show the numerical simulation results of the output optical power and lasing wavelength as functions of the bias, the optical spectra at a fixed bias, the normalized optical field intensity and (minority) carrier density distributions along the cavity, and the small-signal intensity modulation responses for the purely refractive index-coupled uniform-grating DFB, partially gain/loss-coupled in/antiphase DFB, purely gain-coupled DFB, and QWPS DFB, respectively, with those device parameters involved in the model summarized in Table 30.1.

From Figure 30.9, one finds that while all DFB lasers have similar threshold current, the loss-coupled and QWPS DFB lasers have lower slope efficiency. This is expected for loss-coupled DFB lasers as their loss-coupling coefficient introduces an extra contribution equivalent to the non-interband optical modal

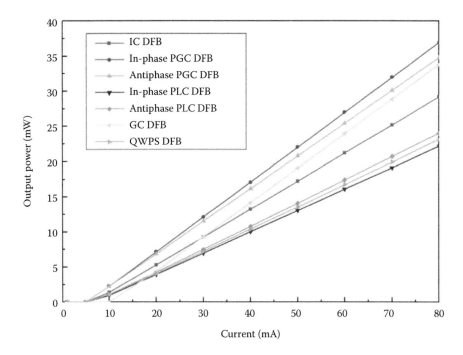

FIGURE 30.9 Output power–bias current curves for different DFB lasers. IC = (purely refractive) index-coupled (uniform grating), PGC = partially gain-coupled, PLC = partially loss-coupled, GC = (purely) gain-coupled, QWPS = quarter-wavelength phase-shifted.

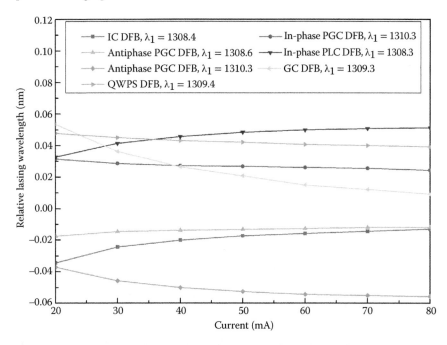

FIGURE 30.10 Relative lasing wavelength shift–bias current curves for different DFB lasers. IC = (purely refractive) index-coupled (uniform grating), PGC = partially gain-coupled, PLC = partially loss-coupled, GC = (purely) gain-coupled, QWPS = quarter-wavelength phase-shifted; the relative lasing wavelength shift is defined as λ-λ_l, with λ indicating the lasing wavelength at any bias current > 20 mA and λ_l the lasing wavelength at bias current = 20 mA for different DFB lasers under comparison.

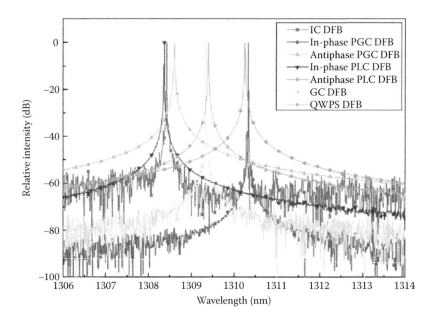

FIGURE 30.11 Lasing spectra for different DFB lasers. IC = (purely refractive) index-coupled (uniform grating), PGC = partially gain-coupled, PLC = partially loss-coupled, GC = (purely) gain-coupled, QWPS = quarter-wavelength phase-shifted.

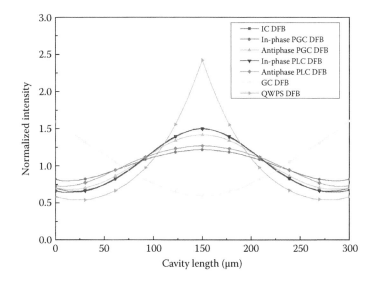

FIGURE 30.12 Normalized optical field intensity distributions for different DFB lasers. IC = (purely refractive) index-coupled (uniform grating), PGC = partially gain-coupled, PLC = partially loss-coupled, GC = (purely) gain-coupled, QWPS = quarter-wavelength phase-shifted.

loss (α_L). As shown in Table 30.1, the assumed non-interband optical modal loss is 10/cm, the normalized loss-coupling coefficient 0.5 would give an extra optical loss around $0.5/L = 0.5/(0.03 \text{ cm}) \sim 17/\text{cm}$. The total optical loss 27/cm indicates a significant increase as compared the original value in 10/cm. Since the slope efficiency is inversely proportional to the optical modal loss, higher loss would certainly cause a lower efficiency. The low slope efficiency of the QWPS DFB laser, however, is not caused by the high

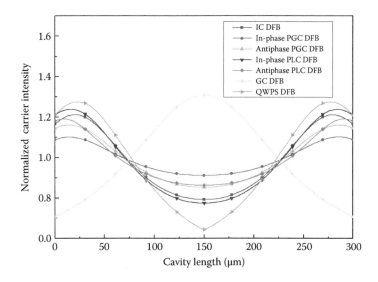

FIGURE 30.13 Normalized (minority) carrier density distributions for different DFB lasers. IC = (purely refractive) index-coupled (uniform grating), PGC = partially gain-coupled, PLC = partially loss-coupled, GC = (purely) gain-coupled, QWPS = quarter-wavelength phase-shifted.

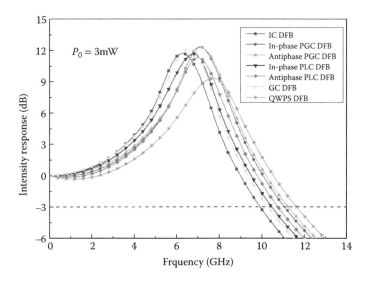

FIGURE 30.14 Small-signal intensity modulation responses for different DFB lasers. IC = (purely refractive) index-coupled (uniform grating), PGC = partially gain-coupled, PLC = partially loss-coupled, GC = (purely) gain-coupled, QWPS = quarter-wavelength phase-shifted.

optical modal loss, but by the low cavity loss. Since this structure has its field intensity mostly concentrated in the middle of the cavity, such a field pattern tends to retain more photons inside the cavity and consequently leads to a low output power by the low field distribution at both facets. This is consistent with our understanding on QWPS DFB as a high Q (quality)-factor laser with low cavity loss. Still because the slope efficiency is determined by the cavity loss over the summation of the cavity loss and the optical modal loss, apparently a low cavity loss will lead to a low slope efficiency once the optical modal loss is a nonvanishing positive value.

TABLE 30.1   Model Parameters

Model Parameter	Value
Cavity length L (μm)	300
Thickness of the quantum well (nm)	5
Number of quantum wells	6
Reflectivity of both laser facets	0, 0
Group index n_g	3.6
Effective index n_{eff}	3.2
Confinement factor Γ	0.06
Ridge width (μm)	2
Normalized index coupling coefficient	2.5
Normalized gain coupling coefficient	0.5
Normalized loss coupling coefficient	0.5
Grating period Λ (nm)	204.7
Gain coefficient a (cm^{-1})	1800
Transparent carrier density N_{tr} (10^{18} cm^{-3})	0.8
Linewidth enhancement factor α_{LEF}	−3
Nonlinear gain saturation factor ε (10^{-2} V^{-2})	9
SRH recombination coefficient A (10^9 s^{-1})	1
Bimolecular and spontaneous emission recombination coefficient B (10^{-10} cm^{-3}s^{-1})	2
Auger recombination coefficient C (10^{-29} cm^{-6} s^{-1})	4
Non-interband modal loss α_L (cm^{-1})	10
Spontaneous emission coupling coefficient γ	10^{-4}
Spontaneous emission over stimulated emission ration_{sp}	1.7
Reference wavelength λ_0 (μm)	1.310

As ideal laser characteristics, the carrier density as well as the averaged gain should be clamped after lasing, but this is true only when the carrier density and optical field intensity both have uniform distributions along the cavity. In DFB lasers, however, the carrier density and optical field intensity have nonuniform distributions in "opposite" shapes as shown in Figure 30.12. Also, following the Schwartz nonequality, one knows the fact that the overlap integration between two functions with different shapes must be smaller than that between two functions in a similar shape. Under higher bias current, stronger LSHB makes the carrier density and optical field intensity distribution shapes more unlike. As a quantity directly in proportion to the overlap integration of the carrier density and optical field intensity distributions along the cavity, the averaged gain will be smaller. To maintain the lasing status, i.e., to balance the cavity loss plus the optical non-interband modal loss, the carrier density will have to increase with the bias current in order to compensate the "lost" averaged gain due to the enhanced LSHB. The increased carrier density will therefore cause a refractive index reduction of the active region following the Kramers–Kronig relation, which is simplified as Equation 30.27 with the negative linewidth enhancement factor (α_{LEF}). The reduction of the active region refractive index brings down the effective index of the waveguide and consequently makes the Bragg wavelength as well as the whole Bragg stop-band shift toward the shorter wavelength side, according to Equation 30.4 ($\lambda_B = 2n_{eff}\Lambda$). This effect is indeed reflected in Figure 30.10 by the blueshift with the increased bias current of the lasing wavelengths of the QWPS DFB and GC DFB lasers, located right in the center of the Bragg stop-band (or at the Bragg wavelength).

Figure 30.10 also shows that, without the thermal effect considered in simulation, the lasing wavelengths of the DFB lasers on the left (blue) side the Bragg stop-band (i.e., the IC DFB, the antiphase PGC DFB, and the in-phase PLC DFB; see Figure 30.11) have redshift toward the longer wavelength side with increased bias current, whereas the lasing wavelengths of the DFB lasers on the right (red) side (i.e., the in-phase PGC DFB and the antiphase PLC DFB; see Figure 30.11) take blueshift toward the shorter wavelength side. This is brought in by the shrinkage of the Bragg stop-band width due to the enhanced LSHB with the increased bias current, which further suggests an effective normalized coupling coefficient reduction or a more opened (lower Q-factor) cavity in accompanying with the stronger LSHB under higher injection level.

Figure 30.11 clearly shows that the QWPS and purely gain-coupled DFB lasers lase at the Bragg wavelength, whereas the in-phase and antiphase partially gain-coupled DFB lasers have their lasing wavelengths at the right (red) and left (blue) side of the Bragg stop-band, respectively. The in-phase and antiphase partially loss-coupled DFB lasers have their lasing wavelengths located the other way around, at the left (blue) and right (red) side of the Bragg stop-band, respectively. Finally, the purely refractive index-coupled DFB laser picks the left (blue) side wavelength rather randomly, with a relatively poor SMSR as expected. All these findings are consistent with previous analysis on these DFB lasers with different grating designs.

From Figure 30.12, one finds that the QWPS DFB laser has its field intensity mostly gathered in the middle of the cavity, as opposed to the purely gain-coupled DFB laser with its field intensity largely concentrated at both edges. The former cavity obviously bears the highest Q-factor or smallest cavity loss, whereas the latter one must correspond to the lowest Q-factor or highest cavity loss. Since the threshold current is proportional to the cavity loss, and the slope efficiency also increases with the cavity loss, the high Q-factor QWPS DFB laser bearing a low cavity loss has a low threshold current and a low slope efficiency. On the contrary, the low Q-factor purely gain-coupled DFB laser bearing a high cavity loss has a high threshold current as well as a high slope efficiency. This result has been confirmed by Figure 30.9 exactly. Other DFB lasers have their field intensities more evenly distributed and have their cavity losses in between. Consequently, they have their threshold current and slope efficiency values in between as shown in Figure 30.9. It is also worth mentioning that the sharp field turning in the middle of the QWPS DFB laser cavity is brought in by the half grating period shift exactly at the same location.

All carrier density distributions take the opposite shape as compared to their optical field intensity distributions. This result is expected as the carriers inside the cavity are consumed to generate photons, indicating a low carrier density wherever the optical field intensity is high, and vice versa.

Simulation result shows that there is no significant difference in terms of the 3 dB small-signal intensity modulation bandwidth among different DFB lasers once they are set to output the same optical power. This is quite different from the result obtained in References [41,42], in which the antiphase partially gain-coupled DFB laser shows a significantly broader 3 dB small-signal intensity modulation bandwidth in comparison with its in-phase counterpart. The inconsistency might come from the different selection on device parameters and further investigation is still needed before a final conclusion can be achieved.

30.3 Coupled-Cavity Single Longitudinal Mode Laser Diodes

30.3.1 A General Optical (Band-Pass) Filtering Model

The conventional FP cavity supports many closely packed longitudinal cavity modes with their spacing given in the form of Equation 30.2. For the edge-emitting laser with its cavity length (L) much longer than the center wavelength (λ_0), the mode spacing is much smaller than the full width at half maximum (FWHM) of the material gain peak. As such, multiple longitudinal modes can lase simultaneously, which makes a typical edge-emitting FP laser operate under the multiple longitudinal mode lasing scheme. Other than the aforementioned grating-assisted methods by introducing the wavelength-selective end reflectors (DBR), or by exploiting various DFB cavities, to achieve the SLM operation, one can still take an alternative approach by inserting an optical (band-pass) filter into the FP cavity to eliminate all other longitudinal cavity modes but leave only one for lasing. This approach is generally known as the coupled-cavity laser that forms the other category of SLM lasers, if we don't count in various external cavity SLM laser configurations.

For a coupled-cavity laser generally presented in the form of an FP cavity with an inserted OBPF shown in Figure 30.15, we have [43]:

$$\begin{bmatrix} A_{out} \\ B_{out} \end{bmatrix} = \frac{1}{T^2} \begin{bmatrix} 1 & -R \\ -R & 1 \end{bmatrix} \begin{bmatrix} e^{j\varphi_2} & 0 \\ 0 & e^{-j\varphi_2} \end{bmatrix} \frac{1}{t_{21}} \begin{bmatrix} t_{12}t_{21} - r_{12}r_{21} & r_{21} \\ -r_{12} & 1 \end{bmatrix} \begin{bmatrix} e^{j\varphi_1} & 0 \\ 0 & e^{-j\varphi_1} \end{bmatrix} \begin{bmatrix} 1 & R \\ R & 1 \end{bmatrix} \begin{bmatrix} A_{in} \\ B_{in} \end{bmatrix}$$
(30.30)

with $A_{in,out}$ and $B_{in,out}$ denoting the optical field amplitude of the forward and backward propagating waves, with their subscript "in" and "out" indicating the assumed input and output port on the left- and right-hand sides, respectively. One should note that the optical field is polarized in the cross-sectional plane perpendicular to the cavity direction. The arrows associated with them in Figure 30.15 indicate their propagation direction, not their polarization direction. Also in Equation 30.30, R and T indicate the amplitude reflectivity and transmissivity of the end facets when looking from inside of the FP cavity, $r_{12,21}$ the amplitude reflectivities when looking from the FP cavity to the OBPF on the left- and right-hand sides, $t_{12,21}$ the amplitude transmissivities from left to right and from right to left when staying inside the FP cavity, respectively. Finally, $\varphi_{1,2}$ are the phase delays of the wave traveling through the two sections ($L_{1,2}$) of the FP cavity separated by the OBPF.

Equation 30.30 can readily be simplified to

$$\begin{bmatrix} A_{out} \\ B_{out} \end{bmatrix} = \frac{1}{t_{21}T^2} \begin{bmatrix} e^{j\varphi_2} & -Re^{-j\varphi_2} \\ -Re^{j\varphi_2} & e^{-j\varphi_2} \end{bmatrix} \begin{bmatrix} \delta & r_{21} \\ -r_{12} & 1 \end{bmatrix} \begin{bmatrix} e^{j\varphi_1} & Re^{j\varphi_1} \\ Re^{-j\varphi_1} & e^{-j\varphi_1} \end{bmatrix} \begin{bmatrix} A_{in} \\ B_{in} \end{bmatrix}$$

$$\equiv \frac{1}{t_{21}T^2} \begin{bmatrix} A_{11} & A_{12} \\ A_{21} & A_{22} \end{bmatrix} \begin{bmatrix} A_{in} \\ B_{in} \end{bmatrix}$$
(30.31)

where

$$\delta \equiv t_{12}t_{21} - r_{12}r_{21} \text{ and } A_{22} \equiv e^{-j(\varphi_2+\varphi_1)} - r_{12}Re^{-j(\varphi_2-\varphi_1)} - r_{21}Re^{j(\varphi_2-\varphi_1)} - \delta R^2 e^{j(\varphi_2+\varphi_1)}$$
(30.32)

The lasing condition is therefore obtained as $A_{in} = B_{out} = 0$, but A_{out} and B_{in} are not zero, which requires $A_{22} = 0$, or:

$$[r_{12}e^{-j(\varphi_2-\varphi_1)} + r_{21}e^{j(\varphi_2-\varphi_1)}]Re^{j(\varphi_2+\varphi_1)} + \delta R^2 e^{2j(\varphi_2+\varphi_1)} = 1$$
(30.33)

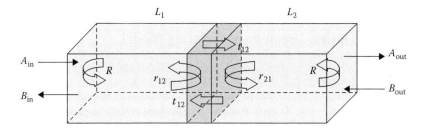

FIGURE 30.15 An illustrative diagram showing the FP cavity with an inserted optical band-pass filter (OBPF), with A_{in} and B_{in}, A_{out} and B_{out} indicating the optical field amplitudes of the forward and backward propagating waves at the input and output port, respectively; r_{12} and t_{12} the amplitude reflectivity and transmissivity from the left section to the inserted OBPF; r_{21} and t_{21} the amplitude reflectivity and transmissivity from the right section to the inserted OBPF; $L_{1,2}$ the lengths of the left and right section, respectively; R the end facet amplitude reflectivity of the FP cavity.

1. FP cavity

   It is obvious that once we remove the OBPF by letting $r_{12} = r_{21} = 0$ and $\delta = 1$, Equation 30.33 reduces to the well-known lasing condition of the FP laser:

   $$R^2 e^{2j(\varphi_2 + \varphi_1)} = 1 \tag{30.34}$$

2. Bidirectional symmetric filter

   Once the OBPF has the bidirectional symmetry, by letting

   $$r_{12} = r_{21} = jre^{j\theta} \tag{30.35}$$

   and

   $$t_{12} = t_{21} = te^{j\theta} \tag{30.36}$$

we have

$$\delta = (t^2 + r^2)e^{2j\theta} \equiv (1 - l)e^{2j\theta} \tag{30.37}$$

where θ and l stand for the single pass phase and loss of the filter, respectively. Hence we find from Equation 30.33:

$$2jrR\cos(\varphi_2 - \varphi_1)e^{j(\varphi_2 + \varphi_1 + \theta)} + (1 - l)R^2 e^{2j(\varphi_2 + \varphi_1 + \theta)} = 1 \tag{30.38}$$

It is apparent that Equation 30.38 can be rewritten as

$$\bar{R}^2 e^{2j(\varphi_2 + \varphi_1 + \theta)} = 1, \text{ with } \bar{R} \equiv \sqrt{\frac{1 - l}{1 - 2rR\cos(\varphi_2 - \varphi_1)e^{j(\varphi_2 + \varphi_1 + \theta + \pi/2)}}} R \tag{30.39}$$

Unlike in Equation 30.34 where the amplitude reflectivity of the end facets R is a constant for a conventional FP laser, a coupled-cavity laser comprising an FP cavity with an inserted bidirectional symmetric filter has a wavelength-dependent effective amplitude reflectivity in its lasing condition, which, conceptually similar to the lasing condition of the DBR laser given as Equation 30.13, provides an extra wavelength selection mechanism to make a conventional FP laser single-moded. For example, if the inserted filter is lossless ($l = 0$), Equation 30.39 shows that the modification on the lasing condition from the original FP cavity is determined by the reflection (r) and length (θ) of the filter and its position inserted inside the FP cavity (φ_1 and φ_2). It is possible that, within the wavelength range set by the gain spectrum bandwidth, there is a single wavelength that makes $1 - 2rR\cos(\varphi_2 - \varphi_1)e^{j(\varphi_2 + \varphi_1 + \theta + \pi/2)} = 1 - 2|r|R$, and $2(\varphi_1 + \varphi_2 + \theta) = 2m\pi$, with m as an integer. Hence $\bar{R}$ will be the maximum with the round trip phase-matching condition satisfied at this wavelength. As a result, the lasing will only happen at this wavelength and such a general coupled-cavity structure supports SLM operation.

In the fabrication of real-world coupled-cavity lasers, the OBPF is usually formed by an equivalent FP etalon. The filter reflectivity and transmissivity in Equations 30.35 and 30.36 can then be derived from Equation 30.30 itself with the inserted filter removed from the FP cavity in Figure 30.15. By doing so, we find that Equation 30.30 reduces to

$$\begin{bmatrix} A'_{out} \\ B'_{out} \end{bmatrix} = \frac{1}{T'^2}\begin{bmatrix} 1 & -R' \\ -R' & 1 \end{bmatrix}\begin{bmatrix} e^{j\varphi'} & 0 \\ 0 & e^{-j\varphi'} \end{bmatrix}\begin{bmatrix} 1 & R' \\ R' & 1 \end{bmatrix}\begin{bmatrix} A'_{in} \\ B'_{in} \end{bmatrix}$$

$$= \frac{1}{T'^2}\begin{bmatrix} e^{j\varphi'}(1 - R'^2 e^{-2j\varphi'}) & 2jR'\sin\varphi' \\ -2jR'\sin\varphi' & e^{-j\varphi'}(1 - R'^2 e^{2j\varphi'}) \end{bmatrix}\begin{bmatrix} A'_{in} \\ B'_{in} \end{bmatrix}$$

with the prime symbols added to the variables and parameters to differentiate them from those being used in Equation 30.30 for the most outside FP cavity in Figure 30.15. In the absence of the incident light coming from the left-hand side, $B'_{out} = 0$, we have

$$R_f \equiv \frac{B'_{in}}{A'_{in}} = j\frac{2R'\sin\varphi' e^{j\varphi'}}{1 - R'^2 e^{2j\varphi'}}, \text{ and } T_f \equiv \frac{A'_{out}}{A'_{in}} = \frac{T'^2 e^{j\varphi'}}{1 - R'^2 e^{2j\varphi'}}$$

Since the common factor $1/(1 - R'^2 e^{2j\varphi'})$ in abovementioned expressions can be expanded in a form of

$$1 + R'^2 e^{2j\varphi'} + (R'^2 e^{2j\varphi'})^2 + \ldots$$

which means that $j2R'\sin\varphi' e^{j\varphi'}$ and $T'^2 e^{j\varphi'}$ are the single pass reflectivity and transmissivity, respectively, Equations 30.35 and 30.36 are valid once we take

$$r = 2R'\sin\varphi' \qquad (30.40)$$

$$t = T'^2 \qquad (30.41)$$

and

$$\theta = \varphi' \qquad (30.42)$$

Equations 30.35 through 30.37 and Equations 30.40 through 30.42 therefore link the model parameters to the design parameters of the physical structure.

30.3.2 Cleaved-Coupled-Cavity (C³) Laser

A typical cleaved-coupled-cavity (C³) laser has a structure shown in Figure 30.16.

Following the general coupled-cavity laser model derived in Section 30.3.1, we have

$$R' = (n_{eff} - 1)/(n_{eff} + 1) \qquad (30.43)$$

$$T'^2 = 4n_{eff}/(n_{eff} + 1)^2 \qquad (30.44)$$

$$\varphi' = 2\pi d/\lambda \qquad (30.45)$$

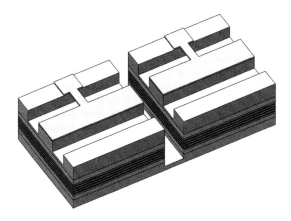

FIGURE 30.16 The C³ laser structure.

with n_{eff} denoted as the laser waveguide effective index and d the air gap spacing between the pair of coupled cavities. By noting

$$\varphi_{1,2} = [2\pi n_{eff}/\lambda - j(g - \alpha_i)]L_{1,2} \tag{30.46}$$

and using Equations 30.40 through 30.42, one can express the lasing condition (Equation 30.39) in terms of the C^3 laser design parameters ($n_{eff}, d, g, \alpha_i, L_{1,2}$). The SLM lasing condition can therefore be found by a proper combination of the two cavity lengths ($L_{1,2}$) and the air gap spacing (d).

A major issue in the fabrication of the C^3 laser is the optical alignment between the two cavities, if the air gap will be formed by cleaving. One can certainly think of using the etching technology to form the air gap, the quality of the facet; however, cannot be guaranteed unless the chemically assisted ion beam etching (CAIBE) is exploited [44]. With CAIBE, the precise control of the air gap spacing becomes difficult, especially when the required spacing is no more than a few micrometers. An elegant approach to form the C^3 laser with a narrow air gap is the microcleaving technique [45]. In this technique, after the air gap section is defined, a selective wet (chemical) etching step is applied first to remove the material (InP) surrounding the active region (InGaAsP or AlGaInAs) inside the air gap, which leaves the latter as a thin bridge hanging over in the middle. The wafer or bar is then soaked inside some liquid and placed inside an ultrasonic cleaning bath for breaking down the active region bridge inside the air gap. The facet formed by this approach has a quality equivalent to the one obtained by the conventional cleaving technique. Since the etching of the air gap won't go very deep after passing though the active region, there is no optical alignment problem as the pair of coupled cavities still sits on top of the same substrate.

Despite its readiness in fabrication [46], the C^3 structure never becomes a popular SLM laser product for its low single-mode yield and the reliability concern. The low yield comes from the sensitive phase-matching condition, which requires precise control on cavity geometrical dimensions, including both cavity lengths and the air gap spacing, as well as the effective index of the waveguide. Since the latter depends not only on the material composition and cross-sectional geometrical dimension design, but also on operating conditions such as the ambient temperature and injection current, it naturally has a very low probability to hit the stable SLM operation condition with a high SMSR over an entire bias range. The reliability concern comes from the extra pair of facets appeared in the C^3 structure, especially when it is difficult to get them protected with dielectric coating layers, for the coating is not only difficult to be applied to the narrow air gap, but also changes the lasing condition.

Figure 30.17 shows the simulated optical spectrum of a typical SLM C^3 laser, with its structural parameter given in Table 30.2, and other parameters the same as those given in Table 30.1.

30.3.3 Etched Slotted Laser

An alternative version of the coupled-cavity structure is the etched slotted SLM laser [47–50] with its typical structure shown in Figure 30.18. The main difference lies in that the coupled cavities in such structure are formed by one or multiple etched slots across the waveguide without passing through the active region. As such, there is no optical alignment issue as the coupled-cavity device still stays as a whole piece. There is no reliability concern either as no extra active region cross section is exposed other than the usual end facets. The major drawback of this structure as compared to the C^3 laser is its much reduced reflectivity (r) of the equivalent OBPF as appeared in Equation 30.39 as the general lasing condition for coupled-cavity lasers. For an individual slot, its reflectivity is given by

$$R' = (n_{eff} - n'_{eff})/(n_{eff} + n'_{eff}) \tag{30.47}$$

with n_{eff} and n'_{eff} as the effective indices of the normal section and the waveguide inside the etched slot, respectively. As compared to the normal section, the etched slot only has its waveguide cladding layer

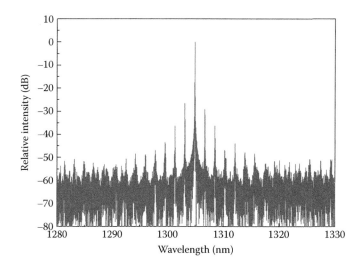

FIGURE 30.17   The simulated lasing spectrum of a typical SLM C^3 laser, with its parameters given in Table 30.2 and Table 30.1, respectively.

TABLE 30.2 C^3 Laser Parameters

Parameter	Value
Cavity lengths L_1, L_2 (μm)	136, 121 [46]
Gap between cavities d (μm)	5
Cleaved facet reflectivity r	0.565

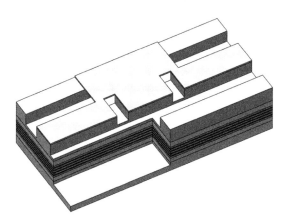

FIGURE 30.18   A typical etched slotted laser structure.

thinned to some extent, one then cannot expect a significant difference between n_{eff} and n'_{eff}. Therefore, R' given by Equation 30.47 is usually much smaller than the corresponding reflectivity in the C^3 laser given by Equation 30.43. From Equations 30.39 and 30.40, one knows that the contrast of the effective reflectivity $\bar{R}$ of the coupled-cavity laser at different wavelength will be reduce with a decreasing reflectivity R', Consequently, its ability on selecting the SLM drops significantly for the etched slotted laser.

To solve the poor mode selection problem, multiple etched slots have been introduced. A cascade of the structure in Figure 30.15 can be exploited to model such coupled-cavity laser with multiple etched slots.

A modified lasing condition similar to Equation 30.39 can also be derived with its effective reflectivity shown as a multiplication of multiple wavelength-dependent factors, with each of the factor corresponding to a single slot. As such, the wavelength spectral contrast of its effective reflectivity will be raised and its mode-selection ability will be improved. However, the multiple slotted coupled-cavity laser usually suffers an even lower single-mode yield, for the associated phase-matching condition becomes more complicated due to the introduction of multiple slots and the chance for it to be satisfied becomes rare.

30.3.4 Discrete Mode Laser

As shown in Figure 30.19, the discrete mode laser [51,52] can be viewed as a specific type of multiple slotted coupled-cavity laser with its slot's spacing properly designed to satisfy the phase-matching condition. As such, its single-mode yield can be significantly improved. With its general design rule given in Reference [52], the discrete mode laser is actually an SLM device between the grating-assisted laser and the coupled-cavity laser. It is apparent that, starting from a pair of coupled FP sections (i.e., the C^3 laser), as one increases the number of coupled FP sections with reduced reflectivities from section to section, and arranges the length of each FP section to satisfy certain phase-matching condition, one readily obtains various of grating-assisted lasers. As a special case, for a design with a uniform unit length from one FP section to another (Λ) selected to match the round trip phase-matching condition $4\pi n_{\mathrm{eff}}\Lambda/\lambda = 2\pi$, one obtains a DFB laser with the first-order Bragg grating, as the uniform unit length design forms a periodic structure (i.e., a grating) with its periodicity (Λ) specified exactly the same as Equation 30.4, the phase-matching condition for the first-order Bragg grating. Therefore, the concept of the discrete mode laser actually bridges the grating-assisted lasers and the coupled-cavity lasers as the two main categories of SLM lasers.

30.3.5 New Aspects in Governing Equations and the Solution Technique

In general, the governing Equations 30.22 through 30.29 can still be used in modeling the coupled-cavity lasers. Since there is no grating involved, π/Λ should be replaced by $2\pi n_{\mathrm{eff}}/\lambda$ with the grating coupling coefficients ($\kappa_{+,-}$) set to zero.

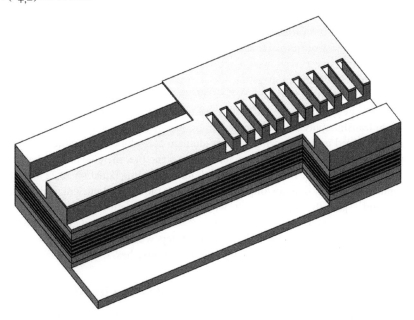

FIGURE 30.19 A typical discrete mode laser structure.

To treat the uniform sections in the coupled-cavity structure, the gain spectral dispersiveness must be considered to make sure that the lasing will happen in the neighborhood of the gain peak, as otherwise there will be no specific wavelength can be referred but the lasing cannot possibly happen at an arbitrary wavelength. Actually, the first term on the right-hand side of Equation 30.22 should be a time-domain convolution between the material gain and the field envelope, rather than be a product of them [53]. This term in its current form in Equation 30.22 is obtained under the assumption that the gain spectral bandwidth is much broader than the field envelope bandwidth, so that the time-domain gain performs as a Dirac δ-function in the convolution, which takes away the integration. For an SLM laser, the latter is usually determined by its linewidth or the modulation signal base bandwidth, whichever is broader, if the laser is under direct modulation. As the gain spectral bandwidth is usually in the range of 60 nm, whereas the field envelope bandwidth of an SLM laser usually is no more than 1 nm, the aforementioned assumption is indeed valid. To restore the gain dispersiveness in the coupled-cavity laser, however, this assumption has to be removed and a special solution technique will have to be introduced in dealing with the time-domain convolution between the material gain and the field envelope. An efficient way to treat the convolution in an initial value problem specified by a set of time-domain differential equations (or more accurately, a mixed initial-boundary value problem specified by a set of time-space domain partial differential equations) is the digital filtering method [16,54]. In this method, the material gain spectral profile is first calculated by the physics-based model within a limited wavelength range [16]. Its shape is then duplicated in the entire wavelength domain to form a periodic function, with nothing changed in its original wavelength range. The time-domain material gain can therefore found by taking the inverse Fourier transform of its spectral function. Since the latter function is turned into a periodic one, the time-domain gain therefore becomes a summation of a set of Dirac δ-functions, weighted by factors obtained as the corresponding coefficients in its Fourier expansion. Consequently, the convolution is reduced to a summation of a set of field envelopes, shifted in time domain and weighted by the said factors. For a given laser structure, these factors just need to be calculated once after the material gain spectral profile is obtained. The digital filtering method is, therefore, very efficient in handling the convolution in conjunction with the time-domain solution techniques, such as the split-step method, for solving Equation 30.22 [38,55].

Yet another problem often appearing in modeling the coupled-cavity laser is the involvement of the passive sections without injection. It is obvious that the material gain, the refractive index change, and the noise contribution in Equation 30.22 all need to be taken away for the passive sections. By noticing that Equation 30.22 become completely decoupled with the rest governing equations, one can solve these in frequency domain only for once to extract the reflectivity and transmissivity of the field envelope for these sections. In the rest (active) sections, the set of governing equations will be solved in time domain as described previously. Upon the propagating waves reach the boundary at the passive section, however, the reflected and transmitted field can be computed by taking the time-domain convolution between the incident field and the inverse Fourier transform of the reflectivity and transmissivity, respectively, as the reflectivity and transmissivity are both defined in frequency domain. Once again, the digital filtering approach can be employed in dealing with the time-domain convolution [56]. Since the reflectivity and transmissivity only need to be calculated once for a given structure, the convolution handled in such a way is very efficient. Generally, the mixed domain method, i.e., treating Equation 30.22 in time domain for active sections and in frequency domain for passive sections, is much more efficient compared with the full time-domain method, especially for passive units with long section lengths, because the frequency-domain treatment links the input and output fields of the passive section directly through transfer functions (i.e., the reflectivity and transmissivity, respectively) in a single step, without any step-by-step marching along the wave propagating direction as required by time-domain approaches.

30.3.6 Performance Comparisons among Coupled-Cavity SLM Lasers

The simulated performance of a few different types coupled-cavity SLM lasers is summarized in Table 30.3. A general tendency is that as the SMSR increases with the number of slots (i.e., the number of coupled

TABLE 30.3 Performance Comparison among Different Coupled-Cavity SLM Lasers

Laser Structure	SMSR (dB)	SLM Yield (SMSR> 15 dB)	SLM Yield (SMSR> 25 dB)
C^3 laser	12–32	60%	20%
Etched slotted laser (2 slots)	6–19	27%	0%
Discrete mode laser (5 slots)	11–29	73%	13%
Discrete mode laser (9 slots)	9–36	60%	40%
Discrete mode laser (19 slots)	7–49 (42, measured value in [52])	47%	33%

Note: The experimentally measured SMSR for a discrete mode laser with 19 slots is given in the brackets.

cavities), the SLM yield decreases. This result agrees with our conclusion drawn by the general OBPF model for the coupled-cavity structures, since the increased slot number enhances the wavelength spectral contrast of the effective reflectivity, thereby enhances the structure's mode selection ability, whereas it simultaneously introduces a more complicated phase-matching condition that is increasingly difficult to be satisfied.

30.4 Recent Development on Single Longitudinal Mode Laser Diode

30.4.1 Open Problems in SLM Lasers

Current SLM lasers are not yet ideal. Grating-assisted lasers generally suffer problems like high fabrication cost and highly sensitive to external feedback. The DBR laser needs the technology to monolithically integrate the active region with the passive grating section in different material compositions. The uniform-grating purely refractive index-coupled DFB laser has a relatively low single-mode yield, whereas the QWPS DFB laser has even poorer immunity to external feedback and wastes half of its output power, and complex-coupled DFB lasers usually have reliability issues. A variety of the grating-assisted laser structures have been proposed, but all as compromised approaches with none of them seeming to be able to solve all the aforementioned problems. Coupled-cavity lasers usually have low fabrication cost and less sensitive to external feedback. However, they all have low single-mode yield and suffer relatively low SMSR. Therefore, it still remains as an open problem on how to obtain reliable SLM lasers with high single-mode yield, high SMSR, high immunity to external feedback, and low fabrication cost.

In the past decade, some progress has been made toward the final solving of the remaining problems in SLM lasers, although most of the recent efforts on laser development have their targets on new applications found in data communication systems and telecommunication access networks with emphases on multiple wavelength accessibility and tunability, and high-speed direct modulation. Effort has also been put on developing SLM lasers with narrow linewidth to meet the strong demand in high-speed long-haul coherent telecommunication systems in advanced modulation-detection schemes with higher spectral efficiency. A thorough description on all these new developments is beyond the scope of this chapter, so that we will only give a brief introduction on the new structures that are closely relevant to SLM lasers.

30.4.2 Bragg Waveguide SLM Laser

A schematic structure of the Bragg waveguide SLM laser [57,58] is shown in Figure 30.20. Unlike the conventional SLM lasers that all utilize their cavity structure to select the single lasing mode, this structure exploits the waveguide itself to eliminate the unwanted modes in an FP cavity.

The working principle of the Bragg waveguide can be understood by a simplified model described in Figure 30.21. By exploiting the effective index method, one reduces the wave vector of an arbitrary plane traveling wave into a two-dimensional plane as shown in Figure 30.21. It can then be decomposed into a pair of orthogonal components, with the one along the waveguide direction (β) as the required wave

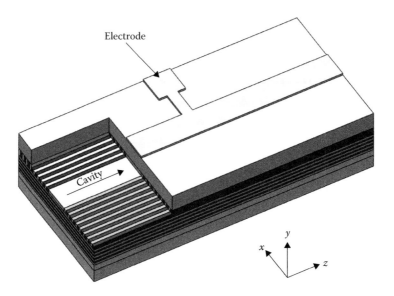

FIGURE 30.20 The Bragg waveguide SLM laser structure.

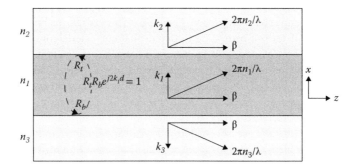

FIGURE 30.21 An illustrative diagram showing the concept of the dielectric waveguide, with $2\pi n_{1,2,3}/\lambda$ indicating the conceptual free-propagating plane wave numbers inside the core, the top cladding, and the bottom cladding layer, respectively, β their corresponding wave vectors' projection along the propagation (z) direction, $k_{1,2,3}$ their corresponding wave vectors' projections along the cross (x) direction in the core, the top cladding, and the bottom cladding layer, respectively, $R_{t,b}$ the amplitude reflectivities between the core and the top cladding layer, and between the core and the bottom cladding layer, respectively; $R_t R_b e^{j2k_1 d} = 1$ gives the resonance condition along the cross direction with d as the core layer thickness, which stands as the necessary and sufficient condition for guided waves in general waveguides.

propagation constant, the other along the cross direction (k_1) as the wave vector needs to be confined. For the wave component that propagates in the cross direction, a round trip travel would make the initial field E_0 become $E_0 R_t R_b e^{j2k_1 d}$, with $R_{t,b}$ denoting the amplitude reflectivity between the core and the top cladding layer, and the core and the bottom cladding layer, respectively, and d the core layer thickness. Therefore, once the resonance happens in the cross direction, i.e.,

$$R_t R_b e^{j2k_1 d} = 1 \tag{30.48}$$

the field distribution along the cross direction remains the same at any cut along the waveguide direction, which means the wave is guided and propagates with β along the waveguide. This is actually a necessary and sufficient condition for a general waveguide, i.e., if and only if the wave resonates in the cross section, it will be guided by the waveguide. Consequently, the guided wave will form a standing wave pattern inside the core area of the waveguide, and will be either zero (for metallic waveguide) or evanescent (for dielectric waveguide) outside of the core. The guided wave's cross-sectional distribution doesn't change as the wave propagates along the waveguide. The only change on the wave as it propagates is its phase, scaled by $\beta z - \omega t$, with z and t indicating the space coordinate along the waveguide (i.e., in the wave propagation direction) and the time variable, respectively.

For the conventional dielectric waveguide that utilizes the total internal reflection effect, one finds that

$$k_{1,2,3} = \sqrt{n_{1,2,3}^2 (2\pi/\lambda)^2 - \beta^2} \tag{30.49}$$

for the wave vector components along the cross direction in the core (with subscript 1), the top cladding (with subscript 2), and the bottom cladding (with subscript 3) layer, respectively, with $n_{1,2,3}$ denoting the refractive indices of these three layers. The propagation constant β, i.e., the wave vector component along the propagation direction (z), must be the same for the same guided wave distributed in different layers, as otherwise the guided wave would be split apart along with its propagating in z. Under the total internal reflection scheme, the refractive indices are chosen in such a way to make k_1 real ($2\pi n_1/\lambda > \beta$) but $k_{2,3}$ imaginary ($2\pi n_{2,3}/\lambda < \beta$). By noticing that $R_{t,b} = (k_1 - k_{2,3})/(k_1 + k_{2,3})$ for the TE wave with its electric field polarized in parallel with, and $R_{t,b} = (1/k_1 - 1/k_{2,3})/(1/k_1 + 1/k_{2,3})$ for the TM wave with its electric field polarized perpendicularly to, the boundary between different layers, respectively, one has $|R_{t,b}| = 1$ as the reflectivity at the boundary always takes the form of $(a - jb)/(a + jb)$. Consequently, the resonant condition (Equation 30.48) in the cross direction is reduced to

$$\varphi_t + \varphi_b + 2k_1 d = 2m\pi \tag{30.50}$$

with $\varphi_{t,b}$ indicating the phase of $R_{t,b}$, i.e.,

$$\varphi_{t,b} = -2\arctan(|k_{2,3}|/k_1) \tag{30.51}$$

for the TE wave and

$$\varphi_{t,b} = -2\arctan(k_1/|k_{2,3}|) \tag{30.52}$$

for the TM wave, respectively. Equations 30.49 through 30.51 or Equation 30.52 defines the dispersion relation of the guided wave, which tells that, for a given waveguide structure and operating wavelength λ below a maximum value known as the cutoff wavelength, one always manage to find at least one real β within the range between $2\pi \max(n_2, n_3)/\lambda$ and $2\pi n_1/\lambda$ as the solution. Therefore, the conventional dielectric waveguide has no wavelength selection ability as the dispersion relation of its guided wave shows a high-pass filter feature that supports the wave propagation with any wavelength shorter than the cutoff value, which also echoes the same statement we made at the beginning of this chapter. It is not surprising for one to reach this conclusion, once one notices that as the total internal reflection condition is satisfied, $|R_t R_b|$ is always 1 regardless of the associated wavelength. The cross-directional resonant condition, as the indication of the wave being guided, is therefore reduced to a phase-matching condition (i.e., the dispersion relation) that is too loose to select a discrete set of wavelengths, not to mention a single wavelength.

If, however, a grating-based reflection is utilized to substitute the total internal reflection, as shown in the structure given by Figure 30.20, it will be possible to make $R_t R_b$ wavelength dependent. Once not only the phase-matching condition in the form of Equation 30.50, but also a wavelength-dependent amplitude condition in the form of $|R_t(\lambda)R_b(\lambda)| = 1$, will jointly be derived from the general resonant condition

(Equation 30.48), it is possible that only a single wavelength can be found for the guided wave to satisfy both constraints imposed on the phase and amplitude.

The Bragg waveguide laser is such a device that exploits the grating reflection, rather than the total internal reflection, in its waveguide in conjunction with the FP cavity to reach the SLM operation. With a structure similar to the DBR laser arranged in the cross direction perpendicular to the waveguide for wave confinement, even a simple FP cavity makes an SLM laser [58]. A unique feature of the Bragg waveguide lies in that it allows its core layer to have a lower refractive index than that of its cladding. As such, the selection on the gain medium to build the active region for the Bragg waveguide laser will be more flexible.

30.4.3 Double-Trench Resonant Tunneling SLM Laser

As concluded by Section 30.3, the concept of the coupled cavity can generally be understood as the insertion of an OBPF into a conventional FP cavity, for purifying the lasing spectrum by eliminating all unwanted FP cavity modes. An accompanying issue is that both the transmitted light passed through the OBPF and the reflected light experienced the complementary optical band-reject filter (OBRF) stay inside the FP cavity, which add up to give no appreciable mode selection mechanism once their gains are the same, since the added spectrum of the complementary OBPF and OBRF turns out to be flat. Although the additionally introduced phase condition by the equivalent filter and any possible gain discrimination between the transmitted and reflected waves will help to eliminate the unwanted FP cavity modes, the single-mode yield and SMSR for such structures are usually low. And the SLM lasing is usually not very stable in the entire laser operating range. For example, the lasing mode often hops or the SMSR deteriorates as the bias current or ambient temperature changes. This can be attributed to the phase-sensitive lasing condition shown as Equation 30.33 or Equation 30.39, since the phase not only varies with any geometrical dimension deviation of the structure in the fabrication process, but also changes with the laser operating condition.

To ensure the SLM operation with an intensified SMSR with high yield, one needs to effectively eliminate the reflected light from the aforementioned OBPF inside the FP cavity. As shown in Figure 30.22, a double-trench structure has been proposed [59] to introduce dual reflections that cancel out the reflected light at a specific wavelength determined by the gap between the two trenches, an effect known as the resonant tunneling. As such, the transmission spectrum of the double trenches resembles that of an OBPF, which selects one of the many cavity modes in a conventional FP laser for single-mode operation. The most important design in this structure, which significantly improves the SLM laser performance, is to make the double trenches slanted so that the reflected waves with unselected wavelengths by the OBPF will escape

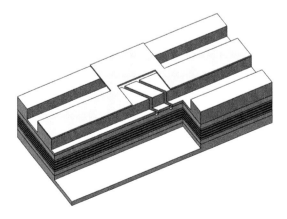

FIGURE 30.22 The double-trench resonant tunneling SLM laser structure.

from the cavity, in order to avoid jeopardizing the single-mode lasing condition, as otherwise the reflected and transmitted light in different wavelengths will compete to each other inside the cavity.

By eliminating the reflection brought in by the OBPF made of the double trenches, i.e., by letting $r_{12}, r_{21} \sim 0, \delta = t_{12}t_{21}$, the lasing condition (Equation 30.33) derived for general coupled-cavity lasers reduces to

$$t_{12}t_{21}R^2e^{2j(\varphi_2+\varphi_1)} = 1 \tag{30.53}$$

which indicates that the lasing will be determined by the transmission spectrum of the inserted OBPF. More specifically, the lasing will happen at the wavelength in the neighborhood of the transmission spectrum ($|t_{12}t_{21}|$) peak. Actually, once the transmission spectrum of the filter resembles that of an OBPF, the lasing condition shown in Equation 30.53 is the same as that of the DBR laser (Equation 30.13).

The simulated and measured lasing spectrum and SLM yield of a typical double-trench resonant tunneling SLM laser are given in Figures 30.23 and 30.24, respectively. The simulation result also shows, and experimental data verifies that, with varying positions of the double trenches inside the FP cavity, the slanted trench pair can always select and lock in one and only one of the many FP modes in the full laser operating range [59]. This can be attributed to the fact that, with the reflection from the double trenches effectively eliminated and consequently with the lasing condition determined by Equation 30.53 rather than Equation 30.33 or Equation 30.39 for the device, its mode selection mechanism is no longer phase sensitive. Hence the precise control of the geometrical dimension in the structure is not required in the fabrication process for such a device, which greatly enhances the single-mode yield, the SMSR, and the single-mode stability in operation, as compared to other existing coupled-cavity lasers.

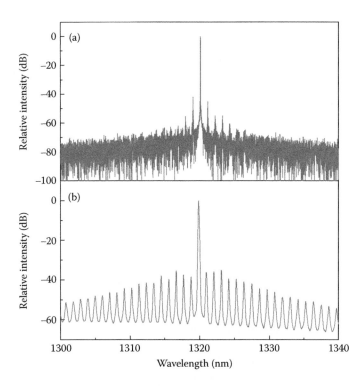

FIGURE 30.23 The (a) simulated and (b) measured lasing spectra of a typical double-trench resonant tunneling SLM laser, with its design parameters given in Reference [59].

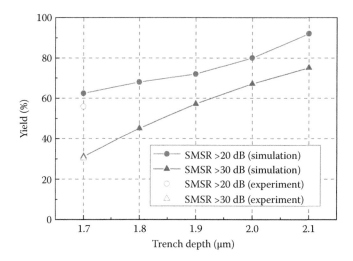

FIGURE 30.24 The SLM yield as a function of the trench depth, with device design parameters given in Reference [59].

30.4.4 Horn Ridge Waveguide (HRW) DFB Laser

With its structure shown in Figure 30.25, the horn ridge waveguide (HRW) DFB laser [60] improves the single-mode yield by breaking the inherent dual-mode degeneracy in the uniform-grating purely refractive index-coupled structure.

In a straight-waveguide purely refractive index-coupled DFB laser with uniform grating, the two longitudinal modes with their respective wavelengths located at each side of the Bragg stopband have the same reflection everywhere along the cavity due to the symmetry in the reflection spectrum, so that they have the degenerated field distribution. The longitudinal modes at the two wavelengths with the degenerated field distribution will then have the same modal gain, which causes either dual-mode operation or SLM lasing with poor SMSR if the degeneracy is somehow removed by the uncontrollable LSHB effect. In the HRW DFB laser, however, an effective chirp to the uniform grating is introduced. As such, the field of the longitudinal mode with the wavelength on the blue side will concentrate in the right section with the wider ridge width, as a consequence of the strong reflection it sees from the left section with the narrow ridge width, where the local Bragg wavelength takes a relative shift toward the shorter wavelength side due to the reduction on the local effective index. On the contrary, the field of the longitudinal mode with the wavelength on the red side will concentrate in the left section with the narrower ridge width, as a consequence of the strong reflection it sees from the right section with the wider ridge width, where the local Bragg wavelength shifts relatively toward the longer wavelength side due to the enhancement on the local effective index. Therefore, the major effect of the effectively chirped grating is to remove the degeneracy of field distributions of the two symmetrical modes on each sides of the Bragg stopband. With the help of such effectively chirped grating, the two longitudinal modes at the Bragg stop-band will take different distributions and have their associated field intensities spatially localized in different sections along the cavity. With the further help of the nonuniform modal gain simultaneously generated by the horn waveguide, the field distribution having a larger overlap with the modal gain turns out to be the only dominant lasing mode. In this particular structure, the modal gain is higher on the wide ridge side due to the larger confinement factor associated. Hence the wavelength at the blue side of the Bragg stop-band, i.e., the shorter wavelength, will be selected to lase, whereas the wavelength at the red side, i.e., the longer wavelength, will be effectively suppressed.

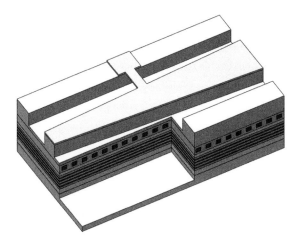

FIGURE 30.25   The horn ridge waveguide (HRW) DFB laser structure.

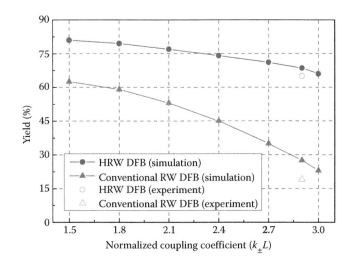

FIGURE 30.26   Comparison on the SLM yield (SMSR >30 dB) between the HRW DFB and the conventional ridge waveguide (RW) DFB with different κ ± L; other device design parameters are given in Reference [60].

Simulation result shows, and experimental result demonstrates, that the SLM yield can be drastically raised (by more than threefolds) with an optimized HRW design, especially for the uniform-grating purely refractive index-coupled DFB laser with high normalized coupling coefficient (κL) [60]. This is further evidenced by the comparison result in Figure 30.26. Finally, measured lasing spectra of a typical HRW DFB laser under different bias current are shown in Figure 30.27.

Acknowledgment

The author would like to express his sincere thankfulness to Cheng Ke for his great help on performing most of the numerical calculations and on drawing all the figures.

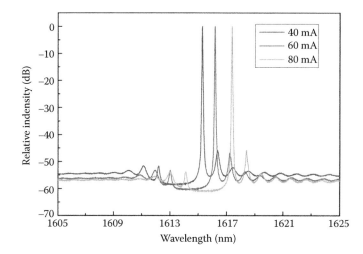

FIGURE 30.27 Measured lasing spectra of a typical HRW DFB laser under different bias current, with its design parameters given in Reference [60].

References

1. K. Ogawa, Analysis of mode partition noise in laser transmission systems, *IEEE Journal of Quantum Electronics*, vol. 18, no. 5, pp. 849–855, May 1982.
2. M. Ahmed and M. Yamada, Theoretical analysis of mode-competition noise in modulated laser diodes and its influence on the noise performance of fibre links, *Journal of Physics D Applied Physics*, vol. 45, no. 40, pp. 4172–4181, Sept. 2012.
3. M. L. Calvo and V. Lakshminarayanan, *Optical Waveguides: From Theory to Applied Technologies*, Boca Raton, FL: CRC Press, Taylor & Francis, ISBN 978-1-57444-698-2, Jan. 2007.
4. D. Marcuse, *Theory of Dielectric Optical Waveguides*, New York, NY: Academic Press, ISBN 0-12-470950-8, 1974.
5. K. Petermann, *Laser Diode Modulation and Noise*, Netherlands: Kluwer Academic Publishers, ISBN 978-0-7923-1204-8, Apr. 1988.
6. B. Mroziewicz, External cavity wavelength tunable semiconductor lasers—a review, *Opto-Electronics Review*, vol. 16, no. 4, pp. 347–366, Sept. 2008.
7. K. Vahala, K. Kyuma, A. Yariv, S. K. Kwong, M. C. Golomb, and K. Y. Lau, Narrow linewidth, single frequency semiconductor laser with a phase conjugate external cavity mirror, *Applied Physics Letters*, vol. 49, no. 23, pp. 1563–1565, Oct. 1986.
8. M. W. Fleming and A. Mooradian, Spectral characteristics of external-cavity controlled semiconductor lasers, *IEEE Journal of Quantum Electronics*, vol. 17, no. 1, pp. 44–59, Jan. 1981.
9. H. Kogelnik and C. V. Shank, Stimulated emission in a periodic structure, *Applied Physics Letters*, vol. 18, no. 4, pp. 152–154, Nov. 1971.
10. S. Wang, Principles of distributed feedback and distributed Bragg-reflector lasers, *IEEE Journal of Quantum Electronics*, vol. 10, no. 4, pp. 413–427, Apr. 1974.
11. J.-W. Mu and W.-P. Huang, Simulation of three-dimensional waveguide discontinuities by a full-vector mode-matching method based on finite-difference schemes, *Optics Express*, vol. 16, no. 22, pp. 18152–18163, Oct. 2008.
12. K. Jiang and W.-P. Huang, Finite-difference-based mode-matching method for 3-D waveguide structures under semivectorial approximation, *Journal of Lightwave Technology*, vol. 23, no. 12, pp. 4239–4248, Dec. 2005.

13. M. Yamada and K. Sakuda, Analysis of almost-periodic distributed feedback slab waveguides via a fundamental matrix approach, *Applied Optics*, vol. 26, no. 16, pp. 3474–3478, Aug. 1987.

14. T. Makino, Transfer-matrix analysis of the intensity and phase noise of multisection DFB semiconductor lasers, *IEEE Journal of Quantum Electronics*, vol. 27, no. 11, pp. 2404–2414, Nov. 1991.

15. G. P. Agrawal and N. K. Dutta, *Semiconductor Lasers*, 2nd ed. New York, NY: Van Nostrand Reinhold, ISBN 0-442-01102-4, 1993.

16. X. Li, *Optoelectronic Devices: Design, Modeling, and Simulation*, Cambridge: Cambridge University Press, ISBN 978-0-521-87510-3, Aug. 2009.

17. W. W. Ng, C. Hong, and A. Yariv, Holographic interference lithography for integrated optics, *IEEE Transactions on Electron Devices*, vol. 25, no. 10, pp. 1193–1200, Oct. 1978.

18. C. J. Armistead, B. R. Butler, S. J. Clements, A. J. Collar, D. J. Moule, S. A. Wheeler, M. J. Fice, and H. Ahmed, DFB ridge waveguide lasers at $\lambda = 1.5$ µm with first-order gratings fabricated using electron beam lithography, *Electronics Letters*, vol. 23, no. 11, pp. 592–593, May 1987.

19. S. Y. Chou, P. R. Krauss, and P. J. Renstrom, Imprint of sub-25 nm vias and trenches in polymers, *Applied Physics Letters*, vol. 67, no. 67, pp. 3114–3116, Sept. 1995.

20. L.-A. Wang, C.-H. Lin, and J.-H. Chen, Fabrication of sub-quarter-micron grating patterns by employing DUV holographic lithography, *Microelectronic Engineering*, vol. 46, no. 1–4, pp. 173–177, Aug. 1999.

21. A. Takemoto, Y. Ohkura, Y. Kawama, Y. Nakajima, T. Kimura, N. Yoshida, S. Kakimoto, and W. Susaki, 1.3-µm distributed feedback laser diode with a grating accurately controlled by a new fabrication technique, *Journal of Lightwave Technology*, vol. 7, no. 12, pp. 2072–2077, Dec. 1989.

22. Y. Abe, K. Kishino, Y. Suematsu, and S. Arai, GaInAsP/InP integrated laser with butt-jointed built-in distributed-Bragg-reflection waveguide, *Electronics Letters*, vol. 17, no. 25, pp. 945–947, Dec. 1981.

23. S. Murata, I. Mito, and K. Kobayashi, Spectral characteristics for 1.5 µm DBR laser with frequency-tuning region, *IEEE Journal of Quantum Electronics*, vol. 23, no. 6, pp. 835–838, June 1987.

24. J. E. Carroll, J. Whiteaway, and D. Plumb, *Distributed Feedback Semiconductor Lasers*, London: Institution of Electrical Engineers, ISBN 0-85296-917-1, Apr. 1998.

25. D. E. Muller, A method for solving algebraic equations using an automatic computer, *Mathematical Tables and Other Aids to Computation*, vol. 10, no. 56, pp. 208–215, Oct. 1956.

26. K. David, G. Morthier, P. Vankwikelberge, R. G. Baets, T. Wolf, and B. Borchert, Gain-coupled DFB lasers versus index-coupled and phase shifted DFB lasers: A comparison based on spatial hole burning corrected yield, *IEEE Journal of Quantum Electronics*, vol. 27, no. 6, pp. 1714–1723, June 1991.

27. H. Su, L. Zhang, A. L. Gray, R. Wang, T. C. Newell, K. J. Malloy, and L. F. Lester, High external feedback resistance of laterally loss-coupled distributed feedback quantum dot semiconductor lasers, *IEEE Photonics Technology Letters*, vol. 15, no. 11, pp. 1504–1506, Nov. 2003.

28. G.-P. Li, T. Makino, R. Moore, N. Puetz, K. Leong, and H. Lu, Partly gain-coupled 1.55 µm strained-layer multiquantum-well DFB lasers, *IEEE Journal of Quantum Electronics*, vol. 29, no. 6, pp. 1736–1742, Jun. 1993.

29. Y. Luo, Y. Nakano, K. Tada, T. Inoue, H. Hosomatsu, and H. Iwaoka, Purely gain-coupled distributed feedback semiconductor lasers, *Applied Physics Letters*, vol. 56, no. 17, pp. 1620–1622, Feb. 1990.

30. Y. Nakano, Y. Luo, and K. Tada, Facet reflection independent, single longitudinal mode oscillation in a GaAlAs/GaAs distributed feedback laser equipped with a gain-coupling mechanism, *Applied Physics Letters*, vol. 55, no. 16, pp. 1606–1608, Aug. 1989.

31. H. A. Haus and C. V. Shank, Antisymmetric taper of distributed feedback lasers, *IEEE Journal of Quantum Electronics*, vol. 12, no. 9, pp. 532–539, Sept. 1976.

32. E. Kapon, A. Hardy, and A. Katzir, The effect of complex coupling coefficients on distributed feedback lasers, *IEEE Journal of Quantum Electronics*, vol. 18, no. 1, pp. 66–71, Jan. 1982.

33. K. Utaka, S. Akiba, K. Sakai, and Y. Matsushima, $\lambda/4$-shifted InGaAsP/InP DFB lasers, *IEEE Journal of Quantum Electronics*, vol. 22, no. 7, pp. 1042–1051, July 1986.

34. S. Akiba, M. Usami, and K. Utaka, 1.5-μmλ/4-shifted InGaAsP/InP DFB lasers, *Journal of Lightwave Technology*, vol. 5, no. 11, pp. 1564–1573, Nov. 1987.

35. S.-L. Chuang, *Physics of Photonic Devices*, 2nd ed. New York, NY: Wiley, ISBN 9781118585658, 2012.

36. A. D. Sadovnikov, X. Li, and W.-P. Huang, A two-dimensional DFB laser model accounting for carrier transport effects, *IEEE Journal of Quantum Electronics*, vol. 31, no. 10, pp. 1856–1862, Oct. 1995.

37. X. Li, A. D. Sadovnikov, W.-P. Huang, and T. Makino, A physics-based three-dimensional model for distributed feedback laser diodes, *IEEE Journal of Quantum Electronics*, vol. 34, no. 9, pp. 1545–1553, Sept. 1998.

38. B. Kim, Y. Chung, and J. Lee, An efficient split-step time-domain dynamic modeling of DFB/DBR laser diodes, *IEEE Journal of Quantum Electronics*, vol. 36, no. 7, pp. 787–794, July 2000.

39. W. H. Press, S. A. Teukolsky, W. T. Vetterling, and B. P. Flannery, *Numerical Recipes: The Art of Scientific Computing*, 3rd ed. Cambridge: Cambridge University Press, ISBN 978-0-511-33555-6, Sept. 2007.

40. X. Li and W.-P. Huang, Simulation of DFB semiconductor lasers incorporating thermal effects, *IEEE Journal of Quantum Electronics*, vol. 31, no. 10, pp. 1848–1855, Oct. 1995.

41. A. J. Lowery and D. Novak, Enhanced maximum intrinsic modulation bandwidth of complex-coupled DFB semiconductor lasers, *Electronics Letters*, vol. 29, no. 5, pp. 461–463, Mar. 1993.

42. L.-M. Zhang and J. E. Carroll, Enhanced AM and FM modulation response of complex coupled DFB lasers, *IEEE Photonics Technology Letters*, vol. 5, no. 5, pp. 506–508, Aug. 1993.

43. S. Hansmann, Transfer matrix analysis of the spectral properties of complex distributed feedback laser structures, *IEEE Journal of Quantum Electronics*, vol. 28, no. 11, pp. 2589–2595, Nov. 1992.

44. C. Youtsey, R. Grundbacher, R. Panepucci, I. Adesida, and C. Caneau, Characterization of chemically assisted ion beam etching of InP, *Journal of Vacuum Science & Technology B*, vol. 12, no. 6, pp. 3317–3321, Dec. 1994.

45. H. Blauvelt, N. Bar Chaim, D. Fekete, S. Margalit, and A. Yariv, AlGaAs lasers with micro-cleaved mirrors suitable for monolithic integration, *Applied Physics Letters*, vol. 40, no. 4, pp. 289–290, Feb. 1982.

46. W. T. Tsang, N. A. Olsson, and R. A. Logan, High-speed direct single-frequency modulation with large tuning rate and frequency excursion in cleaved-coupled-cavity semiconductor lasers, *Applied Physics Letters*, vol. 42, no. 8, pp. 650–652, Feb. 1983.

47. Q.-Y. Lu, W.-H. Guo, D. Byrne, and J. F. Donegan, Design of slotted single-mode lasers suitable for photonic integration, *IEEE Photonics Technology Letters*, vol. 22, no. 11, pp. 787–789, Mar. 2010.

48. Y. Wang, Y.-G. Yang, S. Zhang, L. Wang, and J.-J. He, Narrow linewidth single-mode slotted Fabry–Pérot laser using deep etched trenches, *IEEE Photonics Technology Letters*, vol. 24, no. 14, pp. 1233–1235, May 2012.

49. T.-T. Yu, L. Zou, L. Wang, and J.-J. He, Single-mode and wavelength tunable lasers based on deep-submicron slots fabricated by standard UV-lithography, *Optics Express*, pp. 16291–16299, July 2012.

50. J.-L. Zhao, K. Shi, Y.-L. Yu, and L. P. Barry, Theoretical analysis of tunable three-section slotted Fabry–Perot lasers based on time-domain traveling-wave model, *IEEE Journal of Selected Topics in Quantum Electronics*, vol. 19, no. 5, pp. 1–8, Mar. 2013.

51. S. O'Brien and E. P. O'Reilly, Theory of improved spectral purity in index patterned Fabry–Pérot lasers, *Applied Physics Letters*, vol. 86, no. 20, pp. 201101, May 2005.

52. S. O'Brien, A. Amann, R. Fehse, S. Osborne, E. P. O'Reilly, and J. M. Rondinelli, Spectral manipulation in Fabry–Perot lasers: Perturbative inverse scattering approach, *Journal of the Optical Society of America B*, vol. 23, no. 6, pp. 1046–1056, Jan. 2006.

53. X. Li and J. Park, Time-domain simulation of channel crosstalk and inter-modulation distortion in gain-clamped semiconductor optical amplifiers, *Optics Communications*, vol. 263, no. 2, pp. 219–228, Feb. 2006.

54. W. Li, W.-P. Huang, and X. Li, Digital filter approach for simulation of a complex integrated laser diode based on the traveling-wave model, *IEEE Journal of Quantum Electronics*, vol. 40, no. 5, pp. 473–480, May 2004.
55. Y.-P. Xi, W.-P. Huang, and X. Li, High-order split-step schemes for time-dependent coupled-wave equations, *IEEE Journal of Quantum Electronics*, vol. 43, no. 5, pp. 419–425, May 2007.
56. L. Dong, R.-K. Zhang, D.-L. Wang, S.-Z. Zhao, S. Jiang, Y.-L. Yu, and S.-H. Liu, Modeling widely tunable sampled-grating DBR lasers using traveling-wave model with digital filter approach, *Journal of Lightwave Technology*, vol. 27, no. 15, pp. 3181–3188, Apr. 2009.
57. L. Zhu, A. Scherer, and A. Yariv, Modal gain analysis of transverse Bragg resonance waveguide lasers with and without transverse defects, *IEEE Journal of Quantum Electronics*, vol. 43, no. 10, pp. 934–940, Oct. 2007.
58. Y. Li, Y.-P. Xi, X. Li, and W.-P. Huang, A single-mode laser based on asymmetric Bragg reflection waveguides, *Optics Express*, vol. 17, no. 13, pp. 11179–11186, June 2009.
59. X. Li, Z.-S. Zhu, Y.-P. Xi, L. Han, C. Ke, Y. Pan, and W.-P. Huang, Single-mode Fabry–Perot laser with deeply etched slanted double trenches, *Applied Physics Letters*, vol. 107, no. 9, pp. 091108, Sept. 2015.
60. C. Ke, X. Li, and Y.-P. Xi, A horn ridge waveguide DFB laser for high single longitudinal mode yield, *Journal of Lightwave Technology*, vol. 33, no. 24, pp. 5032–5037, Nov. 2015.

31

Traveling Wave Modeling of Nonlinear Dynamics in Multisection Laser Diodes

Mindaugas
Radziunas

31.1 Introduction

Semiconductor lasers, and a narrow waveguide edge-emitting semiconductor lasers, in particular, are attractive devices for different applications. Among others, these are high-speed all-optical signal processing, optical data storage, thermal and xerographic printing, scanning, directional lighting, secure communications, random number generation, frequency conversion, or various interferometric, spectroscopic, instrumentation, and other quantum-optical experiments.

A typical solitary narrow waveguide (single transversal mode) semiconductor laser exhibits a single-wavelength emission required in different applications. In many cases, however, small fluctuations of the operation conditions impose a significant phase noise which, in turn, causes an unwanted broadening of the emission linewidth. Moreover, the stable performance of the laser can be easily violated by optically reinjected light, and there is a huge number of studies devoted to the analysis of the nonlinear dynamics in lasers with a delayed optical feedback.

A properly designed optical feedback, however, can also play a constructive role when seeking to improve an operation of the solitary laser, or create a new dynamical regime. For example, an external cavity with a diffractive grating can be used for emission linewidth reduction or tuning of the lasing wavelength

[1]. Or, on the contrary, specially designed external cavities allow realizing a chaotic emission usable for cryptography [2] or random number generation [3].

Multisection semiconductor lasers (MSLs) in linear or ring configurations and coupled laser devices provide even more possibilities to tailor laser dynamics for certain applications. For example, a variety of important functionalities of the optical data communications [4], such as pulse generation, clock recovery, and fast switching can be realized by specially designed and differently interconnected MSLs. Several examples of theoretically investigated and experimentally verified dynamic performance of MSLs considered in our previous works are excitability [5], high-frequency mode-beating pulsations [6], and modulation band enhancement [7] in distributed feedback (DFB) lasers with an integrated passive phase tuning section (passive feedback lasers); passive [8] or hybrid [9] harmonic and sub-harmonic mode-locking in lasers with saturable absorber, and pulse broadening in quantum dot (QD) mode-locked lasers [10]; tunable high-frequency pulsations in the detuned grating DFB lasers with an integrated phase tuning section (phase controlled mode-beating lasers) [11,12]; stationary, pulsating, and irregular regimes and their bifurcations in DFB lasers with integrated phase tuning and amplifying sections (active feedback laser) [6,13]; Joule heating–induced transitions between steady states in distributed Bragg reflector (DBR) lasers [14] or external cavity diode lasers (ECDLs) [15].

All these examples confirm the practical importance of modeling, simulations, and analysis of MSLs for designing new devices with a particular dynamical behavior. The most precise models usually are given by 2+1 or 3+1 dimensional systems of partial differential equations (PDEs) [16,17]. The numerical simulations in this case, however, are time consuming, whereas application of analytic methods for the analysis of the nonlinear dynamics is very limited. Unfortunately, numerical simulations of such models are time consuming, whereas an application of analytic methods for the model analysis is very limited. For this reason, we prefer to use simpler approaches which, may be, fail to reproduce a quantitative-, but still allow to get a qualitative agreement between theory and experiments.

For some MSLs, already simple ordinary differential equation (ODE) or delay differential equation (DDE) systems (rate equations) admit a reasonable description of the laser dynamics. An advantage of these models is their simplicity allowing fast numerical simulations and application of advanced analytic methods, such as asymptotic analysis, stability analysis, or numerical continuation and bifurcation analysis. These models, however, usually are based on mean-field approximations, i.e., neglect inhomogeneity of laser parameters and dynamical variables along the laser cavity, take into account only a few fundamental characteristics of the considered lasers, or are suited to describe particular MSL configurations [18–20].

The 1+1-dimensional traveling wave (TW) model considered in this chapter is a compromise between simplicity and precision. It is a first-order PDE system having a single spatial dimension corresponding to the longitudinal (z-) direction along the laser cavity and describing dynamics of the slowly varying optical field amplitudes, polarization functions, and carrier density [21–23]. Comparing to ODE and DDE models mentioned earlier, the TW model is computationally more demanding but still enables an advanced analysis, which is hardly possible in the case of the multidimensional PDE models.

By taking into account or neglecting different physical effects, one can derive a whole hierarchy of TW models of different complexities. The standard part of all such models is a pair of TW equations governing the evolution of the complex forward- and back- propagating field amplitudes, $E^+(z, t)$ and $E^-(z, t)$. These equations originate from the decomposition of the dominant fundamental transverse electric (TE) component of the electromagnetic wave,

$$E(\mathbf{r}, t) = \Phi(x, y) \left[E^+(z, t)e^{-ik_0 z} + E^-(z, t)e^{ik_0 z} \right] e^{i\omega_0 t}.$$

Here, ω_0 is the central reference frequency and k_0 the corresponding wave vector. Whereas the transversal waveguide mode profile, $\Phi(x, y)$, is an eigensolution of the waveguide equation, the related complex eigenvalue of the same problem defines the propagation factor β [24], which determines the evolution of the field amplitudes $E^\pm$. In general, the propagation factor depends on the complex interaction of carriers and photons. In our modeling approach, we apply a phenomenological dependence of this factor on the real carrier

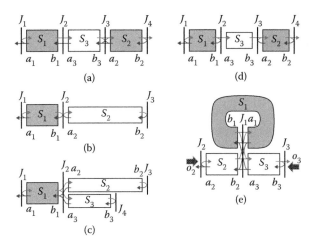

FIGURE 31.1 Schematic representation of five semiconductor laser devices, which can be considered by our modeling approach: (a) three-section laser, (b) laser with a trivial external cavity, (c) laser with a dual external cavity, (d) master–slave laser system, (e) optically injected ring laser with an outcoupling waveguide. Shaded and white frames represent active and passive *sections* (S_*) of the MSL. Thick black segments and thick hatched arrows indicate *junctions* (interfaces of these sections, J_*) and *optical injections* (o_*), respectively. Thin arrows show optical field transmissions and reflections at the interfaces of the laser sections.

density function N, which can represent dynamics of the spatially distributed carrier density, $N(z, t)$, or the section-wise averaged density, $N(t)$. The evolution of N itself is governed by a single or several rate equations.

One can use the TW modeling approach for consideration of various differently interconnected linear and curved, active and passive semiconductor waveguiding parts, taking into account optical injections, field reflections, and transmissions at the interfaces of different laser parts, as well as delayed feedback of the optical fields from the external cavities. For simulation and analysis of the MSLs, we apply our software LDSL-tool [25], which is suited to investigate the longitudinal dynamics of multisection semiconductor lasers. This software allows considering a large variety of MSL devices or coupled laser systems which can be represented by a set of mutually interconnected *sections* and *junctions*, see the schematic representations of several laser devices in Figure 31.1. Besides of numerical integration, LDSL-tool can find longitudinal optical modes and analyze their dynamics [23,26]. In some cases, it locates stable and unstable stationary states of the system [15,27], constructs the reduced ODE models based on a finite number of the optical modes [28], and, together with the software package AUTO [29], performs numerical continuation and bifurcation analysis of these reduced models [7].

In following, we shall introduce a basic TW model for the solitary laser, and present several model extensions allowing to take into account initially neglected physical effects. Next, we shall discuss a possibility to join several laser sections into a single multisection laser or a coupled laser device. For an illustration of the available device complexity, we shall present simulations of a ring laser with four branches of filtered feedback. At the last part of this chapter, we shall briefly introduce the concept of the instantaneous optical modes, discuss the mode analysis, the location and semi analytic continuation of the stationary states, the model reduction, and the numerical bifurcation analysis.

31.2 Basic TW Model in the Solitary Laser

In this section, we formulate the simplest TW model suitable for simulations of a solitary semiconductor laser. Let us consider an edge-emitting narrow-waveguide semiconductor laser (see Figure 31.2a). According to our notations, the "interior" part of this laser is referred as *section* S_1. The longitudinal coordinates

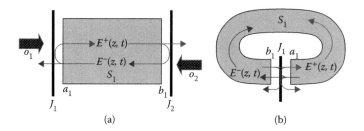

FIGURE 31.2 Schematic representation of the single section lasers in (a) linear and (b) ring configurations. Thin arrows indicate directions of the counter-propagating fields and their reflections/transmissions at the laser facets [J_1 and J_2, panel (a)] or point outcoupling interface [J_1, panel (b)]. Thick hatched arrows represent optically injected fields.

of the section edges and the length of this section are $z = a_1$, $z = b_1$, and $|S_1| = b_1 - a_1$, respectively. The front and the rear laser facets (*junctions* J_1 and J_2), in this case, correspond to the left and the right edges a_1 and b_1 of S_1.

The backbone of the TW model of this laser is the linear system of partial differential (TW) equations describing an evolution of the slowly varying complex amplitudes $E^+(z, t)$ and $E^-(z, t)$ of the counter-propagating optical fields:

$$\begin{cases} \dfrac{n_g}{c_0} \partial_t E^+ + \partial_z E^+ = -i\beta E^+ - i\kappa E^- + F_{sp}^+ \\ \dfrac{n_g}{c_0} \partial_t E^- - \partial_z E^- = -i\beta E^- - i\kappa E^+ + F_{sp}^- \end{cases}, \quad z \in S_1. \tag{31.1}$$

Here, c_0 is the speed of light in vacuum, $F_{sp}^{\pm}$ are the Langevin noise source contributions to the optical fields, and n_g is the group velocity index. The real and the imaginary parts of the complex coefficient κ represent the distributed index and gain/loss coupling of the counter-propagating fields, respectively. κ is nonvanishing in the laser sections containing Bragg grating and is set to zero in the straight sections without the grating. Without an additional scaling of the field functions $E^{\pm}$, $|E(z, t)|^2 = (E, E) = |E^+|^2 + |E^-|^2$ represents the photon density and is proportional to the local field power,

$$P(z, t) = \frac{\sigma c_0}{n_g} \frac{h c_0}{\lambda_0} |E(z, t)|^2.$$

Here, σ is the cross-section area of the active zone, λ_0 the central wavelength, and h the Planck constant.

Active Sections:

The propagation factor β in the TW equations above can be defined as

$$\beta = \delta_0 + \tilde{n}(N) + \frac{i(g(N) - \alpha)}{2}, \tag{31.2}$$

where the peak gain and refractive index change functions $g(N)$ and $\tilde{n}(N)$ are given by the simple linear relations

$$g(N) = \Gamma g' \left(N - N_{tr}\right), \qquad \tilde{n}(N) = \frac{\alpha_H g(N)}{2}. \tag{31.3}$$

Here, N is the carrier density. The parameters δ_0, α, Γ, g', and α_H are the internal field loss, the initial fixed detuning from the central frequency, the confinement factor, the differential gain, and the linewidth enhancement (Henry) factor evaluated at the transparency carrier density N_{tr}, respectively.

To define the evolution of the spatially averaged carrier density $N(t)$, we use a single rate equation

$$\frac{d}{dt}N = \frac{I}{q\sigma|S_1|} - \mathcal{R}(N) - \mathcal{S}(N, E^\pm). \tag{31.4}$$

Here, q is the electron charge, I is the injected current into the active zone of the section, whereas $\mathcal{R}$ and $\mathcal{S}$ are spontaneous and stimulated recombination functions, respectively. We use a cubic spontaneous recombination function,

$$\mathcal{R}(N) = AN + BN^2 + CN^3, \tag{31.5}$$

which can be simplified by assuming vanishing recombination parameters B and C and defining $A = \tau_N^{-1}$, where τ_N denotes the carrier lifetime. Function $\mathcal{S}$ in the carrier rate equation 31.4 represents the spatially averaged stimulated recombination:

$$\mathcal{S}(N, E^\pm) = \frac{c_0}{n_g} g(N)\|E\|_1^2. \tag{31.6}$$

Here, $\|E\|_1^2$ is the spatial average of the local photon density along the section S_1,

$$\|E\|_1^2 = \langle (E, E) \rangle_1, \qquad \langle \eta \rangle_1 = \frac{1}{|S_1|}\int_{S_1} \eta(z)dz, \qquad (\zeta, \xi) = \zeta^{+*}\xi^+ + \zeta^{-*}\xi^-,$$

and * denotes the complex conjugate.

To complete the system, we still need to define the incident forward- and backward-propagating fields at the section edges $z = a_1$ and $z = b_1$, respectively. For the solitary laser, these incident fields can be defined by the following reflection/transmission conditions:

$$E^+(a_1, t) = -r_1^* E^-(a_1, t) + o_1(t), \quad E^-(b_1, t) = r_2 E^+(b_1, t) + o_2(t). \tag{31.7}$$

Here, r_1 and r_2 are the complex field amplitude reflectivity coefficients at the laser facets (junctions J_1 and J_2), whereas complex functions $o_{1,2}(t)$ represent optical injections at these junctions.

One can also use Equations 31.1 through 31.6 for simulations of narrow-waveguide semiconductor ring lasers with the field in- and out-coupling concentrated in the single point of this laser. Figure 31.2b shows a schematic representation of such single-section ring laser device. According to this scheme, we assume that both, "left" and "right" edges a_1 and b_1 of the section S_1, are connected at the single junction J_1. The boundary conditions (Equation 31.7), in this case, should be replaced by the following field transmission-reflection conditions at J_1:

$$E^+(a_1, t) = t_1 E^+(b_1, t) - r_1^* E^-(a_1, t), \quad E^-(b_1, t) = t_1 E^-(a_1, t) + r_1 E^+(b_1, t). \tag{31.8}$$

Here, t_1 is the real field amplitude transmission factor back into the ring section S_1 at the outcoupling point J_1, whereas the complex factor r_1 represents the localized field backscattering at J_1.

To perform simulations of the basic TW model determined by Equations 31.1 through 31.6 and 31.7 or 31.8, one still needs to choose some initial conditions $E^\pm(z, 0)$ and $N(0)$. For the first run of simulations, one can use any small distribution of the optical fields $E^\pm(z, 0)$ and a small positive value of $N(0)$. After some transient, the computed trajectory will be attracted by one of the few regular or irregular attractors of the considered dissipative system. To keep tracing the same attractor during the following parameter continuation calculations, one should better use previously obtained carrier density and field distributions.

Passive Sections:

It is noteworthy that one can also use the TW equations 31.1 for a description of the field propagation in the passive sections, such as gratings, free space between the laser and the external mirror, etc. Here, carriers are absent, do not couple to the emission wavelength (the material gain band of these sections does not support the lasing frequencies), or are just kept at transparency level by an appropriately adjusted bias current. In all such cases, $g(N) = \tilde{n}(N) = 0$, the carrier rate equations 31.4 are decoupled from the field equations 31.1 and, therefore, are irrelevant.

In the case of the passive section S_k containing no grating ($\kappa = 0$), simple analytic relations of the field function values on the both sides of S_k,

$$E^+(b_k, t) = \eta e^{i\varphi/2} E^+(a_k, t - \tau_k), \quad E^-(a_k, t) = \eta e^{i\varphi/2} E^-(b_k, t - \tau_k), \quad \text{where}$$

$$\tau_k = \frac{|S_k| n_g}{c_0}, \quad \eta = e^{-\alpha |S_k|/2}, \quad \varphi' = -2\delta_0 |S_k|, \tag{31.9}$$

can replace the field equations 31.1.

In the case of the passive grating ($\kappa \neq 0$), the analytic solution of the field equations 31.1 in the frequency domain is given by the 2×2-dimensional transfer matrix M [23,26,28],

$$\hat{E}(z, \omega) = M(\beta, \kappa, \omega; z, a_k)\hat{E}(a_k, \omega). \tag{31.10}$$

Here, $\beta = \delta_0 - i\alpha/2$, ω is the relative frequency, $\hat{E}(z, \omega) = \left(\hat{E}^+, \hat{E}^- \right)^T$, T denotes the transpose vector, whereas $\hat{E}^\pm$ are the frequency domain representations of the fields $E^\pm(z, t)$. Within any interval $[z', z]$ where parameters β and κ are constant, the matrix M is defined by

$$M(\beta, \kappa, \omega; z, z') = \begin{pmatrix} \cos \eta(z-z') - \frac{iB}{\eta} \sin \eta(z-z') & -\frac{i\kappa}{\eta} \sin \eta(z-z') \\ \frac{i\kappa}{\eta} \sin \eta(z-z') & \cos \eta(z-z') + \frac{iB}{\eta} \sin \eta(z-z') \end{pmatrix},$$

$$B(\omega) = \beta + \frac{\omega n_g}{c_0}, \quad \eta = \sqrt{B^2 - \kappa^2}. \tag{31.11}$$

Once the parameters β or κ are peacewise constants, i.e., constant within each small subinterval $[z_s, z_{s-1}]$, $z' = z_0 < z_1 < \cdots < z_n = z$, the transfer matrix M is the superposition of the corresponding transfer matrices over these small subintervals:

$$M(\beta, \kappa, \omega; z, z') = M_n \times \cdots \times M_1, \quad M_s = M(\beta(z_{s-1/2}), \kappa(z_{s-1/2}), \omega; z_s, z_{s-1}). \tag{31.12}$$

31.3 Model of Material Gain Dispersion

The relations (Equation 31.3) introduced in Section 31.2 are simple linear approximations of the gain and refractive index functions G and $\tilde{N}$. In general, these functions depend not only on the carrier density N, but also on the optical frequency ω, field intensities $|E^+|^2$ and $|E^-|^2$, and some other physical effects, such as temperature, not considered in our modeling approach. In this section, we introduce the model of the gain dispersion of the semiconductor material, which restricts the gain band in the frequency domain and is the primary optical frequency selection mechanism in Fabry–Pérot (FP) lasers.

Before switching to the modeling of the gain dispersion, let us find out the expression of the laser response $F_l(b_1, \omega)$ to the incident plane wave $e^{i\omega t}$ applied to the right edge of the device (see Figure 31.3a). For this reason, we freeze the propagation factor β and substitute the ansatz $E^\pm(z, t) = \hat{E}^\pm(z, \omega)e^{i\omega t}$ into

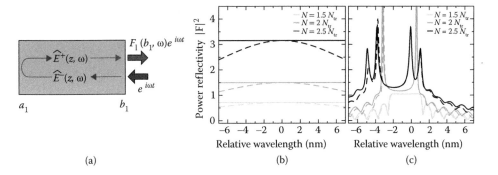

FIGURE 31.3 Laser response to the incident plane wave $e^{i\omega t}$. (a) Schematic representation: thin and thick arrows show field propagation directions and incident/emitted optical fields. (b) Response intensity of the solitary FP laser: parameters $\lambda_0 = 1.57\,\mu\text{m}$, $|S_1| = 0.25\,\text{mm}$, $n_g = 3.6$, $\delta_0 = 0$, $\alpha = 20\,\text{cm}^{-1}$, $\alpha_H = -4$, $\Gamma = 0.15$, $g' = 4 \cdot 10^{-16}\,\text{cm}^2$, $N_{tr} = 10^{18}\,\text{cm}^{-3}$, $r_1 = r_2 = \sqrt{0.3}$, $\kappa = 0$. (c) Response intensity of the solitary DFB laser: parameters are the same as in (b), only $\kappa = 130\,\text{cm}^{-1}$ and $r_1 = r_2 = 0$. Solid and dashed curves in panels (b) and (c) represent models with ($\bar{g} = 100\,\text{cm}^{-1}$, $\bar{\lambda} = 0$, $\bar{\gamma}_\lambda = 40\,\text{nm}$) and without ($\bar{g} = 0$) gain dispersion.

the field equations 31.1. The solution of the resulting system of ODEs within S_1 can be represented by Equation 31.10, where the transfer matrix M is defined in Equation 31.11. The ratio of the outgoing and incident waves at $z = b_1$ together with the (noninjective) boundary condition (Equation 31.7) at $z = a_1$ define the function $F_l(b_1, \omega)$, which shows the laser response dependence on the optical frequency of the injected field[†]. In two simple cases of FP and DFB lasers with vanishing facet reflectivity, the response function is given by

$$F_l(b_1, \omega) = \frac{\widehat{E}^+(b_1,\omega)}{\widehat{E}^-(b_1,\omega)} = \begin{cases} -r_1^* e^{-2i\beta|S_1|} e^{-i2\omega n_g|S_1|/c_0}, & \kappa = 0,\ r_1 \neq 0 \ \text{(FP laser)} \\ \dfrac{\kappa}{i\eta(\omega)\cot[|S_1|\eta(\omega)]-B(\omega)}, & \kappa \neq 0,\ r_1 = 0 \ \text{(DFB laser)} \end{cases}. \qquad (31.13)$$

Figures 31.3b and c show the intensities of these response functions in FP and DFB lasers calculated for different values of carrier densities N. Note also, that the abscissa axis in these figures represents relative wavelengths λ related to the relative frequencies ω by formula $\lambda \approx -\dfrac{\lambda_0^2}{2\pi c_0}\omega$.

The flat laser response curves in Figure 31.3b indicate an absence of frequency selection mechanisms for FP lasers in our fundamental TW model. Thus, this simple model is not suitable for simulations of FP lasers. In contrast, the wavelength selection in DFB lasers is mainly determined by the Bragg grating, and numerical integration of the TW model can provide reliable information. One should note, however, that the index-coupled DFB laser (characterized by a real coupling factor κ) can emit at one of two resonance wavelengths located at both sides of the stopband, see a solid dark gray curve in Figure 31.3c, and the parameter tuning implied jumping between these two resonances can be expected in simulations. The gain dispersion, in this case, can be exploited for the suppression of one of the resonances [12,21].

Lorentzian Approximation of the Material Gain Function:

There are several methods for introduction of the frequency-selective gain dispersion into the time-domain TW model. Many of these approaches use an additional digital filtering of the numerically calculated optical field time series [30–34]. In some cases, these digital filters are equivalent to the numerical schemes obtained by discretization of some additional integrodifferential operators or differential equations. For

---

[†] In the same way, one can also define the response function $F_r(a_1, \omega)$ at the left side of the laser.

the purpose of the analysis of the model equations, it is preferable to introduce the frequency band limiting elements directly into the model equations. For example, the TW model extensions admitting Lorentzian approximation of the material gain dispersion curves can be given by convolution integrals [34] or by an equivalent set of the linear first-order ODEs [21,35]. Another approach to model more sophisticated gain function profiles within the TW modeling frame by including nonlinear polarization equations was used, e.g., in References [22,36,37].

In this chapter, we follow the strategy proposed in References [12,21]. For this reason, we approximate the gain profile in the frequency domain by a Lorentzian with the amplitude $\bar{g}$, the full width at the half maximum $\bar{\gamma} = \frac{2\pi c_0}{\lambda_0^2}\bar{\gamma}_\lambda$, and the detuning of the peak frequency $\bar{\omega} = -\frac{2\pi c_0}{\lambda_0^2}\bar{\lambda}$. Here, $\bar{\gamma}_\lambda$ and $\bar{\lambda}$ are the wavelength representations of the Lorentzian width and its peak position. In the time domain, this approximation is represented by the additional linear operator D in the TW field equations, and a pair of linear differential equations for polarization functions $P^\pm(z, t)$:

$$\frac{n_g}{c_0}\partial_t E^\pm = \mp\partial_z E^\pm - i(\beta - iD)E^\pm - i\kappa E^\mp + F_{sp}^\pm, \quad z \in S_1,$$

$$DE^\pm = \frac{\bar{g}}{2}\left(E^\pm - P^\pm\right), \quad \partial_t P^\pm = \frac{\bar{\gamma}}{2}\left(E^\pm - P^\pm\right) + i\bar{\omega}P^\pm, \quad z \in S_1. \tag{31.14}$$

The introduction of operator D also implies the following modification of the stimulated recombination function S entering the carrier rate equation 31.4:

$$S(N, E^\pm) = \frac{c_0}{n_g}\Re\langle\left(E, [g(N) - 2D]E\right)\rangle_1. \tag{31.15}$$

To understand the impact of the operator D, we consider the laser response function $F_l(b_1, \omega)$ again according to the modified TW equations 31.14. When repeating the procedure described at the beginning of this section, the factor B entering Equation 31.11 takes the form

$$B(\omega) = \beta + \frac{\omega n_g}{c_0} + \chi(\omega), \quad \text{where} \quad \chi(\omega) = \frac{\bar{g}}{2}\frac{(\omega - \bar{\omega})}{\bar{\gamma}/2 + i(\omega - \bar{\omega})}, \quad -iD\widehat{E}(z, \omega) = \chi(\omega)\widehat{E}(z, \omega). \tag{31.16}$$

Thus, an introduction of the linear dispersion operator D implies modifications of both, gain and refractive index change functions. The total gain (twice the imaginary part of $\beta - (\delta_0 - i\alpha/2) + \chi(\omega)$) and the refractive index change function (real part of the same factor), in this case, are given by the expressions

$$G(N, \omega) = g(N) - \frac{\bar{g}(\omega - \bar{\omega})^2}{(\bar{\gamma}/2)^2 + (\omega - \bar{\omega})^2}, \quad \tilde{N}(N, \omega) = \tilde{n}(N) + \frac{\bar{g}}{4}\frac{\bar{\gamma}(\omega - \bar{\omega})}{(\bar{\gamma}/2)^2 + (\omega - \bar{\omega})^2}. \tag{31.17}$$

The dashed curves in Figures 31.3b and c illustrate the impact of the introduced gain dispersion. Whereas these corrections in the case of DFB lasers (panel [c]) are small, for the FP lasers they provide an efficient wavelength selection mechanism.

It is noteworthy that vanishing factor κ in the field equations 31.2 and 31.14 implies the following simple expression of the monochromatic field *transmission* through the laser section:

$$\widehat{E}^+(b_1, \omega) = e^{-iB(\omega)|S_1|}\widehat{E}^+(a_1, \omega), \quad \widehat{E}^-(a_1, \omega) = e^{-iB(\omega)|S_1|}\widehat{E}^-(b_1, \omega),$$

where $B(\omega)$ is defined in Equation 31.16. Thus, the TW equations 31.2 and 31.14 with vanishing functions $g(N)$ and $\tilde{n}(N)$, large Lorentzian amplitude $\bar{g}$, and small Lorentzian width $\bar{\gamma}$ can be effectively used for modeling of the optical filters, i.e., for extracting field frequency components located close to the relative frequency $\bar{\omega}$.

Further Modifications of the Model for Material Gain Dispersion:

The gain peak $g(N)$ and the simple Lorentzian dependence on the optical frequency determined by three fixed parameters $\bar{g}$, $\bar{\gamma}$, and $\bar{\omega}$ define the material gain profile $G(N, \omega)$ in Equation 31.17. To improve fitting of the gain profiles obtained by calculations of microscopic models for various values of N, one can replace these three factors by appropriately selected carrier-dependent functions $\bar{g}(N)$, $\bar{\gamma}(N)$, and $\bar{\omega}(N)$. In the cases, when the gain spectrum has two and more peaks or the asymmetry of the single peak is important, one can also introduce an additional set or several sets of polarization functions $P^{j\pm}(z, t)$. The gain dispersion operator D and corresponding total gain and refractive index functions, in this case, read as

$$D^{(s)}E^{\pm} = \sum_{j=1}^{s} \frac{\bar{g}_j}{2}\left(E^{\pm} - P^{j\pm}\right), \qquad \partial_t P^{j\pm} = \frac{\bar{\gamma}_j}{2}\left(E^{\pm} - P^{j\pm}\right) + i\bar{\omega}_j P^{j\pm}, \quad z \in S_1,$$

$$G(N, \omega) = g(N) + 2\mathfrak{I}\chi^{(s)}(\omega), \quad \tilde{N}(N, \omega) = \tilde{n}(N) + \mathfrak{R}\chi^{(s)}(\omega), \quad \chi^{(s)}(\omega) = \sum_{j=1}^{s} \frac{\bar{g}_j}{2}\frac{(\omega - \bar{\omega}_j)}{\bar{\gamma}_j/2 + i(\omega - \bar{\omega}_j)},$$

where s is the number of polarization function sets. Since the maximal value of $2\mathfrak{I}\chi(\omega)$ is, in general, smaller than zero, one should also correct the function $g(N)$.

Concluding the discussion of this section, we note that a proper numerical resolution of the gain and refractive index functions (Equation 31.17) for the broad frequency band when simulating the time-domain TW mode requires a careful selection of the numerical algorithm and temporal discretization steps. The size of the frequency band that can be represented by calculated discrete time series is inversely proportional to the time step, whereas the precision of the numerical simulations when approaching borders of this band are rapidly degrading. Thus, a suitable time discretization step should ensure that all important frequency regions (Bragg resonances, surrounding of a gain peak frequency, a frequency of optically injected beams, if present) are within the central part of the allowed frequency band.

31.4 Thermal Detuning

Let us switch now to the consideration of the thermal effects. An increase of the bias current implies changes of the device temperature and, consequently, changes in the refractive index and the lasing wavelength. To model these thermal tuning effects in our device, we supplement the propagation factor β from Equation 31.2 with an additional thermal detuning term $\tilde{n}_T$ [14,38]:

$$\beta = \delta_0 + \tilde{n}(N) + \tilde{n}_T(I) + \frac{i(g(N) - \alpha)}{2}, \qquad \tilde{n}_T = \frac{2\pi n_g}{\lambda_0^2}\nu_1^1 I. \tag{31.18}$$

The linear thermal detuning function $\tilde{n}_T(I)$ determines the impact of the injection current I to the refractive index change. The factor ν_1^1 in solitary lasers determines an approximate red shift of the lasing wavelength due to increased bias current:

$$\nu_1^1 \approx \frac{\Delta_\lambda}{\Delta_I},$$

Here, Δ_I is the bias current tuning interval, whereas Δ_λ is the (continuous) lasing wavelength change during this current tuning.

Figures 31.4a and c show the simulated wavelength change with the increased bias current in the solitary FP and DFB lasers, respectively. Here, besides the dominant optical modes shown in white, one can see other slightly excited optical modes which are (almost) equidistant in the FP case (a) or indicate the DFB laser resonance located on the other side of the stopband. The estimated wavelength shift $\frac{\Delta_\lambda}{\Delta_I} \approx 3.14\,\text{nm/A}$ obtained for the FP laser and 3.09 nm/A for the DFB laser slightly differs from the factor $\nu_1^1 = 3.2\,\text{nm/A}$

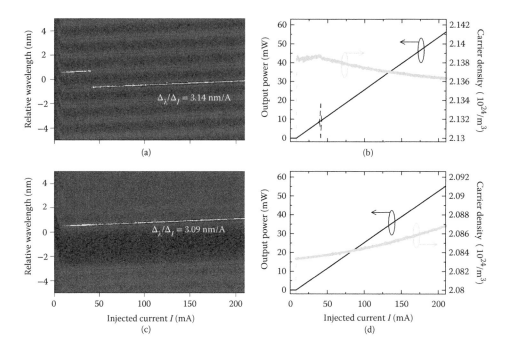

FIGURE 31.4 Mapping of the optical spectra (a), (c), and mean emitted power and carrier density (b), (d) as functions of the increased injected current I in solitary FP (a), (b), and DFB (c), (d) lasers. $\sigma = 2 \cdot 10^{-13} \, \text{m}^2$, $A = 2 \cdot 10^{8}/\text{s}$, $B = 1 \cdot 10^{-16} \, \text{m}^3/\text{s}$, $C = 1.3 \cdot 10^{-41} \, \text{m}^6/\text{s}$, $\nu_1^1 = 3.2 \cdot 10^{-9} \, \text{m/A}$, whereas other parameters are the same as in Figures 31.3b and c in the case of nonvanishing gain dispersion. White lines and light shading in (a) and (c) represent main and side peaks of the calculated optical spectra. Solid black and gray curves in (b) and (d) show time-averaged emission intensity and carrier density, respectively. Dashed black lines indicate minima and maxima of the emission intensity.

used in our simulations. We attribute this slight discrepancy to the additional contribution of the full refractive index change function $\tilde{N}(N, \omega)$ defined in Equation 31.17. Namely, the dependence of the carrier density N and the relative lasing wavelength λ on the bias current I (see gray curves in Figures 31.4b and d and wavelength shifts in Figures 31.4a and c, respectively) implies nonvanishing changes of the function $\tilde{N}(N, \omega) = \tilde{N}(N, -\lambda c_0/\lambda_0^2)$ that counteracts the thermal detuning term $\tilde{n}_T(I)$ and slightly reduces the redshift of the lasing wavelength.

It is noteworthy that an introduction of the thermal detuning term $\tilde{n}_T$ in our still simple TW model of the solitary laser implies, in general, only a continuous tuning of the lasing frequency. Once achieving threshold, the carrier density changes only slightly (gray curves in Figures 31.4b and d), whereas the emitted field intensity increases linearly without a visible saturation (black curves in the same panels), which is still not taken into account in our model. This linear growth of the lasing wavelength can be correctly understood when analyzing TW field Equation 31.1 with the propagation factor β defined by Equation 31.18 and neglected gain dispersion. Due to the transfer matrix formalism (Equations 31.10 and 31.11) and the expression of $\mathcal{B}$ in Equation 31.11, the extension of β by the nonvanishing real term $\tilde{n}_T$ is equivalent to the change of the relative frequency ω by $-\tilde{n}_T c_0/n_\text{g}$, or, alternatively, the change of λ by

$$\frac{-\lambda_0^2}{2\pi c_0} \frac{-\tilde{n}_T c_0}{n_\text{g}} = \nu_1^1 I.$$

The unique, more complicated feature in Figure 31.4 is the transition between two states in the FP laser at $I \approx 40 \, \text{mA}$, see panels (a) and (b). Figure 31.4a shows that in the vicinity of the transition, these two states are determined by two optical modes belonging to the opposite slopes of the wavelength-dependent gain profile with the peak wavelength at $\bar{\lambda} = 0$. Due to the redshift, all optical modes located on the falling (increasing) slope of this gain profile undergo an increase (decrease) of the detuning from the gain peak wavelength and, consequently, a slight rise (fall) in the mode threshold N; see the gray curves in panel (b) for $I < 40$ mA ($I > 40$ mA). A similar increase of the dominant falling-gain-slope mode threshold can also be seen in Figure 31.4d. At the position of the state transition, the wavelengths of two involved modes are symmetric with respect to the gain peak wavelength and, what is more important, their thresholds become equal. Due to a further tuning of the bias current, the previously suppressed mode at the increasing gain profile slope becomes the minimal threshold mode, is amplified and, finally, turns to be the dominant one. See References [14,15,26] for more details on similar and more complex mode transitions.

Modeling of Cross-Talk Heating Effects in Multisection Devices:

In MSLs devices, one can also use a more advanced model for thermal detuning function $\tilde{n}_T$ which takes into account local and nonlocal cross-talk heating effects [38]:

$$\tilde{n}_T|_{z \in S_k} = \tilde{n}_{T,k} = \frac{2\pi n_{g,k}}{\lambda_0^2} \sum_{r=1}^{m} v_k^r I_r. \tag{31.19}$$

Here, m is a number of sections in the considered MSL. The coefficients v_k^r of the linear thermal detuning function $\tilde{n}_T(I)$ determine the impact of the injection currents I_r attributed to the sections S_r on the refractive index change within each laser section S_k.

The effect of the thermal detuning in MSLs is much more complicated than that one of the solitary laser. Besides of the red shift of the lasing wavelength, the MSLs can also exhibit periodically or almost periodically reappearing transitions between different states. The change of mean carrier density in various sections during each such period between state changes can be significant and cannot be explained by simple gain saturation or detuning from the gain peak effects. In some cases, a measured variation of the lasing wavelength with an increase or decrease of the injection current in different laser sections, together with the analysis of the field equations provide good estimates of thermal detuning coefficients including cross-talk effects [14,15,38].

Another well-known effect occurring with the heating of the semiconductor laser is the red shift of the gain peak wavelength [39]. If required, these changes can be accounted by the relation [38]

$$\bar{\lambda}|_{z \in S_k} = \bar{\lambda}_k = \bar{\lambda}_k^0 + \sum_{r=1}^{m} \bar{v}_k^r I_r, \tag{31.20}$$

which is quite similar to the thermal detuning relation (Equation 31.19). Here, $\bar{\lambda}_k^0$ denotes an injection-independent part of the gain peak wavelength in the section S_k, and $\bar{v}_k^r$ are linear thermal gain peak detuning coefficients. When applying these expressions, one should be aware that a proper numerical time-domain resolution of a significant (tens or even hundreds of nanometers) gain peak shift requires very small time and, consequently, space discretization steps.

31.5 Spatially Inhomogeneous Carrier Density

Another important extension of the basic TW model takes into account sectionally inhomogeneous distributions of carrier density N. Namely, in this case, the sectionally averaged carrier density function $N(t)$ is replaced by the spatially distributed function $N(z, t)$, $z \in S_1$. This model extension can be especially important in the situations admitting localization of the high-intensity fields within the laser cavity, which takes place, e.g., during propagation of ultrashort optical pulses in mode-locked lasers, or DBR lasers with a high coupling factor κ. Due to stimulated recombination, the high-intensity fields at these localized regions can significantly deplete the carrier distribution causing a spatial hole burning (SHB) of the carriers [12,40].

To achieve a quantitative description of SHB, we replace simple carrier rate Equation 31.4 by the following equation for spatially distributed carrier density:

$$\partial_t N(z, t) = \mathcal{J}(I, N) - \mathcal{R}(N) - \mathcal{S}(N, E^\pm), \quad z \in S_1,$$

$$\mathcal{J}(I, N(z, t)) = \frac{1}{q\sigma|S_1|} \left[I + \frac{U'_F}{r_S} \left(\langle N \rangle_1 - N(z, t) \right) \right]. \tag{31.21}$$

Here, $\mathcal{J}$ is the inhomogeneous injection current density [12,41,42], N and $\langle N \rangle_1$ are spatially distributed and sectionally averaged carrier densities, U'_F and r_s denote the derivative of the Fermi level separation with respect to N and the series resistivity, whereas $\mathcal{S}$ is the spatially distributed stimulated recombination function,

$$\mathcal{S}(N, E^\pm) = \frac{c_0}{n_g} \Re \left(E, [g(N) - 2D]E \right). \tag{31.22}$$

In the case of the limit $r_s \to 0$, the spatially distributed carrier density, $N(z, t)$, at each position z converges to the sectional average, $N(t) = \langle N \rangle_1$. Since the sectional averaging of the relations 31.21 and 31.22 yields expressions 31.4 and 31.15, the TW models with and without spatial distribution of carriers in this limit case are equivalent.

Figure 31.5 shows some effects occurring due to the SHB of carriers in solitary DFB lasers. The impact of the SHB depends on the injection level. Just above the lasing threshold, the field intensity is small, and the carrier density remains nearly homogeneous, having only a small dip in the center of the laser. With raising injection, this dip increases, but, due to a simultaneous increase of the carrier density at the facets, the mean density remains nearly constant: see only slightly increasing $\langle N \rangle_1$ for $I \leq 90$ mA in panel (b) of the same figure. At these small-to-moderate bias current levels, the carrier density remains symmetric with respect to the laser centrum (dashed curve in panel [d]), and emission at the both facets is the same

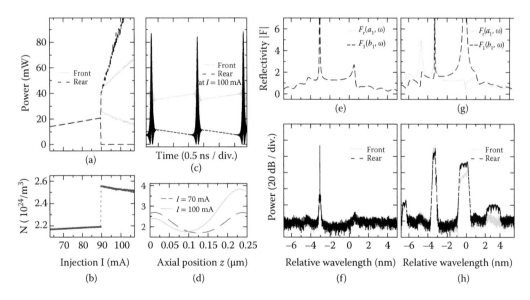

FIGURE 31.5 SHB in solitary DFB laser. (a) Minimal and maximal output power at both facets and (b) sectionally averaged carrier density $\langle N \rangle_1$ as functions of increased bias current. (c) Time trace of the emitted field intensity at both facets for fixed $I = 100$ mA. (d) Time-averaged carrier densities for $I = 70$ mA and $I = 100$ mA. DFB reflectivity spectra (e), (g), and calculated optical spectra (f), (h) for $I = 70$ mA (e), (f), and $I = 100$ mA (g), (h). Solid gray and black dashed curves in panels (a), (c), (e) through (h) represent optical fields at the front (J_1) and rear (J_2) facets of the laser. $U'_F = 10^{-25}$ Vm^3, $r_S = 5 \Omega$, whereas other parameters are the same as in Figures 31.4c and d except $\nu^1_1 = 0$.

(see coinciding solid gray and black dashed curves in panel (a) for $I \leq 90$ mA and panel (f) at $I = 70$ mA). The inhomogeneous carrier density (dashed curve in Figure 31.5d) causes a corresponding longitudinal variation of the index of refraction. The Bragg resonance of the grating thus has not the identical spectral position along the section but varies over a considerable portion of the stopband [12]. As a consequence, the symmetry of the stopband is lost, and the laser preferably operates on the short wavelength side of its stopband (see Figure 31.5f). The preference of the short-wavelength mode is also shown by the laser response functions $F_l(b_1, \omega)$ and $F_r(a_1, \omega)$ (see Figure 31.5e) calculated for spatially distributed carrier density profile (dashed curve in Figure 31.5d) at $I = 70$ mA according to the formulas 31.10 through 31.12 and the algorithm explained in Section 31.3.

At $I \approx 90$ mA, the symmetric solution loses its stability in symmetry breaking pitchfork bifurcation [43]. According to References [12,43], the supercritical pitchfork bifurcation of the stable symmetric steady state in DFB lasers generates a pair of new stable steady states with asymmetric density profiles (each one the mirror of the other). In our case, the pitchfork bifurcation seems to be of the subcritical type. Instead of finding two asymmetric stable steady states, we immediately jump to the pulsating state with different emission at both laser facets and larger mean carrier density (see panels [a] and [b] for $I > 90$ mA, respectively). The spatial distribution of the carrier density, in this case, is strongly asymmetric, see a solid curve in panel (d). This asymmetry together with noncommuting intermediate transfer matrices M_s from Equation 31.12 implies differences in DFB laser response functions estimated at the front (F_r) and the rear (F_l) sides of the laser, see Figure 31.5g. The optical spectra of the emission at the both sides of the laser (Figure 31.5h) also reveal these differences. Like the response functions of the panel (g), the left (thick gray curves) and the right (thin dashed curves) facet emissions have more pronounced contributions at the shorter and longer wavelength sides of the stopband, respectively.

31.6 Nonlinear Gain Saturation

Until now, our phenomenological models for peak gain and refractive index change (Equation 31.3) were taking into account their dependence on the sectionally averaged or local carrier density N. It is known, however, that the high-intensity optical fields saturate the gain function. To account for such saturation, one can introduce the following modifications of the gain and refractive index functions, which should be used for the definition of the propagation factor β in Equations 31.2 or 31.18 and stimulated recombination function S in Equations 31.6, 31.15, or 31.22:

$$g(N, E^\pm) = g(N)\rho_G(E^\pm), \qquad \tilde{n}(N, E^\pm) = \tilde{n}(N)\rho_I(E^\pm),$$

$$\text{where} \quad \rho_j = \begin{cases} \left(1 + \varepsilon_j |E|^2\right)^{-1}, & \text{if } N = N(z,t) \\ \left(1 + \varepsilon_j \|E\|^2\right)^{-1}, & \text{if } N = N(t) \end{cases}, \quad j = G, I.$$

Two different parameters, ε_G and ε_I, separately define the nonlinear gain and refractive index dependence on the local or spatially averaged optical field intensity. A typical assumption $\varepsilon_G = \varepsilon_I$ relates the gain and refractive index functions by the linewidth enhancement factor α_H. Another reasonable assumption $\varepsilon_I = 0$, $\varepsilon_G > 0$ [44] used for modeling of high power amplifiers considers the nonlinear compression of the gain function alone.

The importance of the nonlinear gain compression is best visible in high-power lasers and optical amplifiers showing several Watt emission intensity [44]. Some impact of the gain compression in small-to-moderate (≤ 100 mW) intensity regimes can also be observed when operating in the vicinity of various bifurcations, where a small change of parameters implies qualitative changes of the operating states. We should note, however, that the gain compression, in this case, implies only small shifts of the bifurcation positions, but has no significant impact on the qualitative description of laser dynamics in a large parameter domain. An analysis of simple TW model Equations 31.1 and 31.4 can explain the little influence of the gain compression in these regimes. A nonvanishing gain compression depletes the gain function $g(N)$ what implies a growth of the carrier density needed to reach threshold gain condition g_{th}. In solitary lasers, this growth is given by factor $g_{th}\varepsilon_G|E|^2/(g'\Gamma)$, which for typical gain compression coefficients and small-to-moderate field intensities is not exceeding a few percents of threshold carrier density. Consequently, a similar (up to a few percent) decay of the emission intensity can be observed.

A somehow different situation occurs in semiconductor ring lasers [19,22,23,45], where a proper introduction of nonlinear gain compression is crucial when deciding the type of operation states. In this case, one should distinguish the gain compression implied by co- and counter- propagating fields:

$$g^\pm(N, E^\pm) = g(N)\rho_G^\pm(E^\pm), \qquad \tilde{n}^\pm(N, E^\pm) = \tilde{n}(N)\rho_I^\pm(E^\pm),$$

$$\text{where} \qquad \rho_j^\pm = \begin{cases} \left(1 + \varepsilon_{js}|E^\pm|^2 + \varepsilon_{jc}|E^\mp|^2\right)^{-1}, & \text{if } N = N(z,t) \\ \left(1 + \varepsilon_{js}\langle|E^\pm|^2\rangle + \varepsilon_{jc}\langle|E^\mp|^2\rangle\right)^{-1}, & \text{if } N = N(t) \end{cases}, \quad j \in \{G, I\}, \tag{31.23}$$

whereas parameters ε_{js} and ε_{jc}, $j = G, I$, determine self- and cross-saturation of the gain and refractive index functions. In the ring lasers, usually is assumed that $\varepsilon_{Gc} > \varepsilon_{Gs}$, and $\varepsilon_{Ic} > \varepsilon_{Is}$. A detailed analysis based on the Maxwell–Bloch equations showed that the cross-saturation factor for two resonant modes in the ring cavity is twice larger than the self-saturation one [46].

The generalized functions $g^\pm$ and $\tilde{n}^\pm$ enter the definition of the propagation factor $\beta = \beta^\pm$ and the stimulated recombination function S:

$$\beta^\pm = \delta_0 + \tilde{n}^\pm(N, E^\pm) + \tilde{n}_T(I) + \frac{i(g^\pm(N,E^\pm)-\alpha)}{2},$$

$$S(N, E^\pm) = \begin{cases} \frac{c_0}{n_g}\Re \sum_{\nu=\pm} E^{\nu*}[g^\nu(N, E^\pm) - 2D]E^\nu, & \text{if } N = N(z,t) \\ \frac{c_0}{n_g}\Re \sum_{\nu=\pm} \langle E^{\nu*}[g^\nu(N, E^\pm) - 2D]E^\nu\rangle_1, & \text{if } N = N(t) \end{cases}. \tag{31.24}$$

It is noteworthy that differences in parameters ε_{js} and ε_{jc}, $j = G, I$, imply differences in the propagation factors β^+ and β^- determining the evolution of the fields E^+ and E^-, respectively. These differences are crucial when determining type and stability of operating states in the ring laser, see References [19,23,45] for more details.

To illustrate how an asymmetry of the self- and cross-gain saturation implies different operation states in the ring laser (see Figure 31.2b), we have simulated the TW model equations 31.14, 31.21, 31.8, 31.5, 31.23, 31.24, 31.3 for vanishing ε_{Is} and ε_{Ic}, fixed nonvanishing sum $\varepsilon_{Gc} + \varepsilon_{Gs} = C > 0$ and tuned difference $\varepsilon_{Gc} - \varepsilon_{Gs}$. Figure 31.6 shows the results of these simulations. Solid gray and dashed black curves in all panels of this figure represent clockwise (CW) and counter-clockwise (CCW) propagating field functions $E^-(b_1, t)$ and $E^+(a_1, t)$ at the point scattering source J_1, respectively (see Figure 31.2b). Panel (a) of this figure gives an overview of all obtained states when tuning $\varepsilon_{Gc} - \varepsilon_{Gs}$ from $-C$ (full self-saturation with vanishing ε_{Gc}) up to $+C$ (full cross-saturation with vanishing ε_{Gs}). Panels (b)–(e) of the same figure represent four different observed dynamic regimes. The first three regimes occurring with a consequent increase of the cross-gain saturation are the bidirectional stable stationary state (b), the alternate oscillations (c), and the unidirectional bistable state (d). These three regimes can be observed experimentally and recovered theoretically using a simple two-mode ODE model [19]. An analysis of the TW model performed in Reference [23] has explained the relation between the asymmetry of the gain compression factors, ε_{Gc} and ε_{Gs}, and stability of the bidirectional steady state (regime b) or unidirectional bistable states (regime d). It was also shown, how the difference $\beta^+ - \beta^-$ and localized backscattering r_1 determine the frequency of alternating oscillations (regime c).

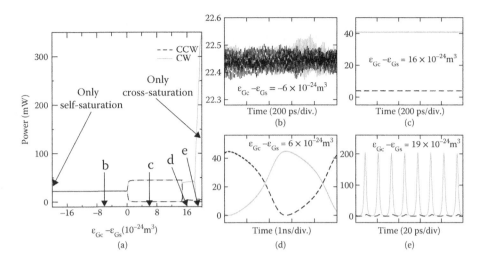

FIGURE 31.6 Dynamic regimes for different contributions of the cross- and self-gain saturations. (a) Maximal and minimal intensities of the optical fields at J_1 for changing values of $\varepsilon_{Gc} - \varepsilon_{Gs}$ but fixed $\varepsilon_{Gs} + \varepsilon_{Gc} = 20 \cdot 10^{-24}$ m^3. (b)–(e) Typical representatives of the observed regimes. Thick gray and dashed black curves indicate clockwise (CW) and counter-clockwise (CCW) propagating fields. Parameters are similar to those of Figures 31.4a and b, only $|S_1| = 1000\,\mu$m, $\alpha_H = -2$, $\alpha = 2$ cm^{-1}, $v_1^1 = \varepsilon_{Is} = \varepsilon_{Ic} = 0$, $I = 100$ mA. U_F' and r_S are the same as in Figure 31.5, whereas the field transmission and localized backscattering parameters at J_1 are $t_1 = \sqrt{0.7}$ and $r_1 = 0.007$, respectively.

The last simulated regime (e) was observed for the dominant cross-gain saturation. Like usual mode-locking pulsations, this regime is characterized by large short pulses occurring with the round-trip period. However, in contrast to the mode-locking observed in multisection ring lasers [47], this state is unidirectional and does not require any fast saturable absorption. Similar mode-locked pulsations in a single-section ring laser were found and discussed theoretically in Reference [45].

31.7 Further Modifications of the TW Model

There exist a vast number of further possible modifications of the TW model for MSLs. Each of these modifications, however, requires a few new not very well-known parameters and, therefore, should be used with the great care. On the other hand, some of these modifications being crucial when analyzing a particular group of MSLs can be irrelevant for simulations and analysis of different type MSLs. Later we present several modifications of the TW model used for investigation of specific types of MSLs.

Multiple Carrier Rate Equations in QD Lasers:

When modeling QD lasers, one should take into account carrier exchange processes between a carrier reservoir (CR) and discrete levels in QDs.

One of the simplest ways to account for all these transitions within the TW modeling frame is provided by the rate equations for the normalized carrier density $N^{cr}(z, t)$ (scaled by the factor Θ_N) within the CR, and occupation probabilities $N^{gs}(z, t)$, $N^{es}(z, t)$ of the ground state (GS) and the first excited state (ES) of QDs, respectively [10,48,49].

To keep the structure of the TW field equations 31.14 unchanged, we neglect the inhomogeneous spectral broadening effect due to QD nonuniformity and consider a simple single-Lorentzian gain spectrum profile, which limits the material gain bandwidth. Besides, we assume that the laser operates at the GS transition only. In this case, the propagation factor β depends on the ground state occupation probability

$N^{gs}(z,t)$ only. The expression of $\beta(N^{gs})$ is equivalent to that one given by Equations 31.2 and 31.3 with spatially distributed occupation probability $N^{gs}(z) \in [0,1]$ and factor $1/2$ instead of the carrier density $N(z,t)$ and transparency carrier density N_{tr}, respectively.

To describe carrier exchange processes between the CR, GS, and ES of the QDs in the active section (see Figure 31.7a), we use the following set of rate equations:

$$
\frac{d}{dt}N^{gs}(z,t) = -\frac{N^{gs}}{\tau_{gs}} + 2\left(\frac{N^{es}(1-N^{gs})}{\tau_{es\to gs}} - \frac{N^{gs}(1-N^{es})}{2\tau_{gs\to es}}\right) - \frac{1}{\theta_E}S(N^{gs},E^{\pm}),
$$

$$
\frac{d}{dt}N^{es}(z,t) = -\frac{N^{es}}{\tau_{es}} - \left(\frac{N^{es}(1-N^{gs})}{\tau_{es\to gs}} - \frac{N^{gs}(1-N^{es})}{2\tau_{gs\to es}}\right) + \left(\frac{N^{cr}(1-N^{es})}{4\tau_{cr\to es}} - \frac{N^{es}}{\tau_{es\to cr}}\right), \qquad (31.25)
$$

$$
\frac{d}{dt}N^{cr}(z,t) = \frac{I}{q|S_1|\theta_I} - \frac{N^{cr}}{\tau_{cr}} - 4\left(\frac{N^{cr}(1-N^{es})}{4\tau_{cr\to es}} - \frac{N^{es}}{\tau_{es\to cr}}\right).
$$

Here, $S(N^{gs},E^{\pm})$ is defined by Equation 31.22, whereas τ_a^{-1} and $\tau_{a\to b}^{-1}$, $a,b \in \{gs,es,cr\}$, denote spontaneous relaxation and transition rates between GS, ES, and CR, respectively. Factors $(1-N^{gs})$ and $(1-N^{es})$ represent the Pauli blocking, factors 2 and 4 account for the spin degeneracy in the QD energy levels. Note that here we neglect direct transitions between CR and GS. θ_I and $\theta_E = \frac{2hc_0\theta_I}{\lambda_0}$ are scaling factors relating the injection current I, the field intensity $|E|^2$, the CR scaling factor θ_N, the differential gain g', and the QD density in the active zone.

In the saturable absorption sections (see Figure 31.7b), there is no pumping, $I = 0$, so that the transitions from CR to ES can be neglected, $\tau_{cr\to es} \to \infty$, and the last of the equations 31.25 can be ignored. The carrier transition from ES to CR can be added to similar spontaneous recombination term: $\bar{\tau}_{es}^{-1} = \tau_{es}^{-1} + \tau_{es\to cr}^{-1}$. Following Reference [48], one can model the carrier transitions in the negatively driven saturable absorber

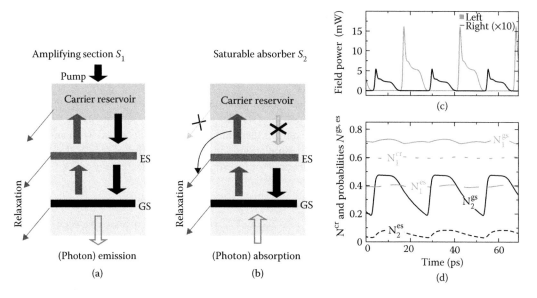

FIGURE 31.7 Schematic representation of carrier transitions in (a) the gain section and (b) the saturable absorption section of the QD MSL. Crossed arrows in panel (b) indicate the transitions which are neglected in the model equations. A sequence of (c) the emitted mode-locking pulses with an enhanced trailing edge plateaux at both sides of the laser and (d) the sectionally averaged carrier functions $N^{gs,es,cr}$ in both laser sections. All parameters and a detail description of the MSL device can be found in References [10,49].

by assuming an exponential decay of $\bar{\tau}_{es}$ with growing negative voltage U, whereas all other relaxation rates remain unchanged.

Figures 31.7c and d presents an example of simulated mode-locked QD laser containing an amplifying and a saturable absorber sections S_1 and S_2. Panel (c) of this figure gives an evidence of strongly asymmetric pulses with a broad trailing edge plateau. Our theoretical analysis has shown that such pulses arise mainly due to noninstant carrier transitions between the CR, ES, and GS of the QD laser shown in panel (d) of the same figure. The presence of these transitions exert a smoothening effect on all spatial/temporal carrier and field intensity distributions and, in turn, imply a broadening of the trailing edge of the pulse. We have also found that an increase of the intradot transition rates leads to a reduction of the filtering effect and, hence, to a growth of the pulse peak intensity and narrowing of the pulse and its trailing edge. More details on our analysis as well as experimental demonstration of such asymmetric pulses can be found in Reference [10,49].

Further modifications of the TW model can be used for more precise simulations of QD lasers. For example, one can improve the model of carrier transitions (Equation 31.25) by separate consideration of electrons and hole densities [50]. To allow a simultaneous radiation on the spectrally well-separated ground and ES, one can introduce another pair of TW equations for optical fields [51]. An inhomogeneous spectral broadening and an accompanying description of the radiation at GS and ES can also be modeled by an introduction of multiple sets of carrier rate and polarization equations representing carrier transitions within the QDs of different size and their impact on the laser emission at different wavelengths [52].

Nonlinear Gain and Refractive Index Functions:

In the earlier discussion, the gain and the refractive index dependence on the carrier density N was modeled by linear functions related to each other by the linewidth enhancement factor α_H. This modeling approach is reasonable for small and slow variation of carrier density N, but can fail once N exhibits some significant changes, see, e.g., Figure 31.5d, where a variation of the spatially distributed $N(z, t)$ was of the order of the mean value of the carrier density. In such situations, one should better use nonlinear peak gain functions, $g(N)$, which can better represent measured or precalculated gain spectra profiles. For this reason, the following logarithmic gain peak function dependence on the carrier density is frequently used:

$$g(N) = \Gamma g' N_{tr} \ln\left(\frac{\max\{N,N^*\}}{N_{tr}}\right), \qquad \tilde{n}(N) = \frac{\alpha_H g(N)}{2}. \tag{31.26}$$

Here, N^* indicates a cutoff carrier density value, which prevents the convergence $g(N) \to -\infty$ with carriers $N \to 0$. These expressions for the gain and index change functions replace the relations 31.3 used in the TW models discussed earlier.

Another issue is related to the linewidth enhancement factor α_H. Initially, this factor was used to relate gain and refractive index functions at a fixed value of N. Such approach implies a rather simple model for propagation factor, β, and can be quite useful when performing an advanced analysis of model equations. In reality, however, the ratio between the gain and refractive index is not a constant, but a function depending on carrier density, temperature, and several other factors not discussed in this chapter. Thus, an experimental estimation of this factor in the semiconductor laser operating at different conditions or using different methods can lead to rather different values of α_H. For this reason, it can be preferable to use separately defined nonlinear peak gain and index change functions $g(N)$ and $\tilde{n}(N)$. These functions depend on the properties of the semiconductor material and the design of the device, and, therefore, should be adjusted individually for each considered laser.

A satisfactory description of these functions for a broad class of semiconductor lasers is given by the logarithmic, and the square-root-like expressions [38]

$$g(N) = \Gamma g' N_{tr} \ln\left(\frac{\max\{N,N^*\}}{N_{tr}}\right), \qquad \tilde{n}(N) = \tilde{n} + \alpha_H \Gamma g' \sqrt{N \cdot N_{tr}}. \tag{31.27}$$

Here, $\tilde{n}$ represents the offset of the refractive index change function, $\Gamma g' = \partial_N g(N_{tr})$, and $\alpha_H = 2\partial_N \tilde{n}(N_{tr}) / \partial_N g(N_{tr})$ is the linewidth enhancement factor evaluated at the transparency carrier density N_{tr}.

We should admit, however, that linear formulas 31.3 with slightly corrected factors g', α_H, N_{tr}, and a proper selection of δ_0 can be used for approximation of nonlinear functions 31.27 not only in the vicinity of N_{tr} but also over a larger range of densities N including the threshold density N_{th}. In many cases, the simplifications of the gain and refractive index functions still imply qualitatively the same results when performing simulations of MSLs with varying parameters [53].

31.8 Multisection Lasers and Coupled Laser Systems

A vast variety of MSLs and coupled laser systems can be represented as a set of differently interconnected laser sections, each characterized by its material and geometry parameters. To distinguish these parameters or functions attributed to different laser sections, we shall use the lower indices. For example, α_k, $g_k(N)$, and a_k denote the field losses, the gain function and the left-edge coordinate of the section S_k. Note also that for unique identification of longitudinal coordinate z within all laser sections, different sections of our device are represented by nonoverlapping intervals (a_k, b_k).

According to our laser device construction, for any edge of all sections S_k, we can attribute a unique junction J_l. On the other hand, each junction has, at least, one section joining it from one or another side, see, e.g., Figure 31.1, where MSLs are represented as sets of laser *sections* mutually interconnected through different *junctions*. To explain the relations between section edges, corresponding junctions, and applied optical injections, we use the following notations in the sequel of this chapter. By l we denote the index of the junction J_l, as well as optical fields and the section edges attributed to this junction. l' (l'') is the vector of length $|l'|$ ($|l''|$) containing indices of the sections connected to J_l by their left (right) edge $a_{l'_j}$ ($b_{l''_j}$); see Figure 31.8a. $l^e = |l'| + |l''| \geq 1$ is a total number of such section edges connected to J_l. By $o_l(t)$ and $E_l^{out}(t)$ we denote the applied optical injection and the recorded emission at the same junction, see solid and dashed thick light gray arrows in Figure 31.8a. When the injection or emission at J_l is absent, the corresponding function is simply set to zero.

General field scattering conditions at the arbitrary junction J_l are defined by the $l^e \times l^e$ dimensional complex field scattering matrix $\mathcal{T}_l$, $l^e \times 1$ dimensional injection distribution matrix $\mathcal{T}_l^i$, and $1 \times (l^e + 1)$

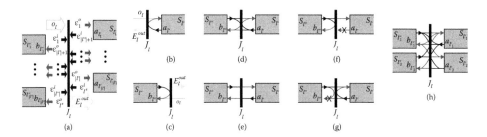

FIGURE 31.8 Schematic representation of different interfaces between the sections at the junction J_l. (a) general configuration, (b) and (c) field reflection and optical injection at laser facets, (d) and (e) reflecting (transmission and reflection) and trivial (transmission only) interface of two section edges, (f) and (g) directionally absorbing interfaces of two section edges for modeling of a master–slave laser system, and (h) transmitting/reflecting interface of four section edges for modeling ring lasers with an outcoupling waveguide. Black segments and gray frames represent junction J_l and all section edges connected to this junction, respectively. Black and dark gray arrows in all diagrams represent all components of the vector fields $\mathcal{E}_l^i$ and $\mathcal{E}_l^o$, respectively. Optical injections o_l and emitted fields E_l^{out} are shown by solid and dashed light gray arrows. Crossed arrows in panels (f) and (g) represent a full absorption of the corresponding fields.

dimensional outcoupling matrix $\mathcal{T}_l^o$:

$$\mathcal{E}_l^o = \mathcal{T}_l \mathcal{E}_l^i + \mathcal{T}_l^i o_l, \quad E_l^{\text{out}} = \mathcal{T}_l^o \begin{pmatrix} \mathcal{E}_l^i \\ o_l \end{pmatrix}, \qquad \text{where}$$

$$\mathcal{E}_l^o = \left(E_{l'_1}^+(a_{l'_1}, t), \ldots, E_{l'_{|l'|}}^+(a_{l'_{|l'|}}, t), E_{l''_1}^-(b_{l''_1}, t), \ldots, E_{l''_{|l''|}}^-(b_{l''_{|l''|}}, t) \right)^T, \tag{31.28}$$

$$\mathcal{E}_l^i = \left(E_{l''_1}^+(b_{l''_1}, t), \ldots, E_{l''_{|l''|}}^+(b_{l''_{|l''|}}, t), E_{l'_1}^-(a_{l'_1}, t), \ldots, E_{l'_{|l'|}}^-(a_{l'_{|l'|}}, t) \right)^T.$$

The vector functions $\mathcal{E}^i$ and $\mathcal{E}^o$ (see black and dark gray arrows in Figure 31.8a) denote the internal optical fields, which are incident into the junction from all adjacent sections and are scattered from the junction back into these sections, respectively.

In most cases, the interfaces between the sections are much simpler. For example, the scattering matrices at the facets of the solitary laser (Figures 31.8b and c as well as junctions J_1 and J_2 in Figure 31.2a) are determined by the boundary conditions (Equation 31.7), i.e.,

$$\begin{cases} \mathcal{T}_l = -r_l^*, \ \mathcal{T}_l^i = 1, \ \mathcal{T}_l^o = (t_l, 0), & \text{single "left" edge } a_{l'}, \ |l'| = 1, \ |l''| = 0 \\ \mathcal{T}_l = r_l, \ \mathcal{T}_l^i = 1, \ \mathcal{T}_l^o = (t_l, 0), & \text{single "right" edge } b_{l''}, \ |l''| = 1, \ |l'| = 0 \end{cases}, \tag{31.29}$$

where r_l and t_l are field reflection and transmission coefficients,

$$|r_l| \le 1, \qquad t_l \le \sqrt{1 - |r_l|^2}. \tag{31.30}$$

Another frequently used case in MSLs is the interface of two adjacent sections (Figures 31.8d through g). At such interfaces, we have no optical injections and field emission, so that we can set $\mathcal{T}_l^i = (0, 0)^T$ and $\mathcal{T}_l^o = (0, 0, 0)$. The scattering of the field at J_l, in this case, is entirely defined by the 2×2 dimensional matrix

$$\mathcal{T}_l = \begin{pmatrix} t_l & -r_l^* \\ r_l & t_l \end{pmatrix}, \tag{31.31}$$

where t_l and r_l satisfy the conditions 31.30 (see Figure 31.8d and, e.g., J_3 in Figure 31.1a). Here, the non-vanishing reflections r_l can appear, e.g., due to different heterostructure of the adjacent sections. In the simplest case of $r_l = 0$ and $t_l = 1$, $\mathcal{T}_l$ is an identity matrix, and the interface admits a full transmission of the optical fields (see Figure 31.8e and J_2 in Figure 31.1a).

When modeling the master–slave laser system (S_1 and S_2 in Figure 31.1d), only one-directional field propagation should be allowed in the air gap between two lasers (section S_3 in the same figure). This effect can be achieved by modification of otherwise standard scattering matrices $\mathcal{T}_l$ (Equation 31.31) at one of the gap section edges. One can model a full absorption of the backward incident beam at the interface of the master laser and the gap section (see Figure 31.8f and J_2 of Figure 31.1d), or prohibit the field backscattering into the air gap at the interface of the slave laser and the gap section (Figure 31.8g and J_3 of Figure 31.1d). Formally, both these situations can be defined by the scattering matrices

$$\mathcal{T}_j = \begin{pmatrix} t_j & 0 \\ r_j & 0 \end{pmatrix} \text{ (master-gap interface)}, \qquad \mathcal{T}_j = \begin{pmatrix} t_j & -r_j^* \\ 0 & 0 \end{pmatrix} \text{ (gap-slave interface)}.$$

More complicated situations occur at the junctions connecting more than two section edges of the MSL. For example, Figure 31.8h and J_1 of Figure 31.1e represent an interface connecting two "left" and two "right" section edges. This situation is used for modeling of a localized coupling of the ring laser (section S_1 in Figure 31.1e) and the outcoupling waveguide (S_2 and S_3 in the same figure).

Similarly to the previously discussed case, the optical injection- and field emission-relevant matrices can be defined by $\mathcal{T}_l^i = (0, 0, 0, 0)^T$ and $\mathcal{T}_l^o = (0, 0, 0, 0, 0)$. By assuming a nonvanishing field reflection r_l at the ring laser part of this junction (section edges $b_{l''}, a_{l'}$ in Figure 31.8h or b_1, a_1 in Figure 31.1e), we model a *localized* linear backscattering of the fields [23,37]. The 4×4 dimensional scattering matrix $\mathcal{T}_l$, in this case, can be defined as

$$\mathcal{T}_l = \begin{pmatrix} t_l & i\tilde{t}_l & -r_l^* & 0 \\ i\tilde{t}_l & t_l & 0 & 0 \\ r_l & 0 & t_l & i\tilde{t}_l \\ 0 & 0 & i\tilde{t}_l & t_l \end{pmatrix}, \qquad t_l^2 + \tilde{t}_l^2 + |r_l|^2 \le 1. \qquad (31.32)$$

Here, t_l is a real field amplitude transmission factor within the same (ring or outcoupling) waveguide and $i\tilde{t}_l$ is an imaginary coefficient representing part of the field amplitude, which is outcoupled from the ring or transmitted into the ring from the external waveguide. It is noteworthy that a proper estimation of the transmission–reflection–outcoupling matrix $\mathcal{T}_l$ in the ring laser case requires some appropriate measurements or an advanced modeling. Such modeling should take into account the curvature of the ring cavity, the length of the coupling regions, the field diffraction, and the overlapping of the lateral modes in the coupling region [54]. Moreover, the coefficients of the scattering matrix are, in general, frequency dependent. In our TW modeling approach, we use constant coefficients describing field scattering at the central reference frequency.

31.9 Simulations of Nontrivial MSL Device

The concept of differently interconnected *sections* and *junctions* allows modeling rather complicated MSLs. One of such nontrivial configurations is a semiconductor ring laser with four separate branches of the filtered optical feedback, see Figure 31.9a. The multichannel feedback scheme of this laser admits a fast switching between steady states determined by the resonances of the ring laser and the wavelengths of the activated filtering channels [55].

The gray-shaded frames in Figure 31.9a represent device sections of different types. Namely, we distinguish here the amplifying sections $S_{A.}$, where the field and carrier dynamics is governed by the full TW model (Equations 31.5, 31.14, 31.21, 31.23, 31.24, 31.26), and two kinds of passive sections, $S_{P.}$ and $S_{F.}$, where gain and refractive index functions are set to zero, allowing to ignore the carrier rate equations at all. The notations of all sections in the section indexes are made according to the cardinal directions "n," "e," "s," and "w".

Almost all parameters of the TW model in all sections of our MSL are the same as in Figure 31.6. A few exceptions are parameters $\alpha_H = -4$ and $\bar{\gamma}_\lambda = 100$ nm. In the passive waveguiding sections, $S_{P.}$ (medium gray), we assume $|S_{Pe}| = |S_{Pw}| = 330 \, \mu\text{m}$, $|S_{Pne}| = |S_{Pnw}| = 50 \, \mu\text{m}$, $|S_{Pse}| = |S_{Psw}| = 2500 \, \mu\text{m}$, and neglect the gain dispersion, $\bar{g} = 0$. In the passive filtering sections, $S_{F.}$ (light gray), we assume $|S_{F.}| = 530 \, \mu\text{m}$ and significantly modify the profile of Lorentzian gain dispersion by setting $\bar{g} = 5 \cdot 10^4 \, \text{m}^{-1}$ and $\bar{\gamma}_\lambda = 4$ nm. The relative peak wavelengths of four filtering branches (sections S_{Fwj} and $S_{Fej}, j = 1, \ldots, 4$) are $\bar{\lambda} = -2, -0.67, 0.67,$ and 2 nm, respectively. Finally, in the amplifying sections (dark gray) within the primary ring laser, $S_{Ajk}, j = n, s$ and $k = e, w$, we use $|S_{A.}| = 380 \, \mu\text{m}$, $\varepsilon_{Gs} = 6 \cdot 10^{-24} \, \text{m}^3$, $\varepsilon_{Gc} = 2\varepsilon_{Gs}$, and the bias currents $I = 26$ mA, which is 1.5 times higher than the lasing threshold in the laser without feedback. In the amplifying sections belonging to the four filtering branches, $S_{Asj}, j = 1, \ldots, 4$, we assume $|S_{A.}| = 190 \, \mu\text{m}$ and $\varepsilon_{Gc} = \varepsilon_{Gs} = 9 \cdot 10^{-24} \, \text{m}^3$. Once the bias current in these sections is zero, $I = 0$ mA,

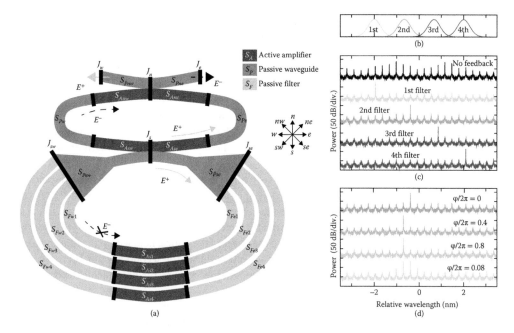

FIGURE 31.9 (a) Scheme of the semiconductor ring laser with four branches of filtered and amplified unidirectional optical feedback. Black segments and gray-shaded frames indicate junctions and different sections of the MSL. Solid light-gray and dashed black arrows show propagation directions and the emission of the fields E^+ and E^-, respectively. (b) Transmission spectra of four filtering branches. Maximal transmission at $I_{Asj} = 10\,\mathrm{mA}, j = 1, \ldots, 4$ is approximately 1.6. (c) Stabilization of the multimode behavior of the ring laser (black) by the single-branch filtered feedback (gray). (d) Dependence of the lasing wavelength on the feedback phase once the second filtering branch is activated.

the feedback branches are efficiently absorbing the optical fields. To activate one of the feedback branches, we set the corresponding injection $I = 10\,\mathrm{mA}$.

The field transmission and reflection conditions at J_n and J_s are given by the conditions 31.32 with $r = 0$, $t = \sqrt{0.8}$, and $\tilde{t} = \sqrt{0.2}$. At J_{se} and J_{sw}, we neglect all possible reflectivity, admit full-field transmission from filtering branches to the passive waveguide sections S_{Psw} or S_{Pse}, equally distribute the intensity of the optical field E^+ propagating from S_{Pse} to the filtering branches, and fully absorb E^- at J_{sw}:

$$E^-(b_{Pse}, t) = \sum_{j=1}^{4} E^-(a_{Fej}, t), \qquad E^+(a_{Fej}, t) = \sqrt{\frac{1}{4}} E^+(b_{Pse}, t), \quad j = 1, \ldots, 4;$$

$$E^+(a_{Psw}, t) = \sum_{j=1}^{4} E^+(b_{Fwj}, t), \qquad E^-(b_{Fwj}, t) = 0, \quad j = 1, \ldots, 4.$$

At J_w and J_e, the fields E^+ and E^- are emitted from our MSL. Here, the field reflection–transmission conditions are given by (Equation 31.29) with the reflectivity factors $r_w = r_e = 0.1$. All other junctions of this MSL are trivial, i.e., the optical fields cross the interfaces according to the relations 31.31 with $r_l = 0$ and $t_l = 1$.

A series of simulations represented in the remaining panels of Figure 31.9 are in good agreement with the experimental results reported in Reference [55]. First of all, panel (b) shows the transmission spectra (modulus of the wavelength-dependent complex transmission function) of the optical fields E^+ propagating through each of four optical feedback branches activated by the injected current into the corresponding amplifying section. For 10 mA injections used in these simulations, the peak amplitude transmission is around 1.6.

Panels (c) and (d) of Figure 31.9 show simulated optical spectra of the emitted field at the "west" facet J_w of the MSL for different operation conditions. An upper black curve in panel (c) represents the optical spectrum in the case of deactivated feedback branches. Multiple significant spectral peaks with the mode separation corresponding to the field round-trip time in the ring laser indicate a multimode lasing of the laser. An optical field E^+ propagating along the filtering branches, however, is not entirely absorbed. For higher ring laser injections, we observed the steady states determined by a single ring resonance mode. In these cases, the amplifier within corresponding filtering branch was optically pumped, and the related peak amplitude transmission was around 0.2. The competition between nominally equivalent unpumped filtering branches, however, does not allow predicting the lasing wavelength of such a steady state.

The four lower spectra in Figure 31.9c represent switching between different optical modes by activation of the corresponding filter and deactivation of the remaining ones. A close inspection of these spectra shows, that whereas the first and the fourth filters select the resonance modes which are closest to the filter peak position, the third, and, especially, the second filter prefers modes admitting smaller optical feedback. We have found, that this mode selection is related to the phase of the optical field within the filtering branch. Figure 31.9d demonstrates, how tuning of the feedback phase within the second filtering branch (realized by variation of the detuning factor $\delta_{0,Fw2}$) implies changes between the resonant modes located within the filtering band.

In conclusion, we have simulated the MSL consisting of 22 sections interconnected at 18 junctions. Our theoretical findings were in a good qualitative agreement with experimental observations of similar ring laser device reported in Reference [55].

31.10 Beyond Numerical Simulations of the TW Model

In the previous sections, we have introduced different modifications of 1+1 dimensional TW model suited for simulations of various MSL devices and coupled laser systems. In the remaining part of this chapter, we introduce the concept of instantaneous optical modes and present several applications of these modes for an advanced analysis of MSLs. In all these cases, we consider MSLs without optical injection and neglect a contribution of Langevin noise term $F_{sp}^{\pm}$, which is of minor importance in the lasers operating well above threshold.

Instantaneous Optical Modes:

The concept of optical modes plays a significant role in understanding laser dynamics in general. They represent the natural oscillations of the electromagnetic field and determine the optical frequency and the lifetime of the photons contained in the given laser cavity. The *instantaneous* optical modes correspond to a fixed *instant* distribution of the propagation factor β [26].

In general, compared to a variation of the optical fields, the changes of the carrier density N are slow. The change of N is mainly determined by the carrier relaxation time which, typically, is measured in nanoseconds (or tens of picoseconds when considering saturable absorbers). On the other hand, picosecond or sub-picosecond time windows are sufficient for significant changes of the photon densities. Since the gain compression for small and moderate field intensities is also small, the propagation factor β experiences only minor modifications in the picosecond range. For this reason, in the remaining part of this chapter we analyze the field equations for the frozen distribution of the propagation factor $\beta(z, t_0)$ at the time instant t_0.

The instantaneous optical modes of MSLs are pairs $(\Omega(\beta), \Theta(\beta, z))$ of complex frequencies Ω and vector-functions $\Theta = (\Theta_E^+, \Theta_E^-, \Theta_P^+, \Theta_P^-)^T$, where imaginary and real parts of $\Omega(\beta)$ are mainly defining the angular frequency and the damping of the mode, whereas $\Theta(\beta, z)$ determines the spatial distribution of the mode.

Complex frequencies Ω and vector-functions $\Theta(\beta, z)$ solve the linear system of algebro-differential equations

$$
\begin{cases}
\frac{d}{dz}\Theta_E^+ = -iB(\Omega)\Theta_E^+ - i\kappa\Theta_E^- \\
\frac{d}{dz}\Theta_E^- = iB(\Omega)\Theta_E^- + i\kappa\Theta_E^+ \\
\Theta_P^\pm(\beta, z) = \frac{\bar\gamma/2}{\bar\gamma/2+i(\Omega-\bar\omega)}\Theta_E^\pm(\beta, z)
\end{cases}
, \quad z \in S_k|_{k=1}^n, \qquad \theta_l^o(\beta) = \mathcal{T}_l \theta_l^i(\beta)\Big|_{l=1}^m, \tag{31.33}
$$

obtained by assuming a nonvarying in time propagation factor, $\partial_t\beta = 0$, and substituting the expressions

$$
E^\pm(z, t) = \Theta_E^\pm(\beta; z)e^{i\Omega(\beta)}, \qquad P^\pm(z, t) = \Theta_P^\pm(\beta; z)e^{i\Omega(\beta)}
$$

into the field equations 31.14 within each of n sections S_k, and boundary conditions 31.28 at each of m junctions J_l. Similarly to the vector functions $\mathcal{E}_l^o$ and $\mathcal{E}_l^i$ in Equation 31.28, complex vectors $\theta_l^i(\beta)$ and $\theta_l^o(\beta)$ in Equation 31.33 represent functions $\Theta_E^\pm(\beta, z)$ at the section edges $z = a_{l_j'}$ or $z = b_{l_j''}$ connected by the junction J_l. The function $B(\Omega)$ entering Equations 31.33 is defined in 31.16.

Each pair of linear ODEs in Equation 31.33 can be solved by the transfer matrix[†] 31.11 with the coefficients nonlinearly depending on still unknown complex frequency Ω. These matrices define $2n$ homogeneous linear equations relating $4n$ components of the complex vector $S = (s_1, \ldots, s_{4n})^T$ representing field functions $\Theta_E^\pm(\beta, z)$ at both edges of all sections S_k. Another $2n$ homogeneous linear equations relating the same complex numbers are given by the field scattering matrices $\mathcal{T}_l$ at all junctions J_l. In such a manner, we build a linear $4n$ dimensional algebraic system

$$
\mathcal{M}(\beta, \kappa; \Omega)S = 0,
$$

determined by a sparse $4n \times 4n$ dimensional matrix $\mathcal{M}$. Nontrivial solutions S (i.e., nontrivial functions Θ of the problem 31.33) are available only for those Ω which are the complex roots of the complex characteristic equation

$$
\det \mathcal{M}(\beta, \kappa; \Omega) = 0. \tag{31.34}
$$

The finite number of these roots can be found using Newton iterations and the homotopy method; see Reference [26] for more details.

It is noteworthy that linear configurations of MSLs admit rather simple expressions of the characteristic equations 31.34 involving the response functions $F_l(z, \Omega)$ and $F_r(z, \Omega)$ defined at some longitudinal position z of the MSL:

$$
\det \mathcal{M}(\beta, \kappa; \Omega) = 0 \qquad \Leftrightarrow \qquad F_l^{-1}(z, \Omega) = F_r(z, \Omega). \tag{31.35}
$$

For example, for the solitary lasers considered in Figure 31.3b and c, $F_l(b_1, \Omega)$ is defined in Equation 31.13, whereas $F_r(b_1, \Omega) = r_2$. In the general case, functions F_l and F_r are defined by a consequent superposition of the sectional transfer matrices $M(\beta, \kappa, \Omega)$ and the left-to-right or right-to-left junction-transfer matrices

$$
\mathcal{T}_{j,22}^{-1}\begin{pmatrix} \det\mathcal{T}_j & \mathcal{T}_{j,12} \\ -\mathcal{T}_{j,21} & 1 \end{pmatrix} \qquad \text{or} \qquad \mathcal{T}_{j,11}^{-1}\begin{pmatrix} 1 & -\mathcal{T}_{j,12} \\ \mathcal{T}_{j,21} & \det\mathcal{T}_j \end{pmatrix};
$$

see References [23,26,28] for more details.

[†] In the case of nonvanishing $\Delta_\beta = \frac{\beta^+-\beta^-}{2}$, the transfer matrix in each section S_k should be constructed for $\bar\beta = \frac{\beta^++\beta^-}{2}$ and later multiplied by the factor $e^{-i(\Delta_\beta)_k|S_k|}$ [23].

The calculated optical field function $\Psi(z, t) = (E^+, E^-, P^+, P^-)^T$ can be represented as a superposition of the suitably normalized vector functions $\Theta(\beta, z)$ which are slowly changing with a variation of the propagation factor $\beta(z, t)$:

$$\Psi(z, t) = \sum_{j=1}^{\infty} f_j(t)\Psi_j(\beta(z, t), z). \tag{31.36}$$

Here, $f_j(t)$ is the complex amplitude of the mode, which can denote the mode contribution to the field emission at the laser facet a_k once normalization of mode functions assumes $\Theta_E^-(\beta, a_k) = 1$. According to our notations, index $_1$ denotes the most significant mode having a largest (instant) amplitude $|f|$ or a lowest damping $\Im\Omega$. An increasing index means a decreasing importance of the mode. This numbering does allow us to achieve good approximations of the field function $\Psi(z, t)$ already by low-dimensional truncated mode expansions 31.36.

Calculation of optical modes and expansion of the field function into the modal components can give a broad understanding of different operating regimes in MSLs and explain parameter change-induced transitions between these states observed in simulations and experiments. We have applied our mode analysis for interpretation of experimental observations in different MSLs. Namely, we have explained a stable operation of ring lasers at alternating oscillation or bi- and unidirectional steady state regimes [23]; almost periodically reappearing state transitions and estimation of thermal tuning parameters in master-oscillator power-amplifier device [38,56], DBR laser [14], or ECDL [15]; and strongly asymmetric pulse shapes in QD mode-locked laser [49]. More theoretical examples of our mode analysis can be found in Reference [26].

Steady States:

Any stationary (rotational wave) state of the MSL is determined by an optical mode with a *real* mode frequency

$$(\Psi(z, t), N(z, t)) = \left(\hat{f}\,\Theta(\hat{\beta}, z)e^{i\hat{\omega}t}, \hat{N}(z)\right), \quad \text{where} \quad \Omega(\hat{\beta}) = \hat{\omega} \in \mathbb{R},$$

and $\hat{\beta}(z)$ is a constant in time spatially distributed propagation factor. Let us consider the TW model with sectionally averaged carrier density and neglected nonlinear gain compression given by Equations 31.4, 31.5, 31.14, 31.18, 31.15, and 31.28. In this case, all steady states are fully defined by a set of $n_a + 2$ real numbers $\left(\hat{\omega}, |\hat{f}|^2, \hat{N}_1, \ldots, \hat{N}_{n_a}\right)$, which are a mode frequency, a mode intensity, and sectionally averaged carrier densities within all n_a "active" sections having nonvanishing functions g and $\tilde{n}$. The rotational invariance of the TW model implies freedom in selection of the phase of the complex mode amplitude $\hat{f}$. The set of these real numbers is a root of a nonlinear algebraic system of one complex characteristic equation and n_a real steady-state carrier rate equations:

$$\det \mathcal{M}\left(\beta(\hat{N}), \kappa; \hat{\omega}\right) = 0,$$
$$\frac{I_r}{q\sigma_r|S_r|} - \mathcal{R}(\hat{N}_r) - |\hat{f}|^2 \frac{c_0}{n_g} G(\hat{N}_r, \hat{\omega})\langle(\Theta_E, \Theta_E)\rangle_r = 0, \quad r = 1, \ldots, n_a. \tag{31.37}$$

Here, the frequency-dependent gain function G is defined in Equation 31.17, whereas the sectional average $\langle(\Theta_E, \Theta_E)\rangle_r$ can be expressed as a function of $\hat{\omega}$ and $\hat{N}$ [28].

In the case of the single active section, $n_a = 1$, the steady state frequency $\hat{\omega}$ and threshold carrier density $\hat{N}_1$ can be directly found from the characteristic equation, whereas the remaining equation determines the value of $|\hat{f}|^2$. Assume that the single active section S_1 of linear MSL is located on the left side of the device (see Figure 31.1b), and the optical fields within the adjacent passive section S_2 are governed by the simple relations 31.9. For an illustration of this situation, we have considered a three section passive dispersive reflector laser consisting of the active DFB, passive DBR, and another passive phase tuning section in between. Due to the relation 31.35, we can replace the complex characteristic equation by a couple of real-valued equations,

$$F_l^{-1}(a_2, \hat{\omega}) = F_r(a_2, \hat{\omega}) = e^{-\alpha_2 |S_2|} e^{i(\varphi - 2\hat{\omega}\tau_2)} F_r(b_2, \hat{\omega}) \qquad \Leftrightarrow$$

$$\begin{cases} \tilde{\mathcal{M}}_\alpha(\hat{N}_1, \hat{\omega}) = e^{\alpha_2 |S_2|} \\ 2\hat{\omega}\tau_2 - \tilde{\mathcal{M}}_\varphi(\hat{N}_1, \hat{\omega}) = \varphi \end{cases}, \text{ where } \tilde{\mathcal{M}}(\hat{N}_1, \hat{\omega}) = \tilde{\mathcal{M}}_\alpha e^{i\tilde{\mathcal{M}}_\varphi} = F_l(a_2, \omega) F_r(b_2, \hat{\omega}) \qquad (31.38)$$

This formulation suggests a simple way for finding the steady states. Namely, each of these equations for fixed parameters α_2 and φ defines one or several curves in frequency ω – carrier threshold N_1 domain, see solid and dashed curves in Figure 31.10. The intersections of these lines determine the steady state pairs $\hat{\omega}, \hat{N}_1$ (hollow bullets in the same figures). It is noteworthy that to any point in the $\omega - N_1$ domain one can attribute a unique triple of loss, phase, and mode power parameters α_2, φ, and $|\hat{f}|^2$. ω and N_1 within the gray shading regions of Figure 31.10 represent the unphysical steady states corresponding to negative damping in the passive section ($\alpha_2 < 0$) and negative mode intensity ($|\hat{f}|^2 < 0$) due to insufficient pumping of the active section.

The fixed level lines of $\tilde{\mathcal{M}}_\alpha$ determined by larger losses $\alpha_2 = 30$ and 40/cm (thin solid-line ellipses located inside of thick solid curves in Figure 31.10) are shrinking toward central points, which are resonances of the solitary DFB laser. Accordingly, the (odd) number of steady states on each ellipse is also reduced. The saddle-node bifurcation that is responsible for creation or annihilation of the steady state pair, occurs at those φ and α_2, where corresponding fixed level lines of $\tilde{\mathcal{M}}_\varphi$ and $\tilde{\mathcal{M}}_\alpha$ become tangent to each other. The last condition formally given by

$$\mathcal{M}_{SN}(\hat{\omega}, \hat{N}_1) \stackrel{def}{=} \partial_\omega \tilde{\mathcal{M}}_\alpha \partial_{N_1} \tilde{\mathcal{M}}_\varphi - \partial_\omega \tilde{\mathcal{M}}_\varphi \partial_{N_1} \tilde{\mathcal{M}}_\alpha = 0$$

is satisfied on the dotted lines of Figure 31.10.

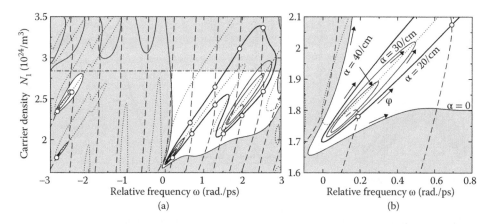

FIGURE 31.10 Stationary states (frequencies and threshold carrier densities) of a three section DFB laser. Panels (a) and (b) show a global overview and a zoomed-in region close to the minimal threshold mode with $\hat{\omega} \approx 0$. Solid curves represent steady states for an arbitrary φ but only a few fixed α_2. Dashed curves indicate steady states for arbitrary α_2 and fixed $\varphi = 0$. Dotted curve shows positions where saddle-node bifurcations hold. Empty bullets: steady states for $\alpha_2 = 20$/cm and $\varphi = 0$. Dash-dotted line indicates maximal N_1 which can be achieved for considered bias current I. Gray shading corresponds to unphysical states. Small arrows in panel (b) indicate directions of the steady state shift along the fixed loss lines for growing φ. All parameters as in Reference [26].

In general, the interpretation of the steady states for large α_2 (small feedback) is in good agreement with the analysis of the external cavity modes in the Lang–Kobayashi (LK) model of lasers with delayed feedback [18,27]. A decrease of α_2 leads to blowing up and collision of different ellipses. This scenario involves multiple modes of the solitary DFB laser and can be no more explained by the LK model.

Mode Approximation Systems:

For some MSL devices, the TW model (Equations 31.2 through 31.5, and 31.14, 31.15, 31.28) with sectionally averaged carrier densities, linear gain and index change functions, and neglected gain compression terms can be reduced to the finite-dimensional system of ODEs describing an evolution of q complex mode amplitudes f and real sectionally averaged carrier densities N within n_a active sections of MSL:

$$
\dot{f}_k = i\Omega_k(N)f_k + \sum_{l=1}^{q}\left(\sum_{r=1}^{n_a} K_{k,l}^r(N)\dot{N}_r\right)f_l, \quad k = 1,\ldots,q;
$$

$$
\dot{N}_r = \frac{I_r}{q|S_r|\sigma_r} - \mathcal{R}(N_r) - \mathfrak{R}\sum_{k,l=1}^{q} L_{k,l}^r(N)f_k^*f_l, \quad r = 1,\ldots,n_a.
$$

(31.39)

This mode approximation (MA) system follows from the substitution of the truncated field expansion (Equation 31.36) into the TW model equations and projection of the resulting field equations onto the linear subspace defined by each of q modes. The nonadjoint nature of the field evolution operator and small but nonvanishing time derivatives of propagation factor $\beta(N)$ imply the appearance of the mode coupling terms $K_{k,l}^r\dot{N}_r$. For the derivation of the MA equations and analytic expressions for carrier and mode frequency Ω-dependent mode-coupling functions, $K_{k,l}^r$ and $L_{k,l}^r$, see Reference [28].

To check the precision of our MA system, we have performed simulations of the TW model and two related MA systems describing the evolution of a mode-locked laser consisting of a saturable absorber and an amplifying section (case of $n_a = 2$). The solid black curve and hollow bullets in Figures 31.11b and c show typical optical spectrum and time trace of the mode-locking pulsations obtained by numerical integrations of the TW model. To determine the most relevant complex mode frequencies $\Omega(N)$

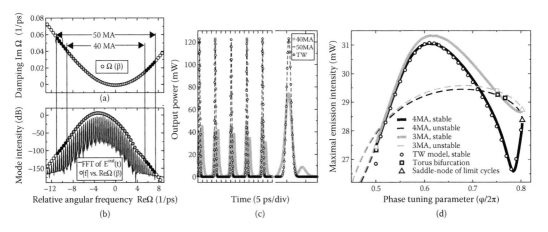

FIGURE 31.11 Calculated (a) complex frequencies Ω, (b) optical spectra as a Fourier-transformed field $E^-(a_1, t)$ (black solid curve) and as a discrete set of mode intensities $|f|^2$ versus $\mathfrak{R}\Omega$ (bullets), and (c) a comparison of the calculated transients of the TW model (bullets) and the reduced ODE systems determined by 40 (solid gray) or 50 (dashed black) optical modes in a two-section mode-locked laser. Full black and small gray bullets in panels (a) and (b) indicate the modes used for the construction of 40MA and 50MA systems. (d) Numerical path following of the stable periodic solution of the TW model (bullets) and periodic orbits of 3MA (gray) and 4MA (black) systems in a three-section phase-controlled mode-beating DFB laser [12,28]. Solid and dashed curves indicate stable and unstable orbits. Empty boxes and triangles denote torus and fold bifurcations, respectively.

(Figure 31.11a) and the coefficients f in the field expansion 31.36 (bullets in panel [b]), we have used the carrier densities $N = (N_1, N_2)$, optical fields E, and polarization functions P obtained as a result of the numerical integration of the TW model. For the construction of the 40MA and 50MA systems, we have used the modes indicated by full black and a bit smaller gray bullets in Figures 31.11a and b. Solid gray and black dashed curves in Figure 31.11c represent the numerical integration of these MA systems. One can see that whereas 40MA system fails to reproduce the stable periodic regime, the 50MA system provides a perfect approximation of the TW model. We note that a significant number of excited optical modes in the example considered above does not allow achieving a low dimensional approximation of the TW model. The number of active modes usually is much smaller in MSLs containing one or more DFB sections. In this case, already three or four appropriately selected optical modes are sufficient for a good approximation of the TW model [7,28].

An integration of the MA system (Equation 31.39) remains a nontrivial task, because for each actual set of carrier densities $N = (N_1, \dots, N_r)$, one should find the corresponding mode frequencies $\Omega_k(N)$, $k = 1, \dots, q$ by solving the characteristic Equation 31.34 numerically. Since the required computation time of the MA systems grows quadratically with the increasing number of modes, one can integrate the TW model faster than the 50MA system. The usefulness of the MA approach starts to be visible when combining our model reduction technique with the numerical continuation and bifurcation analysis tools [29] suited for investigation of nearly arbitrary systems of ODEs. Figure 31.11d presents an example of numerical bifurcation analysis of 3MA (gray) and 4MA (black) systems describing dynamics of the three-section laser consisting of two active DFB sections and a passive phase tuning section in between ($n = 3$ and $n_a = 2$ in this case). Here, solid and dashed curves represent stable and unstable branches of the periodic orbit implied by beating of two closely located resonances, supported by each DFB section. Empty bullets in the same figure represent the continuation of the stable periodic state by direct integration of the TW model. By comparing the bullets and curves, one can see that both MA systems were able to reproduce the stable branch of the periodic orbit, and identify torus and saddle-node bifurcations where this state have lost its stability. The deviation of the solid gray curve from the bullet positions in Figure 31.11d, however, indicates the insufficiency of the 3MA system to reproduce the orbit shape. More detailed analysis of this laser including a continuation of bifurcations in two parameter domain can be found in Reference [28].

31.11 Conclusions

In this chapter, we introduce a hierarchy of TW models describing nonlinear dynamics in individual semiconductor lasers, various MSLs, and coupled laser systems. To simulate these laser devices, we use our software package LDSL-tool, which treats MSLs as a set of differently interconnected laser sections. At the end of the chapter, we introduce several advanced techniques allowing detailed analysis of the model equations. These methods include computation of optical modes, a study of the mode spectra, expansion of electric fields into modal components, a semianalytic location of all steady states of the MSLs, model reduction, numerical continuation, and bifurcation analysis of the reduced system. Altogether, these advanced possibilities of our software tool allow to achieve a thorough understanding of the processes observed both, in the direct integration of model equations and experiments.

References

1. M.W. Fleming and A. Mooradian. Spectral characteristics of external-cavity controlled semiconductor lasers. *IEEE Journal of Quantum Electronics*, 17:44–59, 1981.
2. A. Argyris, D. Syvridis, L. Larger, V. Annovazzi-Lodi, P. Colet, I. Fischer, J. García-Ojalvo, C.R. Mirasso, L. Pesquera, and K.A. Shore. Chaos-based communications at high bit rates using commercial fibre-optic links. *Nature*, 438:343–346, 2005.

3. A. Uchida, K. Amano, M. Inoue, K. Hirano, S. Naito, H. Someya, I. Oowada, T. Kurashige, M. Shiki, S. Yoshimori, K. Yoshimura, and P. Davis. Fast physical random bit generation with chaotic semiconductor lasers. *Nature Photonics*, 2(12):728–732, 2008.

4. E. Murphy. The semiconductor laser: Enabling optical communication. *Nature Photonics*, 4(5):287, 2010.

5. H.-J. Wünsche, O. Brox, M. Radziunas, and F. Henneberger. Excitability of a semiconductor laser by a two-mode homoclinic bifurcation. *Physical Review Letters*, 88(2):023901, 2002.

6. O. Brox, S. Bauer, M. Radziunas, M. Wolfrum, J. Sieber, J. Kreissl, B. Sartorius, and H.-J. Wünsche. High-frequency pulsations in DFB-lasers with amplified feedback. *IEEE Journal of Quantum Electronics*, 39(11):1381–1387, 2003.

7. M. Radziunas, A. Glitzky, U. Bandelow, M. Wolfrum, U. Troppenz, J. Kreissl, and W. Rehbein. Improving the modulation bandwidth in semiconductor lasers by passive feedback. *IEEE Journal of Selected Topics in Quantum Electronics*, 13(1):136–142, 2007.

8. U. Bandelow, M. Radziunas, A. Vladimirov, B. Hüttl, and R. Kaiser. Harmonic mode-locking in monolithic semiconductor lasers: Theory, simulations and experiment. *Optical and Quantum Electronics*, 38:495–512, 2006.

9. R. Arkhipov, A. Pimenov, M. Radziunas, D. Rachinskii, A.G. Vladimirov, D. Arsenijevic, H. Schmeckebier, and D. Bimberg. Hybrid mode locking in semiconductor lasers: Simulations, analysis and experiments. *IEEE Journal of Selected Topics in Quantum Electronics*, 19(4):1100208, 2013.

10. M. Radziunas, A.G. Vladimirov, E.A. Viktorov, G. Fiol, H. Schmeckebier, and D. Bimberg. Pulse broadening in quantum-dot mode-locked semiconductor lasers: Simulation, analysis and experiments. *IEEE Journal of Quantum Electronics*, 47(7):935–943, 2011.

11. M. Möhrle, B. Sartorius, C. Bornholdt, S. Bauer, O. Brox, A. Sigmund, R. Steingrüber, M. Radziunas, and H.-J. Wünsche. Detuned grating multisection-RW-DFB lasers for high-speed optical signal processing. *IEEE Journal of Selected Topics in Quantum Electronics*, 7(2):217–223, 2001.

12. H.-J. Wünsche, M. Radziunas, S. Bauer, O. Brox, and B. Sartorius. Simulation of phase-controlled mode-beating lasers. *IEEE Journal of Selected Topics in Quantum Electronics*, 9(3):857–864, 2003.

13. S. Bauer, O. Brox, J. Kreissl, B. Sartorius, M. Radziunas, J. Sieber, H.-J. Wünsche, and F. Henneberger. Nonlinear dynamics of semiconductor lasers with active optical feedback. *Physical Review E*, 69:016206, 2004.

14. M. Radziunas, K.-H. Hasler, B. Sumpf, T. Quoc Tien, and H. Wenzel. Mode transitions in DBR semiconductor lasers: Experiments, simulations and analysis. *Journal of Physics B: Atomic, Molecular and Optical Physics*, 44:105401, 2011.

15. M. Radziunas, V.Z. Tronciu, E. Luvsandamdin, Ch. Kürbis, A. Wicht, and H. Wenzel. Study of micro-integrated external-cavity diode lasers: Simulations, analysis and experiments. *IEEE Journal of Quantum Electronics*, 51(2):2000408, 2015.

16. O. Hess and T. Kuhn. Spatio-temporal dynamics of semiconductor lasers: Theory, modelling and analysis. *Progress in Quantum Electronics*, 20(2):85–179, 1996.

17. E. Gehrig, O. Hess, and R. Walenstein. Modeling of the performance of high power diode amplifier systems with an optothermal microscopic spatio-temporal theory. *IEEE Journal of Quantum Electronics*, 35:320–331, 2004.

18. R. Lang and K. Kobayashi. External optical feedback effects on semiconductor injection laser properties. *IEEE Journal of Quantum Electronics*, 16:347–355, 1980.

19. M. Sorel, G. Giuliani, A. Scire, R. Miglierina, J.P.R. Laybourn, and S. Donati. Operating regimes of GaAs-AlGaAs semiconductor ring lasers: Experiment and model. *IEEE Journal of Quantum Electronics*, 39:1187–1195, 2003.

20. A.G. Vladimirov and D. Turaev. Model for passive mode-locking in semiconductor lasers. *Physical Review A*, 72:033808, 2005.

21. U. Bandelow, M. Radziunas, J. Sieber, and M. Wolfrum. Impact of gain dispersion on the spatio-temporal dynamics of multisection lasers. *IEEE Journal of Quantum Electronics*, 37:183–188, 2001.

22. J. Javaloyes and S. Balle. Emission directionality of semiconductor ring lasers: A traveling-wave description. *IEEE Journal of Quantum Electronics*, 45:431–438, 2009.

23. M. Radziunas. Longitudinal modes of multisection edge-emitting and ring semiconductor lasers. *Optical and Quantum Electronics*, 47(6):1319–1325, 2015.

24. U. Bandelow, R. Hünlich, and T. Koprucki. Simulation of static and dynamic proper-ties of edge-emitting multiple-quantum-well lasers. *IEEE Journal of Selected Topics in Quantum Electronics*, 9(3):798–806, 2003.

25. LDSL-tool: A software package for simulation and analysis of longitudinal dynamics of multisection semiconductor lasers. Available at www.wias-berlin.de/software/ldsl.

26. M. Radziunas and H.-J. Wünsche. Multisection lasers: Longitudinal modes and their dynamics. In J. Piprek, editor, *Optoelectronic Devices—Advanced Simulation and Analysis*, chapter 5, pp. 121–150. New York, NY: Springer, 2005.

27. M. Radziunas, H.-J. Wünsche, B. Krauskopf, and M. Wolfrum. External cavity modes in Lang-Kobayashi and traveling wave models. In *SPIE Proceeding Series*, 6184:61840X, 2006.

28. M. Radziunas. Numerical bifurcation analysis of the traveling wave model of multisection semicon-ductor lasers. *Physica D*, 213:98–112, 2006.

29. E.J. Doedel, A.R. Champneys, T.F. Fairgrieve, Y.A. Kuznetsov, B. Sandstede, and X. Wang. AUTO97: Continuation and bifurcation software for ordinary differential equations (with HomCont). Tech-nical Report TW-330, Department of Computer Science, K.U. Leuven, Leuven, Belgium, 2001.

30. S. Bischoff, J. Mørk, T. Franck, S.D. Brorson, M. Hofmann, K. Frojdh, and L. Prip. Monolithic col-liding pulse mode-locked semiconductor lasers. *Quantum and Semiclassical Optics*, 9(5):655–674, 1997.

31. M. Kolesik and J. Moloney. A spatial digital filter method for broad-band simulation of semicon-ductor lasers. *IEEE Journal of Quantum Electronics*, 37(7):936–944, 2001.

32. A.J. Lowery. New dynamic semiconductor laser model based on the transmission-line modelling method. *IEE Proceedings J – Optoelectronics*, 134(5):281–289, 1987.

33. D.J. Jones, L. Zhang, J. Carroll, and D. Marcenac. Dynamics of monolithic passively mode-locked semiconductor lasers. *IEEE Journal of Quantum Electronics*, 31(6):1051–1058, 1995.

34. E.A. Avrutin, J.H. Marsh, and E.L. Portnoi. Monolitic and multi-gigaHertz mode-locked semicon-ductor lasers: Constructions, experiments, models and applications. *IEE Proceedings – Optoelectron-ics*, 147(4):251–278, 2000.

35. C.Z. Ning, R.A. Indik, and J.V. Moloney. Effective Bloch equations for semiconductor lasers and amplifiers. *IEEE Journal of Quantum Electronics*, 33(9):1543–1550, 1997.

36. M. Homar, S. Balle, and M. San Miguel. Mode competition in a Fabry-Perot semi-conductor laser: Travelling wave model with asymmetric dynamical gain. *Optics Communications*, 131:380–390, 1996.

37. A. Perez-Serrano, J. Javaloyes, and S. Balle. Longitudinal mode multistability in ring and Fabry-Perot lasers: The effect of spatial hole burning. *Optics Express*, 19:3284–3289, 2011.

38. M. Spreemann, M. Lichtner, M. Radziunas, U. Bandelow, and H. Wenzel. Measurement and simula-tion of distributed-feedback tapered master-oscillators power-amplifiers. *IEEE Journal of Quantum Electronics*, 45(6):609–616, 2009.

39. B. Grote, E.K. Heller, R. Scarmozzino, J. Hader, J.V. Moloney, and S.W. Koch. Fabry-Perot lasers: Temperature and many-body effects. In J. Piprek, editor, *Optoelectronic Devices—Advanced Simula-tion and Analysis*, pp. 27–61. New York, NY: Springer, 2005.

40. H.-J. Wünsche, U. Bandelow, and H. Wenzel. Calculation of combined lateral and longitudinal spa-tial hole burning in $\lambda/4$ shifted DFB lasers. *IEEE Journal of Quantum Electronics*, 29(6):1751–1760, 1993.

41. U. Bandelow, H. Wenzel, and H.-J. Wünsche. Influence of inhomogeneous injection on side-mode suppression in strongly coupled DFB semiconductor lasers. *Electronics Letters*, 28:1324–1325, 1992.

42. P.G. Eliseev, A.G. Glebov, and M. Osinski. Current self-distribution effect in diode lasers: Analytic criterion and numerical study. *IEEE Journal of Selected Topics in Quantum Electronics*, 3:499–506, 1997.

43. R. Schatz. Longitudinal spatial instability in symmetric semiconductor lasers due to spatial hole burning. *IEEE Journal of Quantum Electronics*, 28(6):1443–1449, 1992.

44. V. Tronciu, S. Schwertfeger, M. Radziunas, A. Klehr, U. Bandelow, and H. Wenzel. Amplifications of picosecond laser pulses in tapered semiconductor amplifiers: Numerical simulations versus experiments. *Optics Communications*, 285:2897–2904, 2012.

45. A. Perez-Serrano, J. Javaloyes, and S. Balle. Bichromatic emission and multimode dynamics in bidirectional ring lasers. *Physical Review A*, 81:043817, 2010.

46. E.J. D'Angelo, E. Izaguirre, G.B. Mindlin, L. Gil, and J.R. Tredicce. Spatiotemporal dynamics of lasers in the presence of an imperfect O(2) symmetry. *Physical Review Letters*, 68:3702–3704, 1992.

47. Y. Barbarin, E.A.J.M. Bente, M.J.R. Heck, Y.S. Oei, R. Nötzel, and M.K. Smit. Characterization of a 15 GHz integrated bulk InGaAsP passively modelocked ring laser at 1.53 μm. *Optics Express*, 14(21):9716–9727, 2006.

48. E.A. Viktorov, T. Erneux, P. Mandel, T. Piwonski, G. Madden, J. Pulka, G. Huyet, and J. Houlihan. Recovery time scales in a reversed-biased quantum dot absorber. *Applied Physics Letters*, 94:263502, 2009.

49. M. Radziunas, A.G. Vladimirov, E.A. Viktorov, G. Fiol, H. Schmeckebier, and D. Bimberg. Strong pulse asymmetry in quantum-dot mode-locked semiconductor lasers. *Applied Physics Letters*, 98:031104, 2011.

50. K. Lüdge and E. Schöll. Quantum-dot lasers—desynchronized nonlinear dynamics of electrons and holes. *IEEE Journal of Quantum Electronics*, 45(11):1396–1403, 2009.

51. A. Markus, M. Rossetti, V. Calligari, D. Chek-Al-Kar, J.X. Chen, A. Fiore, and R. Scollo. Two-state switching and dynamics in quantum dot two-section lasers. *Journal of Applied Physics*, 100:113104, 2006.

52. M. Rossetti, P. Bardella, and I. Montrosset. Time-domain travelling-wave model for quantum dot passively mode-locked lasers. *IEEE Journal of Quantum Electronics*, 47(2):139–150, 2011.

53. V.Z. Tronciu, M. Radziunas, Ch. Kürbis, H. Wenzel, and A. Wicht. Numerical and experimental investigations of micro-integrated external cavity diode lasers. *Optical and Quantum Electronics*, 47(6):1459–1464, 2015.

54. T. Krauss and P.J.R. Laybourn. Very low threshold current operation of semiconductor ring lasers. *IEE Proceedings J – Optoelectronics*, 139(6):383–388, 1992.

55. I.V. Ermakov, S. Beri, M. Ashour, J. Danckaert, B. Docter, J. Bolk, X.J.M. Leijtens, and G. Verschaffelt. Semiconductor ring laser with on-chip filtered optical feedback for discrete wavelength tuning. *IEEE Journal of Quantum Electronics*, 48(2):129–136, 2012.

56. M. Radziunas, V.Z. Tronciu, U. Bandelow, M. Lichtner, M. Spreemann, and H. Wenzel. Mode transitions in distributed-feedback tapered master-oscillator power-amplifier: Theory and experiments. *Optical and Quantum Electronics*, 40:1103–1109, 2008.

Mode-Locked Semiconductor Lasers

Eugene Avrutin

and

Julien Javaloyes

32.1 General Principles of Mode Locking, the Important Features of Mode-Locked Semiconductor Lasers, and the Role of Theory and Modeling

In the most general sense, *mode locking* (ML) is a regime of laser operation that involves emitting light in several modes with a time-independent relation between them, i.e., with constant and precisely equidistant frequencies. Usually, the term is used more specifically, referring to what is, rigorously speaking, *amplitude-modulation* (AM) ML, meaning that the phase differences between of adjacent modes are approximately equal. In time domain, this corresponds to the laser's emitting a train of ultrashort optical pulses at a repetition frequency F near the cavity round-trip frequency or its harmonic:

$$F \approx M_{\mathrm{h}} v_{\mathrm{g}}/(2L). \tag{32.1}$$

Here, v_g denotes the group velocity of light in the laser resonator and L is the Fabry–Pérot resonator cavity length. In the case of the ring resonator, $2L$ in Equation 32.1 is substituted by the ring cavity length. The harmonic number M_h corresponds to the number of pulses coexisting in the cavity; in the simplest and most usual case, $M_h = 1$. The pulse duration is then of the order of $2L/(N_M v_g)$, with N_M being the number of lasing modes in the spectrum.

In most cases (some important exceptions will be mentioned below), ML does not occur spontaneously and requires a special laser construction and/or operating conditions. Namely, it is usually achieved either by modulation of the laser net gain at a frequency F or its (sub)harmonic (known as *active* ML) or by exploiting nonlinear properties of the medium to shorten the propagating pulse (known as *passive* ML, PML); in both cases, the pulse shortening mechanism needs to be strong enough to counter the broadening effects of gain saturation and dispersion effectively. PML, in turn, is usually achieved by introducing a saturable absorber (SA) into the laser cavity. The SA both facilitates a self-starting mechanism for ML and, most importantly, plays a crucial role in shortening the duration of the circulating pulses. As a variation of this principle, *refractive index nonlinearities* approximately equivalent in their action to saturable absorption have been intensely studied in the last decades; salient examples are *additive pulse ML* and *Kerr lens ML* in solid-state lasers, see, e.g., Haus (2000) for more detail.

A combination of active and passive methods of ML is known as *hybrid* ML; if the external modulation is in the form of short pulses, the corresponding regime is referred to as *synchronous* ML.

Mode-locked solid-state lasers, often diode pumped, have allowed sub-100 femtosecond pulses to be generated (Brown et al., 2004; Ell et al., 2001; Innerhofer et al., 2003), with peak powers in the range of many kilowatts (partly due to the relatively low repetition rates, typically in the megahertz range or below).

In *semiconductor diode lasers*, the most basic physical mechanisms underlying the generation of short pulses are fundamentally similar to those of other types of lasers, but a number of features are very different, as regards both technology and physics.

From the practical and technological point of view, diode lasers have a number of advantages: They represent the most compact and efficient sources of picosecond and subpicosecond pulses. They are directly electrically pumped, and the bias current can be easily adjusted to determine the pulse duration and the optical power, thus offering, to some extent, electrical control of the characteristics of the output pulses. These lasers also offer the best option for the generation of high-repetition rate trains of pulses, owing to their small cavity size L in Equation 32.1 and hence the large values of F, well into multigigahertz range. Being much cheaper to fabricate and operate than most other types of lasers, ultrafast semiconductor lasers also offer the potential for dramatic cost savings in a number of applications that traditionally use solid-state lasers. The deployment of high-performance ultrafast diode lasers could therefore have a significant economic impact by enabling ultrafast applications to become more profitable and even facilitate the emergence of new applications. At the moment, actual and potential applications of mode-locked lasers include time- and wavelength-multiplexed communications, metrology, biomedical applications, etc.; see Avrutin and Rafailov (2012) for an overview.

From the point of view of physics, which underlies the technology, most of the distinct features of semiconductor lasers, including mode-locked ones, lie in the energy spectrum of semiconductors that consists in continuous bands of energy with relatively high density of states, as opposed to discrete levels in solid-state lasers. Most mode-locked semiconductor lasers operate on fundamental interband transitions (though there has been some work on active ML of intersubband quantum cascade lasers; see, e.g., Revin et al., 2016 and references therein). Semiconductors thus have both a higher gain per unit length (which is one of the reasons of the short cavity length being possible) and a higher nonlinear refractive index than other gain media (the relation between carrier density dependences of gain and refractive index in semiconductor lasers is often quantified via Henry's linewidth enhancement factor α_H that, as discussed below, plays an important role in the theory of ML). The interaction of the pulse with the gain and saturation absorption and the resulting large changes in the nonlinear refractive index lead to significant self-phase modulation, imparting a noticeable *chirp* to the ML pulses, usually up-chirp in the case of passively mode-locked lasers and down-chirp in actively mode-locked ones. This increases the *time-bandwidth product* of the pulse and has been one of the important limitations in obtaining pulse durations of the order of 100 fs *directly* from the diode lasers, with picosecond pulses being the norm. Furthermore, a strong saturation of the gain also results in stabilization of the pulse energy, which limits the average and peak power to substantially lower levels than in vibronic/solid-state lasers. Output average power levels for mode-locked laser diodes are usually between 0.1 and 100 mW, while peak power levels remain between 10 mW and 1 W. Only with additional amplification/compression setups, can the peak power reach the kW level (Kim et al., 2005). Another distinction of semiconductor lasers is that the typical scales of carrier recombination times in semiconductor materials are of the order of hundreds of picoseconds, comparable to the ML repetition time, leading to a rich variety of dynamic instabilities in the laser behavior, some of which, as will be discussed later, determine the *lower* limit for stable ML frequency (and therefore the longest possible laser cavity) for given material parameters.

There has been a large variety of semiconductor laser designs used for ML, from external cavity ones operating at sub-GHz rates to monolithic ones reaching terahertz repetition frequencies (Avrutin et al., 2000), with laser design strongly affecting, not just the laser performance, but the relative importance of the underlying physical effects in determining this performance.

This combination of practical promise, versatility, and scientific challenge has made mode-locked semiconductor lasers an important topic of research for more than two decades; they have arguably attracted considerably more attention than all other methods of ultrashort pulse generation taken together. The most recent years have seen considerable progress in both improving the theoretical understanding of ML in semiconductor lasers and using this understanding to improve their performance in terms of power, pulse duration/chirp, stability, repetition rates accessible, and integrability issues.

This progress has been partly summarized in previous reviews on ML in semiconductor lasers, with some of them (Avrutin et al., 2005) explicitly concentrating on modeling and simulation and others (Avrutin et al., 2000; Avrutin and Rafailov, 2012) paying significant attention to it. Here, we shall partly follow the logic and layout of the previous paper (Avrutin and Rafailov, 2012), but will attempt to present a more modern perspective and cover the recent results by ourselves and other researchers.

32.2 ML Techniques in Laser Diodes: The Main Features

The main advantages of ML over other methods of generating ultrafast pulses by laser diodes are the higher repetition rate pulses and shorter pulse durations. To realize these advantages to the fullest, a variety of ML techniques and device structures have been investigated and optimized (Vasil'ev, 1995). All three main forms of ML—active, passive, and hybrid—have been extensively studied for semiconductor lasers.

Purely *active* ML in a semiconductor laser can be achieved by direct modulation of the gain section current with a frequency very close to the pulse repetition frequency in the cavity or to a subharmonic of this frequency. Alternatively, an electroabsorption segment of a multielement device can be modulated to produce the same effect, or a separate modulation section introduced. The main advantages of active ML techniques are the resultant low jitter (essentially determined by the electrical generator imposing the modulation) and the ability to synchronize the laser output with the modulating electrical signal, which is a fundamental attribute for optical transmission and signal processing applications. However, high repetition frequencies are not readily obtained through directly driven modulation of lasers because fast microwave modulation, particularly of current, becomes progressively more difficult with increase in frequency.

PML of semiconductor lasers typically utilizes an SA region in the laser diode. Upon start-up of laser emission, the laser modes initially oscillate with relative phases that are random; in other words, the temporal radiation pattern consists of irregular bursts. If one of these bursts is energetic enough to provide energy, or more accurately fluence (energy per unit area) of the order of the saturation fluence of the absorber, it will partly bleach the absorption. This means that around the peak of the burst where the intensity is higher, the loss will be smaller, while the low-intensity wings become more attenuated. The pulse generation process is thus initiated by this family of intensity spikes that experience lower losses within the absorber carrier lifetime. The dynamics of absorption and gain play a crucial role in pulse shaping. In steady state, the unsaturated losses are higher than the gain. When the leading edge of the pulse reaches the absorber, the loss saturates more quickly than the gain, which results in a net gain window, as depicted in Figure 32.1. The absorber then recovers from this state of saturation to the initial state of high loss, thus attenuating the trailing edge of the pulse. It is thus easy to understand why the saturation fluence and the recovery time of the absorber are of primary importance in the formation of mode-locked pulses.

In practical terms, the SA can be monolithically integrated into a semiconductor laser by electrically isolating one section of the device (Figure 32.2a). By applying a reverse bias to this section, the carriers photogenerated by the pulses can be more efficiently swept out of the absorber, thus enabling the SA to recover more quickly to its initial state of high loss. An increase in the reverse bias serves to decrease the

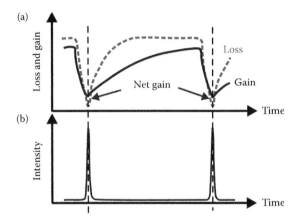

FIGURE 32.1 Schematic illustration of the mechanism of passive mode locking: (a) the loss and gain dynamics that lead to (b) pulse generation.

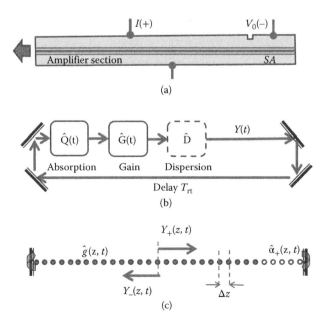

FIGURE 32.2 Schematic of (a) the simplest design of an edge-emitting passively mode locked laser, and its representation in (b) the lumped model of Sections 32.3.1 and 32.3.3, and (c) the traveling-wave model (TWM) of Section 32.3.4.

absorber recovery time, which will have the effect of further shortening the pulses (a faster absorber may act to shorten the trailing as well as the leading edge of the pulse). Alternatively, an SA can also be implemented through ion implantation on one of the facets of the laser, thus increasing the nonradiative recombination (Delpon et al., 1998).

PML provides the shortest pulses achievable by all three techniques, albeit at the expense of somewhat larger pulse jitter and radio frequency (RF) linewidth than in active or hybrid ML. It can be intuitively understood in the following way: Active ML generates a fixed AM of the gain while PML induces a modulation that is proportional to the pulse energy, meaning potentially more efficient mode coupling. Besides, the absence of an RF source simplifies the fabrication and operation considerably. PML also allows for higher pulse repetition rates than those determined solely by the cavity length, by means of harmonic ML ($M_h > 1$ in Equation 32.1; some means of achieving this are considered in more detail below).

Hybrid ML can be achieved by applying RF modulation either to gain or to the SA section. It has been shown, however, that the more efficient method is the latter one, in which case the SA doubles as an electroabsorption modulator. In this case, the pulse generation may be seen as initiated by a modulation provided by the RF signal, while further shaping and shortening are assisted by the SA. This process results in high-quality pulses, synchronized with an external source.

32.3 Theoretical Models of ML in Semiconductor Lasers

Any model of mode-locked laser dynamics should account for pulse shortening by modulation (active/hybrid ML) and/or saturable absorption (passive/hybrid ML) and for pulse broadening by saturable gain and cavity dispersion (including gain/loss dispersion and group velocity/phase dispersion), as discussed above. In addition, if spectral properties are to be accounted for accurately, self-phase modulation needs to be included in the model. In this section, we shall cover general principles of the possible approaches, concentrating on the relatively recent advances in ML theory that have underpinned the significant progress in understanding the details of ML dynamics.

32.3.1 Small-Signal Time-Domain Models and Self-Consistent Pulse Profile

Conceptually the simplest, and historically the oldest, models of mode-locked lasers are *time-domain lumped models* (Figure 32.2b), based on the approximation that the pulsewidth is much smaller than the repetition period, and treating a hypothetic ring laser with unidirectional propagation. The amplification and gain/group velocity dispersion (GVD), which in reality are experienced by the pulse simultaneously, may then be approximately treated in two independent stages. This allows the representation of the distributed amplifier in the model by a lumped *gain element* performing the functions of amplification and self-phase modulation. Mathematically, this element can be described by a nonlinear integral or integro-differential operator acting on the complex pulse shape function (complex slow amplitude) $Y(t)$, t being the *local* time of the pulse. The model was originally designed for solid-state and gas lasers, whose long lengths make for a round-trip time many orders of magnitude longer than the pulse duration, so separate timescales are introduced explicitly for the pulse (the short timescale) and relaxation period between pulses (the long timescale).

The gain operator takes the form

$$\hat{G}Y(t) = \exp\left(\frac{1}{2}(1 - i\alpha_{Hg})G(t)\right) Y(t), \tag{32.2}$$

with α_{Hg} the Henry linewidth enhancement factor in the amplifier and $G(t) = \Gamma \int g(z - v_g t, t)dz$ the total gain integrated over the length of the amplifying region (v_g being the group velocity of light). Further analytical progress can be made by using spatially resolved rate equations for the carrier density N_g in the gain region $\dfrac{\mathrm{d}N_g(z,t)}{\mathrm{d}t} = -g(N_g)\dfrac{P(z,t)}{\hbar\omega A_x} - \dfrac{N_g}{\tau_g}$, where P is the power of light, $\hbar\omega$ the photon energy, A_x the cross section of the optical beam (mode) in the gain section, and τ_g the gain recovery time. A number of approximations are then made. Those involve neglecting dispersion and fast nonlinearities, assuming (which is a safe assumption in semiconductor active media) that $\tau_g \gg \tau_p(\tau_p$ being the pulse duration), so gain relaxation during the pulse can be neglected, and assuming a linear dependence of gain on the carrier density: $g(N_g) \approx \sigma_g(N_g - N_{tr})$, $\sigma_g = \mathrm{d}g/\mathrm{d}N$ being the gain cross section (the derivative of the gain on the carrier density) and N_{tr} the transparency carrier density. With these assumptions, an approximate explicit expression for $G(t)$ on the short timescale commensurate with the pulse duration can be obtained. In the case of $G \ll 1$, it takes the form

$$G(t) = G_- \exp(-U(t)/U_g). \tag{32.3}$$

Here, $U(t) = \int_{-\infty}^{t} P(t')\mathrm{d}t' = v_g\hbar\omega A_X \int_{-\infty}^{t} |Y(t')|^2\mathrm{d}t'$ is the pulse energy up to the time t, $G_- = \Gamma \int g(z, t \to -\infty)\mathrm{d}z$ is the total amplification in the gain element at the time before the arrival of the pulse, and

$$U_g = \dfrac{\hbar\omega A_X}{\sigma_g} \tag{32.4}$$

is the saturation energy of the amplifier.

The SA, if any, is also considered as a lumped element, described, in the simplest case, by an operator similar to Equation 32.3:

$$\hat{Q}_S Y(t) = \exp\left(-\dfrac{1}{2}(1 - i\alpha_{H\alpha})Q_S(t)\right) Y(t), \tag{32.5}$$

with the absorber linewidth enhancement factor $\alpha_{H\alpha}$, and the dimensionless slow saturable absorption $Q_S = \Gamma \int \alpha(z - v_g t, t)\mathrm{d}z$, in the ideal slow absorber approximation and with $Q \ll 1$, given by

$$Q_S(t) = Q_- \exp(-U(t)/U_\alpha). \tag{32.6}$$

Here, again, the total initial absorption $Q_- = \Gamma \int \alpha(z, t \to -\infty)\mathrm{d}z$, and the absorber saturation energy is

$$U_\alpha = \dfrac{\hbar\omega A_{X\alpha}}{\sigma_\alpha}, \tag{32.7}$$

where $\sigma_\alpha = |\mathrm{d}\alpha/\mathrm{d}N|$ is the SA cross section, and, depending on the construction, the cross section of the beam $A_{X\alpha}$ in the SA may be different from that of the amplifying section; as shown below, it is usually advantageous to have $A_{X\alpha} < A_X$.

Equation 32.6 is obtained in the same way as Equation 32.3 by using the spatially resolved rate equation for the carrier density in the absorber region $\dfrac{\mathrm{d}N_\alpha(z,t)}{\mathrm{d}t} = \alpha(N_\alpha)\dfrac{P(z,t)}{\hbar\omega A_{X\alpha}} - \dfrac{N_\alpha}{\tau_\alpha}$, with τ_α the absorber relaxation time and, and making the same assumptions: assuming a linear dependence of the absorption on the SA carrier density $\alpha(N_\alpha) \approx \alpha_0 - \sigma_\alpha N_\alpha$, with α_0 the unsaturated absorption, and neglecting the SA relaxation during the pulse. The latter is known in the theory of ML as *the ideal slow absorber* approximation, meaning that the SA recovery time τ_α, like that of gain τ_g, needs to be much longer than the pulse duration τ_p. However, while the assumption $\tau_g \gg \tau_p$ is readily fulfilled in semiconductor lasers ($\tau_g \sim 1$ ns), the assumption $\tau_\alpha \gg \tau_p$ may be strained ($\tau_\alpha \sim 10$ ps), which may necessitate some modifications to the model, described below, to improve its accuracy.

Absorption in semiconductor SAs tends, in addition, to contain a subpicosecond component that acts as a "fast" SA (recovery time $\ll \tau_p$) even for short pulses ($\tau_p \sim 1$ ps), typically generated by semiconductor lasers. Some lumped time-domain models (Haus and Silberberg, 1985) also include "fast" effects in the SA (SA nonlinearities) as an equivalent fast absorber characterized by an operator $\hat{Q}_F Y(t) = \exp\left(-\frac{1}{2}Q_F(t)\right) Y(t)$ and with an equivalent absorption:

$$Q_F(t) = Q_{iF}(1 - \varepsilon_\alpha |Y(t)|^2). \tag{32.8}$$

Then, the total absorption is

$$\hat{Q}Y(t) = \hat{Q}_S \hat{Q}_F Y(t). \tag{32.9}$$

Gain nonlinearities may, in principle, be included in the same way, although a more accurate account of dynamics may be preferable, particularly in the case of quantum dot (QD) materials.

Finally, in the traditional form of a lumped model, the dispersion of material gain and refractive index, together with any artificial dispersive elements present in the cavity, such as a distributed Bragg reflector (DBR), are combined in a lumped *dispersive element*. In the frequency domain, its effect on the pulse may be written as

$$\hat{D}Y^T(\omega) = e^{i\varphi_0} \left[\frac{1}{1 - i(\omega - \omega_p)/\gamma} + D(\omega - \omega_0)^2 \right] Y^T, \tag{32.10}$$

where Y^T is the Fourier transform of the complex pulse shape $Y(t)$, ω_p and $\gamma \ll \omega_p$ are the peak frequency and the bandwidth of the dispersive element (defined by the gain curve of the amplifier and the frequency selectivity of a grating element, if it is present in the cavity), and ω_0 is the reference frequency as in the analysis of amplifiers. The value of ω_p may change during the pulse (due to gain curve variation with carrier density, most importantly the gain peak shift); this modifies the dispersive operator (Leegwater, 1996), although in the majority of papers on the subject, the effect is not included. ϕ_0 denotes the phase shift introduced by the element and D is the equivalent dispersion (including the GVD of the passive waveguide and the effective dispersion of the external grating element, if any). To rewrite the operator (Equation 32.10) in the time domain, one may expand the first term around the reference frequency ω_0 noting that $|\omega_p - \omega_0| \ll \omega_0$. Then, after a standard transformation, $(\omega - \omega_0)Y^T \div id/dt\, Y$ (Equation 3.7) becomes a differential operator; if the exponential is expanded keeping the first two terms, the operator is reduced to second order.

The dynamics of ML process are then described by cascading the operators and setting:

$$Y_{i+1}(t) = \left(\sqrt{\kappa}\hat{G}\hat{Q}\hat{D} \right) Y_i(t), \tag{32.11}$$

where i is the number of the pulse round-trip (determining the "slow" evolution of the ML pulse), the time t is on the fast timescale commensurate with the pulse duration, and the dimensionless parameter $\kappa < 1$ introduces the total (integrated) unsaturable intensity losses in the cavity, both distributed and due to outcoupling. The model reflects the balance of the main processes affecting the pulse in a mode-locked laser in that the saturable absorption operator $\hat{Q}$ acts to narrow the pulse down, whereas the gain saturation $\hat{G}$ and the dispersion operator $\hat{D}$ act to broaden it. The *stationary* ML equation is thus obtained by writing out the condition that the broadening and narrowing cancel each other, and the shape of the pulse is conserved from one repetition period to the next. In the operator notation introduced above, this means

$$\left(\sqrt{\kappa}\hat{G}\hat{Q}\hat{D} \right) Y(t) = e^{i\delta\psi} Y(t + \delta T), \tag{32.12}$$

where δT is the shift of the pulse or detuning between the repetition period and the round-trip of the "cold" cavity (or its fraction in case of locking at harmonics of the fundamental frequency), and $\delta\psi$ is the optical phase shift induced by the round-trip. In between the pulses, on the slow timescale commensurate with the round-trip time, gain and SA are allowed to recover with their characteristic relaxation times, according to the rate equation for carrier density with $S = 0$. This allows one to calculate the values of gain and saturable absorption at the onset of the pulse, given the pulse energy and repetition period (the only point at which this latter parameter enters a lumped time-domain model).

In the approximation of no dispersion ($\hat{D} = 1$), the broadening of the pulse by gain saturation alone in the lumped model cannot compensate for the shortening by the absorption. The model in this approximation thus predicts the steady output in the form of a series of *infinitely short* (delta function-like) pulses; neither the pulse shape nor the duration can be analyzed in this approximation. However, it is possible to determine the total pulse *energy* and also analyze the stability of the solutions by requiring that net gain both immediately before the pulse and immediately after the pulse is smaller than one that translates into

$$G_- - Q_- - \ln\kappa < 0$$
$$G_+ - Q_+ - \ln\kappa < 0. \tag{32.13}$$

Here G_-, Q_- are the total (integrated) gain and absorption immediately before the pulse, and G_+, Q_+ are the values immediately after the pulse. This is known as New's theory of ML (strictly speaking, G. New's original 1970s paper (New, 1974) related to nonsemiconductor lasers in which $\ln(1/\kappa) \ll 1$; however, Equation 32.13 is also applicable in the generalized version of the theory proposed by Vladimirov et al. (2004) and Vladimirov and Turaev (2005) and covered in more detail below).

Analytical approximations for the *pulse shape and duration* have been originally obtained in the case of weakly nonlinear analysis, that is to say if the pulse energy is smaller than $U_{G,A}$ and the gain and loss (saturable and unsaturable) during one round-trip are small ($\ln(1/\kappa), G, Q \ll 1$). Then, the exponentials in the formulas for the gain and loss operators may be expanded in Taylor series keeping terms up to the second order in Equations 32.3 and 32.6 (weak to moderate saturation of gain during the pulse):

$$\exp\left(-\frac{U(t)}{U_{g,\alpha}}\right) \approx 1 - \frac{U(t)}{U_{g,\alpha}} + \frac{1}{2}\left(\frac{U(t)}{U_{g,\alpha}}\right)^2, \tag{32.14}$$

and to the first order in Equations 32.2 and 32.5 (small gain and loss):

$$\exp\left(\frac{1}{2}(1 - i\alpha_{Hg})G(t)\right) \approx 1 + \frac{1}{2}(1 - i\alpha_{Hg})G(t); \exp\left(-\frac{1}{2}(1 - i\alpha_{H\alpha})Q(t)\right) \approx 1 - \frac{1}{2}(1 - i\alpha_{H\alpha})Q(t) \tag{32.15}$$

(the accuracy of the model can be improved by expanding these equations, too, to the second rather than first order).

Then, following the route pioneered by H. Haus in the first papers on ML in lasers of an arbitrary type (Haus, 1975) and later adapted specifically to diode lasers (Koumans and vanRoijen, 1996; Leegwater, 1996), the ML Equation 32.12 is rewritten as a complex second-order integro-differential equation known as the master equation of ML, which permits an analytical solution of the form

$$Y(t) = Y_0 \exp(i\Delta\omega t)\left(\cosh\frac{t}{\tau_p}\right)^{-1+i\beta}, \tag{32.16}$$

known as the *self-consistent profile* (SCP). The corresponding theoretical approach is known as the SCP, or Haus's ML theory, as applied to semiconductor lasers (in lasers of other kind, for instance, the account for both slow and fast absorbers is typically not necessary). Assembling the terms proportional to the zeroth, first, and second power of $\tanh(t/\tau_p)$ in the ML equation, one obtains three complex, or six real, transcendental algebraic equations (Koumans and vanRoijen, 1996; Leegwater, 1996) for six real variables: pulse amplitude $|Y_0|$, duration measure τ_p, chirp parameter β, optical frequency shift $\Delta\omega = \omega - \omega_0$, repetition period detuning δT, and phase shift arg (Y_0) (which is not a measurable parameter, so in reality there are five meaningful equations). These equations, being nonlinear and transcendental, generally speaking, cannot be solved analytically, but still allow for some insight into the interrelation of pulse parameters. For example, it can be deduced (Leegwater, 1996) that the pulse duration may be considerably shortened by the presence of a fast (instantaneous) component in the saturable absorption and the achievable pulse durations are estimated about 10 times the inverse gain bandwidth, decreasing with increased pulse energy.

By requiring the net small-signal gain before and after the pulse to be negative (Equation 32.13) so that noise oscillations are not amplified, the SCP approach also allows the parameter range of the stable ML regime to be estimated.

Some conclusions from the SCP approach are borne out by more precise models (see below). In particular, it highlights the role of *the gain-to-absorber saturation energy ratio*:

$$s = \frac{U_g}{U_\alpha} = \frac{\sigma_\alpha A_{Xg}}{\sigma_g A_{X\alpha}} \tag{32.17}$$

in the ML laser performance. A minimum value of $s > 1$ is needed to achieve ML at any range of parameters at all and the range of stable ML operation broadens with an increased s. Colliding pulse mode-locked configurations, linear or ring, increase the pulse stability and also lead to shorter pulses by increasing the parameter s.

The SCP model also predicts, correctly, that increasing the dispersion parameter D also increases the parameter range for ML, at the expense of broadening the pulses, and that the slight variation of the frequency F around the estimate (Equation 32.1) shows a minimum in its dependence on current or unsaturated gain.

When applied more quantitatively, however, the SCP model is not too accurate and cannot adequately describe details of pulse shape and spectral features. Indeed, the pulse shape given by the expression (Equation 32.16) is always symmetric, which, in general, needs not, and often is not, the case in practice. Nor are the dynamic regimes of ML faithfully reproduced by the classic SCP model. The reason for this is a large number of approximations involved in the SCP approach, which have been progressively removed by various researchers at the expense of making the model more complex and, in some cases, requiring numerical rather than semianalytical analysis of the pulse profile, even if the model is still lumped.

First, achieving the SCP requires that the relaxation of gain and absorber during the pulse is negligible so that the gain and absorber operators can be written in the form of Equations 32.3 and 32.6. As mentioned above, this is a safe assumption in semiconductor lasers as regards gain media, but not necessarily the SA. The obvious upgrading to the model is then to include the dynamics of the saturable absorption Q by a characteristic recovery time τ_α. If at the same time we abandon the approximation $Q \ll 1$, then the necessary equation will take the form

$$\frac{dQ(t)}{dt} = -X(Q)Q\frac{P(t)}{U_\alpha} - \frac{Q_0 - Q}{\tau_\alpha};$$

$$X(Q) = \frac{1 - \exp(-Q)}{Q}. \tag{32.18}$$

Here, $Q_0 = \alpha_0 - L_\alpha$ (L_α being the absorption region length) is the unsaturated total absorption (at repetition periods $T_{rep} \gg \tau_\alpha$, $Q_- = Q_0$), and X is the geometric factor that stems from averaging the absorption over the length of the absorber area for traveling-wave absorption (in the case of small absorption ($Q_0, Q \ll 1$) treated above, we obtain a constant $X = 1$).

Equation 32.18 is then used with Equation 32.5 instead of Equation 32.6, which it obviously reproduces in the limiting case of $X = 1$, and $\tau_\alpha \gg \tau_p$, or $\tau_\alpha \to \infty$ on the short timescale $t \sim \tau_p$. Unfortunately, even this apparently minor modification to the model means that a closed-form solution in the form of Equation 32.16 is no more possible even with $Q_0, Q \ll 1 (X = 1)$, and the iteration-type procedure (Equation 32.11) has to be repeated numerically until a steady-state profile that satisfies Equation 32.12 is found.

Studies with such a modified SCP model found that even with $X = 1$ in Equation 32.18 (small gain/absorption case) and even with absorber recovery times a few times greater than the pulse duration, the finite τ_α makes some difference to the results, noticeably shortening the pulse, making its shape less symmetric, and affecting boundaries of stable ML regime (Dubbeldam et al., 1997).

32.3.2 Frequency and Time-Frequency Treatment of ML and Dynamic Modal Analysis

An approach conceptually alternative to the time-domain analysis of ML, but to an extent sharing the small-signal nature of the model discussed above, is offered by the technique of modal analysis, static or dynamic (as in Avrutin et al., 2003 and references therein and also in Nomura et al., 2002 and Renaudier et al., 2007). In this approach, instead of analyzing the pulse shape dynamics, a modal decomposition is used and the dynamics of mode amplitudes and phases are analyzed. The advantage of the modal expansion is that the time steps can be much longer than in the spatially distributed models of ML (such as the ones described the following section). Indeed, the stiffness of the dynamics in the modal approach is governed by the temporal evolution of the modal amplitudes, and as PML consists in a steady-state regime for those, one foresees that a modal representation of PML may give rise to smooth solutions—this indeed was shown to be the case (Avrutin et al., 2003). In comparison, spatially distributed models naturally need the time step to be much shorter than the pulsewidth, as discussed in more detail in the following section. Besides, the number of variables can be smaller in the modal analysis, particularly in the case of laser designs with a spectrally selective element where only a few modes are excited, making this approach particularly efficient in analyzing, say, long-scale dynamics of external locking of DBR hybridly mode-locked lasers. It also has the logical advantage of describing steady-state ML as a steady-state solution and, conceptually, allows considering the emergence of stable PML as an order–disorder phase transition in a dissipative system, highlighting the fundamental physical features of PML in addition to its technological implications. Frequency domain analysis can be used as supplementary to time-domain models for some specific problems, as for example the analysis of harmonic operation in a coupled-cavity structure (Yanson et al., 2002), where it actually gives some analytical insight into the modal selectivity of the cavity. Frequency, or time-frequency, modal expansion-based approach to ML is also extremely useful (Kim and Lau, 1993) for analyzing the noise and linewidth properties of the ML signal, as the noise can be seen as exciting higher order *supermodes* (combination of modes) in addition to the fundamental order supermode that is actually realized in ML.

However, the modal expansion approach has a major intrinsic limitation in that it relies on the inherent assumption of weak to modest nonlinearity and modulation, meaning that the results obtained using this method agree reasonably well with time-domain simulations only for the case of high ML frequencies, typically above 100 GHz (short or harmonic cavities) and/or at relatively small currents above threshold. Large-signal instabilities, such as the chaotic leading edge instability, are not predicted accurately, and the accuracy of the frequency-domain models at high amplitudes cannot be guaranteed. In addition, finding the modal structures of complex multisection photonic devices can be cumbersome. Therefore, though the frequency and time-domain analysis of ML originally was introduced approximately simultaneously (reflecting the two major representations of ML, the sequence of periodic pulses and a comb of locked modes), in the context of semiconductor lasers the work on frequency-domain models remains limited. Instead, theoretical progress has been mainly associated with the time-domain models, as they permit a large-signal approach (large modulation of population inversion, or alternatively large nonlinearity) that we cover in the next sections.

32.3.3 Large-Signal Time-Domain Approach and Delay-Differential Equation Model

Both the assumption of weak to moderate pulse saturation during the pulse and, even more so, that of small gain and absorption per pass, as used in small-signal time-domain models discussed in Section 32.3.1, may become even more tenuous in semiconductor lasers than the assumption of an ideally slow absorption—in fact, in edge-emitting lasers, the small-gain assumption is almost always completely inapplicable since at least one of the laser facets is usually uncoated (or even AR-coated to reduce the reflectance to 0.05%– 0.1%) to increase the output power, so the outcoupling losses are by necessity significant, making for large gain per pass even with small saturable absorption. Then, it makes sense to abandon the expansions (Equations 32.14 and 32.15) in the fully numerical procedure and use the full exponential form of Equations 32.2 through 32.8, as well as the more accurate full expression for X in Equation 32.18, thus moving from a small-signal SCP model to a *large-signal iterative model* (see, e.g., Khalfin et al., 1995). This also means that the fast nonlinearities of gain and absorption, and possibly part of the dispersion, may be included directly into the gain and absorber operators.

Even in its large-signal form and with the finite absorber (and gain, if necessary) relaxation time taken into account, the iterative procedure (Equation 32.11) is still somewhat artificial in that it requires a trial pulse shape to start with, and explicitly separates the timescale into the short timescale of the pulse and the long timescale of the repetition period. Moreover, if the time window of the pulse is taken as much smaller than the repetition period (which is the standard thing to do if the repetition period is much longer than the pulse), any instabilities related with secondary pulses arising far away from the main pulse may be missed by the model. In semiconductor lasers, neither of these assumptions is well justified, as the pulse may be only about an order of magnitude shorter than the repetition period, so that the separation of scales is not as justified as in lasers of other types, and the chaotic instabilities with several competing pulse trains are a very real threat.

An elegant solution to these modeling limitations is offered in the form of the most sophisticated and the most realistic of the lumped models of mode-locked lasers. In this form of the lumped approach, the two different scales for pulse analysis are, in general, abandoned, and the iteration procedure (Equation 32.11) is substituted by a *delay* one. In a general form, this procedure may be written as

$$Y(t) = \left(\sqrt{\kappa} \hat{G} \hat{Q} \hat{D} \right) Y(t - T_{RT}), \tag{32.19}$$

where T_{RT} is again the round-trip of the cold cavity, and t is still the local time of the pulse.

A particularly useful form of this model is obtained if the dispersion operator $\hat{D}$ is expanded as a differential one. An efficient form of such an expansion has been derived by Vladimirov et al. (Vladimirov et al., 2004; Vladimirov and Turaev, 2005) who showed that for a bandwidth limiting element with a Lorentzian spectrum similar to Equation 32.10:

$$\hat{D} \cdot Y^T(\omega) = \left[\frac{1}{1 - i(\omega - \omega_p)/\gamma} \right] Y^T \tag{32.20}$$

(i.e., neglecting GVD), assuming without much loss of generality that the peak gain frequency ω_p coincides with one of the laser resonator modes, and taking it as the reference optical frequency, we can rewrite Equation 32.19 as

$$Y(t) = -\gamma^{-1} \frac{\partial Y(t)}{\partial t} + \left(\sqrt{\kappa} \hat{G} \hat{Q} \right) Y(t - T_{RT}). \tag{32.21}$$

Equation 32.21 is a delay-differential one, and the model thus becomes the *delay-differential equation*, or DDE, *model* of ML in semiconductor lasers. The development of this model has been arguably the greatest advance in the theoretical analysis of mode-locked lasers since the original papers by Haus and New (of which it is a rigorous generalization and which it reproduces in limiting cases). It allows a full self-contained treatment of mode-locked operation, including a possibility of some (if by necessity limited) analytical progress with a platform for a full large-signal numerical analysis, which gives a complete, and qualitatively correct (if not necessarily completely accurate), picture of all possible regimes of ML laser dynamics and allows a number of important trends to be identified. Therefore, we shall present it here in some detail, following the original papers (Vladimirov et al., 2004; Vladimirov and Turaev, 2005).

The operators $\hat{G}$ and $\hat{Q}$ can be calculated using Equations 32.2 through 32.6 (in this version of the DDE model, no fast absorption is present); the integrated absorption Q is found in Equation 32.18, and for the integrated gain G, a similar equation is written. Assuming that the pulse in the unidirectional cavity treated by the model passes the absorber before the amplifier, the equation takes the form

$$\frac{dG(t)}{dt} = -\left[\exp\left(G(t)\right) - 1\right] \exp(-Q(t))\frac{P(t)}{U_g} + \frac{G_0 - G(t)}{\tau_g}. \tag{32.22}$$

Here, G_0 is the unsaturated gain determined by the pumping conditions.

Equations 32.21, 32.18, and 32.22 are a closed system suitable for a detailed numerical simulation of both stationary and dynamic behavior of PML. They can also be fairly easily adapted to allow numerical analysis of *hybrid* ML behavior. As shown in Vladimirov and Turaev (2005), the DDE model also allows for significant *analytical* progress, similar to one achieved with classical New's and Haus's models as described above, but for a more general case of large single-pass gain and absorption, more relevant for most semiconductor laser constructions than the classical SCP. In the analytical procedure, the slow absorption and gain approximation have to be reintroduced, and the slow (relaxation of gain and absorption between pulses) and fast (evolution during the pulse) stages of laser dynamics are, as in the traditional SCP model, treated separately. Considering the slow stage results in equations connecting the gain and absorption before and after the pulse, we get

$$G_- = G_0 - (G_0 - G_+)\exp(-T_{RT}/\tau_g), \tag{32.23}$$

and

$$Q_- = Q_0 - (Q_0 - Q_+)\exp(-T_{RT}/\tau_\alpha). \tag{32.24}$$

At the fast stage, as usual in the theory of short pulses in lasers and amplifiers, the relaxation terms are omitted, and so Equations 32.18 and 32.22 take the form

$$\frac{dG(u)}{du} = -\left[\exp\left(G(u)\right) - 1\right]\exp(-Q(u)); \quad \frac{dQ(u)}{du} = -s\left(1 - \exp\left(-Q(u)\right)\right), \tag{32.25}$$

where u is the dimensionless energy within the pulse, $u(t) = U(t)/U_g$, and $U(t) = \int_{-\infty}^{t} P(t')dt' = v_g\hbar\omega A_{Xg}\int_{-\infty}^{t}|Y|^2(t')dt'$. Introducing $u_p = U_p/U_g$, $U_p = U(t \to \infty)$ as the total dimensionless pulse energy (the time of minus infinity on the short timescale meaning the time before the pulse, and plus infinity, covering the entire time of substantial pulse energy, i.e., the entire pulse duration), one can integrate Equation 32.25 to get another set of equations connecting the prepulse and postpulse gain and absorption:

$$Q_+ = Q(u_p) = \ln\left[1 + \exp(-su_p)(\exp(Q_-) - 1)\right], \tag{32.26}$$

$$G_+ = G(u_p) = -\ln\left[1 - \frac{1 - \exp(-G_-)}{\left[\exp(-Q_-)\left(\exp(su_p) - 1\right) + 1\right]^{1/s}}\right]. \tag{32.27}$$

The pulse energy itself may be calculated from Equation 32.21 by taking the modulus square of both sides of the equation and integrating over the pulse. The result can be expressed as

$$\gamma^{-2} v_g \sigma_g \int_{-\infty}^{\infty} \left| \frac{\partial Y}{\partial t} \right|^2 dt + u_p = \kappa \ln \frac{\exp(G_-) - 1}{\exp(G_+) - 1}. \tag{32.28}$$

In general, the integral of the left-hand side cannot be calculated analytically. Two particular cases when this is possible have been analyzed in Vladimirov and Turaev (2005).

The first is the case of a model without spectral filtering when the integral can be set to zero. As noted by Vladimirov and Turaev (2005), this is a fairly crude approximation, as in fact the value of the integral does not disappear even in the limit of infinitely wide gain dispersion curve ($\gamma \to \infty$). Indeed, the integral is over the time of the pulse, and as such roughly proportional to the pulse duration. In the theories of ML, this duration scales as γ^{-1} meaning that $\left| \frac{\partial Y}{\partial t} \right|^2 \propto \gamma^2 Y^2$, so that the integral remains finite as $\gamma \to \infty$. In fact, as mentioned above, the theory with $\gamma \to \infty$ cannot predict the pulse shape or duration, leading to pulses collapsing to a delta-function shape. The total pulse *energy*, however, can be estimated *approximately* by neglecting the integral in the left-hand side of Equation 32.28 and thus obtaining an equation for u_p in the form

$$u_p = \kappa \ln \frac{\exp(G_-) - 1}{\exp(G_+) - 1}. \tag{32.29}$$

Equations 32.23, 32.24, and 32.26 through 32.29 form a closed system of five (nonlinear and transcendental) equations for the five unknowns: $G_\pm$, $Q_\pm$, and u_p. Vladimirov and Turaev (2005) identified this system as the *generalized New's model*, as it does not include spectral filtering (as the original New's model) but, unlike this model, does include arbitrarily large gain and absorption per pass, which are both essential features of diode lasers. The (numerical) solution gives the dependence of pulse energy (though neither duration nor peak power) on pulse parameters, represented by the unsaturated gain (which is related to pumping current) and absorption (which is related to the reverse bias applied to the absorber and the bandgap detuning between the gain and absorber sections). The other fundamental absorber parameter also dependent on the reverse bias, the absorber lifetime, only enters the calculations through the relaxation Equation 32.24 and does not influence the results from this model at all if $\tau_\alpha \ll T_{RT}$ (in which case, obviously, $Q_- \approx Q_0$).

The solution to this nonlinear algebraic equation system can then be substituted into the inequalities (32.13) to analyze the stability boundaries of the ML operating range with respect to the leading-edge and trailing-edge instability. The curves, in general, can only be calculated numerically; however, Vladimirov and Turaev (2005) noted that the leading-edge and trailing-edge instability boundaries met at the codimension-2 point lying on the linear threshold line (G_0-Q_0-$\ln\kappa = 0$). This point can be calculated explicitly as

$$Q_0 = \ln \frac{\kappa(s - 1)}{s\kappa - 1}; \quad G_0 = \ln \frac{s - 1}{s\kappa - 1}. \tag{32.30}$$

This means that the condition $s > 1$ for any range of successful ML to be present, derived in the traditional SCP approach for the case of small gain and loss per period, needs to be generalized in the case of arbitrary losses in the cavity as

$$s\kappa > 1. \tag{32.31}$$

In a more realistic construction, an extra geometric factor could also be required to take into account the fact that the absorber may be saturated by both the forward and reverse propagating wave simultaneously, which is not taken into account by a unidirectional ring laser model on which the approach above is based.

In the case of $G_0, Q_0, \ln \kappa \ll 1$, Equations 32.26 and 32.27 simplify to Equations 32.3 and 32.6, such that the equation for the pulse energy simplifies to $Q_- \frac{\exp(su_p)-1}{s} - G_- \left(\exp(u_p) - 1\right) - u_p \ln \kappa = 0$, which is the equation for u_p featured in the original New's theory of ML.

The second case when full (semi) analytical solution of the DDE model (with relaxation terms during the pulse neglected) is possible is when the dispersion is taken into account, but the saturation of gain and absorption during the pulse is assumed to be small, as in the Haus model of ML (though the gain and absorption themselves are not necessarily small, unlike the case of the traditional Haus model). Vladimirov and Turaev (2005) called this the generalized Haus model. In this case, a steady-state solution is sought in the form similar to Equation 32.12 in our notations, $Y(t + T_{RT}) = e^{-i\delta\psi} Y(t - \delta T)$. Then, from Equation 32.21,

$$\gamma^{-1} \frac{\partial Y(t - \delta T)}{\partial t} + Y(t - \delta T) = F(u(t))Y(t), \tag{32.32}$$

where

$$F(u) = \sqrt{\kappa} \exp\left(G(u)(1 - i\alpha_{Hg}) - Q(u)(1 - i\alpha_{Hq}) - i\delta\psi \right) \tag{32.33}$$

is the "complex net gain," which can be written out explicitly, substituting the expressions (Equations 32.26 and 32.27) (with u instead of u_p) for $G(u)$ and $Q(u)$.

Next, assuming that the single-pass pulse shift is significantly smaller than the pulse duration and that the saturation of both the gain and absorption during the pulse is weak enough ($u(t) < u_p \ll 1/s$)—the latter being the underlying assumption of Haus's theory—both sides of Equation 32.32 can be expanded in Taylor series up to the second-order terms in their respective arguments:

$$Y(t - \delta T) \approx Y(t) - \frac{\partial Y}{\partial t}\delta T + \frac{1}{2}\frac{\partial^2 Y}{\partial^2 t}\delta T^2,$$

and, generalizing the expansions (Equations 32.14 and 32.15) of the original Haus's theory:

$$F(u) \approx F_0 + F_o' u + \frac{1}{2}F_o'' u^2; \; F_0 = F|_{u=0}; \; F_o' = \left.\frac{\partial F}{\partial u}\right|_{u=0}; \; F_o'' = \left.\frac{\partial^2 F}{\partial u^2}\right|_{u=0}. \tag{32.34}$$

Then, the equation governing the pulse evolution is obtained in the form

$$\delta T \left(\frac{\delta T}{2} - \gamma^{-1}\right)\frac{\partial^2 Y}{\partial^2 t} - \left(\gamma^{-1} - \delta T\right)\frac{\partial Y}{\partial t} + \left(F_0 - 1 + F_o' u(t) + \frac{1}{2}F_o'' u^2(t)\right) Y = 0. \tag{32.35}$$

Recalling the definition of $u \propto \int |Y|^2 dt$, one identifies this second-order nonlinear differential equation as the generalization of the master equation of ML in Haus's theory, which admits solutions of the same form (Equation 32.16) as the original master equation. Six equations are then obtained for six real parameters: peak pulse power, duration, time shift δT, optical frequency shift $\Delta\omega$, phase shift per round-trip $\delta\psi$, and the chirp parameter β.

The stability limits in the generalized Haus's form can be obtained by substituting these solutions into the conditions (Equation 32.13). In general, they depend on the linewidth enhancement factors; however, for direct comparison with other models, the case of $\alpha_{Hg} = \alpha_{H\alpha} = 0$ is useful. Results of such analysis, reproduced from Vladimirov and Turaev (2005), are plotted in Figure 32.3. In the plot, the subscript N refers to results from New's model, generalized (solid lines) or standard (dashed lines); and the subscript

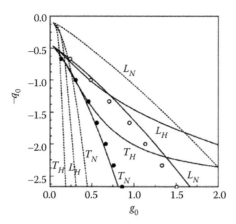

FIGURE 32.3 Stability boundaries of mode locking (ML) with respect to leading (L) and trailing (T) edge instabilities, calculated semianalytically in the DDE approach using traditional and generalized New's (N) and Haus's (H) models. In the calculations, $s = 25$, $T_{RT}/\tau_\alpha = 1.875$, $\tau_\alpha/\tau_g = 0.0133$, $\kappa = 0.1$; $g_0 = (\tau_\alpha/\tau_g)G_0$ as in text; q_0 corresponds to Q_0 in text. (From Vladimirov AG, Turaev D, *Phys Rev A*, 72, 033808, 2005. Reproduced with permission).

H to those from Haus's model (calculated with zero linewidth enhancement factors). The filled/empty dots are the leading/trailing instability boundary calculated by numerical integration of the model. In this numerical integration, the gain and absorber operators are treated on a continuous timescale, without the need to introduce separate timescales for pulse and the free relaxation period as in the iterative procedure. As seen in the figure, standard Haus's and New's models are extremely inaccurate in predicting the instability boundaries of ML in a typical diode laser (with the range predicted by New's model being too wide, and that from Haus's model, too narrow, as noted also in Dubbeldam et al., 1997). The generalized Haus's model gives good agreement within its validity limits at low currents/unsaturated gain values, while the generalized New's model gives very good agreement with numerical simulations at all parameter values (there are some modest deviations that are discussed in more detail below), the reason being that the spectral filtering term neglected in New's approach simply happens to be small in typical diode lasers. Thus, the large-signal nature of the DDE model is proven to be a very important advantage over the classical ML theories.

Apart from allowing some analytical progress in the limiting cases, the DDE model also allows the use of numerical techniques that have been developed for the analysis of DDEs, in particular of numerical packages that allow a full bifurcation analysis of DDEs. Such a study was indeed performed in Vladimirov and Turaev (2005), comprising the full (in)stability analysis of the stationary solution of the DDE. The stationary solution (the steady-state light-current characteristic of the laser) itself is found by seeking the steady-state light output in the form of $Y(t) = Y_{0s}\exp(i\Delta\omega_s t)$. Substituting this into the original Equations 32.21, 32.18, and 32.22 gives the steady-state amplitude and frequency in the parametric form

$$
\begin{aligned}
&\kappa \exp\left(G_s(Y_0) - Q_s(Y_0)\right) - \Delta\omega_s^2 = 0, \\
&\Delta\omega_s \gamma^{-1} + \tan\left[\Delta\omega_s T_{RT} + \left(\alpha_{Hg}G_s(Y_0) - \alpha_{H\alpha}Q_s(Y_0)\right)/2\right] = 0
\end{aligned}
\tag{32.36}
$$

Equation 32.36 is a transcendental trigonometrical equation and thus has an infinite set of formal solutions, corresponding to the cavity modes. The steady-state solution, as usual in the laser theory, is the one with the lowest value of the threshold gain $G_s(Y = 0)$, in other words, the closest to the peak of the gain spectrum. Figure 32.4, after Vladimirov and Turaev (2005), shows the results of a numerical bifurcation analysis of this solution. The line H_1 indicates the *Andronov–Hopf bifurcation* (transition from a steady state to a periodically oscillating solution with an amplitude smoothly increasing from zero as the controlling parameter,

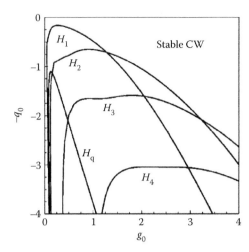

FIGURE 32.4 Bifurcation analysis of the steady-state solutions of the DDE model. Parameters used: $\gamma\tau_\alpha = 33.3$; $\alpha_{Hg,\alpha} = 0$, the rest as in Figure 32.3. The notations H_1, H_2, etc. refer to different harmonic numbers M_h; H_q is the boundary of the Q-switching instability. (From Vladimirov AG, Turaev D, *Phys Rev A*, 72, 033808, 2005. Reproduced with permission).

for example, the unsaturated gain in this case, increases beyond a critical value) corresponding to oscillations at the fundamental ML frequency. ML is predicted for a certain range of conditions regarding the values of the unsaturated gain and absorption, above threshold, whereas at high enough unsaturated gain (or current) and low enough absorption, continuous wave (CW) lasing is expected to be stable. The line H_q indicates the Andronov–Hopf bifurcation corresponding to *passive Q-switching* instability, also known as self-sustained pulsations, which essentially corresponds to the well-known relaxation oscillations in the laser. The positive feedback provided by the SA, which essentially favors pulsed operations, transforms the relaxation oscillations from damped to self-sustained pulsations. The frequency of these oscillations is determined mainly by the unsaturated gain, the gain cross section, the gain relaxation time, and the losses in the cavity, and is typically of the order of 1 GHz, or about an order of magnitude below the ML frequency. Thus, at low frequencies and with high enough amount of saturable absorption in the cavity, the ML pulse train is expected to be modulated by the self-pulsing envelope. The lines H_m, $m > 1$, show the bifurcations corresponding to a solution oscillating at the mth *harmonic* of the fundamental ML frequency. At high enough values of unsaturated absorption, there are ranges of G_0 (or current) in which ML at higher harmonics is predicted to be stable, but ML at fundamental harmonic is not.

These predictions are confirmed by a full numerical integration of the DDE model (Figure 32.5), showing the extrema of the laser intensity time dependence calculated for different values of the pumping parameter $g_0 = (\tau_\alpha/\tau_g)G_0$. For each unsaturated gain, the initial transient is omitted before the start of registering signals. At low values of g_0 (and thus current), the laser exhibits a regime when the ML pulse power is modulated by a passive Q-switching envelope, originally with nearly 100% modulation depth (Figure 32.6a). As the pumping parameter increases, the Q-switching modulation gradually decreases in amplitude and eventually the modulation regime undergoes the backward bifurcation, moving to a stable ML regime (this corresponds to the border of the trailing-edge instability in Figure 32.3). Within the area of stable ML, the fundamental round-trip frequency, a train of short pulses, is observed as in Figure 32.7a, whose amplitude increases with G_0. At higher still pumping, the laser dynamics see areas of harmonic ML at the second and third harmonic of the fundamental ML frequency (see Figure 32.7b and c), separated by narrow areas of unstable operation. Finally, the ML breaks up completely with the onset of chaotic modulation of the pulse power, with multiple pulse trains competing in the cavity, as in Figure 32.6b (the regimes separating fundamental frequency ML and harmonic ML areas are similar). Eventually, the system

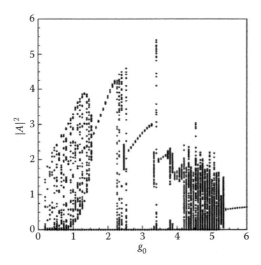

FIGURE 32.5 Bifurcation diagram obtained by direct numerical implementation of a DDE model. $Q_0 = 4$, the other parameters as in Figure 32.3. $|A^2|\propto|Y^2|$ is the normalized output intensity. (From Vladimirov AG, Turaev D, *Phys Rev A*, 72, 033808, 2005. Reproduced with permission).

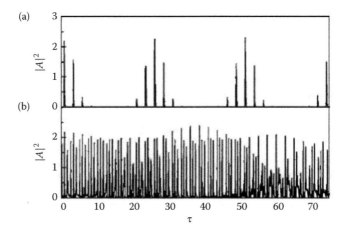

FIGURE 32.6 Illustration of the aperiodic regimes in Figure 32.5: combined mode-locking/Q-switching regime at $G_0 = 50$ (a) and chaotic pulse competition regime at $G_0 = 350$ (b). $|A^2|\propto|Y^2|$ is the normalized output intensity, $\tau = t/\tau_\alpha$ the normalized time. (From Vladimirov AG, Turaev D, *Phys Rev A*, 72, 033808, 2005. Reproduced with permission).

undergoes a transition to CW single-frequency operation in agreement with the bifurcation diagrams of Figure 32.3.

An interesting result obtained in Vladimirov and Turaev (2005) is that, while the conditions (Equation 32.13) of negative net gain before and after the pulse are useful indications of the stability ranges of mode-locked operation, the onset of instabilities in numerical simulations does not coincide with those limits *exactly*. This may be caused in part by the omission of gain dispersion in the analytical study and in part by the neglect of absorber relaxation during the pulse. However, there is also a genuine physical reason for the discrepancy, in that not all small fluctuations in a ML laser were found to grow into full-scale instabilities even if a window of positive gain preceded the ML pulse. Instead, stable ML operation was shown to be possible for a range of parameter (unsaturated gain and absorption) values such that before the pulse,

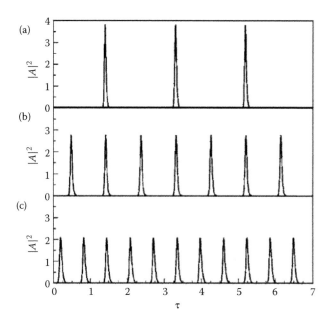

FIGURE 32.7 Illustration of the periodic regimes in Figure 32.5: fundamental frequency ML at $G_0 = 150$ (a) and first and second harmonic ML $G_0 = 225$ (b) and 270 (c). $|A^2|\propto|Y^2|$ is the normalized output intensity, $\tau = t/\tau_\alpha$ the normalized time. (From Vladimirov AG, Turaev D, *Phys Rev A*, 72, 033808, 2005. Reproduced with permission).

the fast absorption had recovered to its unsaturated value, but the slower gain continued recovery, leading to a window of positive net gain preceding the pulse (it may be worth noting that some previous studies, using modifications of Haus's model for semiconductor lasers, indicated the possibility of positive net gain at the *trailing* edge of a stable ML pulse as well; see, e.g., Vladimirov and Turaev, 2005). The possibility of stable ML operation despite a positive net gain window is confirmed by more accurate traveling-wave simulations. One of the consequences of this effect is that the onset of instabilities may be expected to be sensitive to perturbations such as spontaneous noise. The effect of spontaneous emission was indeed studied analytically and numerically in Vladimirov and Turaev (2005), with the noise introduced as a delta-correlated random term in the right-hand side of Equation 32.21. It was concluded that, while the onset of Q-switching oscillations (trailing pulse edge instability) is a dynamic process independent of noise, the onset of the chaotic envelope instability (leading edge instability) is strongly affected by the noise, with an increase in the noise narrowing the window of stable ML. This is fully confirmed by the more complex traveling-wave models (TWMs) described below.

The DDE model, when used as a numerical tool is, not only fully large-signal, but also self-starting: It does not require a trial pulse to start with and can reproduce the emergence of ML pulse train from randomly pulsing light output that is seen as the laser crosses the threshold condition. Thus, the model removes most of the shortcomings traditionally associated with lumped models of mode-locked lasers and presents a relatively simple yet very powerful tool for analyzing the qualitative tendencies of their behavior. As illustrated above, it combines analytical possibilities and numerical methods very naturally within the same framework and, as will be discussed in more detail later, predicts correctly virtually all the dynamic regimes and tendencies observed in a real laser.

An important advantage of the DDE model is that, although strictly speaking derived for the artificial unidirectional ring geometry, it captures enough of the main features of ML to be applicable, with some caution, to predict—at least qualitatively—the phenomena in mode-locked lasers of all types and designs.

An important example is the work presented in Marconi et al. (2014) where the DDE was used to analyze the behavior of mode-locked lasers with *long* delays (cavity round-trip times), comparable to, or even exceeding, the lifetime of carriers in the gain section ($T_{RT} >\sim \tau_g$, unlike the analysis in Vladimirov and Turaev (2005) where the typical situation was $\tau_\alpha < T_{RT} < \tau_g$). Marconi et al. (2014) used a bifurcation analysis similar to that discussed above and presented in Figure 32.3 to analyze the stability of various cavity configurations. The results are shown in Figure 32.8, similar to that presented in Marconi et al. (2014). It was found that the Andronov–Hopf bifurcation mathematically describing the onset of fundamental harmonic ML as discussed above, which is *supercritical* in the case of a short resonator so that ML exists only above its bifurcation point, becomes *subcritical* for a certain value of $T_{RT}/\tau_g > 1$. This means that the (fundamental) ML operation regime can exist below its bifurcation point, coexisting with the CW solution. At longer delays still, the area of stable ML extends below the CW threshold, meaning that ML can coexist with the off solution, thereby implying a bistability between them. Interestingly, during this folding phenomenon, the fundamental ML branch eventually disconnects from the CW solution, meaning that the ML appears for long delays through a saddle-node bifurcation of limit cycles instead of a nascent Andronov–Hopf bifurcation of the CW solution, making this scenario impossible to analyze by any weakly nonlinear analysis such as dynamic modal analysis (Section 32.3.2).

The change in dynamical scenario that occurs in Figure 32.8c has a profound consequence on the mode-locked solutions, as it can be seen in Figure 32.8c and d: The fundamental PML solution becomes stable even in the limit $T_{RT}/\tau_g \gg 1$. Moreover, a very large number of pulsing solutions with different number of pulses per round-trip and different arrangements become stable for the same parameter values. The authors reconstructed analytically some of these solutions, using the generalized New's approximation discussed above, for $T_{RT}/\tau_g = 1$ and restricting the analysis to equally spaced pulses solutions (harmonic PML), as presented in Figure 32.8d. Clearly, all these branches of solutions extend well below the laser threshold, where they stably coexist among them and with the off solution (although the authors noted that New's approximation of neglecting spectral filtering leads to an overestimation of the breadth of the ML region below threshold). This means that the harmonic mode-locked solution of maximal order that exists below threshold becomes fully decomposable, since essentially any pulse of this solution can be set on or off, which the authors confirmed by injecting a digitally modulated sequence of optical pulses into the cavity, which remained stable after a large number of round-trips, meaning that the laser worked as an active version of an optical buffer memory.

The model used in Marconi et al. (2014) was a DDE in its classical form of Vladimirov and Turaev (2005), derived for a hypothetical unidirectional ring laser with large gain and absorption per round-trip (and with the linewidth enhancement factors set to zero as the dynamic effects studied did not significantly depend on them), but the theoretical predictions of the paper were realized experimentally (Marconi et al., 2014)

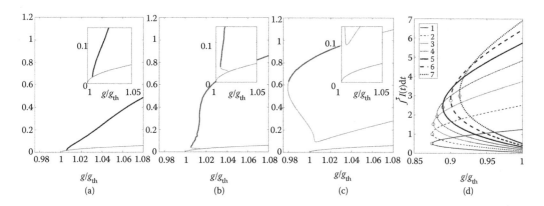

FIGURE 32.8 (a) through (c) Bifurcation diagrams similar to that of Figure 32.3 for different cavity round-trip times: $T_{RT}/\tau_g = 1.2$ (a), 2 (b), 4 (c). In the case of a double mode-locked branch, the upper branch is stable and the lower one unstable. Harmonic ML branches are not shown. (d) Folding of harmonic ML branches with a different number for $T_{RT}/\tau_g = 16$. (Reproduced from Marconi M. et al., *Phys Rev Lett*, 112, 223901, 2014).

using a *vertical external cavity surface-emitting* laser (VECSEL). A laser of this type consists of an amplifying (gain) chip and a semiconductor saturable absorber mirror (SESAM) chip, separated by an unguided free-space propagation path (with collimating optics to direct the beam and control the ratio of the spots over the two facets). Various harmonic regimes were realized with this laser, with the number of pulses between 0 and 19 successfully coexisting at high enough currents. The authors then went on (Marconi et al., 2015) to realize the multistability in the long laser to *generate individually addressable* pulses and sequences of pulses, all very well reproduced by the same DDE model.

This work is an important testament to the generality, power, and versatility of the DDE models: The experiments were stimulated by the theoretical predictions which in turn relied on the analytical capabilities (bifurcation analysis using analytical continuation techniques) unique, among semiconductor laser ML theories, to the DDE approach.

It can be pointed out that VECSELs, in fact, are the one class of mode-locked semiconductor lasers for which a delay-differential model, although of a somewhat different form to that used in Vladimirov and Turaev (2005), may be expected to give a fully *quantitative*, as well as qualitative, description of the behavior of a specific and realistic laser construction. As both the gain chip and SESAM are very short asymmetric resonators, with the gain section "length" much shorter than the spatial pulse duration, the lumped-element formalism is a very natural one for their description. The fact that no integration over length is needed for calculating G and Q allows the use of generic nonlinear $g(N)$ and $\alpha(N)$ dependences with no loss of accuracy, as well as introduction of fast gain and absorption saturation omitted in the original DDE of Vladimirov and Turaev (2005). A model based on an approach of this type was successfully used to analyze the dynamics of external-cavity VECSELs in a simple linear cavity (with the laser and SA chips facing each other and the output being from a partially reflecting mirror located between them). The predicted dynamic regimes, pulse durations, and stability ranges matched the ones previously reported in experimental papers, not only qualitatively, but with a reasonable numerical agreement (Mulet and Balle, 2005).

The mathematical distinction of the DDE model usable for quantitative and rigorous description of realistic VECSELs, such as the one used in Mulet and Balle (2005), from the one reported in Vladimirov and Turaev (2005) is, first, that gain and absorption operators, as well as the rate equations for the carriers, need to be modified to take into account the resonator nature of the amplifier and absorber sections and the short length of their active parts. Second, given the Fabry–Pérot rather than ring nature of the resonator, rather than having a single DDE for the light amplitude with the delay time equal to the cavity round-trip, the model for VECSELs in a linear geometry needs separate equations for the dynamics of light amplitudes in the gain and absorber chips, each of them containing a delayed term with a delay equal to *half* of the cavity round-trip.

With further delayed terms introduced, a model of this type can describe different VECSEL geometries, including a folded (rather than linear) cavity one and a colliding pulse operating VECSEL (Avrutin and Panajotov, in preparation).

More advanced constructions known as MIXCELs (standing for mode-locked integrated external cavity surface emitting laser) with the quantum well (QW) gain and QD absorber layers located in one chip, could be described by a similar, possibly even somewhat simpler, model, with the single chip reflectance operator containing the effects of both the gain and the absorption; the model for such a design would therefore be even closer to that of Vladimirov and Turaev (2005).

In the case of *edge-emitting* lasers, the application of a DDE model to a specific, realistic design is somewhat more tenuous. First of all, the DDE model as studied in Vladimirov and Turaev (2005) does not account for fast gain and saturable absorption nonlinearities due to interband processes. Although it could be possible to include them in the QD case, at least in some approximation, the explicit introduction of fast nonlinearities is not necessarily the best strategy. Instead, separate rate equations for dot and reservoir populations can be used. Second, Equations 32.18 and 32.22 for the gain G and absorption Q integrated over the length of the amplifier/absorber element are only accurate if both the gain and absorption have a simple linear dependence on the carrier densities in the corresponding elements, which is in itself an approximation (or if $G, Q \ll 1$). Third, the geometry of the system analyzed in a DDE model in the form presented in Vladimirov and Turaev (2005) is, as in most lumped models, somewhat artificial in that Equations 32.2 and 32.5 are, strictly speaking, valid only in a hypothetical unidirectional ring cavity. In a real laser, the pulse passes through both the amplifier and absorber twice, not once, with the reflected pulse traveling through areas in which gain or absorption has been already partially saturated by the incident pulse, and possibly partially recovered.

An attempt at taking into account a realistic edge-emitting tandem laser geometry in a large-signal *lumped* iterative (not DDE) model, with an end reflector by introducing averaging of gain/absorption over the corresponding sections, with reflections at facets taken into account, was made in a relatively early paper by Khalfin et al. (1995); the model gave good qualitative predictions of the laser performance; however, such an approach has its own inaccuracies as discussed in Avrutin et al. (2000), since it implicitly assumes pulse duration greater than the round-trip time, which by definition is not the case in mode-locked lasers and the results of Khalfin et al. (1995) were never compared to predictions of more accurate models to ascertain their accuracy.

More recently, a study specifically investigating modeling a Fabry–Pérot edge-emitting laser using a DDE model, importantly with a detailed comparison to the more rigorous and accurate model of a travelling wave type (see the next section), has been presented (Rossetti et al., 2011b). The simulations were performed for the special case of QD mode-locked lasers; however, the results appear to be quite generic. Rossetti et al. (2011b) improved the accuracy of the DDE approach by separating the laser into a number F_s of longitudinal sections (e.g., $F_s = 28$ sections were used in the calculations presented in the paper), some of which belong to the gain region and others to the SA, which results in what the authors termed a multi-section DDE approach. Essentially, in the notations used here (the formalism of Rossetti et al., 2011b was somewhat more complicated because of account for two-level transitions peculiar to QDs), in a multisection DDE model, Equation 32.21 is rewritten with the single product $\sqrt{\kappa}\widehat{G}\widehat{Q}$ substituted by a concatenation of gains and losses in individual segments:

$$Y(t) = -\gamma^{-1}\frac{\partial Y(t)}{\partial t} + \left(\prod_{k=1}^{F_s} \sqrt{\kappa_k}\widehat{G}_k \right) Y(t - T_{\text{RT}}). \tag{32.37}$$

Here, the loss $\sqrt{\kappa_k}$ includes both the distributed losses inside the segment and any lumped scattering/out-coupling loss between the segments k and $k+1$. The complex gain operator $\widehat{G}_k$ for each section is calculated in the way similar to Equation 32.2 if the section is within the gain region and similar to Equation 32.5 if it belongs to an SA, with the length used being the length of the section. The gain (or loss) in each segment is calculated from a local rate equation for population inversion. The use of a number of segments gives the model some longitudinal resolution and allows for a more accurate modeling of the outcoupling losses; although not explicitly done in the model of Rossetti et al. (2011b), it can also include fast gain and absorption saturation in addition to the slow processes described by rate equations. The model, however, still contains a single delay term helping maintain the calculation efficiency that is one of the main advantages of the DDE approach. To account for bidirectional propagation, the Fabry–Pérot resonator had to be represented, somewhat artificially, by an equivalent ring resonator twice the length of the Fabry–Pérot one, with the SA length also doubled, and the distributed losses allocated carefully to represent the loss in the realistic cavity. With these approximations, the laser performance simulated by the multisection DDE approach was in good (though not perfect, e.g., the pulse amplitude was accurate to typically within about 10%) agreement with the traveling wave one as regards the pulse shapes, amplitudes, durations, and stability of the results with respect to a chaotic envelope instability. As in most QD mode-locked lasers, the design simulated did not show Q-switching instability, owing to the large damping of the relaxation oscillations, so it is not certain whether the multisection DDE model would accurately predict its limits in a laser in which Q-switching could occur. It can be said that the multisection DDE approach has some similarity to the decimated traveling wave one considered below, though unlike it still retains the unidirectional ring cavity assumption.

The DDE model also has been used recently to investigate the performance of a number of laser cavity designs and operating regimes more complex than purely PML in a simple tandem cavity, whether monolithic or VECSEL type.

In Arkhipov et al. (2013), the DDE model was used to analyze the performance of *hybridly*, rather than passively, mode-locked lasers. As in the experiments with which the calculations were compared, and as in previous simulations using a TWM (Avrutin et al., 1996), the hybrid ML was implemented by voltage modulation at an approximately resonant frequency applied to the SA. Within the DDE approach, this requires the modification of Equation 32.18 in the form

$$\frac{dQ(t)}{dt} = -\frac{Q_0 - Q}{\tau_\alpha}\left(1 + a_{mod}(1 + F_{mod}(t))\right) - \left(1 - \exp(-Q)\right)\frac{P(t)}{U_\alpha}. \tag{32.38}$$

Here, a_{mod} is the amplitude of absorption modulation, $F_{mod}(t)$ is a periodic function of time, and F_{mod} is a periodic function of time defined so as to vary within the limits $-1 < F_{mod}(t) < 1$. The authors investigated various modulation profiles, the most straightforward of which was sinusoidal modulation $F_{mod}(t) = \cos(2\pi f_{mod}t)$. The modulation frequency could be near resonance with the fundamental ML frequency, with $f_{mod} = f_P + \Delta f_{mod}$, $\Delta f_{mod} \ll 1/T_{rt}$, with $f_P \approx 0.9723/T_{rt}$ being the free-running PML frequency, or at its second harmonic $f_{mod} = 2f_P + \Delta f_{mod}$, or second subharmonic $f_{mod} = f_P/2 + \Delta f_{mod}$.

Note that Equation 32.38 captures an important feature of voltage modulation of an SA: Both the unsaturated absorber and the relaxation time are modulated simultaneously.

The efficient DDE model allowed, first and foremost, for very effective numerical search for the *locking range* of hybrid ML (the range of frequency detunings $\delta f = \Delta f_{mod}T_{rt}$ within which stable hybrid ML was observed for a given modulation amplitude). This was done by long-time (3000 periods, with the last 200 periods stored) direct numerical simulation of the system (Equations 32.18, 32.21, and 32.22). For improved accuracy, the authors used dual characterization of the locking range, using, first, a straightforward bifurcation diagram with the field maxima and minima plotted, and, second, a stroboscopic diagram where, for each considered δf_{mod}, field intensities separated from each other in time by the interval $1/f_{mod}$ were collected. Once locking was achieved and the period of the field intensity time trace became equal to $1/f_{mod}$, all stroboscopic map points had the same value; otherwise, multiple values of the stroboscopic map at a given δf were observed. Figure 32.9 (reproduced from Arkhipov et al., 2013 with permission) illustrates both procedures, as illustrated in Figure 32.9a and b, as well as the calculated *locking tongue* (locking range borders in the coordinates δf, a_{mod}), shown in Figure 32.9c. The latter was in very good qualitative agreement with the observed values. The figure shows the fundamental harmonic locking range (the second harmonic one was very similar as could be expected); however, the simulations also predicted, for the first time and in agreement with the experiments published in the same paper, that a narrower range of *subharmonic* locking was also present due to internal nonlinearities in the laser.

In addition to numerical analysis, the DDE allowed an *analytical* asymptotic model for the calculation of the locking range to be developed, which was in very good qualitative agreement with the numerical simulations up to the modulation amplitude of $a_{mod} \sim 0.5$–0.6. Most importantly, it was shown, in agreement with numerical results, that the magnitude of the locking range was *directly proportional* to the modulation amplitude, a fact also seen in other numerical simulations but only proven analytically in Arkhipov et al. (2013). Explicit, if rather complex, expressions for the proportionality coefficient between the locking range and a_{mod} were derived using perturbative analysis of the periodic ML solution.

Finally, Arkhipov et al. (2013) addressed the issue of the asymmetry of the locking range (it is easier to speed the laser up than to slow down), long known for hybrid ML and shown by both experiments and previous simulations; the authors attributed it to the variation of the absorber relaxation time and hence the average absorption value with modulation.

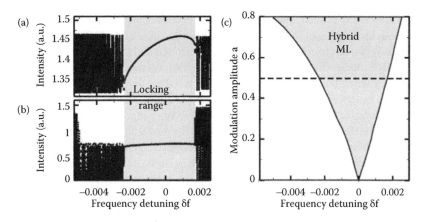

FIGURE 32.9 Numerical DDE estimate of the locking range of a hybrid ML laser. (a) Local maxima of the intensity time trace and (b) 1/fm periodic stroboscopic map points of emitted field intensity time trace at fixed modulation amplitude $a = 0.5$. In the locking range (gray), all ML pulses have the same peak intensity. (c) Locking tongue in the plane of two parameters: frequency detuning δf and modulation amplitude a. (From Arkhipov A. et al., *IEEE J Sel Top Quantum Electron*, 19, 1100208, 2013. Reproduced with permission).

The DDE was also used for modeling the dynamics of mode-locked lasers under the more special conditions of optical injection with a single (Rebrova et al., 2011) or, most recently, dual (Arkhipov et al., 2016c) optical lines, as well as under the condition of external optical feedback (Jaurigue et al., 2015, 2016; Otto et al., 2012). In all of these conditions, the versatility of the DDE allowed analytical insight into operating conditions, as well as direct numerical simulations.

Interestingly, the somewhat abstract nature of the DDE model led to the fact that two very different forms of the model could be used successfully to analyze the situations that in the experiment can be quite similar. In Jaurigue et al. (2015) and Otto et al. (2012), a model with multiple delays (as in the generalized Lang-Kobayashi model of a single-frequency laser with optical feedback from a strong external reflector) was used to analyze ML under external optical feedback, whereas in Arkhipov et al. (2015b), harmonic ML in a compound cavity consisting of an active laser subcavity and a passive one formed by an external reflector (which is very similar to optical feedback) was investigated by modeling both cavities as unidirectional ring ones as in the traditional form of the DDE; mathematically, the compound cavity was represented by two coupled DDEs. Both models showed, and were used to estimate the ranges of, fully or partially rendered harmonic regimes given an integer rational relation between the cavity lengths, and stressed the importance of the subwavelength variations in the cavity length (represented by phase shifts of light amplitude). All of these had been previously independently investigated for Fabry–Pérot laser geometries using frequency-domain and/or traveling wave time-domain models (see the next section), e.g., in Avrutin and Russell (2009). However, the advantage of the DDE model was, first, the possibility of efficient and instructive analysis of the bifurcation diagrams of the laser and, second, the generality as the results apply not just to Fabry–Pérot resonators but to other designs such as ring lasers.

A special version of the DDE model was developed (Viktorov et al., 2006) for analyzing QD mode-locked lasers. Detailed description of the properties of QDs as an active medium is beyond the scope of this chapter; the reader is referred to a specialized monograph (Rafailov et al., 2011) or the relevant chapters in the current handbook. In brief, there are two major (interrelated) features that distinguish the QDs from other semiconductor active media, particularly in the context of ML. The first of these is the complex carrier kinetics which in the case of the gain media involves the relatively slow (~5–10 ps) capture of carriers (electrons and holes) into the QDs and subsequent interlevel relaxation, and in the SA sections, the relatively complex nature of carrier escape involving intermediate levels (Viktorov et al., 2009). The second specific feature of QDs as an active medium is the noticeably nonequilibrium distribution of carriers between dots of different sizes and compositions (which therefore have different energy levels resulting in the inhomogeneous broadening of the laser line), and the possibility of dual-wavelength lasing due to the existence of (at least) two electron levels in each dots, the ground and excited state.

The original DDE model of QD mode-locked lasers (Viktorov et al., 2006) ignored inhomogeneous broadening and the excited-level transitions, and thus captured the first of these characteristic features but not the second. The equation for light field in the model was thus essentially the same as in the standard DDE but the equations for gain and absorber dynamics were changed more significantly to reflect the specific features of QDs. Still, even in this simplified form, the model allowed an explanation for a number of features of QD mode-locked lasers, such, first, as the suppressed Q-switching instability (Viktorov et al., 2006) due to the slow carrier capture in the gain section and, second, the enhanced ML at high temperatures due to faster SA relaxation (Cataluna et al., 2006, 2007). Later, more advanced versions of DDE for QD ML lasers have been developed with inhomogeneous broadening and lasing from different levels taken into account, e.g., Cataluna et al. (2010) and also Rossetti et al. (2011b) already mentioned above.

To summarize, DDE models are a very powerful tool capable of predicting, qualitatively describing, and giving unique analytical insight into all the main features and many of the peculiarities of ML in a semiconductor laser. With some caution, models of this type can be used for quantitative description of the performance of a specific laser design, but the reliability and accuracy of such a procedure may be limited in the case of realistic, multisection edge-emitting laser designs, particularly as they often comprise, in addition to the gain and absorber section, elements such as Bragg mirrors, phase-tuning sections, etc.

For these purposes, TWMs are preferable; they will be considered in the next section.

32.3.4 TWMs: The General Considerations

The most accurate and realistic, though usually the most computationally intensive, approach to simulating edge-emitting mode-locked lasers is offered by *distributed time-domain*, or *TWMs* (shown schematically in Figure 32.2c), which treat the propagation of an optical pulse through a waveguide medium with spatial as well as temporal resolution. The model then starts with decomposing the optical field in the laser cavity into components propagating forward (subscript "+") and backward (subscript "−") in the longitudinal direction (say, z):

$$Y(r, t) = \Phi(x, y) \left(Y_+ \exp(i\beta_{ref}z) + Y_- \exp(-i\beta_{ref}z) \right) \exp(-i\omega_{ref}t), \qquad (32.39)$$

with Φ being the transverse/lateral waveguide mode profile and ω_{ref} and

$$\beta_{ref} = n(\omega_{ref})k_{ref} = n(\omega_{ref})\omega_{ref}/c$$

being the reference optical frequency and the corresponding wave vector, respectively. This results in a reduced equation for slowly varying amplitudes $Y_\pm$, which has the form

$$\pm \frac{\partial Y_\pm}{\partial z} + \frac{1}{v_g} \frac{\partial Y_\pm}{\partial t} = \left(\frac{1}{2}(\widehat{g}_{mod} - \alpha_{int}) + ik_{ref}\widehat{\Delta\eta}_{mod} \right) Y_\pm + iK_{\pm,\mp}Y_\mp + F_{spont}(z, t). \qquad (32.40)$$

The equation is directly solved numerically without the partially analytical integration involved in deriving (Equations 32.2 and 32.5).

The gain and saturable absorption coefficients are most often parametrized as functions of the carrier density and, through the gain and absorption compression coefficients ε_g and ε_α, on the photon densities, which is most efficiently implemented using simple relations:

$$g = \frac{g_{lin}(N)}{1 + \varepsilon_g S}; \quad \alpha = \frac{\alpha_{lin}(N)}{1 + \varepsilon_\alpha S}, \tag{32.41}$$

where $S = |Y_+|^2 + |Y_-|^2$. Note that taking the total intensity in the denominator of Equation 32.41 for both left- and right-traveling waves, although often used, may be an oversimplification for some problems as it assumes identical cross- and self-saturation coefficients between left- and right-traveling waves. A more accurate analysis may be important, in particular, in ring lasers (see Chapter 31 for more details).

The carrier density dependences of the linear gain and absorption $g_{lin}(N)$ and $\alpha_{lin}(N)$ are, in the simplest version of the model, taken in the standard linear $g_{lin}(N) = \sigma_g(N - N_{tr})$ or logarithmic $g_{lin}(N) = \sigma_g N_{tr} \ln \frac{N + N_1}{N_{tr} + N_1}$ forms typical for semiconductor laser modeling in general (Coldren et al., 2012) (N_{tr} being the transparency carrier density and σ_g characterizing the gain cross section near transparency as shown in Sections 3.1 and 3.2; for the absorption, the linear approximation $\alpha_{lin}(N) = \alpha_0 - \sigma_\alpha N$, also as shown in Sections 32.1 and 32.2, is most often used). In more accurate implementations (e.g., Javaloyes and Balle, 2010b), absorption $g_{lin}(N)$ and $\alpha_{lin}(N)$ are calculated microscopically with varying degrees of rigor; some of these implementations will be discussed in more detail in Section 32.5. The compression factors $\varepsilon_{g,\alpha}$ also may be either introduced phenomenologically or calculated microscopically for the two main types of optical nonlinearities in bulk and QW lasers: spectral hole burning and dynamic carrier heating. As pulse duration decreases and particularly for multi-GHz ML, the finite (subpicosecond) relaxation times $\tau_{nl}^{(g,a)}$ of the nonlinearities become important. To take those into account, Equation (32.41) can be substituted by phenomenological relaxation equations: $\frac{d\alpha}{dt} = \frac{1}{\tau_{nl}^{(a)}} \left(\frac{\alpha_{lin}}{1 + \varepsilon_\alpha S} - \alpha \right); \frac{dg}{dt} = \frac{1}{\tau_{nl}^{(g)}} \left(\frac{g_{lin}}{1 + \varepsilon_g S} - g \right)$ (Martins et al., 1995). Some authors choose not to introduce $\varepsilon_{g,\alpha}$ due to carrier heating at all, instead including microscopic analysis of carrier temperature dynamics (Bischoff et al., 1997) and gain-carrier temperature dependence into the model. In QD lasers, with their strong spectral hole-burning effects (nonequilibrium carrier distribution in energy, with the energy levels resonant with the photon energy preferentially depleted), and relatively slow intradot relaxation, kinetic processes are often treated explicitly without introducing $\varepsilon_{g,\alpha}$ (see Section 32.5.1).

The dynamic correction $\Delta\beta = \Delta\eta_{mod} k_{ref}$ to the propagation constant, in bulk and QW lasers, is often approximated as related to the gain variation by means of a single parameter, the Henry's linewidth enhancement factor α_H (with different values used for the gain and absorber sections), e.g., using a relation $\Delta\beta = \Delta\beta_{SPM} = -\alpha_H(g - g_{th})$, the latter parameter being the threshold (or any other reference) value of peak gain. This phenomenological approach, although the simplest and the most traditional, ignores the fact that the spectral dependence of gain and carrier-induced refractive index correction can be different, so the linewidth enhancement factor should be, generally speaking, spectrally and carrier density dependent.

The gain dispersion represented by the operator nature of gain $\widehat{g}_{mod}$ (and to a certain extent GVD, represented by the operator nature of the modal refractive index $\widehat{\Delta\eta}_{mod}$) is very important in determining the *stability range* of ML (unless there is a dispersive element in the laser construction such as a DBR). As mentioned before, in the lumped model, no stable ML with a finite pulsewidth can be simulated in the absence of dispersion. In the distributed model, most models not including gain dispersion cannot predict stable ML either. Some authors (Bischoff et al., 1997) reported stable ML with finite pulsewidths simulated without the dispersion term, but the model of Bischoff et al. (1997) included finite relaxation times of nonlinearities, which may have had a side effect of introducing *effective* dispersion. As regards the numerical implementation of dispersion, in mode-locked laser constructions realized so far, the spectrum of mode-locked lasers, although quite broad, is still usually significantly narrower than that of gain/absorption, meaning that only the top of the gain curve needs to be represented accurately. Therefore, in most studies reported so far, the dispersion has been approximated in frequency domain, i.e., as function of $\Delta\omega = \omega - \omega_{ref}$, as a simple,

Lorentzian curve in complex numbers (similar to Equation 32.10 with $D = 0$), or Equation 32.20, equivalent to approximating the spectral properties of the material by those of an equivalent two-level medium with homogeneous broadening:

$$P_\pm^T(\Delta\omega) = -i\hat{g} \cdot Y_\pm^T(\Delta\omega) = -i\frac{g(N,S)}{1 - i\left(\Delta\omega - \Delta\omega_p(N)\right)/\gamma(N)}Y_\pm^T(\Delta\omega), \tag{32.42}$$

where the superscript T means Fourier transformed variable in frequency domain; $\gamma(N)$ is the gain spectral width parameter as in Equations 32.10 and 32.20; $\Delta\omega_p(N)$ is the spectral shift of the gain peak from the reference frequency, and $g(N,S)$ can be implemented as in Equation 32.41 (for the absorption, the same method can be used). In time domain, this can be implemented numerically by two alternative but largely equivalent methods. The first one, used in a number of papers (e.g., Bandelow et al., 2001), and traceable to early work on pulse generation in lasers with active media with homogeneously broadened gain spectrum (Fleck, 1968), consists of introducing a separate differential equation for gain polarization, as in the DDE approach. With the gain spectrum centered at the reference frequency ($\Delta\Omega'' = 0$), it takes a particularly simple form, similar to that of Equation 32.21 in the DDE approach:

$$\frac{\partial P_\pm(z,t)}{\partial t} = -\gamma\left(P_{f,b}(z,t) - i\frac{g}{2}(N,S)Y_{f,b}(z,t)\right). \tag{32.43}$$

The second method of introducing gain dispersion involves using digital filters of varying complexity (Avrutin et al., 2000, 2005; Heck et al., 2006). In the case of a simple Lorentzian gain, the filter is straightforwardly represented as an infinite impulse response (IIR) one of the form of

$$P_\pm = -i\hat{g}Y_\pm = -i\frac{\gamma}{2}\int_0^\infty g(z,t-\tau)Y_\pm(z,t-\tau)\exp\left(-\tilde{\gamma}_p\tau\right)d\tau; \quad \tilde{\gamma}_p = \gamma - i\Delta\omega_p. \tag{32.44}$$

In practice, the integral requires only storing one iteration in the computer memory: For small integration steps Δt it is easily implemented using a slight generalization of the formula given originally in Schell et al. (1991) as

$$P_\pm(t) = \exp\left(-\tilde{\gamma}_p\Delta t\right)P_\pm(t-\Delta t) - i\left(\gamma/\tilde{\gamma}_p\right)\left(1 - \exp\left(-\tilde{\gamma}_p\Delta t\right)\right)\frac{g(t)}{2}Y_\pm(t).$$

This method of gain dispersion implementation is more tolerant to the simulation time step than the separate differential equation (Equation 32.43), but may be not very accurate if the steps are not small enough (Avrutin et al., 2005; Schell et al., 1991). A more complex, but also somewhat more robust, form of a digital filter implementation of dispersion was described in Carroll et al. (1998) and applied to ML (see, e.g., Jones et al., 1995).

In a recent paper (Javaloyes and Balle, 2010b), a more complex digital filter, representing a more accurate model of the spectra of the complex dielectric permittivity (gain/absorption and refractive index), derived from a microscopic approach and allowing for the realistic fundamental absorption edge spectrum to be modeled, has been implemented; this will be discussed in Section 32.5.

The dispersive nature of the correction $\widehat{\Delta\beta} = k_{ref}\widehat{\Delta\eta}_{mod}$ to the (real part of) the propagation constant is usually less important than gain dispersion. In cases when very short (subpicosecond) pulses may be expected in the simulation, the operator $\widehat{\Delta\beta}$ may include an additional term describing GVD of the structure (Avrutin et al., 1996)

$$\hat{\Delta}\beta Y = k_{\mathrm{ref}}\widehat{\Delta\eta}_{\mathrm{mod}} Y = \hat{\Delta}\beta_{\mathrm{SPM}} Y - \frac{\beta_2}{2}\frac{\partial^2 Y}{\partial t^2}, \tag{32.45}$$

where the first term describes the self-phase modulation effects, and the second, the GVD, with $\beta_2 = \frac{1}{c}\frac{dn_g}{d\omega}$ the first-order GVD coefficient, n_g being the group velocity refractive index of the laser waveguide. Numerical simulations (Avrutin et al., 1996) show that GVD affects the parameters of picosecond pulses significantly for the dispersion values of $dn_g/d\omega \sim 10^{-14}$ s. Thus, this term is usually negligible in most QW lasers (where the GVD magnitude is estimated as $dn_g/d\omega = 10^{-16}$–10^{-15} s) and indeed is omitted in most models of mode-locked laser diodes published to date. In the microscopic or semimicroscopic implementations of TWM (Section 32.5), the dynamic and spectral variation of real, as well as imaginary, part of the dielectric permittivity of the material, and thus the active layer contribution to GVD, is implemented self-consistently. The dispersion (material and waveguide) of the passive waveguide structure is neglected but is believed to be weaker.

The terms containing the forward-back and back-forward propagating coupling constants $K_{\pm,\mp}$ need be included in the model only if a *Bragg grating* is present at the position z (and time t); the constants $K_{\pm}$ and $K_{\mp} = K_{\pm}^{*}$ are, in general, complex due to both refractive index and gain/absorption grating being possible. In the context of a mode-locked laser, accounting for a grating may be needed either if the laser construction contains a DBR section, or to account for *standing wave-induced gratings, or short-scale spatial hole burning* that is important if pulses propagating in opposite directions collide in the active medium (coherent colliding pulse effect). The standing wave-induced grating exists due to the carrier population being increased in the antinodes of the standing wave. Within the SA, where ML pulses are typically engineered to collide, this decreases the local absorption (Martins et al., 1995) and forms an absorption (and possibly refractive index, due to self-phase modulation) grating, to which the fast nonlinearities responding to the standing wave also contribute. The magnitude of the periodic carrier density modulation is then given by the equation (Martins et al., 1995)

$$\frac{d}{dt}N_{\mathrm{grat}}(z,t) = -\frac{N_{\mathrm{grat}}(z,t)}{\tau_{\mathrm{grat}}} + \frac{v_g}{2}(Y_+^*\hat{\alpha}Y_- + Y_-^*\hat{\alpha}Y_+), \tag{32.46}$$

where $\hat{\alpha} = -\hat{g}$ is the saturable absorption operator; the grating relaxation time in QW materials is mainly determined by ambipolar diffusion with the coefficient D_a:

$$\frac{1}{\tau_{\mathrm{grat}}} = \frac{1}{\tau_{\alpha}} + \frac{16\pi^2 D_a n_g^2}{\lambda^2}, \tag{32.47}$$

where λ is the lasing wavelength in vacuum. An estimate gives a value of $\sim$1 ps for the ambipolar diffusion coefficient of 2×10^{-4} m^2/s typical for III–V materials. From the magnitude of carrier density modulation, the coupling in QWs can be estimated as $K_{\mathrm{fb}} \approx i\frac{\partial\alpha}{\partial N}N_{\mathrm{grat}}(z,t)(1 - i\alpha_{\mathrm{H\alpha}})$. Assuming the Henry factor of the absorber is small, as was done in Martins et al. (1995), the grating becomes a purely absorption grating, smoothed down by diffusion. Equation 32.46 is written in terms of SA rather than gain section parameter since in ML lasers designed to utilize the colliding pulse effects, pulses traveling in different directions collide in the SA rather than the gain sections (the former, as will be discussed in the following section, assists ML, the latter impedes it). However, a TWM should contain the grating population in all type of sections; in gain sections, the gain operator $\hat{g} = -\hat{\alpha}$ is used.

The final term in Equation 32.40 is the random noise source that leads to the self-starting of the model and is essential for modeling of noise and pulse jitter. At the laser facets, standard reflection/transmission boundary conditions are imposed on $Y_\pm$; thus, unlike the delay-differential model, the traveling wave one accounts accurately for the laser geometry.

The traveling-wave equations are coupled with coordinate-dependent rate equations for the relevant populations. In the context of QW and bulk lasers:

$$\frac{\mathrm{d}}{\mathrm{d}t}N(z,t) = \frac{J(z,t)}{ed_a} - N\left(BN + \frac{1}{\tau_{nr}} + CN^2\right) - v_g Re\left(Y_+^* \hat{g}Y_+ + Y_=^* \hat{g}Y_-\right), \qquad (32.48)$$

with J/ed as the pumping term, J being the current density, e the elementary charge, d_a the active layer thickness, τ_{nr} the nonradiative recombination rate, and B and C are usually identified as the bimolecular recombination constant and the Auger recombination rate, respectively. Carrier capture dynamics is sometimes taken into account by adding an extra equation for carrier densities in the contact layers, but its significance for most mode-locked lasers (with the exception of QD active media and possibly some specially engineered QW constructions) tends to be modest, except where direct current modulation is involved.

TWMs are very powerful and general and their use is not restricted to ML edge-emitting lasers (see Chapter 31).

The main limitations of TW models are, first, the absence of any analytically solvable cases—the approach is by its very essence numerical, though some analytical insight is given by the modal decomposition of the traveling wave solution (see Chapter 31). Second, there is the fact that, in their traditional form, TWMs pose considerably higher requirements on the computing time and memory compared to delay-differential models. This is mainly due to the fact that, in the traditional implementation of the TWMs, time and space steps are usually related as $\Delta z = v_g \Delta t$, meaning that the computational timescales as $1/\Delta t^2$ and need to be sufficiently short (a typical spatial step being 1–5 μm) to reproduce the pulse characteristics faithfully (the problem is mathematically stiff). However, with the development of computer resources, this limitation has become progressively less important; and the numerical technique of *decimation* has allowed efficient decoupling of time and space steps (see Section 32.5.2). Several commercial or free software simulators of laser diodes include traveling-wave approach of some form as the core of their solver; some of those are directly applicable and have indeed been applied for the analysis of ML lasers; see, e.g., Avrutin and Rafailov (2012), Section 32.5.2 Avrutin and Rafailov (2012), and Chapter 31.

32.4 The Main Predictions of Mode-Locked Laser Theory

In this section, we shall overview some of the main results of ML modeling that can be obtained on a mainly phenomenological level.

32.4.1 Operating Regime Depending on the Operating Point

The most basic result of all the modern ML theories, confirmed by the experiments, is that the dynamics of semiconductor lasers intended for ML can be quite rich and can show, apart from stable ML near the fundamental round-trip frequency, a number of other dynamic regimes. Here, we shall briefly discuss the general trends in their dependence on the laser parameters.

One of the most important features in the dynamic map of operating regimes of a mode-locked laser is the self-sustained pulsations, or passive Q-switching instability at low currents. As shown in Figure 32.5, produced by the DDE model, the range of currents, or unsaturated gain, values in which this regime is observed increases with the amount of saturable absorption in the laser (which, in a given laser construction, either QW or QD, may be varied to some extent with reverse bias, due to electroabsorption). The other model parameter affected by the reverse bias is the *absorber lifetime* τ_a, which is known to decrease approximately exponentially with the reverse bias in QW materials (see Nikolaev and Avrutin, 2003 and references therein) and to some extent in QDs too (Malins et al., 2006). The dependence of the Q-switching range on τ_α is not straightforward; the Q-switching range tends to be broadest at a certain absorber recovery time, of the order of the round-trip time though somewhat longer, as can be seen for example in Figure 32.10. At longer τ_α values, the SP range slightly decreases. However, it also decreases as τ_α is *decreased*, and when τ_α reaches a certain value, of the order of a fraction of the round-trip, the Q-switching instability disappears completely, leaving a broad area of stable ML. This is illustrated in Figure 32.10, which is produced using a TWM and shows approximate borders of different dynamic regimes for a representative laser with a short, relatively broadband (length 50 μm, coupling coefficient $\kappa = 120$ cm^{-1}) DBR section similar to that realized in Bandelow et al. (2006) and with a cavity length designed for the fundamental ML frequency either near 80 GHz (Figure 32.10a) or near 40 GHz as in the experimental study (Figure 32.10b).

The importance of both τ_α and the unsaturated absorption in determining ML properties means, first, that care needs to be taken when interpreting the bias voltage effects on the performance of either QW or QD mode-locked laser, as the unsaturated absorption, the saturable absorption cross section, and the SA recovery time τ_α are all likely to be affected. The effect on the latter is probably the most significant though, as the dependence of τ_α on voltage is quite strong (exponential), while the effect on the unsaturated absorption appears, from measured threshold currents, to be more modest. Second, it means that for the same absorber parameters, longer lasers with longer repetition periods are less likely to suffer from the Q-switching instability, which needs to be kept in mind when analyzing the dynamics of QD lasers (due to the relatively low gain, these often have to be quite long if stable operation at the ground level wavelength band is desired).

The lower current (or unsaturated gain) limit of the self-pulsing instability may be positioned either below or above the low boundary of ML itself, depending on the gain and absorber saturation energies (*s*-parameter) and the absorber recovery time. If the boundary for ML is below that for self-pulsing (which tends to happen in longer lasers, when τ_α is significantly smaller than T_{RT} but not small enough to completely eliminate self-pulsing), then the stable ML range is split in two by the self-pulsing area, with an area of stable ML seen below the Q-switching limit at currents just above the threshold. The area is narrow, however, and the pulse powers generated in this regime are typically rather low. If, on the other hand, the boundary for ML is above that for self-pulsing (which tends to be the case for shorter lasers or longer absorber relaxation time, when $T_{RT} > \tau_\alpha$), then an area of pure self-pulsing, with noisy/chaotic filling of pulses, is seen at small to modest excess currents above threshold, as in Figure 32.5; as the current is increased, the pulses acquire a regular structure and the combined ML/SP regime develops.

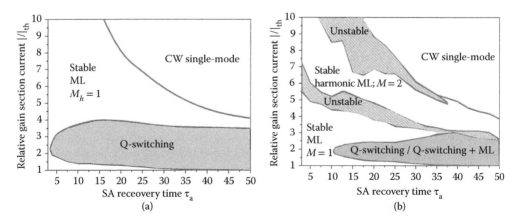

FIGURE 32.10 Schematic diagram of regimes in a generic QW mode-locked laser operating at the repetition rate of 80 GHz (a) and 40 GHz (b), calculated using a phenomenological TWM.

Comparing Figure 32.10a and b, we notice that the dynamics of the laser become richer with the increase of the cavity round-trip time, or, equivalently, of the number of modes in the laser spectrum. Indeed, the dynamic behavior of the short-cavity laser in Figure 32.10a displays only regular deterministic regimes (Q-switching, ML, CW) and only the fundamental frequency ML ($M_h = 1$ in Equation 32.1). In contrast, the dynamics of the longer laser in Figure 32.10b contain a range of currents corresponding to stable *harmonic* ML with $M_h = 2$, as in the DDE model. It is separated from fundamental frequency ML by a band of currents in which the laser shows the second (after Q-switching) main type of instability of ML: the leading edge, or chaotic, instability. As was shown by the DDE model, the onset/lower border of this instability is pushed somewhat toward higher currents by the increased amount of absorption in the laser. In addition, as illustrated in Figure 32.10b, shortening absorber relaxation time also somewhat decreases the risk of this instability.

At high currents, the unstable operating regime gradually evolves into some type of (irregular) quasi-CW operation (few modes present), which for some parameter values, with higher currents still, gives way to single-frequency, stable CW operation, as shown in Figure 32.10 (for longer τ_α) and predicted also by DDE. Whether or not this true single-frequency CW operation is achieved at a practically feasible current depends on the length of the laser and the gain bandwidth; longer lasers (with a repetition frequency $\sim$10 GHz and below) with broader gain spectrum tend to not reach true CW under any realistic pumping current, instead operating in a chaotic quasi-CW regime with a narrow spectrum including only a few modes.

In shorter lasers, as in Figure 32.10, a direct transition from stable ML, fundamental (Figure 32.10a) or harmonic (Figure 32.10b), to CW operation is also possible. This takes the form of a Hopf bifurcation, with the ML pulses acquiring a constant background with increased current and then their amplitude gradually reducing to zero resulting in CW operation.

32.4.2 The Main Parameters That Affect Mode-Locked Laser Behavior

In addition to determining the operating *regime* of the laser as discussed in the previous section, the theoretical models allow also the effects of the operating point on the main parameters of the optical pulse to be analyzed. Here, we shall concentrate mainly on the PML regime, in which case the most important parameters whose effect on the laser behavior the modeling can allow us to investigate are as follows:

32.4.2.1 The Pumping Current

The pulse *amplitude* grows with pumping current, as illustrated in Figure 32.11, calculated by a traveling-wave simulation of the laser of Figure 32.10b. Notice that qualitatively the behavior of the pulse amplitude shown in Figure 32.11a is quite close to the bifurcation behavior seen from Figure 32.5, except that in the TWM, and with the inevitably different set of parameters, only the second rather than third harmonic operation is predicted (it was also noted by some authors (Vladimirov et al., 2009) that the dynamics of an ML laser in the TWM can predict stable or unstable trailing pulses at different time detuning from the prevalent pulse stream, whereas the DDE simulation gives only harmonic operation even in unstable regimes).

As regards the pulse *duration* shown in Figure 32.11b, it has been predicted by early frequency-domain theories (Lau and Paslaski, 1991) to reach a minimum near the area of Q-switching instability; this has been later confirmed by both experiments and time-domain simulations and can be also seen in Figure 32.11b. The pulse duration thus tends to grow with current within the stable ML range *above* the upper boundary of Q-switching. On the other hand, if an area of stable ML below the Q-switching range is observed (which, as discussed above, is sometimes the case in longer resonators), then a *decrease* of pulse duration with current can be expected within this area.

32.4.2.2 The Absorber Relaxation Time

The dependence of the pulse duration on τ_α (and this on the absorber bias) within the stability range is shown in Figure 32.12. As seen in the figure, to achieve stable ML, the absorber relaxation time needs to be

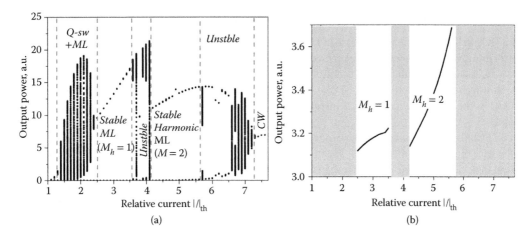

(a)

(b)

FIGURE 32.11 (a) Simulated dependence of the pulse amplitude on the pumping current (bifurcation diagram) for the laser of Figure 32.10b for one value of absorber recovery time, calculated with a phenomenological TWM. (b) Calculated pulse duration for the same laser within the current ranges for stable ML.

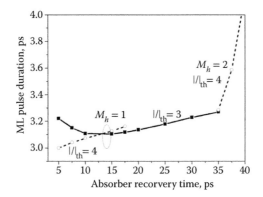

FIGURE 32.12 Typical simulated dependence of pulse duration on the absorber recovery time for two current values: simulations using a phenomenological TMW. Laser design same as in Figure 32.10b.

within a certain range. Values of τ_α above a certain value produce instabilities of either leading or trailing edge type, and may mean switching to harmonic ML. Within the stable ML range, a decrease in τ_α tends to shorten the pulses, due to both the effects of partial absorber relaxation during the pulse and, probably more significantly, to the fact that the slow relaxation of the absorber leads to the absorber being always partially saturated, thus reducing the initial absorption Q_-. This trend can sometimes be reversed for *very* short τ_α, of the order of a few picoseconds, when the absorber may not be saturated efficiently; therefore, there may be an optimum τ_α for shortest pulse generation, though for some laser parameter values this optimum τ_α may be so small as to be technologically unattainable.

The requirement of small τ_α in ML has been recognized from the early days. The methods of reducing τ_α in experiments included, first, ion implantation in early work (Deryagin et al., 1994; Zarrabi et al., 1991), second, choosing QW materials for faster sweepout, e.g., AlGaInAs rather than InGaAsP quaternaries (Green et al., 2011; Hou et al., 2010a,b, 2013) or even QD materials in which SA dynamics is known to be fast (Erneux et al., 2009; Rafailov et al., 2011; Viktorov et al., 2009), third, engineering the QW profile to include steps or oblique rather than vertical walls (Nikolaev and Avrutin, 2004), and fourth, using unitraveling carrier absorbers (Scollo et al., 2005, 2009); see Avrutin and Rafailov (2012) for more detail.

32.4.2.3 The s-Factor, or the Absorber to Gain Saturation Energy Ratio

While it is well known that the increase in s (determined by Equation 32.17) facilitates ML, it is not immediately intuitively clear whether an increased s helps ML stability, as it is known (Kuznetsov, 1985) that the passive Q-switching regime, which is one of the instabilities affecting ML, is also facilitated by an increase in s. However, the results from both DDE and traveling-wave simulations (Bandelow et al., 2006) show that in fact it is the stable ML range that is increased with s at the expense of the Q-switching (or self-pulsing) range.

In VECSELs, the most straightforward way of increasing the s parameter is manipulating the spot ratio of the gain module and the SESAM. For instance, the fast absorption saturation, or the high ratio s, was crucial in achieving the bistability between the off-state and the ML state necessary for low-repetition rate ML and addressable pulse generation studied in Marconi et al. (2014, 2015); this was achieved by placing the SESAM in the Fourier plane of the focusing lens so that the spot area $A_{X\alpha}$ was determined only by the diffraction limit. In *monolithic* diode lasers, a similar strategy can be pursued to some degree by *tapering* the laser waveguide so that the absorber region is narrower than the gain region. Used originally for Q-switching QW lasers in a "bow-tie" construction with the narrow SA in the middle of the cavity and the two amplifier sections tapering outward (Williams et al., 1994), this strategy has later been realized most convincingly in QD mode-locked lasers (Nikitichev et al., 2011a,b; Thompson et al., 2006, 2009); see Avrutin and Rafailov (2012) for an overview of results achieved. More often, monolithic structures use the same waveguide structure in the gain and absorber sections for ease of fabrication and fiber coupling, and so $A_{X\alpha} = A_{Xg}$. The parameter s is then equal to the *absorber-to-gain cross section ratio* $s = \sigma_a / \sigma_g$. QWs, with their sublinear (approximately logarithmic) dependence of gain on carrier density (population inversion) have long been seen as superior to bulk material for ML performance, since the sublinear $g(N)$ helps achieve $\sigma_a / \sigma_g > 1$. Vladimirov et al. (2004) and Vladimirov and Turaev (2005) used their DDE analysis to conclude further that when designing a QW laser for ML purposes, a structure with a *smaller number* of QWs was preferable to one with a larger number—indeed, the smaller number of QWs means a smaller confinement factor, hence a higher threshold carrier density and therefore a smaller dg/dN at threshold due to the sublinear $g(N)$, which in turns gives a higher value of the ratio s. These considerations influenced the choice of structures with just two to three QWs for realizing DBR ML lasers capable of generating very stable pulses about 2-ps long at 40 Gb/s (Bandelow et al., 2006). It may be argued that the same logic also, in part, accounts for the success of QD mode-locked lasers, in which the dependence of gain on the (total) carrier density in the active layer is even more sublinear than in QWs; however, it has to be borne in mind that the concept of total carrier density is somewhat misleading in QDs; a more accurate picture is given by more complex analysis, considering separately the population of the dots themselves and of the reservoir that supplies them with carriers.

32.4.2.4 Gain and Group Velocity Dispersion Parameters

Most models of mode-locked laser operation predict that without *gain* dispersion, stable ML with a finite pulse duration is impossible, and so the gain dispersion, or width of gain curve, represented by the parameter γ in the DDE or ω_L in the TWM, should play an important role in determining the pulsewidth and stability. Within the range of gain dispersion typical in mode-locked semiconductor lasers, which usually corresponds to $\hbar\omega_L$ of the order of tens of meV (or the wavelength range of tens of nm) and does not change too much with operating conditions or construction, gain dispersion is not the most drastic factor limiting the pulsewidth. However, achieving a broad gain spectrum is still desirable. This may be one of the advantages of QD active media, as discussed below.

GVD, like gain dispersion, acts to broaden the pulses in the case of normal dispersion, which is the usual situation in semiconductor lasers. As discussed above, the effect of this parameter is modest in most semiconductor lasers since the pulse durations at which it would become important ($\sim$100 fs) are never achieved; however, with stronger GVD possible in QD lasers, some account for this effect may be necessary.

32.4.2.5 The Gain and Absorber Compression Coefficients

Pulses generated by ML lasers tend to be of picosecond duration. This is below the critical pulsewidth at which the fast gain saturation, rather than the average carrier density dynamics, begins to dominate the pulse amplification and shaping, at least in the gain section; this critical pulse duration has been estimated (Mecozzi and Mork, 1997; Mork and Mecozzi, 1997) to be of the order of $\dfrac{\varepsilon_g}{v_g \dfrac{d_g}{d_N}}$. With typical semiconductor parameters, this estimate gives values of the order of 10 ps. Thus, the gain compression (and, similarly, absorber compression) effects and the coefficients that describe them (if introduced) may be expected to play a significant part in ML properties.

In practice, the effect of nonlinearities is twofold. First, gain compression tends to broaden ML pulses, with absorption compression having the opposite effect. Second, and in some regards more important, an increase in gain compression *stabilizes ML operation*, suppressing the Q-switching instability (the latter can be easily shown by rate equation analysis of Q-switched lasers (Avrutin et al., 1991). Again, fast absorber saturation has the opposite effect.

32.4.2.6 The Self-Phase Modulation in the Gain and Absorber Sections

Within the simplest phenomenological approach, the linewidth enhancement factors have a relatively modest effect on pulse *energy* for a given current and absorption, but a more noticeable one on amplitude and duration. They do not significantly affect the onset of the Q-switching instability (the lower current or unsaturated gain limit of ML stability), but have a stronger effect on the upper limit of ML stability associated with the irregular envelope and pulse competition. Mode-locked behavior is the most stable when the gain and absorber linewidth enhancement factors are not too different from each other. According to the DDE model predictions, the most stable operating point (which also corresponds to the highest pulse amplitude and lowest duration) is for $\alpha_{Hg} = \alpha_{H\alpha}$. Traveling-wave and modal analysis predicts that the best-quality ML is achieved with $\alpha_{Hg} > \alpha_{H\alpha}$, see, e.g., Salvatore et al. (1996); the discrepancy is likely to be caused by the different geometry of the long amplifier and the shorter absorber. The main parameter determined by the linewidth enhancement factors is the *chirp* (dynamic shift of the instantaneous frequency) of the pulse. Passively mode-locked pulses tend to be up-chirped (with the instantaneous optical frequency increasing toward the end of the pulse) when the absorber saturation factor $\alpha_{H\alpha}$ is small and the chirp is mainly caused by α_{Hg}. With a certain combination of α_{Hg} and $\alpha_{H\alpha}$ (typically $\alpha_{Hg} > \alpha_{H\alpha}$), an almost complete compensation of chirp is possible; with $\alpha_{H\alpha} > \alpha_{Hg}$, the pulse is typically down-chirped (Salvatore et al., 1996). As up-chirp is observed more frequently than down-chirp in experiments, one may conclude that typical values of $\alpha_{H\alpha}$ are smaller than α_{Hg}. In active ML, *down-chirp* is typically observed, while hybrid ML allows the chirp to be tuned to some extent, and there is typically a combination of bias and current or voltage modulation amplitude for which the chirp is minimized and close to zero, if only in a very narrow range of operating parameters. In a more detailed semimicroscopic model discussed in Section 32.5, the linewidth enhancement factor is not introduced *explicitly*, with gain and refractive index correction being implemented as imaginary and real part of the same complex dielectric permittivity, respectively, but the tendencies would appear to be general enough to merit being taken into account.

32.4.2.7 The Laser Geometry and Saturable Absorber Location

All the results discussed above could be obtained using either DDE or TWMs, though the latter usually give slightly more realistic predictions (e.g., both types of models predict harmonic operation at high currents, but in the case of multi-GHz PML ($T_{rt} \ll \tau_g$), operation at harmonics above the second one, predicted by the DDE model, is usually not observed experimentally; TWMs tend to predict only fundamental and second harmonic operation, which agrees with the experiment (Bandelow et al., 2006). The effects of a realistic laser cavity geometry (Fabry–Pérot versus ring cavity, absorber position and length, facet reflectances, etc.) on the ML characteristics are one area where the TWMs have an obvious advantage. They are, for example, highly suitable for analyzing the effects of specialized *harmonic* ML designs for high-frequency generation. These fall into two categories.

The first is colliding pulse mode-locking (CPM), including *multiple (MCPM)* (Martins et al., 1995; McDougall et al., 1997) and *asymmetric (ACPM)* (Shimizu et al., 1995, 1997) colliding-pulse ML constructions. These achieve ML at the M_h-th harmonic by positioning one SA (in CPM or ACPM) or several SAs (in MCPM) at fraction(s) M_h'/M_h of the laser cavity length, where $M_h' < M_h$ is an integer, and M' (or at least some of the values of M_h' in case of MCPM) and M are mutually prime. The standard CPM corresponds to $M_h = 2$ (and obviously $M_h' = 1$) with the SA in the center of the cavity. Constructions of this type have produced ML operation at rates of up to 860 GHz (Shimizu et al., 1997) (in that particular case, $\lambda = 1.55$ µm ACPM construction used $M_h' = 5, M_h = 12$). A more detailed review is given in Avrutin et al. (2000). The alternative harmonic ML technique is the use of a *spectrally selective laser cavity*. This may be in the form of a *compound cavity*, which has one or several *intracavity reflectors* (ICRs) positioned at fractions M_h'/M_h of the laser cavity length, rather like the SAs in the MCPM or ACPM technique. The highest ML repetition rates reported to date have been achieved by GaAs/AlGaAs ($\lambda \approx 0.89$ µm) lasers with the ICRs in the shape of deeply etched slots, either single or multiple in a 1-D photonic-bandgap (PBG) mirror arrangement (Yanson et al., 2002). A 608-µm long cavity used ICRs at 1/33 of the cavity lengths, giving $F = 2.1$ THz (similar structures with lower M_h values produced ML at bit rates of the order of hundreds of GHz depending on M_h and L).

TWMs have been long since been used for detailed analysis of the CPM (highlighting the importance of *incoherent* colliding pulse effect, the fact that the SA is saturated simultaneously by pulses propagating in both directions, as described by Equation 32.48, as opposed to coherent CPM effect, the self-induced grating described by Equation 32.46, which in semiconductors is less important than in other active media due to diffusion smoothing of the grating). All studies also predict higher stability and shorter pulses with CPM than with ordinary PML, in good agreement with experiment. A version of TWM was also used to explain and analyze the first realization of MCPM (Martins et al., 1995).

In terms of compound-cavity harmonic ML, TWM can help analyze the effects of the slot reflectances and positions (Yanson et al., 2002; Hou et al., 2010b, 2013) and bias current (Hou et al., 2013), with very good qualitative, and good quantitative, agreement with experiments. A TWM study was also used to propose and analyze the use of an alternative method of harmonic selectivity in a mode-locked laser resonator, a sampled grating reflector (Kim et al., 1999), later successfully realized experimentally, see, e.g., Hou et al. (2014) and references therein.

In principle, qualitative analysis can also be possible with DDE models (in fact, CPM can be analyzed, and ACPM predicted, in VECSEL structures using a DDE model (Avrutin and Panajotov, in preparation)), but this requires a customized model for each design whereas TWMs adapt to new design through a simple change of parameters. Harmonic ML is also one area where the time-frequency domain approaches (Section 32.3.2), despite their limitations, can give some insight into the cavity geometry required for successful harmonic operation (Martins et al., 1995; Yanson et al., 2002); however, TWM is still preferable for full quantitative description of the regime.

In addition to CPM, TWM modeling was also used to highlight another possibility allowing for a better saturation of the SA, which consists in placing the absorber close to antireflection-coated laser facet. This design is termed anticolliding mode-locking (ACML) and is inverse to that used for solid-state lasers where the absorber is usually placed close to a high reflection mirror, leading to self-colliding pulse mode-locking (SCML). In this latter situation, the pulse is allowed to interfere constructively with its own reflection onto the mirror leading to an improved absorber bleaching. For semiconductor laser diodes, however, such interferometric effects are rather weak due to the large value of the carrier diffusion coefficient which, in turn, reduces the coherent population grating created by the pulse self-interference as illustrated by Equation 32.47. The analysis of the ACML (Javaloyes and Balle, 2011) showed that this design leads to a substantial increase in output power, a reduction in amplitude and timing jitter, and an enlargement of the range of currents where stable PML can be obtained. The reason is a consequence of the strong increase of the laser field along the cavity axis and as it propagates toward the anti-reflection facet, yielding a more intense saturation of the absorber, and comparatively a weaker saturation of the gain. These regimes where the field is widely nonuniform along the cavity is one of the cases where a proper analysis demands using a TWM approach.

TWM was also used for analysis and optimization of the geometry of a number of other ML laser designs. Simulations of long-cavity ($F_{rep} \sim 10$ GHz) actively, passively, and hybridly mode-locked DBR lasers predicted that the optimal hybrid and PML performance could be achieved by placing the SA near the facet rather than at the DBR (Hasler et al., 2005), probably largely due to the more efficient absorber saturation as discussed above (Javaloyes and Balle, 2011). TWM studies were also used to predict, in agreement with experiment, that extended cavity laser designs containing a passive section were preferable to all-active ones in terms of pulse duration and chirp (Camacho et al., 1997).

32.5 Microscopic and Semimicroscopic Approaches in Mode-Locked Laser Modeling

32.5.1 The Basics of Microscopic Input in Mode-Locked Laser Simulations

As discussed above, phenomenological models of ML, either of delay-differential or traveling wave types, can be successfully used to predict a number of qualitative tendencies and, to a degree, qualitative parameters of the ML process. However, for detailed comparison with experiments, some microscopic input regarding the gain and SA section material is highly desirable. Some attempts at adding elements of microscopic analysis to models of the type described in the previous section has been undertaken for some time, see, e.g., Bischoff et al. (1997) for a relatively early example. The main challenge, and the main need for microscopic input, arguably, is the accurate modeling of interplay of the *spectral and temporal properties* of the material and laser dynamics. Most importantly, in most QW and, in some cases, QD lasers, the operating wavelength is at the sharp edge of the SA absorption spectrum. As the absorber is saturated, the absorption edge shifts, which lead to a change in the operating point, variation in the absorber saturation parameters affects self-phase modulation, etc. Likewise, due to the absorber selectivity, the operating point can be detuned from the gain peak in the amplifier section. In particular, this makes it difficult to construct a simulated map of regimes in the plane of "gain section bias current/absorber bias voltage," since the voltage affects simultaneously the absorber recovery time (relatively easy to model semimicroscopically as discussed below) and, more importantly, the spectral position of the absorption edge hence the unsaturated absorption and the cross section. This complex interplay of effects is difficult to capture with Equation 32.42 or any of its time-domain implementations, for the SA in particular, and to a degree for the gain section too. A more accurate (and thus ideally microscopically informed) model of gain and, particularly, saturable absorption spectrum as function of population (carrier density), and ideally temperature, is thus required.

The difficulty with this problem is that a realistic microscopic evaluation of gain/absorption in semiconductors is a formidable problem. To do so rigorously requires a many-body quantum mechanical approach (Chow and Koch, 1999), which is a task so complex from the point of view of the various physical effects at work, and so computationally resource-intensive that, at the time of writing, only a handful of research teams worldwide have both the expertise and the computer capacity to attempt it, and to our knowledge only one (the partnership between the Philipps-Universität Marburg and the University of Arizona) is performing this task routinely. The approach has been applied to some mode-locked laser problems, notably single-pass pulse modifications (Bottge et al., 2014) and steady-state ML pulse formation (Kilen et al., 2016) in ultrashort-pulse mode-locked VECSELs, and has informed some work on edge-emitting ML lasers, but to the best of our knowledge not yet been integrated into full analysis of high-bit-rate ML dynamics.

Even in the simpler single-particle approximation ignoring many-body effects, microscopic evaluation of the gain/absorption spectrum in a semiconductor material is, in general, rather nontrivial. For the case of a QW, or, with some reservations, QD active medium, the gain in the single-particle approach is calculated as

$$g(\hbar\omega) = A \sum_{\mu,\nu} \int \left|M_{\mu\nu}^2\right| (E_{e-h})\rho_{\mu\nu}(E_{e-h}) \left(f_e(E_e^{(\mu)}) + f_h(E_h^{(\nu)}) - 1\right) L'\left(\hbar\omega - E_{e-h}\right) dE_{e-h}. \qquad (32.49)$$

Here, A is a proportionality coefficient, E_{e-h} is the electron–hole energy separation, $E_e^{(\mu)}$ and $E_h^{(\nu)}$ are the energies of the electron in the conduction subband (in the case of a QW material) or level (in the case of QD material) μ and the hole in the valence subband/level ν, which are fully defined by the subband/level numbers and energy separation. Furthermore, $M_{\mu\nu}^2$ and $\rho_{\mu\nu}$ are the squared matrix element of the transition and the reduced density of states, respectively (each of which, in general, varies with energy and is, in general, calculated from a transcendental equation in the case of nonparabolic QW subbands), f_e and f_h are the electron and hole distribution functions (which in the case of QW materials are very close to Fermi distributions given by the carrier density N and temperature T, and for QDs, in general, need to be calculated from separate kinetic equations), and L'' is the bell-shaped linewidth broadening function.

For a bulk semiconductor material (rarely used in mode-locked diode lasers), the summation over μ and ν is not necessary as there are no multiple subbands in the valence and conduction bands, and the reduced density of states and the matrix element are simple analytical functions of energy.

Still, the expression for the gain/absorption spectrum, in general, can only be evaluated numerically in all cases. The refractive index modulation, needed to consider self-phase modulation in mode-locked lasers, is then reconstructed using Kramers–Kronig relations.

The results are rather difficult to parametrize and to implement in a time-domain model. Simple linear or logarithmic approximations of the microscopic results, often used in the phenomenological models described in the previous section, capture the carrier density dependence of the gain peak and the saturated absorption, but not the spectrum.

The first step toward simplifying the problem is to approximate the linewidth broadening function L'' (which in the general case of a non-Markoffian phase relaxation characteristic of a highly populated semiconductor has quite a complex shape) by a simple Lorentzian form, which follows from a density matrix analysis with a Markoffian phase relaxation with a constant decay rate γ_T and is thus essentially identical to the function describing homogeneous broadening in a two-level system (Equation 32.42):

$$L'(\Delta\omega) = \text{Re}L(\Delta\omega) = \frac{1}{\pi}\frac{\gamma_T}{\gamma_T^2 + \Delta\omega^2}; L(\Delta\omega) = \frac{1}{\pi}\frac{1}{\gamma_T - i\Delta\omega}. \qquad (32.50)$$

This form of L' is advantageous because the dynamic correction to the refractive index or the real part of the dielectric permittivity ("imaginary gain") is then calculated using the same equation as Equation 32.49 but using the complementary linewidth broadening function $L''(\Delta\omega) = \text{Im}L(\Delta\omega) = \frac{1}{\pi}\frac{\Delta\omega}{\gamma_T^2 + \Delta\omega^2}$, eliminating the need for Kramers–Kronig transformation and allowing both gain and refractive index variation to be implemented in time domain as discussed below.

Two main routes can then be successfully followed for using the results of Equation 32.49 in time-domain analysis required for ML simulations.

The first one consists in using a number of approximations in order to simplify the expression (Equation 32.49) (and its imaginary counterpart for the refractive index variation) to a level that permits the integration in energy to be performed analytically, and thus can be realistically parametrized as a (analytical) function of carrier density N and photon energy $\hbar\omega$. This has been achieved (Balle, 1998; Javaloyes and Balle, 2010a) by considering a QW material with a single, parabolic subband for both electrons and holes (hence constant reduced density of states), ignoring the energy dependence of M^2 (and γ_T) and considering (at least in the original version of the model) a very low (mathematically, zero) temperature. In the *frequency* domain (Balle, 1998), this results in a complex dielectric permittivity correction (with the refractive index correction as the real part and the gain as the imaginary part) in the form

$$\chi(\omega, N) = -\chi_0 \left[2\ln\left(1 - \frac{D}{u-i}\right) - \ln\left(1 - \frac{b}{u-i}\right) \right]. \tag{32.51}$$

Here, χ_0 is a constant proportional to the matrix element M^2 (Balle, 1998) (which in practice can be used as a fitting parameter in the model), $D = \frac{\pi d_{QW} \hbar}{m\gamma_T} N$ is the reduced carrier density, d_{QW} and $m = \left(m_e^{-1} + m_h^{-1}\right)^{-1} \approx m_e$ being the QW thickness and the reduced mass of the electron–hole pair; $b = \frac{\hbar k_m^2}{2m\gamma_T}$, k_m being the maximum wave vector in the first Brillouin zone of the semiconductor crystal; and the frequency of the optical transition is parameterized as $u = \frac{\hbar\omega - E_g}{\hbar\gamma_T}$, where the bandgap E_g includes N- and T-dependent renormalization.

This result was used, among other applications, in the work in DDE modeling of mode-locked VECSELs (Mulet and Balle, 2005) discussed in Section 32.3.3. In such lasers, the operating frequency and the spectral selectivity of the laser cavity are mainly determined by the resonator properties of the gain chip. A full time-domain implementation of the gain spectrum was thus not required; instead, the complex dielectric permittivity was linearized around the spectral point of operation which, as mentioned in Section 32.3.3, was sufficient to achieve very good qualitative, and good quantitative, agreement with previously reported experiments.

In time domain, the general form of polarization has the same form as Equation 32.44

$$P_{\pm} = \int_0^\infty \chi(\tau) Y_{\pm}(t - \tau) d\tau. \tag{32.52}$$

The time-domain version of Equation 32.51 results (Javaloyes and Balle, 2010a) in an integration kernel more complex than the simple exponential of Equation 32.44, specifically of the form

$$\chi_g(\tau, N) = \chi_0 e^{-\left(\gamma_T + i\left(\omega_g - \omega_{ref}\right)\right)\tau} \frac{2e^{-i\gamma_T D\tau} - 1 - e^{-i\gamma_T b\tau}}{\tau}. \tag{32.53}$$

Here, $\omega_g = E_g/\hbar$; the rest of parameters are the same as in Equation 32.51 and the normalized carrier density D is, strictly speaking, evaluated at the time $t - \tau$; however, since the carrier density even in the faster recovering SA, let alone the gain section, does not change noticeably on the timescale of $1/\gamma_T$, using the value at time t is usually accurate enough (Javaloyes and Balle, 2010a). Despite being, strictly speaking, only applicable for very low temperatures such that $\kappa_B T \ll \hbar\gamma_T$, the kernel (Equation 32.53) is a reasonable approximation for the highly degenerate gain medium. Its accuracy can be further improved to include the finite temperature, though this results in a more complex expression including special functions.

For the case of SAs, an alternative kernel expression was derived (Stolarz et al., 2011) with a finite temperature taken into account from the start, but with the Fermi distribution functions substituted by Boltzmann exponentials, which is usually a safe approximation for the weakly degenerate SA material; in this case,

$$\chi_\alpha(\tau, N) = \chi_0 e^{-\left(\gamma_T + i\left(\omega_g - \omega_{ref}\right)\right)\tau} \left\{ D\left[\frac{a_c}{a_c + i\gamma_T\tau} + \frac{a_v}{a_v + i\gamma_T\tau} \right] - \frac{1 - e^{-i\gamma_T b\tau}}{i\gamma_T\tau} \right\}, \tag{32.54}$$

where $a_{c,v} = \frac{m}{m_{c,v}} \frac{\hbar\gamma_T}{\kappa_B T}$ and the rest of the parameters are as in Equation 32.53. Unlike the simple purely Lorentzian kernel of Equation 32.44, neither of the functions (Equations 32.53 and 32.54) allow for a simple iterative solution using only the previous point in time; instead, the integral must in practice go to an integration limit in τ of about $\tau_{\text{lim}} \sim (3\text{--}6)/\gamma_{\text{T}}$, making the number of previous points necessary for the integration $\tau_{\text{lim}}/\Delta t$.

Despite being managed/mitigated in part by the efficient numerical implementation (Javaloyes and Balle, 2010a), this approach is still quite taxing on the computer resources if implemented straightforwardly in a TWM, since the need for numerical integration in the dispersion operator adds to the general limitations of the TWM approach (the stiffness of the problem requiring a small time step, and computational timescaling as $1/\Delta t^2$ due to the $\Delta z = v_g \Delta t$ condition). Its efficiency can, however, be improved drastically with the *decimation* technique discussed in Section 32.5.2.

The semimicroscopical approach in this form was successfully used to model a number of mode-locked laser designs, including a "straightforward" two-section laser with an SA at a facet (Javaloyes and Balle, 2010b), a colliding-pulse mode-locked laser (Tandoi et al., 2013), and a specialist ring laser with an intra-cavity filter for spectral flattening and pulse duration reduction (Moskalenko et al., 2013); we shall consider the results of Tandoi et al. (2013) here in somewhat more detail as an illustration of the possibilities and certain limitations of the model.

While the quantitative agreement with experimental results is still not entirely accurate, microscopic models of this type have allowed a number of new possibilities compared to purely phenomenological approaches. The first of these is a much more meaningful, and directly comparable with experiments, prediction of the map of regimes in the "gain section current/absorber bias" plane. Indeed, in QW materials, the voltage applied to the SA has a double effect. First, it decreases the absorber recombination time. To a good accuracy, this is described with an exponential dependence, e.g., in Tandoi et al. (2013), an expression

$$\tau_{\text{SA}} \approx \tau_0 \exp[-F/(f\, F_0)] \tag{32.55}$$

was used, with $F_0 = (\kappa_{\text{B}} T)/(e d_{\text{QW}})$ being the activation field predicted by the simple thermionic excitation theory (Cavailles et al., 1992), and a correction factor f is introduced heuristically to take into account the interplay between tunneling and thermionic processes (and possibly any well profile distortion). The values $f = 0.5$ and $\tau_0 = 50$ ps were given by fitting the measured $\tau(V_0)$ but agreed well with estimates from more sophisticated theory (Nikolaev and Avrutin, 2003). Second, in materials where the quantum-confined Stark effect (QCSE, absorption shift with voltage) is significant, a voltage variation changes the bandgap, thus affecting the operating point of the laser (unsaturated absorption and saturation cross section), which the microscopic model successfully captures. This tends to be more pronounced in InGaAsP than in aluminum-containing quaternaries. Strictly speaking, QCSE affects also the shape of the absorption spectrum, but this is as yet to be included in the model, since to the best of our knowledge there is as yet no accurate calculation of absorption in a QW with both carriers and electric field present.

Finally, analysis of CW ML behavior requires, in addition to models of absorption spectrum and the recovery time, also a model of the current heating in the absorber. Indeed, the material bandgap that features in the kernels (Equations 32.53 and 32.54) is the renormalized temperature- (and carrier density-) dependent one, and the absorber can get significantly (up to about $10°$) heated by the photocurrent flowing through it. This can be modeled (Tandoi et al., 2013) using a separate equation for the absorber bandgap:

$$\frac{dE_{\text{g}}^{(\text{SA})}}{dt} = -\gamma_{\text{therm}} \left(E_{\text{g}}^{(\text{SA})} - E_{\text{g0}}^{(\text{SA})} + 2\pi R_{\text{th}} j_{\text{SA}} \left(|V_{\text{SA}}| + V_{\text{bi}} \right) \right). \tag{32.56}$$

Here, R_{th} is the thermal resistance of the SA, the photocurrent density can be estimated as $j_{\text{SA}} = e N_{\text{SA}} d_{\text{QW}}/\tau_{\text{SA}}$, and the voltage seen by the active region in the SA is a sum of the built-in potential drop $V_{\text{bi}} = E_{\text{g0}}^{(\text{SA})}/e$ and the applied reverse bias $|V_{\text{SA}}|$. The parameters of the model were fitted in Tandoi et al. (2013) with the experimental measurements, and then the thermal relaxation rate γ_{therm}, which in reality is of the order of inverse microseconds, sped up to 10^8 s^{-1}, still slower than all characteristic times of laser dynamics but within the possibilities of the model.

Several features observed from the experimental results were well reproduced by the model, e.g., the optimal pulses around 600 fs long were found in the case of a SA occupying 4% of the cavity; some deviation from the best experimental results (430 fs) can be attributed to the approximate nature of the fitting of the gain curve with the two-band model (Equation 32.53).

With the microscopically informed gain and SA models as well as the voltage dependence on the SA recovery time, we constructed in Tandoi et al. (2013) maps of major characteristics of laser behavior in the "amplifier section current—absorber voltage" plane (Figure 32.13). The evolution of the ML quality as a function of the current followed the predictions of the DDE and TWM theories, with a minimal bias current to obtain the ML. The inspection of the time traces indicates that the degradation at high currents is due to the leading-edge instability and to the competition with other harmonics ML solutions, where three or four pulses propagate within the cavity. The evolution of the dynamical regimes as a function of the reverse bias was more subtle. In the AlGaAs materials studied in Tandoi et al. (2013), QCSE was weak and only the SA recovery time varied. This was, however, sufficient to capture the essential physical effects at work and to reproduce the output power decrease and the photocurrent density increase with V_{SA}. A faster SA has enough time to recover its full absorption between pulses and, therefore, presents the full amount of its unsaturated absorption to the pulse, which allows increased losses and photocurrent. Additionally, the photocurrent generated in the SA induces a detuning of the SA bandgap, which eventually hinders ML and decreases the output power. This is due to the fact that the SA can efficiently be modulated only within a limited spectral region around the bandgap and that the absorption increases on the blue side of the spectrum.

The alternative approach to including microscopically informed gain and absorption spectra into a mode-locked laser model has been so far implemented mainly in the context of QD lasers. In QD media, carriers are localized in individual dots and so the global carrier–carrier collisions that establish a common quasi Fermi level in a bulk or QW material are not present. QD lasers still show inhomogeneous

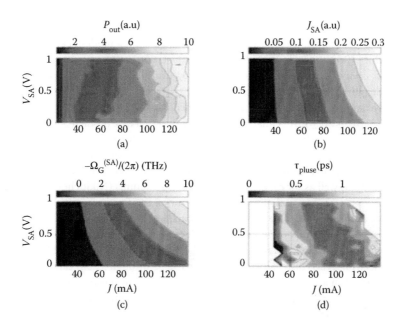

FIGURE 32.13 Parameter maps as a function of the SA reverse bias and of the gain section currently calculated using a TMW with decimation and analytical gain dispersion spectrum, for (a) output power P_{out}, (b) photocurrent density j_{SA} flowing out of the SA section, (c) bandgap of the SA section $-E_g^{(SA)}/h$, and (d) pulsewidth τ_{pulse}. The SA was 4% of the cavity length (optimal case, in agreement with measurements). (From Tandoi G. et al., *IEEE J Sel Top Quantum Electron*, 19, 1100608, 2013. Reproduced with permission).

broadening described mathematically by an equivalent of the reduced density of states in Equation 32.49, but the nature of this broadening is different from the case of bulk or QW materials and stems, not from a continuous energy spectrum of electrons and holes at each point in the material, but from the presence of a large number of dots, with fluctuating dot size and composition, and thus energy levels, in any microscopically large material sample, as already mentioned in Section 32.3.3. The Fermi energy distribution among the carriers occupying the multiple energy levels is established, not by the fast carrier–carrier collisions as in bulk and QW, but by slower processes of capture and escape and is thus only approximate at room temperatures and nonexistent at cryogenic temperatures. The gain thus cannot always be parameterized as a function of the total carrier density N. Instead, the populations of multiple individual levels need to be determined dynamically by solving rate equations including the kinetic processes of capture, interlevel relaxation, and escape, as well as spontaneous and stimulated recombination (see Rafailov et al., 2011 and the corresponding chapter in this handbook for a detailed description). In this case, trying to obtain an integral in Equation 32.49 analytically is problematic (some approximate methods have been discussed in Rafailov et al. (2011) for the room temperature case, but their accuracy has not been tested). Instead, a number of dynamic variables representing population of levels, or slices of the energy spectrum of the carriers in the material (different values of $E_{\text{e-h}}$ in Equation 32.49), at each point in time and space are introduced; the problem thus becomes not two, but (at least) three-dimensional: time, space, and transition energy. To reproduce a spectrum faithfully, the width of the slice should be smaller than, or at least comparable to, the homogeneous broadening γ_T. This approach to QD ML laser modeling is much more complex than the simpler one ignoring the inhomogeneous broadening as presented in Viktorov et al. (2006), but is also much more accurate. Its computational efficiency is comparable to that of using the analytical dielectric permittivity (Equations 32.52 through 32.54). Indeed, on the one hand, the number of variables characterizing carrier polarization (gain and refractive index modulation) at each point is not just the carrier density but the populations of the individual levels, but on the other hand, polarization relaxation of each individual level is Lorentzian (Equation 32.50), so an easy and efficient integration kernel of the type of Equation 32.44 can be implemented.

The resulting model has been used for mapping stability limits in a mode-locked laser and analysis of pulse shape and chirp (Rossetti et al., 2011a,b), as well as the analysis of the effects of laser cavity geometry (absorber location, cavity reflectivities) (Simos et al., 2013; Xu and Montrosset, 2013), in particular finding the optimum SA length (Xu and Montrosset, 2013).

The microscopic approach of this type can be implemented both in the traveling-wave formalism as in the papers above (particularly with decimation to improve the otherwise problematic numerical efficiency) and in a DDE model, particularly the multisection version (MS-DDE) as discussed in Section 32.3.3 (Rossetti et al., 2011b). With inhomogeneous broadening included, an MS-DDE model was used for the analysis of problems including the peculiarities of ML operation involving both ground- and excited-level transitions (Xu et al., 2012), as well as for modeling-tapered laser structures (Nikitichev et al., 2012), the latter in agreement with both a TWM simulation by the same authors and with experiments.

In Pimenov and Vladimirov (2014), a multilevel approach was presented for a TWM of a "generic" inhomogeneously broadened medium, not necessarily QD. It can be foreseen that potentially, it can be used for lasers with QW and bulk active layers, for example for a short-cavity QW laser where transitions from both first and second electron levels (disregarded in Equations 32.52 through 32.54) can become important. In principle, it could allow an arbitrary set of carrier density-dependent spectra, possibly calculated using a many-body approach and/or using realistic QW parameters such as multiple-level transitions, strain, etc., by necessity omitted in Equations 32.52 through 32.54, to be approximated and reproduced in the time-domain model. To a degree, the analytical spectrum approach (Equations 32.52 through 32.54) offers such an approximation possibility as well by treating the effective masses of carriers and matrix elements as fitting parameters, which has the advantages of involving much fewer parameters, but may be somewhat less flexible.

32.5.2 Improving the Numerical Efficiency of TWMs: Decimation/Space–Time Folding

So far, we have considered, on the one hand, the DDE models (Section 32.3.3) which are simple, efficient, instructive, and allow automated bifurcation analysis by numerical continuation and in the simplest cases also analytical insight, but are not always accurate in reproducing the behavior or realistic edge-emitting laser constructions, and, on the other hand, full TWM models (Section 32.3.4) which are potentially very accurate in representing a realistic laser, but rather computationally intensive and incompatible with automated bifurcation analysis.

The numerical technique of decimation presented below, which builds on an earlier model for laser designs where gain and SA dispersion are negligible compared to the effect of a lumped dispersive element (reflector) such as a DBR (Vladimirov et al., 2009) to apply to the general case of significant gain and SA dispersion (Javaloyes and Balle, 2012b), allows the advantages of both models to be combined to a large degree.

The starting point for the technique is a TWM in the general form of Equation 32.40, which can be compactly rewritten (keeping only the deterministic part so far) as

$$\pm \frac{\partial Y_\pm}{\partial z} + \frac{1}{v_g} \frac{\partial Y_\pm}{\partial t} = Z_\pm = P_\pm^t - \alpha_f Y_\pm, \tag{32.57}$$

where $\alpha_f = \alpha_{int}/2$ is the "field" internal loss, and the "total" polarization $P_\pm^t$ includes the dynamic coupling terms:

$$P_\pm^t = \left(\frac{1}{2} \hat{g}_{mod} + i k_{ref} \widehat{\Delta \eta}_{mod} \right) Y_\pm + i \widehat{K}_{\pm,\mp} Y_\mp. \tag{32.58}$$

(Note that the dynamic coupling terms can include dispersion in the same way as the "copropagating" polarization terms, as is indeed the case in the semimicroscopic model used in Javaloyes and Balle (2010a,b), though this is not likely to have a major effect as the dynamic coupling is usually weak anyway.) The formal solution of Equation 32.51 is obtained in the form

$$Y_\pm(z, t) = Y_\pm(z \mp v_g \tau, t - \tau) + \Xi_\pm(z, t); \quad \Xi_\pm(z, t) = \int_0^\tau Z_\pm(z \mp v_g(\tau - \vartheta), t - \tau + \vartheta) d\vartheta. \tag{32.59}$$

Then, assuming that the time interval is sufficiently short for the variation of the field along the traveling coordinate $\xi_\mp = z \mp v_g t$ to be modest (crucially, no such requirement is made to the variations in z and t *separately*—in short-pulse ML, these can be quite strong due to the steep pulse fronts, but the evolution in the pulse shape as it propagates is relatively gentle meaning that variation along $\xi_\mp$ remains gentle), it is possible to approximate $\Xi_\pm(z, t) = \frac{v_g \tau}{2}(Z_\pm(z \mp v_g \tau, t - \tau) + Z_\pm(z, t))$. Then, using a (1,1) Padé approximation $\exp(x) \approx (1 + x/2)/(1 - x/2)$, one obtains a final formula relating the fields and polarizations at the previous and current points:

$$\left(1 + \frac{\alpha_f \tau}{2} \right) Y_\pm(z, t) = \left(1 - \frac{\alpha_f \tau}{2} \right) Y_\pm(z \mp v_g \tau, t - \tau) + \frac{v_g \tau}{2}(P_\pm^t(z \mp v_g \tau, t - \tau) + P_\pm^t(z, t)). \tag{32.60}$$

This formula can be, and is, implemented directly in a numerical solver, in which a laser is separated into a number of segments n_{sect} with a length $v_g \tau_j, j = 1, \ldots, n_{sect}$.

Accurate calculation of the polarization with the spectrum taken into account still requires a small *temporal* integration step Δt (commensurate with $1/\gamma_T$); however, crucially, when using Equation 32.59 as a basis of the numerical procedure, the *spatial* step $v_g \tau_j$ is limited only by the degree of uniformity with the field (roughly speaking, the requirement that $v_g \tau_j \left| g_{\text{mod } j} \right| << 1$, $g_{\text{mod}j}$ being the characteristic value of the modal gain (or absorption) in the section of the laser with the length $v_g \tau_j$). This is a much less stringent requirement, meaning that the space and time steps are essentially decoupled and the computational time therefore scales, not as $1/\Delta t^2$ as in the straightforward TWM, but only approximately as $1/\Delta t$.

From the numerical point of view, it is important to have an integer relation between the time and space steps:

$$\tau_j = n_{dj} \Delta t, \tag{32.61}$$

where the integer number n_{dj} is the *decimation factor* of the spatial section *j*. It is not necessary (nor advisable) for all the sections to have the same decimation factor; e.g., typically, in the SA part of a passively or hybridly mode-locked laser (where the field is less uniform), it is advisable to set the decimation factor smaller than in the more uniform gain sections. For numerical simplicity, identical decimation factors are typically used in all gain sections and in all SA sections, e.g., $n_{d(\text{gain})} = 71$ and $n_{dSA} = 8$ were used in Tandoi et al. (2013). This implies that the computation time required in a model with decimation is only a few percent of the time needed for the same calculation with a straightforward TWM model, which is indeed the case.

Mathematically speaking, a model with decimation is a set of n_{sect} delayed algebraic equations (DAEs) with the delay of $\tau_j = n_{dj} \Delta t$ in each section *j*. This section replaces having to deal with n_{dj} spatial steps which would be the case in a straightforward TWM analysis, which is why the alternative (original) term for the decimation technique is *folding space into time delay*. As the number of the DAEs in a typical simulation is not large (<∼10), the decimated model, unlike a straightforward TWM, is compatible with the numerical tools for bifurcation analysis.

The decimation/space-time folding technique of Tandoi et al. (2013) was not the first attempt at improving the efficiency of the traveling-wave solver. For example, this was attempted, with some degree of success, in the semianalytical TWM technique by Carroll et al. (Jones et al., 1995; Carroll et al., 1998), in which, however, the (homogeneous broadening type) digital filter was explicitly incorporated into the solver and the condition $\Delta z = v_g \Delta t$, eliminated in the decimated TWM, still stood. The decimation technique is thus both much more powerful and more versatile in that it can be used with *any arbitrary method* of calculating the polarization $P^t_\pm$, including either of the two microscopic approaches discussed in Section 32.5.1, or indeed a phenomenological two-level type dispersion representation of Equation 32.44.

However, the decimation technique is both particularly useful and particularly instructive with the analytical kernel technique of Equations 32.54 and 32.57, for which it was originally derived and used (which is the reason we have chosen to place the description of the model within the section on microscopic input). Figure 32.14 illustrates the "stencils" showing the time and space points required for propagating the fields in three major cases: (1) the TWM with a homogeneously broadened line (Figure 32.14a), (2) a straightforwardly implemented TWM with the analytical kernel (Figure 32.14b), and (3) the decimated model with the same analytical kernel and $n_{d(\text{gain})} = 8$, $n_{dSA} = 2$ (Figure 32.14c). The sparsity of the stencil clearly shows how the decimation allows the sophistication of the microscopic model to be retained while drastically improving the efficiency of the calculation.

We note, following Javaloyes and Balle (2012b), that the decimated TWM contains as particular cases the two other major model types discussed in previous sections. Indeed, with $n_{d(\text{gain})} = n_{dSA} = 1$, it reduces to a straightforward TWM of Section 32.3.4. For a unidirectional ring cavity (and particularly for a homogeneously broadened line), it reproduces the DDE approach of Section 32.3.3 (Javaloyes and Balle (2012b) noted also that for a single-frequency laser, their model can, in principle, reduce to a generalized complex rate equations formalism, but that is by definition not relevant for ML). A decimated TWM with an analytical kernel technique for gain/refractive index dispersion has been implemented to include a number of laser geometries including but not restricted to mode-locked lasers, and is available online as free software (MATLAB® toolkit) under GPL license (Javaloyes and Balle, 2012a).

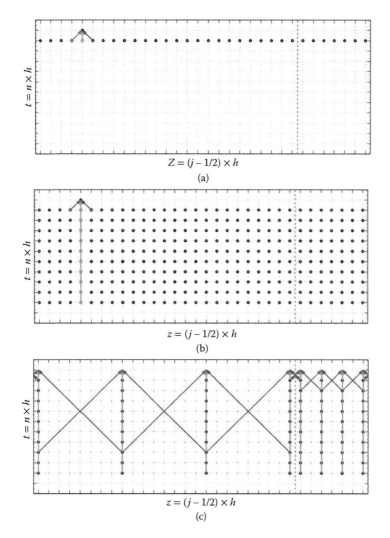

FIGURE 32.14 Calculation "stencils" illustrating field propagation in traveling wave modeling. (a) TWM with homogeneous broadening (Lorentzian gain spectrum), (b) analytical gain spectrum in a straightforward TWM, and (c) analytical gain spectrum in a TWM with decimation.

32.6 Some Novel Problems and Challenges in Mode-Locked Laser Modeling

32.6.1 Coherent Population Effects as a Possible Saturable Absorption Mechanism

One of the most important developments in mode-locked semiconductor laser technology in the recent years has been the direct generation of ultrashort (tens to a couple of hundreds of femtoseconds) pulses that until recently have been the domain of solid-state lasers only. Only one type of semiconductor lasers has been reported to produce such pulses so far, namely an optically pumped VECSEL, with the gain and SESAM chips in a folded (V-shaped) external cavity arrangement (Klopp et al., 2009, 2011; Quarterman et al., 2009; Wilcox et al., 2008). One of the teams that produced the femtosecond ML pulses attributed their results to coherent effects in the SESAM elements, namely the optical (ac) Stark effect. This effect, related to self-induced transparency, has the relaxation rate equal to the inverse of the dephasing (coherence decay) time. In semiconductors, this time is determined by carrier–carrier (and carrier–phonon) collisions and is of the order of 50–100 fs. This belongs to the *fast*, rather than slow, absorber regime, when the SA recovery time is shorter than the pulse duration. Indeed, theoretical analysis (Mihoubi et al., 2008; Wilcox et al., 2008), based on iterative small-signal time domain approach not dissimilar to that covered in Section 32.3.1 for a standard SA, showed that the pulse duration possible with this mechanism is about twice the dephasing time, which agrees well with the experiments. Further developments of this work are ongoing and are likely to rely not only on the experimental progress but also achieving better understanding of the limits and the requirements on the laser model through improved modeling.

There has also been work recently on the theoretical analysis of (so far, hypothetical) coherent ML through self-induced transparency effects in an *edge-emitting* class B (possibly semiconductor) laser geometry (see, e.g., Arkhipov et al., 2015a, 2016a,b), which promises femtosecond pulses but relies on *extending* the phase relaxation time, possibly by working under cryogenic conditions.

32.6.2 Spontaneous ML in Single-Section Lasers

Most of the discussions above concerned passive or hybrid ML constructions including an SA. However, in recent years, a very interesting development in ML laser technology occurred, when several teams have observed—and utilized—ML in single-section lasers without SA sections and without any external modulation either. The ML in this case was not of a pure AM ML type and the laser output only acquired short-pulse characteristics after external chirp compensation.

This effect has been observed in *quantum dash* (see, e.g., Duan et al., 2009; Gosset et al., 2006; Rosales et al., 2011), *QD* (Lu et al., 2011; Renaudier et al., 2005), QW (Sato, 2003; Yang, 2011), and even bulk materials (Yang, 2011), as well as in intersubband, far-infrared quantum cascade lasers (Faist et al., 2016 and references therein) showing that the effect is fairly generic. Further optimization of these lasers may partly depend on the establishment of full understanding of their behavior, yet at the time of writing, a full, universally agreed, theoretical explanation for ML in single-section lasers is still pending, and it is not impossible that different effects can play the main part in different constructions. In the past, some authors (Shore and Yee, 1991; Yee and Shore, 1993) used frequency-domain models with postulated, and differing, self- and cross- nonlinearity coefficients to show that nonlinearities in single-section semiconductor lasers could lead to a steady-state regime with fixed phases; this was predicted to produce, not the AM ML that corresponds to short-pulse emission, but the so-called frequency modulation ML, in which the phases of adjacent modes differ approximately by π, and the outcome is a CW-like regime with periodic carrier frequency oscillation. More recent works (Nomura et al., 2002; Renaudier et al., 2007), also using frequency-domain analysis with microscopically calculated (Nomura et al., 2002) or phenomenological (Renaudier et al., 2007) description of linear and nonlinear gain, predicted a possibility of ML-type signal generation, including AM ML for certain cavity lengths and active layer parameters, in a single-section laser with three modes involved in lasing, due to mode coupling by population pulsations/four-wave mixing effects. This appears to agree with traveling-wave modeling (Section 32.3.4) analysis for the case of a DBR laser without an SA (Bardella and Montrosset, 2005); Bardella and Montrosset (2005) also identified the role of four-wave mixing in their construction. Recently, a microscopic TWM analysis of a *QD* single-section laser has been presented (Gioannini et al., 2015), relying on both the nonequilibrium occupation of multiple levels due to inhomogeneous broadening and an additional phenomenological gain compression coefficient. The authors concluded that the suppressed carrier diffusion and the fast ground state (GS) gain recovery, typical of quantum dashes and QDs, were the mechanisms behind the phase-locking among the laser modes. Time-domain modeling on single-section ML (comb generation) in quantum cascade lasers has also been reported (Tzenov et al., 2016), though it has to be noted that the short upper state lifetime in those lasers makes their dynamics very different to that of other semiconductor lasers.

Further work extending the understanding achieved this far to other active media is desirable and hopefully forthcoming.

32.7 Concluding Remarks

We have attempted to review the most important developments in semiconductor mode-locked laser modeling. The topic is very dynamic and fast developing, so the choice of emphasis in this review was by necessity subjective; we apologize to those authors whose work may not have been given due prominence.

Acknowledgment

J.J. acknowledges financial support project COMBINA (TEC2015-65212-C3-3-P MINECO/FEDER UE)

References

Arkhipov R, Pimenov A, Radziunas M, Rachinskii D, Vladimirov AG, Arsenijevic D, Schmeckebier H, Bimberg D (2013) Hybrid mode locking in semiconductor lasers: Simulations, analysis, and experiments. *IEEE J Sel Top Quantum Electron* 19:1100208.

Arkhipov RM, Amann A, Vladimirov AG (2015b) Pulse repetition-frequency multiplication in a coupled cavity passively mode-locked semiconductor lasers. *Appl Phys B-Lasers Opt* 118:539–548.

Arkhipov RM, Arkhipov MV, Babushkin IV (2015a) On coherent mode-locking in a two-section laser. *Jetp Lett* 101:149–153.

Arkhipov RM, Arkhipov MV, Babushkin I (2016a) Self-starting stable coherent mode-locking in a two-section laser. *Opt Commun* 361:73–78.

Arkhipov RM, Arkhipov MV, Babushkin I, Rosanov NN (2016b) Self-induced transparency mode locking, and area theorem. *Opt Lett* 41:737–740.

Arkhipov RM, Habruseva T, Pimenov A, Radziunas M, Hegarty SP, Huyet G, Vladimirov AG (2016c) Semiconductor mode-locked lasers with coherent dual-mode optical injection: Simulations, analysis, and experiment. *J Opt Soc Am B-Opt Phys* 33:351–359.

Avrutin EA, Arnold JM, Marsh JH (1996) Analysis of dynamics of monolithic passively mode-locked laser diodes under external periodic excitation. *IEE Proc-Optoelectron* 143:81–88.

Avrutin EA, Arnold JM, Marsh JH (2003) Dynamic modal analysis of monolithic mode-locked semiconductor lasers. *IEEE J Sel Top Quantum Electron* 9:844–856.

Avrutin EA, Marsh JH, Portnoi EL (2000) Monolithic and multi-gigahertz mode-locked semiconductor lasers: Constructions, experiments, models and applications. *IEE Proc-Optoelectron* 147:251–278.

Avrutin EA, Nikolaev VV, Gallagher D (2005) Monolithic mode-locked semiconductor lasers. In: *Optoelectronic Devices—Advanced Simulation and Analysis* (Piprek J, ed.), pp. 185–215. New York, NY: Springer.

Avrutin and Panajotov, in preparation.

Avrutin EA, Portnoi EL, Chelnokov AV (1991) Effect of nonlinear amplification on characteristics of quality modulation regime in semiconducting laser with fast saturated absorbers. *Pisma V Zhurnal Tekhnicheskoi Fiziki* 17:49–54.

Avrutin EA, Rafailov EU (2012) Advances in mode-locked semiconductor lasers. In: *Advances in Semiconductor Lasers* (Coleman JJ, Bryce AC, Jagadish C, eds.), San Diego: Academic Press, pp. 93–147.

Avrutin EA, Russell BM (2009) Dynamics and spectra of monolithic mode-locked laser diodes under external optical feedback. *IEEE J Quantum Electron* 45:1456–1464.

Balle S (1998) Simple analytical approximations for the gain and refractive index spectra in quantum-well lasers. *Phys Rev A* 57:1304–1312.

Bandelow U, Radziunas M, Sieber J, Wolfrum M (2001) Impact of gain dispersion on the spatio-temporal dynamics of multisection lasers. *IEEE J Quantum Electron* 37:183–188.

Bandelow U, Radziunas M, Vladimirov A, Huttl B, Kaiser R (2006) 40 GHz mode-locked semiconductor lasers: Theory, simulations and experiment. *Opt Quantum Electron* 38:495–512.

Bardella P, Montrosset I (2005) Analysis of self-pulsating three-section DBR lasers. *IEEE J Sel Top Quantum Electron* 11:361–366.

Bischoff S, Mork J, Franck T, Brorson SD, Hofmann M, Frojdh K, Prip L, Sorensen MP (1997) Monolithic colliding pulse mode-locked semiconductor lasers. *Quantum Semiclass Opt* 9:655–674.

Bottge CN, Hader J, Kilen I, Moloney JV, Koch SW (2014) Ultrafast pulse amplification in mode-locked vertical external-cavity surface-emitting lasers. *Appl Phys Lett* 105:261105.

Brown CTA, Cataluna MA, Lagatsky AA, Rafailov EU, Agate MB, Leburn CG, Sibbett W (2004) Compact laser-diode-based femtosecond sources. *New J Phys* 6:175.

Camacho F, Avrutin EA, Cusumano P, Helmy AS, Bryce AC, Marsh JH (1997) Improvements in mode-locked semiconductor diode lasers using monolithically integrated passive waveguides made by quantum-well intermixing. *IEEE Photonics Technol Lett* 9:1208–1210.

Carroll JE, Whiteaway J, Plumb D (1998) *Distributed Feedback Semiconductor Lasers*, 2nd Edition. Stevenage: IET.

Cataluna MA, Nikitichev DI, Mikroulis S, Simos H, Simos C, Mesaritakis C, Syvridis D, Krestnikov I, Livshits D, Rafailov EU (2010) Dual-wavelength mode-locked quantum-dot laser, via ground and excited state transitions: Experimental and theoretical investigation. *Opt Express* 18:12832–12838.

Cataluna MA, Viktorov EA, Mandel P, Sibbett W, Livshits DA, Kovsh AR, Rafailov EU (2006) Temperature dependence of pulse duration in a mode-locked quantum-dot laser: Experiment and theory. In: *The 19th Annual Meeting of the IEEE Lasers and Electro-Optics Society* (IEEE, ed), pp. 798–799. Montreal, Canada.

Cataluna MA, Viktorov EA, Mandel P, Sibbett W, Livshits DA, Weimert J, Kovsh AR, Rafailov EU (2007) Temperature dependence of pulse duration in a mode-locked quantum-dot laser. *Appl Phys Lett* 90:101102–101103.

Cavailles JA, Miller DAB, Cunningham JE, Wa PLK, Miller A (1992) Simultaneous measurements of electron and hole sweep-out from quantum-wells and modeling of photoinduced field screening dynamics. *IEEE J Quantum Electron* 28:2486–2497.

Chow WW, Koch SW (1999) *Semiconductor Laser Fundamentals. Physics of the Gain Materials*. Berlin, Heidelberg: Springer.

Coldren LA, Corzine SW, Mashanovitch ML (2012) *Diode Lasers and Photonic Integrated Circuits*, 2nd Edition. New York, NY: Wiley.

Delpon EL, Oudar JL, Bouche N, Raj R, Shen A, Stelmakh N, Lourtioz JM (1998) Ultrafast excitonic saturable absorption in ion-implanted InGaAs/InAlAs multiple quantum wells. *Appl Phys Lett* 72:759–761.

Deryagin AG, Kuksenkov DV, Kuchinskii VI, Portnoi EL, Khrushchev IY, Frahm J (1994) Generation of high repetition frequency subpicosecond pulses at 1.535 mu-m by passive mode-locking of InGaAsP/InP laser diode with saturable absorber regions created by ion implantation. In: *14th IEEE International Semiconductor Laser Conference*, pp 107–108. New York, NY: IEEE.

Duan GH, Shen A, Akrout A, Van Dijk F, Lelarge F, Pommereau F, LeGouezigou O, Provost JG, Gariah H, Blache F, Mallecot F, Merghem K, Martinez A, Ramdane A (2009) High performance InP-based quantum dash semiconductor mode-locked lasers for optical communications. *Bell Labs Tech J* 14:63–84.

Dubbeldam JLA, Leegwater JA, Lenstra D (1997) Theory of mode-locked semiconductor lasers with finite absorber relaxation times. *Appl Phys Lett* 70:1938–1940.

Ell R, Morgner U, Kartner FX, Fujimoto JG, Ippen EP, Scheuer V, Angelow G, Tschudi T, Lederer MJ, Boiko A, Luther-Davies B (2001) Generation of 5-fs pulses and octave-spanning spectra directly from a Ti: Sapphire laser. *Opt Lett* 26:373–375.

Erneux T, Viktorov EA, Mandel P, Piwonski T, Huyet G, Houlihan J (2009) The fast recovery dynamics of a quantum dot semiconductor optical amplifier. *Appl Phys Lett* 94:113501.

Faist J, Villares G, Scalari G, Rosch M, Bonzon C, Hugi A, Beck M (2016) Quantum cascade laser frequency combs. *Nanophotonics* 5:272–291.

Fleck JA (1968) Emission of pulse trains by Q-switched lasers. *Phys Rev Lett* 21:131–133.

Gioannini M, Bardella P, Montrosset I (2015) Time-domain traveling-wave analysis of the multimode dynamics of quantum dot Fabry-Perot lasers. *IEEE J Sel Top Quantum Electron* 21:1–11.

Gosset C, Merghem K, Martinez A, Moreau G, Patriarche G, Aubin G, Landreau J, Lelarge E, Ramdane A (2006) Subpicosecond pulse generation at 134 GHz and low radiofrequency spectral linewidth in quantum dash-based Fabry-Perot lasers emitting at 1.5 μm. *Electron Lett* 42:91–92.

Green RP, Haji M, Hou LP, Mezosi G, Dylewicz R, Kelly AE (2011) Fast saturable absorption and 10 GHz wavelength conversion in Al-quaternary multiple quantum wells. *Opt Express* 19:9737–9743.

Hasler KH, Klehr A, Wenzel H, Erbert G (2005) Simulation of high-power pulse generation due to modelocking in long multisection lasers. *IEE Proc-Optoelectron* 152:77–85.

Haus HA (1975) Theory of mode-locking with a slow saturable absorber. *IEEE J Quantum Electron* 11:736–746.

Haus HA (2000) Mode-locking of lasers. *IEEE J Sel Top Quantum Electron* 6:1173–1185.

Haus HA, Silberberg Y (1985) Theory of mode-locking of a laser diode with a multiple-quantum-well structure. *J Opt Soc Am B-Opt Phys* 2:1237–1243.

Heck MJR, Bente E, Barbarin Y, Lenstra D, Smit MK (2006) Simulation and design of integrated femtosecond passively mode-locked semiconductor ring lasers including integrated passive pulse shaping components. *IEEE J Sel Top Quantum Electron* 12:265–276.

Hou LP, Avrutin EA, Haji M, Dylewicz R, Bryce AC, Marsh JH (2013) 160 GHz passively mode-locked AlGaInAs 1.55 mu m strained quantum-well lasers with deeply etched intracavity mirrors. *IEEE J Sel Top Quantum Electron* 19:1100409.

Hou LP, Haji M, Dylewicz R, Stolarz P, Qiu BC, Avrutin EA, Bryce AC (2010b) 160 GHz harmonic mode-locked AlGaInAs 1.55 mu m strained quantum-well compound-cavity laser. *Opt Lett* 35:3991–3993.

Hou LP, Haji M, Marsh JH (2014) Mode-locking and frequency mixing at THz pulse repetition rates in a sampled-grating DBR mode-locked laser. *Opt Express* 22:21690–21700.

Hou LP, Stolarz P, Dylewicz R, Haji M, Javaloyes J, Qiu BC, Bryce C (2010a) 160-GHz passively mode-locked AlGaInAs 1.55-mu m strained quantum-well compound cavity laser. *IEEE Photonics Technol Lett* 22:727–729.

Innerhofer E, Sudmeyer T, Brunner F, Haring R, Aschwanden A, Paschotta R, Honninger C, Kumkar M, Keller U (2003) 60-W average power in 810-fs pulses from a thin-disk Yb: YAG laser. *Opt Lett* 28:367–369.

Jaurigue L, Nikiforov O, Scholl E, Breuer S, Ludge K (2016) Dynamics of a passively mode-locked semiconductor laser subject to dual-cavity optical feedback. *Phys Rev E* 93:022205.

Jaurigue L, Pimenov A, Rachinskii D, Scholl E, Ludge K, Vladimirov AG (2015) Timing jitter of passively-mode-locked semiconductor lasers subject to optical feedback: A semi-analytic approach. *Phys Rev A* 92:053807.

Javaloyes J, Balle S (2010a) Quasiequilibrium time-domain susceptibility of semiconductor quantum wells. *Phys Rev A* 81:062505.

Javaloyes J, Balle S (2010b) Mode-locking in semiconductor Fabry-Perot lasers. *IEEE J Quantum Electron* 46:1023–1030.

Javaloyes J, Balle S (2011) Anticolliding design for monolithic passively mode-locked semiconductor lasers. *Opt Lett* 36:4407–4409.

Javaloyes J, Balle S (2012a) Freetwm: A simulation tool for semiconductor lasers. In: Freetwm: A simulation tool for semiconductor lasers.

Javaloyes J, Balle S (2012b) Multimode dynamics in bidirectional laser cavities by folding space into time delay. *Opt Express* 20:8496–8502.

Jones DJ, Zhang LM, Carroll JE, Marcenac DD (1995) Dynamics of monolithic passively mode-locked semiconductor lasers. *IEEE J Quantum Electron* 31:1051–1058.

Khalfin VB, Arnold JM, Marsh JH (1995) A theoretical-model of synchronization of a mode-locked semiconductor-laser with an external pulse stream. *IEEE J Sel Top Quantum Electron* 1:523–527.

Kilen I, Koch SW, Hader J, Moloney JV (2016) Fully microscopic modeling of mode locking in microcavity lasers. *J Opt Soc Am B-Opt Phys* 33:75–80.

Kim BS, Chung Y, Kim SH (1999) Dynamic analysis of mode-locked sampled-grating distributed Bragg reflector laser diodes. *IEEE J Quantum Electron* 35:1623–1629.

Kim I, Lau KY (1993) Frequency and timing stability of mode-locked semiconductor lasers-passive and active mode locking up to millimeter wave frequencies. *IEEE J Quantum Electron* 29:1081–1090.

Kim K, Lee S, Delfyett PJ (2005) 1.4kW high peak power generation from an all semiconductor mode-locked master oscillator power amplifier system based on eXtreme Chirped Pulse Amplification (X-CPA). *Opt Express* 13:4600–4606.

Klopp P, Griebner U, Zorn M, Klehr A, Liero A, Weyers M, Erbert G (2009) Mode-locked InGaAs-AlGaAs disk laser generating sub-200-fs pulses, pulse picking and amplification by a tapered diode amplifier. *Opt Express* 17:10820–10834.

Klopp P, Griebner U, Zorn M, Weyers M (2011) Pulse repetition rate up to 92 GHz or pulse duration shorter than 110 fs from a mode-locked semiconductor disk laser. *Appl Phys Lett* 98:071103.

Koumans R, vanRoijen R (1996) Theory for passive mode-locking in semiconductor laser structures including the effects of self-phase modulation, dispersion, and pulse collisions. *IEEE J Quantum Electron* 32:478–492.

Kuznetsov M (1985) Pulsations of semiconductor lasers with a proton bombarded segment: Well-developed pulsations. *IEEE J Quantum Electron* 21:587–592.

Lau KY, Paslaski J (1991) Condition for short pulse generation in ultrahigh frequency mode-locking of semiconductor lasers. *IEEE Photonics Technol Lett* 3:974–976.

Leegwater JA (1996) Theory of mode-locked semiconductor lasers. *IEEE J Quantum Electron* 32:1782–1790.

Lu ZG, Liu JR, Poole PJ, Jiao ZJ, Barrios PJ, Poitras D, Caballero J, Zhang XP (2011) Ultra-high repetition rate InAs/InP quantum dot mode-locked lasers. *Opt Commun* 284:2323–2326.

Malins DB, Gomez-Iglesias A, White SJ, Sibbett W, Miller A, Rafailov EU (2006) Ultrafast electroabsorption dynamics in an InAs quantum dot saturable absorber at 1.3 μm. *Appl Phys Lett* 89:171111–171113.

Marconi M, Javaloyes J, Balle S, Giudici M (2014) How lasing localized structures evolve out of passive mode locking. *Phys Rev Lett* 112:223901.

Marconi M, Javaloyes J, Camelin P, Gonzalez DC, Balle S, Giudici M (2015) Control and generation of localized pulses in passively mode-locked semiconductor lasers. *IEEE J Sel Top Quantum Electron* 21:1101210.

Martins JF, Avrutin EA, Ironside CN, Roberts JS (1995) Monolithic multiple colliding pulse mode-locked quantum-well lasers: Experiment and theory. *IEEE J Sel Top Quantum Electron* 1:539–551.

McDougall SD, Vogele B, Stanley CR, Ironside CN (1997) The crucial role of doping for high repetition rate monolithic mode locking of multiple quantum well GaAs/AlGaAs lasers. *Appl Phys Lett* 71:2910–2912.

Mecozzi A, Mork J (1997) Saturation effects in nondegenerate four-wave mixing between short optical pulses in semiconductor laser amplifiers. *IEEE J Sel Top Quantum Electron* 3:1190–1207.

Mihoubi Z, Daniell GJ, Wilcox KG, Tropper AC (2008) *Numerical Model of a Vertical-External-Cavity Surface-Emitting Semiconductor Lasers Mode-Locked by the Optical Stark Effect.* New York, NY: IEEE.

Mork J, Mecozzi A (1997) Theory of nondegenerate four-wave mixing between pulses in a semiconductor waveguide. *IEEE J Quantum Electron* 33:545–555.

Moskalenko V, Javaloyes J, Balle S, Smit M, Bente E (2013) Dynamics of colliding pulse passively semiconductor mode-locked ring lasers with an intra-cavity Mach-Zehnder modulator. In: *2013 Conference on and International Quantum Electronics Conference Lasers and Electro-Optics Europe*. Cleo Europe/Iqec.

Mulet J, Balle S (2005) Mode-locking dynamics in electrically driven vertical-external-cavity surface-emitting lasers. *IEEE J Quantum Electron* 41:1148–1156.

New GHC (1974) Pulse evolution in mode-locked quasicontinuous lasers. *IEEE J Quantum Electron* 10:115–124.

Nikitichev DI, Ding Y, Cataluna MA, Rafailov EU, Drzewietzki L, Breuer S, Elsaesser W, Rossetti M, Bardella P, Xu T, Montrosset I, Krestnikov I, Livshits D, Ruiz M, Tran M, Robert Y, Krakowski M (2012) High peak power and sub-picosecond Fourier-limited pulse generation from passively mode-locked monolithic two-section gain-guided tapered InGaAs quantum-dot lasers. *Laser Phys* 22:715–724.

Nikitichev DI, Ding Y, Ruiz M, Calligaro M, Michel N, Krakowski M, Krestnikov I, Livshits D, Cataluna MA, Rafailov EU (2011a) High-power passively mode-locked tapered InAs/GaAs quantum-dot lasers. *Appl Phys B-Lasers Opt* 103:609–613.

Nikitichev DI, Ruiz M, Ding Y, Tran M, Robert Y, Krakowski M, Rossetti M, Bardella P, Montrosset I, Krestnikov I, Lifshits D, Kataluna MA, Rafailov EU (2011b) Passively mode-locked monolithic two-section gain-guided tapered quantum-dot lasers: II. Record 15 Watt peak power generation. In: *CLEO Europe*, p CB3.4. Munich.

Nikolaev VV, Avrutin EA (2003) Photocarrier escape time in quantum-well light-absorbing devices: Effects of electric field and well parameters. *IEEE J Quantum Electron* 39:1653–1660.

Nikolaev VV, Avrutin EA (2004) Quantum-well design for monolithic optical devices with gain and saturable absorber sections. *IEEE Photonics Technol Lett* 16:24–26.

Nomura Y, Ochi S, Tomita N, Akiyama K, Isu T, Takiguchi T, Higuchi H (2002) Mode locking in Fabry-Perot semiconductor lasers. *Phys Rev A* 65:043807.

Otto C, Ludge K, Vladimirov AG, Wolfrum M, Scholl E (2012) Delay-induced dynamics and jitter reduction of passively mode-locked semiconductor lasers subject to optical feedback. *New J Phys* 14:113033.

Pimenov A, Vladimirov AG (2014) Theoretical analysis of passively mode-locked inhomogeneously broadened lasers. In: *14th International Conference on Numerical Simulation of Optoelectronic Devices* (Piprek J, Javaloyes J, eds.), pp. 151–152. 1–4 September 2014, Palma de Mallorca, Spain.

Quarterman AH, Wilcox KG, Apostolopoulos V, Mihoubi Z, Elsmere SP, Farrer I, Ritchie DA, Tropper A (2009) A passively mode-locked external-cavity semiconductor laser emitting 60-fs pulses. *Nat Photonics* 3:729–731.

Rafailov EU, Cataluna MA, Avrutin EA (2011) *Ultrafast Lasers Based on Quantum Dot Structures*. New York, NY/Berlin: Wiley.

Rebrova N, Huyet G, Rachinskii D, Vladimirov AG (2011) Optically injected mode-locked laser. *Phys Rev E* 83:8.

Renaudier J, Brenot R, Dagens B, Lelarge F, Rousseau B, Poingt F, Legouezigou O, Pommereau F, Accard A, Gallion P, Duan GH (2005) 45 GHz self-pulsation with narrow linewidth in quantum dot Fabry-Perot semiconductor lasers at 1.5 μm. *Electron Lett* 41:1007–1008.

Renaudier J, Duan GH, Landais P, Gallion P (2007) Phase correlation and linewidth reduction of 40 GHz self-pulsation in distributed Bragg reflector semiconductor lasers. *IEEE J Quantum Electron* 43:147–156.

Revin DG, Hemingway M, Wang Y, Cockburn JW, Belyanin A (2016) Active mode locking of quantum cascade lasers in an external ring cavity. *Nat Commun* 7:11440.

Rosales R, Merghem K, Martinez A, Akrout A, Tourrenc JP, Accard A, Lelarge F, Ramdane A (2011) InAs/InP quantum-dot passively mode-locked lasers for 1.55-mu m applications. *IEEE J Sel Top Quantum Electron* 17:1292–1301.

Rossetti M, Bardella P, Montrosset I (2011a) Time-domain travelling-wave model for quantum dot passively mode-locked lasers. *IEEE J Quantum Electron* 47:139–150.

Rossetti M, Bardella P, Montrosset I (2011b) Modeling passive mode-locking in quantum dot lasers: A comparison between a finite-difference traveling-wave model and a delayed differential equation approach. *IEEE J Quantum Electron* 47:569–576.

Salvatore RA, Sanders S, Schrans T, Yariv A (1996) Supermodes of high-repetition-rate passively mode-locked semiconductor lasers. *IEEE J Quantum Electron* 32:941–952.

Sato K (2003) Optical pulse generation using Fabry-Perot lasers under continuous-wave operation. *IEEE J Sel Top Quantum Electron* 9:1288–1293.

Schell M, Weber AG, Scholl E, Bimberg D (1991) Fundamental limits of sub-ps pulse generation by active mode-locking of semiconductor lasers: The spectral gain width and the facet reflectivities. *IEEE J Quantum Electron* 27:1661–1668.

Scollo R, Lobe HJ, Holzman JE, Robin E, Jackel H, Erni D, Vogt W, Gini E (2005) Mode-locked laser diode with an ultrafast integrated uni-traveling carrier saturable absorber. *Opt Lett* 30:2808–2810.

Scollo R, Lohe HJ, Robin F, Erni D, Gini E, Jackel H (2009) Mode-locked InP-based laser diode with a monolithic integrated UTC absorber for subpicosecond pulse generation. *IEEE J Quantum Electron* 45:322–335.

Shimizu T, Ogura I, Yokoyama H (1997) 860 GHz rate asymmetric colliding pulse modelocked diode lasers. *Electron Lett* 33:1868–1869.

Shimizu T, Wang XL, Yokoyama H (1995) Asymmetric colliding-pulse mode-locking in InGaAsP semi-conductor lasers. *Opt Rev* 2:401–403.

Shore KA, Yee WM (1991) Theory of self-locking FM operation in semiconductor lasers. *IEE Proc J Optoelectron* 138:91–96.

Simos H, Rossetti M, Simos C, Mesaritakis C, Xu TH, Bardella P, Montrosset I, Syvridis D (2013) Numerical analysis of passively mode-locked quantum-dot lasers with absorber section at the low-reflectivity output facet. *IEEE J Quantum Electron* 49:3–10.

Stolarz PM, Javaloyes J, Mezosi G, Hou L, Ironside CN, Sorel M, Bryce AC, Balle S (2011) Spectral dynamical behavior in passively mode-locked semiconductor lasers. *IEEE Photonics J* 3:1067–1082.

Tandoi G, Javaloyes J, Avrutin E, Ironside CN, Marsh JH (2013) Subpicosecond colliding pulse mode-locking at 126 GHz in monolithic GaAs/AlGaAs quantum well lasers: Experiments and theory. *IEEE J Sel Top Quantum Electron* 19:1100608.

Thompson MG, Rae A, Sellin RL, Marinelli C, Penty RV, White IH, Kovsh AR, Mikhrin SS, Livshits DA, Krestnikov IL (2006) Subpicosecond high-power mode locking using flared waveguide monolithic quantum-dot lasers. *Appl Phys Lett* 88:133119–133113.

Thompson MG, Rae AR, Xia M, Penty RV, White IH (2009) InGaAs quantum-dot mode-locked laser diodes. *IEEE Journal Sel Top Quantum Electron* 15:661–672.

Tzenov P, Burghoff D, Hu Q, Jirauschek C (2016) Time domain modeling of terahertz quantum cascade lasers for frequency comb generation. *Opt Express* 24:23232–23247.

Vasil'ev P (1995) *Ultrafast Diode Lasers: Fundamentals and Applications.* Boston: Artech House.

Viktorov EA, Erneux T, Mandel P, Piwonski T, Madden G, Pulka J, Huyet G, Houlihan J (2009) Recovery time scales in a reversed-biased quantum dot absorber. *Appl Phys Lett* 94:263502.

Viktorov EA, Mandel P, Vladimirov AG, Bandelow U (2006) Model for mode locking in quantum dot lasers. *Appl Phys Lett* 88:201102–201103.

Vladimirov AG, Pimenov AS, Rachinskii D (2009) Numerical study of dynamical regimes in a monolithic passively mode-locked semiconductor laser. *IEEE J Quantum Electron* 45:462–468.

Vladimirov AG, Turaev D (2005) Model for passive mode locking in semiconductor lasers. *Phys Rev A* 72:033808.

Vladimirov AG, Turaev D, Kozyreff G (2004) Delay differential equations for mode-locked semiconductor lasers. *Opt Lett* 29:1221–1223.

Wilcox KG, Mihoubi Z, Daniell GJ, Elsmere S, Quarterman A, Farrer I, Ritchie DA, Tropper A (2008) Ultrafast optical Stark mode-locked semiconductor laser. *Opt Lett* 33:2797–2799.

Williams KA, Sarma J, White IH, Penty RV, Middlemast I, Ryan T, Laughton FR, Roberts JS (1994) Q-switched bow-tie lasers for high-energy picosecond pulse generation. *Electron Lett* 30:320–321.

Xu TH, Montrosset I (2013) Quantum dot passively mode-locked lasers: Relation between intracavity pulse evolution and mode locking performances. *IEEE J Quantum Electron* 49:65–71.

Xu TH, Rossetti M, Bardella P, Montrosset I (2012) Simulation and analysis of dynamic regimes involving ground and excited state transitions in quantum dot passively mode-locked lasers. *IEEE J Quantum Electron* 48:1193–1202.

Yang WG (2011) Single-section Fabry-Perot mode-locked semiconductor lasers. *Adv Optoelectron* 2011:780373.

Yanson DA, Street MW, McDougall SD, Thayne IG, Marsh JH, Avrutin EA (2002) Ultrafast harmonic mode-locking of monolithic compound-cavity laser diodes incorporating photonic-bandgap reflectors. *IEEE J Quantum Electron* 38:1–11.

Yee WM, Shore KA (1993) Multimode analysis of self-locked FM operation in laser diodes. *IEE Proc J Optoelectron* 140:21–25.

Zarrabi JH, Portnoi EL, Chelnokov AV (1991) Passive-mode locking of a multistripe single quantum-well GaAs-laser diode with an intracavity saturable absorber. *Appl Phys Lett* 59:1526–1528.

33

Quantum Cascade Lasers: Electrothermal Simulation

Song Mei

Yanbing Shi

Olafur Jonasson

and

Irena Knezevic

33.1 Introduction

Quantum cascade lasers (QCLs) are high-power, coherent light sources emitting in the mid-infrared (mid-IR) and terahertz (THz) frequency ranges [1]. QCLs are electronically driven, unipolar devices whose active core consists of tens to hundreds of repetitions of a carefully designed stage. The QCL active core can be considered a superlattice (SL) in which each stage is a multiple quantum well (MQW) heterostructure where confined electronic states with specific energy levels are formed because of quantum confinement. The concept of achieving lasing in semiconductor SLs was first introduced by Kazarinov and Suris [2] in 1971. The first working QCL was demonstrated by Faist et al. [1] two decades later.

QCLs are typically III-V material systems grown on GaAs or InP substrates. Molecular beam epitaxy (MBE) [3] and metal-organic chemical vapor deposition (MOCVD) [4] are the techniques that enable precise growth of thin layers of various III-V alloys. It is also possible to incorporate strain into the structure as long as the total strain in a stage is balanced. Both the precision and the possibility of introducing strain bring great flexibility to the design of the QCL active core, so lasing over a wide range of wavelengths (from 3 to 190 µm) has been achieved. The growth techniques produce high-quality interfaces with atomic-level roughness.

Mid-IR QCLs (wavelength range 3–12 µm) have widespread military and commercial applications. A practical portable detector requires mid-IR QCLs to operate at room temperature (RT) in continuous wave (CW) mode and with high (watt-level) output power. Furthermore, these QCLs must also have high wall-plug efficiency (WPE) (the ratio of emitted optical power to the electrical power pumped in) and long-term reliability under these high-stress operating conditions. As the stress likely stems from excessive nonuniform heating while lasing [5,6], improving device reliability and lifetime goes hand in hand with improving the WPE.

33.1.1 Lasing in QCLs

In QCLs, multiple conduction subbands are formed in the active core by means of quantum confinement. QCLs are unipolar devices, meaning that lasing is achieved through radiative intersubband transitions (transitions between two conduction subbands) instead of radiative interband transitions (transitions between the conduction and valence bands) in traditional quantum well (QW) semiconductor lasers. As a result, electrons do not combine with holes after the radiative transitions and can be used to emit another photon. In order to reuse electrons, the same MQW heterostructure is repeated many times (25–70) in the QCL active core (the so-called cascading scheme).

Figure 33.1 depicts a typical conduction band diagram of two adjacent stages in a QCL under an electric field. Each stage consists of an injector region and an active region. The injector region has several thin wells separated by thin barriers (10–30 Å), so a miniband is formed with multiple subbands that are close in energy and whose associated wavefunctions have high spatial overlap. Typically, the lowest few energy levels in the miniband are referred to as the injector levels. The injector levels collect the electrons that come from the previous stage and inject them into the active region. The active region usually consists of 2–3 wider wells (40–50 Å) separated by thin barriers. Consequently, a minigap forms in the active region between the upper lasing level (3) and the lower lasing level (2). Another important energy level in the active region is the ground state (1). There is a thin barrier (usually the thinnest among all layers) between the injecting region and the active region, called the injection barrier.

By design, the injector levels are close in energy and strongly coupled to the upper lasing level because of the thin injection barrier. The upper and lower lasing levels have large spatial overlap, which allows a radiative transition between the two levels; the wavelength of the emitted light is determined by the energy spacing between these two levels. The lower lasing level overlaps with the ground state for efficient electron extraction. Electron emission of longitudinal optical (LO) phonons is the dominant mechanism for electron extraction, so the energy spacing between the lower lasing level and the ground state is designed to be close to the LO phonon energy to facilitate extraction. With careful design, the electron lifetime in the upper lasing level is longer than in the lower lasing level, so population inversion can be

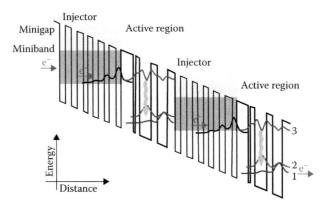

FIGURE 33.1 A typical conduction band diagram of two adjacent QCL stages under an applied electric field. Each stage consists of an injector region and an active region. A miniband is formed in the injector region while a minigap is formed in the active region (between the upper and lower lasing levels). Lasing is associated with a radiative transition from the upper (3) to the lower (2) lasing level. Electrons in the lower lasing level depopulate quickly to the ground level (1) by emission of longitudinal optical phonons.

achieved. After reaching the ground state, electrons tunnel through the injector into the upper lasing level of the next stage, and the process is repeated. Of course, the lasing mechanism description above is idealized. In reality, the efficiency of the radiative transition between the upper and lower lasing levels is very low [7,8].

33.1.2 Recent Developments in High-Power QCLs

In recent years, considerable focus has been placed on improving the WPE and output power of QCLs for RT CW operation. Bai et al. [9] showed 8.4% WPE and 1.3 W output power around 4.6 μm in 2008. Shortly thereafter, Lyakh et al. [10] reported 12.7% WPE and 3 W power at 4.6 μm. Watt-level power with 6% WPE at 3.76 μm and then lower power at 3.39 and 3.56 μm are reported by Bandyopadhyay et al. [11,12]. Bai et al. [13] demonstrated 21% WPE and 5.1 W output power around 4.9 μm in 2011. Much higher WPE and/or output power has been achieved at lower temperatures or at pulsed mode [7,14] near 4.8 μm. A summary of recent developments can be found in review papers [8,15].

While good output powers and WPEs have been achieved, long-term reliability of these devices under RT CW operation remains a critical problem [5,6]. These devices are prone to catastrophic breakdown owing to reasons that are not entirely understood, but are likely related to thermal stress that stems from prolonged high-power operation [5]. This kind of thermal stress is worst in short-wavelength devices that have high strain and high thermal-impedance mismatch between layers [11,12,16,17].

In addition to improved device lifetime, we seek better CW temperature performance (higher characteristic temperatures T_0 and T_1, defined next) [6]. The first aspect is a weaker temperature dependence of the threshold current density. Empirically, the threshold current density (the current density at which the device starts lasing) has an exponential dependence on the operating temperature T: $J_{th} \propto \exp\left(\frac{T}{T_0}\right)$. Higher characteristic temperature T_0 is preferred in QCL design, as it means less variation in J_{th} as the temperature changes.

Another key temperature-dependent parameter is the differential quantum efficiency (also called the slope efficiency or external quantum efficiency), defined as the amount of output optical power dP per unit increase in the pumping current dI: $\eta_d = \frac{dP}{dI} \propto \exp\left(-\frac{T}{T_1}\right)$. The differential quantum efficiency is directly proportional to the WPE (WPE $= \eta_d \eta_f$, where η_f is the feeding efficiency). Therefore, the higher the T_1, the closer the η_d is to unity, and the higher the WPE. Recently, deep-well structures with tapered active regions have demonstrated significant improvements in T_0 and T_1 with respect to the conventional 4.6-μ*m* device [9], underscoring that the suppression of leakage plays a key role in temperature performance [6,18,19]. Still, the microscopic mechanisms and leakage pathways that contribute to these empirical performance parameters remain unclear.

33.1.3 QCL Modeling: An Overview

Under high-power RT CW operation, both electron and phonon systems in QCLs are far away from equilibrium. In such nonequilibrium conditions, both electronic and thermal transport modeling are important for understanding and improving QCL performance.

Electron transport in both mid-IR and THz QCLs has been successfully simulated via semiclassical (rate equations [20–22] and Monte Carlo [23–26]) and quantum techniques (density matrix [27–33], nonequilibrium Green's functions (NEGF) [34–36], and lately Wigner functions [37]). InP-based mid-IR QCLs have been addressed via semiclassical [38] and quantum transport approaches (8.5-μm [39] and 4.6-μm [35,36] devices). There has been a debate whether electron transport in QCLs can be described using semiclassical models, in other words, how much of the current in QCLs is coherent. Theoretical work by Iotti and Rossi [23,40] shows that the steady-state transport in mid-IR QCLs is largely incoherent. Monte Carlo simulation [41] has also been used to correctly predict transport near threshold. However, short-wavelength structures [9] have pronounced coherent features, which cannot be addressed semiclassically

[33]. NEGF simulations accurately and comprehensively capture quantum transport in these devices but are computationally demanding. Density matrix approaches have considerably lower computational overhead than NEGF but are still capable of capturing coherent transport features. A comprehensive review of electron transport modeling was recently written by Jirauschek and Kubis [42].

Electronic simulations that ignore radiative transitions are applicable for modeling QCLs below or near threshold where the interaction between electrons and the laser electromagnetic field can be ignored. Such simulations are useful for predicting quantities such as threshold current density and T_0. However, in order to accurately model QCLs under lasing operations, the effect of the laser field on electronic transport would have to be included. In some cases, the effects of the laser field can be very strong [39], especially for high-WPE devices, where the field-induced current can be dominant [38]. When included in simulations, the laser field is typically modeled either as an additional scattering mechanism [38,43] or as a time-dependent sinusoidal electric field [39,44]. In this work, we ignore the effect of the laser field on electron dynamics.

Thermal transport in QCLs is often described through the heat diffusion equation, which requires accurate thermal conductivity in each region, a challenging task for the active core that contains many interfaces [17,45–48]. It is also very important to include nonequilibrium effects, such as the nonuniform heat generation rate stemming from the nonuniform temperature distribution [47] and the feedback that the nonequilibrium phonon population has on electron transport [26].

In this chapter, we present a multiphysics (coupled electronic and thermal transport) and multiscale (bridging between a single stage and device level) simulation framework that enables the description of QCL performance under far-from-equilibrium conditions [49]. We present the electronic (Section 33.2) and thermal (Section 33.3) transport models, and then bring them together for electrothermal simulation of a real device structure (Section 33.4). We strive to cover the basic ideas while pointing readers to the relevant references for derivation and implementation details.

33.2 Electronic Transport

Depending on the desired accuracy and computational burden, one can model electronic transport in QCLs with varying degrees of complexity. The goal is to determine the modal gain (proportional to the population inversion between the upper and lower lasing levels) under various pumping conditions (current or voltage) and lasing conditions (pulsed wave or CW). A typical electron transport simulator relies on accurately calculated quasibound electronic states and associated energies in the direction of confinement. Electronic wavefunctions and energies are determined by solving the Schrödinger equation or the Schrödinger equation combined with the Poisson's equation in highly doped systems. Section 33.2.3.1 introduces a $\mathbf{k} \cdot \mathbf{p}$ Schrödinger solver coupled with a Poisson solver. More information about other solvers for electronic states can be found in the review paper [42] and references therein.

The simulations of electronic transport fall into two camps depending on how the electron single-particle density matrix is treated. The diagonal elements of the density matrix represent the occupation of the corresponding levels and off-diagonal elements represent the "coherence" between two levels. Transport is semiclassical or incoherent when the off-diagonal coherences are much smaller than the diagonal terms and can be approximated as proportional to the diagonal terms times the transition rates between states [50]. In that case, the explicit calculation of the off-diagonal terms is avoided and only the diagonal elements are tracked, which simplifies the simulation considerably. However, when the off-diagonal terms are appreciable, transport is partially coherent and has to be addressed using quantum transport techniques.

33.2.1 Semiclassical Techniques

Semiclassical approaches assume that electronic transport between stages is largely incoherent "hopping" transport. The key quantities are populations of electronic states that are confined in the QCL growth direction, and electrons transfer between them due to scattering events. The scattering rates can be obtained empirically or more rigorously via Fermi's golden rule. Common semiclassical approaches are the rate equations and ensemble Monte Carlo (EMC), the latter solving a Boltzmann-like transport equation stochastically.

33.2.1.1 Rate Equations

In the rate equation approach [20–22], scattering between relevant states, i.e., the injector level, the upper and lower lasing levels, and the ground state, is captured through transition rates. The rates include all relevant (radiative and nonradiative) scattering mechanisms and can be either empirical parameters or calculated [51,52]. The computational requirements of rate equation models are low, so they are suitable for fast numerical design and optimization of different structures [22].

33.2.1.2 Ensemble Monte Carlo

The heterostructure in the QCL active core is a quasi-two-dimensional (quasi-2D) system, where electrons are free to move in the x–y plane, while confined cross-plane in the z direction; the confinement results in the formation of quasibound states and discrete energy levels corresponding to the bottoms of 2D energy subbands. The electron wavefunctions in 3D are plane waves in the x–y plane and confined wavefunctions in z direction. Electronic transport is captured by a Boltzmann-like semiclassical transport equation [23], which can be solved via the stochastic EMC technique assuming instantaneous hops between states in 3D due to scattering [53]. The simulation explicitly tracks the energy level and in-plane momentum of each particle in the simulation ensemble (typically $\sim 10^5$ particles). Tracking in-plane dynamics makes it more detailed than the rate equation model. The transition rates are generally computed directly from the appropriate interaction Hamiltonians and therefore depend on the energy levels as well as the wavefunction overlaps between different electronic states [42,53,54]. EMC allows us to include nonequilibrium effect into transport, which is covered in more detail in Section 33.2.3.

33.2.2 Quantum Techniques

Density matrix and NEGF are the two most widely used techniques to describe quantum transport in QCLs. Recently, a Wigner function approach was also successfully used to model a SL [37].

33.2.2.1 Density Matrix Approaches

In semiclassical approaches, the central quantity of interest is the distribution function $f_n^{E_k}(t)$, the probability of an electron occupying an eigenstate n and having an in-plane kinetic energy E_k. The quantum-mechanical analogue is the single-electron density matrix, $\rho_{nm}^{E_k}(t)$, where the diagonal elements $\rho_{nn}^{E_k}(t) = f_n^{E_k}(t)$ are occupations and the off-diagonal elements $\rho_{nm}^{E_k}(t)$ are the spatial coherences between states n and m at the in-plane energy E_k. When employing semiclassical methods, off-diagonal matrix elements are assumed to be much smaller than diagonal elements. This approximation may fail in some cases, e.g., when two eigenstates with a large spatial overlap have similar energies. This scenario often arises when modeling THz QCLs [28,32,55], but can also come up in mid-IR QCLs [33]. In these cases, semiclassical models fail.

The density matrix models that have been employed for QCL modeling can be categorized into two groups. The first includes hybrid methods, where transport is treated semiclassically within a region of the device (typically a single stage) while the effects of tunneling between different regions, separated by barriers, are treated quantum mechanically using a density matrix formalism with phenomenological dephasing times [28,31,55]. The second group involves completely quantum-mechanical methods that rely on microscopically derived Markovian master equations that guarantee positivity of the density matrix [32,33]. Both methods are more computationally expensive than their semiclassical counterparts, because the density matrix contains many more elements than its diagonal semiclassical analogue.

33.2.2.2 Nonequilibrium Green's Functions

The NEGF technique (see a good overview in Reference [42]) relies on the relationships between single-particle time-ordered Green's functions and correlation functions [34–36]. The correlation function $G^<_{\alpha,\beta}(k; t_1, t_2)$, often referred to as the lesser Green's function [44,56], is one of the central quantities and can be understood as a two-time generalization of the density matrix, where k refers to the magnitude of in-plane wave vector. The correlation function contains both spatial correlations (terms with $\alpha \neq \beta$) as well as temporal correlations between times t_1 and t_2 (not included in semiclassical or density matrix models). Typically, the potential profile is assumed to be time independent, in which case the correlation function only depends on the time difference $G^<_{\alpha,\beta}(k; t_1, t_2) = G^<_{\alpha,\beta}(k; t_1 - t_2)$. Fourier transform over the time difference into the energy domain gives the energy-resolved correlation function $G^<_{\alpha,\beta}(k, E)$, which is the quantity that is usually solved for numerically [35,44,56]. The main advantages of the NEGF formalism are that it provides spectral (energy-resolved) information and it includes the effects of collisional broadening (the broadening of energy levels due to scattering), which is particularly important when the states are close in energy. These advantages carry a considerable computational cost, so NEGF calculations are much more time-consuming than density matrix approaches [39].

33.2.3 EMC with Nonequilibrium Phonons

Here, we focus on presenting semiclassical modeling of electron transport in QCL structures via EMC [25,26,57]. The solver consists of two parts: a coupled Schrödinger–Poisson solver and a transport kernel. We solve for the electronic states using the coupled Schrödinger–Poisson solver and feed the energy levels and the wavefunctions of the relevant electronic states to the transport kernel. The transport kernel keeps track of the electron momentum, energy, and distribution among subbands. If the electron density inside the device is high, transport kernel will periodically feed the electron distribution back to the Schrödinger–Poisson solver and update the electronic states. This loop is repeated until the electron distribution converges. By doing so, we solve for both the electron transport and the electronic band structure self-consistently.

Since the active QCL core consists of repeated stages, the wavefunctions in any stage can be obtained from the wavefunctions in any other stage by translation in space and energy. This translational symmetry makes it possible to simulate electron transport in only one generic central stage instead of in the whole QCL core [53]. Typically, electronic states in nonadjacent stages have negligible overlap, which also means that the transition rates between them are negligible. As a result, it is sufficient to limit interstage scattering events to only those between adjacent stages.

Figure 33.2 shows a schematic of three adjacent stages under an applied field. We simulate electron transport in the central stage λ, while nearest-neighbor interstage ($\lambda \rightleftarrows \lambda \pm 1$) and intrastage ($\lambda \rightarrow \lambda$) scattering is allowed. Periodic boundary conditions (PBCs) are applied in the simulation, i.e., whenever one electron scatters from the central stage into the next stage (process ②), an electron scatters from the previous stage into the central stage (process ①) and vice versa (process ③ and process ④). PBCs are justified by the cascading scheme.

33.2.3.1 Electronic Bandstructure Solver

We employ the **k·p** method to solve the Schrödinger equation and couple it to a Poisson solver [25,53,57]. The **k·p** method is an efficient way to solve for the electronic band structure near the band edges, where the transport happens in QCLs. The **k·p** method considers the contribution from the conduction band (C), light-hole band (LH), and the spin-orbit split-off band (SO) (the heavy-hole band (HH) decouples from the other three at the band edge) [58]. The contributions from the LH and SO are especially important for narrow-gap materials, such as InP. Moreover, in modern QCLs, strain-balanced structures have been employed to obtain enhanced performance. In these structures, alternate layers are compressively or tensilely strained while the whole structure is strain free, with carefully designed thickness of each

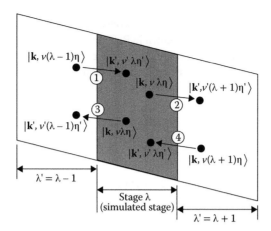

FIGURE 33.2 Schematic of the three simulated stages in a QCL active core under an applied field. Scattering is limited to nearest-neighbor stages and periodic boundary conditions are justified by the cascading scheme; therefore, only three stages are needed in the EMC transport kernel.

layer. The **k·p** method allows for convenient inclusion of the effects of strain on the band structure. The implementation details of the **k·p** solver can be found in Reference [53].

The **k·p** solver can only solve for a finite structure rather than an infinite periodic one. As a result, we need to simulate a finite number of stages and add artificially high barriers to the two ends to confine all the states. If a stage is far enough from the boundaries, the calculated band structure in it should be the same as if we were to solve for the whole periodic structure. Tests have confirmed that three stages, which we also used in EMC, are enough when solving for the electronic states to ensure that the central-stage states are unaffected by the simulation-domain potential boundaries. The states from the central stage are then translated in energy and position to the neighboring stages according to the stage length and the applied electric field.

When we need to solve for electron transport and electronic states self-consistently, it is necessary for the solver to be able to automatically pick out the electronic states belonging to the central stage. One intuitive criterion is to calculate the "center of mass" for each state (the expectation value of the cross-plane coordinate, $\langle z \rangle$) and assign those falling in the central stage to that stage. However, in our three-stage scheme, this method may pick up the states that are too close to the boundary. One can either extend the number of stages in the **k·p** solver to five, so the three stages in the middle are all far from the boundary, or use additional criteria such as that there be more than 50% possibility of finding an electron in the central stage, based on the probability density distribution, or requiring that the location of the probability density peak be in the central stage. Additional criteria requiring strong confinement of states have been explored in [32].

33.2.3.2 Transport Kernel with Nonequilibrium Phonons

The EMC kernel tracks the hopping transitions of electrons between subbands and stages until convergence and outputs the transport information for us to calculate the experimentally relevant quantities such as current and modal gain [53]. In the transport kernel, both electron–electron interactions and electron–LO-phonon interactions are considered. Other scattering processes, such as intervalley scattering, impurity scattering, and interface roughness scattering can be considered under different circumstances [42]. Photon emission is not considered, either. Because EMC tracks individual particles, nonequilibrium electron transport can be automatically captured. (EMC tracks individual simulation particles, each of which might represent thousands of real electrons.)

The most important scattering mechanism in QCLs is electron–LO-phonon scattering, which facilitates the depopulation of the lower lasing level. As shown in Reference [59], phonon confinement has little effect on the electronic transport; therefore, for simplicity, LO phonons are treated as bulklike dispersionless phonons with energy $\hbar\omega_0$. The transition rate between an initial state $\phi_i(z)$ with energy E_i and a final state $\phi_f(z)$ withenergy E_f can be derived from the Fermi's golden rule as follows:

$$\Gamma_{a(-),e(+)} = \frac{e^2\hbar\omega_0 m_f^*}{8\pi^2\hbar^3} \left(\frac{1}{\epsilon_\infty} - \frac{1}{\epsilon_0}\right) \int_0^{2\pi} d\theta \int_{-\infty}^{\infty} dq_z \int_0^{\infty} dE_{kf} N_\mathbf{q} \frac{|\mathcal{I}_{if}(q_z)|^2}{\mathbf{q}_\parallel^2 + q_z^2} \delta(E_f - E_i \mp \hbar\omega_0), \quad (33.1)$$

where e is the electronic charge while ϵ_0 and ϵ_∞ are static and high-frequency electronic permittivities of the material, respectively. The integrals are over the in-plane kinetic energy E_k of the final state and the cross-plane momentum transfer q_z. $\mathbf{q}_\parallel = \mathbf{k}_\parallel' - \mathbf{k}_\parallel$ is the in-plane momentum transfer.

$$|\mathcal{I}_{if}(q_z)|^2 = \left| \int_0^d dz \phi_f^*(z)\phi_i(z)e^{-izq_z} \right|^2 \quad (33.2)$$

is defined as the overlap integral (OI) between the initial and final states, where q_z is the cross-plane momentum transfer. The integration is over the angle between initial and final in-plane momenta $\mathbf{k}_\parallel$ and $\mathbf{k}_\parallel'$ (θ), cross-plane momentum component of the final state (k_{fz}), and the kinetic energy of the final state (E_{kf}). $N_\mathbf{q}$ represents the number of LO phonons with momentum $\mathbf{q} = (\mathbf{q}_\parallel, q_z)$. The expression can be further simplified in the equilibrium case, where $N_\mathbf{q}$ follows the Bose–Einstein distribution [53]. In order to model nonequilibrium phonon effects, we numerically integrated the expression using a phonon number histogram according to both $q_\parallel$ and q_z [54].

According to the uncertainty principle, position and momentum both cannot be determined simultaneously. Since our electrons are all confined in the central stage ($\Delta z'$ is finite), the cross-plane momentum is not exactly conserved during the scattering process ($q_z \neq k_z' - k_z$) [60]. This analysis does not affect the momentum conservation in the x–y plane, because we assume infinite uncertainty in position there. Previously, the cross-plane momentum conservation has been considered through the momentum conservation approximation (MCA) [61,62] and a broadening of q_z according to the well width [60]. The MCA forbids a phonon emitted between subbands i and f to be reabsorbed by another transition between i' and f' if $i \neq i'$ or $f \neq f'$, and thus might underestimate the electron–LO interaction strength [54]. The concept of well width is hard to apply in an MQW structure such as the QCL active core [54]. We observe that the probability of a phonon with cross-plane momentum q_z being involved in an interaction is proportional to the OI in Equation 33.2. Figure 33.3 depicts the typical OIs for both intersubband ($i_1 \to 3$ and $2 \to 1$) and intrasubband ($3 \to 3$) transitions. As a result, in each electron–LO-phonon scattering event, we randomly select a q_z following the distribution from the OI (Figure 33.3). Depending on the mechanism (absorption or emission), a phonon with ($\mathbf{q}_\parallel, q_z$) is removed or added to the histogram according to the 2D density of states (DOS) and the effective simulation area [54]. Once the phonons with a certain momentum are depleted, transitions involving such phonons become forbidden.

In order to couple the EMC solver to the thermal transport solver, we need to keep a detailed log of heat generation during electron transport. In all the relevant scattering events, electron–LO-phonon scattering is the only inelastic mechanism and therefore is the only mechanism that contributes to heat generation. As a result, the total energy emitted and absorbed in the form of LO phonons is recorded during each step of the EMC simulation. The nonequilibrium phonons decay into acoustic longitudinal acoustic (LA) phonons via a three-phonon anharmonic decay process. The formulation and the parameters here follow [63]. The simulation results of EMC including nonequilibrium phonons are shown in Section 33.4.

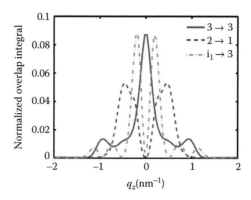

FIGURE 33.3 Normalized overlap integral $|\mathcal{I}_{if}|^2$ from Equation 33.2 versus cross-plane phonon wave vector q_z for several transitions (intersubband $i_1 \rightarrow 3$ and $2 \rightarrow 1$; intrasubband $3 \rightarrow 3$). (Reprinted with permission from Y. B. Shi and I. Knezevic. Nonequilibrium phonon effects in midinfrared quantum cascade lasers. *Journal of Applied Physics*, 116(12):123105, 2014. Copyright 2014, American Institute of Physics.)

33.3 Thermal Transport

The dominant path of heat transfer in a QCL structure is depicted in Figure 33.4. The operating electric field of a typical QCL is high, which means that considerable energy is pumped into the electronic system. These energetic, "hot" electrons relax their energy largely by emitting LO phonons. LO phonons have high energies but flat dispersions, so their group velocities are low and they are poor carriers of heat. An LO phonon decays into two LA phonons via a three-phonon process referred to as anharmonic decay. LA phonons have low energy but high group velocity and are the main carriers of heat in semiconductors [47,63]. If we neglect the diffusion of optical phonons, the flow of energy in a QCL can be described by the following equations:

$$\frac{\partial W_A}{\partial t} = \nabla \cdot (\kappa_A \nabla T_A) + \left.\frac{\partial W_{LO}}{\partial t}\right|_{coll} ; \quad \frac{\partial W_{LO}}{\partial t} = \nabla \cdot (\kappa_A \nabla T_A) + \left.\frac{\partial W_e}{\partial t}\right|_{coll} - \left.\frac{\partial W_{LO}}{\partial t}\right|_{coll}, \quad (33.3)$$

where W_{LO}, W_A, and W_e are the LO phonon, acoustic phonon, and electron energy densities, respectively. κ_A is the thermal conductivity in the system and T_A is the acoustic phonon (lattice) temperature. The term $\nabla \cdot (\kappa_A \nabla T_A)$ describes heat diffusion, governed by acoustic phonons. We have also used the fact that the rate of increase in the LO-phonon energy density equals the difference between the rate of its generation by electron–LO-phonon scattering and the rate of anharmonic decay into LA phonons.

In a nonequilibrium steady state, both the LO and LA energy densities are constant, so

$$-\nabla \cdot (\kappa_A \nabla T_A) = \left.\frac{\partial W_e}{\partial t}\right|_{coll} . \quad (33.4)$$

As described in the previous section, the right-hand side of Equation 33.4 is the heat generation rate Q and can be obtained by recording electron–LO-phonon scattering events in electronic EMC [47,64]:

$$Q = \left.\frac{\partial W_e}{\partial t}\right|_{coll} = \frac{N_{3D}}{N_{sim} t_{sim}} \sum (\hbar\omega_{ems} - \hbar\omega_{abs}), \quad (33.5)$$

where $N_{3D} = N_s/D_{stage}$ is the electron density (N_s is the sheet density and D_{stage} is the length of a single stage) while N_{sim} and t_{sim} are the number of simulation particles and the simulation time, respectively,

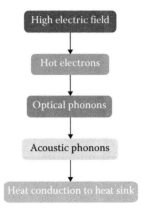

FIGURE 33.4 Flow of energy in a QCL.

and $\hbar\omega_{ems}$ and $\hbar\omega_{abs}$ are the energies of the emitted and absorbed LO phonons, respectively. To solve Equation 33.4, we need information on both the thermal conductivity κ_A and the heat generation rate Q; they are discussed in Subsections 33.3.1 and 33.3.2, respectively.

33.3.1 Thermal Conductivity in a QCL Device

33.3.1.1 Active Core: A III-V Superlattice

The QCL active core is an SL: It contains many identical stages, each with several thin layers made from different materials and separated by heterointerfaces. The thermal conductivity tensor of an SL system reduces to two values: the in-plane thermal conductivity $\kappa_\parallel$ (in-plane heat flow is assumed isotropic) and the cross-plane thermal conductivity $\kappa_\perp$. Experimental results have shown that, in SLs, the thermal conductivity is very anisotropic [65] ($\kappa_\parallel \gg \kappa_\perp$) while both $\kappa_\parallel$ and $\kappa_\perp$ are smaller than the weighted average of the constituent bulk materials [66–70]. Both effects can be attributed to the interfaces between adjacent layers [71,72].

Here, we discuss a semiclassical model for describing the thermal conductivity tensor of III-V SL structures. Note that the model described here is in principle applicable to SLs in other material systems as long as they have high-quality interface and thermal transport is mostly incoherent [48,73–75]. In particular, we focus on thermal transport in III-arsenide-based SLs, as they are most commonly used in mid-IR QCL active cores [48].

Under QCL operation conditions of interest (>77 K, and typically near RT), thermal transport is dominated by acoustic phonons and is governed by the Boltzmann transport equation (BTE). To obtain the thermal conductivity, we solve the phonon BTE with full phonon dispersion in the relaxation-time approximation [48].

33.3.1.2 Twofold Influence of Effective Interface Roughness

To capture both the anisotropic thermal transport and the reduced thermal conductivity in SL systems, we need to observe the twofold influence of the interface. First, it reduces $\kappa_\parallel$ by affecting the acoustic phonon population close to the interfaces [76]. Second, it introduces an interface thermal boundary resistance (ITBR), which is still very difficult to model [65,77]. Common models are the acoustic mismatch model (AMM) and the diffuse mismatch model (DMM) [65,76]; the former assumes a perfectly smooth interface and only considers the acoustic mismatch between the two materials, while the latter assumes complete randomization of momentum after phonons hit the interface. As most III-V-based QCLs are

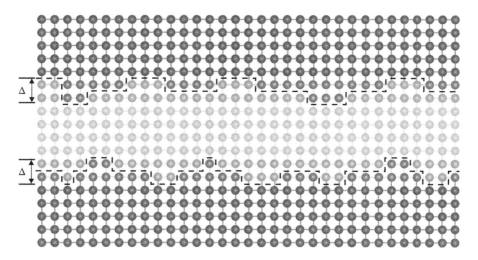

FIGURE 33.5 Even between lattice-matched crystalline materials, there exist nonuniform transition layers that behave as an effective atomic-scale interface roughness with some rms roughness Δ. This effective interface roughness leads to phonon momentum randomization and to interface resistance in cross-plane transport. (From S. Mei and I. Knezevic, *Journal of Applied Physics*, 118, 175101, 2015. With the permission of AIP Publishing.)

grown by MBE or MOCVD, both well-controlled techniques allowing consistent atomic-level precision, neither AMM nor DMM captures the essence of a III-V SL interface. Figure 33.5 shows a schematic of interface roughness in a lattice-matched SL. The jagged dashed boundaries depict transition layers of characteristic thickness Δ between the two materials.

We introduce a simple model that calculates a more realistic ITBR (a key part in calculating $\kappa_\perp$) by interpolating between the AMM and DMM transmission rates using a specularity parameter p_{spec}. The model has a single fitting parameter: the effective interface rms roughness Δ. Since the growth environment is well controlled, using one Δ to describe all the interfaces is justified. We use Δ to calculate a momentum-dependent specularity parameter:

$$p_{\text{spec}}(\vec{q}) = \exp(-4\Delta^2 |\vec{q}|^2 \cos^2 \theta), \qquad (33.6)$$

where $|\vec{q}|$ is the magnitude of the phonon wave vector and θ is the angle between $\vec{q}$ and the normal direction to the interface. Consistent with the twofold impact of interface roughness, Δ affects the thermal conductivity through two channels. Apart from calculating the ITBR, an effective interface scattering rate $\tau_{\text{interface}}^{-1}(\vec{q})$ dependent on the same specularity parameter $p_{\text{spec}}(\vec{q})$ is added to the internal scattering rate to calculate modified $\kappa_\parallel$ (see detailed derivations in [48]). By adjusting only Δ, typically between 1 and 2 Å, the calculated thermal conductivity using this model fits a number of different experiments [66,68,69].

33.3.1.3 $\kappa_\parallel$ and $\kappa_\perp$ of a QCL Active Core

Thermal transport inside the active core of a QCL is usually treated phenomenologically: $\kappa_\parallel$ is typically assumed to be 75% of the weighted average of the bulk thermal conductivities of the constituent materials, while $\kappa_\perp$ is treated as a fitting parameter (constant for all temperatures) to best fit the experimentally measured temperature profile [17,78]. We calculated the thermal conductivity tensor of a QCL active core [78] and showed that the typical assumption is not accurate and that the degree of anisotropy is temperature dependent (Figure 33.6).

The ratio between $\kappa_\parallel$ and the averaged bulk value (inset to Figure 33.6) varies between 45% and 70% over the temperature range of interest. $\kappa_\perp$ has a weak dependence on temperature in keeping with the

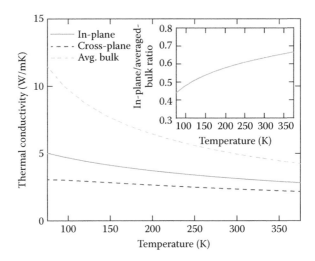

FIGURE 33.6 Thermal conductivity of a typical QCL active region [78] as a function of temperature. A single stage consists of 16 alternating layers of $In_{0.53}Ga_{0.47}As$ and $In_{0.52}Al_{0.48}As$. The solid curve, dashed curve, and dashed-dotted curve show the calculated in-plane, cross-plane, and averaged bulk thermal conductivity, respectively. $\Delta = 1$ Å in the calculations. The inset shows the ratio between the calculated in-plane and the averaged bulk thermal conductivities. (From S. Mei and I. Knezevic, *Journal of Applied Physics*, 118, 175101, 2015. With the permission of AIP Publishing.)

TABLE 33.1 Thermal Conductivity as a Function of Temperature for Materials in a QCL Structure

Materials	Thermal Conductivity (W/mK)
Au	$337 - 600 \times 10^{-4}T$
Si_3N_4	$30 - 1.4 \times 10^{-2}T$
In solder	$93.9 - 6.96 \times 10^{-2}T + 9.86 \times 10^{-5}T^2$

common assumption in simplified models; the weak temperature sensitivity means that ITBR dominates cross-plane thermal transport. These results show that it is important to carefully calculate the thermal conductivity tensor in QCL thermal simulation, and we use this thermal conductivity model in the device-level simulation.

33.3.1.4 Other Materials

The active core is not the only region we need to model in a device-level thermal simulation. Figure 33.7 shows a typical schematic (not to scale) of a QCL device in thermal simulation with a substrate-side mounting configuration [79]. The active core (in this case, consisting of 36 stages and 1.6-µm thick) with width W_{act} is embedded between two cladding layers (4.5-µm-thick GaAs). The waveguide is supported by a substrate (GaAs) with thickness D_{sub}. An insulation layer (Si_3N_4) with thickness D_{ins} is deposited around the waveguide and then etched away from the top to make the contact. Finally, a contact layer (Au) with thickness D_{cont} and a thin layer of solder (D_{sold}) are deposited on top. There is no heat generation in the regions other than the active core. Further, these layers are typically thick enough to be treated as bulk materials. Bulk substrate (GaAs or InP) thermal conductivities are readily obtained for III-V materials from experiment, as well as from relatively simple theoretical models [17,48,78,80] (Table 33.1).

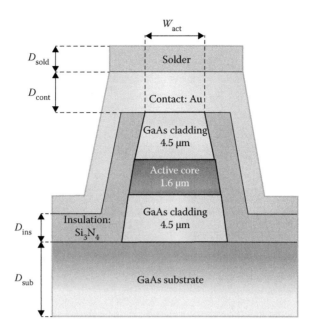

FIGURE 33.7 Schematic of a typical GaAs-based mid-IR QCL structure with a substrate (not to scale).

33.3.2 Device-Level Electrothermal Simulation

33.3.2.1 Device Schematic

The length of a QCL device is much greater than its width, therefore we can assume the length is infinite and carry out a 2D thermal simulation. The schematic of the simulation domain (not to scale) is shown in Figure 33.7. The boundary of the simulation region is delineated, and certain boundary conditions (heat sink at fixed temperature, convective boundary condition, or adiabatic boundary condition) can be applied (independently) to each boundary. Typically, the bottom boundary of the device is connected to a heat sink while other boundaries have the convective boundary condition at the environment temperature (single-device case) or the adiabatic boundary condition (QCL array case). Typical values for the layers thickness are $W_{act} = 15$ μm, $D_{sub} = 50$ μm, $D_{ins} = 0.3$ μm, $D_{cont} = 3$ μm, and $D_{sold} = 1.5$ μm.

We use the finite element method to solve for the temperature distribution. The whole device is divided into different regions according to their materials properties. Each stage of the active region is treated as a single unit with the heat generation rate tabulated in the device table in order to capture the nonuniform behavior among stages. The active core is very small but is also the only region with heat generation, small thermal conductivity, and spatial nonuniformity. To capture the behavior of the active region while saving computational time, we use a nonuniform mesh in the finite element solver to emphasize the active core region. Figure 33.8 shows a mesh generated in the simulation.

33.3.2.2 Simulation Algorithm

It is known that among all the stages in the active core, the temperature T_i and the electric field F_i (i represents the stage index) are not constant [56,78], but we have no *a priori* knowledge of how they depend on the stage index. However, we know that the charge–current continuity equation must hold, and in the steady state $\nabla \cdot \mathbf{J} = 0$; this implies that the current density J must be uniform, as the current flow is essentially in one dimension, along z. This insight is key to bridging the single-stage and device-level simulations.

From Section 33.2.3, we can obtain the heat generation rate Q inside the active core by running the single-stage EMC simulation. Each single-stage EMC is carried out at a specific electric field F and

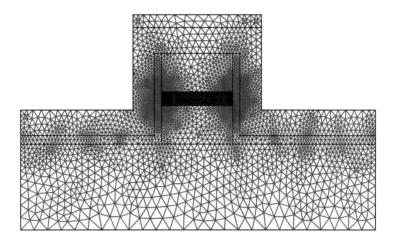

FIGURE 33.8   A typical nonuniform finite-element mesh of the simulated GaAs-based mid-IR QCL structure.

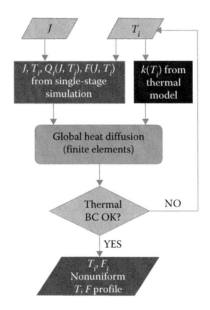

FIGURE 33.9   Flowchart of the device-level thermal simulation. We start by assuming a certain current density J and temperature profile T_i across the whole device. Based on the tabulated information from the single-stage simulation and assumed (J, T_i), we get stage-by-stage profiles for the electric field F_i and the heat generation rate Q_i. An accurate temperature-dependent thermal conductivity model, which includes the boundary resistances of layers, and the temperature profile guess are used as input to the heat diffusion equation, which is then iteratively solved (with updated temperature profile in each step) until the thermal boundary conditions are satisfied.

temperature T and outputs both the current density $J(F, T)$ and the heat generation rate $Q(F, T)$. By sweeping F and T in range of interest, we obtain a table connecting different field and temperature (F, T) to appropriate current density and heat generation rate (J, Q) $[(F, T) \rightarrow (J, Q)]$. However, based on the discussion above, the input in the thermal simulation needs to be the constant parameter J. Therefore, we "flip" the recorded $(F, T) \rightarrow (J, Q)$ table to a so-called device table $(J, T) \rightarrow (Q, F)$, suitable for coupled simulation [49].

Figure 33.9 depicts the flowchart of the device-level electrothermal simulation [49]. Before the simulation, we obtain the device table $[(J, T) \rightarrow (Q, F)]$, as discussed above. We also have to calculate the thermal conductivities ($\kappa_\parallel$ and $\kappa_\perp$) of the active region as a function of temperature and tabulate them, based on the model described in Section 33.3.1. We also need the bulk thermal conductivity of other materials in the device (cladding layer, substrate, insulation, contact, and solder) as a function of temperature. These material properties are standard and already well characterized.

Each device-level thermal simulation is carried out in a certain environment (i.e., for a given set of boundary conditions) and with a certain current density J. At the beginning of the simulation, an initial temperature profile is assigned. With the input from the device table and the thermal conductivity data in each region, we use a finite element method to iteratively solve the heat diffusion equation until convergence. At the end of the simulation, we obtain a thermal map of the whole device. Further, from the temperature T_i in each stage and the injected current density J, we obtain the nonuniform electric field distribution F_i. With the electric field in each stage and given the stage thickness, we can accurately calculate the voltage drop across the device and obtain the current–voltage characteristic. By changing the mounting configuration (W_{act}, D_{sub}, D_{ins}, D_{cont}, D_{sold}) or the boundary conditions, the temperature profile can be changed.

33.4 Device-Level Electrothermal Simulation: An Example

In this section, we present detailed simulation results of a 9-μm GaAs/Al$_{0.45}$Ga$_{0.55}$As mid-IR QCL [79] based on a conventional three-well active region design. The chosen structure has 36 repetitions of the single stage; each stage has 16 layers. Starting from the injection barrier, the layer thicknesses in one stage (in Å) are **46**/19/**11**/54/**11**/48/**28**/34/**17**/30/**18**/28/**20**/30/**26**/30. Here, the barriers (Al$_{0.45}$Ga$_{0.55}$As) are in bold while the wells (GaAs) are in normal font; the underlined layers are doped to a sheet density of $n_{Si} = 3.8 \times 10^{11}$ cm^{-2}. The results at 77 K are shown here.

33.4.1 Electronic Simulation Results

33.4.1.1 Band Structure

Figure 33.10 shows the electronic states of the chosen structure under the designed operating field of 48 kV/cm, calculated from the coupled **k·p**–Poisson solver (see Section 33.2.3.1). The active region states of the central stage are represented by dark solid curves 1, 2, and 3 are the ground state and the lower

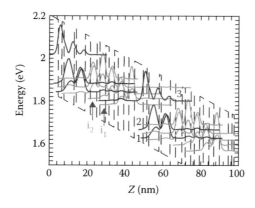

FIGURE 33.10 Energy levels and wavefunction moduli squared of Γ-valley subbands in two adjacent stages of the simulated GaAs/AlGaAs-based structure. The dark solid curves denote the active region states (1, 2, and 3 represent the ground state and the lower and upper lasing levels, respectively). The lighter solid curves represent injector states, with i_1 and i_2 denoting the lowest two. (From Y. B. Shi and I. Knezevic, *Journal of Applied Physics*, 116, 123105, 2014. With the permission of AIP Publishing.)

and upper lasing levels, respectively. Injector states are labeled i_1 and i_2. Other lightly colored solid curves denote the states that form the miniband. (When the electron density in the QCL is high, the electronic bands have to be calculated self-consistently with EMC.)

33.4.1.2 J-F Curve

The current density J versus field F curve, one of the key QCL characteristics at a given temperature, is intuitive to obtain in EMC. After calculating the electronic band structure at a certain field F, the wave-functions, energy levels, and effective masses of each subband and each stage are fed into the EMC solver. In the EMC simulation, we include all the scattering mechanisms described in Section 33.2.3. Since we employ PBCs, the current density J can be extracted from how many electrons cross the stage bound-aries in a certain amount of time in the steady state. The net flow n_{net} of electrons is calculated by subtracting the flow between the central stage and the previous stage ($n_{backward}$) from the flow between the central stage and the next stage ($n_{forward}$) in each time step. The current density is then calculated as follows:

$$J = \frac{en_{net}}{A_{eff}\delta t} = \frac{e(n_{forward} - n_{backward})}{A_{eff}\delta t}, \tag{33.7}$$

where δt is the time interval during which the flow is recorded. A_{eff} is the effective in-plane area of the simulated device. Since doping is the main source of electrons, the area is calculated as follows:

$$A_{eff} = \frac{N_{ele}}{N_s}, \tag{33.8}$$

where N_{ele} is the number of simulated electrons and N_s is the sheet doping density (in cm^{-2}) in the fabricated device. In the current simulation, $N_{ele} = 50,000$ and $N_s = 3.8 \times 10^{11}\ cm^{-2}$.

Because of the stochastic nature of EMC, we need to average the current density over multiple time steps. In practice, one can record the net cumulative number of electrons per unit area that leave a stage over time and obtain a linear fit to this quantity in the steady state; the slope yields the steady-state current density.

From each individual simulation, we extract the current density at a given electric field and temperature. To obtain the J-F curve at that temperature, we sweep the electric field. To demonstrate the importance of including nonequilibrium phonons effects, we carry out the simulation with thermal phonons alone and with both thermal and excess nonequilibrium phonons. Figure 33.11 is the J-F curve for the simulated structure with (filled squares) and without (empty squares) nonequilibrium phonons at 77 K. It can be seen that the current density at a given field considerably increases when nonequilibrium phonons are included and the trend holds up to 60 kV/cm. This difference is prominent at low temperatures (<200 K) and goes away at RT [26].

33.4.1.3 Modal Gain (G_m) and Threshold

We calculate the modal gain as follows [22]:

$$G_m = \frac{4\pi e^2 \langle z_{32} \rangle^2 \Gamma_w \Delta n}{2\varepsilon_0 \underline{n} \gamma_{32} L_p \lambda}, \tag{33.9}$$

where ε_0 is the permittivity of free space. Some constants are obtained from experiment: waveguide con-finement factor $\Gamma_w = 0.31$, stage length $L_p = 45$ nm, optical-mode refractive index $\underline{n} = 3.21$, and full width at half maximum $\gamma_{32}(T_L) \approx 8.68$ meV + 0.045 meV/K $\times T_L$ [26,79]. The dipole matrix element between the upper and lower lasing levels ($\langle z_{32} \rangle = 1.7$ nm) and the emission wavelength ($\lambda = 9$ μm) are

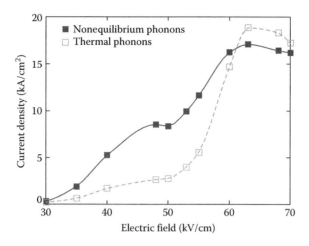

FIGURE 33.11 The current density versus electric field ($J - F$) curve of the simulated device with (filled squares) and without (empty squares) the nonequilibrium phonon effect at 77 K. The inclusion of nonequilibrium phonons considerably increases the current density at a given field up to 60 kV/cm.

also estimated in experiment [79], but we calculate these two terms directly. The dipole matrix element is calculated as follows:

$$\langle z_{32} \rangle = \int_0^d z \varphi_3^*(z) \varphi_2(z) dz . \tag{33.10}$$

The value is slightly different at different fields, as the band structure changes. At 48 kV/cm, the calculated matrix element is $\langle z_{32} \rangle = 1.997$ nm. Similarly, the wavelength of emitted photon also changes at different fields. One can calculate the value from the energy difference between the upper and lower lasing levels. The calculated wavelength at 48 kV/cm is 8.964 µm. $\Delta n = n_{upper} - n_{lower}$ is the population inversion obtained from EMC. Again, due to the randomness of EMC, the population inversion needs to be averaged over a period after the steady state has been reached.

Figure 33.12 shows the modal gain of the device with nonequilibrium (filled squares) and thermal (empty squares) phonons as a function of (1) electric field and (2) current density at 77 K. Horizontal dotted line indicates the total estimated loss in the device, which is used to help find the threshold current density, J_{th}. Lasing threshold is achieved when the modal gain G_m equals the total loss α_{tot}. We consider two sources of loss, mirror (α_m) and waveguide (α_w), so the total loss is $\alpha_{tot} = \alpha_m + \alpha_w$. The intercepts between the total loss line and the G_m versus F (Figure 33.12a) and G_m versus J (Figure 33.12b) curves give the threshold field F_{th} and threshold current density J_{th}, respectively. Like the current density, the modal gain of the device is also considerably higher when nonequilibrium phonons are considered, which leads to a lower F_{th} and a lower J_{th}. The reason for the increased current density and modal gain with nonequilibrium phonons can be attributed to the enhanced injection selectivity and efficiency [26].

33.4.1.4 Heat Generation Rate

The way to obtain the heat generation rate Q is similar to how we get the current density J. We record the cumulative net energy emission as a function of time and fit a straight line to the region where the simulation has reached a steady state. The slope of the line is used in place of $\sum(\hbar\omega_{ems} - \hbar\omega_{abs})/t_{sim}$. Figure 33.13 shows the heat generation rate as a function of electric field at 77 K. The filled squares and the empty squares depict the situation with and without nonequilibrium phonons, respectively.

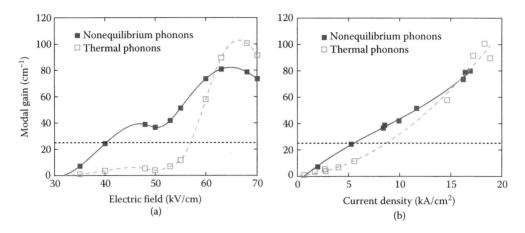

FIGURE 33.12 The modal gain (G_m) of the simulated device with nonequilibrium (filled squares) and thermal (empty squares) phonons as a function of (a) applied electric field and (b) current density J at 77 K. Horizontal dotted line shows the total estimated loss of the device.

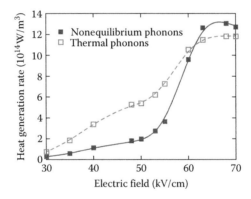

FIGURE 33.13 The heat generation rate of the simulated device as a function of electric field F at 77 K with (filled squares) and without (empty squares) nonequilibrium phonons.

33.4.2 Representative Electrothermal Simulation Results

This section serves to illustrate how the described simulation is implemented in practice and what type of information it provides at the single-stage and device levels.

First, the single-stage-coupled simulation has to be performed at different temperatures, as in Figure 33.14a. We noted the calculated J-F curves show a negative differential conductance region, which is typical for calculations, but generally not observed in experiment. Instead, a flat J-F dependence is typically recorded [30]. At every temperature and field, we also record the heat generation rate, as depicted in Figure 33.14b.

Second, the thermal model for the whole structure is developed. Considering that growth techniques improve over time, structures grown around the same time period should have similar properties. Since the device studied here was built in 2001 [79], we assume the active core should have similar effective rms roughness Δ to other lattice-matched GaAs/AlAs SLs built around the same time [69,70]. From our

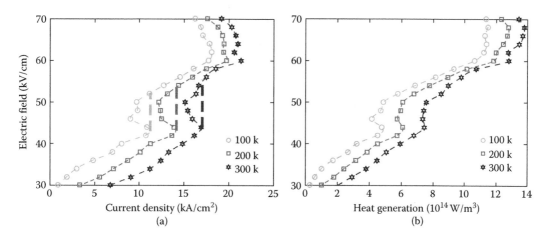

FIGURE 33.14 The field versus current density (a) and heat generation rate versus current density (b) characteristics for the simulated device at 100, 200, and 300 K, as obtained from single-stage simulation with nonequilibrium phonons.

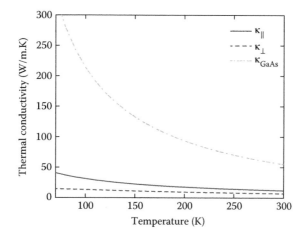

FIGURE 33.15 Calculated in-plane ($\kappa_\parallel$; solid line) and cross-plane ($\kappa_\perp$; dashed line) thermal conductivities of the active core along with the bulk thermal conductivity of the GaAs substrate (dash-dotted line). The effective rms roughness Δ is taken to be 5 Å.

previous simulation work on fitting the SL thermal conductivities [48], we choose an effective rms roughness $\Delta = 5$ Å in this calculation. Figure 33.15 shows the calculated thermal conductivities $\kappa_\parallel$ (solid line) and $\kappa_\perp$ (dashed line) along with the calculated bulk thermal conductivity (dash-dotted line) for the substrate GaAs.

The structure we considered operated in pulsed mode at 77 K. Depending on the duty cycle, the temperature distribution in the device can differ considerably. Figure 33.16 depicts a typical temperature profile across the device, while Figure 33.17 depicts the profile across the active core alone at duty cycles of 100% (essentially CW lasing, if the device achieved it) and 0.01% (as in experiment [79]). Clearly, CW operation would result in dramatic heating of the active region. Finally, Figure 33.18 shows the J–V curve of the entire simulated device at 77 K with duty cycles of 0.01%, 100%, and as observed in experiment [79].

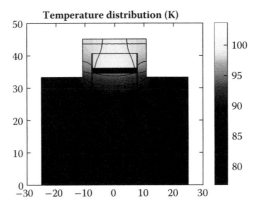

FIGURE 33.16 A typical temperature profile across the structure. At the bottom of the device is a heat sink held at 77 K, while adiabatic boundary conditions are applied elsewhere. The current density is 6 kA/cm² and the duty cycle is 100%.

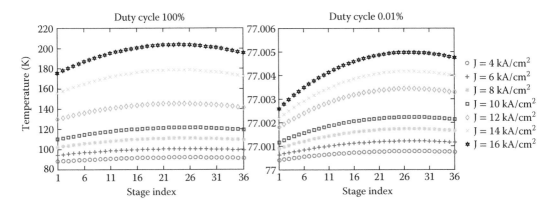

FIGURE 33.17 Temperature profile inside the active region at 100% duty cycle (left) and 0.01% duty cycle (right) for the QCL from [79].

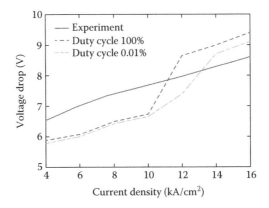

FIGURE 33.18 The current density versus voltage drop for the simulated device in experiment (solid curve) and as calculated at 100% (dashed curve) and 0.01% (dot-dashed curve) duty cycles. The bottom of the device is placed on a heat sink held at 77 K while adiabatic boundary conditions are assumed on the rest of the boundaries (see Section 33.3.2).

33.5 Conclusion

We overviewed electronic and thermal transport simulation of QCLs, as well as recent efforts in device-level electrothermal modeling of these structures, which is appropriate for transport below threshold, where the effects of the optical field are negligible. We specifically focused on mid-IR QCLs in which electronic transport is largely incoherent and can be captured by the EMC technique. The future of QCL modeling, especially for near-RT CW operation, will likely include improvements on several fronts: (1) further development of computationally efficient yet rigorous quantum transport techniques for electronic transport to fully account for coherent transport features that are important in short-wavelength mid-IR devices; (2) a better understanding and better numerical models for describing the role of electron–electron interaction, impurities, and interface roughness on device characteristics; and (3) holistic modeling approaches in which electrons, phonons, and photons are simultaneously and self-consistently captured within a single simulation. The goal of QCL simulation should be nothing less than excellent predictive value of device operation across a range of temperatures and biasing conditions, along with unprecedented insight into the fine details of exciting nonequilibrium physics that underscores the operation of these devices.

Acknowledgment

The authors gratefully acknowledge support by the U.S. Department of Energy, Basic Energy Sciences, Division of Materials Sciences and Engineering, Physical Behavior of Materials Program, Award No. DE-SC0008712. The work was performed using the resources of the UW-Madison Center for High Throughput Computing (CHTC).

References

1. J. Faist, F. Capasso, D. L. Sivco, C. Sirtori, A. L. Hutchinson, and A. Y. Cho. Quantum cascade laser. *Science*, 264(5158):553–556, 1994.
2. R. A. Suris and R. F. Kazarinov. Possibility of the amplification of electromagnetic waves in a semiconductor with a superlattice. *Soviet Physics: Semiconductors*, 5(4):707–709, 1971.
3. K. Y. Cheng. Molecular beam epitaxy technology of III-V compound semiconductors for optoelectronic applications. *Proceedings of the IEEE*, 85(11):1694–1714, 1997.
4. K. H. Goetz, D. Bimberg, H. Jürgensen, J. Selders, A. V. Solomonov, G. F. Glinskii, and M. Razeghi. Optical and crystallographic properties and impurity incorporation of $Ga_xIn_{1-x}As$ ($0.44 < x < 0.49$) grown by liquid phase epitaxy, vapor phase epitaxy, and metal organic chemical vapor deposition. *Journal of Applied Physics*, 54(8):4543–4552, 1983.
5. Q. Zhang, F. Q. Liu, W. Zhang, Q. Lu, L. Wang, L. Li, and Z. Wang. Thermal induced facet destructive feature of quantum cascade lasers. *Applied Physics Letters*, 96(14):141117, 2010.
6. D. Botez, J. C. Shin, J. D. Kirch, C. C. Chang, L. J. Mawst, and T. Earles. Multidimensional conduction-band engineering for maximizing the continuous-wave (CW) wallplug efficiencies of mid-infrared quantum cascade lasers. *IEEE Journal of Selected Topics in Quantum Electronics*, 19(4):1200312–1200312, 2013.
7. P. Q. Liu, A. J. Hoffman, M. D. Escarra, K. J. Franz, J. B. Khurgin, Y. Dikmelik, X. Wang, J.-Y. Fan, and C. F. Gmachl. Highly power-efficient quantum cascade lasers. *Nature Photonics*, 4:95–98, 2010.
8. A. J. Hoffman, Y. Yao, and C. F. Gmachl. Mid-infrared quantum cascade laser. *Nature Photonics*, 6(1749–4885):432–439, 2012.
9. Y. Bai, S. R. Darvish, S. Slivken, W. Zhang, A. Evans, J. Nguyen, and M. Razeghi. Room temperature continuous wave operation of quantum cascade lasers with watt-level optical power. *Applied Physics Letters*, 92(10):101105, 2008.

10. A. Lyakh, R. Maulini, A. Tsekoun, R. Go, C. Pflügl, L. Diehl, Q. J. Wang, F. Capasso, and C. Kumar N. Patel. 3 W continuous-wave room temperature single-facet emission from quantum cascade lasers based on nonresonant extraction design approach. *Applied Physics Letters*, 95(14):141113, 2009.

11. N. Bandyopadhyay, Y. Bai, B. Gokden, A. Myzaferi, S. Tsao, S. Slivken, and M. Razeghi. Watt level performance of quantum cascade lasers in room temperature continuous wave operation at $\lambda \sim$ 3.76 µm. *Applied Physics Letters*, 97(13):131117, 2010.

12. N. Bandyopadhyay, S. Slivken, Y. Bai, and M. Razeghi. High power, continuous wave, room temperature operation of $\lambda \sim 3.4$ µm and $\lambda \sim 3.55$ µm InP-based quantum cascade lasers. *Applied Physics Letters*, 100(21):212104, 2012.

13. Y. Bai, N. Bandyopadhyay, S. Tsao, S. Slivken, and M. Razeghi. Room temperature quantum cascade lasers with 27% wall plug efficiency. *Applied Physics Letters*, 98(18):181102, 2011.

14. Y. Yu, W. Xiaojun, F. Jen-Yu, and C. F. Gmachl. High performance continuum-to-continuum quantum cascade lasers with a broad gain bandwidth of over 400 cm^{-1}. *Applied Physics Letters*, 97(8):081115, 2010.

15. M. Razeghi, Q. Y. Lu, N. Bandyopadhyay, W. Zhou, D. Heydari, Y. Bai, and S. Slivken. Quantum cascade lasers: From tool to product. *Optics Express*, 23(7):8462–8475, 2015.

16. M. Wienold, M. P. Semtsiv, I. Bayrakli, W. T. Masselink, M. Ziegler, K. Kennedy, and R. Hogg. Optical and thermal characteristics of narrow-ridge quantum-cascade lasers. *Journal of Applied Physics*, 103(8):083113, 2008.

17. H. K. Lee and J. S. Yu. Thermal analysis of short wavelength InGaAs/InAlAs quantum cascade lasers. *Solid-State Electronics*, 54(8):769–776, 2010.

18. J. D. Kirch, J. C. Shin, C. C. Chang, L. J. Mawst, D. Botez, and T. Earles. Tapered active-region quantum cascade lasers ($\lambda = 4.8$ µm) for virtual suppression of carrier-leakage currents. *Electronics Letters*, 48(4):234–235, 2012.

19. D. Botez, C. C. Chang, and L. J. Mawst. Temperature sensitivity of the electro-optical characteristics for mid-infrared ($\lambda = 3$–16 µm)-emitting quantum cascade lasers. *Journal of Physical D: Applied Physics*, 49:043001, 2016.

20. D. Indjin, P. Harrison, R. W. Kelsall, and Z. Ikonić. Self-consistent scattering theory of transport and output characteristics of quantum cascade lasers. *Journal of Applied Physics*, 91(11):9019–9026, 2002.

21. D. Indjin, P. Harrison, R. W. Kelsall, and Z. Ikonic. Influence of leakage current on temperature performance of GaAs/AlGaAs quantum cascade lasers. *Applied Physics Letters*, 81(3):400–402, 2002.

22. A. Mircetic, D. Indjin, Z. Ikonic, P. Harrison, V. Milanovic, and R. W. Kelsall. Towards automated design of quantum cascade lasers. *Journal of Applied Physics*, 97(8):084506, 2005.

23. R. Claudia Iotti and F. Rossi. Nature of charge transport in quantum-cascade lasers. *Physical Review Letters*, 87:146603, 2001.

24. H. Callebaut, S. Kumar, B. S. Williams, Q. Hu, and J. L. Reno. Importance of electron-impurity scattering for electron transport in terahertz quantum-cascade lasers. *Applied Physics Letters*, 84(5):645–647, 2004.

25. X. Gao, D. Botez, and I. Knezevic. X-valley leakage in GaAs-based midinfrared quantum cascade lasers: A Monte Carlo study. *Journal of Applied Physics*, 101(6):063101, 2007.

26. Y. B. Shi and I. Knezevic. Nonequilibrium phonon effects in midinfrared quantum cascade lasers. *Journal of Applied Physics*, 116(12):123105, 2014.

27. H. Willenberg, G. H. Döhler, and J. Faist. Intersubband gain in a Bloch oscillator and quantum cascade laser. *Physical Review B*, 67:085315, 2003.

28. S. Kumar and Q. Hu. Coherence of resonant-tunneling transport in terahertz quantum-cascade lasers. *Physical Review B*, 80:245316, 2009.

29. C. Weber, A. Wacker, and A. Knorr. Density-matrix theory of the optical dynamics and transport in quantum cascade structures: The role of coherence. *Physical Review B*, 79:165322, 2009.

30. E. Dupont, S. Fathololoumi, and H. C. Liu. Simplified density-matrix model applied to three-well terahertz quantum cascade lasers. *Physical Review B*, 81:205311, 2010.

31. R. Terazzi and J. Faist. A density matrix model of transport and radiation in quantum cascade lasers. *New Journal of Physics*, 12(3):033045, 2010.

32. O. Jonasson, F. Karimi, and I. Knezevic. Partially coherent electron transport in terahertz quantum cascade lasers based on a Markovian master equation for the density matrix. *Journal of Computational Electronics*, 15:1192, 2016.

33. O. Jonasson, S. Mei, F. Karimi, J. Kirch, D. Botez, L. Mawst, and I. Knezevic. Quantum transport simulation of high-power 4.6-μm quantum cascade lasers. *Photonics*, 3(2):38, 2016.

34. S.-C. Lee and A. Wacker. Nonequilibrium Green's function theory for transport and gain properties of quantum cascade structures. *Physical Review B*, 66:245314, 2002.

35. M. Bugajski, P. Gutowski, P. Karbownik, A. Kolek, G. Hałdaś, K. Pierściński, D. Pierścińska, J. Kubacka-Traczyk, I. Sankowska, A. Trajnerowicz, K. Kosiel, A. Szerling, J. Grzonka, K. Kurzydłowski, T. Slight, and W. Meredith. Mid-IR quantum cascade lasers: Device technology and non-equilibrium Green's function modeling of electro-optical characteristics. *Physica Status Solidi (b)*, 251(6):1144–1157, 2014.

36. A. Kolek, G. Haldas, M. Bugajski, K. Pierscinski, and P. Gutowski. Impact of injector doping on threshold current of mid-infrared quantum cascade laser–non-equilibrium greens function analysis. *IEEE Journal of Selected Topics in Quantum Electronics*, 21(1):124–133, 2015.

37. O. Jonasson and I. Knezevic. Dissipative transport in superlattices within the Wigner function formalism. *Journal of Computational Electronics*, 14:879–887, 2015.

38. A. Matyas, P. Lugli, and C. Jirauschek. Photon-induced carrier transport in high efficiency midinfrared quantum cascade lasers. *Journal of Applied Physics*, 110(1):013108, 2011.

39. M. Lindskog, J. M. Wolf, V. Trinite, V. Liverini, J. Faist, G. Maisons, M. Carras, R. Aidam, R. Ostendorf, and A. Wacker. Comparative analysis of quantum cascade laser modeling based on density matrices and non-equilibrium Green's functions. *Applied Physics Letters*, 105(10):103106, 2014.

40. R. C. Iotti and F. Rossi. Microscopic theory of semiconductor-based optoelectronic devices. *Reports on Progress in Physics*, 68(11):2533, 2005.

41. C. Jirauschek and P. Lugli. Monte-Carlo-based spectral gain analysis for terahertz quantum cascade lasers. *Journal of Applied Physics*, 105(12):123102, 2009.

42. C. Jirauschek and T. Kubis. Modeling techniques for quantum cascade lasers. *Applied Physics Reviews*, 1(1):011307, 2014.

43. C. Jirauschek. Monte Carlo study of carrier-light coupling in terahertz quantum cascade lasers. *Applied Physics Letters*, 96(1), 2010.

44. A. Wacker, M. Lindskog, and D. O. Winge. Nonequilibrium Green's function model for simulation of quantum cascade laser devices under operating conditions. *IEEE Journal of Selected Topics in Quantum Electronics*, 19(5):1–11, 2013.

45. C. A. Evans, D. Indjin, Z. Ikonic, P. Harrison, M. S. Vitiello, V. Spagnolo, and G. Scamarcio. Thermal modeling of terahertz quantum-cascade lasers: Comparison of optical waveguides. *IEEE Journal of Quantum Electronics*, 44(7):680–685, 2008.

46. H. K. Lee and J. S. Yu. Thermal effects in quantum cascade lasers at λ ~ 4.6 μm under pulsed and continuous-wave modes. *Applied Physics B*, 106(3):619–627, 2012.

47. Y. B. Shi, Z. Aksamija, and I. Knezevic. Self-consistent thermal simulation of $GaAs/Al_{0.45}Ga_{0.55}As$ quantum cascade lasers. *Journal of Computational Electronics*, 11(1):144–151, 2012.

48. S. Mei and I. Knezevic. Thermal conductivity of III-V semiconductor superlattices. *Journal of Applied Physics*, 118(17):175101, 2015.

49. Y. B. Shi, S. Mei, O. Jonasson, and I. Knezevic. Modeling quantum cascade lasers: Coupled electron and phonon transport far from equilibrium and across disparate spatial scales. *Fortschritte der Physik – Progress of Physics*, 1600084, 2016. doi:10.1002/prop.201600084

50. S.-C. Lee, F. Banit, M. Woerner, and A. Wacker. Quantum mechanical wavepacket transport in quantum cascade laser structures. *Physical Review B*, 73:245320, 2006.
51. K. Faist, D. Hofstetter, M. Beck, T. Aellen, M. Rochat, and S. Blaser. Bound-to-continuum and two-phonon resonance, quantum-cascade lasers for high duty cycle, high-temperature operation. *IEEE Journal of Quantum Electronics*, 38(6):533–546, 2002.
52. M. Yamanishi, T. Edamura, K. Fujita, N. Akikusa, and H. Kan. Theory of the intrinsic linewidth of quantum-cascade lasers: Hidden reason for the narrow linewidth and line-broadening by thermal photons. *IEEE Journal of Quantum Electronics*, 44(1):12–29, 2008.
53. X. Gao. Monte Carlo Simulation of Electron Dynamics in Quantum Cascade Lasers. PhD thesis, University of Wisconsin–Madison, 2008.
54. Y. Shi. Electrothermal Simulation of Quantum Cascade Lasers. PhD thesis, University of Wisconsin–Madison, 2015.
55. H. Callebaut and Q. Hu. Importance of coherence for electron transport in terahertz quantum cascade lasers. *Journal of Applied Physics*, 98(10):104505, 2005.
56. A. Wacker. Semiconductor superlattices: A model system for nonlinear transport. *Physics Reports*, 357(1):1–111, 2002.
57. X. Gao, D. Botez, and I. Knezevic. X-valley leakage in GaAsAlGaAs quantum cascade lasers. *Applied Physics Letters*, 89(19):191119, 2006.
58. S. L. Chuang. *Physics of Optoelectronic Devices*. New York, NY: Wiley, 1995.
59. X. Gao, D. Botez, and I. Knezevic. Confined phonon scattering in multivalley Monte Carlo simulation of quantum cascade lasers. *Journal of Computational Electronics*, 7(3):209–212, 2008.
60. P. Lugli, P. Bordone, L. Reggiani, M. Rieger, P. Kocevar, and S. M. Goodnick. Monte Carlo studies of nonequilibrium phonon effects in polar semiconductors and quantum wells. I. Laser photoexcitation. *Physical Review B*, 39:7852–7865, 1989.
61. B. K. Ridley. The electron-phonon interaction in quasi-two-dimensional semiconductor quantum-well structures. *Journal of Physics C: Solid State Physics*, 15(28):5899, 1982.
62. J. T. Lü and J. C. Cao. Monte Carlo simulation of hot phonon effects in resonant-phonon-assisted terahertz quantum-cascade lasers. *Applied Physics Letters*, 88(6):061119, 2006.
63. S. Usher and G. P. Srivastava. Theoretical study of the anharmonic decay of nonequilibrium LO phonons in semiconductor structures. *Physical Review B*, 50:14179–14186, 1994.
64. E. Pop, S. Sinha, and K. E. Goodson. Heat generation and transport in nanometer-scale transistors. *Proceedings of the IEEE*, 94(8):1587–1601, 2006.
65. D. G. Cahill, P. V. Braun, G. Chen, D. R. Clarke, S. Fan, K. E. Goodson, P. Keblinski, W. P. King, G. D. Mahan, A. Majumdar, H. J. Maris, S. R. Phillpot, E. Pop, and L. Shi. Nanoscale thermal transport. II. 2003–2012. *Applied Physics Reviews*, 1(1):011305, 2014.
66. T. Yao. Thermal properties of AlAs/GaAs superlattices. *Applied Physics Letters*, 51(22):1798–1800, 1987.
67. G. Chen, C. L. Tien, X. Wu, and J. S. Smith. Thermal diffusivity measurement of GaAs/AlGaAs thin-film structures. *Journal of Heat Transfer*, 116:325–331, 1994.
68. X. Y. Yu, G. Chen, A. Verma, and J. S. Smith. Temperature dependence of thermophysical properties of GaAs/AlAs periodic structure. *Applied Physics Letters*, 67(24):3554–3556, 1995.
69. W. S. Capinski and H. J. Maris. Thermal conductivity of GaAs/AlAs superlattices. *Physica B*, 219–220:699–701, 1996.
70. W. S. Capinski, H. J. Maris, T. Ruf, M. Cardona, K. Ploog, and D. S. Katzer. Thermal conductivity measurements of GaAs/AlAs superlattices using a picosecond optical pump-and-probe technique. *Physical Review B*, 59:8105–8113, 1999.
71. G. Chen. Size and interface effects on thermal conductivity of superlattices and periodic thin-film structures. *Journal of Heat Transfer*, 119(2):220, 1997.
72. G. Chen. Thermal conductivity and ballistic-phonon transport in the cross-plane direction of superlattices. *Physical Review B*, 57:14958–14973, 1998.

73. S. T. Huxtable, A. R. Abramson, C. L. Tien, A. Majumdar, C. LaBounty, X. Fan, G. Zeng, J. E. Bowers, A. Shakouri, and E. T. Croke. Thermal conductivity of Si/SiGe and SiGe/SiGe superlattices. *Applied Physics Letters*, 80(10):1737–1739, 2002.

74. Y. Wang, H. Huang, and X. Ruan. Decomposition of coherent and incoherent phonon conduction in superlattices and random multilayers. *Physical Review B*, 90:165406, 2014.

75. Y. Wang, C. Gu, and X. Ruan. Optimization of the random multilayer structure to break the random-alloy limit of thermal conductivity. *Applied Physics Letter*, 106:073104, 2015.

76. Z. Aksamija and I. Knezevic. Thermal conductivity of $Si1-xGex/Si1-yGey$superlattices: Competition between interfacial and internal scattering. *Physical Review B*, 88:155318, 2013.

77. E. T. Swartz and R. O. Pohl. Thermal boundary resistance. *Reviews of Modern Physics*, 61:605–668, 1989.

78. A. Lops, V. Spagnolo, and G. Scamarcio. Thermal modeling of GaInAs/AlInAs quantum cascade lasers. *Journal of Applied Physics*, 100(4):043109, 2006.

79. H. Page, C. Becker, A. Robertson, G. Glastre, V. Ortiz, and C. Sirtori. 300 k operation of a GaAs-based quantum-cascade laser at $\lambda \approx 9$ μm. *Applied Physics Letters*, 78(22):3529, 2001.

80. V. Spagnolo, A. Lops, G. Scamarcio, M. S. Vitiello, and C. Di Franco. Improved thermal management of mid-IR quantum cascade lasers. *Journal of Applied Physics*, 103(4):043103, 2008.

34

Vertical-Cavity Surface-Emitting Lasers

Tomasz
Czyszanowski

Leszek
Frasunkiewicz

and

Maciej Dems

34.1 Introduction

The history of vertical-cavity surface-emitting lasers (VCSELs) can be traced back almost as far as that of semiconductor lasers themselves. In 1965, Melngailis demonstrated coherent emission parallel to the direction of the current in an n^+pp^+ InSb diode laser.[1] The polished surface on the emission side of InSb and the gold current contact on its opposite side provided sufficient optical feedback to enhance laser emission. In this very early work, a design was proposed for the integration in an array that would provide coherent emission over a large area, with a small beam angle. Unfortunately, this idea was forgotten and not rediscovered until 1979 by Soda et al.,[2] who demonstrated a surface-emitting InP-based diode laser with two metallic mirrors. Further development was initiated by Ogura et al.,[3] who introduced highly reflective distributed Bragg reflectors (DBRs) in 1983.

Unlike typical edge-emitting lasers, in VCSELs the optical cavity is formed between the mirrors above and below the active region (Figure 34.1). The laser light resonates in the vertical direction. Inside the laser structure, the light passes the active region in the vertical direction—i.e., gain is provided over a short distance only and the amplification per photon round-trip is small. Therefore, the mirrors must be highly reflective (over 99% for the emitting mirror and almost 100% for the opposite one) so that the photons make many round-trips before they are emitted. High reflectivity is provided by DBRs composed of stacks of two alternating layers with high refractive index contrast. With quarter-wavelength layer thicknesses, the reflected waves from all DBR interfaces add up constructively, allowing for a total DBR reflectance of over 99%. Such layers are composed of 50 or more semiconductor layers. The current is injected vertically, similar to edge-emitting devices. However, in VCSELs, optical resonance occurs in the vertical direction.

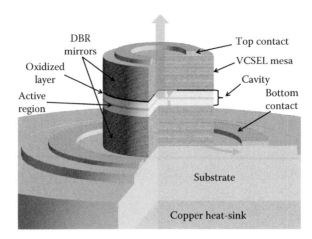

FIGURE 34.1 Intersection of a VCSEL structure showing key construction elements. The arrows inside VCSEL structure illustrate the path of the current flow and vertical arrow the emitted beam.

Optical resonance and current injection taking place in the same direction is the main weakness of these devices. Theoretically, preferential injection of the carriers to the active region may be achieved by placing two electrodes on opposite sides of the laser. However, in such configuration, the stimulated light resonates between the mirrors and no emission occurs, since the metal contacts are highly absorptive. To enable emission, a ring contact is typically used on the top DBR, through which the light is emitted. However, current flows through the shortest path, so maximal stimulated emission overlaps with the top contact, leading to high absorption in the metal and deterioration of the emission beam. To overcome this problem, the carriers must be forced to flow through the central part of the active region. This can be achieved either by selective proton implantation or by selective oxidation of the lateral regions in the VCSEL. Both of these approaches turn the semiconductor layers into electrical insulators. With selective processes, the central part of the VCSEL is conductive and radiative recombination takes place within the central part of the active region. Both methods also allow for cylindrical aperture formation, supporting cylindrical beam emission that cannot be achieved with typical edge-emitting lasers. Such symmetry of the aperture, together with the use of ring contacts, favors the choice of cylindrical geometry for numerical approaches, which are used to simulate the phenomena that take place in VCSELs during laser operation.

34.2 Physical Phenomena

VCSELs are very complex systems due to their multilayered structure (comprising in excess of a hundred layers). Often with nonplanar or buried-type architectures, they feature many heterojunctions, graded layers, quantum wells (QWs), quantum dots (QDs) or quantum wires, superlattices, oxide and oxidized layers, mesa structures, photonic microstructures, and so on. To analyze the operation of VCSELs precisely, therefore, one needs to consider four main classes of phenomena in the device: electrical, thermal, optical, and recombination, which together create a nonlinear network of mutual interrelations. Each class of phenomena is responsible for the following:

1. Optical: Cavity modes, their modal gain and loss; emission wavelengths and optical field distributions within the laser cavity; formation of the output beam
2. Electrical: Current spreading between the top and bottom contacts through the centrally located active region; injection of carriers of both polarities into the active region and their subsequent radiative or nonradiative recombination after radial out-diffusion in the active region; overbarrier carrier leakage

3. Thermal: Generation of heat flux (nonradiative recombination, reabsorption of spontaneous radiation, as well as volume and barrier Joule heating); heat spreading within heat sinks

4. Recombination: Material gain as a result of stimulated recombination (to determine the active region optical gain spectrum); nonradiative monomolecular and Auger recombinations

The main effects induced by nonlinear interaction between the models are thermal focusing and emitted power rollover. The first is related to the dependence of the refractive index on temperature. The heat induced predominantly in the central part of the active region dissipates in all directions, but mostly toward the bottom heat sink. This, in turn, induces different temperatures in the laser cavity. Increasing the temperature of semiconductor materials triggers an increase in their refractive index. The most significant increase of the refractive index is in the central part of the laser, which is heated most intensely. This causes the waveguiding effect, whereby the modes squeeze radially into the central part of the laser. On the one hand, stronger focusing of the modes in the active region increases their interaction with carriers that are injected into the central part of the active region, reducing the current necessary for laser operation (threshold current). On the other hand, if there are more optical modes confined and excited within the central part, this contributes to a broadening of the emission spectrum, since each mode relates to a slightly different emission wavelength. The quality of the output beam also deteriorates, since interference between different modes occurs at various emission angles.

The second effect—emission power rollover—limits the emitted power of the laser, mostly due to thermal effects. The active regions of VCSELs are designed in such a way that the wavelength corresponding to maximal material gain overlaps with the wavelength of the optical resonance induced by the VCSEL cavity. A change in the temperature of the device, due to heat dissipation or ambient temperature, may modify the bandgap and consequently change the wavelength corresponding to the maximum of the material gain. The optical modes do not follow the maximum of the material gain, as is the case in edge-emitting lasers, but follow the optical resonance induced by the cavity. The resonance wavelength also changes with temperature, since the refractive indices of the cavity change. The changes in maximal material gain and of cavity resonance, however, are of different proportions. Typically, the change in the gain spectrum is more sensitive to temperature changes than in the case of optical resonance. Hence, material gain can be insufficient at resonant wavelength to sustain laser operation for higher operational temperatures and the laser ceases to emit stimulated radiation. This effect is manifested as a reduction in emitted power as the injected current increases.

To simulate such complex behaviors, a comprehensive model including electrical, thermal, recombination, and optical submodels is needed. Three of these submodels can be solved in a very similar manner to the optoelectronic devices considered previously in this book. However, special attention must be given to the optical model, since electromagnetic field resonance takes place in the vertical direction, which is orthogonal to the resonance that occurs in typical edge-emitting lasers. In the remainder of this chapter, we have therefore focused on optical simulation.

34.3 Light Confinement in VCSELs

VCSELs inherently emit in single longitudinal mode due to their very short cavities. Controlling the modes in the transversal direction is more complicated. In typical VCSELs made from arsenide-based materials, well-established wet oxidation technology can be used,[4] which produces the waveguide effect through a selectively oxidized Al-rich layer that also plays the role of an electric insulator for improved current funneling to the active region. Another less material dependent technique is proton implantation,[5] which defines highly resistive regions. The waveguide effect is produced by thermal focusing induced by thermal increase of the refractive index in regions where nonradiative recombination and Joule heating occur. Other methods of transverse mode confinement require more sophisticated and expensive technologies. Possible solutions include micro-optical structures, such as surface relief,[6] antiresonant patterning,[7] tunnel junction patterning,[8] photonic crystals[9] and highcontrast gratings.[10] Each of these methods requires a separate analytical approach, since different phenomena are responsible for mode confinement: the waveguide effect, total interior reflections, Bragg reflections or Fano resonance. Since not all those phenomena can be observed in all VCSEL constructions with equal intensity, numerical models should be properly chosen to provide a balance between generalization and efficiency.

34.4 Fundamental Equations

Interaction between a VCSEL and the optical field is governed by Maxwell's equations. The boundary conditions correspond to an optical field that decays to 0 in infinity. These boundary conditions mean that Maxwell's equations turn into an eigenvalue problem. Its solutions are discrete eigenvalues and the corresponding eigenvectors. Eigenvalues of Maxwell's equations are vacuum wavevectors (k_0), corresponding angular frequencies (ω), or wavelengths in a vacuum (λ). Eigenvectors are distributions of electric (**E**) and magnetic (**H**) fields of electromagnetic waves corresponding to particular eigenvalues.

The three-dimensional electric **E** and magnetic **H** field vectors satisfy the following partial differential equations,[11] which correspond to Faraday's and Amper's laws, respectively:

$$\nabla \times \mathbf{E} = -\mu\mu_0 \frac{\partial \mathbf{H}}{\partial t}, \tag{34.1}$$

$$\nabla \times \mathbf{H} = \varepsilon\varepsilon_0 \frac{\partial \mathbf{E}}{\partial t}, \tag{34.2}$$

where μ and μ_0 are the magnetic permittivity of the material and the vacuum, respectively, and ε and ε_0 are analogous dielectric constants.

Let us consider a monochromatic wave only, of which the angular frequency is ω. The time dependency of **E** and **H** is given by the following relations:

$$\mathbf{E} = \mathbf{E} \exp{(i\omega t)}, \tag{34.3}$$

$$\mathbf{H} = \mathbf{H} \exp{(i\omega t)}. \tag{34.4}$$

Embedment of Equations 34.3 and 34.4 to Equations 34.1 and 34.2 results in a time-independent set of equations:

$$\nabla \times \mathbf{E} = -i\omega\mu\mu_0 \mathbf{H}, \tag{34.5}$$

$$\nabla \times \mathbf{H} = i\omega\varepsilon\varepsilon_0 \mathbf{E}. \tag{34.6}$$

We assume that μ is uniform and equal to 1, which is always true for the dielectrics and nonferromagnetic metals typically used in VCSELs. Combining Equations 34.5 and 34.6, we can reduce the set of six equations for all vector components to a set of three equations for the electric field only:

$$\nabla \times \nabla \times \mathbf{E} = k_0^2 n^2 \mathbf{E}, \tag{34.7}$$

where k_0 is the wavenumber of the mode in a vacuum and

$$\sqrt{\varepsilon} = n = n_{\mathrm{re}} + i n_{\mathrm{im}} \tag{34.8}$$

is the complex refractive index. Its imaginary part accounts for the loss and the gain of the medium in which the wave propagates. The relation between the imaginary part of n and losses α is given by

$$n_{\mathrm{im}} = -\frac{\alpha}{2 k_0}. \tag{34.9}$$

Equation 34.9 can be rewritten in an equivalent form:

$$\nabla (\nabla \cdot \mathbf{E}) - \nabla^2 \mathbf{E} = \omega^2 \mu \varepsilon \mathbf{E},$$

$$\nabla (\nabla \cdot \mathbf{E}) - \nabla^2 \mathbf{E} = k_0^2 n^2 \mathbf{E}. \tag{34.10}$$

VCSELs are typically processed to be circular in the epitaxial plane to support symmetric beam emission. This geometry allows the complexity of Equation 34.7 to be reduced by expressing it in a cylindrical coordinate system where the angular dependence of $\mathbf{E}$ can be given in simple analytical terms. However, in recent years, numerous confinement approaches based on photonic microstructures have been proposed. Many of these structures cannot be analyzed using the cylindrical coordinate system and therefore the three-dimensional Cartesian system must be used. We will therefore now show how Equations 34.5, 34.6, and 34.10 can be solved using three approaches of incremental complexity. The first approach is based on a one-dimensional analysis of the optical field along the optical axis. The second assumes that the VCSEL has cylindrical geometry and considers both the scalar optical field in two dimensions along the optical axis and the radius of the VCSEL. The third considers the vectorial optical field in the three-dimensional Cartesian coordinate system.

34.5 VCSEL Structure

Figure 34.2 shows the VCSEL structure with construction details listed in Table 34.1. This structure was used in References [12,13] as a benchmark for comparative analysis of optical numerical models, however it must be stressed here, the structure is significantly simplified in comparison to real-life VCSELs. The structure consists of one wavelength-long GaAs cavity with a 5-nm active GaAs layer at the antinode of the standing wave and a selectively oxidized AlAs layer ~15 nm in thickness. The cavity is sandwiched between 24 and 29.5 pairs of GaAs/AlAs DBRs. In comparison to a real-life VCSEL, there is no vertical carrier confinement that would capture and support optical recombination of the carriers. There is also no doping scheme that would allow efficient injection of the current to the active region.

In what follows, we will be using two coordinate systems: Cartesian and cylindrical (Figure 34.1). These share a z-axis along the optical axis of the device. The Cartesian plane $x - y$ and cylindrical plane $r - \varphi$ are parallel to the epitaxial layers of the VCSEL, and the x-axis relates to $\varphi = 0$.

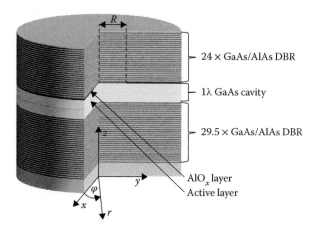

FIGURE 34.2 Intersection of the VCSEL structure with assigned radius (R) of the oxide aperture and definition of Cartesian (x, y, z) and cylindrical (r, φ, z) coordinate systems. DBR = distributed Bragg reflectors.

TABLE 34.1 Refractive Indices and Thicknesses of VCSEL Layers

Refractive Index	Thickness (nm)	Repetition
1.0	∞	–
3.53	69.49	24.5
3.08	79.63	
3.08	63.71	–
2.95 for $r < R$ 1.6 for $r > R$	15.93	–
3.53	136.49	–
3.53 (active)	5.00	–
3.53	136.49	–
3.08	79.63	29.5
3.53	69.49	
3.53	∞	–

Note: The two outer layers are assumed to be infinite.

Vectors **E** and **H** are given in the Cartesian system as

$$\mathbf{E} = \begin{bmatrix} E_x \\ E_y \\ E_z \end{bmatrix}, \quad \mathbf{H} = \begin{bmatrix} H_x \\ H_y \\ H_z \end{bmatrix}, \tag{34.11}$$

and in the cylindrical system as

$$\mathbf{E} = \begin{bmatrix} E_r \\ E_\phi \\ E_z \end{bmatrix}, \quad \mathbf{H} = \begin{bmatrix} H_r \\ H_\varphi \\ H_z \end{bmatrix}. \tag{34.12}$$

34.6 One-Dimensional Scalar Approach

The simplest approach is to consider the distribution of light within the VCSEL along the z-axis only, assuming an electromagnetic field in the form of a plane wave within the device and that the layers are uniform in the lateral direction. This extremely simplified approach allows only the longitudinal modes to be determined. However, it is of tremendous importance in VCSEL wafers designing, as it enables one to determine the size of the cavity and the number of necessary DBRs, as well as the positions of active or highly doped layers.

The assumption of uniform layers corresponds to the Gauss's law relating to electrical fields:

$$\nabla \cdot \varepsilon \mathbf{E} = 0. \tag{34.13}$$

Since ε is constant within any particular layer of the VCSEL, Gauss's law can be reduced to the following form:

$$\nabla \cdot \mathbf{E} = 0, \tag{34.14}$$

which simplifies Equation 34.10:

$$\nabla^2 \mathbf{E} = k_0^2 n^2 \mathbf{E}. \tag{34.15}$$

The operator ∇^2 can be expressed in either coordinate system. However, for the sake of clarity in what follows, we shall express it in the cylindrical system:

$$\nabla^2 \equiv \begin{bmatrix} \partial_r^2 + \frac{1}{r}\partial_r + \frac{1}{r^2}\partial_\varphi^2 + \partial_z^2 & 0 & 0 \\ 0 & \partial_r^2 + \frac{1}{r}\partial_r + \frac{1}{r^2}\partial_\varphi^2 + \partial_z^2 & 0 \\ 0 & 0 & \partial_r^2 + \frac{1}{r}\partial_r + \frac{1}{r^2}\partial_\varphi^2 + \partial_z^2 \end{bmatrix}. \tag{34.16}$$

Substituting Equation 34.16 to Equation 34.15, one obtains three independent equations for each field component of **E**:

$$\left(\partial_r^2 + \frac{1}{r}\partial_r + \frac{1}{r^2}\partial_\varphi^2 + \partial_z^2\right) E_u = k_0^2 n^2 E_u, \tag{34.17}$$

where $u = r, \varphi, z$. An additional assumption with regard to the plane wave imposes $E_r = $ const and $E_\varphi = $ const, which simplifies Equation 34.17 to a one-dimensional expression:

$$\partial_z^2 E_u = k_0^2 n^2 E_u. \tag{34.18}$$

Each of the equations can be solved separately; although by imposing the assumption of layer uniformity, we lose information on the relations between the components of the electric field. Hence, it is useful to assume that when linearly polarized light propagates along the z-axis then

$$E_z = 0$$

and

$E_\varphi = E_r \cot \varphi$ for x polarization or
$E_\varphi = E_r \tan \varphi$ for y polarization.

To solve Equation 34.18, we used the transfer matrix method (TMM),[14] which assumes that within the jth layer the solution of Equation 34.18 can be expressed by the function (Figure 34.3):

$$E_r^j = A_j \exp\left(ik_j z\right) + B_j \exp\left(-ik_j z\right), \tag{34.19}$$

where $k_j = k_0 n_j$ and E_r is continuous and smooth at the interfaces between the consecutive layers of the VCSEL (Figure 34.3). This leads to the following relations:

$$A_j \exp\left(ik_j d_j\right) + B_j \exp\left(-ik_j d_j\right) = A_{j+1} \exp\left(ik_{j+1} 0\right) + B_{j+1} \exp\left(-ik_{j+1} 0\right), \tag{34.20}$$

$$\frac{d}{dz}\left(A_j \exp\left(ik_j z\right) + B_j \exp\left(-ik_j z\right)\right)\Big|_{z=d_j} = \frac{d}{dz}\left(A_{j+1} \exp\left(ik_{j+1} z\right) + B_{j+1} \exp\left(-ik_{j+1} z\right)\right)\Big|_{z=0}, \tag{34.21}$$

at neighboring layers expressed in local coordinates within layers; k_0 is unknown and corresponds to the vacuum wavevector of the mode.

As a consequence, we can write the relation that transforms E_r between the layers as follows:

$$\begin{aligned} A_j \frac{k_{j+1}+k_j}{2k_{j+1}} \exp\left(ik_j d_j\right) + \frac{k_{j+1}-k_j}{2k_{j+1}} B_j \exp\left(-ik_j d_j\right) = A_{j+1} \\ A_j \frac{k_{j+1}-k_j}{2k_{j+1}} \exp\left(ik_j d_j\right) + \frac{k_{j+1}+k_j}{2k_{j+1}} B_j \exp\left(-ik_j d_j\right) = B_{j+1} \end{aligned}, \tag{34.22}$$

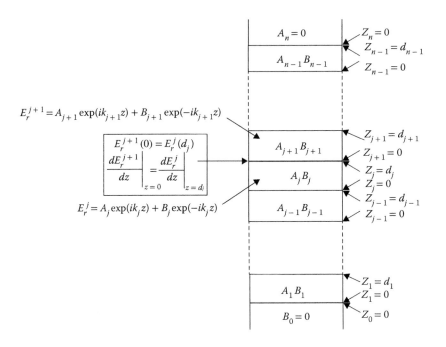

FIGURE 34.3 Schematic view of structure illustrating parameters and variables used in TMM algorithm.

and this can be given in a matrix form:

$$\begin{bmatrix} A_{j+1} \\ B_{j+1} \end{bmatrix} = \begin{bmatrix} \frac{k_{j+1}+k_j}{2k_{j+1}} \exp\left(ik_j d_j\right) & \frac{k_{j+1}-k_j}{2k_{j+1}} \exp\left(-ik_j d_j\right) \\ \frac{k_{j+1}-k_j}{2k_{j+1}} \exp\left(ik_j d_j\right) & \frac{k_{j+1}+k_j}{2k_{j+1}} \exp\left(-ik_j d_j\right) \end{bmatrix} \begin{bmatrix} A_j \\ B_j \end{bmatrix} = \mathbf{M_j} \begin{bmatrix} A_j \\ B_j \end{bmatrix}. \tag{34.23}$$

Matrix $\mathbf{M_j}$ connects the amplitudes of the electric field from layers j and $j + 1$. All layers of the VCSEL can be combined using the recursive relation:

$$\begin{bmatrix} A_n \\ B_n \end{bmatrix} = \prod_{j=0}^{n} \mathbf{M_j} \begin{bmatrix} A_0 \\ B_0 \end{bmatrix} = \mathbf{M} \begin{bmatrix} A_0 \\ B_0 \end{bmatrix}. \tag{34.24}$$

Boundary conditions impose that the electric field decays to infinity, so the exponential elements responsible for infinite values of E in $+\infty$ and $-\infty$ must be eliminated:

$$\begin{bmatrix} 0 \\ B_n \end{bmatrix} = \mathbf{M} \begin{bmatrix} A_0 \\ 0 \end{bmatrix} = \begin{bmatrix} m_{11} & m_{12} \\ m_{21} & m_{22} \end{bmatrix} \begin{bmatrix} A_0 \\ 0 \end{bmatrix}. \tag{34.25}$$

This relation implies that

$$m_{11}\left(k_0\right) = 0, \tag{34.26}$$

which is the condition in the numerical algorithm responsible for finding the longitudinal modes of the VCSEL. k_0, which fulfills the condition (Equation 34.26), corresponds to the longitudinal mode of the VCSEL and is interpreted as the wavenumber of the longitudinal mode in the vacuum. Simple relations give the emitted wavelength as well as the frequency of the mode. The imaginary part of k_0 corresponds to the total losses or gain of the mode. If $\text{im}(k_0) \geq 0$, the mode reaches the threshold and can be observed in the emitted spectrum.

The vertical mode distribution for the structure detailed in Table 34.1, calculated using TMM, is given in Figure 34.4. Mode intensity is defined as follows:

$$I = E_u E_u^*, \tag{34.27}$$

where E^* is a complex conjugate of the electric field of an electromagnetic wave. Figure 34.4 shows typical oscillatory decay of optical field intensity within the top and bottom DBRs. In the simplified example considered here, in which the cavity is uniform, optical intensity is maximal and the amplitude is constant in the cavity. Typically, VCSELs are designed to ensure the coincidence of optical field intensity and the active region to maximize stimulated emission.

Using this simple approach, one can design the dominant wavelength of the VCSEL wafer by tuning the length of the cavity. Figure 34.5 illustrates the dependence of the resonant wavelength and modal gain as functions of detuning the cavity length from its optimal length. The cavity length and DBR stopband were originally designed for 980 nm. Detuning the cavity length and leaving the parameters of the DBRs unchanged shift the emitted wavelength. However, modal gain remains close to its maximal value in the limited range of detuning. Greater detuning induces significant shift of the resonant wavelength, which is more weakly reflected by the DBRs, and modal gain decreases.

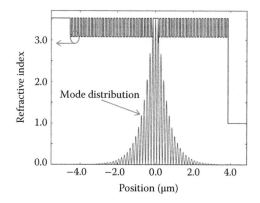

FIGURE 34.4 Distribution of the longitudinal mode and distribution of the refractive index along the optical axis of the VCSEL (Table 34.1).

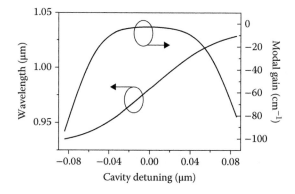

FIGURE 34.5 Modification of the resonant wavelength and modal gain induced by modification of the cavity length.

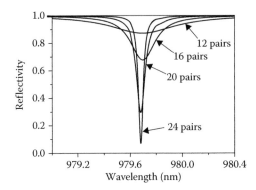

FIGURE 34.6   Power reflectance of the VCSEL wafer as a function of the incident wavelength for different numbers of DBR pairs.

Using the same method, but changing the boundary conditions to allow wave propagation toward the structure, one can calculate the reflectance or transmittance of the VCSEL wafer. In the case of a structure with an unpumped active region (in our calculations, absorption of 3000 cm^{-1} in the active region was assumed), there is a characteristic dip in the reflectance spectrum (Figure 34.6). If the dip is narrow and tends toward 0 reflectance, this indicates that the structure provides strong resonance for incident light that is absorbed by the active region. Such calculations can be compared directly with experimental measurements taken from the reflectance spectrum of the VCSEL wafer.

34.7 Two-Dimensional Scalar Approach

A more comprehensive approach, allowing lateral mode determination but still requiring moderate usage of computational memory and time, assumes that the electric field can be expressed independently using three coordinates: r, φ, and z:

$$E_u = E_u^{\mathrm{rad}}(r)E_u^{\mathrm{azim}}(\varphi)E_u^{ax}(z). \tag{34.28}$$

If the geometry of the device is cylindrical, one can assume the periodic dependence of

$$E_u^{\mathrm{azim}}(\varphi) \sim \sin(m\varphi) \text{ or } E_u^{\mathrm{azim}}(\varphi) \sim \cos(m\varphi), \tag{34.29}$$

which in practice reduces the problem to two dimensions (r, z).

Another assumption that makes use of the one-dimensional solution of (Equation 34.18) axial dependence can be expressed as

$$E_u^{ax} = A \exp\left(ik_{ax}z\right) + B \exp\left(-ik_{ax}z\right), \tag{34.30}$$

where k_{ax} is calculated for each cylinder (Figure 34.7) separately if the lateral distributions of the refractive indices in each cylinder are uniform in the radial direction. Equation 34.17 can then be transformed into

$$\left(\partial_r^2 + \frac{1}{r}\partial_r + k_{ax}^2 - k_0^2 n^2 - \frac{m^2}{r^2}\right) E_u = 0, \tag{34.31}$$

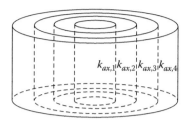

FIGURE 34.7 Schematic arrangement of cylinders. Wavenumbers $k_{ax,i}$ were calculated using one-dimensional TMM.

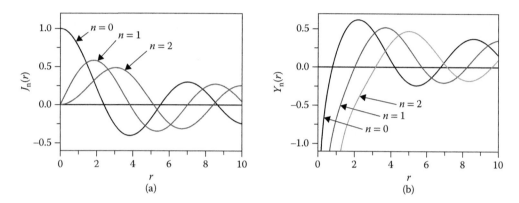

FIGURE 34.8 Bessel functions of the first (a) and second (b) kinds for $n = 0, 1, 2$.

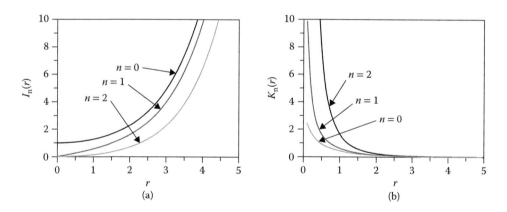

FIGURE 34.9 Modified Bessel function of the first (a) and second (b) kinds for $n = 0, 1, 2$.

which is solved separately for different m. The solutions of these Bessel differential equations are combinations of Bessel functions of the first $J_m(ar)$ and second kind $Y_m(ar)$ (Figure 34.8), where

$$a = k_{ax}^2 - k_0^2 n^2. \tag{34.32}$$

When $a < 0$, it is convenient to use modified Bessel functions of the first $I_m(-ar)$ and second kind $K_m(-ar)$ (Figure 34.9).

The following sums give the solution for an arbitrary layer:

$$E_{u,j}^{\mathrm{rad}} = A_j J_m \left(a_j r \right) + B_j Y_m \left(a_j r \right) \tag{34.33}$$

or

$$E_{u,j}^{\mathrm{rad}} = A_j I_m \left(-a_j r \right) + B_j K_m \left(-a_j r \right). \tag{34.34}$$

Implementing conditions for continuity of E_u^{rad} and continuity of their derivatives at the interfaces between cylinders, one can construct a transfer matrix between the electric fields in separate cylinders. In a similar manner as in Equation 34.24, one can construct a relation combining the fields from the first and last cylinders. The boundary conditions correspond to a finite value for the field at the center of the structure and to a decaying field for $r \to \infty$. The first condition eliminates $Y_m(ar)$ from the solution in the first cylinder and $I_m(-ar)$ in the last, which implies a condition similar to Equation 34.26. The set of k_0's that fulfill this condition correspond to lateral modes. Algorithms based on these approaches are very efficient and can be performed on personal computers.

The intensities of the lateral modes and the corresponding emission wavelengths of the VCSEL (Table 34.2), calculated using the algorithm detailed in this chapter, are presented in Figure 34.10. The set of modes illustrated were calculated assuming $E_u^{\mathrm{azim}}(\varphi) \sim \cos(m\varphi)$. A corresponding set can be calculated for $E_u^{\mathrm{azim}}(\varphi) \sim \sin(m\varphi)$, for which the emitted wavelengths will be exactly the same as for $\sim \cos(m\varphi)$. Superposition of these solutions and their complex conjugation leads to the elimination of angular dependence and only the oval shapes of the modes can be observed. However, in real-life structures in which perfect cylindrical symmetry can be broken by the geometry of the optical aperture or by the crystallographic structure of the layers, separate modes corresponding to both solutions can be observed.

TABLE 34.2 Resonant Wavelengths of Successive Lateral Modes Calculated Using Two-Dimensional Scalar Model

Mode Number	Wavelength (nm)
LP_{01}	979.3025
LP_{11}	978.8136
LP_{21}	978.175
LP_{02}	977.9555
LP_{31}	977.3976
LP_{12}	976.9519
LP_{41}	976.4901
LP_{22}	975.8181
LP_{03}	975.6216
LP_{51}	975.46
LP_{32}	974.5793
LP_{61}	974.3159
LP_{13}	974.2325
LP_{42}	973.3101
LP_{71}	973.0707
LP_{81}	971.7489

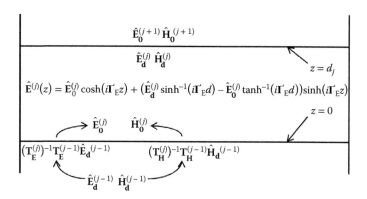

FIGURE 34.10 Definition of symbols for a single, uniform layer. $\hat{\mathbf{E}}_0$ and $\hat{\mathbf{H}}_0$ are virtual fields for the bottom interface of the layer, $\hat{\mathbf{E}}_d$ and $\hat{\mathbf{H}}_d$ are virtual fields for the upper interface, and d_j— is thickness of the layer.

34.8 Three-Dimensional Vectorial Approach

The three-dimensional vectorial approach relies on relations (Equations 34.5 and 34.6). To consider these in general geometries, we use the Cartesian coordinate system and assume that the medium in which field propagates is nonisotropic. The curl operator $\nabla\times$ in the Cartesian coordinate system is defined as follows:

$$\nabla\times \equiv \begin{bmatrix} 0 & -\partial_z & \partial_y \\ \partial_z & 0 & -\partial_x \\ -\partial_y & \partial_x & 0 \end{bmatrix}. \tag{34.35}$$

We can eliminate E_z and H_z:

$$H_z = \begin{bmatrix} -\dfrac{i}{\omega\mu_z\mu_0}\partial_y & \dfrac{i}{\omega\mu_z\mu_0}\partial_x \end{bmatrix} \begin{bmatrix} E_x \\ E_y \end{bmatrix}$$

$$E_z = \begin{bmatrix} -\dfrac{i}{\omega\varepsilon_z\varepsilon_0}\partial_x & \dfrac{i}{\omega\varepsilon_z\varepsilon_0}\partial_y \end{bmatrix} \begin{bmatrix} H_y \\ H_x \end{bmatrix}, \tag{34.36}$$

which leads to the relations:

$$\partial_z\bar{\mathbf{E}} = \begin{bmatrix} -\partial_x\dfrac{i}{\omega\varepsilon_z\varepsilon_0}\partial_x - \omega\mu_y\mu_0 & \partial_x\dfrac{i}{\omega\varepsilon_z\varepsilon_0}\partial_y \\ -\partial_y\dfrac{i}{\omega\varepsilon_z\varepsilon_0}\partial_x & \partial_y\dfrac{i}{\omega\varepsilon_z\varepsilon_0}\partial_y - \omega\mu_x\mu_0 \end{bmatrix} \bar{\mathbf{H}} = \mathbf{R}_E\bar{\mathbf{H}}$$

$$\partial_z\bar{\mathbf{H}} = \begin{bmatrix} -\partial_y\dfrac{i}{\omega\mu_z\mu_0}\partial_y - \omega\varepsilon_x\varepsilon_0 & \partial_y\dfrac{i}{\omega\mu_z\mu_0}\partial_x \\ -\partial_x\dfrac{i}{\omega\mu_z\mu_0}\partial_y & \partial_x\dfrac{i}{\omega\mu_z\mu_0}\partial_x - \omega\varepsilon_y\varepsilon_0 \end{bmatrix} \bar{\mathbf{E}} = \mathbf{R}_H\bar{\mathbf{E}} \tag{34.37}$$

where

$$\bar{\mathbf{H}} = \begin{bmatrix} H_x \\ H_y \end{bmatrix} \text{ and } \bar{\mathbf{E}} = \begin{bmatrix} E_x \\ E_y \end{bmatrix}. \tag{34.38}$$

Equations 34.37 can be transformed into

$$\partial_z^2\bar{\mathbf{E}} = \mathbf{R}_H\mathbf{R}_E\bar{\mathbf{E}}$$
$$\partial_z^2\bar{\mathbf{H}} = \mathbf{R}_E\mathbf{R}_H\bar{\mathbf{H}} \tag{34.39}$$

Both equations can be solved in the base, which simplifies the matrices $\mathbf{R_H R_E}$ and $\mathbf{R_E R_H}$ into diagonal forms. The following analysis leads to the solution of Equation 34.39 and enables the characteristic value of the problem to be found. The solution corresponds to the wavenumber of the mode, as well as to the characteristic vectors, which determine the distribution of the electromagnetic field within the structure.

The equation for the electric field rewritten in the new virtual base (after diagonalization) is

$$\partial_z^2 \hat{\mathbf{E}} = \mathbf{R_H R_E} \hat{\mathbf{E}}, \tag{34.40}$$

where $\hat{\mathbf{E}}$ stand for electric field in the new, virtual base and can be defined as

$$\hat{\mathbf{E}} = \mathbf{T_E^{-1}} \bar{\mathbf{E}}, \tag{34.41}$$

where the matrix $\mathbf{T_E}$ diagonalizes $\mathbf{R_H R_E}$:

$$\mathbf{T_E^{-1} R_H R_E T_E} = \mathbf{\Gamma_E^2}. \tag{34.42}$$

Here $\mathbf{\Gamma_E^2}$ is a diagonal matrix, which allows a solution of Equation 34.39 in the well-known form of a standing wave:

$$\hat{\mathbf{E}}(z) = \mathbf{A} \cosh\left(i\mathbf{\Gamma_E} z\right) + \mathbf{B} \sinh\left(i\mathbf{\Gamma_E} z\right). \tag{34.43}$$

For further derivation, we can use values of the electric field equal to $\hat{\mathbf{E}}_0$ and $\hat{\mathbf{E}}_d$ on the extreme borders of the uniform region along the z direction. The thickness of the considered uniform layer equals d (Figure 34.10).

We can therefore relate the solution to the values of the field at the opposite borders of the uniform region:

$$\hat{\mathbf{E}}(z) = \hat{\mathbf{E}}_0 \cosh\left(i\mathbf{\Gamma_E} z\right) + \left(\hat{\mathbf{E}}_j \sinh^{-1}\left(i\mathbf{\Gamma_E} d\right) - \hat{\mathbf{E}}_0 \tanh^{-1}\left(i\mathbf{\Gamma_E} d\right)\right) \sinh\left(i\mathbf{\Gamma_E} z\right). \tag{34.44}$$

The solution for the magnetic field can be found in a similar manner.

Using the solution for $\hat{\mathbf{E}}$ and $\hat{\mathbf{H}}$ and substituting to Equation 34.37, one can find the relation connecting electric and magnetic fields within the layer:

$$\begin{bmatrix} \hat{\mathbf{H}}_0 \\ -\hat{\mathbf{H}}_d \end{bmatrix} = \begin{bmatrix} \mathbf{y_1} & \mathbf{y_2} \\ \mathbf{y_2} & \mathbf{y_1} \end{bmatrix} \begin{bmatrix} \hat{\mathbf{E}}_0 \\ \hat{\mathbf{E}}_d \end{bmatrix}, \tag{34.45}$$

where

$$\mathbf{y_1} = \left(\mathbf{T_E^{-1} R_H T_H}\right)^{-1} \mathbf{\Gamma_E} \tanh^{-1}\left(i\mathbf{\Gamma_E} d\right), \tag{34.46}$$

$$\mathbf{y_2} = -\left(\mathbf{T_E^{-1} R_H T_H}\right)^{-1} \mathbf{\Gamma_E} \sinh^{-1}\left(i\mathbf{\Gamma_E} d\right). \tag{34.47}$$

The relation between the fields from the ith and $(i-1)$th layers is given in the formulas:

$$\hat{\mathbf{E}}_0^{(i)} = \left(\mathbf{T_E^{(i)}}\right)^{-1} \mathbf{T_E^{(i-1)}} \hat{\mathbf{E}}_d^{(i-1)} = \mathbf{t_E^{(i)}} \hat{\mathbf{E}}_d^{(i-1)}, \tag{34.48}$$

$$\hat{\mathbf{H}}_0^{(i)} = \left(\mathbf{T_H^{(i)}}\right)^{-1} \mathbf{T_H^{(i-1)}} \hat{\mathbf{H}}_d^{(i-1)} = \mathbf{t_H^{(i)}} \hat{\mathbf{H}}_d^{(i-1)}. \tag{34.49}$$

Using Equations 34.45, 34.48, and 34.49, the relationship between the electric and magnetic fields, for the interface separating layers $i + 1$ and i, takes an iterative formula:

$$\hat{\mathbf{H}}_d^{(i)} = \mathbf{Y}^{(i)} \hat{\mathbf{E}}_d^{(i)}, \tag{34.50}$$

$$\mathbf{Y}^{(i)} = -\left(\mathbf{y}_2^{(i)} \left(\mathbf{t}_H^{(i)} \mathbf{Y}^{(i-1)} \left(\mathbf{t}_E^{(i)} \right)^{-1} - \mathbf{y}_1^{(i)} \right)^{-1} \mathbf{y}_2^{(i)} + \mathbf{y}_1^{(i)} \right). \tag{34.51}$$

Propagating by this procedure from the bottom limit of the simulated structure to the upper limit, the characteristic equation can be determined. However, from the numerical point of view, it is more efficient and precise to perform the procedure starting from the bottom limit to the center, and then to start again from the upper limit down to the center. The center does not correspond to the geometrical center but to the plane at which the electromagnetic field reaches or it is close to its maximum, which is termed as the "matching interface." One can write the relation (Equation 34.50) for the top-down and bottom-up algorithms as:

$$\hat{\mathbf{H}}_d^{(m)} = \mathbf{Y}_{up}^{(m)} \hat{\mathbf{E}}_d^{(m)}, \tag{34.52}$$

$$\hat{\mathbf{H}}_d^{(l)} = \mathbf{Y}_{down}^{(l)} \hat{\mathbf{E}}_d^{(l)}. \tag{34.53}$$

m counts layers from the bottom and l from the top.

Since the magnetic field is continuous in the matching interface in the real base, one can write the relation

$$\bar{\mathbf{H}}_d^{(m)} = \bar{\mathbf{H}}_d^{(l)}. \tag{34.54}$$

Transforming back Equation 34.54 to the virtual base:

$$\mathbf{T}_H^{(m)} \hat{\mathbf{H}}_d^{(m)} = \mathbf{T}_H^{(m)} \hat{\mathbf{H}}_d^{(l)}, \tag{34.55}$$

using Equations 34.52 and 34.53:

$$\mathbf{T}_H^{(m)} \mathbf{Y}_{up}^{(m)} \hat{\mathbf{E}}_d^{(m)} = \mathbf{T}_H^{(m)} \mathbf{Y}_{down}^{(l)} \hat{\mathbf{E}}_d^{(l)}, \tag{34.56}$$

and finally using relation (Equation 34.41) to lead us to an equation that includes the fields in the real base:

$$\left(\mathbf{T}_H^{(m)} \mathbf{Y}_{up}^{(m)} \left(\mathbf{T}_E^{(m)} \right)^{-1} - \mathbf{T}_H^{(m)} \mathbf{Y}_{down}^{(l)} \left(\mathbf{T}_E^{(l)} \right)^{-1} \right) \bar{\mathbf{E}} = \mathbf{Y}\bar{\mathbf{E}} = 0. \tag{34.57}$$

Equation 34.57 is also the characteristic value equation. The solution of the equation determines the characteristic values, which correspond to the complex wavenumber. As before, the imaginary part of the characteristic value relates to the modal gain of the propagating wave. Equation 34.57 has a nontrivial solution only if matrix $\mathbf{Y}$ is singular, which imposes the condition that one of the eigenvalues of $\mathbf{Y}$ is 0. This condition is useful in numerical searches for solutions to Equation 34.57.

The intensity distributions of lateral modes calculated using the vectorial approach are very close to those depicted in Figure 34.11. The main difference between the solutions lies in the different number of modes (Table 34.3). In the vectorial solution, there are no modes of shorter wavelengths than LP_{22}. This may be associated with diffraction losses, which are more severe for higher order modes but are not included in the scalar model.

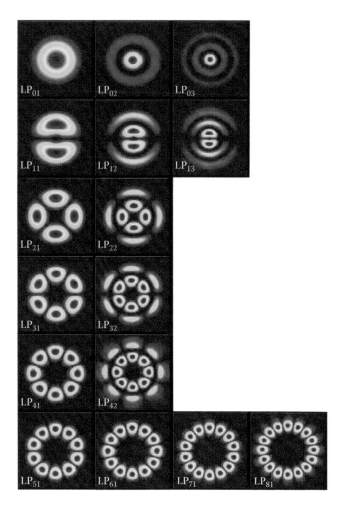

FIGURE 34.11 Distributions of lateral modes in the VCSEL structure detailed in Table 34.1, where $R = 4\ \mu m$, calculated using the scalar two-dimensional approach.

TABLE 34.3 Resonant Wavelengths of Successive Lateral Modes Calculated Using the Three-Dimensional Vectorial Model

Mode Number	Wavelength (nm)
LP_{01}	979.3014
LP_{11}	978.8108
LP_{21}	978.1691
LP_{02}	977.9616
LP_{31}	977.3912
LP_{12}	976.9431
LP_{41}	976.4811
LP_{22}	975.8031

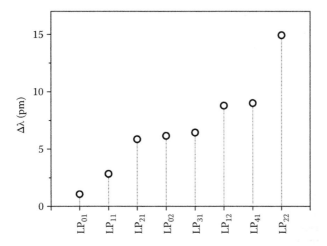

FIGURE 34.12 Difference in resonant wavelengths of successive lateral modes calculated using scalar and vectorial models.

Figure 34.12 illustrates the difference ($\Delta\lambda$) with respect to the wavelength of modes calculated using scalar and vectorial models. The vectorial model provides solutions with shorter wavelengths. However, the difference is on a scale of picometers. The monotonic increase in $\Delta\lambda$ shows that the plane wave assumption used in the scalar model is more accurate in the case of lower order modes.

34.9 Impact of Oxide Aperture Diameter and Position

Finally, we shall analyze the influence of oxide aperture on the emitted wavelength of the VCSEL considered in this study. Although the structure is a very simplified design, the general tendencies presented in Figures 34.13 and 34.14 are applicable to real-life oxide-confined VCSEL structures.

Reducing the aperture (Figure 34.13) size leads to a characteristic blue shift, since a narrower aperture introduces stronger interaction between the mode and the oxide layer, which has a low refractive index. The wider the aperture diameter, the closer the values for the wavelengths determined using each approach.

The location of the oxide layer has a significant effect on the VCSEL parameters (Figure 34.14). If it is positioned at the antinode of the standing wave, the interaction between the optical field and carriers within the active region is intensified, since the modes are strongly confined within the aperture. This configuration reduces the threshold current and broadens the spectrum of emission, as numerous lateral modes are excited. If the oxide layer is in the node position, the number of modes is reduced, which narrows the

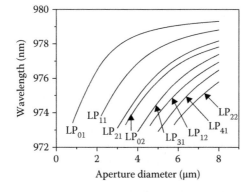

FIGURE 34.13 Dependence of the lateral mode wavelengths on aperture diameter.

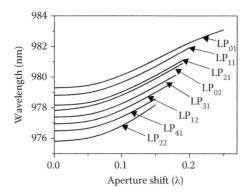

FIGURE 34.14 Dependence of lateral mode wavelengths on aperture shift from the antinode position relative to the wavelength of the standing wave.

emitted spectrum. However, interaction between the mode and the oxide layer is weaker and the threshold current increases. Shifting the oxide layer from the antinode position leads to an increase in the wavelength. This can be explained by weaker interaction between the field and the oxidized layer, which leads to an increase in the effective index of the mode and eventually causes redshift.

34.10 Conclusion

Scalar one- and two-dimensional optical approaches represent very efficient tools for VCSEL designers. Their calculation algorithms, requiring short calculation times and minor memory storage, enable one to perform rapid calculations using even PCs. However, the assumptions that make them so efficient introduce simplifications, which may produce inaccurate results in more sophisticated structures. Most significantly, the basic scalar assumption of plane optical wave propagation is violated in the case of aperture sizes comparable to the radiation wavelength and for higher order modes.

Vectorial methods are much more complicated, since they are based on the solution of six coupled equations instead of only one (in the simplified scalar approach). They give more exact results at the price of significantly longer computational time and a considerably more complicated algorithm. Vectorial methods may not be the optimal approach to modeling a standard VCSEL. However, analysis of VCSELs with in-built refractive index steps in lateral directions or subwavelength patterning certainly requires the use of vectorial methods.

References

1. Melngailis, I. Longitudinal injection plasma laser of InSb. *Appl Phys Lett.* 6, 59–60 (1965).
2. Soda, H. et al. GaInAsP/InP surface emitting injection lasers. *Jap J Appl Phys.* 18, 2329–2330 (1979).
3. Ogura, M. et al. GaAs/Al$_x$Ga$_{1-x}$As multilayer reflector for surface emitting laser diode. *Jap J Appl Phys.* 22, L112–L114 (1983).
4. MacDougal, M. H. et al. Ultralow threshold current vertical-cavity surface-emitting lasers with AlAs oxide-GaAs distributed Bragg reflectors. *IEEE Photon Technol Lett.* 7, 229–231 (1995).
5. Tai, K. et al. Use of implant isolation for fabrication of vertical cavity surface emitting lasers. *Electron Lett.* 25, 1644–1645 (1989).
6. Gadallah, A.-S. et al. High-output-power single-higher-order transverse mode VCSEL with shallow surface relief. *IEEE Photon Technol Lett.* 23, 1040–1042 (2011).

7. Zhou, D. et al. High-power single-mode antiresonant reflecting optical waveguide-type vertical-cavity surface-emitting lasers. *IEEE J Quantum Electron.* 38, 1599–1606 (2002).
8. Kapon, E. et al. Long-wavelength VCSELs: Power-efficient answer. *Nat Photon.* 3, 27–29 (2009).
9. Danner, A. J. et al. Progress in photonic crystal vertical cavity lasers. *IEICE Trans Electron.* E88, 944–950 (2005).
10. Chang-Hasnain, C. J. et al. High-contrast gratings for integrated optoelectronics. *Adv Opt Photon.* 4, 379–440 (2012).
11. Saleh, B. E. A., Teich, M. C. *Fundamentals of Photonics.* New York, NY: John Wiley & Sons (1991).
12. Bienstman, P. R. et al. Comparison of optical VCSEL models on the simulation of oxide-confined devices. *IEEE J Quant Electron.* 37, 1618–1631 (2001).
13. Kuszelewicz, R. *COST 268 Modeling Exercise.* Paris, France: CNET (1998).
14. Bergmann, M. J. et al. Optical-field calculations for lossy multiple-layer $Al_xGa_{1-x}N/In_xGa_{1-x}N$ laser diodes. *J Appl Phys.* 84, 1196–1203 (1998).

VII

Photodetectors and Modulators

35

Photodetector Fundamentals

Prasanta Basu

35.1 Introduction

Photodetectors (PDs) convert light energy into electrical energy, which usually manifests as a photocurrent. They are a kind of transducer. The electrical signal is useful in electronic imaging, providing a means for viewing images much faster than the earlier photographic imaging. PDs also find important applications in present-day fiber-optic communication systems. In this case, PDs receive the stream of weak optical pulses at the far end of the optical fiber originally transmitted by the transmitter. The role of the detector is to convert these optical pulses into electrical pulses, which are then processed for use in a telephone, a computer, or other terminals at the receiving end (Bhattacharyya 1996; Deen and Basu 2012; Kumar and Deen 2014).

Other important applications of PDs include automatic door opener, use in remote control elements in TV, VCR, air conditioners, etc., in CCDs in a video camera, and in astronomy, research, and defense.

This chapter gives an introduction to the working principle of photodetection by semiconductors, the material properties, different types, and their structures, important parameters pertaining to PDs, and materials and makes a list of the application areas.

35.2 Basic Working Principle

The conversion of optical energy into electrical energy by a PD involves three basic processes. First, a bunch of incident photons is absorbed by the semiconducting material to create excess electron–hole pairs (EHPs). The carriers are then transported across the absorption or transit region. In the third stage, the transported carriers are collected to generate photocurrent. In this section, we consider the basic principle, structure, and the process of absorption (Bhattacharyya 1996; Deen and Basu 2012).

35.2.1 Conversion of Light into Current

The process of creation of excess EHPs and their collection is illustrated in Figure 35.1. In Figure 35.1a, a photon of energy $\hbar\omega \geq E_g$, incident on the semiconductor, is absorbed by an electron in the valence band (VB), which then moves up in the conduction band (CB), leaving a hole in the VB. An excess EHP, in excess of the thermal equilibrium values, is therefore created. To collect these carriers, a reverse-biased p-n junction is employed. As shown in Figure 35.1b, the electric field in the depletion layer drives the excess electrons and excess holes in the opposite directions to be collected by the terminals and thus to produce current. The reverse saturation current in the absence of illumination is called the dark current. The photogenerated current is proportional to the flux of incident photons or the power of incident light (Kaiser 2013; Senior and Jamro 2009).

As shown in Figure 35.1b, absorption may also take place outside the depletion layer. The excess EHPs generated therein move by the slower process of diffusion and are collected by the terminals. The diffusion process limits the speed of operation and distorts the electrical pulses.

35.2.2 Absorption in Semiconductors

The process of absorption is now discussed in some detail in Figure 35.2. In Figure 35.2a, a photon of energy $\hbar\omega \geq E_g$ is shown incident on the semiconductor. The photon lifts an electron from the VB to a higher lying state in the CB. The energetic electron then relaxes to the bottom of the CB via phonon emission. A photon thus creates an excess EHP. The pair ultimately recombines to give rise to emission (luminescence). The same processes are illustrated in the E-k diagrams of a direct gap semiconductor in Figure 35.2b and of an indirect gap semiconductor in Figure 35.2c. The photon momentum $\hbar k_{photon}$ is usually negligible and the momentum (**k**) conservation can be written as $\mathbf{k}_e = \mathbf{k}_h + \mathbf{k}_{photon} \approx \mathbf{k}_h$, where $\mathbf{k}_e (\mathbf{k}_h)$ denotes the wave vector for electron (hole). Accordingly, the absorption in a direct gap semiconductor is indicated by a vertical

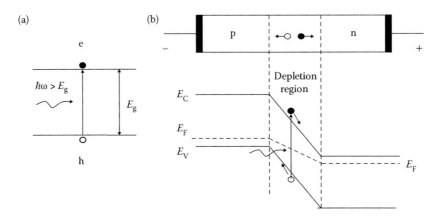

FIGURE 35.1 Conversion of optical energy into electrical energy: (a) absorption of photons and generation of excess EHPs; (b) collection of carriers by using a reverse-biased p-n junction.

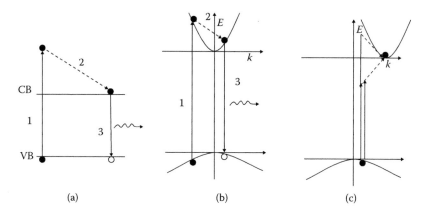

FIGURE 35.2 (a) Absorption (1), relaxation (2), and recombination (3) processes in semiconductors. CB (VB) denotes conduction (valence) band edges; (b) absorption (1), relaxation (2), and recombination (3) processes shown in the E-k diagram of a direct gap semiconductor; (c) the same processes in the E-k diagram of an indirect gap semiconductor. Dashed lines correspond to phonon scattering processes.

line as shown in Figure 35.2b. However, for indirect gap semiconductor, an extra momentum is needed to place the electron in the CB, which is provided by, among other agencies, phonons. In Figure 35.2c, the indirect absorption is indicated by both phonon absorption and emission. The indirect absorption has a very low probability and orders of magnitude being lower than in direct gap semiconductors (Basu 1997).

The variation of light intensity inside a semiconductor, $I(z)$, is governed by the following expression known as the Lambert's law:

$$I(z) = I_0 \exp\left(-\alpha z\right), \tag{35.1}$$

where I_0 is the intensity at the surface ($z = 0$), and α is the absorption coefficient of the material, which is a function of photon energy or wavelength.

The absorption coefficient for direct gap semiconductors is much higher than for indirect gap semiconductors.

35.2.2.1 Absorption in Direct Gap Semiconductors

As mentioned already, the absorption process in a direct gap semiconductor is indicated by a vertical line in the E-k diagram. The absorption coefficient in direct gap semiconductors is calculated by using the first-order time-dependent perturbation theory and the Fermi golden rule. The expression for absorption coefficient is of the form (Basu 1997)

$$\alpha(\hbar\omega) = A\left(\hbar\omega - E_g\right)^{1/2}\left(f_c - f_v\right), \hbar\omega \geq E_g, \tag{35.2}$$

where the prefactor A contains the fundamental constants, material parameters such as permittivity, etc., and $f_{c(v)}$ denotes the Fermi occupation probabilities of electrons in the CB (VB).

35.2.2.2 Absorption in Indirect Gap Semiconductors

Since a photon and a phonon are involved in the absorption process, the second-order perturbation theory is used to calculate the absorption coefficient in indirect gap semiconductors. The expression takes the

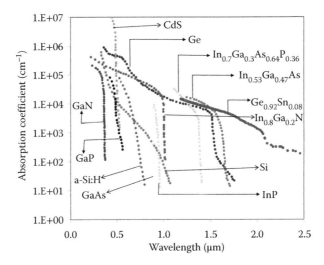

FIGURE 35.3   Absorption spectra of some common semiconductors.

following form at 0 K when a single type of phonon of energy $\hbar\omega_p$ is involved (Basu 1997):

$$\alpha(\hbar\omega) = A_a \left(\hbar\omega - E_g + \hbar\omega_p \right)^2 + A_e \left(\hbar\omega - E_g - \hbar\omega_p \right)^2, \tag{35.3}$$

where the first and second terms correspond, respectively, to the absorption followed by phonon absorption and that followed by phonon emission, and the prefactors A's are distinguished by the corresponding suffixes. There is, in general, a possibility of involvement of different types of phonons. Figure 35.3 shows typical absorption coefficients of some representative semiconductors.

35.2.2.3 Absorption in Quantum Nanostructures

Recent detectors make increasing use of quantum nanostructures such as quantum wells (QWs), quantum wires (QWRs), and quantum dots (QDs), rather than the bulk semiconductors. In QWs, for example, the thickness of the layer of interest is comparable to the de Broglie wavelength of the particle. A QW is realized by sandwiching a thin layer of GaAs between two layers of a higher gap material such as $Al_xGa_{1-x}As$. The typical band diagram of the QW is shown in Figure 35.4a. Due to very small thickness d (<de Broglie wavelength) of GaAs, the motion of electrons and holes is quantized, and assuming that the barrier heights are infinite, the quantized energy levels, or subbands as they are called, are expressed as follows:

$$E_n = \frac{\hbar^2}{2m_{eff}} \left(\frac{n\pi}{d} \right)^2, \quad n = 1, 2, 3, \dots . \tag{35.4}$$

The effective mass, m_{eff}, is to be replaced by the electron effective mass, m_e, heavy hole effective mass, m_{hh}, and light hole effective mass, m_{lh}, in order to calculate the respective subband energies (Basu 1997).

The absorption processes in a QW are shown in Figure 35.4b. Here, transitions can occur in several ways. An electron from a valence subband (both lh and hh) can absorb a photon and can come to one of the subbands in the CB. This is a typical band-to-band absorption; however, the absorption threshold, that is, the photon energy marking the sharp rise in the absorption coefficient, increases from the threshold in bulk material that equals the bandgap (Equation 35.2). In addition, an electron (or a hole) in a lower subband (*i*) can absorb a photon and may move up to a higher subband (*j*) in the same band. This is a representative intersubband transition. The absorption processes in QWs are governed by some selection rules.

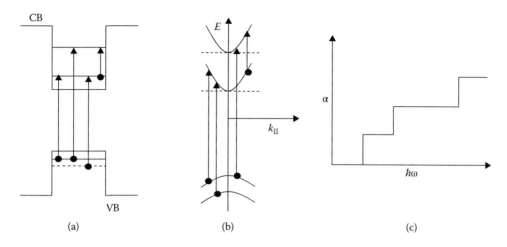

FIGURE 35.4 (a) Band diagram and quantized energy levels or subbands in a QW structure; (b) optical absorption processes in a QW; both band-to-band and intersubband processes are shown; (c) idealized absorption spectra for QWs. α and $\hbar\omega$ denote the absorption coefficient and the photon energy, respectively. k_{II} is the wave vector along the QW layer plane. CB (VB) denotes conduction (valence) band edges.

The idealized band-to-band absorption coefficient in the QW of a direct gap semiconductor takes the following form (Basu 1997):

$$\alpha(\hbar\omega) = K \sum_n H\left(\hbar\omega - E_{nn}\right), \tag{35.5}$$

where the prefactor K contains fundamental constants, effective (reduced) masses of electrons and holes, and the permittivity of the QW material. In Equation 35.5, H is a Heaviside step function, and E_{nn} is the energy difference between nth hole and nth electron subband.

Typical absorption spectra for QWs obtained from Equation 35.5 are shown in Figure 35.4c, which indicates that the spectra are staircase like.

The idea can be extended to the cases of QWRs and QDs, in which the motion of the particles is inhibited along two and three directions, respectively.

The absorption spectra get modified when excitonic effects are included. Excitons are complexes formed by the EHP bound by the mutual Coulomb interaction.

35.3 Important Characteristics

In this section, some of the important characteristics of PDs are discussed.

35.3.1 Wavelength Selectivity

Different materials are used for the detection of photons at different wavelengths. The wavelength range covered by a particular material is determined by the cutoff wavelength, above which the absorption by it is negligibly small. The cutoff wavelength is determined by the expression $\hbar\omega = E_g$, where E_g is the bandgap, from which it is simple to write

$$\lambda_c = \frac{hc}{E_g} = \frac{1.24}{E_g(\text{eV})}\,\mu\text{m}. \tag{35.6}$$

For a bandgap of 1 eV, the cutoff wavelength is 1.24 μm.

35.3.2 Quantum Efficiency

Quantum efficiency (QE), sometimes called the internal quantum efficiency (IQE), is the ratio of the generated EHPs and the number of incident photons. The higher the QE is, larger is the sensitivity of the PD and it is therefore an important figure of merit.

More practical measure of the sensitivity is the external QE (EQE), which is the ratio of EHPs collected at the terminals to the number of incident photons. EQE is less than IQE since a part of the incident photon flux is reflected at the surface and some EHPs are lost by recombination at the surface or by nonradiative processes. EQE is a function of wavelength.

Let I_{pc} be the photocurrent, P_i be the incident power, and $\hbar\omega$ be the photon energy. By definition,

$$\eta = \frac{\text{electron generation rate}}{\text{photon incidence rate}} = \frac{I_{pc}/e}{P_i/\hbar\omega} = \frac{I_{pc}}{P_i} \cdot \frac{\hbar\omega}{e} = \frac{1.24}{\lambda\,(\mu m)} \frac{I_{pc}}{P_i}. \tag{35.7}$$

Equation 35.7 may also be written as a product of three factors as follows:

$$\eta = \zeta(1 - e^{-\alpha d})(1 - R). \tag{35.8}$$

In Equation 35.8, ζ is the fraction of the photogenerated carriers that reach the outer circuit without recombination, the second term within parentheses denotes the fraction of light absorbed in the absorption region of width d, and the last term represents the amount of light transmitted into the semiconductor, R being the reflection coefficient.

35.3.3 Responsivity

Responsivity is defined as the ratio of the primary photocurrent without gain, I_{pc}, and the incident optical power and is expressed as

$$\mathfrak{R} = \frac{I_{pc}}{P_l} = \eta(\lambda) \left[\frac{e}{(hc/\lambda)} \right]. \tag{35.9}$$

In PDs, responsivity, or sensitivity as it is alternately called, may be increased by increasing the absorption region thickness and minimizing reflection from its surface. The latter can be accomplished by coating the surface with an antireflection (AR) coating of refractive index $n_{AR} = \sqrt{n_{air} n_{SC}}$, where the subscripts refer to AR material, air, and semiconductor, respectively.

35.3.4 Response Time

The speed of response of PDs is limited by three main factors, as discussed next.

35.3.4.1 Diffusion Time of Carriers

The time taken by carriers to diffuse a distance d is given by (Streetman and Banerjee 2016)

$$\tau_{diff} = \frac{d^2}{2D}, \tag{35.10}$$

where D is the diffusion constant. Diffusion is a slower process than drift. The carriers that are optically generated within a diffusion length from the depletion layer of a photodiode move to the depletion layer by this slow process. The photogenerated current pulse due to a narrow optical pulse shows "tails" due to the slow diffusion process.

35.3.4.2 Drift Time through Depletion Region

The drift velocity of both electrons and holes in a semiconductor attains a saturation value in the presence of very high fields (Streetman and Banerjee, 2016). Usually, a high electric field exists in the depletion layer of a reverse-biased junction. The longest transit time of carriers in the depletion layer of width W may therefore be written as follows:

$$\tau_{\text{drift}} = \frac{W}{v_{\text{sat}}}, \tag{35.11}$$

where v_{sat} is the saturation velocity of the carrier, which is different for electrons and holes. The maximum speed of operation in a PD is limited by transit time.

35.3.4.3 Junction Capacitance Effect

The electrical equivalent circuit of a PD using a p-i-n diode as an example is shown in Figure 35.5. It includes a photogenerated current source (I_{ph}), the depletion layer capacitance (C_J), a series resistance (R_S) due to bulk resistance and ohmic contacts, a parasitic capacitance (C_P), and a load resistance (R_L). The electrical 3-dB bandwidth may be expressed as

$$f_{\text{3dB}} = \frac{1}{2\pi(C_J + C_P)(R_L + R_S)}. \tag{35.12}$$

35.3.5 Noise Equivalent Power and Detectivity

The term *noise* refers to unwanted electrical signal that masks the signal to be detected. Several noise sources, such as thermal, shot, dark current, and flicker noise, are present in PDs and associated circuits and make it difficult to detect weak signals. The minimum detectable signal corresponds to an rms output signal equal to that generated by noise. The signal-to-noise ratio (SNR) is then unity. The noise equivalent power (NEP), a measure of the minimum detectable signal, is defined as the power of sinusoidally modulated monochromatic optical signal that generates the same rms output power in the ideal noise-free detector as the noise signal produced in a real detector.

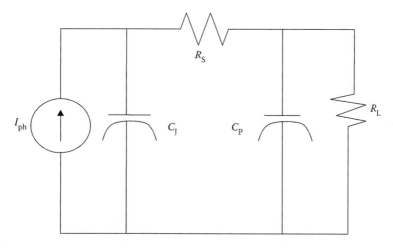

FIGURE 35.5 Electrical equivalent circuit of a photodiode. I_{ph} is the photogenerated current, and C_J and C_p are, respectively, the junction and parasitic capacitances. R_S and R_L are, respectively, the series and load resistances.

A unit NEP^* is defined as

$$NEP^* = \frac{NEP}{(A\Delta f)^{1/2}}, \tag{35.13}$$

where A is the area of the detector and Δf the bandwidth. The commonly used parameter is, however, the specific detectivity D^*, which is the reciprocal of NEP^*.

35.4 Classification of PDs

PDs can be classified in different ways. It may be an intrinsic or an extrinsic type. In the former, the absorption is band to band and the excess EHPs created produce a photocurrent. In the extrinsic types, the photon energy is less than the bandgap energy, and an impurity or deep-level defect in the forbidden gap is involved. An electron from this deep level is lifted by photon absorption process to the CB. Alternately, an electron in the VB may come to the deep level by photon absorption leaving behind a hole in the VB. In both the cases, the excess carriers created produce photocurrent. The absorption may also take place between two confined states in the CB or VB in a QW, QWR, or QD. Such processes give rise to the intersubband transitions.

PDs can be of two types: with or without internal gain. Photoconductors, avalanche PDs (APDs), and phototransistors show internal gain, that is, the primary photocurrent is multiplied. On the other hand, p-n or rather p-i-n PDs do not have any gain.

In another classification scheme, PDs may be classified as having either vertically illuminated or horizontally illuminated structures. In the former type, the absorbing layer thickness is low to reduce the transit time and to be useful for high-speed operation, though sacrificing the QE. However, a high packing density of arrays of PDs can be achieved, also facilitating coupling of the PDs with fibers. The low QE may be increased by enclosing the active layer within a resonant cavity structure, which is realized by using multilayer dielectrics (Unlu et al. 2001).

The simplest example of a horizontally illuminated PD is an edge-illuminated detector. However, due to diffraction, light spreads outside the high-field region and slow diffusion process leads to tails in the impulse response. The problem may be solved by using a waveguide structure, confining the light into the high-field region covering the optical waveguide. The speed of the devices is limited by the capacitance due to a thin intrinsic absorbing layer. A traveling wave detector may solve the limitations due to capacitance.

Another way of classifying the PDs is based on the type of absorbing layer. Various possibilities exist that include (1) bulk semiconductor, (2) QWs/multiple QWs (MQWs), (3) QWRs, (4) QDs, (5) strained quantum nanostructures, and (6) quantum nanostructures of different band alignments (see Basu 1997) such as type I, type II, type III, etc., as the active absorbing region.

35.5 Different PD Configurations

In this section, we shall discuss the structure and working principles of photoconductors, p-i-n and APDs, and PDs involving intersubband transitions. PDs relying on extrinsic processes will not be addressed. Detailed discussions about different types are given in many text books and reviews (see, for example, Bhattacharyya 1996; Bowers and Burrus 1987; Bowers and Wey 1995; Deen and Basu 2012; Kaiser 2013; Senior and Jamro 2009)

35.5.1 Photoconductive Detectors

The simplest form of PD is a photoconductor, in which the conductivity of the semiconductor is changed by absorption of photons. The changed photoconductivity produces photocurrent proportional to the incident optical power. In this section, the structure, principle, and noise analysis of such photoconductive detectors are presented.

35.5.1.1 Structure and Analysis

Consider a slab of semiconductor of length L, width W, and thickness d as shown in Figure 35.6. A voltage V is applied across the length and the slab is illuminated by light at the top. The current I flowing through the photoconductor may be expressed as

$$I = eWd(n\mu_e + p\mu_h)\left(\frac{V}{L}\right) = eWd\left(\frac{V}{L}\right)[(n_0\mu_e + p_0\mu_h) + (\Delta n\mu_e + \Delta p\mu_h)] = I_D + I_{ph}, \qquad (35.14)$$

where I_D is the dark current, I_{ph} is the photocurrent, n_0 and p_0 are the equilibrium electron and hole concentrations, respectively, Δn (Δp) is the excess electron (hole) concentration, and μ_e (μ_h) is the electron (hole) mobility.

The average generation rate r_g within the photoconductor is expressed as follows:

$$r_g = \frac{\eta I_0 WL}{\hbar\omega WLd} = \frac{\eta I_0}{\hbar\omega d}, \qquad (35.15)$$

where η is the efficiency, I_0 the light intensity at the surface, and $\hbar\omega$ the photon energy. Under steady state, the generation rate equals the recombination rate so that $r_g = r_r = \Delta n/\tau = \Delta p/\tau$, where τ is the recombination lifetime. Thus, one obtains

$$\Delta n = \Delta p = r_g \tau. \qquad (35.16)$$

Substituting Equation 35.16 in Equation 35.14, the photocurrent is obtained as follows:

$$I_{ph} = \frac{Wd}{L} r_g \tau e(\mu_e + \mu_h)V. \qquad (35.17)$$

The photoconductive gain, defined as the ratio of the rate of flow of electrons from the device to the rate of generation of EHPs within the device, takes the following form:

$$G = \frac{I_{ph}}{e} \frac{1}{r_g WLd} = \frac{\tau(\mu_e + \mu_h)V}{L^2}. \qquad (35.18)$$

Equation 35.17 has been used to arrive at the last equality in Equation 35.18. The variation in gain with voltage V may be understood if we assume that $\mu_e \gg \mu_h$, which is usually the case. The transit time of electrons through the device is $\tau_{tre} = L^2/\mu_e V$ and from Equation 35.18 $G = \tau/\tau_{tre}$. At low bias, $\tau < \tau_{tre}$ and the gain is <1. With larger V, electrons move faster, $\tau > \tau_{tre}$, and the gain becomes larger than unity.

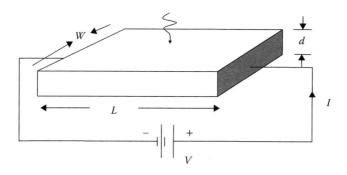

FIGURE 35.6 Schematic of a photoconductive PD. L, W, and d are the length, width, and depth of the photoconductor, respectively. V is the applied voltage and I is the current.

At still larger bias, the velocities of both electrons and holes are saturated. The gain is again reduced below unity. By following these arguments, it is easily concluded that higher values of gain may be achieved by increasing V, or by decreasing L, or by increasing the recombination lifetime.

35.5.1.2 Noise in Photoconductors

There are three main sources of noise in a photoconductive detector. The first is thermal or Johnson noise that arises due to random motion of carriers in the device. The mean squared noise current due to a conductor having finite resistance R is given as follows:

$$\langle i_{\text{th}}^2 \rangle = \frac{4k_{\text{B}}TB}{R}, \tag{35.19}$$

where B is the bandwidth of the device. In developing the noise equivalent circuit, the noisy photoconductor is replaced by a noiseless photoconductor and an ideal noise current or voltage generator.

Fluctuations in generation and recombination rates give rise to fluctuations in current and hence generate noise. The mean squared generation-recombination (GR) current is expressed as follows:

$$\langle i_{\text{GR}}^2 \rangle = \frac{4eGI_0B}{1 + \omega^2\tau^2}, \tag{35.20}$$

where I_0 is the steady photogenerated current. GR noise is also termed as shot noise. Since shot noise decreases as ω^{-2}, the thermal noise is the dominating noise source at high frequencies.

The third kind of noise, the flicker noise, has its origin in the presence of surface and interface defects and traps in the bulk material, and the mean squared current is expressed as:

$$\langle i_f^2 \rangle \propto \frac{1}{f}. \tag{35.21}$$

Equation 35.21 indicates that flicker noise is of importance at very low frequencies, less than 1 KHz.

In the noise equivalent circuit, the noise current sources appear in parallel. Assuming that the noise sources are uncorrelated and also that flicker noise is negligible, one may write for the SNR as

$$\frac{S}{N} = \frac{i_{\text{ph}}^2}{\langle i_{\text{th}}^2 \rangle + \langle i_{\text{GR}}^2 \rangle}. \tag{35.22}$$

Let us assume that the optical signal incident on the detector is sinusoidally modulated by a modulating signal of angular frequency ω_{m}, and its power variation has the form $P_1 \exp(j\omega_{\text{m}}t)$. The rms optical power is $P_1/\sqrt{2}$. The rms photocurrent is expressed as

$$i_{\text{ph}} = \frac{eP_1\eta}{\sqrt{2}\hbar\omega} \frac{\tau}{\tau_{\text{tr}}} \frac{1}{\left(1 + \omega_{\text{m}}^2\tau^2\right)^{1/2}}. \tag{35.23}$$

In arriving at Equation 35.23, the expression for responsivity, Equation 35.9, has been used to relate photocurrent with power and the gain is replaced by the ratio τ/τ_{tr}. The SNR as given by Equation 35.22 may now be written by using Equations 35.21 through 35.23 as

$$\frac{S}{N} = \frac{\eta P_1^2}{8B\hbar\omega P_0} \frac{1}{1 + \frac{k_BT}{eG}\left(1 + \omega_{\text{m}}^2\tau^2\right)\frac{1}{RI_0}}. \tag{35.24}$$

35.5.2 p-n Photodiode

The simplest PD without internal gain is a reverse-biased p-n junction, shown schematically in Figure 35.7 along with the electric field profile. Both depletion and diffusion regions are shown. Photons are absorbed in the absorption region ($\sim 1/\alpha(\lambda)$). In case of weak absorption, the absorption region may cover the whole device, a part of which may contain the depletion region. Note that EHPs are generated in both the depletion and diffusion regions. While the electric field in the depletion region sweeps the photogenerated carriers quickly, the EHPs created in the diffusion region move slowly and thus limit the response of the PD. To improve the response, it is to be ensured that all photons are absorbed in the depletion region. Therefore, its width is made as large as possible by using lower doping.

35.5.3 p-i-n Photodiode

An efficient way to overcome the limitations of a p-n PD is to use a p-i-n configuration. Near the absorption edge, the weak absorption coefficient makes the absorption region quite long. The depletion region must match this, and to achieve this, the n region is low-doped to make it intrinsic (i). To make a low-resistance contact, a heavily doped n layer (n$^+$) is added. The resulting structure, a p-i-n structure, is shown in Figure 35.8 including the electric field profile, depletion and absorption regions. As shown, in this structure, the EHPs created are subject to the depletion layer field.

The speed of response of the p-i-n photodiode depends on (1) the transit time of carriers across the depletion layer as given by Equation 35.11 and (2) the RC time constant, where C is the junction capacitance.

The junction capacitance may be used either by using a mesa structure or by using a planar geometry with selective diffusion in the contact region (Bandyopadhyay and Deen 2001). The capacitance may also be reduced by increasing the width of the i layer. This increases the absorption and hence the QE, but increases the transit time, thereby decreasing the speed of response. To have high values of both QE and

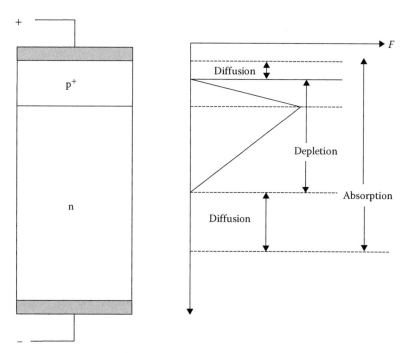

FIGURE 35.7 p-n PD structure (left) and electric field profile across the depth of the device (right). The depletion, diffusion, and absorption regions are shown.

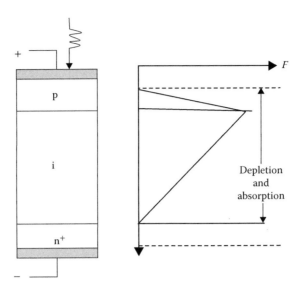

FIGURE 35.8 The p-i-n photodiode structure (left) and absorption and depletion regions and electric field (F) profile in the device (right).

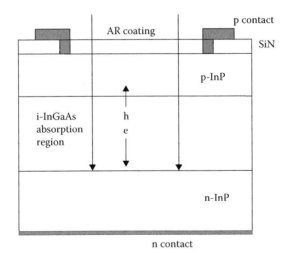

FIGURE 35.9 A cross-sectional view of a vertically illuminated p-i-n photodiode using p-InP/i-InGaAs/n-InP and SiN as antireflection (AR) coating.

speed (or bandwidth), the usual methods are (1) to employ an edge-illuminated structure or (2) to use a resonant-cavity-enhanced (RCE) structure.

A typical structure for a vertically illuminated p-i-n PD is shown in Figure 35.9. The n and p layers may be InP and the i layer is an undoped (or unintentionally doped) InGaAs layer. Since InP is transparent to 1.3–1.55 μm, the incident light is absorbed in the InGaAs layer, in which the electric field appears. Thus, there is no recombination or diffusion of the carriers. This leads to a fast response. The area of the absorbing layer must be optimized for giving high-speed operation (Bandyopadhyay and Deen 2001; Beling and Campbell 2009).

Edge-illuminated and RCE structures are presented in subsections 35.6.1 and 35.6.2, respectively.

35.5.4 Avalanche Photodiodes

APDs possess internal gain and as such are used to detect very weak optical signals. The working principle of an APD is discussed first and then some of the structures for improved performance are presented. The noise performance of APDs is discussed next.

35.5.4.1 Principles

The internal gain arises due to impact ionization in the presence of a high electric field in the depletion layer. A primary electron gets enough kinetic energy to break a covalent bond, thereby creating one secondary EHP. The primary and secondary carriers are accelerated by the depletion field and may again create tertiary EHPs. The multiplication process gives rise to an increased photocurrent. A primary hole can also initiate the ionization process. By this process, one initial electron or a hole may generate M extra EHPs, where M is the multiplication gain of the PD. In general, the ionization coefficients for electrons (α) and for holes (β) are different. The structure of an APD is shown in Figure 35.10, in which the electric field profile is also included.

In order to cause impact ionization, the carriers must possess a minimum energy, called the ionization threshold, which must exceed the bandgap energy. The ionization thresholds for electrons and holes are different. This leads to different values of ionization coefficients α and β, for electrons and holes, respectively. The ionization coefficient is the reciprocal of the average distance traversed by a carrier in the direction of the electric field to create an EHP. In its motion, a carrier may lose or gain energy via carrier–phonon interaction, which becomes more dominant with rise in temperature, thereby reducing the ionization probability. The breakdown voltage in a p-n junction therefore increases with temperature.

Defects in the device can lead to microplasma effect, in which breakdown at a lower voltage than in the remaining part of the junction occurs in a small area. This needs careful device processing and fabrication.

One of the important performance parameters for a PD used in optical communication is the maximum bit rate of operation. The detector must convert faithfully the incident optical pulses into electrical pulses. This is governed by how quickly the detector current responds to a pulse of photons. At low bias, carriers may be generated in the undepleted region and the response of an APD to an optical pulse may produce

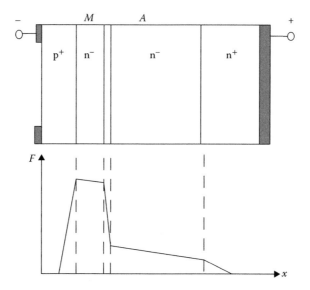

FIGURE 35.10 Structure of an APD (top) and the electric field (F) profile within it (bottom). M and A denote the multiplication and absorption regions, respectively.

a diffusion tail in the current pulse as in a p-i-n diode. When the bias is high enough to fully deplete the junction, the following three factors contribute to the response time:

1. Transit time of the carriers across the absorption layer of width W, given by Equation 35.11
2. The time taken by the carriers to create EHPs by avalanche multiplication
3. The RC time constant, where R is the device and load resistance and C is the junction capacitance

For low gain, the response time is determined by the transit time and the RC time constant. These two limit the value of bandwidth. When the gain is large, avalanche multiplication time becomes dominant and the bandwidth is reduced. APDs are therefore characterized by a constant gain bandwidth product.

35.5.4.2 Separate Absorption Grading and Multiplication and Separate Absorption, Grading, Charge Sheet, and Multiplication APDs

PDs for optical communication at a wavelength of 1.55 μm have $In_{0.53}Ga_{0.47}As$ as the absorbing layer. The usual p-n junction shows a high dark current due to the low bandgap of InGaAs. Furthermore, at high reverse field in the depletion region, the breakdown may be due to tunneling as opposed to the avalanche process. Therefore, the p-n junction is made using InP. The design involves separate absorption and multiplication (SAM). In an SAM structure, a sufficiently high electric field exists to cause avalanche multiplication in InP, and at the same time the field is low enough to prevent tunneling in the lower bandgap InGaAs layer in which optical absorption takes place. However, the photogenerated holes in the InGaAs layer while flowing toward InP layers find a band offset ΔE_v at the InGaAs/InP interface. Holes are thus trapped at the heterointerface, giving rise to a slower response. Incorporation of one or more layers of intermediate gap InGaAsP in between the absorption and multiplication regions eliminates this problem. The structure is called a separate absorption grading and multiplication (SAGM) structure.

The presence of quaternary and ternary layers in SAGM structures leads to another problem. The ionization rate of holes is higher than that of electrons in InP, but is lower in both the InGaAs and InGaAsP layers. This may give rise to unwanted multiplication in the ternary and quaternary layers. As the ionization ratio k approaches unity there, higher noise and lower speed are expected. The doping and thickness of the multiplication layer should be accurately controlled in order to have the increased gain–bandwidth product. A planar SAGM APD that avoids premature edge breakdown employs a partial charge sheet layer in between the multiplication layer and the grading region. The structure is known as a separate absorption, grading, charge sheet, and multiplication (SAGCM) structure. The structure consists of a lightly (or unintentionally) doped wide gap multiplication region where the field is high and an adjacent doped charge layer or a field control region. The presence of the charge (C) layer offers an extra degree of freedom such that the electric field profiles in the active region and in the periphery of the device can be controlled independently.

A cross-sectional view of an SAGCM APD using InP multiplication, InP charge, InGaAsP grading, and InGaAs absorption layer is shown in Figure 35.11. APDs used in telecommunications are reviewed by Campbell (2007).

35.5.4.3 Noise in APDs

Three important noise sources are operational in p-i-n PDs and in APDs. They are (1) thermal noise, (2) dark current noise, and (3) shot noise. In addition, APDs are characterized by an excess noise due to the random nature of the avalanche multiplication process. The statistical variation of multiplication is responsible for multiplication excess noise. The excess noise factor F depends on the ratio $k = \alpha/\beta$ and is expressed as

$$F = kM + (1 - k)\left(2 - \frac{1}{M}\right) = M\left\{1 - \left[(1 - k)\left(\frac{M - 1}{M}\right)^2\right]\right\}. \tag{35.25}$$

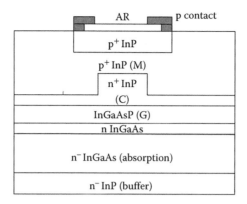

FIGURE 35.11 Cross-sectional view of an SAGCM APD. Antireflection (AR), multiplication (M), charge sheet (C), and graded (G) regions are shown.

The excess noise factor therefore reduces with large value of k. The total dark current in an APD is given by

$$I_d = I_s + I_{dm}M, \tag{35.26}$$

where I_s is the unmultiplied surface current, and I_{dm} is the unmultiplied dark current. Let I_{ph0} be the primary DC photocurrent, I_{bk} the photo-induced background current, and B the bandwidth. The mean square shot noise current is then

$$\langle I_M \rangle^2 = 2e \left[I_s + \left(I_{dm} + I_{bk} + I_{ph0} \right) M^2 F \right] B. \tag{35.27}$$

The SNR of APD is written as

$$\text{SNR} = \frac{M^2 I_{ph0}^2}{\langle I_M \rangle^2 + \left(4k_B TB/R \right)} = \frac{M^2 I_{ph0}^2}{2e \left[I_s + \left(I_{dm} + I_{bk} + I_{ph0} \right) M^2 F \right] B + \left(4k_B TB/R \right)}. \tag{35.28}$$

In the above expression, the thermal noise contribution $\left(4k_B TB/R \right)$ has been included. Note that by putting $M = 1$ and $F = 1$, we may obtain the SNR for p-i-n PD.

It is easy to assess the relative contributions of the shot noise (first term in the denominator of Equation 35.28) and of the thermal noise term (second term in the denominator), by dividing the numerator and denominator by M^2. The shot noise increases with M as does F. However, the thermal noise decreases with M as M^{-2}. It appears therefore that there exists an optimum value of M for which SNR can be a maximum. The variation of M is effected by change of reverse bias and may be expressed by the following empirical expression:

$$M \approx \frac{1}{\left\{ 1 - \left(\frac{V_a - I_{ph0}R}{V_B} \right)^n \right\}}, \tag{35.29}$$

where V_a is the bias voltage, R the effective resistance, V_B the breakdown voltage, and n a constant that depends on the design of the device.

35.5.5 Schottky Barrier PD

In a Schottky barrier (SB) PD, the rectifying junction is formed between a metal and a semiconductor, rather than by a p-n junction. A schematic structure of an SB PD is shown in Figure 35.12, which indicates that the active or absorbing layer is quite narrow. This leads to short transit time and hence higher bandwidth, but to a lower QE.

35.5.6 Metal-Semiconductor-Metal PD

A metal-semiconductor-metal (MSM) PD is shown schematically in Figure 35.13. The schematic layer structure consists of a thin undoped layer grown on a semi-insulating substrate. Metal electrodes are then deposited on the active layer as interdigitated fingers. Each set of fingers forms a Schottky contact with the semiconductor and is connected to a large pad for connection to the external circuit. Both contacts in the MSM PD are on the same side of the substrate. Since the active layer thickness is small, the QE is less than in a p-n PD. Moreover, the reflections from the metal and semiconductor surfaces lead to smaller responsivity. However, the small transit time and very low capacitance greatly enhance the bandwidth.

A contact between a metal and a semiconductor may be either ohmic or rectifying. To obtain a rectifying contact, a certain combination of metal and a given semiconductor is needed. In such a junction, potential barrier develops due to the difference between the work functions of the metal and semiconductor. Another difference between a p-n junction and an SB diode is that both electrons and holes contribute to the current in a p-n junction, whereas SB diodes are majority carrier devices. The MSM PD consists of two junctions,

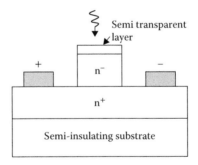

FIGURE 35.12 Schematic diagram of an SB PD. Shaded regions represent metallic contacts.

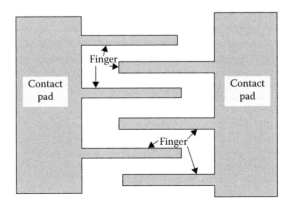

FIGURE 35.13 Top view of an MSM PD showing fingers and contact pads for connection to external circuits.

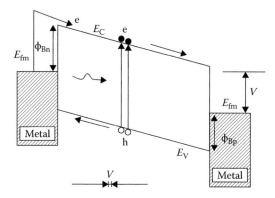

FIGURE 35.14 Band diagram of an MSM PD showing directions of flow of electrons and holes. E_c and E_v are conduction and valence band edges, respectively, E_{fm} is the Fermi level of the metal, ϕ_{Bn} and ϕ_{Bp} are the barrier heights at the n and p sides, respectively, and V is the applied bias between two metallic contacts. The PD is represented by two rectifying contacts shown at the bottom.

both of which may be rectifying or one junction ohmic and the other rectifying. The band diagram of an MSM PD is shown in Figure 35.14, in which the potential barrier ϕ_b is due to the work function difference. The two electrodes are connected serially, back-to-back, and the device contains two diodes as shown in Figure 35.14.

The dark current in the PD is due to the flow of thermally generated EHPs and by carriers from the metal surmounting the potential barrier ϕ_b. Under illumination, excess EHPs are generated and flow of electrons and holes is in opposite directions under the influence of the external bias V, as shown in Figure 35.14.

As stated already in connection with SB PDs, MSM PDs make a compromise between QE and bandwidth. The use of multiple fingers increases the QE. However, to increase the bandwidth, the effective absorption layer thickness needs to be reduced. Introduction of a highly doped layer at a certain depth may restrict the electric field within a certain absorption region and may provide a solution to the bandwidth problem.

Current improved technology may produce very thin fingers with narrow spacings between fingers. This reduces the transit time, but the speed of response is governed by the RC time constant. The following simple analytical expression for the capacitance may be used to calculate the time constant (Averine et al. 2001):

$$C = 0.226NL\varepsilon_0 \left(\varepsilon_S + 1\right) \left(6.5\theta^2 + 1.08\theta + 2.37\right). \tag{35.30}$$

In Equation 35.30, L is the length of a finger, ε_0 is the permittivity of free space, ε_S is the relative permittivity of the semiconductor, and $\theta = (D + t)/t$ is the finger-to-period ratio, where D is the width of the metal finger and t is the gap spacing between the metal fingers.

The EQE is given by

$$\eta = (1 - R)\frac{t}{t + D}\left[1 - \exp(-\alpha d)\right], \tag{35.31}$$

where R is the reflection coefficient, α the absorption coefficient, and d the thickness of the absorption layer of the semiconductor. The QE may be increased by replacing the metal electrodes by transparent conducting indium tin oxide (ITO), thereby reducing R. It can be increased by reducing the finger width at the cost of increased resistance however.

35.5.7 Phototransistors

Phototransistors may be a bipolar junction transistor (BJT) or metal-oxide-semiconductor field-effect transistor (MOSFET). The BJT may be either p-n-p or n-p-n type. It may be used in two-terminal (2T) configuration, with electrical contacts to the emitter and collector with the base floating, or in three-terminal (3T) configuration as in a normal BJT.

We first consider the 2T configuration. The structure is akin to a heterojunction bipolar transistor in which the emitter region is a higher bandgap material. The emitter is transparent to the incident light and thus acts as a window layer; the incident light falls directly on the base and is primarily absorbed in the base, base-collector depletion region, and the collector region. The heterojunction also increases the current gain. The base-collector junction is reverse biased so that the transistor acts as a reverse-biased PD.

Under no illumination the usual collector saturation current flows. However, when light is incident, the photogenerated electrons cross the depletion region in the base-collector junction and are collected at the collector terminal. The photogenerated holes travel in the opposite direction to the emitter terminal, thus reducing the emitter injection efficiency and the current gain.

The incident light is absorbed in the base and the base-collector depletion layer. The collector current for the (2T) configuration is given by (Chand et al. 1985):

$$I_c = I_{ph}(1 + h_{fe}),\tag{35.32}$$

where I_{ph} is the photogenerated current and h_{fe} the common emitter current gain of the BJT. A dark current also contributes to the collector current.

The currently used phototransistors are heterojunction bipolar transistors with larger gap and smaller gap semiconductors forming the emitter and base layers, respectively [see Basu et al. (2015) and Chang et al. (2016)] for work on GeSn-based heterojunction phototransistors and references therein for work on III-V compounds and alloy-based devices).

As stated already, a p-i-n diode is formed by the base, collector, and subcollector of a heterojunction phototransistor (HPT). Under optical illumination in a 2T-HPT, the photocurrent generated within the p-i-n diode is injected across the emitter-base junction and then amplified through transistor action. In this way, both the photocurrent in the p-i-n structure and the dc current gain determine the final collector photocurrent and the optical gain. When the incident optical power is low, the dc current gain is low, thereby making the optical gain low. This situation is improved in a 3T-HPT, in which a dc bias is applied to the base. The bias shifts the operating point to a higher collector current, making the dc current gain and optical gain higher.

A detailed analysis of the 3T-HPT has been presented by Frimel and Roenker (1997a,b) using a thermionic emission-diffusion model. The theory is too lengthy to be included here.

Useful references for 3T-HPTs for III–V compound-based materials as well as recently developed GeSn alloys may be found in Basu et al. (2015) and Chang et al. (2016).

In an MOS phototransistor, light incident at the source-gate region changes the surface potential under the gate oxide. This change causes a change in the current flowing through the channel.

35.5.8 QWIP and QDIP

It has been mentioned in Section 35.4 that the absorbing layer in a PD may contain quantum nanostructures. QWs and MQWs, in particular, are incorporated in the intrinsic (i) layer of a p-i-n PD, an APD, or in the base of a hetero phototransistor. The absorption in MQWs is from a valence subband to a conduction subband.

Quantum well infrared photodetectors (QWIPs) have QWs as the absorbing region, but the transition occurs from one subband to another in the same band, as shown in Figure 35.11. The intersubband

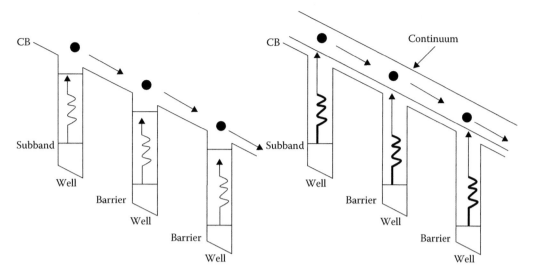

FIGURE 35.15 Intersubband absorption in QWIPs: (a) a bound-to-bound transition and (b) a bound-to-continuum absorption. The movement of electrons (solid balls) is shown in the figures.

transition energy is lower than the energy involved in band-to-band transitions in a QW. The QWIPs therefore work at the mid-infrared range (3–20 μm). Equation 35.4 may be used to calculate the energy difference between two subbands corresponding to the wavelength to be detected (see Equation 35.6). The width of the QW is then calculated from the value of the energy difference.

The types of transitions involved for CB are illustrated in Figure 35.15. The MQWs form the intrinsic layer of a p-i-n structure, in which the reverse bias causes the band bending. As shown in Figure 35.15a, the transition takes place from the lowest subband to the upper subband that lies close to the CB edge in the barrier layer. The electron lifted to the upper subband tunnels to the continuum states in the barrier and is drifted by the built-in electric field. In Figure 35.15b, an electron from the lone lowest subband absorbs the MIR photon and goes to the continuum states in the barrier and is then transported (see Schneider and Liu 2007).

Instead of QWs, multiple stacks of QDs can act as the absorbing layer, and the PD is called a quantum dot infrared photodetector (QDIP) (Barve et al. 2010).

35.5.9 Single-Photon Detectors

Single-photon emitters and detectors are needed for quantum information processing systems. Semiconductor-based single-photon detectors are APDs that operate with reverse bias exceeding the breakdown voltage. In this regime, the gain is enormously high. An incoming photon creates a charge, the number of which is multiplied (avalanche process) until it saturates at a current typically limited by an external circuit. In order that the single-photon APD should respond to a subsequent optical pulse, the saturated avalanche current must be terminated by lowering the bias voltage below the breakdown voltage (Eisaman et al. 2011).

35.6 Improved Devices

The PD structures such as p-i-n, APDs, etc., may show improved performances if some structural or materials modifications are introduced, some of which are discussed briefly in this section.

35.6.1 Waveguide PDs

In all the earlier discussions, vertically illuminated PD structures have been considered. For good PD performance, both the QE and speed of operation must be as high as possible. The absorbing layer in vertically illuminated devices must be very thin in order to reduce transit time; however, the thin layer cannot absorb the incident photons, thereby reducing the QE. To achieve high QE and high speed, edge-illuminating devices are used (Beling and Campbell 2014). The advantage of using waveguide structure has been pointed out in Section 35.4. The waveguide confines the otherwise spreading light into the high-field region. (Beling and Campbell 2014).

The waveguide photodetector (WPD) structure employs waveguides with absorbing layers either embedded in the waveguide or grown adjacent to the waveguide. The edge-coupled structure has been realized in p-i-n, APD, MSM, or uni traveling wave configuration.

The simplest type of edge-illuminated WPD structure is shown in Figure 35.16. The incident light falls directly into the edge of the absorbing layer. Here the direction of propagation of light is normal to the direction of transit of the photogenerated carriers. Note that the incident light may spread outside the absorbing region, making the optical confinement factor low and effective absorption less than in the bulk. However, the absorption length is quite large (~mm) so that responsivity and QE are enhanced. At the same time, the photogenerated carriers traverse a very short distance equal to the thickness of the absorbing layer making the transit time low. The increase in the capacitance due to the thin depletion layer may be arrested by reducing the area of the device. The state-of-the art technology for WPDs is discussed by Beling and Campbell (2014).

One drawback of the edge-coupled WPD is the misalignment of the optical field with the waveguide region. Use of lensed fiber or additional optics can increase the coupling efficiency. A useful way to improve coupling efficiency or to avoid the misalignment problem is to employ the evanescent waveguide structure as shown in Figure 35.17. In this configuration, the absorbing layer is placed adjacent to the waveguide core. The optical coupling is enhanced by inserting a transition layer between the waveguide and the detector. Since the waveguide mode has a smaller cross-section than that of a single-mode fiber, the input coupling efficiency is increased by using different approaches, for example, using a spot-size converter.

35.6.2 RCE Structure

The performance of a PD can be enhanced by placing the active device structure inside a Fabry–Pérot cavity. The cavity acts as a resonant structure providing wavelength selectivity and a large enhancement of the optical field, which in turn allows thinner PD layer. The device shows faster response and increased QE at the resonant wavelength. The wavelength selectivity and high-speed response make the devices the right candidates for wavelength division multiplexing applications (Unlu et al. 2001).

Photonic crystal structures have been employed to realize nanoscale PDs offering ultralow capacitance, thereby increasing the bandwidth and enabling ultralow power operation (Nozaki et al. 2013) (Figure 35.16).

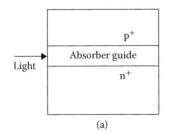
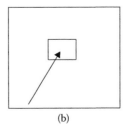

(a)   (b)

FIGURE 35.16 Schematic view of a WPD: (a) side view and (b) front view.

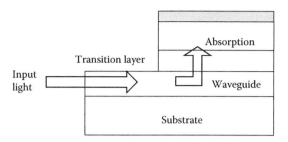

FIGURE 35.17 Schematic of an evanescently coupled WPD. Light from the waveguides couples into the absorption region via the transition layer.

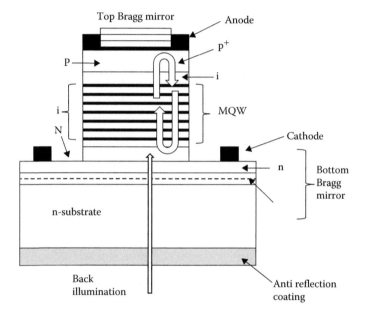

FIGURE 35.18 Schematic diagram of a RCE PD. MQW, multiple quantum well.

35.6.3 Incorporation of Heterojunctions and Quantum Nanostructures

The active materials for detection are usually grown on a suitable substrate leading to heterojunctions. However, heterojunctions and MQWs and stacks of QDs are also used in the active regions of PDs. The heterostructures extend or control the range of wavelength to be detected. For example, in Ge_xSi_{1-x}/Si heterostructures or MQWs, the absorption takes place in the lower gap alloy material, and the heterobarrier prevents leakage of carriers, thus reducing the dark current.

35.7 Materials

The wavelength range covered by a PD depends on the active material, the bandgap of which determines the maximum wavelength (cutoff wavelength). Table 35.1 includes a few important materials and the corresponding wavelength range.

TABLE 35.1   Photodetector Materials and the Wavelength Range Covered

Material	Bandgap (eV)	Wavelength Range (μm)	Substrate
GaAs	1.42	<0.87	GaAs
$In_{0.53}Ga_{0.47}As$	0.81	1.3–1.6	InP
Si	1.12	0.2–1.12	Si
Ge	0.66	<1.8	Si
$Si_{1-x}Ge_x$	1.12–0.66	<1.8	Si
$Si_{1-x}Ge_x$/Si QWs	1.12–0.66	2–12	Si
$Ge_{1-x}Sn_x$ ($x > 0.08$)	0.58– ~ 0	1.2–2.2	Si
GaAlAsSb/GaSb	0.60	0.75–1.9	GaSb
InGaAlAs/InGaAs/InP	–	0.5–2.1	InP
HgCdTe/CdTe	–	0.8–2.0	CdTe
$Ga_{1-x}Al_xN$	0.38–0.22	0.36–<0.2	Sapphire

Acknowledgments

The author is thankful to Dr. Rikmantra Basu of NIT Delhi, and Dr. Bratati Mukhopadhyay and Vedatrayee Chakraborty of Institute of Radio Physics and Electronics for their help in preparation of the manuscript.

References

Averine S. V., Chan Y. C., and Lam Y. L. (2001) Geometry optimization of interdigitated Schottky-barrier metal-semiconductor-metal photodiode structures. *Solid State Electron.*, vol. 45, 441–446.

Bandyopadhyay A. and Deen M. J. (2001) Photodetectors for optical fiber communications, Ch. 5 in *Photodetectors and Fiber Optics*, Nalwa H. S (ed) San Diego, CA: Academic Press

Barve A J., Lee S. J., Noh S. K., and Krishna S. (2010) Review of current progress in quantum dot infrared photodetectors. *Laser Photon. Rev.*, vol. 6, 738–750.

Basu P. K. (1997) *Theory of Optical Processes in Semiconductors: Bulk and Microstructures.* Oxford, UK: Oxford University Press.

Basu R., Chakraborty V., Mukhopadhyay B., and Basu P. K. (2015) Predicted performance of Ge/GeSn hetero-phototransistors on Si substrate at 1.55 μm. *Opt. Quant. Electron.*, vol. 47, no. 2., 387–399.

Beling A. and Campbell J. C. (2009) InP-based high-speed photodetectors. *J. Lightwave Technol*, vol. 27, no. 3, 343–355.

Beling A. and Campbell J. C. (2014) High-speed photodiodes. *IEEE J. Sel. Top. Quantum Electron*, vol. 20, no. 6, 3804507.

Bhattacharyya P. (1996) *Semiconductor Optoelectronic Devices*, 2nd edn. Upper Saddle Point, NJ: Prentice-Hall.

Bowers J. E. and Burrus, C. A., Jr. (1987) Ultrawide-band long-wavelength p-i-n photodetectors. *J. Lightwave Technol.*, vol. 5, no. 10, 1339–1350.

Bowers J. E. and Wey Y. G. (1995) High speed photodetectors, Ch. 17 in *Handbook of Optics, Fundamentals, Techniques and Design*, Vol. 1, Bass M (ed.). Pennsylvania, NY: McGraw-Hill.

Campbell J. C. (2007) Recent advances in telecommunications avalanche photodiodes. *J. Lightwave Technol*, vol. 25, no. 1, 109–121.

Chand N., Houston P. A., and Robson P. N. (1985) Gain of a heterojunction bipolar phototransistor. *IEEE Trans. Electron Dev*, vol. 32, no. 3, 622–627.

Chang G. E., Basu Rikmantra, Mukhopadhyay B. and Basu P. K. (2016) Design and modeling of GeSn-based heterojunction phototransistors for communication applications. *IEEE J. Sel. Topi. Quantum Electronics*, vol. 22, no. 6, 1–9.

Deen M. J. and Basu P. K. (2012) *Silicon Photonics: Fundamentals and Applications*, Chichester: John Wiley.

Eisaman M. D., Fan J., Migdall A., and Polyakov S. V. (2011) Single-photon sources and detectors. *Rev. Sci. Instruments*, vol. 82, 071101.

Frimel S. M. and Roenker K. P. (1997a) A thermionic-field-diffusion model for Npn bipolar heterojunction phototransistors. *J. Appl. Phys.*, vol. 82, 1427–1437.

Frimel S. M. and Roenker K. P. (1997b) Gummel-Poon model for Npn heterojunction bipolar phototransistor. *J. Appl. Phys.*, vol. 82, 3581–3592.

Kaiser G. (2013) *Optical Fiber Communication*, 5th edn. New Delhi; McGraw Hill Education (India) Pvt. Ltd.

Kumar S. and Deen M. J. (2014) *Fiber Optic Communications: Fundamentals and Applications*. Hoboken, NJ: Wiley.

Nozaki K., Matsuo S., Takeda K., Sato T., Kuramochi E., and Notomi M. (2013) InGaAs nano-photodetectors based on photonic crystal waveguide including ultracompact buried heterostructure. *Opt. Express*, vol. 21, no. 16, 19022–19028.

Schneider H. and Liu H. C. (2007) *Quantum Well Infrared Photodetectors: Physics and Applications*. Berlin: Spriner Verlag.

Senior M. and Jamro M. Y. (2009) *Optical Fiber Communications: Principles and Practice*. Harlow, England: Pearson Prentice-Hall.

Streetman B. G. and Banerjee S.K. *Solid State Electronic Devices*, 5th edn. Harlow, England: Pearson Education Ltd.

Unlu M. S., Ulu G., and Gokkavas M. (2001) Resonant cavity enhanced photodetectors, Ch. 2 in *Photodetectors and Fiber Optics*, Nalwa H. S (ed) San Diego, CA: Academic Press

36

P-N Junction Photodiodes

Weida Hu

36.1 Introduction

Semiconductor p-n junction photodiodes have been available for half a century. By applying an electric field across the p-n junction, the light-generated carriers create an electrical current flow, thus the p-n junction photodiodes can convert optical signals into electrical signals (Shi and Li, 2014). Sensing applications in infrared, visible, and ultraviolet spectral bands promote theoretical research of these photodiodes. Numerical simulations, containing structural details such as layer thicknesses, doping profiles, and trap concentrations, provide key insights into device design and reliability degradation mechanisms, which could effectively reduce costly and time-consuming testing when developing and characterizing a new semiconductor device or technology. However, simulations of p-n junction photodiodes involve many important physical models, which play a decisive role in the calculations of accurate results. This chapter covers some basic equations and models used in the simulations of p-n junction photodiodes. Another key point is the accurate knowledge of device material parameters. Because of the complexity of an actual device structure, different device and material parameters are needed for various designs and structures. A number of these characteristic input parameters required for accurate modeling when building a basic theoretical framework are explained in this chapter.

First, this chapter presents several basic equations and models in detail. Second, principles and characteristics of several common optical generation models are reviewed and compared. Then, a number of important characteristic input parameters are introduced. Several p-n junction photodiode simulation examples are listed in the end of this chapter.

Figure 36.1 presents the general process simulation of p-n junction. After building the device structure, the device model could be defined by adding mesh generation and doping for different region. Different physical models are added in the optical/electrical simulation step, and then the desired information could be achieved. Figure 36.2 examples the results of some of the important characteristics of p-n junction.

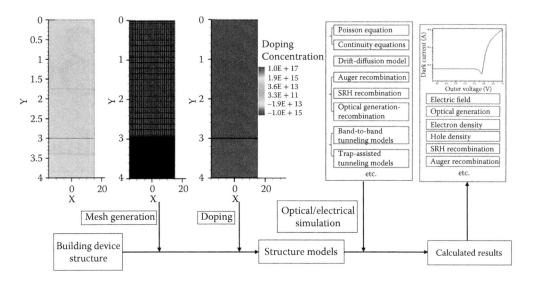

FIGURE 36.1   The flow chart of numerical simulations of p-n junction photodiodes.

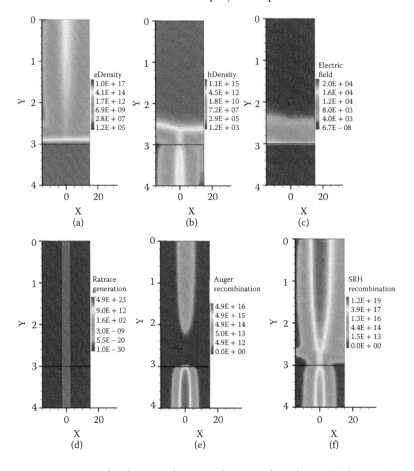

FIGURE 36.2   Important output/plots for a two-dimensional numerical simulation: (a) electron density (eDensity), (b) hole density (hDensity), (c) electric field, (d) raytrace generation, (e) Auger recombination, and (f) Shockley–Read–Hall (SRH) recombination.

36.2 Basic Equations and Models

Physical phenomena in semiconductor devices are very complicated and, for different devices and materials, models of different levels of complexity are needed in simulations. Different partial differential equations and models determine different microscopic physical mechanisms in a simulation, and can depend on different materials and device structures. This chapter introduces a number of widely used basic equations and models.

Depending on the p-n junction photodiodes under investigation and the level of modeling accuracy required, there are three governing equations for charge transport in semiconductor devices: *the Poisson equation* and *the electron and hole continuity equations*. The Poisson equation is as follows:

$$\nabla \varepsilon \cdot \nabla \psi = -q \left(p - n + N_{D^+} - N_{A^-} \right) \tag{36.1}$$

where ε is the electrical permittivity, ψ is the electrostatic potential, q is the elementary electronic charge, n and p are the electron and hole densities, N_{D^+} is the number of ionized donors, and N_{A^-} is the number of ionized acceptors. The electron and hole continuity equations are written as:

$$\frac{1}{q} \nabla \vec{J_n} + \left(G^{opt} - R \right) = 0 \tag{36.2}$$

$$\frac{1}{q} \nabla \vec{J_p} - \left(G^{opt} - R \right) = 0 \tag{36.3}$$

where R is the net electron–hole recombination rate, $\vec{J_n}$ the electron current density, and $\vec{J_p}$ is the hole current density.

Otherwise, *the drift-diffusion model* is widely used for the simulation of carrier transport in p-n junction photodiodes and is defined by the basic semiconductor equations, where current densities for electrons and holes are given by

$$\vec{J_n} = -nq\mu_n \nabla \phi_n \tag{36.4}$$

$$\vec{J_p} = -pq\mu_p \nabla \phi_p \tag{36.5}$$

where μ_n and μ_p are the electron and hole mobilities, ϕ_n and ϕ_p are the electron and hole quasi-Fermi potentials, respectively.

Additionally, *the thermodynamic model*, which is defined by a basic set of partial differential equations, and the lattice heat flow equations have been used in the simulations. The relations are generalized to include the temperature gradient as a driving term:

$$\vec{J_n} = -nq\mu_n \left(\nabla \phi_n + P_n \nabla T \right) \tag{36.6}$$

$$\vec{J_p} = -pq\mu_p \left(\nabla \phi_p + P_p \nabla T \right) \tag{36.7}$$

where P_n and P_p are the absolute thermoelectric powers.

36.3 Optical Generation Models

In the numerical simulation of the photoelectric detector, different structures and devices have different requirements for optical absorption models, including photoelectric incident position, power density, incident angle, reflection, refraction, and so on. Therefore, it is very important to select suitable optical

TABLE 36.1 Comparison of Some Common Optical Generation Models

	Optical Beam	Raytracing	Transfer Matrix	FDTD	Beam Propagation
Layered media	×	√	√	√	√
Anisotropy on horizontal direction	×	√	×	√	√
Geometrical optics	×	√	×	√	√
Diffraction	×	×	×	√	×
Reflection	×	√	√	√	√
Oblique incidence light	×	√	×	√	√

Note: FDTD = finite-difference time-domain.

generation models for different devices. The application conditions of some common optical generation models are listed in Table 36.1, and the characteristics of each model are introduced in this chapter.

Optical beam method is a simple optical absorption simulation method using Beer's law, and is only applicable to very simple optical process. Raytracing is capable of simulating a wide variety of optical effects such as reflection, refraction, and scattering, but its accuracy is severely limited, and cannot handle complex optical problems such as diffraction. Transfer matrix is very suitable for layered media, but requires media to be transversely isotropic. The FDTD method belongs in the general class of grid-based differential numerical modeling methods (finite difference methods). The time-dependent Maxwell's equations (in partial differential form) are discretized using central-difference approximations to the space and time partial derivatives, and could deal with more complex optical effects such as diffraction. Beam propagation method (BPM) is a quick and easy method of solving for fields in integrated optical devices, and a higher accuracy could be achieved. However, the BPM relies on the slowly varying envelope approximation, and is inaccurate for the modeling of discretely or fastly varying structures. What's more, BPM has some problems when dealing with scattering and diffraction problems.

36.3.1 Optical Beam Absorption

The optical beam absorption method supports the simulation of photogeneration using Beer's law. Multiple vertical photon beams can be defined to represent the incident light. The following equations describe useful relations for the photogeneration problem:

$$E_{\text{ph}} = \frac{hc}{\lambda} \tag{36.8}$$

$$J_0 = \frac{P_0}{E_{\text{ph}}} \tag{36.9}$$

where J_0 denotes the optical beam intensity (number of photons that cross an area of 1 cm^2 per 1 s) incident on the semiconductor device, P_0 is the incident wave power per area (W/cm^2), λ is the wavelength (cm), h is Planck's constant (J $\cdot$ s), c is the speed of light in a vacuum (cm/s), and E_{ph} is the photon energy that is approximately equal to $\frac{1.24}{\lambda\,(\mu m)}$ in eV.

The optical beam absorption model computes the optical generation rate along the z-axis taking into account that the absorption coefficient varies along the propagation direction of the beam according to

$$G^{\text{opt}}(z, t) = J_0 F_t(t) F_{xy} \cdot \alpha(\lambda, z) \cdot \exp\left(-\left|\int_{z_0}^{z} \alpha(\lambda, z)\,dz\right|\right) \tag{36.10}$$

where t is the time, $F_t(t)$ is the beam time behavior function: for a Gaussian pulse, it is equal to 1 for t in $[t_{min}, t_{max}]$ and shows a Gaussian distribution decay outside the interval with the standard deviation σ_t; F_{xy} is equal to 1 inside the illumination window and zero otherwise; z_0 is the coordinate of the semiconductor surface; and $\alpha(\lambda, z)$ is the nonuniform absorption coefficient along the z-axis.

The optical beam absorption method supports a wide range of window shapes as well as arbitrary beam time behavior functions. It is also not limited to beams propagating along the z-axis.

36.3.2 Raytracing

The optical generation method of raytracing supports the simulation of photogeneration in two dimensions (2Ds) and three dimensions (3Ds) for arbitrarily shaped structures. The calculation of refraction, transmission, and reflection follows geometric optics. A plane wave can be partitioned and each partition is represented by a 1D ray of light. Raytracing can approximate the behavior of a plane wave on a device by following such rays.

The optical generation along a ray, when the propagation is thought to be in the z-axis for a nonuniform absorption coefficient $\alpha(\lambda, x, y, z)$, can be written as a photon beam generation:

$$G^{opt}(z, t) = J\left(x, y, z_0\right) \alpha\left(\lambda, z\right) \cdot \exp\left(-\left|\int_{z_0}^{z} \alpha\left(\lambda, z\right) dz\right|\right) \qquad (36.11)$$

where $J(x, y, z_0)$ is the beam spatial variation of intensity over a window where rays enter the device, and z_0 is the position along the ray where absorption begins.

When a ray passes through a material boundary, the raytracer uses a recursive algorithm: It starts with a source ray and builds a binary tree that tracks the transmission and reflection of the ray. A reflection/transmission process occurs at interfaces with refractive index differences.

An incident ray impinges on the interface of two different refractive index (n_1 and n_2) regions, resulting in a reflected ray and a transmitted ray. The angles involved in raytracing are governed by Snell's law using

$$n_1 \sin\left(\theta_1\right) = n_2 \sin\left(\theta_2\right) \qquad (36.12)$$

Absorption of photons occurs when there is an imaginary component (extinction coefficient), κ, to the complex refractive index. To convert the absorption coefficient to the necessary units, the following formula is used for power/intensity absorption:

$$\alpha\left(\lambda\right)\left[cm^{-1}\right] = \frac{4\pi\kappa}{\lambda} \qquad (36.13)$$

In the complex refractive index model, the refractive index is defined element wise. In each element, the intensity of the ray is reduced by an exponential factor defined by $\exp(-\alpha L)$ where L is the length of the ray in the element. Therefore, the photon absorption rate in each element is

$$G^{opt}\left(x, y, z, t\right) = I\left(x, y, z\right)\left(1 - e^{(-\alpha L)}\right) \qquad (36.14)$$

where $I(x, y, z)$ is the rate intensity of the ray in the element. After all of the photon absorptions in the elements have been computed, the values are interpolated onto the neighboring vertices and are divided by its sustaining volume to obtain the final absorption rate. The absorption of photons occurs in all materials with a positive extinction coefficient for raytracing. Depending on the quantum yield, a fraction of this value is added to the carrier continuity equation as a generation rate so that correct accounting of particles is maintained.

36.3.3 Transfer Matrix Method

The propagation of plane waves through layered media could be calculated by using a transfer matrix approach.

In the underlying model of the optical carrier generation rate, monochromatic plane waves with arbitrary angles of incidence and polarization states penetrating a number of planar, parallel layers are assumed. Each layer must be homogeneous, isotropic, and optically linear. In this case, the amplitudes of forward and backward running waves $A_j^\pm$ and $B_j^\pm$ in each layer shown in Figure 36.3 are calculated with the help of transfer matrices.

These matrices are functions of the complex wave impedances Z_j given by $Z_j = n_j \cdot \cos\theta_j$ in the case of E polarization (TE) and by $Z_j = n_j/\cos\theta_j$ in the case of H polarization (TM). Here, n_j denotes the complex index of refraction and θ_j is the complex counterpart of the angle of refraction ($n_0 \cdot \sin\theta_0 = n_j \cdot \sin\theta_j$).

The transfer matrix of the interface between layers j and $j + 1$ is defined by

$$T_{j,j+1} = \frac{1}{2Z_j} \cdot \begin{bmatrix} Z_j + Z_{j+1} & Z_j - Z_{j+1} \\ Z_j - Z_{j+1} & Z_j + Z_{j+1} \end{bmatrix} \tag{36.15}$$

The propagation of the plane waves through layer j can be described by the transfer matrix:

$$T_j\left(d_j\right) = \begin{bmatrix} \exp\left(2\pi i n_j \cos\theta_j \frac{d_j}{\lambda}\right) & 0 \\ 0 & \exp(-2\pi i n_j \cos\theta_j \frac{d_j}{\lambda}) \end{bmatrix} \tag{36.16}$$

with the thickness d_j of layer j, and the wavelength λ of the incident light. The transfer matrices connect the amplitudes shown in Figure 36.3 as follows:

$$\begin{bmatrix} B_j^+ \\ A_j^+ \end{bmatrix} = T_{j,j+1} \begin{bmatrix} A_{j+1}^- \\ B_{j+1}^- \end{bmatrix} \tag{36.17}$$

$$\begin{bmatrix} A_j^- \\ B_j^- \end{bmatrix} = T_j\left(d_j\right) \begin{bmatrix} B_j^+ \\ A_j^+ \end{bmatrix} \tag{36.18}$$

It is assumed that there is no backward running wave behind the layered medium, and the intensity of the incident radiation is known. Therefore, the amplitudes $A_j^\pm$ and $B_j^\pm$ at each interface can be calculated with appropriate products of transfer matrices.

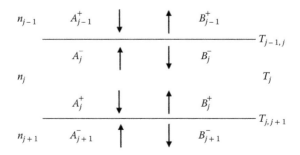

FIGURE 36.3 Wave amplitudes in a layered medium and transfer matrices connecting them.

For both cases of polarization, the intensity in layer j at a distance d from the upper interface $(j, j+1)$ is given by

$$I_{T(E,TM)}(d) = \frac{\Re(Z_j)}{\Re(Z_0)} \cdot \left\| T_j(d) \cdot \begin{pmatrix} A_j^- \\ B_j^- \end{pmatrix} \right\|^2 \tag{36.19}$$

with the proper wave impedances, $\Re(Z_j)$ means the real part of Z_j. If δ is the angle between the vector of the electric field and the plane of incidence, the intensities should be added according to

$$I(d) = aI_{TM}(d) + I_{TE}(d) \tag{36.20}$$

where $I_{TM} = (1 - a)I(d)$ and $I_{TE} = aI(d)$ with $a = \cos^2\delta$.

36.3.4 FDTD Method

Light propagation through the sub-wavelength microstructure (such as nanoparticles, nanometal grating, photonic crystals, photo-trapping structures, and diffractive microlenses)-enhanced photodetectors is simulated by FDTD method based on a rigorous vector solution of Maxwell's equations. The computation area of the FDTD method is discretized into grid space where the electric field vector and the magnetic field vector are respectively arranged in the space and time step by 1/2, and the differential form of Maxwell's equations is rewritten as the central difference form with two-order accuracy.

In the calculation, these correspond six time varying scalar equations in each Yee's grid (i.e., corresponding to the three spatial components of the electric and magnetic fields, as shown in Figure 36.4), along with the propagation of the electromagnetic field, the electromagnetic field is constantly updated in the set period until convergence. At this time, the corresponding electromagnetic field (light) has a stable distribution in the whole calculation area. At this point, the absorbed power density and the distribution of carrier density are calculated by assuming that each of the photons absorbed could excite an electron–hole pair.

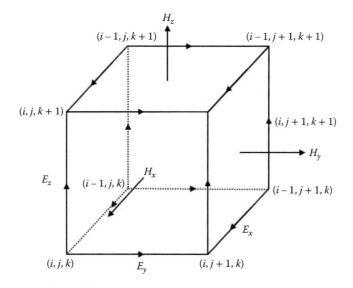

FIGURE 36.4 Structure of 3D Yee grid showing one cell of primary grid.

The differential forms of Maxwell equations are

$$\frac{\partial H}{\partial t} = -\frac{1}{\mu}\nabla \times E - \frac{\rho}{\mu}H \tag{36.21}$$

$$\frac{\partial E}{\partial t} = \frac{1}{\varepsilon}\nabla \times H - \frac{\sigma}{\varepsilon}E \tag{36.22}$$

Maxwell's curl equations can be discretized by the central difference method to achieve second-order accuracy. Equations 36.21 and 36.22 could be rewritten as formulas of electromagnetic field component:

$$\frac{H_x\big|_{i,j,k}^{n+1/2} + H_x\big|_{i,j,k}^{n-1/2}}{\Delta t} = \frac{1}{\mu_{i,j,k}}\left(\frac{E_y\big|_{i,j,k+1/2}^{n} - E_y\big|_{i,j,k-1/2}^{n}}{\Delta z} - \frac{E_z\big|_{i,j+1/2,k}^{n} - E_z\big|_{i,j-1/2,k}^{n}}{\Delta y} - \rho_{i,j,k}\cdot H_x\big|_{i,j,k}^{n}\right) \tag{36.23}$$

$$\frac{H_y\big|_{i,j,k}^{n+1/2} + H_y\big|_{i,j,k}^{n-1/2}}{\Delta t} = \frac{1}{\mu_{i,j,k}}\left(\frac{E_z\big|_{i+1/2,j,k}^{n} - E_z\big|_{i-1/2,j,k}^{n}}{\Delta x} - \frac{E_x\big|_{i,j,k+1/2}^{n} - E_x\big|_{i,j,k-1/2}^{n}}{\Delta z} - \rho_{i,j,k}\cdot H_y\big|_{i,j,k}^{n}\right) \tag{36.24}$$

$$\frac{H_z\big|_{i,j,k}^{n+1/2} + H_z\big|_{i,j,k}^{n-1/2}}{\Delta t} = \frac{1}{\mu_{i,j,k}}\left(\frac{E_x\big|_{i,j+1/2,k}^{n} - E_x\big|_{i,j-1/2,k}^{n}}{\Delta y} - \frac{E_y\big|_{i+1/2,j,k}^{n} - E_y\big|_{i-1/2,j,k}^{n}}{\Delta x} - \rho_{i,j,k}\cdot H_z\big|_{i,j,k}^{n}\right) \tag{36.25}$$

$$\frac{E_x\big|_{i,j,k}^{n+1} + E_x\big|_{i,j,k}^{n}}{\Delta t} = \frac{1}{\varepsilon_{i,j,k}}\left(\frac{H_z\big|_{i,j+1/2,k}^{n+1/2} - H_z\big|_{i,j-1/2,k}^{n+1/2}}{\Delta y} - \frac{H_y\big|_{i,j,k+1/2}^{n+1/2} - H_y\big|_{i,j,k-1/2}^{n+1/2}}{\Delta z} - \sigma_{i,j,k}\cdot E_x\big|_{i,j,k}^{n+1/2}\right) \tag{36.26}$$

$$\frac{E_y\big|_{i,j,k}^{n+1} + E_y\big|_{i,j,k}^{n}}{\Delta t} = \frac{1}{\varepsilon_{i,j,k}}\left(\frac{H_x\big|_{i,j,k+1/2}^{n+1/2} - H_x\big|_{i,j,k-1/2}^{n+1/2}}{\Delta z} - \frac{H_z\big|_{i+1/2,j,k}^{n+1/2} - H_z\big|_{i-1/2,j,k}^{n+1/2}}{\Delta x} - \sigma_{i,j,k}\cdot E_y\big|_{i,j,k}^{n+1/2}\right) \tag{36.27}$$

$$\frac{E_z\big|_{i,j,k}^{n+1} + E_z\big|_{i,j,k}^{n}}{\Delta t} = \frac{1}{\varepsilon_{i,j,k}}\left(\frac{H_y\big|_{i+1/2,j,k}^{n+1/2} - H_y\big|_{i-1/2,j,k}^{n+1/2}}{\Delta x} - \frac{H_x\big|_{i,j+1/2,k}^{n+1/2} - H_x\big|_{i,j-1/2,k}^{n+1/2}}{\Delta y} - \sigma_{i,j,k}\cdot E_z\big|_{i,j,k}^{n+1/2}\right) \tag{36.28}$$

Therefore, the electromagnetic field component can be solved by the Equations 36.23 through 36.28 for any given grid node at any moment.

For more detailed introduction of FDTD, see Chapter 52.

36.3.5 Beam Propagation Method

The BPM can be applied to find the light propagation and penetration into devices such as photodetectors. Its efficiency and relative accuracy are much better than the optical beam absorption method. The BPM solver is available for both 2D and 3D device geometries, where its computational efficiency compared with a full-wave approach becomes particularly apparent in 3Ds.

The BPM is based on the fast Fourier transform (FFT) and is a variant of the FFT BPM, which was developed by Feit and Fleck (1978).

The solution of the scalar Helmholtz equation is

$$\left[\nabla_t^2 + \frac{\partial^2}{\partial z^2} + k_0^2 n^2 (x, y, z) \right] \phi (x, y, z) = 0 \tag{36.29}$$

At $z + \Delta z$ with $\nabla_t^2 = \frac{\partial^2}{\partial x^2} + \frac{\partial^2}{\partial y^2}$ and $n(x, y, z)$ being the complex refractive index can be written as:

$$\frac{\partial}{\partial z} \phi (x, y, z + \Delta z) = \mp i \sqrt{k_0^2 n^2 + \nabla_t^2} \phi (x, y, z) = \pm i \xi \phi (x, y, z) \tag{36.30}$$

In the paraxial approximation, the operator ξ reduces to

$$\xi = \sqrt{k_0^2 n_0^2 + \nabla_t^2} + k_0 \delta n \tag{36.31}$$

where n_0 is taken as a constant reference refractive index in every transverse plane. By expressing the field ϕ at z as a spatial Fourier decomposition of plane waves, the solution to Equation 36.29 for forward-propagating waves reads

$$\phi (x, y, z + \Delta z) = \frac{1}{(2\pi)^2} \exp \left(i k_0 \delta n \Delta z \right) \int\limits_{-\infty}^{+\infty} d\vec{k_t} \exp \left(i \vec{k_t} \cdot \vec{r} \right) \exp \left(i \sqrt{k_0^2 n_0^2 \to \vec{k_t^2}} \Delta z \right) \tilde{\phi} \left(\vec{k_t}, z \right) \tag{36.32}$$

where $\tilde{\phi}$ denotes the transverse spatial Fourier transform. As can be seen from Equation 36.32, each Fourier component experiences a phase shift, which represents the propagation in a medium characterized by the reference refractive index n_0. The phase-shifted Fourier wave is then inverse transformed and given an additional phase shift to account for the refractive index inhomogeneity at each $(x, y, z + \Delta z)$ position. In the numeric implementation of Equation 36.32, an FFT algorithm is used to compute the forward and inverse Fourier transform.

For more detailed introduction of BPM, see Chapter Chapter 4.

36.4 Characteristic Input Parameters

The basic physical mechanism involved in numerical simulations of p-n junction photodiodes is very complicated and, depending on devices and use environments, different material parameters should be chosen for appropriate physical models. Some are related to bandgap such as effective mass, and another part of the parameters contributes to noise such as all of the recombination rate, while the absorption coefficient is dominant for the light absorption properties of device. In addition, the minority carrier lifetime of each region significantly influence the carrier transport. This chapter introduces some typical material parameters and related physical models (Table 36.2).

36.4.1 Minority Carrier Lifetimes

The minority carrier lifetime refers to the average survival time of minority carriers, and can be calculated as:

$$\tau = \frac{\Delta n}{U} \tag{36.33}$$

where Δn is the nonequilibrium carrier concentration and U is the carrier recombination rate.

TABLE 36.2   Parameters That Can Be Specified from
Parameter File

Parameters	Units
Room-temperature bandgap, E_g	eV
Dielectric constant, ε_r	1
Density of states in conduction band, N_C	cm^{-3}
Density of states in valance band, N_V	cm^{-3}
Electron mobility, μ_e	cm^2/V·s
Hole mobility, μ_h	cm^2/V·s
Electron saturation velocity, v_{Se}	cm/s
Hole saturation velocity, v_{Sh}	cm/s
Electron effective mass (relative), m_e	1
Hole effective mass (relative), m_h	1
Electron Auger coefficient, R_{Ae}	cm^6/s
Hole Auger coefficient, R_{Ah}	cm^6/s
Electron SRH lifetime, τ_{Se}	s
Hole SRH lifetime, τ_{Sh}	s
Radiative coefficient, R_R	cm^3/s
Absorption spectra, A_1	cm^{-1}
Absorption spectra, A_2	cm^{-1}
Absorption spectra, E_1	eV
Absorption spectra, E_2	eV
Absorption spectra, P	1

Note: SRH, Shockley–Read–Hall.

The electron concentration n_0 and the hole concentration p_0 under the thermal equilibrium state fit the function:

$$n_0 p_0 = n_i^2 \tag{36.34}$$

where n_i is the intrinsic carrier concentration.

The generation-recombination rate of radiative recombination model is given by

$$U_R = r(np - n_i^2) \tag{36.35}$$

where r is the electron–hole recombination probability and n, p are the electron and hole concentration, respectively. The minority carrier lifetime is given by

$$\tau = \frac{1}{r(n_0 + p_0 + \Delta p)} \tag{36.36}$$

To the SRH recombination model, the generation-recombination rate can be written as

$$U_{SRH} = \frac{N_t r_n r_p (np - n_i^2)}{r_n \left(n + n_1\right) + r_p (p + p_1)} \tag{36.37}$$

with

$$n_1 = n_0 \exp\left(\frac{E_t - E_f}{k_0 T}\right) \tag{36.38}$$

$$p_1 = p_0 \exp\left(-\frac{E_t - E_f}{k_0 T}\right) \tag{36.39}$$

where r_n and r_p are the capture coefficient of electron and hole, N_t is the concentration of recombination center, E_t is the recombination center energy level, E_f is the Fermi energy level, and T is the lattice temperature. The SRH minority carrier lifetime can be calculated by

$$\tau = \frac{r_n\left(n + n_1\right) + r_p(p + p_1)}{N_t r_n r_p(n_0 + p_0 + \Delta p)} \tag{36.40}$$

When the surface recombination model is considered, a revised formula is used based on the SRH recombination model:

$$U_{\text{Surf}} = \frac{s_n s_p(np - n_i^2)}{s_p\left(n + n_1\right) + s_n(p + p_1)} \tag{36.41}$$

where s_n and s_p are the surface recombination velocity of electron and hole.

The rate of Auger recombination is given by

$$U_A = (\gamma_n n + \gamma_p p)(np - n_i^2) \tag{36.42}$$

where γ_n and γ_p are the auger recombination coefficient of electron and hole. The Auger recombination minority carrier lifetime can be written by

$$\tau = \frac{1}{\left(n_0 + p_0 + \Delta p\right)\left(\gamma_n n + \gamma_p p\right)} \tag{36.43}$$

36.4.2 Interface Traps and Fixed Charges

The following expressions for trap concentration versus the energy (E) define different types of distributions:

N_0 for $E = E_0$ Single-energy level

N_0 for $E_0 - 0.5E_s < E < E_0 + 0.5E_s$ Uniform distribution

$N_0\exp\left(-\left|\dfrac{E - E_0}{E_S}\right|\right)$ Exponential distribution

$N_0\exp\left[-\dfrac{(E - E_0)^2}{2E_S^2}\right]$ Gaussian distribution

$$\begin{cases} N_0 & \text{for} & E = E_1 \\ N_2 & \text{for} & E = E_2 \\ \cdots & & \cdots \\ N_m & \text{for} & E = E_m \end{cases} \quad \text{User-defined table distribution} \tag{36.44}$$

Trapped electron and hole concentrations on one energy level are related to occupation probabilities for electrons (r_n) and holes (r_p) as follows:

$$\begin{aligned} n_{\text{Dt}} = N_{\text{Dt}} r_n \quad & n_t = N_{\text{Et}} r_n \\ p_{\text{At}} = N_{\text{At}} r_p \quad & p_t = N_{\text{Ht}} r_p \end{aligned} \tag{36.45}$$

where n_{Dt} is the electron concentration of the donor trap level, p_{At} is the hole concentration of the acceptor trap level, n_t is the electron concentration of the neutral electron trap level, and p_t is the hole concentration of the neutral hole trap level, N_{Dt} is the donor trap concentration, N_{At} is the acceptor trap concentration,

N_{Et} is the neutral electron trap concentration, N_{Ht} is the neutral hole trap concentration. It's obvious that $r_n = 1 - r_p$.

In consideration of the presence of traps, the Poisson equation is revised:

$$\nabla\varepsilon \cdot \nabla\psi = -q\left(p - n + N_{D^+} - N_{A^-} + \sum_{E_t} N_{Dt} - n_{Dt} - \sum_{E_t} N_{At} - p_{At} + \sum_{E_t} p_t - \sum_{E_t} n_t\right) \quad (36.46)$$

where E_t means trap energy levels.

The balance of the carrier flow to and from a trap level gives the following expression for the electron concentration of the trap level:

$$\frac{dn_t}{dt} = v_{th}^n \sigma_n N_{Et}\left[\frac{n_1}{g_n}r_n - n(1 - r_n)\right] + v_{th}^p \sigma_p N_{Et}\left[\frac{p_1}{g_p}(1 - r_n) - pr_n\right] \quad (36.47)$$

where v_{th}^n and v_{th}^p are the electron and hole thermal velocities, σ_n and σ_p are the electron and hole capture cross sections, and g_n and g_p are the electron and hole degeneracy factors, respectively.

In the steady state, the trap-assisted recombination process can be analyzed by standard SRH recombination; however, to the transient case, the recombination processes for electrons and holes are different, and the SRH form cannot be applied. The differential equations for the electron-related occupation probability at each energy level are

$$\frac{dr_n}{dt} + \left[v_{th}^n \sigma_n\left(n + \frac{n_1}{g_n}\right) + v_{th}^p \sigma_p\left(p + \frac{p_1}{g_p}\right)\right]r_n = v_{th}^n \sigma_n n + v_{th}^p \sigma_p \frac{p_1}{g_p} \quad (36.48)$$

This equation is solved self-consistently with the transport and Poisson equations.

The trap total recombination term in the continuity equations is the sum for all levels:

$$R_{trap} = \sum_{E_t} R_{Dt} + \sum_{E_t} R_{Et} + \sum_{E_t} R_{At} + \sum_{E_t} R_{Ht} \quad (36.49)$$

At nonheterointerface vertices, bulk traps are considered only from the region with the lowest bandgap. In cases where the bulk traps from other regions are important, assume appropriate value and solve nonlinear Poisson equation by changing the numeric damping of the trap charge, which properly considers bulk traps from all adjacent regions.

36.4.3 Absorption Coefficients

A widely used absorption coefficient model that fits all the materials is

$$\alpha\left(E_{ph}\right) = \begin{cases} \alpha_1 \exp\left[\frac{\left(E_{ph}-E_1\right)}{E_2}\right] & E_{ph} < E_1 \\[2mm] \alpha_1 + \alpha_2 \exp\left[\frac{\left(E_{ph}-E_1\right)}{E_2}\right]^p & E_{ph} \geq E_1 \end{cases} \quad (36.50)$$

where E_{ph} is the photon energy, and $\alpha_1, \alpha_2, E_1, E_2$, and p could be specified in the parameter file.

The abovementioned parameters can be obtained by fitting the data acquired in the experiment. A fitting result (Wang et al., 2015) for GaN and AlGaN are shown in Figures 36.5 and 36.6.

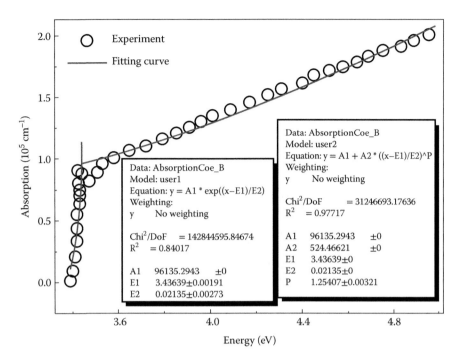

FIGURE 36.5 Fitting results of absorption spectrum for GaN at room temperature.

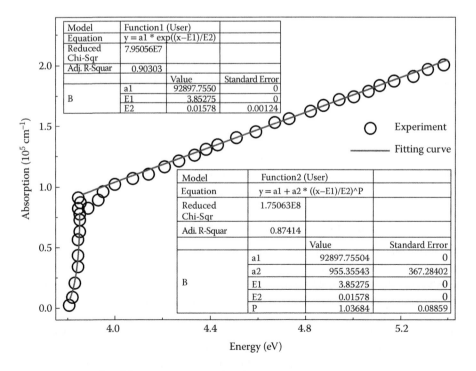

FIGURE 36.6 Fitting results of absorption spectrum for $Al_{0.15}Ga_{0.85}N$ at room temperature.

An absorption model (Rajkanan et al., 1979) is implemented for silicon and materials with similar bandgap structure:

$$\alpha(T) = \sum_{i=1}^{2} \sum_{j=1}^{2} C_i A_j \left\{ \frac{\left[hf - E_{gj}(T) + E_{pi} \right]^2}{e^{-\frac{E_{pi}}{kT}} - 1} + \frac{\left[hf - E_{gj}(T) - E_{pi} \right]^2}{1 - e^{-\frac{E_{pi}}{kT}}} \right\} + A_d \left[hf - E_{gd}(T) \right]^{1/2} \quad (36.51)$$

where

$$E_g(T) = E_g(0) - \frac{\beta T^2}{T + \gamma} \quad (36.52)$$

and T is the temperature. Parameter values used in this formula can be changed for some materials with bandgap structures similar to silicon or for silicon itself.

Instead, the complex refractive index model is adopted to replace the absorption coefficient model in some numerical simulation commercial software. The complex refractive index $\tilde{n}$ can be written as

$$\tilde{n} = n + i \cdot k \quad (36.53)$$

where n refers to the refractive index, and k to the extinction coefficient. The numerical conversion of the extinction coefficient k and the absorption coefficient α is

$$\alpha = \frac{4\pi k}{\lambda} \quad (36.54)$$

For specific extinction coefficients of different materials, one can refer to the website http://refractiveindex. info/.

36.5 Simulation Examples of P-N Junction Photodiodes

In the following section, examples of some widely used p-n junction photodiodes are introduced with the calculation results and parameters shown graphically. First, examples about visible photodiodes using different materials and structure are given. Second, photodiodes simulations applied to ultraviolet and infrared wave band are introduced. Finally, we modeled a two-color heterojunction photodiode, in which more complex mechanism and structure are included. All the absorption models involved in examples are based on Equation 36.50.

36.5.1 Physical Mechanism and Performance Parameters

In order to obtain comprehensive understanding of photoresponse (including dark current) mechanisms of p-n junction photodiodes and improve the device performance, these key parameters, such as electron field (when the effect of electric field intensity becomes dominant, the dark current will be increased and the photoresponse starts reducing); different kinds of junction types (see Figure 36.7a and b), which have a great effect on technological requirements, surface leakage current, and so on; dark current (see Figure 36.7c through f), which will influence our device performance; as well as responsivity and quantum efficiency (QE) should be taken into consideration. In addition, surface leakage current is another consideration of p-n junction photodiodes. A detailed description of the surface leakage current of HgCdTe infrared photodiode is shown in Figure 36.8.

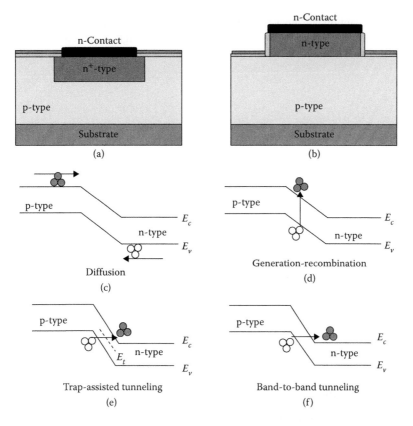

FIGURE 36.7 Schematic structures of two types of main structure of p-n junction for (a) planar junction type, (b) mesa junction type, and the main dark current components of the bulk p-n junction: (c) diffusion, (d) generation-recombination, (e) trap-assisted tunneling, and (f) direct band-to-band tunneling.

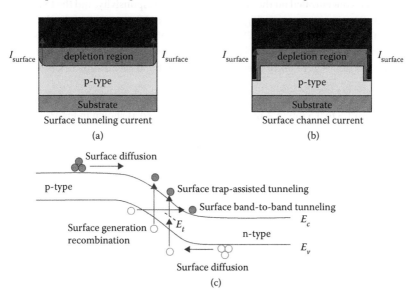

FIGURE 36.8 Surface leakage current of HgCdTe infrared photodiode: (a) surface tunneling current, (b) surface channel current, (c) the main dark current components of the surface-induced p-n junction surface diffusion current, surface generation-recombination, surface trap-assisted tunneling, and surface direct band-to-band tunneling.

The responsivity, quantum efficiency (QE) and cross-talk are index parameters measuring the ability of the device to acquire the target radiation signal. Responsivity is the light current formed by unit radiation power. The calculation formula is

$$R = \frac{I_0}{P_{\text{opt}}} \tag{36.55}$$

where R is the responsivity, I_0 is the optical current under zero bias, and P_{opt} is the power of incident light.

QE means the efficiency of the device with which the incident photons of light are converted to current:

$$\eta = \left(\frac{I_p}{q}\right)\left(\frac{P_{\text{opt}}}{h\nu}\right)^{-1} = \left(\frac{I_0}{P_{\text{opt}}}\right)h\nu = R\frac{1.24}{\lambda} = \frac{1.24R}{\lambda} \tag{36.56}$$

where λ is the wavelength of incident light and I_p is the light excitation current when the wavelength is λ (corresponding to the photon energy $h\nu$) and the power of incident light is P_{opt}.

Cross-talk is when the radiation signal in a unit produces a response signal R_0, the adjacent unit also appears in the response signal R_1. Cross-talk can be used to measure the image quality detection. The cross-talk is defined by:

$$C = \frac{R_1}{R_0} \times 100\% \tag{36.57}$$

36.5.2 Infrared Photodiodes

36.5.2.1 InP/In$_{0.53}$Ga$_{0.47}$As/InP PIN Photodiode

For an InP/In$_{0.53}$Ga$_{0.47}$As/InP PIN photodiode (Wang et al., 2008), the thickness of the absorption layer (T_{abs}) as a key structural parameter can determine, to a large extent, the magnitude of responsivity. The following discussion is concentrated on the effect of T_{abs} on the responsivity, and the following simulations are performed only at room temperature. Figure 36.9 shows the 2D structure of the device. The structure consists of a 0.5-μm heavily doped n-type InP buffer layer, a lightly doped n-type InGaAs absorption layer, and a 0.2-μm heavily doped p-type InP cap layer. The thickness of the InGaAs absorption layer (T_{abs}) is set as 1.5, 3.5, and 5.5 μm. Figure 36.10 presents the energy band of InP/InGaAs PIN device structure, respectively.

Here, the thickness of the InGaAs absorption layer is set as 1.5 μm. Vertical distributions of electric field with varying T_{abs} from 1.5 to 5.5 μm in 2-μm steps are investigated as shown in Figure 36.11; the dark and

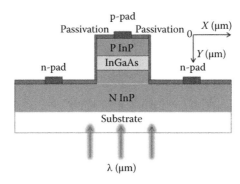

FIGURE 36.9 Schematic cross section of InP/InGaAs PIN photodiode.

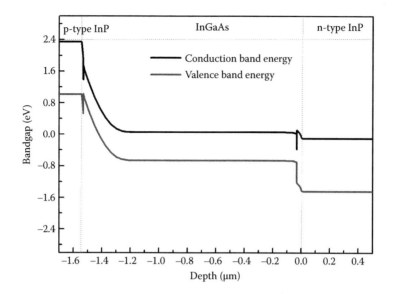

FIGURE 36.10 Energy band of the InP/InGaAs PIN photodiode.

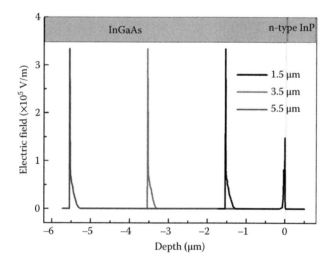

FIGURE 36.11 Vertical distributions of electric field with T_{abs} increasing from 1.5 to 5.5 μm in 2-μm steps.

light current characteristics with the absorption layer (T_{abs}) increasing from 1.5 to 5.5 μm in 1-μm steps are shown in Figure 36.12; the responsivity and QE, as a function of T_{abs} for different incident wavelengths, are shown in Figure 36.13. For clarity, the fitting parameters used in the simulation are listed in Table 36.3.

36.5.2.2 HgCdTe-Based P-N Junction Photodiode

Another example is the HgCdTe-based p-n junction photodiode infrared detector (Hu et al., 2008, 2010; Liang et al., 2014; Qiu and Hu, 2015; Yin et al., 2009; Wang et al., 2011a). The thickness of the n-region of the device has been chosen as a key structural parameter. This section discusses the effect of the thickness of the n-region on responsivity and QE of the device. The following simulations are performed at 77 K. Figure 36.14 presents a schematic diagram of the device where the white region represents the depletion region which is a space charge layer of a PN junction with positive charges in the n-type region and negative

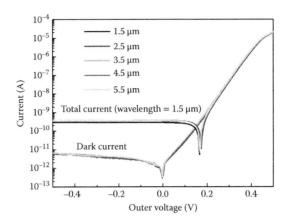

FIGURE 36.12 Dark and light current characteristics with the absorption layer (T_{abs}) increasing from 1.5 to 5.5 µm in 1-µm steps.

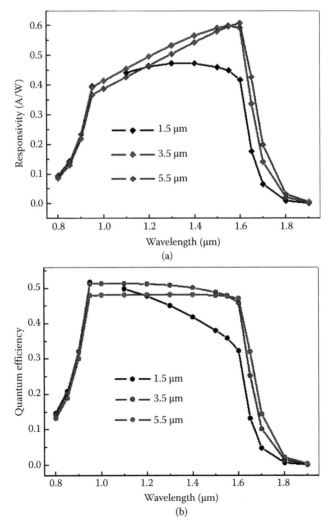

FIGURE 36.13 (a) Responsivity and (b) quantum efficiency for InP/InGaAs/InP PIN photodiode with the absorption layer (T_{abs}) increasing from 1.5 to 5.5 µm in 2-µm steps.

TABLE 36.3   InP/InGaAs/InP PIN Photodiode Simulation Parameter List

Parameters	Units	$In_{0.53}Ga_{0.47}As$	InP
Room-temperature bandgap, E_g	eV	0.78	1.34
Dielectric constant, ε_r	1	13.9	12.4
Density of states in conduction band, N_C	cm^{-3}	2.75×10^{17}	5.66×10^{17}
Density of states in valance band, N_V	cm^{-3}	7.62×10^{18}	2.03×10^{19}
Electron mobility, μ_e	cm^2/V·s	12,000	4,730
Hole mobility, μ_h	cm^2/V·s	450	151
Electron saturation velocity, v_{Se}	cm/s	8.9×10^6	2.6×10^7
Hole saturation velocity, v_{Sh}	cm/s	8.9×10^6	2.6×10^7
Electron effective mass (relative), m_e	1	0.0489	0.08
Hole effective mass (relative), m_h	1	0.45	0.861
Electron Auger coefficient, R_{Ae}	cm^6/s	3.2×10^{-28}	3.7×10^{-31}
Hole Auger coefficient, R_{Ah}	cm^6/s	3.2×10^{-28}	8.7×10^{-30}
Electron SRH lifetime, τ_{Se}	s	1×10^{-6}	1×10^{-9}
Hole SRH lifetime, τ_{Sh}	s	1×10^{-6}	1×10^{-11}
Radiative coefficient, R_R	cm^3/s	1.43×10^{-10}	2×10^{-11}
Absorption spectra, A_1	cm^{-1}	6687.08	6045.319
Absorption spectra, A_2	cm^{-1}	618.99	4008.099
Absorption spectra, E_1	eV	0.776	1.363
Absorption spectra, E_2	eV	0.012	0.013
Absorption spectra, P	1	1.013	0.65

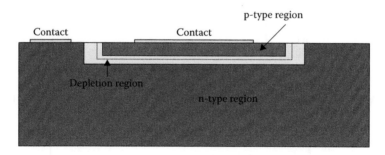

FIGURE 36.14   Schematic cross section of HgCdTe p-n junction photodiode.

charges in the p-type region. Figure 36.15 shows the energy band of the device with different thicknesses (d). Vertical distributions of electric field with the thickness of device (d) increasing from 3 to 7 in 2-µm steps under reverse bias of 200 mV are shown in Figure 36.16. Dark current and photocurrent with varying d from 3 to 7 µmin 2-µm steps under reverse bias are shown in Figure 36.17. The vertical distributions of the electric field of the device with three different conditions are very close. The responsivity and QE, as a function of d, for different incident wavelengths are shown in Figure 36.18. The responsivity and QE of the device with thickness of 5 µm are greater than those devices with thickness of 3 µm, because more photons can be absorbed by the material as the absorber region becomes thicker. However, the responsivity and QE of the device with thickness of 7 µm is smaller than that of the device with the thickness of 5 µm, because the recombination rate of the device with thickness of 7 µm increases, causing that photon-generated carriers to recombine before they arrive at the contacts. For clarity, the fitting parameters used in the simulations are listed in Table 36.4.

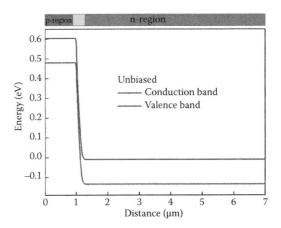

FIGURE 36.15 Energy band of HgCdTe p-n junction photodiode.

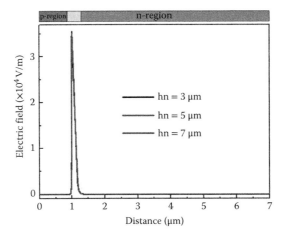

FIGURE 36.16 Vertical distributions of electric field with the height of n-region (hn) increasing from 3 to 7 μm in 2-μm steps.

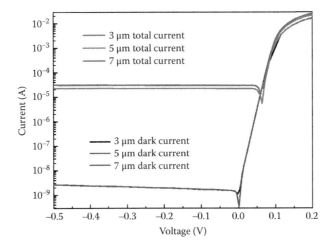

FIGURE 36.17 Dark and light current characteristics with the height of n-region increasing from 3 to 7 μm in 2-μm steps.

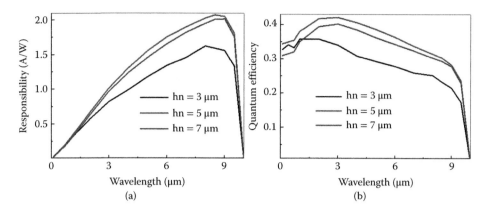

FIGURE 36.18 (a) Responsivity and (b) quantum efficiency for HgCdTe PN photodiode with the height of n-region (hn) increasing from 3 to 7 μm in 2-μm steps.

TABLE 36.4 List of the Key Parameters Used in the Simulation

Parameters	Units	HgCdTe
Bandgap	eV	0.12
Dielectric constant	1	17.4
Effective conduction band density of states	cm^{-3}	0.64×10^{19}
Effective valence band density of states	cm^{-3}	0.70×10^{19}
Electron mobility	cm^2/V·s	6,1367
Hole mobility	cm^2/V·s	610
Temperature exponent	1	0
Electron effective mass (relative)	1	1.09
Hole effective mass (relative)	1	1.15
Electron Auger coefficient	cm^6/s	2.3×10^{-25}
Hole Auger coefficient	cm^6/s	4.66×10^{-26}
Electron SRH lifetime	s	1×10^{-7}
Hole SRH lifetime	s	1×10^{-7}
Radiative recombination coefficient	cm^3/s	3.47×10^{-17}

36.5.3 Visible Photodiodes

36.5.3.1 Si P-N Junction Photodiode

In the Si p-n junction photodiode, the influence of p-region thickness, operating temperature, and doping is calculated. Figures 36.19 and 36.20 show a schematic diagram and the energy band of a silicon p-n junction device structure, respectively. Vertical distributions of electric field with varying doping concentration are investigated as shown in Figure 36.21. Figure 36.22 shows the light and dark current distribution as a function of different operating temperatures. The responsivity and QE as a function of incident wavelengths for different p-region absorption layer are shown in Figure 36.23. Parameters not described are all according to those given as (default) in parameter Table 36.5.

36.5.4 Ultraviolet Photodiodes

36.5.4.1 GaN/AlGaN PIN Photodiode

The structure of a visible blind GaN/AlGaN PIN photodiode (Wang et al., 2011b, 2014) consists of three epitaxial layers grown on a 2-inch-diameter sapphire substrate. The first layer is a 0.6-μm n$^+$-doped AlGaN

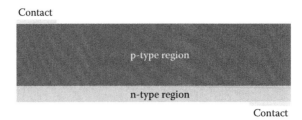

FIGURE 36.19 Schematic cross section of Si p-n junction photodetector.

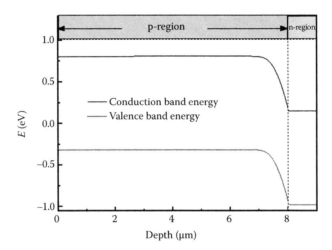

FIGURE 36.20 Energy band of Si p-n junction photodetector.

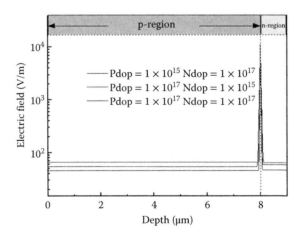

FIGURE 36.21 Vertical distributions of electric field with temperature increasing from 280 to 320 K in 20-K steps.

layer (n-layer) with a doping concentration of 3×10^{18} cm^{-3}. This layer is followed by a 0.1-μm unintentionally doped GaN absorption layer with an electron concentration of 5×10^{15} cm^{-3}, and a 0.15 μm p^{+}-doped GaN layer (p-layer) with a doping concentration of 3×10^{17} cm^{-3}. For simplicity, the thicknesses of the absorption layer and n-layer are defined as d_{abs} and d_{n}, respectively. The schematic cross section of

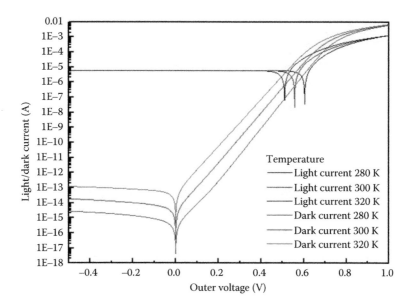

FIGURE 36.22 Dark and light current characteristics with temperature increasing from 280 to 320 K in 20-K steps.

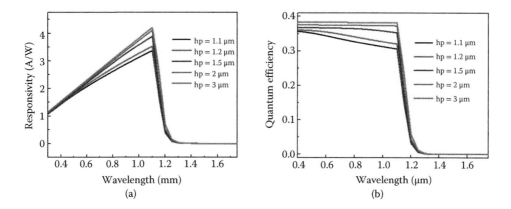

FIGURE 36.23 (a) Responsivity and (b) quantum efficiency for Si PN photodiode with the height of p-region (hp) under 1.1, 1.2, 1.5, 2, and 3 μm.

GaN/AlGaN PIN photodiode is shown in Figure 36.24. Figure 36.25 shows energy band as a function of vertical distance at equilibrium state, and the d_{abs} is set as 0.1 μm. Figure 36.26 describes responsivity versus the absorption layers (d_{abs}) with incident wavelengths of 0.30, 0.32, 0.34, and 0.36 μm, respectively, and electric field profile in the vertical direction with the absorption layers (d_{abs}) under 0.05, 0.2, 0.4, 0.6, and 0.8 μm.

The dark and light current characteristics with a bias voltage are shown in Figure 36.27. The absorption layer (d_{abs}) is set as 0.2 μm. Figure 36.28 shows responsivity and QE for GaN/AlGaN/GaN PIN photodiode with the absorption layer (d_{abs}) under 0.2, 0.3, 0.4, 0.6, and 1.1 μm. For clarity, the fitting parameters used in the simulation are listed in Table 36.6.

TABLE 36.5 List of the Key Parameters Used in the Simulation

Parameters	Units	Silicon
Bandgap	eV	1.124
Temperature (default)	K	300
Dielectric constant		11.7
p-type doping level (default)	cm^{-3}	10^{15}
n-type doping level (default)	cm^{-3}	10^{17}
p-type layer thickness (default)	μm	8
n-type layer thickness	μm	1
Lattice heat capacity	J/K·cm^3	1.63
Electron affinity	eV	4.05
Electron's lifetime	s	10^{-5}
Hole's lifetime	s	10^{-6}
Electron effective mass		1.09
Hole effective mass		1.15
Electron mobility	cm^2/V·s	1.42×10^3
Hole mobility	cm^2/V·s	4.70×10^2

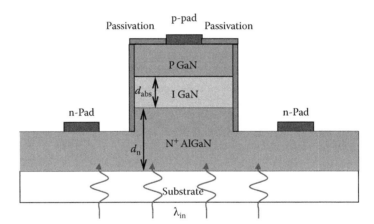

FIGURE 36.24 Schematic cross section of GaN/AlGaN/GaN PIN photodiode.

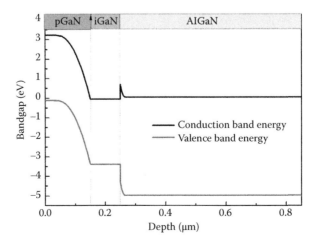

FIGURE 36.25 Energy band of GaN/AlGaN/GaN PIN photodiode.

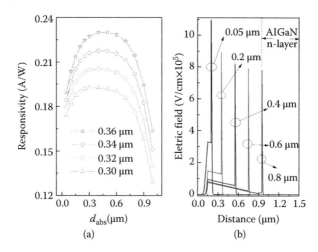

FIGURE 36.26 (a) Responsivity versus the absorption layers (d_{abs}) with incident wavelengths of 0.30, 0.32, 0.34, and 0.36 μm, respectively and (b) electric field profile in the vertical direction with the absorption layers (d_{abs}) changing from 0.05 to 0.8 μm.

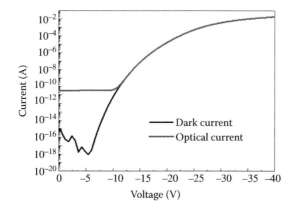

FIGURE 36.27 Dark and light current characteristics of GaN/AlGaN/GaN PIN photodiode.

36.5.5 Two-Color Heterojunction Photodiodes

36.5.5.1 Long-Wavelength/Mid-Wavelength Two-Color HgCdTe

There are two important parameters for the long-wavelength/mid-wavelength (LW/MW) two-color HgCdTe infrared focal plane array detector (Hu et al., 2009, 2010, 2011, 2012, 2013, 2014; Qiu et al., 2016): (1) absorption layer thickness and (2) opening dimensions of the groove (the length of the groove top, L_{top}; the length of the groove bottom, L_{bot}). In the simulation, the electrical cross-talk of both LW to MW ($C_{\lambda e\text{-}LW\text{-}to\text{-}MW}$) and MW to LW ($C_{\lambda e\text{-}MW\text{-}to\text{-}LW}$) contributes to the total spectral cross-talk dependent on thickness and band energy offset of the barrier layer (as shown in Figure 36.29). Figure 36.30 shows the spectral photoresponse for the HgCdTe two-color infrared detector with the proposed structure from the previous numerical simulations. Figure 36.31 shows the QE and cross-talk as a function of the thickness of the MW layer for the grooved n_1^+-p_1-P-p_2-n_2^+ HgCdTe two-color infrared detector. It is found that a minimum MW-to-LW cross-talk and maximum MW QE can be achieved with an MW layer of approximately 7 μm. To further optimize the groove structure, the QEs with different L_{top} and L_{bot} are calculated at cutoff wavelengths. It indicates that the MW QE at the MW cutoff wavelength of 4.8 μm increases with L_{bot} (no

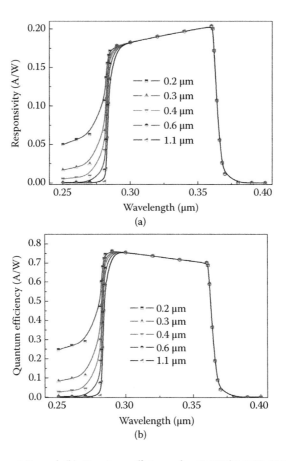

FIGURE 36.28 (a) Responsivity and (b) Quantum efficiency for GaN/AlGaN/GaN PIN photodiode with the absorption layer (Tabs) under 0.2,0.3,0.4,0.6, and 1.1 μm.

TABLE 36.6 List of the Key Parameters Used in the Simulation

Parameters	Units	$Al_{0.3}Ga_{0.7}N$	GaN
Band gap	eV	4.29	3.47
Dielectric constant	1	9.2	9.5
Effective conduction band density of states	cm^{-3}	3.09×10^{18}	2.65×10^{18}
Effective valence band density of states	cm^{-3}	1.03×10^{20}	2.5×10^{19}
Electron mobility	$cm^2/V{\cdot}s$	800	1010
Hole mobility	$cm^2/V{\cdot}s$	18	20
Saturation velocity	cm/s	1.5×10^{17}	1.5×10^{17}
Temperature exponent	1	0	0
Electron effective mass (relative)	1	0.25	0.222
Hole effective mass (relative)	1	2.2	1
Electron Auger coefficient	cm^6/s	2.1×10^{-29}	3.2×10^{-29}
Hole Auger coefficient	cm^6/s	6.5×10^{-30}	8.7×10^{-30}
Electron SRH lifetime	s	1×10^{-10}	1×10^{-9}
Hole SRH lifetime	s	1×10^{-10}	1×10^{-9}
Radiative recombination coefficient	cm^3/s	2×10^{-10}	2×10^{-10}

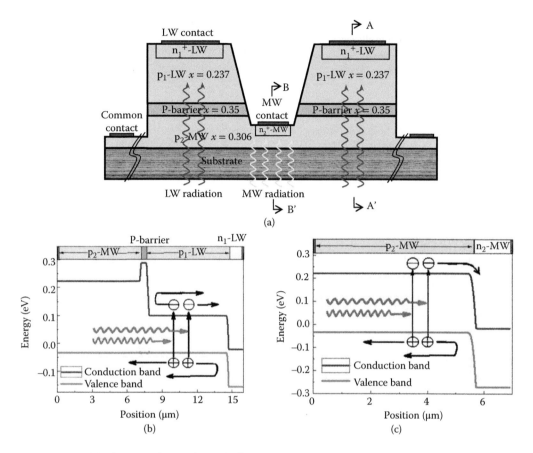

FIGURE 36.29 (a) Schematic of grooved HgCdTe two-color infrared detector, (b) equilibrium energy band diagram cut at A-A′, and (c) equilibrium energy band diagram cut at B-B′.

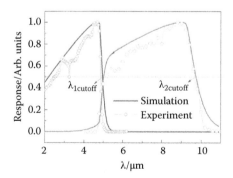

FIGURE 36.30 Experimental (dotted) and simulated (solid) spectral photoresponse of LW/MW HgCdTe two-color infrared detector with cutoff wavelengths of $\lambda_{1cutoff}$ = 4.8 μm and $\lambda_{2cutoff}$ = 9.7 μm for the LW and MW diodes respectively. A voltage of 0.01 V is used in simulation and experiment.

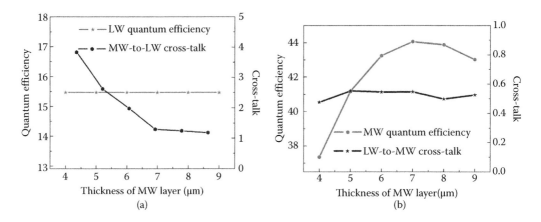

FIGURE 36.31 (a) LW quantum efficiency (QE) and MW-to-LW cross-talk and (b) MW QE and LW-to-MW cross-talk as a function of the thickness of MW layer for the infrared detector. The QE is calculated at cutoff wavelength.

TABLE 36.7   Material and Structural Parameters of the Two-color HgCdTe Photodiode at 77 K

Parameters	Value of LW (Units)	Value of MW (Units)	Value of Barrier (Units)
Cd molar fraction (x)	0.232	0.309	0.35
Doping density of n-region	1×10^{17} (cm^{-3})	1×10^{17} (cm^{-3})	–
Doping density of p-region	8×10^{15} (cm^{-3})	8×10^{15} (cm^{-3})	8×10^{15} (cm^{-3})
Thickness of n-region	1.2 (μm)	1.2 (μm)	–
Thickness of p-region	9 (μm)	5.8 (μm)	0.8 (μm)

dependence on L_{top}). When $L_{bot} \geq 14$ μm, the MW QE saturates at a maximum value as the photogenerated carriers with the diffuse length of ~10 μm are efficiently collected by the MW contact. The specific details of the detector are listed in Table 36.7.

36.6 Summary

With the rapid development of computational capabilities and algorithm, numerical simulation becomes a very important tool in both qualitative and quantitative studies of p-n junction photodiodes. In this chapter, physical equations, models, and parameters that play important roles in the numerical simulation have been introduced. First, for the construction of basic theoretical framework, the well-known Poisson and continuity equations are introduced. Meanwhile, the widespread used drift-diffusion model and thermodynamic model is also described. The results show the effects of the important device parameters such as structure size, doping concentration, and incident wavelength on the device performance such as dark current, photoresponsivity, QE, and so on. Second, we introduced and compared the characteristics of several common optical generation models in detail, which could achieve optoelectronic simulations of different precision and structural requirements. Third, the characteristic input parameters and widely used models in the simulations of typical p-n junction photodiodes are discussed and introduced. Different input parameters and models can be selected according to different optoelectronic materials, device structure, and microscopic physical mechanism. Finally, we introduced some key parameters and measures of diode simulation, and simulation examples of some typical p-n junction photodiodes are given. In addition, some important results are numerically extracted, and the parameter tables used in the simulations are provided.

References

Feit MD and Fleck JA (1978) Light propagation in graded-index optical fibers, *Applied Optics*, 17: 3990–3998.

Hu WD, Chen XS, Ye ZH, Chen YG, Yin F, Zhang B, and Lu W (2012) Polarity inversion and coupling of laser beam induced current in As-doped long-wavelength HgCdTe infrared detector pixel arrays: Experiment and simulation, *Applied Physics Letters*, 101: 181108.

Hu WD, Chen XS, Ye ZH, Feng AL, Yin F, Zhang B, Liao L, and Lu W (2013) Dependence of ion-implant-induced LBIC novel characteristic on excitation intensity for long-wavelength HgCdTe-based photovoltaic infrared detector pixel arrays, *IEEE Journal of Selected Topics in Quantum Electronics*, 19: 4100107.

Hu WD, Chen XS, Ye ZH, Lin C, Yin F, Zhang J, Li ZF, and Lu W (2010a) Accurate simulation of temperature-dependence of dark current in HgCdTe infrared detector assisted by analytical modeling, *Journal of Electronic Materials*, 39: 981–985.

Hu WD, Chen XS, Ye ZH, and Lu W (2010b) An improvement on short-wavelength photoresponse for heterostructure HgCdTe two-color infrared detector, *Semiconductor Science and Technology*, 25: 045028.

Hu WD, Chen XS, Ye ZH, and Lu W (2011) A hybrid surface passivation on HgCdTe long wave infrared detector with in-situ CdTe deposition and high-density hydrogen plasma modification, *Applied Physics Letters*, 99: 091101.

Hu WD, Chen XS, Yin F, Ye ZH, Lin C, Hu XN, Li ZF, and Lu W (2009) Numerical analysis of two-color HgCdTe infrared photovoltaic heterostructure detector, *Optical and Quantum Electronics*, 41: 699–704.

Hu WD, Chen XS, Yin F, Ye ZH, Lin C, Hu XN, Quan ZJ, Li ZF, and Lu W (2008) Simulation and design consideration of photoresponse for HgCdTe infrared photodiodes, *Optical and Quantum Electronics*, 40: 1255–1260.

Hu WD, Ye ZH, Liao L, Chen HL, Chen L, Ding RJ, He L, Chen XS, and Lu W (2014) A 128 × 128 long-wavelength/mid-wavelength two-color HgCdTe infrared focal plane array detector with ultra-low spectral crosstalk, *Optics Letters*, 39: 5130–5133.

Liang J, Hu WD, Ye ZH, Liao L, Li ZF, Chen XS, and Lu W (2014) Improved performance of HgCdTe infrared detector focal plane arrays by modulating light field based on photonic crystal structure, *Journal of Applied Physics*, 115:184504.

Qiu WC and Hu WD (2015) Laser beam induced current microscopy and photocurrent mapping for junction characterization of infrared photodetectors, *Science China-Physics Mechanics & Astronomy*, 58 (2): 1–13.

Qiu WC, Hu WD, Lin C, Chen XS, and Lu W (2016) Surface leakage current in 12.5 μm long-wavelength HgCdTe infrared photodiode arrays, *Optics Letters*, 41 (4): 828–831.

Rajkanan K, Singh R, and Shewchun J (1979) Absorption coefficient of silicon for solar cell calculations, *Solid-State Electronics*, 22: 793–795.

Shi M and Li ML (2014) *Semiconductor Devices Physics and Technology*, Soochow, China: Soochow University Press.

Wang XD, Chen XY, Hou LW, Wang BB, Xie W, and Pan M (2015) Role of n-type AlGaN layer in photoresponse mechanism for separate absorption and multiplication (SAM) GaN/AlGaN avalanche photodiode, *Optical and Quantum Electronics*, 47: 1357–1365.

Wang J, Chen XS, Hu WD, Wang L, Lu W, Xu FQ, Zhao J, Shi YL, and Ji RB (2011a) Amorphous HgCdTe infrared photoconductive detector with high detectivity above 200 K, *Applied Physics Letters*, 99: 113508.

Wang XD, Hu WD, Chen XS, Lu W, Tang HJ, Li T, and Gong HM (2008) Dark current simulation of InP/In0.53Ga0.47As/InP p-i-n photodiode, *Optical and Quantum Electronics*, 40: 1261–1266.

Wang XD, Hu WD, Chen XS, Xu JT, Li XY, and Lu W (2011b) Photoresponse study of visible blind GaN/AlGaN p-i-n ultraviolet photodetector, *Optical and Quantum Electronics*, 42: 755–764.

Wang XD, Hu WD, Pan M, Hou LW, Xie W, Xiu JT, Li XY, Chen XS, and Lu W (2014) Study of gain and photoresponse characteristics for back-illuminated separate absorption and multiplication GaN avalanche photodiodes, *Journal of Applied Physics*, 115:013103.

Yin F, Hu WD, Zhang B, Li ZF, Hu XN, Chen XS, and Lu W (2009) Simulation of laser beam induced current for HgCdTe photodiodes with leakage current, *Optical and Quantum Electronics*, 41: 805–810.

37

Quantum Well Infrared Photodetectors

37.1 Introduction

Quantum well infrared photodetectors (QWIPs) are based on optical transitions among quantized states in the conduction band of the quantum wells (QWs). They are widely used in long-wavelength infrared (LWIR) detection such as in gas sensing (Hinnrichs and Guptab, 2008) and in Earth observation from space (Jhabvala et al., 2011). The popularity of QWIPs is gained from the advantages in material quality and uniformity, detector sensitivity and stability, and array availability and cost. Another unique advantage of this technology is its predictable optoelectronic properties. In the absence of the extrinsic effects imposed by impurities and surfaces, the observed photocurrent and dark current of a QWIP are intrinsic to the detector structure and are well understood from fundamental physics. More specifically, its dark current is found to agree with the familiar thermionic emission (TE) process, while the photocurrent can be quantitatively predicted from first principle calculations on QW energy levels and electromagnetic (EM) field distributions inside the detector volume. With the known dark and photocurrents, the detection characteristics and performance of a QWIP array placed in the focal plane of an optical lens system can readily be predicted with very few system parameters. But because of the arbitrariness of the QW materials and the detector geometries, the associated physical equations can only be solved by numerical means. In this chapter, we describe the theoretical framework and numerical modeling of QWIPs.

The role of photocurrent I_p and dark current I_d in affecting the sensitivity of a QWIP focal plane array (FPA) has been analyzed (Choi et al., 1998). Here, it suffices to note that while the signal S of a detector is directly proportional to I_p, the noise N is directly proportional to $(I_p + I_d)^{1/2}$, which can be derived from the Poisson statistics of charge collection. The signal-to-noise ratio of a QWIP is therefore proportional to

$$\left(\frac{S}{N}\right)_{\text{detector}} \propto \frac{I_{\text{p}}}{\sqrt{I_{\text{p}} + I_{\text{d}}}} = \left(\frac{I_{\text{p}}}{1 + I_{\text{d}}/I_{\text{p}}}\right)^{1/2} = \left(\frac{I_{\text{p}}}{1 + 1/r}\right)^{1/2}, \tag{37.1}$$

where r is the photocurrent to dark current ratio ($I_{\text{p}}/I_{\text{d}}$). From Equation 37.1, it is obvious that the S/N ratio is large only when I_{p} and r are large. In this chapter, we focus on the modeling of I_{p} and I_{d} in QWIP design and optimization.

37.2 Theoretical Model of QWIPs

A QWIP uses a multiple quantum well (MQW) structure to detect light (Levine et al., 1987). This structure is realized by growing alternate material layers of different bandgaps on a suitable substrate. The MQW creates a series of energy subbands, $E_1, E_2, \ldots$, in each QW unit. The QWs are doped n-type with a doping density N_{D} so that there are free electrons in the ground subband E_1 with a Fermi energy E_{F}. The band structure of a typical QWIP made of $Al_xGa_{1-x}As/In_yGa_{1-y}As$ is shown in Figure 37.1. In Figure 37.1, x is the aluminum mole fraction, y is the indium mole fraction, W is the well width, and B is the barrier thickness. An inter-subband absorption brings an electron from E_1 to E_n where the conductivity across the layers is higher. A photocurrent is thus created.

37.2.1 Dark Current

The dark current I_{d} of a QWIP can be qualitatively separated into three main components. They are direct tunneling (DI) between adjacent ground states, thermally assisted tunneling (TAT) near the top of the barrier, and TE over the barrier. These processes are indicated in Figure 37.2.

The TAT current originates from the finite thermal spreading of in-plane electron energy in E_1. Upon a scattering process, the in-plane energy combines with the out-of-plane energy, with which the electron tunnels more efficiently through the QW barrier. Therefore, the barrier transmission depends on the total energy rather than its energy components. Although each current component differs in details, such as the densities of initial and final states, the major factor in determining their magnitudes is the impedance of the QW barrier. By adopting an energy-dependent tunneling probability $\gamma(E, V)$ for the barrier, the total

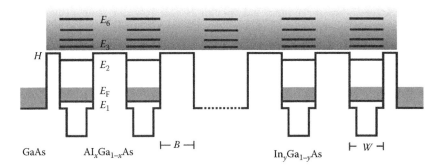

FIGURE 37.1 A typical energy diagram of an n-type QWIP along the z axis. E_1 to E_6 are the resonant QW states. The shaded region is the global conduction band.

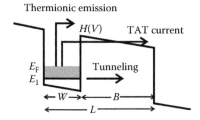

FIGURE 37.2 Different dark current transport processes in a QWIP. (TAT, thermally assisted tunneling.)

dark current can be obtained by summing over all electron energies (Levine et al., 1990):

$$J_d(V, T) = \int_{E_1}^{\infty} e\rho_d(E, T) v_d(V) dE,$$

$$= \int_{E_1}^{\infty} e\left[\frac{g_{2D}}{L} f(E)\gamma(E, V)\right] v_d(V) dE, \tag{37.2}$$

$$= \int_{E_1}^{\infty} e\frac{m^*}{\pi\hbar^2 L}\frac{\gamma(E, V) v_d(V)}{1 + \exp\left(\frac{E - E_F - E_1}{kT}\right)} dE.$$

In Equation 37.2, $J_d(V, T)$ is the total dark current density under a bias of V per period at an operating temperature T, E the electron total energy, $\rho_d(E)$ the three-dimensional (3D) electron density per electron-volt, $v_d(V)$ the drift velocity, g_{2D} the 2D density of states, L the QW period length, $f(E)$ the Fermi–Dirac distribution, $\gamma(E, V)$ the tunneling probability of an electron with E under V, and m^* the electron effective mass in the well. The value of γ can be obtained from the usual Wentzel–Kramers–Brillouin (WKB) approximation (Choi, 1997) of a trapezoidal barrier with a barrier height H, a barrier thickness B, and an electron effective mass m_b^* in the barrier. Specifically,

$$\gamma(E, V) = \exp\left\{\frac{-4B}{3e\hbar V}(2m_b^*)\left[(H(V) - E)^{3/2} - (H(V) - eV - E)^{3/2}\right]\right\} \quad \text{for } E < H(V) - eV,$$

$$= \exp\left\{\frac{-4B}{3e\hbar V}(2m_b^*)(H(V) - E)^{3/2}\right\} \quad \text{for } H(V) > E > H(V) - eV,$$

$$= 1 \text{ for } E > H(V), \tag{37.3}$$

where H is generally a function of V because there is a potential drop across the well as depicted in Figure 37.2, which depends on the W/L ratio. The drift velocity was experimentally found to be linearly proportional to the electric field (V/L) with a low-field mobility μ initially and then becomes saturated at a constant velocity v_{sat} when the field is high. Therefore, $v_d(V)$ can be represented as follows (Levine et al., 1990):

$$v_d(V) = \frac{\mu V/L}{\sqrt{1 + \left(\mu V/L v_{sat}\right)^2}}. \tag{37.4}$$

To fit the measured J_d, one can adjust H, μ, and v_{sat} in accounting for uncertainties in the material composition and quality.

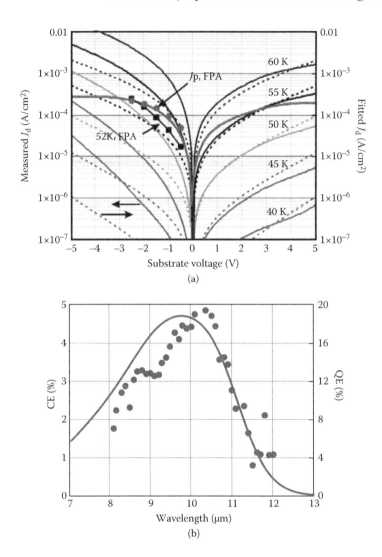

FIGURE 37.3   (a) The measured J_d (solid curves), fitted J_d (dashed curves), and measured J_p (thick solid curves) of a test detector are shown. The symbols are measured from a focal plane array (FPA). (b) The predicted and measured quantum efficiency (QE) and conversion efficiency (CE) of the FPA at −3 V based on angled sidewall optical coupling are shown.

Figure 37.3 shows the typical current–voltage (I–V) characteristics of a QWIP. The substrate voltage refers to the bias applied on the substrate while the top contact is grounded. They are measured from a QWIP with a cutoff wavelength $\lambda_c = 11.2$ μm (Choi et al., 2011). The QWIP is made up of 60 periods of 700 Å $Al_{0.19}Ga_{0.81}As$ and 50 Å GaAs. The wells are doped with N_D of 1.7×10^{18} cm^{-3}. By adopting a constant $H = 191$ meV, $\mu = 400$ cm^2/V · s, and $v_{sat} = 1 \times 10^7$ cm/s, the dark current I–V characteristics under positive bias can be explained in a large range of temperatures. In this example, H is independent of V because the W/L ratio ($= 50/750 = 0.067$) is very small. The adopted H is consistent with the expected band offset from the barrier composition, and the values of μ and v_{sat} are nearly the same for all experimental LWIR detectors. This consistency shows that Equation 37.2 is a satisfactory dark current model for QWIPs. According to the symmetrical band structure shown in Figure 37.1, the I–V characteristics of a QWIP should be symmetric with respect to bias polarity. The higher measured dark current shown in Figure 37.3

under the negative bias is attributed to dopant migration during the material growth, which lowers the barrier height in the barrier region immediately after the doping.

37.2.2 Photoconductive Gain

The photocurrent I_p of a QWIP can be expressed as $I_p = A\rho_p ev_d$, where A is the detector area, ρ_p the photoelectron density, and v_d the electron drift velocity. The parameter ρ_p can be further expressed as $G\tau/(A\ell)$, where G is the total photoelectron generation rate in the entire detector volume $A\ell$, ℓ the detector total thickness, and τ the average recombination lifetime of a photoelectron. After photoexcitation, a photoelectron only retains its energy in a period τ before it recombines into one of the QWs. Therefore, $G\tau/(A\ell)$ represents the time averaged photoelectron density. If we define the quantum efficiency (QE) η as the fraction of incident photon flux Φ being absorbed by the detector and converted into photoelectrons, the generation rate G will be $\eta\Phi$. Hence, I_p is given by

$$
\begin{aligned}
I_p &= A\rho_p ev_d, \\
&= A\left(\frac{G\tau}{A\ell}\right)e\left(\frac{\ell}{\tau_{tr}}\right) = \left(\frac{\eta\Phi\tau}{\ell}\right)e\left(\frac{\ell}{\tau_{tr}}\right) = \eta\Phi\left(\frac{\tau}{\tau_{tr}}\right)e, \\
&= \eta\Phi ge,
\end{aligned}
\tag{37.5}
$$

where τ_{tr} is the time taken by a photoelectron to travel across the entire detector thickness ℓ, and $g = \tau/\tau_{tr} = \tau v_d/\ell$ is defined as the photoconductive gain. Equation 37.5 indicates that the measured number of photoelectrons (I_p/e) is g times the number of photons absorbed ($\eta\Phi$), resulted in a gain. Since the lifetime τ of a photoelectron is usually shorter than the transit time τ_{tr} in a QWIP, g is usually less than unity. Its photocurrent will then be lower than that of a photodiode with a gain of unity for the same QE. The value of g is usually deduced from noise measurements and theoretically fitted to the same functional form of v_d with τ as a fitting parameter. Experimentally, we found $\tau = 25$ ps for most LWIR QWIPs.

37.2.3 Photocurrent

Expanding the last line of Equation 37.5, the photocurrent is given by

$$
\begin{aligned}
I_p(V, T_B) &= eg(V)\int_{\lambda_1}^{\lambda_2}\eta(V, \lambda)\Phi(T_B, \lambda)d\lambda, \\
&= eg(V)\int_{\lambda_1}^{\lambda_2}\eta(V, \lambda)\Omega AL(T_B, \lambda)d\lambda, \\
&= \frac{\pi}{4F^2 + 1}Aeg(V)\int_{\lambda_1}^{\lambda_2}\eta(V, \lambda)L(T_B, \lambda)d\lambda.
\end{aligned}
\tag{37.6}
$$

In Equation 37.6, Ω is the solid angle of the light cone sustained at the detector toward the lens. $L(T_B, \lambda)$ is the photon spectral radiance of the scene, which has a unit of number of photons arrived per unit time per unit detector area per unit solid angle per unit wavelength. Expressing Ω in terms of the usual specification for a lens system, the f-number F, which is the ratio between the focal length f and the lens diameter D, $\Omega = \pi\sin^2(\theta) = \pi(D/2)^2/[f^2 + (D/2)^2] = \pi/(4F^2 + 1)$, where θ is the half-cone angle.

The parameter $g(V)$ is the photoconductive gain at V, λ_1 and λ_2 are the lower and upper wavelengths encompassing the detector absorption spectrum, and $\eta(V, \lambda)$ is the detection QE spectrum. $\eta(V, \lambda)$ can be decomposed into $\eta_{abs}(\lambda)\gamma(V)$, where η_{abs} is the fraction of incident photons being absorbed and $\gamma(V)$ is the tunneling probability of the photoelectron out of the barrier. For the typical QWIP designs, the upper subbands are above the barrier and thus $\gamma(V)$ is close to unity. Therefore, we usually assume $\eta(V, \lambda) \approx \eta_{abs}(\lambda)$, and label $\eta_{abs}(\lambda)$ simply as $\eta(\lambda)$ unless the effect of γ is apparent.

37.2.4 Absorption Coefficient

The magnitude of the photocurrent I_p in Equation 37.6 depends on the QE spectrum $\eta(\lambda)$. In QWIPs, the optical transitions are generated from the transition dipole moment between two envelope eigenfunctions in the z-direction. The optical electric field E of the radiation must be oriented in this direction to drive the transitions. When the optical field is polarized in the x–y plane, the dipole moment is zero due to conservation of momentum, and therefore, there is no absorption. Therefore, the optical absorption of the material depends on the direction of light propagation, and the absorption coefficient α is anisotropic. The parameter α is defined as $(-1/\Phi)(d\Phi/dx)$, where Φ is the photon flux and x is the traveling distance. In general, when the light is traveling at an angle θ relative to the normal of the layer surface, only the vertical electric component $E_z = E \cdot \sin(\theta)$ in the transverse magnetic (TM) mode is able to initiate the transitions. The transverse electric mode, on the other hand, is totally uncoupled. Therefore, for an unpolarized light, only half of its intensity is being absorbed, and α of the remaining half varies as $\sin^2\theta$. When θ is zero under the usual detection orientation, i.e., under normal incidence, α will be zero unless there is an optical coupling structure to convert the incident electric polarization from horizontal to vertical. Overall, $\eta(\lambda)$ of a QWIP depends on both the material intrinsic absorption properties and the detector physical structure.

For a parallel propagating light with vertical polarization, the absorption coefficient $\alpha(\hbar\omega)$ depends on the oscillator strength between the ground state and one of the upper states and the number of free carriers in the ground state. Quantitatively, $\alpha(\hbar\omega)$ is given by (Choi, 1997)

$$\alpha(\hbar\omega) = \frac{N_s}{L} \frac{\pi e^2 \hbar}{2m^* \sqrt{\varepsilon_r}\, \varepsilon_0 c} \sum_{n=2}^{\infty} f_n \rho_n(\hbar\omega + E_1), \tag{37.7}$$

where $N_s = N_D W$ is the 2D doping density per QW period, L the length of a QW period, m^* the effective electron mass of individual layers, ε_r the real part of the dielectric constant, f_n the oscillator strength between E_1 and E_n, and ρ_n the normalized density of states due to the presence of line broadening mechanisms. The oscillator strength f_n is given by

$$f_n = \frac{2\hbar^2}{m^* \left(E_n - E_1\right)} \left| \langle \psi_n \left| \frac{\partial}{\partial z} \right| \psi_1 \rangle \right|^2, \tag{37.8}$$

where ψ_n is the eigenfunction of E_n. The overall absorption lineshape depends mostly on the energy distribution of E_n and the respective f_n with a weaker dependence on the individual ρ_n. We assume a Gaussian shape for ρ_n, i.e.,

$$\rho_n(E) = \frac{1}{\sqrt{2\pi}\sigma_n} \exp\left(-\frac{\left(E - E_n\right)^2}{2\sigma_n^2}\right), \tag{37.9}$$

where σ_n is the line broadening parameter. In the next section, we show modeling procedures for $\alpha(\hbar\omega)$.

37.3 Material Absorption Modeling

The eigenfunction ψ_n and eigenvalue E_n of an MQW can be obtained from the transfer matrix method (TMM) (Choi et al., 2002). In this method, ψ_j in the jth material layer is written as two counter-propagating plane wave states:

$$\psi_j = A_j\,e^{i\,k_j\,z} + B_j\,e^{-i\,k_j\,z}, \tag{37.10}$$

where A_j and B_j are two constants to be determined, and k_j is the wavevector given by

$$k_j(E) = \frac{\sqrt{2\,m_j^*(E)}}{\hbar}\sqrt{E - \Delta E_j}, \tag{37.11}$$

where $m_j^*(E)$ is the energy-dependent effective mass, E and ΔE_j are the electron energy and the conduction band offset measured from the GaAs conduction band edge, respectively. The constants A_j and B_j relate to A_{j+1} and B_{j+1} of the adjacent $(j+1)$th layer through the boundary conditions that both ψ and $(1/m^*)d\psi/dz$ are continuous across the interface:

$$
\begin{bmatrix} A_j \\ B_j \end{bmatrix} = \frac{1}{2}\begin{bmatrix} (1+\gamma_{j,j+1})\,e^{i\,(k_{j+1}-k_j)\,d_{j,j+1}} & (1-\gamma_{j,j+1})\,e^{-i\,(k_{j+1}+k_j)\,d_{j,j+1}} \\ (1-\gamma_{j,j+1})\,e^{i\,(k_{j+1}+k_j)\,d_{j,j+1}} & (1+\gamma_{j,j+1})\,e^{-i\,(k_{j+1}-k_j)\,d_{j,j+1}} \end{bmatrix} \begin{bmatrix} A_{j+1} \\ B_{j+1} \end{bmatrix}
$$

$$
= \frac{1}{2}\,M_{j,j+1}\begin{bmatrix} A_{j+1} \\ B_{j+1} \end{bmatrix}, \tag{37.12}
$$

where $\gamma_{j,j+1} = (m_j^*k_{j+1})/(m_{j+1}^*k_j)$ and $d_{j,j+1}$ is the spatial coordinate of the interface. The energy level structure of a multilayer material system can be obtained from the maxima of the global transmission coefficient T_G of a plane wave through the material. When the incident energy of the wave coincides with one of E_n's, strong resonant transmission occurs and T_G attains a local maximum. Assuming the plane wave incident from the left, the coefficients A_1 and B_1 for the first material layer on the left and the coefficients A_p and B_p of the last pth layer are connected by

$$\begin{bmatrix} A_1 \\ B_1 \end{bmatrix} = \frac{1}{2^{p-1}}\,M_{1,p}\begin{bmatrix} A_p \\ B_p \end{bmatrix}, \tag{37.13}$$

where $M_{1,p} = M_{1,2}M_{2,3}\ldots M_{p-1,p}$, and $M_{j,j+1}$ is defined in Equation 37.12. By assuming that there is no wave traveling to the left in the last period, one can set $A_p = 1$ and $B_p = 0$, with which all the coefficients in the layers are determined. The value of T_G at a given energy E is then equal to

$$T_G(E) = \frac{1}{|A_1(E)|^2}\frac{v_p(E)}{v_1(E)} = \frac{2^{2p-2}}{|a_{11}(E)|^2}\frac{m_1^*(E)}{m_p^*(E)}\frac{k_p(E)}{k_1(E)}, \tag{37.14}$$

where v is the electron group velocity and a_{11} the first diagonal element of $M_{1,p}$. The local maxima in $T_G(E)$ determine the locations of E_n. In general, these local maxima are less than unity unless the material structure is symmetrical. Calculating $A_n(E)$ and $B_n(E)$ at $E = E_n$ will yield ψ_n.

37.3.1 Material Parameters

The TMM calculation can be handled efficiently by a personal computer. The material parameters required for the InGaAs/AlGaAs material system had been collected and summarized (Shi and Goldys, 1999).

To begin with, the unstrained bandgap of the *i*th layer in the unit of meV is

$$E_{g,j}^u = 1424 + 1186\, x_j + 370\, x_j^2 - 1499\, y_j + 429\, y_j^2, \tag{37.15}$$

where x_j and y_j are the Al and In molar ratios in the *j*th layer, respectively. (In the present calculation, x and y both cannot be nonzero in a given layer, i.e., AlInGaAs quaternary compound is not considered.) Since the material layers are under strain, the values of bandgaps are modified. Assuming the strain is accommodated exclusively in the InGaAs well layers, which are usually much thinner than the GaAs/AlGaAs layers, the modified bandgap for $y_j > 0$ is

$$E_{g,j}^s = E_{g,j}^u + 2a\left(1 - \frac{c_{12}}{c_{11}}\right)\varepsilon_{11}, \tag{37.16}$$

where $a = -8700 + 2700y_j$ meV is the deformation potential, $c_{12}/c_{11} = (5.32 - 0.79y_j)/(11.81 - 3.48y_j)$ is the ratio of the elastic constants, and $\varepsilon_{11} = -0.405y_j/(5.6533 - 0.405y_j)$ is the hydrostatic component of the strain induced by the adjacent GaAs/AlGaAs material layers. Therefore, in the present assumption, the bandgap of InGaAs is widened while that of GaAs and AlGaAs are unchanged. The band offset ΔE_j is taken as

$$\begin{aligned}\Delta E_j &= \beta_x\left(E_{g,j}^u - 1424\right) \quad \text{if} \quad x_j > 0 \\ &= \beta_y\left(E_{g,j}^s - 1424\right) \quad \text{if} \quad y_j > 0,\end{aligned} \tag{37.17}$$

where β_x is the band offset parameter for the GaAs/AlGaAs interface and β_y is for the InGaAs/GaAs interface. From fitting detectors with different wavelengths, we notice that β_x increases from 0.60 to 0.77 when x decreases from 0.4 to 0.16, and β_y is 0.63 when $y < 0.34$. The effective mass at the conduction band edge is given by

$$m_j^* = 0.0665\left(1 + 1.256\, x_j - 0.579\, y_j\right). \tag{37.18}$$

Taking the band nonparabolicity into account, the effective mass $m_j^*(E)$ can be expressed in the form of

$$m_j^*(E) = m_j^*\left(1 + \frac{E - \Delta E_j}{E_{g,j}}\right), \tag{37.19}$$

where $E_{g,j}$ is the strain modified bandgap. With all these material parameters, the input parameters in Equation 37.14 are known. Since m^* varies from layer to layer, strictly speaking, Equations 37.7 and 37.8 need to be evaluated individually in each layer. For simplicity, we assume m^* to be the effective mass at the GaAs conduction edge since optical transitions mostly occur in the GaAs layers.

37.3.2 Numerical Examples

In this section, we present some examples and compare them to experiments. To measure the α lineshape, infrared (IR) light is incident on a 45° polished facet at the edge of a substrate (referred to as the edge coupling method), so that the optical electric field pointing at the *z*-direction is a factor of $1/\sqrt{2}$ of that of the TM component. The measured quantity is the responsivity (R) or the conversion efficiency ($CE \equiv \eta g$). The values of R and CE are related by

$$R(\lambda) = \frac{I_p}{P(\lambda)} = \frac{e\eta g\Phi(\lambda)}{\frac{hc}{\lambda}\Phi(\lambda)} = \frac{\lambda}{\frac{hc}{e}}\eta g = \frac{\lambda}{1.24 \times 10^{-6}}CE(\lambda) \quad \left[\frac{\text{A}}{\text{W}}\right], \tag{37.20}$$

where $P(\lambda)$ is the incident optical power. Since in the weak absorption limit, η is directly proportional to α (see Section 37.4.4 for details on this coupling scheme), we can compare either the experimental $CE(\lambda)$ lineshape to the theoretical $\alpha(\lambda)$ or $R(\lambda)$ lineshape to $\lambda\alpha(\lambda)$. In Figure 37.4, we show the calculated $f_n(\lambda)$ of a typical QWIP, labeled as Det. A, which is made of 21 periods of 500 Å $Al_{0.27}Ga_{0.73}As$/46 Å GaAs QWs. The TMM calculation was performed on a single QW unit. The result will be the same if the calculation is performed on the entire MQW when E_1 of individual wells are not coupled (Choi et al., 2002), and in this case, the total f is independent of B.

Substituting the f spectrum into Equation 37.7, $\alpha(\lambda)$ can be obtained. Figure 37.5 shows the calculated $\alpha(\lambda)$ for Det. A along with three other GaAs/AlGaAs detectors having different cutoff wavelength λ_c and doping density N_D. Figure 37.5 also shows the measured CE lineshape of each detector in arbitrary units. To fit the measurements, the values of β_x needs to vary from 0.665 to 0.715 for the nominal x from 0.27

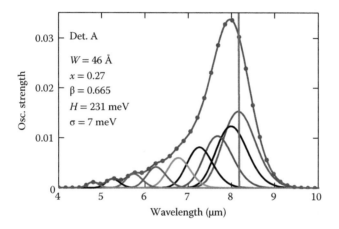

FIGURE 37.4 The calculated oscillator strengths f_n of a QWIP with $W = 46$ Å, $B = 500$ Å, $x = 0.27$, and $\sigma = 7$ meV. The curve with circles is the total oscillator strength. The vertical line shows λ at which the final state energy $= H - E_1$.

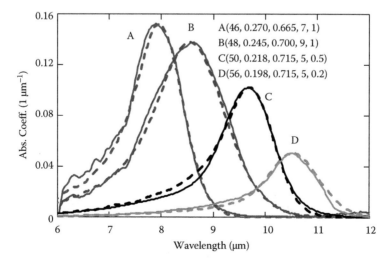

FIGURE 37.5 The measured CE lineshapes (solid curves) from four materials. The dashed curves are the calculated α assuming $(W, x, \beta_x, \sigma, N_D)$ shown in the figure, where N_D is in 10^{18} cm^{-3}. The barrier thickness is 500 Å for all detectors except 600 Å for Det. D.

to 0.198. The broadening parameter σ ranges from 5 to 9 meV for the first upper state, and it increases by about 0.5 meV for each successive upper state. With the two fitting parameters, β_x and σ, the intrinsic absorption lineshape can be satisfactorily explained.

The bandwidths shown in Figure 37.5 are rather limited. To broaden the bandwidth, one approach is to use superlattices (SLs) to create miniband-to-miniband (M–M) transitions. When the barrier thickness B is small, E_n states among different wells are coupled to form minibands M_n. Since the degenerate E_1 states develop into a range of occupied states in M_1, the absorption can cover a wider spectral range. This bandwidth can further be expanded if the QWs in the SL have different well widths (Choi et al., 2002).

To cover the entire 8- to 14-μm LWIR window, we designed a structure (Li et al., 2006) that contains seven SLs, and they are separated by 600-Å $Al_{0.17}Ga_{0.83}As$ barriers. Each SL contains seven 30-Å $Al_{0.23}Ga_{0.77}As$ barriers and four 70-Å and four 75-Å wells alternately placed together. The doping is 4×10^{17} cm^{-3} in the wells while the barriers are undoped. Since the eight miniband ground states within M_1 have different energies, their electron populations are also different. In this example, the first state in M_1 is at 39.4 meV and the last state is at 55.7 meV, while the common Fermi energy for the present doping is at 57.8 meV. Therefore, the electron population in the first state is 8.8 times larger than the last state, resulting in a significant modification of the lineshape from that of MQWs. Calculating Equation 37.7 with a different effective 3D doping density (N_s/L) for each miniband ground state according to the different Fermi levels, the spectral lineshape of the SL can be obtained. Figure 37.6 shows the calculated lineshape for the present SL. It agrees well with the measurement using a constant σ of 7 meV. The spectral responsivity is seen to cover the entire LWIR window. Following this material design approach, a QWIP with any cut-on and cutoff wavelengths can be designed.

By using InGaAs/AlGaAs QWs to increase the combined barrier height, a mid-wavelength IR QWIP can be developed (Choi et al., 2002). The illustrated structure contains eight $In_{0.34}Ga_{0.66}As/Al_{0.4}Ga_{0.6}As$ SL units separated by 500-Å $Al_{0.4}Ga_{0.6}As$ blocking barriers. Each unit consists of five wells and four barriers. The wells are made of 5-Å GaAs/25 Å InGaAs/5Å GaAs layers, and the barriers are made of 30-Å AlGaAs layers. There are top and bottom GaAs contact layers. All SL layers are doped to 2×10^{18} cm^{-3}. The modeled responsivity curve matches the experiment if $\sigma = 27.5$ meV as shown in Figure 37.7. This much larger σ is usually observed in the MWIR regime.

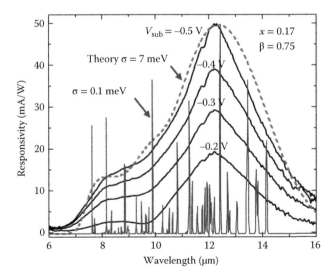

FIGURE 37.6 The measured responsivity from an SL structure with alternate well widths (solid curves). The fitted spectral lineshape of $\lambda\alpha(\lambda)$ with $\sigma = 7$ meV is shown as the dashed curve. The curve with $\sigma = 0.1$ meV is to reveal the spectral distribution of individual M-M transitions.

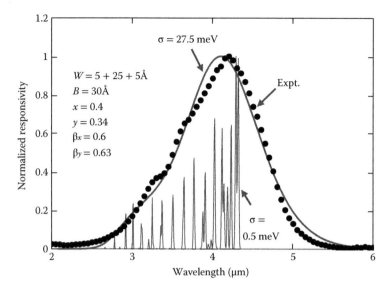

FIGURE 37.7 The measured normalized responsivity (circles) and the calculated $\lambda\alpha(\lambda)$ with $\sigma = 27.5$ meV (solid curve) for the combined transitions in the SL with blocking barriers. The curve with $\sigma = 0.5$ meV is to reveal the spectral distribution of individual M-M transitions.

37.4 Detector QE Modeling

Although $\alpha(\lambda)$ for parallel propagating light can be modeled accurately, the detector QE under normal incidence is still far from definite because it also depends on the optical coupling scheme. The search for an optimum coupling structure for a given application occupies the bulk of QWIP research to date. To yield the theoretical QE, one has to know the EM field distribution inside the detector volume in the presence of a polarization-dependent absorption coefficient. This task can only be solved by employing 3D numerical EM modeling. Since this capability was not readily available in the past, the common QWIP structures are not optimized.

With the advent of computer hardware and software technologies, one can now perform realistic and sophisticated EM modeling using commercial software on a personal computer. These advances have greatly facilitated the QWIP development. There are many numerical methods in solving EM problems. The more common ones are finite element method (FEM), finite-difference time-domain method, finite boundary method, and modal transmission-line (MTL) method, etc. In this chapter, we describe using FEM to model the theoretical QE of QWIPs.

FEM solves Maxwell's equation,

$$\nabla \times \left(\mu_r^{-1} \nabla \times E\right) - k_0^2 \varepsilon_{rc} E = 0, \tag{37.21}$$

via trial basis functions under a user-defined boundary condition. In Equation 37.21, μ_r and $\varepsilon_{rc} = \varepsilon_r - j\varepsilon_i$ are the relative permeability and relative permittivity, respectively, and $k_0 = 2\pi/\lambda$ is the free-space wavevector. Through variational principles, FEM transforms Equation 37.21 into a large set of linear equations, which can be solved by matrix multiplications (Sadiku, 2001). To yield a valid solution, it is important to present a correct boundary condition in the model.

The present modeling is performed in the RF Module of a commercial EM solver, COMSOL Multiphysics version 3.5 (COMSOL AB, Nov 2008). The modeling procedures involve selecting the EM analysis mode, building the 2D or 3D detector geometry, defining constants, variables, and functions, inputting subdomain properties, selecting appropriate boundary conditions, building mesh structures, setting solver parameters, performing computation, and using postprocessing to yield the required information.

To model the QE of a QWIP, we note that η can be expressed as (Choi et al., 2012, 2013a)

$$\eta(\lambda) = \frac{1}{P_0} \int_V dI(\mathbf{r}) = \frac{1}{P_0} \int_V \alpha(\lambda) I(\mathbf{r}, \lambda) d^3 r$$

$$= \frac{\alpha(\lambda)}{A \frac{c\varepsilon_0}{2} E_0^2} \int_V \frac{n(\lambda) c\varepsilon_0}{2} \left| E_z(\mathbf{r}, \lambda) \right|^2 d^3 r = \frac{n(\lambda)\alpha(\lambda)}{A E_0^2} \int_V \left| E_z(\mathbf{r}) \right|^2 d^3 r, \qquad (37.22)$$

where P_0 is the incident optical power, V the detector active volume, I the optical intensity associated with E_z, α the absorption coefficient, A the detector area, and n the material refractive index. Equation 37.22 states that the QE of a QWIP can be calculated from the volume integral of $|E_z|^2$ in the presence of a finite α. If E_z in Equation 37.22 is replaced by the total E, it will be applicable to detectors with isotropic absorption as well. Since E_0 and E_z are linearly proportional to each other, E_0 can be set arbitrarily. Besides E_0, the only free input parameter in Equation 37.22 is the wavelength-dependent $\alpha(\lambda)$ (if $n(\lambda)$ of the material is known accurately). For a known $\alpha(\lambda)$ from Equation 37.7, there will be no free parameters, and the value of $\eta(\lambda)$ for a given combination of material and geometry can be uniquely and unambiguously determined.

37.4.1 Modeling Validation and Phase Coherence

Since numerical EM modeling involves many steps, assumptions, and idealizations of materials and structures, it is critical to validate the model through theoretical and experimental means. One important assumption in the model is the complete phase coherence of radiation. While it is true for radio waves, it is generally not the case in the IR wavelengths and the rationale for EM modeling needs further explanation. If the optical field distribution in a detector indeed depends on the degree of coherence of light, the present modeling clearly will not be relevant. It follows that a QWIP will not have a definite QE, and its value depends on the light source characteristics. For a blackbody source, it is well known that the radiation is incoherent and its coherence length ℓ_c is only about $0.8\lambda_{max}$, where λ_{max} is the free-space wavelength at the intensity maximum (Donges, 1998). At a blackbody temperature $T_B = 300$ K, λ_{max} is 10 µm, and ℓ_c is 8 µm in free space and 2.7 µm in GaAs. Such a short ℓ_c would suggest phase coherence exists only in a very small detector volume and coherent EM modeling would not be meaningful.

The abovementioned picture is certainly true when considering all of the radiations from the blackbody. Consistent with this picture, it also follows that the modeled coherent QE at λ_{max} will not be valid for all the blackbody radiations, which is also certainly the case. Nonetheless, when all the incoherent wave packets from the blackbody are Fourier transformed into the plane wave components, which happen to assume the Planck distribution, each individual wavelength component is in fact phase coherent. (By the definition of a plane wave with a definite λ, its coherence length is infinite.) Since the detector absorbs one photon with a definite λ at a time, not a wave packet with a broad spectrum at a time, the detector is interacting with each Fourier component individually, and because of this specific absorption feature, the spectrally dispersed radiation appears to be completely coherent to the detector at the individual wavelength component basis. Following this reasoning, there will only be one QE value for each λ, and it is the value evaluated under phase coherent condition. The modeled QE is thus an intrinsic property of the detector independent of the light source. Consequently, the detector maintains the same optical properties when detecting objects with different coherence lengths caused by, for example, different scene temperatures.

For experimental verification, when one of the Fourier components is separated out from a blackbody by an instrument, either using a grating or a filter, its coherence length $\ell_c = \lambda^2 / \Delta\lambda$ becomes finite, where $\Delta\lambda$ is bandwidth of the instrument (Fowles, 1975). It is more convenient to use a light source with a long ℓ_c, i.e., a small $\Delta\lambda$, so that no QE convolution is need. However, even if $\Delta\lambda$ is substantial in comparison with the variation of modeled $\eta(\lambda)$, one can still make comparison by characterizing the spectral power density $P(\lambda)$ of the source around the nominal incident wavelength λ_0. The weighted theoretical $\eta(\lambda_0) = \int \eta(\lambda - \lambda_0)P(\lambda - \lambda_0)d\lambda / \int P(\lambda - \lambda_0)d\lambda$ can then be compared to experiment.

It is interesting to mention that when $\Delta\lambda$ of the instrument is broader than the detector absorption bandwidth Γ, the detector can only sense part of each wave packet emitted by the instrument, with which the coherence length appears to be longer at λ^2/Γ instead of $\lambda^2/\Delta\lambda$ for the original wave packet. Therefore, a detector with a given Γ will not be able to perceive the true phase coherence of a source if the source ℓ_c is shorter than λ^2/Γ. In the following, we present examples of modeling validation.

37.4.2 Planar Solar Cell

A planar GaAs solar cell provides a simple test for the modeling since it lacks any $x-y$ spatial variations. Its material structure includes an active GaAs absorbing layer, a front antireflection coating, and a back metal reflector as shown in Figure 37.8 (Grandidier et al., 2012). With the known complex refractive indices of GaAs, and assuming 100% transmission for the antireflection (AR)-coating and 100% reflection for the reflector, a simple classical model for QE using ray-tracing technique is represented by the dash curve in Figure 37.9. On the other hand, using the complex refractive indices from all the constituent materials, the 2D modeling based on Equation 37.22 with isotropic absorption yields the solid curve. It coincides

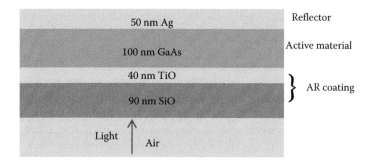

FIGURE 37.8 The modeled layer structure of a planar solar cell.

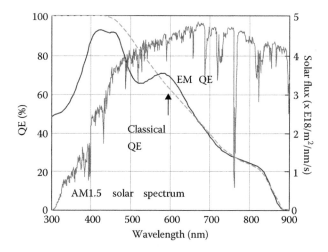

FIGURE 37.9 The quantum efficiency (QE) obtained by raytracing (dashed curve) and FEM (solid curve). The figure also shows the solar flux under 1-sun. (EM, electromagnetic.)

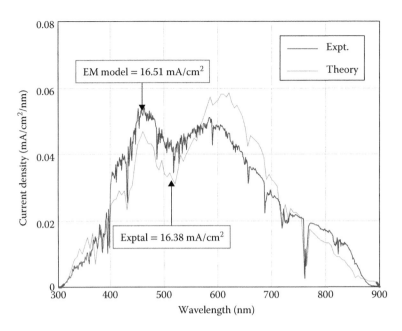

FIGURE 37.10 The measured and modeled photocurrent density under 1-sun without fitting parameters. (EM, electromagnetic.)

with the classical solution when the coatings in the modeling are effective. At the same time, the predicted photocurrent spectrum in Figure 37.10 is very similar to the measured spectrum (Grandidier et al., 2012) without any fitting parameters. In particular, the model accurately predicts two photocurrent peaks at 430 and 600 nm, which are absent in the simplified classical model. Furthermore, the model predicts the integrated photocurrent to be 16.51 mA/cm^2 under 1-sun condition, which is in excellent agreement with the measured 16.38 mA/cm^2. Therefore, without knowing the actual material properties in the experimental study, the agreement between theory and experiment is satisfactory for this device structure.

37.4.3 InAs/GaInSb IR Photodiode

When the detector material is fabricated into detector pixels, it acquires spatial variations. In this example, we show an IR photodiode in Figure 37.11 with a linear dimension of 22 μm in a 25-μm pitch array. It consists of a 2-μm thick p-InAs/Ga$_{0.75}$In$_{0.25}$Sb strained SL (p-SLS2) layer and a 0.5-μm thick wider bandgap n-SLS1 layer to form a photodiode. The photodiode is sandwiched between a 0.5-μm thick wider bandgap p-contact layer (p-SLS3) and a 2-μm thick n-InAs$_{0.9}$Sb$_{0.1}$ contact layer. On top of the p-contact, there is a 12-μm wide gold layer for ohmic contact. Light is incident from the n-contact layer. The detector is assumed to be surrounded by a semi-infinite epoxy material with $n = 1.5$ and with negligible IR absorption.

Since there is no AR-coating in this detector, Fabry–Pérot (FP) oscillations arise from optical interference between the top and bottom material interfaces. Because of the inhomogeneous phase change along the top surface, the FP oscillations are irregular. In Figure 37.12, we show the theoretical α spectrum of the p-SLS2 layer and the deduced α spectrum of n-InAs$_{0.9}$Sb$_{0.1}$ contact layer from open literature. The latter spectrum is obtained by shifting the experimental n-InAs spectrum by $\lambda = 0.7$ μm in accounting for the bandgap reduction (Miles et al., 1990). We further assume α of the n-SLS1 and p-SLS3 to be the same as that of p-SLS2 but at the shorter wavelengths (Dixon et al, 1961). The $n(\lambda)$ value in Equation 37.22 is taken to be a constant of 3.5, which is the average between 3.4 for InAs and 3.7 for GaSb at 10 μm. All IR absorption is assumed to be isotropic. Since the absorption of all the layers except the p-SLS2 layer will be dissipated as heat and no photocurrents will be generated in these layers, their absorption reduces the QE of the p-SLS2 layer in IR detection.

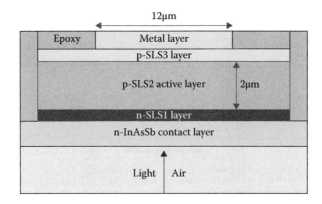

FIGURE 37.11 The material layer structure of a SLS photodiode. The pixel pitch is 25 μm.

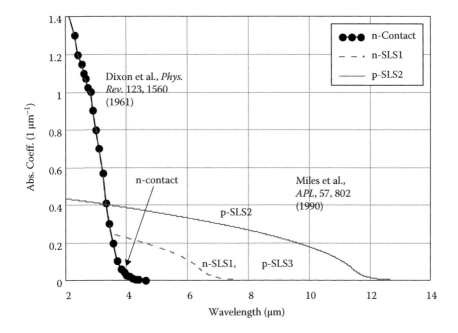

FIGURE 37.12 The theoretical and experimental α of individual layers obtained from open literature.

With these material and detector structures, the detector QE spectrum is modeled and the result is shown in Figure 37.13. Without adjusting any input parameters, the theoretical prediction matches satisfactorily with the experiment both in magnitude and in spectral lineshape in the entire spectral range. This example shows the validity of the modeling in the presence of spatial variations.

37.4.4 Edge Coupling for QWIPs

In Section 37.3.2, we assumed optical coupling efficiency using a 45° facet to be λ-independent in order to employ this coupling scheme for material characterization. It is worth checking its validity (Choi et al., 2012). Figure 37.14 shows the detector geometry and the calculated E fields with $E_0 = 377$ V/m. The detector has a 3-μm thick active QWIP layer, a 1-μm top contact layer, and a 1-μm gold layer. The detector is 100-μm wide. The value of α is assumed to be 0.15 μm^{-1}, and it is λ independent.

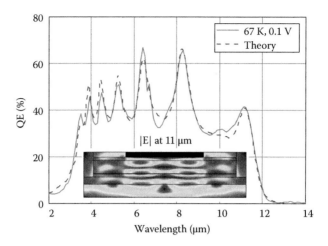

FIGURE 37.13 The calculated and the measured quantum efficiency (QE) of the p-SLS2 layer. The experimental data are provided by M. Jhabvala at NASA Goddard Space Flight Center. The insert shows the color plot of the $|E|$ distribution at $\lambda = 11$ μm.

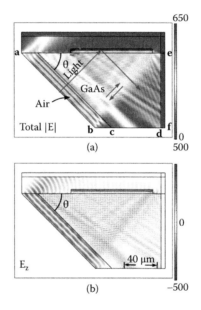

FIGURE 37.14 (a) The detector geometry at wedge angle θ of $45°$. The arrows show the typical light paths. (b) The corresponding E_z is shown.

The 2D modeling with anisotropic absorption shows that QE at $\theta = 45°$ as well as other angles are nearly independent of λ as shown in Figure 37.15, justifying the experimental approach. The small oscillations are due to FP oscillations established in one of the detector corners shown in Figure 37.14a, but they will not be noticeable in the presence of edge roughness. In Figure 37.15, we also show the classical QE calculated from

$$\eta_c(\theta) = \frac{4n}{(1+n)^2} \frac{\cos\theta}{2} \left[1 - \exp\left(-\alpha \sin^2\theta \cdot \frac{2\ell}{\cos\theta} \right) \right], \tag{37.23}$$

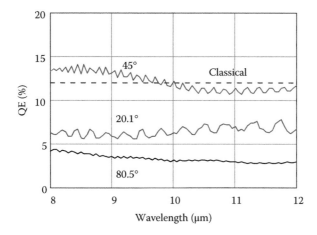

FIGURE 37.15 The calculated quantum efficiencies (QEs) at three different wedge angles. The dashed lines show the classical value at $\theta = 45°$.

where $\ell = 3$ μm is the active layer thickness and $\theta = 45°$ is the angle of the polished wedge, which is also the angle of incidence measured from the detector normal. In Equation 37.23, $4n/(1 + n)^2$ is the transmission coefficient of the GaAs substrate. The term $\cos(\theta)/2$ accounts for the smaller projected detector area in the direction of light and the fact that only the TM mode is coupled. The factor $\alpha \cdot \sin^2(\theta)$ is the absorption coefficient under oblique incidence, and $2\ell/\cos(\theta)$ is the total optical path length inside the detector accounting for the internal reflection at the top of the detector. In this classical model, if α is constant respect to the wavelength, η_c will also be wavelength independent. In Figure 37.15, η_c is shown to agree with the mean value from EM modeling. Therefore, EM modeling indeed supports the edge coupling approach in material characterization.

37.4.5 Quantum Grid Infrared Photodetector

Equipped with a quantitative model, we can revisit some of the QWIP coupling structures used in the past. Figure 37.16a shows the cross section of a structure known as the quantum grid infrared photodetector (QGIP) (Choi et al., 2004). It uses a metal layer on top of long and linear grid lines to scatter light incident from the bottom. The width w of the grid line acts as a half-wave dipole antenna and its dimension determines the detection wavelength. This detector geometry can also be modeled by 2D MTL method (Tamir and Zhang, 1996). Figure 37.16b compares the MTL solution and the present FEM solution for the same QGIP structure, assuming the same constant ε_i of $9.722 + i$. The close agreement in the QE spectrum between these two methods and the agreement between the methods and the experiment (Choi et al., 2004; not shown here) confirm the validity of both EM models. The E_z field distribution under the individual metal grid lines shows the expected scattering pattern from a dipole antenna.

37.4.6 Plasmonic-Enhanced QWIP

The structures described earlier were modeled in 2D. In this example, we perform 3D modeling as the structure cannot be reduced to 2D. The present structure, shown in Figure 37.17, uses surface plasmon to couple incident light into a QWIP (Wu et al., 2010). Surface plasmon is the resonant oscillation of conduction electrons at the interface between a metal and a semiconductor under optical excitation. The detector in this case contains a 400-Å thick gold film deposited on a 0.528-μm-thick InGaAs/InP active material.

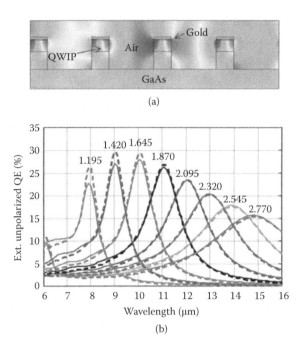

(a)

(b)

FIGURE 37.16 (a) The QGIP structure and the E_z distribution for $w = 1.87\ \mu m$ and $\lambda = 11.1\ \mu m$. (b) The quantum efficiency (QE) spectra calculated based on FEM (solid curves) and MTL (dashed curves) having the same detector parameters. The numbers are w in microns. (QWIP, quantum well infrared photodetectors.)

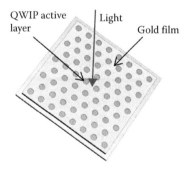

FIGURE 37.17 The 3D perspective of a plasmonic-enhanced quantum well infrared photodetectors (QWIP). Light is shining through the perforated gold layer from the top.

The gold layer is perforated with circular holes in a hexagonal lattice with spacing of 2.9 μm, and the hole diameter is 1.4 μm. The light incident on the gold layer excites surface plasmons in the opposite side with a resonant wavelength determined by the hole pattern. The resonant surface plasmons then create an intense vertical field at the gold/semiconductor interface. The active material has a low doping, with which the peak α is calculated to be 0.05 μm^{-1}. The modeled QE assuming this constant α for E_z is shown in Figure 37.18. It has a narrow peak at $\lambda = 8.3\ \mu m$. Meanwhile, from the measured responsivity spectrum (Wu et al., 2010) and the estimated gain of 8.5 from a similar detector structure (Eker et al., 2010), the experimental peak QE is deduced to be 12.6%, which agrees with the modeling to within 10%, and the two spectra have similar lineshapes.

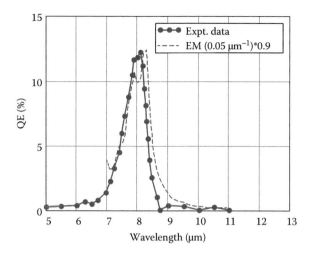

FIGURE 37.18 The measured (solid curve) and modeled (dashed curve) quantum efficiency (QE) of the plasmonic-enhanced QWIP. (EM, electromagnetic.)

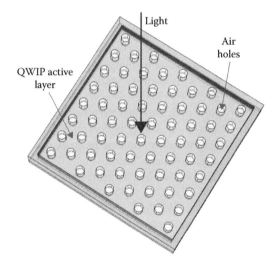

FIGURE 37.19 The 3D perspective of the PCS-QWIP. (QWIP, quantum well infrared photodetectors.)

37.4.7 Photonic Crystal Slab QWIP

To show another 3D optical structure, Figure 37.19 shows a photonic crystal slab QWIP (PCS-QWIP) studied (Kalchmair et al., 2011). The PCS is made of QWIP material, which is filled with periodic holes. These holes create a periodic modulation of refractive index in the x- and y-directions. Together with refractive index difference between GaAs and air in the z-direction, the PCS creates a series of characteristic photonic bands and bandgaps for the parallel propagating light. The normal incident light excites the stationary photonic eigenstates at the Γ-point of these bands in strong and sharp resonances. The present detector consists of an array of holes with spacing of 3.1 μm and diameter of 1.24 μm etched through the active and bottom contact layers. There is no metal layer in the optical path. The active material thickness is 1.5 μm, and the bottom contact thickness is 0.5 μm. The PCS is suspended in the air at a height t_{air} of 2.0 μm above the GaAs substrate.

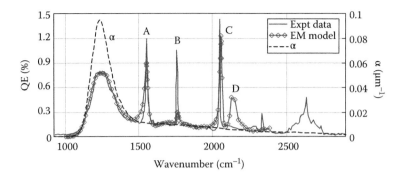

FIGURE 37.20 The measured photocurrent spectrum in arbitrary unit (solid curve) and the calculated QE spectrum based on the displayed α spectrum. (EM, electromagnetic.)

The measured photocurrent spectrum is shown in Figure 37.20. Based on the material α spectrum shown in Figure 37.20, which is deduced from the edge-coupled detector, the QE spectrum is modeled and it indeed contains the characteristic sharp peaks. However, the calculated peaks do not align exactly with the measurement. To obtain a better alignment, the theoretical hole spacing is reduced slightly from 3.1 to 2.9 μm as shown in Figure 37.20. This discrepancy could be due to the assumption of an average GaAs refractive index n to be 3.239 over a wide range of wavelengths. In reality, the value of n varies from 3.34 at 4 μm to 3.04 at 11 μm. Furthermore, the magnitudes of these peaks depend weakly on t_{air}. To obtain a larger peak at position B in Figure 37.20, t_{air} is adjusted from 2.0 to 0.9 μm. A different t_{air} could be due to the sagging of the PCS in the air at the operating temperature. From the modeling, the sharpness of these peaks is caused by two factors. One is the nearly symmetrical detector structure, both in vertical and horizontal directions, which induces strong resonances. The second is the weak material absorption at the peak wavelengths, which introduces only small damping effects on the resonances. The close match of the main peak and some of the side peaks lends support to the present modeling approach.

The large responsivity peaks A to C in the absorption tail illustrate that a large relative photoresponse can be obtained by using resonance in spite of a small intrinsic material absorption. Besides the examples presented earlier, we have also modeled other coupling structures, and the agreements between theory and experiment are similar to those presented (Choi et al., 2013a). They include air resonant cavities, linear and cross gratings, random-gratings, corrugated-QWIPs (C-QWIPs), etc. After studying all these examples, we found that the present FEM is highly accurate when comparing with other theoretical models. It also agrees well with experiments in both magnitudes and spectral lineshapes when both modeling and experimental uncertainties are taken into account.

37.5 EM Design of Resonator-QWIPs

The present FEM solves Maxwell equations numerically and thus should be able to optimize any detectors mentioned earlier as well as exploring new ideas. Among the many possible mechanisms and geometries, one structure that can yield a large QE is a photon trap. The function of a photon trap is to capture and hold incident light inside the detector until it is absorbed by the detector.

37.5.1 Resonator-QWIP Design

One realization of a photon trap is the resonator-QWIP (R-QWIP) as shown in Figure 37.21 (Choi et al., 2013a). It consists of an active layer, a thin GaAs substrate layer, and a number of diffractive elements (DEs) made of the same GaAs material on top. The topmost surface is then covered with a metal layer, and the detector is encapsulated with a low refractive index material.

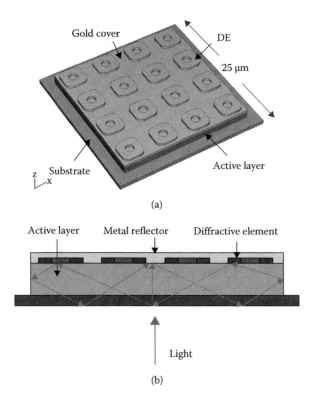

FIGURE 37.21 (a) A 3D perspective of an R-QWIP. Light is incident from the bottom. (b) The cross-section and one possible optical path.

The function of the DEs is to diffract incident light such that the subsequent angles of incidence at all detector boundaries are larger than the critical angle for total internal reflection (~18° for GaAs/air interface). If this is the case, the light will be totally confined in the detector. To account for interference effects, the size and shape of the detector volume are adjusted such that the scattered optical paths form a constructive interference pattern inside the detector. Under this condition, the newly incident light will be able to reinforce the light already under circulation, and the optical energy can be accumulated and stored in the detector as in a resonator. Therefore, by designing the detector into a resonator with a diffractive surface, i.e., an R-QWIP structure, an effective photon trap can be obtained.

To demonstrate the light storage capability, we first set α to zero so that there is no optical absorption in the detector. We then adjust the detector dimensions such that the intensity enhancement ratio (r_{ie}) given by

$$r_{ie} = \frac{1}{\frac{c\varepsilon_0}{2}E_0^2} \frac{nc\varepsilon_0}{2} \frac{1}{V} \int_V |E|^2 d^3r \qquad (37.24)$$

reaches a maximum. The detector dimensions include the shape, size, and spacing of the DEs, the DE distribution pattern, the material layer thicknesses, the detector areal dimensions, etc. The quantity r_{ie} compares the average intensity built up inside the active volume relative to the incident intensity in the air. Figure 37.22 shows that r_{ie} can attain a value of 20 at 9.6 μm with the optimized structure shown in Figure 37.21a. For QWIPs, only the vertical intensity initiates absorption. The corresponding r_{ie} in Figure 37.22 for the vertical polarization component is 11. Such a large vertical r_{ie} produces a QE of 71% when α becomes finite and equals to 0.2 μm^{-1}, despite the very thin active thickness of 1.2 μm is adopted.

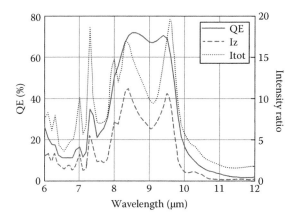

FIGURE 37.22 The calculated r_{ie} for the total intensity (dotted curve) and the vertical intensity (dashed curve) assuming $\alpha = 0$. The figure also shows the calculated QE (solid curve) if $\alpha = 0.2\ \mu m^{-1}$. The detector dimension is 22 μm. (DE, diffractive element.)

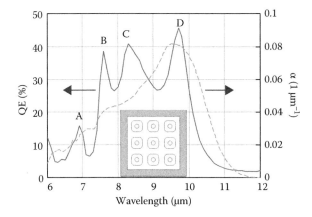

FIGURE 37.23 The measured α spectrum and the modeled quantum efficiency (QE) of an R-QWIP if $\alpha = 0.08\ \mu m^{-1}$. The pixel size is 20 μm, and the pixel pitch is 25 μm. The insert shows the DE pattern used for optical coupling.

37.5.2 R-QWIP Realization

We have conducted experimental research into R-QWIPs, and found QE as high as 70% in materials with doping larger than $1 \times 10^{18}\ cm^{-3}$ (Choi et al., 2013b, 2014, 2015). In this section, we present a more recent R-QWIP to show its advantages. The material consists of 19 QW periods. Each period is made of 50-Å GaAs/60 Å $Al_{0.22}Ga_{0.78}As$/50-Å GaAs/500-Å $Al_{0.22}Ga_{0.78}As$ double-barrier (DW) structure and each well is doped to $N_D = 0.5 \times 10^{18}\ cm^{-3}$. The adoption of the DW structure is to widen the absorption width, and the theoretical peak α is 0.08 μm^{-1} for this N_D. The measured α lineshape based on edge coupling is shown in Figure 37.23. Its bandwidth of 3 μm is about twice that of the single well design. This experimental lineshape and a theoretical peak value of 0.08 μm^{-1} is used for QE modeling. A matching resonator structure for this spectral range is shown in the insert of Figure 37.23. Assuming a λ-independent α of 0.08 μm^{-1} and $n = 3.05$, the resonator shows four resonant peaks (A, B, C, and D) in Figure 37.23, and the highest peak value is 45%.

The measured I–V characteristics of the R-QWIP averaged over 1600 pixels are shown in Figure 37.24a, and the QE spectral lineshape measured at −1.1 V averaged over 1000 pixels is shown in Figure 37.24b. Figure 37.24b also shows the calculated spectral lineshape using the material $\alpha(\lambda)$. It is seen that the relative

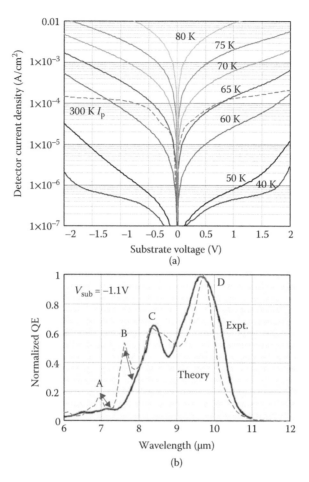

FIGURE 37.24 (a) The average *I–V* characteristics of 1600 R-QWIP pixels. (b) The measured average quantum efficiency (QE) lineshape of 1000 pixels in an FPA at −1.1 V and the modeled lineshape.

magnitudes of the experimental C and D peaks are as predicted while those of A and B peaks are much lower. The lower peaks also shift to longer wavelengths. This wavelength shift is expected when a constant n of 3.05 suitable only for 10 μm is used in modeling, while the actual n is known to increase toward shorter wavelengths. In addition to the less accurate n, the E_z distributions shown in Figure 37.25 at λ for A and B peaks indicate that they are higher order resonant modes that require higher precisions in the global structural dimensions. With the presence of processing nonuniformity, these resonances will be reduced.

For the QE magnitude, the main peak at 9.7 μm is measured to be 20% at −2 V and 37% at +2 V. The QE at positive bias is thus consistent with the modeling in this case. The polarity asymmetry is usually observed in R-QWIPs and is related to the highly localized optical intensity created in these detector structures as shown in Figure 37.26. For the two main peaks C and D, E_z is much stronger at the top of the active layer, where the QW layers are more active under positive bias. Therefore, this polarity produces a larger detection QE (Choi et al., 2014). In our simple dark current model described in Section 2.1, we assumed a linear potential drop in the MQW structure. In reality, the potential drop is slightly curved with potential drop near the cathode larger than the anode (Choi, 1997). This nonlinearity makes $\gamma(V)$ in $\eta_{abs}(\lambda)\gamma(V)$ larger near the top contact region than the substrate under positive bias. While the change of $\gamma(V)$ along z does not affect the average I_d with uniformly excited thermal electrons, it does promote larger I_p under positive bias when the optical intensity is localized. Therefore, R-QWIPs usually prefer positive bias operation. For future modeling efforts, the inclusion of $n(\lambda)$ and $\gamma(z)$ are clearly needed.

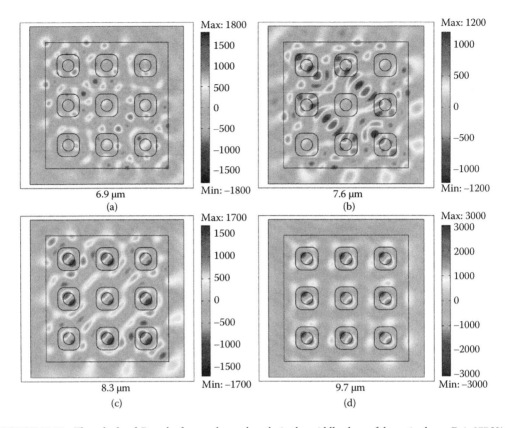

FIGURE 37.25   The calculated E_z at the four peak wavelengths in the middle plane of the active layer. E_0 is 377 V/m. Note the change of scale in the plots.

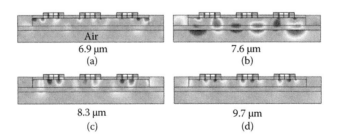

FIGURE 37.26 E_z at the center cross-section of the R-QWIP with distribution in (a), (b), (c), and (d) corresponding to peak A, B, C, and D, respectively. The value of E_z decreases exponentially from the top toward the substrate in (a), (c), and (d).

The advantages of the R-QWIPs can be seen by comparing the corrugated-QWIP performance (C-QWIP) in Figure 37.3a and the R-QWIP performance in Figure 37.24a. The C-QWIP uses angled sidewalls to couple normal incident light (Choi et al., 2011, 2012). The cutoff wavelength λ_c of these two materials are similar, which is 11.2 µm for the C-QWIP and 10.6 µm for the R-QWIP. The C-QWIP shows $J_d = 9 \times 10^{-4}$ A/cm² at 3 V and 60 K and $J_p = 1.5 \times 10^{-4}$ A/cm² at 3 V. For the R-QWIP, $J_d = 2.5 \times 10^{-5}$ A/cm² at 1 V and 60 K and $J_p = 1.5 \times 10^{-4}$ A/cm² at 1 V. While the difference in the dark currents of the two detectors is a factor of 36, the photocurrents are the same. Accounting for the longer λ_c for the C-QWIP, the R-QWIP improves the I_p/I_d ratio by approximately a factor of 10. Figure 37.27 shows an IR image taken by the present R-QWIP FPA, which shows its excellent sensitivity and uniformity. The FPA operating condition is as follows: temperature = 61 K, applied bias = −1.1 V, $F/\# = 2.5$, signal integration

FIGURE 37.27   This image was taken by the R-QWIP FPA with 1024 × 1024 pixels.

time = 3.06 ms, and signal processing = 1-point background subtraction. Under this operating condition, the pixel operability is 99.5% and thermal sensitivity is 45 mK. Even higher sensitivity is expected at a lower temperature or at positive bias.

37.6 Conclusion

In this chapter, we have discussed the theoretical methods in modeling the optoelectronic properties of QWIPs. We found that analytical methods can be used to explain the detector dark current and photoconductive gain. The intrinsic material absorption properties can be calculated from the TMM, and the detector QE can be solved by performing finite-element EM modeling. With these methods, all of the relevant optoelectronic properties of a QWIP can be predicted with very few empirical parameters. This ability greatly facilitates QWIP FPA production, since it allows more consistent and capable products be designed and manufactured.

Due to limited space, some of the advantages of EM modeling have not been discussed. For example, EM modeling can be used to design structures that use thinner materials without antireflection coatings or optical filters to reduce manufacturing cost and improve uniformity; structures that are more immune to fabrication nonuniformity and have less pixel spatial and spectral cross-talk; structures that have special functionality such as extremely narrowband detection, two-color detection, broadband detection, and polarization detection; and structures that are very small in size but are still sensitive so that ultra-high-density FPAs can be produced. For the last benefit, the model predicts that a QE of 56% can be obtained with an absorption coefficient of 0.2 μm^{-1} and pixel size of 5×5 μm^2. Its realization will enable the production of very high-resolution FPAs with much smaller system size, weight, power, and cost. Besides their usefulness to QWIPs, EM modeling and the photon trap concept are also applicable to other sensors and energy convertors in the IR and other nearby wavelengths. Therefore, one can foresee the increasing vital role played by EM modeling in advancing different optoelectronic technologies in the future.

References

Choi KK (1997) *The Physics of Quantum Well Infrared Photodetectors*. River Edge, NJ: World Scientific.

Choi KK (2012) Electromagnetic modeling of edge coupled quantum well infrared photodetectors. *J. Appl. Phys.* 111:124507.

Choi KK, Bandara SV, Gunapala SD, Liu WK, and Fastenau JM (2002) Detection wavelength of InGaAs/Al-GaAs quantum wells and superlattices. *J. Appl. Phys.* 91:551–564.

Choi KK, Chen CR, Goldberg AC, Chang WH, and Tsui DC (1998) Performance of corrugated quantum well infrared photodetectors. *Proc. SPIE* 3379:441–452.

Choi KK, Dang G, Little JW, Leung KM, and Tamir T (2004) Quantum grid infrared spectrometer. *Appl. Phys. Lett.* 84:4439–4441.

Choi KK, Jhabvala, MD, Forrai D, Sun J, and Endres D (2011) Corrugated quantum well infrared photodetectors for far infrared detection. *Opt. Eng.* 50:061005-1-6.

Choi KK, Jhabvala MD, Forrai DP, Waczynski A, Sun J, and Jones R (2012) Electromagnetic modeling of quantum well infrared photodetectors. *IEEE J. Quant. Electron.* 48:384–393.

Choi KK, Jhabvala MD, Forrai DP, Waczynski A, Sun J, and Jones R (2013a) Electromagnetic modeling and design of quantum well infrared photodetectors. *IEEE J. Sel. Top. Quant. Electron.* 19:1–10.

Choi KK, Jhabvala MD, Sun J, Jhabvala CA, Waczynski A, and Olver K (2013b) Resonator-quantum well infrared photodetectors. *Appl. Phys. Lett.* 103:201113.

Choi KK, Jhabvala MD, Sun J, Jhabvala CA, Waczynski A, and Olver K (2014) Resonator-QWIPs and FPAs. *Proc. SPIE* 9070:907037.

Choi KK, Sun J, and Olver K (2015) Resonator-QWIP FPA development. *Proc. SPIE* 9451:94512K.

Dixon JR and Ellis JM (1961) Optical Properties of n-type indium arsenide in the fundamental absorption edge region. *Phys. Rev.* 123:1560–1566.

Donges A (1998) The coherence length of black-body radiation. *Eur. J. Phys.* 19:245–249.

Eker SU, Arslan Y, Onuk AE, and Besikci C (2010) High conversion efficiency InP/InGaAs strained quantum well infrared photodetector focal plane array with 9.7 μm cut-off for high-speed thermal imaging. *IEEE J. Quantum Electron.* 46:164–168.

Fowles GR (1975) *Introduction to Modern Optics*, 2nd Edition. New York, NY: Holt, Rinehart, and Winston.

Grandidier J, Callahan DM, Munday JN, and Atwater HA (2012) Gallium arsenide solar cell absorption enhancement using whispering gallery modes of dielectric nanospheres. *IEEE J. Photovolt.* 2:123–128.

Hinnrichs M and Guptab N (2008) Comparison of QWIP to HgCdTe detectors for gas imaging. *Proc. SPIE* 6940:69401Q.

Jhabvala MD et al. (2011) Performance of the QWIP focal plane arrays for NASA's Landsat Data Continuity Mission. *Proc. SPIE* 8012:80120Q.

Kalchmair S, Detz, H, Cole GD, Andrews AM, Klang P, Nobile M, Gansch R, Ostermaier C, Schrenk W, and Strasser G (2011) Photonic crystal slab quantum well infrared photodetector. *Appl. Phys. Lett.* 98:011105.

Levine BF, Bethea CG, Hasnain G, Shen VO, Pelve E, Abbott RR, and Hsieh SJ (1990) High sensitivity low dark current 10 μm GaAs quantum well infrared photodetectors. *Appl. Phys. Lett.* 56:851–863.

Levine BF, Choi KK, Bethea CG, Walker J, and Malik RJ (1987) A new 10 micron infrared detector using inter-subband absorption in resonant tunneling GaAs/GaAlAs superlattices. *Appl. Phys. Lett.* 50:1092–1094.

Li J, Choi KK, Klem JF, Reno JL, and Tsui DC (2006) High Gain, broadband InGaAs/InGaAsP quantum well infrared photodetectors. *Appl. Phys. Lett.* 89:081128.

Miles RH, Chow DH, Schulman JN, and McGill TC (1990) Infrared optical characterization of InAs/Ga$_{1-x}$InxSb superlattices. *Appl. Phys. Lett.* 57:801–803.

Sadiku MNO (2001) Numerical techniques in electromagnetics, 2nd Edition. Boca Raton, FL: CRS Press.

Shi JJ and Goldys EM (1999) Intersubband optical absorption in strained double barrier quantum well infrared photodetectors. *IEEE Trans. Elect. Dev.* 46:83–88.

Tamir T and Zhang S (1996) Modal transmission-line theory of multilayered grating structures. *J. Lightwave Technol.* 14:914–927.

Wu W, Bonakdar A, and Mosheni H (2010) Plasmonic enhanced quantum well infrared photodetector with high detectivity. *Appl. Phys. Lett.* 96:161107.

38

Optical Modulators

Dominic F. G.
Gallagher

and

Dmitry Labukhin

38.1 Introduction

The modulation of an optical signal is a key requirement of modern optical communications. Even when the transmitter is a laser diode, which can itself be directly modulated, there are advantages of using an unmodulated laser and an external modulator on the output.

Optical modulators can be constructed using many different physical effects. The main ones of interest are as follows:

1. Linear electro-optic, or Pockel's effect—an electrostatic field is used to modulate the refractive index of certain crystals.
2. The plasma effect—the alteration of the free carrier density in a semiconductor can change both the absorption and the refractive index.
3. The Franz–Keldysh (FK) effect and the related quantum-confined Stark effect (QCSE)—where an electrostatic field changes the band structure of a semiconductor, altering both the absorption spectrum and the refractive index.

In choosing a modulator, the following factors should be taken into account:

1. Do you want a phase modulator or an absorption modulator?
2. The modulation bandwidth needed.
3. Its energy efficiency, typically measured in pJ/bit.
4. The optical bandwidth required—some designs are based on resonant structures and can modulate only a narrow range of wavelengths, others can work over very wide optical bandwidths.
5. Size.

There is no such thing as a pure phase modulator or pure absorption modulator. The Kramers–Kronig relationship states that the dispersion spectrum (real part of the refractive index) can be known given the absorption spectrum over all frequencies, and vice versa:

$$\varepsilon'(\omega) = 1 + \frac{2}{\pi} P \int_0^\infty \frac{\omega' \varepsilon''(\omega')}{\omega'^2 - \omega^2} d\omega' \tag{38.1}$$

$$\varepsilon''(\omega) = -\frac{2\omega}{\pi} P \int_0^\infty \frac{\varepsilon'(\omega') - 1}{\omega'^2 - \omega^2} d\omega' \tag{38.2}$$

where the dielectric constant $\varepsilon = \varepsilon' + i \cdot \varepsilon''$, and P is the *Cauchy principal value*. The Kramers–Kronig relationships tell us that the largest change in real index occurs near a change in absorption. Thus the most efficient phase modulators are likely to have absorption at nearby wavelengths.

38.2 QCSE-Based Modulators

38.2.1 Figures of Merit

Design of an electroabsorption modulator (EAM) requires accurate computation of its key characteristics. Among most important figures of merit are (a) on/off ratio defined as the output light power ratio between on and off states of the EAM; (b) the change in absorption coefficient per unit applied voltage; and (c) maximum data transmission speed. Calculating (a) and (b) is based on accurate modeling of absorption spectrum as a function of applied voltage, which will be the main focus of this chapter. Computing (c) involves modeling carrier dynamics. A detailed account of the latter can be found in [1–3].

38.2.2 Structure

Let us consider a typical traveling-wave EAM shown in Figure 38.1. This type of structure is used in this chapter as an example. Although other configurations have been reported as well (e.g., VCSEL-type EAM [4]), the traveling-wave type is the most common and the general principles of its operation can be extended to others with some modifications.

In a traveling-wave configuration, the light propagation dimension (z in Figure 38.1) is much greater than light- and carrier-confining dimensions (x and y in Figure 38.1) and, therefore, z can be separated from the x and y for computational efficiency. This means that we can split the device along axis z into a finite

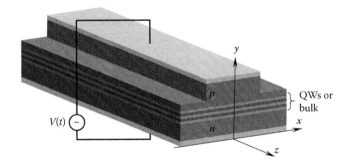

FIGURE 38.1 Traveling-wave EAM structure. The light signal propagates in the z-direction. Light confinement in the x–y plane can be realized via various configurations, the rigde waveguide shown in the figure being just one example. Epitaxial layers are grown in the y-direction. Top and bottom layers are metallic contacts. QWs and barriers comprise the active region. Next to the metallic contact layers are the substrate (n-doped) and cladding (p-doped). (QW=quantum well.)

number of short segments and assume optical properties within each segment to be z-independent, thus reducing the problem of modeling the optical properties to two transversal dimensions, x and y. Further-more, in many cases, when the variation of the optical field and carrier density in one transversal direction (e.g., y in Figure 38.1) is much greater than in the other, the problem becomes one-dimensional. Then the calculated optical properties, namely, *material absorption* and *refractive index*, are used as input to an opti-cal mode solver to compute the *effective index* and *modal absorption* of a particular waveguide mode that carries the optical signal. Finally, the result is passed as input to a one-dimensional z-propagation algo-rithm to obtain the device's characteristics. The first stage of this sequence, modeling material absorption and refractive index, is the focus of this chapter.

The optical absorption in EAMs is caused by several processes, among which are carrier transitions between conduction and valence band, free-carrier absorption (FCA), intraband transitions, photon–phonon interaction, etc. By far, the strongest contribution within the relevant spectral bandwidth is made by the first of these. It can be controlled by electrical bias applied across the epitaxial structure via one of the two main physical mechanisms: the FK effect in bulk layer structures or the QCSE in quantum wells (QWs). It is the latter one that is used in the majority of EAMs due to its higher efficiency. However, the FK effect is briefly examined here for the benefit of general understanding of these two phenomena.

38.2.3 Franz–Keldysh Effect

The Franz–Keldysh (FK) effect is the change in photon absorption spectrum induced by reverse bias applied to a bulk semiconductor structure (Figure 38.1). In an unbiased bulk layer, the electron and hole states are represented by space-harmonic functions. The probability of a photon-absorbing transition between two states is proportional to the overlap integral between their respective wavefunctions. When electric field F is applied across spatial direction y, the electron and hole wavefunctions change their shape and become Airy functions as shown in Figure 38.2b.

As the overlap integral between them changes, the probability of the photon absorption changes accord-ingly. Furthermore, the minimum absorption frequency falls below the bandgap, as can be seen from Figure 38.2b. This enables the control of the transmitted power of the light traveling through the EAM with applied voltage at a given wavelength (Figure 38.3). However, absorption change per applied voltage is relatively low and QW structures are preferred due to the higher efficiency of the QCSE.

38.2.4 Quantum-Confined Stark Effect

The QCSE is a counterpart of the FK effect in QW structures. It has, however, an additional feature: the presence of thin peaks in the absorption spectrum caused by the excitonic transitions. Indeed, unlike the bulk structures, QWs facilitate strong electron and hole confinement (see Figure 38.4), which causes

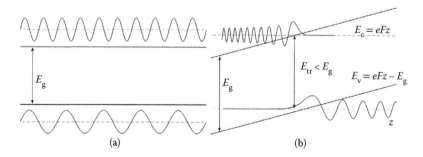

FIGURE 38.2 Electron and hole wavefunction in (a) an unbiased bulk semiconductor with bandgap E_g and when (b) a constant electric field F is applied across spatial direction y.

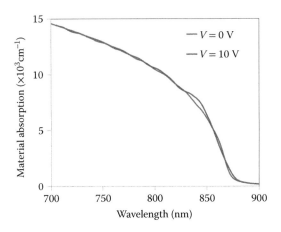

FIGURE 38.3   Change in absorption spectrum in 70-nm bulk AlGaAs under reverse bias due to the FK effect.

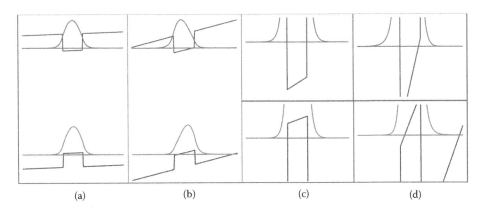

(a)   (b)   (c)   (d)

FIGURE 38.4   QW potential, electron and hole wavefunctions (a) without and (b) with a reverse bias. Plots (c) and (d) are (a) and (b), respectively, zoomed in to demonstrate the redshift in the energy levels.

the two carrier types to form bound states, *excitons*, similar to those between a proton and an electron in a hydrogen atom. These states introduce new discrete energy levels in addition to those of the QWs: but whereas the contribution to the absorption spectrum from the QW transitions has a staircase-like shape, the exciton contribution appears as peaks in the absorption spectrum.

Like in bulk layers, applying a reverse bias to the QW potential creates a double effect (see Figure 38.4). First, the electron and hole wavefunctions start "escaping" through the lower potential barrier, which decreases their overlap and, consequently, the probability of photon absorption. Second, their energy levels move closer to one another, which results in the redshift of the respective absorption peaks.

It can be seen from Figure 38.5 that if the absorption is measured at wavelength λ_0 near the exciton peak, the net effect of the two phenomena is the *increase* in absorption. This enables the control of the amount of light passing through the EAM by varying electric bias applied to the device. The material absorption change is much higher than in the case of the bulk structure. However, the material absorption is factored by the optical confinement factor, which is much smaller in QWs. To overcome this problem, multiple quantum wells (MQWs) are normally used to increase the efficiency of the device.

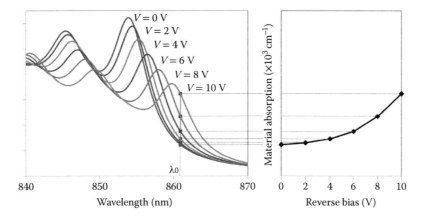

FIGURE 38.5 Change in material absorption spectrum in a 10-nm single QW AlGaAs EAM under reverse bias due to the QCSE effect. The peaks represent exciton's positions.

38.2.5 Modeling Absorption Spectrum

Modeling of the absorption spectrum as a function of voltage bias consists of several stages. Two most important effects contributing to the absorption is the carrier transitions between (a) energy levels created by QW confinement and (b) exciton binding energy levels. Photon absorption rate is proportional to the probability of the transition between an occupied state n in the valence band (hole) and an unoccupied state m in the conduction band (electron) multiplied by the Fermi population level $\alpha_{mn} \propto P_{mn}(f_m - f_n)$. The total absorption is given by the double sum over all relevant transitions mn, while each probability P_{mn} is determined from the overlap between the respective electron and hole wavefunctions. Therefore, the central part of the modeling is solving the Schrödinger equation in the QW potential and with an excitonic term added to its Hamiltonian. The QW potential, in its turn, is created by the juxtaposition of epitaxial layers with different bandgaps and applied electric field. It can be calculated using the Poisson-drift-diffusion model. In the case of EAMs, the coupling between the Schrödinger and Poisson equations can be neglected, which significantly reduces computational load.

38.2.6 Poisson-Drift-Diffusion Model

The Poisson-drift-diffusion model is based on three first-order differential equations: the carrier continuity equations for electrons and holes and the Gauss's law (see Chapter 3.1 in [5]),

$$\frac{\partial n}{\partial t} = G_n - R_n - \nabla \cdot (\mu_n n \nabla \phi - D_n \nabla n)$$

$$\frac{\partial p}{\partial t} = G_p - R_p - \nabla \cdot (\mu_p p \nabla \phi - D_p \nabla p) \tag{38.3}$$

$$0 = \nabla \cdot \varepsilon \nabla \phi + q_e(p - n + N_D + N_A) \tag{38.4}$$

where $n = n(\boldsymbol{\rho}_{xy})$, $p = p(\boldsymbol{\rho}_{xy})$ are the electron and hole densities; $\boldsymbol{\rho}_{xy} = \{x, y\}$ is the spatial coordinate; $G_{n,p}(\boldsymbol{\rho}_{xy})$ and $R_{n,p}(\boldsymbol{\rho}_{xy})$ are generation and recombination rates, respectively; $\mu_{n,p}(\boldsymbol{\rho}_{xy})$ is electron and hole mobilities; $D_{n,p}(\boldsymbol{\rho}_{xy})$ is the electron and hole diffusion coefficients; $\varphi(\boldsymbol{\rho}_{xy})$ is the static field potential; $\varepsilon(\boldsymbol{\rho}_{xy})$ is the electric permittivity; $N_{D,A}(\boldsymbol{\rho}_{xy})$ is the donor and acceptor doping levels; and q_e is the electron charge.

Since in most EAMs the carrier density is very low, the generation and recombination terms can be discarded. With the reverse bias, the currents are also assumed negligible and, therefore, for the steady-state solution, the time derivatives are set to zero:

$$\nabla \cdot (\mu_n n \nabla \phi - D_n \nabla n) = 0$$

$$\nabla \cdot (\mu_p p \nabla \phi - D_p \nabla p) = 0 \tag{38.5}$$

$$\nabla \cdot \varepsilon \nabla \phi + q_e (p - n + N_D + N_A) = 0$$

Equation 38.5 is solved with respect to $\{n, p, \varphi\}$, whereas $\mu_{n,p}$, $D_{n,p}$, ε, and $N_{D,A}$ are usually assumed to be independent of $\{n, p, \varphi\}$ and are precomputed from the epitaxial layer composition. The potential φ is then used to find the spatial profile of the band edges:

$$E_{c,v}(\boldsymbol{\rho}_{xy}) = E_{c0,v0}(\boldsymbol{\rho}_{xy}) - \phi(\boldsymbol{\rho}_{xy}) \tag{38.6}$$

In addition to static field potential, the Poisson-drift-diffusion model calculates the distribution of carriers and, consequently, the distribution of charge in the PIN-junction as a function of applied bias. This is essential for calculating the device's characteristics in an electric circuit, such as capacitance, resistance, and, eventually, the modulation response.

38.2.7 Schrödinger Equation for QW States and Excitons

The next step after finding the band edge potentials $E_{c,v}(\boldsymbol{\rho}_{xy})$ is to solve the Schrödinger equation in order to find energy levels and spatial distributions of electron and hole wavefunctions. There exist several methods to do so, with different tradeoffs between accuracy and computational efficiency. The general form of the Schrödinger equation is

$$\hat{H}\Psi = E\Psi \tag{38.7}$$

The Hamiltonian operator $\hat{H}$ in the presence of excitonic effect is split into electron, hole, and exciton parts [6]:

$$\hat{H} = \hat{H}^e + \hat{H}^h + \hat{H}^{exc}$$

$$\hat{H}^e = \hat{H}^e_{kin} + E_c(\boldsymbol{r}_e)$$

$$\hat{H}^h = \hat{H}^h_{LK} + E_v(\boldsymbol{r}_h) \tag{38.8}$$

where $\hat{H}^e_{kin} \propto \nabla^2_e$ is the kinetic part of the electron Hamiltonian, $E_{c,v}(\boldsymbol{r}_{e,h})$ is the QW potentials with applied bias for conduction and valence bands, respectively, as calculated by solving Equations 38.5 and 38.6. Note that the spatial coordinates for electrons and holes are different. $\hat{H}^h_{LK}$ is the Luttinger–Kohn Hamiltonian for holes. The latter can be a 4 × 4 matrix operator, in case if only heavy and light holes are taken into account [6]. When highly strained structures are considered, the influence of the spin-orbit split-off band becomes significant and it needs to be included in $\hat{H}^h_{LK}$, increasing its size to a 6×6 matrix operator [29,30]. The exciton part, $\hat{H}^{exc}$, is due to the Coulomb interaction between electron and holes. The additional term in the Hamiltonian of the system is, therefore,

$$\hat{H}^{exc} = \frac{q_e}{4\pi\varepsilon_0\varepsilon} \frac{1}{|r_e - r_h|} \tag{38.9}$$

As the first step, the energies and the wavefunctions of the conduction and valence bands without the exciton contribution are obtained by solving the Schrödinger equation with respective Hamiltonians:

$$\hat{H}^e \psi_m^e(\mathbf{k}, \mathbf{r}) = E_m^e \psi_m^e(\mathbf{k}, \mathbf{r})$$
$$\hat{H}_{LK}^h \psi_n^h(\mathbf{k}, \mathbf{r}) = E_n^h \psi_n^h(\mathbf{k}, \mathbf{r})$$

(38.10)

It is assumed that the scale of the spatial variation of parameters in the x- and z-directions is insignificant in comparison to that in y-direction. The solution of these equations is, therefore, obtained by separating their spatial dependence: the wavefunction is assumed to be harmonic in the x- and z-directions and (Equation 38.10) is reduced to a one-dimensional eigenvalue problem with a y-dependent Hamiltonian only (note that in some references quoted here the perpendicular axis is denoted z, not y).

The one-dimensional Schrödinger equation can be solved using a finite-difference method or a so-called "shooting" method. In the case of the effective mass approximation (parabolic dispersion relation $E(\mathbf{k}_{xz})$), it only needs to be solved for $\mathbf{k}_{xz} = 0$, where $\mathbf{k}_{xz}$ is an in-plane wave vector (see, e.g., [7]).

If nonparabolicity of the dispersion curve needs to be taken into account, Equation 38.10 must be solved for a range of $\mathbf{k}_{xz}$ in the vicinity of zero. This can present a considerable computational challenge. The axial approximation removes the dependency of $\hat{H}_{LK}^h$ on azimuthal angle θ and the two-dimensional vector $\mathbf{k}_{xz}$ is replaced with its magnitude k, which significantly reduces computational time (see, e.g., [6]).

After the energies and wavefunctions in Equation 38.10 are found, the second step is to compute the exciton transition energy levels and wavefunctions with the full Hamiltonian:

$$\hat{H}\, \Psi^X(\mathbf{r}_e, \mathbf{r}_h) = E^X \Psi^X(\mathbf{r}_e, \mathbf{r}_h)$$

(38.11)

Here X denotes different exciton states, such as $1s$, $2s$, $2p$, etc., called so by analogy with the atomic states. Two main methods of solving this problem are the momentum-space method [6] and the variation method [7].

In the momentum-space method, the exciton function is expanded in terms of conduction and valence wavefunctions $\psi_n^e(\mathbf{k}_{xz}, y_e)$ and $\psi_m^h(\mathbf{k}_{xz}, y_h)$ (found by solving Equation 38.10) in the $\mathbf{k}_{xz}$ space. As a result, Equation 38.11 is reduced to an eigenvalue problem with the integral operator on the right-hand side (see Equations 38.18 and 38.19 in [6]). Similarly to Equation 38.10, the axial approximation can be applied to full Hamiltonian $\hat{H}$: vector $\mathbf{k}_{xz}$ is replaced with its absolute value k, and the dependence of the wavefunctions on the azimuthal angle is confined to the phase term only: $\varphi(\mathbf{k}_{xz}) \rightarrow \varphi(k)e^{j\theta l}$. As mentioned earlier, the exciton states with $l = 0, \pm 1, \pm 2$, and ± 3, are s, p, d, and f.

The exact mathematical details can be found in [6], but it is important to mention here that the final equation will contain a quadruple integral: over the spatial domain of the electron wavefunctions y_e, over that of the hole wavefunctions y_h spatial domain, and over the momentum angle of the in-plane vector $\mathbf{k}$; and over its magnitude k. The method was demonstrated to be in excellent agreement with the experimental results [6]. However, computing the quadruple integral for large structures (on the order of tens of nanometers in y-direction) can become prohibitively slow.

The variational method [7] is used to increase computational efficiency, albeit at the cost of accuracy. The exciton function in Equation 38.11 is expanded in terms of conduction and valence wavefunctions in real space $\boldsymbol{\rho}$, not $\mathbf{k}$. Only the $1s$ exciton state is taken into account (so that the θ dependence is dropped, $\boldsymbol{\rho} \rightarrow \rho$):

$$\Psi^{1s}(\rho, y_e, y_h) = \phi_{mn}^{1s}(\rho)\psi_m^e(y_e)\psi_n^h(y_h)$$

(38.12)

where $\phi_{mn}^{1s}(\rho)$ is a $1s$ trial function with variational parameter λ:

$$\phi_{mn}^{1s}(\rho) = \sqrt{\frac{2}{\pi}}\frac{e^{-\rho/\lambda}}{\lambda}$$

(38.13)

38.2.8 Permittivity Spectrum

Equations 38.7 through 38.13 serve to calculate the spectral positions of the exciton and QW transitions. The next step is to calculate the strength and the shape of the absorption $\alpha(\omega)$ as a function of light frequency ω due to these transitions and the corresponding change in the refractive index $n_{re}(\omega)$, both making part of the complex permittivity $\varepsilon(\omega) = [n_{re}(\omega) + i\alpha(\omega)c_0/(2\omega)]^2$, where ω is the angular frequency of light and c_0 is the speed of light in vacuum. The expression for total complex permittivity can be written as

$$\varepsilon(\omega) = \varepsilon^{opt} + \sum_{m(e)} \sum_{n(h)} (\varepsilon_{mn}^{QW}(\omega) + \varepsilon_{mn}^{exc}(\omega)) \tag{38.14}$$

where ε_{opt} is the permittivity for low optical frequencies, $\varepsilon_{mn}^{QW}(\omega)$ and $\varepsilon_{mn}^{exc}(\omega)$ is the permittivity change due to the QW and exciton transition between hole state n and electron state m. Whereas the QW contribution is a staircase-like function with steps located at energies $(E_m - E_n)$, the exciton transition is a delta function found at energies $(E_m - E_n - E_X)$. The exact formulae for the momentum-space and variational methods can be found in [6,7], respectively.

Two things must be pointed out. First, as mentioned earlier, the contribution both to the QW "steps" and exciton peaks is proportional to the overlap integral between the respective hole and electron functions:

$$\varepsilon_{mn}^{QW,exc}(\omega) \sim \int \psi_m^e(y)\psi_n^{h*}(y)dy \tag{38.15}$$

Second, due to the finite lifetime of the exciton states, these peaks have a spectral width $\simeq 1/\tau_{exc}$. Similarly, the steps of the staircase-like function of the QW contribution are smoothed out by a QW linewidth function, whose width is inversely proportional to the lifetime of the QW states. Mathematically, it is expressed as a convolution integral between $\varepsilon_{mn}^{QW,exc}(\omega)$ and respective linewidth functions $L_{mn}^{QW,exc}(\omega)$. The latter are complex-valued functions satisfying Kramers–Kronig relations (Equations 38.1 and 38.2) and they contribute to both the real and imaginary parts of the complex permittivity. Strictly speaking, each electron–hole pair should have its own $L_{mn}^{QW,exc}(\omega)$, but often in practice the same function is used for all pairs mn. There have been a few publications dedicated to calculating the shape of $L^{QW,exc}(\omega)$ (see, e.g., [9]). Still in many cases, a simple Lorenzian shape with an experimentally measured spectral width is used in calculations with satisfying accuracy [6].

38.2.9 Numerical Results

Below, experimental and modeling results are shown for several structures. Figure 38.6 shows the comparison between experimental results obtained for an SiGe modulator [10] and those simulated using the variational method. The variational method demonstrates excellent agreement in positions and shapes of the primary excitonic peaks and their dependence on the applied electric bias. It is not as accurate, however, in modeling higher energy spectra.

Furthermore, experimental results for an AlGaAs EAM, reported in [6], are compared with curves computed by the variational method and the momentum-space method (Figure 38.7).

It can be seen that the latter captures the general shape of the curve much more accurately, especially at higher energies. It can be argued, however, that the working frequency of the EAM is below the bandgap (to the left of the excitonic peak at zero bias) and it is, therefore, the most important spectral area to model. Moreover, the behavior in this area is largely dependent on the correct modeling of the exciton shape, not covered by either of the methods (see [9]). Therefore, in many cases, the variational method is preferred as a computationally fast alternative of the momentum-space method, especially for large MQW structures.

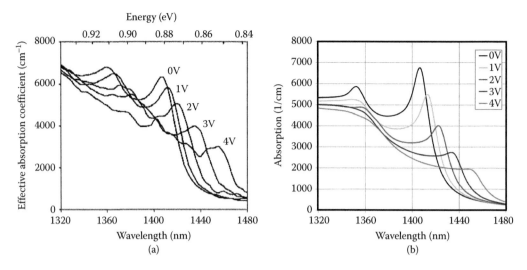

FIGURE 38.6 SiGe modulator. Transverse electric (TE) material absorption spectra for various values of the bias voltage: (left) experimental data from [10] (see detailed description of the structure in Section II of [10]) and (right) variational method by Photon Design Harold EAM.

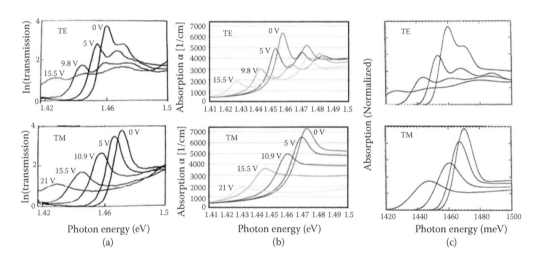

FIGURE 38.7 AlGaAs modulator [6] ([9.4 nm AlGaAs]/$Al_{0.3}Ga_{0.7}As$ QW, see detailed description of the structure in [31]). Absorption spectra for various values of the bias voltage: (a) experimental data from; (b) simulated curves using variational method for (top) TE polarization and (bottom) TM polarization; and (c) simulated curves using momentum-space method.

38.3 Electro-Optic Effect Modulators

The linear electro-optic effect, otherwise known as the Pockel's effect is present in crystals that do not have an inversion symmetry. This means they need an anisotropic dielectric tensor. Popular materials exploiting the Pockel's effect include $LiNbO_3$, $LiTaO_3$, and β-barium borate (BBO). The Pockel's effect is effectively instantaneous, once the electrostatic field has been established, making it suitable for exceedingly high-speed modulators.

LiNbO$_3$ has been a very popular material for electrorefractive modulators for many years because of its relatively large electro-optic coefficient [11,12], i.e., variation of refractive index with an applied electrostatic field. The rest of this section focuses on LiNbO$_3$ not because it is the only material available, but because of its dominant industrial position in photonics.

LiNbO$_3$ is strongly anisotropic implying a nontrivial dielectric tensor ε_{ij}. Since the tensor is symmetric, the double indexing conventionally simplified as

$$\begin{pmatrix} D_x \\ D_y \\ D_z \end{pmatrix} = \begin{pmatrix} \varepsilon_1 & \varepsilon_6 & \varepsilon_5 \\ \varepsilon_6 & \varepsilon_2 & \varepsilon_4 \\ \varepsilon_5 & \varepsilon_4 & \varepsilon_3 \end{pmatrix} \begin{pmatrix} E_x \\ E_y \\ E_z \end{pmatrix} \tag{38.16}$$

In addition, the variation of each ε_{ij} with applied electrostatic field depends on the direction of the electrostatic field E, leading to a second-rank tensor:

$$\Delta \left(\frac{1}{\varepsilon_i} \right) = \sigma_{ik} E_k \tag{38.17}$$

This implies that different optical polarizations will experience different modulations. In LiNbO$_3$, only a few of the electro-optic tensor coefficients are significant:

$$\sigma_{13} = 8.6 \text{ pm/V}, \sigma_{22} = 3.4 \text{ pm/V}, \sigma_{33} = 30.8 \text{ pm/V}, \sigma_{51} = 28 \text{ pm/V} \tag{38.18}$$

Unlike for example, silicon modulators, there is no macroscopic movement of charge except that associated with the capacitance of the dielectric: $C = \epsilon_r \epsilon_0 A/d$; d is thickness, A is area. This leads to a very high-speed modulator capable of working to 30 GHz or more [13]. In fact, the change in refractive index is near instantaneous once the electric field has been established and the speed of an LiNbO$_3$ modulator is limited almost entirely by it capacitance and the speed of the electrical circuit driving it.

As the electro-optic tensor coefficients given earlier imply, the effect is anisotropic in both the direction of the applied field and in the modulation of the optical dielectric tensor. The crystal must be carefully oriented to obtain the desired effect. The two favored orientations are denoted x-cut and z-cut. An x-cut crystal will create a strong modulation of a TE-like polarized optical mode in response to an electrostatic field in the same transverse axis as the optical one, exploiting the σ_{33} coefficient. An x-cut crystal requires a vertical electrostatic field to modulate the same TE-like polarized optical mode and exploits the σ_{51} coefficient.

38.3.1 The Mach–Zehnder Modulator

The change in refractive index can be converted to a change in signal amplitude using a Mach–Zehnder interferometer design (Figure 38.8). With sufficient voltage, a 180°phase shift between the two optical paths is created by the time the waveguides combine, causing full cancellation of the signal.

A modulator can be made using a field between an electrode placed on top of the LiNb0$_3$ chip and a ground plane underneath. However, since the chip is rather thick, this leads to a low field strength, so typically both electrodes are placed on top of the chip either side of the waveguide, as shown in Figure 38.9.

The design in Figure 38.9 has one big flaw. The electrodes are very close to the optical mode; this causes substantial optical losses especially in the TM-like polarization. To cure this, a cap layer can be deposited on the LiNbO$_3$ before the electrodes are deposited. The buffer has another advantage in that it alters the velocity of the electrical signal along the modulator, allowing it to better match the velocity of the optical signal. This substantially improves the modulation bandwidth.

The disadvantage of the buffer is a reduction in electric field strength in the vicinity of the optical mode, requiring a longer modulator to achieve the same phase change. In Figure 38.10, you can see the voltage contours of a design with a 100-nm SiO$_2$ buffer. Notice that a lot of potential is dropped across the buffer layer. This typically leads to a ~30% drop in modulator efficiency.

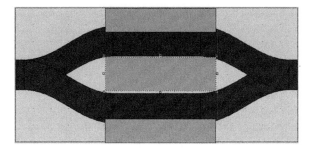

FIGURE 38.8 An LiNbO$_3$ modulator based on a Mach–Zehnder interferometer. A 3-dB splitter divides the input signal between two arms. One arm is given a positive electric field and the other arm a negative field, creating an increase in mode effective index in one branch and a similar decrease in effective index in the other.

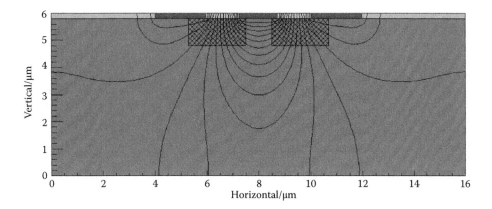

FIGURE 38.9 Cross section of a typical x-cut LiNbO$_3$ Mach–Zehnder modulator showing the Ti-diffused waveguides (hashed), three electrodes on top of the structure and electric potential contour lines. Typically, the outer electrodes are grounded and the central one carries the modulating signal. The opposite electrostatic field direction across each waveguide creates a natural push-pull effect for the Mach–Zehnder modulator. Note how the strongest field is generated in the middle of the waveguides.

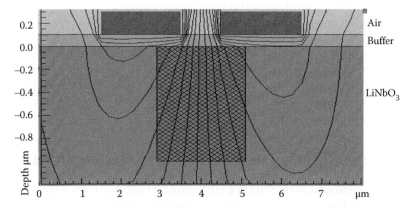

FIGURE 38.10 An x-cut LiNbO$_3$ modulator cross section with a cap layer creating an offstand between the electrodes and the optical mode. But note the strong field drop across the cap layer. This will reduce the electric field in the vicinity of the optical mode.

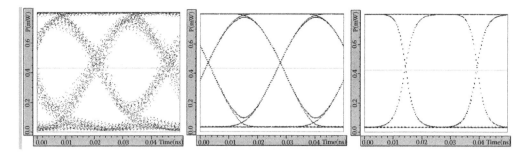

FIGURE 38.11 Eye diagrams of a traveling-wave LiNbO$_3$ MZI modulator at 40 Gb/s, computed with an optical circuit simulator PicWave [14] that includes traveling-wave time domain models for both the optical field and the electrodes: (left) with an unmatched electrode contacted a few millimeters from the end of the 20-mm-long modulator; (middle) contacted at the end; and (right) with a velocity-matched electrode.

38.3.2 Traveling-Wave Design

Because the change in effective index with voltage is typically very small, ~0.0001 per volt, the modulator has to be very long. This means that both the optical signal and the electrical signal will take a significant time to travel along the device. In order to maximize the modulation bandwidth, it is advantageous to match the velocity of the optical and electrical signals by appropriate design of the electrodes. Figure 38.11 shows some simulations of a traveling-wave LiNbO$_3$ Mach–Zehnder modulator under 40 Gb/s NRZ data stream. The figures show "eye diagrams" of the modulation response. To create an eye diagram the circuit is driven by a random bit sequence, and response of each clock period is superimposed. A clear open "eye" indicates a good signal.

38.4 Plasma-Effect Modulators

The plasma effect is the name given to the interaction of free electrons with an optical field. A plasma naturally occurs in a metal, which is why all metals are opaque. In order to make a modulator, one needs a way to control the carrier density, which is not realistic in a metal. In a semiconductor, the free-carrier population can be readily controlled by either injecting carriers into a diode (forward bias) or by lightly doping a region of the semiconductor and using a gate electrode or a reverse-biased junction to create an electrostatic field to deplete the carriers from the lightly doped region.

This section focuses on silicon-based plasma-effect modulators. Silicon is becoming a popular platform for manufacturing photonic components, largely because of the benefits of access to the huge electronic silicon manufacturing industry, its mature foundries and tools. Although silicon is not capable of generating light efficiently, i.e., making a laser, it is capable of absorbing light or modulating its phase by controlling the plasma effect in the optical path. Silicon optical modulators are now capable of reaching tens of gigahertz modulation speed, nearly rivalling the best LiNbO$_3$ modulators and in a much smaller volume, with much lower power requirements.

38.5 Si Electroabsorption versus Electrorefractive Effect

Silicon can act as an optical modulator either by a variation of the optical absorption in the silicon [15]—an "electroabsorptive modulator" or more popularly by modulation of the refractive index [16–19]—a so-called "electrorefractive modulator". In fact this distinction is somewhat arbitrary, since all changes in the real part of the refractive index will cause a change in the imaginary part, and vice versa. As discussed in Introduction, the real and imaginary changes do not generally happen at the same wavelength and the relationship is described by the so Kramers–Kronig relationship. At a wavelength of 1.55 μm and for low doping densities, the real and imaginary changes are given approximately by [17]

$$\Delta n = -8.8x10^{-22}\Delta N - 8.6x10^{-18}(\Delta P)^{0.8}$$

$$\Delta\alpha = +8.5\times10^{-18}\Delta N + 6\times10^{-22}\Delta P$$

A big disadvantage of an electrorefractive modulator is that usually we want to modulate the amplitude of an optical signal and to create an amplitude modulation out of such a modulator generally requires interference effects in for example, a Mach–Zehnder interferometer [18–20] or a ring resonator [16,17, 21,22]. Both of these have a strong wavelength-dependent response, especially the ring resonator. Despite this, silicon electrorefractive modulators have arguably had the greater impact due to their lower power and higher modulation bandwidths.

Some of the earliest silicon modulator designs are based on carrier injection in a PIN diode [20–22]. Here a forward current is injected into the diode and the free charge carriers in the intrinsic region cause a change in refractive index. Figure 38.12 shows the cross section of a typical carrier-injection PIN diode modulator. The effect is only weakly dependent on wavelength, though once included in a Mach–Zehnder or resonant cavity structure, this advantage is lost. The effect is however only weakly dependent on temperature unlike QCSE or other bandgap-dependent effects.

An alternative Si modulator structure reverse biases a pn (or PIN) junction [18,19,23,24]. The reverse bias causes carrier depletion in the junction, and an associated alteration of the local refractive index. Figure 38.13 shows a possible geometry, one with a vertical junction [8]. In general, it is preferable to move the p-n junction to one side so most modulation is done by p-type carriers. This is because the p-carriers create less absorption at wavelengths of interest for a given modulation depth, since the absorption peak is moved further out.

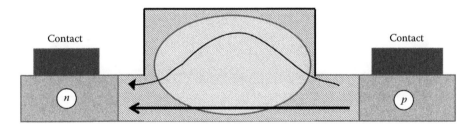

FIGURE 38.12 A PIN diode type silicon optical modulator. The silicon is etched to form a rib waveguide in the intrinsic region. Forward current interacts with the optical mode in the rib, changing its velocity or effective index.

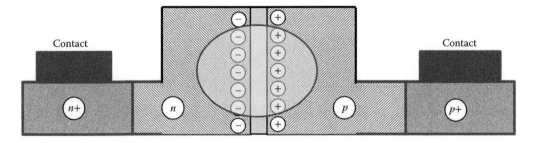

FIGURE 38.13 A reverse-bias type Si optical modulator. In these designs, the reverse bias caused electrostatic charge to build up at the p-n junction. These charges change the local refractive index.

Geometries similar to the former with asymmetric p-n junction have achieved a Vπ of 10 V in a 1-mm waveguide and bandwidths in excess of 50 Gb/s for a 1-mm long asymmetric Mach–Zehnder interferometer [19]. An experimentally measured eye diagram from this modulator is shown in Figure 38.14.

Figure 38.15 shows a recent design with a horizontal junction corresponding to a capacitor device [23]. The device is forward biased to cause carrier accumulation across the capacitor. This device achieved a modulation capability of 10 Gb/s at a length of 625 μm. Because the p and n carriers are close to each other, this design does not avoid the higher absorption of the n-type carriers.

Although most Si optical modulators are primarily electrorefractive, some electroabsorptive modulators have been demonstrated. In this case the free carriers are increased to a sufficient density that substantial FCA occurs. FCA is caused by carriers close to the band edge absorbing a photon and being excited to a state further up the conduction band (electrons) or down into the valence band (holes). The excited carriers then rapidly lose their energy by nonradiative transitions. An interesting design was recently proposed based on a Schottky diode [25]. The structure is predicted to reach a 3-dB bandwidth at 7 GHz for a 500-μm-long traveling-wave design. This is equivalent to a data rate of over 15 Gb/s using NRZ coding. Other groups have demonstrated EAMs using SiGe QWs—the QCSE effect (see Section 38.2) or by depositing graphene on the silicon [15,26,27].

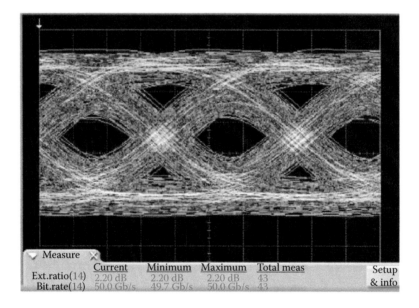

FIGURE 38.14 Experimental measurement of 50 Gb/s data stream in a depletion-type Silicon modulator in an asymmetric Mach–Zehnder interferometer configuration.

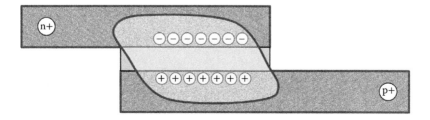

FIGURE 38.15 A capacitor-like structure also works by charge build up at the interface.

38.6 Resonant Modulators

In previous paragraphs, we outlined the use of traveling-wave designs such as the Mach–Zehnder interferometer to implement a modulator. Because the change in refractive index tends to be small for most modulator effects, this leads to long devices—hundreds of microns to tens of millimeter. This is not practical for many applications and so many have looked into resonant structures to reduce the modulator size. The principle is simple—in a resonant cavity, the light will remain trapped, bouncing or circulating around for a time much greater than L/c, where L is some characteristic dimension of the device. In this section, we focus on a popular structure—the ring resonator [16,17,21,22]—but many other designs are possible, for example, a photonic crystal cavity [28].

A resonator is popularly characterized by its "Q" factor, which can be defined as the photon lifetime divided by the optical period. A ring resonator has a roundtrip time of

$$t_{\text{round}} = \frac{2\pi R}{c}\mu \tag{38.19}$$

where μ is the characteristic refractive index, R is the ring radius. The photon lifetime of a cavity for optical frequency f is Q/f. This means that a ring resonator can increase the time spent in the ring by a factor of

$$\frac{Q\lambda_0}{2\pi R\mu} \tag{38.20}$$

This gives the modulator more time to change the phase of the signal, and allows the modulator to be shrunk be a corresponding factor. The disadvantage is that the ring resonator optical bandwidth is reduced to f/Q, making it sensitive to small changes in input wavelength.

A typical resonator configuration is shown in Figure 38.16 (left). This design has only one tangential waveguide and will act as a phase modulator. If a second tangential waveguide is added (Figure 38.16, right) then the modulator will direct the light into this when in resonance.

The high Q lowers the modulation speed of the device as well as its optical bandwidth. Figure 38.17 shows a PicWave [14] simulation of an eye diagram for a 20-μm-diameter silicon ring resonator modulator driven in carrier depletion mode with a 20 Gb/s NRZ bit pattern. Notice the overshoot caused by the high Q of the cavity, in this case over 6000. In practice, this ringing would be filtered out by a low-pass filter on the detector.

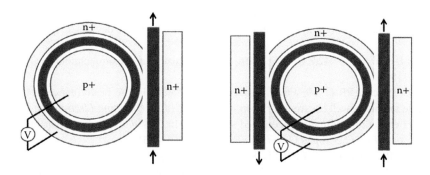

FIGURE 38.16 (Left) A silicon ring resonator modulator. The waveguides in blue (dark) are surrounded by electrodes imposing a reverse bias on the p-n junction (depletion-mode modulator). This design is a phase modulator. (Right) An amplitude modulator.

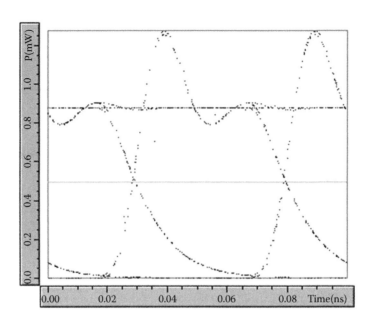

FIGURE 38.17 Eye diagram of a 20-μm-diameter silicon ring resonator modulator. The overshoot is caused by the high Q of the cavity.

References

1. Lefevre KR, Anwar AFM (1997) Electron escape time from single quantum wells. *IEEE J Quantum Electron*. 33: 187–191.
2. Hojfeldt S (2002) Modeling of carrier dynamics in electroabsorption modulators. PhD Thesis. Technical University of Denmark
3. Nikolaev V, Avrutin E (2003) Photocarrier escape time in quantum-well light-absorbing devices: Effects of electric field and well parameters. *IEEE J Quantum Electron*. 16: 24–26.
4. van Eisden J et al. (2007) Modulation properties of VCSEL with intracavity modulator. *Proc SPIE*. 6484: 64840A.
5. Piprek J (2003) *Semiconductor Optoelectronic Devices: Introduction to Physics and Simulations*. Academic Press London
6. Chao CYP, Chuang SL (1993) Momentum-space solution of exciton excited states and heavy-hole-light-hole mixing in quantum wells. *Phys Rev B*. 48: 8210–8221.
7. Mares PJ, Chuang SL (1993) Modeling of self-electro-optic-effect devices. *J Appl Phys*. 74: 1388–1397.
8. Alloatti L et al. (2011) 42.7 Gbit/s electro-optic modulator in silicon technology. *Opt Express*. 19: 11841–11851.
9. Cho HS, Prucnal PR (1989) Effect of parameter variation in the performance of GaAs/AlGaAs multiple-quantum-well electroabsorption modulator. *IEEE J Quantum Electron*. 25: 1682–1690.
10. Kuo YH et al. (2006) Quantum-confined stark effect in Ge–SiGe quantum wells on Si for optical modulators. *IEEE J Sel Topics Quantum Electron*. 12: 1503–1513.
11. Alferness RC, Joyner CH, Buhl LL, Korotky SK (1983) High-speed travelling-wave directional coupler switch/modulator for l = 1.32 μm. *IEEE J Quantum Electron*. 19: 1339–1341.
12. Kubota K, Noda J, Mikami O (1980) Traveling wave optical modulator using a directional coupler LiNbO₃ waveguide. *IEEE J Quantum Electron*. 16: 754–760.

13. Kondo J et al. (2002) 40-Gb/s X-cut LiNbO$_3$ optical modulator with two-step back-slot structure. *J Lightwave Technol.* 20: 2110–2114.
14. PicWave - optoelectronic simulator, ver. 5. www.photond.com/products/picwave.htm.
15. Hu YT et al. (2014) Broadband 10 Gb/s graphene electro-absorption modulator on silicon for chip-level optical interconnects. In Proceedings of 2014 IEEE Electron Devices Meeting (IEDM), San Francisco, CA.
16. Dong P et al. (2009) Low Vpp, ultralow-energy, compact, high-speed silicon electro-optic modulator. *Opt Express.* 17: 22484–22490.
17. Lipson M (2006) Compact electro-optic modulators on a silicon chip. *IEEE J Sel Topics Quantum Electron.* 12: 1520–1526.
18. Liu A et al. (2007) High-speed optical modulation based on carrier depletion in a silicon waveguide. *Opt Express.* 15: 660–668.
19. Thompson D et al. (2012) 50-Gb/s silicon optical modulator. *IEEE Phot Tech Lett.* 24: 234–236.
20. Zhou GR et al. (2008) Effect of carrier lifetime on forward-biased silicon Mach-Zehnder modulators. *Opt Express.* 16: 5218–5226.
21. Manipatruni S, Xu Q, Schmidt B, Shakya I, Lipson M (2007) High speed carrier injection 18 Gb/s silicon micro-ring electro-optic modulator. In Proceedings of 2007 Lasers and Electro-Optics Society, Lake Buena Vista, FL (LEOS 2007), 537–538.
22. Xu Q, Manipatruni S, Schmidt B, Shakya J, Lipson M (2007) 12.5 Gbit/s carrier-injection-based silicon microring silicon modulators. *Opt Express.* 15: 430–436.
23. Abraham A, Olivier S, Marris-Morini D, Vivien L (2014) Evaluation of the performances of a silicon optical modulator based on a silicon-oxide-silicon capacitor. In Proceedings of 2014 IEEE 11th International Conference on Group IV Photonics, 3–4.
24. Gardes FY, Tsakmakidis KL, Thomson D, Reed GT, Mashanovich GZ, Hess O (2007) Micrometer size polarisation independent depletion-type photonic modulator in Silicon on insulator. *Opt Express.* 15: 5879–5884.
25. Jeong U, Han D, Lee D, Lee K, Kim J, Park J (2014) A broadband silicon electro-absorption modulator (EAM) using a Schottky diode. *Proc SPIE.* 8988: 89881M.
26. Liu M, Yin X, Zhang X (2012) Double-layer graphene optical modulator. *Nano Lett.* 12(3): 1482–1485.
27. Phare CT, Lee YD, Cardenas J, Lipson M (2015) Graphene electro-optic modulator with 30 GHz bandwidth. *Nat Photonics.* 9: 511–514.
28. Tanabe T, Nishiguchi K, Kuramochi E, Notomi M (2009) Low power and fast electro-optic silicon modulator with lateral p-i-n embedded photonic crystal nanocavity. *Opt Express.* 17: 22505–22513.
29. Chao CYP, Chuang SL (1992) Spin-orbit-coupling effects on the valence-band structure of strained semiconductor quantum wells. *Phys Rev B.* 46: 4110–4122.
30. Lever L et al. (2010) Design of Ge-SiGe quantum-confined stark effect electroabsorption heterostructures for CMOS compatible photonics. *J Lightwave Technol.* 28: 3273–3281.
31. Weiner JS et al. (1985) Strong polarization-sensitive electroabsorption in GaAs/AlGaAs quantum well waveguides. *Appl Phys Lett.* 43: 1148–1150.

VIII

Solar Cells

39

Solar Cell Fundamentals

Matthias Müller

39.1 Introduction

Solar cells convert the energy of the sun light to electricity. It is an optical to electrical energy conversion. In principle, they make use of the photovoltaic (PV) effect, i.e., the generation of a potential difference between two electrodes under illumination. It was firstly observed by A. E. Becquerel (Becquerel, 1839) and practically implemented in a solar cell at Bell Laboratories in the United States (Chapin et al., 1954). Today it is believed that PV energy generation will be a part of the world's base of energy supply established in the twenty-first century. Conservative estimates by the International Energy Agency (IEA) show a share of the world electrical energy generation of about 1% nowadays to about 16% in 2050 (IEA, 2014). This would correspond to a yearly energy generation of about 6300 TWh compared to slightly more than 200 TWh today.

Looking back on the past decades, we can see that the development in solar research and industry was very fast and economically successful, which enabled such a bright perspective for solar electricity. The price for solar modules reduced with increasing manufactured PV module capacity following a power-law relation: the so called PV learning curve could be established which is shown in Figure 39.1.

For example, the price of solar modules decreased by about 80% between the years 2008 and 2015. About 1000 GW module shipments may be reached in 2024 for 10% market growth, or even in 2021 for 30% market growth, which corresponded to the market development in the past 15 years. Those cost reductions

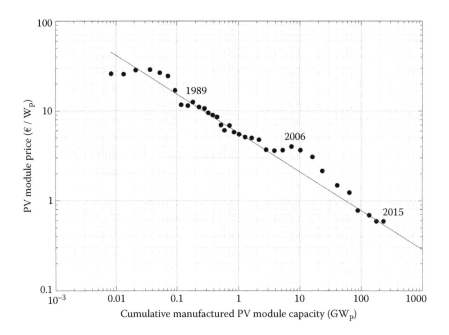

FIGURE 39.1 PV learning curve (Fraunhofer ISE, 2013; IEA, 2014; pvXchange, 2016) and a fit of a power law model to the years 1990–2000 and 2012–2015. In between the years 2000–2010, there was a shortage on raw silicon which slowed down the price reduction.

result largely from economies of scale and technological improvements. In 2015, the limit of 200 GW-shipped solar modules was exceeded, and the same volume is expected to be the future world production capacity to maintain the global solar energy supply in the twenty-first century starting from 2025 (IEA, 2014).

The current solar cell production is dominated by wafer-based crystalline silicon (c-Si) solar cells, which make about 90% of the market (ITRPV, 2015) and only 10% are thin-film technologies based mainly on cadmium telluride (CdTe) and copper indium gallium diselenide (CIGS). That is why the author focuses on the c-Si technology within this chapter.

Looking closely on c-Si, the shares of different wafer technologies that are applied nowadays are shown in the upper graph of Figure 39.2 (ITRPV, 2015), where mainly boron-doped p-type material is used (95%). The cast silicon materials multicrystalline (mc) and high-performance multicrystalline (HPmc-Si) currently make up to over 65%, whereas monocrystalline, mainly Czochralski (Cz), make 35%.

The share of different solar cell technologies is shown in the lower graph of Figure 39.2, where the main product currently is the full-area back-surface-field (BSF) solar cell (Mandelkorn and Lamneck, 1972), whereas the passivated emitter and rear cell (PERC) technology (Blakers et al., 1989) shows increasing market shares in the near future (ITRPV, 2015). The passivated emitter and rear totally diffused solar cell (PERT) (Zhao et al., 1999), the Si-heterojunction (SHJ) solar cell (Tanaka et al., 1992) and the back contacted solar cells—mainly interdigitated back contact solar cells (IBC) (Lammert and Schwartz, 1977)—are expected to play a minor role in the coming years but may be candidates for future high-efficiency solar cells with higher market shares (ITRPV, 2015). The structure of these different solar cells is shown in Figure 39.3.

This chapter gives a general overview about the working principle and basic models of solar cells. The view will be extended to the energy yield delivered by solar cells under real world conditions, i.e., the view will be extended to solar cells working in solar modules and the corresponding cell demands. Examples will

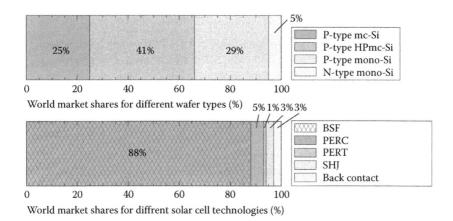

FIGURE 39.2 Share of different silicon wafer technologies for solar cell application and share of different solar cell technologies. [BSF, back-surface-field; PERT, passivated emitter and rear totally diffused; PERC, passivated emitter and rear cell; SHJ, Si-heterojunction. (ITRPV, 2015)]

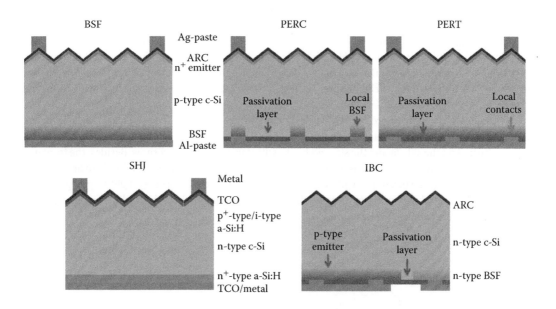

FIGURE 39.3 Schematic of typical solar cell structures in their two-dimensional cross section representation where antireflection coating (ARC) and transparent conductive oxide (TCO) are used. (BSF, back-surface-field; PERC, passivated emitter and rear cell; PERT, passivated emitter and rear totally diffused; SHJ, Si-heterojunction; IBC; interdigitated back contact solar cell.)

be presented considering only nonconcentrated crystalline silicon-based solar cells for 1-sun application. For additional information and explanations which are not given in this chapter, the author would like to refer to the notable literature by Wuerfel (2005), Green (1982), and Goetzberger et al. (1998). Additionally, the author would like to refer to the free online resources for PV engineers and scientists PVEducation (PVEDUCATION.ORG, 2016) and PV Lighthouse (PV Lighthouse, 2016).

39.2 The PV System

The solar cell is only a part, even so the functional device, of a full PV system. Such systems may be grid-connected as shown in Figure 39.4, or stand-alone systems, which means that all produced electricity is consumed locally. Those stand-alone PV systems are good options for rural areas of developing countries and special applications (e.g., parking meter). However, in comparison, grid-connected systems produce nowadays and will produce in the future a higher share of electricity. A grid-connected system consists of solar modules, which contain the solar cells. These modules are typically connected in series (called module string) to an inverter. Those inverters first do Maximum Power Point Tracking (MPPT) of the module strings and secondly convert the direct current produced by solar cells in alternating current. The inverter is connected to the load and to the electricity grid via an AC isolation switch and via an import/export electricity meter (compare Figure 39.4).

A current trend for grid-connected systems is the combination with storage systems. It increases energy self-consumption, as it is expected that during noon an excess of solar energy is produced. This excess energy can then be stored and used during times without irradiation. The trend will be enhanced, as the price for electricity is decreasing more and more at noon time when more solar power plants are connected to the grid. So the storage system will help for grid stabilization as PV electricity is a fluctuating energy challenging the electricity grid in the future. Those battery PV systems are controlled by an energy management system (EMS) which regulates production, load, storage, and grid connection (compare Figure 39.4).

Looking in more detail at the structure of a solar module as shown in Figure 39.5, one can see that solar cells are basically embedded between glass and plastics.

A typical solar module contains 60 solar cells on an area of approximately 1.7 m^2. They are connected in series by tabbing ribbons to strings of 20 cells while bus ribbons interconnect the three cell strings. Those cell strings are connected in parallel by a bypass diode which accounts for lower power losses during shading conditions. Finally, the bus ribbons are connected to the junction box, where cables with connectors go out.

Solar modules are made to produce electricity for at least 20–30 years; hence all used materials have to withstand changing environmental conditions. Solar glass, that in most cases is low-iron rolled glass or float glass, is situated on the side facing the sky and has to withstand rain, hail and snow. Besides, the front glass guarantees mechanical stability against wind and snow load. It may be manufactured with an antireflection coating (ARC) and/or textured, i.e., with a structured front surface for improved light trapping. The rear side is typically covered with polyethylene terephthalate (PET) polymer backsheet which features moisture resistance and electrical isolation. If replaced by another glass to produce a glass-glass module, the solar

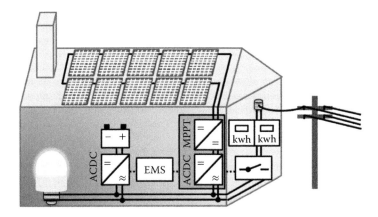

FIGURE 39.4 Schematic of a PV system with solar modules, cables, inverter, battery energy managements system (EMS), battery, load, AC isolation switch, and import/export electricity meter.

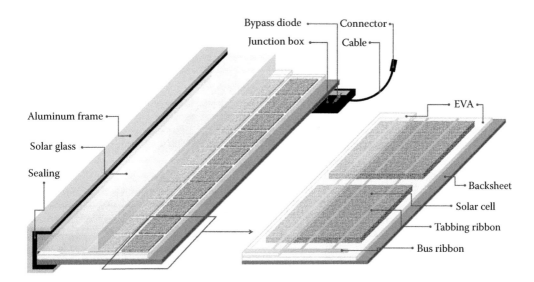

FIGURE 39.5 Structure of a solar module. EVA, ethylenvinylacetate.

module lifetime increases by about 5–10 years. Finally, the solar cells are embedded between the solar glass and the backsheet in Ethylenvinylacetate (EVA) as an encapsulant. It is a polymer which cross-links at about 150°C during manufacturing. Solar modules are sealed at the edges with silicone and an aluminum frame.

39.3 The Solar Spectrum

Our sun is basically a fusion power plant which emits electromagnetic radiation from its surface, called photosphere, into the solar system. It has a spectrum similar to a blackbody with a temperature of 5777 K, which is shown in Figure 39.6.

The light intensity reaching the planet earth is reduced by a factor of 2.17×10^{-5} due to geometrical considerations. The factor is the square of the suns radius divided by the square of the distance sun–earth. The resulting spectrum above the earth's atmosphere is shown in Figure 39.6 denoted as AM0, where AM stands for air mass index and the "zero" denotes that the light doesn't travel through the earth's atmosphere. This extraterrestrial solar spectrum AM0 has an irradiance of 1366.1 W/m^2 as defined within (ASTM E490-00a, 2014), which is called the solar constant. The latest measured value is 1360.8 W/m^2 (Kopp and Lean, 2011). This spectrum is relevant for solar cell power measurements for space applications.

Light traveling through the atmosphere faces absorption, scattering, and reflection effects which change the spectrum and the irradiance as shown schematically in Figure 39.7.

Absorption in the earth's atmosphere takes place in its molecules: ozone, oxygen, water vapor, carbon dioxide, nitrogen dioxide. Scattering in the atmosphere takes place at particles that are smaller than the wavelength of the light (Rayleigh scattering, e.g., N_2, O_2 molecules) and which are larger than the wavelength of the light (Mie scattering, e.g., aerosol particles, cloud droplets). The third mechanism is reflection of the light in the atmosphere, mainly at the clouds, and finally at the ground. This introduces a direct, a diffuse, and a ground-reflected component of irradiance on the module plane. The diffuse component describes all light coming from any region of the sky, mainly Rayleigh-scattered light which we see as blue sky on sunny days.

The standard spectrum AM1.5g is the result of light passing the earth's atmosphere approximately on a very sunny day and it is shown in Figure 39.6. The air mass index AM defines the direct optical path length through the atmosphere of the earth. Thus, AM1.5 is one and a half times the path length; "g" stands for

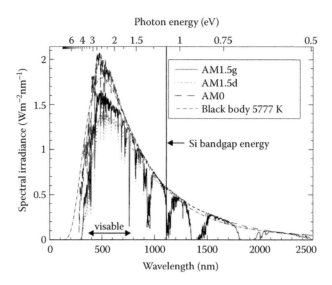

FIGURE 39.6 Comparison of different spectra: blackbody spectrum with a temperature of 5777 K, extraterrestrial spectrum AM0 (ASTM E490-00a, 2014), reference spectra AM1.5g and AM1.5d (ASTM G173-03, 2012) for global and direct solar spectral irradiance at AM1.5, respectively.

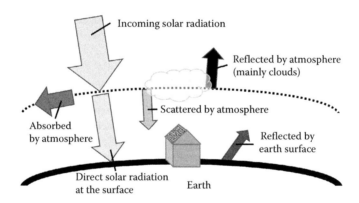

FIGURE 39.7 Schematic of solar radiation traveling through the earth's atmosphere.

global, i.e., with direct and diffuse light. The spectrum AM1.5d, which is also shown in Figure 39.6, is only the direct component of the AM1.5g and is used to measure solar cells for concentrated PV (CPV) applications.

39.4 Real Operating Conditions of Non-concentrated Solar Cells and Modules

The output power of a solar system, typically expressed in "Watt-peak" (W_P), and the efficiency of a solar system or solar components like solar cells or solar modules are determined by a standardized measurement of the current–voltage (I-V) characteristic under the "Standard Test Conditions" (STCs), i.e., under a light intensity of 1000 W/m^2 (one-sun), an ambient temperature of 25°C and the standardized spectrum AM1.5g (ASTM G173-03, 2012) which may altogether correspond approximately to a very sunny summer day in the morning on a 37° sun facing tilted surface with low ambient temperature.

Unfortunately, those environmental conditions do not occur very often. During the year and day-time deviations from the reference efficiency under STC are observed and the efficiency of the solar system changes due to a severe number of effects:

- Efficiency decrease due to module heat generation above 25°C
- Efficiency decrease due to low light/partial load condition (e.g., cloudy weather)
- Reflection losses due to incident light under increased angle of incidence (AOI) between the module and the sun
- Decreased or increased efficiency due to the change of the spectrum of the incident light
- Mismatch loss due to interconnection of cells in a module and solar modules in a PV system with different I-V characteristics
- Losses due to soiling and snow
- Losses due to shading
- Losses due to interconnection and (bypass) diode failures
- MPPT failure of the MPP tracker
- Conversion losses and self-consumption of the inverter (stand-by loss)
- Energy yield loss due to differences in real, measured and labeled module power
- Degradation, i.e., efficiency decrease over time due to environmental stress load

The effect of efficiency degradation of solar modules happens mainly due to mechanical stress and moisture resulting in corrosion, discoloration, delamination, breakage, and cracking cells (Jordan and Kurtz, 2013), but also due to solar cell degradation resulting from the formation of defects within the silicon wafers under illumination, e.g., the boron-oxygen-related defect (Schmidt et al., 1997). Another degradation mechanism may occur in solar modules under high potential (Pingel et al., 2010), which is called potential-induced degradation (PID). PID results in a low shunt resistance of the module and can decrease the output power down to zero. Nevertheless, the solar module lifetime exceeds 20–25 years maintaining typically at least 80% of its initial efficiency or for a glass-glass module even more than 30 years. Thus a solar module is one of the longest lasting products for electricity generation that does not need maintenance.

A parameter that considers all the real world influences on a PV system is the performance ratio *PR*, as defined by

$$PR = \frac{\text{real energy yield}}{\text{ideal energy yield}} = \frac{W_{SP}}{H_{sun} \times A_P \times \eta_{STC}} \tag{39.1}$$

PR is defined as the actual energy yield of a PV system W_{SP} related to the incoming sun irradiation H_{sun} on the array area A_P converted by the solar system efficiency under STC η_{STC} over a time interval of months/years. Also a wide spread parameter for PV system assessment is the specific or final system yield Y_f in kWh/kWp, as defined by Equation 39.2, which allows to compare the energy production for different systems at similar locations. Typical values are in the range of 500–2500 kWh/kWp.

$$Y_f = \frac{W_{SP}}{P_{syst}} \tag{39.2}$$

39.4.1 Irradiation

The frequency with which STC applies in real world is negligible. For a better understanding of which environmental conditions are more typical for a solar system, exemplarily three geographical locations are chosen for further visualization within this chapter, which are situated in middle-Europe (Halle, Germany), southern-Europe (Almeria, Spain) and in tropical climate close to the Equator (Brazzaville, Republic of the Congo).

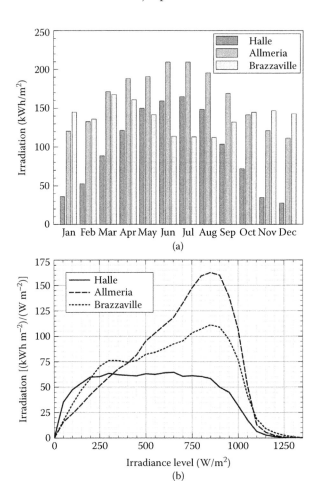

FIGURE 39.8   (a) Comparison of different irradiation conditions (Meteotest, 2008) per month, and lower graph versus in-plane irradiance level. (b) A histogram in steps of 50 W/m^2 per irradiance level bin where the frequency is multiplied by the mean irradiance of each bin.

An overview about the applying in-plane irradiation conditions, i.e., irradiation on the tilted surface, are shown in the upper graph of Figure 39.8 per month and in the lower graph versus irradiance level.

The geometric considerations concerning sun light hitting a tilted surface may be found in references (Duffie and Beckman, 2013; Quaschning, 2005). The yearly solar energy yield, i.e., in-plane irradiation, is 1159 kWh/m^2 for Halle, 1961 kWh/m^2 for Almeria and 1657 kWh/m^2 for Brazzaville for tilted surfaces having tilted angles of 30° for Halle and Almeria, and 10° for Brazzaville, respectively.

Meteorological data may be found worldwide at NASA Surface Meteorology and Solar Energy Data Set (NASA, 2016) and using the software Meteonorm (Meteotest, 2008), for EU at PVGIS EU Joint Research Centre (JRC, 2016), and for US at National Renewable Energy Laboratory (NREL, 2016).

39.4.2 Temperature

The illumination of solar modules does not only generate electricity, but also heat. The main reasons are the heat generation of the solar cells due to thermalization of charge carriers and parasitic absorption within the module of photons with higher wavelength than the bandgap of the semiconductor. A typical solar cell operating temperature increases linearly with the in-plane irradiance G_I above ambient temperature.

As commercial solar modules have a very similar structure as shown in Figure 39.5, they have comparable thermal properties and heat in a similar way. The thermal properties are measured in a standardized way determining the "Nominal Operating Cell Temperature" (NOCT) (IEC 61215, 2005) of the solar module under 800 W/m^2 light intensity, 20°C ambient temperature, AM1.5g spectrum, and 1 m/s wind velocity, which results in typical NOCT = 45°C ± 3°C.

For a better understanding of the system efficiency losses due to different operating temperatures deviating from 25°C (STC), an annual average operating temperature T_{ann} is calculated from the meteorological data set shown in Figure 39.8. The solar cell operating temperature T_{Cell} is calculated by (Ross and Smokler, 1986)

$$T_{Cell} = T_{amb} + (NOCT - 20°C)\frac{G_I}{800\frac{W}{m^2}}$$ (39.3)

This model represents a standard roof top PV system installation. For an open-rack mounting condition lower T_{Cell} can be expected whereas for building integrated installations higher T_{Cell} is expected (Garcia and Balenzategui, 2004).

Assuming that a PV system only depends on irradiation and different ambient temperatures, the annual average (irradiance weighted) cell operating temperature T_{ann} is calculated by

$$T_{ann} = \frac{\int_{1a} P(t) \times T_{Cell}(t)dt}{\int_{1a} P(t)dt} = \frac{1}{W_{SP}}\int_{1a} P(t) \times T_{Cell}(t)dt$$ (39.4)

Exemplarily, the monthly average operating temperature is calculated and shown in Figure 39.9, assuming a *NOCT* of 45°C and applying Equations 39.3 and 39.4 to the meteorological data sets.

The annual average cell operating temperature for the selected locations are those of 32.0°C for Halle, 42.9°C for Almeria, and 47.7°C for Brazzaville, respectively. Hence Almeria and Brazzaville are close to the *NOCT* = 45°C, which confirms that NOCT is a relevant environmental condition for many locations. Assuming that the relative temperature coefficient of the efficiency of a typical c-Si solar cell is −0.4%$_{rel}$/°C, an energy yield loss compared to STC may be calculated based on the annual average cell operating temperatures. For Halle (32.0°C − 25.0°C) × −0.4$\frac{\%_{rel}}{°C}$ = −2.8%$_{rel}$ may be expected, for Almeria −7.2%$_{rel}$, and for Brazzaville −9.1%$_{rel}$, respectively.

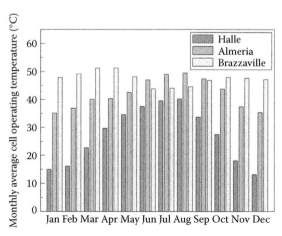

FIGURE 39.9 Monthly average (irradiance weighted) cell operating temperature per month for the locations Halle, Almeria, and Brazzaville.

39.5 Solar Cell Optics

In the field and under real operating conditions, solar cell's optical behavior is mainly determined by the optics of a solar module. In more detail, it strongly depends on the optics of the first medium and its surface structure facing the air (Winter et al., 2014), which is typically the solar glass. During year and day time, sun light appears under different oblique incidence to the module. An angle of incidence (AOI) θ can be defined, which is the angle between the sun's rays and the module normal. Light reaching the module's surface at an angle is spread out over a larger area, thus reduces the (direct) irradiance on the tilted plane following a cosine relation, as shown in Figure 39.10, described by:

$$G_{\text{direct}}(\theta) = G_{\text{direct}}(0^\circ) \times \cos(\theta) \tag{39.5}$$

The first interface air–glass may be simply described by the Fresnel equations and Beer–Lambert–Bouguer and Snell's laws (Duffie and Beckman, 2013), which are in general wavelength dependent. The angle of refraction θ_r is therefore

$$\theta_r = \arcsin\left(\frac{n_1}{n_2}\sin(\theta)\right) \tag{39.6}$$

where $n_1 = 1$ and $n_2 = 1.52$ are the refractive indices of air and solar glass, respectively. The angle-dependent transmission of such system is approximated by Equation 39.7 considering both reflective losses at the interface and absorption losses within the glass (Duffie and Beckman, 2013).

$$\left(A_{\text{opt}}T_{\text{opt}}\right)(\theta) = A_{\text{opt}}\left(1 - R_{\text{opt}}\right) = e^{-(KL/\cos\theta_r)}\left[1 - \frac{1}{2}\left(\frac{\left(\sin\left(\theta_r - \theta\right)\right)^2}{\left(\sin\left(\theta_r + \theta\right)\right)^2} + \frac{\left(\tan\left(\theta_r - \theta\right)\right)^2}{\left(\tan\left(\theta_r + \theta\right)\right)^2}\right)\right] \tag{39.7}$$

where $K = 4\ \text{m}^{-1}$ may be assumed for the extinction coefficient and $L = 2$ mm for the thickness of the glass (Duffie and Beckman, 2013).

The calculation of the product of optical absorption and transmission through the solar glass ($A_{\text{opt}}T_{\text{opt}}$) approximates here the angle-dependent losses of the solar module, which is defined as the incident angle modifier (IAM). However, *IAM* accounts for all angular and spectral losses due to increased absorption

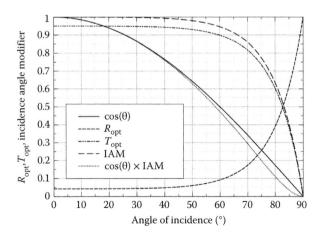

FIGURE 39.10 Cosine, optical reflection, and transmission ($n_1 = 1$ for air and $n_2 = 1.52$ for glass), incidence angle modifier (IAM), angle-dependent losses of a solar module versus angle of incidence.

within the EVA and for geometrically different, wavelength-dependent light propagation under oblique incident, i.e., multiple reflections at the cell or the backsheet. *IAM*, also shown in Figure 39.10, is used to correct the incoming direct irradiance G_{direct} on the tilted module plane as derived in Equation 39.8. The indirect irradiance component may be corrected in a similar approach.

$$G_{\text{direct}}(\theta) = G_{\text{direct}} \times \cos\theta \times IAM(\theta) = G_{\text{direct}} \times \cos\theta \times \frac{\left(A_{\text{opt}} T_{\text{opt}}\right)(\theta)}{\left(A_{\text{opt}} T_{\text{opt}}\right)(0°)} \tag{39.8}$$

Calculations for the angular-dependent irradiation losses over 1 year using Equations 39.5 through 39.8 and the meteorological data sets, as shown in Figure 39.8, result in $-1.3\%_{\text{rel}}$ for Halle, $-1.3\%_{\text{rel}}$ for Almeria, and $-0.9\%_{\text{rel}}$ for Brazzaville.

Latest developments of solar glass lead more and more to the introduction of ARCs and partially deeply structured (textured) glass, which improves the module optics.

The same concepts are also applied for solar cell optic optimization. ARCs for solar cells, as drawn in Figure 39.11, are typically applied as a single layer of silicon-nitride SiN_X films of about 75 nm thickness and a refractive index of about 2, whereas its optimum is slightly lower against air and slightly higher if optimized for the use in a module.

Those thin film layers may be well modeled by the transfer matrix method, which considers oblique incidences, absorbing media, coherent light, and even assemblies of thin films (e.g., double ARC) (Macleod, 2010). The second optimization strategy is to apply a texture on the front side of the solar cell, which increases the chance that reflected light reaches again the solar cell surface a second or even more times as shown in Figure 39.11. Most relevant textures for industrial application are random pyramids for monocrystalline (Cz) cells and isotexture, which forms bowl-like structures, for mc cells. Both are produced by wet chemical etching. Additional benefits of texturing arise because light is coupled into the silicon under an angle. First, this increases the chance of a photon to be absorbed close to the surface, second the path length for absorption is increased within the semiconductor and thirdly such photons may reach the rear side under the critical angle of total internal reflection. This internal light trapping may be even improved further if the rear side has properties of a back reflector, as reflected light of wavelengths close to the bandgap energy gets the chance to be absorbed on the way there and back (at least double length) within the semiconductor. Those back reflectors are, e.g., full area Aluminum layers for a BSF solar cell or dielectric layer stacks for PERC solar cells. Note that semiconductors having a direct bandgap (e.g., GaAs, CIGS, CdTe) may be fabricated with much smaller absorber thickness compared to, e.g., silicon with an indirect bandgap, and are called thin film solar cells.

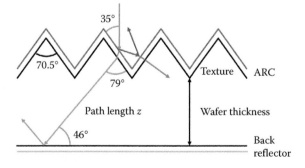

FIGURE 39.11 Schematic of light trapping within a solar cell with antireflection coating (ARC) and a random pyramid texture. Note that the schematic is not to scale, as a typical side length of such pyramid is about 5 μm and a wafer thickness is about 200μm.

39.5.1 Photogenerated Current

A way to model light-induced photocurrent of solar cells, as shown in Figure 39.11, is to use ray-tracing, which is a method to calculate the path of a photon through the solar cell considering its geometry. The light propagation may be calculated at surfaces and interfaces as necessary, e.g., using Fresnel equations. Raytracing software for solar cells are, e.g., Sunrays (Brendel, 1994), TCAD Sentaurus Device (Synopsys, 2016) and for solar modules, e.g., Daidalos (Holst et al., 2013). Also wide spread in the PV community is the usage of the simulation tool PC1D (Rover et al., 1985), which can also analytically calculate solar cell optics with texture and ARC.

The absorption of photons within the active semiconductor leads to the creation of electron–hole pairs, i.e., to a current generation. This photogenerated current density j_{gen} is not influenced by front contact shading and recombination losses and hence can be treated as the maximum current potentially extracted from the solar cell described by

$$j_{gen} = -q \int_0^\infty A_{opt} \left(E_{ph} \right) dj_{ph}(E_{ph}) \tag{39.9}$$

where q is the elementary charge, E_{ph} is the photon energy and j_{ph} is the photon flux. Assuming $A_{opt}(E_{ph} \geq E_g) = 1$, Equation 39.9 solves to

$$j_{gen} = -q \int_{E_g}^\infty dj_{ph}(E_{ph}) \tag{39.10}$$

The optical absorption in silicon is described by its absorption coefficient α which is shown in Figure 39.12.

The maximal achievable photogenerated current density for c-Si is about 46 mA/cm^2 for AM1.5g reference spectrum. Photons with more energy than the bandgap energy E_g quickly loose this "extra" energy to the lattice atoms in a process called thermalization. Photons with lower energy than E_g are not able to excite any electron from valence to conduction band and are not absorbed under the creation of an electron–hole pair.

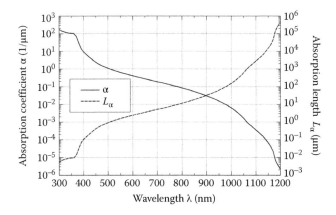

FIGURE 39.12 Absorption coefficient α and absorption length $L_\alpha = 1/\alpha$ for silicon at 300 K.

39.6 Conversion into Electrical Energy

The conversion of the optical energy of the sun light into electrical energy can be regarded in general in terms of particles in a two-level energy system. The particles are excited from their ground state to an excited state by illumination and carry the energy of the photon in the excited state. In general, the excited particles need to be extracted from the absorber into an external electrical circuit; otherwise recombination takes place which is the reverse reaction to excitation, also called generation. This is realized by the usage of selective contacts which only allow the excited particle to flow out at one side of the absorber, preventing the "overflow" of the absorber, which is the driving force. The excited particles flow into the external electrical circuit to perform work and then flow back into the absorber as relaxed particles through the other contact.

A possible realization for selective contacts is to create oppositely situated highly n- and p-type doped regions in the absorbing semiconductor for the respective contact, which is shown in the schematic of an illuminated semiconductor-based solar cell in Figure 39.13. Under illumination-free electrons and holes (excess charge carriers), introduced as excited particles, are generated in the conduction band and valence band, respectively. Thus the quasi-Fermi-levels of electrons E_{Fn} and holes E_{Fp} split in the semiconductor describing the energy distribution of electrons and holes by Fermi–Dirac statistics within the conduction and valence band, respectively. The Fermi levels of the majority carriers join at the surfaces at the respective metal contacts because the surface states exchange electrons very easily with the metal and maintain the same potential. The opposing doping at each contact results in three main effects. First, low minority carrier densities occur at the contacts due to the law of mass action and hence lead to almost zero contact recombination (selective contact). In consequence, only one kind of charge carrier is extracted. Second, it can result in ohmic contact behavior with sufficiently high doping close to the contact, which leads to low resistive losses at the contacts. Third, the creation of p-type and n-type regions in the semiconductor leads to the creation of a p-n junction with a SCR and a built-in potential, also called diffusion voltage, which is typically thought to be a necessity for solar cells, but it is not the "engine" of a solar cell (Wuerfel, 2005). Both carrier types are continuously created under illumination, and, e.g., under short-circuit condition electrons and holes flow out at their respective selective contacts as the reverse direction is blocked for them. Thus an excess of charge carriers is produced and the majority of excess charge carriers do not recombine before they diffuse along their concentration gradients. This creates a charge current through the solar cell. As charge current flows are the result of potential gradients, the behavior of solar cells is described below in this section in terms of potentials in more detail. Under an external forward bias, the difference of E_{Fn} in the n-type region and E_{Fp} in the p-type region correspond to the external voltage V, so that

$$E_{Fn} - E_{Fp} = qV \qquad (39.11)$$

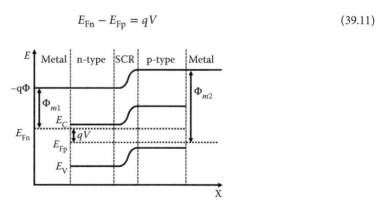

FIGURE 39.13 Schematic of the band diagram of an illuminated solar cell with two selective contacts with their respective metal work functions Φ_m separated by a p-n junction with a space-charge region (SCR) and the electrostatic potential Φ.

where q is the elementary charge. The forward bias reduces the built-in potential of the p-n junction. The driving forces of charge currents in a solar cell are the gradients of the quasi-Fermi-levels E_{Fn}/dx and E_{Fp}/dx, which are the gradients of the electrochemical potentials for electrons and holes, respectively. The electrochemical potential of electrons, which is constant in thermal equilibrium, is determined by the chemical potential $E_C - E_{Fn}$ and the electrostatic potential $-q\Phi$:

$$E_R - E_{Fn} = (E_C - E_{Fn}) - q\phi \tag{39.12}$$

where E_R is the reference potential. The absolute value of E_R is not relevant as only gradients of the potentials are of interest. The chemical potential describes the diffusion aspect of carrier transport due to a gradient in the carrier concentration whereas the electrostatic potential describes the force acting on charge carriers due to coulomb interaction and it is the drift aspect of charge carriers transport due to an electric field. Those two transport mechanisms have to be considered together, because diffusing mobile electrons and holes inherently carry a charge.

39.7 Modeling of Ideal *I–V* Characteristics

A PV solar cell can in first-order approximation be described in an equivalent circuit model as a photocurrent generator in parallel with a diode. It corresponds to the single diode equation which will be derived below, in a similar way to the reference (Green, 2003). In order to derive the model, the following assumptions are made: infinite thickness of the absorber, negligible contact resistances, zero contact recombination, zero surface recombination, low injection condition, i.e., the excess charge carrier density is much smaller than the doping density ($\Delta n \ll N_{dop}$). Under open-circuit condition, there is not a charge carrier flow (no current) into the external circuit resulting in the recombination of all generated charge carriers in the solar cell. Thus the diode equation can be determined from the net recombination rate. The continuity equation for electrons in one dimension under steady-state condition is

$$q\frac{\partial n}{\partial t} = 0 = \nabla \cdot j_n + q\left(G_n - U_n\right) = \nabla \cdot j_n + qG - q\frac{\Delta n}{\tau_n} \tag{39.13}$$

where n is the electron carrier density, j_n is the electron current density, G_n is the generation rate, U_n is the recombination rate and τ_n can be interpreted as the average time after which an excess minority carrier recombines. Solving Equation 39.13 for j_n leads to

$$j_n = q\int_0^{L_n} \left[G_n - q\frac{\Delta n}{\tau_n}\right] dx = qG_n L_n + q\int_0^{L_n} \frac{\Delta n}{\tau_n} dx \tag{39.14}$$

where L_n is the diffusion lengths for electrons. The relation between carrier lifetime and diffusion length for electrons is determined by the electron diffusion coefficient D_n:

$$\tau_n = \frac{L_n^2}{D_n} \tag{39.15}$$

Respective equations are derived for holes as minority carriers relevant for the n-type region. The excess charge carrier density Δn may be derived from the law of mass action at all dopant and injection conditions from

$$np = \left(N_D + \Delta n\right)\left(N_A + \Delta n\right) = n_{i, eff}^2 e^{\left(\frac{E_{Fn} - E_{Fp}}{k_B T}\right)} = n_{i, eff}^2 e^{\left(\frac{qV}{k_B T}\right)} \tag{39.16}$$

where p is the hole concentration, N_D is the donor concentration, N_A is the acceptor concentration, k_B is the Boltzmann constant, T is the temperature and $n_{i,eff}$ is the effective intrinsic carrier density (Altermatt et al., 2003). The latter depends on the intrinsic carrier concentration n_i and the bandgap narrowing (BGN) ΔE_g, which is temperature-, doping- and injection-dependent (Schenk, 1998) and is defined by

$$n_{i,eff}^2 = n_i^2 e^{\left(\frac{\Delta E_g}{k_B T}\right)} \tag{39.17}$$

For a p-type semiconductor under low-level injection, i.e., $\Delta n \ll N_A$, Equation 39.16 reduces to

$$\Delta n \approx \frac{n_{i,eff}^2}{N_A} \left(e^{\left(\frac{qV}{k_B T}\right)} - 1\right) \tag{39.18}$$

Substituting Equations 39.15 and 39.18 in Equation 39.14 for electrons and holes and integrating over x from L_n to L_p computes to the total current density:

$$j(V) = j_n + j_p = qG\left(L_n + L_p\right) + \left(\frac{qn_{i,eff}D_p}{N_D L_p} + \frac{qn_{i,eff}D_n}{N_A L_n}\right)\left(e^{\left(\frac{qV}{k_B T}\right)} - 1\right) \tag{39.19}$$

This can be simplified to the single diode equation

$$j(V) = j_{SC} + \left(j_{0e} + j_{0b}\right)\left(e^{\left(\frac{qV}{k_B T}\right)} - 1\right) = j_{SC} + j_0\left(e^{\left(\frac{qV}{k_B T}\right)} - 1\right) \tag{39.20}$$

where j_{SC} is the short-circuit current density, which differs from the photogenerated current density j_{gen}, as not all generated charge carriers are collected because of their finite diffusion lengths. The carriers recombine before they reach the p-n junction. The saturation current density j_0 can be separated in the contribution of the n-type region called emitter saturation current density j_{0e} and in the contribution of the p-type region called base saturation current density j_{0b}. A more realistic saturation current density of the base for a finite thickness of the solar cell and a non-zero surface recombination is defined by (Goetzberger et al., 1998)

$$j_{0b} = \frac{qn_{i,eff}D_n}{N_A L_n}G_{Fn} \tag{39.21}$$

where G_{Fn} is the geometry factor considering the effective surface recombination velocity of the electrons S_n and the thickness of the base (p-type region) W_p:

$$G_{Fn} = \frac{\cosh\left(\frac{W_p}{L_n}\right) + \frac{D_n}{S_n L_n}\sinh\left(\frac{W_p}{L_n}\right)}{\frac{D_n}{S_n L_n}\cosh\left(\frac{W_p}{L_n}\right) + \sinh\left(\frac{W_p}{L_n}\right)} \tag{39.22}$$

In Figure 39.14, an ideal solar cell I-V and power characteristic are shown, which are calculated using Equation 39.20 assuming $j_{SC} = -40$ mA/cm^2, $j_0 = 100$ fA/cm^2, and $T = 25°C$.

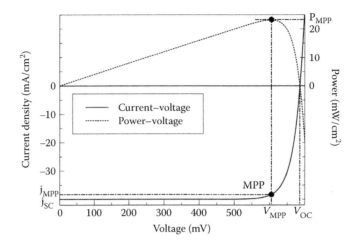

FIGURE 39.14　Example of solar cell current–voltage (I–V) and power characteristics. (MPP, maximum power point.)

39.7.1 I–V Parameters and Quantum Efficiency

In Figure 39.14, important output parameters of I–V and P–V characteristics can be extracted as there are: short-circuit current density $j_{SC}(V = 0 \text{ V})$, open-circuit voltage V_{OC} ($j(V_{OC}) = 0 \text{ mA/cm}^2$) and the maximum power point (MPP) ($P_{MPP}(V) = P_{max}$) with the current density and voltage at MPP j_{MPP} and V_{MPP}, respectively.

The short-circuit current density j_{SC} is the part of photogenerated current density j_{ph} which is collected by the p-n junction due to limited carrier diffusion length within the emitter and the base. A measure of this difference is the internal quantum efficiency (IQE). It is the relation between extracted charge carriers from the solar cell and generated charge carriers in the solar cell. The wavelength-dependent IQE(λ) can be derived from the measurable external quantum efficiency EQE(λ):

$$\text{IQE}(\lambda) = \frac{\text{extracted charge carriers}}{\text{generated charge carriers}} = \frac{\text{EQE}(\lambda)}{(1 - R_{opt}(\lambda) - A_{opt,par}(\lambda) - T_{opt}(\lambda))} \quad (39.23)$$

The EQE also considers optical losses due to reflection, transmission, and parasitic absorption $A_{opt,par}$ in, e.g., dielectric layers at front and rear side. It can be defined by the ratio:

$$\text{EQE}(\lambda) = \frac{\text{extracted charge carriers}}{\text{incident photons}} \quad (39.24)$$

The fill factor (FF) and the efficiency η are derived from the I–V characteristic. The FF is defined by

$$\text{FF} = \frac{j_{MPP} V_{MPP}}{j_{SC} V_{OC}} = \frac{P_{MPP}}{j_{SC} V_{OC}} \quad (39.25)$$

FF is always below 1 and is introduced to describe nonideal losses in the solar cell arising, e.g., due to series resistance losses as discussed in Section 39.8. The efficiency is calculated considering the irradiance G_I and the cell area A_{Cell}:

$$\eta = \frac{P_{MPP}}{G_I A_{Cell}} = \frac{\text{FF} j_{SC} V_{OC}}{G_I A_{Cell}} \quad (39.26)$$

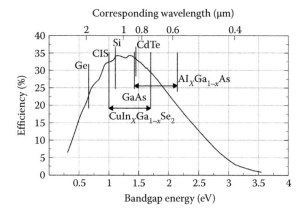

FIGURE 39.15 Dependence of a maximal achievable efficiency on the bandgap energy under an AM1.5g spectrum (1 sun) and a cell operating temperature of 300 K. Optimal bandgap energies of semiconductors are in the range of 1.0–1.5 eV. Established semiconductor absorber materials for solar cells are silicon (Si), Gallium-Arsenide (GaAs), Aluminum-Gallium-Arsenide ($Al_xGa_{1-x}As$), Germanium (Ge), CdTe, Copper-Indium-Diselenide (CIS), Copper-Indium-Gallium-Diselenide (CIGS).

39.7.2 Maximum Efficiency of Ideal Single Junction Solar Cells

For ideal single-junction solar cells, a maximal reachable efficiency may be calculated depending on their bandgap energy E_g following the reference (Wuerfel, 2005) which is very similar to the so-called Shockley–Queisser limit calculation (Shockley and Queisser, 1961). Two loss mechanisms are solely considered. First, nonabsorption losses due to photons with energy lower than the bandgap energy of the semiconductor are considered by Equation 39.10. Second, radiative recombination losses of the semiconductor occur assuming a one-sided emission of only photons of a blackbody spectrum with a temperature of 300 K higher than the bandgap energy, which is described by

$$j_{0,\text{rad}} = q j_{\text{ph,emission}}^{300\text{K}} = q \int_{E_g}^{\infty} dj_{\text{ph}}^{300\text{K}}(E_{\text{ph}}) \tag{39.27}$$

Equation 39.27 is derived in Wuerfel (2005) assuming a constant distribution of charge carriers over the absorber thickness, large diffusion lengths, and perfected optics where all photons with the bandgap energy or lower are reabsorbed. Using the single diode model described by Equation 39.20 and substituting 39.26 and 39.27 computes a bandgap energy-dependent maximal efficiency $\eta(E_g)$ for a single p-n junction which is shown in Figure 39.15 for an AM1.5g reference spectrum.

39.8 Nonideal *I–V* Characteristics

39.8.1 Series and Parallel Resistance

Solar cells are large diodes. They suffer under real operation from resistance losses. One differentiates two resistances in solar cells: series resistance r_S and parallel resistance r_P. Note that the small letters indicate area-weighted quantities in $\Omega \cdot cm^2$ similar to the current density j in mA/cm^2. Series resistance losses arise from current transport in the front metal busbars and metal fingers as well as in front contacts, emitter, base, rear contact, and rear metallization which is visualized by a sketch of a BSF solar cell in Figure 39.16.

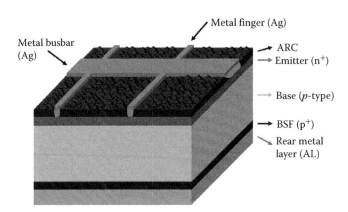

FIGURE 39.16 Sketch of back-surface-field (BSF) solar cell with an indicated random pyramid texture. The screen-printed front and rear metallization consist mainly of silver and aluminum, respectively. During solar cell fabrication, the p-type boron-doped wafer is phosphorus-diffused creating a highly n-type doped emitter. The BSF is formed by an alloying process during a fast firing step while creating the front and rear contact at the same time.

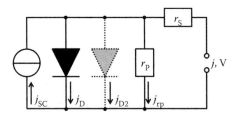

FIGURE 39.17 Equivalent circuit diagram of the single diode and double diode (dotted second diode) model of a solar cell.

As all these parts are included, r_s is called lumped series resistance. Parallel resistance losses, also called shunt resistance losses, arise when the p-n junction is short-circuited, e.g., by grown-in defects of the material (Breitenstein et al., 2004). A corresponding single diode equivalent circuit model is set up as shown in Figure 39.17.

It is described by the implicit Equation 39.28 and examples for analytically calculated I-V characteristics are shown in Figure 39.18.

$$j(V) = j_{SC} + j_D + j_{rp} = j_{SC} + j_0 \left(e^{\left(\frac{q(V+j(V)r_S)}{k_B T} \right)} - 1 \right) + \frac{V + j(V)\, r_S}{r_P} \tag{39.28}$$

39.8.2 Recombination

As r_S cannot be zero and r_P cannot be infinite, both reduce the FF and thus also the efficiency compared to the ideal case. The consideration of series and parallel resistance is in most cases not sufficient to fully describe I–V characteristics. Additional influences which are the root causes for deviations from the ideal I–V curve are the following:

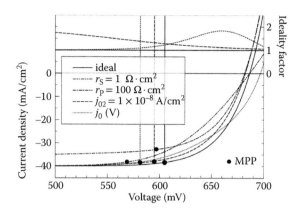

FIGURE 39.18 Solar cell *I–V* characteristics: ideal, suffering r_S, suffering r_P, suffering nonidealities modeled by the double diode model (compare Equation 39.30) and by a voltage-dependent saturation current density $j_0(V)$ in the voltage range from 500 mV to 700 mV, which is the relevant range for maximum power point (MPP). For details on $j_0(V)$ please refer to the text in Section 39.8.2.

- Distributed series resistance losses (Araújo et al., 1986; Fong et al., 2011) leading to a voltage-dependent change of current flow pattern in solar cells (Altermatt et al., 1996) which results in a voltage-dependent series resistance $r_S(V)$
- Injection-dependent lifetimes, i.e., voltage-dependent recombination within the volume, e.g., due to the boron-oxygen-related defect (Schmidt et al., 2001), or injection-dependent surface recombination velocities, e.g., at thermal oxides (Aberle et al., 1993)
- Edge recombination (Grove and Fitzgerald, 1966; Kuehn et al., 2000)
- Recombination in the space-charge region (SCR) (Wuerfel, 2005)

An injection-dependent lifetime within the absorber, i.e., voltage-dependent recombination, may be modeled by a voltage-dependent saturation current density $j_0(V)$ (Müller, 2016) which is typically experienced for Shockley–Read–Hall (SRH) recombination with a single defect with asymmetric capture cross sections, as, e.g., for the boron-oxygen-related defect (Schmidt et al., 1997). A $j_0(V)$ may be derived by Equation 39.29 due to a change of the injection density Δn in the solar cell with voltage as described by Equation 39.29. Assuming an excess carrier lifetime for a SRH midgap defect with the electron lifetime parameter $\tau_{n0} = 100$ µs and the hole lifetime parameter $\tau_{p0} = \tau_{n0} \times 10 = 1000$ µs for the parameters $j_{0e} = 10$ fA/cm², $N_A = 7.2 \times 10^{15}$ cm⁻³, $W_A = 160$ µm, a corresponding *I–V* curve is calculated with the single diode Equation 39.20 and is shown in Figure 39.18.

$$j_0(V) = j_0(\Delta n) = j_{0e} + j_{0b}(\Delta n) = j_{0e} + \frac{qW_A n_{i,eff}^2(N_A)}{N_A \tau_{eff}(\Delta n)} \tag{39.29}$$

Recombination in the SCR can be described by the double diode model which is given by

$$j(V) = j_{SC} + j_D + j_{D2} + j_{rp} = j_{SC} + j_0\left(e^{\left(\frac{q(V+jr_S)}{k_B T}\right)} - 1\right) + j_{02}\left(e^{\left(\frac{q(V+jr_S)}{2k_B T}\right)} - 1\right) + \frac{V + jr_S}{r_P} \tag{39.30}$$

Its corresponding equivalent circuit model is also shown in Figure 39.17 and an example for an *I–V* characteristic is shown in Figure 39.18. The second diode is analytically derived with an ideality factor $n = 2$ (Wuerfel, 2005) for recombination via defects in the space-charge region, compared to $n = 1$ for the first diode. However, SCR recombination is typically not the main contribution to nonideal behavior, i.e., the saturation current density of the second diode j_{02} is mainly a fit parameter, when applying the double diode model to fit dark I-V characteristics of solar cells. For crystalline wafer-based silicon solar cells, the root cause is typically different and a fitted j_{02} accounts for all effects listed earlier.

Another possibility for the description of nonideal I–V characteristics is the usage of a voltage-dependent ideality factor $n(V)$ (Rhoderick and Williams, 1988), which can be calculated from dark or light I–V characteristics by Equation 39.31 if $V > 3k_B T/q$:

$$\frac{1}{n} = \frac{k_B T}{q} \frac{d\left(\ln j\right)}{dV} \tag{39.31}$$

Exemplarily deriving the voltage-dependent ideality factor $n(V)$, the double diode model is now compared to the voltage-dependent saturation current density $j_0(V)$ using Equation 39.29, which is shown in in the upper part of Figure 39.18. The second diode model has as expected an ideality factor of $n = 2$ for voltages below 400 mV and it decreases toward $n = 1$ almost at V_{OC}. At MPP it is in-between at $n = 1.33$ for the shown example. For $j_0(V)$ the ideality factor increases in this example from $n = 1$ to $n = 1.1$ at MPP to a maximum of about $n = 1.83$ and then decreases again toward $n = 1$ while it is $n = 1.25$ at V_{OC}.

Thus, in general, nonideal I–V behavior can be expected for any real solar cell and the $n(V)$ evaluation may give hints for its origin (McIntosh, 2001). When analyzing I–V characteristics, it is necessary to firstly derive a voltage-dependent series resistance with preferable a multiple-light-level determination method (Fong et al., 2011) because current flow patterns within a solar cell remain similar and injection-dependent effects are reduced when changing the illumination only by 10%–20%. In a second step, a series resistance free I–V curve is derived for the $n(V)$ evaluation (Steingrube et al., 2011).

As an alternative and quicker approach to analyze I–V characteristics, the FF can be evaluated in terms of series resistance induced and recombination-induced losses. The FF is compared to the pseudo fill factor (pFF) described by Equation 39.32 which accounts only for recombination driven FF changes if r_P is sufficiently high ($r_P > 10\,k\Omega \cdot cm^2$).

$$pFF = FF - \Delta FF_{rs} = FF - \frac{j_{MPP}^2 r_S(V_{MPP})}{j_{SC} V_{OC}} \tag{39.32}$$

39.9 Numerical Simulation and Models for Crystalline Silicon Solar Cells

The diode model-based approaches are very often not sufficient to describe solar cells; therefore numerical device simulations are carried out. The advantage of numerical simulations compared to analytical solutions is an increased accuracy and provides a deeper insight into device physics and solar cell losses. Analytical solutions are restricted to particular assumptions such as constant photogeneration, constant doping or low injection conditions, which often only approximate real device behavior. Typical problems that demand numerical solar cell simulation are the following:

- Local line front contacts desire a two-dimensional simulation
- Local point rear contacts desire a three-dimensional simulation
- Inhomogeneous distributed wafer quality in case of mc silicon
- Injection-dependent SRH recombination in the bulk
- Injection-dependent SRH recombination at surfaces
- Consideration of Auger recombination in the emitter or BSF with their doping profiles

39.9.1 Numerical Simulation

The physics for the numerical simulation of solar cells are implemented in various simulators. Most developed and widely used simulators are PC1D (Rover et al., 1985), PC1Dmod (Haug et al., 2015) for one-dimensional problems and TCAD Sentaurus Device (Synopsys, 2016), Atlas (Silvaco, 2016), Quokka (Fell, 2013), and AFORS-HET (Varache et al., 2015) for two and three-dimensional simulations. An in-depth introduction into numerical device simulation of crystalline silicon solar cells can be found in Altermatt (2011) which is also used as the source of the most important facts and models discussed below in this section. An update will be given below for the intrinsic recombination model and the incomplete ionization model.

For numerical solar cell simulation a coupled set of differential equations, which are the basic equations of semiconductor device physics, have to be solved by an iterative procedure. This set consists of the Poisson Equation 39.33, which describes the relationship between the electrostatic potential Φ and the charge distribution, the continuity equations for electrons and holes (Equations 39.13, 39.34, 39.35), which conserve the quantity charge carrier, and the current density equations in the drift-diffusion approach for electrons (Equation 39.36) and holes (Equation 39.37) in one dimension:

$$\nabla \cdot (\varepsilon \nabla \Phi) = -q \left(p - n + N_{\text{don}}^+ - N_{\text{acc}}^- \right) \tag{39.33}$$

$$q \frac{\partial n}{\partial t} = \nabla \cdot j_{\text{n}} + q \left(G - U \right) \tag{39.34}$$

$$q \frac{\partial p}{\partial t} = \nabla \cdot j_{\text{p}} + q \left(G - U \right) \tag{39.35}$$

$$j_{\text{n}} = -q \mu_{\text{n}} n \nabla \Phi + q D_{\text{n}} \nabla n \tag{39.36}$$

$$j_{\text{p}} = -q \mu_{\text{p}} p \nabla \Phi - q D_{\text{p}} \nabla p \tag{39.37}$$

In Equations 39.33 through 39.37, ε is the permittivity of the material, p is the hole carrier density, N_{D}^+ is the ionized donor concentration and N_{A}^- is the ionized acceptor concentration, μ_{n} and μ_{p} are the mobilities for electrons and holes, respectively and D_{p} is the hole diffusion coefficient. The mobilities and diffusion coefficients are related through the Einstein relationships $D_{\text{n}} = \frac{k_{\text{B}}T}{q} \mu_{\text{n}}$ and $D_{\text{p}} = \frac{k_{\text{B}}T}{q} \mu_{\text{p}}$ for a nondegenerated semiconductor. For a degenerated semiconductor (e.g., doping densities higher than 1×10^{19} cm^{-3}), Fermi–Dirac statistics are valid and the electron transport equation is extended by the additional term $-nk_{\text{B}}T\nabla \left(\ln \left(\frac{n}{N_{\text{C}}} e^{(E_{\text{C}} - E_{\text{Fn}})/(k_{\text{B}}T)} \right) \right)$ and the hole transport equation is extended by $-pk_{\text{B}}T\nabla \left(\ln \left(\frac{p}{N_{\text{V}}} e^{(E_{\text{Fp}} - E_{\text{V}})/(k_{\text{B}}T)} \right) \right)$, respectively. N_{C} and N_{V} are the effective density of states for electrons in the conduction band and for holes in the valence band, respectively. At those doping densities higher than 1×10^{19} cm^{-3}, Boltzmann statistics overestimates the product of electron and hole density and may lead to a distorted loss analysis between Auger recombination losses in the highly doped region and surface recombination. Thus Fermi–Dirac statistics should be used, especially as the widely used PC1D software is updated using Fermi–Dirac statistics nowadays (Haug et al., 2015).

39.9.2 Models for Crystalline Silicon Solar Cells

A doping and carrier concentration-dependent as well as temperature-dependent mobility model (Klaassen, 1992a,b) is used to calculate Figure 39.19.

For highly doped semiconductors, the bandgap decreases, which is called bandgap narrowing (BGN). It occurs due to carrier-dopant and carrier–carrier interactions, which may be considered for silicon using a BGN model (Schenk, 1998) which is used to calculate Figure 39.20.

The effective intrinsic carrier concentration $n_{\text{i,eff}}$ changes correspondingly with the doping density (Altermatt et al., 2003) using Equation 39.17 and $n_{\text{i,eff}}$ is shown in Figure 39.21.

For medium-doped semiconductors in the range 1×10^{18} cm^{-3} to 1×10^{19} cm^{-3} the effect of incomplete ionization occurs, which is caused by the Fermi-level approaching the donor or acceptor level, resulting

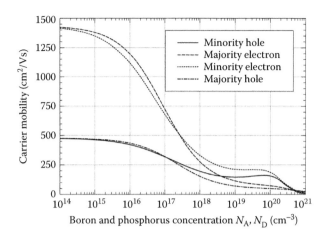

FIGURE 39.19 Klaassen's parametrization of the minority and majority carrier mobility in boron- and phosphorus-doped silicon at 300 K.

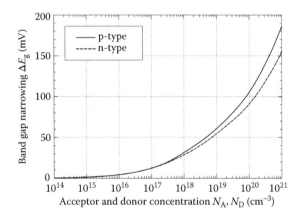

FIGURE 39.20 Bandgap narrowing in p- and n-type silicon calculated with the Schenk model at 300 K.

in the restriction of electrons or holes to localized donor or acceptor states, which prevents free electron or hole movement, respectively. The consideration of incomplete ionization is recommended for the BSF solar cell and a parametrization may be found in reference (Steinkemper et al., 2015).

In order to include the geometrical properties of the solar cell and to decrease the simulation time geometrical considerations have to be performed to derive a well-suited geometry of the device for simulations. Most solar cells have an area of about 15.6×15.6 cm^2 and are about 150–200 µm thick. Smallest features are, e.g., front contacts with a width down to 10 µm. However, it is not necessary to simulate the complete solar cell because its structure is highly symmetric. The simulation domain can be reduced to a geometrically irreducible *standard domain* as shown in Figure 39.22.

This domain is very often two-dimensional where its width is half the spacing between the front metal fingers. Note that the front surface is typically textured, which it is not accounted for in the flat device domain due to the fact that, e.g., random pyramids would result in a huge three-dimensional domain. For simulation, the device domain is discretized on a mesh as shown in Figure 39.22. The mesh should be fine

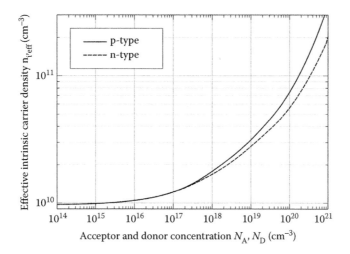

FIGURE 39.21 Influence of the doping density on the effective intrinsic carrier concentration in p- and n-type silicon at 300 K.

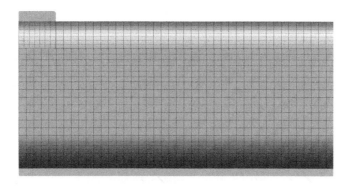

FIGURE 39.22 Geometrically irreducible two-dimensional *standard domain* for a BSF solar cell. Mesh refinements near the front surface, in diffused layers and close to contacts are indicated.

enough where any input or output parameter varies strongly with distance, which is for solar cells close to the front surface where the blue part of the sunlight is absorbed within nanometers, and where a front diffusion layer is typically situated. Besides near metal contacts the mesh should be refined. The generation of charge carriers as described in Section 39.4 is typically considered by an one-dimensional generation profile $G(x)$. The reverse process of intrinsic recombination may be included by a combined model for the radiative recombination and the Auger recombination (Richter et al., 2012), which is used to calculate Figure 39.23.

At surfaces, surface recombination occurs via surface states within the bandgap and additionally fixed charges are present which lead to local band bending (Kingston and Neustadter, 1955).

The most interesting result of a simulation is the sweep of an $I-V$ curve which is performed changing the applied voltage at the contacts as boundary condition. For a full solar cell $I-V$ characteristic, the standard domain $I-V$ curves can be interconnected in a circuit by ohmic resistances, in, e.g., a SPICE network simulation, which represent the front metal fingers. A sketch of such a network is shown in Figure 39.24.

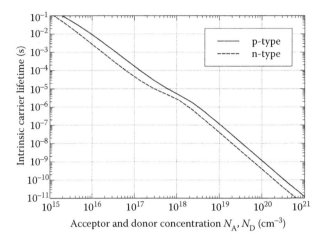

FIGURE 39.23 Influence of the doping density on radiative and Auger recombination in silicon (intrinsic recombination) in p- and n-type silicon at 300 K.

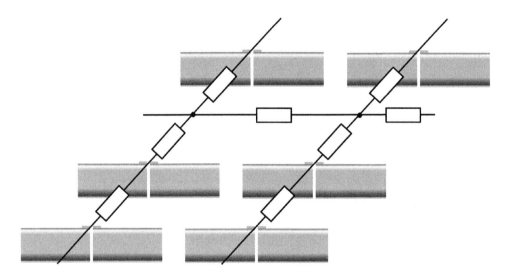

FIGURE 39.24 Sketch of a network of resistances and standard domains which allow simulating a full solar cell I–V characteristic by device and circuit simulation. Compare Figure 39.16 to the sketch of the same solar cell part.

39.10 Real Working Behavior of Solar Cells

Under real operating conditions, main changes of solar cell performance arise due to its intensity and temperature behavior.

39.10.1 Light Intensity Behavior

The intensity dependence is mainly governed by the decrease of irradiance G which results in a decrease of the short-circuit current density j_{SC}. For typical crystalline silicon solar cells under nonconcentrated sunlight conditions, the output power is mainly determined by the linear relationship $G \propto j_{SC}$. Deviations may occur for irradiance conditions below 0.2 suns, where injection-dependent effects, e.g., injection-dependent carrier lifetime, become apparent. The general change of the open-circuit voltage V_{OC} with irradiance G may be derived from the ideal-behavior solving the single diode Equation 39.20 to V_{OC} for $j(V_{OC}) = 0$ mA/cm^2:

$$V_{OC} = \frac{k_B T}{q} \ln \left(1 + \frac{j_{SC}}{j_0} \right) \tag{39.38}$$

The change of *FF* with irradiance G can be approximated by Equation 39.39 (Green, 1982) for the ideal-behavior, which depends on V_{OC}, and is called *ideal fill factor FF$_0$*.

$$FF_0 (v_{OC}) = \frac{v_{OC} - \ln (v_{OC} + 0.72)}{v_{OC} + 1}, \text{ with } v_{OC} = \frac{q}{k_B T} V_{OC} \tag{39.39}$$

An example of a calculated *I-V* characteristic under low illumination using the single diode Equation 39.20 is shown in Figure 39.25, assuming $j_{SC} = 20$ mA/cm^2 and $j_0 = 100$ fA/cm^2. The almost halved j_{SC} results in nearly half of the maximum power.

The so-called low light behavior of a solar cell is shown in Figure 39.26 using the single diode equivalent circuit model (Equation 39.28) with j_{SC} (STC) $= -40$ mA/cm^2 and $j_0 = 100$ fA/cm^2. Additionally, the relative efficiency normalized to its STC value is calculated and compared:

$$\eta_{rel} (G, T) = \frac{\eta (G, T)}{\eta(1000 \frac{W}{m^2}, 25°C)} \tag{39.40}$$

For the ideal-behavior ($r_S = 0 \ \Omega \cdot$ cm^2) in Figure 39.26, it is deduced that the efficiency decreases with decreasing irradiance, which is mainly due to the V_{OC} decrease and only slightly due to the *FF* decrease. The variation of the lumped series resistance in Figure 39.26 shows that its influence governs

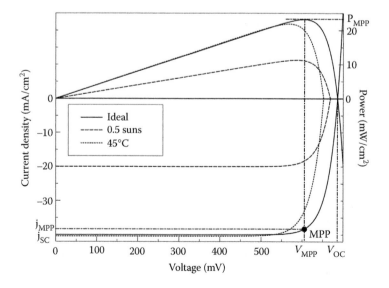

FIGURE 39.25 Comparison of solar cell I-V and power characteristics under STC, under low illumination of 0.5 suns, and under increased temperature of $T = 45°C$. (MPP, maximum power point.)

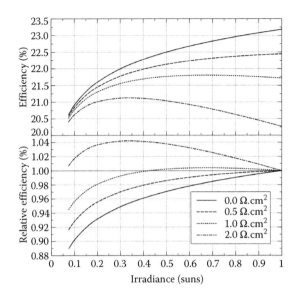

FIGURE 39.26 Relative efficiency versus irradiance for different lumped series resistances r_s, where $r_S = 0\ \Omega \cdot cm^2$ represents the ideal behavior, whereas $r_S = 0.5\ \Omega \cdot cm^2$ is a typical value for c-Si solar cells and r_S between 1 and $2\ \Omega \cdot cm^2$ is typical for solar modules due to the cell interconnection, e.g., the tabbing ribbons cause additional series resistances.

within the irradiance range from 200 W/m^2 to 1000 W/m^2. In this range, the current in the solar cell is sufficiently high and η saturates or even reduces toward STC with increasing r_S. This results in an increasing relative efficiency under low light condition for solar cells and modules with high r_S (compare lower part of Figure 39.26), which seems beneficial under real operating condition. However, the series resistance solely limits the STC efficiency (compare upper part of Figure 39.26) and the solar module will stay behind its potential. In conclusion, r_S should be low for maximum energy yield.

39.10.2 Temperature Behavior

The temperature behavior is determined by several effects (Dupré et al., 2015). It increases the effective intrinsic carrier density $n_{i,eff}$ (Misiakos and Tsamakis, 1993), which decreases the quasi-Fermi-level splitting and thus at higher temperature the voltage at the same injection density is reduced (compare Equations 39.17 and 39.18). A detrimental effect, but smaller, is the increase of the bandgap with increasing temperature, which results in a slight increase in absorbed photon current density. Thus j_{SC} is slightly increased. The change of FF with temperature depends on a variety of influences, e.g., temperature-dependent conductivity of the front and rear metallization, as well as front and rear contacts. For silicon, the FF typically decreases with higher temperature. Surprisingly, the temperature behavior of j_{SC}, V_{OC} and η is typically linearly under relevant operating conditions in the range of 0°C–75°C. Therefore, it is expressed in terms of their absolute temperature coefficients $TC_{abs}(j_{SC})$, $TC_{abs}(V_{OC})$ and $TC_{abs}(\eta)$ or more often in terms of their relative counterparts $TC_{rel}(j_{SC})$, $TC_{rel}(V_{OC})$ and $TC_{rel}(\eta)$ normalized to STC similar to Equation 39.40. The relevant temperature coefficient of the efficiency $TC_{rel}(\eta)$ is in the range of −0.5%/K to −0.25%/K for crystalline silicon solar cells. An example of a calculated I-V curve at 45°C using the single diode model is shown in Figure 39.25 assuming a $j_{SC}(45°C) = 40.4$ mA/cm^2 and j_0 (45°C) = 191 fA/cm^2. This corresponds to a $TK_{rel}(j_{SC}) = 0.05\%/K$, i.e., $\Delta j_{SC}\,(\Delta T) = j_{SC}\,(25°C) \times TC_{rel}\left(j_{SC}\right) \times \Delta T$, and a j_0 increase which is approximated by the proportionality to $T^3 \times e^{-\frac{E_g(T)}{k_B T}}$ (Wolf et al., 1977). The increased temperature leads to decreased V_{OC} and thus to slightly lower P_{MPP}.

39.10.3 Energy Yield

The real working behavior of solar cells and solar modules, respectively, is regarded in terms of their delivered energy yield below to complete the information given in this chapter.

The main influences on solar cell performance under real operating conditions are changing irradiance, cell temperature, angle of incidence, and spectral variations compared to the STC condition. The spectral variation is below 1% for c-Si solar cells (Huld et al., 2011) and hence will not be considered in the further discussion. For a typical solar module, the energy yield is calculated for the three locations introduced in Section 39.3 using the Equations 39.3, 39.5 through 39.8, 39.26 and 39.28 assuming c-Si solar module relevant parameters $j_0 = 100$ fA/cm^2, $r_S = 1\Omega \cdot cm^2$, $r_P = \infty$, $TC_{rel}(\eta) = 0.4\%/°C$, $K = 4$ m^{-1} and $L = 2$ mm and $NOCT = 45°C$. The specific yields per month are shown in the lower part of Figure 39.27 using Equation 39.2 and the specific yields over 1 year are $Y_f = 1120$ kWh/kWp for Halle, $Y_f = 1823$ kWh/kWp for Almeria, and $Y_f = 1504$ kWh/kWp for Brazzaville.

In a first-order approximation, the specific yield follows the available solar irradiation as shown in Figure 39.8 which results, e.g., in low Y_f for winter months in Halle or similar Y_f over the whole year for Brazzaville near the equator. To separate the irradiation influence, the performance ratio has been introduced in Equation 39.1 and here PR is shown in the upper part of Figure 39.27. It is mostly below 100% due to increased cell operating temperature above 25°C in summer months for all locations and due to the warmer climate in Almeria and Brazzaville. Performance ratios over 1 year are $PR = 96.6\%$ for Halle, $PR = 92.9\%$ for Almeria and $PR = 90.8\%$ for Brazzaville, which are very similar to the estimated temperature losses derived from the annual average cell operating temperature in Section 39.3.

The calculations are carried out assuming no additional cell-to-module efficiency changes while module series resistance and optics under oblique incidence are included. Additional PR losses arise in reality due to losses in the inverter (up to 5%), due to cell-to-cell and module-to-module mismatches (about 1%), due to shading (depends on location), thus good PR are in the range of 85%–90% in the first years. Additionally, efficiency degradation occurs over time of about 0.5%$_{rel}$/year (Jordan and Kurtz, 2013), which further decreases the PR of about 6.25% over the system lifetime of about 25 years.

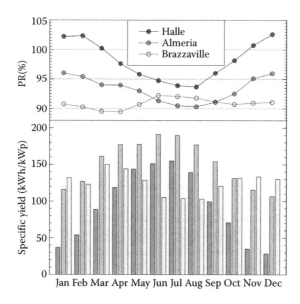

FIGURE 39.27 Specific yield (lower graph) and performance ratio (upper graph) per month for the three locations Halle, Almeria, and Brazzaville.

39.11 Conclusion and Outlook

PV systems became already an integral part of our world in terms of energy supply and also in our land-scapes. The levelized cost of electricity (LCOE) of PV is approaching that of conventional electricity power plants (IEA, 2014). Thus, a major share of electricity is expected to be PV electricity in the next decades. However, it is a fluctuating energy resource and storage is getting more and more important.

Wafer-based silicon solar cells became the market leading technology and at the moment a change is unlikely to happen. The competitive thin-film solar cells on CIGS and CdTe absorber material benefit from reduced material consumption, but are still built in solar module with front glass for long lifetime which nowadays lowers significantly the inherent cost advantage. Organic and dye-sensitized (inorganic) thin-film absorbers are not showing a high efficiency potential yet (Green et al., 2015).

New solar cell structures have to be designed right away in cost competitive solar modules with a long lifetime. For example, the recently emerging perovskite-based solar cells (Green et al., 2014) have shown an incredible efficiency increases in the past 5 years, but they still lack stability against UV light and will face similar module cost disadvantage as thin-films solar cells do. Efficiency matters, but costs and stability have to be competitive in order to gain economic success.

Besides the single-junction solar cells for 1 sun illumination, there are several other developing technologies and approaches. Concentrating photovoltaics (CPV) is close to enter the market. CPV uses optical elements to increase the irradiance on a smaller solar cell size for increased efficiency and reduced material consumption. However, for higher concentration (>20 suns) the PV system needs to track the sun, which results in higher complexity and higher maintenance requirements. Increased efficiency of solar cells may be reached by so-called third generation PVs (Green, 2003). Examples are multijunction solar cells which are stacks of semiconductor absorber materials of different bandgap energies reducing thermalization losses which have shown great potential (Green et al., 2015), whereas hot carrier solar cells, multiband solar cells, and photon up- and down-converting materials are still in focus of basic research.

Acknowledgments

I would like to thank Eric Schneiderlöchner, Pietro Altermatt, Byungsul Min, and Julia Müller for their great support.

References

Aberle A G, Robinson S J, Wang A, Zhao J, Wenham S R and Green M A (1993) High-efficiency silicon solar cells: Fill factor limitations and non-ideal diode behaviour due to voltage-dependent rear surface recombination velocity. *Prog. Photovolt. Res. Appl.* 1(2):133–143.

Altermatt P P (2011) Models for numerical device simulations of crystalline silicon solar cells—A review. *J. Comput. Electron.* 10(3):314–330.

Altermatt P P, Heiser G, Aberle A G, Wang A, Zhao J, Robinson S J, Bowden S and Green M A (1996) Spatially resolved analysis and minimization of resistive losses in high-efficiency Si solar cells. *Prog. Photovolt Res. Appl.* 4:399–414.

Altermatt P P, Schenk A, Geelhaar F and Heiser G. (2003) Reassessment of the intrinsic carrier density in crystalline silicon in view of band-gap narrowing. *J. Appl. Phys.* 93(3):1598–1604.

Araújo G L, Cuevas A and Ruiz J M (1986) The effect of distributed series resistance on the dark and illuminated current-voltage characteristics of solar cells. *IEEE Trans. Electron Dev.* 33(3):391–401.

ASTM E490-00a (2014) *Standard Solar Constant and Zero Air Mass Solar Spectral Irradiance Tables.* West Conshohocken, PA: ASTM International. www.astm.org.

ASTM G173-03 (2012) *Standard Tables for Reference Solar Spectral Irradiances: Direct Normal and Hemispherical on 37° Tilted Surface.* West Conshohocken, PA: ASTM International. www.astm.org.

Becquerel A E (1839) Recherches sur les effets de la radiation chimique de la lumière solair au moyen des courants électriques. *C. R. Acad. Sci.* 9:561–567.

Blakers A, Wang A, Milne A, Zhao J and Green M A (1989) 22.8% efficient silicon solar cell. *Appl. Phys. Lett.* 55:1363–1365.

Breitenstein O, Rakotoniaina J P, Al Rifai M H and Werner M (2004) Shunt types in crystalline silicon solar cells. *Prog. Photovolt Res. Appl.* 12(7):529–538.

Brendel R (1994) Sunrays: A versatile ray tracing program for the photovoltaic community. In: *Proceedings of the 12th European Photovoltaic Solar Energy Conference*, Amsterdam, Netherlands, pp. 1339–1342.

Chapin D M, Fuller C S and Pearson G O (1954) A new silicon p–n junction photocell for converting solar radiation into electrical power. *J. Appl. Phys.* 25:676–677.

Duffie J A and Beckman W A (2013) *Solar Engineering of Thermal Processes* (4th edition). Hoboken, NJ: John Wiley & Sons Inc.

Dupré O, Vaillon R and Green M A (2015) Physics of the temperature coefficients of solar cells. *Sol. Energ. Mat. Sol. Cells.* 140:92–100.

Fell A (2013) A free and fast three-dimensional/two-dimensional solar cell simulator featuring conductive boundary and quasi-neutrality approximations. *IEEE Trans. Electron Dev.* 60(2):733–738.

Fong K C, McIntosh K R and Blakers A W (2011) Accurate series resistance measurement of solar cells. *Prog. Photovolt Res. Appl.* 21(4):490–499.

Fraunhofer ISE (2013) *Photovoltaics Report*, 7 November 2013. http://www.ise.fraunhofer.de/en/downloads-englisch/pdf-files-englisch/photovoltaics-report-slides.pdf

Garcia M C A and Balenzategui J L (2004). Estimation of photovoltaic module yearly temperature and performance based on Nominal Operation Cell Temperature calculations. *Renew. Energy* 29:1997–2010.

Goetzberger A, Knobloch J and Voß B (1998) *Crystalline Silicon Solar Cells*. Chichester: John Wiley & Sons Ltd.

Green M (1982) *Solar Cells: Operating Principles, Technology, and System Applications*. Englewood Cliffs, NJ: Prentice-Hall, Inc.

Green M A (2003) *Third Generation Photovoltaics–Advanced Solar Energy Conversion*. Berlin Heidelberg: Springer-Verlag.

Green M A, Emery K, Hishikawa Y, Warta W and Dunlop E D (2015). Solar cell efficiency tables (Version 45). *Prog. Photovolt Res. Appl.* 23(1):1–9.

Green M A, Ho-Baillie A and Snaith H J (2014) The emergence of perovskite solar cells. *Nat. Photonics.* 8(7):506–514.

Grove A S and Fitzgerald D J (1966) Surface effects on p-n junctions: Characteristics of surface space-charge regions under non-equilibrium conditions. *Solid State Electron.* 9:783–806.

Haug H, Greulich J, Kimmerle A and Marstein E S (2015) PC1Dmod 6.1—State-of-the-art models in a well-known interface for improved simulation of Si solar cells. *Sol. Energ. Mat. Sol. Cells.* 142:47–53.

Holst H, Winter M, Vogt M R. Bothe K, Kontges M, Brendel R and Altermatt P (2013) Application of a new ray tracing framework to the analysis of extended regions in Si solar cell modules. *Energy Procedia* 38:86–93.

Huld T, Friesen G, Skoczek A, Kenny R P, Sample T, Field M and Dunlop E D (2011) A power-rating model for crystalline silicon PV modules. *Sol. Energ. Mat. Sol. Cells.* 95(12):3359–3369.

IEC 61215 (2005) *Crystalline Silicon Terrestrial Photovoltaic (PV) Modules—Design Qualification and Type Approval*. Geneva, Switzerland: International Electrotechnical Commission (IEC). http://www.iec.ch.

International Energy Agency (IEA) (2014) *Technology Roadmap: Solar Photovoltaic Energy* (2014 edition). Accessed January 12, 2016. https://www.iea.org/media/freepublications/technologyroadmaps/solar/TechnologyRoadmapSolarPhotovoltaicEnergy_2014edition.pdf

ITRPV (July 2015) *International Technology Roadmap for Photovoltaic: Results 2014*, 6th edition. Accessed January 2016. http://www.itrpv.net/.cm4all/iproc.php/Reports%20downloads/ITRPV_Roadmap_2015_Rev1_July_150722.pdf?cdp=a

Joint Research Center (JRC) (2016) Photovoltaic Geographical Information System (PVGIS). Accessed 18 January 2016. http://re.jrc.ec.europa.eu/pvgis/

Jordan D C and Kurtz S R (2013) Photovoltaic degradation rates—An analytical review. *Prog. Photovolt Res. Appl.* 21(1):12–29.

Kingston R H and Neustadter S F (1955) Calculation of the space charge, electric field, and free carrier concentration at the surface of a semiconductor. *J. Appl. Phys.* 26(6):718–720.

Klaassen D B M (1992a) A unified mobility model for device simulation—I. Model equations and concentration dependence. *Solid State Electron.* 35(7):953–959.

Klaassen D B M (1992b) A unified mobility model for device simulation—II. Temperature dependence of carrier mobility and lifetime. *Solid State Electron.* 35(7):961–967.

Kopp G and Lean J L (2011) A new, lower value of the total solar irradiance: Evidence and climate significance. *Geophys. Res. Lett.* 38:L01706.

Kuehn R, Fath P and Bucher E (2000) Effects of pn junctions bordering on surfaces investigated by means of 2D-modeling. In: *Proceedings of the 28th IEEE Photovoltaic Specialists Conference*, Anchorage, AK, pp. 116–119.

Lammert M and Schwartz R (1977) The interdigitated back contact solar cell: A silicon solar cell for use in concentrated sunlight. *IEEE Trans. Electron Dev.* 24(4):337–342.

Macleod H A (2010) *Thin-Film Optical Filters* (4th edition). Boca Raton, FL: CRC Press, Taylor & Francis.

Mandelkorn J and Lamneck J (1972) Simplified fabrication of back surface electric field silicon cell and novel characteristics of such cells. In *Proceeding 9th IEEE Photovoltaic Specialists Conference*, Silver Springs, MD.

McIntosh K R (2001) *Lumps, Humps and Bumps: Three Detrimental Effects in the Current-Voltage Curve of Silicon Solar Cell.* PhD thesis, University of New South Wales, Sydney, Australia.

Meteotest (2008) *Meteonorm Version 6.1.0.0.* Accessed January 12, 2016. www.meteotest.com.

Misiakos K and Tsamakis D (1993) Accurate measurements of the silicon intrinsic carrier density from 78 to 340 K. *J. Appl. Phys.* 74(5):3293–3297.

Müller M (2016) Reporting effective lifetimes at solar cell relevant injection densities. *Energy Proc.* 92:138–144.

National Aeronautics and Space Administration (NASA) (2016) Surface meteorology and Solar Energy Accessed January 18, 2016. https://eosweb.larc.nasa.gov/sse/.

National Renewable Energy Laboratory (NREL) (2016) *Renewable Resource Data Center (RReDC).* Accessed January 18, 2016. http://www.nrel.gov/rredc/.

Pingel S, Frank O, Winkler M, Daryan S, Geipel T, Hoehne H and Berghold J (2010) Potential induced degradation of solar cells and panels. In: *Proceedings 35th IEEE Photovoltaic Specialists Conference*, Honolulu, HI, 2010, pp. 2817–2822.

PV Lighthouse (2016) PV Lighthouse. Accessed 12 January 2016. https://www.pvlighthouse.com.au

PVEDUCATION.ORG (2016) PVEducation Accessed January 12, 2016. http://pveducation.org

pvXchange Trading GmbH (2016) Market price data. Accessed January 12, 2016. http://www.pvxchange.com/priceindex/Default.aspx?langTag=de-DE

Quaschning V (2005) *Understanding Renewable Energy Systems.* London, UK: Earthscan.

Rhoderick E H and Williams R H (1988) *Metal-Semiconductor Contacts* (2nd edition). New York, NY: Oxford University Press.

Richter A, Glunz S W, Werner F, Schmidt J and Cuevas A (2012) Improved quantitative description of Auger recombination in crystalline silicon. *Phys. Rev. B.* 86(16):165202.

Ross R G Jr and Smokler M I (1986). Flat-Plate Solar Array Project–Final Report, Vol. VI Engineering Sciences and Reliability, JPL Pub. No. 86–31, http://ntrs.nasa.gov/search.jsp?R=19870011218.

Rover D T, Basore P A, Thorson G M (1985) PC-1D version 2: Enhanced numerical solar cell modelling. In: *Proceedings of the 18th IEEE Photovoltaic Specialists Conference*, Las Vegas, NV, p. 703.

Schenk A (1998) Finite-temperature full random-phase approximation model of band gap narrowing for silicon device simulation. *J. Appl. Phys.* 84(7):3684–3695.

Schmidt J, Aberle A G and Hezel R (1997) Investigation of carrier lifetime instabilities in Cz-grown silicon. In: *Proceedings of the 26th IEEE Photovoltaic Specialists Conference*, New York, NY, pp. 13–18.

Schmidt J, Cuevas A, Rein S and Glunz S W (2001) Impact of light-induced recombination centres on the current-voltage characteristic of czochralski silicon solar cells. *Prog. Photovolt Res. Appl.* 9(4):249–255.

Shockley W and Queisser H J (1961) Detailed balance limit of efficiency of p-n junction solar cells. *J. Appl. Phys.* 32(3):510–519.

Silvaco (2016) *Atlas*. Santa Clara, CA: Silvaco. http://www.silvaco.com.

Steingrube S, Wagner H, Hannebauer H, Gatz S, Chen R, Dunham S T, Dullweber T, Altermatt P P, Brendel R (2011) Loss analysis and improvements of industrially fabricated Cz-Si solar cells by means of process and device simulations. *Energy Procedia* 8:263–268.

Steinkemper H, Rauer M, Altermatt P P, Heinz F D, Schmiga C and Hermle M (2015) Adapted parameterization of incomplete ionization in aluminum-doped silicon and impact on numerical device simulation. *J. Appl. Phys.* 117(7):074504.

Synopsys (2016) *TCAD Sentaurus Device version 2015.06*. Mountain View, CA: Synopsys. http://www.synopsys.com.

Tanaka M, Taguchi M, Matsuyama T, Sawada T, Tsuda S, Nakano S, Hanafusa H and Kuwano Y (1992) Development of new a-Si/c-Si heterojunction solar cells: ACJ-HIT (artificially constructed junction-heterojunction with intrinsic thin-layer). *Jpn. J. Appl. Phys.* 31:3518–3522.

Varache R, Leendertz C, Gueunier-Farret M E, Haschke J, Muñoz D and Korte L (2015) Investigation of selective junctions using a newly developed tunnel current model for solar cell applications. *Sol. Energ. Mat. Sol. Cells.* 141:14–23.

Winter M, Vogt M R, Holst H, Altermatt P P (2014) Combining structures on different length scales in ray tracing: Analysis of optical losses in solar cell modules. In: *Proceedings of the 14th International Conference on Numerical Simulation of Optoelectronic Devices*, Palma de Mallorca, Spain.

Wolf M, Noel G T and Stirn R J (1977) Investigation of the double exponential in the current-voltage characteristics of silicon solar cells. *IEEE Trans. Electron Dev.* 24(4):419–428.

Wuerfel P (2005) *Physics of Solar Cells*. Weinheim, Germany: Wiley-VCH Verlag GmbH. KGaA.

Zhao J, Wang A and Green M A (1999) 24.5% efficiency silicon PERT cells on MCZ substrates and 24.7% efficiency PERL cells on FZ substrates. *Prog. Photovolt Res. Appl.* 7:474–474.

40

Multijunction Solar Cells

Matthew Wilkins

and

Karin Hinzer

40.1 Introduction

Multijunction solar cells (MJSCs) currently hold the records for greatest power conversion efficiency of any solar cell, under one-sun conditions (Spectrolab, 38.8% [1]) and under concentration (Soitec/Fraunhofer, 46.0% at 508 suns). MJSCs based on III–V materials such as GaAs and GaInP find applications in aerospace applications where size and weight take precedence over cost, and in terrestrial concentrating photovoltaic (CPV) systems where concentrating optics focus the incident sunlight onto very small ($< 5 \times 5$ mm) cells. MJSCs based on material systems, such as amorphous silicon/silicon and II-VI compounds, and those using III–Vs grown on alternative substrates, have potential to combine high efficiency with lower costs suitable for nonconcentrating terrestrial applications.

Single-junction solar cells have two major loss mechanisms: thermalization and transmission. Photons of energy less than the semiconductor bandgap are transmitted through the material without being absorbed (transmission), while photons of energy in excess of the bandgap convert that excess energy to heat. Single junctions, then, must be designed to minimize transmission (requiring a low bandgap E_g), and minimizing thermalization (requiring a large E_g). For typical solar spectra, this leads to an optimal bandgap of ~1.3–1.4 eV [2,3].

MJSCs consist of a stack of n-p semiconductor diodes of different materials, each chosen to absorb a different part of the solar spectrum. As light passes through the stack, the short wavelengths are absorbed in the top junction which has a large E_g, and longer wavelengths are absorbed in each of the subsequent junctions. This arrangement minimizes the impact of transmission and thermalization losses by providing materials that are capable of absorbing the vast majority of photons within the solar spectrum, and by ensuring that each photon is absorbed in a junction with E_g similar to the photon energy. Typically, the junctions are all grown epitaxially in a monolithic device, with all junctions connected in series; this implies that the current through all junctions must be the same. Less commonly, some devices are designed such that separate contacts are made to each junction so that they can each be operated independently.

Since MJSCs are more expensive to produce than single-junction solar cells, terrestrial applications of MJSCs typically involve concentrating optics that focus light from a large aperture onto a small area of solar cell. In these systems, the sun's intensity is concentrated by a factor of 10 to 2000. Besides the cost savings from using smaller cells, there is also a significant increase in cell efficiency under concentration. We can write the efficiency η as

$$\eta = \frac{P_{max}}{XE_{1-sun}} = \frac{J_{SC} V_{OC} FF}{XE_{1-sun}}, \qquad (40.1)$$

where E_{1-sun} is the incident light intensity under standard 1-sun conditions, X is the concentration factor, P_{max} is the output electrical power density at the optimal voltage bias, J_{SC} is the short-circuit current density, V_{OC} is the open-circuit voltage, and FF is the fill factor.

A simple equivalent circuit model for the series-connected triple-junction solar cell is shown in Figure 40.1b; each junction i is modeled with a current source providing the 1-sun photocurrent for that

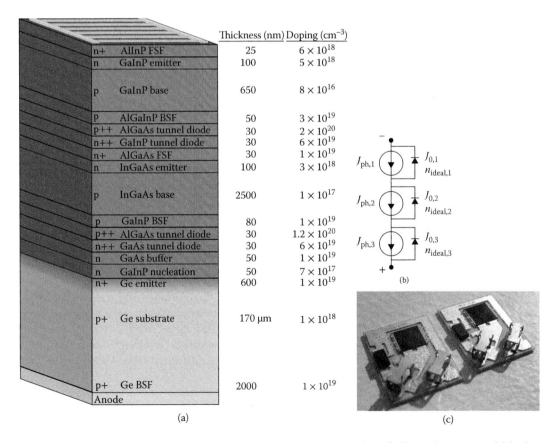

(a)   (b)   (c)

FIGURE 40.1 (a) Layer structure of a triple-junction GaInP/(In)GaAs/Ge solar cell. (b) Simple circuit model for the triple-junction cell. (c) Two 5 × 5-mm MJSCs mounted on ceramic carriers.

subcell, $J_{ph,i}$. A parallel diode provides the exponential diode behavior of each cell, so the device current density

$$J = XJ_{ph,i} - J_{0,i}\left(\exp\frac{qV_i}{n_{ideal,i}kT} - 1\right) \quad \forall\, i, \tag{40.2}$$

where kT/q is the thermal voltage, $n_{ideal,i}$ is the diode ideality factor of each subcell and $J_{0,i}$ is the saturation current density. It is useful to model each junction with two parallel diodes with different ideality factors, and also add parasitic series and shunt resistances. The diodes effectively model the Shockley–Read–Hall (SRH), radiative, and Auger recombination mechanisms, which are active in each junction.

If all three photocurrents $J_{ph,i}$ are equal, then the device's short-circuit current density $J_{SC} = J_{ph,i}$ and the current density through the diodes is zero. If they are not equal, then $J_{SC} = \min(J_{ph,i})$, and any excess photocurrent is dissipated through the diodes.

Under open-circuit conditions, the current at the device terminals, and at the interconnections between subcells, is zero. Therefore, all photogenerated current recombines via the diodes.

The short-circuit current density J_{SC} increases linearly with concentration, but there is also an increase in open-circuit voltage V_{OC}, which comes about due to the higher concentration of carriers in the semiconductor and larger separation of the quasi-Fermi levels. Under a concentration factor X, the open-circuit voltage is [4]

$$V_{OC} = \frac{kT}{q}\left[\sum_i n_{ideal,i}\ln\left(\frac{J_{ph,i}}{J_{0,i}}\right) + \ln X\sum_i n_{ideal,i}\right]. \tag{40.3}$$

If we neglect effects due to parasitic resistances, the fill factor also increases with concentration; consequently, MJSCs generally gain in efficiency as concentration is increased. Of course, this relation does not extend to infinitely high concentration; eventually the assumptions used to derive the diode-like behavior of Equation 40.2 break down, while resistive terms become increasingly important. In such cases, a more complete physics-based device simulation can be useful to study these limits.

The most common commercially available MJSC is a lattice-matched design with triple junctions consisting of $Ga_{0.49}In_{0.51}P$, $In_{0.01}Ga_{0.99}As$, and Ge, respectively [5]. The GaInP and (In)GaAs junctions are clad with a front surface field and a back surface field of a larger bandgap, which helps to confine minority carriers within the junction region. The triple *n-p* junctions (or subcells) are connected in series with *p*++/*n*++ tunnel diodes, which provide an optically transparent, ohmic connection between the *p*-type base of one subcell and the *n*-type emitter of the next subcell. The structure is grown by metal-organic chemical vapor deposition (MOCVD) on a germanium substrate, with the germanium junction being formed by diffusion of group V dopants into the top surface of the substrate during the initial growth of epitaxial material [6,7]. The remaining subcell and tunnel-diode layers are formed by conventional MOCVD growth, requiring ~4 μm of epitaxial material in ~20 layers. A typical layer structure is shown in Figure 40.1a, and some typical 5 × 5-mm cells are shown in Figure 40.1c. This basic structure will be used as an example throughout this chapter.

MJSCs have been simulated with good results using a variety of approaches including lumped-parameter models [2], distributed parameter models [8], and drift-diffusion-based simulators in 1D, 2D, and 3D. In this chapter, we focus primarily on drift-diffusion-based methods. The simulation results shown here were generated using the Sentaurus TCAD package from Synopsys, Inc.; example simulation projects for MJSC devices are available from most of the major simulation software vendors [9–11] and should provide a good starting point.

One of the greatest challenges in simulating these devices is managing the large number of material parameters that are required for each of the materials; this is particularly difficult for ternary and quaternary materials such as AlGaAs and InGaAs for example, where the material composition may be subject to optimization and so material parameters are needed for a wide variety of compositions.

40.2 Device Model and Simulation Workflow

The process of setting up and running a simulation begins by defining each of the material regions (Figure 40.2). In the case of MJSC, this will be a simple structure of uniform layers for the most part; the layers are defined in Figure 40.1a. The exception in our example structure is the germanium subcell, which is formed by diffusion of dopants rather than by epitaxial growth of the emitter. For the germanium subcell, we model the *p*-type substrate as a single region, with a smoothly varying "error function" profile of *n*-type doping added to the emitter area. The back surface field at the bottom of the substrate is also modeled in the same way.

Each of the regions is discretized with a mesh, material parameters are assigned to each region, and ohmic contact boundary conditions are defined at the cathode and anode.

From this point, the process differs for QE or for *J–V* curve calculations. For a QE calculation, additional "virtual contacts" are be defined between each junction, enabling the current from each junction to be extracted separately. An optical calculation is done to determine the optical generation rate as a function of position within the device at the starting wavelength. The Poisson equation is solved to find the potential profile at equilibrium, and then the coupled Poisson and electron and hole continuity equations

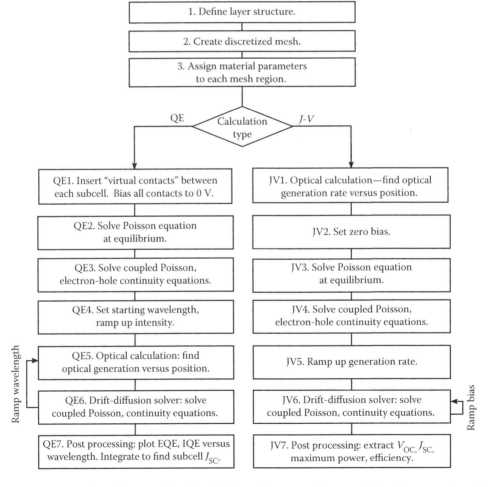

FIGURE 40.2 Typical simulation workflow for calculation of quantum efficiency (QE) and current–voltage (*J–V*) characteristics of multijunciton solar cells.

are solved as the incident optical intensity (and the resulting optical generation rate) is ramped up to the desired value. Finally, the incident wavelength is ramped over the range of interest, with the optical problem being recalculated at each wavelength, and the coupled Poisson-continuity problem is also solved at each wavelength to find the short-circuit current of each junction. This wavelength and short-circuit current data are postprocessed to produce plots of QE.

In the case of a $J–V$ curve calculation, the optical problem is only solved once, but with an appropriate broadband spectrum. The simulation then starts with a Poisson solution at zero bias. Intensity is gradually ramped up as was the case for the QE calculation, and then the voltage bias is ramped to produce a $J–V$ curve. Postprocessing is done to identify the maximum power point and extract key parameters, such as open-circuit voltage (V_{OC}), short-circuit current (J_{SC}), fill factor (FF), and power conversion efficiency (η).

40.3 Mesh Discretization

Depending on the physical phenomena which are being studied and the symmetry of the device, the region of interest can be discretized with a one-dimensional mesh [12], a 2D mesh representing a small symmetry element such as a half-gridline pitch [13], an entire half of the device, or a 3D model of a portion of the device [14]. The mesh must resolve steps in potential at material heterointerfaces, so ideally mesh vertices along the interfaces should be duplicated with a boundary condition relating the solution variables on each side of the interface (Figure 40.3).

In the direction normal to the interfaces, the minimum required mesh density is typically driven by a need to resolve potential gradients near the interfaces and $p–n$ junctions. Tunnel diodes in particular have depletions regions of perhaps 10 nm in thickness, and these must be resolved in order for a nonlocal tunneling model to work properly. Far from the interfaces and junctions, the minimum mesh density is driven by the need to resolve standing waves in the optical intensity profile. In the lateral direction, mesh elements can usually be allowed to grow quite large, except at the edges of gridlines and at the device perimeter.

For many studies, a 1D model or a 2D half-gridline pitch option is sufficient to study the phenomena of interest; however, certain phenomena will not be accurately represented. Table 40.1 summarizes the

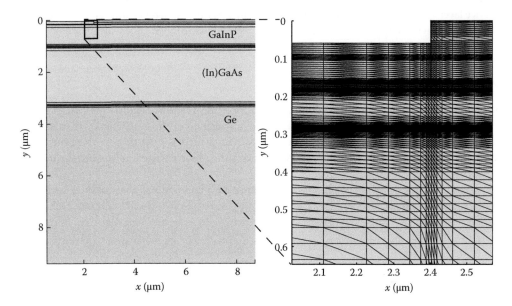

FIGURE 40.3 (a) A 2D, half-grid pitch mesh for the triple-junction solar cell. The full simulation region extends to 60×170 μm. (b) Detail of the area near the top contact gridline.

TABLE 40.1 Limitations of Reduced Model Geometries.

	1D	2D Half-Grid Pitch	2D Half-Device	3D Distributed Parameter Model Calibrated Using 2D Simulations	3D
p-n junctions and tunnel diodes	•	•	•	•	•
ARC/window interface		•	•	•	•
Emitter sheet resistance		•	•	•	•
Shading due to gridlines		•	•	•	•
Perimeter recombination			Partial	•	•
Grid ohmic losses				•	•
Nonuniform bias due to grid resistance				•	•
Nonuniform illumination			•	•	•
Busbar regions			Partial	•	•

ARC, anti-reflection coating

limitations of various reduced-dimensional representations. The full 2D and 3D options may be very computationally intensive and will likely be impractical using standard desktop computer hardware; however Létay et al. have demonstrated a method of addressing this by modeling small portions of the device representing the perimeter and the center of the device, linked through an external circuit simulator [13]. Alternatively, a 2D half-grid pitch model can be used to characterize the illuminated *I–V* performance of a small cell element, and that information can be used to calibrate the parameters of a distributed circuit model [15].

40.4 Optics

Several types of solvers exist which can calculate the carrier generation rate as a function of position within the device layer stack for a given illumination. Raytracing and finite-difference time domain methods [14] can be used, but here we focus on the transfer matrix method (TMM), which is widely used in simulating MJSCs because they have a planar structure and minimal surface texture.

40.4.1 Transfer Matrix Method

In a 2D or 3D simulation including gridlines, we can apply the TMM to calculate the optical generation rate as a function of depth within the layer stack. The propagation of electromagnetic waves through the layer stack is represented by a series of matrices. For each layer there is a matrix describing the reflections from the top interface, and a second matrix representing the phase change and attenuation through the thickness of the layer.

For a stack of N layers, we define phasors representing the amplitude and phase of the y-component of the electric field for downward- and upward-propagating waves at the top and bottom of each layer. The vector

$$\begin{bmatrix} E^+_{m,\text{top}} \\ E^-_{m,\text{top}} \end{bmatrix} \tag{40.4}$$

contains the phasors describing the electric field at the top of layer m. The symbol $+$ refers to downward propagation (into the cell) and $-$ refers to upward propagation (reflected waves). Layer 0 is the medium above the cell, and layer $N + 1$ is the medium below the cell, which represents the opaque back contact, or air in the case of a bifacial solar cell.

Note that for a plane wave of wavelength λ_0 with an angle of incidence θ_0 with respect to the surface normal, the angle of propagation within any layer m can be found from Snell's equation, $n_0 \sin \theta_0 = n_m \sin \theta_m$. The y-component of the wave vector is then

$$k_{y,m} = \frac{2\pi}{\lambda_0} \frac{n_m}{n_0} \cos \theta_m. \tag{40.5}$$

The phasors at the top and bottom interfaces of a layer are related by a propagation matrix $\mathbf{P}_m$,

$$\begin{bmatrix} E_{m,\text{top}}^+ \\ E_{m,\text{top}}^- \end{bmatrix} = \begin{bmatrix} e^{-jk_{y,m}d_m} & 0 \\ 0 & e^{jk_{y,m}d_m} \end{bmatrix} \begin{bmatrix} E_{m,\text{bot}}^+ \\ E_{m,\text{bot}}^- \end{bmatrix}, \tag{40.6}$$

where j is $\sqrt{-1}$, and d_m is the thickness of layer m.

The phasors at bottom of one layer and the top of the next are are related by

$$\begin{bmatrix} E_{m-1,\text{bot}}^+ \\ E_{m-1,\text{bot}}^- \end{bmatrix} = \frac{1}{t_{m-1,m}} \begin{bmatrix} 1 & r_{m-1,m} \\ r_{m-1,m} & 1 \end{bmatrix} \begin{bmatrix} E_{m,\text{top}}^+ \\ E_{m,\text{top}}^- \end{bmatrix}, \tag{40.7}$$

where $t_{m-1,m}$ and $r_{m-1,m}$ are the Fresnel coefficients for reflection and transmission at the interface. We denote this transmission matrix $\mathbf{T}_{m-1,m}$.

A transfer matrix for the entire layer stack can then be assembled as

$$M = \prod_{m=1}^{N} \left(\mathbf{T}_{m-1,m} \mathbf{P}_m \right) \mathbf{T}_{N,N+1}. \tag{40.8}$$

And finally, we set the electric field incident on the top of the cell, $E_{0,\text{bot}}^+$, to any convenient value and write

$$\begin{bmatrix} E_{0,\text{bot}}^+ \\ E_{0,\text{bot}}^- \end{bmatrix} = M \begin{bmatrix} E_{N+1,\text{top}}^+ \\ 0 \end{bmatrix}. \tag{40.9}$$

This can be solved for the reflected field, $E_{0,\text{bot}}^-$, and from that point the total electric field at any point in the structure can be found straightforwardly by assembling a transfer matrix to relate the fields at the top surface with the fields at the point of interest.

This type of calculation is often done assuming that unpolarized light is incident perpendicular to the device, but is also valid for other angles of incidence. The calculation is done as a 1D problem and the resulting carrier generation rate profile is applied to all regions that are not shaded by the grid (Figure 40.4b). Where a broad spectrum illumination is being studied, the TMM is solved repeatedly at each wavelength and the carrier generation rate is summed, weighted according to the incident spectral irradiance [16]. Similarly, illumination from a wide range of angles can be modeled by summing the optical generation calculated from many TMM calculations.

40.4.2 Complex Refractive Index

Solar cell simulations require data on the complex refractive index (refractive index n and extinction coefficient k) over a wide range of wavelengths. This can be a challenge as published data are often limited to a certain range of wavelengths [17]; it may be necessary to measure properties of some materials

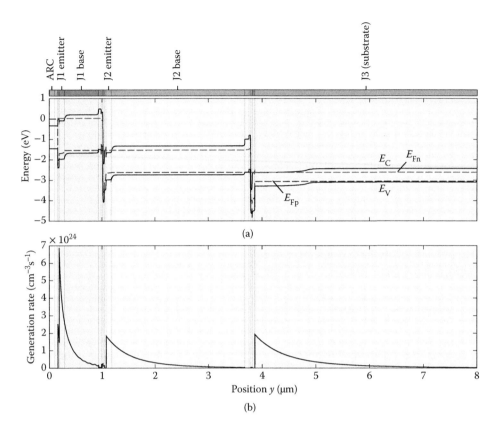

FIGURE 40.4 (a) Simulated band diagram at maximum power point for the triple-junction solar cell. (b) Profile of optical generation rate through the depth of the triple-junction solar cell, as calculated using the TMM. Illumination is broadband (AM1.5D spectrum). (ARC, antireflection coating.)

using ellipsometry or to combine data from several studies to yield a useable dataset. There can be further difficulties due to the use of ternary and quaternary alloys where data are only available for specific compositions.

In some cases, research has been done to model the dielectric function of a material in terms of a model dielectric function (MDF) consisting of a series of Lorentzian oscillators and other terms representing the various band-to-band transitions [18–20]. If the oscillator parameters are estimated as a function of alloy composition, this can be an effective method to generate n,k data for arbitrary alloy compositions. For use in simulating solar cells, care must be taken because these MDF formulations usually yield excessively large k values at photon energies below the bandgap.

40.4.3 Absorption and Quantum Yield

Once the electric field and optical intensity as a function of position within the structure is calculated using the TMM (or any other method), the local absorbed power density is the negative derivative of the optical intensity, $A = -\nabla(\frac{1}{2}\epsilon|\mathbf{E}|^2)$, where ϵ is the dielectric permittivity. If we assume that the quantum yield is unity (i.e., all absorbed photons result in generated carriers), then the local carrier generation rate is

$$G = -\frac{\lambda}{hc}\nabla(\frac{1}{2}\epsilon|\mathbf{E}|^2). \tag{40.10}$$

The assumption of unity quantum yield generally works very well in III–V heterostructures, but in some materials such as highly doped silicon, free-carrier absorption is significant and will result in a quantum yield less than unity.

The resulting optical generation profile is plotted in Figure 40.4b, for the case of broadband illumination with the AM1.5D spectrum as defined by ASTM G173-03. We can note that within each subcell, there is strong carrier generation near the top of the junction, which then decays quickly through the thickness of the junction. This is to be expected since each subcell is designed to absorb nearly all of the incident light of energy greater than its bandgap, so the generation decays nearly to zero at the bottom of each junction.

40.5 Drift-Diffusion Calculation

Having determined the optical generation rate as a function of position, we then proceed to simulation of the electrical device performance. The electrostatic potential within the device follows the Poisson equation,

$$\nabla \cdot (\epsilon \nabla \phi + \vec{P}) = -q(p - n - N_A + N_D), \tag{40.11}$$

where ϕ is the electrostatic potential, $\vec{P}$ is the polarization which is zero in semiconductors with a zincblende crystal structure, p and n are the free hole and electron concentrations respectively, and N_A and N_D are the acceptor and donor concentrations. ϵ is the permittivity of the material.

The electron and hole concentrations also follow the continuity equations,

$$\nabla \cdot \vec{J}_n = qR_{net} + q\frac{\partial n}{\partial t},$$
$$-\nabla \cdot \vec{J}_p = qR_{net} + q\frac{\partial p}{\partial t}. \tag{40.12}$$

$\vec{J}_n$ and $\vec{J}_p$ are the electron and hole current densities; R_{net} is the net recombination rate, i.e., the sum of contributions from all recombination and generation processes. At a minimum this will include SRH recombination, Auger recombination, and radiative recombination, as well as the optical generation calculated using the TMM. It can also include additional terms such as band-to-band tunneling and coupled generation, as we discuss below. For studies of solar cells, we are generally interested in steady-state solutions so the time derivaties are zero. The current densities $\vec{J}_n$ and $\vec{J}_p$ can be further expanded into drift and diffusion terms, which depend on the gradient of the local electrical potential and the gradients of the carrier concentrations, the same set of unknowns as in Equation 40.11. Using the Einstein relation, the current densities are [21]

$$\vec{J}_n = -nq\mu_n \nabla E_{Fn},$$
$$\vec{J}_p = -pq\mu_p \nabla E_{Fp}, \tag{40.13}$$

where μ_n and μ_p are the electron and hole mobilities and E_{Fn} and E_{Fn} are the electron and hole quasi-Fermi levels. Relating the conduction band energy and valence band energy E_C and E_V to the material properties and the electrostatic potential, we have

$$E_C = -\chi - q(\phi - \phi_{ref}),$$
$$E_V = -\chi - E_{g,eff} - q(\phi - \phi_{ref}), \tag{40.14}$$

where χ is the electron affinity of the material, $E_{g,eff}$ is the effective bandgap (including the effect of bandgap narrowing), and ϕ_{ref} is an arbitrarily chosen reference potential.

Using Fermi–Dirac statistics for the electron and hole concentrations and making the assumptions that the carrier populations and the crystal are at a constant temperature throughout the device and that the conduction and valence bands have parabolic band structures, the electron and hole quasi-Fermi levels are [21]

$$E_{Fn} = kT\mathcal{F}_{1/2}^{-1}\left(\frac{n}{N_C}\right) + E_C,$$

$$E_{Fp} = -kT\mathcal{F}_{1/2}^{-1}\left(\frac{p}{N_V}\right) + E_V,$$

(40.15)

where k is the Boltzmann constant, T is the device temperature, $\mathcal{F}_{1/2}^{-1}$ is the inverse Fermi-Dirac integral of order 1/2, and N_C and N_V are the effective densities of states in the conduction and valence bands, respectively. By substituting Equations 40.13 through 40.15 into Equation 40.12, we arrive at a formulation of the three coupled differential equations for steady-state problems, stated solely in terms of the material properties and functions of the three variables n, p, and ϕ:

$$\nabla \cdot \epsilon \nabla \phi = -q(p - n - N_D - N_A),$$

(40.16)

$$\nabla \cdot \left[n\mu_n \nabla \left(kT\mathcal{F}_{1/2}^{-1}\left(\frac{n}{N_C}\right) - \chi - q(\phi - \phi_{ref}) \right) \right] = -R_{net},$$

(40.17)

$$\nabla \cdot \left[p\mu_p \nabla \left(kT\mathcal{F}_{1/2}^{-1}\left(\frac{p}{N_V}\right) + \chi + E_{g,eff} + q(\phi - \phi_{ref}) \right) \right] = -R_{net}.$$

Typically, we need to solve the coupled set of Equations 40.16 and 40.17 to find the local potential and carrier concentrations throughout the device. Once those quantities are known, then the local electron and hole current densities and the quasi-Fermi levels are also known, and the voltage and current at each electrode can also be found. We are dealing with heterostructures with abrupt changes in the valence and conduction band energies and other quantities at each material interface. To handle these discontinuities, the coupled equations are solved separately in each material region with a set of boundary conditions relating the solution variables on each side of the interfaces. The boundary conditions allow for thermionic emission across or tunneling through small potential barriers at the interfaces [22,23].

It can be difficult to achieve convergence of the coupled Poisson and continuity equations, so we typically begin first with a solution of the Poisson equation under equilibrium conditions (i.e., with all currents and net recombination/generation equal to zero). This provides an initial guess for the solution of the coupled Poisson and continuity equations, again under equilibrium. Next, we gradually ramp up the illumination intensity, leading to a solution at nonequilibrium, and then sweep either the bias at the contacts (to produce a J–V curve) or the illumination wavelength (to produce a QE plot). In the case of QE, the optical generation must be recalculated at each wavelength.

In order to provide physically accurate results, the set of physical models and material parameters must be carefully chosen. Given the wide range of materials and compositions that are present in a single device, managing this database of material parameters is probably the single most difficult part of a MJSC device simulation. Altermatt provides an excellent guide to the parameters needed to do good simulations of silicon solar cells [24]; similar studies are needed for each of the materials in the MJSC structure. Wherever possible, it is advisable to validate simulations of single junctions of each material before attempting to simulate a multijunction device.

40.5.1 Band Structure and Intrinsic Carrier Concentration

While material bandgaps are often well known to a high degree of accuracy, Altermatt makes the point that it is more important to have a self-consistent set of parameters (i.e., bandgap, effective masses, and doping-induced bandgap narrowing) that lead to an accurate value of the effective intrinsic carrier concentration, than to have any one of these individually correct.

If Fermi–Dirac statistics are being used in the simulation, then the experiments used to determine these parameters must also have been based on the same statistics. For the III–V materials commonly used in MJSCs, [25] is a good starting point.

Ternary materials in the $(Al,Ga)_{0.5}In_{0.5}P$ system can have varying degrees of Cu:Pt ordering, where Al or Ga atoms and In atoms are found on separate planes of the crystal structure [26]. This means that a range of bandgap values are possible for the same material composition. For example, $Ga_{0.5}In_{0.5}P$ can have a bandgap in the range between 1.8 and 1.9 eV.

In Tables 40.2 through 40.4, we provide the material parameters used for the drift-diffusion-based simulation of the triple-junction solar cell as presented in this chapter. The effective bandgap, $E_{g,eff}$ at a temperature T is described using a Varshni model [25] for the temperature dependence,

$$E_{g,eff} = E_{g,ref} - \frac{\alpha T^2}{T + \beta} + \frac{\alpha T_{ref}^2}{T_{ref} + \beta} - \Delta E_g, \tag{40.18}$$

where $E_{g,ref}$ is the bandgap at a reference temperature T_{ref}, and α and β are the Varshni parameters. ΔE_g is a doping-dependent bandgap narrowing term, which we specify using a table of data points. Similarly, the electron affinity χ (difference between the vacuum energy level and the conduction band energy) is described by

$$\chi = \chi_{ref} + \frac{\alpha T^2}{T + \beta} - \frac{\alpha T_{ref}^2}{T_{ref} + \beta} + \Delta E_g \cdot (\Delta\chi/\Delta E_g), \tag{40.19}$$

where the ratio $\Delta\chi / \Delta E_g$ is specified as a constant.

We calculate the conduction band and valence band densities of states, N_C and N_V, from the electron and hole effective masses m_n and m_p [21],

$$N_C = 2.5094 \times 10^{19} \left(\frac{m_n}{m_0} \cdot \frac{T}{300\ K} \right)^{3/2} cm^{-3}, \tag{40.20}$$

$$N_V = 2.5094 \times 10^{19} \left(\frac{m_p}{m_0} \cdot \frac{T}{300\ K} \right)^{3/2} cm^{-3}. \tag{40.21}$$

The effective intrinsic carrier density is then calculated from the band edge densities of states and the effective bandgap,

$$n_{i,eff} = \sqrt{N_C N_V} \exp\frac{-E_{g,eff}}{2kT}. \tag{40.22}$$

Carrier mobilities μ_n and μ_p are adjusted for temperature- and doping-dependent effects. One of two parameterizations, the "Masetti model" [27] or the "Arora model" [28] are used depending on what data is available and/or which model provides the best fit to data. The Masetti model takes the form

$$\mu = \mu_{min1} \exp\left(\frac{-P_c}{N_A + N_D} \right) + \frac{\mu_{max}(T/300\ K)^\zeta - \mu_{min2}}{1 + [(N_A + N_D)/C_r]^\alpha} - \frac{\mu_1}{1 + [C_s/(N_A + N_D)]^\beta}, \tag{40.23}$$

where N_D and N_A are the concentrations of donors and acceptors respectively. The remaining parameters are listed in the parameter tables.

TABLE 40.2   Table of Material Parameters for the GaInP/GaInAs/Ge Solar Cell Example.

	Symbol	Units	$Ga_{0.49}In_{0.51}P$	$In_{0.01}Ga_{0.99}As$	Germanium
	ϵ/ϵ_0		12.005	12.91	16.0
			[29]	[29]	[30,31]
	n, k		[20]	[32]	[33]
Band gap	$E_{g,0}$	eV	1.88	1.40	0.744
	α	eV/K	1.82×10^{-4}	5.405×10^{-4}	4.77×10^{-4}
	β	K	81	204	235
	T_{ref}	K	300	300	0
	χ_0	eV	3.924	4.07	3.960
	ΔE_g	eV	[29]	[34]	[29]
	$\Delta\chi/\Delta E_g$		0.5	0.5	0.5
			[35]	[25]	[31]
Density of states	m_n/m_0		0.08515	0.06553	0.5438
	$N_{C,300}$	cm^{-3}	6.235×10^{17}	4.210×10^{17}	1.0063×10^{19}
	m_p/m_0		0.7125	0.5236	0.3406
	$N_{V,300}$	cm^{-3}	1.5093×10^{19}	9.509×10^{18}	4.9883×10^{18}
			[29,35]	[25,35]	[29,35]
Mobility (Masetti model)	μ_{max}	cm^2/V.s		(9400, 491.5)	
	ζ			(2.1, 2.2)	
	μ_{min1}	cm^2/V.s		(500.0, 20.0)	
	μ_{min2}	cm^2/V.s		(500.0, 20.0)	
	μ_1	cm^2/V.s		(0, 0)	
	P_c	cm^{-3}		(0, 0)	
	C_r	cm^{-3}		$(6 \times 10^{16}, 1.48 \times 10^{17})$	
	C_s	cm^{-3}		(0, 0)	
	α			(0.394, 0.38)	
	β			(0, 0)	
				[36]	
Mobility (Arora model)	A_{min}	cm^2/V.s	(400.0, 15.0)		(850.0, 300.0)
	α_m		(0, 0)		(0, 0)
	A_d	cm^2/V.s	(3900, 135)		(2950, 1500)
	α_d		(0, 0)		(0, 0)
	A_N	cm^{-3}	$(2 \times 10^{16}, 1.5 \times 10^{17})$		$(2.6 \times 10^{17}, 1 \times 10^{17})$
	α_N		(1.955, 1.47)		(0, 0)
	A_a		(0.7, 0.8)		(0.56, 1.0)
	α_a		(0, 0)		(0, 0)
			[36]		[31]
Recombination	B_{rad}	cm^3/s	1×10^{-10}	2×10^{-10}	6.4×10^{-14}
	τ_{max}	s	$(2 \times 10^{-8}, 2 \times 10^{-8})$	$(1 \times 10^{-7}, 2 \times 10^{-8})$	(0.001, 0.001)
	τ_{min}	s	(0, 0)	(0, 0)	(0, 0)
	N_{ref}	cm^{-3}	$(1 \times 10^{19}, 1 \times 10^{19})$	$(1 \times 10^{16}, 2 \times 10^{18})$	$(1 \times 10^{16}, 1 \times 10^{16})$
	γ		(1.0, 1.0)	(1.0, 3.0)	(1.0, 1.0)
	T_α	K	(0, 0)	(0, 0)	(0, 0)
	T_{coeff}		(0, 0)	(0, 0)	(0, 0)
	A_A	cm^{6}s^{-1}	$(3 \times 10^{-30}, 3 \times 10^{-30})$	$(5 \times 10^{-30}, 5 \times 10^{-30})$	$(1 \times 10^{-30}, 1 \times 10^{-30})$
	B_A	cm^{6}s^{-1}	(0, 0)	(0, 0)	(0, 0)
	C_A	cm^{6}s^{-1}	(0, 0)	(0, 0)	(0, 0)
	H		(0, 0)	(0, 0)	(0, 0)
	N_0	cm^{-3}	$(1 \times 10^{18}, 1 \times 10^{18})$	$(1 \times 10^{18}, 1 \times 10^{18})$	$(1 \times 10^{18}, 1 \times 10^{18})$
			[35]	[35]	[35]
Nonlocal tunneling	g		(0.21, 0.4)	(0.21, 0.4)	
	m_t/m_0		(0.24, 0.48)	(0.085, 0.34)	
			[37]	[38]	

Note: Paired values in parentheses are specified for electrons and for holes, respectively.

TABLE 40.3 Table of Material Parameters for the GaInP/GaInAs/Ge Solar Cell Example

	Symbol	Units	$Al_{0.5}In_{0.5}P$	$Al_{0.3}Ga_{0.7}As$	$Al_{0.2}Ga_{0.8}As$
	ϵ/ϵ_0		11.355	12.055	12.34
			[29]	[29]	[29]
	n, k		[20]	[19]	[19]
Band gap	$E_{g,0}$	eV	2.382	1.7976	1.6726
	α	eV/K	1.76×10^{-4}	4.98×10^{-4}	5.12×10^{-4}
	β	K	134.2	142.8	163.2
	T_{ref}	K	300	300	300
	χ_0	eV	3.75	3.756	3.877
	ΔE_g	eV	[29]	[34]	[34]
	$\Delta\chi/\Delta E_g$		0.5	0.5	0.5
			[25]	[25,35]	[25,35]
Density of states	m_n/m_0		0.23	0.0879	0.0796
	$N_{C,300}$	cm^{-3}	2.768×10^{18}	6.540×10^{17}	5.636×10^{17}
	m_p/m_0		0.36	0.5842	0.5593
	$N_{V,300}$	cm^{-3}	5.4203×10^{18}	1.1206×10^{19}	1.0496×10^{19}
			[25]	[25,35]	[25,35]
Mobility (Masetti model)	μ_{max}	cm^2/V.s	(150.0, 180.0)	(3721.1, 240)	(5896.6, 308.21)
	ζ		(1.0, 1.0)	(1.74, 1.83)	(1.81, 1.9)
	μ_{min1}	cm^2/V.s	(0, 0)	(195.09, 5.0)	(313.6, 8.8571)
	μ_{min2}	cm^2/V.s	(0, 0)	(195.09, 5.0)	(313.6, 8.8571)
	μ_1	cm^2/V.s	(0, 0)	(0, 0)	(0, 0)
	P_c	cm^{-3}	(0, 0)	(0, 0)	(0, 0)
	C_r	cm^{-3}	$(5 \times 10^{17}, 2.75 \times 10^{17})$	$(1.16 \times 10^{17}, 1.00 \times 10^{17})$	$(9.33 \times 10^{16}, 1.07 \times 10^{17})$
	C_s	cm^{-3}	(0, 0)	(0, 0)	(0, 0)
	α		(0.436, 0.397)	(0.5758, 0.324)	(0.5152, 0.33425)
	β		(0, 0)	(0, 0)	(0, 0)
			[36]	[36]	[36]
Mobility (Arora model)	A_{min}	cm^2/V.s			
	α_m				
	A_d	cm^2/V.s			
	α_d				
	A_N	cm^{-3}			
	α_N				
	A_a				
	α_a				
Recombination	B_{rad}	cm^3/s	1×10^{-10}	1×10^{-10}	1×10^{-10}
	τ_{max}	s	$(2.5 \times 10^{-8}, 2.5 \times 10^{-8})$	$(1 \times 10^{-6}, 1 \times 10^{-8})$	$(1 \times 10^{-6}, 1 \times 10^{-8})$
	τ_{min}	s	$(5 \times 10^{-10}, 5 \times 10^{-10})$	(0, 0)	(0, 0)
	N_{ref}	cm^{-3}	$(1 \times 10^{16}, 1 \times 10^{16})$	$(1 \times 10^{16}, 2 \times 10^{18})$	$(1 \times 10^{16}, 2 \times 10^{18})$
	γ		(1.0, 1.0)	(1.0, 3.0)	(1.0, 3.0)
	T_α	K	(0, 0)	(0, 0)	(0, 0)
	T_{coeff}		(0, 0)	(0, 0)	(0, 0)
	A_A	cm^{6}s^{-1}	$(3 \times 10^{-30}, 3 \times 10^{-30})$	$(1 \times 10^{-30}, 1 \times 10^{-30})$	$(1 \times 10^{-30}\ 1 \times 10^{-30})$
	B_A	cm^{6}s^{-1}	(0, 0)	(0, 0)	(0, 0)
	C_A	cm^{6}s^{-1}	(0, 0)	(0, 0)	(0, 0)
	H		(0, 0)	(0, 0)	(0, 0)
	N_0	cm^{-3}	$(1 \times 10^{18}, 1 \times 10^{18})$	$(1 \times 10^{18}, 1 \times 10^{18})$	$(1 \times 10^{18}, 1 \times 10^{18})$
			[35]	[35]	[35]
Nonlocal tunneling	g			(0.21, 0.4)	
	m_t/m_0			(0.09, 0.37)	
				[38]	

Note: Paired values in parentheses are specified for electrons and for holes, respectively.

TABLE 40.4   Table of Material Parameters for the GaInP/GaInAs/Ge Solar Cell Example.

	Symbol	Units	$Al_{0.25}Ga_{0.25}In_{0.5}P$	MgF	TiOx
	ϵ/ϵ_0		11.84	5.3	6
			[29]	[39]	[40]
	n, k		[20]	[41]	[42,43]
Band gap	$E_{g,0}$	eV	1.984		
	α	eV/K	1.76×10^{-4}		
	β	K	134.2		
	T_{ref}	K	300		
	χ_0	eV	3.997		
	ΔE_g	eV	[29]		
	$\Delta\chi/\Delta E_g$		0.5		
			[25]		
Density of states	m_n/m_0		0.1214		
	$N_{C\,300}$	cm^{-3}	1.061×10^{18}		
	m_p/m_0		0.6244		
	$N_{V\,300}$	cm^{-3}	1.24×10^{19}		
			[29,35]		
Mobility (Masetti model)	μ_{max}	cm^2/V.s			
	ζ				
	μ_{min1}	cm^2/V.s			
	μ_{min2}	cm^2/V.s			
	μ_1	cm^2/V.s			
	P_c	cm^{-3}			
	C_r	cm^{-3}			
	C_s	cm^{-3}			
	α				
	β				
Mobility (Arora model)	A_{min}	cm^2/V.s	(400.0, 15.0)		
	α_m		(0, 0)		
	A_d	cm^2/V.s	(3900, 135)		
	α_d		(0, 0)		
	A_N	cm^{-3}	$(2 \times 10^{16}, 1.5 \times 10^{17})$		
	α_N		(1.955, 1.47)		
	A_a		(0.7, 0.8)		
	α_a		(0, 0)		
			[36]		
Recombination	B_{rad}	cm^3/s	1×10^{-10}		
	τ_{max}	s	$(1 \times 10^{-9}, 1 \times 10^{-9})$		
	τ_{min}	s	(0, 0)		
	N_{ref}	cm^{-3}	$(1 \times 10^{19}, 1 \times 10^{19})$		
	γ		(1.0, 1.0)		
	T_α	K	(0, 0)		
	T_{coeff}		(0, 0)		
	A_A	cm^{6}s^{-1}	$(3 \times 10^{-30}, 3 \times 10^{-30})$		
	B_A	cm^{6}s^{-1}	(0, 0)		
	C_A	cm^{6}s^{-1}	(0, 0)		
	H		(0, 0)		
	N_0	cm^{-3}	$(1 \times 10^{18}, 1 \times 10^{18})$		
			[35]		
Nonlocal tunneling	g				
	m_t/m_0				

Note: Paired values in parentheses are specified for electrons and for holes, respectively.

TABLE 40.5 Table of Interface Parameters for the GaInP/GaInAs/Ge Solar Cell Example.

	Symbol	Units	TiOx/AlInP	AlInP/GaInP	GaAs/AlGaAs	GaInP/Ge
Interface	S_0	cm/s	0	$(2 \times 10^5, 2 \times 10^5)$	(200, 200)	$(1 \times 10^5, \ 1 \times 10^5)$
recombination			fitted	[44]/fit to data	[44]	[6]/fit to data

Note: Paired values in parentheses are specified for electrons and for holes respectively.

The Arora model for mobility uses a set of temperature-corrected constants,

$$N_0 = A_N \left(\frac{T}{300 \text{ K}} \right)^{\alpha_N} ; A^* = A_a \left(\frac{T}{300 \text{ K}} \right)^{\alpha_a} \tag{40.24}$$

$$\mu_{\min} = A_{\min} \left(\frac{T}{300 \text{ K}} \right)^{\alpha_m} ; \mu_d = A_d \left(\frac{T}{300 \text{ K}} \right)^{\alpha_d} . \tag{40.25}$$

Mobility is then calculated with

$$\mu = \mu_{\min} + \frac{\mu_d}{1 + \left[(N_A + N_D)/N_0 \right]^{A^*}} . \tag{40.26}$$

40.5.2 Recombination, Lifetimes, and Surface/Interface Recombination Velocities

Having defined some of the basic material properties, we can now look at the recombination rates, each of which have their own material parameters. Here we present the rates as they are implemented in the Sentaurus software [21]; similar formulas can be found in texts on device simulation such as Palankovski [31]. The radiative recombination rate can be written as

$$R_{\text{rad}} = B_{\text{rad}}(np - \gamma_n \gamma_p n_{\text{i,eff}}^2), \tag{40.27}$$

where γ_n and γ_p are correction factors which are needed for simulations using Fermi–Dirac statistics, and are defined by

$$\gamma_n = \frac{n}{N_C} \exp \frac{E_C - E_{Fn}}{kT}, \tag{40.28}$$

$$\gamma_p = \frac{p}{N_V} \exp \frac{E_{Fp} - E_V}{kT} . \tag{40.29}$$

The SRH recombination rate is somewhat more complicated. We begin by finding the doping-dependent minority carrier lifetimes for the SRH process,

$$\tau_{\text{SRH}} = \tau_{\min} + \frac{\tau_{\max} - \tau_{\min}}{1 + \left(\frac{N_A + N_D}{N_{\text{ref}}} \right)^\gamma} . \tag{40.30}$$

This calculation will be done separately using different parameters for the electron and hole minority carrier lifetimes. Then we can write the expression for the SRH recombination rate,

$$R_{\text{SRH}} = \frac{np - \gamma_n \gamma_p n_{\text{i,eff}}^2}{\tau_{\text{SRH},p}(n + \gamma_n n_1) + \tau_{\text{SRH},n}(p + \gamma_p p_1)}, \tag{40.31}$$

where n_1 and p_1 are

$$n_1 = n_{i,eff}\exp\left(\frac{E_{trap}}{kT}\right); \quad p_1 = n_{i,eff}\exp\left(\frac{-E_{trap}}{kT}\right).$$ (40.32)

Note that we have not included a temperature dependence of the lifetimes, or any enhancement of the SRH process due to electric field, but these could be significant in some devices. E_{trap} is the position of the trap level relative to the intrinsic level; we generally take it to be zero.

Finally, we write the Auger recombination rate,

$$R_{Auger} = (C_n n + C_p p)(np - n_{i,eff}^2).$$ (40.33)

Due to the long nonradiative lifetimes in the bulk of many III–V materials and the relatively thin layers, interface recombination can be a significant factor in the performance of these devices. The values of surface recombination velocity can vary widely due to various growing conditions, and so this along with the SRH lifetime can be treated as a "fitting parameter" rather than a known material property [44]. The surface recombination rate is treated with an SRH model,

$$R_{surf} = \frac{np - \gamma_n\gamma_p n_{i,eff}^2}{(n + \gamma_n n_1)/S_p + (p + \gamma_p p_1)/S_n},$$ (40.34)

where S_n and S_p are the electron and hole surface recombination velocities.

40.5.3 Tunnel Diodes

The tunnel diodes that allow for the flow of carriers between subcells in MJSCs operate via band-to-band tunneling across a degenerately doped p++/n++ junction [45]. The depletion width in this case will be extremely narrow (~5 nm), with carrier populations at similar energy levels on both sides of the junction (Figure 40.5a). When a small bias is applied, the free electrons become aligned in energy with holes on the opposite side, and can tunnel across the junction, creating a net current.

A typical J–V curve of a tunnel diode is shown in Figure 40.5b. Under normal conditions, the tunnel diode should remain in the ohmic region, where large currents can be conducted with minimal potential difference across the tunnel diode. This requires that the peak tunneling current density should be significantly greater than the solar cell's short-circuit current density J_{SC}. If the bias voltage is increased beyond the ohmic region, the free electrons and holes are no longer aligned in energy and the tunneling current decreases. Eventually, the diode will be forward biased and will conduct via normal mechanisms.

Many drift-diffusion-based simulators allow for modeling band-to-band tunneling using approaches derived from the work of Kane et al. [46] and others [47,48]. These models can reproduce all of the behavior of a tunnel junction [37,38,49] including the ohmic region, tunneling peak and excess current region of the tunnel diode J–V curve [30].

Tunnel diodes typically have minimal impact on the performance of MJSCs, except in pathological cases or where the device is operated at very high current density. For many studies, a simplistic band-to-band tunneling model based only on local quantities will be sufficient. For example, in Synopsys Sentaurus, there is a model that adds a band-to-band recombination term to the continuity equation following the work of Hurkx [47],

$$R^{bb} = A\frac{np - n_{i,eff}^2}{(n + n_{i,eff})(p + n_{i,eff})}\left(\frac{|F|}{1V/cm}\right)^\sigma\exp\left(-\frac{F_0}{|F|}\right),$$ (40.35)

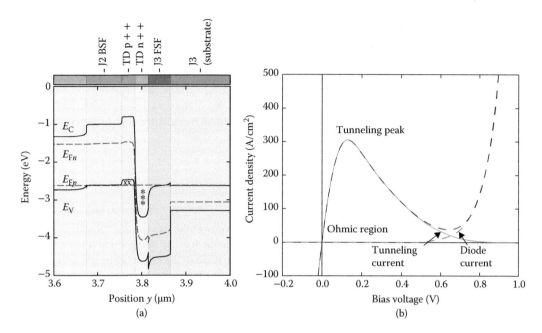

FIGURE 40.5 (a) Detail of the bottom tunnel-diode portion of the cross-sectional band diagram. The tunnel diode is short circuited. "xx" symbols represent holes and "*" symbols represent electrons that are available to tunnel across the junction into the opposite band. (b) *J–V* curve of a tunnel diode. Contributions from band-to-band tunneling and the diode diffusion current are shown.

where n and p are the electron and hole concentrations, $n_{i,\text{eff}}$ is the effective intrinsic carrier concentration, and F is the local electric field strength. σ has a value of 2 for direct transitions and 5/2 for indirect transitions. A and F_0 are material-dependent parameters that must be determined experimentally. This type of model should be adequate to give good results in cases where tunnel diodes are not dominating the behavior of an MJSC, but because it does not consider the variation of the band profile and material properties across the tunneling path, it will not be physically accurate.

A more accurate, but also more computationally intensive, approach uses a nonlocal band-to-band tunneling model combined with trap-assisted tunneling [50]. A "nonlocal" model is one where the tunneling current at a given position is calculated based on quantities at other locations, such as the potential profile across the tunneling path, and carriers are effectively removed from one location in the structure and reinserted at a different location. The use of this type of nonlocal model is described in detail in [38] and [51]. Implementation in Sentaurus is shown in Reference [52]. Figure 40.5b was simulated using nonlocal models for band-to-band tunneling as well as SRH recombination enhanced by trap-assisted tunneling. The individual contributions from each of these two mechanisms is shown in the figure. Some typical values of the tunneling masses m_t and effective Richardson constants g are included in Tables 40.2 and 40.3. Readers should refer to references [38,50–52] for interpretation of these parameters.

It is our experience that, while it it very possible to fit tunnel-diode parameters to a measured *J–V* curve, the resulting model tends not to be predictive for varying tunnel junction designs (i.e., varying doping levels or material compositions). The peak tunneling current of devices can vary by several orders of magnitude for relatively small changes in doping or material composition.

40.6 Current–Voltage Calculation

A *J–V* curve calculation is performed by first calculating the optical generation as a function of position for the appropriate illumination conditions, then solving the drift-diffusion problem while ramping bias at the contacts. It can also be useful to generate J–V curves for individual subcells. Additional "virtual" ohmic contact boundary conditions are defined at locations in between the subcells in order to bias a subcell individually. The current through a given subcell is then the sum of currents at all contacts above it (Figure 40.6).

It's very informative to study the subcell dark J–V curves on a log-current axis, as the contributions from various physical processes can be distinguished [2]. In Figure 40.6c, the dark J–V curve of the (In)GaAs subcell is plotted along with the recombination currents due to SRH, radiative, and Auger mechanisms. The recombination currents are calculated as

$$J_i = \frac{q}{A} \int_V R_i \, dV, \tag{40.36}$$

where q is the electron charge, R is the recombination rate per unit volume, A is the cell area and i refers to the recombination process in question. The integral is evaluated over the volume of a subcell. Comparing simulated dark-current curves against measured single-junction devices can be a good first step to validating multijunction simulations. When plotted on a log axis, the slope of the curve is directly related to the diode ideality factor. In particular, the SRH mechanism usually produces a region where the ideality factor $n_{\text{ideal}} \simeq 2$; this gives an excellent method to verify that the SRH current is modeled accurately independent of other factors. In contrast, fitting the SRH lifetimes based on V_{OC} measured at high intensity (1 sun or greater) would not be effective as the cell is not strongly dominated by the SRH process at that level of bias.

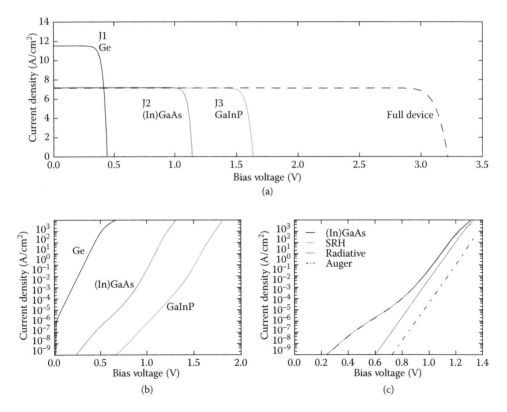

FIGURE 40.6 (a) J–V curves of the subcells and the triple-junction device under 500 suns, AM1.5D illumination. (b) Dark current curves of each of the subcells. (c) Dark J–V curve of (In)GaAs with the contributions from individual processes.

40.7 Quantum Efficiency Calculation

Experimentally, QE of multijunction cells is measured by using bias lights to ensure that the subcell of interest is limiting the overall device current, as described in Reference [53]. In this way, it is possible to probe each of the subcells individually. This experiment can also be replicated in simulation by adding bias illumination sources, or simply adding a fixed amount of carrier generation to the non-limiting subcells, in addition to a monochromatic source, which is scanned in wavelength over the range of interest. The bias light intensity must be adjusted such that the short-circuit currents of the non-probed junctions are greater than the current produced by the probed junction at any wavelength; this requires some iteration to find the correct bias intensities.

An alternative approach that can be implemented in simulation is to simply define ohmic "virtual contacts" to the front surface field and back surface field layers of each subcell. This provides a more direct control over the voltage bias being applied to each subcell than the light-biasing method, but should generally yield identical results. It also has advantages over the light-biasing method in that there is no iteration required to find the correct setup, and where the light-biasing method requires a separate sweep over wavelength for each junction, with the "virtual contact" method the short-circuit currents and QE plots of all triple junctions can be collected in one wavelength sweep. The simulated QE for the triple-junction solar cell, calculated using this method, is presented in Figure 40.7. The simulation will output short-circuit current density as a function of wavelength λ for subcell i, $J_{SC,i}$. From the optical calculation, we also know the reflectivity from the top of the layer stack, $R(\lambda)$ and transmissivity through the layer stack, $T(\lambda)$. The external quantum efficiency (EQE) of a given subcell is defined as the ratio of short-circuit current density to incident photon flux density $\Phi(\lambda)$,

$$EQE_i(\lambda) = \frac{J_{SC,i}(\lambda)}{q\Phi(\lambda)}, \tag{40.37}$$

or in terms of the optical intensity $E(\lambda)$,

$$EQE_i(\lambda) = \frac{hc}{q\lambda}\frac{J_{SC,i}(\lambda)}{E(\lambda)}, \tag{40.38}$$

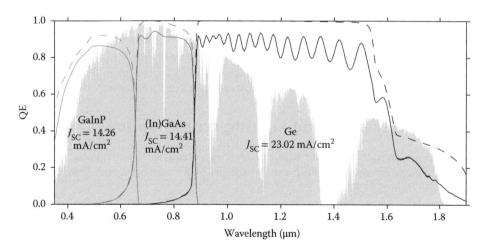

FIGURE 40.7 External (solid) and internal (dashed) quantum efficiency (EQE and IQE) of the triple-junction solar cell. The normalized AM1.5D solar spectral irradiance is shown filled in gray.

where h is Planck's constant, c is the speed of light and q is the electron charge. Internal quantum efficiency (IQE) is given by

$$IQE_i(\lambda) = \frac{EQE_i(\lambda)}{1 - R(\lambda) - T(\lambda)}. \qquad (40.39)$$

QE curves as calculated in software often show Fabry–Pérot type oscillations at long wavelengths which do not appear in physical measurements. Frequently these oscillations are physically correct for a perfectly collimated beam, but in an experimental setup with a finite beam divergence and monochromator resolution these fine details cannot be resolved. In order to make meaningful comparisons between measured and simulated data, it can therefore be useful to convolve the simulated QE with a spectral function approximating the monochromator lineshape, thus giving a representation of the simulated QE as it would be measured by an instrument. We can write this effective QE as

$$EQE_{\text{eff}}(\lambda) = \int_{-\infty}^{\infty} EQE(\lambda + l)w(l)dl, \qquad (40.40)$$

where $EQE(\lambda)$ is the simulated QE, $w(\lambda)$ is a function approximating the spectral lineshape of the monochromator which is centered at $\lambda = 0$ and integrates to unity, and l is a variable of integration.

Besides calculating the experimentally measurable quantities R, T, EQE, and IQE, simulation also provides the opportunity for introspection – examining quantities, which would not be measurable. We can integrate various recombination rates over the volume of the device to quantify each of the loss mechanisms and explain any nonideality in the simulation. For example, integrating the SRH recombination rate over the volume of subcell i, we find a loss component

$$L_{\text{SRH},i}(\lambda) = \frac{1}{A\Phi(\lambda)} \int_{V_i} R_{\text{SRH}}(\lambda)dV. \qquad (40.41)$$

Similarly for interface recombination at the interfaces above and below subcell i,

$$L_{\text{surf},i}(\lambda) = \frac{1}{A\Phi(\lambda)} \left(\int_{A_{\text{em},i/\text{fsf},i}} R_{\text{surf}}dA + \int_{A_{\text{base},i/\text{bsf},i}} R_{\text{surf}}dA \right), \qquad (40.42)$$

where the integrals are evaluated over the area of the front surface field/emitter and base/back surface field interfaces of the subcell in question. In general, it is possible to account for all EQE losses in this way, in which case the sum of all subcell EQEs and all loss components should be unity (Figure 40.8). Just as the EQE can be integrated with incident spectrum to give J_{SC} of each subcell, the loss components can be integrated to indicate the amount of photocurrent lost via a particular mechanism.

40.8 Photon Recycling and Luminescent Coupling

Some MJSCs show significant effects due to photon recycling, which can enhance the voltage of individual subcells, and luminescent coupling which can redistribute photocurrent between junctions and reduce sensitivity to spectrum [54,55].

To understand this, we can modify the circuit model of Figure 40.1b with two diodes per subcell, one of which represents the nonradiative recombination current and one the radiative current, $J_{\text{rad},i}$ (Figure 40.9). A portion of the radiatively emitted current results in generation within other junctions, according to

$$J_{\text{LC},i} = \sum_j K_{i,j} J_{\text{rad},j}, \qquad (40.43)$$

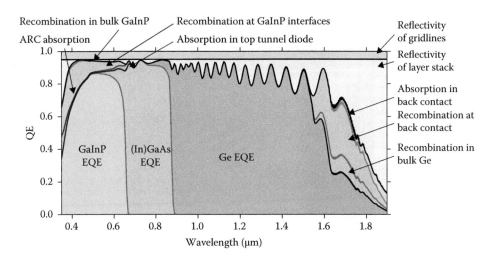

FIGURE 40.8 Quantum efficiency plot showing the breakdown of losses. (ARC, antireflection coating; EQE, external quantum efficiency.)

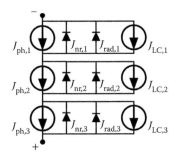

FIGURE 40.9 Equivalent circuit model of a triple-junction cell, modified to distinguish radiative and nonradiative diode currents and to include coupled generation.

where $K_{i,j}$ is a junction-to-junction coupling coefficient. If any of the photocurrents $J_{ph,i}$ are in excess of the device current J, that excess current will recombine through the nonradiative and radiative diodes in Figure 40.9. At sufficiently high bias, GaAs and GaInP junctions have predominantly (>90%) radiative recombination. These junctions are thick enough that the self-coupling coefficient $K_{i,i}$ can also be >90%, meaning that carriers can cycle repeatedly through the process of radiative recomination, photon emission, and reabsorption within the same junction. This process is called "photon recycling" and can be responsible for an increase in carrier concentrations.

A related term, "luminescent coupling," refers to the coupling between different junctions. If one of the bottom junctions is current limiting (i.e., it has the samllest J_{ph} of any subcell), then the excess current in the upper junctions can be radiatively coupled into the bottom junction, allowing a J_{SC} greater than the photocurrent in the limiting junction.

Luminescent coupling can also be observed in experimental measurements of multijunction cell QE. When bias lights are applied to the top two subcells, some of that bias light is coupled into the subcell that is being probed, interfering with the measurement [56].

These effects can be modeled in a drift-diffusion device simulator with the addition of a coupled generation term to the continuity equation [57,58], or through a process of successive approximations to the optical generation function [59].

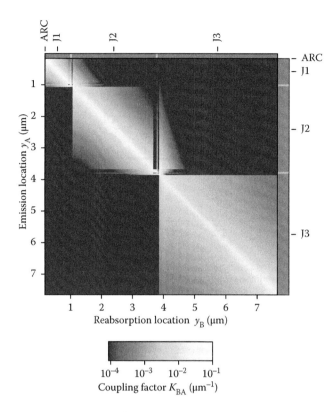

FIGURE 40.10 Luminescent coupling matrix for the triple-junction GaInP/(In)GaAs/Ge example structure. (ARC, antireflection coating.)

We discuss the first option here, and begin by modifying the process flow outlined in Figure 40.2 and insert a step after the assignment of material parameters for calculation of the coupling matrix, which gives the probability that a photon emitted from a position A will be reabsorbed at another position, B (shown in Figure 40.10). Naturally, photons have a great probability of being reabsorbed near the point where they are emitted (i.e., along the main diagonal of the matrix), or in layers of smaller bandgap such as a lower subcell. Indeed, the figure shows significant absorption in the GaAs junction (J2) due to emissions from J1, and also absorption in J3 from emissions in J2. This coupled transfer is unidirectional; there is no coupling of emitted photons from the lower subcells into the upper ones. Also, it should be noted that Figure 40.10 indicates only the efficiency with which emitted photons are reabsorbed. Some layers may not emit strongly at all depending on the local carrier concentration and the ratio of radiative and nonradiative lifetimes.

Once the coupling matrix has been calculated, the simulation can proceed as normal with a modification to the electron and hole continuity equations to add the coupled generation term which is a function of radiative recombination rates at all positions within the device. While the simulator used here, Synopsys Sentaurus, does not directly provide a method to add this term, it does provide a facility to define custom generation/recombination terms through a C++ programming interface. The custom C++ code is executed at each iteration of the nonlinear equation solver, and has access to quantities such as carrier concentrations (and by extension, recombination rates) at all mesh vertices within the simulation domain. The technique is described in detail in the appendix to Reference [58].

The result of this calculation can be seen in Figure 40.11; the "plateau" in QE between 650 and 880 nm is due to light that is radiatively coupled from the (In)GaAs junction into the germanium junction, and is an undesirable effect as it interferes with measuring the actual germanium photocurrent $J_{\mathrm{ph,3}}$.

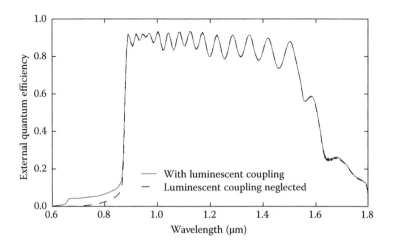

FIGURE 40.11 Calculated quantum efficiency of the germanium junction including coupled generation from light emitted by the (In)GaAs junction.

In multijunction phototransducers, which are devices closely related to MJSCs but are designed to operate under single-wavelength illumination, the luminescent coupling causes a significant broadening of the wavelength response. This broadening effect has also been modeled using this technique[58,60].

In contrast with the circuit-model representation, this treatment of luminescent coupling using a drift-diffusion simulation provides some advantages. The radiative emission rate, and the resulting coupled generation, is calculated as a function of position and reflects local changes in carrier density. The rates of emission from the emitter, the space charge region, and the base will each vary depending on the bias conditions, and the simulation will handle these changes appropriately. Areas that are shaded by gridlines and busbars will also be treated correctly. Furthermore, it is straightforward to consider coupling between all layer combinations in the device structure, not just those layers that comprise the photovoltaic junctions.

40.9 Summary

MJSCs are relatively complex devices, with typically 25 or more heterointerfaces and a number of photovoltaic junctions and tunnel diodes. With sufficient calibration, models can be developed that provide a very accurate representation of the device performance and can indicate the relative importance of various loss mechanisms. Once the model is calibrated, a detailed design of experiments can be done to pinpoint optimal design for a given requirement. In order to do accurate simulations, an extensive set of material parameters must be collected for each of the semiconductor materials. In particular, the complex refractive index, the minority carrier lifetimes, and the tunneling parameters tend to be difficult to find in the literature and need to be found through experiment.

In many cases, lateral current spreading and uneven bias across the surface of the cell have significant impacts on efficiency, which cannot be represented in a simple equivalent circuit model. In concentrating systems that have chromatic aberrations, the illumination of each subcell will have a different, nonuniform distribution over the the cell aperture. This leads to local mismatch in the subcell current densities and is an interesting problem to study through device simulation. Simulations based on drift-diffusion enable the device to be simulated including transport across all heterojunctions, without any presupposition of diode-like behavior, and hence this type of model can be used to study the effects of extreme operating conditions or manufacturing errors where the diode model is not applicable.

References

1. M. A. Green, K. Emery, Y. Hishikawa,W. Warta, and E. D. Dunlop. Solar cell efficiency tables (version 47). *Prog. Photovolt. Res. Appl.*, 24(1):3–11, 2016.
2. J. Nelson. *Physics of Solar Cells*. London: Imperial College Press, 2003.
3. C. H. Henry. Limiting efficiencies of ideal single and multiple energy gap terrestrial solar cells. *J. Appl. Phys.*, 51(8):4494, 1980.
4. R. R. King, D. Bhusari, A. Boca, D. Larrabee, X. Liu, W. Hong, C. M. Fetzer, D. C. Law, and N. H. Karam. Band gap-voltage offset and energy production in next-generation multijunction solar cells. *Prog. Photovolt. Res. Appl.*, 19:797–812, 2011.
5. R. R. King, N. H. Karam, J. H. Ermer, N. Haddad, P. Colter, T. Isshiki, H. Yoon, H. L. Cotal, D. E. Joslin, D. D. Krut, R. Sudharsanan, K. Edmondson, B. T. Cavicchi, and D. R. Lillington. Next-generation, high-efficiency III-V multijunction solar cells. In *Conference Record 28th IEEE Photovoltaic Specialists Conference.*, pages 998–1001, Anchorage, AK, 2000. IEEE.
6. D. J. Friedman and J. M. Olson. Analysis of Ge junctions for GaInP/GaAs/Ge three-junction solar cells. *Prog. Photovolt. Res. Appl.*, 9(3):179–189, 2001.
7. D. J. Friedman, J. M. Olson, S. Ward, T. Moriarty, K. Emery, S. Kurtz, A. Duda, R. R. King, H. L. Cotal, D. R. Lillington, J. H. Ermer, and N. H. Karam. Ge concentrator cells for IIIV multijunction devices. In *Conference Record 28th IEEE Photovoltaic Specialists Conference*, Anchorage, AK, pages 965–967. IEEE, 2000.
8. I. García, C. Algora, I. Rey-Stolle, and B. Galiana. Study of non-uniform light profiles on high concentration III-V solar cells using quasi-3D distributed models. In *2008 33rd IEEE Photovolatic Specialists Conference*, San Diego, CA, pages 1–6. IEEE, May 2008.
9. Synopsys Inc. Simulation of a GaAs / GaInP Dual-Junction Solar Cell, Synopsys Inc., Mountain View, CA, 2012.
10. Solarex17.in: EQE of III-V Tandem Cell. Silvaco Inc., Santa Clara, CA, 2015. http://www.silvaco.com/examples/tcad/section44/example17/index.html
11. Z. Q. Li, Y. G. Liao, and Z. M. S. Li, Modelling of multi-junction solar cells by Crosslight APSYS, High and Low Concentration for Solar Electric Applications, *Proceedings of the SPIE*, vol. 6339, pp. 633909–15, San Diego, CA, 2006.
12. W. E. McMahon, J. M. Olson, J. F. Geisz, and D. J. Friedman. An examination of 1D solar cell model limitations using 3D SPICE modeling. In *38th IEEE Photovoltaic Specialists Conference.*, pp. 002088–002091. Austin, TX, Jun 2012.
13. G. Létay, M. Hermle, and A. W. Bett. Simulating single-junction GaAs solar cells including photon recycling. *Prog. Photovolt. Res. Appl.*, 14(8):683–696, 2006.
14. T. Rahman and K. Fobelets. Efficient tool flow for 3D photovoltaic modelling. *Comput. Phys. Commun.*, 193:124–130, 2015.
15. P. Sharma, A. W. Walker, J. F. Wheeldon, K. Hinzer, and H. Schriemer. Enhanced efficiencies for high concentration, multijunction PV systems by optimizing grid spacing under nonuniform illumination. *Int. J. Photoenergy*, 2014:582083, 2014.
16. ASTM International. ASTM G173-03: Standard Tables for Reference Solar Spectral Irradiances: Direct Normal and Hemispherical on 37 degree Tilted Surface. Technical report, ASTM International, 2012.
17. D. E. Aspnes, S. M. Kelso, R. A. Logan, and R. Bhat. Optical properties of AlxGa1−xAs. *J. Appl. Phys.*, 60(2):754, 1986.
18. T. Kim, T. Ghong, Y. Kim, S. Kim, D. Aspnes, T. Mori, T. Yao, and B. Koo. Dielectric functions of InxGa1−xAs alloys. *Phys. Rev. B.*, 68(11):115323, 2003.
19. A. B. Djurisic, A. D. Rakic, P. C. K. Kwok, E. H. Li, M. L. Majewski, and J. M. Elazar. Modeling the optical constants of Al(x)Ga(1-x)As alloys. *J. Appl. Phys.*, 86(1):445–451, 1999.

20. M. Schubert, J. A. Woollam, G. Leibiger, B. Rheinlander, I. Pietzonka, T. Sab, and V. Gottschalch. Isotropic dielectric functions of highly disordered AlxGa1$-x$InP (0 <= x <= 1) lattice matched to GaAs. *J. Appl. Phys.*, 86(4):2025–2033, 1999.

21. Sentaurus Device User Guide, version K-2015. Synopsys Inc., Mountain View, CA. 1446p, 2015.

22. D. Schroeder. *Modelling of Interface Carrier Transport for Device Simulation*. Wien: Springer-Verlag, 1994.

23. K. Horio and H. Yanai. Numerical modeling of heterojunctions including the thermionic emission mechanism at the heterojunction interface. *IEEE Trans. Electron Dev.*, 37(4):1093–1098, 1990.

24. P. P. Altermatt. Models for numerical device simulations of crystalline silicon solar cells—a review. *J. Comput. Electron.*, 10(3):314–330, 2011. Multi-Junction Solar Cells 40–43

25. I. Vurgaftman, J. R. Meyer, and L. R. Ram-Mohan. Band parameters for III-V compound semiconductors and their alloys. *J. Appl. Phys.*, 89(11):5815, 2001.

26. D. H. Levi. Effects of ordering on the optical properties of GaInP$_2$. *Proc. SPIE.*, 5530:326–337, 2004.

27. G. Masetti, M. Severi, and S. Solmi. Modeling of carrier mobility against carrier concentration in arsenic-doped, phosphorus-doped, and boron-doped silicon. *IEEE Trans. Electron Dev.*, 30(7):764–769, 1983.

28. N. D. Arora, J. R. Hauser, and D. J. Roulston. Electron and hole mobilities in silicon as a function of concentration and temperature. *IEEE Trans. Electron Dev.*, 29(2):292–295, 1982.

29. J. Piprek. *Semiconductor Optoelectronic Devices: Introduction to Physics and Simulation*. San Diego, CA: Academic Press, 2003.

30. S. M. Sze and K. Ng Kwok. *Physics of Semiconductor Devices*,   3rd edition. Hoboken, NJ: Wiley-Interscience, 2007.

31. V. Palankovski and R. Quay. *Analysis and Simulation of Heterostructure Devices*. Vienna: Springer-Verlag, 2004.

32. S. Adachi. *Optical Constants of Crystalline and Amorphous Semiconductors*. Boston, MA: Springer, 1999.

33. E. D. Palik, editor. *Handbook of Optical Constants of Solids*. Cambridge, MA: Academic Press, 1998.

34. E. F. Schubert. *Physical Foundations of Solid-State Devices*. Rensselaer Polytechnic Institute, Troy, NY, 273 pages, 2006.

35. M. Levinshtein, S. Rumyantsev, and M. Shur, editors. *Handbook Series on Semiconductor Parameters*, vol. 2: Ternary And Quaternary III-V Compounds. Singapore: World Scientific, 1999.

36. M. Sotoodeh, A. H. Khalid, and A. A. Rezazadeh. Empirical low-field mobility model for III-V compounds applicable in device simulation codes. *J. Appl. Phys.*, 87(6):2890, 2000.

37. J. F. Wheeldon, C. E. Valdivia, A. W. Walker, G. Kolhatkar, A. Jaouad, A. Turala, B. Riel, D. Masson, N. Puetz, S. Fafard, R. Arès, V. Aimez, T. J. Hall, and K. Hinzer. Performance comparison of AlGaAs, GaAs and InGaP tunnel junctions for concentrated multijunction solar cells. *Prog. Photovolt. Res. Appl.*, 19(4):442–452, 2010. 40–44 Book title goes here.

38. A. W. Walker, O. Thériault, M. M. Wilkins, J. F. Wheeldon, and K. Hinzer. Tunnel-junction- limited multijunction solar cell performance over concentration. *IEEE J. Sel. Top. Quantum Electron.*, 19(5):4000508, 2013.

39. J. Fontanella, C. Andeen, and D. Schuele. Low-frequency dielectric constants of α-quartz, sapphire, MgF2, and MgO. *J. Appl. Phys.*, 45(7):2852–2854, 1974.

40. D. R. Lide, editor. *CRC Handbook of Chemistry and Physics*, 86th edition. Boca Raton, FL: CRC Press, 2005.

41. J. M. Siqueiros, R. Machorro, and L. E. Regalado. Determination of the optical constants of MgF(2) and ZnS from spectrophotometric measurements and the classical oscillator method. *Appl. Opt.*, 27(12):2549–2553, 1988.

42. J. R. Devore. Refractive indices of rutile and sphalerite. *J. Opt. Soc. Am.*, 41(6):416, 1951.

43. J. Kischkat, S. Peters, B. Gruska, M. Semtsiv, M. Chashnikova, M. Klinkmüller, O. Fedosenko, S. Machulik, A. Aleksandrova, G. Monastyrskyi, Y. Flores, and W. T. Masselink. Mid-infrared optical properties of thin films of aluminum oxide, titanium dioxide, silicon dioxide, aluminum nitride, and silicon nitride. *Appl. Opt.*, 51(28):6789–6798, 2012.

44. S. R. Kurtz, J. M. Olson, D. J. Friedman, J. F. Geisz, and A. E. Kibbler, Passivation of interfaces in high-efficiency photovoltaic devices, In *MRS Online Proceedings Library Archive*, vol. 573, pp. 95–106, San Francisco, CA, 1999.

45. W. Guter and A. W. Bett. IV-Characterization of Devices Consisting of Solar Cells and Tunnel Diodes. In *4th World Conference Photovoltaic Energy Conversion.*, pages 749–752. Waikoloa, HI, 2006.

46. E. O. Kane. Theory of tunneling. *J. Appl. Phys.*, 32(1):83, 1961.

47. G. A. M. Hurkx, D. B. M. Klaassen, and M. P. G. Knuvers. A new recombination model for device simulation including tunneling. *IEEE Trans. Electron Dev.*, 39(2):331–338, 1992.

48. A. Ajoy. Complex bandstructure of direct bandgap III-V semiconductors: application to tunneling. In M. Katiyar, B. Mazhari, and Y. N. Mohapatra, editors, *International Workshop on Physics of Semiconductor Devices*, Kanpur, India, 2012.

49. J. F. Wheeldon, C. E. Valdivia, A. Walker, G. Kolhatkar, T. J. Hall, K. Hinzer, D. Masson, S. Fafard, A. Jaouad, A. Turala, R. Ares, and V. Aimez. AlGaAs tunnel junction for high efficiency multi-junction solar cells: Simulation and measurement of temperature-dependent operation. In *34th IEEE Photovoltaic Specialists Conference*, Philadelphia, PA, pages 000106–000111, 2009.

50. M. Baudrit and C. Algora. Tunnel diode modeling, including nonlocal trap-assisted tunneling: A focus on III–V multijunction solar cell simulation. *IEEE Trans. Electron Dev.*, 57(10):2564–2571, 2010.

51. M. Hermle, G. Létay, S. P. Philipps, and A. W. Bett. Numerical simulation of tunnel diodes for multi-junction solar cells. *Prog. Photovolt. Res. Appl.*, 16:409–418, 2008.

52. Synopsys Inc. Simulation of GaAs Tunnel Diode for Multijunction Solar Cells, Synopsys, Inc., Mountain View, CA, 9 pages, 2012.

53. K. Emery, M. Meusel, R. Beckert, A. W Bett, and W. Warta. Procedures for evaluating multijunction concentrators. In *Conference Record 28th IEEE Photovoltaic Specialists Conference.*, pages 1126–1130. Anchorage, AK, 2000.

54. D. J. Friedman, J. F. Geisz, and M. A. Steiner. Effect of luminescent coupling on the optimal design of multijunction solar cells. *IEEE J. Photovolt.*, 4(3):986–990, 2014.

55. D. J. Friedman, J. F. Geisz, and M. A. Steiner. Analysis of multijunction solar cell current voltage characteristics in the presence of luminescent coupling. *IEEE J. Photovolt.*, 3(4):1429–1436, 2013.

56. S. H. Lim, J.-J. Li, E. H. Steenbergen, and Y.-H. Zhang. Luminescence coupling effects on multijunction solar cell external quantum efficiency measurement. *Prog. Photovolt. Res. Appl.*, 21(3):344–350, 2013.

57. M. Wilkins, A. M. Gabr, A. H. Trojnar, H. Schriemer, and K. Hinzer. Effects of luminescent coupling in single- and 4-junction dilute nitride solar cells. In *Proceeding 40th Photovoltaic Specialists Conference*, pages 6–9, Denver, CO, 2014.

58. M. Wilkins, C. E. Valdivia, A. M. Gabr, D. Masson, S. Fafard, and K. Hinzer. Luminescent coupling in planar opto-electronic devices. *J. Appl. Phys.*, 118(14):143102, 2015.

59. A. W. Walker, O. Hohn, D. N. Micha, L. Wagner, H. Helmers, A. W. Bett, and F. Dimroth. Impact of photon recycling and luminescence coupling on III - V photovoltaic devices. In *Proc. SPIE*, vol. 9358, San Francisco, CA, pages 11–13, 2015.

60. M. Wilkins, C. E. Valdivia, S. Chahal, M. Ishigaki, D. P. Masson, S. Fafard, and K. Hinzer. Performance impact of luminescent coupling on monolithic 12-junction phototransducers for 12 V photonic power systems. In *Proc. of SPIE*, vol. 9743, San Francisco, CA, 2016.

41

Nanostructure Solar Cells

Urs Aeberhard

41.1 Introduction: Nanostructures in Photovoltaics

The operation of solar cell (SC) devices is governed by a large variety of solid-state properties associated with the interplay of optical, electronic, and vibrational degrees of freedom of the component materials. Light-matter interaction provides the elementary optoelectronic processes of photogeneration and radiative recombination, while coupling of charge carriers to lattice vibrations leads to energy conversion losses due to thermalization, dissipative transport, and nonradiative recombination.

In the presence of spatial inhomogeneities, the optoelectronic material properties change if the characteristic length scale of the spatial variation is reduced below a certain threshold. The critical length scale depends on the physical degree of freedom (optical, electronic, vibrational) and corresponds to the wavelength of the associated wave (electromagnetic [EM] wave, de Broglie wave, sound wave). In this *sub-wavelength* regime, the physics is governed by the corresponding wave equations (Schrödinger, Maxwell, harmonic lattice dynamics), with the peculiar features such as interference or nonlocality. Along with the states, the interaction of different degrees of freedom is strongly modified in the presence of confinement, mainly due to the effect of localization and symmetry breaking. This tunability of physical properties via variation of material composition, size and shape of structures below a characteristic length scale—termed *nanostructures* (NSs)[†]—provides the basis for the vast field of nanotechnological applications, and the relation between configurational parameters and target NS functionality is a major topic of past and current research. Almost since the advent of semiconductor NS fabrication technologies, such as molecular beam epitaxy (MBE) or metal-organic chemical vapor deposition (MOCVD), but also lithographic processes,

---

[†] While the term *nano* is used for all of the above degrees of freedom (nanophotonics, -electronics, -phononics) to distinguish the confinement regime from bulk behavior, the characteristic confinement length varies strongly (0.1 μm to 1 cm for photons, 0.5 to 10 nm for electrons, 5 nm to 1 cm for phonons).

the peculiar physical properties of the NS produced in this way have also been utilized for the design of novel SC devices with the potential of increased photovoltaic (PV) conversion efficiency. In later years, new NS fabrication techniques such as nanoimprinting, catalytic vapor deposition, or solution-based processes have enlarged the zoo of NSs and NS architectures proposed for SC with either enhanced efficiency or the potential for substantial cost reduction.

A common way to characterize NSs is according to the dimensionality: The material without confinement, i.e., featuring extended states in all three spatial dimensions, is referred to as *bulk*. For electronic confinement in one, two, and three spatial dimensions, the resulting 2D, 1D, and 0D NSs are called *quantum well* (QW), *quantum wire* (QWR), and *quantum dot* (QD), respectively. At length scales where only optical confinement is present, the corresponding objects are usually called thin films, nanowires, and nanoparticles, respectively. On the other hand, the functionalities of nanostructure solar cell (NSSC) components can be categorized according to the physical degree of freedom that is to be affected: *Optical* functionalities are used for the purpose of light trapping for absorption enhancement via engineering of the EM field strength inside the absorber; the (opto-)*electronic* functionalities, which are related to the tuning of (opto-)electronic properties such as bandgaps, absorption coefficients, mobilities, and carrier relaxation rates via electronic structure engineering; and *vibrational* functionalities, which affect the dissipation of carrier energy due to coupling to lattice vibrations, for instance, via inhibition of suitable phonon modes. As explained earlier, all of the functionalities emerge as the consequence of the nanoscale dimension of the structures that leads to quantization and confinement of the photonic, electronic, and vibrational states and results in physical properties that can deviate considerably from the bulk behavior.

The main applications of NS in photovoltaics are in the so-called second and third generations of PV devices [1], where NS are used in the implementation of SC concepts aiming at either lower cost through strongly reduced material usage and cheap production technology or at increased energy conversion efficiencies through enhanced spectrum utilization and/or reduced thermalization losses. NS with optical functionality have been instrumental in the realization of thin-film SC (second generation) based on low absorption materials, such as thin-film silicon devices [2] or ultrathin absorbers [3]. Relevant representatives of such NS applications for light-trapping include plasmonic metal nanoparticles [4] and nanostructured metal films [5], dielectric scatterers [6] and diffraction gratings [7], engineering of the optical density of states (DOS) via photonic crystals [8], light focusing antenna effects in nanowires and nanorods [9], as well as spectral conversion via QDs [10]. This wide field of optical NS for PV applications is covered by a number of topical reviews [11–14].

The electronic functionality of NS concerns first of all the tuning of optical transitions and nonradiative charge carrier dynamics in high-efficiency SC architectures for concentrator and space applications (third generation). Since the early nineties, multi-QW (MQW—multiple decoupled QW) and QW superlattice (QWSL—periodic and coupled QW) structures have been implemented in different III-V semiconductor materials and have been investigated for application in single-junction SC [15] and multijunction (MJ) PV devices [16], and, more recently, also for hot-carrier SC (HCSC) [17]. The practical relevance of these structures lies mainly in the adjustability of the absorption edge to the wavelength required for efficiency optimization, in the case where suitable bulk materials are not available. Similar bandgap engineering features are provided by multi-QD (MQD) or QD superlattice (QDSL) structures implemented in III-V semiconductors [18–21] or silicon alloys [22], with the advantage of improved strain relaxation mechanisms. However, while QWSCs based on strain-balanced components have enabled efficiencies comparable to the bulk limit [23], severe recombination losses did not allow thus far a similar achievement in QDSC [24]. On the other hand, QD structures were considered to be particularly useful for the implementation of a number of specific third-generation PV concepts: for intermediate band SC (IBSC) due to the possibility to obtain an energetically separated intermediate state or band from the ground state of the QD [25]; for multiple exciton generation (MEG) via impact ionization (providing increased current) due to enhanced electron–electron interaction as a consequence of wave function localization [26]; and for the extraction of hot carriers in HCSC (providing increased voltage) due to inhibited electron–phonon scattering resulting from the low density and spectral sparsity of electronic and vibrational modes [27]. These effects have been demonstrated in dedicated experimental setups [28–33], but they remain to be shown in working SC devices.

Besides these highly ordered epitaxial QD structures, there is a whole class of QD materials made as colloids from solution, such as, e.g., PbSe and PbS [34], which are arranged into different SC architectures, ranging from QD-sensitized TiO_2 electrodes to hybrid organic–inorganic devices and closely packed QD solids with carrier-selective contacting schemes [35–38]. While present efficiencies are still low (11.3%), these approaches promise low-cost manufacturing, and there is still the prospect of MEG exploitation as observed in the constituent particles [39].

For the sake of completeness, a word is due here on the case of QWRs. Even though they offer a perfect combination of size tunability in transverse dimensions with extended states in transport direction, there are almost no experimental structures including 1D NS with electronic functionality. The reasons are twofold. On the one hand, it is very challenging to produce wires with diameters small enough such as to exhibit electronic confinement. On the other hand, it is not possible to achieve at the same time optical and electronic confinement: At the dimensions suitable for optical confinement, the absorber is electronically bulk-like, and no guided optical modes can exist in the tiny dimensions required for electronic confinement.

As compared to the optical and electronic features emerging at the nanoscale, the vibrational functionalities of NSs have received the least attention so far. Most potential applications consider exploitation of reduced carrier cooling by engineering the phonon modes in NSs such as QWSL [40] and QDSL [22,27]. The issue of confined phonons is discussed also in the context of fast relaxation of hot photogenerated excitons in isolated QD, one of the detrimental factors suppressing MEG, which was thought to be reduced in QD. Recent experiments and simulations emphasize the influence of softened surface modes enabling efficient multiphonon relaxation [41].

All of the above functionalities might be relevant in a single SC device, for instance a HCSC based on a III-V QWSL structure (similar to [40]) with absorption enhancement induced by plasmonic nanoparticles or a cavity effect. In such a case, deviations from bulk physics need to be considered for photons, electrons, and phonons at the same time. Furthermore, while the single functionality of an NS component (e.g., enhancement of interband absorption due to wave function localization) might be beneficial for a certain aspect of device operation (e.g., photogeneration), it may at the same time have a detrimental impact on another aspect (e.g., recombination and transport). Indeed, in experimental implementations of advanced PV concepts, NSs are often found to introduce electronic and optical losses, e.g., due to high defect density at NS interfaces and parasitic absorption into localized states with insufficient carrier extraction efficiency. In the idealized models considered for the determination of efficiency bounds such as the famous Shockley–Queisser (SQ) limit [42], such losses are usually neglected. This explains to some extent the large discrepancy between theoretical ideal case predictions and experimentally achieved results. However, for a realistic assessment of the technological potential of a given NSSC concept, the detrimental aspects associated with the NS need to be considered and quantified. This underlines the importance of advanced modeling approaches reaching beyond bulk semiconductor physics for this category of PV devices, which will be in the focus of the remainder of this chapter.

Space does not permit a detailed discussion of the modeling approaches for all of the NS used in photovoltaics as outlined above. Since nanophotonic aspects of optoelectronic device modeling are covered elsewhere in this book, NS with purely optical functionality—including the important case of the nanowire SC, where the absorber is at the same time its own light-trapping structure—will not be addressed here. The other important class of NSSC that will not be covered comprises the colloidal QD SC architectures. In these devices, the PV processes are strongly influenced by the surface chemistry, such as, e.g., the properties associated with different ligand molecules, but also by strong carrier localization, large exciton binding energies, and the considerable fluctuations in size and position. As a consequence, mobility is usually limited by thermally activated hopping processes [43], resulting in poor transport properties. The disorder-induced smearing of the band edge has also unfavorable impact on the achievable open-circuit voltages, even at the radiative limit and under the assumption of perfect transport [44]. Overall, the situation is

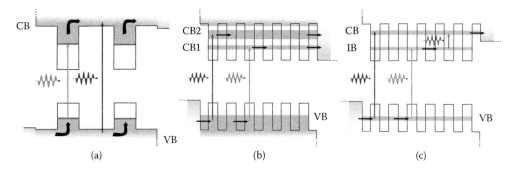

FIGURE 41.1 Common architectures of NSSC: (a) multi-QW or QD structures, where the confined states of distinct NS are not coupled, and photogenerated carriers therefore need first to escape to extended (quasi-)continuum states in order to contribute to current; (b) superlattices of QW or QD, where miniband formation or sequential tunneling enables photocarrier transport and extraction via partially localized NS states; (c) in contrast to (b), the IBSC configuration features noncontacted NS states, which, however, can still be delocalized, but, contrary to (a), should be only radiatively coupled to the contacted continuum states by which carriers are extracted. CB, conduction band; VB, valence band; IB, intermediate band.

closer to the case of organic and hybrid SC, for which a whole range of dedicated simulation approaches exist [45].

The main focus of this chapter, however, will be on the modeling of *regimented* QW and QD architectures as utilized for the implementation of high-efficiency SC concepts using *inorganic* semiconductor materials. Among these structures, three main categories can be distinguished according to the participation of the NS states in the transport of charge carriers. The first category includes devices where the NS states are localized in transport direction and need to be coupled to another set of extended (i.e., bulk-like or unconfined) states to enable charge carrier flow. Typical representatives are the MQW and MQD SC (Figure 41.1a). The second category consists of architectures where both generation-recombination and transport processes are mediated by NS states, which requires a finite degree of wave function delocalization in transport direction related to the coupling of NS, as in QDSL devices with large confinement barriers (Figure 41.1b). Finally, there are structures where coupling-induced delocalization exists between both confined and unconfined states, but only the continuum states are connected to contacts, as in some QD-IBSC implementations (Figure 41.1c).

In Section 41.2, the main physical mechanisms governing the PV device operation are discussed for the different types of NS architectures with regard to their theoretical description and modelization. In Section 41.3, existing methods are reviewed to obtain the device characteristics of NS-based SCs under consideration of the modeling requirements resulting from Section 41.2 and the associated implementation challenges, establishing a simulation hierarchy ranging from global detailed balance approaches for limiting efficiency calculations to microscopic quantum-kinetic theories for the local charge carrier dynamics in nonclassical device regions. In Section 41.4, the simulation of a prototypical NSSC in the form of a single QW photodiode is considered in the light of the approaches discussed in Section 41.3.

41.2 Physical Mechanisms of NSSC Device Operation

In the following, the different stages of the PV energy conversion process in NSSC—from the incident photon flux to the extracted charge current—are considered with focus on the modeling requirements imposed by the presence of NS components.

41.2.1 Photogeneration of Electron–Hole Pairs

The structures that are of interest in NS PV architectures are small compared to the coherence length of sunlight. The spectral *photon flux* $\Phi_\gamma(\mathbf{r}, \hbar\omega)$ at energy $\hbar\omega$, with $\hbar$ the reduced Planck constant and ω the frequency of the light, and position $\mathbf{r}$ inside the absorber of the SC under consideration is therefore well described by the propagation of coherent EM radiation in a dispersive medium, as provided by the solution of (macroscopic) Maxwell's equations. In the most general case, the linear response[†] of the medium to the transverse electrical field $\mathcal{E}^t$ is characterized by the complex microscopic dielectric tensor $\vec{\varepsilon}(\mathbf{r}, \mathbf{r}', \hbar\omega)$, which considers the electronic excitations in dependence of the energy and polarization of the incident radiation field and of the available electronic states. The corresponding local volume rate $\mathcal{G}$ for the generation of electron–hole pairs by photon absorption is obtained from the dissipated EM energy as follows [46]:

$$\mathcal{G}(\mathbf{r}, \hbar\omega) = \eta_{gen}(\mathbf{r}, \hbar\omega)(2\hbar\mathcal{V})^{-1} \sum_{\mu,\nu} \int d^3r' \; \mathfrak{I}\left\{\varepsilon_0\varepsilon_{\mu\nu}(\mathbf{r}, \mathbf{r}', \hbar\omega)\mathcal{E}_\mu^{t*}(\mathbf{r}, \hbar\omega)\mathcal{E}_\nu^t(\mathbf{r}', \hbar\omega)\right\}, \tag{41.1}$$

where $\mathfrak{I}$ denotes the imaginary part, ε_0 is the vacuum permittivity, η_{gen} is the generation efficiency ($\eta_{gen} = 1$ means that each photon that is absorbed generates exactly one electron-hole pair), and $\mathcal{V}$ is the absorbing volume. In the above expression, the time-harmonic monochromatic Ansatz $\mathcal{E}^t(\mathbf{r}, t) = \mathfrak{R}\{\mathcal{E}^t(\mathbf{r}, \hbar\omega)\exp(-i\omega t)\}$ was used. While the interaction with the light is inherently nonlocal from the electronic point of view, an averaged local version of the dielectric tensor can be used if the EM field varies only slowly over the length scale of electronic inhomogeneities. For an isotropic medium and unit generation efficiency, Equation 41.1 can then be rewritten as

$$\mathcal{G}(\mathbf{r}, \hbar\omega) = (2\hbar)^{-1}\varepsilon_0\mathfrak{I}\varepsilon(\mathbf{r}, \hbar\omega)|\mathcal{E}^t(\mathbf{r}, \hbar\omega)|^2 \approx \alpha(\mathbf{r}, \hbar\omega)\Phi(\mathbf{r}, \hbar\omega), \tag{41.2}$$

where

$$\alpha(\mathbf{r}, \hbar\omega) \equiv \frac{\omega}{n_r c_0}\mathfrak{I}\varepsilon(\mathbf{r}, \hbar\omega) \tag{41.3}$$

denotes the (locally defined) *absorption coefficient* that reflects the decay of the optical intensity due to all optical excitations possible at a given photon energy, weighted by their respective strength (c_0: speed of light in vacuum, n_r: refractive index of the background).

41.2.1.1 Photon Flux

In the general time-harmonic case, the photon flux is given by the Poynting vector $\mathbf{S}$ of the transverse EM fields $(\mathcal{E}^t, \mathcal{B}^t)$:

$$\mathbf{S}(\mathbf{r}, \hbar\omega) = (2\mu_0)^{-1}\left\{\mathcal{E}^t(\mathbf{r}, \hbar\omega) \times \mathcal{B}^t(\mathbf{r}, \hbar\omega)\right\} \equiv \Phi(\mathbf{r}, \hbar\omega)\hbar\omega\,\hat{\mathbf{n}}(\mathbf{r}, \hbar\omega), \tag{41.4}$$

where μ_0 is the vacuum permeability and $\hat{\mathbf{n}}$ is the unit vector in the direction of propagation. The evolution of the flux inside the absorber is related to the EM dissipation due to light-matter coupling via the optical conservation law:

$$\nabla \cdot \mathbf{S}(\mathbf{r}, \hbar\omega) = -\frac{1}{2}\omega\varepsilon_0\,\mathfrak{I}\left\{\varepsilon(\mathbf{r}, \hbar\omega)[\mathcal{E}^t(\mathbf{r}, \hbar\omega)]^* \cdot \mathcal{E}^t(\mathbf{r}, \hbar\omega)\right\}, \tag{41.5}$$

[†] This is appropriate for standard solar illumination conditions, i.e., the AM1.5g spectrum.

where the transverse electrical field is given by Maxwell's equations acquiring the form

$$\mu_0^{-1} \nabla \times \nabla \times \boldsymbol{\mathcal{E}}^{\mathrm{t}}(\mathbf{r}, \hbar\omega) = \omega^2 \varepsilon_0 \varepsilon(\mathbf{r}, \hbar\omega) \boldsymbol{\mathcal{E}}^{\mathrm{t}}(\mathbf{r}, \hbar\omega), \tag{41.6}$$

$$\nabla \cdot \varepsilon_0 \varepsilon(\mathbf{r}, \hbar\omega) \boldsymbol{\mathcal{E}}^{\mathrm{t}}(\mathbf{r}, \hbar\omega) = 0. \tag{41.7}$$

The *absorptance* of the absorbing volume is then given by the ratio $a \equiv P_{\mathrm{abs}}/P_{\mathrm{in}}$ of absorbed and incident EM power, with $P_{\mathrm{abs}} = \int_{\mathcal{V}} d\mathcal{V}(-\nabla \cdot \mathbf{S})$ and $P_{\mathrm{in}} = \int_S d\mathbf{S} \cdot \mathbf{S}$, where S is the surface of the absorbing volume. In the absence of optical confinement—for instance, in optically thick absorbers—the decay of the flux can be written in terms of the absorption coefficient via the *Lambert–Beer* (LB) law, e.g., for propagation along the z-direction:

$$\Phi_{\mathrm{LB}}(z, \hbar\omega) = \Phi_0^{\swarrow}(\hbar\omega) \exp\left(-\int_0^z dz'\, \alpha(z', \hbar\omega) \right), \tag{41.8}$$

with $\Phi_0^{\swarrow}$ denoting the incident photon flux at $z=0$. This defines the absorptance $a_{\mathrm{LB}}(z, \hbar\omega) \equiv 1 - \Phi_{\mathrm{LB}}(z, \hbar\omega)/\Phi_0^{\swarrow}(\hbar\omega)$ after an optical path of length z. For planar systems featuring optical elements such as antireflection coatings (ARC) or a back reflector, as frequently used in thin-film PV devices due to the requirement of absorption enhancement via light trapping, the required solution of Maxwell's equations can be performed using a transfer-matrix method (TMM) [47], which relates the coefficients of incident, reflected, and transmitted plane wave components for piecewise constant dielectric function. In the case of 2-D/3-D nanophotonic or plasmonic light trapping structures, Maxwell's equations (Equation 41.7) have to be solved by advanced numerical approaches such as the finite-difference time-domain (FDTD) method or the finite element method (FEM).

41.2.1.2 Absorption Coefficient

Equations 41.2 and 41.3 can be used to determine the absorption coefficient from the complex dielectric function or the linear *susceptibility* $\chi(\omega) = \varepsilon(\omega) - \varepsilon_b$ ($\varepsilon_b \equiv n_r^2$: background dielectric constant) or, equivalently, from the net absorption rate $r_{\mathrm{abs}} = \mathcal{G}/\eta_{\mathrm{gen}}$ (net means after subtraction of the rate for stimulated emission). For weak perturbation, the net volume rate of uncorrelated direct electronic transitions between states v and c by absorption of photons of energy $E_\gamma = \hbar\omega$ is given by *Fermi's Golden Rule* (FGR):

$$r_{v \to c}^{\mathrm{abs}}(E_\gamma) = \frac{2\pi}{\hbar\mathcal{V}} |\langle c|\hat{H}_{e\gamma}|v\rangle|^2 \delta(\varepsilon_c - \varepsilon_v - E_\gamma)(f_v - f_c), \tag{41.9}$$

with $\hat{H}_{e\gamma}$ the electron–photon interaction Hamiltonian, $\varepsilon_{c,v}$ the energies of single electron and hole states, and $f_{c,v}$ the occupation of these states. The electron–photon Hamiltonian has the general form

$$\hat{H}_{e\gamma}(\mathbf{r}, t) = \frac{e}{2m_0} \left\{ \hat{\mathbf{A}}(\mathbf{r}, t) \cdot \hat{\mathbf{p}} + \hat{\mathbf{p}} \cdot \hat{\mathbf{A}}(\mathbf{r}, t) \right\} + \frac{e^2 [A(\mathbf{r}, t)]^2}{2m_0} \approx \frac{e}{m_0} \hat{\mathbf{A}}(\mathbf{r}, t) \cdot \hat{\mathbf{p}}, \tag{41.10}$$

where $\hat{\mathbf{A}}$ is the EM vector potential, $\hat{\mathbf{p}} = -i\hbar\nabla$ is the momentum operator, and for the last expression (minimal coupling), the assumption of weak vector potential $\hat{\mathbf{A}}$ was made and Coulomb gauge ($\nabla \cdot \mathbf{A} = 0$) was used. In the standard *semiclassical* picture of light-matter interaction in semiconductors, the light is treated as a classical wave. Using for the incident field, a monochromatic time-harmonic plane wave expansion:

$$\mathbf{A}(\mathbf{r}, t) = \frac{A_0}{2} \boldsymbol{\epsilon} \left\{ e^{i(\mathbf{q} \cdot \mathbf{r} - \omega t)} + c.c. \right\} \tag{41.11}$$

of the vector potential, with amplitude $A_0/2$, wave vector $\mathbf{q} = \frac{\omega}{c}\hat{\mathbf{n}}$, and polarization $\boldsymbol{\epsilon}$, together with the minimal coupling form of $H_{e\gamma}$ in the FGR rate (Equation 41.9) yields

$$r^{abs}_{v\to c}(E_\gamma) = \frac{2\pi}{\hbar V}\frac{e^2 A_0^2}{4m_0^2}\mathcal{M}^2_{cv}\delta(\varepsilon_{cv} - E_\gamma)(f_v - f_c),\tag{41.12}$$

$$\mathcal{M}_{cv} \equiv |\langle c|e^{i\mathbf{q}\cdot\mathbf{r}}\boldsymbol{\epsilon}\cdot\hat{\mathbf{p}}|v\rangle|,\tag{41.13}$$

where $\varepsilon_{cv} \equiv \varepsilon_c - \varepsilon_v$. If $\mathbf{q}\cdot\mathbf{r} \ll 1$ (i.e., the wavelength of the light is much larger than the characteristic length scale of the absorbing structure), the phase factor $e^{i\mathbf{q}\cdot\mathbf{r}}$ in the matrix element can be neglected, which corresponds to the *dipole approximation*, meaning that contributions of higher order electric and magnetic multipoles are not considered. In this approximation, the coupling via the momentum operator $\hat{\mathbf{p}}$ can be replaced by a coupling via the dipole operator $\hat{\mathbf{d}} = -e\hat{\mathbf{r}}$ ($\hat{\mathbf{r}}$: position operator) due to the relation $(e/m_0)\mathbf{A}\cdot\hat{\mathbf{p}} \approx -\hat{\mathbf{d}}\cdot\boldsymbol{\mathcal{E}}^t$ [48]. The photon flux associated with Equation 41.11 amounts to

$$\Phi_0(\hbar\omega) = \frac{n_r c_0 \varepsilon_0 \omega A_0^2}{2\hbar},\tag{41.14}$$

which, together with Equation 41.12, provides the general FGR absorption coefficient:

$$\alpha^{FGR}(E_\gamma) = \frac{\pi\hbar}{n_r c_0 \varepsilon_0 E_\gamma V}\left(\frac{e}{m_0}\right)^2 \sum_{v,c} \mathcal{M}^2_{cv}\delta(\varepsilon_{cv} - E_\gamma)(f_v - f_c),\tag{41.15}$$

where the sum is over all initial and final states.

In Equation 41.15, the occupation of the initial and final states is assumed to be known, and in the case of interband transitions, $f_v = 1$ and $f_c = 0$ are often assumed. However, this is no longer appropriate in the case of transitions between partially filled subbands, as for instance in the case ot the IBSC concept. Thus, in general, the occupation needs to be determined by a microscopic treatment of the charge carrier dynamics under illumination in the framework of the semiconductor Bloch equations [48–50] or nonequilibrium Green's functions [51].

The main quantities determining the generation of electron-hole pairs are, thus, the local value of the transverse EM field on the one hand, and the local value of the dipole or momentum matrix elements (MMEs) and of the density of occupied initial and empty final states of the optical transitions, respectively, on the other hand. While the spatial variation of the transverse EM fields is small over the extent of semiconductor NS in the range of few nanometers, the electronic quantities reflect the impact of symmetry breaking, wave function localization, and energy quantization associated with reduced dimensionality. A prominent example is silicon QDs, where the contribution of direct transitions—absent in bulk—and an increase in oscillator strength due to larger overlap of electron and hole wave functions (in this case equivalent to larger MME) result in a strongly increased absorption [52,53].

Considering Equation 41.12, evaluation of the absorption in semiconductor NSs starts with the computation of the wave functions $\psi_{c,v}$ and energies $\varepsilon_{c,v}$ of the single-particle states $\{|v\rangle, |c\rangle\}$ participating in the optical transitions. There is a vast amount of literature on that topic, which is also covered by a number of chapters in this volume, and will thus not be treated explicitly here. With the electronic structure information, idealized absorption spectra can be obtained from Equation 41.12. However, for a realistic assessment of the PV response, the experimental spectra should be reproduced as closely as possible, which requires consideration of the actual effects that have an impact on the line shape, such as finite fields, excitonic contributions, lifetime broadening due to coupling to phonons and hybridization with continuum states, compositional disorder, and finite-size distributions. Various approaches with increasing degree of sophistication can be found in the literature.

For **QW**, excitons and field effects (i.e., electroabsorption) were included in [54] using a variational ansatz for the exciton wave functions. The consistent consideration of the bound-to-bound state and the bound-to-quasibound state transitions including excitonic and field effects was achieved using an exciton Green's function approach [55]. To describe lineshape broadening beyond phenomenological approaches, microscopic quantum-kinetic theories such as the density matrix formalism, including the different scattering mechanisms (electron–phonon, electron-electron, etc.) can be used [48–50]. Recently, excitonic absorption enhancement for QWs under consideration of the electron–phonon interaction was formulated in a general nonequilibrium Green's function framework [56]. In practice, however, some degree of fitting is used to reproduce the experimental spectra. For QWSC, a semiempirical approach was introduced by Paxman et al. [57], based on earlier work on QW optical modulators [58,59], in which the field-dependent absorption coefficient is expressed as

$$\alpha(E_\gamma, \mathcal{E}^z) = \sum_{c,v} \alpha_{cv} M_{cv}(\mathcal{E}^z) \Big[\sum_l r_{cv}^l \mathcal{L}[E_\gamma, \varepsilon_{cv}(\mathcal{E}^z) - B_{cv}^l] + \int_{\varepsilon_{cv}(\mathcal{E}^z)}^{\infty} dE' \mathcal{L}(E_\gamma, E') \Big], \qquad (41.16)$$

where $\mathcal{E}^z$ is the longitudinal electric field in growth direction, $M_{cv} = \int_L dz \psi_c^*(z, \mathcal{E}^z) \psi_v(z, \mathcal{E}^z)$ (L: normalization length) is the overlap matrix element of the field-dependent single-particle electron and hole wave functions, r_{cv}^l (in units of energy) and B_{cv}^l are the relative oscillator strength and binding energy of the *l*th exciton, respectively, and $\mathcal{L}(E_\gamma, E') = \Gamma/(2\pi[(E_\gamma - E')^2 + \Gamma^2/4])$ is a Lorenzian lineshape function to include homogeneous broadening characterized by Γ. The parameter α_{cv} corresponds to the absorption coefficient at the onset of the continuum absorption, and, in practice, is used together with B and r to fit the experimental spectra. For the absorption at photon energies above the bulk bandgap of the barrier material, the bulk absorption coefficient of the well material is used, neglecting any field effects. A similar semiempirical approach was used in [60], based on [61].

For **QD**, the single-particle absorption spectra can be obtained using similar density matrix formalisms for the linear susceptibility [62] or the equivalent FGR, with straightforward extension from single QD to QDSL [63,64]; in the case of significant excitonic effects, the latter can be treated by means of the configuration interaction (CI) method [65]. While lineshape broadening due to dephasing via electron–phonon interactions can in principle be included in quantum-kinetic approaches, the finite-size distribution of QD ensembles requires additional convolution with some distribution function representing inhomogeneous broadening, commonly of the Gaussian type [66]. In self-assembled QD formed on a wetting layer (WL), the WL states may also contribute to the subgap absorption [67]. For practical device simulation, a common approach is the use of an effective medium absorption coefficient [68]:

$$\bar{\alpha}_{\text{eff}} = w_{\text{QD}} \bar{\alpha}_{\text{QD}} + w_{\text{WL}} \bar{\alpha}_{\text{WL}} + w_{\text{bulk(QD)}} \bar{\alpha}_{\text{bulk(QD)}} + w_{\text{barrier}} \bar{\alpha}_{\text{barrier}}, \qquad (41.17)$$

which contains the contribution of the confined states in the QD, the WL, the bulk absorption of the QD material for continuum states, and the barrier absorption, weighted according to the relative volume fractions and a given distribution function reflecting the variations in size and shape. The general expression for the absorption of an ensemble of QD configurations C with distribution f using these simple approaches reads:

$$\bar{\alpha}_{\text{QD}}(E_\gamma) = \int dC f(C) \alpha_C(E_\gamma), \qquad (41.18)$$

$$\alpha_C(E_\gamma) = \frac{\pi \hbar}{n_r c_0 \varepsilon_0 E_\gamma \mathcal{V}_{\text{QD}}} \left(\frac{e}{m_0}\right)^2 \sum_{c,v} \bar{p}_{cv}^2(C) \frac{\Gamma}{\left(\varepsilon_{cv}(C) - E_\gamma\right)^2 + \Gamma^2/4} (f_v - f_c), \qquad (41.19)$$

where $\bar{p}_{cv}$ is the polarization-averaged MME and Γ is the homogeneous broadening. The WL absorption is given by the expression for QW (usually neglecting excitonic absorption), where consideration of thickness fluctuations results in an additional absorption tail. Since the absorbing volume of a QD is not always well defined, especially in the case of quasibound states, it is common to write the QD contribution absorption in terms of an *absorption cross-section* σ_{QD} per QD and of the QD concentration N_{QD} via $\alpha_{QD} = N_{QD}\sigma_{QD}$.

41.2.2 Charge Separation and Extraction

In general, the extraction of photogenerated charge carriers represents one of the most critical aspects in NS-based PV devices. Indeed, while in some devices based on QW or QD, the tailoring of the absorption via band structure engineering could be successfully demonstrated [69,70], the efficiency of carrier collection has until now remained below the values found for the bulk counterparts, in some cases on a dramatically poor level, especially in devices where transport proceeds via states with increased degree of localization. There are several reasons for the observed performance issue, which are associated with different stages in the extraction process, such as exciton dissociation, carrier escape and capture between localized and extended states, mobility issues associated with scattering, and the virtual absence of true miniband formation in realistic situations, which shall be discussed below.

41.2.2.1 Exciton Dissociation

Since electrons and holes have to be extracted via separate contacts, the dissociation of the photogenerated excitons and the subsequent charge separation need to occur at some point in the device. In bipolar bulk SC, exciton binding energies are usually in the range of a few meV, resulting in thermal dissociation immediately after generation, and the generation can thus safely be described via (Equation 41.2) in terms of noninteracting electron-hole pairs. In isolated NS on the other hand, spatial confinement enhances the electron-hole interaction, and exciton binding energies can amount to multiples of the thermal energy $k_B T$ (k_B: Boltzmann constant, T: temperature), in which case rapid thermal dissociation is no longer possible, and the dissociation process needs to be considered explicitly [53]. Inefficient exciton dissociation can result in a transport behavior dominated by exciton diffusion, similar to the situation in organic SC devices, where a specially designed bulk–heterojunction interface is required for exciton dissociation. As a SC based on exciton diffusion is not likely to reach very high efficiencies, the description of excitons in the simulation of efficient NSSC devices is primarily focused on the excitonic enhancement of optical transitions close to the effective band edge, as described above in the treatment of the absorption coefficient.

41.2.2.2 Charge Carrier Escape and Capture

Charge carriers that are generated in localized states of electronically decoupled NS need to be transferred to extended states in order to contribute to current, which normally amounts to a sequence of scattering events involving confined (2D—QW/0D—QD) subband states followed by scattering from a bound or quasibound state to a (3D) continuum state. While this *escape* process constitutes a loss mechanism in light-emitting devices, which is sought to be suppressed in optimized designs, the opposite applies in NSSC architectures. Hence, understanding the escape process via identification of dominant and limiting mechanisms and design of corresponding models has been an important focus in NSSC research from the beginning. On the other hand, any escape channel enables its inverse capture process, which removes photocarriers from contacted states and increases dark current flow.

In general, while the microscopic processes underlying the carrier dynamics are always related to the fundamental coupling between the electronic, vibrational, and optical degrees of freedom, there are several mechanisms by which photogenerated charge carriers can escape from confined NS states, which are commonly characterized by respective time constants τ_{esc}^i, with total escape rate proportional to $\tau_{esc}^{-1} = \sum_i (\tau_{esc}^i)^{-1}$. For **QW** at room temperature—the relevant temperature regime for terrestrial SC operation—and with shallow confinement, the dominant escape mechanism was identified as *thermionic emission* [71], which describes the current due to charge carriers that occupy the part of the energy distribution located above the barrier edge. In the presence of a rectifying built-in field, the escape process is affected by the latter through the modified height (with respect to the center of the QW) and triangular shape of the barrier, giving rise to escape enhancement due to *tunneling*. For the common assumption of a constant field, the effective barrier for escape in direction of and against the field is given by $V_{B\mp}(\mathcal{E}^z) = \Delta E_b \mp q\mathcal{E}^z L_w/2$, where ΔE_b is the offset of bulk band edges and L_w is the QW width. A general *phenomenological* expression for the corresponding escape time is then given by [71]

$$\tau_{\text{esc}}^{-1} = \frac{1}{L_w} \int_0^\infty \tilde{n}(E)\mathcal{T}(E)v(E)dE, \quad \tilde{n}(E) = \frac{n(E)}{\int_0^\infty n(E)dE}, \tag{41.20}$$

where $\mathcal{T}(E)$ is the probability function for transmission through the barriers, $v(E)$ is the carrier velocity, and $n(E) = g(E)f(E)$ is the spectral carrier density at an energy E measured from the center of the bottom and assuming a given distribution function $f(E)$ for charge carriers in the QW. The actual form of Equation 41.20 depends on the specific description of the spectral quantities $(\mathcal{T}, v, g, f)$. The simplest assumption for $\mathcal{T}$ is a step function with unit transmission above the effective barrier edge and zero below, which neglects tunneling contributions. Tunneling can be included on the level of the Wentzel–Kramers–Brillouin (WKB) theory via [72]

$$\mathcal{T}_{\text{WKB}}(E) = \exp\left(-\frac{2}{\hbar}\int_0^{L_B} dz\sqrt{2m_B^*[V_B(z)-E]}\right), \tag{41.21}$$

where m_B^* is the effective mass of carriers in the barrier material, L_B is the width of the barrier, and $V_B(z)$ is the spatial profile of the barrier potential. More accurate approaches for determination of the transmission function are based on analytical or numerical solution of the Schrödinger equation using the exact barrier potential, e.g., by means of an electronic transfer matrix formalism [71]. For free carriers, the velocity is related to the (kinetic) carrier energy via the momentum, e.g., $v_z = \hbar k_z/m_w^* = \sqrt{2E_z/m_w^*}$ for propagation in the z-direction. In basic approximation, the thermal velocity v_{th} is used, corresponding to $E_z = k_B T$. The exact form of the DOS g reflects the transition of dimensionality from 2D to 3D at energies close to the barrier edge and can be captured in advanced descriptions based on scattering states in a real space basis representation, which do not discriminate between bound, quasibound, and continuum states, such as the Green's function approaches [73,74]. In most implementations, however, a combination of 2D square well DOS and 3D bulk DOS is used [71]. Finally, the occupation function f needs to encode the nonequilibrium steady-state carrier distribution under illumination and bias voltage. In general, this is achieved by using Fermi or Boltzmann statistics together with a quasi-Fermi level (QFL) for the carriers in the QW. In the case of Boltzmann statistics, the QFL drops out of the expression for the escape time due to the normalization to the total carrier density [75]. The advantage of the above approach is that it covers any combination of thermionic, direct tunneling, and thermally assisted tunneling escape. The drawback is that it requires assumptions on the carrier distribution in the QW and it does not contain any information on the microscopic mechanisms that lead to the carrier population at a certain energy. Such mechanisms include, e.g., the very fast electron-electron scattering that establishes a hot-carrier quasi-equilibrium at a fs timescale. But even the slower scattering of electrons with longitudinal optical (LO) phonons was shown to produce a hot-carrier population via subband scattering that enables fast thermally assisted carrier escape [76–78]. This latter result, while based on simple application of FGR to electron-LO-phonon scattering between electronic subbands, was confirmed by the state-of-the-art quantum-kinetic simulation of the phonon-assisted tunneling escape [79].

In **QD** devices, in addition to continuum (3D) and confined (0D) QD states, the 2D states of the WL play an important role in the dynamics of carrier transitions. Experimentally, similar thermionic and tunneling escape processes were identified [80–82]. However, thermal escape is much less efficient than in QW, and larger fields are required to enable tunneling escape. In terms of microscopic mechanisms, evidence for both Auger scattering [83] and (multi)phonon absorption [84] was found. While the carrier-carrier scattering depends strongly on carrier density and decreases quickly with QD size due to shrinking Coulomb (Cb) matrix elements—and is, therefore, less prominent in QW—the phonon processes depend critically on the energetic spacing of the states involved in the transitions, which produces strong resonances in the size dependence of the scattering mechanism. However, the lifetime broadening of QD states due to dynamical processes and delocalization increasing with energy alleviate the impact of the discrete DOS. Finally, in addition to the nonradiative escape processes, QDs offer the possibility of photon-assisted or purely radiative escape via subband transitions as described in Section 41.2.1. In QW, the subband absorption is strongly suppressed for in-plane polarization; hence, special photon management is required to scatter light into large angles, preferentially with coupling to guided modes.

The inverse process to escape is carrier *capture*. In SCs, this process is relevant in MQW and MQD devices, where photogenerated carriers that escaped to the current-carrying continuum states are recaptured into confined states on their way to the contacts. Since on the other hand, this process is instrumental for light-emitting devices, its theoretical description and numerical simulation have received considerable attention and are extensively covered in the literature. For **QW**, traditional models consider the FGR rate for transitions from 3D bulk states to confined states via the emission of LO-phonons [85–95] or carrier-carrier scattering [96,97], in combination with population dynamics from rate equations [98,99] or Monte Carlo simulations [100–104], while more recent approaches are based on quantum-kinetic theory [105]. At the relatively low densities of injected carriers at the operating point of a SC (which is below the bias regime of LED operation), relaxation due to emission of phonons was found to dominate over the effects of carrier-carrier scattering. In the case of **QD** devices, capture from the WL and subband relaxation were evaluated in the FGR framework for Auger processes [106,107], including the effects of line broadening induced by the carrier dynamics [108], and for the emission of single phonons [109,110] and multiple phonons [111], with extension to a polaron picture with finite phonon lifetime [112] indicated due to the strong coupling of QD states to phonons. The mechanisms were also combined in unified theories on semiclassical [113] and quantum-kinetic [114] levels.

41.2.2.3 Transport to and Extraction at Carrier-Selective Contacts

The choice of the picture for charge transport is primarily dictated by the kind of NS device architecture under consideration. In the second and third types of NSSC devices displayed in Figure 41.1, the NS states are involved not only in the absorption, but also in the carrier transport. These states thus need to be to some degree extended, which amounts to the requirement of coupling between individual NS. Indeed, most of the concepts of this kind are based on the formation of minibands in superlattices of QWs or QDs. For that situation, transport may be described as band-like, with a low-field Bloch-type electron *mobility* $\mu = \sigma/(\rho e)$ defined in terms of the conductivity tensor σ, and in the diffusive limit is obtained from the linearized Boltzmann transport equation (BTE) [115]:

$$\mu_{\alpha\beta} = e \left[\sum_{n,\mathbf{K}} \tau_{n\mathbf{K}} v_{n\mathbf{K}}^{\alpha} v_{n\mathbf{K}}^{\beta} \left(-\frac{\partial f_{n\mathbf{K}}^0}{\varepsilon_{n\mathbf{K}}} \right) \right] \Big/ \sum_{n,\mathbf{K}} f_{n\mathbf{K}}^0. \tag{41.22}$$

In Equation 41.22, n is the band index, $\mathbf{K}$ the superlattice wave vector, $\tau_{n\mathbf{K}}$ is the relaxation time, $\varepsilon_{n\mathbf{K}}$ is the miniband dispersion, $f_{n\mathbf{K}}^0$ is the equilibrium Fermi-Dirac electron distribution function, and $v_{n\mathbf{K}}^{\alpha} = \hbar^{-1} \partial_{K_\alpha} \varepsilon_{n\mathbf{K}}$ is the band velocity. The relaxation time depends on the scattering processes that are present in the device and can be computed—in some approximation—from the corresponding FGR formalism [116]:

$$\tau_{n\mathbf{K}}^{-1} = \sum_{n'\mathbf{K}'} \left(\frac{1 - f_{n'\mathbf{K}'}^0}{1 - f_{n\mathbf{K}}^0} \right) \{1 - \cos(\theta_{\mathbf{K}\mathbf{K}'})\} P_{\mathbf{K}\mathbf{K}'}^{nn'} \tag{41.23}$$

where $\cos(\theta_{\mathbf{KK'}}) = (\mathbf{v}_{n\mathbf{K}} \cdot \mathbf{v}_{n'\mathbf{K'}})/(|\mathbf{v}_{n\mathbf{K}}||\mathbf{v}_{n'\mathbf{K'}}|)$ and $P^{nn'}_{\mathbf{KK'}}$ is the FGR rate for scattering from the occupied state $|n\mathbf{K}\rangle$ to the empty state $|n'\mathbf{K'}\rangle$. However, even in the ideal case of a perfect miniband, the associated Bloch mobility may be critically low, as in the case of silicon QDs in a dielectric matrix material [117]. In realistic situations, due to the presence of built-in fields and any kind of spatial, configurational and compositional disorder, the NS states usually show a high degree of localization such that tunneling between NS is restricted to nearest neighbors. In addition to inducing localization, these effects lead to a misalignment of energy levels in adjacent NS, with the formation of Wannier-Stark ladders in the extreme case of very strong fields [118], in which situation transport is only possible via an inelastic scattering process and is best described by hopping between localized states. In coupled QD systems, due to the sparsity of the DOS, the presence of charge on the NS can lead to pronounced shifts in the energy level structure [53], inhibiting transport (Coulomb blockade regime). In devices where charge transport acquires molecular character, such as in colloidal QD solids, the hopping process involves even a change in configuration and may be phenomenologically described by the Marcus theory [44]. The bulk picture of charge carrier transport in band states does thus in general not provide a proper description of the real situation, which in addition to the remaining coherence and nonlocality effects should include all the localization effects and scattering mechanisms required to overcome energetic misalignment. The modification of the electronic structure due to the presence of NS also affects the scattering processes responsible for the limitation of mobility and for carrier relaxation, such as electron–phonon interaction, which becomes especially relevant in devices where this form of energy dissipation is sought to be suppressed, as in the hot-carrier concept, and in the case where extraction distances are on the order of the mean free path and transport approaches the ballistic regime.

The final step of the PV energy conversion consists in the extraction of the charge carriers at carrier-selective contacts. Carrier selectivity—which is an essential requirement on any PV device architecture [119]—can be described on a phenomenological level via the surface recombination velocities for majority and minority carriers used in the boundary conditions for carrier density/current in macroscopic transport equations. Microscopically, transport across the boundary to the contact depends again on the electronic structure of the interface and on the prevailing scattering processes. Since those can be largely engineered using NS, the latter also play an important role in the implementation of energy-selective [22] and carrier-selective [120] contacts.

41.2.3 Recombination

On the way to or from the contacts, the interaction of optically or electronically injected charge carriers with the EM and vibrational fields can induce different types of *recombination* (the mutual annihilation of an electron and a hole), which are usually classified in *radiative*, *Auger*, and *defect-mediated* processes. Like for the generation process, a proper consideration of NS effects on the recombination processes requires a careful consideration of the wave functions and local DOS together with the occupation of these states in general nonequilibrium conditions.

41.2.3.1 Radiative

Even in ideal, defect-free absorber materials, the principle of *detailed balance* dictates the presence of radiative recombination by spontaneous and stimulated photon emission as the inverse process to carrier generation by photon absorption. Conventionally, the local volume rate of radiative recombination is related to the charge carrier densities and the local optical material constants (α, n_r) as follows [121]:

$$r^{\mathrm{em}}(\mathbf{r}) = B(\mathbf{r})\{\rho_e(\mathbf{r})\rho_h(\mathbf{r}) - \rho_e^0(\mathbf{r})\rho_h^0(\mathbf{r})\}, \tag{41.24}$$

$$B(\mathbf{r}) = \rho_i^{-2} \int dE_\gamma \, \alpha(\mathbf{r}, E_\gamma) \bar{\Phi}_{bb}(E_\gamma), \tag{41.25}$$

where ρ^0 and ρ_i denote equilibrium and intrinsic carrier densities, respectively, and $\bar{\Phi}_{bb} = 4\pi\Phi_{bb}$ is the angle-integrated blackbody radiation flux, with

$$\Phi_{bb}(E_\gamma) = \frac{E_\gamma^2 n_r^2}{4\pi^3 \hbar^3 c_0^2} \left\{ \exp\left(\frac{E_\gamma}{k_B T}\right) - 1 \right\}^{-1}. \tag{41.26}$$

In Equation 41.25, the assumption of emission in an optically homogeneous medium is made, with isotropic photon DOS. In thin-film SCs with nanophotonic light trapping, this is often not appropriate, and the radiative recombination should be modeled starting from the FGR for spontaneous emission under consideration of the actual density of final photon states. For NS, the emission is enhanced as compared to the bulk, due to carrier localization [122] resulting in maximized density product in Equation 41.25 and to larger overlap of electron and hole wave functions in Equation 41.13. A further point to note in the case of NSSC is the dependence of the absorption coefficient on the operating conditions, mainly the bias voltage via the effect of the built-in field on the local electronic structure, which should be reflected in the quantities used in Equation 41.25.

41.2.3.2 Auger

In the regime of large optical injection, relevant for the operation under high optical concentration, the PV performance of a defect-free absorber is limited by the interband Auger recombination mechanism [123], which as an intrinsic effect is unavoidable and needs thus to be considered for a realistic estimate of the limiting energy conversion efficiency. In this process originating in carrier-carrier scattering, electrons recombine with holes by giving up the excess energy to excite either another electron or another hole. In the semiclassical bulk continuum formulation, this rate takes the form [124]

$$r^{\text{Aug}}(\mathbf{r}) = \left[C_e(\mathbf{r})\rho_e(\mathbf{r}) + C_h(\mathbf{r})\rho_h(\mathbf{r}) \right] \left\{ \rho_e(\mathbf{r})\rho_h(\mathbf{r}) - \rho_e^0(\mathbf{r})\rho_h^0(\mathbf{r}) \right\}, \tag{41.27}$$

where $C_{e/h}$ are the Auger coefficients for electron/hole excitation. For NS, some or all of the band indices are replaced by the labels of subbands (QW) or discrete levels (QD). The Auger coefficients can be computed microscopically by application of perturbation theory for the carrier-carrier interaction under consideration of the appropriate dimensionality of the (screened) Cb potential. For QW, FGR-type approaches were used to assess direct [125–129] and phonon-assisted [130] Auger processes, and the latter was also investigated using a Green's function technique [131]. For QD, similar FGR models were developed for colloidal QD [132] and applied to self-assembled III-V single QD [133] and QD arrays [134]. Recently, the complete dynamics of excitons in colloidal QD was also simulated on an *ab initio* level using a combination of density functional theory and molecular dynamics [135]. While in QW, the magnitude of the Auger recombination rate was found to be similar to bulk, it is strongly enhanced in QD due to the effects of wave function localization and associated relaxation of momentum selection rules.

41.2.3.3 Defect Mediated

While radiative and Auger losses are unavoidable and thus present even in structures without imperfections, recombination in real-world devices is usually dominated by defect-mediated processes where the carrier energy is dissipated as heat and is lost for the conversion process. Indeed, there is an inherent increase in surface area associated with the presence of NS, which are thus likely to act as centers of nonradiative recombination. In some cases, it is possible to reduce the NS-related defects via specially engineered strain-balancing techniques [136] or via passivation [137], but in general, the insertion of NS leads to losses in the open-circuit voltage of the SC. With regard to the modeling, the role of lower dimensional NS in the context of defect-mediated recombination has received little attention so far, and models are usually based on the conventional bulk Shockley–Read–Hall (SRH) theory [138,139], with modification to relate the recombination to interface or surface defects rather than volume defects, e.g., for QW [60]

$$r^{\text{SRH}}(\mathbf{r}) = \frac{\rho_e(\mathbf{r})\rho_h(\mathbf{r}) - \rho_e^0(\mathbf{r})\rho_h^0(\mathbf{r})}{(\rho_e(\mathbf{r}) + \rho_e^t(\mathbf{r}))\tau_e + (\rho_h(\mathbf{r}) + \rho_h^t(\mathbf{r}))\tau_h}, \tag{41.28}$$

where $\tau_s^{-1} = \sigma_s v_{th} N_S / L_w$ ($s = e, h$) with σ_s the carrier capture cross-section, v_{th} the thermal velocity, N_S the sheet density of interface defects, L_w the QW width, and ρ^t the carrier density for a Fermi level at the trap energy. The capture cross-sections depend on the local electronic structure of the NS and defect states, as well as on the phonon modes coupling to those states in the multiphonon relaxation mechanism [124], and can be obtained based on microscopic theories on phenomenological [140] and fully *ab initio* [141] levels.

41.3 Modeling of NSSC Device Characteristics

The basic function of a SC consists in the conversion of the solar energy flux impinging on its surface into electrical power. A central requirement on any theory suitable for the modeling of SC device operation is thus the ability to predict the PV energy conversion efficiency as defined via the ratio $\eta_{\text{PV}} \equiv P_{el}^{\swarrow} / P_{\gamma}^{\swarrow}$ of generated electrical power to the incident radiative power, with

$$P_{el}^{\swarrow} = \max_V \{I(V) \cdot V\}|_{T_C}, \tag{41.29}$$

$$P_{\gamma}^{\swarrow} = \int dE_{\gamma} \int_S \int_{\Omega} d\mathbf{\Omega} \cdot d\mathbf{S} \, \Phi_{\gamma}^{\swarrow}(\mathbf{r}_S, \Omega, E_{\gamma}) E_{\gamma}, \tag{41.30}$$

where V is the terminal voltage (i.e., related to the separation of chemical potentials at the contacts), $I(V)$ is the charge current extracted from the device at this voltage, T_C is the temperature of the SC, E_{γ} is the photon energy, and $\Phi_{\gamma}^{\swarrow}$ denotes the incident spectral photon flux impinging on a cell of (outer) surface S under a solid angle of incidence $\Omega = (\theta, \varphi)$. Hence, any PV model needs to provide I as a function of V and $\Phi_{\gamma}^{\swarrow}$ at given T_C.

As outlined in Sections 41.1 and 41.2, depending on the functionality of the NS within the SC device, some or all of the fundamental PV processes are affected by the peculiar physical properties of the NS component, with major implications for the requirements on a suitable simulation approach. The general approach reflects the multiscale and multiphysics scheme depicted in Figure 41.2: Starting from the NS states, the dynamics giving the occupation of these states is evaluated and the resulting generation and recombination rates are used to obtain the device characteristics. Three levels of sophistication that can be identified in the hierarchy of theoretical descriptions of SC device operation providing the current–voltage characteristics $I(V, \Phi_{\gamma}^{\swarrow})$, depending on the explicity of consideration of the various physical mechanisms, are: *detailed balance* theories providing efficiency limits of idealized systems, *semiclassical* transport models with detailed balance generation-recombination rates, and *quantum-kinetic* theories with consistent and fully microscopic treatment of charge carrier dynamics and transport [142]. These three approaches shall now be discussed in some more detail.

41.3.1 Global Detailed Balance Theories for Ideal Systems

On the most basic level, NS properties are considered only in generation and recombination at the radiative limit, while transport is assumed to be ideal. This means that from a device point of view, the system is treated as a black box, and there is no information on geometry or any spatial inhomogeneity in the electronic system that would depart from effective bulk properties. The most prominent examples of this

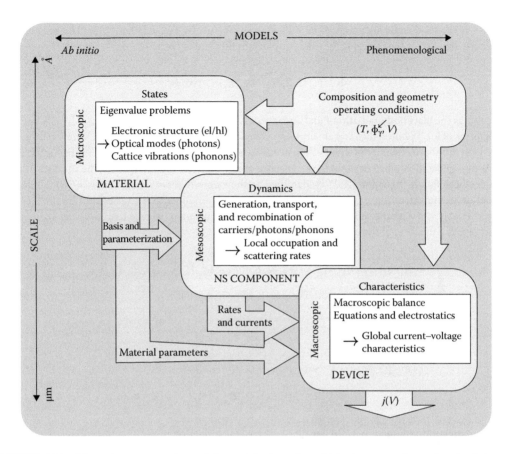

FIGURE 41.2 Schematic representation of the multiscale and multiphysics approach to the simulation of nanostructure-based solar cell devices.

category are the detailed balance models [143] used to obtain an upper bound to the energy conversion efficiency, such as the SQ limit. In this approach, the current–voltage characteristics are obtained by equating the terminal current with the difference of the number of photons per unit time entering and leaving the absorber volume at given terminal bias voltage [143,144]:

$$I(V, \Phi_\gamma^{\swarrow})|_{T_C} = -q \int dE_\gamma \int_S \int_\Omega d\Omega \cdot dS \left\{ \Phi_{abs}(\mathbf{r}_S, \Omega, E_\gamma) - \Phi_{em}(\mathbf{r}_S, \Omega, E_\gamma) \right\}|_{V, \Phi_\gamma^{\swarrow}, T_C}. \tag{41.31}$$

At the radiative limit, where the recombination is determined by the emission, and assuming unit generation efficiency, the electronic equivalent to the above balance relation is obtained from the integration of the charge continuity equation over the absorber volume $\mathcal{V}$ (e/h: electrons/holes):

$$I(V, \Phi_\gamma^{\swarrow})|_{T_C} = \mp q \int_\mathcal{V} d^3\mathbf{r} \left\{ \mathcal{G}_{e/h}(\mathbf{r}) - \mathcal{R}_{e/h}(\mathbf{r}) \right\}|_{V, \Phi_\gamma^{\swarrow}, T_C}. \tag{41.32}$$

The absorbed photon flux is derived in terms of the absorptance—or, synonymously, the absorptivity—a of the device, and of the reflection coefficient r for incident light, via

$$\Phi_{abs}(\mathbf{r}_S, \Omega, E_\gamma) = [1 - r(\mathbf{r}_S, \Omega, E_\gamma)]a(\mathbf{r}_S, \Omega', E_\gamma)\left\{\Phi_\gamma^{\swarrow}(\mathbf{r}_S, \Omega, E_\gamma) + \Phi_{\gamma 0}(E_\gamma, \Omega, T_C)\right\}, \tag{41.33}$$

where Ω' is the solid angle after refraction, related to the former by Fresnel's law [145]. The second term in the curly brackets is the isotropic thermal equilibrium flux and is given by

$$\Phi_{\gamma 0}(E_\gamma, \Omega, T_C) = \Phi_{bb}(E_\gamma, T_C)\cos\theta, \tag{41.34}$$

where Φ_{bb} is the blackbody radiation flux (Equation 41.26) introduced in Section 41.2.3. The generation rate in terms of local photon flux and absorption coefficient is given in Equation 41.2. However, for equivalence with the above result, photon recycling needs to be considered in the evaluation of the local photon flux. At unit generation efficiency and in the absence of nonradiative recombination, the product in front of the incident flux in Equation 41.33 is identified as the *external quantum efficiency* (EQE). Under the assumption of perfect transport, i.e., infinite mobility, which amounts to flat QFLs split by the terminal voltage, i.e., $\Delta\mu = qV$, and of an optically isotropic medium, detailed balance then provides the emitted photon flux as a function of the absorptance [144,146]:

$$\Phi_{em}(\mathbf{r}_S, \Omega, E_\gamma)|_V = [1 - r(\mathbf{r}_S, \Omega, E_\gamma)]a(\mathbf{r}_S, \Omega', E_\gamma)\tilde{\Phi}_{bb}(\Omega, E_\gamma, T_C, V) \tag{41.35}$$

$$\equiv EQE(\mathbf{r}_S, \Omega, E_\gamma)\tilde{\Phi}_{bb}(\Omega, E_\gamma, V), \tag{41.36}$$

where:

$$\tilde{\Phi}_{bb}(\Omega, E_\gamma, T_C, V) = \Phi_{bb}(E_\gamma - qV, T_C)\cos\theta \tag{41.37}$$

$$\approx \Phi_{\gamma 0}(E_\gamma, \Omega, T_C)e^{qV/(k_B T_C)} \qquad (E_\gamma \gg qV). \tag{41.38}$$

The approximation in Equation 41.38 permits a compact formulation of the current–voltage characteristics:

$$I(V, \Phi_\gamma^\swarrow)|_{T_C} \approx -q\int dE_\gamma \int_S \int_\Omega d\Omega \cdot d\mathbf{S}\, EQE(\mathbf{r}_S, \Omega, E_\gamma)$$

$$\times \left\{\Phi_\gamma^\swarrow(\mathbf{r}_S, \Omega, E_\gamma) - \Phi_{\gamma 0}(E_\gamma, \Omega, T_C)\left[e^{qV/(k_B T_C)} - 1\right]\right\}. \tag{41.39}$$

Finally, the local recombination rate for the determination of the radiative dark current via Equation 41.32 is given by Equation 41.25.

In the SQ limit, the absorptance is a unit step function with threshold energy at the value of the bandgap E_g, which is then the only electronic parameter entering the calculation of the current–voltage characteristics. This simplifies the expressions for the absorbed and emitted photon fluxes considerably. In a first step beyond the SQ analysis, the effect of incomplete absorption with non–abrupt onset as well as the impact of light-trapping structures can be considered by using in Equations 41.33 and 41.35 the proper absorptance computed by optical models for the light propagation as outlined in Section 41.2. It is common to neglect the bias dependence of the photocurrent also in situations where a realistic absorptance is considered, which results in current–voltage characteristics given by the *superposition* of photocurrent I_γ and dark current I_d:

$$I(V, \Phi_\gamma^\swarrow) = I_\gamma(\Phi_\gamma^\swarrow) + I_d(V). \tag{41.40}$$

However, this tends to be inappropriate in nanoscale absorbers due to the strong impact of built-in fields and finite state filling on the local absorption coefficient [147,148].

Due to its formal simplicity, the detailed balance analysis of the radiative limit is applied to most novel NS PV devices in a first assessment of the performance potential. On the other hand, the restrictive assumptions made are seldom met in realistic devices, and the upper efficiency bound obtained in this way may therefore be far from a realistic estimate.

Within the above setting of idealized conditions, an alternative starting point for the device characteristics is a modification of the ideal diode equation [149]:

$$I(V, \Phi_\gamma^{\checkmark}) = I_0 \left(e^{qV/(k_B T_C)} - 1 \right) - I_G(\Phi_\gamma^{\checkmark}) + I_R(V), \qquad (41.41)$$

which is formally equivalent to Equation 41.39, but where the generation and recombination currents I_G and I_R are composed of the contributions of the bulk "baseline" cell and of the NS components. The NS contributions are related to those of the baseline using correction factors for quantities such as geometry (fraction of NS material), oscillator strength, and DOS (for radiative transitions). It needs to be noted, however, that such modifications still have to comply with the detailed balance relations between absorption and emission as given above. Furthermore, even in its original formulation devised for the radiative limit, the theory goes beyond that limit in considering in the reverse saturation current I_0 the standard nonradiative minority carrier diffusion contribution of an ideal p-(i-)n diode.

41.3.1.1 Application to QWSC

Soon after the proposal of the QWSC concept, the detailed balance approach was used to show that under the idealized conditions of that theory, the QWSC could not provide an efficiency in excess of that of an ideal gap bulk cell, since both photocurrent generation and radiative recombination are based on the same modified absorptivity [150,151]. An SQ-type detailed balance model was then used to demonstrate that QWSC efficiency could still exceed the bulk limit if variations of the quasi-Fermi level splitting (QFLS) between bulk and QW material were allowed for [152]. This was shown to be equivalent to the detailed balance picture of the intermediate band concept, with the difficulty of the QW featuring suppressed sub-band absorption and fast population equilibration due the continuous electronic DOS [153]. Indeed, the issue of QFLS variation in QWSC is still a matter of debate and ongoing research. In more recent QWSC developments, the detailed balance approach was used to model tandem configurations and the impact of directional emission in strained QW [23]. Following the original work on the ideal diode QWSC model [149], a number of variations and refinements of this semianalytical approach were devised for MQWSC [154–161] and QWSLSC [159,162–164].

41.3.1.2 Application to QDSC

For QD intermediate band SC (QDIBSC), the original detailed balance IBSC model [165] was adapted to account for the specific absorption characteristics of QD structures [166,167] and was implemented for specific material systems using a Kronig-Penney model for supracrystals of InAsN/GaAsSb QD [168], InGaN/GaN QD [169], and InGaN/InN [170], as well as a $\mathbf{k} \cdot \mathbf{p}$ model for isolated InAs/GaAs QD [171]. The ideal diode theory was applied to theoretically investigate an InAs/GaAs MQD concept with AlGaAs fences to prevent capture of photogenerated carriers [172].

41.3.2 Hybrid Models: Semiclassical Transport with Detailed Balance Rates

The main shortcoming of global detailed balance approaches and ideal diode theories is the lack of a realistic consideration of photocarrier transport, i.e., the assumption of unit collection efficiency, which is often inappropriate in NSSC devices. In order to account for the effects of finite mobility, actual transport equations have to be solved for the charge carriers. The main difficulty in that undertaking resides in the consideration of the contribution of NS states with a higher degree of localization. To obtain the self-consistent occupation of confined NS states (c) and bulk host states (b), separate but coupled rate equations need to be formulated for the densities of carriers occupying different types of states. Under the standard assumption of continuum (i.e., no atomistic information) and complete thermalization (i.e., occupation is characterized by a local QFL), the current–voltage characteristics are obtained from the *drift-diffusion* (DD) charge current for carrier species $s \in \{e, h\}$:

$$\mathbf{J}_{s,i}(\mathbf{r}) = -q\{\mu_{s,i}(\mathbf{r})\rho_{s,i}(\mathbf{r})\nabla\phi(\mathbf{r}) \mp D_{s,i}(\mathbf{r})\nabla\rho_{s,i}(\mathbf{r})\}, \qquad i = b, c \qquad (41.42)$$

where μ denotes the mobility, $\rho_{e/h}$ denotes the charge carrier densities, D is the diffusion coefficient, ϕ is the electrostatic potential, q is the elementary charge, and the upper (lower) sign is for electrons (holes). In this general formulation, no restriction is made on the charge transport in NS states, which allows for application to all three NS architectures introduced in Section 41.1. The charge carrier densities and the electrostatic potential that determine the charge currents in Equation 41.42 are obtained from the coupled solution of the steady-state continuity equations:

$$\mp\frac{1}{q}\nabla \cdot \mathbf{J}_{s,i}(\mathbf{r}) = \sum_{j \neq i} \{\mathcal{G}_{s,j \to i}(\mathbf{r}) - \mathcal{R}_{s,i \to j}(\mathbf{r})\}, \qquad (41.43)$$

where $\mathcal{G}/\mathcal{R}$ is the carrier (volume) generation/recombination rate due to transitions between (sub)bands and the upper (lower) sign applies again to electrons (holes), and Poisson's equation for the total charge density:

$$\epsilon_0 \nabla \cdot \{\varepsilon(\mathbf{r})\nabla\phi(\mathbf{r})\} = -q\left\{ \sum_i \rho_{h,i}(\mathbf{r}) - \sum_j \rho_{e,j}(\mathbf{r}) + N_{\text{dop}}(\mathbf{r}) \right\}, \qquad (41.44)$$

where ε is the static dielectric function and $q \cdot N_{\text{dop}}$ is the net charge density due to ionized dopants. The boundary conditions used in the solution of the coupled equations are the same as in the bulk case, reflecting the nature of the physical contact in terms of (Schottky-)barriers and surface recombination. In Equation 41.43, the generation rate $\mathcal{G}_{s,j \to i} = \mathcal{G}_{s,j \to i}[\Phi_\gamma(\mathbf{r}), \alpha_{j \to i}(\mathbf{r}, E_\gamma)]$ $(i = b, c)$ is a functional of the local photon flux and the local absorption coefficient as given by Equation 41.2 and the recombination rate $\mathcal{R}_{s,i \to j} = \mathcal{R}_{s,i \to j}[\rho_{e,i}, \rho_{h,j}, \ldots]$ depends on the carrier densities and on recombination mechanism (radiative, Auger, defect-mediated)–specific parameters. In the case where, due to strong localization as for isolated QW or QD absorbers, direct transport between NS states can be neglected, the current term in Equation 41.43 is absent for these states, and exchange of carriers between localized states always proceeds via scattering to and from extended states. The coupling of the equations for localized and extended states is then provided on the level of a pure rate equation, where the rates are subject to the detailed balance condition and are commonly expressed via the density in the initial state and an associated lifetime depending on the scattering process, e.g.,

$$\mathcal{R}^x_{e,i \to j} = \frac{\rho_{e,i}}{\tau^x_{e,i \to j}}, \qquad \text{x: scattering mechanism.} \qquad (41.45)$$

The lifetimes τ (which are local quantities) are either used as fitting parameters to reproduce experimental characteristics, or derived from an appropriate description for the microscopic mechanisms of the scattering process responsible for the coupling as reviewed in Section 41.2, e.g., via FGR, based on the solution of the Schrödinger-Poisson problem for the NS states and energies. For consistency, the macroscopic material parameters based on microscopic information, such as absorption coefficients and mobilities used in the generation term and the current expressions, should be computed on the basis of the same solutions of the microscopic equations for the electronic structure. It needs to be emphasized at this point that the lifetimes themselves still depend on the occupation of the states, e.g., for the radiative recombination, $\tau^{\text{em}}_e(\mathbf{r}) \approx [B(\mathbf{r})\rho_h(\mathbf{r})]^{-1}$, where the very small equilibrium term in (Equation 41.24) is neglected.

In the above formulation of the charge current, the local electronic structure does not appear as a parameter that could be determined microscopically. If the Einstein relation between mobility and diffusion coefficient, i.e., $D = \mu k_B T/q$, and the Boltzmann approximation for the carrier distribution function are valid, an equivalent formulation of the charge carrier densities and currents in terms of effective DOS $\mathcal{N}$ and QFL E_F can be used:

$$\rho_{s,i}(\mathbf{r}) = \int dE\, D_{s,i}(\mathbf{r}, E) f_{s,i}(E) \approx \mathcal{N}_{s,i} \exp\left(\frac{\pm E_{F_{s,i}}(\mathbf{r}) \mp \varepsilon_{s,i}(\mathbf{r})}{k_B T} \right), \tag{41.46}$$

$$\mathcal{N}_{e,i} \equiv \int_{\varepsilon_{e,i}}^{\infty} dE\, D_{e,i}(\mathbf{r}, E) \exp\left(\frac{\varepsilon_{e,i}(\mathbf{r}) - E}{k_B T} \right), \tag{41.47}$$

$$\mathcal{N}_{h,i} \equiv \int_{-\infty}^{\varepsilon_{h,i}} dE\, D_{h,i}(\mathbf{r}, E) \exp\left(\frac{E - \varepsilon_{h,i}(\mathbf{r})}{k_B T} \right), \tag{41.48}$$

$$\mathbf{J}_{s,i}(\mathbf{r}) = \pm \mu_{s,i}(\mathbf{r}) \rho_{s,i}(\mathbf{r}) \nabla E_{F_{s,i}}(\mathbf{r}), \tag{41.49}$$

with upper (lower) sign applying to electrons (holes) and D denoting the DOS for carrier species s and band i, with $\varepsilon_{s,i}$ the minimum (maximum) energy associated ("band edge"). This approach, in addition to the implicit consideration of the microscopic DOS for NS states, provides the possibility to study the emergence of distinct QFL for the NS populations, which is instrumental in PV concepts such as the IBSC.

The *hybrid* (macroscopic-microscopic) approach outlined above can be used to reproduce experimental device characteristics with remarkable accuracy. However, being a macroscopic and local model, any situation requiring energy resolution, nonlocality, or coherence cannot be described properly, e.g., nonthermalized carrier distributions or resonant tunneling. To include such processes in a consistent description, transition to a truly microscopic picture of carrier transport is indicated.

41.3.2.1 Applications to QWSC

Semiclassical DD Poisson models with detailed balance rates for generation and recombination were applied to III-V MQWSC [60,173,174] and to Si-SiO$_x$ QWSLSC [175]. Simplifications include rate equations with semianalytical solutions for current contributions [176,177] and the analytical solution of diffusion equation for excess minority carriers in neutral layers [57,154,178].

41.3.2.2 Applications to QDSC

The above mentioned DD Poisson approach with local rates was implemented in a comprehensive fashion for MQDSC [68,179,180] and QD-IBSC [181–183]. Again, there are also approximations in the form of analytical solutions of the diffusion equation [184–186], of simple rate equations for the population of bulk and NS states [187], or a combination of the two [188].

41.3.3 Fully Microscopic Quantum-Kinetic Models

In situations where the NSs are also involved in transport and the latter is not entirely band-like or purely diffusive, a consistent microscopic picture of the PV processes can be formulated by means of nonequilibrium quantum statistical mechanics using the nonequilibrium Green's function formalism (NEGF) [189]. In the NEGF picture, the device current density $\mathbf{J} = \mathbf{J}_n + \mathbf{J}_p$ is given in terms of the charge carrier Green's function components as follows:

$$\mathbf{J}_{n/p}(\mathbf{r}) = \pm \lim_{\mathbf{r}' \to \mathbf{r}} \frac{e\hbar}{m_0} \left(\nabla_{\mathbf{r}} - \nabla_{\mathbf{r}'} \right) \int \frac{dE}{2\pi} G^{\lessgtr}(\mathbf{r}, \mathbf{r}'; E). \tag{41.50}$$

The Green's functions in Equation 41.50 are determined by the steady-state Keldysh equation:

$$G^{\lessgtr}(\mathbf{r}, \mathbf{r}'; E) = \int d\mathbf{r}_1 \int d\mathbf{r}_2 G^R(\mathbf{r}, \mathbf{r}_1; E) \Sigma^{\lessgtr}(\mathbf{r}_1, \mathbf{r}_2; E) G^A(\mathbf{r}_2, \mathbf{r}'; E), \tag{41.51}$$

with retarded and advanced components resulting from the Dyson equation:

$$\int d\mathbf{r}_1 \left[\{G_0^{R/A}\}^{-1}(\mathbf{r}, \mathbf{r}_1; E) - \Sigma^{R/A}(\mathbf{r}, \mathbf{r}_1; E) \right] G^{R/A}(\mathbf{r}_1, \mathbf{r}'; E) = \delta(\mathbf{r} - \mathbf{r}'). \tag{41.52}$$

While the retarded and advanced Green's functions $G^{R/A}$ are related to the charge carrier DOS, the correlation functions $G^{\lessgtr}$ additionally contain information on the (nonequilibrium) occupation of these states. The noninteracting system is described by G_0. The self-energies Σ, on the other hand, are scattering functions describing the renormalization of the Green's functions due to coupling to the environment in the form of interactions with photons (mediating photogeneration and radiative recombination), phonons (mediating relaxation and indirect transitions), and other carriers (relevant for excitonic and Auger processes). An additional self-energy term describes injection and extraction of carriers as contacts with arbitrary chemical potential, enabling the treatment of an open nonequilibrium system. The computation of the NEGF is self-consistently coupled to the Poisson equation (Equation 41.44) for the mean-field electrostatic potential ϕ—entering the equation for G_0—via the charge carrier density:

$$\rho_{e/h}(\mathbf{r}) = \mp i \int \frac{dE}{2\pi} G^{\lessgtr}(\mathbf{r}, \mathbf{r}; E). \tag{41.53}$$

For an in-depth review of the general NEGF formalism and its application to NS PV, the reader is referred to [189] and references therein.

In contrast to the semiclassical models presented in Section 41.3.2, the microscopic approach provides insight into the physical mechanisms underlying the PV device operation, as those are properly captured by the formalism. On the downside, the method is extremely demanding from a computational point of view and applications are therefore limited to functional NSSC components of mesoscopic extension. An important direction of future research will thus consist in the combination of quantum-kinetic models for local carrier dynamics in NS regions with semiclassical models for the global characteristics of the extended device.

41.3.3.1 Applications to Nanowire SCs

As discussed in Section 41.1, QWRs are not in the focus of efficient solar energy harvesting systems. However, the low dimensionality allows for an efficient numerical treatment of the system. Indeed, the earliest application of NEGF for PV energy conversion considered quasi-1D carbon nanotube systems with tight-binding (TB) band structure [190–192]. More recently, a $\mathbf{k} \cdot \mathbf{p}$-NEGF approach was formulated to study photocurrents in GaAs nanowires [193].

41.3.3.2 Applications to QWSC

Single and double QW structures in III-V material systems were studied in [194,195] using a simple TB band structure and in [79,196] within an effective mass approach, with focus on the modification of absorption and emission spectra by the complex potential profile and on the escape of carriers from confined to extended, current-carrying states. Photocarrier transport regimes from ballistic extraction to phonon-mediated sequential tunneling in SL absorbers were investigated for Si-SiO$_x$ [197] (including phonon-mediated indirect transitions) and InAlGaAs-InGaAs [198].

41.3.3.3 Applications to QDSC

The main challenge for the implementation of NEGF models for QDSC is the complexity of the electronic structure with respect to 3D spatial variation, which prevents the use of the slab models conveniently applied to planar NSs. A simple approach is to use a coarse-grained model with a basis of localized QD orbitals—e.g., the Wannier functions of a regular QDSL—in the spirit of the TB approximation in atomic systems. This method was used to formulate a NEGF model for Si QDSL absorbers embedded in the SiO_x-SiC matrix material [199], and to study the impact of inter-QD and QD-contact coupling on the PV performance of generic QD arrays [200]. Within such an approach, however, it is not possible to study on a microscopic level, similar to that achieved in MQWSC, the carrier escape and capture processes that are instrumental in the operation of MQDSC.

41.4 Case Study: Single Quantum Well p-i-n SC

In this section, the different elements and levels of sophistication in the simulation of NSSC shall be illustrated by consideration of a prototypical device architecture, for which a single quantum well (SQW) p-i-n photodiode is chosen. The device as displayed in Figure 41.3 consists of a 100-nm thin bulk GaAs baseline SC with heavily doped emitter and base components ($N_A = N_D = 10^{18}$ cm^{-3}) of 20-nm extension. In the center of the intrinsic region, a slab of 10-nm InGaAs is inserted, which forms a type-I QW. For the band offsets, $\Delta E_{C/V} = 0.15/0.10$ eV is chosen. The contacts are designed to be perfectly carrier selective, which corresponds to the assumption of ohmic boundary conditions for majorities and infinite barriers for minorities. The light is assumed to be incident from the p-side, whereas a gold reflector is applied to the n-side, which, however, is not considered in the electronic simulation.

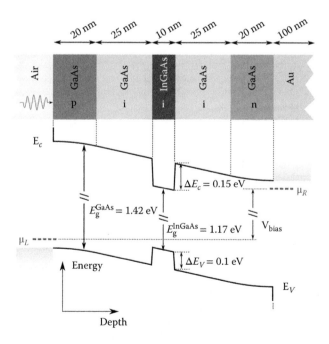

FIGURE 41.3 Single GaAs-InGaAs quantum well p-i-n solar cell architecture used as a test case for the simulation. The contacts are designed to be perfectly selective, i.e., with ohmic boundary conditions for majorities and infinite barriers for minorities. The gold reflector is used only in the optical simulation.

41.4.1 States

In the first step of the simulation, the electronic and optical device states need to be determined. For the description of the electronic structure of the component semiconductor bulk materials, a basic combination of two decoupled single-band effective mass approximation (EMA) models is used, with $m_{e/h}^{*\text{GaAs}} = 0.067/0.1 \, m_0$ and $m_{e/h}^{*\text{InGaAs}} = 0.055/0.22 \, m_0$. Under the standard thin-film assumption of a transverse continuum due to very large spatial extension perpendicular to the growth direction, the wave functions (WF) can be written as the decomposition

$$\Psi_{n\alpha\mathbf{k}_{\parallel}}(\mathbf{r}) = \frac{e^{i\mathbf{k}_{\parallel}\mathbf{r}_{\parallel}}}{\sqrt{\mathcal{A}}} \psi_{n\alpha\mathbf{k}_{\parallel}}(z)u_{n0}(\mathbf{r}), \tag{41.54}$$

where n is the bulk Bloch band index, $\parallel$ denotes transverse quantities, $\mathcal{A}$ is the transverse cross-section area, ψ is the envelope function in the growth direction, and u_{n0} is the periodic part of the Bloch function at the Γ point. The 1D EMA Schrödinger equation for the envelope functions of the QW states is solved numerically for closed-system (Dirichlet) boundary conditions and at vanishing field. The wave functions $\psi_{c/v}(z)$ (z: depth) and eigenenergies $\varepsilon_{c/v}$ of the resulting NS states at $\mathbf{k}_{\parallel} = 0$ are displayed in Figure 41.4a. There are two electron and three hole states with energies below the band edge of the host material, exhibiting the expected symmetry properties and increasing barrier penetration as a function of energy. For comparison, the local density of states (LDOS) at $\mathbf{k}_{\parallel} = 0$ is computed using the NEGF formalism for identical EMA Hamiltonian and at flat potential (Figure 41.4b), but with open boundary conditions, which, however, does not affect either shape or energies of the confined states. While the solutions of the square well potential are readily available, the situation in the device at operating conditions is quite different, as demonstrated by Figure 41.4c, where the LDOS is shown for a realistic band profile at an operating forward bias voltage

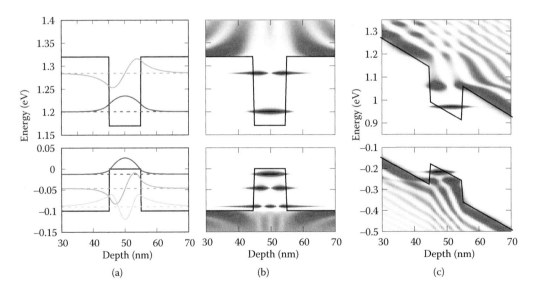

(a)   (b)   (c)

FIGURE 41.4 (a) Wave functions and energies of the eigenstates in the 10-nm InGaAs QW embedded in a GaAs host as obtained by the solution of the 1D single band effective mass Schrödinger equation along the growth direction. (b) One dimensional 1D LDOS ($\mathbf{k}_{\parallel} = 0$) obtained using the NEGF formalism with the same effective mass Hamiltonian and at flat band conditions. (c) 1D LDOS of the same SQW, but in the p-i-n device configuration close to operating conditions at an applied forward bias voltage $V = 0.8$ V. The most remarkable feature is the unbinding of the higher lying QW states.

of $V = 0.8$ V. Most remarkably, the higher lying QW states lose their confinement in the large built-in field and turn into quasibound resonances.

The optical modes of the device are computed under consideration of the absorbing components and of the reflector using an in-house developed wave propagation solver based on the TMM. As will be shown, this is more appropriate than simple consideration of LB law, since both slab and reflector form an optical cavity, which can lead to nontrivial spatial variation of the optical intensity inside the absorber. Since the states of the quasicontinuum in the QW material are not well described by the chosen approach with closed boundaries, the optical response will be computed only for energies below the host gap. The determination of the optical absorption strength based on the computed electronic states thus forms the next step in the simulation procedure.

41.4.2 Dynamics

The absorption coefficient is evaluated following the FGR procedure outlined in Section 41.2, using the states computed in Section 41.4.1. For energies below the host gap, assuming fully occupied valence bands and empty conduction bands, and neglecting the variation of states and energies with transverse momentum, the absorption coefficient of a square well (SW) potential of width L_w reads

$$\alpha_{SW}(E_\gamma) = f \cdot E_\gamma^{-1} \sum_{i,j} |M_{c_i v_j}|^2 \Theta(E_\gamma - \varepsilon_{c_i v_j}), \tag{41.55}$$

with Θ the unit step function and

$$f = \frac{e^2 m_r^* P_{cv}^2}{m_0^2 \hbar n_r c_0 \varepsilon_0 L_w}, \tag{41.56}$$

where m_r^* is the reduced effective mass, $P_{cv} = \sqrt{E_P m_0/6}$ is the bulk MME for Kane energy $E_P = 26.9$ eV, $n_r = 3.7$ is the refractive index of the bulk material, and c_0 and ε_0 are speed of light and permittivity in vacuum, respectively. The overlap matrix elements are

$$M_{c_i v_j} = \int_L dz\, \psi_{c_i}^*(z)\psi_{v_j}(z), \tag{41.57}$$

where $L\ (> L_w)$ is the normalization length of the wave functions, introduce the optical selection rules according to the WF symmetry. As can be verified in Figure 41.5, displaying the above FGR absorption coefficient (full line), only two of the six transitions between the three hole and two electron levels are not suppressed by symmetry. The validity of the selection rules in the absorption coefficient computed via NEGF (dotted line), while evident, is a nontrivial result, which requires careful consideration of the nonlocality in the Green's function picture of electron-photon interaction. In the absence of a symmetric potential, which is not found in the real operating device, the selection rules are strongly relaxed, and the absorption coefficient exhibits a shape that interpolates between 2D and 3D DOS (dashed line).

Based on this absorption coefficient, the emission coefficient B of the radiative recombination rate introduced in Section 41.2 is determined using the Van Roosbroeck–Shockley (VRS) relation (Equation 41.25). These two quantities are the main ingredients for the assessment of the device characteristics at the radiative limit.

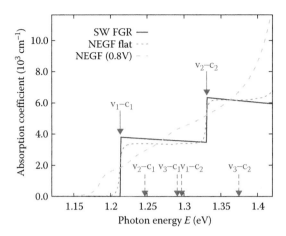

FIGURE 41.5 Absorption coefficient as computed within the Fermi's golden rule approach based on the solution of the single-band effective mass Schrödinger equation for the square well potential (SW FGR, full line), together with the NEGF absorption coefficients (spatially averaged over the QW region) for flat band (dotted) and the real potential at $V = 0.8$ V (dashed). While the selection rules are properly reflected in the NEGF approach, the real spectra feature subband tails and deviate from purely 2-D character due to transitions between quasibound states as a consequence of the large built-in fields present in the intrinsic region of the p-i-n diode.

41.4.3 Characteristics

First, the current–voltage characteristics are evaluated on the level of a detailed balance analysis, i.e., photocurrent and radiative dark current are computed from the absorptance as explained in Section 41.3. To this end, the absorptance is determined both from the LB law (a_{LB}) and from TMM (a_{TMM}) using the previously computed absorption coefficient α_{SW}. The photocurrent density can be evaluated at normal incidence. Dark current, on the other hand, should in principle be evaluated under consideration of the finite opening angle of the loss cone, as determined by the condition of total internal reflection. In the case of emission from high to low refractive index medium, which applies in the situation under consideration here ($n_r^{GaAs} = 3.6$ to $n_r^{air} = 1$), the angular dependence may be dropped, i.e., $a(\Omega, E_\gamma) \equiv a(\theta, E_\gamma) \approx a(E_\gamma)$, where the first equality is due to the planar geometry. The angular integration in Equation 41.39 thus provides just a factor of π, and the current–voltage characteristics can be written as

$$J(V) = J_{SC} - J_{0r}\left(e^{qV/(k_B T)} - 1\right),\qquad(41.58)$$

with

$$J_{SC} = -q\int dE_\gamma a(E_\gamma)\Phi_\gamma^{\swarrow}(E_\gamma),\qquad(41.59)$$

$$J_{0r} = -q\int dE_\gamma a(E_\gamma)\pi\Phi_{bb}(E_\gamma).\qquad(41.60)$$

For simplicity, a monochromatic incident photon flux at $E_\gamma = 1.3$ eV and intensity of $I_\gamma \equiv \Phi_\gamma^{\swarrow}(E_\gamma)E_\gamma = 0.1$ kW/cm^2 is used for the evaluation of the characteristics. Figure 41.6 displays the JV curves obtained in this way using the absorptances a_{LB} and a_{TMM} based on α_{SW}, which shows that in this unoptimized situation, the reflector does not increase the absorptance of the QW significantly. Comparison with a 1D numerical DD-Poisson-TMM solver (*ASA*-TU Delft) using the same absorption coefficient α_{SW} and corresponding VRS emission coefficient, but otherwise the transport properties of bulk GaAs ($\mu_{n/p} = 400/40$ cm^2/Vs, $\mathcal{N}_{C/V} = 4.35/7.93 \times 10^{17}$ cm^{-3}, $\varepsilon_r = 13.18$, $S_{min/maj} = 0/10^8$ m/s), reveals—as to be expected—coincidence of the photocurrent with that from a_{TMM}, but also a significant underestimation

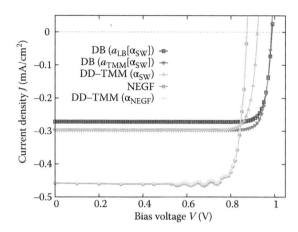

FIGURE 41.6 Current–voltage characteristics as provided by different simulation approaches: detailed balance (DB) currents based on the absorptance from Lambert–Beer's (LB) law or the TMM method, both using the square well potential (SW) absorption coefficient; the semiclassical drift-diffusion-Poisson-TMM model (DD-TMM) for the SW absorption coefficient; the fully microscopic NEGF formalism, revealing a bias-dependent structure in the level of photogeneration. The NEGF characteristics can be reproduced by using in the DD-TMM model the optical data as computed by NEGF, which hints at close to ideal transport in terms of carrier escape from the QW.

of the recombination due to the neglectance of the proper loss cone. The insensitivity to transport is a consequence of the high mobility of GaAs in combination with the short extraction distance and the absence of fast recombination, which results in the flat Fermi levels assumed in the global detailed balance approach. In the full NEGF simulation, on the other hand, both photocurrent and dark current are increased as compared to the characteristics based on the SW absorptance. Remarkably, the NEGF characteristics can be recovered almost completely in the DD-TMM model if local absorption and emission coefficients are extracted from NEGF. This hints at almost perfect transport in terms of unit carrier escape probability (implicitly assumed in the semiclassical model), in consistency with experimental observation for QW structures of similar depth and operated at room temperature. Deviations between the results of the two models are expected in the case of deep wells with presence of fast recombination processes competing with the slow carrier escape: While the NEGF picture considers that a fraction of the photo-generated carriers recombines in the QW prior to escape to the continuum, the semiclassical bulk model assumes the whole generation in bulk states, in which the carriers diffuse quickly from the highly absorbing and—due to detailed balance—emitting QW locations to regions with lower optical coupling strength. That regime, however, is currently not accessible due to the lack of a suitable NEGF picture of nonradiative recombination.

41.5 Summary

The use of NSs in photovoltaics is manifold, and many applications require dedicated models to assess their impact on the PV device performance. There is, however, a common rationale in the consideration of NS in PV device models, according to the device architecture and the NS functionality, which in general amounts to the application of a multiscale and multiphysics scheme. Hence, NS device simulation generally involves many different levels of modeling and simulation, from the determination of NS states to their consideration in the dynamics of carrier generation, transport, and recombination and the final synthesis in the propagation of NS properties to features in the global device characteristics. Reliable consideration of NS effects hence demands a derivation of the key PV material parameters (α, μ, τ) based on a consistent microscopic picture. In the case of ideal transport, an assessment in terms of the optical properties and detailed balance relations between absorption and emission gives already a good estimate of the device performance. For less ideal systems, the individual loss mechanisms need to be considered in more detail, as provided by semiclassical models on device level, down to the microscopic picture of the fundamental physical mechanisms as delivered by quantum-kinetic approaches in the case where nonclassical effects are relevant, such as transport away from diffusive and band-like regimes.

References

1. M. A. Green. Third generation photovoltaics: Ultra-high conversion efficiency at low cost. *Prog. Photovolt. Res. Appl.*, 9:123, 2001.
2. J. Müller, B. Rech, J. Springer, and M. Vanecek. TCO and light trapping in silicon thin film solar cells. *Sol. Energy*, 77(6):917–930, 2004.
3. S. Liu, D. Ding, S. R. Johnson, and Y.-H. Zhang. Approaching single-junction theoretical limit using ultra-thin GaAs solar cells with optimal optical designs. In *38th IEEE Photovoltaic Specialists Conference (PVSC)*, pages 002082–002087, 2012.
4. S. Pillai, K. R. Catchpole, T. Trupke, and M. A. Green. Surface plasmon enhanced silicon solar cells. *J. Appl. Phys.*, 101(9):093105, 2007.
5. I. Massiot, C. Colin, N. Péré-Laperne, P. Roca i Cabarrocas, C. Sauvan, P. Lalanne, J.-L. Pelouard, and S. Collin. Nanopatterned front contact for broadband absorption in ultra-thin amorphous silicon solar cells. *Appl. Phys. Lett.*, 101(16):163901, 2012.
6. J. R. Nagel and M. A. Scarpulla. Enhanced absorption in optically thin solar cells by scattering from embedded dielectric nanoparticles. *Opt. Express*, 18(S2):A139–A146, 2010.
7. C. Heine and R. H. Morf. Submicrometer gratings for solar energy applications. *Appl. Opt.*, 34(14):2476–2482, 1995.
8. P. Bermel, C. Luo, L. Zeng, L. C. Kimerling, and J. D. Joannopoulos. Improving thin-film crystalline silicon solar cell efficiencies with photonic crystals. *Opt. Express*, 15(25):16986–17000, 2007.
9. L. Cao, P. Fan, A. P. Vasudev, J. S. White, Z. Yu, W. Cai, J. A. Schuller, S. Fan, and M. L. Brongersma. Semiconductor nanowire optical antenna solar absorbers. *Nano Lett.*, 10(2):439–445, 2010.
10. K. W. J. Barnham, I. I. Ballard, J. G. Connolly, B. G. Kluftinger N. Ekins-Daukes, J. Nelson, C. Rohr, and M. Mazzer. Recent results on quantum well solar cells. *J. Mater. Sci. Mater. Electron.*, 11:531, 2000.
11. K. R. Catchpole and A. Polman. Plasmonic solar cells. *Opt. Express*, 16(26):21793–21800, 2008.
12. H. A. Atwater and A. Polman. Plasmonics for improved photovoltaic devices. *Nat. Mater.*, 9(3):205–213, 2010.
13. K. R. Catchpole, S. Mokkapati, F. Beck, E.-C. Wang, A. McKinley, A. Basch, and J. Lee. Plasmonics and nanophotonics for photovoltaics. *MRS Bull.*, 36:461–467, 2011.
14. S. Mokkapati and K. R. Catchpole. Nanophotonic light trapping in solar cells. *J. Appl. Phys.*, 112(10):101101, 2012.
15. K. W. J. Barnham and G. Duggan. A new approach to high-efficiency multi-band-gap solar cells. *J. Appl. Phys.*, 67:3490–3493, 1990.
16. D. B. Bushnell, K. W. J. Barnham, J. P. Connolly, M. Mazzer, N. J. Ekins-Daukes, J. S. Roberts, G. Hill, R. Airey, and L. Nasi. Effect of barrier composition and well number on the dark current of quantum well solar cells. In *Proceeding of 3rd World Conference on Photovoltaic Energy Conversion*, Osaka, Japan, page 2709, 2003.
17. L. C. Hirst, M. Fuhrer, D. J. Farrell, A. Le Bris, J.-F. Guillemoles, M. J. Y. Tayebjee, R. Clady, T. W. Schmidt, M. Sugiyama, Y. Wang, H. Fujii, and N. J. Ekins-Daukes. InGaAs/GaAsP quantum wells for hot carrier solar cells. *Proc. SPIE.*, 8256:82560X, 2012.
18. R. Oshima, A. Takata, and Y. Okada. Strain-compensated InAs/GaNAs quantum dots for use in high-efficiency solar cells. *Appl. Phys. Lett.*, 93(8):083111, 2008.
19. S. M. Hubbard, C. D. Cress, C. G. Bailey, R. P. Raffaelle, S. G. Bailey, and D. M. Wilt. Effect of strain compensation on quantum dot enhanced GaAs solar cells. *Appl. Phys. Lett.*, 92(12):123512, 2008.

20. Y. Okada, R. Oshima, and A. Takata. Characteristics of InAs/GaNAs strain-compensated quantum dot solar cell. *J. Appl. Phys.*, 106(2):024306, 2009.

21. C. G. Bailey, D. V. Forbes, R. P. Raffaelle, and S. M. Hubbard. Near 1 V open circuit voltage InAs/GaAs quantum dot solar cells. *Appl. Phys. Lett.*, 98(16):163105, 2011.

22. G. Conibeer, M. Green, E.-C. Cho, D. König, Y.-H Cho, T. Fangsuwannarak, G. Scardera, E. Pink, Y. Huang, T. Puzzer, S. Huang, D. Song, C. Flynn, S. Park, X. Hao, and D. Mansfield. Silicon quantum dot nanostructures for tandem photovoltaic cells. *Thin Solid Films*, 516(20):6748–6756, 2008.

23. J. G. J. Adams, B. C. Browne, I. M. Ballard, J. P. Connolly, N. L. A. Chan, A. Ioannides, W. Elder, P. N. Stavrinou, K. W. J. Barnham, and N. J. Ekins-Daukes. Recent results for single-junction and tandem quantum well solar cells. *Prog. Photovolt. Res. Appl.*, 19(7):865–877, 2011.

24. D. Alonso-Álvarez, A. G. Taboada, J. M. Ripalda, B. Alen, Y. González, L. González, J. M. García, F. Briones, A. Martí, A. Luque, A. M. Sánchez, and S. I. Molina. Carrier recombination effects in strain compensated quantum dot stacks embedded in solar cells. *Appl. Phys. Lett.*, 93(12):123114, 2008.

25. A. Martí, L. Cuadra, and A. Luque. Partial filling of a quantum dot intermediate band for solar cells. *IEEE Trans. Electron Dev.*, 48(10):2394–2399, 2001.

26. A. J. Nozik. Quantum dot solar cells. *Physica E*, 14(1–2):PII S1386–9477(02)00374–0, 2002.

27. R. Patterson, M. Kirkengen, B. Puthen Veettil, D. König, M. A. Green, and G. Conibeer. Phonon lifetimes in model quantum dot superlattice systems with applications to the hot carrier solar cell. *Sol. Energy Mater. Sol. Cells*, 94(11):1931–1935, 2010.

28. V. I. Klimov. Detailed-balance power conversion limits of nanocrystal-quantum-dot solar cells in the presence of carrier multiplication. *Appl. Phys. Lett.*, 89(12):123118, 2006.

29. O. E. Semonin, J. M. Luther, S. Choi, H.-Y. Chen, J. Gao, A. J. Nozik, and M. C. Beard. Peak external photocurrent quantum efficiency exceeding 100% via MEG in a quantum dot solar cell. *Science*, 334(6062):1530–1533, 2011.

30. A. Martí, E. Antolín, C. R. Stanley, C. D. Farmer, N. López, P. Diaz, E. Cánovas, P. G. Linares, and A. Luque. Production of photocurrent due to intermediate-to-conduction-band transitions: A demonstration of a key operating principle of the intermediate-band solar cell. *Phys. Rev. Lett.*, 97(24):247701, 2006.

31. P. G. Linares, A. Martí, E. Antolín, and A. Luque. III-V compound semiconductor screening for implementing quantum dot intermediate band solar cells. *J. Appl. Phys.*, 109(1):014313, 2011.

32. L. C. Hirst, R. J. Walters, M. F. Fuehrer, and N. J. Ekins-Daukes. Experimental demonstration of hot-carrier photo-current in an InGaAs quantum well solar cell. *Appl. Phys. Lett.*, 104(23):231115, 2014.

33. J. A. R. Dimmock, S. Day, M. Kauer, K. Smith, and J. Heffernan. Demonstration of a hot-carrier photovoltaic cell. *Prog. Photovolt. Res. Appl.*, 22(2):151–160, 2014.

34. F. W. Wise. Lead salt quantum dots: The limit of strong quantum confinement. *Acc. Chem. Res.*, 33(11):773–780, 2000.

35. A. J. Nozik, M. C. Beard, J. M. Luther, M. Law, R. J. Ellingson, and J. C. Johnson. Semiconductor quantum dots and quantum dot arrays and applications of multiple exciton generation to third-generation photovoltaic solar cells. *Chem. Rev.*, 110(11):6873–6890, 2010.

36. Y. Kanai, Z. Wu, and J. C. Grossman. Charge separation in nanoscale photovoltaic materials: Recent insights from first-principles electronic structure theory. *J. Mater. Chem.*, 20(6):1053, 2010.

37. A. Ruland, C. Schulz-Drost, V. Sgobba, and D. M. Guldi. Enhancing photocurrent efficiencies by resonance energy transfer in CdTe quantum dot multilayers: Towards rainbow solar cells. *Adv. Mater.*, 23(39):4573–7, 2011.

38. I. J. Kramer and E. H. Sargent. The architecture of colloidal quantum dot solar cells: Materials to devices. *Chem. Rev.*, 114(1):863–882, 2014.

39. R. J. Ellingson, M. C. Beard, J. C. Johnson, P. Yu, O. I. Micic, A. J. Nozik, A. Shabaev, and A. L. Efros. Highly efficient multiple exciton generation in colloidal PbSe and PbS quantum dots. *Nano Lett.*, 5(5):865–871, 2005.

40. D. König, Y. Takeda, and B. Puthen-Veettil. Technology-compatible hot carrier solar cell with energy selective hot carrier absorber and carrier-selective contacts. *Appl. Phys. Lett.*, 101(15), 2012.

41. D. Bozyigit, N. Yazdani, M. Yarema, O. Yarema, W. M. M. Lin, S. Volk, K. Vuttivorakulchai, M. Luisier, F. Juranyi, and V. Wood. Soft surfaces of nanomaterials enable strong phonon interactions. *Nature*, 531(7596):618–622, 2016.

42. W. Shockley and H. J. Queisser. Detailed balance limit of efficiency of p-n junction solar cells. *J. Appl. Phys.*, 32(3):510–519, 1961.

43. R. E. Chandler, A. J. Houtepen, J. Nelson, and D. Vanmaekelbergh. Electron transport in quantum dot solids: Monte Carlo simulations of the effects of shell filling, Coulomb repulsions, and site disorder. *Phys. Rev. B*, 75(8):085325, 2007.

44. P. Guyot-Sionnest. Electrical transport in colloidal quantum dot films. *J. Phys. Chem. Lett.*, 3(9):1169–1175, 2012.

45. D. Beljonne and J. Cornil, editors. *Multiscale Modelling of Organic and Hybrid Photovoltaics*. Berlin: Springer, Berlin, Heidelberg, 2014.

46. J. D Jackson. *Classical Electrodynamics*. Hoboken, NJ: John Wiley & Sons, 1998.

47. P. Yeh, A. Yariv, and C. S. Hong. Electromagnetic propagation in periodic stratified media.1. General theory. *J. Opt. Soc. Am.*, 67(4):423–438, 1977.

48. W. Schäfer and M. Wegener. *Semiconductor Optics and Transport Phenomena*. Berlin: Springer, 2002.

49. H. Haug and A. P. Jauho. *Quantum Kinetics in Transport and Optics of Semiconductors*. Berlin: Springer, 1996.

50. H. Haug and S. W. Koch. *Quantum Theory of the Optical and Electronic Properties of Semiconductors*. Singapore: World Scientific, 2004.

51. U. Aeberhard. Quantum-kinetic theory of photocurrent generation via direct and phonon-mediated optical transitions. *Phys. Rev. B*, 84(3):035454, 2011.

52. G. Allan and C. Delerue. Efficient intraband optical transitions in Si nanocrystals. *Phys. Rev. B*, 66(23):233303, 2002.

53. J.-W. Luo, P. Stradins, and A. Zunger. Matrix-embedded silicon quantum dots for photovoltaic applications: A theoretical study of critical factors. *Energy Environ. Sci.*, 4(7):2546, 2011.

54. D. A. B. Miller, D. S. Chemla, T. C. Damen, A. C. Gossard, W. Wiegmann, T. H. Wood, and C. A. Burrus. Electric field dependence of optical absorption near the band gap of quantum-well structures. *Phys. Rev. B*, 32:1043–1060, 1985.

55. S.-L. Chuang, S. Schmitt-Rink, D. A. B. Miller, and D. S. Chemla. Exciton Green's-function approach to optical absorption in a quantum well with an applied electric field. *Phys. Rev. B*, 43:1500–1509, 1991.

56. U. Aeberhard. Nonequilibrium Green's function theory of coherent excitonic effects in the photocurrent response of semiconductor nanostructures. *Phys. Rev. B*, 86:115317, 2012.

57. M. Paxman, J. Nelson, B. Braun, J. Connolly, K. W. J. Barnham, C. T. Foxon, and J. S. Roberts. Modeling the spectral response of the quantum well solar cell. *J. Appl. Phys.*, 74:614, 1993.

58. P. J. Stevens, M. Whitehead, G. Parry, and K. Woodbridge. Computer modeling of the electric field dependent absorption spectrum of multiple quantum well material. *IEEE J. Quant. Electron.*, 24:2007, 1988.

59. D. Atkinson, G. Parry, and E. J. Austin. Modelling of electroabsorption in coupled quantum wells with applications to voltage optical modulation. *Semicond. Sci. Technol.*, 5:516, 1990.

60. S. M. Ramey and R. Khoie. Modeling of multiple-quantum-well solar cells including capture, escape, and recombination of photoexcited carriers in quantum wells. *IEEE Trans. Electron Dev.*, 50:1179, 2003.

61. G. Lengyel, K. W. Jelley, and R. W. H. Engelmann. A semi-empirical model for electroabsorption in GaAs/AlGaAs multiple quantum well modulator structures. *IEEE J. Quant. Electron.*, 26(2):296–304, 1990.

62. S. Schmitt-Rink, D. A. B. Miller, and D. S. Chemla. Theory of the linear and nonlinear optical properties of semiconductor microcrystallites. *Phys. Rev. B*, 35:8113–8125, 1987.

63. D. L. Nika, E. P. Pokatilov, Q. Shao, and A. A. Balandin. Charge-carrier states and light absorption in ordered quantum dot superlattices. *Phys. Rev. B*, 76(12):125417, 2007.

64. S. Tomić, T. S. Jones, and N. M. Harrison. Absorption characteristics of a quantum dot array induced intermediate band: Implications for solar cell design. *Appl. Phys. Lett.*, 93(26):263105, 2008.

65. Y. Z. Hu, M. Lindberg, and S. W. Koch. Theory of optically excited intrinsic semiconductor quantum dots. *Phys. Rev. B*, 42(3):1713–1723, 1990.

66. W.-Y. Wu, J. N. Schulman, T. Y. Hsu, and U. Efron. Effect of size nonuniformity on the absorption spectrum of a semiconductor quantum dot system. *Appl. Phys. Lett.*, 51(10):710–712, 1987.

67. A. Vasanelli, R. Ferreira, and G. Bastard. Continuous absorption background and decoherence in quantum dots. *Phys. Rev. Lett.*, 89:216804, 2002.

68. A. W. Walker, O. Thériault, J. F. Wheeldon, and K. Hinzer. The effects of absorption and recombination on quantum dot multijunction solar cell efficiency. *IEEE J. Photovolt.*, 3(3):1118–1124, 2013.

69. R. Riölver, B. Berghoff, D. Bätzner, B. Spangenberg, H. Kurz, M. Schmidt, and B. Stegemann. Si/SiO2 multiple quantum wells for all silicon tandem cells: Conductivity and photocurrent measurements. *Thin Solid Films*, 516(20):6763–6766, 2008.

70. M. Zacharias, J. Heitmann, R. Scholz, U. Kahler, M. Schmidt, and J. Bläsing. Size-controlled highly luminescent silicon nanocrystals: A SiO/SiO2 superlattice approach. *Appl. Phys. Lett.*, 80(4):661–663, 2002.

71. J. Nelson, M. Paxman, K. W. J. Barnham, J. S. Roberts, and C. Button. Steady state carrier escape rates from single quantum wells. *IEEE J. Quant. Electron.*, 29:1460–1468, 1993.

72. O. Y. Raisky, W. B. Wang, R. R. Alfano, C. L. Reynolds Jr., and V. Swaminathan. Investigation of photoluminescence and photocurrent in InGaAsP/InP strained multiple quantum well heterostructures. *J. Appl. Phys.*, 81:339, 1997.

73. K. R. Lefebvre and A. F. M. Anwar. Electron and hole escape times in single quantum wells. *J. Appl. Phys.*, 80(6):3595–3597, 1996.

74. K. R. Lefebvre and A. F. M. Anwar. Electron escape time from single quantum wells. *IEEE J. Quant. Electron.*, 33(2):187–191, 1997.

75. H. Schneider and K. V. Klitzing. Thermionic emission and Gaussian transport of holes in a GaAs/AlxGa1−xAs multiple-quantum-well structure. *Phys. Rev. B*, 38(9):6160, 1988.

76. A. Larsson, P. A. Andrekson, S. T. Eng, and A. Yariv. Tunable superlattice p-i-n photodetectors: characteristics, theory, and application. *IEEE J. Quant. Electron.*, 24(5):787, 1988.

77. D. J. Moss, T. Ido, and H. Sano. Calculation of photogenerated carrier escape rates from GaAs/AlxGa1−xAs quantum wells. *IEEE J. Quant. Electron.*, 30(4):1015, 1994.

78. C.-Y. Tsai, L. F. Eastman, Y.-H. Lo, and C.-Y. Tsai. Breakdown of thermionic emission theory for quantum wells. *Appl. Phys. Lett.*, 65(4):469–471, 1994.

79. U. Aeberhard. Microscopic theory and numerical simulation of quantum well solar cells. *Proc. SPIE*, 7597:759702, 2010.

80. P. N. Brounkov, A. A. Suvorova, M. V. Maximov, A. F. Tsatsul'nikov, A. E. Zhukov, A. Yu Egorov, A. R. Kovsh, S. G. Konnikov, T. Ihn, S. T. Stoddart, L. Eaves, and P. C. Main. Electron escape from self-assembled InAs/GaAs quantum dot stacks. *Physica B*, 249–251:267–270, 1998.

81. C. M. A. Kapteyn, F. Heinrichsdorff, O. Stier, R. Heitz, M. Grundmann, N. D. Zakharov, D. Bimberg, and P. Werner. Electron escape from InAs quantum dots. *Phys. Rev. B*, 60:14265–14268, 1999.

82. P. W. Fry, J. J. Finley, L. R. Wilson, A. Lemaitre, D. J. Mowbray, M. S. Skolnick, M. Hopkinson, G. Hill, and J. C. Clark. Electric-field-dependent carrier capture and escape in self-assembled InAs/GaAs quantum dots. *Appl. Phys. Lett.*, 77(26):4344–4346, 2000.

83. S. Marcinkevius and R. Leon. Carrier capture and escape in InxGa1−xAs/GaAs quantum dots: Effects of intermixing. *Phys. Rev. B*, 59:4630–4633, 1999.

84. M. De Giorgi, C. Lingk, G. von Plessen, J. Feldmann, S. De Rinaldis, A. Passaseo, M. De Vittorio, R. Cingolani, and M. Lomascolo. Capture and thermal re-emission of carriers in long-wavelength InGaAs/GaAs quantum dots. *Appl. Phys. Lett.*, 79(24):3968–3970, 2001.

85. J. A. Brum and G. Bastard. Resonant carrier capture by semiconductor quantum wells. *Phys. Rev. B*, 33:1420–1423, 1986.

86. P. W. M. Blom, J. E. M. Haverkort, and J. H. Wolter. Optimization of barrier thickness for efficient carrier capture in graded-index and separate-confinement multiple quantum well lasers. *Appl. Phys. Lett.*, 58(24):2767–2769, 1991.

87. S. C. Kan, D. Vassilovski, T. C. Wu, and K. Y. Lau. Quantum capture and escape in quantum-well lasers—Implications on direct modulation bandwidth limitations. *IEEE Photonic. Technol. Lett.*, 4(5):428–431, 1992.

88. P. Sotirelis, P. von Allmen, and K. Hess. Electron intersubband relaxation in doped quantum wells. *Phys. Rev. B*, 47:12744–12753, 1993.

89. P. Sotirelis and K. Hess. Electron capture in GaAs quantum wells. *Phys. Rev. B*, 49:7543–7547, 1994.

90. B. K. Ridley. Space-charge-mediated capture of electrons and holes in a quantum well. *Phys. Rev. B*, 50:1717–1724, 1994.

91. C.-Y. Tsai, L. F. Eastman, Y.-H. Lo, and C.-Y. Tsai. Carrier capture and escape in multisubband quantum well lasers. *IEEE Photonic. Technol. Lett.*, 6(9):1088–1090, 1994.

92. L. Thibaudeau and B. Vinter. Phonon-assisted carrier capture into a quantum well in an electric field. *Appl. Phys. Lett.*, 65(16):2039–2041, 1994.

93. D. Bradt, Y. M. Sirenko, and V. Mitin. Inelastic and elastic mechanisms of electron capture to a quantum well. *Semicond. Sci. Technol.*, 10(3):260, 1995.

94. K. Kalna and M. Mosko. Electron capture in quantum wells via scattering by electrons, holes, and optical phonons. *Phys. Rev. B*, 54:17730–17737, 1996.

95. M. Mosko and K. Kalna. Carrier capture into a GaAs quantum well with a separate confinement region: Comment on quantum and classical aspects. *Semicond. Sci. Technol.*, 14(9):790, 1999.

96. J. A. Brum and G. Bastard. Direct and indirect carrier capture by semiconductor quantum wells. *Superlattice. Microst.*, 3(1):51–55, 1987.

97. P. W. M. Blom, C. Smit, J. E. M. Haverkort, and J. H. Wolter. Carrier capture into a semiconductor quantum well. *Phys. Rev. B*, 47(4):2072–2081, 1993.

98. C.-Y. Tsai, C.-Y. Tsai, Y.-H. Lo, and L. F. Eastman. Carrier dc and ac capture and escape times in quantum-well lasers. *IEEE Photonic. Technol. Lett.*, 7(6):599–601, 1995.

99. C.-Y. Tsai, C.-Y. Tsai, Y.-H. Lo, R. M. Spencer, and L. F. Eastman. Nonlinear gain coefficients in semiconductor quantum-well lasers: Effects of carrier diffusion, capture, and escape. *IEEE J. Sel. Top. Quant. Electron.*, 1(2):316–330, 1995.

100. Y. Lam and J. Singh. Monte Carlo studies on the well-width dependence of carrier capture time in graded-index separate confinement heterostructure quantum well laser structures. *Appl. Phys. Lett.*, 63(14):1874–1876, 1993.

101. L. Davis, Y. L. Lam, Y. C. Chen, J. Singh, and P. K. Bhattacharya. Carrier capture and relaxation in narrow quantum wells. *IEEE J. Quant. Electron.*, 30(11):2560–2564, 1994.

102. L. F. Register and K. Hess. Simulation of carrier capture in quantum well lasers due to strong inelastic scattering. *Superlattice. Microst.*, 18(3):223–228, 1995.

103. G. C. Crow and R. A. Abram. Monte Carlo simulations of capture into quantum well structures. *Semicond. Sci. Technol.*, 14(1):1, 1999.

104. M. Ryzhii and V. Ryzhii. Monte Carlo modeling of electron transport and capture processes in AlGaAs/GaAs multiple quantum well infrared photodetectors. *Jpn. J. Appl. Phys.*, 38(10R):5922, 1999.

105. J. Hader, J. V. Moloney, and S. W. Koch. Structural dependence of carrier capture time in semiconductor quantum-well lasers. *Appl. Phys. Lett.*, 85(3):369–371, 2004.

106. A. V. Uskov, J. McInerney, F. Adler, H. Schweizer, and M. H. Pilkuhn. Auger carrier capture kinetics in self-assembled quantum dot structures. *Appl. Phys. Lett.*, 72(1):58–60, 1998.

107. I. Magnusdottir, S. Bischoff, A. V. Uskov, and J. Mørk. Geometry dependence of Auger carrier capture rates into cone-shaped self-assembled quantum dots. *Phys. Rev. B*, 67:205326, 2003.

108. A. V. Uskov, I. Magnusdottir, B. Tromborg, J. Mørk, and R. Lang. Line broadening caused by Coulomb carrier-carrier correlations and dynamics of carrier capture and emission in quantum dots. *Appl. Phys. Lett.*, 79(11):1679–1681, 2001.

109. R. Ferreira and G. Bastard. Phonon-assisted capture and intradot Auger relaxation in quantum dots. *Appl. Phys. Lett.*, 74(19):2818–2820, 1999.

110. J. M. Miloszewski, M. S. Wartak, S. G. Wallace, and S. Fafard. Theoretical investigation of carrier capture and escape processes in cylindrical quantum dots. *J. Appl. Phys.*, 114(15):154311, 2013.

111. J.-Z. Zhang and I. Galbraith. Rapid hot-electron capture in self-assembled quantum dots via phonon processes. *Appl. Phys. Lett.*, 89(15):153119, 2006.

112. I. Magnusdottir, A. V. Uskov, R. Ferreira, G. Bastard, J. Mørk, and B. Tromborg. Influence of quasibound states on the carrier capture in quantum dots. *Appl. Phys. Lett.*, 81(23):4318–4320, 2002.

113. T. R. Nielsen, P. Gartner, and F. Jahnke. Many-body theory of carrier capture and relaxation in semiconductor quantum-dot lasers. *Phys. Rev. B*, 69:235314, 2004.

114. J. Seebeck, T. R. Nielsen, P. Gartner, and F. Jahnke. Polarons in semiconductor quantum dots and their role in the quantum kinetics of carrier relaxation. *Phys. Rev. B*, 71:125327, 2005.

115. G. D. Mahan. *Many-Particle Physics*, 2nd ed. New York, NY: Plenum, 1990.

116. T. Gunst, T. Markussen, K. Stokbro, and M. Brandbyge. First-principles method for electron-phonon coupling and electron mobility: Applications to two-dimensional materials. *Phys. Rev. B*, 93:035414, 2016.

117. C.-W. Jiang and M. A. Green. Silicon quantum dot superlattices: Modeling of energy bands, densities of states, and mobilities for silicon tandem solar cell applications. *J. Appl. Phys.*, 99(11):114902, 2006.

118. A. Wacker. Semiconductor superlattices: A model system for nonlinear transport. *Phys. Rep.*, 357:1–111, 2002.

119. P. Würfel. *Physics of Solar Cells: From Basic Principles to Advanced Concepts*. Weinheim: Wiley-VCH, 2009.

120. I. Mora-Sero, L. Bertoluzzi, V. Gonzalez-Pedro, S. Gimenez, F. Fabregat-Santiago, K. W. Kemp, E. H. Sargent, and J. Bisquert. Selective contacts drive charge extraction in quantum dot solids via asymmetry in carrier transfer kinetics. *Nat. Commun.*, 4:2272, 2013.

121. W. Van Roosbroeck and W. Shockley. Photon-radiative recombination of electrons and holes in germanium. *Phys. Rev.*, 94:1558, 1954.

122. E. O. Göbel, H. Jung, J. Kuhl, and K. Ploog. Recombination enhancement due to carrier localization in quantum well structures. *Phys. Rev. Lett.*, 51(17):1588–1591, 1983.

123. A. R. Beattie and P. T. Landsberg. Auger effect in semiconductors. *Proc. R. Soc. London Ser. A*, 249:16, 1959.

124. P. T. Landsberg. *Recombination in Semiconductors*. Cambridge: Cambridge University Press, 1991.

125. L. Chiu and A. Yariv. Auger recombination in quantum-well InGaAsP heterostructure lasers. *IEEE J. Quant. Electron.*, 18(10):1406–1409, 1982.

126. C. Smith, R. A. Abram, and M. G. Burt. Auger recombination in a quantum well heterostructure. *J. Phys. C Solid State Phys.*, 16(5):L171, 1983.

127. N. K. Dutta. Calculation of Auger rates in a quantum well structure and its application to InGaAsP quantum well lasers. *J. Appl. Phys.*, 54(3):1236–1245, 1983.

128. A. Sugimura. Auger recombination effect on threshold current of InGaAsP quantum well lasers. *IEEE J. Quant. Electron.*, 19(6):932–941, 1983.

129. P. K. Basu. Auger recombination rate in quantum well lasers: Modification by electronelectron interaction in quasi two dimensions. *J. Appl. Phys.*, 56(11):3344–3346, 1984.

130. A. Haug. Phonon-assisted auger recombination in quantum well semiconductors. *Appl. Phys. A*, 51:354, 1990.

131. M. Takeshima. Phonon-assisted Auger recombination in a quasi-two-dimensional structure semiconductor. *Phys. Rev. B*, 30(6):3302–3308, 1984.

132. L.-W. Wang, M. Califano, A. Zunger, and A. Franceschetti. Pseudopotential theory of auger processes in CdSe quantum dots. *Phys. Rev. Lett.*, 91:056404, 2003.

133. G. A. Narvaez, G. Bester, and A. Zunger. Carrier relaxation mechanisms in self-assembled (In,Ga)As/GaAs quantum dots: Efficient $P \rightarrow S$ Auger relaxation of electrons. *Phys. Rev. B*, 74:075403, 2006.

134. S. Tomić. Intermediate-band solar cells: Influence of band formation on dynamical processes in InAs/GaAs quantum dot arrays. *Phys. Rev. B*, 82(19):195321, 2010.

135. K. Hyeon-Deuk and O. V. Prezhdo. Time-domain ab initio study of Auger and phonon-assisted auger processes in a semiconductor quantum dot. *Nano Lett.*, 11(4):1845–1850, 2011.

136. N. J. Ekins-Daukes, K. W. J. Barnham, J. P. Connolly, J. S. Roberts, J. C. Clark, G. Hill, and M. Mazzer. Strain-balanced GaAsP/InGaAs quantum well solar cells. *Appl. Phys. Lett.*, 75(26):4195–4197, 1999.

137. K. Ding, U. Aeberhard, O. Astakhov, U. Breuer, M. Beigmohamadi, S. Suckow, B. Berghoff, W. Beyer, F. Finger, R. Carius, and U. Rau. Defect passivation by hydrogen reincorporation for silicon quantum dots in SiC/SiO$x$hetero-superlattice. *J. Non-Cryst. Solids*, 358(17):2145–2149, 2012.

138. W. Shockley and W. T. Read. Statistics of the recombinations of holes and electrons. *Phys. Rev.*, 87:835–842, 1952.

139. R. N. Hall. Electron-hole recombination in germanium. *Phys. Rev.*, 87:387, 1952.

140. A. Schenk. An improved approach to the Shockley-Read-Hall recombination in inhomogeneous fields of space-charge regions. *J. Appl. Phys.*, 71(7):3339–3349, 1992.

141. A. Alkauskas, Q. Yan, and C. G. Van de Walle. First-principles theory of nonradiative carrier capture via multiphonon emission. *Phys. Rev. B*, 90:075202, 2014.

142. U. Aeberhard. Simulation of nanostructure-based high-efficiency solar cells: Challenges, existing approaches, and future directions. *IEEE J. Sel. Top. Quant. Electron*, 19(5):4000411, 2013.

143. G. L. Araújo and A. Martí. Absolute limiting efficiencies for photovoltaic energy conversion. *Sol. Energy Mater. Sol. Cells*, 33:213, 1994.

144. U. Rau. Reciprocity relation between photovoltaic quantum efficiency and electroluminescent emission of solar cells. *Phys. Rev. B*, 76(8):085303, 2007.

145. M. Born and E. Wolf. *Principles of Optics*, 7th ed. Cambridge: Cambridge University Press, 1999.

146. P. Würfel. The chemical potential of radiation. *J. Phys. C Solid State Phys.*, 15:3967, 1982.

147. U. Aeberhard. Simulation of ultra-thin solar cells beyond the limits of the semi-classical bulk picture. *IEEE J. Photovolt.*, 6(3):654–660, 2016.

148. U. Aeberhard. Impact of built-in fields and contact configuration on the characteristics of ultra-thin GaAs solar cells. *Appl. Phys. Lett.*, 109(3), 2016.

149. N. G. Anderson. Ideal theory of quantum well solar cells. *J. Appl. Phys.*, 78:1861, 1995.

150. G. L. Araújo, A. Martí, F. W. Ragay, and J. H. Wolter. Efficiency of multiple quantum well solar cells. In *Proceeding of 12th European Photovoltaic Solar Energy Conference*, Amsterdam, The Netherlands, page 1481, 1994.

151. G. L. Araújo and A. Martí. Electroluminescence coupling in multiple quantum well diodes and solar cells. *Appl. Phys. Lett.*, 66(7):894–895, 1995.

152. S. P. Bremner, R. Corkish, and C. B. Honsberg. Detailed balance efficiency limits with quasi-Fermi level variations. *IEEE Trans. Electron Dev.*, 46(10):1932–1939, 1999.

153. A. Luque, A. Martí, and L Cuadra. Thermodynamic consistency of sub-bandgap absorbing solar cell proposals. *IEEE Trans. Electron Dev.*, 48:2118, 2001.

154. J. P. Connolly, J. Nelson, K. W. J. Barnham, I. Ballard, C. Roberts, J. S. Roberts, and C. T. Foxon. Simulating multiple quantum well solar cells. In *Proceeding of 28th IEEE Photovoltaic Specialists Conference*, Anchorage, Alaska, USA, page 1304, 2000.

155. J. C. Rimada and L. Hernández. Modelling of ideal AlGaAs quantum well solar cells. *Microelectron. J.*, 32(9):719–723, 2001.

156. S. J. Lade and A. Zahedi. A revised ideal model for AlGaAs/GaAs quantum well solar cells. *Microelectron. J.*, 35(5):401–410, 2004.

157. J. C. Rimada, L. Hernandez, J. P. Connolly, and K. W. J. Barnham. Quantum and conversion efficiency calculation of AlGaAs/GaAs multiple quantum well solar cells. *Phys. Status Solidi B*, 242(9):1842–1845, 2005.

158. J. C. Rimada, L. Hernández, J. P. Connolly, and K. W J Barnham. Conversion efficiency enhancement of AlGaAs quantum well solar cells. *Microelectron. J.*, 38(4–5):513–518, 2007.

159. M. Courel, J. C. Rimada, and L. Hernández. GaAs/GaInNAs quantum well and superlattice solar cell. *Appl. Phys. Lett.*, 100(7):073508, 2012.

160. C. I. Cabrera, J. C. Rimada, M. Courel, L. Hernandez, J. P. Connolly, A. Enciso, and D. A. Contreras-Solorio. Modeling multiple quantum well and superlattice solar cells. *Nat. Resour.*, 4(03):235–245, 2013.

161. A. Chatterjee, A. Kumar Biswas, and A. Sinha. An analytical study of the various current components of an AlGaAs/GaAs multiple quantum well solar cell. *Physica E*, 72:128–133, 2015.

162. M. Courel, J. C. Rimada, and L. Hernandez. An approach to high efficiencies using GaAs/GaInNAs multiple quantum well and superlattice solar cell. *J. Appl. Phys.*, 112(5):054511, 2012.

163. A. Alemu and A. Freundlich. Resonant thermotunneling design for high-performance single-junction quantum-well solar cells. *IEEE J. Photovolt.*, 2(3):256–260, 2012.

164. M. Courel, J. C. Rimada, and L. Hernández. AlGaAs/GaAs superlattice solar cells. *Prog. Photovolt. Res. Appl.*, 21(3):276–282, 2013.

165. A. Luque and A. Martí. Increasing the efficiency of ideal solar cells by photon induced transitions at intermediate levels. *Phys. Rev. Lett.*, 78(26):5014–5017, 1997.

166. Z. Wang, T. P. White, and K. R. Catchpole. Plasmonic near-field enhancement for planar ultra-thin photovoltaics. *IEEE Photon. J.*, 5(5):8400608, 2013.

167. W. Hu, M. Maksudur Rahman, M.-Y. Lee, Y. Li, and S. Samukawa. Simulation study of type-II Ge/Si quantum dot for solar cell applications. *J. Appl. Phys.*, 114(12):124509, 2013.

168. Q. Shao, A. A. Balandin, A. I. Fedoseyev, and M. Turowski. Intermediate-band solar cells based on quantum dot supracrystals. *Appl. Phys. Lett.*, 91(16):163503, 2007.

169. Q. Deng, X. Wang, H. Xiao, C. Wang, H. Yin, H. Chen, Q. Hou, D. Lin, J. Li, Z. Wang, and X. Hou. An investigation on $In_xGa_{1-x}N/GaN$ multiple quantum well solar cells. *J. Phys. D Appl. Phys.*, 44(26):265103, 2011.

170. W. Wei, Q. Zhang, S. Zhao, and Y. Zhang. Two intermediate bands solar cells of InGaN/InN quantum dot supracrystals. *Appl. Phys. A*, 116(3):1009–1016, 2014.

171. A. Mellor, A. Luque, I. Tobias, and A. Marti. Realistic detailed balance study of the quantum efficiency of quantum dot solar cells. *Adv. Funct. Mater.*, 24(3):339–345, 2014.

172. G. Wei, K.-T. Shiu, N. C. Giebink, and S. R. Forrest. Thermodynamic limits of quantum photovoltaic cell efficiency. *Appl. Phys. Lett.*, 91(22), 2007.

173. J. Nelson, I. Ballard, K. Barnham, J. P. Connolly, J. S. Roberts, and M. Pate. Effect of quantum well location on single quantum well p-i-n photodiode dark currents. *J. Appl. Phys.*, 86:5898, 1999.

174. P. Kailuweit, R. Kellenbenz, S. P. Philipps, W. Guter, A. W. Bett, and F. Dimroth. Numerical simulation and modeling of GaAs quantum-well solar cells. *J. Appl. Phys.*, 107(6):064317, 2010.

175. T. Kirchartz, K. Seino, J.-M. Wagner, U. Rau, and F. Bechstedt. Efficiency limits of Si/SiO2 quantum well solar cells from first-principles calculations. *J. Appl. Phys.*, 105(10):104511, 2009.

176. C.-Y. Tsai and C.-Y. Tsai. Effects of carrier escape and capture processes on quantum well solar cells: A theoretical investigation. *IET Optoelectron.*, 3(6):300–304, 2009.

177. O. Kengradomying, S. Jiang, Q. Wang, N. Vogiatzis, and J. M. Rorison. Modelling escape and capture processes in GaInNAs quantum well solar cells. *Phys. Status Solidi C Curr. Top. Solid State Phys.*, 10(4):585–588, 2013.

178. J. Nelson, K. W. J Barnham, J. P Connolly, G. Haarpaintner, C. Button, and J. Roberts. Quantum well solar cell dark currents. In *Proceeding of 12th EU PVSEC*, Amsterdam, The Netherlands, 1994.

179. M. Gioannini, A. P. Cedola, N. Di Santo, F. Bertazzi, and F. Cappelluti. Simulation of quantum dot solar cells including carrier intersubband dynamics and transport. *IEEE J. Photovolt.*, 3(4):1271–1278, 2013.

180. K. Driscoll, M. F. Bennett, S. J. Polly, D. V. Forbes, and S. M. Hubbard. Effect of quantum dot position and background doping on the performance of quantum dot enhanced GaAs solar cells. *Appl. Phys. Lett.*, 104(2):023119, 2014.

181. K. Yoshida, Y. Okada, and N. Sano. Self-consistent simulation of intermediate band solar cells: Effect of occupation rates on device characteristics. *Appl. Phys. Lett.*, 97(13):133503, 2010.

182. Y.-X. Gu, X.-G. Yang, H.-M. Ji, P.-F. Xu, and T. Yang. Theoretical study of the effects of InAs/GaAs quantum dot layer's position in i-region on current-voltage characteristic in intermediate band solar cells. *Appl. Phys. Lett.*, 101(8):081118, 2012.

183. T. Sogabe, T. Kaizu, Y. Okada, and S. Tomić. Theoretical analysis of GaAs/AlGaAs quantum dots in quantum wire array for intermediate band solar cell. *J. Renew. Sustain. Ener.*, 6(1):011206, 2014.

184. V. Aroutiounian, S. Petrosyan, A. Khachatryan, and K. Touryan. Quantum dot solar cells. *J. Appl. Phys.*, 89(4):2268, 2001.

185. A. Martí, L. Cuadra, and A. Luque. Quasi-drift diffusion model for the quantum dot intermediate band solar cell. *IEEE Trans. Electron Dev.*, 49(9):1632–1639, 2002.

186. C.-C. Lin, M.-H. Tan, C.-P. Tsai, K.-Y. Chuang, and T. S. Lay. Numerical study of quantum-dot-embedded solar cells. *IEEE J. Sel. Top. Quant. Electron.*, 19(5), 2013.

187. V. Aroutiounian, S. Petrosyan, and A. Khachatryan. Studies of the photocurrent in quantum dot solar cells by the application of a new theoretical model. *Sol. Energy Mater. Sol. Cells*, 89(2–3):165–173, 2005.

188. S. M. Willis, J. A. R. Dimmock, F. Tutu, H. Y. Liu, M. G. Peinado, H. E. Assender, A. A. R. Watt, and I. R. Sellers. Defect mediated extraction in InAs/GaAs quantum dot solar cells. *Sol. Energy Mater. Sol. Cells*, 102:142–147, 2012.

189. U. Aeberhard. Theory and simulation of quantum photovoltaic devices based on the non-equilibrium Green's function formalism. *J. Comput. Electron.*, 10:394–413, 2011.

190. D. A. Stewart and F. Leonard. Photocurrents in nanotube junctions. *Phys. Rev. Lett.*, 93:107401, 2004.

191. D. A. Stewart and F. Leonard. Energy conversion efficiency in nanotube optoelectronics. *Nano Lett.*, 5:219, 2005.

192. J. Guo, M. A. Alam, and Y. Yoon. Theoretical investigation on photoconductivity of single intrinsic carbon nanotubes. *Appl. Phys. Lett.*, 88(13):133111, 2006.

193. A. Buin, A. Verma, and S. Saini. Optoelectronic response calculations in the framework of k.p coupled to non-equilibrium Green's functions for 1D systems in the ballistic limit. *J. Appl. Phys.*, 114(3):033111, 2013.

194. U. Aeberhard and R. H. Morf. Microscopic nonequilibrium theory of quantum well solar cells. *Phys. Rev. B*, 77:125343, 2008.

195. U. Aeberhard. Spectral properties of photogenerated carriers in quantum well solar cells. *Sol. Energy Mater. Sol. Cells*, 94(11):1897–1902, 2010.

196. N. Cavassilas, F. Michelini, and M. Bescond. Theoretical comparison of multiple quantum wells and thick-layer designs in InGaN/GaN solar cells. *Appl. Phys. Lett.*, 105(6), 2014.

197. U. Aeberhard. Theory and simulation of photogeneration and transport in Si-SiOx superlattice absorbers. *Nanoscale Res. Lett.*, 6:242, 2011.

198. U. Aeberhard. Simulation of nanostructure-based and ultra-thin film solar cell devices beyond the classical picture. *J. Photon. Energy*, 4(1):042099, 2014.

199. U. Aeberhard. Effective microscopic theory of quantum dot superlattice solar cells. *Opt. Quant. Electron.*, 44:133–140, 2012.

200. A. Berbezier and U. Aeberhard. Impact of nanostructure configuration on the photovoltaic performance of quantum-dot arrays. *Phys. Rev. Applied*, 4:044008, 2015

42

Nanowire Solar Cells: Electro-Optical Performance

Bernd Witzigmann

42.1 Introduction

Nanowire arrays show unique optical and electronic features for photovoltaic applications. It has been demonstrated that they exhibit intrinsic optical antireflection properties [1,2]. Moreover, they also can act as optical microconcentrators [3] that focus light to an active region. Nanowires have a large aspect ratio, improving the electronic properties by combining a long optical absorption path along the wire axis and a short lateral carrier extraction [4]. Short carrier extraction paths are beneficial for the carrier collection efficiency. These mechanisms are illustrated in Figure 42.1. On the left, a conventional bulk solar cell collects light in the vertical direction, and the photogenerated carriers move to the respective contacts causing the photo current. A nanostructured solar cell (Figure 42.1a) can act as an optical concentrator in order to focus the incoming light to a small area. Moreover, the carrier extraction path length to the contacts is reduced, which minimizes the probability of carrier loss by unwanted recombination.

Finally, multijunction cells can be realized with strain-relaxed axial heterointerfaces. This allows optimum bandgap combinations [5] and low-defect densities at the same time. As a potential drawback, the large surface to volume ratio can lead to enhanced surface recombination. Epitaxial passivation techniques have been developed in order to suppress surface recombination [6].

In this chapter, a detailed simulation study of the electro-optical characteristics of nanowire array solar cells is presented. An efficiency analysis is derived that treats solar cells with nanostructures on a broader basis using the Shockley–Queisser framework [7]. As addition, the modification of optical density of states (DOS) and wave optics effects are included. Using this framework, the numerical simulation of nanostructured solar cells is discussed. The simulation model consists of the three-dimensional solution of the vectorial Helmholtz equation for solar illumination and a drift-diffusion/Poisson equation system for

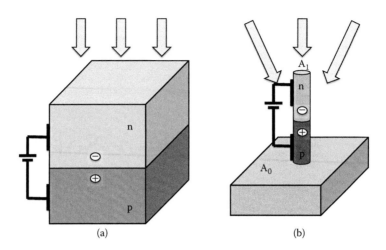

FIGURE 42.1   (a) Bulk solar cell with p-n junction. (b) Solar cell structured as a nanowire.

carrier extraction. In the results section, it is first shown how a nanowire array can act as an electromagnetic broadband absorber. Second, the choice of material composition and dopant placement is discussed with respect to efficiency in single-junction designs. Surface recombination is discussed for both radial and axial p-n junction arrangement.

42.2 Efficiency of Nanostructured Solar Cells

In this section, the efficiency of solar cells that contain nanostructures is discussed. Nanostructures can modify the optical or the electronic properties. As a basic theory, the established derivation by Shockley and Queisser [7] is used, which assumes detailed balance in the radiative limit, i.e., nonradiative recombination processes are not included. The main extension of Shockley and Queisser's classical model is to include the fact that light is viewed as electromagnetic waves and their modes in resonant structures. This leads to a modal dispersion and a strong local variation of the electromagnetic power. Moreover, resonant structures modify the optical DOS, and those enter the absorption as well as the emission properties of a cell, which is based on the Purcell effect [8].

The focus will be put on semiconductor materials, although the derivation can be transferred to other materials in a straightforward manner.

42.2.1 Solar Cell Current Density

The electronic current density of the solar cell can be easily obtained from current continuity in the steady state. It is derived from Maxwell's equations and builds on the mass conservation principle of charged particles:

$$\nabla \vec{J} = q(G - R), \tag{42.1}$$

with $\vec{J}$ being the local current density, G and R representing generation and recombination rates, respectively, and q being the elementary charge. In the integral form, the continuity equation reads

$$\int \vec{J} d\vec{A} = q \int (G - R) \, dV. \tag{42.2}$$

The surface area integration on the left hand side is carried out over the contacts (where carriers can leave or enter the solar cell), and the right hand side integration includes the solar cell volume.

42.2.2 Thermal Equilibrium

In thermal equilibrium, all of the emission and absorption processes of the solar cell can be equated, as there is no net current at the contacts. The left hand side of Equation 42.2 is zero, therefore

$$\int G dV = \int R dV = \int G^0 dV = \int R^0 dV. \tag{42.3}$$

The local generation rate per volume is

$$G^0 = \int \alpha d j_{phot}, \tag{42.4}$$

with α being the absorption coefficient (a material property) and j_{phot} is the photon current density (in units $1/m^2 s$), which in thermal equilibrium is basically a blackbody radiation at the given reservoir temperature T_0. In order to relate the photon current density to the blackbody radiation (which is an energy density per spectral width), the following relation is used:

$$d j_{phot} = d E_{phot} \frac{c}{n} \frac{1}{\hbar \omega} \tag{42.5}$$

and

$$d E_{phot} = D_{opt} f_{phot} \hbar \omega d \hbar \omega. \tag{42.6}$$

The optical DOS D_{opt} depends on the dielectric surrounding of the solar cell and f_{phot} is the photon distribution function. Within an electromagnetic theory, D_{opt} can be calculated as the electromagnetic response of a local dipole excitation [9]:

$$D_{opt}(\omega, \vec{x}) = \frac{1}{2\pi} \sum_m \int_{\vec{k}} d\vec{k} \delta(\omega - \omega_m) |\vec{e}_d \vec{E}_{m,\vec{k}}(\vec{r})|^2, \tag{42.7}$$

where $\vec{E}$ is the electric field as obtained from the solution of Maxwell's equations, $\vec{e}_d$ is the dipole unit source, and ω_m the eigenfrequencies of the system.

The photon distribution function in equilibrium is

$$f_{phot} = \frac{1}{\exp\left(\frac{\hbar \omega}{k_B T_0}\right) - 1}. \tag{42.8}$$

The generation rate, therefore, can be expressed as

$$G^0(\vec{x}) = R^0(\vec{x}) = \int \alpha(\vec{x}, \hbar\omega) \frac{c}{n_r} D_{opt}^{sc}(\vec{x}, \hbar\omega) f_{phot}(T_0) d\hbar\omega, \tag{42.9}$$

which is equal to the recombination in equilibrium.

42.2.3 Nonequilibrium

In the nonequilibrium case, the solar cell is illuminated by solar radiation power, which in our case is represented by an external blackbody source at the temperature T_s. This leads to an energy density outside of the solar cell of

$$d E_{phot}^s = D_{opt}^{vac} f_{phot}(T_s) \hbar\omega d\hbar\omega. \tag{42.10}$$

The optical DOS here is the vacuum DOS, and the solar temperature T_s enters the Bose–Einstein photon distribution function. In principle, other illumination sources can be used, which then enter the calculation, e.g., as tabulated energy density. The external photon source generates additional electron–hole pairs ΔG inside the solar cell, and therefore

$$G^s = \int \alpha a \frac{c}{n_r^{\text{vac}}} D_{\text{opt}}^{\text{vac}} f_{\text{phot}}(T_s) \, d\hbar\omega. \tag{42.11}$$

Here, a is the relative absorptivity, the amount of electromagnetic solar power entering the solar cell. It can be calculated using electromagnetic theory by

$$a(\vec{x}) = \frac{\langle \int Re\left\{\nabla\vec{S}\right\} dV \rangle}{\int Re\left\{\vec{S}\right\} d\vec{A}}, \tag{42.12}$$

where $\vec{S}$ is the Poynting vector, and the integration in the denominator is performed over a closed surface around the solar cell. This is especially important for nonplanar devices where the solar energy is captured not only from the surface, but from an angle of incidence larger than the half space π. The $\langle \rangle$ symbol in the numerator denotes an averaging over a volume that contains several optical wavelengths so that an inclusion into a drift-diffusion carrier transport framework is possible. The absorptivity $a(\vec{x})$ is the local relative electromagnetic energy entering in the solar cell, and in the ray optics approximation often is expressed as $(1 - r)\exp(-\alpha x)$, where the first factor is the back-reflected power from the air–top surface interface, and the second factor is the diminishing optical intensity inside the device due to internal absorption (Beer–Lambert law).

In the nonequilibrium case, the generation in the cell is increased, which in turn leads to an increased radiative recombination rate. Assuming that the radiative recombination is proportional to the product of electron and hole density, one gets

$$R^s = R^0 \frac{np}{n_i^2} = \frac{N_c N_v}{n_i^2} \exp\left(\frac{\eta_c \eta_v}{k_B T_0}\right) G^0 = G^0 \exp\left(\frac{\Delta E_F}{k_B T_0}\right), \tag{42.13}$$

with the following nomenclature: The intrinsic density is n_i, the effective state densities are N_c or N_v, the reduced Fermi energies are $\eta_c = E_F^n - E^c$ and $\eta_v = -E_F^p + E^v$, and ΔE_F is the quasi-Fermi level separation.

With these quantities, the current densities of the solar cell can be calculated. Using the current continuity equation, one gets

$$\int \vec{J} d\vec{A} = q \int (G^s - R^s) dV = q \int \left(G^s - G^0 \exp(\frac{\Delta E_F}{k_B T_0})\right) dV = I_{\text{gen}} - I_{\text{rec}}. \tag{42.14}$$

The quasi-Fermi level separation can be approximated by the voltage at the contacts, and after integration of the left hand side, the current of the solar cell becomes

$$I = I_{\text{gen}} - I_{\text{sat}} \exp\left(\frac{qV}{k_B T_0}\right). \tag{42.15}$$

The currents introduced here are the total generation and recombination currents I_{gen} and I_{rec}, and the reverse saturation current I_{sat}. They can be evaluated from

$$I_{\text{sat}} = q \int \int \alpha \frac{c}{n_r^{\text{sc}}} D_{\text{opt}}^{\text{sc}} f_{\text{phot}}(T_0) \, d\hbar\omega dV \tag{42.16}$$

and

$$I_{gen} = q \int \int \alpha a \frac{c}{n_r^{vac}} D_{opt}^{vac} f_{phot}(T_s) d\hbar\omega dV. \tag{42.17}$$

The short-circuit current of a solar cell is the current under illumination with zero voltage at the contacts, and therefore accounts to

$$I_{sc} = I_{gen} - I_{sat}. \tag{42.18}$$

With this definition, the current–voltage relationship can be written in the more conventional Shockley form by replacing I_{gen} in Equation 42.15:

$$I = I_{sat}\left(1 - \exp(\frac{qV}{k_B T_0})\right) + I_{sc}. \tag{42.19}$$

The open-circuit voltage is the voltage where the net current flow at the contacts is zero, and it can be evaluated from the relations obtained here as

$$V_{oc} = \frac{k_B T_0}{q} \log\left(\frac{I_{gen}}{I_{sat}}\right) = \frac{k_B T_0}{q} \log\left(1 + \frac{I_{sc}}{I_{sat}}\right). \tag{42.20}$$

42.2.4 Efficiency Evaluation

From the current versus voltage characteristics, the electro-optical efficiency of a solar cell can be obtained from

$$\eta_{sc} = \max\left(\frac{IV}{P_{inc}}\right), \tag{42.21}$$

where the denominator is the total, spectrally integrated incident solar power. Obtaining a high efficiency translates into maximizing the open-circuit voltage V_{oc} and the short-circuit current I_{sc} as defined in Section 42.2.3, which maximizes the product IV. A maximum short-circuit current is achieved by capturing the incident broadband solar radiation and absorbing the photons in the solar cell. According to Equation 42.17, this requires a large absorption coefficient α, which is a material property, and a good absorptivity a, which is the ability to capture light into the solar cell, an electromagnetic property. In bulk solar cells, the absorptivity a can be maximized by broadband antireflection coatings. If the solar cell thickness is in the order of a few optical wavelengths, waveguiding or even evanescent wave effects can help maximize a. It will be shown in Section 42.2.4.2 how this is applied to nanowire solar cells.

The evaluation of the open-circuit voltage involves a maximum optical generation current I_{gen} and a minimum reverse saturation current I_{sat}, which is a challenge to fulfill. As can be seen from Equation 42.16, the major ingredient to I_{sat} is the optical DOS of the solar cell.

42.2.4.1 Bulk Solar Cell

First, the efficiency evaluation is shown for a bulk solar cell. In principle, all current commercial technologies fall into this category, also the thin-film designs. For a homogeneous medium, the optical DOS is

$$D_{opt}^{hom} = \left(\frac{n_r}{c}\right)^3 \frac{1}{4\pi\hbar^3}(\hbar\omega)^2 \Omega, \tag{42.22}$$

where Ω is the solid angle of the radiation emission and absorption, respectively.

For the bulk solar cell, the solid angle of illumination is $\Omega = \pi$, and therefore

$$D_{\text{opt}}^{\text{vac}} = \left(\frac{1}{c}\right)^3 \frac{1}{4\hbar^3}(\hbar\omega)^2. \tag{42.23}$$

The solid angle of emission at any location inside the solar cell is $\Omega = 4\pi$, leading to

$$D_{\text{opt}}^{\text{sc}} = \left(\frac{n_r}{c}\right)^3 \frac{1}{\hbar^3}(\hbar\omega)^2. \tag{42.24}$$

For the absorptivity a, a simple ray optics model can be applied, which leads to

$$a = (1 - r)(1 - \exp(-\alpha d)) \approx 1, \tag{42.25}$$

with r being the reflectivity at the air-semiconductor interface, and d the cell thickness. By design, one can realize a good antireflective coating and sufficient cell thickness, leading to the relation $a \approx 1$. The short-circuit current can therefore be maximized by engineering the cell.

The open-circuit voltage contains the ratio of $\frac{I_{\text{gen}}}{I_{\text{sat}}}$, which is approximately proportional to the ratio $\frac{D_{\text{opt}}^{\text{vac}} n_r^{\text{sc}}}{D_{\text{opt}}^{\text{sc}} n_r^{\text{vac}}}$. This results in a factor $\frac{1}{4n_r^2}$ in the short-circuit current calculation. The physics behind this factor is the light intensity concentration. The factor of 4 in the denominator can be reduced by efficient light management, as it comes from the photon collection from the angle π but photon emission into an angle of 4π [10]. Introducing a dielectric environment that allows emission into an angle π only (by, e.g., photonic crystals or conventional mirrors) increases the open-circuit voltage. Therefore, light management is an important physical mechanism for advanced solar cell design [10,11].

42.2.4.2 Optical Microconcentrators

Waveguides, resonators, or optical antennas can influence the optical absorption behavior of a solar cell. As first case, an optical concentrator that acts as absorber is taken that focuses the incoming solar radiation to the absorber area A_1, whereas the illuminated (large) area is A_0 (for illustration, see Figure 42.1). This leads to an absorptivity $a \propto \frac{A_0}{A_1} = f$, as the absorber concentrates the energy flux density to the area A_1. As a consequence, the short-circuit current is obtained with a much smaller volume of absorption compared to a nonconcentrated case (see integral in Equation 42.18). At the same time, the reverse saturation current (Equation 42.16) is also integrated over a smaller volume compared to a bulk solar cell. The ratio of $\frac{I_{\text{gen}}}{I_{\text{sat}}}$, and therefore the open circuit voltage, increases. However, two assumptions need to be made: First, the reverse saturation current density needs to be independent of the local carrier density, which is, within the Shockley theory, a valid assumption. Second, the optical concentrator modifies the optical DOS of the absorber. This needs to be taken into account in the open-circuit voltage calculation. Therefore, the open-circuit voltage is only increased if the following relation holds:

$$\frac{D_{\text{opt}}^{\text{vac}}}{D_{\text{opt}}^{\text{mc}}} f > \frac{D_{\text{opt}}^{\text{vac}}}{D_{\text{opt}}^{\text{bulk}}}, \tag{42.26}$$

or

$$\frac{D_{\text{opt}}^{\text{bulk}}}{D_{\text{opt}}^{\text{mc}}} f > 1, \tag{42.27}$$

where $D_{\text{opt}}^{\text{bulk}}$ is the optical DOS for the bulk cell and $D_{\text{opt}}^{\text{mc}}$ is the optical DOS for the micro concentrated cell, respectively; their ratio can be interpreted as the Purcell factor. This result shows an interesting design opportunity for high-efficiency nanostructured solar cells: If optical elements are introduced that focus the solar electromagnetic energy to a small volume, the open-circuit voltage, and therefore the efficiency, can be increased compared to a bulk or thin-film solar cell. However, the Purcell factor, or the ratio of optical DOS, should not be increased at the same rate, as it compensates the concentration effect.

42.2.4.3 Electronic Nanostructures

In the field of lasers and light-emitting diodes, electronic nanostructures such as quantum wells and quantum dots have increased the performance considerably [12]. In solar cells, the application of electronic nanostructures has not been used as a standard feature to improve the cell efficiency. In general, the introduction of nanostructures alters the electronic band structure. It leads to the following consequences:

- The effective bandgap, i.e., the absorption edge, is shifted to the lowest subband energy difference.
- The absorption coefficient increases, as the electronic DOS is higher at the band edge compared to a bulk material, and in addition, the overlap of the envelope wave functions enters the absorption coefficient.
- The absorption coefficient becomes anisotropic due to the anisotropy of the underlying carrier wave functions and DOS.

As a drawback, electronic nanostructures typically have dimensions around 10 nm, which is much smaller than the optical wavelength or a typical absorption path length. Therefore, in order to obtain efficient absorption, multiple nanometer-sized regions need to be stacked in the direction of the light path or efficient focusing of the light onto the absorbing nanostructure needs to be done. Fundamentally, the anisotropy of the absorption and emission can be used to restrict the absorptivity α to the solid angles of the solar radiation, and to reduce the reverse saturation current by a maximum theoretical factor of 4 (see explanation in Section 42.2.4.1) [13]. In addition, the increase of the absorption coefficient leads to smaller active regions, which increases the ratio of short-circuit current to reverse saturation current as outlined in Section 42.2.4.2.

Finally, some advanced concepts have been pursued with electronic nanostructures such as intermediate band solar cells [14] or hot-carrier contacts [15]. The basic idea is to introduce additional energy levels with nanostructures in order to overcome the single bandgap limitation of the semiconductor. Such structures require nonequilibrium carrier transport models and are not discussed here.

42.3 Computational Models

In the previous sections, some fundamental properties of nanostructured solar cells have been described within the semianalytical framework of the Shockley–Queisser model. In general, the incoming light generates electron–hole pairs by absorption in the semiconductor. In Section 42.2, it has been shown that the efficiency of a solar cell depends on the solar optical generation $G^s(\vec{x})$, the generation in equilibrium $G^0(\vec{x})$, and the recombination under illumination $R^s(\vec{x})$ (see Equation 42.14). In principle, those are all local quantities. In nanostructured solar cells, both light intensity and carrier density can be controlled on the scale of tens of nanometers by optical waveguides, plasmonic subwavelength structures, or quantum wells or dots. This introduces a strong local variation of the physical quantities contributing to the cell efficiency. Computational models can capture the local variations, and a correct determination of the efficiency is feasible.

42.3.1 Optical Models

The governing optical equations for solar cells with structures similar or smaller than the optical wavelength are Maxwell's equations. The solar illumination is modeled as a broadband current source, and the resulting electromagnetic field contributes to the local carrier generation rate. There is a multitude of models to solve Maxwell's equations either analytically, semianalytically, or numerically.

A popular analytical solution is the plane wave ansatz, which after some simplifications leads to the transfer matrix method (TMM). Using TMM, light propagation in multilayer structures can be evaluated or the even simpler Beer–Lambert law can be obtained:

$$I(x) = I_0 e^{-\alpha x}, \tag{42.28}$$

with the intensity I.

Most semianalytical methods expand the solution of Maxwell's equations in modes. For nanowire solar cells, which are essentially cylinders, the modes of a cylindrical optical fiber can be used, and the solution is then represented as superposition of the fiber modes [16].

Numerical methods essentially discretize the simulation domain and solve the respective equations only on parts of the domain, e.g., vertices or edges. The finite-difference time-domain (FDTD) method [17] is a popular method that discretizes space and time using explicit finite differences. With sources placed in the simulation domain, the propagation of electromagnetic waves can be computed iteratively in space and time. The FDTD method is straightforward to implement and can be adapted for parallel computing in a natural manner. The original FDTD kernels [17] require rectangular tensor product meshes, which can be restrictive for complex geometries. Moreover, the solutions are only second-order accurate in space and time, which requires many grid points per wavelength for accurate solutions, in particular for optical frequencies. A rather elegant approach to overcome these limitations is the discontinuous Galerkin time-domain method [18]. It combines explicit time stepping with a local finite-element space discretization, and therefore powerful high-order accuracy and flexible meshes can be used. The local elements are connected with constraints for the numerical flux in order to ensure consistency and convergence to the exact solution. For solar cell simulations, which use the broadband solar spectrum as source, the time-domain methods allow the use of spectral broadband sources, with only one simulation run for the entire spectrum. Dispersive materials need special treatment in the time domain, and therefore additional computational resources.

A second class of numerical methods solve Maxwell's equations in the frequency domain. Here, monochromatic harmonic waves replace the time derivatives with an imaginary frequency factor, and any multifrequency solution can be built as superposition in the linear case. Maxwell's equations can be reduced to the Helmholtz equation, which for the electric field $\vec{E}$ reads

$$\left(\Delta - \omega^2 \mu \epsilon \right) \vec{E} = -j\mu\vec{J}, \tag{42.29}$$

where μ and ϵ are the respective permeability and permittivity, ω is the frequency, and $\vec{J}$ is a current source. This equation is valid for homogeneous isotropic materials, and the permittivity is in general a dispersive complex quantity in optics. This equation can be solved by a finite-element method in three dimensions, with sophisticated meshes, and an approximation to the exact solution. In its homogeneous form (without current sources), it yields an eigenvalue problem that give the eigenfrequencies and eigenvectors of a given geometry. For complex permittivities, the eigenvalue problem becomes complex as well, and the imaginary part of the eigenfrequency gives the quality factor of the eigenmode [19,20].

In order to evaluate the efficiency of a solar cell beyond the analytical Shockley–Queisser limit, the framework presented in Section 42.2 as efficiency analysis for nanostructured solar cells requires evaluation of the classical electromagnetic field for the solar cell geometry in the presence of a complex refractive index. This enters both the absorptivity a (see Equation 42.12) and the optical DOS D_{opt} (Equation 42.7).

42.3.2 Electronic Models

Any electronic model describes the dynamics of the charge carriers inside the solar cell, i.e., the photogenerated carriers and the intrinsic carriers under applied bias. In a generalized description, solar cell operation means that the photogenerated carriers thermally relax to the band edges and are subsequently driven to the contacts creating the photocurrent. In the radiative (i.e., Shockley–Queisser) limit, the carriers can recombine radiatively on their way to the contacts, and of course, the resulting photons can be reabsorbed. This gives an upper limit to the photocurrent. However, carriers can be lost by nonradiative recombination, which decreases the efficiency further. An excellent description of such an efficiency analysis can be found in [21]. In order to understand local features of the carrier transport, such as surface recombination effects, or defect recombination related to specific doping species or materials, a space-resolved carrier transport model is needed. In nanostructures, additional physical properties are potentially relevant, e.g., the thermal relaxation process is different from bulk processes, which can lead to ballistic or semiballistic carrier transport. If these heated carriers can be detected by a contact without thermal relaxation, a higher efficiency is achieved as the thermalization energy loss can be avoided [22].

Therefore, electronic properties of nanostructures require simulations by quantum transport models in order to include effects such as carrier tunneling, phase coherence (and carrier localization), and ballistic transport. These models have been successfully applied to nanowire devices by several groups [23] and established methods are the nonequilibrium Green's functions (NEGF) method or the Wigner function formalism [24]. The NEGF method has been applied to quantum solar cells [25,26]; however, these methods in general are computationally demanding, as they discretize both real and phase space. Therefore, simulation domains of only a few nanometers can be accounted for. In the future, the so called multiscale approaches might be feasible that combine computationally intense quantum transport methods with efficient semiclassical methods; several approaches have been made, with still many open questions [27].

In the following sections, the carrier transport in nanowire solar cells is analyzed. It will turn out that these solar cells have dimensions with diameters larger than 50 nm and lengths more than 1000 nm. In this regime, semiclassical models still have a decent validity, therefore the established and efficient drift-diffusion model will be used. It comprises the self-consistent solution of the current continuity equations for electrons and holes with the Poisson equation for the electrostatic potential.

42.4 Light Capture and Emission

Figure 42.2 shows the arrangement of the nanowire array for a solar cell. The wire diameter is d, the lattice constant is a, and the wire height is L. The first question is how such an array can incouple and absorb light. Solar radiation spans from the ultraviolet to the infrared wavelength range, and the solar cell therefore needs to be a broadband absorber. It has been reported in the literature that an array of nanowires can act as a good broadband absorber [16,28]. For highest efficiency with a single bandgap material, the classical Shockley–Queisser analysis predicts a band gap of the semiconductor material between $E_g \approx 1.3$ eV and 1.4 eV [11]. This bandgap optimum can be explained by the balance of having minimum carrier thermalization losses from high-energy photons on the one hand and minimizing the losses due

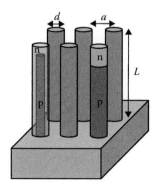

FIGURE 42.2 Illustration of an ordered nanowire array solar cell. The p-n junction can be arranged in a radial (left) or axial (right) manner. The periodicity can be hexagonal or cubic.

to the optical transparency of the semiconductor for low-energy photons. Indium phosphide (InP) with $E_g = 1.34eV$ matches this range very well; therefore, the following analysis will use this material. The findings are applicable to any other material as well.

42.4.1 Absorptivity

Using a three-dimensional finite-element solution of the complex vectorial Maxwell equations, the wire arrangement as shown in Figure 42.2 is illuminated from the top side with a linearly polarized sheet source. The source matches the power spectrum of AM1.5D solar power and is placed at a distance above the wires, which is beyond the evanescent mode decay length. The space between the wires is vacuum, and both the wire and the substrate material is InP, with the complex refractive indices following the dispersion relation given in [12]. The simulation is carried out for a quarter of the wire only, with periodic boundary conditions on the cut planes in order to represent an infinite periodicity of the array. Perfectly matched layer absorbing boundary conditions are placed at the top and the bottom of the simulation domain.

From the electromagnetic fields, the relative local absorptivity $a(\vec{x}, \lambda)$ is calculated according to Equation 42.12, which becomes, after integration of the wire volume, the spectral absorptivity of the entire wire $\tilde{a}(\lambda)$. The optical absorption in the substrate is not included in the integration and does not contribute to the efficiency of the cell. Figure 42.3 shows the spectral absorptivity for different wire periodicity and diameters; the wire height is constant at $L = 2$ µm. A geometric fill factor is defined that describes the volume of active material in a nanostructured cell relative to a bulk material volume. This fill factor can be expressed as

$$f_{nw} = \frac{\pi d^2}{4a^2}. \tag{42.30}$$

The highest fill factor can be achieved with $d = a$ and a fill factor of $f_{nw} \approx 0.79$, i.e., when the nanowires touch. For the simulations shown in Figure 42.3, combinations of periodicity and diameter have been

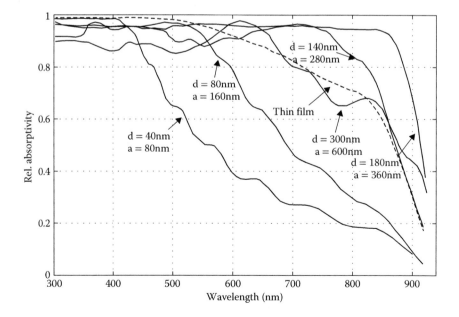

FIGURE 42.3 Spectral relative absorptivity of an InP nanowire array. The wire diameter is labeled d, and the wire distance is a.

chosen so that the material volume fill factor is identical at $f_{nw} = 0.196$. The arrangement with the smallest wire diameter of $d = 40$ nm and distance of $a = 80$ nm shows a substantial reduction of absorptivity above a wavelength of 500 nm. Nanowires with larger diameter and larger array pitch show a better long wavelength absorptivity, and the case of $d = 180$ nm and $a = 360$ nm has an absorptivity of more than 90% up to the bandgap wavelength of InP. The dashed curve shown in Figure 42.3 is the absorptivity of a homogeneous InP film with a thickness of $2\mu m \times 0.196 = 392$ nm, which is the same material volume as the nanowire array. This thin film, however, is covered by an ideal antireflective coating in order to eliminate reflections (which is not added to the wire array case). The decrease of absorptivity at wavelengths larger than 600 nm comes from the insufficient thickness of the InP layer.

42.4.2 Modal Analysis

In order to understand the findings of Section 42.4.1, the spectral behavior of the absorptivity of the nanowire array can be analyzed by a modal description of the electromagnetic field distribution. In [16], it has been shown that the absorptivity obtained from the three-dimensional calculation can be constructed from a separation of the power absorptivity parallel to the nanowire axis according to the Beer–Lambert law. The effective attenuation coefficient is obtained from the two-dimensional mode calculation of a cross-section of the wire array perpendicular to the wire axis. This assumes an infinite extension of the wires along their axes, and the calculation gives the eigenmodes with their complex eigenvalues $n_m^{\text{eff}} = n_m' + in_m''$, with m as the mode index. The real part of the eigenvalue describes the coupling efficiency of the electromagnetic incident TEM wave to the wire array and is inserted in the separation ansatz for the calculation of the reflected power. The imaginary part is the effective absorptivity that enters the Beer–Lambert expression. For an illustration, see Figure 42.4. The optical power at the interface between vacuum and the nanowire array top can be expressed as

$$P_m(z = 0) = P_0 \frac{4n_0 n_m'}{(n_0 + n_m')^2} \frac{\Gamma_m}{\sum_i \Gamma_i}, \tag{42.31}$$

with n_0 being the vacuum refractive index, P_0 the source power, and Γ_m is the overlap integral of the electric field of the incoming TEM mode with the mth mode of the wire array cross-section. The absorbed power in the nanowire material is then

$$P_{\text{abs}}(\lambda) = \sum_i P_i(z = 0) \left(1 - \exp(\frac{4\pi}{\lambda} n_i'' L)\right). \tag{42.32}$$

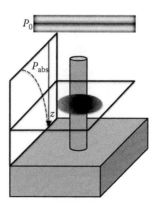

FIGURE 42.4 Illustration of the electromagnetic separation approach for a nanowire array. The incoming plane wave power P_0 is separated into two-dimensional cross-sectional modes and an exponential power decay in z-direction following a Beer–Lambert law.

Using this separation ansatz, the spectral absorptivity curves of the three-dimensional calculation can be reproduced with excellent accuracy, as shown in [28].

This procedure identifies the type of modes that couple the linearly polarized free-space electromagnetic field to the nanowire array. The unit cell of the array consists of the wire itself (which resembles a cylindrical fiber) and the periodic boundary to the neighboring wires that can be viewed as the Neumann boundary condition at the border of the unit cell.

The broadband absorptivity for the nanowire array can be described by only very few modes. For the case of $d = 180$ nm and $a = 360$ nm, only two modes are necessary to create the absorptivity characteristic as shown in Figure 42.3. These modes are plotted in Figure 42.5 with their electric field mode profile and their relative incoupling efficiency. The latter describes the ability of the wire array to act as an antireflective structure.The effect of relative incoupling in Figure 42.5 (bottom figure) is a combination of the confinement factor Γ and the real part of the effective index of the in-plane mode in the nanowire array n' (see Equation 42.31). For incoupling of 90%, the effective index therefore is close to 1 (i.e., the surrounding air), which for a fiber mode, is the regime of weak confinement. Still, most of the electromagnetic power couples into the wire array, and the field distribution as shown in in Figure 42.5 (e.g., mode A at 500 nm) shows a strong field between the wires, which is also absorbed according to the Beer–Lambert law. Between 300 nm and 700 nm, mode A couples close to 90% into the wire, and the electric field analysis shows that the electric field distribution resembles an fiber mode. Finally, the electric field distribution between the wires resembles a fundamental TEM mode, meaning parallel arrangement of the electric field vector. At 750 nm, mode A reaches its cutoff wavelength, and most of the light couples into mode B, which is of type HE_{11}. For further detail, the reader is referred to the optical fiber standard dispersion relations and eigenmodes, which are similar to the case of the sparse wire array.

It is surprising that only two modes are responsible for the broadband absorptivity of the electromagnetic power. This modal description helps understand the characteristics: If the wire spacing is increased from the ideal distance, the overlap integral of the wire modes decreases, and therefore the absorptivity decreases. If the wire spacing is decreased, the effective index of the wire array n' increases, and therefore the relative incoupling decreases. If, at the same distance, the wire diameter is modified, similar effects occur; an increase of the wire diameter moves the fiber mode dispersion relation to the multimode regime, where the effective indices are larger (a standard optical fiber dispersion relation illustrates this). A decrease of the wire diameter will decrease the mode number to a point where only the fundamental mode can be used.

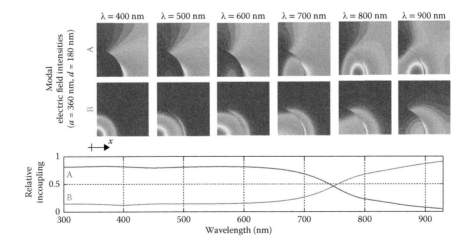

FIGURE 42.5 Top: Electric field distribution in the nanowire array for mode A and mode B, for different wavelengths. Bottom: Relative incoupling P_A/P_0 and P_B/P_0 (see Equation 42.31). (Reprinted from J. Kupec et al., *Optics Express*, 18, 27589–27605, 2010. With permission.)

As a rule of thumb, therefore, the wire distance and diameters should be chosen so that in the dispersion relation of the cross-sectional eigenvalue problem, the bandwidth of operation can span the fundamental and a few higher order modes. Moreover, the wire length should be chosen so that the imaginary part of the eigenvalue ensures full absorption of the optical power according to the Beer–Lambert law.

The modal analysis of the wire array can also be applied for solar radiation under an incident angle different from vertical illumination. For moderate incident angles (below 15°), the results do not change significantly.

42.4.3 Shockley–Queisser Efficiency of a Nanowire Array

Having calculated the relative absorptivity from electromagnetics, the Shockley–Queisser efficiency can subsequently be evaluated from Equation 42.21. Carrier transport effects are not included at this level of description and will be described later. Also, as approximation, the optical DOS in the semiconductor D_{opt}^{sc} is taken to be a bulk DOS, for simplicity. This neglects the Purcell effect for the radiative emission, and therefore we get an upper limit for the solar cell efficiency. The impact of the modified optical DOS will be discussed in Section 42.4.4.

Figure 42.6 shows the calculated efficiency of the InP nanowire array with a square wire arrangement and a wire length of $L = 2\,\mu m$. The efficiency is plotted versus the volumetric fill factor. For small wire diameters of below 100 nm, the efficiency is highest for high fill factors and decreases for larger wire distances. Here, mostly the fundamental HE_{11} mode is responsible for the absorptivity; however, the overlap with the wire is too small for efficient absorption. For larger wire distance, the electromagnetic energy therefore is lost in the substrate. For thicker wires with diameters larger than 100 nm, the most dense array with highest fill factor shows the same efficiency as the thinner wires; however, with increasing wire distance, the efficiency increases up to values of 30% for the diameter of $d = 180$ nm. This increase can be explained by a reduced effective index of the wire array with lower fill factor, and therefore less back reflection from the vacuum–wire array interface. The electromagnetic energy is concentrated to the wires by the modal light capture, and the decrease of efficiency below fill factors of 10% is due to the decreasing coupling factor. As a result, a maximum efficiency of 30% can be reached with a fill factor of approximately 20%, a wire diameter of $d = 180$ nm, and a wire separation of $a = 360$ nm, without an additional anti-reflective coating. Hence, the wire array acts as a microconcentrator element as well as an active absorber at the same time. An InP bulk cell with ideal broadband antireflective coating gives a detailed balance efficiency of $\eta_{ideal} = 31\%$ for comparison.

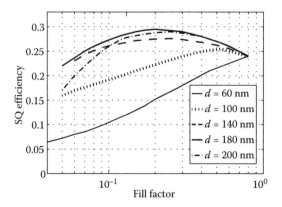

FIGURE 42.6 Efficiency of an InP nanowire array based on the Shockley–Queisser analysis in the radiative limit for different wire diameters *d*. An infinitely thick, perfect antireflective InP bulk cell has an efficiency of approximately 32%.

42.4.4 Purcell Effect

The nanowire array acts as a microconcentrator, and therefore, as discussed in Section 42.2.4.2, the open-circuit voltage can be increased compared to a bulk solar cell, if the concentration factor outweighs the increase in optical DOS. The reciprocity theorem states that any lens or concentrator not only increases the collection angle of the solar cell, but also the emission angle. In addition, the spontaneous emission rate (which in the bulk case is a material property) can be altered if the dielectric surrounding of the emission volume is modified, which changes the recombination rates. For the InP nanowire model array, the question is now how large the Purcell factor is and if it substantially alters the efficiency considerations described up to now.

In order to calculate the optical DOS, dipole sources were placed in the nanowire array, with vertical (z-) and horizontal (xy-) polarization. As position, the cross-section of the nanowire was sampled at a height of $z_0 = 1.9$ μm, where the absorptivity is large. A wavelength sample of $\lambda = 700, 800$, and 900 nm was used, which is close to the spontaneous emission maximum in InP. A description similar to Equation 42.7 has been used in order to calculate the local optical density of states (LDOS) (for a detailed procedure, please see [19]). In Figure 42.7, the LDOS is plotted, normalized to the free-spaced DOS. It basically describes the enhancement or suppression factor for the spontaneous emission rate; a number larger than 1 would mean that spontaneous emission is enhanced. From the calculation, the z-polarized normalized LDOS reaches values around 2–3, close to the wire surface, where the maximum absorption takes place (see Figure 42.5 for comparison). This is a photonic crystal effect, as the z-polarized dipole emits predominantly in the horizontal direction. The xy-LDOS is 0.5–0.8, close to unity. As a consequence, the averaged LDOS over all directions leads to a normalized LDOS very close to unity, which according to Equation 42.27 allows the full benefit of the optical microconcentration effect to enter the open-circuit voltage calculation.

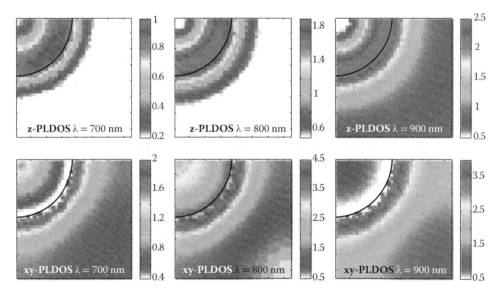

FIGURE 42.7 Local optical density of states (LDOS) relative to the bulk density of states in a nanowire array with diameter of $d = 180$ nm and pitch $a = 360$ nm. The top row is from an emitter polarized parallel to the wire axis, the bottom row for a dipole emitter with horizontal polarization. (Reprinted from J. Kupec et al., *Optics Express*, 18, 27589–27605, 2010. With permission)

42.5 Electronic Properties

42.5.1 Radial versus Axial Junction Arrangement

The nanowire geometry by nature has a high aspect ratio of length to diameter, and the p-n junction can be arranged in radial (i.e., vertical) or axial (i.e., horizontal) manner generating large or small quasi-neutral regions. Both implementations are shown in Figure 42.2 schematically and also have been demonstrated experimentally by epitaxial growth techniques. This p-n junction placement introduces another important aspect to nanostructured solar cell design. Assuming an abrupt junction, and the depletion approximation, the p-n junction can be separated into three distinct regions: the depleted region, and the quasi-neutral p- and n-regions. Photogenerated carriers in the depleted region are separated by the electric field and constitute a local photo current. In the quasi-neutral regions, the carriers diffuse, ideally to the contacts, if a local gradient is present. The cell current at the contacts is reduced if photogenerated carriers recombine on their way to the contact. This can be internal radiative or nonradiative recombination, consisting of spontaneous emission, Auger recombination, and Shockley–Read–Hall recombination. As a general guideline for minimizing the influence of nonradiative mechanisms, the recombination limited carrier lifetime needs to be much larger than the transit time of the carriers to the contacts. This principle is used in the thin-film solar cell technology.

In the analysis in Section 42.2.3, only the radiative recombination was included in the detailed balance limit. The analysis here is also done in the detailed balance limit as well, as it gives an upper boundary for the efficiency.

As example, the InP nanowire geometry is used, with a diameter of $d = 180$ nm and a wire height of 2 μm. The top n-region has a thickness of 100 nm, and the core diameter of the core-shell implementation is $d = 120$ nm. As p-doping, an active concentration of $p = 1 \times 10^{18} cm^{-3}$ and n-doping of $n = 5 \times 10^{18} cm^{-3}$ is used. The junction arrangement of the respective axial and core-shell structure is shown in Figure 42.2. The space and frequency resolved absorptivity profile has been calculated solving the vectorial Helmholtz equation under AM1.5D illumination; an illustration can be seen in Figure 42.8 after integration over frequency.

The wire array fill factor is 20%, which gives a maximum Shockley–Queisser efficiency. After spectral integration of the absorptivity, a local optical generation rate can be evaluated. The electronic simulation has been carried out only for a single wire by solving the continuity equations for holes and electrons coupled to the Poisson equation for the electrostatic potential. This means that any interaction between the

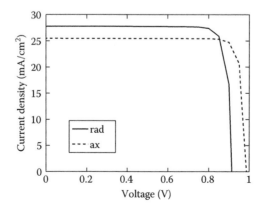

FIGURE 42.8 Current density versus voltage characteristics of axial and radial junction nanowire array solar cells made of InP, illuminated with the AM15.D solar spectrum. (ax, axial; rad, radial.)

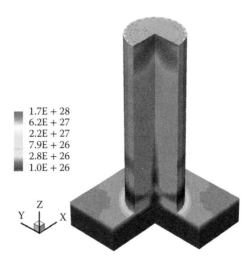

FIGURE 42.9 Optical generation rate profile in a nanowire array (only single nanowire shown) with a solar AM1.5D illumination spectrum (in units 1/s).

wires has been neglected for the carrier transport. Figure 42.9 shows the current versus voltage characteristics for the two designs. The axial p-n junction has a lower short-circuit current than the radial junction; however, it also shows a larger open-circuit voltage. The resulting maximum efficiencies are therefore close: 22% for the axial cell and 21% for the radial cell. In the axial wire, the photocurrent is generated from two effects. For one, the electric field in the depletion zone leads to drift current of the photogenerated carriers. This drift current removes carriers from the depletion zone, and in turn results in a diffusion current of the photogenerated carriers as excess minority carriers toward the junction. For the doping values in this wire, the depletion region is less than 15-nm thick at zero voltage, which would yield almost no photocurrent from direct drift. However, the simulations show that 600 nm of the quasi-neutral p-region contributes to the electron photocurrent by diffusion, and the entire quasi-neutral top n-region of 100 nm contributes to the hole photocurrent. The thickness of these diffusion regions depends critically on the minority carrier lifetime, and therefore impurity densities in the quasi-neutral regions need to be minimized. A more detailed analysis of this effect has been shown in [4]. In our simulation of the axial structure (with only radiative recombination present), a total of 700 nm of the wire height therefore can be viewed as the active region for photocurrent generation [29], which is 35% of the total length of the wire. In contrast, the radial junction contains a depletion region along the entire wire, and the charge separation due to drift and diffusion currents takes place in the entire wire volume. This leads to a higher photocurrent compared to the axial wire. The difference is not as large as the active volume ratio though, as the optical generation rate decreases exponentially from top to bottom, and the high generation top region is an active region for both structures. In [4], it has been found that the efficiency of a radial nanowire solar cell is much less prone to degradation if the minority carrier lifetime decreases, which can be directly explained by the increased depleted volume in the radial cell. It should be noted that the n-doped shell thickness of 30 nm for the radial wire has been chosen since it is larger than the depletion region so that a vertical photocurrent to the n-contact can be established.

The active region analysis explains the increased photocurrent for the radial structure, but it also can explain the reduction in open-circuit voltage compared to the axial device. The open-circuit voltage is given by Equation 42.20, with the fraction $\frac{I_{gen}}{I_{sat}}$ as contribution. Including the above described carrier transport effects, the volume integration for the calculation of the optical generation current I_{gen} and the reverse saturation current I_{sat} should only be done for the active volume. In the case of the axial junction, it is the depletion volume plus the diffusion zones in the quasi-neutral regions, which, in our specific case, account to the top third of the wire. The radial junction wire collects photocarriers in the entire wire volume, and therefore the integration volume is the entire wire. As stated, this maximizes the optical generation current, and at the same time, also increases the reverse saturation current. Toward the bottom of the radial junction wire, the integration adds the same amount of reverse saturation current as in the upper region; however, the optical generation contribution decreases to a negligible amount. This is due to the exponential decay of the relative absorptivity a (see Equation 42.12). As result, the open-circuit voltage decreases if the wire is too long. This is plotted in Figure 42.10, where the wire array efficiency of a radial junction cell has been calculated for different wire lengths. For wire lengths below 2 μm, the solar spectrum is not fully absorbed within the wire (especially the long wavelengths), which limits the short-circuit current. For lengths above 2 μm, the reverse saturation current increases more strongly than the photocurrent, which decreases the open-circuit voltage. It should be noted that the values for the wire length and resulting efficiencies are specific to the material, dimensions, and doping concentrations that have been chosen for this example; however, the physical principles are valid for the general case of nanowire solar cells. For silicon nanowires, the optical absorption coefficient is smaller, and therefore the optimum wire length would be larger.

42.5.2 Surface Recombination

Up to now, all the calculations have been presented with radiative recombination only, excluding any non-radiative processes. In this section, the impact of surface recombination is analyzed. Nanostructures show a large surface to volume ratio, and any surface effect has a pronounced impact on the characteristic behavior of a device. For solar cells, surface recombination contributes to the nonradiative recombination mechanisms and decreases the cell efficiency, as discussed in the previous section. Here, the goal is to study the impact of surface recombination in the cell efficiency within the framework of the carrier continuity equation and the Poisson equation. The surface recombination rate is calculated via a Shockley–Read–Hall recombination with 2D trap densities [30]:

$$R_s = \frac{n_s p_s - n_i^2}{1/v_p(n_s + n_1) + 1/v_n(p_s + p_1)}, \tag{42.33}$$

where n_i is the intrinsic carrier concentration, v_n and v_p represent the surface recombination velocities, and n_1 and p_1 represent the surface trap densities. This model is used in the simulation procedure. The surface carrier densities are n_s and p_s, respectively.

The radial and axial structures have been simulated with constant surface recombination velocity of $v_s = 10^4$ cm/s, which is a realistic value for InP [31], and surface trap densities n_1 and p_1 set to a value for midgap states . Figure 42.11 shows the current–voltage characteristics with (dashed lines) and without surface recombination (solid lines). Turning on surface recombination, the maximum efficiency drops

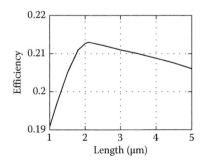

FIGURE 42.10   Cell efficiency of a radial junction nanowire array versus nanowire length.

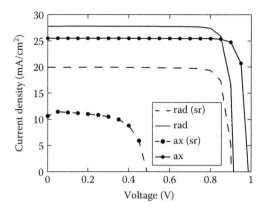

FIGURE 42.11   Current versus voltage relationship of a radial (rad) and axial (ax) junction nanowire array with and without surface recombination (sr).

from 21% to 15.5% for the radial structure and from 22% to 3.5% for the axial structure. It can be shown that the maximum surface recombination rate is achieved for the condition $v_n n_s = v_p p_s$ [32]. Assuming similar recombination velocities for electrons and holes for simplicity, the maximum surface recombination is hence located where electron and hole densities are both high at the same location. For a p-n junction, this occurs in the depletion region. It explains why the axial junction solar cell shows a substantial decrease in efficiency; in this geometry, the depletion region is exposed to the surface. This can be avoided by adding a passivation layer to the surface. For the radial junction, the depletion region thickness is smaller than the outer n-doped shell thickness, which makes the surface a quasi-neutral region. In that respect, the radial junction arrangement is vastly superior and much less prone to degradation from surface recombination effects.

42.6 Summary

Photovoltaic cells have become a fast growing technology for electrical power generation. Conversion efficiency and cost have been major drivers for the development. In that context, the use of nanostructures in solar cells has not lead to a major performance advantage up to now, and most implementations rely on classical p-n junction electronics and classical ray optics. In this chapter, the use of periodic nanowire arrays and their impact on solar cell efficiency has been analyzed. An efficiency analysis is presented that includes the nature of wave optics based on electromagnetic propagation and the optical DOS. The former leads to resonance and interference effects, which in turn causes localization of the electromagnetic energy and spectral dispersion. As main result, the efficiency of a bulk cell with ideal antireflective coating can be achieved with wires of only 20% of material without antireflective coating. In addition, efficient light management improves the solar cell efficiency via a modification of the optical DOS. This leads to a higher open-circuit voltage compared to a classical solar cell.

Acknowledgements

This work has been supported by the EU FP7 grant AMON-RA (grant agreement FP7-214814-1). The author acknowledges the contribution of J. Kupec, S. Yu on the numerical simulation results.

Appendix 42.A: Material Parameters

The InP material parameters used in the simulations are listed here. The refractive index dispersion is shown in Figure 42A.1.

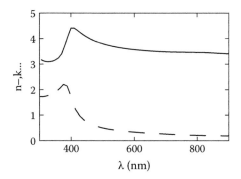

FIGURE 42A.1 InP refractive index n and extinction k versus wavelength (From S. L. Chuang, *Physics of Photonic Devices. Wiley Series in Pure and Applied Optics*, 2009. Copyright Wiley-VCH Verlag GmbH & Co. KGaA).

As mobility model for the electronic simulations, the Arora model has been used:

$$\mu = \mu_{min} + \frac{\mu_D}{1 + (N_i/N_0)^A},\tag{42A.1}$$

with N_i being the impurity density, and the remaining parameters have been chosen as [33].

	$\mu_{min}(m^2/V \cdot s)$	$\mu_D(m^2/V \cdot s)$	$N_0(m^{-3})$	A
el	0.040	0.520	3.00e23	0.47
hl	0.001	0.017	4.87e23	0.62

The intrinsic carrier density is $n_i = 1.2 \times 10 \ m^{14}m^{-3}$, the bandgap energy is $E_G = 1.34$ eV, and the radiative recombination coefficient is $1.2 \times 10 \ m^{-16}m^3/s$.

References

1. L. Tsakalakos, J. Balch, J. Fronheiser, M.-Y. Shih, S. F. Leboeuf, M. Pietrzykowski, P. J. Codella, B. A. Korevaar, O. Sulima, J. Rand, and A. Davuluru. Strong broadband optical absorption in silicon nanowire films. *Instrumentation*, 1:1–10, 2007.
2. E. Garnett and P. Yang. Light trapping in silicon nanowire solar cells. *Nano Letters*, 10:1082–1087, 2010.
3. J. Wallentin, N. Anttu, D. Asoli, M. Huffman, I. Berg, M. H. Magnusson, G. Siefer, P. Fuss-Kailuweit, F. Dimroth, B. Witzig-Mann, H. Q. Xu, L. Samuelson, K. Deppert, and M. T. Borgstrm. InP nanowire array solar cells achieving 13.8% exceeding the ray optics limit. *Science*, 339(6123):1057–1060, 2013.
4. B. Kayes, H. Atwater, and N. Lewis. Comparison of the device physics principles of planar and radial pn junction nanorod solar cells. *Journal of Applied Physics*, 97(11):114302, 2005.
5. C. Boecklin, R. G. Veprek, S. Steiger, and B. Witzigmann. Computational study of an InGaN/GaN nanocolumn light-emitting diode. *Physical Review B*, 81:155306, 2010.
6. G. Mariani, A. C. Scofield, C.-H. Hung, and D. L. Huffaker. GaAs nanopillar-array solar cells employing in situ surface passivation. *Nature Communications*, 4:1497, 2013.

7. W. Shockley and H. J. Queisser. Detailed balance limit of efficiency of pn junction solar cells. *Journal of Applied Physics*, 32(3):510–519, 1961.
8. E. M. Purcell, H. C Torrey, and R. V Pound. Resonance absorption by nuclear magnetic moments in a solid. *Physical Review*, 69(1–2):37–38, 1946.
9. Ivan S. Nikolaev, Willem L. Vos, and A. Femius Koenderink. Accurate calculation of the local density of optical states in inverse-opal photonic crystals. *Journal Optical Society of America B*, 26(5):987–996, 2009.
10. E. Yablonovitch. Statistical ray optics. *Journal of the Optical Society of America*, 72(7):899–907, 1982.
11. J. Nelson. *The Physics of Solar Cells*. London: Imperial College Press, 2003.
12. S. L. Chuang. *Physics of Photonic Devices. Wiley Series in Pure and Applied Optics*, 2nd edition. New York, NY: Wiley, 2009.
13. K. Barnham and G. Duggan. A new approach to high-efficiency multi-band-gap solar cells. *Journal of Applied Physics*, 67:3490–3493, 1990.
14. A. Luque and A. Marti. Increasing the efficiency of ideal solar cells by photon induced transitions at intermediate levels. *Physical Review Letters*, 78:5014–5017, 1997.
15. R. T. Ross and A. J. Nozik. Efficiency of hot-carrier solar energy converters. *Journal of Applied Physics*, 53(5):3813–3818, 1982.
16. J. Kupec and B. Witzigmann. Dispersion, wave propagation and efficiency analysis of nanowire solar cells. *Optics Express*, 17(12):10399–10410, 2009.
17. A. Taove and S. C. Hagness. *Computational Electrodynamics: The Finite-Difference Time-Domain Method*, 3rd edition. Norwood, MA: Artech House, 2005.
18. B. Cockburn and C.-W. Shu. The Runge-Kutta discontinuous Galerkin method for conservation laws v: Multidimensional systems. *Journal of Computational Physics*, 141:199–224, 1998.
19. F. Roemer, B. Witzigmann, O. Chinellato, and P. Arbenz. Investigation of the Purcell effect in photonic crystal cavities with a 3D finite element nanowire solar cells: Electro-optical performance Maxwell solver. *Optical and Quantum Electronics*, 39(4–6):341–352, 2007.
20. F. Roemer and B. Witzigmann. Spectral and spatial properties of the spontaneous emission enhancement in photonic crystal cavities. *Journal Optical Society of America B*, 25(1):31–39, 2008.
21. O. D. Miller, E. Yablonovitch, and S. R. Kurtz. Strong internal and external luminescence as solar cells approach the Shockley–Queisser limit. *IEEE Journal of Photovoltaics*, 2(3):303–311, 2012.
22. D. Knig, K. Casalenuovo, Y. Takeda, G. Conibeer, J. F. Guillemoles, R. Patterson, L. M. Huang, and M.A. Green. Hot carrier solar cells: Principles, materials and design. *Physica E: Low-Dimensional Systems and Nanostructures*, 42(10):2862–2866, 2010.
23. M. Luisier and G. Klimeck. Atomistic full-band simulations of silicon nanowire transistors: Effects of electron-phonon scattering. *Physical Review B*, 80:155430, 2009.
24. W. Frensley. Wigner-function model of resonant-tunneling semiconductor device. *Physical Review B*, 36(3):1570–1580, 1987.
25. U. Aeberhard and R. Morf. Microscopic nonequilibrium theory of quantum well solar cells. *Physical Review B*, 77(12):125343, 2008.
26. U. Aeberhard. Simulation of nanostructure-based and ultra-thin film solar cell devices beyond the classical picture. *Journal of Photonics for Energy*, 4(1):042099, 2014.
27. M. Auf der Maur, G. Penazzi, G. Romano, F. Sacconi, A. Pecchia, and A. Di Carlo. The multiscale paradigm in electronic device simulation. *IEEE Transactions on Electron Devices*, 58(5):1425–1432, 2011.
28. J. Kupec, L. R. Stoop, and B. Witzigmann. Light absorption and emission in nanowire array solar cells. *Optics Express*, 18(26):27589–27605, 2010.
29. S. Yu, J. Kupec, and B. Witzigmann. Efficiency analysis of III-V axial and core-shell nanowire solar cells. *Journal of Computational and Theoretical Nanoscience*, 9(5):1–8, 2012.
30. F. Roemer and B. Witzigmann. Modelling surface effects in nanowire optoelectronics. *Journal of Computational Electronics*, 11(4):431–439, 2012.

31. S. Bothra, S. Tyaqi, S. K. Ghandhi, and J. M. Borrego. Surface recombination velocity and lifetime in InP. *Solid-State Electronics*, 34:47–50, 1991.

32. P. P. Altermatt, A. G. Aberle, J. Zhao, A. Wang, and G. Heiser. A numerical model of p-n junctions bordering on surfaces. *Solar Energy Materials and Solar Cells*, 74:165–174, 2002.

33. M. Sotoodeh, A. H. Khalid, and A. A. Rezazadeh. Empirical low-field mobility model for III–V compounds applicable in device simulation codes. *Journal of Applied Physics*, 87(6), 2000.

Thin-Film Solar Cells

43.1 Introduction

In this chapter, we review simulation models for thin-film solar cells, and especially the specific challenges encountered in modeling of thin-film cells based on emerging technologies. From a technological point of view, the numerous currently available photovoltaic (PV) technologies can be roughly subdivided into two classes. One is based on conventional (mono-)crystalline semiconductor technology, to which belong the first generation of solar cells such as monocrystalline silicon or gallium arsenide p-n junction cells, and the other is based on thin-film technology [1]. The term *thin-film* in this context does not primarily refer to the thickness of the devices, but to the type of production technologies used, although their physical thickness is typically small in the range of few tens of micrometer to well below 1 μm [19,90].

Thin-film technology is characterized by several advantages over conventional technology [4]. First of all, material deposition techniques are often more simple. Depending on the material system, there is a vast diversity in techniques. Many of these are low-temperature methods, when compared to standard crystalline semiconductor technology, and many of them are easily scalable to large-scale production and large device areas. In particular, solar cells based on organic polymers, e.g., can be produced from liquid solutions with processes similar to printing technology, which opens a way to very low-cost large-scale roll-to-roll production [5]. Overall, simpler and lower cost production methods and often cheaper, more abundant, or environmentally unproblematic materials are the main driving factors for development and optimization of thin-film PV concepts. Still, these technologies have to compete with costs of traditional PV, other renewable energy sources, and most of all with that of fossil energy sources in order to penetrate the energy market. In 2014, thin-film technologies accounted for roughly 9% of the global PV module production [6].

The "classical" thin-film solar cell technologies include the second generation cells like amorphous and microcrystalline silicon (a-Si, μc-Si), the chalcogenide alloy copper indium gallium diselenide (CIGS), and cadmium telluride (CdTe), which together count up for almost all of the current thin-film solar cell market.

Although the record efficiency map regularly published by the National Renewable Energy Laboratory (NREL) [7] groups only these technologies under the term "thin-film," we assume a broader interpretation by including and actually focusing on emerging concepts. These concepts are organic-based (bulk) heterojunction cells (organic photovoltaics [OPV]), dye solar cells (DSCs), and Perovskite-based cells. Recent reviews on the state of the art of these emerging technologies, also known as third-generation cells [8], can be found in [9–11].

Modeling of thin-film solar cells requires three essential ingredients, namely a model for the optical absorption, i.e., carrier or charge generation, for the charge separation, and for electronic transport. While the details of the latter depend mostly on the material system and technology of a given cell, the optical absorption is particularly critical for thin-film devices compared to devices employing thick absorbers. This is because they typically consist of several layers of different materials with thicknesses comparable to or smaller than the optical wavelength. Therefore, the optical absorption profile can be significantly determined by interference effects, so that the calculation of the optical absorption requires basically the explicit solution of the full Maxwell equations. Even if the optical absorption in some planar devices with thicker absorber layers, like DSCs, may be calculated with simplified approaches, the inclusion of light management–improving measures like scattering elements or back contact gratings usually requires a scheme based on the solution of Maxwell's equations, too.

Most of the models described in literature are using for the optical modeling constant (but material and wavelength dependent) bulk absorption coefficients, such that the optical absorption can be calculated once, providing the carrier generation profile for the electronic transport simulation. Analytical models for carrier transport can be found under suitable simplifying assumptions, as described e.g., in [1], and such models are very useful to understand basic features of solar cell behavior. Here, however, we are interested in physics-based simulation beyond analytical approximations, which provides more detailed insight into device operation and which, in particular, can be used for device optimization. Moreover, analytical models can typically be formulated only for one-dimensional (1D) problems, while numerical simulations can treat higher dimensions. This allows for simulation-based optimization of the device geometry, e.g., related to contacting schemes, which is hardly possible using only 1D simulations.

The chapter is organized as follows. First, we give a short overview on the basic features of the different thin-film technologies. Then we review the basic physics-based modeling approaches typically used for thin-film solar cells, both for optical absorption and electronic transport. Then we will describe critical details and modeling issues for the different third-generation thin-film concepts. More advanced concepts based on low-dimensional structures, hot carriers or using plasmonic effects, which generally require more complex simulation models, are not covered in this chapter.

43.2 Technology Overview

In the following sections, we shortly review the main characteristics of different thin-film technologies. For more in-depth discussions and further references, we refer, e.g., to [2,12].

43.2.1 Silicon-Based Cells

The currently most important silicon thin-film technologies are based on amorphous (a-Si) and microcrystalline (μc-Si) form. In contrast to crystalline silicon, which forms a regular tetrahedral configuration in a diamond lattice such that every silicon atom has four neighboring atoms, amorphous silicon does not show the same regular structure. The tetrahedral coordination of the atoms is mostly maintained, but random variations in bond angles and random appearance of dangling bonds, i.e., atoms having less than four neighbors, destroys the regular lattice and the associated periodicity.

The dangling bonds in amorphous silicon lead to a large number of defect states inside the bandgap, which adversely influence its electronic properties. In particular, the material can hardly be doped.

However, passivation of at least part of these dangling bonds by hydrogen atoms, known as *hydrogenation*, has been found to be necessary and sufficient to improve considerably material quality and allow it to be doped [13]. The hydrogenated form of a-Si is known as a-Si:H.

μc-Si (or more correctly μc-Si:H) is a more complex material, consisting of amorphous phases and conglomerates of silicon nanocrystals. Their microstructure depends largely on deposition methods and underlying layers, and the presence of grain boundaries between crystalline domains will influence further electronic properties.

As expected, the electronic properties of such disordered materials are worse than those of crystalline silicon. In particular, carrier mobilities are much lower. However, both materials are good enough to build working solar cells.

Both a-Si and μc-Si can be produced by a plasma-enhanced chemical vapor deposition (PECVD) technique introduced around 1970 [13,14]. The advantages of this technique over crystalline silicon deposition methods is that the required substrate temperature is significantly lower ($<600\,°C$) and that it can be employed for large-area deposition. The former allows the use of cheap substrate materials such as plastic, while the latter allows to process large modules at once. Moreover, a-Si and μc-Si have two material intrinsic advantages. First, the loss of long-range order effectively leads to a transition from an indirect bandgap to a direct one, resulting in higher optical absorption coefficient. This allows to use thinner absorber layers without loss of absorption and thus to a cost reduction. Second, because of the different electronic transport properties these cells actually perform better at higher temperatures, contrary to crystalline silicon solar cells.

Typically, a-Si and μc-Si solar cells are built as p-i-n structures as shown in Figure 43.1a rather than pn-junctions, i.e., with an intrinsic layer between the n- and p-doped contact layers, since the carrier diffusion length in doped material are rather short due to the relatively poor transport properties. The transparent conductive oxide (TCO) layer in Figure 43.1a is e.g., indium tin oxide (ITO) or SnO_x. This layer is common to all thin-film solar cell technologies and serves to achieve a sufficiently low front contact resistivity without optical absorption. Cell thickness is usually rather small (around 400 nm for a-Si, few μm for μc-Si) in order to reduce recombination losses. Current best research cell efficiency is around 10% for a-Si:H and 12% for μc-Si single junction cells, compared to 25% of single-crystal silicon cells [15].

One of the most important problems in hydrogenated silicon cells at current is a light-induced performance degradation by prolonged illumination, known as the Wronski–Staebler effect, which is believed to be related to light-induced diffusion of the hydrogen atoms [16].

A recent review on thin-film Si solar technology can be found, e.g., in [17].

43.2.2 CdTe-Based Cells

CdTe is a very suitable material for PV applications. It has a direct bandgap of $E_g \approx 1.45$ eV, which is in the optimum range of values, and it has a high optical absorption coefficient $>10^4$ cm^{-1} over the whole solar spectrum, allowing for thin absorbers.

Both monocrystalline and polycrystalline CdTe cells can be produced. Polycrystalline CdTe can be deposited by a vast range of techniques, including vacuum techniques such as sublimation, evaporation, atomic layer deposition, chemical vapor deposition, but also nonvacuum techniques based on solution processes. Low-temperature processes are especially interesting since they allow for deposition on cheap and flexible substrates and for large-scale roll-to-roll process technology. Deposition is followed by a $CdCl_2$ treatment to improve material quality.

Some disadvantages of CdTe are the inability to tune the bandgap by alloying, and the presence of native defects to an amount that makes it difficult to achieve high-doping levels. The latter is amended by smart contacting schemes. At the front contact, an n-doped CdS buffer layer is introduced before the p-type CdTe absorber layer, which also acts as a buffer layer for CdTe deposition. The back contact is challenging because of the impossibility to achieve high p-doping density and since CdTe has a high electron affinity leading to high Schottky barriers for usual contact metals. Several solutions exist, e.g., suitable combinations of

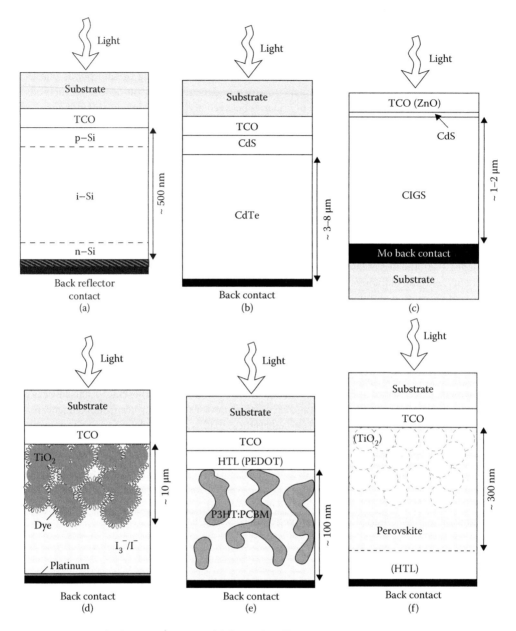

FIGURE 43.1 Typical schematic structures of different thin-film solar cells. (a) Amorphous silicon, with intrinsic absorber layer (i-Si) and doped contact layers (n/p-Si); (b) cadmium telluride (CdTe); (c) copper indium gallium selenide (CIGS); (d) dye solar cell (DSCs): TiO_2 nanoparticles are in the order of 10 nm diameter; (e) organic solar cell: the characteristic dimension of the absorber blend is in the order of 10 nm; and (f) perovskite cell, in some versions including a mesoporous TiO_2 scaffold as used in DSCs. Typical absorber thicknesses are indicated in the figures. TCO is the transparent conducting oxide, like indium tin oxide (ITO) or zinc oxide (ZnO), HTL is the hole transport layer.

buffer layers with metallization or particular surface treatments. Transport and recombination properties in CdTe films depend much on deposition technology and on the properties of the polycrystallinity of the film. The best research cell efficiency reported so far is 21% [15]. A typical CdTe solar cell scheme is shown in Figure 43.1b.

CdTe thin-film solar cells have been commercialized successfully and in 2014 they covered roughly 44% of the market share of all thin-film technologies [6]. It has to be noted that the presence of cadmium in CdTe has induced some doubts about the environmental compatibility of this technology. Recent reviews can be found in [18,19].

43.2.3 Copper Indium Gallium Selenide–Based Cells

Chalcopyrite-based solar cells is the second most important thin-film technology nowadays with a market share of 39% [6]. The most relevant compounds of this material class for PV applications are $CuInSe_2$, $CuInS_2$, and $CuGaSe_2$ having a bandgap of 1.0, 1.5, and 1.7 eV, respectively. These compounds can be easily alloyed, providing an extended range of bandgaps and lattice constants. Typically, the absorber consists of a $Cu(In,Ga)(S,Se)_2$ alloy, giving the name "CIGS," or of $CInS_2$ which has a gap of 1.5 eV. The highest cell efficiency reported in 2015 was around 21%, higher than multicrystalline Si cells [15]. Beside significantly higher efficiency than a-Si thin-film cells, CIGS cells show better stability and have no light-induced degradation.

A typical CIGS cell is shown in Figure 43.1c [20]. The nontransparent back contact is typically molybdenum sputtered onto a suitable substrate. Then follows a polycrystalline p-type CIGS absorber layer of few micrometer thickness. A heterointerface with a very thin n-type CdS is subsequently formed, followed by a TCO-like heavily doped ZnO as front contact and window layer.

Like CdTe and a-Si, CIGS cells can be produced on flexible substrates, allowing for specialized application fields. Reviews can be found, e.g., in [21,22].

43.2.4 Dye-Sensitized Solar Cells

A dye-sensitized solar cell (DSC) is an electrochemical system where the absorption and charge transport components are spatially separated [23]. The standard device is made of a mesoporous oxide, usually titanium dioxide (TiO_2), composed of nanoparticles of 15–20 nm in diameter. The surface of the mesoporous oxide is covered by a monolayer of molecules, the dye, conventionally organometallic complexes like the N719 [24] based on ruthenium. Finally, the oxide is sourrounded by a liquid electrolyte containing a redox pair. The most common one is triiodide/iodide pair (I_3^-/I^-), but many other redox pairs have been tried. In particular, cobalt-based redox pairs [25] have shown interesting properties. The contacts also are made of a glass with a transparent conducting oxide layer like ITO or fluorine-doped tin oxide (FTO) on the TiO_2 side, the anode, while the cathode is a thin layer of platinum deposited on a metallic substrate. In the device, the dye is the active material which absorbs light, while the redox pair in the electrolyte and the TiO_2 are the hole and electron transporting materials, respectively. The mesoporous structure of the TiO_2 is necessary in order to increase the surface for the dye and thus enhance dramatically the absorption efficiency, since necessarily only a monolayer of dye molecules can be used in order to efficiently extract the electrons and regerenate the dye.

The photoconversion in a DSC is far more complex than in conventional PV devices. After the absorption of the photon by the dye, the electron is transferred to the TiO_2 by an extremely fast charge transfer. Then the electron percolates in the mesoporous TiO_2 until it is collected by the ITO or FTO at the anode. The oxidized dye is regenerated by an electrochemical process with the redox pair. In a standard DSC this means that iodide is consumed and triiodide is generated nearby of the dye. The process generates an excess of triiodide with respect to the equilibrium concentration of the I_3^-/I^- couple in the bulk electrolyte. This concentration gradient moves triiodide toward the cathode where it is transformed back into iodide by taking two electrons from the catalyst, the platinum, closing the circuit. The platinum takes electrons directly from the external circuit.

There are several advantages of a DSC: it is a thin-film device, with a thickness in the range of tens of micrometer; it is quite cheap as the most expensive components (platinum and dye) are used in tiny amounts (few nanometer in thickness and a single molecular layer, respectively); the decoupling of the absorption and transport components strongly reduces recombination processes as electron and hole pairs are immediately separated on different materials at the very beginning of the photoconversion process; and fabrication processes are very cheap.

The main drawback of the device is the liquid component that can easily leak out from microcracks in the encapsulation material or be polluted by the external environment. Moreover, the most common redox pair, triiodide/iodide, is highly corrosive leading to problems in the metals used to make the external contacts and degradation of the dye layer due to desorption of molecules from the surface. All these effects are obviously detrimental for the life span of the device. In order to correct these problems, an alternative DSC has been proposed [26], where the liquid electrolyte for hole transport is substituted by an organic semiconductor (e.g., Spiro-OMeTAD or P3HT [27]). The advantage is that the new hole transporter is more stable and noncorrosive; however, the charge transport in these organic semiconductors is worse leading to higher recombinations at the interface. Moreover, the intermixing between the organic material and the mesoporous TiO_2 is less efficient. DSCs have reached 12% of confirmed efficiency [15]. Recent reviews can be found in [28,29].

43.2.5 Organic Solar Cells

OPV cells are classified as organic because their active layer consists entirely of carbon-based materials, such as polymers or small organic molecules. A major advantage of OPVs is their compatibility with techniques suitable for large-scale and low-cost fabrication, such as roll-to-roll processing. The ability to chemically engineer organic materials allows to create a variety of compounds suitable for OPV applications. Especially due to the high absorption coefficient of specific polymers, very thin layers of 50–300 nm provide sufficient absorption and make OPVs a favorable choice for thin-film solar cells. Additional features, such as mechanical flexibility and low weight, open up the possibility for a variety of applications.

A peculiarity of the organic materials used for OPVs is their low permittivity ($\varepsilon_r \approx 3-4$), which leads to a weak screening of electrostatic interactions. On absorption of a photon, an electron–hole pair is created. Due to the strong Coulomb interaction between the electron and the hole, the pair forms an exciton with binding energies of several hundred millielectron volts (meV). The thermal energy of 25 meV at room temperature is not large enough to split the exciton and separate the charges. Therefore, the concept of a heterojunction composed of a (electron) donor, usually a polymer or small molecule, and an (electron) acceptor was introduced. The most prominent acceptor material is a derivative of the Buckminsterfullerene, [6,6]-phenyl-C61 butyric acid methyl ester (PCBM), because its large electronegativity has shown major improvement in exciton dissociation [30].

The first organic solar cell with efficiency of >1% was developed by Tang [31] and consisted of a planar heterojunction between donor and acceptor materials. Planar architectures have the disadvantage of a low exciton splitting efficiency, since excitons typically have a very small diffusion length and thus only excitons created nearby the junction can diffuse to the interface and become separated before recombination. A boost in device performance was achieved by introducing the concept of a bulk heterojunction (BHJ) [32], which is an intermixed morphology between donor and acceptor materials, as indicated schematically in Figure 43.1c. It provides a trade-off between efficient exciton splitting and good charge transport. The spatial structure of the BHJ is crucial for device performance and is still object of intense study.

The pioneering organic devices nowadays are based on the polymer/fullerene bulk-heterojunction concept with efficiencies of up to 10.8% [15,33]. Although much progress has been achieved in the last decade, higher efficiencies are needed in order for organic solar cells to be commercially competitive with established (inorganic) technologies. Since the experimental investigation of OPVs during operation is difficult, simulations can give valuable insights into the still not fully understood charge generation and transport processes.

More details and reviews on OPV can be found, e.g., in [9,34,35].

43.2.6 Perovskites-Based Cells

Since the first attempts in 2009 and a breakthrough in 2012 regarding efficiency and stability [36], the organometal lead halide perovskite $CH_3NH_3PbX_3$ (with X = Cl, Br, or I) has attracted great attention as absorber material in PV devices. Originally, perovskites have been used as alternative dyes in DSCs, but then perovskite-based solar cells started forming a PV technology on their own. Structurally, these perovskites consist of an ionic cage built from PbX in a tetragonal structure, with an organic methylammonium (CH_3NH_3) molecule at the center [37].

One of the most astonishing properties of perovskite-based solar cells is the fact that perovskite layers can be very easily deposited with simple solution processes, and at the same time easily reaching high photoconversion efficiencies. In fact, recent perovskite cells show efficiencies comparable to the best CdTe and CIGS cells, and thus much higher than all other emerging PV concepts [15]. Such a good performance is due to good absorption properties and good, balanced electron and hole transport properties.

A further useful property of methylammonium lead halide perovskites ($MAPbX_3$) is the ability to tune the bandgap by alloying $MAPbI_3$ with $MAPbBr_3$, which allows to extend the absorption spectra up to 800 nm.

Since perovskite solar cells originally emerged from the DSC technology, their structure often resembles a DSC. Usually, a perovskite layer of few hundred nanometer thickness is deposited on a TiO_2 layer, with or without mesoporous phase, followed by a hole transport layer (HTL), as shown in Figure 43.1f. However, the latter may even be omitted [38].

Substantially, perovskite technology is "dirty," cheap but performing at the same time, reaching efficiencies as high as 20% at cell level [15]. However, it struggles with several open issues. Most importantly, stability is still very poor [39], which is mostly due to the sensitivity of the $MAPbI_3$ to humidity [40]. Second, often the measured current–voltage characteristics show pronounced hysteresis, which is nowadays believed to be due to migration of charged defects, especially lead vacancies, in the perovskite. This poses difficulties for the accurate experimental evaluation of cell efficiencies. Finally, toxicity of lead halide perovskites may be an issue. In fact, lead compounds are toxic, and therefore often banned by law and certainly unpopular in industry. Identification of lead-free perovskites, however, did not yet lead to satisfactory candidates. It seems, that lead-based halides provide best performance.

Recent reviews on perovskite-based solar cells can be found in [37,38,40,41].

43.3 Optical Modeling

The calculation of the optical field distribution in a solar cell is a critical modeling step for any solar cell, as it provides the carrier generation profile, which is the most important factor determining the photocurrent. The generation of carriers is related to the optical absorption of a material for a given wavelength λ and can be written as [42,43]

$$G(\lambda; x) = \alpha(\lambda; x)I(\lambda; x) \tag{43.1}$$

where G is the carrier generation rate per wavelength interval at a given incident free-space wavelength λ at position x, usually given in units of $cm^{-3}s^{-1}nm^{-1}$; α is the optical absorption coefficient at wavelength λ, with units cm^{-1}; and $I(\lambda; x)$ is the local light intensity given as photon flux in $cm^{-2}s^{-1}nm^{-1}$. In Equation 43.1, we assume that every absorbed photon produces a carrier, which may be an unbound electron–hole pair or an exciton. The total generation profile is then obtained by summing the contributions of all relevant wavelengths present in the incident light. The knowledge of the local absorption requires the knowledge of the local wavelength-dependent absorption coefficient, which is usually taken as a material constant, and of the local intensity spectrum. The latter is related to the local strength of the electromagnetic field, which can be written as [44]

$$I(\lambda; x) = \frac{\lambda \varepsilon_0 n}{2h} |E(\lambda; x)|^2 \tag{43.2}$$

where ε_0 is the vacuum permittivity, n the refractive index, h Planck's constant, E the amplitude of the electric field, and λ the free space wavelength. Therefore, the knowledge of the electromagnetic field distribution in a solar cell device allows to calculate the local generation rate G.

While the local electromagnetic field is accessible only via simulation, it is possible to both calculate and measure the absorption spectrum α. The usual way to calculate α is by using Fermi's golden rule and evaluating the optical transition matrix elements between the involved electronic states [1,45]. In this approximation, the transition probability between two states i and f with different energies E_i and E_f is given by

$$P_{i \to f} = \frac{2\pi}{\hbar} |\langle i| H' |f\rangle|^2 \delta(E_f - E_i \pm \hbar\omega), \tag{43.3}$$

where H' is the Hamiltonian describing the light-matter interaction, treated as a perturbation, and $\hbar\omega$ is the photon energy. $\langle i| H' |f\rangle$ in Equation 43.2 is the matrix element coupling initial and final states i and f, and its calculation requires an explicit representation of these states in some suitable basis. The delta function is an expression of the conservation of energy, and the minus sign describes photon absorption, while a positive sign describes photon emission, i.e., radiative recombination. The net rate of photogeneration is then given by the product of transition probability with the number of occupied initial and available final states, and by subtracting emission from absorption. The occupation of the states is given by the carriers' Fermi–Dirac distribution functions f_i and f_f. Dividing the net rate by the incident photon flux leads to the absorption coefficient [45]

$$\alpha(\hbar\omega) = \frac{\pi e^2}{n_r c \varepsilon_0 m_0^2 \omega V} \sum_{\substack{\mathbf{k}_i, \mathbf{k}_f, \\ \sigma_i, \sigma_f}} |\hat{e} \cdot \mathbf{p}_{fi}|^2 \delta(E_f - E_i - \hbar\omega)(f_i - f_f) \tag{43.4}$$

Here, e is the elementary charge, n_r is the relative dielectric constant, ε_0 is the permittivity, m_0 is the electron mass, V is the normalization volume, and the optical matrix element has been substituted by the momentum matrix element applying the dipole approximation [45]. The summation has to be done over the crystal momentum $\mathbf{k}$ and all quantum numbers σ of both initial and final states. For absorption and at not too high optical excitation, one can usually assume all initial states as occupied and all final states as empty, i.e., $f_i \approx 1$ and $f_f \approx 0$.

Radiative recombination is obtained in a similar way; however, in that case only transitions between occupied electron and hole states are relevant. Therefore, the summation can be restricted to states with energies near the band extrema. For absorption, however, all transitions are relevant, and therefore the electronic states have to be known with sufficient accuracy for all energy bands and in the whole reciprocal k-space (or more appropriately in the full first Brillouin zone). This makes calculation of absorption spectra more difficult, so that usually measured spectra are used.

In principle, the local electric field strength at all relevant wavelengths can be obtained by solving Maxwell's equations using one of the standard methods such as finite difference time domain (FDTD) [46], finite elements (FEM) [47], or rigorous coupled-wave analysis (RCWA) [48], but this is not always necessary. For a homogeneous thick absorber, in the absence of coherent effects, the Beer–Lambert approximation can be used [1]. It allows to calculate the local optical intensity simply considering the absorption along the optical path, starting from Equation 43.1. Assume a monochromatic light beam with wavelength λ with intensity I traveling in a material with absorption coefficient α. When it travels an infinitesimal distance dx, the intensity of the beam reduces by αIdx, since a part of the photons are absorbed. We can therefore deduce a continuity law for the light intensity given by

$$\frac{\mathrm{d}I}{\mathrm{d}x} = -\alpha I \tag{43.5}$$

The solution to this is an exponential function, describing exponential decay of the intensity known as Beer–Lambert law, $I(x) = I_0 e^{-\alpha x}$, where I_0 is the intensity of the incident beam (after subtraction of reflection losses, i.e., immediately inside the device) and x the distance from the point of incidence. In the case of spatially inhomogeneous absorption, the result is easily generalized to

$$I(\lambda; x) = I_0(\lambda) e^{-\int\limits_0^x \alpha(\lambda; x)\, \mathrm{d}x} \tag{43.6}$$

Equation 43.6 together with Equation 43.1 are used succesfully for the simulation of solar cells where layer thicknesses are larger than the optical wavelength. However, since this model is based on a ray optics description of light, it will break down whenever effects like interference or diffraction are important. In that case a more rigorous approach is needed to calculate the optical intensity profile.

Under the assumption of a flat, planar layered structure, a more precise model can be formulated, without increasing much the computational efforts. This model is based on a 1D approximation along the optical path of a plane wave, including sequential reflections at material interfaces and transmissions in homogeneous portions of the device, and is known as the *Transfer Matrix Method* (TMM) [49,50]. Today, this is probably the most used model for one-dimensional optical modeling of thin-film devices. The model has been first applied to organic thin-film solar cells in the late 1990s [51–53], and since then it has become a standard tool [54–60].

The basic principle of the TMM is shown schematically in Figure 43.2. To apply the method, we assume that the device can be sliced into layers of homogeneous properties with flat, parallel interfaces. We do not include a possible thick substrate that should be treated with an incoherent method (details on this can be found, e.g., in [44]). Further, we assume a monochromatic optical field written as $E(x) \exp(i\omega t)$, where $E(x)$ is a complex local field amplitude, where x is the coordinate along the stack, with $x = 0$ being the point of incidence. For simplicity, we assume a transverse electromagnetic (TEM) field with orthogonal incidence here, and in the following we drop the harmonic time dependence.

The amplitudes of the electric and magnetic field at each position can be found by using two relations derived from Maxwell's equations. The first relates the fields at the left interface of a layer to that at the right interface by means of the free propagation in a homogeneous medium, while the second relates the fields at the left and the right of each interface by means of continuity conditions. Maxwell's equations in

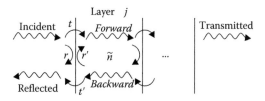

FIGURE 43.2 Scheme of the transfer matrix method. The device is assumed to be sliced in homogeneous layers, and monochromatic light is incident from the left. We decompose the optical field in a forward and backward traveling component. At each interface between two layers, these two components undergo reflection and transmission, r and t, while inside the layer the waves propagate freely with propagation constant according to the complex refractive index $\tilde{n}$.

homogeneous form (without source terms) are given by [49,61]

$$\nabla \times \mathbf{E} = -\frac{\partial \mathbf{B}}{\partial t}, \quad \nabla \cdot \mathbf{D} = 0 \tag{43.7}$$

$$\nabla \times \mathbf{H} = \frac{\partial \mathbf{D}}{\partial t}, \quad \nabla \cdot \mathbf{B} = 0 \tag{43.8}$$

$$\mathbf{B} = \mu \mathbf{H}, \quad \mathbf{D} = \varepsilon \mathbf{E} \tag{43.9}$$

Here, $\mathbf{E}$, $\mathbf{H}$, $\mathbf{D}$, and $\mathbf{B}$ are the electric field, the magnetic field (or magnetizing field, magnetic field strength), the electric displacement, and the magnetic flux density (or magnetic induction, magnetic field), respectively. The two electric fields and magnetic fields are related by the constitutive relations (Equations 43.7 through 43.9), where ε and μ are the electric permittivity and the magnetic permeability, respectively. The latter is usually equal to the vacuum permeability μ_0 for the materials of interest in this context. Since we are interested in time harmonic fields, Equations 43.7 through 43.9 are conveniently transformed to Fourier space, leading to

$$\nabla \times \mathbf{E} = -i\omega \mathbf{B} \tag{43.10}$$

$$\nabla \times \mathbf{H} = i\omega \mathbf{D} \tag{43.11}$$

while the other expressions remain the same. Note that the fields are now complex quantities. Moreover, the permittivity, which we write as product of vacuum permittivity ε_0 and relative permittivity $\tilde{\varepsilon}$, is frequency dependent (dispersive) and complex in case of lossy, i.e., absorbing materials.

In a homogeneous medium, the solution of Equations 43.7 through 43.9, or equivalently Equations 43.10 and 43.11, is found to be a propagating wave $E(x) = E_0 \exp(\pm ikx)$, where k is the propagation constant (in our case the component along x). The latter is related to the relative permittivity $\tilde{\varepsilon}$, or, equivalently, to the refractive index $\tilde{n} = \sqrt{\tilde{\varepsilon}}$ by [43]

$$k = \frac{2\pi}{\lambda} = \frac{\omega}{c} = \frac{\omega\sqrt{\tilde{\varepsilon}}}{c_0} = \frac{2\pi}{\lambda_0}\tilde{n} \tag{43.12}$$

where c_0 and λ_0 are the vacuum light speed and the free space wavelength, respectively; $\omega = 2\pi f$ is the angular frequency, and we have used the expression for the light speed $c = 1/\sqrt{\varepsilon\mu}$. Note that the refractive index $\tilde{n} = n_0 - i\kappa$ is a complex quantity, where the imaginary part is related to absorption.

It is convenient to decompose the electric field into *forward* and *backward* travelling components E^+ and E^-, as indicated in the figure. This is written as

$$E(x) = E^+(x) + E^-(x) \tag{43.13}$$

where once again all amplitudes are complex quantities.

We can now relate the field amplitudes at the left and right interface of the layer j by

$$E_{j;R}^+ = E_{j;L}^+ e^{-ik_j t_j} \tag{43.14}$$

$$E_{j;R}^- = E_{j;L}^- e^{ik_j t_j} \tag{43.15}$$

where k_j and t_j are the propagation constant and thickness of layer j, respectively. This can be put into matrix form as

$$\begin{pmatrix} E^+ \\ E^- \end{pmatrix}_{j;R} = \mathbf{M}_j \begin{pmatrix} E^+ \\ E^- \end{pmatrix}_{j;L}, \quad \mathbf{M}_j = \begin{pmatrix} e^{-ik_j t_j} & 0 \\ 0 & e^{ik_j t_j} \end{pmatrix} \tag{43.16}$$

where we have defined the transmission matrix of layer j as $\mathbf{M}_j$. Comparing Equation 43.16 with Equation 43.6 and considering Equation 43.2, the relation between $\mathrm{Im}(k)$, commonly called *extinction coefficient*, and the absorption coefficient α can be immediately found as

$$\alpha = \frac{4\pi}{\lambda}\kappa \tag{43.17}$$

Based on Maxwell's equations, one can derive the boundary conditions for the in-plane and orthogonal components of the electric and magnetic fields at planar material interfaces (orthogonal to the x-axis) as shown in Figure 43.3, which are well known and given by

$$\mathbf{n}_{lm} \times (\mathbf{E}_m - \mathbf{E}_l) = 0 \tag{43.18}$$

$$\mathbf{n}_{lm} \times (\mathbf{H}_m - \mathbf{H}_l) = 0 \tag{43.19}$$

i.e., the tangential components of $\mathbf{E}$ and $\mathbf{H}$ need to be continuous. In the above equations, $\mathbf{n}_{lm}$ denotes the normal on the interface between layers l and m. The orthogonal components of $\mathbf{D}$ and $\mathbf{B}$ have to be continuous, which is not relevant in our case of normal incidence.

Furthermore, using Equation 43.10 we can express the magnetic field in terms of the electric field as

$$B^\pm = \pm\frac{\tilde{n}}{c_0}E^\pm \tag{43.20}$$

which is obtained by evaluating the curl of the electric field and using Equation 43.12. With this we can rewrite Equations 43.18 and 43.19 in terms of the forward and backward traveling electric field components as

$$E_l^+ + E_l^- = E_m^+ + E_m^- \tag{43.21}$$

$$\tilde{n}_l(E_l^+ - E_l^-) = \tilde{n}_m(E_m^+ - E_m^-) \tag{43.22}$$

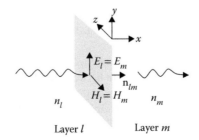

FIGURE 43.3 A material interface and the respective boundary conditions for the in-plane field components.

which can be rearranged such as to provide the transfer matrix of the material interface

$$\begin{pmatrix} E^+ \\ E^- \end{pmatrix}_{l,R} = \frac{\tilde{n}_l + \tilde{n}_m}{2\tilde{n}_l} \underbrace{\begin{pmatrix} 1 & \frac{\tilde{n}_l - \tilde{n}_m}{\tilde{n}_l + \tilde{n}_m} \\ \frac{\tilde{n}_l - \tilde{n}_m}{\tilde{n}_l + \tilde{n}_m} & 1 \end{pmatrix}}_{\mathbf{I}_{lm}} \begin{pmatrix} E^+ \\ E^- \end{pmatrix}_{m,L} \tag{43.23}$$

In the expressions for the elements of the transfer matrix $\mathbf{I}_{lm}$, we can identify the reflection and transmission coefficients of the interface between adjacent layers l and $m = l + 1$ with respect to the amplitudes

$$r = r_{lm} = \frac{E^-_{l;R}}{E^+_{l;R}} = \frac{n_l - n_m}{n_l + n_m} \tag{43.24}$$

$$t = t_{lm} = \frac{E^+_{m;L}}{E^+_{l;R}} = \frac{2n_l}{n_l + n_m} = 1 + r_{lm} \tag{43.25}$$

$$r' = r_{ml} = \frac{E^+_{m;L}}{E^-_{m;L}} = -r_{lm} \tag{43.26}$$

$$t' = t_{ml} = \frac{E^-_{l;R}}{E^-_{m;L}} = \frac{1 - r^2_{lm}}{t_{lm}} \tag{43.27}$$

so that we can write

$$\mathbf{I}_{lm} = \frac{1}{t_{lm}} \begin{pmatrix} 1 & r_{lm} \\ r_{lm} & 1 \end{pmatrix} \tag{43.28}$$

Combining Equations 43.16 and 43.28 we obtain the transfer matrix for the fields at the left interface between layers l and $m = l + 1$ as

$$\begin{pmatrix} E^+ \\ E^- \end{pmatrix}_l = \mathbf{M}_l \mathbf{I}_{lm} \begin{pmatrix} E^+ \\ E^- \end{pmatrix}_m \tag{43.29}$$

and finally the transfer matrix $\mathbf{T}$ relating the fields to the left and to the right of stack as a product

$$\mathbf{T} = \prod_l \mathbf{M}_l \mathbf{I}_{l,l+1} \tag{43.30}$$

Note that TMM can be implemented in several ways, and that it is inherently numerically unstable for larger layer thicknesses [62]. Stabilization is possible, e.g., by resorting to alternative representations of the transfer matrix [63].

As an example for the importance of interference effects in layered structures, we present a comparison of the local field strength and generation rate between results of a TMM calculation and those of an incoherent calcuation. For the latter, we considered the reflectance and transmittance at each material interface and summed up the intensities of the forward and backward traveling waves. The test structure, similar to the one used in [64], is shown in Figure 43.4a. For simplicity, we assumed total reflection at the aluminum back contact, and we did not consider the glass substrate. The results have been obtained for a wavelength of 500 nm. Figure 43.4b and c presents the field and generation profiles for two different

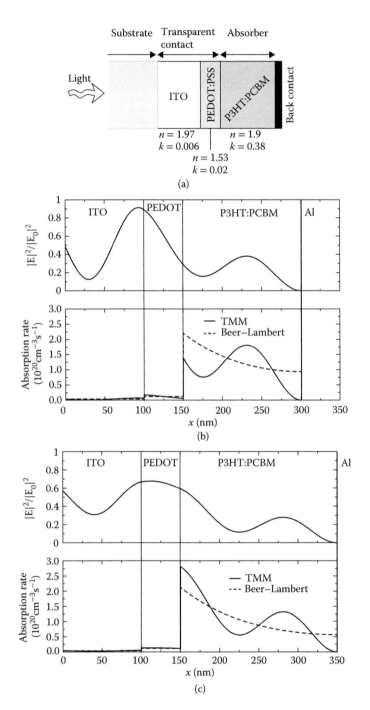

FIGURE 43.4 Structure for comparison of coherent and incoherent optical field calculation. The parameters used in the simulation are given in (a). In (b) and (c) the optical field intensity and absorption profiles for two different absorber thicknesses is shown. The local absorption rate considerably deviates from the Beer–Lambert prediction, in particular for thin layers. For thick layers Beer–Lambert behavior is recovered. (ITO, indium tin oxide; PCBM, [6,6]-phenyl-C61 butyric acid methyl ester; TMM, transfer matrix method.)

absorber thicknesses, showing that the profiles differ significantly from what would be expected from the Beer–Lambert model, and that it depends critically on layer thicknesses. In particular, absorption maxima can be located deeply inside the layer instead of near the interface. When the layers are becoming optically thick, the local absorption rate mediated over a wavelength converges to the result obtained by the Beer–Lambert law.

Figure 43.5 shows another example, borrowed from Reference [53], where the obtained results have been compared with experimental data. In this case, the donor–acceptor heterojunction is formed by the

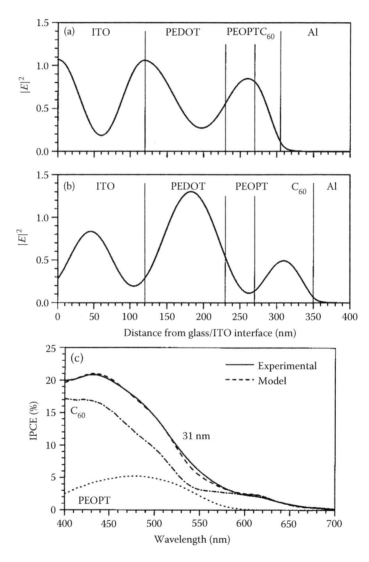

FIGURE 43.5 Field intensity profiles in an organic heterojunction solar cell for two different thicknesses of the C_{60} layer (a, b). The quality and accuracy of the optical simulation is proven by the comparison of experimental and simulated quantum efficiency in (c). Note that the simulation provides the information on where absorption happens, which cannot be obtained from the measurement. For example, in (c) the wavelength resolved contributions to the total absorption in the C_{60} and PEDOT layers is reported for an optimized C_{60} layer thickness. (Reproduced with permission from Leif A.A. Pettersson et al. Modeling photocurrent action spectra of photovoltaic devices based on organic thin films. *Journal of Applied Physics*, 86(1):487—496, 1999 Copyright 1999, American Institute of Physics Publishing LLC.)

PEOPT/C_{60} interface. To achieve maximum efficiency, a peak in field intensity should be obtained near this interface. Figure 43.3 presents the intensity profile at a wavelength of 460 nm for a C_{60} layer thickness of 35 and 80 nm, respectively, indicating that device optimization can benefit substantially from optical simulations. Note that the field profile is optimized in this way for a single wavelength only. Although the field profile cannot be measured directly experimentally, the accuracy of the optical simulation can be assessed by measuring the incident photocurrent efficiency (IPCE) (also denoted as external quantum efficiency, EQE), which is the product of quantum absorptivity and the current collection efficiency at each wavelength and gives the ratio of the number of collected electrons to the number of incident photons [43]. This quantity has been simulated in [53] by combining optical and current transport simulations, and the comparison between measurement and simulation is presented in Figure 43.3 for 31-nm thick C_{60} layer, showing excellent agreement. This example also shows how a simulation can give accurate insight into internal device behavior that is inaccessible by measurement alone, opening the way to efficient device optimization.

While 1D TMM-based modeling is a powerful tool to describe the optical field in a planar layered structure, it is less suitable to treat devices where light management techniques are employed, such as (possibly random) front or back surface texturing. Such approaches are becoming more and more important in order to approach the theoretical limits of light-to-electricity conversion, especially for technologies that are already optimized from an electrical point of view [65]. For organic solar cells, e.g., there is interest in grating structures on the back contact in order to modulate the electric field profile and thus to optimize optical generation and efficiency. Plasmonic effects are also studied for improving solar cell performance [66]. To accurately model texturing, gratings, or plasmonic effects, usually full 3D Maxwell solvers are employed [67,68]. Unfortunately, such calculations are computationally quite expensive, so that approximate models formulated in one dimension are also of interest [69]. Special approaches have been devised also for other cases, like thick layers with incoherent coupling [70].

Also note that from Figure 43.1 it might seem that TMM is not suited for devices based on mesoporous or blended materials, since the materials are not really piecewise homogeneous and there are apparently no planar interfaces. However, since the characteristic dimensions are in the order of at most tens of nanometers, these composite materials can be described with sufficient accuracy as effective materials so that TMM can be applied without particular restrictions.

43.4 Transport Modeling

While optical calculations are important for the simulation of solar cells, taken alone they can only provide upper limits for cell efficiency. For a comprehensive view on cell operation, the transport of the generated carriers and their collection at the electrical contacts has to be taken into account. This section reviews the two carrier transport models that are used most often for solar cell simulations, namely the drift-diffusion model and kinetic Monte Carlo (kMC).

43.4.1 Drift-Diffusion Model

Modeling of electronic transport in thin-film solar cells is mostly based on the drift-diffusion model, which has been the work-horse of physics-based electronic device simulation since decades. A thorough introduction, derivation of the model and implementation hints can be found in different textbooks such as [71], and a discussion of theoretical and mathematical aspects can be found, e.g., in [72]. A more detailed description is provided in Chapter 50.

Since the drift-diffusion model is well known and is described in numerous books and articles, we will provide only a short generic description and concentrate more on the features important for thin-film solar cell simulation.

For a "normal" semiconductor where only electrons and holes are mobile carriers, the drift-diffusion model consists of a set of three continuity equations for the electric displacement $D = -\varepsilon\nabla\varphi$ and the two particle currents J_n and J_p:

$$\nabla(\varepsilon\nabla\varphi) = q(n - p - C) \tag{43.31}$$

$$\frac{\partial n}{\partial t} - \frac{1}{q}\nabla J_n = -R, \qquad J_n = q(D_n\nabla n - \mu_n n\nabla\varphi) \tag{43.32}$$

$$\frac{\partial p}{\partial t} + \frac{1}{q}\nabla J_p = R, \qquad J_p = -q(D_p\nabla p + \mu_p p\nabla\varphi) \tag{43.33}$$

This nonlinear system of partial differential equations is also known as the *basic semiconductor device equations* or the van Roosbroeck system, who was the first to formulate this set of equations in the framework of electronic transport in semiconductors [73]. The first of the above equations (Equation 43.31) is the Poisson equation, relating the electrostatic potential φ with the total charge density $\rho = q(p - n + C)$ where n and p are the electron and hole densities, respectively, while ε is the permittivity. C denotes the net density of fixed charges such as dopants or traps. The other two equations are the continuity equations for the electron and hole currents, where the constitutive equations for the currents are given on the right (which are also known as *drift-diffusion equations*) where q is the electron's charge, $D_{n,p}$ are the electron and hole diffusion coefficients, and $\mu_{n,p}$ are the electron and hole mobilities. R is the net recombination rate, including all processes which lead to destruction or generation of electron–hole pairs.

The three most important recombination models are the trap-assisted (Shockley–Read–Hall [SRH]) recombination, bimolecular (direct, radiative) recombination, and Auger recombination, given by [71,74]

$$R_{\text{SRH}} = \frac{np - n_i^2}{\tau_p(n + n_t) + \tau_n(p + p_t)} \qquad \text{SRH} \tag{43.34}$$

$$R_{\text{rad}} = B(np - n_i^2) \qquad \text{Radiative} \tag{43.35}$$

$$R_{\text{auger}} = (C_n n + C_p p)(np - n_i^2) \qquad \text{Auger} \tag{43.36}$$

Here, n_i^2 is the square of the intrinsic carrier density, given by the product of the equilibrium densities n_0 and p_0. n_t and p_t are parameters depending on the energetic position inside the gap of the trap. Note that Equation 43.34 is valid for a single discrete trap level and in the absence of an electric field, however, it can be generalized to other cases [75,76]. The other parameters in the expressions are mostly material related. The relative importance of the different recombination mechanisms depends on the operating conditions and on material quality. SRH recombination is important at low carrier densities and in presence of defects, while Auger recombination usually becomes dominant only at high carrier densities.

Written in standard form (Equations 43.31 through 43.33), the particle currents are seen to be decomposed into a diffusion term proportional to the gradient of the carrier density and a drift component proportional to the conductivity times the electric field. This is an intuitive description of particle transport, whenever the system is dominated by scattering, that is when the particles undergo Brownian motion with some thermal velocity, and the external perturbation (electric field or density gradient) induces only a small nonzero mean carrier velocity leading to a net current flow.

The system of Equations 43.31 through 43.33, supplemented with suitable boundary conditions, is usually solved numerically, although under certain circumstances analytic solutions exist in one dimensions. For implementation details and numerical methods, we refer to [71] and the vast existing literature on the topic.

The device equations (Equations 43.31 through 43.33) are still the most important model for transport in semiconductor devices, since they can be solved much faster than any other transport model and therefore represent a good compromise between computational efficiency and physical accuracy. While the equations allow easy interpretation, and can be solved relatively easily, they are based on a number of assumptions that are important when applying them to systems other than crystalline semiconductors. The other important point to note is that they give a good description of device behavior only when good models for the quantities appearing as parameters in Equations 43.31 through 43.33 can be provided. In particular, the models for the recombination and generation processes and for the carrier mobilities, in combination with accurate material-dependent parameters, are important in order to reach satisfactory results with the drift-diffusion model. Most often this implies careful parameter fitting against experimental data, careful checking whether the fitted parameters provide a predictive model or not.

In the following, we discuss some of the basic assumptions underlying the semiconductor device equations (Equation 43.31 through 43.33) and possible consequences when applying them to the materials of interest in this chapter. A discussion of technology-specific issues regarding the basic equations or the parameter models is deferred to Section 43.5.

The standard way to derive the drift-diffusion equations is to start from the semiclassical Boltzmann transport equation [72]. These derivations include implicitly or explicitly assumptions on the distribution function such that the carrier density takes the form of the well-known Boltzmann approximation, i.e.,

$$n = N_c e^{\frac{E_{F,n}-E_c}{k_B T}}, \quad p = N_v e^{-\frac{E_{F,p}-E_v}{k_B T}} \tag{43.37}$$

Here, N_c and N_v are the effective density of states (DOS) in the conduction and valence band, $E_{F,n}$ and $E_{F,p}$ are the electron and hole quasi Fermi levels, and E_c and E_v are the conduction and valence band edges. The latter can be expressed as $E_{c,v} = E_{c,v}^{(0)} - q\varphi$, where $E_{c,v}^{(0)}$ are material band edges in the absence of an electric field, identifiable with electron affinity and ionization energy, k_B is the Boltzmann constant, and T the temperature.

The diffusion constants and mobilities in Equations 43.31 through 43.33 are not independent, which can be easily seen when substituting Equation 43.27 into Equations 43.31 through 43.33 and solving for equilibrium, i.e., zero current under no external bias nor carrier generation. It is found that the two quantities are related by a so-called Einstein relation [74]:

$$D_n = \frac{k_B T}{q}\mu_n, \quad D_p = \frac{k_B T}{q}\mu_p \tag{43.38}$$

which in actual implementations of the drift-diffusion model is usually used to obtain the diffusion coefficients from the mobilities.

The expressions (Equation 43.37) provide an adequate description of the carrier densities for most nondegenerate crystalline semiconductors, i.e., at not too high carrier densities. This is because the same expressions are obtained when calculating the carrier density from the DOS and the Fermi–Dirac distribution, and making the Boltzmann approximation for small densities. Generally, the density, e.g., of electrons is found under the assumption of a local equilibrium distribution by occupying the available states, given by the DOS, with a local quasi Fermi level and integrating over all energies, i.e.,

$$n(\mathbf{r}) = \int g(E) f\left(\frac{E - E_{F,n}(\mathbf{r})}{k_B T}\right) dE \tag{43.39}$$

where $f(x)$ is the Fermi–Dirac distribution and $g(E)$ the DOS. The DOS of a crystalline semiconductor at the relevant energies, i.e., near the band extrema, is obtained from the energy dispersion, which is approximately parabolic in this range, as shown schematically in Figure 43.6. Details on the calculation of the carrier densities can be found, e.g., in [42,71,74,77].

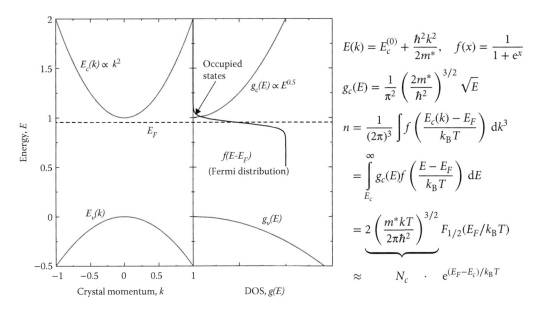

FIGURE 43.6 Parabolic approximation to the energy dispersion $E(k)$ of a simple semiconductor and the DOS resulting from this dispersion, obtained by setting $1/(2\pi)^3 \int dk^3 = \int g(E)dE$. On the right, some of the expressions are given, and the calculation of the density is sketched ($m^* = \hbar^2(\partial^2 E(k)/\partial k^2)^{-1}$ is the effective mass, E_F is the quasi Fermi level).

The key point is that the derivations are based on the assumption that the carriers are moving freely in a periodic potential of the underlying crystal lattice, which allows to define the momentum k of the carriers and which leads to the parabolic dispersion valid at least in the vicinity of the band extrema. While this is a good approximation for crystalline semiconductors, it does not hold in many materials of interest in thin-film PV technology. In particular, whenever the periodicity of the crystal structure is lost, the DOS will become substantially different. In an amorphous semiconductor such as a-Si, e.g., the crystalline lattice structure is maintained only locally and only approximately: a-Si is mostly tetrahedrally bonded, but with a statistical distribution of bond angles and with certain number of silicon atoms with three or five neighbors. This modifies the DOS by washing out the band edge and appearance of approximately exponential band tails, known as Urbach tails [78], and appearance of localized states in the gap even in absence of impurity atoms, as shown schematically in Figure 43.7b. Similarly, organic semiconductors are usually characterized by a certain energetic disorder in the molecular energy levels, and the definition of carrier momentum is largely meaningless. In this case, the DOS is completely different from the "standard" semiconductor bulk DOS, and is often well described by a Gaussian distribution with a certain width σ characterizing the statistical scattering of the molecular levels [79], shown in Figure 43.7c.

A consequence of a DOS, which differs from that resulting from a parabolic dispersion, is that the Einstein relation relating the diffusion constants and the mobilities needs to be generalized. This general form can be easily found from Equations 43.31 through 43.33 as

$$D_n = n \left(\frac{\partial n}{\partial \varphi}\right)^{-1} \mu_n, \quad D_p = -p \left(\frac{\partial p}{\partial \varphi}\right)^{-1} \mu_p \tag{43.40}$$

An accurate description of semiconductor transport therefore requires to take the functional form of the DOS into account, although it does not appear explicitly in the basic semiconductor equations. Beside the formal modification of the Einstein relation, the DOS also influences some directly measurable quantities.

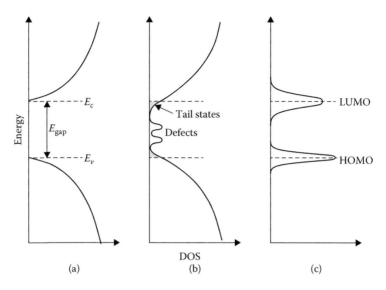

FIGURE 43.7 Sketch of different density of states (DOS): (a) crystalline semiconductor with parabolic band dispersion, (b) amorphous semiconductor with band tails and defect states, (c) disordered semiconductor, e.g., organic, with the DOS formed from energetic and spatial disorder of molecular orbital energies. In the last case, the bands are usually denoted as HOMO (highest occupied molecular orbital) and LUMO (lowest unoccupied molecular orbital).

Assume, e.g., a bulk absorber in flat band condition subject to homogeneous optical carrier generation, so that there is no current flow, and assume that only radiative recombination is relevant. In this case, all generated electron–hole pairs have to recombine radiatively according to Equation 43.35, i.e., $G = Bnp$. The splitting of the electron and hole quasi Fermi levels induced by the illumination is a measure for the open-circuit voltage V_{oc} that can be expected. Using Equation 43.37, an analytic formula can be easily found as [80]

$$V_{oc} = \frac{1}{q}(E_{F,n} - E_{F,p}) = \frac{1}{q}E_g + \frac{k_B T}{q} \ln\left(\frac{G}{BN_c N_v}\right) \tag{43.41}$$

and apparently the V_{oc} should vary by a constant rate of $2.3 k_B T/q$ per decade in intensity. A different result is obtained if the same quantity is calculated using a different expression for the DOS, e.g., using a Gaussian DOS (GDOS) as typical for organic semiconductors and given by

$$g(E) = \frac{N_0}{\sigma\sqrt{2\pi}} \exp\left(-\frac{1}{2}\frac{(E - E_0)^2}{2\sigma^2}\right) \tag{43.42}$$

Here N_0 is the total density, σ the width (variance) of the Gaussian, and E_0 the center of Gaussian. Using approximations of the formulas for the densities [81], and assuming the same parameters for the electron and hole DOS, we can obtain the open-circuit voltage for sufficiently large carrier densities in this case as

$$V_{oc} = \frac{1}{q}E_g - 2\frac{\sigma\sqrt{2}}{qH(\sigma)} \text{erfc}^{-1}\left(2\sqrt{\frac{G}{BN_0^2}}\right) \tag{43.43}$$

Here, E_g is defined as the difference of the center energies of the electron and hole DOS, and $H(\sigma) = k_B T \sqrt{2}/\sigma \cdot \text{erfc}^{-1}(\exp(-\frac{1}{2}\sigma^2/(k_B T)^2))$. Figure 43.8 compares the derivative of V_{oc} with respect to the logarithm of the generation rate for a parabolic DOS and a GDOS, showing that for typical Gaussian widths

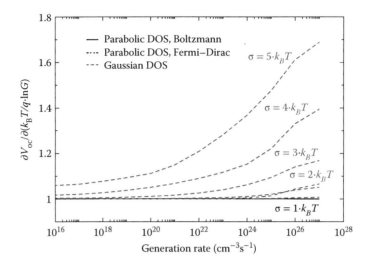

FIGURE 43.8 Variation of the bulk quasi Fermi level splitting due to a homogeneous generation rate. Deviation from the exponential form of the carrier density leads to considerable deviation from the expected rate of change. For the Gaussian density of states (DOS), $N_0 = 10^{21}$ cm^{-3} has been used, for the parabolic case $N_c = N_v = 10^{19}$ cm^{-3}. The recombination rate constant was $B = 10^{-10}$ cm$^3 \cdot$s^{-1}.

the DOS influences the rate even at moderate generation rates. The effect becomes stronger with increasing intensity. The measurable effect of energetic disorder is described, e.g., in [82].

Note that a further modification of the form (Equations 43.31 through 43.33) of the drift-diffusion equations is needed whenever the material parameters in the formula for the carrier densities (i.e., effective DOS or the band edges, see Figure 43.6), or generally the DOS, change spatially. This can be seen easily when evaluating the gradient of the densities, which then leads to additional terms that can be interpreted as effective potentials. These corrections are simple in the case of the standard exponential form (Equation 43.37) of the carrier densities, but they become more complicated in the degenerate case and for other DOS. It is therefore sometimes preferred to write the drift-diffusion model in a different but equivalent form, derived from thermodynamics [61,83]:

$$\nabla(\varepsilon\nabla\varphi) = q(n - p - C) \tag{43.44}$$

$$\frac{\partial n}{\partial t} - \frac{1}{q}\nabla J_n = -R, \qquad J_n = -q\mu_n n\nabla\phi_n \tag{43.45}$$

$$\frac{\partial p}{\partial t} + \frac{1}{q}\nabla J_p = -R, \qquad J_p = -q\mu_p p\nabla\phi_p \tag{43.46}$$

Here, $\phi_{\{n,p\}}$ are the electrochemical potentials of the electrons and holes, respectively, related to the quasi Fermi levels by $E_{F,\{n,p\}} = -q\phi_{\{n,p\}}$. While in Equations 43.31 through 43.33, the dependent variables are the electrostatic potential and the carrier densities, in (Equations 43.44 through 43.46) they are the electrostatic and the electrochemical potentials, and the particle currents are driven by the gradient of the electrochemical potential. The advantage of this formulation with respect to the one in Equation 43.31 through 43.33 is that it is independent on the functional form of the density, and the Einstein relation is not needed explicitly. For the rest, it has the same limitations of applicability as Equations 43.31 through 43.33.

From a technical point of view, note that the decoupled carrier transport equations in Equations 43.44 through 43.46 are quasilinear, with the differential operator being proportional to the carrier density, while in Equations 43.31 through 43.33 the transport equations are linear. This makes the numerical solution procedure more involved.

If we are using the thermodynamic formulation (Equations 43.44 through 43.46) for simulation of solar cells, then we need an adequate model for the DOS or, respectively, the carrier density for all the materials involved, and we need models for the carrier mobilities and the recombination rates. Especially, the mobility is a critical parameter. In a crystalline semiconductor, the mobility is related to the scattering the carriers are subject to. So it depends strongly on temperature and defect or doping densities, and usually mobility increases with decreasing temperature and decreasing defect densities, since the associated scattering rates decrease. Mobility also depends on the electric field, typically by decreasing with increasing field due to a saturation of the mean drift velocity. In fact, the proportionality between drift velocity and electric field, given by $v = \mu_n E$, holds only at low field strength [84]. In noncrystalline materials such as amorphous or organic semiconductors, these dependencies can be radically different. For example, mobility may increase with temperature because carriers can escape more easily local energy minima or overcome small energy barriers. Or it may increase with electric field because the field reduces the effective energy barriers hindering carrier flow. In some cases, such behavior is a manifestation of an underlying microscopic transport that deviates from a diffusive one.

In disordered organic materials, e.g., the deviation from a periodic crystalline structure makes the concept of band-transport inapplicable. Instead, charges are spatially localized and charge transport occurs by transitions between localized states, a process termed as *hopping*. In reality, disordered systems show both disordered and (semi-)crystalline regions so that charge transport is always an interplay between hopping and band conduction. Higher contributions of hopping transport usually lead to lower charge mobilities. Localized states can be imagined as local energy valleys in which charges get confined. Transport can either take place by surmounting the energy barrier to a neighboring localized state by absorbing phonons, or by tunneling through the barrier. The concept of phonon-assisted hopping was introduced by Mott [85]. In 1960, Miller and Abrahams described a hopping model between two localized states consisting of both a contribution due to thermal activation and tunneling. After their model, the hopping process between the states i and j with energy difference $\Delta E = E_j - E_i$ and distance r_{ij} is of the form [86]

$$a_{i \to j} = a_0 \exp\left(-2\gamma r_{ij}\right) \begin{cases} \exp\left(-\frac{\Delta E}{k_B T}\right) & \text{if } \Delta E > 0 \\ 1 & \text{if } \Delta E \leq 0 \end{cases} \tag{43.47}$$

where a_0 is given by a typical phonon frequency in the system and γ is the inverse localization constant. The formula combines the contribution of two different processes to hopping: the first exponential describes a tunneling contribution through a barrier given by the distance r_{ij} of the localized states, and the latter term stands for the thermally activated contribution with the relation between barrier height ΔE_{ij} and the thermal energy $k_B T$.

Another model that takes into account strong coupling between charges and lattice is the polaron model. In organic materials with low permittivities, a charge on a molecule can considerably deform the structure of the molecule. The complex of charge and the distortion of the lattice it induces is known as a quasiparticle called *polaron*. Marcus [87] developed a model for the transition of polarons:

$$a_{i \to j} = \frac{|J_{ij}|^2}{\hbar} \sqrt{\frac{\pi}{\lambda k_B T}} \exp\left(-\frac{(\Delta G_{ij} - \lambda)^2}{4\lambda k_B T}\right) \tag{43.48}$$

where J_{ij} is the transfer integral for electron and hole transfer, ΔG_{ij} is the free energy difference between state, i and j and λ the reorganization energy, which accounts for the reconfiguration of the molecule due to the deformation by the charge.

In order to derive a mobility model for disordered materials, a description for the distribution of available states must be available. The spatial irregularity of these systems leads to fluctuations in the intermolecular interactions. A successful model to represent these energetic fluctuations is the GDOS as given in

Equation 43.42, in which the energy distribution of the localized states is spread by a Gaussian distribution function around the HOMO/LUMO energy levels of the organic material. The GDOS in combination with the Miller–Abrahams hopping was applied by Bässler [79] by Monte Carlo simulations to study the effect of temperature and electric field on the charge mobility. He found a non-Arrhenius temperature dependence $\mu(T) = \mu_0 \exp\left(-(t\hat{\sigma})^2\right)$ with $\hat{\sigma} = \sigma/k_B T$, and, Poole–Frenkel type behavior $\mu(F) = \mu_0 \exp\left(\gamma\sqrt{F}\right)$ for the dependence on the electric field F for a limited range of electric fields. To include the influence of charge carrier densities on the mobility, Pasveer et al. [88] parametrized the results based on a direct solution of the Master equation in a system with a GDOS. They found as expression for the hopping mobility:

$$\mu(T, \rho, F) = \mu(T, \rho)f(T, F) \text{ with} \tag{43.49}$$

$$\mu(T, \rho) = \mu_0 c_1 \exp\left[-c_2\left(\frac{\sigma}{k_B T}\right)^2\right] \exp\left[\frac{1}{2k_B T}\left(\frac{\sigma^2}{k_B T} - \sigma\right)(2\rho a^3)^\delta\right] \tag{43.50}$$

$$f(T, F) = \exp\left\{0.44\left[\left(\frac{\sigma}{k_B T}\right)^{\frac{3}{2}} - 2.2\right]\left[\sqrt{1 + 0.8\left(\frac{eFa}{\sigma}\right)^2} - 1\right]\right\} \tag{43.51}$$

where ρ is the charge carrier density, either electrons or holes, F is the electric field, $k_B T$ the thermal energy, $a = N_0^{-\frac{1}{3}}$ the distance between localized states, σ the energetic disorder in the GDOS, $c_1 = 1.8 \cdot 10^{-9}$, $c_2 = 0.42$, $\mu_0 = a^2 \nu_0 e/\sigma$, with ν_0 the prefactor in the Miller–Abrahams formula, and

$$\delta = 2\frac{\log(s^2 - s) - \log(\log(4))}{s^2} \text{ with } s = \frac{\sigma}{k_B T} \tag{43.52}$$

This parametrization can be used in drift-diffusion models to account for the hopping nature of transport.

43.4.2 Kinetic Monte Carlo

In this section, we give an overview of the kinetic Monte Carlo (kMC) method [89,90] and how it can be applied to set up a model for organic solar cells.

Monte Carlo methods offer a simple yet powerful tool to solve real world problems by the use of random numbers. In the hierarchy of transport models shown in Figure 43.9, kMC bridges the gap between macroscopic models such as drift-diffusion and molecular dynamics (MD) and quantum mechanical models.

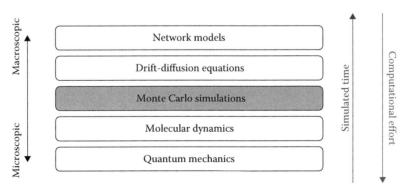

FIGURE 43.9 Hierarchy of transport models. Kinetic Monte Carlo simulations bridge the gap between macroscopic and microscopic modeling techniques.

The idea for the use of random numbers in an algorithm came from Fermi [91] and was further pursued by Ulam and Metropolis in the 1940s studying neutron diffusion in fissionable materials. They developed the most used method of this kind, the Metropolis algorithm [92], during a study about the calculation of state equations for a system of interacting particles described by a hard sphere cutoff potential.

In the following decades, Monte Carlo methods have proven to be a versatile tool and much progress in the development of calculations for a variety of areas, ranging from surface adsorption processes, radiation damage annealing, and statistical physics [93–95], has been achieved. Especially, the development of a Monte Carlo algorithm class that models physical systems as dynamically evolving from state to state by a certain set of transition rates obtained large interest as it allowed to study time-dependent properties of physical systems. This class of algorithms is today commonly known as *kinetic Monte Carlo*. It has received great benefit in the growing computational power of modern processors and provides a flexible tool to simulate dynamical behavior of physical systems, e.g., transport processes in inorganic semiconductor devices [96]. Major contributions to the modern kMC algorithm, as it is described here, have been made by Bortz et al. in 1975 [95] and Gillespie in 1976 [89] and 1977 [97].

Compared to MD simulations, which include the interactions of single atoms with all dynamics of the molecules like vibrations, occurring on very small time scales (fs), kMC characterizes the state of a system by a set of more macroscopic variables which do not include underlying dynamics like fast vibrational modes. These motions on smaller time scales can be ignored as long as the current state can still be assigned to a long-time state. The dynamical evolution of the system is only seen by transitions between such long-time states, called events. Such a system is called an infrequent-event system [90]. Figure 43.10 illustrates this concept. By ignoring the underlying fast motions and only considering the transitions between long-time states, a sufficient amount of processes can be neglected. Thus, much larger overall simulation times of the order of milliseconds to seconds can be reached.

The above mentioned concept can be applied in particular to organic solids, where particles such as electrons, holes and excitons reside in localized states and transport is determined by hopping from one localized state to another. The time scale of these hopping events is much longer than other processes, which do not change the spatial localization of the carriers.

Compared to drift-diffusion, kMC has the advantages of being able to capture the discrete nature of the problem by modeling individual particles at their molecular positions, and of providing a physically more accurate description of hopping transport without having to resort to effective continuous media models. Moreover, it can be much easier and physically more sound to treat blended materials with nanoscale morphology in kMC.

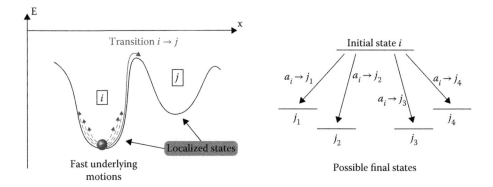

FIGURE 43.10 Transitions from long-time state to long-time state with underlying fast modes in an infrequent-event system (left). State-to-state transitions reduced to the needs for kinetic Monte Carlo (right).

43.4.2.1 kMC Algorithm

The essential concept of the kMC method is the characterization of the considered system by a set of system-specific long-time states. A state is, e.g., defined by the position of all particles in the system at a certain time. Under dynamical evolution, it is possible for the system to change its state by certain system-specific transitions, thus motion of particles from one state to another. The task for an accurate model is to identify all the important processes that take place in the system. By considering processes that happen on long time scales and neglecting those which act on very small time scales, one is able to accomplish a tradeoff between accuracy and sufficiently long simulation times. It is of importance to carefully choose the implemented processes in order not to lose essential parts of the system dynamics.

Let the system be in a certain state i. Depending on the current system setup, a set of several different states j exists to which a transition can take place. This is indicated in Figure 43.10 (right). The transitions from beginning states i to possible end states j can be characterized by rate constants a_{ij} given in units of s^{-1}. Rate constants are a measure for the probability of the corresponding transition from $i \rightarrow j$. An important property must be that the rates are memoryless, meaning that they are only dependent on the setup of the current state and are not dependent on how the system evolved into it. This property is known from the theory of Markov chains [98] and allows, if the set of rate constants is known, to determine the dynamical evolution of the system.

As known from statistical physics [99], stochastical processes characterized by discrete transitions from state to state under temporal evolution can phenomenologically be described by the master equation

$$\frac{dp_i(t)}{dt} = -\sum_j a_{ij} p_i(t) + \sum_j a_{ji} p_j(t) \tag{43.53}$$

The probability $p_i(t)$ to find the system in state i at a certain time t changes in a way that is determined by the rate constants a_{ij} and the probabilities of only the initial state $p_i(t)$ and the possible final states $p_j(t)$. Finding solutions of the master equation, be it by analytical or numerical methods, is often complex and not feasible. The kMC method is based on a stochastical framework and provides a numerical approach to obtain the dynamic system evolution based on the time-dependent propagations from state to state. By choosing a transition pathway through a chain of subsequent states, one possible dynamic system evolution is obtained. Such a walk through a pathway of transitions forms a Markov chain. Averaging over a large number of such Markov chains provides an equivalent system behavior as described by the master equation. Therefore, the kMC algorithm is essentially emulating the master equation.

For a detailed derivation of the associated mathematical framework, the reader is referred to the original work of Gillespie [89]. In the following, we present only the equations necessary for a possible implementation of the algorithm. The algorithm as proposed by Gillespie relies on the ability to characterize the system events by a set of transitions

$$\{\mu\} \text{ with } \mu = 1, 2, \dots, M \tag{43.54}$$

where M is the total number of transitions. μ is equal to a transition from state i to one of the possible states j in that configuration: $i \rightarrow j_\mu$. Then a_μ is the rate value describing process μ and $\{a_\mu\}$ is the corresponding set of rates. All transitions must describe distinct processes and their number M can vary over time, depending on which events are enabled at a certain step in the system evolution. It is essential to consider all major processes in order to be able to accurately reproduce the behavior of the system. The main task is now to develop a procedure to simulate the time-dependent evolution of the system under consideration of the given transitions. This makes it necessary to derive an algorithm that is able to choose which transition will be carried out and (for the time dependency) at what time span, all based on the drawing of random numbers.

At any given time, the set of $\{a_\mu\}$ characterizes the possible transitions. They are functions of the current system setup and will change over time, as described later.

Instead of finding a solution to the master equation, the kMC method is based on a function called the probability density function (PDF). The PDF is the suitable function to select a transition and its corresponding time step based on random numbers. The PDF is defined as

$$P(\tau, \mu)\, d\tau = \text{probability at time } t \text{ for the next transition}$$

$$\text{to occur in time interval } (t + \tau,\, t + \tau + d\tau) \qquad (43.55)$$

$$\textit{and being a transition of type } R_\mu$$

where $P(\tau, \mu)$ is the PDF. This function can be interpreted as follows: it is a joint PDF of the continuous time variable τ ($0 \le \tau < \infty$) and the integer variable μ ($\mu = 1, 2, \ldots, M$) characterizing the transition. At a certain time t, the PDF represents the probability for transition μ to occur within an infinitesimal time span of $d\tau$ after no other event has taken place for a time interval τ. Hence, $P(\tau, \mu)$ can be seen as the independent probabilities to have no transition in the time interval $(t,\, t + \tau)$ *and* to have a transition of type R_μ immediately afterward in the infinitesimal interval $(t + \tau,\, t + \tau + d\tau)$. In common terms, the transition μ needs time τ before the next transition occurs.

Using the abbreviation

$$a = \sum_{\nu=1}^{M} a_\nu \qquad (43.56)$$

as the total transition probability per unit time, the total expression for the PDF is given by

$$P(\tau, \mu) = a_\mu \cdot \exp\left(-\sum_{\nu=1}^{M} a_\nu \tau\right) = a_\mu \exp(-a\tau) \qquad (43.57)$$

This function, only dependent on the set of events a_μ, is the starting point to choose a random number pair (τ, μ) that characterizes the next transition and time step in the dynamic evolution of the system. The PDF is properly normalized, meaning that every pair (τ, μ) with $\tau \in (0 \le \tau < \infty)$ and $\mu = 1, 2, \ldots, M$ has its weighted contribution to the PDF and that it is guaranteed to obtain an associated value for the PDF by randomly choosing one pair:

$$\int_{t=0}^{\infty} d\tau \sum_{\nu=1}^{M} P(\tau, \nu) = \sum_{\nu=1}^{M} a_\nu \int_{t=0}^{\infty} d\tau \exp(-a\tau) = 1 \qquad (43.58)$$

Now, the fundamental Monte Carlo step can be defined: by choosing a random pair of the variables (τ, μ) according to the PDF $P(\tau, \mu)$ after Equation 43.57, a weighting based on the magnitude of the various transitions and the exponential decrease in time is performed. For the execution of this step, two approaches were suggested by Gillespie: the *direct method* and the *first reaction method*. In this context, the former method is used. The direct method is most efficient in the way it utilizes random numbers to determine a pair of transition and a time step, while the first reaction method can save computational power at the cost of accuracy.

In the direct method, the PDF is divided into two separate PDFs, each of which is itself normalized and only dependent on either the time τ or the transition μ:

$$P(\tau, \mu) = P_1(\tau) \cdot P_2(\mu|\tau) \tag{43.59}$$

Here, the first factor represents the probability density that any of the possible transitions takes place in the time interval $(t + \tau, t + \tau + d\tau)$, irrespective of which one. Thus, $P_1(\tau)$ is the sum over all transitions. With Equation 43.57 follows

$$P_1(\tau) = \sum_{\nu=1}^{M} P(\tau, \nu) = \sum_{\nu=1}^{M} a_\nu \exp(-a\tau) = a \cdot \exp(-a\tau) \tag{43.60}$$

From this and Equation 43.59, $P_2(\mu|\tau)$ is easily derived as

$$P_2(\mu|\tau) = \frac{P(\tau, \mu)}{P_1(\tau)} = \frac{a_\mu}{a} \tag{43.61}$$

and $P_2(\mu|\tau) \, d\tau$ represents the probability that a reaction μ occurs, given the condition that a time step of τ has been chosen before. Both P_1 and P_2 are properly normalized PDFs on their own.

A pair of random numbers (τ, μ) according to the distributions (Equations 43.60 and 43.61) can be produced as follows, using (pseudo-)random number generator with uniform distribution available in all modern programming languages. Every normalized PDF $P(x)$ has an associated cumulative distribution function, defined by

$$F_1(x) = \int_{-\infty}^{x} P_1(x') \, dx' \qquad \text{for a continuous } P_1(x) \tag{43.62}$$

$$F_2(i) = \sum_{i'=1}^{i} P_2(i') \qquad \text{for a discrete } P_2(i) \tag{43.63}$$

Because $\lim_{x \to -\infty} F_1(x) = 0$ and $\lim_{x \to \infty} F_1(x) = 1$ for the continuous case, and $F_2(0) = 0$ and $F_2(M) = 1$ for the discrete case, both functions are monotone and limited. Thus, there exists an inverse function F^{-1} for each of them. It is made sure that every value from the output range has a corresponding value x or i in the domain range. The concept of the inversion method is to draw two uniform random numbers $r_1, r_2 \in (0, 1)$, and calculate $x = F_1^{-1}(r)$ and i from $F_2^{-1}(i)$. This way the obtained values for x and i are distributed according to the PDF from which F is derived.

Using this inversion method and the PDF $P_1(\tau) = a \cdot \exp(-a\tau)$, the corresponding cumulative probability function is[†]

$$F_1(\tau) = \int_0^\tau a \cdot \exp(-a\tau') \, d\tau' = 1 - \exp(-a\tau) \tag{43.64}$$

The continuous and exponentially distributed time variable τ, which describes the time passed before one transition (or, for a more vivid interpretation: the time "needed" for one transition), is then

---

[†] Note that $P_1(t)$ is only defined for $t \geq 0$.

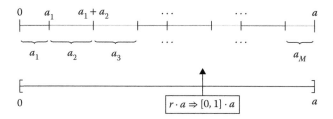

FIGURE 43.11 Drawing a discrete transition by a continuous random variable $r \in [0, 1]$. All enabled rate constants are added to yield the total rate a. Multiplication of a random number in the unit interval with the total rate corresponds to one specific rate process, weighted by its magnitude. Rates are added until the next addition will exceed $r \cdot a$. The last rate in the addition is chosen.

given by the inverse function of F_1. Hence,[†]

$$\tau = F_1^{-1}(r_1) = \frac{1}{a} \ln \left(\frac{1}{r_1} \right) \qquad \text{with } a = \sum_{\nu=1}^{M} a_\nu \text{ and } r_1 \in (0, 1) \tag{43.65}$$

To choose a transition μ from the set of events $\{a_\mu\}$ ($\mu = 1, 2, ..., M$) by a random number r_2, the cumulative probability must satisfy

$$F_2(\mu - 1) < r_2 \le F_2(\mu) \tag{43.66}$$

That is, $r_2 \in (0, 1)$ corresponds to one plateau in the discrete histogram of F_2. Substituting Equation 43.61 in Equation 43.63 yields an expression for $F_2(\tau|\mu)$:

$$F_2(\tau|\mu) = \sum_{\nu=1}^{\mu} \frac{a_\nu}{a} \tag{43.67}$$

With this, the abovementioned condition reads

$$\sum_{\nu=1}^{\mu-1} \frac{a_\nu}{a} < r_2 \le \sum_{\nu=1}^{\mu} \frac{a_\nu}{a} \tag{43.68}$$

or, multiplying by a,

$$\sum_{\nu=1}^{\mu-1} a_\nu < r_2 \cdot a \le \sum_{\nu=1}^{\mu} a_\nu \quad \text{with } a = \sum_{\nu=1}^{M} a_\nu \text{ and } r_2 \in (0, 1) \tag{43.69}$$

This represents a sum of all transition probabilities, up to an index for which the sum is larger than the random variable r_2. This index is set to be the chosen transition μ. The process to derive μ computationally from a random number is illustrated in Figure 43.11.

The essence of the Monte Carlo method is condensed in the two equations (Equations 43.65 and 43.69) to calculate a pair of random numbers (τ, μ), while the rest of the method is fairly straightforward and can be reduced to only the following few, simple steps, shown in a flowchart in Figure 43.12:

[†] The term $1 - r$ has been replaced by r. Since we are dealing with probabilities, this is valid for uniform random numbers in the unit interval.

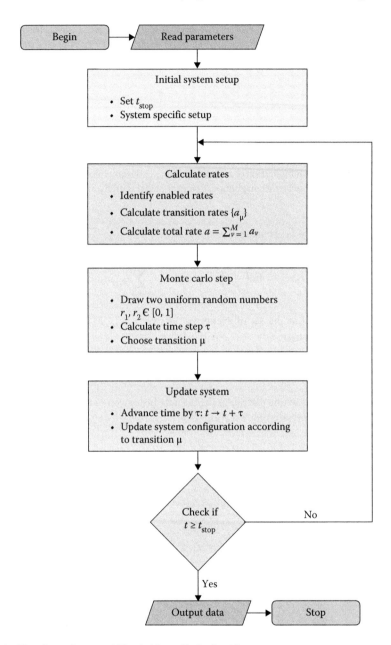

FIGURE 43.12 Flowchart of a general kinetic Monte Carlo algorithm.

0 – Initialization

In the beginning, system-specific parameters are defined in the program. This includes all parameters that must be known in order to set up the system and to calculate the transition rates. Variables to store the occurrence of relevant processes during the simulation are set up here. The simulation time t_{stop} is defined. It determines how long the dynamic evolution should be simulated. The starting time is set to $t = 0$.

1 – Calculation of transition rates

The kMC method is reliant on physical models for all processes that can occur in the system. All physics enters the model in terms of the event rates that are calculated in this step. Depending on the current state, only some transitions may be activated. Conditions must be met in order to make sure that rates for processes that cannot occur in a certain system state are deactivated. This is realized by setting the value for a deactivated rate $a_\mu = 0$ so that it is not considered in the selection a transition after Equation 43.69. After all rates have been calculated and stored, they are summed up to provide the total rate a.

2 – Monte Carlo step

As described in Section 43.4.2.1, a random pair (τ, μ) is generated according to the PDF $P(\tau, \mu)$ from the list of available (and activated) transitions $\{a_\mu\}$. This accounts for the stochastically correct time step τ for the selection of a transition μ.

3 – Update system

To update the system the time t is advanced by τ. Based on which transition μ has been chosen, the system configuration needs to be updated accordingly. How every update is specifically depends on the physical process.

After updating the system, a check is performed whether the simulation time has reached the specified stopping time t_{stop}. If $t < t_{\text{stop}}$, the simulation will continue by jumping back to step 1 and recalculate the event rates. This way, the system update on the behavior of other processes is reflected. The steps 1–3 represent the so called *Monte Carlo loop*. They will be executed as long as termination condition is not yet fulfilled.

4 – Termination and data output

Through the successive time advance t eventually becomes larger than the stopping time and the Monte Carlo loop is terminated. At last the variables tracking selected events can be evaluated.

Overall, the kMC method offers a straightforward algorithm to simulate the time-dependent evolution of a system. It is only reliant on (1) the assumption that all physical processes can be determined by rate expressions, and (2) that a (pseudo-)random number generator with a sufficiently small correlation between two subsequent numbers and a large period is provided.

43.4.3 Example: Transport through 1D Disordered System by kMC

To illustrate an actual implementation of the kMC algorithm, we provide a short manual on how to model charge transport in a 1D, energetically disordered system, e.g., an organic polymer. We will go through the steps presented in the previous section and in Figure 43.12: the initial system setup, the simulation loop, and the evaluation of quantities.

43.4.3.1 System Setup and Starting Conditions

In the beginning, the available states that the system can occupy must be defined. The setup is illustrated in Figure 43.14. Here, we consider a linear chain of N equidistant nodes indexed by $i = 1, \ldots, N$. The position of each node is quantified by $x_i = i \cdot L$ where L is the fixed distance between all nodes. Then, energy values are assigned to each node accounting for the influence of an applied electric field F, which is assumed to be constant leading to a linear energy drop across the system, and an energetic fluctuation $r(\mu, \sigma)$ drawn from a GDOS distribution with mean μ and standard deviation σ as given in Equation 43.42. This makes up for an energy landscape of the form $E_i = qFx_i + r_{\text{GDOS}}(\mu = 0, \sigma)$. The disorder is picked independently of the position. A typical configuration for E_i is depicted in Figure 43.14.

Next, initial conditions must be set. The simulation time is set to an initial time $t = 0$. It will be updated during each simulation step. The termination condition is defined by the particle position: if the particle reaches the last node N, the simulation loop stops. We allow one electron of charge $q = -e$ at a time to be in the system. In the beginning it is set to the first node x_1.

A model for the transition rates needs to be defined. We allow the particle to move only to its immediately neighboring sites to the left or to the right. The transitions are described by hopping rates according to the Miller–Abrahams formula (Equation 43.49). There are then only two distinct rates: a_l for hopping to the left and a_r for hopping to the right (Figure 43.13). They depend on the relative energy difference $\Delta E = E_j - E_i$ of the current node i of the particle and the final node $i \pm 1$. Typical values are $a_0 = 10^{11}$ s^{-1}, $\gamma = 2$nm, and $r_{ij} = L = 1$ nm.

43.4.3.2 Simulation Loop

After system setup, the simulation loop is executed until the termination condition is reached. It consists of the following steps:

1. Check the energy landscape to the right and to the left of the current particle position $E_{i\pm1}$.
2. Calculate the hopping rates $a_l(E_{i-1} - E_i)$ and $a_r(E_{i+1} - E_i)$.
3. Pick two distinct random numbers r_1, r_2 uniformly from the interval $(0, 1)$.
4. Calculate the total rate $a = a_l + a_r$. Then, determine the transition (hopping left or right): If $r_1 \cdot a \leq a_l$, pick a_l. Otherwise, if $a_l < r_1 \cdot a \leq a_l + a_r = a$, pick a_r.
5. Calculate timestep: $\tau = -\frac{r_2}{a}$.
6. Update system according to the transition picked, i.e., change the particle position to the node on the right or on the left.
7. Advance time $t \to t + \tau$.
8. Check termination condition. If still true, go to step 1. Otherwise, exit loop.

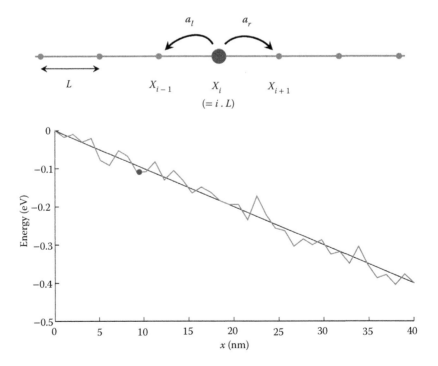

FIGURE 43.13 Setup to model hopping in a 1D system. Top: Spatial discretization of the problem in equidistant nodes. Bottom: Energetic configuration of the nodes. The setup shows the energetic landscape for $N = 40$ nodes comprised of a linear drop of the electric field $F = 10^5$ V/cm and a Gaussian energetic disorder with $\sigma = 30$ meV. The particle is allowed to hop to neighboring nodes under the influence of the energetic configuration.

43.4.3.3 Evaluation

After termination, desired quantities can be evaluated. In this case, since the time that the particle needed to hop through the disordered landscape t_{sim} is now known, the effective velocity $v = \frac{N \cdot L}{t_{sim}}$ can be calculated. From v and the electric field F, the effective transport mobility can be extracted as $\mu = \frac{v}{F}$.

43.4.3.4 Further Info

The general system setup and transition rates can be extended to 2D or 3D (see Figure 43.15). More particle types (holes, excitons) and more effects than hopping, such as injection, recombination, photogeneration of excitons can be introduced to obtain a more realistic model. The principle of choosing transition and timestep stays the same.

43.5 Specific Modeling Issues

In this section, we provide further details on the implementation of models specific to some of the emerging PV technologies, and we cover generic modeling issues.

43.5.1 Dye Solar Cells

DSCs have been extensively simulated using analytical models or numerical ones at the level of drift-diffusion. Most simulations have been performed using 1D models [100–102] and an effective material approximation (see Section 43.5.4 for an extensive description of this approximation). Only few simulations going beyond the effective medium model and trying to include the morphology [103] or to go toward 2D and 3D simulations [104,105] have been reported so far.

The main issues concerning simulations of a DSC are for sure related to the complex morphology of the mesoporous material and the chemical reactions at the interfaces. The latter include both the dye regeneration processes as well as all the recombination events where electrons in the TiO_2 are transferred to reduce ionized dyes or triiodide ions. The main issue in modeling is that all these processes are in general complex electrochemical rates where the charge transfer occurs in several steps. They depend on the effect of the electrolyte, the reorganization energy of the environment, temperature, and voltage [106]. Moreover, different dyes or electrolytes will lead to different kinetics and thus different recombination rates. This leads to recombination rates that formally are completely different from the standard formulas given in Equations 43.34 through 43.36.

Beside that, boundary conditions at the contacts also are very uncommon with respect to more conventional PV devices. While the anode with the ITO or FTO can be approximated by a standard ohmic or Schottky contact, the cathode at the platinum is an electrochemical contact which requires to take into account the formation of an ion accumulation layer. The standard boundary condition is a Butler–Volmer (BV) equation, which is generally a Robin-type boundary condition, since it connects the electron transfer rate with the electrochemical potential at the contact. A BV equation [107] describes the dependency of the electrical current at an electrode j on the electrode potential when both anodic and cathodic reactions are involved:

$$j = j_0 \cdot \left\{ \exp\left[\frac{\alpha_A z e}{k_B T} \eta \right] - \exp\left[\frac{\alpha_C z e}{k_B T} \eta \right] \right\} \tag{43.70}$$

where j_0 is the exchange current density prefactor, $\eta = E - E_{eq}$ the activation overpotential given by the difference in electrode potential and equilibrium potential, α_A and α_C are charge transfer coefficients for anode and cathode, respectively, z is the number of electrons participating in the reaction at the electrode, e

is the elementary charge and $k_B T$ the thermal energy. A Robin-type boundary condition in general is a combination of Dirichlet boundary conditions and Neumann boundary conditions. Imposed on a differential equation of function f, it reads in a general form

$$a f + b \frac{\partial f}{\partial n} = g \qquad (43.71)$$

where a and b are nonzero constants and g is a function defined on the boundary of the problem. Depending on whether the limiting process is the ionic diffusion of the redox pair or the charge transfer between platinum and triiodide ions, we can have different shapes of the BV equation. Finally, electron transport in the TiO_2 is mediated by trapping and detrapping events, a physical situation captured by the multitrapping model, where the mobility becomes density dependent [108,109]. For solid state DSCs, the simulation issues are more similar to OPVs as the electrolyte is replaced by an organic semiconductor.

43.5.2 Organic Photovoltaics

Modeling of OPV devices incurs several important differences when compared to inorganic devices. First of all, as mentioned already in section 43.4.1, the DOS is fundamentally different as it does not result from electronic states in a periodic potential, but from a both energetically and spatially disordered energy landscape of molecular levels as shown schematically in Figure 43.14. This has to be considered when using or implementing a drift-diffusion based simulation tool, since the standard drift-diffusion model has been derived for carriers with a parabolic DOS. As a further consequence of this energy disorder, transport cannot be described adequately by carriers moving freely in conduction and valence bands, but rather by hopping between energy levels on different molecules [86,110–112], as described in Section 43.4.1. Fortunately, on a macroscopic level, i.e. on a scale much larger than the molecular structure, transport can still be described in terms of drift and diffusion, if a suitable model for the carrier mobilities can be found. Such a model is described in [88], where the mobility has been parametrized based on numerical results of the Master equation.

Due to the different DOSs, the models for transport across interfaces and in particular for injection at the contacts also are different from their counterparts employed in standard drift-diffusion simulations of inorganic semiconductors. The properties of a metal-semiconductor contact, known as Schottky contact in the inorganic case, are controlled by the enetrgetic alignment of the metal work function with the semiconductor band edges, by the semiconductor's Fermi level, presence of interface states, and by the semiconductor DOS. The boundary condition on the particle current e.g. for the electrons at the metal-semiconductor contact can be written as $J_n = q v_R (n - n_0)$, where n_0 is the electron density at the contacts in equilibrium and v_R is a recombination velocity, with units cm/s. The latter quantity depends on the exact

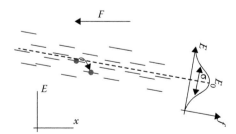

FIGURE 43.14 Schematic representation of the energetic disorder in an organic semiconductor. Each level represents a molecular energy level, which are disordered in energy and in space, following approximately a Gaussian distribution. Transport is driven by hopping between levels, described by energy difference and distance-dependent coefficients that depend also on the external electric field F.

transport mechanism and on the semiconductor DOS, and is therefore very different between inorganic and organic semiconductors [113].

A further difference of OPVs is their low permittivity of $3 \sim 4$ compared to inorganic materials with $\varepsilon_r \approx 10$. This leads to strong local Coulomb interaction which induces the formation of excitons even at room temperature. Since only electrons and holes can be collected at the contacts, the photo-generated excitons need to dissociate first. For this they need to reach the donor/acceptor interface, which is however possible only within their diffusion length of approximately 10 nm. For this reason, OPVs employ a blended donor/acceptor absorber, forming a BHJ. In combination with the hopping transport of charges in spatially and energetically disordered materials, these factors are the key differences to inorganics and make up for most of the challgenges associated with the modeling of OPVs. Since excitonic effects are so important in OPVs, several authors have proposed drift-diffusion-based models including explicitly exciton generation and dissociation [114–116].

The main disadvantage of state-of-the-art drift-diffusion models is the simplification of the BHJ morphology to a 1D effective medium. More details on the background and modeling techniques of the morphology can be found in Section 43.5.4. Since no interfaces exist in the effective medium, all interface effects between the donor/acceptor boundaries, such as exciton splitting and electron–hole recombination, are assumed to be bulk processes and evenly distributed across the photoactive layer. An absorbed photon directly creates and electron–hole pair and exciton diffusion and loss processes cannot be considered. Also, no charge accumulation effects at the donor/acceptor interface can be observed due to the lack of the blend morphology.

Most of these limitations do not apply to kMC simulations. A real 3D blend morphology, explicit treatment of exciton diffusion and splitting, and other interface processes, can easily be implemented. Also, the hopping transport in energetically and spatially disordered materials is naturally representable on a discrete lattice. However, since kMC is a statistical method, it often has to deal with unfeasible simulation times in order to achieve a reasonable accuracy for the desired quantities. This is especially critical when one has to sample very slow processes in presence of very fast events. Specifically, the evaluation of Coulomb interaction is expensive, and since particles are modeled explicitly and not by continuous quantities, the kMC algorithm is very demanding, in particular for a large number of particles. Large particle numbers are induced by a high energetic disorder, small injection barriers ("small barrier problem" in kMC [117]), or large device dimensions/thicknesses and limit the choice of parameters where kMC can be applied to certain ranges. Procedures to parallelize the algorithm and the Coulomb interaction need to be applied [118] to model larger systems and/or get better statistics in the results.

43.5.3 Perovskites

Perovskite, based devices pose two types of issues related to simulation. On the one hand, the fundamental working principle has not been completely unveiled yet, in particular the role of the perovskite in carrier transport. Therefore, it is not clear yet if the perovskite should be used merely as sensitizing absorber similar to the dye in a DSC, or if it should be used in a configuration more similar to an inorganic cell where the absorber acts also as electron and hole transporter. Partly, such questions arise because the development of perovskite PV is driven by the DSC and OPV comunities, with different views on device structure and functioning, and so also the point of view for modeling varies. From a fundamental point of view, being a crystalline material, perovskite seems to be more similar to an inorganic material with well-defined band structure [119], and it would seem that standard simulation models used for inorganic semiconductors should apply.

There are several issues encountered experimentally, where device simulation could give a strong contribution for a better understanding. One open question is whether ferroelectricity and ferroelectric domains exist in perovskite layers and whether they play a role on device functioning. While a fundamental answer to such questions based on density functional theory or MD is not easy, device-level simulation can at least provide inside into the qualitative effects that could be expected in presence of spatially varying polarization fields. This requires, however, simulation models that can include such effects. In particular, realistic domains have to be generated numerically, and the drift-diffusion model has to include the variable polarization field. Attempts in this direction have been made [120].

Another challenge for modeling is the hysteresis typically observed experimentally when measuring perovskite solar cell current–voltage characteristics [40,121]. Currently, the most probable cause of such hysteresis is believed to be migration of halide ion vacancies [122,123]. However, modeling of this effect in a standard device simulator is not easy, since the simulation model assumes transport of electrons and holes for a given fixed spatial distribution of defects. Therefore, self-consistent modeling of electronic transport and ion migration requires an extension of the standard drift-diffusion model with further mobile species, which is currently not available in standard device simulation software.

43.5.4 Modeling of Morphology and Grain Effects

Many device architectures for OPVs, (solid-state-) DSSCs and perovskite solar cells have regions with complex intermixed morphologies between two different materials where important processes are assumed to be controlled by the shape of the intermixing, especially mobilities through the morphologies and interface processes such as charge recombination and charge generation. For the sake of simplicity and low computational effort, the 3D blend is often simplified and treated as an effective medium, hiding the interface processes and 3D morphology effects in effective models for charge generation and recombination, and the mobility. An effective medium can be defined such that its band edges are given by the lower and higher values of the conduction and valence band edges of the two blended materials, respectively. Parameters like recombination rate constants and mobilities are then usually fitted to reproduce experimental data like current–voltage curves, or experimentally measured mobilities. It is yet to be investigated under which conditions an effective medium treatment is valid and which models are accurate for this description. Alternatively to the effective medium, the mobility can be derived from microscopic simulations considering the 3D blend morphology, like kMC. A comparison between the effective medium approach and a real blend kMC simulation in BHJ OPVs shows, however, that it might be difficult to include all interface effects in a blend into mobility and recombination models within an effective medium [124], especially in low-ε_r materials. A common basis in as many parameters as possible needs to be established for a comparison: the MO levels, the generation profile, and the injection processes, to name a few. By extracting electron and hole mobilities from the 3D simulations, the effect of the blend mobility can effectively be plugged into the effective medium model. It is then tried to obtain an agreement between the two models by tuning the recombination ratio. No general agreement could be obtained, which might indicate that it is not possible to merge all parameters in effective models accurately. It has been observed that, for instance, the charge density distribution within the 3D blend cannot be recovered by an effective model, since local accumulations at an interface are not present in the effective medium where both phases are assumed to be everywhere.

Thus, performing simulations on a realistic 3D blend is important to get a detailed picture about the processes in the blend. For a drift-diffusion model, complex 3D meshes are not easy to generate, increase considerably computation time and are prone to cause problems related to convergence. To avoid these problems for 3D blends, the kMC method is a suitable choice. The kMC algorithm does not have issues with convergence, even for complex morphologies. However, due to the popularity and establishment of drift-diffusion models, more effort should be made to include morphology effects in drift-diffusion-based simulation models. An example in this direction is given, e.g., in [125].

In kMC models, a blend can be created using a spin-exchange algorithm based on the work of Watkins et al. [111] which allows to control the grade of intermixing between two phases. However, the structure is based on a discrete lattice and is not suitable as basis for a DD simulation for which a high-quality finite element mesh is needed. A smoothing algorithm can be applied in order to achieve a mesh suitable for DD simulations. An illustration of a discrete BHJ lattice and a successively smoothed mesh is depicted in Figure 43.15. Further tests with an implementation of the 3D blend morphology as a mesh for DD simulations need to be performed to increase the common basis of the two models and get more information about where the two models differ and when the effective medium is appropriate.

Apart from nanometer-scale morphology effects, grain boundaries in multicrystalline materials can also affect optoelectronic device properties. Grain dimensions in such materials can range from nanometers to micrometers. A correct description of the effects of the grain boundaries, which can be related to both interface recombination and transport, may require to include the grain interfaces explicitly in the simulation model, especially if there is a limited number of interfaces along the carrier transport path such that an effective medium approximation may become questionable. Grain boundaries in organic materials have been studied related mostly to organic thin-film transistors [126,127]. Similar studies have been performed on polycrystalline silicon [128], and the effect of grain boundaries on carrier collection and recombination has been in CIGS devices has been studied in [129] and [130].

Note that also kMC does not automatically include the effect of grain boundaries, which should be included ad hoc with parameters based on *ab initio* or MD simulations [131].

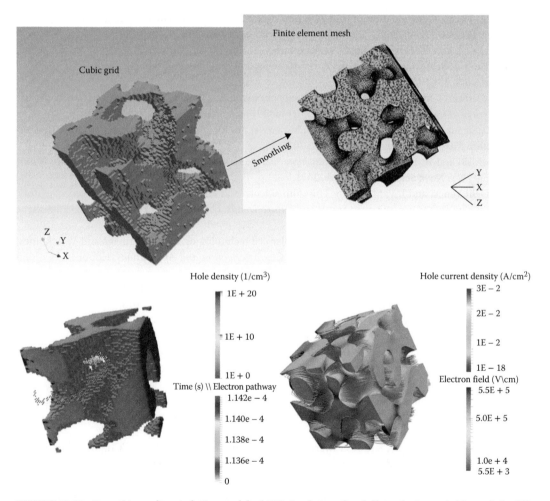

FIGURE 43.15 Smoothing a discrete lattice used for kMC simulations (top left) to obtain a suitable mesh for DD simulations (top right). Exemplary results of the kMC method (bottom left) highlighting the advantage to track single charge carrier trajectories and the DD model to extract, e.g., directed streams such as current densities through the morphology (bottom right).

References

1. J. Nelson. *The Physics of Solar Cells*. London: Imperial College Press, 2003.
2. J. Poortmans and V. Arkhipov, editors. *Thin Film Solar Cells—Fabrication, Characterization and Applications*. Hoboken, NJ: John Wiley & Sons, 2006.
3. K.L. Chopra, P.D. Paulson, and V. Dutta. Thin-film solar cells: An overview. *Progress in Photovoltaics: Research and Applications*, 12(2–3):69–92, 2004.
4. A. Jäger-Waldau. Status and perspectives of thin film photovoltaics. In A. Bosio and A. Romeo, editors, *Thin Film Solar Cells—Current Status and Future Trends*. Hauppauge, NY: Nova Science Publishers, 2011.
5. T.M. Brown, A. Reale, and A. Di Carlo. Organic and hybrid solar cells. In A. Bosio and A. Romeo, editors, *Thin Film Solar Cells—Current Status and Future Trends*. Hauppauge, NY: Nova Science Publishers, 2011.
6. Fraunhofer Institute for Solar Energy (ISE). *Photovoltaics Report*, 2015. Available at http://www.ise.fraunhofer.de/.
7. http://www.nrel.gov/ncpv.
8. M.A. Green. Consolidation of thin-film photovoltaic technology: The coming decade of opportunity. *Progress Photovoltaics: Research and Applications*, 14:383–392, 2006.
9. L. Dou, J. You, Z. Hong, Z. Xu, G. Li, R.A. Street, Y. Yang, L. Dou, J. You, Z. Hong, Z. Xu, G. Li, and Y. Yang. 25th Anniversary Article: A decade of organic/polymeric photovoltaic research. *Advanced Materials*, 25:6642–6671, 2013.
10. W. Cao and J. Xue. Recent progress in organic photovoltaics: Device architecture and optical design. *Energy & Environmental Science*, 7:2123–2144, 2014.
11. S.D. Stranks and H.J. Snaith. Metal-halide perovskites for photovoltaic and lightemitting devices. *Nature Nanotechnology*, 10(5):391–402, 2015.
12. A. Bosio and A. Romeo, editors. *Thin Film Solar Cells—Current Status and Future Trends*. Hauppauge, NY: Nova Science Publishers, 2011.
13. P. Delli Veneri and L.V. Mercaldo. Thin film silicon solar cells. In A. Bosio and A. Romeo, editors, *Thin Film Solar Cells—Current Status and Future Trends*. Hauppauge, NY: Nova Science Publishers, 2011.
14. M. Zeman. Advanced amorphous silicon solar cell technologies. In J. Poortmans and V. Arkhipov, editors, *Thin Film Solar Cells—Fabrication, Characterization and Applications*. Hoboken, NJ: John Wiley & Sons, 2006.
15. M.A. Green, K. Emery, Y. Hishikawa, W. Warta, and E.D. Dunlop. Solar cell efficiency tables (version 46). *Progress in Photovoltaics: Research and Applications*, 23(7):805–812, 2015.
16. H. Fritzsche. Development in understanding and controlling the Staebler-Wronski effect in a-Si:H. *Annual Review of Materials Science*, 31:47–79, 2001.
17. E.A. Schiff, S. Hegedus, and X. Deng. Amorphous silicon-based solar cells. In A. Luque and S. Hegedus, editors, *Handbook of Photovoltaic Science and Engineering*, pp. 487–545. Hoboken, NJ: John Wiley and Sons, 2011.
18. A. Bosio, A. Romeo, D. Menossi, S. Mazzamuto, and N. Romeo. Review: The secondgeneration of CdTe and CuInGaSe2 thin film PV modules. *Crystal Research and Technology*, 46(8):857–864, 2011.
19. S. Girish Kumar and K.S.R. Koteswara Rao. Physics and chemistry of CdTe/CdS thin film heterojunction photovoltaic devices: Fundamental and critical aspects. *Energy and Environmental Science*, 7:45–102, 2014.
20. R. Klenk and M.C. Lux-Steiner. Chalcopyrite based solar cells. In J. Poortmans and V. Arkhipov, editors, *Thin Film Solar Cells—Fabrication, Characterization and Applications*. Hoboken, NJ: John Wiley & Sons, 2006.

21. P. Reinhard, A. Chirila, P. Blosch, F. Pianezzi, S. Nishiwaki, S. Buechelers, and A.N. Tiwari. Review of progress toward 20% efficiency exible cigs solar cells and manufacturing issues of solar modules. *IEEE Journal of Photovoltaics,* 3(1):572–580, 2013.

22. T. Unold and C.A. Kaufmann. *Chalcopyrite Thin-Film Materials and Solar Cells,* Volume 1. Amsterdam: Elsevier, 2012.

23. B. O'Regan and M. Grätzel. A low-cost, high-efficiency solar cell based on dye-sensitized colloidal TiO2 films. *Nature,* 353:737–740, 1991.

24. K. Kalyanasundaram, editor. *Dye Sensitized Solar Cells.* Lausanne: EPFL Press, 2010.

25. S. Mathew, A. Yella, P. Gao, R. Humphry-Baker, B.F.E. Curchod, N. Ashari-Astani, I. Tavernelli, U. Rothlisberger, M.K. Nazeeruddin, and M. Grtzel. Dye-sensitized solar cells with 13% efficiency achieved through the molecular engineering of porphyrin sensitizers. *Nature Chemistry,* 6(3):242–247, 2014.

26. U. Bach, D. Lupo, P. Comte, J.E. Moser, F. Weissrtel, J. Salbeck, H. Spreitzer, and M. Grtzel. Solid-state dye-sensitized mesoporous TiO_2 solar cells with high photon-to-electron conversion efficiencies. *Nature,* 395(6702):583–585, 1998.

27. F. Matteocci, G. Mincuzzi, F. Giordano, A. Capasso, E. Artuso, C. Barolo, G. Viscardi, T.M. Brown, A. Reale, and A. Di Carlo. Blocking layer optimisation of poly(3-hexylthiopene) based solid state dye sensitized solar cells. *Organic Electronics,* 14(7):1882–1890, 2013.

28. A. Hagfeldt, G. Boschloo, L. Sun, L. Kloo, and H. Pettersson. Dye-sensitized solar cells. *Chemical Reviews,* 110:6595–6663, 2010.

29. A. Hinsch. Status of the dye solar cell technology (DSC) as a guideline for further research. In *28th European PV Solar Energy Conference and Exhibition,* Paris, France, 2013.

30. N.S. Sariciftci, L. Smilowitz, A.J. Heeger, and F. Wudl. Photoinduced electron transfer from a conducting polymer to buckminsterfullerene. *Science,* 258(5087):1474–1476, 1992.

31. C.W. Tang. Two-layer organic photovoltaic cell. *Applied Physics Letters,* 48(2):183–185, 1986.

32. G. Yu, J. Gao, J.C. Hummelen, F. Wudl, and A.J. Heeger. Polymer photovoltaic cells: Enhanced efficiencies via a network of internal donor-acceptor heterojunctions. *Science,* 270(5243):1789–1790, 1995.

33. Y. Liu, J. Zhao, Z. Li, C. Mu, W. Ma, H. Hu, K. Jiang, H. Lin, H. Ade, and H. Yan. Aggregation and morphology control enables multiple cases of high-efficiency polymer solar cells. *Nature Communications,* 5:5293, 2014.

34. C. Deibel and V. Dyakonov. Polymer-fullerene bulk heterojunction solar cells. *Reports on Progress in Physics,* 73(9):096401, 2010.

35. G. Li, R. Zhu, and Y. Yang. Polymer solar cells. *Nature Photonics,* 6(3):153–161, 2012.

36. H.-S. Kim, C.-R. Lee, J.-H. Im, K.-B. Lee, T. Moehl, A. Marchioro, S.-J. Moon, R. Humphry-Baker, J.-H. Yum, J.E. Moser, M. Grtzel, and N.-G. Park. Lead iodide perovskite sensitized all-solid-state submicron thin film mesoscopic solar cell with efficiency exceeding 9%. *Scientific Reports,* 2:591, 2012.

37. H.S. Jung and N.-G. Park. to Perovskite solar cells: From materials devices. *Small,* 11(1):10–25, 2015.

38. P. Gao, M. Grätzel, and M.K. Nazeeruddin. Organohalide lead perovskites for photovoltaic applications. *Energy and Environmental Science,* 7(8):2448–2463, 2014.

39. D. Wang, M. Wright, N.K. Elumalai, and A. Uddin. Stability of perovskite solar cells. *Solar Energy Materials and Solar Cells,* 147:255–275, 2016.

40. M. Grätzel. The light and shade of perovskite solar cells. *Nature Materials,* 13(9):838–842, 2014.

41. P.P. Boix, K. Nonomura, N. Mathews, and S.G. Mhaisalkar. Current progress and future perspectives for organic/inorganic perovskite solar cells. *Materials Today,* 17(1):16–23, 2014.

42. P. Würfel. *Physics of Solar Cells.* Weinheim: Wiley-VCH Verlag, 2009.

43. G.P. Smestad. *Optoelectronics of Solar Cells.* SPIE Press, 2002.

44. N.-K. Persson and O. Inganäs. Simulations of optical processes in organic photovoltaic devices. In S.-S. Sung and N.S. Sariciftci, editors, *Organic Photovoltaics—Mechanisms, Materials and Devices.* Hoboken, NJ: Taylor & Francis, 2005.

45. S.L. Chuang. *Physics of Optoelectronic Devices. Wiley Series in Pure and Applied Optics*, 1st edition. Hoboken, NJ: John Wiley & Sons, 1995.

46. A. Taove and S.C. Hagness. *Computational Electrodynamics: The Finite-Difference Time-Domain Method.* Norwood, MA: Artech House Publishers, 2005.

47. J. Jin. *The Finite Element Method in Electromagnetics.* Hoboken, NJ: Wiley, 2002.

48. I. Semenikhin, M. Zanuccoli, M. Benzi, V. Vyurkov, E. Sangiorgi, and C. Fiegna. Computational efficient RCWA method for simulation of thin film solar cells. *Optical and Quantum Electronics*, 44(3):149–154, 2012.

49. M. Born and E. Wolf. *Principles of Optics.* New York, NY: Cambridge University Press, 1999.

50. B. Harbecke. Coherent and incoherent reflection and transmission of multilayer structures. *Applied Physics B Photophysics and Laser Chemistry*, 39(3):165–170, 1986.

51. J.J.M. Halls, K. Pichler, R.H. Friend, S.C. Moratti, and A.B. Holmes. Exciton diffusion and dissociation in a poly(p-phenylenevinylene)/C_{60} heterojunction photovoltaic cell. *Applied Physics Letters*, 68(22):3120–3122, 1996.

52. J.J.M. Halls. *Photoconductive Properties of Conjugated Polymers.* PhD dissertation, Cambridge: St. John's College, 1997.

53. L.A.A. Pettersson, L.S. Roman, and O. Inganäs. Modeling photocurrent action spectra of photovoltaic devices based on organic thin films. *Journal of Applied Physics*, 86(1):487–496, 1999.

54. L.A.A. Pettersson, L.S. Roman, and O. Inganäs. Quantum efficiency of exciton-to-charge generation in organic photovoltaic devices. *Journal of Applied Physics*, 89(10):5564–5569, 2001.

55. N.-K. Persson, M. Schubert, and O. Inganäs. Optical modelling of a layered photovoltaic device with a polyuorene derivative/fullerene as the active layer. *Solar Energy Materials and Solar Cells*, 83(23):169–186, 2004.

56. H. Hoppe, N. Arnold, N.S. Sariciftci, and D. Meissner. Modeling the optical absorption within conjugated polymer/fullerene-based bulk-heterojunction organic solar cells. *Solar Energy Materials and Solar Cells*, 80(1):105–113, 2003.

57. D.P. Gruber, G. Meinhardt, and W. Papousek. Spatial distribution of light absorption in organic photovoltaic devices. *Solar Energy*, 79(6):697–704, 2005.

58. F. Monestier, J.-J. Simon, P. Torchio, L. Escoubas, F Flory, S. Bailly, R. de Bettignies, S. Guillerez, and C. Defranoux. Modeling the short-circuit current density of polymer solar cells based on p3HT:PCBM blend. *Solar Energy Materials and Solar Cells*, 91(5):405–410, 2007.

59. R. Husermann, E. Knapp, M. Moos, N.A. Reinke, T. Flatz, and B. Ruhstaller. Coupled optoelectronic simulation of organic bulk-heterojunction solar cells: Parameter extraction and sensitivity analysis. *Journal of Applied Physics*, 106(10):104507, 2009.

60. G.F. Burkhard, E.T. Hoke, and M.D. McGehee. Accounting for interference, scattering, and electrode absorption to make accurate internal quantum efficiency measurements in organic and other thin solar cells. *Advanced Materials*, 22(30):3293–3297, 2010.

61. L.D. Landau, E.M. Lifshitz, and L.P. Pitaevskii. *Electrodynamics of Continuous Media.* Amsterdam: Elsevier, 1993.

62. H.-Y.D. Yang. A spectral recursive transformation method for electromagnetic waves in generalized anisotropic layered media. *IEEE Transactions on Antennas and Propagation*, 45(3):520–526, 1997.

63. J. Ning and E.L. Tan. Hybrid matrix method for stable analysis of electromagnetic waves in stratified bianisotropic media. *IEEE Microwave and Wireless Components Letters*, 18(10):653–655, 2008.

64. A.H. Fallahpour, A. Gagliardi, F. Santoni, D. Gentilini, A. Zampetti, M. Auf der Maur, and A. Di Carlo. Modeling and simulation of energetically disordered organic solar cells. *Journal of Applied Physics*, 116(18), 2014.

65. M. Zeman, O. Isabella, K. Jger, R. Santbergen, S. Solntsev, M. Topic, and J. Krc. Advanced light management approaches for thin-film silicon solar cells. *Energy Procedia*, 15:189–199, 2012.

66. S. In, D.R. Mason, H. Lee, M. Jung, C. Lee, and N. Park. Enhanced light trapping and power conversion efficiency in ultrathin plasmonic organic solar cells: A coupled optical-electrical multiphysics study on the effect of nanoparticle geometry. *ACS Photonics*, 2(1):78–85, 2015.

67. M. Peters, B. Bläsi, S.W. Glunz, A.G. Aberle, J. Luther, and C. Battaglia. Optical simulation of silicon thin-film solar cells. *Energy Procedia*, 15:212–219, 2012.

68. A.H. Fallahpour, G. Ulisse, M. Auf der Maur, A. Di Carlo, and F. Brunetti. 3-D simulation and optimization of organic solar cell with periodic back contact grating electrode. *IEEE Journal of Photovoltaics*, 5(2):591–596, 2015.

69. M. Zeman, O. Isabella, S. Solntsev, and K. Jäger. Modelling of thin-film silicon solar cells. *Solar Energy Materials and Solar Cells*, 119:94–111, 2013.

70. N. Tucher, J. Eisenlohr, P. Kiefel, O. Höhn, H. Hauser, M. Peters, C. Müller, J.C. Goldschmidt, and B. Bläsi. 3D optical simulation formalism optos for textured silicon solar cells. *Optics Express*, 23(24):A1720–A1734, 2015.

71. S. Selberherr. *Analysis and Simulation of Semiconductor Devices*, 1st edition. New York, NY: Springer-Verlag Wien, 1984.

72. P.A. Markowich, C.A. Ringhofer, and C. Schmeiser. *Semiconductor Equations*, 1st edition. New York, NY: Springer-Verlag Wien, 1990.

73. W.V. van Roosbroeck. Theory of the flow of electrons and holes in germanium and other semiconductors. *Bell System Technical Journal*, 29:560–607, 1955.

74. S.M. Sze. *Semiconductor Devices: Physics and Technology*. New York, NY: John Wiley & Sons, 1985.

75. A. Schenk. An improved approach to the Shockley-Read-Hall recombination in inhomogeneous fields of spacecharge regions. *Journal of Applied Physics*, 71(7):3339–3349, 1992.

76. T. Goudon, V. Miljanovic, and C. Schmeiser. On the Shockley-Read-Hall model: Generation-recombination in semiconductors. *SIAM Journal on Applied Mathematics*, 67(4):1183–1201, 2007.

77. E. O'Reilly. *Quantum Theory of Solids*. London: Taylor & Francis, 2002.

78. S. John, C. Soukoulis, M.H. Cohen, and E.N. Economou. Theory of electron band tails and the urbach optical-absorption edge. *Physical Review Letters*, 57:1777–1780, 1986.

79. H. Bässler. Charge transport in disordered organic photoconductors: A Monte Carlo simulation study. *Physica Status Solidi (B)*, 175(1):15–56, 1993.

80. L.J.A. Koster, V.D. Mihailetchi, R. Ramaker, and P.W.M. Blom. Light intensity dependence of open-circuit voltage of polymer: Fullerene solar cells. *Applied Physics Letters*, 86(12):123509, 2005.

81. G. Paasch and S. Scheinert. Charge carrier density of organics with Gaussian density of states: Analytical approximation for the Gauss-Fermi integral. *Journal of Applied Physics*, 107(10):104501, 2010.

82. A. Zampetti, A.H. Fallahpour, M. Dianetti, L. Salamandra, F. Santoni, A. Gagliardi, M. Auf der Maur, F. Brunetti, A. Reale, T.M. Brown, and A. Di Carlo. Influence of the interface material layers and semiconductor energetic disorder on the open circuit voltage in polymer solar cells. *Journal of Polymer Science Part B: Polymer Physics*, 53(10):690–699, 2015.

83. G.K. Wachutka. Rigorous thermodynamic treatment of heat generation and conduction in semiconductor device modeling. *IEEE Transactions on Computer-Aided Design*, 11:1141–1149, 1990.

84. N.A. Zakhleniuk. Nonequilibrium drift-diffusion transport in semiconductors in presence of strong inhomogeneous electric fields. *Applied Physics Letters*, 89(25):252112, 2006.

85. N.F. Mott. On the transition to metallic conduction in semiconductors. *Canadian Journal of Physics*, 34(12A):1356–1368, 1956.

86. A. Miller and E. Abrahams. Impurity conduction at low concentrations. *Physical Review*, 120(3):745, 1960.

87. R.A. Marcus. Chemical and electrochemical electron-transfer theory. *Annual Review of Physical Chemistry*, 15(1):155–196, 1964.

88. W.F. Pasveer, J. Cottaar, C. Tanase, R. Coehoorn, P.A. Bobbert, P.W.M. Blom, D.M. De Leeuw, and M.A.J. Michels. Unified description of charge-carrier mobilities in disordered semiconducting polymers. *Physical Review Letters*, 94(20):206601, 2005.

89. D.T. Gillespie. A general method for numerically simulating the stochastic time evolution of coupled chemical reactions. *Journal of Computational Physics*, 22(4):403–434, 1976.

90. A.F. Voter. Introduction to the kinetic Monte Carlo Method. *Radiation Effects*, 235:1–23, 2005.

91. H.L. Anderson. Scientific uses of the maniac. *Journal of Statistical Physics*, 43(5–6):731–748, 1986.

92. N. Metropolis, A.W. Rosenbluth, M.N. Rosenbluth, A.H. Teller, and E. Teller. Equation of state calculations by fast computing machines. *The Journal of Chemical Physics*, 21(6):1087, 1953.

93. J.R. Beeler (Jr). Displacement spikes in cubic metals. i. α-Iron, copper, and tungsten. *Physical Review*, 150(2):470, 1966.

94. E.S. Hood, B.H. Toby, and W.H. Weinberg. Precursor-mediated molecular chemisorption and thermal desorption: The interrelationships among energetics, kinetics, and adsorbate lattice structure. *Physical Review Letters*, 55(22):2437, 1985.

95. A.B. Bortz, M.H. Kalos, and J.L. Lebowitz. A new algorithm for Monte Carlo simulation of Ising spin systems. *Journal of Computational Physics*, 17(1):10–18, 1975.

96. C. Jacoboni and P. Lugli. *The Monte Carlo Method for Semiconductor Device Simulation. Computational Microelectronics*. Vienna: Springer, 1989.

97. D.T. Gillespie. Exact stochastic simulation of coupled chemical reactions. *The Journal of Physical Chemistry*, 81(25):2340–2361, 1977.

98. J.R. Norris. *Markov Chains*. Number 2008. Cambridge: Cambridge University Press, 1998.

99. F. Schwabl. *Statistical Mechanics. Advanced Texts in Physics*. Berlin: Springer, 2006.

100. J. Ferber, R. Stangl, and J. Luther. An electrical model of the dye sensitized solar cell. *Solar Energy Materials and Solar Cells*, 53:29–54, 1998.

101. P.R.F. Barnes and B.C. O'Regan. Electron recombination kinetics and the analysis of collection efficiency and diffusion length measurements in dye sensitized solar cells. *The Journal of Physical Chemistry C*, 114(44):19134–19140, 2010.

102. A. Gagliardi, S. Mastroianni, D. Gentilini, F. Giordano, A. Reale, T.M. Brown, and A.D. Carlo. Multiscale modeling of dye solar cells and comparison with experimental data. *IEEE Journal of Selected Topics in Quantum Electronics*, 16(6):1611–1618, 2010.

103. A. Gagliardi, M. Auf der Maur, D. Gentilini, F. Di Fonzo, A. Abrusci, H.J. Snaith, G. Divitini, C. Ducati, and A. Di Carlo. The real TiO_2/HTM interface of solid-state dye solar cells: Role of trapped states from a multiscale modelling perspective. *Nanoscale*, 7(3):1136–1144, 2015.

104. A. Gagliardi, M. Auf der Maur, D. Gentilini, and A. Di Carlo. Simulation of dye solar cells: Through and beyond one dimension. *Journal of Computational Electronics*, 10(4):424–436, 2011.

105. A. Gagliardi, M. Auf der Maur, and A. Di Carlo. Theoretical investigation of a dye solar cell wrapped around an optical fiber. *IEEE Journal of Quantum Electronics*, 47(9):1214–1221, 2011.

106. J. Nissfolk, K. Fredin, A. Hagfeldt, and G. Boschloo. Recombination and transport processes in dye-sensitized solar cells investigated under working conditions. *The Journal of Physical Chemistry B*, 110(36):17715–17718, 2006.

107. U. Landau. Current distribution in electrochemical cells: Analytical and numerical modeling. In *Modern Aspects of Electrochemistry No. 44: Modelling and Numerical Simulations II*, pp. 451–501. New York, NY: Springer New York, 2009.

108. J. Bisquert and V.S. Vikhrenko. Interpretation of the time constants measured by kinetic techniques in nanostructured semiconductor electrodes and dye-sensitized solar cells. *Journal of Physical Chemistry B*, 108(7):2313–2322, 2004.

109. D. Gentilini, A. Gagliardi, M. Auf der Maur, L. Vesce, D. D'Ercole, T.M. Brown, A. Reale, and A. Di Carlo. Correlation between cell performance and physical transport parameters in dye solar cells. *Journal of Physical Chemistry C*, 116(1):1151–1157, 2012.

110. S.D. Baranovskii. Theoretical description of charge transport in disordered organic semiconductors. *Physica Status Solidi (B)*, 251(3):487–525, 2014.

111. P.K. Watkins, A.B. Walker, and G.L.B. Verschoor. Dynamical Monte Carlo modelling of organic solar cells: The dependence of internal quantum efficiency on morphology. *Nano Letters*, 5(9):1814–1818, 2005.

112. M. Casalegno, G. Raos, and R. Po. Methodological assessment of kinetic Monte Carlo simulations of organic photovoltaic devices: The treatment of electrostatic interactions. *Journal of Chemical Physics*, 132(9):094705, 2010.

113. F. Santoni, A. Gagliardi, M. Auf der Maur, and A. Di Carlo. The relevance of correct injection model to simulate electrical properties of organic semiconductors. *Organic Electronics*, 15(7):1557–1570, 2014.

114. I. Kamohara, M. Townsend, and B. Cottle. Simulation of heterojunction organic thin film devices and exciton diffusion analysis in stacked-hetero device. *Journal of Applied Physics*, 97(1):014501, 2005.

115. Il-S. Park, S.-R. Park, D.-Y. Shin, J.-S. Oh, W.-J. Song, and J.-H. Yoon. Modeling and simulation of electronic and excitonic emission properties in organic host/guest systems. *Organic Electronics*, 11(2):218–226, 2010.

116. B. Ruhstaller, T. Beierlein, H. Riel, S. Karg, J.C. Scott, and W. Riess. Simulating electronic and optical processes in multilayer organic light-emitting devices. *IEEE Journal of Selected Topics in Quantum Electronics*, 9(3):723–731, 2003.

117. D. Kipp and V. Ganesan. A kinetic Monte Carlo model with improved charge injection model for the photocurrent characteristics of organic solar cells. *Journal of Applied Physics*, 113(23):234502, 2013.

118. N.J. van der Kaap and L.J.A. Koster. Massively parallel kinetic monte carlo simulations of charge carrier transport in organic semiconductors. *Journal of Computational Physics*, 307:321–332, 2016.

119. L. Pedesseau, J.-M. Jancu, A. Rolland, E. Deleporte, C. Katan, and J. Even. Electronic properties of 2D and 3D hybrid organic/inorganic perovskites for optoelectronic and photovoltaic applications. *Optical and Quantum Electronics*, 46(10):1225–1232, 2014.

120. A. Pecchia, D. Gentilini, D. Rossi, M. Auf der Maur, and A. Di Carlo. Role of ferroelectric nanodomains in the transport properties of perovskite solar cells. *Nano Letters*, 16(2):988–992, 2016.

121. D. Liu, M.K. Gangishetty, and T.L. Kelly. Effect of CH3NH3PbI3 thickness on device efficiency in planar heterojunction perovskite solar cells. *Journal of Materials Chemistry A*, 2:19873–19881, 2014.

122. S. Meloni, T. Moehl, W. Tress, M. Franckeviius, M. Saliba, Y.H. Lee, P. Gao, M.K. Nazeeruddin, S.M. Zakeeruddin, U. Rothlisberger, and M. Graetzel. Ionic polarization-induced current-voltage hysteresis in $CH_3NH_3PbX_3$ perovskite solar cells. *Nature Communications*, 7, 2016, Article number: 10334. doi:10.1038/ncomms10334.

123. J.M. Azpiroz, E. Mosconi, J. Bisquert, and F. De Angelis. Defect migration in methylammonium lead iodide and its role in perovskite solar cell operation. *Energy and Environmental Science*, 8(7):2118–2127, 2015.

124. T. Albes, P. Lugli, and A. Gagliardi. Investigation of the blend morphology in bulk-heterojunction organic solar cells. *IEEE Transactions on Nanotechnology*, 15(2):281–288, 2016.

125. H.K. Kodali and B. Ganapathysubramanian. A computational framework to investigate charge transport in heterogeneous organic photovoltaic devices. *Computer Methods in Applied Mechanics and Engineering*, 247:113–129, 2012.

126. A. Bolognesi, M. Berliocchi, M. Manenti, A. Di Carlo, P. Lugli, K. Lmimouni, and C. Dufour. Effects of grain boundaries, field-dependent mobility, and interface trap states on the electrical characteristics of pentacene TFT. *IEEE Transactions on Electron Devices*, 51(12):1997–2003, 2004.

127. Y. Hu, L. Wang, Q. Qi, D. Li, and C. Jiang. Charge transport model based on single-layered grains and grain boundaries for polycrystalline pentacene thin-film transistors. *Journal of Physical Chemistry C*, 115(47):23568–23573, 2011.

128. M. Kimura, S. Inoue, T. Shimoda, and T. Sameshima. Device simulation of carrier transport through grain boundaries in lightly doped polysilicon films and dependence on dopant density. *Japanese Journal of Applied Physics, Part 1: Regular Papers and Short Notes and Review Papers*, 40(9A):5237–5243, 2001.

129. W.K. Metzger and M. Gloeckler. The impact of charged grain boundaries on thin-film solar cells and characterization. *Journal of Applied Physics*, 98(6):063701, 2005.

130. K. Taretto and U. Rau. Numerical simulation of carrier collection and recombination at grain boundaries in Cu(In,Ga)Se$_2$ solar cells. *Journal of Applied Physics*, 103(9), 2008.

131. H. Kobayashi and Y. Tokita. Modeling of hole transport across grain boundaries in organic semiconductors for mesoscale simulations. *Applied Physics Express*, 8(5):051602, 2015.

Novel Applications

44

Electroluminescent Refrigerators[*]

Kuan-Chen Lee

and

Shun-Tung Yen

44.1 Introduction

Solid-state optical refrigeration, which is featured by carrying the internal thermal energy away through light emission, has progressed rapidly after the first demonstration of photoluminescent (PL) refrigeration in 1995 (Epstein et al., 1995). To date, it has been shown that an Yb-doped $LiYF_4$ cooler can reach a temperature of 114 K and give a cooling power of 750 mW at room temperature (Melgaard et al., 2014). Further improving this kind of rare-earth-doped coolers to an even lower cooling temperature will be a difficult task. This is because the cooling process involves the thermalization of dopants and becomes inefficient when the thermal energy is comparable to the energy difference between discrete dopant levels in the ground-state manifold (Sheik-Bahae and Epstein, 2007). Different to the doped coolers, a semiconductor luminescent refrigerator is not restricted by such limitation and is expected to have lower operation temperature and higher cooling power. Moreover, it can be directly integrated with other semiconductor devices. These attractive features have stimulated extensive research on semiconductor luminescent refrigeration (Sheik-Bahae and Epstein, 2007, 2009). However, the required external efficiency for cooling is as high as nearly unity such that the realization of semiconductor PL refrigeration was not achieved until recently by Zhang et al. (2013) on CdS mircorod.

Semiconductor luminescent refrigeration using electrical pumping scheme, on the other hand, receives growing attention because of its looser requirement on the external efficiency compared to semiconductor PL refrigeration. Despite several works on the feasibility of electroluminescent (EL) refrigeration several decades ago (Berdahl, 1985; Dousmanis et al., 1964), the experimental verification of EL cooling has been reported just recently. In 2012, Santhanam et al. (2012) biased a light-emitting diode (LED) under several tens of microvolts and showed that the output emission power can exceed twice of the applied electrical

---

[*] This chapter is reproduced with permission from Yen and Lee (2010), and Lee and Yen (2012).

input power. This positive difference of the output over the input powers implies that the LED is capable of extracting thermal energy from the environments, as a refrigerator, to keep heat flow balanced. However, the cooling power is smaller than nanowatts so that the cooling phenomenon is not directly observed. It has been known for a while that the cooling power density of an EL cooler can be significantly high if the cooler is biased at a voltage of $\sim E_{\mathrm{g}}/q$, where E_{g} is the bandgap energy of the active layer and q the elementary charge. Extensive studies have been made to investigate the cooling feasibility and capability of EL refrigeration in this bias region. Mal'shukov and Chao studied the influence of the Auger recombination on the cooling capability of a GaAs-based refrigerator. They showed that a net cooling power density of several watts per square centimeter is achievable (Mal'shukov and Chao, 2001). Wang et al. performed a detailed self-consistent calculation with various recombination mechanisms accounted. Their analysis showed an impressive cooling efficiency of 35% for a cooler of p-GaAs active layer (Wang et al., 2005). Oksanen and Tulkki (2010) proposed a thermal heat pump, which consists of an EL cooler integrated directly with an absorbing diode, to deal with the problem of low light extraction efficiency encountered in semiconductor luminescent refrigeration. Other theoretical works, such as on heat transfer and cooling temperature (Han et al., 2007; Heikkilä et al., 2010; Lin et al., 2012; Piprek and Li, 2016), the design of refrigerator structure (Oksanen and Tulkki, 2010; Yen and Lee, 2010), the influence of the active materials (Yu et al., 2007), light extraction efficiency (Yen and Lee, 2010), and the photon recycling effect (Heikkilä et al., 2009; Lee and Yen, 2012), has also been conducted to provide a deep insight into EL refrigeration. Experimental works on the feasibility (Xue et al., 2015) and the application (Liu et al., 2013) of EL refrigeration are also being conducted. Although the high-cooling-power EL refrigeration has not been achieved, a thorough understanding of its operation mechanism still plays a necessary role in its realization.

In the following sections, we begin with the fundamental concept of EL refrigeration. The self-consistent calculation generally used for analyzing EL refrigeration and the operation principle of EL refrigeration is introduced in Section 44.2. Calculated results together with the analysis of various structural and optical effects on EL refrigeration are given in Section 44.3. Finally, a conclusion is given in Section 44.4.

44.2 Theoretical Methods

44.2.1 Fundamentals of EL Refrigeration

In EL refrigeration, electrons and holes are mediated as refrigerants and are intended to be injected into the active region for radiative recombination, which generates photons with thermal energy carried away. Recently, studies indicate that the extracted thermal energy comes from the thermalization of carriers during transport, similar to the Peltier effect in thermoelectric (TE) cooling (Santhanam et al., 2012; Oksanen and Tulkki, 2015; Xue et al., 2015). This process can be seen during the cooling cycle of EL refrigeration as illustrated in Figure 44.1a. At terminal contacts, carriers are injected into the device under an external bias V and flow all the way to the active layer. In the process of carrier transport, local cooling occurs in the place where the external bias is not large enough for carriers to overcome the potential barriers. This situation can be seen, for instance, in the cladding layer near the active layer. In such regions, only carriers of sufficiently high energy can surmount the barrier and move forward as shown in the inset circle. This causes high-energy empty states that are then occupied by carriers through thermalization, i.e., drawing phonons from the lattice matrix. The redistribution of carriers progresses rapidly and hence the quasi-thermal equilibrium is achieved as a result of much short phonon relaxation time. In the active layer, the injected electron–hole pairs recombine through various mechanisms. A heating energy of qV is generated if the electron–hole pair recombines nonradiatively. Contrary, for each photon generated from radiative recombination escaping out of the device, the internal energy of the whole system is reduced by $\hbar\omega - qV$. Here, $\hbar\omega$ is the photon energy, which comes from the combination of the thermal energy from the lattice matrix and the electric potential energy from the external bias. These three processes, i.e., electrical excitation from external bias, thermalization during transport, and relaxation through radiative recombination, complete the cooling cycle of EL refrigeration. Figure 44.1b shows the system view of an ideal EL cooler

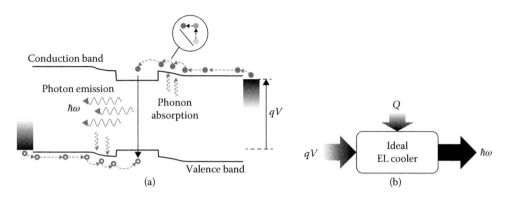

FIGURE 44.1 (a) The illustration of cooling cycling for EL refrigeration. The inset circle shows the thermalization process of carriers, which happens in the cladding layer near the active region. This is similar to the Peltier effect in thermoelectric cooling and attributed to EL refrigeration. (b) The system view of an ideal EL cooler. Net cooling occurs when $Q > 0$; otherwise, the device generates heat.

for a single cooling cycle, in which the parasitic effects are omitted. In this figure, an energy flow per cycle $Q = \hbar\omega - qV$ is required to be drawn from the ambient environment for compensating the internal energy loss of the system. If $Q > 0$, heat flows into the device and net cooling occurs; otherwise, the device generates heat. In this chapter, the cooling power, one of the most important parameters for coolers, is obtained in a similar way as Q described earlier, for which the input and the output powers are calculated from a self-consistent calculation based on one-dimensional drift-diffusion model with various recombination mechanisms accounted. In such a picture, the cooler is assumed to be in direct contact with a heat reservoir and the thermal resistance inside the cooler is neglected so that the temperature of the whole device is kept at a constant value of 300 K. In reality, a temperature gradient between the cooler and its environment exists due to a finite thermal conductance. The temperature distribution may also be nonuniform inside the cooler because of a nonuniform distribution of TE heat exchange (Yu et al., 2007). To account these effects, the heat balance equation between the cooler and its environment and the heat transfer equation inside the cooler have to be considered. Interested readers can refer to the works of Ashcroft and Mermin (1976), Han et al. (2007), Heikkilä et al. (2010), Lin et al. (2012), and Piprek and Li (2016) for more information. The geometric effects, such as current spreading and the detailed optical processes, are not accounted either. The consideration of these effects relies on the detailed information about the device geometry and a full three-dimensional simulation, which is beyond the scope of this chapter. The goal of this chapter is to provide a comprehensive view on how various structural and optical components affect the performance of an EL cooler without losing accuracy.

44.2.2 Device Structure and Main Equations

The device structure used in this work is composed of seven layers, as shown in Figure 44.2. They are, in order, a 100-nm p-doped $Al_{0.25}Ga_{0.75}As$ cladding layer (with an acceptor concentration of $N_a = 1 \times 10^{18}$ cm^{-3}), a 50-nm p-doped $Al_xGa_{1-x}As$ carrier blocking layer ($N_a = 1 \times 10^{18}$ cm^{-3}), a 50-nm undoped $Al_{0.25}Ga_{0.75}As$ spacer, an undoped GaAs active layer of thickness $L = 100$ nm, a 50-nm undoped $Al_{0.25}Ga_{0.75}As$ spacer layer, a 50-nm n-doped $Al_xGa_{1-x}As$ carrier blocking layer (with a donor concentration of $N_d = 1 \times 10^{18}$ cm^{-3}), and a 100-nm n-doped $Al_{0.25}Ga_{0.75}As$ cladding layer ($N_d = 1 \times 10^{18}$ cm^{-3}). Figure 44.2 also demonstrates the calculated energy band profiles, including the band edges of the conduction band E_c and the valence band E_v, the electron quasi-Fermi level E_{Fc}, and the hole quasi-Fermi level E_{Fv}, along the growth direction of z under a forward bias of $V = 1.4$ V. The Al content x is 0.4 in the carrier blocking layer. For the drift-diffusion model, the central problem is to find out the electrostatic potential Φ

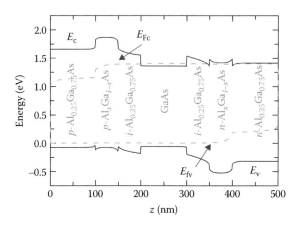

FIGURE 44.2 The energy band profiles, E_c and E_v, and the quasi-Fermi level profiles, E_{Fc} and E_{Fv}, along the growth direction z of the conduction and the valence bands, respectively, for the heterostructure under consideration. The profiles are obtained under a forward bias of 1.4 V.

and the quasi-Fermi levels for electrons E_{Fc}, and for holes E_{Fv} by solving three coupled equations (Horio and Yanai, 1990; Yen and Lee, 2010). These are the one-dimensional Poisson equation in the steady state:

$$\frac{d}{dz}\epsilon\frac{d}{dz}\Phi = -q\left(p - n + N_d^+ - N_a^-\right), \tag{44.1}$$

and the one-dimensional continuity equations for the electron and the hole currents:

$$\frac{d}{dz}J_n = q\left(R - G\right),$$

$$\frac{d}{dz}J_p = -q\left(R - G\right), \tag{44.2}$$

where ϵ is the electric permittivity; p, n, N_d^+, and N_a^- are the concentrations of the holes, electrons, ionized donors, and ionized acceptors, respectively; J_n (J_p) is the current densities due to the electron (hole) flow; R is the electron–hole recombination rate minus the intrinsic generation rate due to thermal excitation; and G is the electron–hole generation rate due to the photon recycling effect to be described. It is noted that all of the aforementioned p, n, N_d^+, N_a^-, J_n, J_p, R, and G are functions of Φ, E_{Fc}, and E_{Fv}. The first six ones can be expressed by the following formulas:

$$n = N_c F_{1/2}\left(\eta_n\right), \eta_n \equiv \left(E_{Fc} - E_c\right)/k_B T, \tag{44.3}$$

$$p = N_v F_{1/2}\left(\eta_p\right), \eta_p \equiv \left(E_v - E_{Fv}\right)/k_B T, \tag{44.4}$$

$$N_d^+ = \frac{N_d}{\left\{1 + 2\exp\left[\left(E_{Fc} - E_d\right)/k_B T\right]\right\}}, \tag{44.5}$$

$$N_a^- = \frac{N_a}{\left\{1 + 4\exp\left[\left(E_a - E_{Fv}\right)/k_B T\right]\right\}}, \tag{44.6}$$

$$J_n = -q\mu_n n d\Phi/dz + qD_n dn/dz, \tag{44.7}$$

$$J_p = -q\mu_p p d\Phi/dz - qD_p dp/dz, \tag{44.8}$$

where N_c (N_v) is the effective density of states for the conduction (valence) band, F_j the Fermi–Dirac integral of order j, k_B the Boltzmann constant, T the temperature, E_d (E_a) the donor (acceptor) level, μ_n (μ_p) the electron (hole) mobility, and D_n (D_p) the diffusivity for the electrons and the holes.

44.2.3 Recombination and Generation Processes

For the recombination processes in the device, the Shockley–Read–Hall (SRH), the radiative, and the Auger recombinations are considered in the high-quality GaAs active layer, while only the SRH recombination process is assumed to be dominant in the other layers containing aluminum. Therefore, the recombination rate is

$$
R = \begin{cases} R_{SRH} + R_{rad} + R_{Aug} & z \text{ in the active layer,} \\ R_{SRH} & \text{otherwise,} \end{cases}
\tag{44.9}
$$

where R_{SRH}, R_{rad}, and R_{Aug} are the recombination rates due to the SRH, the radiative, and the Auger recombination processes, respectively. In the calculation of R_{rad}, we consider the spontaneous emission contributed from the transition from the conduction band (c) to both the heavy-hole (hh) and the light-hole (lh) bands and express R_{rad} as

$$
R_{rad} = \int r(\hbar\omega)\, d\hbar\omega, \quad r = r_{hh} + r_{lh},
\tag{44.10}
$$

where $\hbar$ is the reduced Planck constant, ω the optical angular frequency, and r the total spontaneous emission rate that consists of r_{hh} and r_{lh} attributed to the transitions of c→hh and c→lh, respectively. With the parabolic band approximation, r_{hh} can be expressed as (Chuang, 1995)

$$
r_{hh}(\hbar\omega) = C_{hh} \frac{\hbar\omega \sqrt{\hbar\omega - E_g}\,\Theta(\hbar\omega - E_g)}{\left[1 + e^{-\eta_n} \exp\left(\frac{m_r}{m_c}\frac{\hbar\omega - E_g}{k_B T}\right)\right]\left[1 + e^{-\eta_p} \exp\left(\frac{m_r}{m_{hh}}\frac{\hbar\omega - E_g}{k_B T}\right)\right]},
\tag{44.11}
$$

where

$$
C_{hh} = \frac{q^2 n_r m_r^{3/2} E_P}{3\sqrt{2}\pi^3 \epsilon_0 m_0 \hbar^5 c^3}.
\tag{44.12}
$$

ϵ_0, m_0, and c are the electric permittivity, the electron mass, and the light speed, *in vacuo*, respectively; n_r is the refractive index; E_P is the optical matrix parameter; $m_r = m_c m_{hh}/(m_c + m_{hh})$ is the reduced mass; and Θ is the unit step function. For r_{lh}, it is simply obtained by replacing the role of m_{hh} in Equations 44.11 and 44.12 with the effective mass of the light hole band, m_{lh}.

For the Auger recombination rate, the CHHS and CHCC processes are considered. Therefore, R_{Aug} is

$$
R_{Aug} = \left(C_p p + C_n n\right)\left(np - n_i^2\right),
\tag{44.13}
$$

where C_p (C_n) is the Auger coefficient corresponding to the CHHS (CHCC) process and n_i the intrinsic carrier density. For the SRH recombination, we use

$$
R_{SRH} = \frac{pn - n_i^2}{\tau_n\left(p + n_i\right) + \tau_p\left(n + n_i\right)},
\tag{44.14}
$$

where τ_n (τ_p) is the minority carrier lifetime when $p \gg n, n_i$ ($n \gg p, n_i$).

For the generation rate G, the carriers generated by photon recycling are accounted in the active layer. Photon recycling is a process involving the absorption of spontaneously emitted photons by the active material itself and hence strongly related to the distribution of optical field in the device. A detailed consideration of the recycling process relies on the information about the device geometry, which is out of the scope for the present work. In this work, the generation rate G is directly connected to R_{rad} by

$$G = \frac{1}{L} \eta_{pr} \left(1 - \eta_{xp} \right) \int_{z \in \text{ active layer}} R_{rad} dz, \tag{44.15}$$

where η_{xp} is the light extraction efficiency, the probability for the spontaneously generated photons emitting out of the device, and η_{pr} is the photon recycling efficiency, the ratio of trapped photons eventually being absorbed by the active layer. Therefore, Equation 44.15 implies that the photon recycling process generates carriers uniformly in the active layer, which results from the absorption of a portion of spontaneously generated photons trapped in the device. The assumption of a uniform carrier generation rate is justifiable for an active layer thickness no longer than the inverse of the absorption coefficient.

44.2.4 Boundary Conditions

To obtain an unambiguous distribution of Φ, E_{Fc}, and E_{Fv}, the boundary conditions for Equations 44.1 and 44.2 have to be specified. For the boundary condition of Φ, it is simply assumed that the electrostatic potential Φ and the displacement field $-\epsilon d\Phi/dz$ are continuous at interfaces. The boundary conditions of E_{Fc} and E_{Fv} are, however, complicated and classified into two cases according to the magnitude of band offsets. For the case of a band offset smaller than $2k_B T$, the quasi-Fermi levels are determined from the continuity of drift-diffusion currents at the both sides of the interface under the assumption of a continuous quasi-Fermi level. For the other case of a larger band offset, the drift-diffusion model is no longer applicable for carriers' transport between layers. In this case, the thermionic process has to be taken into account, which allows quasi-Fermi levels to split at the interface. The thermionic emission current density due to the electron flow at the interface z_i with Fermi–Dirac distribution accounted can be expressed as[†]

$$J_n \left(z_i \right) = A_n \left(z_i^+ \right) T^2 F_1 \left[\xi_n \left(z_i^+ \right) \right] - A_n \left(z_i^- \right) T^2 F_1 \left[\xi_n \left(z_i^- \right) \right], \tag{44.16}$$

where $z_i^\pm = \lim_{\varepsilon \to 0} \left(z_i \pm \varepsilon \right)$, $A_n \left(z_i^\pm \right)$ is the effective Richardson constant of electrons at $z_i^\pm$, and $\xi_n \left(z_i^\pm \right)$ is the energy difference, normalized by $k_B T$, from the quasi-Fermi levels at $z_i^\pm$ to the larger one of the conduction band edges at the interface. By matching the drift-diffusion current to the thermionic current at interfaces, we can write the boundary condition for E_{Fc} as

$$\begin{aligned} J_n \left(z_i \right) &= J_n \left(z_i^\pm \right), \quad \text{for } \left| E_c \left(z_i^+ \right) - E_c \left(z_i^- \right) \right| \geq 2k_B T, \\ J_n \left(z_i^+ \right) &= J_n \left(z_i^- \right), \quad \text{otherwise,} \end{aligned} \tag{44.17}$$

with $J_n \left(z_i^\pm \right)$ being the drift-diffusion current at $z_i^\pm$. The boundary condition for E_{Fv} is determined by the continuity of J_p at the interface, similar to that for E_{Fc} just described above.

At the end of surfaces on which the metal contacts are to be made, the Fermi levels are assumed to be pinned at a fixed position from the band edges such that the carrier densities equal to the values of those

---

[†] It is noted that, at a low temperature, the contribution of the tunneling current through an abrupt heterojunction may not be neglected. In such a situation, Equation 44.16 has to be modified. Interested readers can refer to Yang et al. (1993) for detailed expression.

in thermal equilibrium. The Fermi level positions at the surface are thus determined by the charge neutral condition, $p + N_d^+ = n + N_A^-$, with $E_{Fc} = E_{Fv}$.

With the boundary conditions being specified, the differential Equations 44.1 and 44.2 can be solved steadily by the Newton–Raphson iteration scheme for each applied voltage V. The external electrical work qV done on each of injected carriers gives the Fermi level difference between two surfaces $E_{Fc}|_{z\,\text{at cathode}} - E_{Fc}|_{z\,\text{at anode}} = qV$. Parameters used in this work are listed in Table 44.1, except for mobility, which is obtained from an empirical model (Sotoodeh et al., 2000). The band offset ratio ($\Delta E_c / \Delta E_v$) is set as 0.6:0.4 for GaAs/Al$_x$Ga$_{1-x}$As interfaces. The binding energies of donors and acceptors are set at $k_B T$ for simplicity.

44.2.5 Output Quantities and Models for Steady-State Analysis

Once the solutions Φ, E_{Fc}, and E_{Fv} are obtained, the quantities of interest, including various current components, power densities, and efficiencies, can be calculated and used for further analysis. Figure 44.3 shows the working principle of an EL refrigerator. In the steady state, electrons and holes are injected

TABLE 44.1 Material Parameters of GaAs and AlGaAs at 300 K Used for Calculation

Parameter	Symbol	Unit	GaAs	Al$_x$Ga$_{1-x}$As ($x < 0.45$)
Bandgap	E_g	eV		$1.424 + 1.247x$
	m_c			$0.063 + 0.083x$
Effective mass	m_{hh}	m_0		$0.51 + 0.25x$
	m_{lh}			$0.082 + 0.068x$
Effective density of states	N_c	cm^{-3}		$2.5 \times 10^{19}(m_c/m_0)^{3/2}$
	N_v			$2.5 \times 10^{19}[(m_{hh}/m_0)^{3/2} + (m_{lh}/m_0)^{3/2}]$
Effective Richardson constant	A_n	A/cm^2/K^2		$120 \times (m_c/m_0)$
	A_p			$120 \times (m_{hh} + m_{lh})/m_0$
Electric permittivity	ϵ	ϵ_0		$12.9 - 2.84x$
Refractive index	n_r	–	3.6	–
Optical matrix parameter	E_p	eV	25.7	–
SRH recombination lifetime	τ_n	s	1.3×10^{-6}	1×10^{-8}
	τ_p		1.2×10^{-6}	1×10^{-8}
Auger coefficient	C_n	cm^6/s	1.0×10^{-31}	–
	C_p		1.2×10^{-30}	–

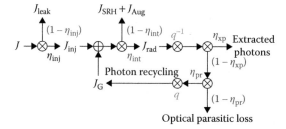

FIGURE 44.3 A diagram illustrating of the operation of an electroluminescent refrigerator with photon recycling as a feedback. It indicates the relationships among various currents. Symbol + stands for an addition operation that sums its inputs, indicated by inward arrows with the tips pointing at the symbol, and gives its output, indicated by an outward arrow with the tip pointing away from the symbol. Symbol × stands for multiplication operations, which give their outputs, indicated by outward arrows, by multiplying their inputs, indicated by inward arrows, by the factors along with the individual outward arrows.

from the electrodes under a forward bias, leading to a total current J in the device. Most of the electrons and holes are intended to be injected into the active region, recombine therein, and result in a major current component, called the injection current J_{inj}. The injection efficiency η_{inj} is defined as the ratio J_{inj}/J. The other part of the small current, called the leakage current J_{leak}, arises from the SRH recombination outside the active region $J'_{SRH} = \int_{z \notin active\ layer} R_{SRH}dz$ and the recombination current at the device surfaces $J_{sr} = J_n|_{z\ at\ anode} + J_p|_{z\ at\ cathode}$. In this figure, we see that there are two means of injecting carriers into the active region. One is by the external electric injection and the other is by the photon recycling effect, which result in two effective current components J_{inj} and J_G, respectively. As a result, the sum $J_{inj} + J_G$ accounts for the total effective injection rate of electrons and holes to the active region. In the steady-state condition, the injection rate is balanced by the recombination rate in the active layer. Hence,

$$J_{inj} + J_G = J_{SRH} + J_{rad} + J_{Aug}, \tag{44.18}$$

where J_{SRH}, J_{rad}, and J_{Aug} are the current components in the active region due to the SRH, the radiative, and the Auger recombinations, respectively, with $J_i = q \int_{z \in active\ region} R_i dz$, for $i =$ SRH, rad, and Aug. The internal quantum efficiency η_{int} is naturally defined as the probability for electron–hole pairs recombining through the radiative recombination:

$$\eta_{int} = \frac{J_{rad}}{J_{SRH} + J_{rad} + J_{Aug}}. \tag{44.19}$$

Owing to the radiative recombination, photons are generated with a rate J_{rad}/q. Since only a ratio η_{xp} of generated photons can escape out of the device, the emission flux is $\eta_{xp}J_{rad}/q$ and the corresponding emission power is:

$$P_{rad} = \eta_{xp} \int \int r\hbar\omega d\hbar\omega dz \equiv \eta_{xp} \langle \hbar\omega \rangle J_{rad}/q, \tag{44.20}$$

where $\langle \hbar\omega \rangle$ is the average photon energy. In addition to these emitted photons, the others are trapped in the device, either being parasitically absorbed, causing heat, or being recycled, reincarnating into electron–hole pairs in the active region. Figure 44.3 shows that the photon recycling effect is a feedback process, supplying a portion of electron–hole pairs required for recombinations. From Figure 44.3, we can simply express J_G as $J_G = \eta_{pr} (1 - \eta_{xp}) J_{rad}$, which is equivalent to Equation 44.15 with the equality $J_G = qGL$.

In conventional optoelectronic devices, the external quantum efficiency η_{ext} is defined as the ratio of the photon flux emitted out of the devices to the carrier flux injected into the devices, i.e., $\eta_{ext} = \eta_{xp}J_{rad}/J$. Using Equation 44.18 and the definitions of various efficiencies, η_{ext} with photon recycling effect accounted can be expressed in terms of the other efficiencies:

$$\eta_{ext} = \frac{\eta_{xp}\eta_{int}\eta_{inj}}{1 - \eta_{int}\eta_{pr} (1 - \eta_{xp})}. \tag{44.21}$$

In the absence of photon recycling ($\eta_{pr} = 0$), η_{ext} returns to the conventional form of $\eta_{ext} = \eta_{xp}\eta_{int}\eta_{inj}$.

For the purpose of refrigeration, the cooling power and the cooling efficiency are two important parameters of interests. Following the concepts shown in Figure 44.1b, we define the cooling power as the power difference of the output and the input powers, which is required to keep the cooler operating steadily at 300 K:

$$P_c = P_{rad} - JV = \eta_c JV, \tag{44.22}$$

where $\eta_c \equiv P_c/JV$ is known as the cooling efficiency. The notion of EL refrigeration is to set the devices in the cooling mode, which requires a positive cooling power or cooling efficiency. Using Equation 44.20 and the definition of η_{ext}, one can rewrite the cooling efficiency as

$$\eta_c = \frac{\eta_{ext}}{\eta_{ext,cr}} - 1, \tag{44.23}$$

where

$$\eta_{ext,cr} \equiv \frac{qV}{\langle \hbar\omega \rangle}, \tag{44.24}$$

is the critical external efficiency for the devices to act in the cooling mode, which is consistent with the result from Figure 44.1b if the external efficiency of the cooler is considered. From Equation 44.21, the critical extraction efficiency for cooling is:

$$\eta_{xp,cr} = \frac{1 - \eta_{pr}\eta_{int}}{\eta_{int}\left(\eta_{inj}/\eta_{ext,cr} - \eta_{pr}\right)}, \tag{44.25}$$

for $\eta_{inj}/\eta_{ext,cr} > \eta_{pr}$.

44.3 Results and Discussion

In this section, we present the results simulated by the aforementioned self-consistent calculation to investigate various structural and optical effects on the cooling feasibility and capability of GaAs/AlGaAs EL refrigerators. In the discussion of the structural effects, we have assumed an unity light extraction efficiency $(\eta_{xp} = 1)$.

44.3.1 Influence of Device Structures on EL Refrigeration

44.3.1.1 Leakage Current and Carrier Blocking Layer

We first study the cooling feasibility of a simple double heterostructure (DH) EL cooler by plotting the internal, the external, and the critical efficiencies shown in Figure 44.4 as functions of the applied voltage

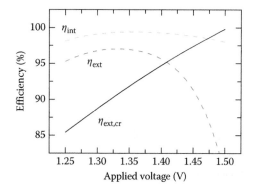

FIGURE 44.4 The internal efficiency η_{int}, the external efficiency η_{ext}, and the critical external efficiency $\eta_{ext,cr}$ as functions of the applied voltage V for the heterostructure with an active layer of width $L = 0.1\ \mu m$ and carrier blocking layers of Al composition $x = 0.25$, which has no effect on blocking carriers.

V for $\eta_{xp} = 1$. The Al content x in the blocking layer is 0.25, which is identical with that in the cladding and the spacer layers so that there is no blocking effect. As can be seen, while the critical external efficiency $\eta_{ext,cr}$ increases linearly with the applied voltage V, the internal efficiency η_{int} first increases to a maximum at $V = 1.36$ V and then decreases. The variation of η_{int} on V reveals a more important role played by the SRH and the Auger recombinations at a lower and a higher voltage, respectively. Similar to η_{int}, the external efficiency η_{ext} reaches to a maximum at $V = 1.33$ V and then droops more severely than η_{int} as V further increases, crossing with $\eta_{ext,cr}$ at the critical voltage $V_{cr} = 1.41$ V, where $\eta_{ext} = \eta_{ext,cr}$. We see that the DH EL cooler can act in the cooling mode for $V < 1.41$ V.

It is not difficult to find that the deterioration of η_{ext} is caused by a rapid reduction in the injection efficiency η_{inj} since $\eta_{inj} = \eta_{ext}/\eta_{int}$ for $\eta_{xp} = 1$. This implies that V_{cr} can be improved if the leakage current can be alleviated. In our calculation, it is shown that the strong surface recombination dominates in the leakage current as a result of a high minority carrier density near the terminal contacts. This situation becomes worse in particular for the devices of shorter cladding layers as in this work. The high minority carrier density can be reduced significantly by carrier blocking layers. In Figure 44.5a, we show the electron density in the center of the p-cladding layer as a function of V for $x = 0.25, 0.3, 0.35,$ and 0.4. As can be seen, the electron density decreases monotonically with increasing x. This indicates that the energy barrier provided by the blocking layer can effectively block the minority carriers, preventing them from flowing over the carrier blocking layer. This leads to a lower minority carrier density in the cladding layer and hence a lower surface recombination current. The low surface recombination current is confirmed by the injection efficiency as plotted in Figure 44.5b. Evidently, the injection efficiency is improved significantly by increasing the barrier height when $x \leq 0.3$. The improvement however becomes insignificant as x increases from 0.35 to 0.4, in contrast to the result from Figure 44.5a This is because the leakage current across the blocking layer is almost eliminated when $x = 0.35$. In this case, $\eta_{inj} \approx 1$ and the residue of the leakage current is mostly due to the SRH recombination in the space layer. Contrarily, for $x \leq 0.3$, the injection efficiency decreases rapidly at high bias as a result of J_{sr}.

In Figure 44.6, we show the cooling power as a function of the applied voltage for EL cooler of various Al content in the blocking layer. As expected, the critical voltage increases from 1.41 V for $x = 0.25$ to 1.47 V for $x = 0.35$ and 0.4. Moreover, the cooling power also increases from 7.5 to 20 W/cm^2 for $x = 0.25$ to 0.35. The improvement of cooling power comes from an improved η_{inj} and hence η_{ext} as indicated by Equations 44.22 and 44.23, i.e., $P_c = JV\left(\eta_{ext}/\eta_{ext,cr} - 1\right)$. Further increasing the Al content to 0.4 does not help in improving the cooling power, which is consistent with the variation of the injection efficiency as shown in Figure 44.5b.

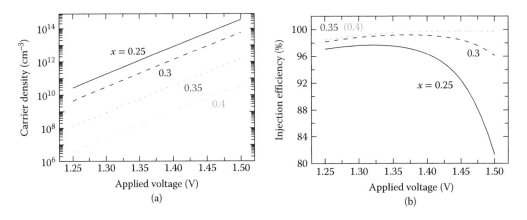

FIGURE 44.5 (a) The minority carrier (electron) density in the center of the *p*-cladding layer and (b) the injection efficiency as functions of the applied voltage V for various aluminum content $x = 0.25, 0.3, 0.35,$ and 0.4.

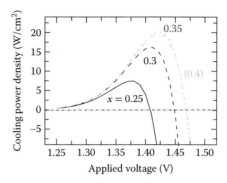

FIGURE 44.6 The ideal cooling power density P_c^0 as a function of the applied voltage V for various Al compositions of $x = 0.25$, 0.3, 0.35, and 0.4 in the carrier blocking layer. The extraction efficiency is set at unity.

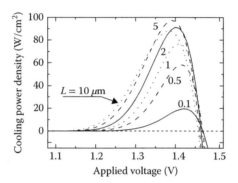

FIGURE 44.7 The ideal cooling power density P_c^0 as a function of V for $L = 0.1$, 0.5, 1, 2, 5, and 10 μm.

44.3.1.2 Active Layer Thickness

The cooling power density of 20 W/cm² seems to be the limiting value for the device of active layer thickness $L = 0.1$ μm. To further enhance it, one can increase the active layer thickness. Intuitively, a wider active layer can contain more electron–hole pairs to recombine therein, leading to an enhanced J_{rad} and hence more effective cooling cycles. Figure 44.7 shows the cooling power density P_c as a function of V for the structures with the same Al content ($x = 0.4$) in the carrier blocking layer but various active layer thickness $L = 0.1$, 0.5, 1, 2, 5, and 10 μm; here, $\eta_{xp} = 1$. As can be seen, the maximum cooling power $P_{c,max}$ can be improved about five times up to as high as 97 W/cm² by increasing L from 0.1 to 5 μm. However, this improvement is only prominent for $L \leq 5$ μm. Further increasing the active layer thickness does not enhance but degrades the maximum cooling power due to the remarked nonuniformity of carrier distribution, as L is comparable to or greater than the carrier diffusion length. From our calculation, we find that the average carrier density in the active layer is lower for a larger L; this difference in the average carrier density between different Ls is particularly remarkable when V is high. This renders the cooling power for $L = 10$ μm larger than that for $L = 5$ μm at a low V but smaller at a high V.

In Figure 44.8, we show the dependences of various efficiencies on the active layer thickness by plotting the injection efficiency η_{inj} and the internal efficiency η_{int} in Figure 44.8a, the product $\eta_{inj}\eta_{int}$ and the critical external efficiency $\eta_{ext,cr}$ in Figure 44.8b, and the cooling efficiency η_c in Figure 44.8c as functions of the applied voltage V for the active layer thickness $L = 0.1$, 1, and 10 μm and $\eta_{xp} = 1$. The Al content of the blocking layer is $x = 0.4$. Since J_{inj} is enhanced by increasing L, the η_{inj} is also enhanced. For $L \geq 1$ μm, the injection efficiency is nearly 100%, almost independent of the applied voltage. As been described, the internal efficiency η_{int} reaches to a maximum at about 1.37 V. It declines more profoundly as V decreases

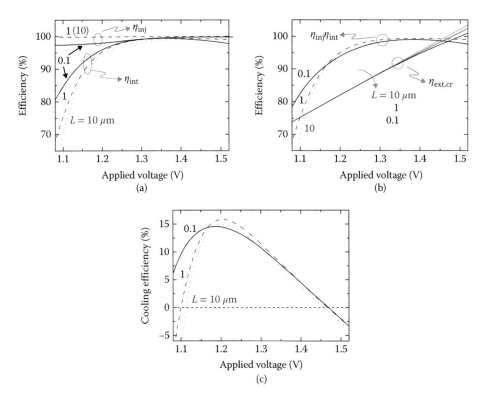

FIGURE 44.8 (a) The injection efficiency η_{inj} and the internal efficiency η_{int}, (b) the production of $\eta_{inj}\eta_{int}$ and the critical extraction efficiency $\eta_{ext,cr}$, and (c) the ideal cooling efficiency η_c^0 as functions of the applied voltage for $L = 0.1$, 1, and 10 μm.

than increases, implying a more important SRH recombination at low V than the Auger recombination at high V. Such a lowering of η_{int} at low V is more prominent for a larger L because of a lower carrier density in the active layer. On the other side, at high V, the Auger recombination is less important for a larger L due also to a lower carrier density. The ratio of $\eta_{ext,cr}/(\eta_{inj}\eta_{int})$ is the critical extraction efficiency $\eta_{xp,cr}$, which gives the design margin for η_{xp}. For instance, at $V = 1.2$ V, $\eta_{inj}\eta_{int} = 94\%$ and $\eta_{ext,cr} = 82\%$, as can be found in Figure 44.8b, give $\eta_{xp,cr} = 87\%$. Evidently, the margin becomes tighter when bias V deviates from 1.2 V. If an EL cooler is improperly biased at a voltage where $\eta_{xp,cr} \geq 1$, the ideal cooling efficiency will become negative (Figure 44.8c). In this case, it is impossible for the device to act in the cooling mode. At $V = 1.39$ V where the maximum cooling power occurs (see Figure 44.7), $\eta_{xp,cr} = 96\%$ for $L = 10$ μm. In such a bias condition, the corresponding cooling efficiency is only about 4%. The cooling efficiency can be improved by biasing the cooler at $V = 1.2$ V at which it can be as high as 15% at the cost of the cooling power density.

44.3.1.3 Doping in Active Layer

In Figure 44.9, we show the influences of the doping types and concentrations on EL refrigeration by plotting the ideal cooling power as a function of the applied voltage V for the active regions which are undoped, doped with $N_d = 1 \times 10^{18}$ cm^{-3}, with $N_a = 1 \times 10^{18}$ cm^{-3}, and with $N_a = 5 \times 10^{18}$ cm^{-3}. Here, the thickness of the active layer is $L = 1$ μm and the Al content of the blocking layers is $x = 0.4$. We find that intentionally doping into the active region does not improve the cooling power. The degradation caused by p-type doping is mainly due to the enhanced Auger recombination. The n-type doping suppresses the radiative recombination and also the Auger recombination, leading to a slightly enhanced cooling power at high V but a reduced cooling power at low V.

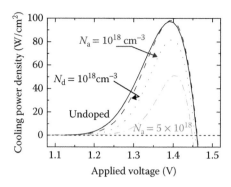

FIGURE 44.9 The ideal cooling power density P_c^0 as a function of the applied voltage for active layers with different doping types and concentrations, including undoped, doped with $N_d = 1 \times 10^{18}$ cm^{-3}, with $N_a = 1 \times 10^{18}$ cm^{-3}, and with $N_a = 5 \times 10^{18}$ cm^{-3}. Here, the active layer thickness is $L = 1$ μm, the Al content of the blocking layer is $x = 0.4$, and the light extraction efficiency is set at unity.

44.3.2 Influence of Optical Processes on EL Refrigeration

44.3.2.1 Influence of Light Extraction

Light extraction efficiency influences significantly the performance of an EL refrigerator in the absence of photon recycling. This is because the photons trapped within the device turn out to be totally absorbed and transformed into thermal energy. In this subsection, we study the influence of the light extraction efficiency on EL refrigeration for the worst case of zero photon recycling efficiency ($\eta_{pr} = 0$). Here, the active layer thickness L is set at 1 μm and the Al content $x = 0.4$ is used for the carrier blocking layer. Figure 44.10a shows the curves of the cooling power P_c as a function of the applied voltage V for five values of extraction efficiency $\eta_{xp} = 0.8, 0.85, 0.9, 0.95$, and 1. The curves exhibit a peak for $\eta_{xp} = 0.9, 0.95$, and 1, but no peak for $\eta_{xp} = 0.8$ and 0.85. As expected, the height of the peak diminishes drastically from 79 to 2 W/cm^2 as η_{xp} reduces slightly from unity to 0.9. The voltage at which the curve peaks moves from 1.41 V for $\eta_{xp} = 1$ to 1.28 V for $\eta_{xp} = 0.9$. The cooling power falls rapidly from the peak value to negative as the voltage increases.

These behaviors can be understood from Equation 44.23, i.e., $\eta_c = \eta_{ext}/\eta_{ext,cr} - 1$, with the help of the relationships between the curves of the corresponding external efficiencies η_{ext} and the critical external efficiencies $\eta_{ext,cr}$ shown in Figure 44.10b. The curves of $\eta_{ext,cr}$ for the different η_{xp} values merge into a single line, since $\eta_{ext,cr} = qV/\langle \hbar\omega \rangle$ is nearly proportional to V and independent of η_{xp}. The curves of η_{ext} for the different η_{xp} values behave similarly and separate almost equidistantly from their neighbors. As V increases, the η_{ext} value increases and then almost reaches saturation at a level $\eta_{ext} \approx \eta_{xp}$, with a slight decline in the high-voltage region. These features can be explained by the simple relation $\eta_{ext} = \eta_{xp}\eta_{int}\eta_{inj}$, where the injection efficiency η_{inj} is almost unity for the device containing carrier blocking layers. Therefore, the variation of η_{ext} with V for various η_{xp} simply follows the variation of the internal efficiency η_{int} multiplied by a scaling factor of η_{xp}. In Figure 44.10b, we see that there exist operation regions of refrigeration for $\eta_{xp} = 0.9, 0.95$, and 1, but no operation region for $\eta_{xp} = 0.8$ and 0.85, where the operation region of refrigeration is a range of voltage in which the device operates in the cooling modes ($\eta_c > 0$ or $\eta_{ext} > \eta_{ext,cr}$). This explains the feature shown in Figure 44.10a that only the three curves for $\eta_{xp} = 0.9, 0.95$, and 1 exhibit a peak of positive P_c. Figure 44.10c shows the cooling efficiency η_c as a function of V for $\eta_{xp} = 0.85, 0.9, 0.95$, and 1. We see that the peaks for η_c-V curves fall at $V = 1.2$ V. The cooling efficiencies decline when the operation voltage deviates from 1.2 V. Different to the peak voltage for η_c-V curves, the peaks of P_c-V curves in Figure 44.10a occur at voltages near the critical voltages, for which the cooling efficiency is small. This is why a slight increase of V may cause the drastic reduction of P_c from the peak value and turn the operation of the device from the cooling mode to the heating mode.

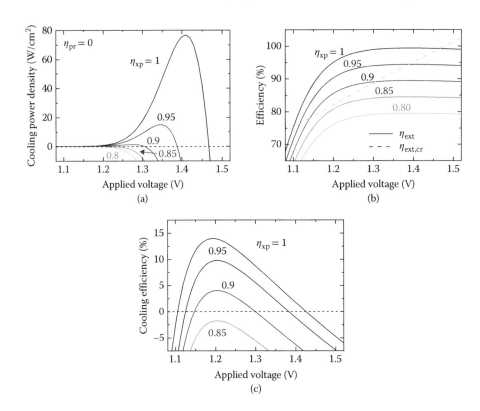

FIGURE 44.10   (a) The cooling power density P_c, (b) the external quantum efficiency η_{ext} and the critical extraction efficiency $\eta_{ext,cr}$, and (c) the cooling efficiency η_c as functions of the applied voltage for various extraction efficiency $\eta_{xp} = 0.8, 0.85, 0.9, 0.95$, and 1. The photon recycling efficiency is $\eta_{pr} = 0$.

44.3.2.2 Influence of Photon Recycling

Actually, most of the trapped photons can be reabsorbed by the high-quality GaAs active region and reincarnated into electron–hole pairs rather than into the thermal energy if the device structure is well designed. The photon recycling, as will be described, alleviates considerably the requirement of extraction efficiency for EL refrigeration. In this subsection, we demonstrate the photon recycling effect by examining another extreme case of unity recycling efficiency $\eta_{pr} = 1$, i.e., all of the trapped photons are recycled, turning into electron–hole pairs in the active region.

Figure 44.11a shows the cooling power density P_c as a function of V for five values of the extraction efficiency $\eta_{xp} = 0.2, 0.4, 0.6, 0.8$, and 1. All the curves in the figure behave as an asymmetrical peak of positive P_c, even for the low extraction efficiency of 0.2. The height of the peak now diminishes gently as η_{xp} reduces. These peaks occur almost at the same voltage ($V = 1.41$ V) except for $\eta_{xp} = 0.2$, for which the curve peaks at 1.39 V. Similarly, these behaviors can also be understood from Equation 44.23 and the relations between the corresponding η_{ext} and $\eta_{ext,cr}$ shown in Figure 44.11b. Comparing with those in Figure 44.10b, the external efficiencies η_{ext} shown in Figure 44.11b still vary with V in the manner that they first rise, reach a maximum, and then slightly decline. The difference is that η_{ext} is now much less sensitive to η_{xp}, especially in the voltage range where the internal efficiency η_{int} is near unity, as plotted in Figure 44.11c together with η_{inj} for various η_{xp}. In the voltage range between 1.3 and 1.5 V, where η_{int} is close to unity, the external efficiency η_{ext} can be approximated by

$$\eta_{ext} \approx \eta_{inj} \left[1 - \frac{1}{\eta_{xp}} \left(1 - \eta_{int} \right) \right]. \tag{44.26}$$

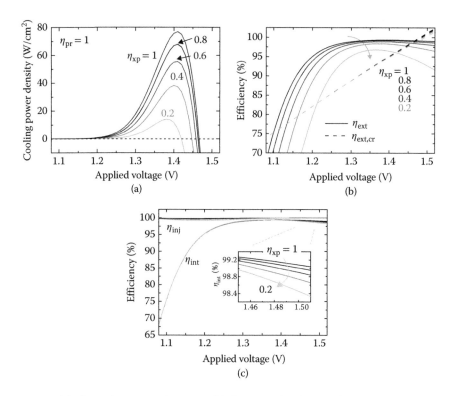

FIGURE 44.11 (a) The cooling power density P_c, (b) the external quantum efficiency η_{ext} and the critical external efficiency $\eta_{ext,cr}$, and (c) the injection efficiency η_{inj} and the internal efficiency η_{int} as functions of V for various extraction efficiency $\eta_{xp} = 0.2, 0.4, 0.6, 0.8$, and 1. The inset of (c) shows η_{int} at high V. The photon recycling efficiency is set at unity.

Equation 44.26 explains that, as $1 - \eta_{int}$ is small, η_{ext} is insensitive to η_{xp} and has a value close to η_{inj}, which is nearly unity.

In principle, the variation of η_{ext} basically follows the variation of η_{int} modified by photon recycling. This is because the only important loss is the nonradiative recombination in the active region when we set $\eta_{pr} = 1$ and $\eta_{inj} \approx 1$. For $\eta_{xp} = 1$, all the generated photons are emitted out of the device and there is no photon recycling, regardless of the value of η_{pr}, as can also be seen from Equation 44.21. The photon recycling effect manifests when η_{xp} deviates from unity. It causes an effective feedback current J_G and boosts the carrier concentration in the active region. The increasing carrier concentration degrades η_{int} and hence η_{ext} in the high-voltage region. This effect is particularly important for a low η_{xp} (see Figure 44.11b and the inset in Figure 44.11c). The lines of $\eta_{ext,cr}$ shown in Figure 44.11b do not merge as completely as those for the case of $\eta_{pr} = 0$ shown in Figure 44.10b. This is a consequence of the photon recycling effect, which causes a higher average photon energy $\langle \hbar\omega \rangle$ and, hence, a lower $\eta_{ext,cr}$ for a lower η_{xp}, due to a higher carrier concentration. The condition that $P_c = 0$ occurs at a critical voltage of about 1.47 V for $\eta_{xp} = 0.4$, 0.6, 0.8, and 1. The voltages are close to each other because of the crowding of the η_{ext}-V curves in the high-voltage region, which cross the $\eta_{ext,cr}$-V curves in the vicinity of 1.47 V. This implies that the P_c-V curves peak almost at the same voltage. The voltages for the peaks of the η_c-V curves, which can be estimated from Figure 44.11b, move from 1.2 to 1.3 V as η_{xp} decreases from 1 to 0.2.

Under the condition that $\eta_{pr} = 1$ and $\eta_{inj} \approx 1$, the critical extraction efficiency $\eta_{xp,cr}$ for the device to operate in the cooling mode can be written as

$$\eta_{xp,cr} \approx \frac{\left(1/\eta_{int}\right) - 1}{\left(1/\eta_{ext,cr}\right) - 1}. \tag{44.27}$$

The above expression implies that $\eta_{xp,cr}$ can be very small when η_{int} approaches unity, while $\eta_{ext,cr}$ does not. An EL refrigerator having a very low η_{xp} could operate in the cooling mode if there were no optical parasitic loss ($\eta_{pr} = 1$) and no current leakage ($\eta_{inj} = 1$).

The aforementioned discussion reveals that a high internal efficiency is essential to the photon recycling process. This implies that the quality of the materials making up the device is critical for refrigeration. If the material quality is not as high as in the study and gives $\eta_{int} < \eta_{ext,cr}$, it is impossible for the device to operate in the cooling mode as a result of the inequality $\eta_{ext} < \eta_{int}$ (see Equation 44.21), regardless of the values of the other efficiencies.

We now know that the internal efficiency η_{int} is an important factor in determining η_{ext} and, hence, to the performance of the device. It is influenced by photon recycling through the feedback generation of electron–hole pairs in the active region. To further gain insight into this effect, we set $\eta_{pr} = 1$ and investigate the average carrier density in the active region (Figure 44.12) as a function of η_{xp} when the device is biased at $V = 1.2$ and 1.4 V. As shown in Figure 44.12a, the average carrier density is low ($\approx 3 \times 10^{16}$ cm^{-3}) and nearly independent of η_{xp} for the device biased at 1.2 V, but is more than 5×10^{17} cm^{-3} and increases as η_{xp} reduces for the device biased at 1.4 V. This difference arises from the difference in the electric potential energy of the electrons (holes) in the active region relative to that in the n-type (p-type) cladding layer, from which the electrons (holes) are injected. This situation can be seen in Figure 44.12b, which shows the conduction band profiles at the n-doping side as functions of z for $V = 1.2$ and 1.4 V. When the device is biased at a low voltage, such as $V = 1.2$ V, the electric potential energy of the electrons is higher in the active layer than in the n-type cladding layer. This drop in electric potential energy causes the excess carriers, due to optical generation, to leak out of the active region, leading to an effective current J_G opposite the injection current J_{inj}. Consequently, accumulation of carriers in the active region is negligible, and the carrier density is basically independent of the photon recycling (and, hence, η_{int}). The drop in electric potential energy diminishes as the bias increases and, eventually, the situation reverses. At $V = 1.4$ V, the electric potential energy of the electrons is lower in the active region than in the n-type cladding layer. The potential profile now forms a potential well for the carriers and prohibits most of the excess carriers in the active region from leaking out. As a result, the optically generated carriers are mostly accumulated within the active region, recombined therein, and contribute an additional current, known as J_G. The carrier accumulation depends on the photon recycling and becomes particularly important for low η_{xp}, as shown in Figure 44.12a.

The variation of recombination current densities (J_{SRH}, J_{rad}, and J_{Aug}) on the extraction efficiency also follows the variation of carrier density such that they are invariant on η_{xp} at $V = 1.2$ V but increase considerably at $V = 1.4$ V for a decreasing η_{xp} (Lee and Yen, 2012). The feedback current J_G, which compensates J_{inj} for the total recombination in the active region, increases with decreasing η_{xp} for both

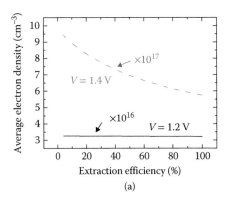
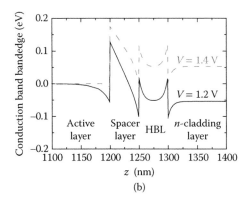

FIGURE 44.12 (a) The average electron densities in the active region for $V = 1.2$ and 1.4 V as functions of the extraction efficiency and (b) the conduction band profiles near the n-doping side as functions of z for $V = 1.2$ and 1.4 V. (HBL, hole-blocking layer.)

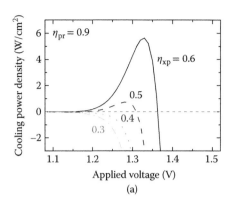 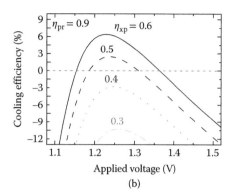

FIGURE 44.13 (a) The cooling power density P_c and (b) the cooling efficiency η_c as functions of applied voltage for various light extraction efficiencies $\eta_{xp} = 0.3, 0.4, 0.5$, and 0.6. The recycling efficiency is set at a practical value of 0.9.

applied voltages because $J_G = \eta_{pr}(1 - \eta_{xp})J_{rad}$. This leads to a significant reduction in J_{inj} and hence improves external efficiency for a low η_{xp}.

For a more practical case, we show the cooling power P_c in Figure 44.13a and the cooling efficiency η_c in Figure 44.13b as functions of the applied voltage V. Since $L = 1$ μm, η_{xp} is at most 60%. We therefore consider $\eta_{xp} = 0.3, 0.4, 0.5$, and 0.6 and set $\eta_{pr} = 0.9$, which is potentially achievable for a high-quality GaAs material. The resulting P_c-V relationship changes with η_{xp} to a degree milder than for $\eta_{pr} = 0$ (Figure 44.10a) but more severe than for $\eta_{pr} = 1$ (Figure 44.11a). The peak cooling power is now 5.6 W/cm² at $V = 1.33$ V for $\eta_{xp} = 0.6$, which is only 10% for the case of $\eta_{pr} = 1$ (Figure 44.11a). At the voltage of 1.33 V, the cooling efficiency η_c is small (2%) for $\eta_{xp} = 0.6$ and falls negative as η_{xp} reduces to 0.5, associated with the severe degradation of P_c from 5.6 to -3 W/cm². The situation can be cured by biasing the device at a lower voltage for an improved η_c. For example, at $V = 1.24$ V, η_c is 6.4% for $\eta_{xp} = 0.6$ and remains positive (about 2%) for $\eta_{xp} = 0.5$. The device can operate in the cooling mode for $\eta_{xp} > 0.45$. However, the cooling power is sacrificed. It is now only 0.9 W/cm² for $\eta_{xp} = 0.6$ and 0.3 W/cm² for $\eta_{xp} = 0.5$.

In the end of this chapter, we remark on the choices of active materials and operation temperatures for EL refrigeration. From Equation 44.24, the critical external efficiency of an EL cooler operating at E_g/q for maximum cooling power can be approximated by $\eta_{ext,cr} \approx E_g/\left(E_g + 2k_B T\right)$. Evidently, the external efficiency required for cooling is alleviated with decreasing bandgap energy and an increasing operating temperature. Indeed, recent research has shown that operating a 450-nm InGaN LED at 615 K is capable of producing a negative Peltier heating power inside the device (Xue et al., 2015). Moreover, theoretical study also confirms that elevating the operating temperature can improve the cooling power and efficiency for a GaAs EL cooler (Heikkilä et al., 2010). These improvements, as can be seen from the approximation of $\eta_{ext,cr}$, are strongly related to the ratio $E_g/2k_B T$. A smaller value of $E_g/2k_B T$ implies a fewer number of effective cooling cycles required for compensating a single parasitic loss. This means that the EL refrigeration will be more likely achieved for a cooler with an active material of smaller bandgap energy and operating in a higher temperature if the internal efficiency is high.

44.4 Summary

We discussed the influences of device structures and optical processes on the feasibility and capability of GaAs-based EL refrigerators by self-consistent calculation. For EL devices with short cladding layers, the leakage current, which arises from the minority carriers flowing toward the opposite electrodes, deteriorates the injection efficiency significantly at high bias. The degradation of injection efficiency can be cured by employing carrier blocking layers of $Al_xGa_{1-x}As$ ($x \geq 0.35$), leading to an injection efficiency of nearly unity. The maximum cooling power density generally increases with the active layer thickness because of an increasing total radiative recombination rate. A GaAs active layer with thickness of 5 μm can give a high internal efficiency of nearly 100% and a limiting cooling power density of 97 W/cm^2. Further increasing the active layer thickness causes a nonuniform distribution of carriers, reducing the radiative power and also the maximum cooling power. Doping in the active layer does not help significantly in improving the cooling power density. Contrarily, a heavily *p*-doped active layer has much lower cooling power density than an undoped one owing to the enhanced Auger recombination.

The cooling capability depends strongly on the external efficiency, and hence the extraction efficiency, due to a small ratio of the extracted thermal energy to the work done on each injected electron–hole pair. For a GaAs EL cooler to act in the cooling mode, an external efficiency of 82% is needed at 1.2 V. This corresponds to an extraction efficiency of 87% if the trapped photons are all absorbed parasitically. This harsh requirement of extraction efficiency can be alleviated by photon recycling process provided that the internal efficiency is sufficiently high. The photon recycling behaves as an additional means of generating carriers, improving the external efficiency by reducing the driving currents. Therefore, for an ideal case of unity recycling efficiency, an EL device can act in the cooling mode even if the extraction efficiency is only 20%. For a more practical case of 90% recycling efficiency, our calculation shows that a maximum cooling power of 5.6 W/cm^2 and a maximum cooling efficiency of 6.4% can be achieved for an EL cooler of 1-μm GaAs active layer. The limiting extraction efficiency for this cooler is found to be 45%, which is achievable for the state-of-the-art technology. This reveals a good possibility of experimentally realizing EL refrigeration for high-cooling-power application.

References

Ashcroft NW, Mermin ND (1976) *Solid State Physics*. Fort Worth, TX: Harcourt Brace College.

Berdahl P (1985) Radiant refrigeration by semiconductor diodes. *J. Appl. Phys.* 58:1369–1374.

Chuang SL (1995) *Physics of Optoelectronic Devices*. New York, NY: Wiley.

Dousmanis GC, Mueller CW, Nelson H, Petzinger KG (1964) Evidence of refrigerating action by means of photon emission in semiconductor diodes. *Phys. Rev.* 133:A316–A318.

Epstein RI, Buchwald MI, Edwards BC, Gosnell TR, Mungan CE (1995) Observation of laser-induced fluorescent cooling of a solid. *Nature* 377:500–503.

Han P, Jin KJ, Ren SR, Zhou YL, Lu HB (2007) Numerical analysis of optothermionic refrigeration in semiconductor triple-well structure. *J. Appl. Phys.* 102:114501.

Heikkilä O, Oksanen J, Tulkki J (2009) Ultimate limit and temperature dependency of light-emitting diode efficiency. *J. Appl. Phys.* 105:093119.

Heikkilä O, Oksanen J, Tulkki J (2010) The challenge of unity wall plug efficiency: The effects of internal heating on the efficiency of light emitting diodes. *J. Appl. Phys.* 107:033105.

Horio K, Yanai H (1990) Numeral modeling of heterojunctions including the thermionic emission mechanism at heterojunction interface. *IEEE Tran. Electron. Devices* 37:1093–1098.

Lee KC, Yen ST (2012) Photon recycling effect on electroluminescent refrigeration. *J. Appl. Phys.* 111:014511.

Lin FR, Lee KC, Yen ST (2012) The ultimate cooling temperature of semiconductor electroluminescent refrigeration. In *the 12th International Conference on Numerical Simulation of Optoelectronic Devices (NUSOD 2012)*, Shanghai, China, ThPD2.

Liu X, Zhao G, Zhang Y, Deppe DG (2013) Semiconductor laser monolithically pumped with a light emitting diode operating in the thermoelectrophotonic regime. *Appl. Phys. Lett.* 102:081116.

Mal'shukov AG, Chao KA (2001) Opto-thermionic refrigeration in semiconductor heterostructures. *Phys. Rev. Lett.* 86:5570–5573.

Melgaard S, Seletskiy D, Polyak V, Asmerom Y, Shiek-Bahae M (2014) Identification of parasitic loss in Yb:YLF and prospects for optical refrigeration down to 80 K. *Opt. Express* 22:7756–7764.

Oksanen J, Tulkki J (2010) Thermophotonic heat pump—A theoretical model and numerical simulations. *J. Appl. Phys.* 107:093106.

Oksanen J, Tulkki J (2015) LEDs feed on waste heat. *Nat. Photonics* 9:782–783.

Piprek J, Li Z-M (2016) Electroluminescent cooling mechanisms in InGaN/GaN light-emitting diodes. *Opt. Quant. Electron.* 48:472.

Santhanam P, Gray DJ Jr., Ram RJ (2012) Thermoelectrically pumped light-emitting diodes operating above unity efficiency. *Phys. Rev. Lett.* 108:097403.

Sheik-Bahae M, Epstein RI (2007) Optical refrigeration. *Nat. Photonics* 4:693–699.

Sheik-Bahae M, Epstein RI (2009) Laser cooling of solids. *Laser Photon. Rev.* 3:67–83.

Sotoodeh M, Khalid AH, Rezazadeh AA (2000) Empirical low-field mobility model for III-V compounds applicable in device simulation codes. *J. Appl. Phys.* 87:2890–2900.

Wang JB, Ding D, Yu SQ, Johnson SR, Zhang YH (2005) Electroluminescence cooling in semiconductors. In 2005 *Conference on Lasers and Electro-Optics/Quantum Electronics and Laser Science and Photonic Applications Systems Technologies, QELS'05,* Baltimore, MD, 655–657.

Xue J, Zhao Y, Oh SH, Herrington WF, Speck JS, Denbaars ST, Nakamura S, Ram RJ (2015) Thermally enhanced blue light-emitting diode. *Appl. Phys. Lett.* 107:121109.

Yang K, East JR, Haddad GI (1993) Numerical modeling of abrupt heterojunctions using a thermionic-field emission boundary condition. *Solid State Electron.* 36:321–330

Yen ST, Lee KC (2010) Analysis of heterostructures for electroluminescent refrigeration and light emitting without heat generation. *J. Appl. Phys.* 107:054513.

Yu SQ, Wang JB, Ding D, Johnson SR, Valsileska D, Zhang YH (2007) Impact of electronic density of states on electroluminescence refrigeration. *Solid State Electron.* 51:1387–1390.

Zhang J, Li D, Chen R, Xiong Q (2013) Laser cooling of a semiconductor by 40 Kelvin. *Nature* 493:504–508.

45

Photonic Crystal Laser Diodes

Maciej Dems

45.1 Photonic Crystals Basics

45.1.1 Optical Periodic Structures

Photonic crystals are periodic structures designed to affect electromagnetic waves in a similar manner to how solid-state crystals affect electrons. The simplest, one-dimensional form of a photonic crystal is a periodic multilayer film (Bragg mirror). Such films were already being studied 120 years ago, in 1887, by Rayleigh [1]. They continued to be the subject of intensive research, resulting in the development of thin-film optics: the science and technology of fabricating dielectric multilayer mirrors, filters, polarizers, antireflection coatings, and so on. However, it took a whole century before the technology was expanded into two and three dimensions by Yablonovitch [2] and John [3] in 1987. Since then, the term *photonic crystal* has become widely used and the subject is studied all over the world.

The physics of photonic crystals has much in common with solid-state physics, as both concern the solution of wave equations in a periodic medium. Hence, concepts known from solid-state physics can also be used in photonics. In particular, for any regular lattice, it is possible to define a *reciprocal lattice* in the Fourier space, i.e., the space of the wavevectors. Such reciprocal lattices are an important tool for the analysis of light propagation in photonic crystals. Because of the Bloch theorem, any eigenmode in periodic medium can be represented as

$$\mathbf{E}(\mathbf{r}) = \mathbf{E_k}(\mathbf{r}) \exp(-i\,\mathbf{k} \cdot \mathbf{r}), \tag{45.1}$$

where $\mathbf{k}$ is a wavevector and $\mathbf{E_k}$ is a periodic function with the same periodicity as the refractive index, i.e.,

$$n_R(\mathbf{r} + \mathbf{a}_i) = n_R(\mathbf{r}), \tag{45.2}$$

$$\mathbf{E_k}(\mathbf{r} + \mathbf{a}_i) = \mathbf{E_k}(\mathbf{r}). \tag{45.3}$$

Vectors $\mathbf{a}_i$ are referred to as the base vectors of the periodic lattice and their number is equal to the number of dimensions (in Equations 45.2 and 45.3).

A reciprocal lattice has vectors $\mathbf{b}_j$ defined in the space of wavevectors. They fulfill the condition

$$\mathbf{a}_i \cdot \mathbf{b}_j = 2\pi\delta_{ij},$$

where δ_{ij} is Kronecker's delta. For three 3D lattices, $\mathbf{b}_j$ can be expressed directly as

$$\mathbf{b}_1 = 2\pi\frac{\mathbf{a}_2 \times \mathbf{a}_3}{a_1 a_2 a_3}, \qquad \mathbf{b}_2 = 2\pi\frac{\mathbf{a}_3 \times \mathbf{a}_1}{a_1 a_2 a_3}, \text{ and} \qquad \mathbf{b}_3 = 2\pi\frac{\mathbf{a}_1 \times \mathbf{a}_2}{a_1 a_2 a_3},$$

and for 2D lattices as

$$\mathbf{b}_1 = 2\pi\frac{\mathbf{a}_2 \times \hat{\mathbf{n}}}{a_1 a_2}, \qquad \mathbf{b}_2 = 2\pi\frac{\hat{\mathbf{n}} \times \mathbf{a}_1}{a_1 a_2},$$

where $\hat{\mathbf{n}}$ is the unit vector perpendicular to the lattice plane. These vectors define the periodicity of the wavevector $\mathbf{k}$ from Equation 45.1. This periodicity enables the modal analysis of PCs with the vectors $\mathbf{k}$ located within a limited area, called the first Brillouin zone [4].

In the case of two-dimensional photonic crystals, two lattice configurations are particularly common. These are the square lattice and the hexagonal lattice, shown in Figure 45.1a and c. Their reciprocal lattices have the same properties as corresponding physical lattices and are depicted in Figure 45.1b and d. Each of these lattices possesses several high symmetry points, which are important for determination of the photonic band structure. The Γ point is always defined for wavevector $\mathbf{k} = \mathbf{0}$. Others are located at the edges of the first Brillouin zone. For the square lattice, these are X for $\mathbf{k} = \frac{\pi}{a}\hat{\mathbf{x}}$ and M for $\mathbf{k} = \frac{\pi}{a}\hat{\mathbf{x}} + \frac{\pi}{a}\hat{\mathbf{y}}$[†]. In the case of the hexagonal lattice, the high symmetry points are M ($\mathbf{k} = \frac{\pi}{a}\hat{\mathbf{x}} + \frac{\pi}{a\sqrt{3}}\hat{\mathbf{y}}$) and K ($\mathbf{k} = \frac{4\pi}{3a}\hat{\mathbf{x}}$). These positions are given for the points marked in Figure 45.1, although there are other points of each kind on the other edges and corners. In practice, this changes little, as the eigenmodes may be subjected to any symmetry transformation that does not change the distribution of the refractive index. Hence, the triangles made by the adjacent symmetry points are the *irreducible zone*, into which all the possible wavevectors can

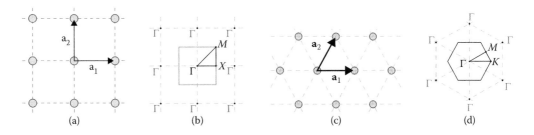

FIGURE 45.1 Arrangements of periodic cells and base vectors $\mathbf{a}_1$ and $\mathbf{a}_2$ of two-dimensional photonic crystals for (a) rectangular and (c) triangular lattices and reciprocal lattices of (b) rectangular and (d) triangular two-dimensional photonic crystals. The important high-symmetry points in the middle and at the edges of the first Brillouin zone (marked with solid lines) are shown. The periodicity of this lattice allows each point outside of the first Brillouin zone to be mapped within it. All the lattice nodes correspond to the Γ point. The triangles formed by the high-symmetry points and shown in the figures are irreducible zones. Each point outside this zone can be mapped to one inside it by periodicity and symmetry transformations.

[†] $\hat{\mathbf{x}}$ and $\hat{\mathbf{y}}$ are unit tensors of the Cartesian coordinate system.

be transformed using periodicity and symmetry transformations. As a consequence, analysis of the Bloch modes of a photonic crystal can be limited to this irreducible zone. In order to determine the bandgap, it is even sufficient to compute the frequencies for the wavevectors at the edges of that zone.

45.1.2 Photonic Bandgap

One of the most important properties of photonic crystals is the *photonic bandgap* (PBG), i.e., a range of frequencies in which light cannot propagate through the crystal [5]. If a light wave of such frequency were to propagate into the photonic crystal, it would be almost 100% reflected. This effect is well known in one-dimensional periodic stacks and is commonly used in distributed Bragg reflectors (DBRs, see Chapter 34). Figure 45.2a shows the reflectivity spectrum of a typical quarter-wavelength GaAs/AlAs DBR. The clearly visible high-reflectivity range between 927 nm and 1039 nm is the bandgap. The same effect in two dimensions can be seen in Figure 45.2b, which shows the optical bands as a horizontal wavevector to normalized frequency relation and the density of optical states within an infinite crystal. The bandgap is the region with zero density of states. In this frequency range, there is no band for any wavevector.

Two- and three-dimensional photonic crystals can be considered a multi-dimensional equivalent of DBRs. Hence, one of their major applications is to confine light. This is what they are most commonly used for in vertical-cavity surface-emitting lasers (VCSELs).

In order to confine light using a photonic crystal, one must create a *defect*: a localized perturbation of the periodic structure. Such perturbations can be either linear, in which case we have a waveguide, or isolated, in which case it makes a cavity. Examples of two-dimensional photonic crystal waveguides are shown in Figure 45.3. Because of the reflective properties of the photonic crystal within the bandgap, the waveguides can be bended with sharp angles and still effectively transmit light. In lasers, such waveguides can be coupled with edge-emitting lasers. In the case of surface-emitting lasers, the light is confined within a two-dimensional photonic crystal cavity. The most common configurations of such cavities are shown in Figure 45.4.

45.1.3 Photonic Crystals in Lasers

Photonic crystals are seldom used in edge-emitting lasers. One design of such a laser is presented in [6]. It is a W1 or W3 linear waveguide filled with an active material. Such lasers are reported to be intrinsically

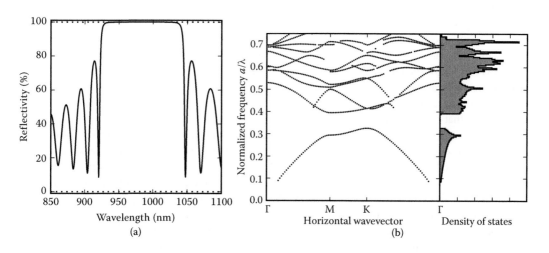

FIGURE 45.2 (a) Reflectivity of a quarter-wavelength one-dimensional stack and (b) band diagram and density of states (right histogram) of a two-dimensional photonic crystal. A photonic bandgap is visible in both cases as the frequency/wavelength region with very high reflectivity or vanishing density of states.

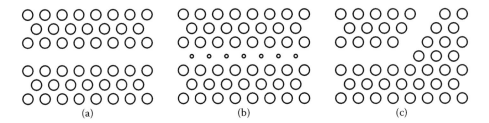

FIGURE 45.3 Examples of photonic crystal waveguides: (a) simple W1 waveguide with one row of missing holes, (b) linear waveguide with smaller holes, and (c) bent waveguide.

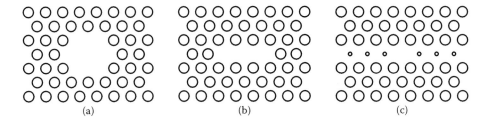

FIGURE 45.4 Examples of photonic crystal cavities: (a) cavity formed by removing seven holes, (b) L3 cavity with shifted edge holes, and (c) birefringent cavity formed by one missing hole and a row of smaller holes.

single-mode, in contrast to classical first-order DFB lasers. Depending on the configuration, the emitting mode has its wavevector either at the Γ or K point of the hexagonal photonic crystal lattice. The emission spectrum has two peaks, corresponding to the edges of the so-called mini-stop-band, which appears when two modes cross (as is the case at both Γ and K points). If emission takes place at frequencies that do not fall within the bandgap of the photonic crystal, only one of these modes is confined to the waveguide. The other spreads infinitely into the photonic crystal, and therefore suffers significantly larger losses. This is the main cause of single-mode operation.

In contrast to edge-emitting lasers, photonic crystals are much more commonly used in surface-emitting lasers. The first experimental designs were published at the end of the nineties. Similarly to photonic-crystal edge-emitting lasers (PC-EELs), such lasers were at first pumped with an optical beam and their design was not very complicated. The simplest surface-emitting photonic crystal laser is a PC membrane with a defect inside. The greatest advantage of such a design is its simplicity and the ease of fabricating a photonic crystal with a small lattice constant. For a typical hexagonal lattice of air holes, etched in a semiconductor material, the pitch must be around $0.3\,\lambda$ if the frequency of the lasing mode c/λ is to be inside the bandgap. Because of the Purcell effect, the small size of the cavity helps to achieve a high spontaneous emission rate and a high Q-factor that can reach up to 10^4 [7].

In most designs of photonic crystal surface-emitting lasers [8–12], the light is not confined by the bandgap, but through the effective index of the defect area, which is higher than that of the surrounding photonic crystal. The role of the photonic crystal is to provide single-mode emission on either a higher order [9] or a fundamental mode [12]. Single-mode emission is one of the most promising applications of photonic crystals in the area of VCSEL design. This is especially true for high-power lasers [13], where classical designs fail because of the necessity of using broad area cavities.

The origin of a single-mode emission in PC-VCSELs can be attributed to one of the two factors. The first is a significant increase of modal losses for the higher-order modes, caused by light scattering. By carefully choosing the etching depth of the holes in the top DBR and their diameter, one can maximize this difference while keeping optical losses for the fundamental mode reasonably low [14,15]. The other factor is a cutoff similar to that found in so-called *endlessly single-mode* fibers [16], i.e., fibers that remain single-mode for any particular wavelength. The origin of this phenomena lies in the dependence of the PC effective index on the wavelength. The number of modes in a classical fiber depends on the parameter V_{eff}, which is defined as

$$V_{\text{eff}} = \frac{2\pi r}{\lambda} \left(n^2_{\text{core}} - n^2_{\text{clad}}\right)^{1/2},$$ (45.4)

where r is the radius of the core, λ is the wavelength, and n_{core} and n_{clad} are the effective indexes of the fiber core and the cladding, respectively. The fiber is single-mode if $V_{\text{eff}} < 2.405$. It is possible to design a photonic crystal fiber in which the V_{eff} does not exceed 2.405 even for $\lambda \to 0$. Because the VCSEL cavity can be approximated as a waveguide (although many phenomena are disregarded in such an approximation), Equation 45.4 can be used in this case. In practice, this means that single-mode operation is achieved if the diameters and depths of the holes in the photonic crystal are sufficiently small.

45.2 Fully Vectorial Method for Modeling VCSELs with Subwavelength Structures

Unlike classical oxide-confined VCSELs, discussed in Chapter 34, VCSELs incorporating subwavelength structures cannot be modeled using either the effective frequency method or any other approximate scalar method. The reason for this is the fact that such methods do not adequately consider light scattering from the subwavelength structures. Hence, in order to correctly model the optical properties of such devices, one needs to apply more exact vector methods, such as the finite-difference time-domain method [17], plane-wave expansion [18], the finite element method (FEM) [19], the coupled mode model (CMM) [20], the plane-wave admittance method (PWRT) [21], or plane-wave reflection transfer [22]. The greatest drawback of all these models is the increased computation time and computer memory usage. This cost is, however, necessary to bear, as only the rigorous solution of Maxwell equations provided by these methods allows light scattering from subwavelength structures to be included in the model.

In this chapter, we illustrate the fundamental optical properties of surface-emitting lasers incorporating subwavelength periodic structures, called photonic crystals. First, we need to sketch the principles of the rigorous vector methods used to obtain the results presented later in the chapter. Instead of describing the details of all popular vector models (some of which are discussed elsewhere in this book), we focus on the two that have proved to be the most useful for our purposes: the plane-wave admittance method and PWRT.

45.2.1 Plane-Wave Admittance Method

The plane-wave admittance method is an extension of the vector approach presented in Chapter 34. It uses the same procedure—admittance transfer—to find the eigenmodes of a resonant cavity. However, in this method, the electromagnetic field is expanded in the plane-wave Fourier basis. This basis is particularly useful for periodic structures, as it has natural periodic boundary conditions. Hence, modeling of infinite two-dimensional photonic crystals is straightforward and requires little effort. However, to analyze localized modes in lasers and correctly consider emitted radiation, one needs to introduce absorbing perfectly matched layers at the edges of the computational domain.

As the exact computation procedure was described in Chapter 34, it need not be repeated here. Instead, this section presents a derivation of the matrix form of Maxwell equations using the plane-wave expansion approach. Such matrices can be entered straightforwardly into computer memory for effective computations.

To start the derivation, consider an anisotropic linear dielectric or semiconductor material without any free charges ($\rho = 0$) or currents ($\mathbf{j} = 0$). In such a material, the time-independent Maxwell equations read

$$\nabla \times \mathbf{E} = -i\omega\mu\mu_0\mathbf{H}, \tag{45.5}$$

$$\nabla \times \mathbf{H} = i\omega\varepsilon\varepsilon_0\mathbf{E}, \tag{45.6}$$

where $i^2 = -1$, ω is the angular frequency of light, and physical observables are the real parts of $\mathbf{E}$ and $\mathbf{H}$.

Now consider an anisotropic material in Cartesian coordinates. Assuming that both permittivity and permeability of the material can be represented as diagonal tensors, the Equations 45.5 and 45.6 can be expanded as

$$-\partial_z E_y + \partial_y E_z = -i\omega\mu_x\mu_0 H_x, \tag{45.7}$$

$$\partial_x E_z - \partial_z E_x = -i\omega\mu_y\mu_0 H_y, \tag{45.8}$$

$$-\partial_y E_x + \partial_x E_y = -i\omega\mu_z\mu_0 H_z, \tag{45.9}$$

$$-\partial_z H_y + \partial_y H_z = i\omega\varepsilon_x\varepsilon_0 E_x, \tag{45.10}$$

$$\partial_x H_z - \partial_z H_x = i\omega\varepsilon_y\varepsilon_0 E_y, \tag{45.11}$$

$$-\partial_y H_x + \partial_x H_y = i\omega\varepsilon_z\varepsilon_0 E_z, \tag{45.12}$$

where ∂_ξ means a partial derivative in the ξ direction. These equations can be transformed into the form already introduced in Chapter 34:

$$\partial_z \begin{bmatrix} -E_y \\ E_x \end{bmatrix} = -i\frac{\eta_0}{k_0} \begin{bmatrix} \partial_y \varepsilon_z^{-1} \partial_y + \mu_x k_0^2 & -\partial_y \varepsilon_z^{-1} \partial_x \\ -\partial_x \varepsilon_z^{-1} \partial_y & \partial_x \varepsilon_z^{-1} \partial_x + \mu_y k_0^2 \end{bmatrix} \begin{bmatrix} H_x \\ H_y \end{bmatrix}, \tag{45.13}$$

$$\partial_z \begin{bmatrix} H_x \\ H_y \end{bmatrix} = -i\frac{1}{\eta_0 k_0} \begin{bmatrix} \partial_x \mu_z^{-1} \partial_x + \varepsilon_y k_0^2 & \partial_x \mu_z^{-1} \partial_y \\ \partial_y \mu_z^{-1} \partial_x & \partial_y \mu_z^{-1} \partial_y + \varepsilon_x k_0^2 \end{bmatrix} \begin{bmatrix} -E_y \\ E_x \end{bmatrix}, \tag{45.14}$$

where $k_0 = \omega/c = \omega\left(\mu_0\varepsilon_0\right)^{1/2}$.

Now, consider an electromagnetic field in a structure exhibiting two-dimensional periodicity, i.e.,

$$\varepsilon(\mathbf{r}) = \varepsilon(\mathbf{r} + \mathbf{a}_i) \qquad (i = 1, 2), \tag{45.15}$$

$$\mu(\mathbf{r}) = \mu(\mathbf{r} + \mathbf{a}_i) \qquad (i = 1, 2), \tag{45.16}$$

with $\mathbf{a}_1$ and $\mathbf{a}_2$ being elementary lattice vectors in the xy plane. Because of their periodicity, the solutions of the Maxwell equations obey the Bloch theorem, i.e., both the electric and magnetic fields can be represented as

$$\mathbf{E}(\mathbf{r}) = \bar{\mathbf{E}}(\mathbf{r}) \exp(-i\,\mathbf{k}_{xy} \cdot \mathbf{r}), \tag{45.17}$$

$$\mathbf{H}(\mathbf{r}) = \bar{\mathbf{H}}(\mathbf{r}) \exp(-i\,\mathbf{k}_{xy} \cdot \mathbf{r}), \tag{45.18}$$

where $\bar{\mathbf{E}}(\mathbf{r})$, $\bar{\mathbf{H}}(\mathbf{r})$ are periodic functions and $\mathbf{k}_{xy}$ is a projection of the mode wavevector into the xy plane. Furthermore, the functions $\bar{\mathbf{E}}(\mathbf{r})$ and $\bar{\mathbf{H}}(\mathbf{r})$ can be expanded into the two-dimensional

Fourier (plane-wave) series as

$$\bar{\mathbf{E}}(\mathbf{r}) = \mathbf{E}^{\mathbf{g}} \exp(i\,\mathbf{g}\cdot\mathbf{r}) = \mathbf{E}^{\mathbf{g}} \left| \varphi_{\mathbf{g}} \right\rangle, \tag{45.19}$$

$$\bar{\mathbf{H}}(\mathbf{r}) = \mathbf{H}^{\mathbf{g}} \exp(i\,\mathbf{g}\cdot\mathbf{r}) = \mathbf{H}^{\mathbf{g}} \left| \varphi_{\mathbf{g}} \right\rangle, \tag{45.20}$$

where the Einstein summation convention[†] is used and $\mathbf{g}$ is the reciprocal lattice vector:

$$\mathbf{g} = l_1\mathbf{b}_1 + l_2\mathbf{b}_2, \tag{45.21}$$

$$\mathbf{a}_i \cdot \mathbf{b}_j = 2\pi\delta_{ij}, \tag{45.22}$$

with l_1 and l_2 being arbitrary integers and δ_{ij} the Kronecker's delta. As we perform Fourier expansion only in the xy plane, the coefficients $\mathbf{E}^{\mathbf{g}}$ and $\mathbf{H}^{\mathbf{g}}$ are functions of z.

Using the Bloch theorem together with Equations 45.19 and 45.20, we can represent both fields as

$$\mathbf{E}(\mathbf{r}) = \mathbf{E}^{\mathbf{g}} \left| \varphi_{\mathbf{g}-\mathbf{k}_{xy}} \right\rangle, \tag{45.23}$$

$$\mathbf{H}(\mathbf{r}) = \mathbf{H}^{\mathbf{g}} \left| \varphi_{\mathbf{g}-\mathbf{k}_{xy}} \right\rangle. \tag{45.24}$$

Now consider Equations 45.13 and 45.14. As the Fourier basis is orthonormal, i.e.,

$$\langle \varphi_{\mathbf{g}} | \varphi_{\mathbf{g}'} \rangle = \delta_{\mathbf{g}\mathbf{g}'}, \tag{45.25}$$

we can introduce Equations 45.23 and 45.24 into Equations 45.13 and 45.14 and left-multiply them by $\langle \varphi_{\mathbf{g}-\mathbf{k}_{xy}} |$ to obtain

$$\partial_z \begin{bmatrix} E_y^{\mathbf{g}} \\ E_x^{\mathbf{g}} \end{bmatrix} = -i\frac{\eta_0}{k_0} \left\langle \varphi_{\mathbf{g}-\mathbf{k}_{xy}} \left| \begin{array}{cc} \partial_y \varepsilon_z^{-1}\partial_y + \mu_x k_0^2 & -\partial_y \varepsilon_z^{-1}\partial_x \\ -\partial_x \varepsilon_z^{-1}\partial_y & \partial_x \varepsilon_z^{-1}\partial_x + \mu_y k_0^2 \end{array} \right| \varphi_{\mathbf{g}'-\mathbf{k}_{xy}} \right\rangle \begin{bmatrix} H_x^{\mathbf{g}'} \\ H_y^{\mathbf{g}'} \end{bmatrix}, \tag{45.26}$$

$$\partial_z \begin{bmatrix} H_x^{\mathbf{g}'} \\ H_y^{\mathbf{g}'} \end{bmatrix} = -\frac{i}{\eta_0 k_0} \left\langle \varphi_{\mathbf{g}-\mathbf{k}_{xy}} \left| \begin{array}{cc} \partial_x \mu_z^{-1}\partial_x + \varepsilon_y k_0^2 & \partial_x \mu_z^{-1}\partial_y \\ \partial_y \mu_z^{-1}\partial_x & \partial_y \mu_z^{-1}\partial_y + \varepsilon_x k_0^2 \end{array} \right| \varphi_{\mathbf{g}'-\mathbf{k}_{xy}} \right\rangle \begin{bmatrix} E_y^{\mathbf{g}} \\ E_x^{\mathbf{g}} \end{bmatrix}, \tag{45.27}$$

where, in order to simplify notation, we represent $E_y(\mathbf{r})$ as

$$E_y(\mathbf{r}) = -E_y^{\mathbf{g}} \left| \varphi_{\mathbf{g}-\mathbf{k}_{xy}} \right\rangle \tag{45.28}$$

instead of directly using Equation 45.23.

The matrices in the above equations can be computed easily when both permittivity and permeability are expanded in the Fourier basis. For this purpose, the analyzed layer is assumed to be invariant in the

---

[†] By this convention, if an index is repeated twice in any product, we perform summation over this index. We use this convention in this whole chapter, unless explicitly stated otherwise.

z-direction. Then, we can write

$$\varepsilon_i(\mathbf{r}) = \varepsilon_i^{\mathbf{g}} \left| \varphi_{\mathbf{g}} \right\rangle \quad (i = x, y), \tag{45.29}$$

$$\mu_i(\mathbf{r}) = \mu_i^{\mathbf{g}} \left| \varphi_{\mathbf{g}} \right\rangle \quad (i = x, y), \tag{45.30}$$

$$\varepsilon_z^{-1}(\mathbf{r}) = \kappa^{\mathbf{g}} \left| \varphi_{\mathbf{g}} \right\rangle, \tag{45.31}$$

$$\mu_z^{-1}(\mathbf{r}) = \gamma^{\mathbf{g}} \left| \varphi_{\mathbf{g}} \right\rangle. \tag{45.32}$$

Substituting this into Equations 45.26 and 45.27, we finally obtain the matrix form of the equations in the plane-wave basis:

$$\partial_z \begin{bmatrix} E_y^{\mathbf{g}} \\ E_x^{\mathbf{g}} \end{bmatrix} = -i \frac{\eta_0}{k_0} \begin{bmatrix} -(g_y - k_y)(g_y' - k_y)\kappa^{\mathbf{g}-\mathbf{g}'} + k_0^2 \mu_x^{\mathbf{g}-\mathbf{g}'} & (g_y - k_y)(g_x' - k_x)\kappa^{\mathbf{g}-\mathbf{g}'} \\ (g_x - k_x)(g_y' - k_y)\kappa^{\mathbf{g}-\mathbf{g}'} & -(g_x - k_x)(g_x' - k_x)\kappa^{\mathbf{g}-\mathbf{g}'} + k_0^2 \mu_y^{\mathbf{g}-\mathbf{g}'} \end{bmatrix} \begin{bmatrix} H_x^{\mathbf{g}'} \\ H_y^{\mathbf{g}'} \end{bmatrix}, \tag{45.33}$$

$$\partial_z \begin{bmatrix} H_x^{\mathbf{g}} \\ H_y^{\mathbf{g}} \end{bmatrix} = -\frac{i}{\eta_0 k_0} \begin{bmatrix} -(g_x - k_x)(g_x' - k_x)\gamma^{\mathbf{g}-\mathbf{g}'} + k_0^2 \varepsilon_y^{\mathbf{g}-\mathbf{g}'} & -(g_x' - k_x)(g_y' - k_y)\gamma^{\mathbf{g}-\mathbf{g}'} \\ -(g_y - k_y)(g_x' - k_x)\gamma^{\mathbf{g}-\mathbf{g}'} & -(g_y - k_y)(g_y' - k_y)\gamma^{\mathbf{g}-\mathbf{g}'} + k_0^2 \varepsilon_x^{\mathbf{g}-\mathbf{g}'} \end{bmatrix} \begin{bmatrix} E_y^{\mathbf{g}'} \\ E_x^{\mathbf{g}'} \end{bmatrix}, \tag{45.34}$$

where k_x, k_y, g_x, and g_y are the corresponding components of the mode wavevector and the reciprocal lattice vector $\mathbf{g}$, respectively. These equations can be introduced into computer memory for further computations by truncating the Fourier basis $\{\mathbf{g}\}$ at some point. The introduced error will depend on two factors—the distribution of the electromagnetic field and the distribution of the material parameters. The former, as a result of simulation, cannot be influenced directly, although the latter can be convoluted with a Gaussian window to improve convergence.

The vertical components of electric and magnetic fields can also be expanded in the Fourier basis as

$$E_z^{\mathbf{g}} = \frac{\eta_0}{k_0} \kappa^{\mathbf{g}-\mathbf{g}'} \left[-(g_y' - k_y) H_x^{\mathbf{g}'} + (g_x' - k_x) H_y^{\mathbf{g}'} \right], \tag{45.35}$$

$$H_z^{\mathbf{g}} = \frac{1}{k_0 \eta_0} \gamma^{\mathbf{g}-\mathbf{g}'} \left[(g_x' - k_x) E_y^{\mathbf{g}'} + (g_y' - k_y) E_x^{\mathbf{g}'} \right]. \tag{45.36}$$

Once Equations 45.33 and 45.34 have been derived and represented in their numerical forms, we can use them to determine the eigenmodes of the analyzed structure. For this purpose, we first show their solution for a single z-invariant layer and then present a numerically stable technique for combining fields in all layers and deriving an eigenmode condition for the whole structure. The procedure described below was originally used with the method of lines [23] and is also applied to the plane-wave admittance method.

Further in this section, we will use the short representation of the vectors and matrices introduced in Equations 45.33 and 45.34. In particular, we will name

$$\hat{\mathbf{E}} = \begin{bmatrix} E_y^{\mathbf{g}} \\ E_x^{\mathbf{g}} \end{bmatrix},$$

$$\hat{\mathbf{H}} = \begin{bmatrix} H_x^{\mathbf{g}} \\ H_y^{\mathbf{g}} \end{bmatrix},$$

$$\mathbf{R_E} = \frac{1}{\eta_0 k_0} \begin{bmatrix} -(g_x - k_x)(g'_x - k_x)\gamma^{\mathbf{g}-\mathbf{g}'} + k_0^2 \varepsilon_y^{\mathbf{g}-\mathbf{g}'} & -(g'_x - k_x)(g'_y - k_y)\gamma^{\mathbf{g}-\mathbf{g}'} \\ -(g_y - k_y)(g'_x - k_x)\gamma^{\mathbf{g}-\mathbf{g}'} & -(g_y - k_y)(g'_y - k_y)\gamma^{\mathbf{g}-\mathbf{g}'} + k_0^2 \varepsilon_x^{\mathbf{g}-\mathbf{g}'} \end{bmatrix},$$

$$\mathbf{R_H} = \frac{\eta_0}{k_0} \begin{bmatrix} -(g_y - k_y)(g'_y - k_y)\kappa^{\mathbf{g}-\mathbf{g}'} + k_0^2 \mu_x^{\mathbf{g}-\mathbf{g}'} & (g_y - k_y)(g'_x - k_x)\kappa^{\mathbf{g}-\mathbf{g}'} \\ (g_x - k_x)(g'_y - k_y)\kappa^{\mathbf{g}-\mathbf{g}'} & -(g_x - k_x)(g'_x - k_x)\kappa^{\mathbf{g}-\mathbf{g}'} + k_0^2 \mu_y^{\mathbf{g}-\mathbf{g}'} \end{bmatrix}.$$

It is worth noting that the whole procedure described below can be applied with any correct representation of the vectors and matrices above. Thus, it is applicable not only to plane-wave expansion, but also to any orthogonal or non-orthogonal basis. In particular, the representation of $\hat{\mathbf{E}}$, $\hat{\mathbf{H}}$, $\mathbf{R_E}$, and $\mathbf{R_H}$ by finite-difference approximation is used in the method of lines.

Take Equations 45.33 and 45.34 and represent them in their short form:

$$\partial_z \hat{\mathbf{E}} = -i\mathbf{R_H}\hat{\mathbf{H}}, \tag{45.37}$$

$$\partial_z \hat{\mathbf{H}} = -i\mathbf{R_E}\hat{\mathbf{E}}. \tag{45.38}$$

By taking the z-derivative of one of these equations and substituting the other into it, one can write two second-order equations with the fields decoupled:

$$\partial_z^2 \hat{\mathbf{E}} = -\mathbf{Q_E}\hat{\mathbf{E}}, \tag{45.39}$$

$$\partial_z^2 \hat{\mathbf{H}} = -\mathbf{Q_H}\hat{\mathbf{H}}, \tag{45.40}$$

where $\mathbf{Q_E} = \mathbf{R_H}\mathbf{R_E}$ and $\mathbf{Q_H} = \mathbf{R_E}\mathbf{R_H}$.

The eigenmodes can be found using the procedure described in Chapter 34 by constructing and solving equation $\mathbf{Y}\bar{\mathbf{E}} = 0$.

45.2.2 Plane-Wave Reflection Transfer

PWRT differs from the plane-wave admittance method in the way the diagonalized Equations 45.33 and 45.34 are solved. In this method, the electric field in each layer uniform in z-direction is chosen to have the form

$$\tilde{\mathbf{E}} = \exp(-i\Gamma z)\tilde{\mathbf{F}} + \exp(i\Gamma z)\tilde{\mathbf{B}}, \tag{45.41}$$

where $\tilde{\mathbf{F}}$ denotes the amplitude of the forward propagating wave and $\tilde{\mathbf{B}}$ the amplitude of the backward propagating wave. The reflection transfer means an iterative determination of the reflection matrix $\mathbf{R}$, which is defined in each layer as a relation between $\tilde{\mathbf{F}}$ and $\tilde{\mathbf{B}}$ such that

$$\tilde{\mathbf{F}} = \mathbf{R}\,\tilde{\mathbf{B}}. \tag{45.42}$$

We assume that in the first layer, which has infinite thickness, there is no in-going field (we will introduce any optional incident field in the last layer). Thus, $\tilde{\mathbf{F}} \equiv 0$ and so $\mathbf{R} = \mathbf{0}$. The reflection matrices in the subsequent layers can be determined in an iterative process using the requirement that the electric and magnetic fields are continuous at layer interfaces.

By careful selection of units in the magnetic field, it is possible to show that in the diagonalized domain it takes the form of

$$\tilde{\mathbf{H}} = \exp(-i\Gamma z)\tilde{\mathbf{F}} - \exp(i\Gamma z)\tilde{\mathbf{B}}, \tag{45.43}$$

which can be derived from Equation 45.41. The requirement of field continuity between the n-th and $(n + 1)$th layers can be expressed as

$$\mathbf{T_E}^n \left\{ \boldsymbol{\varphi}^n \tilde{\mathbf{F}}^n + \left(\boldsymbol{\varphi}^n\right)^{-1} \tilde{\mathbf{B}}^n \right\} = \mathbf{T_E}^{n+1} \left\{ \tilde{\mathbf{F}}^{n+1} + \tilde{\mathbf{B}}^{n+1} \right\}, \tag{45.44}$$

$$\mathbf{T_H}^n \left\{ \boldsymbol{\varphi}^n \tilde{\mathbf{F}}^n - \left(\boldsymbol{\varphi}^n\right)^{-1} \tilde{\mathbf{B}}^n \right\} = \mathbf{T_H}^{n+1} \left\{ \tilde{\mathbf{F}}^{n+1} - \tilde{\mathbf{B}}^{n+1} \right\}, \tag{45.45}$$

where $\boldsymbol{\varphi}^n = \exp\left(-i\boldsymbol{\Gamma} d_n\right)$ and d_n is the thickness of nth layer. Using the relation (Equation 45.42), we obtain

$$\tilde{\mathbf{F}}^{n+1} + \tilde{\mathbf{B}}^{n+1} = \mathbb{E}^n \left(\boldsymbol{\varphi}^n\right)^{-1} \tilde{\mathbf{B}}^n, \tag{45.46}$$

$$\tilde{\mathbf{F}}^{n+1} - \tilde{\mathbf{B}}^{n+1} = \mathbb{H}^n \left(\boldsymbol{\varphi}^n\right)^{-1} \tilde{\mathbf{B}}^n, \tag{45.47}$$

where

$$\mathbb{E}^n = \left(\mathbf{T_E}^{n+1}\right)^{-1} \mathbf{T_E}^n \left\{ \boldsymbol{\varphi}^n \mathbf{R}^n \boldsymbol{\varphi}^n + \mathbf{I} \right\}, \tag{45.48}$$

$$\mathbb{H}^n = \left(\mathbf{T_H}^{n+1}\right)^{-1} \mathbf{T_H}^n \left\{ \boldsymbol{\varphi}^n \mathbf{R}^n \boldsymbol{\varphi}^n - \mathbf{I} \right\}. \tag{45.49}$$

This, after simple algebra, gives the reflection matrix for the $(n + 1)$th layer in the form

$$\mathbf{R}^{n+1} = \left[\mathbb{E}^n + \mathbb{H}^n\right] \left[\mathbb{E}^n - \mathbb{H}^n\right]^{-1}. \tag{45.50}$$

In contrast to plane-wave admittance transfer, we need not assume the electric field at the boundaries is equal to zero and, hence, there is no need to use absorbing perfectly matched layers as vertical boundary conditions. Instead, at the edge of the structure, we can assume the $\mathbf{R}^0$ matrix to be equal to 0. This means that in the outermost layer, there is no forward propagating field. This field corresponds to the wave coming into the structure. As the analyzed device is a laser resonant cavity, it only emits light; hence, $\tilde{\mathbf{F}}^0 = \mathbf{0}$ and $\tilde{\mathbf{B}}^0 \neq \mathbf{0}$. Then, similarly to PWAM, we can perform iterative progressions of the reflection matrix from both top and bottom boundaries, up to a selected layer. In this, the following relations hold:

$$\tilde{\mathbf{F}}^{\text{top}} = \tilde{\mathbf{B}}^{\text{bottom}}, \tag{45.51}$$

$$\tilde{\mathbf{B}}^{\text{top}} = \tilde{\mathbf{F}}^{\text{bottom}}. \tag{45.52}$$

Using this condition, it is possible to construct a matching matrix $\mathbf{M}$ such that

$$\mathbf{M}\tilde{\mathbf{F}}^{\text{bottom}} = \left[\mathbf{R}^{\text{top}} - \left(\mathbf{R}^{\text{bottom}}\right)^{-1}\right] \tilde{\mathbf{F}}^{\text{bottom}} = \mathbf{0}.$$

In other words, this matrix must be singular, which is the condition for finding the eigenmodes.

45.3 VCSELs with Photonic Crystals for Light Confinement

In VCSELs, photonic crystals are used to provide effective light confinement. By etching a regular lattice of holes in either the DBRs or the VCSEL cavity (or both), together with intentional lattice defects, it is possible to mold the optical mode distribution. In particular, one can laterally confine the fundamental laser mode within a photonic crystal defect without adding any more lowindex oxidation layers. There are several advantages of this approach: First, it allows the fabrication of VCSELs with well-defined cavities in material systems where selective oxidation is either impossible or difficult to achieve. Second, nonuniform patterns in the periodic lattice can be tuned to interact differently with fundamental and high-order modes, providing increased mode selectivity and allowing single-mode VCSEL operation at higher currents (see Section 45.1.3).

To illustrate the exclusive optical properties of PC-VCSELs, consider a basic optical-only VCSEL structure made of GaAs, as shown in Figure 45.5a. It contains 24 GaAs/AlAs pairs of top DBRs, 29 pairs of GaAs/AlAs bottom DBRs, and a very short single-wavelength GaAs cavity. The exact layer structure is listed in Table 45.1. The photonic crystal is based on a triangular lattice (of pitch L) and takes the form of three hexagons surrounding a small cavity in the form of a missing PC hole. Holes with diameter d are etched either in the top DBRs (with varying etching depths), throughout the whole structure, or only inside the cavity.

We now investigate the impact of photonic crystal parameters on the properties of the fundamental mode in the VCSEL in question. Due to its Gaussian-like shape, this mode is the most preferred in most applications. Furthermore, VCSELs with incorporated photonic crystals are known for their high mode selectivity [24], which makes it possible to tune them into a single-mode source with high power emission.

Our results are depicted in Figures 45.6 through 45.9. They express typical tendencies observed in PC-VCSELs. Figure 45.6 shows that increasing the size of the holes results in a decrease in the resonant wavelength, which can be explained by two factors. The first is the decrease in the area of the photonic crystal defect, which forms the cavity (corresponding to a decrease in the aperture of a classical VCSEL).

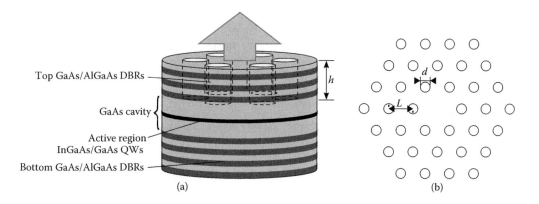

(a)   (b)

FIGURE 45.5 Analyzed VCSEL structure: (a) schematic three-dimensional view, (b) PC structure. (DBRs, distributed Bragg reflectors; QWs, quantum wells.)

TABLE 45.1 Details of the Layer Structure of the Analyzed VCSEL.

	Thickness [nm]	Material	Refractive index	
		Air	1.00	
Top DBR	69.40	GaAs	3.53	
24 pairs	79.55	AlGaAs	3.08	
	121.71	GaAs	3.53	
Cavity	$3 \times 8.00 + 2 \times 5.00$	QW/GaAs	$\begin{cases} 3.56 + jn_g & \text{for } r < a \\ 3.56 - 0.01j & \text{for } r \geq a \end{cases}$	3.53
	121.71	GaAs	3.53	
Bottom DBR	79.55	AlGaAs	3.08	
29 pairs	69.40	GaAs	3.53	
Substrate		GaAs	3.53	

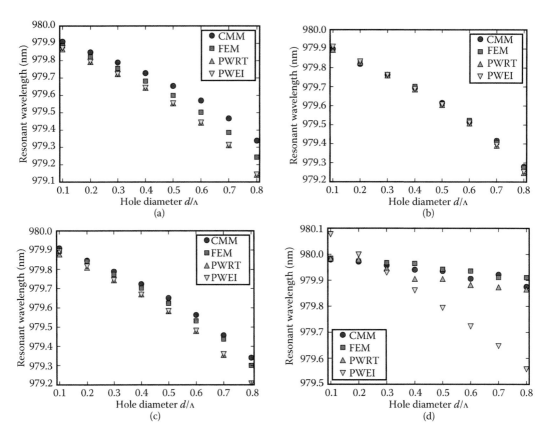

FIGURE 45.6 Resonant wavelengths of PC-VCSELs with photonic crystals etched in different parts of the structure ((a) whole structure, (b) cavity only, (c) 24 pairs of the top DBR, (d) 10 pairs of the top DBR) as a function of hole diameters d. The photonic crystal pitch Λ is 4 μm in all the plots.

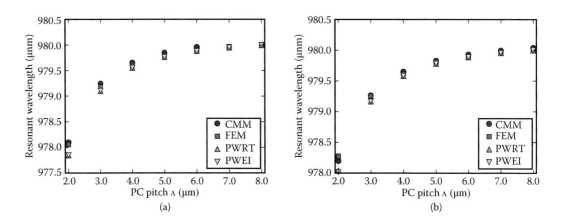

FIGURE 45.7 Resonant wavelengths of PC-VCSELs with photonic crystals etched in different parts of the structure ((a) whole structure, (b) cavity only, (c) 24 pairs of the top DBR, (d) 10 pairs of the top DBR) as a function of the photonic crystal pitch Λ. The diameter of the etched holes d is proportional to the pitch and equals 0.5 Λ.

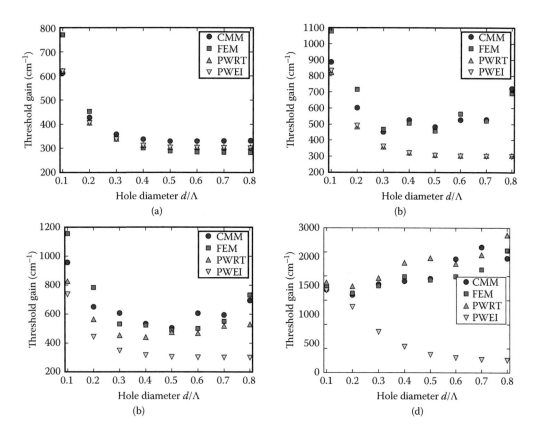

FIGURE 45.8 Threshold gains of PC-VCSELs with photonic crystals etched in different parts of the structure ((a) whole structure, (b) cavity only, (c) 24 pairs of the top DBR, (d) 10 pairs of the top DBR) as a function of the hole diameters d. The photonic crystal pitch Λ is 4 μm in all the plots.

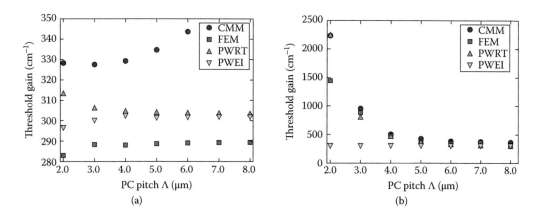

FIGURE 45.9 Threshold gains of PC-VCSELs with photonic crystals etched in different parts of the structure ((a) whole structure, (b) cavity only, (c) 24 pairs of the top DBR, (d) 10 pairs of the top DBR) as a function of the photonic crystal pitch Λ. The diameter of the etched holes d is proportional to the pitch and equals 0.5 Λ.

The other factor is the increase of the effective refractive index contrast between the cavity and the photonic crystal surrounding it. Larger holes mean a larger average air-to-semiconductor ratio, which results in a lower effective index outside the cavity.

The opposite tendency is shown in Figure 45.7. Increasing the photonic crystal pitch results in a rapid increase in the resonant wavelength, as the cavity area increases in proportion to Λ. Due to the constant d/Λ ratio, the effective index contrast remains unchanged. The change is fastest for smaller pitches, where the light confinement is high. With increasing Λ, it increases asymptotically to 980 nm, which is the resonant wavelength of the presented VCSEL with no lateral confinement at all.

Much more interesting is the behavior of the threshold gain. In each case, it is large for small holes (Figure 45.8) because of the weak confinement and large overlap of the mode with the absorbing areas of the active region outside the gain aperture. As the size of the holes increases, the mode is increasingly better confined and therefore the threshold decreases. However, with shallow etching (Figure 45.8b through d), scattering starts to play an important role, and with larger holes the threshold starts increasing again. On the other hand, the photonic crystal pitch does not influence the threshold gain significantly (Figure 45.9), except when it drops too strongly (below 4 μm) in structures with holes etched only in the DBR (Figure 45.9b), in which case light scattering starts to play an important role.

In each case, the same optical properties were computed using four different numerical methods: PWRT presented above, the CMM [20], the FEM [25], and the plane-wave effective index (PWEI) [26]. These methods have already been rigorously compared [27]. Here, we show them to emphasize that the three-dimensional structure of a PC-VCSEL is complex to analyze numerically, even nowadays. Practically all of the methods have shown behavior qualitatively different to the others in at least one case. This shows how important it is to calibrate numerical models and to choose the right one, which will yield correct results for a particular case.

In most circumstances, however, numerical analysis allows some conclusions to be made regarding the optical properties of PC-VCSELs. The results depicted in Figures 45.6 and 45.8 show computed resonant wavelengths and threshold gains as functions of the hole diameters when the photonic crystal pitch Λ equals 4 μm. A pitch of 4 μm is a typical value and has been kept fixed for all etchings. Figures 45.7 and 45.9 show the results obtained from different sizes of the photonic crystal lattice constant, with holes etched through the whole cavity and all layers of the top DBR.

In most cases, the PC-VCSEL resonant wavelength is predicted consistently by all numerical methods. The only exception is with large, shallow etched holes, as can be seen in Figure 45.6d. The differences between the vector methods (CMM, FEM, and PWRT) and the scalar method (PWEI), which gives a too small wavelength, are clearly noticeable. The origin of these discrepancies is the significant light scattering at the bottoms of the holes, which is correctly considered by the vector methods, but neglected by the scalar method.

Unsurprisingly, greater differences are observable in the threshold gain data. The methods show good agreement only in case of the $\Lambda = 4$ μm wholly etched structure. All other situations require more detailed analysis. Figure 45.8c shows the threshold gains for holes etched only in the top DBR. Due to their complicated arrangement, significant scattering occurs at their edges, which is reflected in an increase in the threshold gain for $d/\Lambda > 0.4$. This effect is not predicted by PWEI, which—as a variation of the effective index method—is incapable of properly considering scattering losses. PWEI even more significantly underestimates the threshold gain for shallow holes (Figure 45.6d), where scattering is much stronger due to the weaker mode confinement.

45.4 Subwavelength Gratings in VCSELs

In the first part of this chapter, we described the two-dimensional photonic crystals used to provide light confinement of the optical mode. However, periodic subwavelength structures can offer much more. In what follows, we will look at specific one-dimensional photonic crystals: subwavelength gratings with high refractive index contrast (referred to as high-contrast gratings or HCGs). Such structures may effectively replace DBR mirrors in VCSELs, as they can provide very high reflectivity, while being significantly thinner than a multilayer DBR stack.

Because the light emitted by most VCSELs is linearly polarized, one-dimensional linear gratings can be efficiently applied. Furthermore, they ensure stable polarization of the emitted light, which otherwise might suddenly change at higher output powers [28]. For the same reason, circular gratings are much less suitable as mirror replacements for VCSELs, as they cannot provide uniform reflectivity for linearly polarized light.

In order to investigate the properties of VCSELs with subwavelength gratings, we first need to understand the behavior of grating mirrors. For this reason, in the next subsection, we will outline the basic physical principles behind the high reflectivity of HCGs. Next, we will study their reflectivity in various configurations. After that, we will move on to the determination of resonant modes in VCSEL cavities with subwavelength grating.

45.4.1 High-Contrast Subwavelength Grating Theory

Despite their very slight thicknesses—usually only a few hundreds of nanometers—HCGs can be effective mirrors with reflectivity comparable or exceeding that of thick multilayered DBRs. This behavior can be explained in two ways: either as an effect of the Fano resonance between the incident wave and the excited mode propagating inside the grating [29] or as an effect of destructive interference between transmitted and leaky modes [30]. Below, we briefly summarize the latter explanation.

Consider a grating suspended in air. It can be seen as a periodic array of waveguides, propagating light in a direction perpendicular to the grating plane (Figure 45.10). Assume an incident plane wave propagating in the z-direction and polarized parallel to the grating bars (with the nonzero E_y component). At the incidence side of the grating (region 1 in Figure 45.10), the electric field can be described as the single zero-order incident plane wave and multiple orders of the reflected wave:

$$E_1(x, z) = E_I \exp\left(-ik_0 z\right) + \sum_n E_R^n \cos\left(2n\pi x/L\right) \exp\left(+i\sqrt{k_0^2 - (2n\pi/L)^2}\, z\right), \quad (45.53)$$

where E_I is the amplitude of the incident field, E_R^n is the horizontal distribution of the reflected modes of different orders, $k_0 = \omega/c$ is the normalized frequency, and L is the grating period. Inside the grating, the field is the sum of forward and backward propagating waveguide array modes:

$$E_2(x, z) = \sum_m \left[E_F^m(x) \exp\left(-i\beta_m z\right) + E_B^m(x) \exp\left(+i\beta_m z\right)\right], \quad (45.54)$$

where β_m is the propagation constant of the mth mode and E_F^m and E_B^m are amplitudes of the forward and backward propagating modes, respectively. Finally, in the last section, the field consists only of the

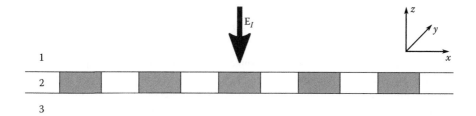

FIGURE 45.10 Grating as a periodic waveguide array.

transmitted modes with amplitudes E_T^n:

$$E_3(x, z) = \sum_n E_T^n \cos{(2n\pi x/L)} \exp{\left(-i\sqrt{k_0^2 - (2n\pi/L)^2} z \right)}.$$

For subwavelength gratings $L < \lambda = 2\pi/k_0$, only the zero orders of both the reflected and transmitted modes propagate and the other modes are evanescent. Although they do not transport any energy, they are very important for ensuring proper boundary conditions for Maxwell equations. By applying these boundary conditions to both grating boundaries, it is possible to determine the amplitudes of both the reflected and transmitted modes (E_R^n and E_T^n) and the modes inside the grating (E_F^m and E_B^m). As only one transmitted mode (E_T^0) propagates energy, it is possible to tune the grating parameters in such a way that $E_T^0 = 0$. This corresponds to the zero energy transmission equivalent, as there is no gain inside the grating, up to 100% reflectivity.

For gratings operating in a near-subwavelength regime, the periodic waveguide array supports only two even modes with propagation constants β_0 and β_2 [31]. In such a case, E_T^0 can be derived analytically as

$$E_T^0 = \frac{1}{L} \int_0^L \left\{ E_F^0(x) + E_B^0(x) \right\} dx + \frac{1}{L} \int_0^L \left\{ E_F^2(x) + E_B^2(x) \right\} dx.$$

In other words, E_T^0 is a sum of the spatial averages of both waveguide array modes. If these averages are antimatched, i.e.,

$$\int_0^L \left\{ E_F^0(x) + E_B^0(x) \right\} dx = -\int_0^L \left\{ E_F^2(x) + E_B^2(x) \right\} dx,$$

there is no energy transmission through the grating. This corresponds to the destructive interference of these modes at the far edge of the grating.

For a fuller derivation and discussion of the modal properties of subwavelength gratings, we refer the reader to [31], where the interaction between gratings and light waves is described in detail.

45.4.2 Modeling of Subwavelength Grating Reflectivity

The reflectivity of subwavelength HCGs can be studied using the PWRT method presented at the beginning of this chapter. This method is mathematically equivalent to the popular rigorous coupled-wave analysis (RCWA) [32]. Both methods rely on expansion of the electromagnetic field into a complex Fourier series and both require the analyzed structure to be split into a set of layers, each of which is uniform in the surface normal direction (see Figure 45.11). PWRT, which was used to obtain all the results presented in this chapter, is applied in the same way as for cavity analysis, as described in Section 45.2.2 with the last step omitted. Instead, after determining the reflection matrix $\bar{\mathbf{R}}$, this is multiplied by a predefined incidence vector $\mathbf{F}$ to calculate the vector of reflected diffraction orders $\mathbf{B}$:

$$\mathbf{B} = \bar{\mathbf{R}} \, \mathbf{F}. \tag{45.55}$$

In most cases, the incident vector has only one nonzero component, which corresponds to the incident plane wave of the zero-order. It is possible to construct more complex incidence vectors, e.g., one forming a Gaussian beam. However, for subwavelength gratings the reflection of the zero-order plane wave is dominant and the results for a Gaussian beam do not differ significantly from those for a plane wave.

Figure 45.11 shows a typical single-period cell of a subwavelength grating. It is comprised of parallel bars made of a high-refractive-index material (usually Si or GaAs), surrounded on both sides by low-index

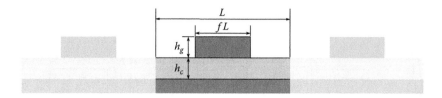

FIGURE 45.11 Schematic diagram of the computational domain for grating reflection computation. The structure is separated into a number of layers uniform in the perpendicular direction and only one grating period is analyzed.

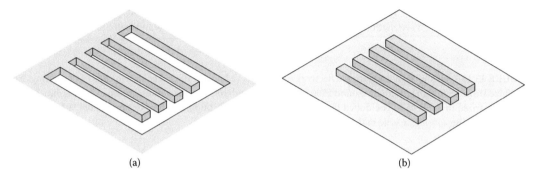

(a)   (b)

FIGURE 45.12 (a) High-contrast grating membrane suspended in air and (b) similar grating located on low-index cladding (e.g., SiO$_2$). The former structure is more difficult to fabricate; however, it allows the membrane to be shifted using piezoelectric forces for VCSEL cavity tuning.

layers. These layers can be either the air on both sides (in which case we speak of grating membranes; see Figure 45.12a) or one of the layers may be some kind of low-index solid (usually an oxide, such as amorphous SiO$_2$). The former are significantly more difficult to fabricate, as the process involves etching a sacrificial layer first and then dissolving it. The resulting grating is suspended in the air. A common problem is then buckling of the stripes, so some additional stress-relaxing etching may be necessary. The advantage of this solution is the fact that the membrane can be placed on a piezoelectric element and then moved freely, allowing for cavity tuning. On the other hand, the latter (Figure 45.12b) is usually much easier to fabricate and the bars are permanently attached to the cladding. The lack of tunability is compensated by the durable mount and no problems with buckling. If the refractive index of the cladding is low enough, there are no significant differences in the optical properties of either type of grating.

Although the gratings are very thin compared to multilayered DBRs, they can provide very high reflectivity, which in an ideal situation can even reach 100%. Due to their strongly anisotropic geometry, the gratings' reflective properties are qualitatively different, depending on the incident light polarization. For an ideal uniform grating, we can distinguish two distinct polarizations: one with an electrical field vector parallel to the grating bars (transverse electric, TE) and the other perpendicular to it (transverse magnetic, TM). Depending on the grating design, any of these polarizations can be strongly reflected, while the other is reflected only moderately. This effect can be used to design VCSELs with strong anisotropy and fixed polarization of the emitted light.

Sample reflectivities computed for a Si membrane suspended in air are shown in Figure 45.13a and b. These figures show reflectivity for both light polarizations with two different gratings. Their exact dimensions are summarized in Table 45.2. Both version were designed to provide high reflectivity at a wavelength of 1500 nm. As can be seen in Figure 45.13, both TE and TM polarization can be strongly reflected, with almost zero transmission. If two such peaks coincide, we observe a wide reflectivity bandwidth. This is clearly visible in Figure 45.13a, where the wide reflectivity area has a small dip in the middle, indicating that it consists of two independent peaks.

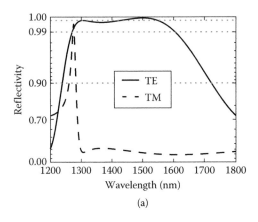
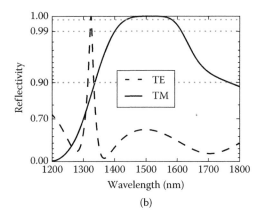

(a)   (b)

FIGURE 45.13 Silicon-on-silica grating reflectivity for TE and TM polarization. The dimensions of the gratings are optimized to provide high reflectivity around 1500 nm for (a) TE and (b) TM polarization. The exact grating dimensions are summarized in Table 45.2. The dotted lines show reflectivity levels important for laser operation: 90%, 99%, and 99.8%. Notice the non-linear scale in the plots.

TABLE 45.2 Dimensions for Gratings with Reflectivities Shown in Figure 45.13a and b.

	(a)	(b)
Grating period L [μm]	1.00	0.64
Grating fill-factor f	0.30	0.64
Grating thickness h_g [μm]	0.22	0.42
Air-gap/cladding thickness h_c [μm]	0.83	0.83

Note: The meaning of the parameters is explained in Figure 45.11. The refractive index of grating bars and substrate is 3.48 (silicon) and of the cladding is 1.47 (silica).

Due to the finite thickness of the grating, it can rather be considered as an array of cavities for the two waveguide array modes. The negative interference described in Section 45.4.1 corresponds to the resonance thickness of these cavities. If we plot grating reflectivity as a function of wavelength and grating thickness (Figure 45.14a and b), we can notice two overlapping periodic patterns, corresponding to the resonance conditions of both waveguide array modes (with propagation constants β_0 and β_2), giving a characteristic chessboard pattern.

Similar calculations can be performed for an HCG located on cladding (Figure 45.12b). With carefully chosen dimensions, it can reflect both TE and TM polarizations equally well. A sample reflectivity spectrum and reflectivity map for TM light polarization is shown in Figure 45.15. The grating parameters are as follows: $L = 0.70\,\mu m$, $h_g = 0.46\,\mu m$, $f = 0.75$, and $h_c = 0.83\,\mu m$. The cladding material is silica with a refractive index of 1.47. Such gratings are usually easier to manufacture and show the same patterns as air membranes.

45.4.3 Cavities with Subwavelength Grating Mirrors

The grating mirrors presented in Section 45.4.2 can be used to replace multi-layered Bragg mirrors as reflectors in VCSELs. Typical structures for such lasers—both with an HCG membrane suspended in air and an HCG grating located on a low-index cladding—are presented in Figure 45.16. For manufacturing feasibility, the bottom mirrors are classical DBRs and the top ones are replaced with HCGs. In the case of the

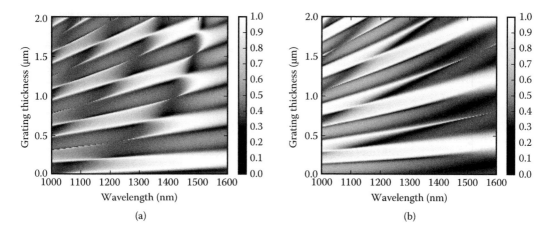

FIGURE 45.14 Reflectivity maps for gratings optimized for (a) TE and (b) TM polarizations as a function of the wavelength and grating thickness.

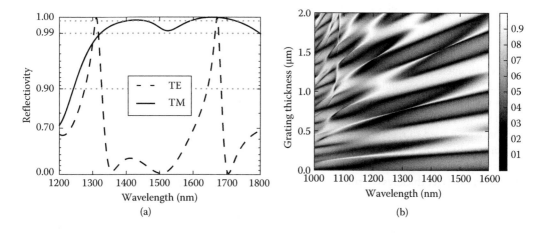

FIGURE 45.15 (a) Reflectivity spectrum and (b) TM reflectivity map for various thicknesses of a grating located on dielectric cladding (Figure 45.12b).

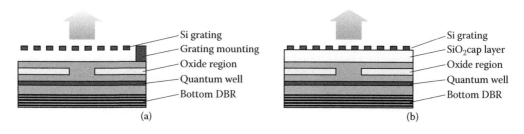

FIGURE 45.16 Typical structure of an HCG-VCSEL with (a) grating membrane suspended in air and (b) grating layer located on a low-index cladding.

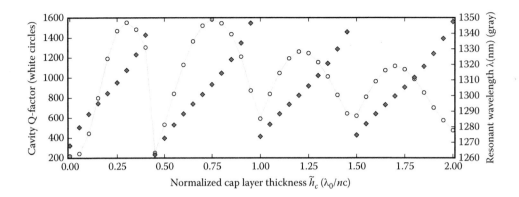

FIGURE 45.17   Cavity Q-factor (light circles) and resonant wavelength (gray diamonds) as functions of the normalized cap layer thickness.

design shown in Figure 45.16a, the membrane can be shifted by piezoelectric forces, effectively changing the cavity length and enabling the emitted wavelength to be tuned. Such a tunable HCG VCSEL has been demonstrated by [33].

In this section, we will analyze an HCG-VCSEL with the grating located on a low-index cladding layer (Figure 45.16b). It consists of a 2-λ GaAs/AlGaAs cavity[†] with a quantum well in one anti-node and an oxide aperture in the second. The bottom mirror is constructed from 29 pairs of GaAs/AlGaAs quarter-wavelength DBRs. At the top, there is an Si HCG separated by a SiO$_2$ cap layer that provides low index on both sides of the HCG. The cavity and the bottom DBR are designed for a wavelength of 1.3 µm and the HCG parameters are as follows (see Figure 45.11 for the meaning of the symbols): $L = 0.567$ µm, $h_g = 0.373$ µm, $f = 0.65$, and h_c varies. We are considering the cold-cavity properties of the laser, thus there is no gain in the quantum well. The oxide aperture is 4 µm and there are a total of seven grating periods within the aperture window. As the computations have to be fully three-dimensional, 891 plane waves have been used to represent the lateral electromagnetic field distribution in each layer: 11 in the direction parallel to the grating and 81 in the direction perpendicular to the grating.

Figure 45.17 shows the computed resonant wavelength of the emitted light and the cavity Q-factor[‡] as functions of cap layer thickness normalized to the target cavity wavelength λ_0 of 1.3 µm. This normalized thickness is computed as $\tilde{h}_c = h_c n_c / \lambda_0$. As can be seen, both the resonant wavelength and the cavity Q-factor depend periodically on $\tilde{h}_c$, with a period of $\lambda/2$ and the highest Q-factor corresponding to the quarter-wavelength cap thickness (i.e., $\tilde{h}_c = 1/4 + k/2$ with k being an arbitrary integer positive number). However, as the cap thickness increases over $\tilde{h}_c = 1.0$, the consequent maxima correspond to a lower and lower Q-factor. This can be attributed to the fact that the finite size of the grating causes an increase in scattering losses, as the divergent beam is not fully back-reflected to the active region.

45.4.4 Manufacturing Sensitivity of High-Contrast Grating Mirrors

The results in Figure 45.17 show that HCGs can be effectively used to replace DBR mirrors in VCSELs. However, the question arises whether they are similarly robust. The need to etch a precise lithographic pattern can make them prone to manufacturing errors. We therefore investigated the reflectivity of a grating mirror with an imperfect periodic grating. We performed simulations of multiple grating periods in which grating bar thicknesses and positions were randomly varied with a specified standard deviation σ. Due

---

[†] 2-λ means that the optical cavity length is exactly two wavelengths.

[‡] Q-factor is a dimensionless rate of energy loss relative to the stored energy of the resonator. It can be determined from the equation $Q = \frac{1}{2}\mathrm{Re}(\tilde{\omega})/\mathrm{Im}(\tilde{\omega})$, where $\tilde{\omega}$ is the complex eigen-frequency of the resonant mode in the cavity.

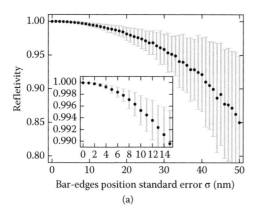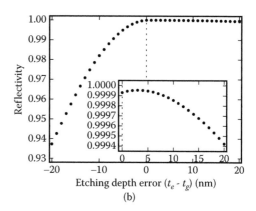

FIGURE 45.18 (a) Computed reflectivity for an HCG mirror with the position of the bar edges varied randomly using different standard deviations. The vertical bars show the uncertainties of the results. The inset shows the magnified part of the full figure. (b) Dependence of reflectivity on the difference between etching depth (h_e) and the thickness of the Si layer (h_g, see Figure 45.11). Negative values mean too shallow holes while positive values designate holes that are too deep. The latter are shown magnified in the inset.

to the randomness introduced, we repeated our calculations 1000 times for each value of σ. The results are presented in Figure 45.18a. Their uncertainties (shown with gray vertical lines) are computed using Student's t-distribution for a confidence interval of 90%.

As can be seen, even a small variation causes a significant drop in reflectivity. As the precision of typical photo-lithography is around 10 nm, mirror reflectivity can drop by 0.5% for $\sigma = 10$ nm. Larger manufacturing errors can make the mirrors unusable in VCSELs, as reflectivity can drop well below 90%. The largest allowable error, for which the mirrors reflect more than 99% of the incident light in at least half of the cases, is $\sigma = 15$ nm. The stronger requirement of reflectivity exceeding 99.8% is fulfilled if the error is not greater than $\sigma = 8$ nm. This clearly shows that the main physical phenomenon behind the properties of HCG mirrors is Bragg reflection, in which perfect periodicity plays a crucial role. This contrasts with most modern designs for photonic crystal VCSELs, in which the most important factor is the reduced effective index of the photonic crystal area [8].

Another test we have performed is an analysis of the precision of etching depth. In the ideal case, the Si bars should rest fully separated on a SiO_2 layer. In other words, the spaces between the bars should be etched to a depth exactly equal to the thickness of the Si layer. In reality, this may not always be the case, as the etching can be slightly deeper or slightly shallower. However, locally one can assume uniform etching conditions, whereby the depths of adjacent etched areas should be the same.

In order to investigate the impact of etching depth precision on the performance of HCG mirrors, we computed their reflectivity for various etching depths h_e (h_e is the depth of etching spaces between the bars; in an ideal situation it should be equal to the thickness of the high-index grating layer h_g, see Figure 45.11). Incident light is the same as in the previous case, i.e., the polarization is perpendicular to the bars and the wavelength is 1.3 μm. The results are shown in Figure 45.18b. As one can see, too shallow etching causes a strong decrease in reflectivity. On the other hand, too deep etching does not influence the results significantly. Furthermore, etching 3 nm of the SiO_2 layer can increase reflectivity slightly, although this effect is so weak that it is negligible in practice. The main conclusion is that it is better to etch the space around the bars a little more than desired in order to provide optimum mirror reflectivity even in the case of manufacturing imprecisions.

Similar analysis can be performed for HCG VCSELs, as described in Section 45.4.3. As the reflectivity of HCGs drops with manufacturing imprecisions, so should the VCSEL cavity Q-factor. A numerical analysis of this effect is shown in Figure 45.19. The computation procedure is identical to that used for generating Figure 45.18. However, each point was averaged over 100 simulations instead of 1000 (due to the larger

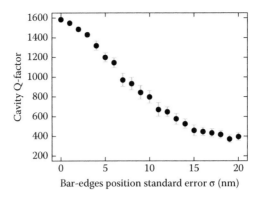

FIGURE 45.19 HCG-VCSEL cavity Q-factor with an imperfect mirror. The edges of the grating bars are varied randomly using different standard deviations. The vertical bars show the uncertainties of the results.

numerical effort necessary to determine the proper VCSEL modes). The resonant wavelength is obviously unaffected, as it is determined by the cavity itself and not the quality of the mirror. We can notice that even a 10 nm error can lead to a twofold reduction in the Q-factor. This must be taken into consideration in designs for devices that are to be produced on an industrial scale. Furthermore, the analysis presented does not consider the effects of the surface roughness of the bars, which to some extent is inevitable in the lithography process.

45.5 Summary

Periodic structures have been used in laser diodes for a long time. Their most basic form are one-dimensional distributed Bragg reflectors, commonly used in VCSELs. However, two- or three-dimensional photonic crystals allow further tuning of laser properties, e.g., to achieve a single-mode light emission. One-dimensional subwavelength high-contrast gratings can make efficient broadband laser mirrors with reduced size and wavelength tuning ability.

The downside of such structures is the increased laser complexity, which requires more advanced numerical methods for the device optical modeling. In particular, only vectorial methods like the finite-difference time-domain method, the plane-wave expansion, the finite element method, the coupled mode model, the plane-wave admittance method, or the plane-wave reflection transfer can yield proper optical characteristics of PC-VCSELs.

References

1. Strutt, J. W. (1887). On the maintenance of vibrations by forces of double frequency, and on the propagation of waves through a medium endowed with a periodic structure. *Phil. Mag.* 27, 145–159.
2. Yablonovitch, E. (1987). Inhibited spontaneous emission in solid-state physics and electronics. *Phys. Rev. Lett.* 58, 2059–2062.
3. John, S. (1987). Strong localization of photons in certain disordered dielectric superlattices. *Phys. Rev. Lett.* 58, 2486–2489.
4. Kittel, C. (2004). *Introduction to Solid State Physics* (8th ed.). Hoboken, NJ: Wiley.
5. Joannopoulos, J. D., R. D. Meade, and J. N. Winn (1995). *Photonic Crystals: Molding the Flow of Light.* Princeton, NY: Princeton University Press.
6. Checoury, X., P. Boucaud, J.-M. Lourtioz, F. Pommereau, C. Cuisin, E. Derouin, O. Drisse, L. Legouezigou, O. L. Legouezigou, F. Lelarge, F. Poingt, G.-H. Duan, S. Bonnefont, D. Mulin, J.

Valentin, O. Gauthier-Lafaye, F. Lozes-Dupuy, and A. Talneau (2005). Distributed feedback-like laser emission in photonic crystal waveguides on InP substrate. *IEEE J. Sel. Top. Quant. Electron.* 11, 1180–1186.

7. Painter, O., J. Vuckovic, and A. Scherer (1999). Defect modes of a two-dimensional photonic crystal in an optically thin dielectric slab. *J. Opt. Soc. Am. B.* 16, 275–285.

8. Danner, A. J., J. J. Raftery Jr., L. P. O., and K. D. Choquette (2006). Single mode photonic crystal vertical cavity lasers. *Appl. Phys. Lett.* 88, 091114.

9. Danner, A. J., J. J. Raftery Jr., N. Yokouchi, and K. D. Choquette (2004). Transverse modes of photonic crystal vertical-cavity lasers. *Appl. Phys. Lett.* 84, 1031–1033.

10. Leisher, P. O., A. J. Danner, and K. D. Choquette (2006). Single-mode 1.3 μm photonic crystal vertical-cavity surface-emitting laser. *IEEE Photon. Technol. Lett.* 18, 2156–2158.

11. Song, D. S., S. H. Kim, H. G. Park, C. K. Kim, and Y. H. Lee (2002). Single-fundamental mode photonic-crystal vertical-cavity surface-emitting lasers. *Appl. Phys. Lett.* 80, 3901–3903.

12. Yokouchi, N., A. J. Danner, and K. D. Choquette (2003). Two-dimensional photonic crystal confined vertical-cavity surface-emitting lasers. *IEEE J. Sel. Top. Quant. Electron.* 9, 1439–1445.

13. Liu, H., M. Yan, P. Shum, H. Ghafouri-Shiraz, and D. Liu (2004). Design and analysis of anti-resonant reflecting photonic crystal VCSEL lasers. *Opt. Express.* 12, 4269–4274.

14. Czyszanowski, T., M. Dems, and K. Panajotov (2007a). Impact of the hole depth on the modal behaviour of long wavelength photonic crystal VCSELs. *J. Phys. D: Appl. Phys.* 40, 2732–2735.

15. Czyszanowski, T., M. Dems, R. P. Sarzała, W. Nakwaski, and K. Panajotov (2011). Precise lateral mode control in photonic crystal vertical-cavity surface-emitting lasers. *IEEE J. Quant. Electron.* 47, 1291–1296.

16. Birks, T. A., J. C. Knight, and P. S. J. Russell (1997). Endlessly single-mode photonic crystal fiber. *Opt. Lett.* 22, 961–963.

17. Taflove, A. and S. C. Hagness (2000). *Computational Electrodynamics: The Finite-Difference Time-Domain Method* (2nd ed.). Boston, MA: Artec House.

18. Johnson, S. G. and J. D. Joannopoulos (2001). Block iterative frequency-domain methods for Maxwell's equations in a planewave basis. *Opt. Express.* 8, 173–190.

19. Nyakas, P. (2007). Full-vectorial three-dimensional finite element optical simulation of vertical-cavity surface-emitting lasers. *IEEE J. Lightwave Techn.* 25, 2427–2434.

20. Bava, G. P., P. Debernardi, and L. Fratta (2001). Three-dimensional model for vectorial fields in vertical-cavity surface-emitting lasers. *Phys Rev. A.* 63, 23816.

21. Dems, M., R. Kotynski, and K. Panajotov (2005). Plane-wave admittance method—A novel approach for determining the electromagnetic modes in photonic structures. *Opt. Express.* 13, 3196–3207.

22. Dems, M. (2011). Modelling of high-contrast grating mirrors. The impact of imperfections on their performance in VCSELs. *Opto-Electr. Rev.* 19, 340–345.

23. Conradi, O., S. F. Helfert, and R. Pregla (2001). Comprehensive modeling of vertical-cavity laser-diodes by the method of lines. *IEEE J. Quant. Electron.* 37, 928–935.

24. Czyszanowski, T., M. Dems, and K. Panajotov (2007b). Single mode condition and modes discrimination in photonic-crystal 1.3 μm AlInGaAs/InP VCSEL. *Opt. Express.* 15, 5604–5609.

25. Gedney, S. D. (1996). An anisotropic perfectly matched layer-absorbing medium for the truncation of FDTD lattices. *IEEE T. Antenn. Propag.* 44, 1630–1639.

26. Dems, M. (2009). Semi-vectorial method based on effective index for VCSEL analysis. *J. Opt. Soc. Am. B.* 26, 792–796.

27. Dems, M., I.-S. Chung, P. Nyakas, S. Bischoff, and K. Panajotov (2010). Numerical methods for modeling photonic-crystal VCSELs. *Opt. Express.* 18, 16042–16054.

28. Panajotov, K., B. Ryvkin, J. Danckaert, M. Peeters, H. Thienpont, and I. Veretennicoff (1998). Polarization switching in VCSEL's due to thermal lensing. *IEEE Photon. Technol. Lett.* 10, 6–8.

29. Fan, S., W. Suh, and J. Joannopoulos (2003). Temporal coupled-mode theory for the Fano resonance in optical resonators. *J. Opt. Soc. Am. A.* 20 (3), 569–572.

30. Karagodsky, V., F. G. Sedgwick, and C. J. Chang-Hasnain (2010). Theoretical analysis of subwavelength high contrast grating reflectors. *Opt. Express.* 18, 16973–16988.

31. Chang-Hasnain, C. J. and W. Yang (2012). High-contrast gratings for integrated optoelectronics. *Adv. Opt. Photon.* 4, 379–440.

32. Rosenblatt, D., A. Sharon, and A. A. Friesem (1997). Resonant grating waveguide structures. *IEEE J. Quant. Electron.* 33, 2038–2059.

33. Huang, M. C. Y., Y. Zhou, and C. J. Chang-Hasnain (2008). A nanoelectromechanical tunable laser. *Nat. Photon.* 2, 180–184.

46

Single-Photon Sources

Niels Gregersen

Dara P. S.
McCutcheon

and

Jesper Mørk

46.1 Introduction

The single-photon source (SPS) is a solid-state device capable of emitting single photons on demand. It is a key component within the emerging field of quantum information processing, where the information is encoded on single photons. In these applications, the quantum mechanical properties of the single photon are exploited, e.g., to enable secure communication where undetected eavesdropping is impossible or to perform quantum calculations taking advantage of massive parallelization.

This tutorial chapter reviews the characteristic figure of merits of the deterministic SPS based on a quantum emitter in a semiconductor solid-state system. The theory and the modeling methods used to describe single-photon emission are discussed. Photon extraction strategies based on cavity and waveguide quantum electrodynamics (QED) are presented, and finally we discuss decoherence mechanisms due to the solid-state environment.

46.1.1 Characteristics

For use in quantum information technology, the SPS must meet three basic performance requirements: pure single-photon emission, indistinguishability, and high efficiency, as discussed separately in the three following subsections.

46.1.1.1 Pure Single-Photon Emission

The most fundamental characteristic of the SPS is that it should emit only one photon at a time. This ability is tested in the so-called Hanbury, Brown, and Twiss (HBT) experiment illustrated in Figure 46.1.

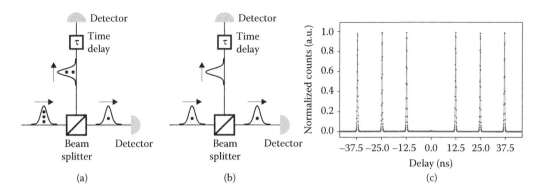

FIGURE 46.1 The Hanbury, Brown, and Twiss experiment illustrated for input pulses containing (a) multiple photons and (b) single photons. (c) The $g^{(2)}(\tau)$ autocorrelation function for a micropillar SPS. (Reprinted with permission from Ding, X. et al., *Physical Review Letters*, 116, 020401, 2016. Copyright 2016 by the American Physical Society.)

Here an optical pulse ideally containing only one photon is directed toward a beam splitter and is either reflected or transmitted toward two detectors. A variable time delay τ is implemented in one of the two output paths, and it is adjusted such that the photons from either of the two ports of the beam splitter arrive simultaneously at the two detectors for zero time delay. The single-photon emission ability is now characterized by measuring the autocorrelation function $g^{(2)}(\tau)$. The experiment is carried out by sending in light pulses from the SPS and by measuring the number of simultaneous detection events in the two detectors as function of time delay.

If the input pulse contains multiple photons as depicted in Figure 46.1a, some photons may be reflected while others are transmitted leading to coincidence clicks in the two detectors at zero time delay such that $g^{(2)}(\tau = 0) \neq 0$. However, if the pulse contains only a single photon as illustrated in Figure 46.1b, this photon will be either transmitted or reflected and no coincidence events are measured. An example of such a HBT measurement is given in Figure 46.1c (Ding et al., 2016), where we observe coincidence counts as function of delay for a SPS generating pulses every 12 ns. The peaks in the measurement correspond to integer values of the pulse time separation, where photons from different pulses may lead to coincidence clicks. However, at zero time delay, the correlation counts are strongly suppressed corresponding to an ideal single-photon emission with $g^{(2)}(\tau = 0) = 0$.

46.1.1.2 Indistinguishability

A second requirement for the SPS is that the emitted photons should be quantum mechanically indistinguishable, meaning that the emitted photons should have the same wavelength, phase variation across the pulse, polarization, optical mode profile, etc. The single-photon indistinguishability is characterized in the Hong–Ou–Mandel (HOM) experiment sketched in Figure 46.2 (Hong et al., 1987).

In this experiment, two single-photon pulses are inbound on a 50–50 beam splitter, where the two pulses are either transmitted or reflected into the output paths. A variable time delay τ is implemented in one of the two output paths adjusted such that the photons reach the detectors simultaneously at zero time delay. The experiment then takes place by sending in single-photon pulses along the input paths and by measuring the correlation function $g^{(2)}(\tau)$ describing the simultaneous detection events for this setup as function of time delay.

The impinging photons may leave the beam splitter either in separate output paths or together as illustrated in Figure 46.2a and b, respectively. In the case where the input photons are distinguishable, the output photons may leave the beam splitter along separate paths leading to $g^{(2)}(\tau = 0) \neq 0$. However, when the input photons are completely indistinguishable and arrive simultaneously at the beam splitter, they always exit the beam splitter along the same output path as depicted in Figure 46.2b, resulting in a $g^{(2)}(\tau = 0) = 0$. This phenomenon is known as the HOM effect and can be understood from a quantum-optical consideration of the probabilities of the four possible output states sketched in Figure 46.2c. The

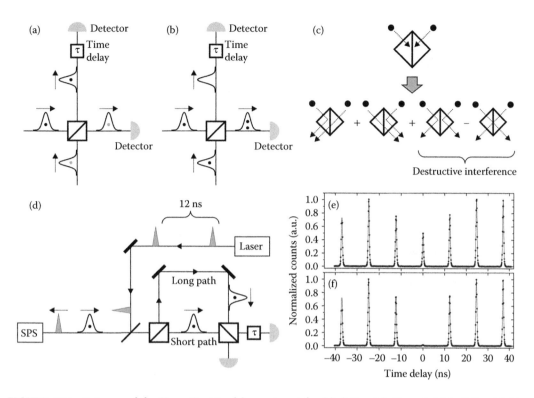

FIGURE 46.2 Outcome of the Hong–Ou–Mandel experiment for (a) distinguishable and (b) indistinguishable photons. (c) Various possible output states. Sketch of a typical experimental setup. Measured coincidence counts in (e) cross and (f) parallel polarization. (Reprinted figure with permission from Ding, X. et al., *Physical Review Letters*, 116, 020401, 2016. Copyright 2016 by the American Physical Society.)

probability amplitude of the photons leaving the beam splitter along separate output paths consists of two contributions corresponding to the photons being both either reflected or transmitted at the beam splitter, and for indistinguishable photons these two contributions add up destructively (Gerry and Knight, 2008).

In most HOM experiments, the impinging photons are generated by the same source using the setup sketched in Figure 46.2d. A laser generates classical pulses at a typical repetition rate of 80 MHz. The pulses are transmitted to the SPS, and each laser pulse triggers a single-photon emission event leading to a train of single photons with 12.5-ns time separation inbound on the first beam splitter. Two subsequent photons in the train can propagate along the short and the long paths in various configurations. The long and short path lengths are adjusted such that a first photon taking the long path and a second photon taking the short path will arrive simultaneously at the beam splitter. Typical HOM measurements for indistinguishable photons having orthogonal or parallel polarization are illustrated in Figure 46.2e and f, respectively. For orthogonal polarizations, the photons are distinguishable and $g^{(2)}(\tau = 0) \neq 0$, whereas for parallel polarizations the HOM effect with $g^{(2)}(\tau = 0) = 0$ is observed. Again, $g^{(2)}(\tau)$ has peaks for time delays equal to an integer times the pulse separation. These peaks correspond to coincidence events for photon pairs that do not arrive simultaneously at the beam splitter and are thus not subject to the HOM effect. The indistinguishability is then quantified as one minus the normalized area of the $\tau = 0$ peak and should ideally equal unity.

46.1.1.3 Efficiency

The third important characteristic of the SPS is its efficiency or brightness (Gregersen et al., 2013). The ideal SPS should be deterministic and provide single photons on demand. It should act as a "single-photon gun" such that every trigger releases exactly one photon into the collection optics. The efficiency ε is then defined as the number of photons detected by the collection optics per trigger. Within the QED design approaches discussed in Sections 46.3 and 46.4, the efficiency is characterized using two governing parameters as illustrated in Figure 46.3. The first is the spontaneous emission factor β describing the fraction of light coupled from the emitter into the optical mode of interest. The β factor can be written as

$$\beta = \frac{\Gamma_M}{\Gamma_M + \Gamma_{Rad}}, \tag{46.1}$$

where Γ_M is the spontaneous emission rate into the optical mode and Γ_{Rad} is the emission rate into background radiation modes. One may also add a term Γ_{NonRad} to the denominator in the definition (46.1) to take into account nonradiative recombination channels of the quantum emitter. Whereas Γ_M and Γ_{Rad} can be modified by tailoring the photonic environment, the parameter Γ_{NonRad} represents inherent properties of the emitter that generally cannot easily be controlled.

The second governing parameter is the outcoupling efficiency γ describing the fraction of light transmitted from the optical mode to the first objective lens. In this simple picture, the efficiency is given by $\varepsilon = \beta\gamma$. In a multiphoton interference experiment involving j photons, the total success probability of the experiment will scale as ε^j. For scalable optical quantum information technologies involving a large number of photons, a single-photon generation efficiency close to unity is thus required.

46.1.2 Single-Photon Generation Schemes

A major workhorse for generating single photons has been the spontaneous parametric down-conversion (SPDC) process (Brendel et al., 1999). Let us recall that in a nonlinear isotropic material, the polarization **P** is related to the electric field **E** as

$$\mathbf{P} = \varepsilon_0 \left(\chi^{(1)}\mathbf{E} + \chi^{(2)}\mathbf{E}^2 + \chi^{(3)}\mathbf{E}^3 + \dots \right), \tag{46.2}$$

where $\chi^{(n)}$ is the nth-order susceptibility. In the SPDC scheme, a crystal featuring a $\chi^{(2)}$ nonlinearity such as lithium niobate is illuminated with a strong laser pump pulse with an energy $\hbar\omega_p$ as sketched in Figure 46.4. Provided that proper phase-matching conditions are met, the $\chi^{(2)}$ nonlinearity then enables conversion of a pump photon to two single photons, the signal and the idler photons of energies $\hbar\omega_s$ and $\hbar\omega_i$, respectively, such that $\hbar\omega_p = \hbar\omega_s + \hbar\omega_i$. The advantage of the SPDC source is its simplicity and the high

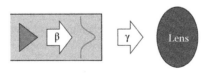

FIGURE 46.3 Efficiency model for the quantum electrodynamics approach. The single-photon emitter is represented by the triangle.

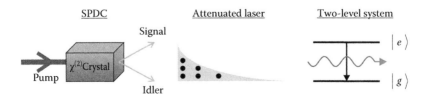

FIGURE 46.4 Major single-photon generation schemes.

degree of indistinguishability of the signal and idler photons. Its main drawback is that the downconversion process is inherently probabilistic, leading to efficiencies below 0.01. While the efficiency may be improved by cranking up the pump power, this occurs at the cost of significant increase of the probability of multiphoton emission. The probabilistic nature thus limits SPDC to experiments involving only a few photons.

Another commercially important source is the attenuated laser SPS (Buttler et al., 2000). A pulse generated using a classical laser propagates through an absorbing material and is attenuated to the single-photon level. The advantage of the attenuated laser is its simplicity, and similar to the SPDC process its main drawback is the probabilistic nature of the attenuation leading to either low single-photon generation efficiency or high probability of multiphoton events. Additionally, the generated photons are not indistinguishable, preventing their use in many quantum optics and quantum information experiments. However, the attenuated laser remains an important device in commercial products for quantum key distribution, where clean single-photon emission and indistinguishability are not required.

The quantum-dot emitter (Shields, 2007) has recently emerged as a solution, overcoming the probabilistic limitations of the SPDC and attenuated laser strategies. It consists of a two-level system as shown in Figure 46.4, where an electron can occupy either a ground state $|g\rangle$ or an excited state $|e\rangle$ characterized by energies E_g and E_e, respectively. The system is initially in the ground state and is then excited using an optical or electrical pulse, after which the electron returns to the ground state by the spontaneous emission of a single photon of energy $\hbar\omega_0 = E_e - E_g$. A main advantage of the two-level emitter is that the upper level can contain only one excitation at a time, thereby ensuring the single-photon nature of the emitted light, combined with the deterministic nature of the spontaneous emission event and the possibility of structuring the medium to realize highly directional emission, potentially allowing for unity efficiency.

Any solid-state system featuring discretized energy levels can in principle be employed as a SPS. The InAs quantum dot (QD) grown in a GaAs host material as shown in Figure 46.5a represents a mature platform for a deterministic SPS. The QD consists of a pyramidal-shaped island of InAs atoms surrounded by GaAs. Due to the lower bandgap of InAs, the electron wavefunction will be fully confined in all three dimensions leading to discretized energy levels in the conduction and valence bands depicted in Figure 46.5b. The emission of InAs QD ensembles features strong inhomogeneous broadening with typical emission wavelength between 900 and 950 nm. The lowest energy s state in the conduction band is typically used as the excited state. Due to the spin degree of freedom, the s state supports a biexciton leading to two-photon emission. However, the InAs QD features a fine-structure splitting arising from the exciton binding energy of a few nanometers allowing for the selection of the excitonic line using spectral filtering thus enabling single-photon emission at cryogenic temperatures. The indistinguishability of the emitted photons depends greatly on the charge environment and the excitation scheme employed, as discussed in detail in Section 46.6.

Other two-level systems including diamond nitrogen-vacancy centers (Babinec et al., 2010) and organic molecules (Siyushev et al., 2014) have been pursued for single-photon generation. These systems feature the advantages of room temperature operation and very long spin coherence times. However, the maturity

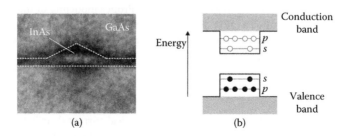

FIGURE 46.5 Scanning electron micrograph (a) and energy level diagram (b) of an InAs quantum dot.

FIGURE 46.6   Major photon extraction strategies. (Reprinted from Gregersen, N et al., *IEEE Journal of Selected Topics in Quantum Electronics* © 2013 IEEE and Reprinted by permission from Macmillan Publishers Ltd: Claudon, J et al., *Nature Photonics* 4(3): 174–177, copyright 2010.)

of III–V semiconductor cleanroom fabrication has made the InAs QD the leading platform for highly efficient SPSs, and for this reason we focus in this tutorial on the InAs QD.

46.1.3 Photon Extraction Strategies

While the self-assembled InAs QD features an internal quantum efficiency η of almost unity, a major challenge in single-photon engineering lies in the efficient extraction of the emitted photon (Gregersen et al., 2013). Figure 46.6 summarizes the main SPS photon extraction strategies. The simplest approach consists of a planar sample with a QD in a bulk material. The main advantages of this approach are its simplicity as no additional structuring is needed after QD growth as well as the absence of fluctuating surface charges near the QD as discussed further in Section 46.6. The disadvantage is the poor extraction efficiency. The symmetry of the system leads to light emission in all directions limiting the efficiency to only a few percent.

The first strategy proposed to improve the efficiency and control the light emission was based on cavity QED, where the QD was placed in a microcavity (Gérard, 2003). Here the Purcell effect ensures a preferential single-photon emission from the QD into the cavity mode. This preferential coupling requires a careful spectral alignment between the QD emission line and the cavity line as discussed in detail in Section 46.3. This approach has led to efficiencies of 0.8 (Gazzano et al., 2013) as well as an indistinguishability of 0.99 from the micropillar cavity SPS (Ding et al., 2016; Somaschi et al., 2016), representing the current state of the art in highly efficient sources of indistinguishable photons.

A second design strategy is based on waveguide QED, where the QD is positioned inside a waveguide (Claudon et al., 2013). The preferential coupling to the waveguide mode is ensured by a dielectric screening effect (Bleuse et al., 2011). The main advantage of this nonresonant approach is that no spectral alignment is required. While an efficiency of 0.75 has been demonstrated for the photonic nanowire SPS, the demonstration of highly indistinguishable photons from a waveguide-based SPS is still lacking.

46.2 Modeling of Spontaneous Emission

A major challenge in engineering deterministic SPSs lies in optimizing the photon extraction efficiency and the indistinguishability to achieve values close to unity. Proper modeling of the light emission from a two-level system generally requires a QED description of the spontaneous emission process. However, SPSs generally operate in the weak coupling regime, and here it turns out that the spontaneous emission rate can be calculated in a purely classical manner (Novotny and Hecht, 2012).

46.2.1 Classical Dipole

We first consider the electromagnetic field generated by a classical current distribution $\mathbf{j}$ emitting light at the angular frequency ω_0. We thus work in the frequency domain with harmonic time dependence such that $\mathbf{E}(\mathbf{r}, t) = \mathbf{E}(\mathbf{r})e^{-i\omega_0 t}$ and $\mathbf{j}(\mathbf{r}, t) = \mathbf{j}(\mathbf{r})e^{-i\omega_0 t}$. Furthermore, we assume nonmagnetic materials and the absence of free charges for simplicity. The classical Maxwell's equations are then combined to yield the second-order wave equation for the electric field:

$$\nabla \times \nabla \times \mathbf{E}(\mathbf{r}) - \varepsilon_r(\mathbf{r})k_0^2\mathbf{E}(\mathbf{r}) = i\omega_0\mu_0\mathbf{j}(\mathbf{r}), \tag{46.3}$$

where $k_0 = \omega_0/c$ is the free-space wavenumber and ε_r is the dielectric constant.

Equation 46.3 can be solved in an elegant manner by introducing the dyadic Green's function $\overline{\overline{G}}$ (Novotny and Hecht, 2012). It is defined as the solution to

$$\nabla \times \nabla \times \overline{\overline{G}}(\mathbf{r}, \mathbf{r}') - \varepsilon_r(\mathbf{r})k_0^2\overline{\overline{G}}(\mathbf{r}, \mathbf{r}') = \overline{\overline{I}}\delta\left(\mathbf{r} - \mathbf{r}'\right), \tag{46.4}$$

where $\overline{\overline{I}}$ is the unit dyad. $\overline{\overline{G}}$ and $\overline{\overline{I}}$ are both 3×3 tensors and we stress that $\overline{\overline{G}}(\mathbf{r}, \mathbf{r}')$ depends on both $\mathbf{r}$ and $\mathbf{r}'$. Now, using this definition of the dyadic Green's function, we note that

$$\mathbf{E}(\mathbf{r}) = i\omega_0\mu_0 \int_V \overline{\overline{G}}(\mathbf{r}, \mathbf{r}')\mathbf{j}(\mathbf{r}')\, d\mathbf{r}' \tag{46.5}$$

is a solution to Equation 46.3. The Green's function $\overline{\overline{G}}(\mathbf{r}, \mathbf{r}')$ can thus be interpreted physically as the field generated at the position $\mathbf{r}$ by a current at the position $\mathbf{r}'$. Once the Green's function has been calculated, the entire field profile can be determined from Equation 46.5.

The standard procedure for computing the power P emitted by the current distribution is to integrate the time average of the normal component of the Poynting vector over the surface S of volume V enclosing the current distribution as

$$P = \frac{1}{2} \int_S Re\left(\mathbf{E} \times \mathbf{H}^*\right) \cdot \mathbf{n}_S dS, \tag{46.6}$$

where $\mathbf{n}_S$ is a unit vector normal to the surface S. Now, the average energy dissipation rate W inside the volume is given by

$$W = -\frac{1}{2} \int_V Re\left(\mathbf{j}^* \cdot \mathbf{E}\right) d\mathbf{r}, \tag{46.7}$$

and according to Poynting's theorem, we have $P = W$.

We now consider the current distribution generated by a point dipole $\mathbf{d}$ positioned at $\mathbf{r}_0$, as illustrated in Figure 46.7, such that

$$\mathbf{j}(\mathbf{r}) = -i\omega_0\mathbf{d}\delta\left(\mathbf{r} - \mathbf{r}_0\right). \tag{46.8}$$

By inserting Equation 46.8 into Equation 46.5 and subsequently Equation 46.5 into Equation 46.7, we obtain an expression for the power P in terms of the dipole moment given by

$$P = \frac{\omega_0^3\mu_0 |\mathbf{d}|^2}{2} Im\left(\mathbf{n}_d \cdot \overline{\overline{G}}(\mathbf{r}_0, \mathbf{r}_0) \cdot \mathbf{n}_d\right), \tag{46.9}$$

where $\mathbf{n}_d = \mathbf{d}/|\mathbf{d}|$ is the orientation of the dipole moment. The power P emitted by the point dipole can thus be obtained by evaluating the imaginary part of the Green's function at the position of the dipole.

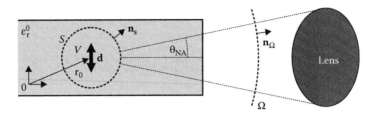

FIGURE 46.7   Light emission from a classical dipole **d** placed at the position $\mathbf{r}_0$ inside a volume V in a material of dielectric constant ε_r^0. The volume is limited by the surface S having normal unit vector $\mathbf{n}_S$. The entire structure is enclosed in a sphere Ω with normal unit vector $\mathbf{n}_\Omega$.

46.2.2 Spontaneous Emission

In the weak coupling regime, the total spontaneous emission rate Γ of a two-level system at the position $\mathbf{r}_0$, with transition energy $\hbar\omega_0$ and with dipole moment **d**, is derived from Fermi's golden rule as

$$\Gamma = \frac{\pi\omega_0}{\hbar\varepsilon_0}|\mathbf{d}|^2\,\rho_L\left(\mathbf{n}_d,\mathbf{r}_0,\omega_0\right). \tag{46.10}$$

Here, $\rho_L\left(\mathbf{n}_d,\mathbf{r}_0,\omega_0\right)$ is the local density of states (LDOS) defined as

$$\rho_L\left(\mathbf{n}_d,\mathbf{r}_0,\omega_0\right) = \int\left|\mathbf{n}_d\cdot\widetilde{\mathbf{E}}_\alpha(\mathbf{r}_0)\right|^2\delta(\omega_0-\omega_\alpha)d\alpha, \tag{46.11}$$

where $\widetilde{\mathbf{E}}_\alpha$ are the so-called normal modes solutions to the wave equation in the absence of free currents and charges:

$$\nabla\times\nabla\times\widetilde{\mathbf{E}}_\alpha(\mathbf{r}) - \varepsilon(\mathbf{r})\left(\frac{\omega_\alpha}{c}\right)^2\widetilde{\mathbf{E}}_\alpha(\mathbf{r}) = 0, \tag{46.12}$$

and the index α is a short notation for all appropriate discrete and continuous parameters describing a mode. The normal modes are normalized such that

$$\int\varepsilon_r(\mathbf{r})\widetilde{\mathbf{E}}_\alpha^*(\mathbf{r})\cdot\widetilde{\mathbf{E}}_{\alpha'}(\mathbf{r})d\mathbf{r} = \delta(\alpha-\alpha'). \tag{46.13}$$

We now write the dyadic Green's function as an expansion in normal functions. Referring the reader interested in the derivation to Novotny and Hecht (2012), we simply state the result:

$$Im\left(\overline{\overline{G}}(\mathbf{r}_0,\mathbf{r}_0,\omega_0)\right) = \frac{\pi c^2}{2\omega_0}\int\widetilde{\mathbf{E}}_\alpha^*(\mathbf{r}_0)\widetilde{\mathbf{E}}_\alpha(\mathbf{r}_0)\delta(\omega_0-\omega_\alpha)d\alpha, \tag{46.14}$$

where the integrand contains the outer product of the normal modes. Comparing this expression to Equation 46.11, we observe that the imaginary part of the Green's function is closely related to the LDOS. Indeed, outer multiplications of the dyad with the dipole unit vector and using the definition (46.11) lead to

$$\rho_L\left(\mathbf{n}_d,\mathbf{r}_0,\omega_0\right) = \frac{2\omega_0}{\pi c^2}Im\left(\mathbf{n}_d\cdot\overline{\overline{G}}(\mathbf{r}_0,\mathbf{r}_0,\omega_0)\cdot\mathbf{n}_d\right), \tag{46.15}$$

and we observe that, except for physical constants, the LDOS is identical to the average power emitted by a classical dipole as given in Equation 46.9. This means that we obtain the very important relationship

$$\frac{\Gamma}{\Gamma_{\text{Bulk}}} = \frac{P}{P_{\text{Bulk}}}, \tag{46.16}$$

which states that the spontaneous emission rate, Γ, of a two-level system embedded in a medium with dielectric constant ε_r^0 normalized to its value, Γ_{Bulk}, in a bulk material of dielectric constant ε_r^0 is equal to the power dissipation, P, from a classical dipole normalized to the rate, P_{Bulk}, in the bulk material.

46.2.3 Simulation Techniques

Analytic expressions for the Green's function are only available in a limited number of geometries and generally we rely on numerical simulation techniques to compute the electric field. When using spatial discretization methods such as the finite-difference time-domain technique and the finite-elements method (Lavrinenko et al., 2014), the field is expanded on a grid as sketched in Figure 46.8a and either Maxwell's equations or a vectorial wave equation are directly solved to give the optical field.

Alternatively, one may use a modal method (Lavrinenko et al., 2014). Here, the geometry is sliced into layers uniform along a propagation axis as illustrated in Figure 46.8b and the eigenmodes of each layer are then computed. The optical field is then expanded on the eigenmodes, and the fields at each side of a layer interface are connected using the scattering matrix formalism. This allows for direct access to scattering coefficients and the possibility of studying the various parts of the SPS individually.

As we have seen, the QD can be modeled as a classical point dipole placed inside the geometry under study. The electric field profile generated by the dipole is then computed, and the classical power P can then be obtained from Equation 46.7. We note that, for the current source defined by Equation 46.8, the real part of the electric field diverges at the position of the dipole, and an evaluation of the power P by integrating over a small sphere using Equation 46.6 is thus often preferred for numerical stability. The power collected by the lens of numerical aperture NA can be evaluated by computing the electromagnetic field on the surface of a large sphere Ω using a standard near-field to far-field transformation (Balanis, 1989) and integrating the Poynting vector over the unit solid angle subject to the condition $\theta < \theta_{\text{NA}}$, where $\theta_{\text{NA}} = \text{asin(NA)}$. The efficiency ε is then given as

$$\varepsilon = \frac{\frac{1}{2} \int\limits_{\theta < \theta_{\text{NA}}} \text{Re}(\mathbf{E} \times \mathbf{H}^*) \cdot \mathbf{n}_\Omega d\Omega}{P}. \tag{46.17}$$

While Equation 46.17 is exact, this direct procedure does not give complete insight into the governing physics. For this reason, the efficiency in the QED SPS designs is instead treated using an elements-splitting

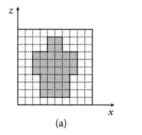
(a)

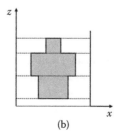
(b)

FIGURE 46.8 Discretization schemes for (a) the finite-difference time-domain technique and for (b) the modal method.

approach where β, Γ_M, Γ_{Rad}, and γ are analyzed and optimized separately and the efficiency is computed using $\varepsilon = \beta\gamma$. The rates Γ_M and Γ_{Rad} are available from Equation 46.11 as

$$\Gamma_M = \frac{\pi\omega_0}{\hbar\varepsilon_0} |\mathbf{d}|^2 \left| \mathbf{n}_d \cdot \tilde{\mathbf{E}}_M(\mathbf{r}_0) \right|^2 \tag{46.18}$$

$$\Gamma_{Rad} = \frac{\pi\omega_0}{\hbar\varepsilon_0} |\mathbf{d}|^2 \int\limits_{\alpha \neq M} \left| \mathbf{n}_d \cdot \tilde{\mathbf{E}}_\alpha(\mathbf{r}_0) \right|^2 \delta(\omega_0 - \omega_\alpha)d\alpha, \tag{46.19}$$

and we observe that their evaluation requires access to the electric field profiles of the optical mode of interest as well as the radiation modes.

When performing numerical simulations of SPS designs, the advantages of the spatial discretization methods are their maturity, their availability in commercial user-friendly software packages (Lumerical, n.d.; JCMWave, n.d.), and their ability to exploit parallelization as required in the treatment of large geometries. Their major weakness is that they do not give direct access to the mode profiles $\tilde{\mathbf{E}}_\alpha(\mathbf{r})$. While the total power P can be obtained from Equation 46.6 or Equation 46.7, the individual rates Γ_M and Γ_{Rad} are not directly available and β thus cannot be directly evaluated. On the other hand, in the modal expansion methods, the mode profiles $\tilde{\mathbf{E}}_\alpha(\mathbf{r})$ are directly available, and this represents a major advantage. However, the drawbacks of the modal method are the lack of user-friendly software packages as well as its difficulty in handling large 3-D geometries due to the lack of eigenvalue solvers capable of exploiting parallelization.

46.3 Cavity QED Approach

Inspection of Equation 46.1 reveals two options for optimizing β. The first is to increase Γ_M using cavity QED effects in the weak coupling regime. Here, the emitter is placed inside a cavity with quality factor Q, mode volume V, and resonance wavelength $\lambda_r = 2\pi c/\omega_r$ as illustrated in Figure 46.9a. If the emitter is placed at a maximum of the optical field profile and the emitter and cavity lines are spectrally aligned such that $\omega_r = \omega_0$, the spontaneous emission rate Γ_M into the cavity mode is enhanced by the factor F_P known as the Purcell factor (Purcell, 1946). It describes the rate Γ_M relative to the rate Γ_{Bulk} for an emitter in a bulk medium and is given by (Gérard, 2003)

$$F_P = \frac{\Gamma_M}{\Gamma_{Bulk}} = \frac{3}{4\pi^2} \frac{Q}{V} \left(\frac{\lambda_r}{n} \right)^3, \tag{46.20}$$

where n is the refractive index of the material enclosing the emitter. The mode volume is computed as[†]

$$V = \frac{\int \varepsilon_r(\mathbf{r}) \left| \tilde{\mathbf{E}}_M(\mathbf{r}) \right|^2 d\mathbf{r}}{\varepsilon_r(\mathbf{r}_{max}) \left| \tilde{\mathbf{E}}_M(\mathbf{r}_{max}) \right|^2}, \tag{46.21}$$

where the denominator is evaluated at the position $\mathbf{r}_{max}$ of a maximum of the electric field intensity profile.

The expression (Equation 46.20) holds provided the emitter line is narrow compared to the cavity linewidth. If this is not the case, an integration over the electronic density-of-states should be performed,

---

[†] When integrated over all space, the integral in Equation 46.21 diverges. To remedy this, it is possible to define a generalized mode volume (Kristensen et al., 2012) that converges in an unambiguous way. However, in high-Q cavities, one can usually employ Equation 46.21 and limit the integration to the device geometry with negligible error.

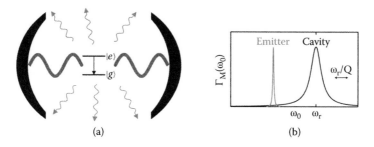

FIGURE 46.9 (a) Cavity-enhanced spontaneous emission. (b) Cavity emission rate Γ_M as a function of frequency.

which generally leads to a reduction in the enhancement. The β factor can be written in terms of the Purcell factor as

$$\beta = \frac{F_P}{F_P + (\Gamma_{Rad}/\Gamma_{Bulk})} \approx \frac{F_P}{F_P + 1}. \tag{46.22}$$

For micropillars with diameters in the few micrometer range, the spontaneous emission into radiation modes is approximately equal to the rate in a bulk material, and we can make the approximation $\Gamma_{Rad} = \Gamma_{Bulk}$ leading to the second equality in Equation 46.22.

We observe from Equations 46.20 through 46.22 that a high-efficiency cavity-based SPS requires a large Q factor combined with a low mode volume, and the optical design challenge for the cavity-based SPS is to maximize the ratio Q/V while maintaining a high outcoupling coefficient γ.

The expression (46.20) assumes perfect spectral alignment between the emitter and the cavity line. If they are detuned, a correction factor given by a Lorentzian lineshape function is introduced, such that (Lorke et al., 2013)

$$\frac{\Gamma_M(\omega_0)}{\Gamma_{Bulk}} = \frac{3}{4\pi^2} \frac{Q}{V} \left(\frac{\lambda_r}{n}\right)^3 \frac{(\delta\omega_r)^2}{(\delta\omega_r)^2 + (\omega_0 - \omega_r)^2}, \tag{46.23}$$

where $\delta\omega_r = \omega_r/Q$ is the cavity linewidth. The cavity emission rate Γ_M is plotted in Figure 46.9b as function of frequency and we observe that for an emitter-cavity misalignment of only a few linewidths $\delta\omega_r$, the enhancement is lost. For the cavity design approach to work, it is thus absolutely necessary to control the spectral alignment.

46.3.1 Micropillar

The AlGaAs microcavity pillar geometry (Reitzenstein and Forchel, 2010) illustrated in Figure 46.10a is one of the most popular platforms for constructing a high-efficiency SPS. It is a vertical structure consisting of a GaAs cavity with an embedded InAs QD surrounded by GaAs/AlAs distributed Bragg reflectors (DBRs). The optical cavity mode profile is confined vertically by the DBRs and laterally by total internal reflection. The emitted light is collected by an objective lens from above.

The thicknesses of the cavity and the DBR layers should be chosen as λ_d/n_{eff} and $\lambda_d/(4n_{eff})$, respectively, where λ_d is the design wavelength and n_{eff} is the effective index of the fundamental mode. These thicknesses ensure that the resonance is centered on the DBR stopband for maximum reflection (Coldren et al., 2012).

In the ideal structure, light escapes only through the bottom and top DBRs. The light emitted into the substrate is lost and the reflectivity of the bottom DBR should thus be significantly higher than that of the top DBR to obtain a high efficiency. The transmission γ to the lens for a micropillar of infinite diameter is presented in Figure 46.10b, and we observe that $\gamma > 0.99$ requires approximately 18 extra layer pairs in the bottom DBR.

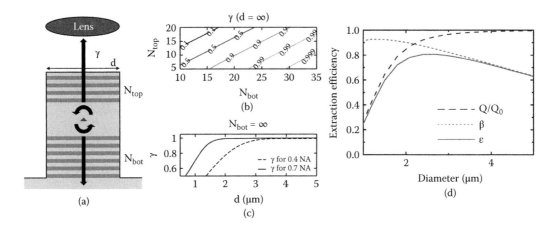

FIGURE 46.10 (a) The micropillar geometry. Transmission γ as function of DBR layer pairs (b) and as function of the pillar diameter (c). (d) Extraction efficiency as function of the pillar diameter. (Reprinted from Gregersen, N. et al., *IEEE Journal of Selected Topics in Quantum Electronics* © 2013 IEEE and reprinted by permission from Macmillan Publishers Ltd: Gazzano, O. et al., *Nature Communications* 4: 1425, copyright 2013.)

After the numbers of layer pairs are fixed, the pillar diameter should be chosen. The transmission γ as function of the pillar diameter is shown in Figure 46.10c for a perfectly reflecting bottom DBR. The far-field emission pattern is related to the Fourier transform of the near-field profile. For small diameters, the far-field emission pattern is broad and the transmission is low. For larger diameters, the beam divergence is reduced and the transmission improves. We observe that a pillar diameter above ~2 μm leads to near-unity transmission for a 0.7 NA lens.

In the optimization of the extraction efficiency, two additional mechanisms should be considered. While the ideal structure emits light only along the vertical axis, fabricated pillar structures suffer from sidewall imperfections leading to in-plane scattering of light. The amount of light lost can be estimated by measuring the Q factor and comparing with the quality factor Q_0 of a planar sample without microstructuring. The light lost due to imperfections is given by $1 - Q/Q_0$ and the predicted efficiency should then be corrected by the fraction Q/Q_0 such that $\varepsilon = \beta\gamma Q/Q_0$. The dashed curve in Figure 46.10d shows a fit to experimentally measured Q/Q_0 ratios for the pillar geometry described in Gazzano et al. (2013). As the diameter increases, the relative overlap of the mode profile with the imperfections decreases, and Q thus increases with diameter toward the planar value.

While a large pillar diameter optimizes γ and Q/Q_0 toward unity, it also increases the mode volume V leading to a reduced F_P. When choosing the diameter, it is thus necessary to consider the magnitude of β as function of diameter as given by Equation 46.22. The calculated ratio $\beta = F_P/(F_P + 1)$ is displayed in Figure 46.10d as the dotted curve and it decreases with the diameter due to the increase in V. The product $\varepsilon = \beta Q/Q_0$ is shown in the figure as the full curve, and we observe that a maximum of ~ 0.8 is obtained for $d \sim 2.5$ μm. This estimation of the efficiency assumes a transmission γ of unity, which is a good approximation when using a 0.7 NA lens.

This micropillar SPS reported in Gazzano et al. (2013) indeed features an efficiency of 0.79. The simultaneous spectral and spatial alignment was achieved using an *in situ* fabrication method pioneered by the group.

46.4 Waveguide QED Approach

The second option for optimizing β as defined by Equation 46.1 is to decrease Γ_{Rad} and this represents the waveguide QED approach. Two main geometries taking advantage of this approach are the photonic nanowire presented in Section 46.4.1 and the photonic crystal waveguide discussed in Section 46.4.2.

Assuming that the QD is positioned at a maximum of the guided mode, the spontaneous emission rate to the guided mode in an infinite uniform waveguide is given by (Claudon et al., 2013)

$$\frac{\Gamma_M}{\Gamma_{Bulk}} = \frac{3}{4\pi A} \left(\frac{\lambda}{n}\right)^2 \frac{n_g}{n}. \tag{46.24}$$

Here, n_g is the group index and

$$A = \frac{\int \varepsilon_r(\mathbf{r}_\perp) |\mathbf{E}(\mathbf{r}_\perp)|^2 \, d\mathbf{r}_\perp}{\varepsilon_r(\mathbf{r}_{max}) |\mathbf{E}(\mathbf{r}_{max})|^2}, \tag{46.25}$$

where the integration is over the lateral plane normal to the waveguide axis. In the case of the periodic waveguide of periodicity a, the rate is given by

$$\frac{\Gamma_M}{\Gamma_{Bulk}} = \frac{3a}{4\pi V} \left(\frac{\lambda}{n}\right)^2 \frac{n_g}{n}, \tag{46.26}$$

where the integration (46.21) of the mode volume V is limited to a single periodic cell.

The waveguide QED approach relies on significant suppression of the spontaneous emission into radiation modes, and this rate Γ_{Rad} should be quantified. The suppression occurs for waveguide diameters on the order of λ/n, and Γ_{Rad} can no longer be approximated to its value in a bulk material. Instead, Γ_{Rad} should be computed using full 3-D numerical simulations. Whereas the calculation of Γ_M is usually straightforward, a precise computation of Γ_{Rad} is generally quite demanding because the radiation modes propagate away from the device and perfect absorbing boundary conditions are required to avoid artificial reflections from the boundary walls. It is often determined indirectly by first computing the total spontaneous emission rate Γ_T, which for most methods is more accessible, and then subtracting Γ_M such that $\Gamma_{Rad} = \Gamma_T - \Gamma_M$.

46.4.1 Photonic Nanowire

The photonic nanowire (Claudon et al., 2013) is a vertical GaAs cylinder with an embedded QD sitting on a substrate. The geometry is illustrated in Figure 46.11a. Light is guided in the nanowire by total internal reflection at the nanowire-air interface. Single photons emitted by the QD are predominantly coupled to the fundamental HE_{11} waveguide mode as will be discussed in the following. The emission is again collected from an objective lens above the nanowire.

Similar to the micropillar SPS, a bottom mirror is required to avoid loss of light into the substrate. Furthermore, the small nanowire diameter leads to a highly divergent far-field emission pattern, and a top taper strategy is required to reduce this divergence. The reader is referred to (Claudon et al., 2013; Gregersen et al., 2013) for more details on the mirror and the taper strategies. In this tutorial, we will simply study spontaneous emission in the infinite nanowire of Figure 46.11b.

The physical origin of the strong suppression of the spontaneous emission into radiation modes can be understood from Figure 46.11c and d. We first consider a radiation mode defined from the illumination of an infinite photonic nanowire by a plane wave $E_x(\mathbf{r}) = E_0 \exp(ik_0 y)$ propagating along the y axis. The corresponding field intensity profile is shown in Figure 46.11c for a nanowire with diameter $d \ll \lambda$. Outside the nanowire we observe some field enhancement; however, inside the nanowire the field is strongly suppressed. This field suppression is due to a dielectric screening effect (Bleuse et al., 2011) illustrated in Figure 46.11d, which displays the refractive index profile $n(x)$ and the normalized electric field component $E_x(x)/E_0$ in along the $y = 0$ axis. Now, at the nanowire-air interface, continuity is dictated by Maxwell's equations for the component $D_x(x) = n(x)^2 E_x(x)$ of the displacement field normal to the boundary.

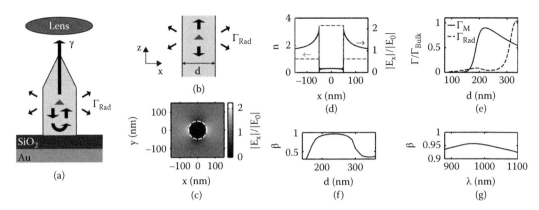

FIGURE 46.11 (a) The full photonic nanowire SPS geometry. (b) The infinite nanowire geometry. (c) The normalized E_x field amplitude of a plane wave propagating along the y axis and scattered by a nanowire of diameter $d = 100$ nm. The nanowire boundary is represented by the dashed white line. (d) Normalized E_x field amplitude (solid) and refractive index profile (dashed) at the position $y = 0$. (e) Spontaneous emission rates into the fundamental waveguide mode (solid) and into radiation modes (dashed) as function of nanowire diameter d. Spontaneous emission β factor (f) as function of d and (g) as function of λ for $d = 240$ nm. Other parameters: $\lambda = 950$ nm and $n_{GaAs} = 3.46$ unless otherwise noted. (Reprinted from Gregersen, N. et al., *IEEE Journal of Selected Topics in Quantum Electronics* © 2013 IEEE.)

Since the refractive index of GaAs is large, around 3.5, the E_x component inside the nanowire must be correspondingly suppressed. Indeed, an analytic expression for the spontaneous emission inside the nanowire is available in the electrostatic limit, where $d \ll \lambda$, and is given by

$$\frac{\Gamma}{\Gamma_{Bulk}} = \frac{\Gamma_{Rad}}{\Gamma_{Bulk}} = \frac{4}{n\left(n^2 + 1\right)^2}. \tag{46.27}$$

We observe that for GaAs the suppression is on the order of 150. The expression (46.27) is valid for a dipole oriented in the plane normal to the axis, which is the case for InAs QDs. However, for a dipole oriented along the z axis, the relevant boundary condition for the radiation modes at the nanowire-air interface is simply continuity of the E_z component. The mechanism ensuring suppression of the electric field is no longer present, and Γ_{Rad} thus remains substantial even when $d \ll \lambda$.

The normalized spontaneous emission rates Γ_M and Γ_{Rad} are shown in Figure 46.11e for finite nanowire diameters for an on-axis radially oriented dipole. For $d < 150$ nm, both rates are low. As d increases toward 200 nm, the nanowire becomes large enough to support a well-confined fundamental mode and Γ_M increases toward unity. However, for $d < 300$ nm, Γ_{Rad} remains below 0.1 Γ_{Bulk} due to the dielectric screening effect discussed above. The corresponding β factor is presented in Figure 46.11f, and we observe a maximum β of ~0.95 for a diameter of ~240 nm.

Importantly, this screening effect is not a resonant effect. The β factor as a function of wavelength is presented in Figure 46.11g, and we observe that $\beta > 0.93$ is obtained in a 100-nm range. The waveguide QED is thus broadband, and no spectral alignment between an emitter line and a narrow cavity is required. This represents a significant advantage in the fabrication and a major asset of the waveguide QED approach.

46.4.2 Photonic Crystal Waveguide

For integrated quantum photonics, in-plane emission of the emitted light into a planar waveguide is desired. A platform allowing for the efficient in-plane emission of single photons is the photonic crystal (PhC) membrane with an embedded QD (Yao, Manga Rao, and Hughes, 2010) as illustrated in Figure 46.12a. By

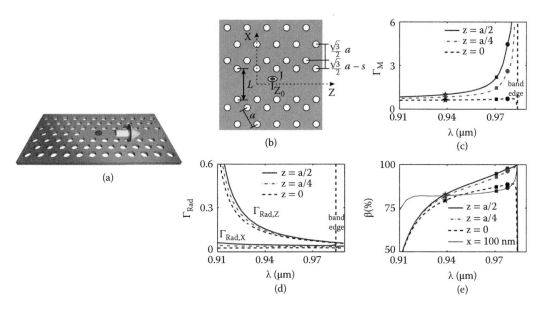

FIGURE 46.12 (a) The PhC waveguide SPS. (b) Geometry of the infinite periodic PhC waveguide. (c) Normalized SE decay rates Γ_M as function of λ. Normalized SE decay rates (d) into the radiation modes for the two in-plane dipole moment orientations X and Z. (e) Associated β factors. The thin solid curve is obtained for an off-axis dipole location ($x = 100$ nm and $z = a/2$). Circles, squares, and stars are reference marks locating the fundamental Bloch-mode group velocities equal to $c/20$, $c/10$, and $c/5$, respectively, where c is the speed of light. (Reprinted with permission from Lecamp, G. et al., *Physical Review Letters*, 99, 023902, 2007. Copyright 2007 by the American Physical Society.)

removing a row of holes, a waveguide can be formed. The waveguide confines light in the plane using the PhC bandgap and vertically using total internal reflection.

In the following, we study the spontaneous emission of the infinite periodic PhC waveguide illustrated in Figure 46.12b. In this tutorial, let us simply consider a dipole positioned in the center of the waveguide $x = 0$ at the position $z = a/2$ with hole displacement $s = 0$. The corresponding results are displayed in Figure 46.12c through 46.12e as full curves. The spontaneous emission rate Γ_M into the fundamental PhC waveguide mode for a dipole oriented along the x axis is shown in Figure 46.12c. For short λ, the rate is similar to that in a bulk material; however, near the band edge Γ_M diverges. This strong enhancement is due to the slow-light effect occurring near the band edge, where the group refractive index diverges (Baba, 2008).

The emission into radiation modes is illustrated in Figure 46.12d for dipoles aligned along the x and the z axes. $\Gamma_{Rad,X}$ is strongly suppressed in the entire wavelength regime due to the same dielectric screening effect discussed for the photonic nanowire. As for the photonic nanowire, this suppression does not take place for the dipole oriented along the z axis and $\Gamma_{Rad,Z}$ is small only near the band edge.

The average spontaneous emission β factor for dipoles oriented along the x and z axes can be written as

$$\beta = \frac{\Gamma_M}{\Gamma_M + \Gamma_{Rad,X} + \Gamma_{Rad,Z}} \tag{46.28}$$

and is depicted in Figure 46.12e. We observe that β approaches unity near the band edge thanks to the suppression of Γ_{Rad} combined with the enhanced Γ_M. The β factor remains larger than 0.75 over a 50 nm wavelength range, an effect predominantly due to the broadband suppression of the spontaneous emission into radiation modes. By displacing the QD x position, the range can be increased as illustrated by the thin solid curve for $x = 100$ nm in Figure 46.12e.

46.5 Planar Dielectric Antenna Approach

A final SPS design discussed in this tutorial is the planar dielectric antenna (PDA) illustrated in Figure 46.13a. This design approach (Chen, Götzinger, and Sandoghdar, 2011) differs conceptually from the previous QED strategies. Whereas the quantum emitter in the QED approaches couples to a specific optical cavity or waveguide mode with the coupling coefficient β, the quantum emitter in the PDA design instead couples to a continuous range of radiation modes and the collection efficiency ε is computed by integrating the contributions from this entire range.

We consider the geometry illustrated in the inset of Figure 46.13b, where a quantum emitter with horizontal electric dipole (HED) or vertical electric dipole (VED) moment orientation is placed in a material of refractive index n_2. In this geometry, light will be collected by an objective lens placed below the emitter.

Without any precautions, light is emitted in all directions due to symmetry, and it is necessary to implement a top mirror to reflect light propagating upward back toward the bottom. While a gold mirror can provide a high broadband reflectivity, coupling to surface plasmons reduces the reflectivity for certain angles of incidence (Chen, Götzinger, and Sandoghdar, 2011). However, the high reflectivity can be restored by introducing an intermediate low-index layer of index $n_3 < n_2$.

With the introduction of the metal mirror, the light is now propagating downward with a highly divergent far-field pattern. To reduce this divergence, a solid immersion lens of index $n_1 > n_2$ is introduced and the downward propagating field is now limited to a cone of polar angle $\theta = \sin^{-1}(n_2/n_1)$. The power density and the collection efficiency as function of the polar angle θ is shown in Figure 46.13b for the two dipole orientations. We observe that a collection efficiency above 0.99 is obtained for an angle $\theta = 55°$. The wavelength dependence on the efficiency is presented in Figure 46.13c, and we observe that ε > 0.99 is maintained over a wavelength range of several hundred nanometers for both emitter orientations. The inset of Figure 46.13c shows the influence of the spacer thickness s on the collection efficiency. As s decreases, the increased coupling of the vertical components of the electric field generated by a VED with surface plasmons reduces the efficiency. However, the field generated by the HED has much weaker vertical components; the coupling to surface plasmons is reduced and the efficiency for this dipole orientation thus remains high.

Whereas the PDA design features record-high broadband efficiency, the condition $n_1 > n_2$ imposes constraints on the choice of material for the medium 1 if the emitter is, e.g., a QD in GaAs with index $n_2 = 3.5$. This can be remedied by introducing an intermediate low-index buffer layer between media 1 and 2. However, an important currently unresolved issue for the PDA design remains, namely the highly non-Gaussian far-field emission pattern illustrated in Figure 46.13b, which makes efficient coupling of the emitted photons to the Gaussian mode profile of a single-mode fiber a challenge.

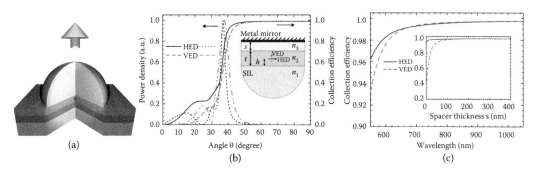

(a)   (b)   (c)

FIGURE 46.13   (a) Planar dielectric antenna. (b) Power density (left axis) and collection efficiency (right axis) versus emission/collection angle θ at a wavelength of 637 nm. The inset shows the schematics of the device with vertically and horizontally oriented dipole emitters. (c) Collection efficiency as a function of wavelength. The inset plots the collection efficiency versus spacer thickness s at 637 nm. (Reprinted from Chen, X,-W. et al., *Optics Letters* 36, 3545–3547, 2011. With permission of Optical Society of America.)

46.6 Decoherence Effects

As discussed above, as well as producing single photons and having a high extraction efficiency, an ideal SPS would produce photons that are quantum mechanically indistinguishable. With reference to Figure 46.2, when two indistinguishable photons are incident on a 50–50 beam splitter they emerge in the same output arm, which provides the fundamental quantum interference necessary for quantum information applications. For photons that are not completely indistinguishable, there remains a nonzero probability that a single photon emerges in both output arms. Based on these arguments, the indistinguishability of a SPS can be quantified by Kaer et al. (2013):

$$I = \frac{\int_0^\infty \int_0^\infty \left|\left\langle \widehat{A}^\dagger(t+\tau)\widehat{A}(t)\right\rangle\right|^2 dt d\tau}{\int_0^\infty \int_0^\infty \left\langle \widehat{A}^\dagger(t)\widehat{A}(t)\right\rangle \left\langle \widehat{A}^\dagger(t+\tau)\widehat{A}(t+\tau)\right\rangle dt d\tau}, \tag{46.29}$$

where $I = 1$ for perfectly indistinguishable photons and $I = 0$ for distinguishable photons. The operators $\widehat{A}^\dagger$ and $\widehat{A}$ are the creation and annihilation operators for the electric field incident on the beam splitter. Further details of the derivation leading to Equation 46.29 can be found in Unsleber et al. (2015). The challenge in modeling the indistinguishability of photons emitted from a semiconductor-based SPS lies in (a) relating the correlation function and expectation values in Equation 46.29 to operators pertaining to the quantum emitter, and (b) ensuring that the model of the emitter itself captures all important processes that affect the photon indistinguishability.

46.6.1 Quantum Emitter Master Equation

To gain some insight, we first consider a QD emitting in a bulk material or a very low-Q cavity. In this case, provided the emitter is sufficiently far from the beam splitter optics, the field operators in Equation 46.29 can be replaced with operators describing the emitter. We write

$$\widehat{A}(t) \sim \widehat{\sigma}(t), \tag{46.30}$$

where $\widehat{\sigma} = |g\rangle\langle e|$ annihilates an excitation of the emitter (Kiraz et al., 2004). The proportionality constants that we have omitted are independent of time, and with reference to Equation 46.29 we see that they will cancel when the indistinguishability is calculated.

Our task now is to calculate the correlation function in the numerator of Equation 46.29, which can be achieved with help from the quantum regression theorem. To begin, we must first write down a master equation describing the dynamics of the quantum emitter, which takes the form of an equation of motion for the density operator $\widehat{\rho}$ (Carmichael, 1999). This can be done to various levels of rigor, but in order to capture the most important qualitative features it suffices to write

$$\frac{d}{dt}\widehat{\rho} = \Gamma_{\mathrm{Bulk}} L_{\widehat{\sigma}}(\widehat{\rho}) + 2\gamma^* L_{\widehat{\sigma}^\dagger\widehat{\sigma}}(\widehat{\rho}), \tag{46.31}$$

where the Lindblad operators are defined to satisfy $L_{\widehat{O}}(\widehat{\rho}) = \widehat{O}\widehat{\rho}\widehat{O}^\dagger - \left(\widehat{O}^\dagger\widehat{O}\widehat{\rho} + \widehat{\rho}\widehat{O}^\dagger\widehat{O}\right)/2$ and Γ_{Bulk} is the bulk spontaneous emission rate. The first term on the right side of Equation 46.31 describes the decay of the excited state while the second term describes the decoherence of the two-level system. The quantity γ^* is the dephasing or decoherence rate, which we will discuss further in the following sections. Equation 46.31

allows us to establish equations of the motion for expectation values of the emitter. The dipole expectation value, e.g., obeys

$$\frac{d}{dt}\langle\hat{\sigma}\rangle = -\frac{1}{2}\left(\Gamma_{\text{Bulk}} + 2\gamma^*\right)\langle\hat{\sigma}\rangle. \tag{46.32}$$

Armed with Equation 46.31, we can now invoke the quantum regression theorem, which states that the correlation function in Equation 46.29 obeys the same equation of motion as the dipole operator in Equation 46.32 but with respect to τ, namely $\frac{d}{d\tau}\langle\hat{\sigma}^\dagger(t+\tau)\hat{\sigma}(t)\rangle = -\frac{1}{2}\left(\Gamma_{\text{Bulk}} + 2\gamma^*\right)\langle\hat{\sigma}^\dagger(t+\tau)\hat{\sigma}(t)\rangle$. The initial condition is found by setting $\tau = 0$, from which we find $\langle\hat{\sigma}^\dagger(t)\hat{\sigma}(t)\rangle = e^{-\Gamma_{\text{Bulk}}t}$ if we assume the emitter is initially in its excited state at time $t = 0$. This allows us to find the correlation function, and inserting this solution into Equation 46.29 we find that the indistinguishability is given by Kaer et al. (2013):

$$I = \frac{\Gamma_{\text{Bulk}}}{\Gamma_{\text{Bulk}} + 2\gamma^*}. \tag{46.33}$$

This expression demonstrates that for our simplistic model, the indistinguishability depends solely on the ratio, $\Gamma_{\text{Bulk}}/\gamma^*$ that we should seek to maximize.

46.6.2 Emitter-Cavity Master Equation

So far we have considered a QD emitting into a bulk medium. However, using similar techniques to those above it is also possible to calculate the indistinguishability of photons emitted from a coupled emitter-cavity system. The calculation is performed by replacing the master equation in Equation 46.31 with the more complicated expression (Kaer et al., 2013)

$$\frac{d}{dt}\hat{\rho} = -i\left[\hat{H}, \hat{\rho}\right] + \Gamma_{\text{Bulk}}L_{\hat{\sigma}}(\hat{\rho}) + 2\gamma^* L_{\hat{\sigma}^\dagger\hat{\sigma}}(\hat{\rho}) + \kappa L_{\hat{a}}(\hat{\rho}). \tag{46.34}$$

Here, $\hat{H} = g\left(\hat{\sigma}\hat{a}^\dagger + \hat{a}\hat{\sigma}^\dagger\right)$ describes the interaction between the emitter and the cavity with g the emitter-cavity coupling strength and $\hat{a}$ the annihilation operator of the cavity mode. The last term in Equation 46.34 describes the outcoupling of light from the cavity with κ the cavity mode decay rate. Broadly speaking, provided the losses in the system are weaker than the emitter-cavity coupling strength, the cavity mode can be adiabatically eliminated from this equation of motion. For details of this procedure, we refer the reader to Kaer et al. (2013). This results in a master equation describing the emitter only, which is equal to Equation 46.31 but with Γ_{Bulk} replaced with $\Gamma = \Gamma_{\text{Bulk}} + \Gamma_{\text{M}}$ with $\Gamma_{\text{M}} = \Gamma_{\text{Bulk}}F_{\text{P}}$ the emission rate into the cavity, and the Purcell factor now given by

$$F_{\text{P}} = \frac{4g^2}{\kappa\Gamma_{\text{Bulk}}}. \tag{46.35}$$

We note that the Purcell factor given here is in fact equivalent to that given in Equation 46.20. The apparent difference is a matter of notation, with Equation 46.35 being preferred in the present context since it makes explicit the dependence on the emitter-cavity coupling strength g. With this modification, the indistinguishability measure for a weakly coupled emitter-cavity system becomes

$$I = \frac{\left(F_{\text{P}} + 1\right)\Gamma_{\text{Bulk}}}{\left(F_{\text{P}} + 1\right)\Gamma_{\text{Bulk}} + 2\gamma^*}. \tag{46.36}$$

This expression encompasses the two main strategies used to increase photon indistinguishability. First, as one might expect, eliminating sources of decoherence, and thus reducing γ^*, serves to increase I. Methods to achieve this are briefly discussed in the following section. Second, placing the emitter in a photonic structure such as a micropillar can give rise to a Purcell enhancement of the spontaneous emission rate, which will also increase the indistinguishability. In either case, the strategy amounts to extracting the single photon on a timescale faster than the decoherence rate. We note, however, that these simple arguments only apply as long as the dephasing rate γ^* and the Purcell factor are independent parameters. As we will see below, dephasing caused by coupling to phonons can increase as the Purcell factor is increased, meaning that increasing photon indistinguishability by increasing the spontaneous emission rate may not always be advisable.

46.6.3 Spectral Wandering

Having introduced the basic theory behind photon indistinguishability, we shall now discuss two main mechanisms of dephasing in semiconductor emitters. We begin with spectral wandering, which refers to the process by which the spectral emission line of a QD moves (wanders) over time (Berthelot et al., 2006). In such a case, successively emitted photons may have different frequencies and therefore cannot be considered indistinguishable. Many processes in semiconductor systems can give rise to spectral wandering, with the amplitude and frequency distribution being dependent on the nature and timescale of the process. The most common process in charge-neutral QDs is spectral diffusion caused by the temporary trapping of charges in nearby sites, which induces a change of the electrostatic environment of the QD.

In order to model spectral diffusion, it is typically sufficient to include a phenomenological dephasing term in the master equation describing the QD degrees of freedom as in Equation 46.31 above. As we have seen, dephasing of this form can be combated by increasing the Purcell effect. Recalling also that the efficiency of a source increases with the Purcell effect, we conclude that in the case of simple phenomenological dephasing, such as that caused by spectral wandering, an increased Purcell factor is always advantageous.

46.6.4 Coupling to Longitudinal Acoustic Phonons

Coupling to phonons, present in any semiconductor sample, represents an additional source of decoherence. It has been demonstrated that the dominant phonon-mediated dephasing mechanism is due to coupling of the QD exciton to longitudinal acoustic phonons via the deformation potential (Ramsay et al., 2010). Phonons deform the QD structure and give rise to fluctuations in the exciton energy, which in turn cause a dephasing of superpositions of the ground and excited state. As such phonons are sensitive to the QD dressed state energy difference, which itself depends on the strength of any QD-cavity coupling g. Therefore, in contrast to spectral wandering, phonons cannot be accurately included by the addition of a simple phenomenological dephasing rate that is independent of other parameters.

For these reasons, correctly incorporating phonon coupling into the description of a QD-based SPS is particularly challenging. One approach is to include the exciton-phonon coupling at a Hamiltonian level, where one essentially takes a large but finite number of harmonic oscillators to model the phonon environment. The dynamics describing the complete system can then be solved numerically, giving results that are exact to within a set numerical tolerance (Kaer et al., 2012, 2013). This method, however, requires a large amount of computing power and can obscure the underlying physical processes. Another less numerically demanding method involves deriving a time-local master equation describing the emitter-cavity system using projection operators. This results in more tractable expression from which the indistinguishability can be calculated, though is valid only to second order in the exciton–phonon coupling strength. A comparison of these methods and a simple phenomenological approach can be found in Kaer et al. (2013).

Here we will adopt a simple model that manages to capture the salient qualitative features while being simple to implement. We replace the constant dephasing rate γ^* in Equation 46.34 with the expression (Ramsay et al., 2010)

$$\gamma_{\mathrm{ph}} = \frac{\pi}{4} J(2g) \coth\left(\frac{g}{k_{\mathrm{B}} T}\right), \tag{46.37}$$

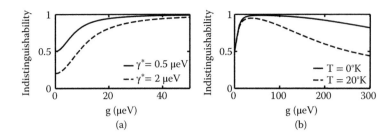

FIGURE 46.14 (a) Photon indistinguishability as a function of emitter-cavity coupling strength for a simple constant dephasing rate of $\gamma^* = 0.5\,\mu\text{eV}$ (solid) and $\gamma^* = 2\,\mu\text{eV}$ (dashed). (b) A similar plot where the dephasing rate is composed of a constant value of $\gamma^* = 0.5\,\mu\text{eV}$ plus a phonon-induced contribution given by Equation 46.37 for temperatures of $T = 0$ K (solid) and $T = 20$ K (dashed). Other parameters are $\Gamma_{\text{Bulk}} = 1\,\mu\text{eV}$, $\kappa = 100\,\mu\text{eV}$, $\eta = 0.032$ ps^2 and $\omega_c = 0.95\,\mu\text{eV}$.

where T is the sample temperature and the function $J(\omega)$ is the exciton-phonon spectral density. For typical GaAs QDs, the spectral density takes the form

$$J(\omega) = \eta\omega^3 e^{-\left(\frac{\omega}{\omega_c}\right)^2},\tag{46.38}$$

where η is a quantity describing the overall strength of the interaction and can be determined from material parameters, while ω_c is a characteristic cutoff frequency that is inversely proportional to the linear dimension of the QD size (Ramsay et al., 2010). Using Equation 46.37 in Equation 46.34 and again invoking the quantum regression theorem, Equation 46.29 can be evaluated to explore the indistinguishability of photons emitted from a coupled QD-cavity system subject to dephasing caused by phonons.

One final consideration is the correct operator to insert into Equation 46.29. Since in the majority of this chapter we are interested in the emission via the cavity mode, if this mode cannot be adiabatically eliminated it is more appropriate to use the cavity operator $\hat{a}$ rather than the emitter operator in Equation 46.29. In doing so, we are explicitly considering the indistinguishability of photons emitted into the cavity mode of interest.

In Figure 46.14a, we show the photon indistinguishability as a function of the emitter-cavity coupling strength g for a constant dephasing rate of $\gamma^* = 0.5\,\mu\text{eV}$ (solid) and $\gamma^* = 2\,\mu\text{eV}$ (dashed). For $g = 0$, the indistinguishability is given by its value in bulk (Equation 46.33). As g is increased, the coupling of the emitter to the cavity gives rise to an enhancement of the Purcell effect that accelerates the spontaneous emission, thus increasing the indistinguishability. Figure 46.14b shows a similar plot where we now include a constant dephasing rate $\gamma^* = 0.5\,\mu\text{eV}$ and also a phonon contribution given by Equation 46.37. For small values of g, we again observe an increase in I due to the Purcell enhancement. However, as g is increased further, the phonon contribution to the dephasing increases, which serves to reduce the indistinguishability. These plots demonstrate the subtle interplay between the beneficial effects of the Purcell enhancement and the detrimental effects of increased coupling to phonons. The reader is referred to (Kaer et al., 2012, 2013; Kaer and Mørk, 2014) for a more extensive discussion of these effects.

46.7 Summary

In this chapter, we have presented the key figures of merit for the single-photon source. For applications in quantum information processing, clean single-photon emission, efficient photon extraction, and indistinguishable photon emission are required. Sources based on spontaneous parametric down-conversion or attenuation are simple to implement but suffer inherently from low efficiency. We have presented the two-level system with focus on the semiconductor quantum dot as an attractive platform for realizing a deterministic highly efficient single-photon source. Design strategies for controlling the light emission and ensuring a near-unity coupling efficiency based on cavity and waveguide quantum electrodynamics approaches as well as the planar dielectric antenna have been presented. We have discussed the theory and major numerical simulation techniques for modeling the light emission. Finally, we have discussed the influence of several decoherence processes arising from the solid-state environment including simple pure dephasing mechanisms such as charge fluctuations and the more advanced phonon-induced decoherence process due to interaction with longitudinal acoustic phonons.

While, in recent years, tremendous progress has been witnessed towards achieving efficient and pure emission of indistinguishable photons, the simultaneous realization of near-unity efficiency and indistinguishability remains elusive. Fundamental challenges persist both on the design level and on the fabrication level making single-photon sources an interesting research topic most likely for many years to come.

References

Baba, T. 2008. Slow light in photonic crystals. *Nature Photonics* 2 (8): 465–473. doi:10.1038/nphoton.2008.146.

Babinec, T. M., B. J. M. Hausmann, M. Khan, Y. Zhang, J. R. Maze, P. R. Hemmer, and M. Lončar. 2010. A diamond nanowire single-photon source. *Nature Nanotechnology* 5 (3): 195–199. doi:10.1038/nnano.2010.6.

Balanis, C. A. 1989. *Advanced Engineering Electromagnetics*. 1st ed. Hoboken, NJ: Wiley.

Berthelot, A., I. Favero, G. Cassabois, C. Voisin, C. Delalande, P. Roussignol, R. Ferreira, and J-M. Gérard. 2006. Unconventional motional narrowing in the optical spectrum of a semiconductor quantum dot. *Nature Physics* 2 (11): 759–764. doi:10.1038/nphys433. http://arxiv.org/abs/cond-mat/0610346/nhttp://www.nature.com/doifinder/10.1038/nphys433.

Bleuse, J., J. Claudon, M. Creasey, N. S. Malik, J-M. Gérard, I. Maksymov, J-P. Hugonin, and P. Lalanne. 2011. Inhibition, enhancement, and control of spontaneous emission in photonic nanowires. *Physical Review Letters* 106 (10): 103601. doi:10.1103/PhysRevLett.106.103601.

Brendel, J., N. Gisin, W. Tittel, and H. Zbinden. 1999. Pulsed energy-time entangled twin-photon source for quantum communication. *Physical Review Letters* 82 (12): 2594–2597. doi:10.1103/PhysRevLett.82.2594. http://journals.aps.org/prl/abstract/10.1103/PhysRevLett.82.2594.

Buttler, W. T., R. J. Hughes, S. K. Lamoreaux, G. L. Morgan, J. E. Nordholt, and C. G. Peterson. 2000. Daylight quantum key distribution over 1.6 Km. *Physical Review Letters* 84 (24): 5652–5655.

Carmichael, H. J. 1999. *Statistical Methods in Quantum Optics*. 1st ed. Berlin-Heidelberg: Springer.

Chen, X-W., S. Götzinger, and V. Sandoghdar. 2011. 99% efficiency in collecting photons from a single emitter. *Optics Letters* 36 (18): 3545–3547. doi:10.1364/OL.36.003545. http://ol.osa.org/abstract.cfm?URI=ol-36-18-3545.

Claudon, J., J. Bleuse, N. S. Malik, M. Bazin, P. Jaffrennou, N. Gregersen, C. Sauvan, P. Lalanne, and J-M. Gérard. 2010. A highly efficient single-photon source based on a quantum dot in a photonic nanowire. *Nature Photonics* 4 (3): 174–177. doi:10.1038/nphoton.2009.287. http://dx.doi.org/10.1038/nphoton.2009.287.

Claudon, J., N. Gregersen, P. Lalanne, and J-M. Gérard. 2013. Harnessing light with photonic nanowires: Fundamentals and applications to quantum optics. *ChemPhysChem* 14 (11): 2393–2402. doi:10.1002/cphc.201300033.

Coldren, L. A., S. W. Corzine, and M. L. Mashanovitch. 2012. *Diode Lasers and Photonic Integrated Circuits*. 2nd ed. Hoboken, NJ: Wiley.

Ding, X., Y. He, Z.-C. Duan, N. Gregersen, M.-C. Chen, S. Unsleber, S. Maier, et al. 2016. On-demand single photons with high extraction efficiency and near-unity indistinguishability from a resonantly driven quantum dot in a micropillar. *Physical Review Letters* 116 (2): 020401. doi:10.1103/PhysRevLett.116.020401. http://arxiv.org/abs/1601.00284.

Gazzano, O., S. Michaelis de Vasconcellos, C. Arnold, A. Nowak, E. Galopin, I. Sagnes, L. Lanco, A. Lemaître, and P. Senellart. 2013. Bright solid-state sources of indistinguishable single photons. *Nature Communications* 4: 1425. doi:10.1038/ncomms2434. http://www.ncbi.nlm.nih.gov/pubmed/23385570.

Gérard, J.-M. 2003. Solid-state cavity-quantum electrodynamics with self-assembled quantum dots. In *Single Quantum Dots, Topics in Applied Physics*, vol. 90, edited by P. Michler, 269–315. Berlin, Heidelberg: Springer-Verlag. doi:10.1007/978-3-540-39180-77. http://www.springerlink.com/index/4nn94hfbm1gpmyxq.pdf/nhttp://www.springerlink.com/content/4NN94HFBM1GPMYXQ.

Gerry, C. C. and P. L. Knight. 2008. *Introductory Quantum Optics*. New York, NY: Cambridge University Press.

Gregersen, N., P. Kaer, and J. Mørk. 2013. Modeling and design of high-efficiency single-photon sources. *IEEE Journal of Selected Topics in Quantum Electronics* 19 (5): 9000516. doi:10.1109/JSTQE.2013.2255265. http://ieeexplore.ieee.org/lpdocs/epic03/wrapper.htm?arnumber=6493378.

Hong, C. K., Z. Y. Ou, and L. Mandel. 1987. Measurement of subpicosecond time intervals between two photons by interference. *Physical Review Letters* 59 (18): 2044–2046. doi:10.1103/PhysRevLett.59.2044. http://journals.aps.org.globalproxy.cvt.dk/prl/abstract/10.1103/PhysRevLett.59.2044.

JCMWave. n.d. JCMSuite. http://www.jcmwave.com.

Kaer, P., N. Gregersen, and J. Mørk. 2013. The role of phonon scattering in the indistinguishability of photons emitted from semiconductor cavity QED systems. *New Journal of Physics* 15: 035027. doi:10.1088/1367-2630/15/3/035027.

Kaer, P. and J. Mørk. 2014. Decoherence in semiconductor cavity QED systems due to phonon couplings. *Physical Review B* 90 (3): 035312. doi:10.1103/PhysRevB.90.035312. http://journals.aps.org/prb/abstract/10.1103/PhysRevB.90.035312.

Kaer, P., T. R. Nielsen, P. Lodahl, A.-P. Jauho, and J. Mørk. 2012. Microscopic theory of phonon-induced effects on semiconductor quantum dot decay dynamics in cavity QED. *Physical Review B* 86 (8): 085302. doi:10.1103/PhysRevB.86.085302.

Kiraz, A., M. Atatüre, and A. Imamoğlu. 2004. Quantum-dot single-photon sources: Prospects for applications in linear optics quantum-information processing. *Physical Review A* 69 (3): 032305. doi:10.1103/PhysRevA.69.032305. http://journals.aps.org/pra/abstract/10.1103/PhysRevA.69.032305.

Kristensen, P. T., C. Van Vlack, and S. Hughes. 2012. Generalized effective mode volume for leaky optical cavities. *Optics Letters* 37 (10): 1649–1651. doi:10.1364/OL.37.001649.

Lavrinenko, A. V., J. Lægsgaard, N. Gregersen, F. Schmidt, and T. Søndergaard. 2014. *Numerical Methods in Photonics*. Boca Raton, FL: CRC Press. http://www.crcpress.com/product/isbn/9781466563889.

Lecamp, G., P. Lalanne, and J. P. Hugonin. 2007. Very large spontaneous-emission β factors in photonic-crystal waveguides. *Physical Review Letters* 99 (2): 023902. doi:10.1103/PhysRevLett.99.023902.

Lorke, M., T. Suhr, N. Gregersen, and J. Mørk. 2013. Theory of nanolaser devices: Rate equation analysis versus microscopic theory. *Physical Review B* 87 (20): 205310. doi:10.1103/PhysRevB.87.205310.

Lumerical. n.d. FDTD Solutions. http://www.lumerical.com/.

Novotny, L. and B. Hecht. 2012. *Principles of Nano-Optics*. 2nd ed. New York, NY: Cambridge University Press.

Purcell, E. M. 1946. Spontaneous emission probabilities at radio frequencies. *Physical Review* 69: 681.

Ramsay, A. J., A. V. Gopal, E. M. Gauger, A. Nazir, B. W. Lovett, A. M. Fox, and M. S. Skolnick. 2010. Damping of exciton rabi rotations by acoustic phonons in optically excited InGaAs/GaAs quantum dots. *Physical Review Letters* 104 (1): 017402. doi:10.1103/PhysRevLett.104.017402. http://journals.aps.org/prl/abstract/10.1103/PhysRevLett.104.017402.

Reitzenstein, S. and A. Forchel. 2010. Quantum dot micropillars. *Journal of Physics D: Applied Physics* 43 (3): 033001. doi:10.1088/0022-3727/43/3/033001.

Shields, A. J. 2007. Semiconductor quantum light sources. *Nature Photonics* 1 (4): 215–223. doi:10.1038/nphoton.2007.46. http://www.nature.com/nphoton/journal/v1/n4/full/nphoton.2007.46.html.

Siyushev, P., G. Stein, J. Wrachtrup, and I. Gerhardt. 2014. Molecular photons interfaced with alkali atoms. *Nature* 509 (7498): 66–70. doi:10.1038/nature13191. http://www.ncbi.nlm.nih.gov/pubmed/24784217.

Somaschi, N., V. Giesz, L. De Santis, J. C. Loredo, M. P. Almeida, G. Hornecker, S. L. Portalupi, et al. 2016. Near-optimal single-photon sources in the solid state. *Nature Photonics* 10 (5): 340–345. doi:10.1038/nphoton.2016.23. http://dx.doi.org/10.1038/nphoton.2016.23.

Unsleber, S., D. P. S. McCutcheon, M. Dambach, M. Lermer, N. Gregersen, S. Höfling, J. Mørk, C. Schneider, and M. Kamp. 2015. Two-photon interference from a quantum dot microcavity: Persistent pure dephasing and suppression of time jitter. *Physical Review B* 91 (7): 075413. doi:10.1103/PhysRevB.91.075413.

Yao, P., V. S. C. Manga Rao, and S. Hughes. 2010. On-chip single photon sources using planar photonic crystals and single quantum dots. *Laser & Photonics Reviews* 4 (4): 499–516. doi:10.1002/lpor.200810081. http://doi.wiley.com/10.1002/lpor.200810081.

47

Nanoplasmonic Lasers and Spasers

A Freddie Page

and

Ortwin Hess

47.1 Introduction

Semiconductor-based electronics has, in recent years, enjoyed a miniaturization trend down to structure sizes on nanometer scales. However, corresponding photonic devices, such as, in particular, lasers, have not been able to follow suit with a similarly dramatic pace [1]. This is because the generation of photons in small volumes appeared challenging due to the diffraction limit, which apparently seemed to set the minimum wavelength of light to be approximately inversely proportional to its frequency. Considering typical semiconductor band gaps with associated frequencies, this would imply that we would have a lower limit of optical devices and lasers of typically around 1 µm. Yet the availability of nanoscale lasers would be a true starting point for on-chip optical circuits, with applications in telecommunications, data storage, optical computing and sensing. So, can we shrink lasers to be much smaller than the wavelength of the light they emit, particularly down to the nanoscale?

Let us recall that lasers operate on the basis of two features: feedback and gain. In general, a cavity provides feedback by supporting resonant lasing modes (closed loops where light constructively interferes with itself on the round trip). On each loop, the gain medium amplifies the light, and when the amplification is greater than the threshold of energy lost (to dissipation in the material or by escaping the cavity), the system can enter a lasing regime. The physical size of a laser is thus dependent on the cavity supporting lasing modes and this, in turn, is limited by diffraction. Metallic cavities, on the other hand, allow for the circumvention of this diffraction limit by confining light in the subwavelength surface plasmon modes [2,3]. On a metal-insulator interface, light couples to the oscillations of surface charges as *surface plasmon polariton* (SPP) modes. But how does this help? There are two primary features of SPPs that make them useful in nano-optics applications: First, SPPs break the diffraction limit, i.e., they have wavevectors larger than what is generally allowed for a photon of a given frequency. Second, SPPs have or are associated with large field enhancements at the material interface. This combination of shorter wavelengths and higher field densities then significantly enhances the interaction of light with the electronic systems of active media

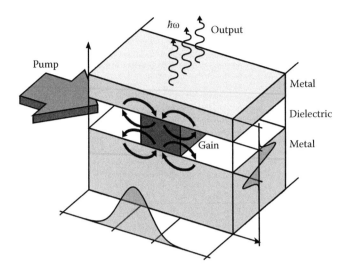

FIGURE 47.1 Nanoplasmonic laser illustration. Light is trapped in a metallic structure such that it forms a closed path (black arrows) that forms in overlap with a gain medium (dark box). In this setup, the gain is pumped from the side and emits from the top. A field strength profile is also shown superimposed on the structure.

[4,5], particularly those interactions that are usually weak. Having subwavelength confinement is, however, frequently traded off to a certain degree with not insignificant dissipative material losses in the metallic components, which requires a corresponding round-trip gain in order to be compensated [6].

Lasers that operate below the diffraction limit on nanometer scales by using metallic cavities are hence called nanoplasmonic lasers. A schematic illustration is given in Figure 47.1. Here there are two options for the light emission: Generated light can be channeled into photonic modes, as in a conventional laser, but can also generate surface plasmons, in which case the laser is more aptly described as a *spaser* for "surface plasmon amplification by stimulated emission of radiation" [7]. Indeed, the possibility to make nanoplasmonic lasers has been explored in various configurations such as, nanorod [8], double-fishnet [9], whispering gallery [10], and stopped light lasers [11].

This chapter will outline a range of techniques that enable the design, optimization, and simulation of nanoplasmonic lasers while tracking the example of a stopped light laser to outline concepts.

47.2 Models for Metals and Gain

47.2.1 Drude–Lorentz Model

In the following, we will chart our theoretical basis for describing the optical response of both metals (which provide subwavelength confinement) and semiconductor gain media (which provides the amplification in a nanoplasmonic laser), involving a combination of a Drude–Lorentz model and a multilevel Maxwell–Bloch theory, respectively, in the harmonic oscillator form. The Drude–Lorentz model describes the oscillations of charges in response to an applied electric field, while a Maxwell–Bloch approach is used to describe the behavior of dipole transitions such as in a four-level system model of a semiconductor.

In its most general terms, the Drude–Lorentz model relates the time evolution of the polarization of a material, **P**, to a driving electric field, **E**, by a second-order differential equation:

$$\frac{\partial^2 \mathbf{P}}{\partial t^2} + \gamma_L \frac{\partial \mathbf{P}}{\partial t} + \omega_L^2 \mathbf{P} = \omega_p^2 \varepsilon_0 \mathbf{E} \, , \tag{47.1}$$

where γ_{L} is a damping term, ω_{L} is the restoring force frequency, and ω_{p} is the plasma frequency—the coupling strength to the driving $\mathbf{E}$ field. It is a causal response that captures the behavior of most resonance phenomena. In the frequency domain, it is expressed as

$$\tilde{\mathbf{P}} = \varepsilon_0 \frac{\omega_{\mathrm{p}}^2}{\omega_{\mathrm{L}}^2 - \omega(\omega + i\gamma_{\mathrm{L}})} \tilde{\mathbf{E}} \,, \tag{47.2}$$

where the Drude–Lorentz susceptibility can be extracted:

$$\chi_{\mathrm{L}}(\omega) = \frac{\omega_{\mathrm{p}}^2}{\omega_{\mathrm{L}}^2 - \omega(\omega + i\gamma_{\mathrm{L}})} \,. \tag{47.3}$$

For $\gamma_{\mathrm{L}} > 0$, this obeys the Kramers–Kronig relations relating real and imaginary parts as required by causality.

The Drude–Lorentz model describes noble metals well, and one may explain the origins of each of the terms with a simple model for a free electron gas. For each electron, let the displacement from its nucleus be given as $\mathbf{x}(t)$. A Newtonian force equation can be built assuming the electron acts as a free particle under the Lorentz force law of an external field $\mathbf{E}$, i.e.,

$$m^* \frac{\partial^2 \mathbf{x}}{\partial t^2} = -e\mathbf{E} \,, \tag{47.4}$$

where m^* is the effective mass of an electron and e is its charge. Additional forces can be included, such as a collision term proportional to the velocity and inversely to an average collision time $F_{\mathrm{col}} = -\frac{\partial \mathbf{x}}{\partial t}/\tau$. In this free electron picture, there is no restoring force as for metals, electrons are not bound to their lattice sites. The macroscopic polarization can be related to these electron displacements, as each displacement induces an electronic dipole moment, and these can be averaged to become the polarization by multiplication by the electron number density, $\mathbf{P} = -en\mathbf{x}$. A comparison of terms allows for the associating:

$$\omega_{\mathrm{p}}^2 \to \frac{ne^2}{m^*} \,, \ \gamma \to \tau^{-1} \,, \ \omega_{\mathrm{L}} \to 0 \,. \tag{47.5}$$

This is the Drude model, and with an additional static background permittivity $\varepsilon_{\mathrm{bg}}$, it describes metals well.

Real materials typically have multiple resonances, which can be incorporated by adding multiple Drude–Lorentz resonances:

$$\chi = \sum_i \frac{\omega_{\mathrm{p}i}^2}{\omega_{\mathrm{L}i}^2 - \omega(\omega + i\gamma_{\mathrm{L}i})} \,. \tag{47.6}$$

These can have a variety of origins, i.e., in addition to the plasma mode discussed, rotational modes, electronic transitions, etc., and indeed the magnetic response, χ_{m}, can also be described as a sum of Drude–Lorentz resonances.

47.2.2 Multilevel Maxwell–Bloch Theory

A Drude–Lorentz (harmonic oscillator) type system can also model the electronic transitions between two levels of a semiconductor system, with an energy difference $\hbar\omega_{\mathrm{e}}$. In this picture, there is, however, a field-dependent occupation density of upper and lower levels. Light interacts with this *two-level system* by stimulated emission and absorption processes. When the lower level is more occupied than the upper level, photons are absorbed and electromagnetic energy is dissipated within the system; however, when the reverse is true, photons can be emitted, stimulated by other photons, adding gain to the system. This dissipative/amplifying response is accompanied by a corresponding refractive response. The strength of the resonance is proportional to the *inversion density* of the two levels, i.e., how much more the higher level is occupied than the lower one, $\Delta N = N_2 - N_1$, such that a layer with emitters embedded may be represented by

$$\varepsilon = \varepsilon_{bg} + \frac{-\Delta N \omega_{pe}^2}{\omega_e^2 - \omega(\omega + i\gamma_e)}, \tag{47.7}$$

$$\omega_{pe}^2 = \sqrt{\varepsilon_{bg}} \gamma_e \sigma_e c, \tag{47.8}$$

where ε_{bg} is the permittivity of the layer hosting the resonance, σ_e is the emission cross-section, and γ_e is the width of the resonance [6]. Note that for negative inversion, the emitter becomes an absorber as there is a higher density of emitters in the lower state. In time, the inversion varies as light is emitted or absorbed. The auxiliary equations to take this into account are

$$\frac{\partial N_1}{\partial t} = -\frac{1}{\hbar \omega_e} \left(\frac{\partial \mathbf{P}_e}{\partial t} + \frac{\gamma_e}{2} \mathbf{P}_e \right) \cdot \mathbf{E} + \frac{N_2}{\tau_{21}}, \tag{47.9}$$

$$\frac{\partial N_2}{\partial t} = +\frac{1}{\hbar \omega_e} \left(\frac{\partial \mathbf{P}_e}{\partial t} + \frac{\gamma_e}{2} \mathbf{P}_e \right) \cdot \mathbf{E} - \frac{N_2}{\tau_{21}}, \tag{47.10}$$

with the first terms representing stimulated emission, and the final ones to account for non-radiative recombination at a rate τ_{21}^{-1}.

The two-level system is appropriate for modeling a single electronic transition; however, there is no way in which a two-level system can reach a state of inversion by relying solely on the processes of spontaneous emission and absorption—the best that can be achieved is equal occupation of the levels. A *four-level system*, on the other hand, can be constructed such that there can be a dynamically maintained population inversion between two of its levels, from where stimulated emission can occur. The construction contains two two-level systems, labeled e and a for emission and absorption; e with energy levels 1 and 2 and energy gap $\hbar \omega_e$, in-between a with energy levels 0 and 3 and greater energy gap $\hbar \omega_a$, as depicted in Figure 47.2. The two upper levels, 2 and 3, are coupled by a fast non-radiative relaxation channel, with rate τ_{32}^{-1}, as are the two lower levels, 0 and 1 (τ_{10}^{-1}). This has the effect of rapidly depleting the first and third levels shortly

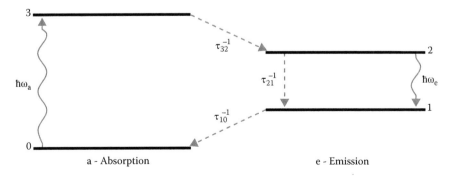

FIGURE 47.2 Schematic of the four-level system. A four-level system comprises two two-level systems, a and e, with active transitions (0 ↔ 3) and (1 ↔ 2), respectively, which are coupled by nonradioactive transition rates τ_{32}^{-1} and τ_{10}^{-1}. In this scheme, the emission subsystem permits radiative transitions at an energy of $\hbar \omega$ and has a slow nonradiative recombination rate τ_{21}^{-1}.

after they become occupied. Between the levels of the emission two-level subsystem, there is additionally a (slow) nonradiative channel (τ_{21}^{-1}).

The key point of a four-level system is that electrons that are pumped from levels 0 and 3, will quickly decay to level 2, leading to an inversion density of level 2 over level 1, which is available for stimulated emission. This allows the optical pumping between levels 0 and 3 to generate inversion between levels 1 and 2.

Four-level systems are incorporated into time-domain simulations using the differential equation for the polarization given in Equation 47.1 for each transition:

$$\frac{\partial^2 \mathbf{P}_a}{\partial t^2} + \gamma_a \frac{\partial \mathbf{P}_a}{\partial t} + \omega_a^2 \mathbf{P}_a = -\Delta N \omega_{pa}^2 \varepsilon_0 \mathbf{E}, \tag{47.11}$$

$$\frac{\partial^2 \mathbf{P}_e}{\partial t^2} + \gamma_e \frac{\partial \mathbf{P}_e}{\partial t} + \omega_e^2 \mathbf{P}_e = -\Delta N \omega_{pe}^2 \varepsilon_0 \mathbf{E}. \tag{47.12}$$

Here the polarization has been split off into a two parts $\mathbf{P}_a$ and $\mathbf{P}_e$ that are connected with the radiative resonances, which are added to the total polarization. The corresponding level occupation densities update with the auxiliary equations [6]

$$\frac{\partial N_3}{\partial t} = \left(\frac{\partial \mathbf{P}_a}{\partial t} + \frac{\gamma_a}{2} \mathbf{P}_a \right) \cdot \mathbf{E} - \frac{N_3}{\tau_{32}}, \tag{47.13}$$

$$\frac{\partial N_2}{\partial t} = \frac{N_3}{\tau_{32}} + \frac{1}{\hbar \omega_e} \left(\frac{\partial \mathbf{P}_e}{\partial t} + \frac{\gamma_e}{2} \mathbf{P}_e \right) \cdot \mathbf{E} - \frac{N_2}{\tau_{21}}, \tag{47.14}$$

$$\frac{\partial N_1}{\partial t} = \frac{N_2}{\tau_{21}} - \frac{1}{\hbar \omega_e} \left(\frac{\partial \mathbf{P}_e}{\partial t} + \frac{\gamma_e}{2} \mathbf{P}_e \right) \cdot \mathbf{E} - \frac{N_1}{\tau_{10}}, \tag{47.15}$$

$$\frac{\partial N_0}{\partial t} = \frac{N_1}{\tau_{10}} - \left(\frac{\partial \mathbf{P}_a}{\partial t} + \frac{\gamma_a}{2} \mathbf{P}_a \right) \cdot \mathbf{E}, \tag{47.16}$$

which encode both the stimulated terms and nonradiative recombinations.

47.3 Transfer Matrix Method

The geometry of the perhaps simplest types of nanoplasmonic lasers is planar waveguide structures. Here, planar layers of heterogeneous materials are stacked to form a waveguide in the center. The modes of such structures depend strongly on the width and material of each layer and can be calculated using the transfer matrix method (TMM), a linear, frequency-domain analysis that shall be used to design and optimize structures before simulation in time domain.

Isotropic homogeneous materials support transverse plane wave solutions in bulk. These waves have the dispersion relation

$$\varepsilon \mu \frac{\omega^2}{c^2} - k^2 = 0, \tag{47.17}$$

and the relative amplitudes of the electric and magnetic fields set by

$$E = \sqrt{\frac{\mu}{\varepsilon}} Z_0 H, \tag{47.18}$$

where Z_0 is the impedance of free space. When two materials meet at a planar interface, the solutions in each space must be matched together, and the prescription here is that the E and H field components parallel to the interface are continuous. Each interface couples a forward traveling and a backward traveling wave in each layer together.

Let us consider a stack with an axis parallel to the z-direction and waves traveling in a plane perpendicular to the y-direction. There are two independent polarization solutions, *transverse magnetic* (TM) where H is perpendicular to the plane of incidence (parallel to y) with E parallel to the plane, and *transverse electric* (TE) where E is perpendicular to the plane of incidence. TM equations shall be considered herein as this is the character of the SPP solutions. The corresponding TE equations can be recovered by making the substitutions: $\varepsilon \to \mu, \mu \to \varepsilon, E \to -Z_0H, Z_0H \to E$.

The system can be solved by performing a change of basis between forward and backward propagating waves and electric and magnetic field components, i.e.,

$$\begin{pmatrix} E_{xi} \\ Z_0 H_{yi} \end{pmatrix} = \underbrace{\begin{pmatrix} u_i & -u_i \\ 1 & 1 \end{pmatrix}}_{\mathbf{U}} \begin{pmatrix} A_i^+ \\ A_i^- \end{pmatrix}, \tag{47.19}$$

and translating the forward and backward field components within a layer:

$$\begin{pmatrix} A_i^+ \\ A_i^- \end{pmatrix}\Bigg|_z = \underbrace{\begin{pmatrix} e^{ik_{zi}(z-z')} & 0 \\ 0 & e^{-ik_{zi}(z-z')} \end{pmatrix}}_{\mathbf{T}} \begin{pmatrix} A_i^+ \\ A_i^- \end{pmatrix}\Bigg|_{z'}, \tag{47.20}$$

with $u_i = ck_{zi}/(\varepsilon_{xi}\omega)$. These are the matrices of the TMM. The matrices, $\mathbf{U}$ and $\mathbf{T}$, are assigned such that $\mathbf{U}$ converts from the basis of forward and backward traveling waves to electric and magnetic field components, and $\mathbf{T}$ propagates forward and backward traveling waves from a point z' to a point z.

For a stack of N planar slabs on a substrate as in Figure 47.3, each with its own material properties, we demonstrate how the TMM can determine the bound modes. Let the top interface of the ith slab be at z_i, vector components with that label, e.g., $A_i^\pm$, be measured from there, i.e., just below the top interface in that

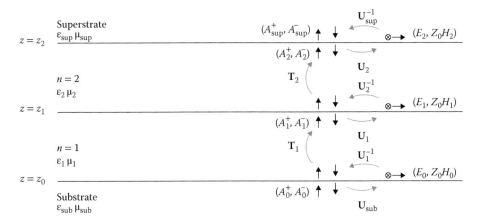

FIGURE 47.3 Schematic of the transfer matrix method. Fields (black arrows) are operated on by matrix transformations (gray arrows). The forward and backward components at the top of the substrate layer (A_0^+, A_0^-) are transferred through the stack to forward and backward components at the bottom of the superstrate layer $(A_{\text{sup}}^+, A_{\text{sup}}^-)$. Fields move between bases of forward and backward waves ($\uparrow \downarrow$) and electric and magnetic components ($\otimes \to$).

layer. Starting in the forward and backward basis, one may use the transfer matrices to propagate through the stack from the one layer to the next:

$$\begin{pmatrix} A_{i+1}^+ \\ A_{i+1}^- \end{pmatrix} = \mathbf{T}_{i+1} \mathbf{U}_{i+1}^{-1} \mathbf{U}_i \begin{pmatrix} A_i^+ \\ A_i^- \end{pmatrix} . \tag{47.21}$$

The exception to this being the superstrate, the $(N+1)$th layer, which by definition does not have a top interface. As such, $E_{\text{sup}}^\pm$ is defined specially as the forward and backward components just above the top interface of the Nth layer, i.e.,

$$\begin{pmatrix} A_{\text{sup}}^+ \\ A_{\text{sup}}^- \end{pmatrix} = \mathbf{U}_{\text{sup}}^{-1} \mathbf{U}_N \begin{pmatrix} A_N^+ \\ A_N^- \end{pmatrix} . \tag{47.22}$$

The full transfer matrix $\mathcal{M}$ will map forward and backward traveling waves in the substrate to the superstrate:

$$\begin{pmatrix} A_{\text{sup}}^+ \\ A_{\text{sup}}^- \end{pmatrix} = \mathcal{M} \begin{pmatrix} A_{\text{sub}}^+ \\ A_{\text{sub}}^- \end{pmatrix} , \tag{47.23}$$

and is given as the product of each layer hop:

$$\mathcal{M} = \mathbf{U}_{\text{sup}}^{-1} \mathbf{U}_N \cdots \mathbf{T}_1 \mathbf{U}_1^{-1} \mathbf{U}_{\text{sub}} . \tag{47.24}$$

Terms with similar indices can be grouped to give a matrix for each layer:

$$\mathbf{M}_i = \mathbf{U}_i \mathbf{T}_i \mathbf{U}_i^{-1} , \tag{47.25}$$

which maps electric and magnetic fields at the bottom of the layer through to the top. As such, the transfer matrix for the whole stack is

$$\mathcal{M} = \mathbf{U}_{\text{sup}}^{-1} \left(\prod_{i=N}^{1} \mathbf{M}_i \right) \mathbf{U}_{\text{sub}} , \tag{47.26}$$

noting the reverse order on the product.

The eigenmodes of the system that have non-zero fields for a zero driving field, i.e.,

$$\begin{pmatrix} A_{\text{sup}}^+ \\ 0 \end{pmatrix} = \mathcal{M} \begin{pmatrix} 0 \\ A_{\text{sub}}^- \end{pmatrix} , \tag{47.27}$$

with the prescription that $\text{Im}\, k_z > 0$ in both the superstrate and substrate such that fields on both sides decay away to infinity. This yields the condition that $\mathcal{M}_{22} = 0$ for bound modes. For dispersive media, this is satisfied for either one of the frequency ω or wavevector q being a complex number, with the other remaining real. In general, analytical dispersion relation solutions do not exist, but rather solutions for $\mathcal{M}_{22}(\mathbf{q}, \omega) = 0$ are found implicitly, using a complex root-finding algorithm.

To illustrate the method, let's consider the SPP solution. Consider a metal substrate with a planar interface to a dielectric layer. In this case, the transfer matrix is given as

$$\mathcal{M} = \mathbf{U}_{\text{sup}}^{-1}\mathbf{U}_{\text{sub}} \tag{47.28}$$

$$= \frac{1}{2}\begin{pmatrix} \frac{\varepsilon_{\text{bg}}\omega}{ck_{z,\text{air}}} & 1 \\ -\frac{\varepsilon_{\text{bg}}\omega}{ck_{z,\text{air}}} & 1 \end{pmatrix}\begin{pmatrix} \frac{ck_{z,\text{metal}}}{\varepsilon(\omega)\omega} & -\frac{ck_{z,\text{air}}}{\varepsilon(\omega)\omega} \\ 1 & 1 \end{pmatrix}, \tag{47.29}$$

which gives the SPP condition:

$$\mathcal{M}_{22} = \frac{1}{2}\left(\frac{\varepsilon_{\text{bg}}\omega}{ck_{z,\text{air}}}\frac{ck_{z,\text{metal}}}{\varepsilon(\omega)\omega} + 1\right) = 0, \tag{47.30}$$

$$\frac{\varepsilon_{\text{bg}}}{\varepsilon(\omega)} + \frac{k_{z,\text{air}}}{k_{z,\text{metal}}} = 0 . \tag{47.31}$$

One can see that for this equation to have solutions, given $\varepsilon_{\text{bg}} > 0$ and $\text{Im}\, k_z > 0$, it must hold that $\text{Re}\,\varepsilon(\omega) < 0$, which is characteristic of metals for frequencies below their plasma frequency.

For a lossless Drude metal, i.e., $\varepsilon(\omega) = \varepsilon_\infty - (\omega_p/\omega)^2$, an analytical solution exists:

$$q^2 = \varepsilon_{\text{bg}}\left(\frac{\omega}{c}\right)^2\frac{\varepsilon_\infty\omega^2 - (\varepsilon_{\text{bg}} + \varepsilon_\infty)\omega_{\text{sp}}^2}{(\varepsilon_{\text{bg}} + \varepsilon_\infty)(\omega^2 - \omega_{\text{sp}}^2)}, \tag{47.32}$$

where $\omega_{\text{sp}} = \omega_p/\sqrt{\varepsilon_{\text{bg}} + \varepsilon_\infty}$ is the surface plasmon frequency.

47.4 Evolutionary Algorithm

For a planar stratified structure whose properties can be determined by a TMM, one may wish to optimize some aspect of its response to electromagnetic radiation. This may be in the dispersion relation of bound modes or in the transmission and reflection spectra. For example, to prescribe the frequency of a bound mode, or a reflection peak; or set whether there is only a single mode or many modes within a frequency range; or control the amplitude and phase response to an incoming wavepacket. These properties are entirely determined by the composition of the structure; however, it is often not obvious how changes in the structure lead to changes in the response.

The structures to be considered can be described by a finite number of numeric or discrete parameters, i.e., for each layer: the thickness of the layer, the material model to use, and the parameters of that model. This is well suited to optimization using an *evolutionary algorithm* (EA) [12].

In brief, an EA operates by storing multiple copies of these parameter sets, perturbing the values of the parameters randomly, ordering the sets by how the system they describe performs against a fitness metric, and keeping only the best variants to the next round.

More fully, in the language of an EA, this would be to say that a *population* of *chromosomes* is kept, and on each iteration of the algorithm, new chromosomes are generated from *mutations* of the current ones; the next-generation population is filled by *selection* of these chromosomes by a *fitness function*. The goal of the EA is to find a chromosome that extremizes the fitness function globally, and as such an EA is most suitable when the evaluation of the fitness function can be performed relatively quickly as a large number of members of the state space must be evaluated and compared.

A chromosome is simply the set of parameters that describes a system, i.e., a list of numbers or discrete option values. They are atomic entities, independent of other chromosomes. Each chromosome may have its fitness function evaluated to determine how suitable its corresponding physical system is to a problem.

Each iteration of the EA holds a set of chromosomes, ordered by their fitness function, known as a *population*. The first-generation population is either seeded with randomly generated structures, or structures that have been tuned by hand, perhaps by approximative theoretical methods. Populations of subsequent generations are constructed by first taking a small quantile of the fittest chromosomes from the previous iteration, with the remaining places filled with *mutations* of chromosomes selected from the previous set. The progress of the algorithm can be tracked by examining the fittest chromosome at the end of each iteration. At the end of each iteration, a *terminating condition* is evaluated, which determines if the algorithm should stop, returning the fittest chromosome, e.g., if the best structure in the population is within a defined tolerance of the target output, or if a set number of iterations have elapsed. If the terminating condition is not met, then the algorithm continues for another iteration. The operation of the EA is described as a flowchart in Figure 47.4.

The mutation process generates a new chromosome by randomly perturbing the values of parameters of a parent chromosome. It is by mutation that the state space of chromosomes is mapped.

For the layered structures considered here, the mutation process will perturb both the properties of the layers themselves and the relationship between layers. Numeric parameters, such as layer thicknesses and those used in material models, are perturbed by a Gaussian random variable, and the discrete parameters, such as the layer model, can be randomly selected from a list. Typically, only one or few parameters will be changed in each mutation. The parameters of the layers themselves do not sit in isolation. The number of layers in these structures is variable and mutation allows for adding or removing of layers or indeed swapping the order of layers and shifting the boundary between adjacent layers.

Whereas the EA is widely applicable to problems in nano-optics, the fitness function is not general and has to be constructed for each problem based on the type and kind of characteristics that are to be selected for. As an example, the EA can be applied to the case of stopped-light lasing. Here structures are desired that support a bound mode with two points of zero group velocity (ZGV) in their dispersion. An SL structure is deemed better if the ratio of the separations of the wavevectors and frequencies of the ZGV points, the band index $|c/v_b| = (k_2 - k_1)/(\omega_2 - \omega_1)$, is higher. This ensures that a wide range of wavevectors can be held at a single frequency, which reduces the group velocity for the whole band and the wave dispersion that will occur for a wavepacket made from Fourier components in the band. Figure 47.5 shows an example of

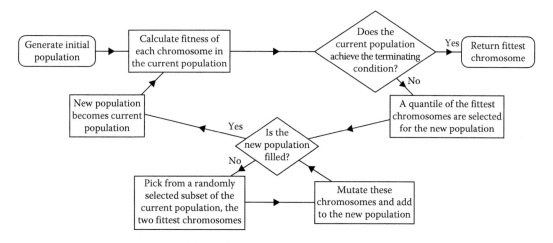

FIGURE 47.4 Evolutionary algorithm flowchart.

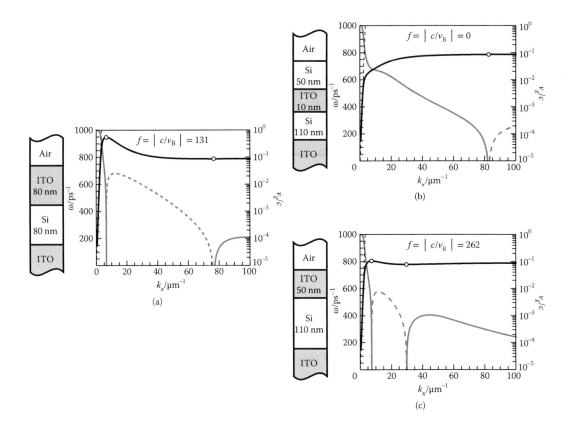

FIGURE 47.5 Diagrams of planar plasmonic nanostructures and corresponding dispersion relations (black solid curve). The group velocity is also shown (gray solid, positive; gray dashed, negative) as well as zero group velocity points (circles). An evolutionary algorithm, using *band index* as a metric, given a. as a starting chromosome may produce b. and c. as child chromosomes. b. has a metric of zero and is thrown out, whereas c. may stay to the next generation as it has a higher metric than its parent.

potential SL structures, their dispersion relations, and group velocity. Each structure is labeled by its band index, and for a parent structure, two possible child structures are presented.

47.5 Finite-Difference Time-Domain Method

When describing processes such as gain, a frequency-domain analysis can only bring insight to a certain level as it has limitations to describe nonlinearities. When plasmons or phonons are emitted, electrons are demoted from higher energetic states to lower ones. This depletes the population inversion, reducing the available gain over time, leading to a nonlinear dependence in the field strength. In addition, spatially resolved effects, such as spatial hole burning, cannot be considered in TMM that assumes uniformity in the direction of propagation, whereas the level of inversion can vary both in this direction and perpendicular to the stacking. Depending on the mode formed, some regions may host higher field densities that can deplete the local gain. Thus a dynamic, spatially resolved, time-domain simulation is required to capture all the aspects of a nanoplasmonic lasing device. Here we will discuss and use the *finite-difference time-domain* (FDTD) method [13,14].

In FDTD, the goal is to solve Maxwell's equations for successive instants in time on a discretized lattice. This is approached by focusing on the curl equations and making the time derivatives of the fields the subject. Let us assume that there are no external sources:

$$\frac{\partial \mathbf{E}}{\partial t} = \frac{1}{\varepsilon_0}\left(\nabla \times \mathbf{H} - \frac{\partial \mathbf{P}}{\partial t}\right) \tag{47.33}$$

$$\frac{\partial \mathbf{H}}{\partial t} = -\frac{1}{\mu_0}\left(\nabla \times \mathbf{E} - \frac{\partial \mathbf{M}}{\partial t}\right), \tag{47.34}$$

where $\mathbf{P}$ and $\mathbf{M}$ are the polarization and magnetization, respectively. Now, space and time are discretized as $(t, x, y, z) \to (n\Delta_t, i\Delta_x, j\Delta_x, k\Delta_x)$ and derivatives are turned into finite differences, i.e.,

$$\frac{\partial \mathbf{E}}{\partial t}(\mathbf{x}, t) \to \frac{\mathbf{E}^{n+1}_{i,j,k} - \mathbf{E}^n_{i,j,k}}{\Delta_t}. \tag{47.35}$$

Strictly, the above is a central difference and approximates the time derivative at a time $(n+1/2)\Delta_t$. Therefore, the corresponding right-hand side terms must be evaluated at this time. The curl operator is handled similarly:

$$\nabla \times \mathbf{E}(\mathbf{x}, t)|_x \to \frac{E_z|^n_{i,j+1,k} - E_z|^n_{i,j,k}}{\Delta_x} - \frac{E_y|^n_{i,j,k+1} - E_y|^n_{i,j,k}}{\Delta_x}. \tag{47.36}$$

In order to make best use of this central difference, the lattice is arranged as a Yee grid [13], where fields are stored separately by component and $\mathbf{E}$ and $\mathbf{P}$ fields are specified on integer time coordinates, whereas $\mathbf{H}$ and $\mathbf{M}$ are specified on half-integer time coordinates. The lattice coordinates are more complicated; for $\mathbf{E}$, components are stored at half-integer coordinates in their own direction and at integer coordinates otherwise. This is the opposite for the $\mathbf{H}$ field, as illustrated in Figure 47.6. From here, one can write the *update equations*:

$$E_x|^{n+1}_{i+\frac{1}{2},j,k} = E_x|^n_{i+\frac{1}{2},j,k} - \varepsilon_0^{-1}P_x|^{n+1}_{i+\frac{1}{2},j,k} + \varepsilon_0^{-1}P_x|^n_{i+\frac{1}{2},j,k} \tag{47.37}$$

$$+ \frac{c\Delta_t}{\Delta_x}\left(Z_0H_z|^{n+\frac{1}{2}}_{i+\frac{1}{2},j+\frac{1}{2},k} - Z_0H_z|^{n+\frac{1}{2}}_{i+\frac{1}{2},j-\frac{1}{2},k}\right)$$

$$- \frac{c\Delta_t}{\Delta_x}\left(Z_0H_y|^{n+\frac{1}{2}}_{i+\frac{1}{2},j,k+\frac{1}{2}} - Z_0H_y|^{n+\frac{1}{2}}_{i+\frac{1}{2},j,k-\frac{1}{2}}\right)$$

$$Z_0H_x|^{n+\frac{1}{2}}_{i,j+\frac{1}{2},k+\frac{1}{2}} = Z_0H_x|^{n-\frac{1}{2}}_{i,j+\frac{1}{2},k+\frac{1}{2}} + cM_x|^{n+\frac{1}{2}}_{i,j+\frac{1}{2},k+\frac{1}{2}} - cM_x|^{n-\frac{1}{2}}_{i,j+\frac{1}{2},k+\frac{1}{2}} \tag{47.38}$$

$$+ \frac{c\Delta_t}{\Delta_x}\left(E_z|^n_{i,j+1,k+\frac{1}{2}} - E_z|^n_{i,j,k+\frac{1}{2}}\right)$$

$$- \frac{c\Delta_t}{\Delta_x}\left(E_y|^n_{i,j+\frac{1}{2},k+1} - E_y|^n_{i,j+\frac{1}{2},k}\right),$$

with corresponding equations for the omitted components.

With the updated equations of the primary fields determined, it now remains to be described how the auxiliary fields are updated. In general, the constitutive equations are functionals of all past values of all fields, though this would prove tricky to implement directly. In practice, material responses can be cast in terms of a finite difference if they can be expressed in terms of a temporal differential equation. This lends the Drude–Lorentz model well to inclusion within FDTD by converting its auxiliary equations into finite differences.

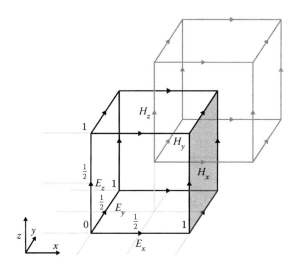

FIGURE 47.6 Schematic of FDTD Yee cell. Electric (black) and magnetic (gray) field grids. They are shifted half a step from each other, forming an interlocking link. Each field line passes through the center of a square formed from lines of the other field, e.g., H_x here passes through the center of the highlighted E-field face, where the finite difference $\nabla \times \mathbf{E}$ is defined.

The FDTD algorithm proceeds by iterating over, and updating in order, the $\mathbf{P}$ field, $\mathbf{E}$ field, $\mathbf{M}$ field, and $\mathbf{H}$ field over all space for each timestep.

Of course, there are limitations to FDTD. Its staggered grid structure means that material interfaces are not sharply defined, and any curvature on an interface will also experience a staircasing effect that can lead to spurious hotspots especially in plasmonic structures where geometric effects play a key role. The grid is also regular, which is to say it cannot be adaptively refined in areas of interest, rather all points must be simulated to the same resolution. This adds to the numerical load, as the smallest wavelength component of the fields, as well as the smallest geometrical features must be provided with sufficient resolution. This is again especially relevant in plasmonics, where field confinement is a feature.

Because of its nature as a full-wave time-domain solver, FDTD can often be the final say on a theoretical investigation. Short of doing a real-world experiment, FDTD numerical simulations find their utility as a heavy-duty tool. That is, the Maxwell's equations are solved at every discretized point in space, so it is often slower and more cumbersome, but can as a result of the inherent thoroughness produce results for questions that other techniques are unable to.

47.6 Loss Compensation

The high field enhancements afforded by plasmon modes supported by negative permittivity in metals must be traded off against Ohmic losses in the optical frequencies. Loss compensation is the incorporation of gain elements into a system in order to offset the loss in a particular frequency range while retaining desired properties from the lossy structure.

The properties one might wish to retain, e.g., negative effective index [6], stopped light points [11], or hyperbolic dispersion [15], etc., are frequency dependent. Causality will influence the position and width of the frequency range where losses can be compensated and desirable properties are to be found simultaneously [16,17], and adding gain to a system will necessarily change the refractive properties in accordance with the Kramers–Kronig relations.

The induced change in the mode structure by the addition of gain can be analyzed in a frequency-domain analysis, modeling gain as a four-level system in frequency domain with constant electron occupation densities. Using the TMM, this would mean replacing a part or whole of a dielectric layer with an equivalent Drude–Lorentz model with properties that map to the dielectric layer for equilibrium carrier occupation densities.

To illustrate this, the SL structure as optimized previously is modified by replacing the dielectric layer with a quantum emitter. The emission frequency is set to match either one of the ZGV points, and the other parameters are set representing the inclusion of realistic laser dye molecules [18]. The resulting dispersion and loss relations are shown in Figure 47.7 for inversion densities ΔN varying between 0 and N, with a fixed emitter density, and excitation about ZGV1 and ZGV2. The first point to note is that, even on full inversion, the addition of gain does not significantly change the dispersion, with the maximum shift in frequency being around 0.6%. The presence of ZGV points is preserved, though they may drift slightly, i.e., ZGV2 moves slightly right with emission about ZGV1. There is no change at the frequency that is being excited because the permittivity change of a Lorentzian is zero at the resonant frequency. Adding gain has the side effect of making the structure a slightly better SL structure as the band velocity is marginally reduced.

The key change though is in the loss. As the inversion density increases, the loss decreases for wavevectors where the dispersion is within the gain width. Initially the plasmonic loss is reduced as the inversion increases, and then for an inversion of around $\Delta N/N \approx 0.4$, some wavevectors become undamped, and even eventually experience gain. Plasmons sitting in these modes will grow exponentially in amplitude while small enough to remain in the small-signal gain regime. As ZGV2 is flatter than ZGV1, when the

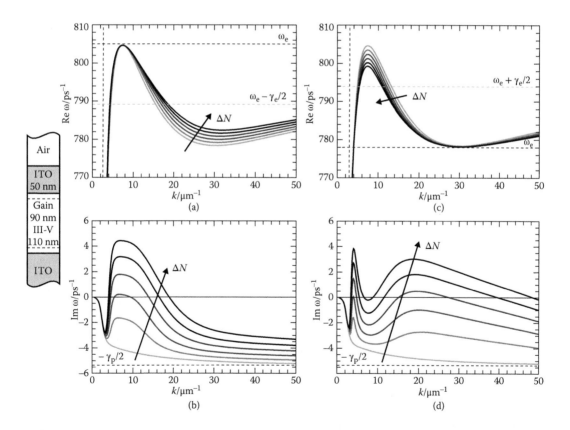

FIGURE 47.7 Perturbation of dispersion relation and loss with Lorentzian gain. The change of mode shape plotted as Lorentzian gain is introduced to the system. For subfigures on the left, emission is about ZGV1, and on the right about ZGV2. a and b show a zoom in of how the dispersion relation changes with increasing inversion density, while c and d show the corresponding loss/gain of the mode.

emission is about this point, a wider range of frequencies fall within the gain width, leading to a larger range of k values that can become undamped.

An analysis of loss compensation can also be performed in the time domain. This has been done in Reference [6] where the structure was a negative refractive index double-fishnet metamaterial. Pump-probe spectroscopy was performed in FDTD in order to retrieve effective parameters using realistic parameters of rhodamine 800 dye molecules as a four-level system. It was shown that the imaginary part of the effective refractive index, which encodes the loss, transitioned from positive to zero as pumping increased while retaining a negative real part of the effective index between wavelengths of 706 and 716 nm. The results are shown in Figure 47.8.

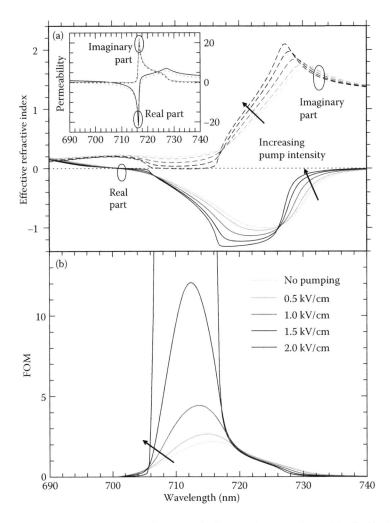

FIGURE 47.8   (a) Real and imaginary part of the retrieved effective refractive indices of the double-fishnet structure for different pump amplitudes. The peak electric field amplitude of the pump increases in steps of 0.5 kV/cm from no pumping (lightest) to a maximum of 2.0 kV/cm (darkest). The inset shows the real and imaginary part of the effective permeability (black and gray line, respectively) and the result of the Kramers–Kronig relation (dotted lines) for the highest peak electric field amplitude of 2.0 kV/cm . (b) The figures of merit (FOM) for the same pumping amplitudes. (Reproduced from Sebastian Wuestner et al., *Phys. Rev. Lett.*, 105,127401, 2010.)

47.7 Lasing Dynamics

In order to transition from loss compensation to lasing, photons emitted by the gain medium must feedback and stimulate on average more emission than photons that are lost on the round trip. Photons can be lost by escaping the cavity to radiation modes, absorbed incoherently by dissipative channels, or absorption by the gain medium.

Initially, for a cold cavity with no lasing mode field built up, an inverted gain system may spontaneously emit into cavity modes. This spontaneous emission may trigger further stimulated emission in a process known as *amplified spontaneous emission* (ASE), which can be added to the semiclassical FDTD description by including spatially resolved dynamic Langevin noise to the system, which accounts for the dissipative reservoirs feeding back stochastically on the system. The noise couples to the four-level system and induces incoherent transitions, which then become amplified, allowing for the triggering of the lasing regime [8].

As a device enters a lasing regime, characteristic relaxation oscillations can be seen where inversion initially builds up and spontaneous emission events are induced. The fields then are amplified as they stimulate further emission, growing the field in the mode coherently. From here on the emitted fields start to grow exponentially, and when they are of sufficient strength will deplete the inversion density. This reduction of inversion feeds back by decreasing the available gain, leading to a decrease in field energy as the energy in the field is lost to dissipative processes in the metal layers. The decrease in energy density allows for the inversion to rebuild. This continues in an oscillatory manner with the energy density lagging behind the inversion by 90°. The amplitude of the inversion and energy oscillations decrease with each cycle until a stable steady value for both is reached. In contrast, the control structure, albeit entering a regime of relaxation oscillations, is more erratic and the oscillations do not settle to a steady state. Instead, the oscillations continue with a factor of 4 between the peak energy density and the trough. An example of these relaxation oscillations in shown for the SL laser set in Figure 47.9 IN contrast to a "control structure" where the waveguide is not of high enough quality for the device to enter the lasing regime. Relaxation oscillations are accompanied by a characteristic decrease of the width of the power spectrum, from a wide peak in ASE to a sharp one in the lasing regime. This observation is usually taken as a crucial evidence for lasing in experimental setups.

47.8 Stopped-Light Lasing

Stopped-light lasing differs from traditional nanoplasmonic lasers in that there is no cavity with predefined modes. Whereas in the structures presented, light is trapped in a waveguide mode, it is free to propagate in one direction, albeit at ZGV. This means that there is a continuum of planewave modes for the gain to emit into, which get selected on the basis of the overlap with the gain media. The feedback in the system can be understood by considering the spatially resolved, cycle-averaged Poynting vector, energy density, and inversion that are plotted in Figure 47.10 for a structure with a gain width of 1000 nm. The energy is concentrated on metal-dielectric interfaces, strongest on the lower interface, and is localized around the gain medium despite there being no cavity along the horizontal direction.

The Poynting vector shows how energy circulates in the structure. There are four energy flux vortices, one in each corner of the structure; as described in the figure caption, these have energy moving out of the gain region into the metal layers and into it in dielectric layers. Considering the x component of the Poynting flux Figure 47.10a, the forward and backward flows are in exact balance. The balanced counter-propagation of energy in the negative-permittivity (metal) layers against the dielectric layer is the basis of the (subwavelength) feedback in the system.

The inversion is shown in Figure 47.10d; there are areas of spatial hole burning where the energy density is highest. The spatial modulation seen is explained when considering the field profile and its formation.

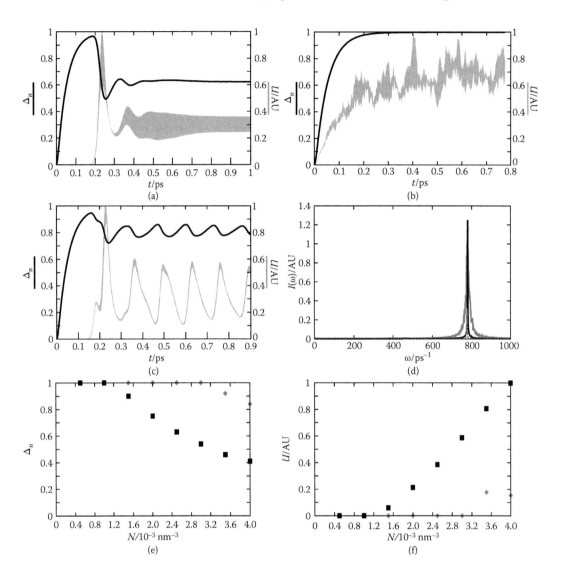

FIGURE 47.9 Lasing onset with increasing gain density. a, b, c. Energy density and mean inversion for (a) the stopped light structure with a gain density $N = 0.002\ \text{nm}^{-3}$, (b) the stopped light structure with a gain density $N = 0.001\ \text{nm}^{-3}$, and (c) the control structure with a gain density $N = 0.004\ \text{nm}^{-3}$. (d) Power spectra for SL structure at $N = 0.002\ \text{nm}^{-3}$ (black), SL structure at $N = 0.001\ \text{nm}^{-3}$ (×5000 dashed gray), and control structure at $N = 0.004\ \text{nm}^{-3}$ (solid gray). (e) Steady-state inversion of SL (black) and control (gray) structures with varying gain density. (f) Steady-state cycle-averaged energy density with varying gain density.

47.9 Summary

In this chapter, we have explored techniques to design, optimize, and simulate a nanoplasmonic laser. We started by defining harmonic oscillator type Drude–Lorentz and Maxwell–Bloch models, a universal type of models for metals, semiconductors, and gain media. From a simple model of forces in time domain, a causal frequency-domain response function was derived. From here, the four-level system was introduced as two dipole transitions in the frequency domain and as a dynamic system with self-consistently evolving

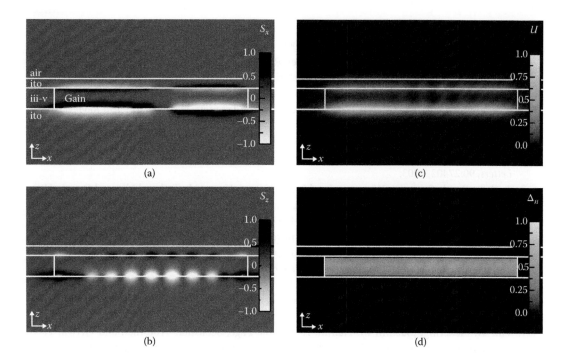

FIGURE 47.10 Energy flux and density of the lasing mode. The lasing mode of a stopped-light laser structure with a width 1000 nm. (a,b) The x and z components of the cycle-averaged Poynting vector, respectively, with positive values in black and negative in white. Focusing on the bottom right lobe of the flux in panel a, energy flows into the gain region in the dielectric layers and out of it in the metal layer. For the same lobe in panel b, energy moves downward out of the gain region toward the center, and upward into the region at the edge. The counter-clockwise energy vortex in this corner is mirrored in the other corners. (c) The cycle-averaged energy density and (d) inversion plots. The inversion is the complement of the energy density, being highest in areas with low energy density and depleted where the mode sits.

electron occupation levels in the time domain. The TMM was next introduced as an exact frequency-domain tool to analyze planar structures. EA are suited to optimize systems that can be characterized by a range of numeric or discrete parameters and perform characterized by a fitness metric. The EA approach was introduced to and demonstrated in the case of the TMM and applied to designing stopped light laser structures. Next, FDTD, the time-domain numerical method, was discussed and used to simulate active hybrid metallo-dielectric structures with nonlinearities and time-dependant dispersion such as a lasing system. This allowed for the simulation of lasing dynamics where loss compensation, ASE, and relaxation oscillations could be captured. Finally, some specifics of the recently established stopped light laser principle were explored as an example system of a designed and optimized nanoplasmonic laser.

References

1. Jennifer A. Dionne, Luke A. Sweatlock, Matthew T. Sheldon, A. Paul Alivisatos, and Harry A. Atwater. Silicon-based plasmonics for on-chip photonics. *IEEE Journal of Selected Topics in Quantum Electronics*, 16(1):295–306, 2010.
2. Anatoly V. Zayats and Igor I. Smolyaninov. Near-field photonics: Surface plasmon polaritons and localized surface plasmons. *Journal of Optics A: Pure and Applied Optics*, 5(4):S16, 2003.
3. Dmitri K. Gramotnev and Sergey I. Bozhevolnyi. Plasmonics beyond the diffraction limit. *Nature Photonics*, 4(2):83–91, 2010.

4. Kevin F. MacDonald, Zsolt L. Samson, Mark I. Stockman, and Nikolay I. Zheludev. Ultrafast active plasmonics. *Nature Photonics*, 3(1):55–58, 2009.

5. Ortwin Hess, John B. Pendry, Stefan A. Maier, Rupert Oulton, Joachim M. Hamm, and Kosmas L. Tsakmakidis. Active nanoplasmonic metamaterials. *Nature Materials*, 11(7):573–584, 2012.

6. Sebastian Wuestner, Andreas Pusch, Kosmas L. Tsakmakidis, Joachim M. Hamm, and Ortwin Hess. Overcoming losses with gain in a negative refractive index meta-material. *Physical Review Letters*, 105(12):127401, 2010.

7. David J. Bergman and Mark I. Stockman. Surface plasmon amplification by stimulated emission of radiation: Quantum generation of coherent surface plasmons in nanosystems. *Physical Review Letters*, 90:27402, 2003.

8. Andreas Pusch, Sebastian Wuestner, Joachim M. Hamm, Kosmas L. Tsakmakidis, and Ortwin Hess. Coherent amplification and noise in gain-enhanced nanoplasmonic metamaterials: A Maxwell-Bloch Langevin approach. *ACS Nano*, 6(3):2420–2431, 2012.

9. Sebastian Wuestner, Joachim M. Hamm, Andreas Pusch, Fabian Renn, Kosmas L. Tsakmakidis, and Ortwin Hess. Control and dynamic competition of bright and dark lasing states in active nanoplasmonic metamaterials. *Physical Review B*, 85(20):201406, 2012.

10. Bumki Min, Eric Ostby, Volker Sorger, Erick Ulin-Avila, Lan Yang, Xiang Zhang, and Kerry Vahala. High-Q surface-plasmon-polariton whispering-gallery microcavity. *Nature*, 457(7228):455–458, 2009.

11. Tim W. Pickering, Joachim M. Hamm, A. Freddie Page, Sebastian Wuestner, and Ortwin Hess. Cavity-free plasmonic nanolasing enabled by dispersionless stopped light. *Nature Communications*, 5:4972, 2014.

12. William M. Spears. *Evolutionary Algorithms: The Role of Mutation and Recombination. Natural Computing Series*. Heidelberg: Springer Berlin Heidelberg, 2000.

13. Kane Yee. Numerical solution of initial boundary value problems involving Maxwell's equations in isotropic media. *IEEE Transactions on Antennas and Propagation*, 14:302–307, 1966.

14. Allen Taflove. *Computational Electrodynamics: The Finite-Difference Time-Domain Method. Antennas and Propagation Library*. Norwood, MA: Artech House, 1995.

15. Xingjie Ni, Satoshi Ishii, Mark D. Thoreson, Vladimir M. Shalaev, Seunghoon Han, Sangyoon Lee, and Alexander V. Kildishev. Loss-compensated and active hyperbolic metamaterials. *Optics Express*, 19(25):25242–25254, 2011.

16. Johannes Skaar. Fresnel equations and the refractive index of active media. *Physical Review E*, 73(2):26605, 2006.

17. Paul Kinsler and Martin McCall. Causality-based criteria for a negative refractive index must be used with care. *Physical Review Letters*, 101(16):1–4, 2008.

18. P. Sperber, W. Spangler, Bernd Meier, and Alfons Penzkofer. Experimental and theoretical investigation of tunable picosecond pulse generation in longitudinally pumped dye laser generators and amplifiers. *Optical and Quantum Electronics*, 20(5):395–431, 1988.

48

Quantum-Dot Nanolasers

Christopher Gies

Michael Lorke

Frank Jahnke

and

Weng W. Chow

48.1 Going Nano: A New Era of Small Lasers

The impact of semiconductor lasers on our daily lives is immense. They perform work every time someone gathers information for the internet, makes a telephone call, prints an article or pays for an item at a store. There are over 2.5 billion VCSELs[†] in use at our homes or offices, and roughly 100 million are produced each year to keep up with demand. VCSEL technology is now considered conventional, and research has progressed to smaller devices, modulated at higher speeds and producing output with greater spectral stability. This chapter focuses on two of the advances. One is the reduction of optical cavity volume by as much as two orders of magnitude, making a transition from microlasers (such as VCSELs) to nanolasers. Two is a change in active medium from quantum wells (QWs) to quantum dots (QDs), bringing about, e.g., devices operating with very few (tens) of emitters. These developments ushered in a new era for semiconductor device physics, one where quantum optical and many-body electron interaction effects dominate. The underlying motivation is the control of spontaneous emission [1]. The typical laser mitigates the randomness (noise) caused by spontaneous emission by overwhelming it with stimulated emission. Here, we are speaking of actually quieting the spontaneous emission noise, both spatially and temporally. In the former, we use nanocavities to inhibit spontaneous emission in undesirable directions. With the latter, we use very

---

[†] Vertical cavity surface emitting lasers, cf. Chapter 34.

few QDs (ideally only one) to control the timing of photon emission, thereby improving photon statistics beyond the limit described by the Poisson distribution. Devices incorporating the two advances are being fabricated and experiments are being performed. The results are both promising and exciting from the device and physics aspects, respectively. They also lead to many new questions and renewed interest in some old ones: What is lasing and where is the threshold? Why is there not a phonon bottleneck? What is the homogeneous width of a QD transition [2–4]? What is the inhomogeneous broadening in my samples? Is there really thresholdless lasing [5–7]? Are QD lasers better than QW lasers [8]? Why do QD lasers not show modulation speeds as advertised? This chapter describes a theoretical framework capable of addressing all the abovementioned questions. The building blocks come from quantum electrodynamics (QED), which is the quantum theory of the interaction of light with matter, and carrier interactions described by many-body theory.

The fundamental concepts of lasing are often discussed in terms of rate equations for material excitations and cavity photons that may well be familiar to the reader. We begin this chapter with an excursion to show profit and limitations of laser rate equations for understanding the behavior of nanolasers, and at the same time establish a connection to more sophisticated semiconductor laser models. In Section 48.3, we give a short overview of the key aspects in modeling nanolasers and the characterization of their emission properties. In Section 48.4, we introduce a microscopic laser model for solid-state emitters that goes beyond the rate equations and that provides access to the quantum-statistical properties of the emission. A central aspect of device performance is determined by the efficiency of carrier scattering processes following excitation. Section 48.5 provides an overview of the underlying mechanisms of carrier–carrier and carrier-phonon interaction and explains methods to incorporate the dynamics into semiconductor laser models within different levels of approximation. The underlying formalism allows one to address a wide variety of effects present in semiconductor nanolasers, and we provide a few examples in Sections 48.6 and 48.7, where QD lasing and radiative coupling between individual emitters are discussed. Section 48.8 gives an outlook toward systems with only a few QDs inside the cavity, where the individual electronic many-particle properties of each emitter matter, and where device properties are dominated by quantum and correlation effects.

48.2 Connecting to Earlier Work: Laser Rate Equations and the β Factor

Lasing operation takes place when pumping of the active material provides sufficient gain to compensate the losses. It is thus determined by the interplay of the excitation and emission dynamics of the gain material, taking into account the feedback provided by the radiation field inside the resonator. A basic understanding of lasing action can be obtained from laser rate equations in the form of two coupled equations for the gain (in this case, the number of excitons N in the QD ensemble) and the intracavity photon number n:

$$\frac{\mathrm{d}}{\mathrm{d}t} N = P - \gamma_l (n+1)N - \gamma_{\mathrm{nl}} N,$$

$$\frac{\mathrm{d}}{\mathrm{d}t} n = -\kappa n + \gamma_l (n+1)N. \tag{48.1}$$

Here, P and κ are the pump and photon loss rates, γ_l the spontaneous emission rate into the laser mode, γ_{nl} the rate of radiative recombination into other modes, and $\gamma_{\mathrm{sp}} = \gamma_l + \gamma_{\mathrm{nl}}$ the total spontaneous emission rate. The bracket $(n+1)$ contains both the stimulated and spontaneous emission contributions. The ratio

$$\beta = \frac{\gamma_l}{\gamma_{\mathrm{sp}}} \tag{48.2}$$

defines the β-factor illustrated in Figure 48.1. It is a key characteristic quantity of any laser device as it defines device performance in the spontaneous-emission regime (i.e., below threshold) and, more importantly, the threshold current. The rate equations (Equation 48.1) are often expressed in terms of β:

$$\frac{d}{dt} N = P - \beta\gamma_{sp}(n+1)N - (1-\beta)\gamma_{sp}N,$$

$$\frac{d}{dt} n = -\kappa n + \beta\gamma_{sp}(n+1)N. \tag{48.3}$$

In conventional lasers, $\beta \approx 10^{-5}$ [9], so that the majority of spontaneously emitted photons are lost. Once the system starts lasing at sufficiently high excitation, stimulated emission completely overwhelms spontaneous emission, so that the fraction of photons lost into nonlasing modes becomes insignificant. This behavior can be directly traced by accessing both emission channels separately. In micropillar lasers, the fundamental mode often used as laser mode has a far-field emission in vertical direction of the pillar, whereas losses occur laterally through the pillar side walls. The input–output curve (photon number versus pump rate) collected separately from both directions, together with a schematic of the experiment, is shown in Figure 48.2 [10]. As long as spontaneous emission is the dominating emission process, both emission intensities increase linearly with pump with a fixed ratio determined by the β factor. At higher excitation powers, stimulated emission from the laser mode sets in and photon emission from the laser mode increases nonlinearly and more strongly than the emission into loss channels. The resulting increase of the normalized ratio r of both emission channels therefore gives direct account of the onset of stimulated emission. Note that the S shape of the intensity curve for the lateral emission is due to stray light from the laser mode that is collected in lateral direction. Ideally, a clamped intensity is expected.

When continuous excitation is considered, the rate equations are solved in the steady state. Input–output curves obtained from the rate equations (Equations 48.1) for different values of β are shown in Figure 48.3. For $\beta \ll 1$, a sudden increase in emission intensity is visible that allows one to identify the lasing threshold. With increasing coupling efficiency β, the threshold shifts to lower pump rates, offering the prospect of devices with reduced threshold currents. At the same time, the intensity jump decreases until the "thresholdless" case is obtained for $\beta = 1$. In this regime an identification of the transition into lasing is impossible from the emission intensity alone. A characterization of the threshold requires going beyond the rate equation approximation to study statistical and coherence properties of the emission. One of the attractiveness of nanolasers is that they are capable of operating in a regime close to $\beta = 1$ due to their small effective mode volumes and large cavity lifetimes that cause strong funneling of spontaneously emitted photons into a single resonator mode [1]. At the same time, the photonic density of states is reduced away from the cavity-mode frequency, which further suppresses emission into nonlasing modes and thereby also

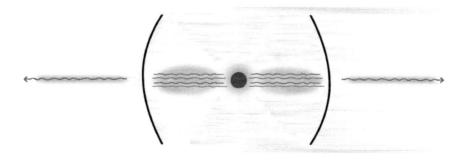

FIGURE 48.1 Spontaneous emission of the emitter into the laser mode (black lines) and into nonlasing modes (gray lines). The ratio of the spontaneous emission into the laser mode and the total spontaneous emission is described by the β factor.

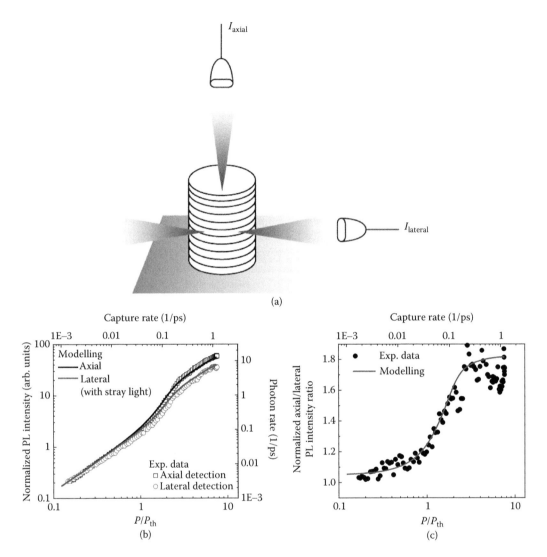

FIGURE 48.2 (a) Sketch of an experiment for detecting laser-mode emission in axial direction, and leaky mode emission in lateral direction separately. (b) Input–output characteristics of a quantum dot (QD) micropillar laser. Emission intensity into the laser mode (axial direction) and into nonlasing modes (lateral direction) are shown separately. (c) Their ratio *r* increases at the onset of stimulated emission, providing a direct visualization of the transition into the lasing regime. The results have been obtained from the semiconductor laser theory introduced in Section 48.4. Note that in lateral emission some scattered light from the axial emission is collected, which is responsible for the unclamped intensity behavior. [PL, photoluminescence. (Figures b and c adapted from Musiał, A. et al., *Physical Review B*, 91, 205310, 2015.)]

increases the β factor. While cavities operate at a specific narrow frequency window, structures that cause a broadband enhancement of spontaneous emission without using a cavity, such as photonic trumpets [11], are based on the very same concept.

The rate equations (Equation 48.1) have further drawbacks for modeling QD-based nanolasers. Most importantly, they do not provide information on laser field coherence and photon correlations. The rate equations are obtained by neglecting all correlations between light and matter degrees of freedom. Due to the enhanced interaction inside the microresonator, photon, and carrier-photon correlations have been demonstrated to play an important role in nanolasers, and this is where new approaches are required

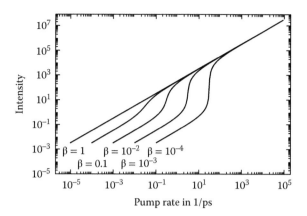

FIGURE 48.3 Laser input–output characteristics as obtained from rate equations. Shown are steady-state results at each pump value, parameterized by the β factor and the cavity loss rate. With increasing β, the threshold shifts to lower pumping and the laser threshold becomes increasingly hard to identify, and additional indicators beyond the reach of the rate-equation description are needed.

to understand and control these properties for new devices. In Section 48.4, we introduce a systematic approach to include semiconductor and correlation effects in extended laser models and discuss effects in nanolasers that set these systems apart from conventional laser devices.

48.3 Key Aspects of Nanolasers

Nanolasers use resonators that confine light in a volume of the light wavelength cube or smaller. The free spectral range, i.e., the energetic distance between resonator modes, grows with decreasing size of the cavity mode volume. Consequently, there is increased potential that the material gain overlaps only with one, or at most, a very few modes [12] of the resonator, as illustrated in the top right panel of Figure 48.4. This sets them apart from other lasers, such as edge-emitting lasers and the typical VCSELs [13,14], where optical modes lie more densely and emission is typically multimode (left panel). A nano-resonator may be a disk or pillar, employing distributed-Bragg reflectors [15], total internal reflection at semiconductor-air interfaces [16], or photonic crystal cavities [17].

It is a great advantage of small resonators that spontaneous emission enhancement $\propto Q/V$ can be used to strongly enhance the light–matter interaction and to favor emission into the single mode and thereby attain high values of the β factor that can approach unity. Here, V is the effective mode volume, and $Q = E/\Delta E$, with $E = \hbar\omega$ the mode energy and ΔE the mode linewidth, is the quality factor of the mode. The spontaneous-emission enhancement is direly needed to make up for the limited amount of gain material that can be placed inside the cavity. This so-called Purcell enhancement [18] can be directly observed in the temporal decay of the emission signal following short excitation. This is shown in Figure 48.5 [19] for micropillar nanolasers of different diameter and, therefore, different effective mode volume V.

As mentioned in Section 48.2, additional indicators are required to identify lasing in high-β devices due to the vanishing kink in the input–output curve. The most established method is to identify the statistical fluctuations of the emitted light. In the thermal regime below threshold, photons prefer to come in bunches. Lasing, on the other hand, originates from a coherent state of the light field, and photons arrive randomly in time. A transition from bunching to close to Poissonian emission is therefore a strong indicator for crossing the threshold to lasing. State-of-the-art streak-camera measurements give a direct visual account of this behavior: Figure 48.6 [20] shows temporally resolved single photon detection events for a system that has been excited by a short laser pulse. Counting statistics can be used to identify single photons, photon pairs, or bunches of more photons as indicated by the boxes. In practice, measurements are repeated thousands

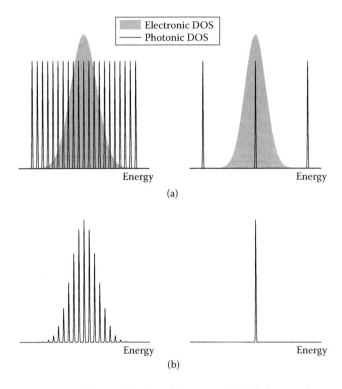

(a)

(b)

FIGURE 48.4 (a) Illustration of the photonic (black) and electronic (shaded) density of states of a typical laser (left) and a nanolaser (right) with the latter having a much larger free spectral range (energetic spacing between photon modes). The QD gain in the right panel is inhomogeneously broadened due to variations in composition or dimension, see Section 48.8. (b) The resulting spontaneous emission spectra are determined by the overlap of the photonic and electronic density of states. On the other hand, with the proper pillar, disk or photonic crystal design, a nanolaser can be single mode both below and above the lasing threshold. (DOS, density of states.)

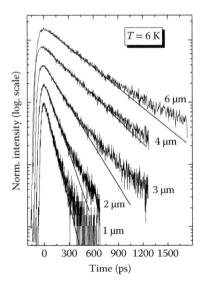

FIGURE 48.5 Time-resolved photoluminescence spectra following a *weak* excitation pulse. A decrease of the spontaneous emission time is observed with decreasing diameter of the micropillar cavities, reflecting the Purcell-enhancement $\propto Q/V$. The weak excitation ensures that the effect is not related to stimulated emission. (Reprinted from Schwab, M. et al., *Physical Review B*, 74, 045323, 2006. With permission.)

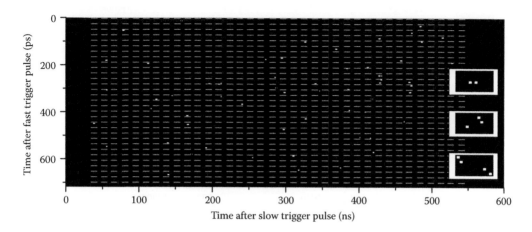

FIGURE 48.6 Streak-camera measurements provide impressive account of statistical fluctuations of the emitted light on the level of single photons. Single- and multiphoton emission events can be detected within a time-bin size of 2 ps (see magnification in the boxes) and provide insight into the quantum-mechanical state of the light field. To perform counting statistics in order to construct correlation functions like the ones shown in Figure 48.7, ten thousands of such measurements are performed. (Reprinted from Aßmann, M. et al., *Science* 325, 297–300, 2009. With permission).

of times. From the collected data, photon correlation functions $g^{(n)}$ can be obtained. They are of particular relevance, as they allow one to characterize the state of the quantum-mechanical light field as given by the distribution function p_n, which describes the *photon statistics* in terms of the probability of finding the system in the photon number state $|n\rangle$. Photon correlation functions are connected to the moments of the photon statistics:

$$\langle n^i \rangle = \sum_{k=0}^{\infty} k^i p_k. \tag{48.4}$$

As an example, the second-order photon-correlation function (sometimes referred to as photon autocorrelation function) is given by

$$g^{(2)} = \frac{\langle n^2 \rangle - \langle n \rangle}{\langle n \rangle^2}. \tag{48.5}$$

Correlation functions of order n take on a value of $n!$ if the light field is in a thermal state, and at all orders they possess a value of 1 for a perfectly coherent state [21]. For a micropillar QD nanolaser, this has been experimentally observed for $g^{(2)}, g^{(3)}$ and $g^{(4)}$ as function of pump rate, shown in Figure 48.7.

More commonly used than a streak camera is a Hanbury Brown and Twiss type measurement [22] schematically shown in Figure 48.8, where the bunching behavior of photons is quantified by the simultaneous registration of two photons at the two detector arms, providing access to the second-order correlation function $g^{(2)}$. For detectors, avalanche photodiodes (APDs) are used. Their time resolution is typically in the range of hundreds of picoseconds. Whether this resolution is sufficient depends on the type of measurement performed and the coherence time τ_{coh} of the emission signal. On the timescale of the latter, $g^{(2)}(\tau)$ goes to unity with respect to the delay time τ between two detection events, and the actual $g^{(2)}(0)$ at $\tau = 0$ can no longer be resolved if the detector's time resolution exceeds the coherence time, leading to the effect that $g^{(2)}(0)$ seems to take on a value of 1 even for thermal light [23].

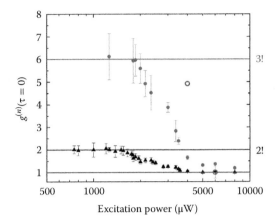

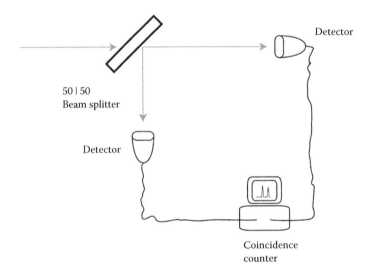

FIGURE 48.7 Photon autocorrelation functions $g^{(2)}(0)$ (triangles), $g^{(3)}(0)$ (dots), and $g^{(4)}(0)$ (open circles) as function of excitation power obtained from streak-camera measurements like the ones shown in the previous figure. For a typical LED to laser transition, all correlation functions show a transition from their thermal values $g^{(n)} = n!$ to that of a coherent field, $g^{(n)} = 1$. (Courtesy of Marc Aßmann, TU Dortmund.)

FIGURE 48.8 Illustration of a Hanbury Brown and Twiss measurement. A beam splitter distributes the emission between two detectors. Equal-time detection of two or more photons provides access to the second-order photon autocorrelation function $g^{(2)}(\tau = 0)$. A delay line in one arm can be used to measure $g^{(2)}(\tau \neq 0)$.

Another accessible indicator for coherent emission is a linewidth narrowing, or equivalently, an increase of the coherence time. Experimentally, the linewidth can be determined from spectra or Michelson interferometry measurements [21]. In the steady state $t = t_{ss}$, the coherence time τ_{coh} is related to the first-order coherence of the emission

$$\tau_{coh} = \int\limits_{-\infty}^{\infty} d\tau \, |g^{(1)}(t_{ss}, \tau)|^2, \tag{48.6}$$

which requires the calculation of the two-time quantity in the integral by means of the quantum-regression theorem [24,25]. We will not go into more detail here and refer the reader to [26].

The criteria for lasing in nanolaser devices are a subject of much discussion [5,27]. Ideally, a combination of several criteria is used to identify the lasing regime and the threshold. These include the following:

1. Coherent emission approaching Poissonian photon statistics p_n.
2. A value close to 1 in the second-order photon correlation function $g^{(2)}(0)$. Note that this is a weaker criterion than 1., as $g^{(2)}(0)$ is only linked to the first two moments of the full photon distribution function p_n, c.f. Equation 48.5.
3. An increase in the coherence time at the onset of lasing, when phase-coherent stimulated emission begins to dominate over spontaneous emission.
4. Cavity-photon-induced feedback of the gain material providing gain instead of absorption.
5. A mean intracavity photon number above 1.
6. An S-shaped nonlinear region in the input–output characteristics at threshold.
7. Carrier dynamics developing a hole-burning effect in the carrier-population functions at the energy of the laser transition, cf. Figure 48.13.

All of these criteria have their justifications, but on their own, neither may suffice to conclusively identify lasing. For example, coherent emission (1) is a prerequisite of lasing, but it may be realized without amplification (4 and 5) or originate from an external coherent drive. An S-shaped jump (6) in the input–output characteristic can originate from a variety of effects, such as a transition between multiexciton configurations feeding the laser mode [28], background contributions [29], or spectral wandering. On its own it is not a clear indicator for lasing. Moreover, lasing can take place in the absence of a nonlinear emission regime, e.g., if spontaneous emission losses are small ($\beta \to 1$), or saturation effects set in before the S shape can fully develop, which is particularly relevant under pulsed excitation [30].

48.4 Essential Physics I: Quantum-Optical Semiconductor Laser Theory

Different approaches can be taken in formulating a laser theory, including semiclassical methods, where the radiation field is classical and the system dynamics under the influence of the light–matter interaction is described by Maxwell–Bloch equations [31,32]. In the following, we introduce an extended laser model that accounts for semiconductor-specific effects, such as carrier scattering and dephasing, and that provides access to the statistical properties of the emission. The model is obtained in the framework of a correlation-expansion technique for the quantized light field interacting with carriers occupying the discrete QD conduction- and valence states. The method is versatile, as it allows one to include various aspects of semiconductor and correlation effects on a required level in a systematic and consistent manner. At the same time, the rate equations are contained as limiting case, as has been shown in [33].

In contrast to semiclassical approaches, using a quantized light field introduces the possibility to investigate emission properties beyond the classical regime. In fact, nonclassical effects, such as antibunched light emission around the threshold region, have been demonstrated in QD nanolasers [34]. In semiclassical models, the electromagnetic field drives a coherent polarization $\langle v_\nu^\dagger c_\nu \rangle$ across the bandgap of the semiconductor medium, as illustrated in Figure 48.9. The operators c_ν ($c_\nu^\dagger$) annihilate (create) a conduction-band carrier in the state $|\nu\rangle$, the operators v_ν ($v_\nu^\dagger$) are the equivalent for valence carriers. The resulting dynamics is familiar from atomic two-level models, where a resonant field drives Rabi oscillations. In a fully quantum-mechanical approach, the electromagnetic field is described by modes that are occupied by field quanta, the photons. If the modes are energetically well separated, like in a nanolaser, where the cavity possesses a large free spectral range, it suffices to consider the interaction of the active medium with photons in a single mode q. The coherent polarization $\langle v_\nu^\dagger c_\nu \rangle$ is then replaced by a photon-assisted polarization $\langle b_q^\dagger v_\nu^\dagger c_\nu \rangle$ that describes a carrier deexcitation (or excitation via $\langle b_q c_\nu^\dagger v_\nu \rangle$) by simultaneous emission (absorption)

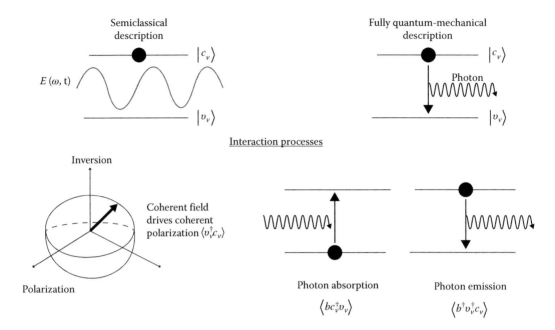

FIGURE 48.9 Illustration of the fundamental light–matter interaction in a semiclassical (left) and fully quantum-mechanical description (right). Semiclassically, matter degrees of freedom are described quantum-mechanically (e.g., by the states $|c_\nu\rangle$ and $|v_\nu\rangle$). The interaction with a classical coherent field $E(\omega, t)$ drives a coherent polarization $\langle v_\nu^\dagger c_\nu\rangle$. The polarization and population dynamics of often described in terms of the rotation of a vector on a unit sphere (Bloch sphere), as illustrated in the bottom left. By quantizing the light field, field quanta (photons) are introduced, and the fundamental interaction processes become the electronic (de)excitations via the simultaneous absorption or emission of a photon (bottom right). The semiclassical coherent polarization is replaced by the photon-assisted polarization $\langle b^\dagger v_\nu^\dagger c_\nu\rangle$.

of a photon into the mode q; the Bose operators b_q $(b_q^\dagger)$ annihilate (create) a photon in mode q. The quantum-mechanical and semiclassical pictures are schematically compared in Figure 48.9.

The fully quantum-mechanical laser model is obtained by calculating the dynamical equations for carrier populations $f_\nu^c = \langle c_\nu^\dagger c_\nu\rangle$, $f_\nu^v = \langle v_\nu^\dagger v_\nu\rangle$ in the conduction and valence states of the QD, and the expectation value of the photon number operator in the laser mode q, for which we drop the index q in the following. These equations are obtained from Heisenberg's equations of motion for operators $A(t)$, $\dot{A} = i[H, A]/\hbar$. In the Hamiltonian, we consider the free contributions of QD single-particle states (first and second term in Equation 48.8) and the electromagnetic field (third term in Equation (48.8)), and the light–matter interaction,

$$H = H^{\text{free}} + H^{\text{LM}}, \tag{48.7}$$

with

$$H^{\text{free}} = \sum_\nu \varepsilon_\nu^c c_\nu^\dagger c_\nu + \sum_\nu \varepsilon_\nu^v v_\nu^\dagger v_\nu + \hbar\omega\left(b^\dagger b + \frac{1}{2}\right) \tag{48.8}$$

and

$$H_{\text{LM}} = -i\sum_{\alpha\nu}\left(g_{\alpha\nu}\, c_\alpha^\dagger v_\nu b + g_{\alpha\nu}\, v_\alpha^\dagger c_\nu b\right) + \text{h.c.} \tag{48.9}$$

Here, $\hbar\omega$ is the mode energy, and $g_{\alpha\nu}$ is the light–matter coupling strength that is determined by the overlap of the wave functions of the QD single-particle states $|\alpha\rangle$ and $|\nu\rangle$ and the mode function of the confined electromagnetic field [33]. The sums run over the single-particle states of the individual QD emitters, and the index is to be understood to contain both, QD label and single-particle state. The interaction Hamiltonian contains the elementary processes of electron–hole recombination and generation via emission and absorption of a photon, including the ones schematically shown in the bottom right of Figure 48.9. Note that the counter-rotating terms in Equation (48.9) are often neglected in the rotating wave approximation.

48.4.1 Dynamical Laser Equations

For the dynamical evolution of the photon number $\langle b^\dagger b \rangle$ in the given mode and the electron and hole populations $f_\nu^e = \langle c_\nu^\dagger c_\nu \rangle$, $f_\nu^h = 1 - \langle v_\nu^\dagger v_\nu \rangle$, the contribution of the light–matter interaction H_{LM} in the Heisenberg equations of motion leads to [33]

$$\left(\hbar \frac{\mathrm{d}}{\mathrm{d}t} + \kappa \right) \langle b^\dagger b \rangle = 2 \operatorname{Re} \sum_{\nu'} |g_{\nu'}|^2 \langle b^\dagger v_{\nu'}^\dagger c_{\nu'} \rangle, \tag{48.10}$$

$$\hbar \frac{\mathrm{d}}{\mathrm{d}t} f_\nu^{e,h} = -2 \operatorname{Re} |g_\nu|^2 \langle b^\dagger v_\nu^\dagger c_\nu \rangle. \tag{48.11}$$

Here, we have scaled $\langle b^\dagger v_\nu^\dagger c_\nu \rangle \rightarrow g_{q\nu} \langle b^\dagger v_\nu^\dagger c_\nu \rangle$ to have the modulus of the coupling matrix elements appear and have further assumed that g is diagonal in the equal-envelope approximation. The finite lifetime of the cavity mode is introduced by considering a complex mode energy, $\tilde\omega = \omega - i\kappa$, where κ is directly connected to the Q-factor of the laser mode, $Q = \hbar\omega/2\kappa$. The dynamics of the photon number is determined by the abovementioned photon-assisted polarization $\langle b^\dagger v_\nu^\dagger c_\nu \rangle$ that describes the expectation value for a correlated event, where a photon in the mode q is created in connection with an interband transition of an electron from a conduction to a valence state. The sum over ν involves all possible interband transitions from various QDs.

An important component of the carrier dynamics is the recombination into nonlasing modes giving rise to the β factor discussed in Section 48.2. For the fraction $1 - \beta$ of carriers emitting into nonlasing modes, a term proportional to the spontaneous emission rate can be added to the carrier population equations to account for the losses [33]:

$$\left. \frac{\mathrm{d}}{\mathrm{d}t} f_\nu^{e,h} \right|_{\mathrm{nl}} = \gamma_{\mathrm{sp}} (1 - \beta) f_\nu^e f_\nu^h. \tag{48.12}$$

In contrast to the rate equations (Equation 48.3), where this contribution is typically accounted for by the β coefficient, the spontaneous emission losses in a semiconductor are determined by both electron and hole populations. The dynamical equation for the cavity-photon-assisted polarization is given by

$$\left(\hbar \frac{\mathrm{d}}{\mathrm{d}t} + \kappa + \Gamma + i \left(\tilde\varepsilon_\nu^e + \tilde\varepsilon_\nu^h - \hbar\omega \right) \right) \langle b^\dagger v_\nu^\dagger c_\nu \rangle =$$

$$f_\nu^e f_\nu^h - \left(1 - f_\nu^e - f_\nu^h \right) \langle b^\dagger b \rangle + \frac{1}{g_\nu} \sum_\alpha g_\alpha C_{\alpha\nu\nu\alpha}^{\mathrm{x}} + \delta \langle b^\dagger b c_\nu^\dagger c_\nu \rangle - \delta \langle b^\dagger b v_\nu^\dagger v_\nu \rangle. \tag{48.13}$$

The free evolution of $\langle b^\dagger v_\nu^\dagger c_\nu \rangle$ is determined by the detuning of the QD transitions from the cavity laser mode. This oscillatory term drops out if the transition is in perfect resonance with the mode. Only accounting for identical resonant QDs is a common and well-justified practice in ensembles of many (>100 emitters), as the summation over all QDs becomes a mere prefactor of the QD number, which is a significant simplification. In a semiconductor, the source term of spontaneous emission is described by an expectation value of four carrier operators $\langle c_\alpha^\dagger v_\alpha v_\nu^\dagger c_\nu \rangle$, see [35]. The Hartree–Fock factorization of this source term leads to $f_\nu^e f_\nu^h$, which appears as the first term on the right hand side of Equation 48.13. Corrections to this factorization are included in $C_{\alpha'\nu\nu\alpha}^x = \delta\langle c_{\alpha'}^\dagger v_\nu^\dagger c_\nu v_\alpha\rangle$. Dephasing is what destroys a coherent polarization. In semiconductors, dephasing arises from carrier scattering processes, excitation of carriers, and from the interaction with lattice vibrations that are responsible for the homogeneous linewidth. These contributions are here summarized in a phenomenological constant Γ.

The correlation functions $\delta\langle b_q^\dagger b_q c_\nu^\dagger c_\nu\rangle$ and $\delta\langle b_q^\dagger b_q v_\nu^\dagger v_\nu\rangle$ introduce carrier-photon correlations that are neglected on the level of rate equations. Their time evolution is given by [33]

$$\left(\hbar\frac{d}{dt} + 2\kappa\right)\delta\langle b^\dagger b c_\nu^\dagger c_\nu\rangle = -2\,|g_\nu|^2\,\mathrm{Re}\left[\delta\langle b^\dagger b^\dagger b v_\nu^\dagger c_\nu\rangle + \left(\langle b^\dagger b\rangle + f_\nu^e\right)\langle b^\dagger v_\nu^\dagger c_\nu\rangle\right], \tag{48.14}$$

$$\left(\hbar\frac{d}{dt} + 2\kappa\right)\delta\langle b^\dagger b v_\nu^\dagger v_\nu\rangle = 2\,|g_\nu|^2\,\mathrm{Re}\left[\delta\langle b^\dagger b^\dagger b v_\nu^\dagger c_\nu\rangle - +\left(\langle b^\dagger b\rangle + f_\nu^h\right)\langle b^\dagger v_\nu^\dagger c_\nu\rangle\right] \tag{48.15}$$

which couples to

$$\left(\hbar\frac{d}{dt} + 3\kappa + \Gamma + i\left(\tilde{\varepsilon}_\nu^e + \tilde{\varepsilon}_\nu^h - \hbar\omega\right)\right)\delta\langle b^\dagger b^\dagger b v_\nu^\dagger c_\nu\rangle = -2\,|g_\nu|^2\,\langle b^\dagger v_\nu^\dagger c_\nu\rangle^2 + 2\left[\langle b^\dagger b\rangle + f_\nu^h - f_\nu^e\right]\times$$

$$\left[\delta\langle b^\dagger b c_\nu^\dagger c_\nu\rangle - \delta\langle b^\dagger b v_\nu^\dagger v_\nu\rangle\right] - \left(1 - f_\nu^e - f_\nu^h\right)\delta\langle b^\dagger b^\dagger bb\rangle \tag{48.16}$$

and

$$\left(\hbar\frac{d}{dt} + 4\kappa\right)\delta\langle b^\dagger b^\dagger bb\rangle = 4\,|g_\nu|^2\sum_{\nu'}\delta\langle b^\dagger b^\dagger b v_{\nu'}^\dagger c_{\nu'}\rangle. \tag{48.17}$$

On this level of approximation, we obtain a closed system of coupled equations that go beyond the rate equations by containing carrier-photon and photon–photon correlation functions that provide access to the second-order photon-correlation function

$$g^{(2)}(\tau = 0) = 2 + \frac{\delta\langle b^\dagger b^\dagger bb\rangle}{\langle b^\dagger b\rangle^2}. \tag{48.18}$$

48.4.2 Pump Process for Continuous-Wave-Excited Lasers

The earlier equations describe the light–matter interaction of a semiconductor system in terms of carrier populations and the polarization-analogue for the quantized light field, i.e., the photon-assisted polarization. An important component is yet missing, namely the interaction of charge carriers with the environment that allows for some sort of excitation mechanism. Pumping can be either done by current injection or by optical excitation. In both cases, electrons and holes are typically created in the continuum states of the wetting layer (WL) or barrier material, from where they are captured into the discrete QD states. This incoherent capture is one example of carrier scattering processes that, at this stage, are still missing from our description. While the quantitative modeling of these many-body processes is discussed in detail in Section 48.5, for continuous-wave (CW) excitation it is often sufficient to consider only the dynamics of carriers in the localized QD states. In its most simple form, two localized states are used for electrons and holes each, which closely resembles a four-level laser scheme. This choice is physically motivated by the electronic structure of typical QDs, which, depending on their size, consists of several confined states. We refer to the lowest (highest) and second-lowest (second highest) electron (hole) state as s- and p-states, respectively, as shown in Figure 48.10. In this reduced level scheme, the p-states are used for carrier pumping at rate P, which approximates the combined effect of carrier excitation in higher

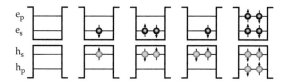

FIGURE 48.10 In a QD with several confined single-particle states for electrons and holes (the s- and p-states are shown), a multitude of many-particle configurations can be realized by occupying these states with electrons and holes. These many-particle configurations contain the neutral, charged, and multiexciton states, and dark states that differ in energy due to the Coulomb interaction between carriers.

states and the successive capture into the QD states:

$$\frac{d}{dt} f_p^e \bigg|_{Pump} = P(1 - f_p^e - f_p^h) \tag{48.19}$$

$$\frac{d}{dt} f_p^h \bigg|_{Pump} = P(1 - f_p^e - f_p^h). \tag{48.20}$$

The blocking factors in the brackets ensure that the occupation remains bounded by unity. From their generation in the QD *p*-states, a phenomenological set of equations distributes carriers into the QD *s*-states:

$$\frac{d}{dt} f_s^e \bigg|_{scatt} = \gamma_{p \to s}^{r,e} f_p^e (1 - f_s^e) - \gamma_{s \to p}^{r,e} f_s^e (1 - f_p^e) \tag{48.21}$$

$$\frac{d}{dt} f_s^h \bigg|_{scatt} = \gamma_{p \to s}^{r,h} f_p^h (1 - f_s^h) - \gamma_{s \to p}^{r,h} f_s^h (1 - f_p^h) \tag{48.22}$$

$$\frac{d}{dt} f_p^e \bigg|_{scatt} = -\gamma_{p \to s}^{r,e} f_p^e (1 - f_s^e) + \gamma_{s \to p}^{r,e} f_s^e (1 - f_p^e) \tag{48.23}$$

$$\frac{d}{dt} f_p^h \bigg|_{scatt} = -\gamma_{p \to s}^{r,h} f_p^h (1 - f_s^h) + \gamma_{s \to p}^{r,h} f_s^h (1 - f_p^h). \tag{48.24}$$

Equations 48.21 through 48.24 are motivated in the more general concepts of carrier kinetics in the following section. Together, Equations 48.10–48.24 serve as an accessible and complete model for QD-based lasers, and we discuss exemplary results in Section 48.6.

48.5 Essential Physics II: Carrier Kinetics in QD Systems

As we overviewed in Section 48.2, rate equations can provide important insights for the theoretical understanding of lasers. They can be used to describe input–output curves [36–39], and, combined with Langevin approaches [40,41] even the photon statistics and intensity noise.

An important reason why carrier dynamics is essential for the understanding of laser operation is the excitation mechanism. To see this, let us assume that carrier populations in lasers are given by Fermi–Dirac distribution functions (often a reasonable approximation). This implies that states with below-average kinetic energy are Pauli-blocked for the excitation mechanism. This is illustrated in Figure 48.11, where states with above-average kinetic energy are available for the excitation, while highly occupied states (shaded region, applied to 77K) are not. Therefore, it is important to understand how carriers, injected at the green-shaded area, end up in the spectral window from where they recombine. Accountable for this are two mechanisms that we discuss in this section. The carrier–carrier Coulomb interaction redistributes carriers between different states and leads to an equilibration towards a Fermi–Dirac distribution [42–46]. But as every carrier scattered downward in energy is partnered by a scattering partner that is scattered upward (right sketch in Figure 48.12), the Coulomb scattering is not able to dissipate energy.

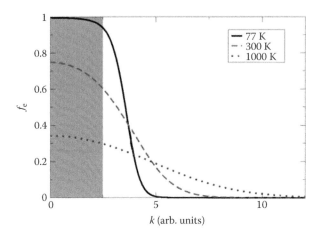

FIGURE 48.11 Fermi–Dirac distribution for electrons in a QW at different temperatures and a carrier density of $5 \times 10^{11} \text{cm}^{-2}$, the shaded background indicates the availability for carrier injection for the low-T distribution.

FIGURE 48.12 Conceptual sketch of carrier scattering processes. Left: carrier–phonon interaction, right: carrier–carrier Coulomb interaction.

This leads to a hot Fermi–Dirac function such as depicted in Figure 48.11, that has significantly different properties regarding inversion and gain. The cooling toward the lattice temperature is provided by the carrier-phonon interaction [47–51], that provides for the cooling by taking up the energy of the hot carriers and distributing it toward crystal lattice (left sketch in Figure 48.12). In a laser under CW-excitation conditions the energetic difference from excitation to emission energy also leads to heating and often the carrier-phonon interaction is not strong enough to provide complete cooling towards the lattice temperature. Therefore, a dynamical equilibrium is reached that may consist of Fermi–Dirac distributions with several 100K above the lattice temperature [52,53].

As already outlined in Section 48.4, it is generally necessary to set up equations of motion not for the total carrier density N but for individual carrier populations of different single particle states $f_\alpha^c = \langle c_\alpha^\dagger c_\alpha \rangle$ to describe these effects. Such so-called "kinetic equations" together with the laser equations, Equations 48.10 through 48.20, enable us to treat excitation kinetics and turn-on dynamics of semiconductor QD-based laser systems and will be the topic of this section.

The question of what timescales and by which processes the carriers approach (quasi-) equilibrium conditions (if at all) influences the design of laser devices, as the answer to these questions determines, e.g., the dynamical response and turn-on delay [53–57]. While rate equations can also describe the turn-on dynamics for conventional laser devices, where carrier dynamics takes place on a fs to ps timescale, whereas the laser turn-on happens on the order of 10 ns, the situation is very much different in nanocavities with large Purcell enhancement, i.e., the enhancement of the spontaneous emission rate due to the presence of the cavity, where the turn-on delay can be on the order of 10 ps [53,58] and thus on the same timescale as the carrier scattering. The use of population function both for QD and WL states also allows one to describe effects such as spectral hole burning, or the difference between optical and injection pumping, that are unavailable to a theoretical model that tracks only the dynamics of the total carrier density N.

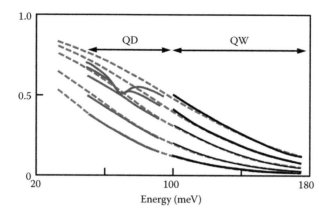

FIGURE 48.13 Carrier populations in a QD+WL system under laser operation for increasing pump intensity. The gray lines describe the population of QD carriers (subject to inhomogeneous broadening), the black lines show the populations of the WL states. For high pump rates the "hole" burnt into the population function by laser operation is clearly visible. The shaded-dashed lines are corresponding Fermi–Dirac functions. (From Chow, W. W. et al., *Light: Science & Applications*, 3, e201, 2014. With permission.)

An example of such carrier population functions are shown in Figure 48.13. The seemingly continuous features of the discrete QD single-particle state populations originate from the large number of emitters in the inhomogeneously broadened ensemble, cf. Section 48.8.

The simplest equation for such state-dependent carrier dynamics is given by

$$\frac{d}{dt}f_\alpha = (1 - f_\alpha)\Gamma_\alpha^{in} - f_\alpha\Gamma_\alpha^{out}, \tag{48.25}$$

where Γ_α^{in} and Γ_α^{out} give the in- (out-) scattering rates into (out of) the state α. A more detailed version of Eq. (48.25) has the structure

$$\frac{d}{dt}f_\alpha = \sum_\beta f_\beta(1 - f_\alpha)\gamma_{\beta\to\alpha} - f_\alpha(1 - f_\beta)\gamma_{\alpha\to\beta} \tag{48.26}$$

and describes that the population of state α is enlarged by scattering from β to α and reduced by scattering from α to β. The Pauli-blocking terms $f_\beta(1 - f_\alpha)$ and $f_\alpha(1 - f_\beta)$ occur as scattering has to be proportional to the occupancy f of the initial and to the non-occupancy $(1 - f)$ of the final state. Under continuous excitation (CW laser), it may suffice to restrict the description to the s-shell and p-shell states of the QD and to use phenomenological scattering rates $\gamma_{\beta\to\alpha}^{e,h}$, in which case we recover Equation 48.21 through 48.24 in Section 48.4.

An often used approximation that is generally of relevance if not only QD states are discussed but also the continuous states of the WL or barrier are taken into consideration is the so-called relaxation time approximation

$$\frac{d}{dt}f_\alpha = \frac{f_\alpha - F_\alpha(N, T)}{\tau}, \tag{48.27}$$

which describes the relaxation of a nonequilibrium carrier population f_α toward a Fermi function $F_\alpha(N, T)$ at temperature T and carrier density N on a time-scale τ. While this approximation often works well, it also necessitates to determine the time-scale τ and the carrier temperature T. The last point is not trivial as in laser devices the carriers typically posses a higher temperature than the lattice [52,53]. To answer these questions and also to investigate situations where the relaxation time approximation (Equation 48.27) is not applicable we have to take the microscopic origin of the scattering process into account. In the following, we discuss the dominant physical mechanisms behind the carrier dynamics, which are carrier-phonon scattering and carrier–carrier Coulomb scattering.

48.5.1 Carrier–Phonon Interaction

As discussed earlier, carrier temperature is an important factor for laser operation as it controls the population inversion. Hence the optical gain depends strongly on the temperature of the carrier system [59,60]. As the carrier–carrier Coulomb interaction cannot dissipate energy, the source for thermalization of the carriers toward the lattice temperature is the interaction with phonons. To investigate this interaction mechanism, we first have to determine which phonon modes are responsible for the carrier dynamics. A main difference between different phonon branches is whether they are longitudinal or transversal in nature. One can show [61] that due to their ability to form polarizations, mainly longitudinal phonons couple to the carrier system and that for most situations the interaction with transversal phonons is negligible. This leaves us with two phonon branches that are most important for the carrier-phonon interaction. These are the longitudinal optical (LO) phonons and longitudinal acoustic (LA) phonons. The speed of propagation of an acoustic phonon, which is also the speed of sound in the lattice, is given by the slope of the acoustic dispersion relation, $\frac{\partial \omega_k}{\partial k}$. At low values of k, the dispersion relation is almost linear, and the speed of sound is approximately independent of the phonon wavenumber.

Even though we are interested in laser structures, where the active material is composed of nanostructures (QDs), it is a good approximation to take the phonon modes and interaction potentials of the surrounding bulk material. This is the case, as we are mostly interested in long-wavelength phonons where the phonon wavelength is much larger than the size of the nanostructure we are investigating.

The interaction of carriers with LO phonons can be described by the Fröhlich coupling [61]. As this coupling mechanism is Coulombic in nature, the corresponding interaction matrix elements drop off rapidly with increasing momentum transfer $\mathbf{q}$. Therefore a often used approximation consists of treating the LO phonons as dispersionless with constant frequency $\hbar\omega_\mathbf{q} = \hbar\omega_{LO}$ [62]. For acoustic phonons, the main interaction mechanisms are the deformation potential coupling and the piezoelectric coupling [62]. While these play an important role in the line broadening of QD lines in the low temperature and low carrier-density regime, they do not provide an efficient source for carrier scattering.

The interaction matrix elements for the interaction of carriers with LO-phonons are [60,63]

$$|M_{\alpha\beta}|^2 = \frac{M_{LO}^2}{e^2/\varepsilon_0} V_{\alpha\beta\alpha\beta},$$
(48.28)

containing Coulomb matrix elements $V_{\alpha\beta\alpha\beta}$ (see Equation 48.33 and Reference [64]), the elementary charge e and the vacuum permittivity ε_0. The prefactor $M_{LO} = 4\pi\alpha\frac{\hbar}{\sqrt{2m}}(\hbar\omega_{LO})^{\frac{3}{2}}$ includes the polar coupling strength α and the reduced mass m. The matrix elements are given here in a way that explicitly shows the Coulombic nature of the carrier-LO-phonon coupling.

A first approach toward carrier scattering is given by the Boltzmann equation [40,61]. For carrier-phonon scattering, this is given by Equation 48.26 with

$$\gamma_{\beta\rightarrow\alpha} = \sum_q M_{\beta\alpha}(q)\Big((1 + n_q)\delta(\epsilon_\beta - \epsilon_\alpha - \hbar\omega_{LO}) + n_q\delta(\epsilon_\beta - \epsilon_\alpha + \hbar\omega_{LO})\Big).$$
(48.29)

The terms proportional to n_q and $(1 + n_q)$ describe phonon absorption and emission processes. In QD systems, additional effects exist that are not described within the framework of a Boltzmann equation, but are closely connected to carrier scattering, as they stem from the same microscopic interaction mechanism. These are quasi-particles and non-Markovian effects. The main reason is that the carrier-phonon interaction leads to the formation of a new quasi-particle. The quasi-particle obtained by dressing the carriers with the carrier-phonon interaction—the polaron—describes the lattice distortion accompanying the electron in its motion. In the framework of Green's functions (GFs), polarons are described using the retarded GF Kadanoff–Baym equation [63]:

$$\left[i\hbar \frac{\partial}{\partial \tau} - e_\alpha^a \right] G_\alpha^{a,R}(\tau) = \delta(\tau) + \int d\tau' \ \Sigma_\alpha^{a,R}(\tau - \tau') \ G_\alpha^{a,R}(\tau'). \tag{48.30}$$

The corresponding retarded selfenergy in random-phase approximation (RPA) [62] is given by

$$\Sigma_\alpha^{a,R}(\tau) = i\hbar \sum_\beta |M_{\alpha\beta}|^2 G_\beta^{a,R}(\tau) \ d^<(-\tau), \tag{48.31}$$

where the phonon propagators $d^{\gtrless}$ contain the phonon frequency and the phonon population [63]. The spectral function $\hat{G}_\alpha(\hbar\omega) = 2\text{Im}(G_\alpha^{a,R}(\hbar\omega))$, which follows from the retarded GF, can be seen as a generalization of the single particle energy, i.e., a free particle has a spectral function of the form $\hat{G}_\alpha(\hbar\omega) = \delta(\epsilon_\alpha - \hbar\omega)$. In contrast, the Polaronic spectral functions but also posses a finite spectral width that is connected to the lifetime of the respective quasi-particle.

Using the quasi-particles as described by the polaron retarded GF, we can formulate a quantum-kinetic equation for polaron scattering:

$$\frac{\partial f_\alpha^a(t)}{\partial t} = 2\,\text{Re} \sum_\beta \int_{-\infty}^{t} dt' \ |M_{\alpha\beta}|^2 G_\beta^{a,R}(t, t') \left[G_\alpha^{a,R}(t, t') \right]^*$$

$$* \left\{ \left[f_\beta^a(t')(1 - f_\alpha^a(t')) \right] d^>(t', t) - \left[f_\alpha^a(t')(1 - f_\beta^a(t')) \right] d^<(t', t) \right\}, \tag{48.32}$$

which contains scattering by phonon emission and absorption processes just as in the Boltzmann equation. Effects included here that go beyond the Boltzmann treatment are a weakening of the exact energy conservation due to the polaron properties (finite spectral width) and due to non-Markovian effects described by the explicit dependence of the time change of f_α^a at time t on all earlier times.

Results from such a quantum-kinetic treatment are shown in Figure 48.14. The QD level spacing is chosen to be 10% higher than the LO-phonon energy of 36meV. Therefore, a Boltzmann scattering equation predicts a complete inhibition of scattering due to the δ-functions in Equation (48.29). In contrast, the quantum-kinetic treatment shows that, due to quasi-particle effects (dotted line), scattering channels open up. The solid line shows the combined influence of quasi-particles and non-Markovian effects, that are caused by the time integrals in Equation (48.32). The inhibition of scattering as given by a Boltzmann equation, was predicted as a "phonon-bottleneck" early on [65,66], however, fast-carrier scattering was observed e.g., in Reference [67], showing the relevance of polaronic effects. For details see References [63,68,69].

48.5.2 Coulomb Interaction

The most important carrier scattering mechanism in the regime of high excitation densities, relevant for laser operation, is the carrier–carrier Coulomb interaction. It leads to a fast redistribution of carriers from

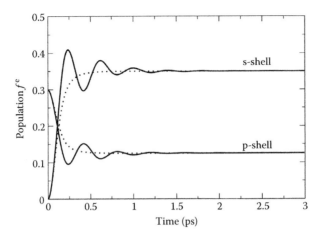

FIGURE 48.14   Redistribution of carrier occupations due to scattering of electrons from p-shell to s-shell in a QD via the carrier phonon interaction. In a Boltzmann-equation approach the occupations would remain constant due to a complete lack of scattering processes as described in the text (Reprinted from Seebeck, J. et al., *Physical Review B*, 71, 125327, 2005. With permission).

the spectral window where excitation takes place toward the spectral window from where they recombine. While Coulomb scattering cannot dissipate energy and thermalize the system (see Figure 48.12), it leads to equilibration toward a hot Fermi–Dirac function.

To determine the strength of the Coulomb interaction in nanostructures, we need to determine matrix elements of the Coulomb interaction in the single-particle basis given by, e.g., tight-binding descriptions of QDs. Using such single-particle wavefunctions, we can construct the Coulomb matrix elements

$$V_{\alpha\beta\gamma\delta} = \frac{1}{A} \sum_{\mathbf{q}} V_{\mathbf{q}} \langle \alpha | e^{-i\mathbf{q}\cdot\mathbf{r}} | \delta \rangle \langle \beta | e^{+i\mathbf{q}\cdot\mathbf{r}} | \gamma \rangle, \tag{48.33}$$

consisting of overlap integrals between single-particle wavefunctions and the Coulomb potential $V_{\mathbf{q}}$ [64]. A closer analysis of these Matrix elements also points towards a source for the efficiency of Coulomb scattering in nanostructures. Even though the available phase space of scattering partners is reduced significantly in QDs, compared to QW or bulk systems, this is partially balanced by enhanced interaction matrix elements due to the strong localization of the QD wave functions.

Like for the carrier-phonon scattering a first description of carrier–carrier scattering can be given by a Boltzmann-type equation

$$\frac{\mathrm{d}}{\mathrm{d}t} f_{\alpha}(t) = \frac{2\pi}{\hbar} \sum_{\beta\gamma\delta} \left(\left| W_{\alpha\gamma\delta\beta} \right|^2 - W_{\alpha\gamma\delta\beta} W^*_{\alpha\gamma\beta\delta} \right) \delta(\epsilon_{\alpha} - \epsilon_{\beta} + \epsilon_{\gamma} - \epsilon_{\delta})$$

$$\left[f_{\beta}(t)(1 - f_{\alpha}(t))f_{\delta}(t)(1 - f_{\gamma}(t)) - f_{\alpha}(t)(1 - f_{\beta}(t))f_{\gamma}(t)(1 - f_{\delta}(t)) \right], \tag{48.34}$$

which contains screened interaction matrix elements $W_{\alpha\gamma\delta\beta}$ and an energy conserving δ-function. A feature of the carrier–carrier Coulomb scattering in QDs is that, while most scattering channels become extremely inefficient at low carrier density due to the population factors in Equation (48.34), some some scattering channels like electron–hole scattering are possible at very low carrier densities [70].

Results for the carrier–carrier Coulomb scattering in a QD laser under optical cw excitation are presented in Figure 48.15, where the carrier population functions is shown as a function of energy for different times. Discrete symbols denote the QD populations for s-shell and p-shell and the lines the continuous

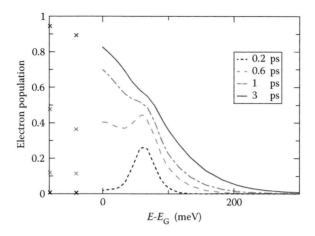

FIGURE 48.15 Carrier populations as a function of energy for electrons in a QD system under laser operation for different times. Symbols denote QD populations, lines WL populations. (Reprinted from Lorke, M. *Applied Physics Letters*, 99, 151110, 2011. With permission.)

WL states that are optically excited. After the initial Gaussian population profile has been excited by the pump, an ultra-fast redistribution is observed, that immediately starts to populate the QD states. It should be noted that, contrary to an often assumed situation [56,71], the timescales between the relaxation dynamics in the WL and the dynamics of the QD populations are not decoupled. On one hand, this is due to the efficiency of the initial in-scattering into the QD, as the QD states are initially empty. On the other hand, the relaxation within the WL is significantly slower than in a pure QW, as the capture into the QD states is most efficient for the WL states with low quasi-momenta. Therefore, these states are constantly depleted during the early stage of the kinetics, which slows down the relaxation of the WL distribution itself toward a quasi-equilibrium distribution [53] Furthermore, after the initial relaxation that lasts about 3 ps, a heating of the carrier population is observed on a timescale of 100 ps, as the pumping into the WL injects carriers at a higher energy than the emission energy.

These results also show (compared to those of the preceding section) that Coulomb and phonon scattering act on the same timescale in QD systems.

As for the phonons, quantum-kinetic effects lead to quasi-particle formation and non-Markovian effects. The corresponding quantum-kinetic equations can be written as

$$
\frac{\partial f_\alpha^a(t)}{\partial t} = \frac{2}{\hbar} \mathrm{Re} \sum_{\beta\gamma\delta} \int_{-\infty}^{t} dt' \left(\left| W_{\alpha\gamma\delta\beta} \right|^2 - W_{\alpha\gamma\delta\beta} W_{\alpha\gamma\beta\delta}^* \right) G_\beta^{a,R}(t,t') \left[G_\alpha^{a,R}(t,t') \right]^* * G_\delta^{a,R}(t,t') \left[G_\gamma^{a,R}(t,t') \right]^*
$$

$$
* \left[f_\beta(t')(1 - f_\alpha(t')) f_\delta(t')(1 - f_\gamma(t')) - f_\alpha(t')(1 - f_\beta(t')) f_\gamma(t')(1 - f_\delta(t')) \right]. \quad (48.35)
$$

As for the Boltzmann-type equation for carrier scattering, this equation contains direct and exchange scattering, proportional to $\left| W_{\alpha\gamma\delta\beta} \right|^2$ and $W_{\alpha\gamma\delta\beta} W_{\alpha\gamma\beta\delta}^*$, respectively. The energy conserving δ-function is generalized by the product of retarded GFs just as for the carrier-phonon interaction and via the explicit time dependence on earlier times memory effects are included. While these equations have not been evaluated in detail due to their numerical complexity, it is expected that the included quasi-particle and non-Markovian effects would influence several properties of nanolasers. On one hand, large signal modulation and turn-on delay are expected to be influenced by non-Markovian effects. On the other hand, quasi-particle effects can, as for the carrier-phonon interaction, alter the scattering rates influencing also the modulation response and threshold current. In the photon-assisted polarization, Equation 48.13 exist analogous expressions to the scattering contributions given in Equation 48.35 that lead to excitation-enhanced dephasing

$$\frac{d}{dt}\langle b^{\dagger} v_{\alpha}^{\dagger} c_{\alpha}\rangle(t) = \int_{-\infty}^{t} dt'(-\Gamma_{\alpha}^{DD}(t,t'))\langle b^{\dagger} v_{\alpha}^{\dagger} c_{\alpha}\rangle(t') + \sum_{\beta} \Gamma_{\alpha\beta}^{OD}(t,t')\langle b^{\dagger} v_{\beta}^{\dagger} c_{\beta}\rangle(t'), \qquad (48.36)$$

where Γ^{DD} and Γ^{OD} consist of interaction matrix elements W, retarded GFs $G^{a,R}$ and population functions f_{α} in a similar way as they appear in Equation 48.35 for the carrier scattering. We have investigated the influence of quasi-particle and non-Markovian effects on the quasi-classical analogues of these quantities in detail to study optical gain spectra of QD systems [8,60,72–75] and we expect an influence of these contributions on the turn-on behavior and the modulation dynamics [38].

48.6 Connecting to Experiments: Transition to Lasing and the β Factor

As first example, we discuss characteristic input–output curves and statistical properties of a Purcell-enhanced QD nanolaser device. In Figure 48.16 results from the semiconductor laser model are shown for various values of the β factor and fixed number of QD emitters. For β = 1, all spontaneous emission is

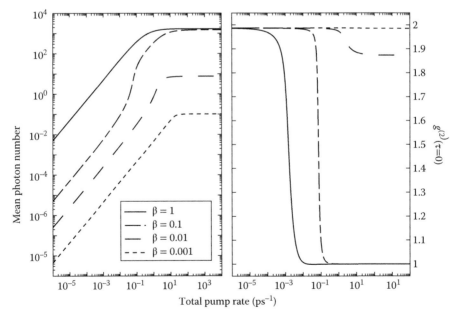

FIGURE 48.16 Input–output curves (left) and autocorrelation function $g^{(2)}(0)$ (right) for a nanolaser with 300 QD emitters. Curves are shown for different values of the β factor. For β = 1, a visible threshold is absent in the input–output curve, whereas $g^{(2)}(0)$ shows a clear transition from thermal to coherent emission. Note that the jump from below to above threshold is larger than in the results obtained from rate equations (Figure 48.3) and does here no longer scale with $1/\beta$. For β = 0.1, a soft kink appears in the input–output curve. For smaller values of β, the limited number of QD emitters cannot provide sufficient gain to reach lasing before saturation is reached. A cavity-Q factor of 32,000 and a cavity-enhanced spontaneous emission rate of $\tau_{sp} = 50$ ps has been used.

directed into the laser mode. In contrast to the rate-equation result that we have discussed in the context of Figure 48.3, saturation is visible at strong pumping. This effect is important in QD nanolasers, since the small size of the resonator naturally limits the amount of gain material that can be brought into the cavity. A monotonous increase of the output intensity with pumping, as given by the rate equations (Equation 48.1), would be unphysical. As a consequence, for smaller values of β, the gain can become insufficient to reach lasing before saturation sets in. In the example in Figure 48.16, this is visible for $\beta = 0.01$, where the threshold is reached, but the emission remains largely thermal with $g^{(2)}(0) \approx 1.9$, and for $\beta = 0.001$, where the QD emission saturates long before the laser threshold at a low mean intracavity photon number of 0.1. Access to the photon autocorrelation function $g^{(2)}(0)$ provides an important tool in judging the emission properties far beyond what can be inferred from the emission intensity alone and is a significant advancement over the rate equations.

Deeper insight into cavity-QED effects in nanolasers can be obtained from the emission and correlation *dynamics*. The top panel in Figure 48.17 shows time resolved the emission pulse (solid curves) and autocorrelation function $g^{(2)}(t, \tau = 0)$ (symbols) following picosecond-short optical excitation of the nanolaser. Also here the high time resolution has been realized by using a streak-camera setup. Following a sufficiently strong excitation pulse, the statistical properties of the emission pulse change as function of time from thermal to coherent and back to thermal. Coherent emission is reached around the peak region of the emission pulse and forms a plateau when the excitation level is above the threshold region. The data agrees well with time-dependent results of the laser model (bottom panel), which are readily available from the time evolution of the coupled differential equations presented in Section 48.4. In contrast to conventional measurements that integrate over the emission pulse, here it is possible to track the degree of coherence during the pulse and to provide an understanding of the coherence it contains. In their combination, state-of-the art measurements and microscopic models can reveal new aspects of solid-state light sources, reflecting the emission dynamics and the dynamics of quantum-mechanical correlations down to the few-photon level. The explicit treatment of dynamical properties of nanolasers, such as carrier dynamics, turn-on delay, and modulation response requires to go beyond the simple approach in Equations 48.21 through 48.24 for the carrier scattering and use the more involved methods of Section 48.5 due to the sensitivity of these quantities [55].

48.7 Interemitter Coupling Effects and Superradiance in Nanolasers

When several emitters couple to a common light field, the interaction introduces correlations between them. The eigenstates of the interacting system are then no longer given by those of the individual systems, but by so-called Dicke states of the collective system. In 1954, Dicke has shown [76] that from certain collective states, emission can be enhanced or suppressed, which is referred to as superradiance and subradiance. In particular, for the Dicke state of the half inverted system of N two-level emitters, the recombination rate is proportional to $N(N - 1)$ [77]. This quadratic behavior with emitter number has become one of the hallmarks of superradiance. Since its discovery by Dicke, superradiance has been extensively and continuously studied in a variety of systems. Most commonly, however, superradiance is associated with temporal modifications of the emission in spatially extended systems, such as clouds of atoms [78] or, when it comes to semiconductor systems, excitons in extended systems [79].

In an ensemble of QDs that are embedded in a microcavity, depending on the inhomogeneous broadening of the QDs and the cavity Q factor, a fraction of the QDs overlaps both spectrally and spatially with the mode. The common light field these resonant solid-state emitters are subjected to provides a mechanism to form inter-emitter correlations that can lead to sub- and superradiant effects in QD nanolasers. Laser models typically assume that emitters act individually and neglect such effects. In the following, we outline how to use the formalism introduced in Section 48.4 to formulate a laser theory that includes inter-emitter coupling in an approximative way and give a quantitative discussion of its effects. As it turns out, their

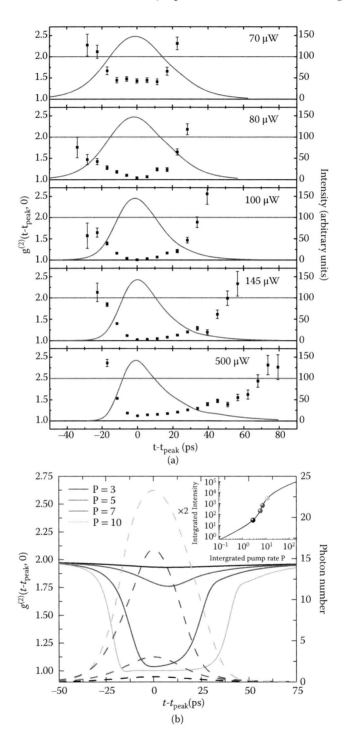

FIGURE 48.17 Time-resolved measurements (a) and theoretical results (b) for emission intensity and photon autocorrelation function following picos-second short optical excitation of a QD-micropillar laser. (Reprinted from Aßmann, M. et al., *Physical Review B* 81, 165314, 2010. With permission).

impact on both stationary and dynamical laser properties can be quite significant. A detailed description of the formalism is found in [80].

To this end, we introduce a formulation for operators that act on the electronic QD states in terms of the many-particle configurations $|i\rangle$, where i refers to configurations like ground state, exciton, trion, biexciton, various dark states, etc. These configuration states, some of which are shown in Figure 48.10, form the basis of the single-QD subspaces of the Hilbert space. Operators Q_{ij}^{α} are introduced that describe either transitions from configuration j to configuration i for $i \neq j$, or the probability that configuration i is realized for $i = j$ in QD α. This approach has advantages over descriptions that use single-particle operators c_v and v_v, like we did in Section 48.4. In doing so, many-electron configurations of one emitter are addressed by a single operator Q rather than many single-particle creation and annihilation operators. It further simplifies the systematic approximation scheme used to truncate the hierarchy of equations of motion with respect to correlations involving *different* QDs. Interemitter correlations are described by a hierarchy of expectation values $\langle Q_{ij}^{\alpha} Q_{kl}^{\beta} \rangle$, $\langle Q_{ij}^{\alpha} Q_{kl}^{\beta} Q_{mn}^{\gamma} \rangle$, ... that involve transitions in QDs α and β, or QDs α, β and γ, etc. The cluster expansion approach facilitates a truncation at a certain order of interemitter coupling, e.g. on the level of pair-correlations between emitters given by $\langle Q_{ij}^{\alpha} Q_{kl}^{\beta} \rangle$ and illustrated in Figure 48.18. All quantum-mechanical expectation values containing electronic QD operators acting on more than two emitters are then neglected. On this level, which is the lowest order that contains interemitter coupling effects, a system of N identical QDs with four confined states each is described by about 300 coupled differential equations, also including photon correlations up to the fourth order providing access to the autocorrelation function $g^{(2)}(0)$. The generation of the equations of motion can be assisted by using computer algebra [80]. While the formalism may appear to involve a great deal of effort, it is highly efficient in the numerical evaluation, and computations for many emitters can typically be performed on a single workstation.

The impact of interemitter coupling on the steady-state properties of a continuously driven QD nanolaser can be seen in Figure 48.19. In the top panel, input–output curves are shown that have been obtained from two separate calculations, one including and one omitting QD–QD correlations. C_F in the bottom panel visualizes the difference in the output intensity I, i.e.,

$$C_F = \frac{I_{\text{rad. coupled QDs}}}{I_{\text{independent QDs}}} - 1. \tag{48.37}$$

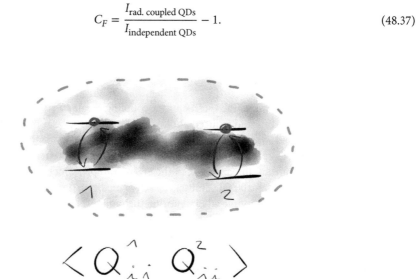

FIGURE 48.18 Illustration of photon-mediated inter-emitter dipole correlations between two emitters.

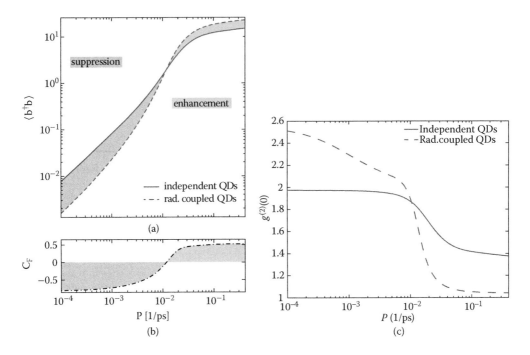

FIGURE 48.19 (a) Input–output curve for a QD nanolaser with typical specifications: 100 resonant emitters, cavity loss rate $\kappa = 0.05/\text{ps}$ (corresponding to $Q = 20,000$), carrier relaxation rate $\gamma_r = 0.05/\text{ps}$, and spontaneous losses into leaky modes $\gamma_{\text{spont}} = 0.01/\text{ps}$. (b) The cooperativity factor visualizes the difference caused by the radiative coupling. (Adapted from Leymann, H. A. M. et al., *Physical Review Applied*, 4, 044018, 2015. With permission.) (c) Corresponding autocorrelation functions. At low excitation, superthermal bunching is a signature of the subradiant emission regime. At high excitation, the threshold to lasing is not fully crossed if emitters act individually and radiative coupling is neglected. Dipole–dipole correlations between distant emitters create additional coherence in the superradiant regime.

The possibility to "switch" the coupling on an off is a particular advantage of the equation-of-motion method and allows to directly assess the impact of the radiative coupling. The following effects are observed:

1. In the low-excitation regime, the inter–emitter coupling creates a subradiant state with a reduced spontaneous emission rate and reduced photon output. Once the laser threshold is crossed, a super-radiant state is formed which enhances the laser output *in addition* to stimulated emission provided by photons in the cavity. In sum, the effect on the input–output curve over the whole excitation regime is quite significant. In particular, in the given example, which has been calculated for a typical microcavity laser with 100 QD emitters as active medium, one would underestimate the true β factor by one order of magnitude when ignoring the radiative coupling

2. The input/output characteristics exhibit a different and untypical slope below threshold.

3. In the subradiant regime (low emission intensity), dipole correlations between pairs of QDs cause an increased probability to emit photons synchronously, leading to super-thermal bunching behavior in the statistical properties of the emission, as shown in the right panel of Figure 48.19.

4. Due to inter-emitter coupling, fewer QDs are required to reach the lasing threshold. In small systems, where achieving sufficient gain due to the limited number of emitters in the cavity has always been an issue, the presence of radiative coupling can explain why lasing is more easily reached than expected from conventional laser theories that do not include QD–QD correlations.

5. Coherent emission with $g^{(2)}(0) = 1$ is achieved at lower intracavity mean photon number in the presence of interemitter coupling. This may be interpreted as an increase in the "coherence per photon" due to the alignment of dipole correlations between emitters.

Interestingly, these theoretical predictions are hard to verify experimentally. A strong indicator for the presence of a dipole-correlated phase is the super-thermal photon bunching [81], which has recently been observed in time-resolved studies using a streak-camera setup in combination with microscopic theory [82]. In the same study, highly accelerated spontaneous emission and subradiant photon trapping have been demonstrated. In their sum, the observation of these criteria give a convincing account of the presence of strong inter-emitter coupling effects in a typical QD nanolaser.

48.8 Nonresonant Coupling and Lasing from Multiexciton States in Few-QD Systems

An inherent property of self-assembled QDs that are grown in the so-called *Stranski–Krastanov* mode [83] is inhomogeneous broadening. It refers to inhomogeneity in size, shape and material composition from one QD to the next, causing differences in the confinement potential and with it the electronic single-particle states. While the typical linewidth of QD transitions is tens to hundreds of μeV depending on temperature [2,84,85], the line of an ensemble of self-assembled QDs consists of the spectral lines of the individual emitters, which for many emitters then appears as a single line with a typical broadening of 20–50 meV. An illustrative demonstration of this behavior is given in Figure 48.20 [86]. Inhomogeneous broadening is generally seen as a weak point of self-assembled QDs, as it takes away some of the advantage of the narrow and well-defined transition energies that is characteristic for the single emitter. On the other hand, in cavity-QED it simplifies the task of creating spectral overlap between emitters and a cavity mode (cf. Figure 48.4) and compensates for spectral wandering, since the ensemble broadening is large compared to the cavity linewidth (typically 50–500 μeV). The influence of detuning of emitters in the ensemble and a cavity mode is greatly reduced by nonresonant coupling mechanisms discussed later. The influence of inhomogeneous broadening on QD laser properties has been studied in the framework of a QD laser theory in [27].

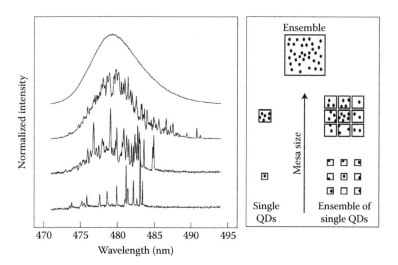

FIGURE 48.20 Experimental results illustrating inhomogeneous broadening in a quantum dot (QD) ensemble. With decreasing mesa size, the QD ensemble consists of fewer and fewer emitters (illustrated right). The broad inhomogenous emission spectrum for many emitters (top left) is revealed to consist of discrete lines as the number of emitters is reduced (bottom left). (Reprinted from Scheibner, M. et al., *Nature Physics*, 3, 106–110, 2007. With permission.)

In this section, we look into what happens if the resonator contains only a few emitters, and the gain spectrum consists of a multitude of separate sharp lines. Such a situation is depicted in the left panel of Figure 48.21 [87]. Then, the emission properties are determined by the interplay of various many-particle configurations of each QD emitter, and their contribution to the photon production depends on the relative spectral position to the cavity mode. While at first it seems contradictory that lasing can be achieved in such a system, strong emission enhancement can drive even a single emitter close to the regime of lasing. At the same time, nonresonant coupling mechanisms haven been identified to provide means for a detuned QD to emit photons into the cavity mode. Nonresonant cavity feeding plays a significant role in nanolasers with ensembles of emitters, but a quantification of its effect is nearly impossible to obtain in such a system. The few-emitter limit, however, offers the unique possibility to study non-resonant mode coupling in a highly controllable environment. In fact, the underlying physical process has long been elusive. Intensive research on this topic has identified three mechanisms that are responsible for the effect:

1. Accoustic phonons can bridge small energy gaps in the range of 1 meV between a detuned emitter resonance and the cavity mode [88–93]. The efficiency is determined by the imbalance between phonon-assisted cavity-photon emission, and the reverse process. Low temperature favors this asymmetry, because low phonon population makes the emission of a phonon more likely than its absorption. Therefore, phonon-mediated nonresonant coupling is more efficient for coupling blue-detuned emitter transitions to the mode [94].

2. At higher carrier densities, interactions with the quasi-continuous WL states provide means for Auger-like scattering processes that allow to bridge larger energy gaps of up to 10 meV [95–99]. While their efficiency is small in comparison to the ultrafast intraband relaxation [53,64,100–102], Auger-coupling is of central importance in the high-excitation regime of nanolasers.

3. QDs possess a rich electronic structure that gives rise to a multitude of multiexciton states, as illustrated in Figure 48.10. The Coulomb interaction energetically separates transitions from different multi-exciton states. This energetic separation, and the possibility of interband recombinations from higher confined states (e.g., the p shell) covers an even larger energetic window, as has been shown

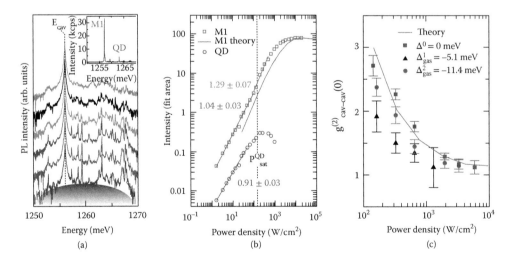

FIGURE 48.21 (a) Spectra of a few-quantum dot (QD) photonic-crystal cavity laser for excitation energies ranging from 0.14 to 5.9 kW/cm². (b) Input–output curves at the cavity-mode energy (squares) and the QD exciton (circles), the latter showing saturation while the cavity-mode emission continues to increase. (c) Second-order photon correlation function at the cavity mode for three different detunings of the cavity-mode energy. Irrespective of the detuning between QD transitions and mode, lasing is achieved at sufficiently high excitation. PL, photoluminescence. (Reprinted from Lichtmannecker, S. et al., *Scienti. Rep.*, 7, 2017).

in [99,103,104]. Thus, even if the QD exciton line, which dominates the spectrum at low excitation, is detuned from the cavity mode, a higher transition, e.g., from the biexciton, may be in perfect resonance with the mode and lead to efficient photon production at higher excitation.

There are several approaches to include the non-resonant coupling of detuned emitters in a laser theory, which superficially all have the same effect of producing additional photons in the resonator mode. In their origin they differ fundamentally, which is reflected in the saturation behavior and in the statistical properties of the emission. On the one hand, a "direct" cavity feeding is often used in the literature [105–108], where photons are directly generated in the cavity mode by an inverse cavity loss term. Such a feeding mechanism, acting together with the cavity loss, gives rise to a thermal photonic state with a temperature defined by the cavity-pump-to-loss ratio. This overlooks the fact that the background photons are generated through exciting an active medium, and hence are expected to have a coherent component [89]. On the other hand, detuned emitters can be explicitly included in the calculation, and their coupling to a detuned cavity mode via phonon or Auger coupling is facilitated by an effective system-bath coupling (Lindblad) term. This approach has the advantage that properties of the gain medium are preserved. The resulting cavity feeding is strongly non-linear with pump rate and exhibits saturation if the number of detuned emitters is limited. Furthermore, the overall emission can become coherent if the background contributions drive the laser above the threshold [89].

As an example, the left part of Figure 48.21 [109] shows the spectrum of a few-QD photonic-crystal nanocavity laser. Only few discrete QD lines are found in the vicinity of the mode (M1). The input–output characteristics is that of a nanolaser with a slight kink, hinting at a β-factor close to unity (middle panel), and coherent emission with $g^{(2)}(0) = 1$ being reached (right panel). Remarkably, the performance of the device is insensitive to tuning of the cavity-mode energy, which can be realized by gas deposition on the photonic crystal cavity. Evidence for this is found in the autocorrelation function in the right panel, which shows a clear transition to lasing with $g^{(2)}(0) = 1$ *independently* of the exact cavity position. In a system with few discrete emission channels this provides strong evidence for the importance of nonresonant coupling effects. Similar results have been reported in [110], to which the authors referred to as "self-tuning gain" effect. Moreover, in the spontaneous emission regime the coupling of various multiexcitonic emission channels from different emitters inside the cavity leads to super-thermal photon bunching with $g^{(2)}(0) > 2$. A density-matrix theory including quantum-mechanical correlations between these transitions, shown as green curve in the right panel, is in good agreement with the measured data. Neither rate equations, nor conventional laser models can unravel the intricate quantum-mechanical nature that governs the laser properties of few-emitter cavity-QED systems.

48.9 Outlook

With present-day nanolaser devices, we are closer than ever to reaching fundamental limits of light–matter interaction in semiconductor systems. Electronic excitations can be generated and used on the level of single excitons [111,112]. Brought into interaction with a single mode of microresonator, semiconductor nanolasers operate at the transition to the regime of quantum optics, where cavity-QED effects, such as photon blockade, photon-antibunching, and vacuum Rabi oscillations can coexist with lasing in different excitation regimes of the same device [106,113,114]. This offers fascinating prospects both from a fundamental and an applied point of view.

Acknowledgments

The authors would like to thank Joachim Piprek for the kind invitation to write this chapter. We also thank Andreas Beuthner for his help in creating the schematic figures. Christopher Gies and Frank Jahnke acknowledge financial support from the German Science Foundation (DFG).

Sandia National Laboratories is a multimission laboratory managed and operated by National Technology and Engineering Solutions of Sandia, LLC., a wholly owned subsidiary of Honeywell International, Inc., for the U.S. Department of Energy's National Nuclear Security Administration under contract DE-NA-0003525. WWC thanks the hospitality of the Technical University Berlin and travel support from the German Research Foundation via the collaborative research center 787.

References

1. Vahala, K. J. (2003). Optical microcavities. *Nature 424* (6950), 839–846.

2. Borri, P., W. Langbein, S. Schneider, U. Woggon, R. L. Sellin, D. Ouyang, and D. Bimberg (2002). Relaxation and dephasing of multiexcitons in semiconductor quantum dots. *Physical Review Letters 89* (18), 187401.

3. Borri, P., W. Langbein, U. Woggon, V. Stavarache, D. Reuter, and A. D. Wieck (2005). Exciton dephasing via phonon interactions in InAs quantum dots: Dependence on quantum confinement. *Physical Review B 71* (11), 115328.

4. Florian, M., A. Steinhoff, C. Gies, and F. Jahnke (2016). Scattering-induced dephasing of many-particle transitions in semiconductor quantum dots. *Applied Physics B 122* (1), 1–7.

5. Blood, P. (2013). The laser diode: 50 years on. *IEEE Journal of Selected Topics in Quantum Electronics 19* (4), 1503201–1503201.

6. Ning, C. (2013). What is laser threshold? *IEEE Journal of Selected Topics in Quantum Electronics 19* (4), 1503604–1503604.

7. Rice, P. R. and H. J. Carmichael (1994). Photon statistics of a cavity-QED laser: A comment on the laser–phase-transition analogy. *Physical Review A 50* (5), 4318–4329.

8. Chow, W. W., M. Lorke, and F. Jahnke (2011). Will quantum dots replace quantum wells as the active medium of choice in future semiconductor lasers? *IEEE Journal of Selected Topics in Quantum Electronics 17*, 1349–1355.

9. Yokoyama, H. and S. D. Brorson (1989). Rate equation analysis of microcavity lasers. *Journal of Applied Physical 66*, 4801.

10. Musiał, A., C. Hopfmann, T. Heindel, C. Gies, M. Florian, H. A. M. Leymann, A. Foerster, C. Schneider, F. Jahnke, S. Höfling, M. Kamp, and S. Reitzenstein (2015). Correlations between axial and lateral emission of coupled quantum dot micropillar cavities. *Physical Review B 91* (20), 205310.

11. Munsch, M., N. S. Malik, E. Dupuy, A. Delga, J. Bleuse, J.-M. Gérard, J. Claudon, N. Gregersen, and J. Mørk (2013). Dielectric GaAs antenna ensuring an efficient broadband coupling between an InAs quantum dot and a Gaussian optical beam. *Physical Review Letters 110* (17), 177402.

12. Leymann, H. A. M., C. Hopfmann, F. Albert, A. Foerster, M. Khanbekyan, C. Schneider, S. Höfling, A. Forchel, M. Kamp, J. Wiersig, and S. Reitzenstein (2013). Intensity fluctuations in bimodal micropillar lasers enhanced by quantum-dot gain competition. *Physical Review A 87* (5), 053819.

13. Agrawal, G. P. and N. K. Dutta (1993). *Semiconductor Laser*. Norwell, MA: Kluwer.

14. Chuang, S. L. (1995). *Physics of Optoelectronic Devices*. New York, NY: Wiley.

15. Lohmeyer, H., K. Sebald, C. Kruse, R. Kröger, J. Gutowski, H. Hommel, J. Wiersig, N. Baer, and F. Jahnke (2006). Confined optical modes in monolithic II-VI pillar microcavities. *Applied Physics Letters 88* (5), 051101.

16. McCall, S. L., A. F. J. Levi, R. E. Slusher, S. J. Pearton, and R. A. Logan (1992). Whispering gallery mode microdisk lasers. *Applied Physics Letters 60* (3), 289–291.

17. Joannopoulos, J. D., P. R. Villeneuve, and S. Fan (1997). Photonic crystals: Putting a new twist on light. *Nature 386* (6621), 143–149.

18. Purcell, E. M. (1946). Spontaneous emission probability at radio frequencies. *Physical Review 69*, 681.

19. Schwab, M., H. Kurtze, T. Auer, T. Berstermann, M. Bayer, J. Wiersig, N. Baer, C. Gies, F. Jahnke, J. P. Reithmaier, A. Forchel, M. Benyoucef, and P. Michler (2006). Radiative emission dynamics of quantum dots in a single cavity micropillar. *Physical Review B 74* (4), 045323.

20. Aßmann, M., F. Veit, M. Bayer, M. v. d. Poel, and J. M. Hvam (2009). Higher-order photon bunching in a semiconductor microcavity. *Science 325* (5938), 297–300. PMID: 19608912.

21. Loudon, R. (2000). *The Quantum Theory of Light*. Oxford, UK: Oxford University Press.

22. Hanbury Brown, R. and R. Q. Twiss (1956). Correlation between photons in two coherent beams of light. *Nature 177*, 27–29.

23. Ulrich, S. M., C. Gies, J. Wiersig, S. Reitzenstein, C. Hofmann, A. Löffer, A. Forchel, F. Jahnke, and P. Michler (2007). Photon statistics of semiconductor microcavity lasers. *Physical Review Letters 98*, 043906.

24. Gardiner, C. and P. Zoller (2004). *Quantum Noise A Handbook of Markovian and Non-Markovian Quantum Stochastic Methods with Applications to Quantum Optics*. Heidelberg, Berlin, New York: Springer Verlag.

25. Gies, C., M. Florian, P. Gartner, and F. Jahnke (2012). Modelling single quantum dots in microcavities. In F. Jahnke (Ed.), *Quantum Optics with Semiconductor Nanostructures*. Philadelphia, PA: Woodhead Publishing Limited.

26. Ates, S., C. Gies, S. M. Ulrich, J. Wiersig, S. Reitzenstein, A. Löffler, A. Forchel, F. Jahnke, and P. Michler (2008). Influence of the spontaneous optical emission factor β on the first-order coherence of a semiconductor microcavity laser. *Physical Review B 78* (15), 155319.

27. Chow, W. W., F. Jahnke, and C. Gies (2014). Emission properties of nanolasers during the transition to lasing. *Light: Science & Applications 3* (8), e201.

28. Ritter, S., P. Gartner, C. Gies, and F. Jahnke (2010). Emission properties and photon statistics of a single quantum dot laser. *Optics Express 18* (10), 9909–9921.

29. Reitzenstein, S., C. Böckler, A. Bazhenov, A. Gorbunov, A. Löffer, M. Kamp, V. D. Kulakovskii, and A. Forchel (2008). Single quantum dot controlled lasing effects in high-q micropillar cavities. *Optics Express 16* (7), 4848–4857.

30. Gies, C., J. Wiersig, and F. Jahnke (2008). Output characteristics of pulsed and continuous-wave-excited quantum-dot microcavity lasers. *Physical Review Letters 101* (6), 067401.

31. Chow, W. and S. Koch (2005). Theory of semiconductor quantum-dot laser dynamics. *IEEE Journal of Quantum Electronics 41* (4), 495–505.

32. Kolarczik, M., N. Owschimikow, J. Korn, B. Lingnau, Y. Kaptan, D. Bimberg, E. Schöll, K. Lüdge, and U. Woggon (2013). Quantum coherence induces pulse shape modification in a semiconductor optical amplifier at room temperature. *Nature Communications 4*, 2953.

33. Gies, C., J. Wiersig, M. Lorke, and F. Jahnke (2007). Semiconductor model for quantum-dot-based microcavity lasers. *Physical Review A 75* (1), 013803.

34. Wiersig, J., C. Gies, F. Jahnke, M. Aßmann, T. Bestermann, M. Bayer, C. Kistner, S. Reitzenstein, S. H. A. Forchel, C. Kruse, J. Kalden, and D. Hommel (2009). Direct observation of correlations between individual photon emission events of a microcavity laser. *Nature 460*, 245.

35. Baer, N., C. Gies, J. Wiersig, and F. Jahnke (2006). Luminescence of a semiconductor quantum dot system. *European Physical Journal B 50*, 411.

36. Gregersen, N., T. Suhr, M. Lorke, and J. Mørk (2012). Quantum-dot nano-cavity lasers with Purcell-enhanced stimulated emission. *Applied Physics Letters 100* (13), 131107.

37. Lermer, M., N. Gregersen, M. Lorke, E. Schild, P. Gold, J. Mørk, C. Schneider, A. Forchel, S. Reitzenstein, S. Hing, and M. Kamp (2013). High beta lasing in micropillar cavities with adiabatic layer design. *Applied Physics Letters 102* (5), 052114.

38. Lorke, M., T. Suhr, N. Gregersen, and J. Mørk (2013). Theory of nanolaser devices: Rate equation analysis versus microscopic theory. *Physical Review B 87*, 205310.

39. Suhr, T., N. Gregersen, M. Lorke, and J. Mørk (2011). Modulation response of quantum dot nanolightemitting-diodes exploiting Purcell-enhanced spontaneous emission. *Applied Physics Letters 98* (21), 211109.

40. Chow, W. and S. Koch (1999). *Semiconductor-Laser Fundamentals* (1st ed.). Berlin: Springer-Verlag.

41. Coldren, L. and S. Corzine (1995). *Diode Lasers and Photonic Integrated Circuits.* New York, NY: Wiley.

42. Binder, R., H. Köhler, M. Bonitz, and N. Kwong (1997). Green's function description of momentum orientation relaxation of photoexcited electron plasmas in semiconductors. *Physical Review B 55*, 5110.

43. Bonitz, M., D. Kremp, D. Scott, R. Binder, W. Kraeft, and H. Köhler (1996). Numerical analysis of nonmarkovian effects in charge-carrier scattering: One-time kinetic equations. *Journal of Physics: Condensed Mattter 8*, 6057.

44. ElSayed, K., S. Schuster, H. Haug, G. Herzel, and K. Henneberger (1994). Subpicosecond plasmon response: Buildup of screening. *Physical Review B 49*, 7337.

45. Henneberger, K. (1988). Resonant laser excitation and electron-hole kinetics of a semiconductor: I. nonequilibrium green's function treatment and fundamental equations. *Physica A 150*, 419.

46. Jahnke, F. and S. W. Koch (1995). Many-body theory for semiconductor microcavity lasers. *Physical Review A 52*, 1712.

47. Bányai, L., P. Gartner, and H. Haug (1998). Self-consistent RPA retarded polaron Green function for quantum kinetics. *European Physical Journal B 1*, 209.

48. Gartner, P., L. Bányai, and H. Haug (1999). Two-time electron-LO-phonon quantum kinetics and the generalized Kadanoff-Baym approximation. *Physical Review B 60*, 14234.

49. Gartner, P., L. Bányai, and H. Haug (2002). Self-consistent RPA for the intermediate-coupling polaron. *Physical Review B 66*, 75205.

50. Hartmann, M. and W. Schäfer (1992). Real time approach to relaxation and dephasing processes in semiconductors. *Physical Status Solidi (B) 173*, 165.

51. Haug, H. (1992). Interband quantum kinetics with LO-phonon scattering in a laser-pulse-excited semiconductor I. Theory. *Physical Status Solidi (B) 173*, 139.

52. Jahnke, F. and S. W. Koch (1993). Theory of carrier heating through injection pumping and lasing in semiconductor microcavity lasers. *Optics Letters 18* (17), 1438–1440.

53. Lorke, M., T. R. Nielsen, and J. Mørk (2011). Switch-on dynamics of nanocavity laser devices. *Applied Physics Letters 99* (15), 151110.

54. Lingnau, B., K. Lüdge, E. Schöll, and W. W. Chow (2010). Many-body and nonequilibrium effects on relaxation oscillations in a quantum-dot microcavity laser. *Applied Physics Letters 97*, 111102.

55. Lorke, M., T. R. Nielsen, and J. Mørk (2010). Influence of carrier dynamics on the modulation bandwidth of quantum-dot based nanocavity devices. *Applied Physics Letters 97* (21), 211106.

56. Luedge, K., M. J. P. Bormann, E. Malic, P. Hoevel, M. Kuntz, D. Bimberg, A. Knorr, and E. Schoell (2008). Turn-on dynamics and modulation response in semiconductor quantum dot lasers. *Physical Review B 78*, 035316.

57. Luedge, K. and E. Scholl (2009). Quantum-dot lasers—Desynchronized nonlinear dynamics of electrons and holes. *IEEE Journal of Quantum Electronics 45* (11), 1396–1403.

58. Aßmann, M., F. Veit, M. Bayer, C. Gies, F. Jahnke, S. Reitzenstein, S. Höfling, L. Worschech, and A. Forchel (2010). Ultrafast tracking of second-order photon correlations in the emission of quantum-dot microresonator lasers. *Physical Review B 81* (16), 165314.

59. Lorke, M., W. Chow, T. R. Nielsen, J. Seebeck, P. Gartner, and F. Jahnke (2006). Anomaly in the excitation dependence of optical gain in semiconductor quantum dots. *Physical Review B 74*, 035334.

60. Lorke, M., T. Nielsen, J. Seebeck, P. Gartner, and F. Jahnke (2006). Influence of carrier-carrier and carrier-phonon correlations on optical absorption and gain in quantum-dot systems. *Physical Review B 73*, 85324.

61. Schäfer, W. and M. Wegener (2002). *Semiconductor Optics and Transport Phenomena* (1st ed.). Berlin: Springer-Verlag.

62. Mahan, G. (1990). *Many-Particle Physics*. New York, NY: Plenum Press.

63. Seebeck, J., T. Nielsen, P. Gartner, and F. Jahnke (2005). Polarons in semiconductor quantum dots and their role in the quantum kinetics of carrier relaxation. *Physical Review B 71*, 125327.

64. Nielsen, T. R., P. Gartner, and F. Jahnke (2004). Many-body theory of carrier capture and relaxation in semiconductor quantum-dot lasers. *Physical Review B 69* (23), 235314.

65. Benisty, H. (1995). Reduced electron-phonon relaxation rates in quantum-box systems: Theoretical analysis. *Physical Review B 51*, 13281.

66. Inoshita, T. and H. Sakaki (1996). Electron-phonon interaction and the so-called phonon bottleneck effect in semiconductor quantum dots. *Physica B 227*, 373.

67. Kurtze, H., J. Seebeck, P. Gartner, D. R. Yakovlev, D. Reuter, A. D. Wieck, M. Bayer, and F. Jahnke (2009). Carrier relaxation dynamics in self-assembled semiconductor quantum dots. *Physical Review B 80*, 235319.

68. Schuh, K., F. Jahnke, and M. Lorke (2011). Rapid adiabatic passage in quantum dots: Influence of scattering and dephasing. *Applied Physics Letters 99* (1), 011105.

69. Seebeck, J., T. Nielsen, P. Gartner, and F. Jahnke (2006). Quantum kinetic theory of phonon-assisted carrier transitions in nitride-based quantum-dot systems. *European Physical Journal B 49*, 167.

70. Seebeck, J., M. Lorke, P. Gartner, and F. Jahnke (2009). Carrier-carrier and carrier-phonon scattering in the low-density and low-temperature regime for resonantly pumped semiconductor quantum dots. *Physica Status Solidi (C) 6*, 488.

71. Uskov, A. V., C. Meuer, H. Schmeckebier, and D. Bimberg (2011). Auger capture induced carrier heating in quantum dot lasers and amplifiers. *Applied Physics Express 4* (2), 022202.

72. Goldmann, E., M. Lorke, T. Frauenheim, and F. Jahnke (2014). Negative differential gain in quantum dot systems: Interplay of structural properties and many-body effects. *Applied Physics Letters 104* (24), 242108.

73. Lorke, M., W. Chow, and F. Jahnke (2007). Reduction of optical gain in semiconductor quantum dots. *Proceedings of SPIE 6468*, 646818.

74. Lorke, M., F. Jahnke, and W. Chow (2007). Excitation dependence of gain and carrier induced refractive index changes in quantum dots. *Applied Physical Letters 90*, 51112.

75. Lorke, M., J. Seebeck, T. Nielsen, P. Gartner, and F. Jahnke (2006). Excitation dependence of the homogeneous linewidths in quantum dots. *Physical Status Solidi (C) 3*, 2393–2396.

76. Dicke, R. H. (1954). Coherence in spontaneous radiation processes. *Physical Review 93*, 99–110.

77. Mandel, L. and E. Wolf (1995). *Optical Coherence and Quantum Optics*. New York, NY: Cambridge University Press.

78. Bohnet, J. G., Z. Chen, J. M. Weiner, D. Meiser, M. J. Holland, and J. K. Thompson (2012). A steady-state superradiant laser with less than one intracavity photon. *Nature 484* (7392), 78–81.

79. Timothy Noe Ii, G., J.-H. Kim, J. Lee, Y. Wang, A. K. Wójcik, S. A. McGill, D. H. Reitze, A. A. Belyanin, and J. Kono (2012). Giant superfluorescent bursts from a semiconductor magneto-plasma. *Nature Physics 8* (3), 219–224.

80. Leymann, H. A. M., A. Foerster, F. Jahnke, J. Wiersig, and C. Gies (2015). Sub—And superradiance in nanolasers. *Physical Review Applied 4* (4), 044018.

81. Auffèves, A., D. Gerace, S. Portolan, A. Drezet, and M. F. Santos (2011). Few emitters in a cavity: From cooperative emission to individualization. *New Journal of Physics 13* (9), 093020.

82. Jahnke, F., C. Gies, M. Aßmann, M. Bayer, H.A.M. Leymann, A. Foerster, J. Wiersig, C. Schneider, M. Kamp, and S. Höfling (2016). Giant photon bunching, superradiant pulse emission and excitation trapping in quantum-dot nanolasers. *Nature Communications 7*, 11540.

83. Bimberg, D., M. Grundmann, and N. N. Ledentsov (1998). *Quantum Dot Heterostructures*. Hoboken, NJ: John Wiley & Sons.

84. Bayer, M. and A. Forchel (2002). Temperature dependence of the exciton homogeneous linewidth in In0.60Ga0.40As/GaAs self-assembled quantum dots. *Physical Review B 65* (4), 041308.

85. Moody, G., R. Singh, H. Li, I. A. Akimov, M. Bayer, D. Reuter, A. D. Wieck, and S. T. Cundiff (2013). Fifth-order nonlinear optical response of excitonic states in an InAs quantum dot ensemble measured with two-dimensional spectroscopy. *Physical Review B 87* (4), 045313.

86. Scheibner, M., T. Schmidt, L. Worschech, A. Forchel, G. Bacher, T. Passow, and D. Hommel (2007). Superradiance of quantum dots. *Nature Physics 3* (2), 106–110.

87. Lichtmannecker, S., M. Florian, T. Reichert, M. Blauth, M. Bichler, et al. (2017). A few-emitter solid-state multi-exciton laser. *Scientific Reports 7* (7420). doi: 10.1038/s41598-017-07097-9

88. Ates, S., S. M. Ulrich, A. Ulhaq, S. Reitzenstein, A. Löffler, S. Höfling, A. Forchel, and P. Michler (2009). Non-resonant dot—cavity coupling and its potential for resonant single-quantum-dot spectroscopy. *Nature Photonics 3* (12), 724–728.

89. Florian, M., P. Gartner, C. Gies, and F. Jahnke (2013). Phonon-mediated off-resonant coupling effects in semiconductor quantum-dot lasers. *New Journal of Physics 15* (3), 035019.

90. Hohenester, U., A. Laucht, M. Kaniber, N. Hauke, A. Neumann, A. Mohtashami, M. Seliger, M. Bichler, and J. J. Finley (2009). Phonon-assisted transitions from quantum dot excitons to cavity photons. *Physical Review B 80* (20), 201311.

91. Majumdar, A., E. D. Kim, Y. Gong, M. Bajcsy, and J. Vučković (2011). Phonon mediated off-resonant quantum dot–cavity coupling under resonant excitation of the quantum dot. *Physical Review B 84* (8), 085309.

92. Press, D., S. Götzinger, S. Reitzenstein, C. Hofmann, A. Löffler, M. Kamp, A. Forchel, and Y. Yamamoto (2007). Photon antibunching from a single quantum-dot-microcavity system in the strong coupling regime. *Physical Review Letters 98* (11), 117402.

93. Tarel, G. and V. Savona (2009). Emission spectrum of a quantum dot embedded in a nanocavity. *Physica Status Solidi (C) 6* (4), 902–905.

94. Roy, C. and S. Hughes (2011). Influence of electron–acoustic-phonon scattering on intensity power broadening in a coherently driven quantum-dot–cavity system. *Physical Review X 1* (2), 021009.

95. Chauvin, N., C. Zinoni, M. Francardi, A. Gerardino, L. Balet, B. Alloing, L. H. Li, and A. Fiore (2009). Controlling the charge environment of single quantum dots in a photonic-crystal cavity. *Physical Review B 80* (24), 241306.

96. Florian, M., P. Gartner, A. Steinhoff, C. Gies, and F. Jahnke (2014). Coulomb-assisted cavity feeding in nonresonant optical emission from a quantum dot. *Physical Review B 89* (16), 161302(R).

97. Karrai, K., R. J. Warburton, C. Schulhauser, A. Högele, B. Urbaszek, E. J. McGhee, A. O. Govorov, J. M. Garcia, B. D. Gerardot, and P. M. Petroff (2004). Hybridization of electronic states in quantum dots through photon emission. *Nature 427* (6970), 135–138.

98. Settnes, M., P. Kaer, A. Moelbjerg, and J. Mork (2013). Auger processes mediating the nonresonant optical emission from a semiconductor quantum dot embedded inside an optical cavity. *Physical Review Letters 111* (6), 067403.

99. Winger, M., T. Volz, G. Tarel, S. Portolan, A. Badolato, K. J. Hennessy, E. L. Hu, A. Beveratos, J. Finley, V. Savona, and A. Imamoğlu (2009). Explanation of photon correlations in the far-off-resonance optical emission from a quantum-dot–cavity system. *Physical Review Letters 103* (20), 207403.

100. Bockelmann, U. and T. Egeler (1992). Electron relaxation in quantum dots by means of auger processes. *Physical Review B 46*, 15574.

101. Efros, A. L., V. A. Kharchenko, and M. Rosen (1995). Breaking of the phonon bottleneck in nanometer quantum dots role of auger-like processes. *Solid State Communications 93*, 281.

102. Uskov, A. V., F. Adler, H. Schweizer, and M. H. Pilkuhn (1997). Auger carrier relaxation in self-assembled quantum dots by collisions with two-dimensional carriers. *Journal of Applied Physics 81*, 7895.

103. Dekel, E., D. Gershoni, E. Ehrenfreund, D. Spektor, J. M. Garcia, and P. M. Petroff (1998). Multiexciton spectroscopy of a single self-assembled quantum dot. *Physical Review Letters 80* (22), 4991–4994.

104. Laucht, A., M. Kaniber, A. Mohtashami, N. Hauke, M. Bichler, and J. J. Finley (2010). Temporal monitoring of nonresonant feeding of semiconductor nanocavity modes by quantum dot multiexciton transitions. *Physical Review B 81*, 241302.

105. del Valle, E., F. P. Laussy, and C. Tejedor (2009). Luminescence spectra of quantum dots in microcavities. II. Fermions. *Physical Review B 79*, 235326.

106. Nomura, M., N. Kumagai, S. Iwamoto, Y. Ota, and Y. Arakawa (2010). Laser oscillation in a strongly coupled single-quantum-dot–nanocavity system. *Nature Physics 6*, 279–283.

107. Valle, E. d. and F. P. Laussy (2011). Effective cavity pumping from weakly coupled quantum dots. *Superlattices and Microstructures 49* (3), 241–245.

108. Yao, P., P. K. Pathak, E. Illes, S. Hughes, S. Münch, S. Reitzenstein, P. Franeck, A. Löffer, T. Heindel, S. Höfling, L. Worschech, and A. Forchel (2010). Nonlinear photoluminescence spectra from a quantum-dot cavity system: Interplay of pump-induced stimulated emission and anharmonic cavity QED. *Physical Review B 81* (3), 033309.

109. Lichtmannecker, S., M. Florian, T. Reichert, M. Blauth, M. Bichler, F. Jahnke, J. J. Finley, C. Gies, and M. Kaniber (2016). A few-emitter solid-state multi-exciton laser. *ArXiv:1602.03998 [condmat]*.

110. Strauf, S., K. Hennessy, M. T. Rakher, Y.-S. Choi, A. Badolato, L. C. Andreani, E. L. Hu, P. M. Petroff, and D. Bouwmeester (2006). Self-tuned quantum dot gain in photonic crystal lasers. *Physical Review Letters 96*, 127404.

111. Mlynek, J. A., A. A. Abdumalikov, C. Eichler, and A. Wallraff (2014). Observation of Dicke superradiance for two artificial atoms in a cavity with high decay rate. *Nature Communications 5*, 5186.

112. Müller, M., S. Bounouar, K. D. Jöns, M. Glässl, and P. Michler (2014). On-demand generation of indistinguishable polarization-entangled photon pairs. *Nature Photonics 8* (3), 224–228.

113. Muñoz, C. S., E. del Valle, A. G. Tudela, K. Müller, S. Lichtmannecker, M. Kaniber, C. Tejedor, J. J. Finley, and F. P. Laussy (2014). Emitters of n-photon bundles. *Nature Photonics 8* (7), 550–555.

114. Gies, C. et al., Phys. Rev. A **96**, 023806 (2017).

Nonlinear Dynamics in Quantum Photonic Structures

Gabriela Slavcheva

and

Mirella Koleva

49.1 Introduction

Quantum optics unites physical optics and the quantum field theory of light. The foundations of the latter were laid by Planck, who succeeded to explain theoretically the black-body radiation spectrum by postulating that the energy of a harmonic oscillator is quantized. In 1927, Dirac developed this theory by solving the long-standing problem of the wave-particle duality and unifying the wave and particle aspects of light by quantization of the electromagnetic radiation. Thus, according to Dirac's single-particle interpretation, the light emitted from a spontaneously emitting source with low photon density is described as single-photon emission events.

A fundamental consequence of light-field quantization is the appearance of vacuum (zero-point energy) fluctuations, or quantum noise, which has no classical analogue and stems from the Heisenberg uncertainty relations. However, some of the greatest successes of quantum theory—such as Planck's black-body spectrum and Einstein's theory of the photoelectric effect—do not actually require a full quantum-mechanical treatment and can instead be explained using semiclassical theory, whereby matter is quantized but light remains a classical wave. Only in the 1960s, motivated by the development of lasers, did researchers start to examine effects which truly required a quantum theory of light. For instance, light quantization is required to explain spontaneous emission, irreversible decay, and the origin of the laser spectral linewidth. The demand to develop further the quantum theory of light in order to explain these effects was met by the pioneering work of Glauber [1,2].

Vacuum quantum fluctuations are necessary to account for phenomena, such as the Lamb shift: a radiative correction to the energy spectrum of hydrogen, as well as for the attractive force between two uncharged conductive plates in a vacuum placed a few nanometres apart, known as the Casimir effect. Some examples of effects which the semiclassical approach fails to explain are quantum beat phenomena in a three-level atom, two-photon interferometry and production of "entangled" states of light, resonance fluorescence (quantum jumps) in the emission of atoms, and the so-called "nonclassical" states of light.

The classification of light states as classical and nonclassical stems from the statistical properties of radiation and, more specifically, from the second-order coherent properties of the light emitter. A crucial step in elucidating these properties was the correlation experiment of Hanbury Brown and Twiss [3], which was designed to measure the coincidence of individual photons at two detectors placed equidistantly from a half-silvered mirror splitting a beam of oncoming low-intensity light. Using the second-order correlation function, $g^{(2)}(\tau)$, quantifying the synchronicity with which photons arrive at the detectors, it was then possible to discern the manner in which different types of light sources emitted light. This method revealed that classical light sources—such as a thermal black-body radiation source or a lamp—exhibited Bose–Einstein statistics manifested by photon "bunching", i.e., an increased likelihood of the photons to arrive at the detectors at the same time, as though they had "bunched together" while traveling from the source to the detectors. On the other hand, a stable coherent light source (e.g., a laser) exhibits Poissonian statistics, which is manifested by a constant unity second-order correlation function, $g^{(2)}(\tau)$, meaning that photons are not intercorrelated and are equally likely to impinge on the detectors with a delay of any length with respect to each other. However, most interestingly of all, nonclassical states of light are described by sub-Poissonian statistics (a diminished likelihood of coincident arrival) called "antibunching": the photons are "diluted" as they travel and eventually emitted one by one. Antibunching is characteristic of resonant fluorescence from single atoms, molecules, or semiconductor nanocrystals and quantum dots (QDs), and is a signature of a single-photon emitter. For perfectly antibunched light, the second-order correlation function vanishes for coincident detections. This is irreconcilable with the classical wave optics and is considered a hallmark of pure quantum behavior.

Photon-number (Fock) states are states of light containing a specific number of photons. The single-photon state is a special case of a Fock state and is a squeezed-light state [4], a quantum state of light resulting from the Heisenberg relations with, e.g., reduced amplitude fluctuations at the expense of enhanced fluctuations of the phase. An ideal amplitude-squeezed source delivers a stream of photons at regular time intervals, which is the precise realization of a single-photon source [5].

Another quantum-mechanical property is entanglement. Quantum systems consisting of interacting subsystems become highly correlated and their individual constituents become "entangled". The global quantum state of an entangled system—for e.g., one consisting of two particles (e.g., a photon pair)—remains correlated even when the two particles are separated by large distances from each other and unable to interact or communicate in any known way. The most famous example of an entangled state is the Einstein–Podolsky–Rosen–Bohm [6] state, whereby two spin 1/2 particles reside in a state with zero total spin angular momentum: $|\Psi\rangle = \frac{1}{\sqrt{2}}(|\uparrow\rangle \otimes |\downarrow\rangle - |\downarrow\rangle \otimes |\uparrow\rangle)$. Notably, this state cannot be represented as a simple product of the pure states of the individual particles.

Many of these quantum effects have been observed only in carefully controlled atomic systems; however, with the advent of quantum photonics (the applied device-oriented field of quantum optics), recent advances in semiconductor technology have enabled the fabrication of solid-state devices that emit single photons, opening new avenues for nonclassical light state generation. For instance, the use of radiative cascades from QD excited states is a principal method for generation of quantum entanglement for applications of QDs as polarization-entangled sources. The exchange interaction and the fine structure splitting of the energy levels involved in radiative cascades from QD excited states are of fundamental importance, since they determine the polarization and entanglement of the emitted photons.

Quantum photonics has attracted great interest in recent years. This is due to its potential to revolutionize science and day-to-day life alike through enabling the implementation of faster algorithms, secure transmission of information, vast increase in data storage, and execution of more accurate measurements. One of the major challenges of modeling quantum photonic devices is working out a way of efficient simulation and control of realistic, "open" quantum systems (i.e., allowing for energy to flow in and out of the system) and devices for applications in quantum technologies. For optical quantum information processing to advance beyond demonstration experiments, the development of on-chip capabilities will be essential. Solid-state qubit architectures have emerged as a most promising physical implementation of quantum photonic circuits, taking advantage of the state-of-the-art semiconductor technologies. Reliable, integrated sources of nonclassical light, as well as realization of high-fidelity on-chip readout of solid-state qubits, are a critical requirement for next-generation quantum technologies [7].

Recently, significant progress has been made toward the achievement of high-efficiency solid-state single-photon sources [8–12]. Among these sources, QDs are particularly promising since they can be embedded in devices enabling electrical injection while maintaining efficient light emission. Furthermore, the dot doping—or the number and type of the charge carriers within the dot—can be controlled by applying electrical bias in a Schottky-type structure. A quantum bit (qubit) is a linear superposition of basis states $|0\rangle$ (ground level) and $|1\rangle$ (excited level) in a two-level system, given by: $|\Psi\rangle = \alpha |0\rangle + \beta |1\rangle$, with α and β being complex numbers such that $\alpha^2 + \beta^2 = 1$. In a classical computer, the information is encoded in digital bit states 0 and 1. By contrast, in a quantum computer, the state of each bit is permitted to be any quantum-mechanical state of a qubit (manipulated using qubit rotations, also known as gate operations), leading to a massive parallelism and an exponential speed-up of the quantum computations.

The $|0\rangle$ and $|1\rangle$ states of a qubit could be encoded in the spin-up ($|\uparrow\rangle$) and spin-down ($|\downarrow\rangle$) states of a charge carrier. The spin-up and spin-down states of a single electron or hole confined in a charged semiconductor QD are a particularly attractive qubit candidate [13]. This arises from several advantages intrinsic to this quantum system. The first advantage is that the atomic-like electronic structure suppresses coupling of the spin to the solid-state quantum environment and thus impedes the spin decoherence processes. This leads to a very long spin coherence time (denoted T_2)—that is, the time within which the spin loses its phase information—and thus a long-lived quantum state, which is required for performing a large number of spin manipulations. Another important advantage is optical convertibility—i.e., the ability to switch through transfer of angular momentum between a "flying" qubit encoded in a photon and the stationary matter qubit of the spin confined in a QD. This enables the transport of qubits over large distances and facilitates the realization of scalable solid-state qubit architectures for optical quantum computing. Furthermore, this promising platform allows integrating the operations of optical initialization, manipulation, and readout of a spin qubit on a chip. By employing a combination of optical orientation [14] and quantum coherent control [15] techniques, using ultrashort polarized light pulses one can address individual carrier spins in semiconductor QDs, initialize and manipulate them coherently through optically excited states (charged excitons), and detect the resulting spin qubit [16]. For quantum technologies applications, control of both QD position and emission energy is crucial. For instance, deterministic site control of QD location is indispensable for multiqubit operations and scalability. Therefore, new simulation tools are needed to determine the allowed tolerances in the dot positions and emission energies to ensure observation of the desired quantum-optical effects.

Quantum coherent control represents a universal approach for predictable manipulation of the properties of quantum systems, such as atoms and molecules. This technique has most recently been applied to QD solid-state systems. Coherent control in semiconductor nanostructures allows for coherent manipulation of the carrier wavefunctions on a time scale shorter than typical dephasing times. This is a prerequisite for successful implementation of ultrafast optical switching and quantum information processing. Quantum coherent control requires use of ultrashort pulses considerably shorter than the characteristic relaxation times in matter ($\tau \ll T_1, T_2$). This is equivalent to the photon-dipole coupling rate exceeding all dissipative rates in the system, a general condition for achieving the strong-coupling regime. The ultrashort pulses are usually characterized by high field amplitudes and, consequently, lead to nonlinear optical effects, such as coherent pulse propagation and self-induced transparency (SIT) [17]. The phenomenon of SIT is observable above a critical power threshold for a given pulse width: a high-intensity, ultrashort pulse propagating through a medium composed of an ensemble of resonant quantum two-level absorbers, whose relaxation times greatly exceed the pulse duration, travels at a reduced speed and unchanged shape with anomalously low energy loss. The absorbers are driven into the excited state by absorbing ultrashort pulse energy; by reradiating this energy into the pulse, they return to the ground state. Thus, the optical energy is carried through the medium not by the electromagnetic field, but by a coupled light-matter polariton wave. The polariton is a mixed light-matter quasiparticle resulting from the strong coupling of the optical wave to the medium's polarization. As a result, the pulse travels as a solitary wave, known as a SIT-soliton. This soliton is localized both in space and time, in contrast with the well-known nonlinear optics solitons which result from the interplay between nonlinearity and the medium's dispersion and/or diffraction. The condition of SIT is predicted by the remarkable pulse-area theorem (PAT) [18] which establishes a general criterion for stable, ultrashort pulse propagation in attenuating or amplifying media based on an integral quantity— namely, the pulse area. This phenomenon can be preserved to a great extent in solid-state systems—e.g., semiconductors—and has been experimentally demonstrated by picosecond pulse propagation in QD waveguides [19].

When the laser pulse's temporal width becomes comparable with the optical period (i.e., when the pulse envelope contains only a few optical cycles of the carrier wave), a transition to a qualitatively new "extreme nonlinear" regime of strong light-matter interactions is induced in which the electric field itself, rather than intensity envelope, drives the interaction. The majority of the current analytical approaches are based on the standard, slowly varying amplitude approximation (SVEA) and rotating-wave approximation (RWA). However, a number of works have demonstrated the limitations of these approximations, and new phenomena have been predicted on the basis of a non-perturbative approach (see, e.g., [20], and references therein). This requires development of new, non-perturbative theoretical and computational methods beyond the above approximations as shown in Section 49.2, which could account properly for the ultrafast carrier-wave dynamics.

The earlier high-intensity (i.e., high photon density) limit allows to achieve the strong-coupling regime in free space. However, in the opposite, low-intensity (i.e., few-photon) limit, the light-matter coupling of a single photon to a single atom in free space is weak, and despite attempts by several groups [21–23] to improve it, the experimental realization of high-fidelity excited-state preparation remains challenging. In order to improve the performance of single-photon sources, it is necessary to incorporate the emitters in cavities. The cavity helps to select a single mode out of the infinite number of radiation field modes coupled to the emitter. As the emission is preferentially directed into a single mode, owing to the Purcell effect the spontaneous emission lifetime is reduced for emitters located at the antinode of the electric-field intensity in cavities with a small mode volume and high quality factor. Faster emission is desired because it makes the single-photon source more deterministic and reliable.

If the exciton-light coupling in a semiconductor microcavity is weak compared to all dissipative rates in the system, irreversible decay occurs—that is, energy cannot be alternately passed from light to matter. Depending on whether the quantum wells inside the semiconductor microcavity are situated near the node or antinode of the excitatory electric field, either an inhibited or enhanced spontaneous emission takes place. The cavity then operates in the weak coupling regime, whereby the dynamics of the light-matter interaction is dominated by incoherent processes and damping. Conversely, in the strong coupling limit, no irreversible decay takes place. Instead, a sequence of coherent reabsorption and reemission processes occurs and the energy oscillates between the exciton and photon modes, a phenomenon known as optical Rabi oscillations. The spectral signature of strong coupling is an anticrossing between the dot line (i.e., the spectral linewidth of the QD) and the cavity mode. The eigenmodes of the system are mixed entangled light-matter states, giving rise to quasiparticles known as microcavity exciton–polaritons which exhibit bosonic properties. Considerable scientific effort worldwide has been invested in exploration and control of a strongly coupled single QD in a cavity. The latter is considered a solid-state analogue of the atom-cavity system for which many quantum-optical effects have been demonstrated. Some examples include cavity field quantization [24] and photon blockade [25]. Embedding QDs in confined geometries—such as planar [26], micropillar [27], microdisk [28], microsphere [29], nanowire [30], or photonic crystal nanocavities [31]—provides a means for controlling the light-matter interactions. However, unlike isolated atoms, QDs are embedded in a solid-state environment, so confined charge carriers may experience additional interactions with phonons and other neighboring states and charges. These interactions may result in a departure from the behavior of an idealized two-level quantum system, but can also be exploited as additional control measures.

Cavity quantum electrodynamics (CQED), traditionally applied to atom-cavity systems, has gained new momentum after recent reports of quantum-optical experiments in solid-state systems demonstrating nonclassical light behavior, such as nonclassical photon statistics, nonlocality, and quantum entanglement [32,33]. These cutting-edge studies put to the test the very foundations of quantum mechanics and open up new avenues for on-demand generation and exploitation of various nonclassical states of light. So far, close to perfect single-photon interaction with a QD has been achieved in a cavity under the strong coupling regime [34,35].

A major goal of the simulation of quantum-photonic devices is to find an optimum design for combining photonic and solid-state devices on a chip. Integrated optical technologies for quantum computing based on linear optics [36] have been under active development in the recent decade. However, nonlinear optical interactions "on a chip" offer new, unexplored functionalities, and may play a key role in future technologies. In this respect, a major theoretical and computational challenge is the development of new, tractable models of the nonlinear quantum dynamics in realistic devices with complex geometries in multiple dimensions. This chapter reviews the evolution of the original methodology into our fully non-perturbative, mesoscopic, Maxwell's curl-pseudospin equation-based model of the optically induced spatiotemporal dynamics in quantum-photonic devices. This chapter contains further details about its numerical implementation and some concrete applications.

As we shall show in Section 49.2, our theoretical and computational approach differs substantially from the methods described in Chapter 46 of this book. In that chapter, a number of different approaches for calculation of the spontaneous emission have been employed. In the weak-coupling regime, a classical dipole model is developed whereby the dyadic Green's function is calculated and the spontaneous emission rate is found from Fermi's Golden Rule. Analytical expressions for Green's function are obtainable only for a limited number of device geometries, and more generally one has to resort to numerical methods (e.g., the finite-element method) to compute the electric field. This entails expanding the fields in the eigenmodes in different layers of the specific geometry under consideration. Another approach for modeling cavity-emitter systems is CQED, which is based on the Jaynes–Cummings Hamiltonian, describing the interaction of a two-level system strongly coupled to a single mode of an electromagnetic field. Two approximations are made: (1) for a very low Q cavity, the optical field operators can be replaced with material excitation operators; the dissipation in the system (spontaneous emission rate and dephasing) is taken into account through Lindblad operators. Using the quantum regression theorem that is valid for Markovian processes, an analytical expression for the correlation function is obtained; (2) for a coupled cavity-emitter system, a more complicated master equation is written with an additional dissipation term describing the cavity photon loss. However, for losses weaker than the cavity-emitter coupling, the cavity can be adiabatically eliminated and this results in a master equation describing the emitter only but with modified spontaneous emission rates. By contrast, we use the full master equation for the system density matrix which is valid for all regimes: weak, intermediate, and strong coupling. In addition, in our case the longitudinal and transverse dissipation are taken into account not through Lindblad operators, but by considering the dynamics of population transfer and decoherence between the discrete levels. Another difference between our model presented in Section 49.2 and the model of Gregersen et al. is that in our case, the cavity-emitter (Rabi) coupling is not a model parameter, but is calculated from the initial pulse amplitude, duration, and dipole moment of the specific optical transition, as in the actual experiment. Furthermore, the cavity loss is not a phenomenological parameter either, as it is calculated numerically through the perfectly transparent boundary conditions of the specific cavity geometry. The master equation is solved self-consistently with the vector Maxwell equations for the pulse propagation.

Our approach, predominantly using quantum-optical description based on the Liouville equation of motion for the density matrix, is quite distinctive from the many-body theoretical methods applied in Chapter 48. For instance, the developed by the aforementioned authors quantum-optical semiconductor laser theory beyond the semiclassical approach is based on a correlation-expansion and truncation technique for the quantized light field interacting with carriers in the single-particle picture (electrons and holes) occupying the discrete QD levels. Dynamical equations for the carrier populations in the conduction and valence states of the QD, for the expectation value of the photon number operator of the laser mode and the photon-assisted polarization are derived using Heisenberg operator equations of motion for the creation and annihilation operators. These coupled equations contain carrier–photon and photon–photon correlations, providing access to the second-order correlation function, $g^{(2)}$. The interaction with the environment is accounted through quantum kinetic equations, beyond the Boltzmann equation approach, for the carrier dynamics due to carrier–phonon and carrier-Coulomb interactions. Many-body theory is applied to calculate the corresponding matrix elements. In the former case, effects such as polaron scattering and non-Markovian dynamics are shown for QD systems. This is a very thorough many-body quantized electromagnetic field approach; however, it does not take into account the specific macroscopic boundary conditions of the particular device geometry considered, as the light-matter interaction is described by a single laser mode coupled to the quantum system. An extension for multiple emitters coupled to a single common light mode is proposed, taking advantage of the many-particle picture description of the QD states, such as excitons, trions, multiexcitons, etc., to describe interemitter correlations and radiative coupling. This is somewhat similar to our approach in that it considers many-particle states (e.g., excitons); however, our approach is extended in that it also considers multiple cavity electromagnetic field modes. The cluster expansion approach in Chapter 48 by Gies et al. is in essence a generation of equations of motion for the photon correlations up to a given order of truncation of the expansion and thus providing access to $g^{(2)}$. The dissipation phenomena within this approach—such as the carrier–phonon and carrier–carrier Coulomb scattering—are calculated from a microscopic many-body model (compared to the Lindblad terms and our approach, where they are taken as phenomenological parameters), which is an advantage, as the theory is self-consistent.

In Section 49.2, we will give an overview of the strategy we have used to construct a model for describing coherent light-matter interactions in semiconductor nanostructures.

49.2 Maxwell-Pseudospin Method

49.2.1 Pseudospin Equation

In what follows, we will consider either resonant or near-resonant laser optical field interactions with atom-like discrete-level quantum systems, such as QDs. The optical field we consider is monochromatic and (near-)resonant with a given transition between a pair of atomic levels, thus effectively mimicking a two-level atom. As the two-level atom is conceptually equivalent to a half-spin particle in a magnetic field, the dynamics of the laser-atom interaction is governed by the same Bloch spin vector equation of motion [37] developed for describing nuclear magnetic resonance and the precession of a classical gyromagnet (or its quantum analogue, the spin vector, **S**) in a constant magnetic field, **B**:

$$\frac{d\mathbf{S}}{dt} = f\mathbf{B} \times \mathbf{S}. \tag{49.1}$$

Here f guarantees the constant length of the spin vector, thus confining the spin dynamics to the surface of a Bloch sphere. (Note, however, that this equation is valid only for systems with equally spaced energy levels). The formal analogy between a spin and a nonspin (pseudospin) system was pointed out for the first time by Feynman and coworkers [38], showing that when coherent processes are involved in a two-level system, it is sufficient to consider a real three-vector, rather than the complex probability amplitudes in the Schrödinger equation or the complex density-matrix elements [37]. This real-vector representation provides an elegant and intuitive geometrical framework for comprehending the system dynamics in terms of rotations of a real state vector in the Hilbert space. Furthermore, in contrast to the wavefunction formalism, it describes both pure and mixed quantum states. Attempts to extend the formalism analogously for (N > 2)-level quantum systems fail to preserve the simple vector form of the basic equation of motion for the real state pseudospin (coherence, or Bloch) vector. The only known exception to this is the special case of excitation of a spin-J system in constant magnetic field (thus yielding a $(N = 2J + 1)$-level system), due to the equal spacing of the energy levels in this system [39]. Preserving this simple form of the vector equation has a number of advantages: (1) it accounts for the intrinsic symmetry of the underlying Hilbert space of the system and therefore is an exact description, independent of the strength, number, or time dependence of the external forces acting on the system; (2) similar to the two-level system, the dynamical evolution can be illustrated as a rotation in the real physical space of a real coherence vector.

The case of a three-level system is of particular interest, since it provides a useful framework for study-ing such phenomena as two-photon coherence, resonance Raman scattering, three-level echoes, three-level superradiance, coherent multistep photoionization/photodissociation, coherent population trapping, and electromagnetically induced transparency (EIT). A solution to this problem has been given by Elgin [40] by invoking the invariance of the state vector under rotations of the SU(3) transformation group. The dynam-ical evolution of a three-level system can be expressed in terms of an eight-dimensional real coherence vector, taking advantage of the group-theory and Gell-Mann's SU(3) generators, which had been developed for the quark triplet in the domain of high-energy physics [41].

A general solution for an N-level quantum system with arbitrary level spacing has been provided by Hioe and Eberly [42]. The starting point is the dynamical evolution of a quantum system described by the Liouville equation for the density operator in the Schrödinger picture

$$i\hbar \frac{\partial \hat{\rho}}{\partial t} = \left[\hat{H}, \hat{\rho}\right], \tag{49.2}$$

where $\hat{H}$ is the total system Hamiltonian:

$$\hat{H} = \hat{H}_0 + \hat{H}_{\text{int}}(t), \tag{49.3}$$

with $\hat{H}_0$ being the unperturbed Hamiltonian of an N-level system: a diagonal matrix with eigenenergies of each level, $\hbar\omega_k$ ($k = 1, ..., N$), along the main diagonal, and $\hat{H}_{int}$ is a time-dependent perturbation (not necessarily small).

Hioe and Eberly showed that the presence of unitary group generators in the time evolution of an N-level quantum system permits to describe the evolution in terms of the rotations of a real coherence vector. Expanding the system Hamiltonian and the density matrix operators in terms of λ-generators of the SU(N) Lie algebra, $\hat{\lambda}_k$, they derived the pseudospin equation of motion in the Heisenberg picture, depicting the time evolution of the real pseudospin vector, **S**, as a generalized rotation in $N^2 - 1$ Hilbert space:

$$\dot{S}_i = f_{ijk}\gamma_j S_k, \quad i, j, k = 1, 2, ..., N^2 - 1, \tag{49.4}$$

where the dot stands for the time derivative, γ_j is the torque vector, and $f_{ijk} = \frac{1}{4}i\left[Tr\left(\hat{\lambda}_i\hat{\lambda}_k\hat{\lambda}_j\right) - Tr\left(\hat{\lambda}_i\hat{\lambda}_j\hat{\lambda}_k\right)\right]$, $i, j, k = 1, 2, ..., N^2 - 1$, is the fully antisymmetric tensor of the structure constants of the SU(N) group, whose asymmetry guarantees the constant length of the pseudospin vector, $|\mathbf{S}|$, in the (N^2-1)-dimensional Hilbert space. The coherence vector is expressed as the trace of the product of the density and the SU(N) group generators, whereas the torque vector is expressed as the trace of the product of the system Hamiltonian operators and the SU(N) group generators

$$S_j(t) = Tr\left(\hat{\rho}(t).\hat{\lambda}_j\right); \quad \gamma_j(t) = \frac{1}{\hbar}Tr\left(\hat{H}(t).\hat{\lambda}_j\right). \tag{49.5}$$

The SU(N) group generators obey the group property

$$\left[\hat{\lambda}_j, \hat{\lambda}_k\right] = 2if_{jkl}\hat{\lambda}_l \tag{49.6}$$

and orthogonality relations, $Tr\left(\hat{\lambda}_j.\hat{\lambda}_k\right) = \delta_{jk}$, however their choice is not unique.

For the simplest case of a two-level system, $N = 2$, the λ-generators are simply the Pauli matrices $\hat{\sigma}_x$, $\hat{\sigma}_y$, and $\hat{\sigma}_z$. This is obtained from Equation 49.4 when f_{ijk} is replaced by the Levi-Civita symbol, ε_{ijk}. For a dipole-coupling interaction $\hat{H}_{int} = e\mathbf{E}.\hat{\mathbf{r}}$, where **r** is the local displacement operator and e is the electron charge, and a linearly polarized dipole optical transition ($\mathbf{E} \parallel \mathbf{r}$ with an optical selection rule for the z-axis projection of the total angular momentum, $\Delta J_z = 0$) the optical Bloch equations are easily recovered:

$$\dot{S}_1 = \omega_0 S_2$$
$$\dot{\mathbf{S}} = \boldsymbol{\gamma} \times \mathbf{S} \quad \dot{S}_2 = -\omega_0 S_1 + 2\Omega_R S_3 \tag{49.7}$$
$$\dot{S}_3 = -2\Omega_R S_2,$$

where S_i ($i = 1, 2, 3$) is the real pseudospin three-vector, and the torque vector is given by $\boldsymbol{\gamma} = \left(2\Omega_R, 0, \omega_0\right)$ from Equation 49.5. Here we have defined the Rabi frequency, $\Omega_R = \frac{\wp}{\hbar}E$, with $\wp = \langle i|e\hat{r}|j\rangle$, transition dipole matrix element of $i \rightarrow j$, and ω_0 is the atomic transition resonant frequency. In case of near-resonant excitation, ω_0 should be replaced by $\Delta\omega = \omega_0 - \omega$, where ω is the excitation frequency [43].

In the general case of an electric field vector noncollinear with the local displacement (dipole moment) vector, the optical transition is excited by circularly or elliptically polarized light obeying the selection rule $\Delta J_z = \pm 1$, where the sign "+/−" corresponds to right/left circularly polarized light. The dipole matrix elements are complex vectors yielding complex Rabi frequencies along the x- and y-direction in the plane determined by **E** and **r** (or equivalently, along the dipole moment vector): $\Omega_x = \frac{\wp}{\hbar}E_x$; $\Omega_y = \frac{\wp}{\hbar}E_y$. The torque vector is given by $\boldsymbol{\gamma} = \left(\Omega_x, \Omega_y, \omega_0 \right)$ and the resulting pseudospin equations read (see also [37])

$$\dot{\mathbf{S}} = \boldsymbol{\gamma} \times \mathbf{S} \quad \begin{aligned} \dot{S}_1 &= -\omega_0 S_2 - \Omega_y S_3 \\ \dot{S}_2 &= \omega_0 S_1 + \Omega_x S_3 \\ \dot{S}_3 &= \Omega_y S_1 - \Omega_x S_2 . \end{aligned} \tag{49.8}$$

The relationship between the real-state coherence vector components and the density-matrix components for the general N-level case is given by Equation 49.5, and for a two-level system reads

$$\begin{aligned} S_1 &= \hat{\rho}_{12} + \hat{\rho}_{21} = 2\text{Re}\left(\hat{\rho}_{12} \right) \\ S_2 &= i \left(\hat{\rho}_{12} - \hat{\rho}_{21} \right) = -2\text{Im}\left(\hat{\rho}_{12} \right) = 2\text{Im}\left(\hat{\rho}_{21} \right) \\ S_3 &= \hat{\rho}_{22} - \hat{\rho}_{21} . \end{aligned} \tag{49.9}$$

The physical meaning of the first two components of the pseudospin vector are dispersive (in-phase) and absorptive (in-quadrature) parts of the dipole polarization, and the third component is the population difference between the two levels. Since the level occupation probability is conserved, i.e., $\rho_{11} + \rho_{22} = 1$, it can be easily shown that the length of the coherence vector is unity: $|S|^2 = S_1^2 + S_2^2 + S_3^2 = 1$, and the evolution of the pseudospin is "inscribed" by the pseudospin vector on the surface of the Bloch sphere.

Up until now, we have only considered coherent processes; however, in an open quantum system there is always energy dissipation. For a two-level system, it is easy to define phenomenological longitudinal relaxation time (also known as population relaxation time), T_1, and transverse relaxation time (also known as polarization-decay or dephasing time), T_2, leading to the well-known optical Bloch equations for a two-level system with damping for e.g., linearly polarized transitions:

$$\begin{aligned} \frac{\partial S_1}{\partial t} &= \omega_o S_2 - \frac{1}{T_2} S_1 \\ \frac{\partial S_2}{\partial t} &= -\omega_o S_1 + 2\Omega_x S_3 - \frac{1}{T_2} S_2 \\ \frac{\partial S_3}{\partial t} &= -2\Omega_x S_2 - \frac{1}{T_1} \left(S_3 - S_{30} \right) , \end{aligned} \tag{49.10}$$

where $S_{30} = \pm 1$ is the initial population profile; "–" and "+" correspond, respectively, to all of the population in the ground and the excited state. Upon optical excitation, the former describes absorption, while the latter describes optical gain. However, it is not so easy to generalize the dissipation dynamics for a system with an arbitrary number of levels (N-level system), because each pair of levels within this system is characterized by their individual population transfer rates and dephasing rates. We show how to do this generalization next. The time evolution of the two-level quantum system is described by the generalized rotation of the real three-vector, $\mathbf{S}$, in the three-dimensional Hilbert space; at all times the tip of $\mathbf{S}$ is just touching the surface of the Bloch sphere, shown in Figure 49.1.

49.2.2 Dissipative Dynamics and Master Pseudospin Equation

To include energy relaxation and decoherence processes in the system dynamics, we adopt an unconventional approach. We introduce longitudinal relaxation in the Liouville equation for the density matrix by considering first the diagonal population components of the Liouville equation, ρ_{ii}, and taking into account population transfer between all allowed dipole optical transitions in the system with corresponding rates.

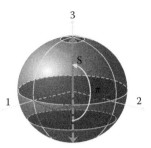

FIGURE 49.1 Geometrical representation of the time evolution of a two-level quantum system by generalized rotation of the real state pseudospin vector, **S**, in the 3D Hilbert space, drawing a trajectory on the $|\mathbf{S}| = const.$ sphere surface. A one-qubit rotation occurs when the vector is rotated through π from state 0 (spin down, dashed arrow) to state 1 (spin up, solid arrow along $\hat{3}$-axis), or vice versa.

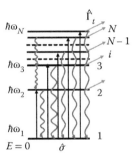

FIGURE 49.2 Schematic of an N-level quantum system with energy zero at the ground state. [Vertical solid arrows, coherent optical excitation; wavy lines, population relaxation $\hat{\sigma}$ (spontaneous emission); tilted arrows, dephasing $\hat{\Gamma}_t$ (decoherence).]

For each Liouville equation we can construct a longitudinal damping matrix of rank N: $\hat{\Gamma}_i$, $i = 1, ..., N$, whose elements are the population transfer rates between all pairs of levels. This allows us to introduce a diagonal matrix, $\hat{\sigma} = diag(Tr(\hat{\Gamma}_i.\hat{\rho}))$, which accounts for the longitudinal population relaxation. Similarly, we introduce an off-diagonal matrix, $\hat{\Gamma}_t$, of rank N with zero main diagonal components to account for the dephasing rates of all levels (Figure 49.2). Adding the relaxation and dephasing terms in the Liouville equation thus leads to the following master equation:

$$\frac{\partial \hat{\rho}}{\partial t} = \frac{1}{\hbar}\left[\hat{\rho}, \hat{H}\right] + \hat{\sigma} - \hat{\Gamma}_t.\hat{\rho}. \tag{49.11}$$

Using Equation 49.5, we can now rewrite the density-matrix equations in terms of the pseudospin vector, **S** [44]:

$$\frac{\partial S_j}{\partial t} = \begin{cases} f_{jkl}\Gamma_k S_l + \frac{1}{2}Tr\left(\hat{\sigma}.\hat{\lambda}_j\right) - \frac{1}{T_j}\left(S_j - S_{je}\right), & j = 1, 2, ...N\,(N-1) \\ f_{jkl}\Gamma_k S_l + \frac{1}{2}Tr\left(\hat{\sigma}.\hat{\lambda}_j\right), & j = N\,(N-1)+1, ..., N^2 - 1, \end{cases} \tag{49.12}$$

where we have defined an equilibrium coherence vector, $\mathbf{S_E} = (S_{1e}, S_{2e}, ..., S_{(N^2-1)e})$, and $T_j, j = 1, 2...,$ $N^2 - 1$ are phenomenologically introduced nonuniform decay times describing the relaxation of the components of the real state vector towards their equilibrium values, S_{je}. Because of the incoherent processes involved in maintaining the system at a definite level of excitation, the equilibrium system polarization is zero. This means that the subset of $\mathbf{S_E}$ containing the polarization components of the equilibrium coherence vector is zero, i.e., $\mathbf{S_E^*} = (S_{1e}, S_{2e}, ..., S_{N(N-1)e}) = 0$. The second term in the second line of the Equation 49.12 containing the trace of the density matrix through $\hat{\sigma}$ can be rewritten in terms of nonuniform relaxation times, $T_j = T_{N(N-1)+1}, ..., T_{N^2-1}$, describing the relaxation of the population components of the real coherence vector toward their equilibrium values. Therefore, the total equilibrium coherence vector is of the form: $S_E = (0, 0, ..., 0, S_{N(N-1)+1}, ..., S_{N^2-1})$. We shall refer to Equation 49.12 as the master pseudospin equations.

49.2.3 Vector Maxwell Equations and Coupling to Pseudospin Equations

The optical wave propagation is described by the vector Maxwell equations:

$$\frac{\partial \mathbf{H}}{\partial t} = -\frac{1}{\mu} \nabla \times \mathbf{E}$$

$$\frac{\partial \mathbf{E}}{\partial t} = \frac{1}{\varepsilon} \nabla \times \mathbf{H} - \frac{1}{\varepsilon} \frac{\partial \mathbf{P}}{\partial t}, \tag{49.13}$$

where $\mathbf{P}$ is the macroscopic medium polarization vector given by

$$\mathbf{P} = -eN_a \, Tr\,(\hat{\rho}.\hat{r}), \tag{49.14}$$

with N_a signifying the density of the ensemble of resonant dipoles in the medium. Starting with a phenomenologically constructed dipole-coupling interaction Hamiltonian with coherent Rabi frequency terms describing optical transitions between pairs of levels, one can decompose the dipole moment operator, and represent it as a linear combination of λ-generators, whose coefficients are the corresponding Cartesian dipole moment vector components. Then, using Equation 49.14, we can find a relationship between the macroscopic polarization components, $P_{x,y,z}$, and the coherence vector components, S_j. To illustrate this procedure, suppose we have a transverse electromagnetic wave propagating along the z-axis exciting a given $i \rightarrow j (i < j)$ transition in an N-level system. Let the wave be circularly polarized in the plane perpendicular to z, with in-plane electric-field components E_x and E_y. Then, the (i,j) term of the interaction Hamiltonian will contain a complex coherent term $\hbar(\Omega_x \pm i\Omega_y)$, where the sign "+/−" depends on the helicity of the pulse, σ^+ or σ^-. (The sign convention is such that "+" and "−" correspond to, respectively, clockwise and counterclockwise rotation of the electric-field vector, when viewed from a direction opposite to that of the propagation of the pulse.) Note that, for the case of a linearly polarized excitation, the Rabi frequency will be real. The in-plane dipole moment vector can be decomposed into two Cartesian components: $er = e(\hat{r}_x \vec{e}_x + \hat{r}_y \vec{e}_y)$, where $\vec{e}_x$ and $\vec{e}_y$ are the unit vectors along the x- and y-axes. The coefficients $\hat{r}_x$ and $\hat{r}_y$ are $\hat{\lambda}$-generators; $\hat{r}_x$ contains unity at positions (i,j) and (j,i), whereas $\hat{r}_y$ contains the imaginary number, $-i$, at position (i,j) and i at position (j,i):

$$\mathbf{r} = r_0 \left\{ \begin{pmatrix} 0 & 0 & 0 & \cdot & \cdot & 0 & 0 & 0 \\ \cdot & \cdot & \cdot & & \cdot & \cdot & \cdot \\ 0 & 0 & 0 & \cdot & \cdot & 1 & \cdot & 0 \\ 0 & 0 & 0 & \cdot & & \cdot & 0 \\ \cdot & \cdot & \cdot & & \cdot & \cdot & \cdot \\ 0 & \cdot & 1 & 0 & \cdot & \cdot & 0 \\ \cdot & \cdot & \cdot & & \cdot & \cdot & \cdot \\ 0 & 0 & 0 & \cdot & \cdot & 0 & 0 & 0 \end{pmatrix} \vec{e}_x + \begin{pmatrix} 0 & 0 & 0 & \cdot & \cdot & 0 & 0 & 0 \\ \cdot & \cdot & \cdot & & \cdot & \cdot & \cdot \\ 0 & 0 & 0 & \cdot & \cdot & -i & \cdot & 0 \\ 0 & 0 & 0 & \cdot & & \cdot & 0 \\ \cdot & \cdot & \cdot & & \cdot & \cdot & \cdot \\ 0 & \cdot & i & 0 & \cdot & \cdot & 0 \\ \cdot & \cdot & \cdot & & \cdot & \cdot & \cdot \\ 0 & 0 & 0 & \cdot & \cdot & 0 & 0 & 0 \end{pmatrix} \vec{e}_y \right\}, \tag{49.15}$$

where r_0 is the dipole length scale. The following general formula can be used to obtain the corresponding $\hat{\lambda}$-generators, depending on the number of discrete levels of the system, N:

$$\hat{r}_x = \hat{\lambda}_{j-i+\sum_{d=1}^{i-1}(N-d)}$$

$$\hat{r}_y = \hat{\lambda}_{j-i+\sum_{d=1}^{i-1}(N-d)+\frac{N(N-1)}{2}}. \tag{49.16}$$

In fact, this linear combination reproduces the phenomenologically constructed dipole coupling interaction Hamiltonian, $\hat{H}_{int} = -e\mathbf{r}.\mathbf{E}$, when the dipole moment and the decomposed electric-field vector $(\vec{E} = E_x\vec{e}_x + E_y\vec{e}_y)$ are substituted in

$$\hat{H}_{int} = \hbar \begin{pmatrix} 0 & 0 & 0 & \cdot & \cdot & 0 & 0 & 0 \\ \cdot & \cdot & \cdot & \cdot & \cdot & \cdot & \cdot & \cdot \\ 0 & 0 & 0 & \cdot & \cdot & \Omega_x - i\Omega_y & \cdot & 0 \\ 0 & 0 & 0 & \cdot & \cdot & \cdot & \cdot & 0 \\ \cdot & \cdot & \cdot & \cdot & \cdot & \cdot & \cdot & \cdot \\ 0 & \cdot & \Omega_x + i\Omega_y & 0 & \cdot & \cdot & \cdot & 0 \\ \cdot & \cdot & \cdot & \cdot & \cdot & \cdot & \cdot & \cdot \\ 0 & 0 & 0 & \cdot & \cdot & 0 & 0 & 0 \end{pmatrix}. \tag{49.17}$$

The macroscopic polarization vector components, P_x and P_y, are then obtained from Equation 49.14, giving

$$P_x = -\wp N_a Tr(\hat{\rho}.\hat{r}_x)$$

$$P_y = -\wp N_a Tr(\hat{\rho}.\hat{r}_y). \tag{49.18}$$

Now we can write down the vector Maxwell equations for a circularly polarized optical wave exciting the $i \to j$ transition of an N-level system, and thus inducing dipole polarizations P_x and P_y along the x- and y-directions, respectively:

$$\frac{\partial H_x(z,t)}{\partial t} = \frac{1}{\mu}\frac{\partial E_y(z,t)}{\partial z}$$

$$\frac{\partial H_y(z,t)}{\partial t} = -\frac{1}{\mu}\frac{\partial E_x(z,t)}{\partial z}$$

$$\frac{\partial E_x(z,t)}{\partial t} = -\frac{1}{\varepsilon}\frac{\partial H_y(z,t)}{\partial z} - \frac{1}{\varepsilon}\frac{\partial P_x(z,t)}{\partial t} \tag{49.19}$$

$$\frac{\partial E_y(z,t)}{\partial t} = \frac{1}{\varepsilon}\frac{\partial H_x(z,t)}{\partial z} - \frac{1}{\varepsilon}\frac{\partial P_y(z,t)}{\partial t}.$$

For the special case of a transverse optical wave linearly polarized along x and propagating in the z-direction, the above system (Equation 49.19) is reduced to

$$\frac{\partial H_y(z,t)}{\partial t} = -\frac{1}{\mu}\frac{\partial E_x(z,t)}{\partial z}$$

$$\frac{\partial E_x(z,t)}{\partial t} = -\frac{1}{\varepsilon}\frac{\partial H_y(z,t)}{\partial z} - \frac{1}{\varepsilon}\frac{\partial P_x(z,t)}{\partial t}. \tag{49.20}$$

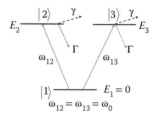

FIGURE 49.3 Energy-level scheme of a V-type three-level system. $\omega_{jk} = (E_j - E_k)/\hbar$, with energy zero chosen at the ground state $|1\rangle$. (Γ, population relaxation rate; γ, dephasing rate.)

We solve Maxwell's curl equations, (Equations 49.19 or 49.20), and the pseudospin equations (Equation 49.12) self-consistently, having established a link between them via the medium's polarization (Equation 49.18). We shall refer to this coupled system of first-order differential equations as the coherent Maxwell's curl-pseudospin model.

49.2.4 Coherent Maxwell-Pseudospin Equations in 2D

Description of nonlinear pulse propagation, interaction, and localization phenomena in multiple dimensions requires dipole coupling to at least a 2D resonant medium, whereby the electric field of a single-photon excitation is coupled to at least two distinct dipole optical transitions. We have previously shown that as a minimum requirement, it is sufficient to consider a degenerate three-level ensemble of dipoles in which two of the allowed dipole optical transitions are excited by each of the two components of the electric-field vector in the waveguide plane [45]. The pseudospin geometrical picture thus allows for adequate modeling of the interaction of an ultrashort laser pulse with a medium in two spatial dimensions. Therefore, we consider a V-type degenerate three-level system of resonant dipoles shown in Figure 49.3, for which a linearly polarized monochromatic electromagnetic wave induces polarizations (dipoles) in two orthogonal directions in the waveguide plane. Two cases arise: a transverse electric (TE) wave, for which $\mathbf{E} = (E_x, 0, 0)$ and $\mathbf{H} = (0, H_y, H_z)$, and transverse magnetic (TM) wave, for which $\mathbf{E} = (0, E_y, E_z)$ and $\mathbf{H} = (H_x, 0, 0)$. The respective vector Maxwell equations are

$$\frac{\partial H_y}{\partial t} = -\frac{1}{\mu}\frac{\partial E_x}{\partial z}$$

$$\frac{\partial H_z}{\partial t} = \frac{1}{\mu}\frac{\partial E_x}{\partial y}$$

$$\frac{\partial E_x}{\partial t} = \frac{1}{\varepsilon}\left(\frac{\partial H_z}{\partial y} - \frac{\partial H_y}{\partial z}\right) - \frac{1}{\varepsilon}\frac{\partial P_x}{\partial t}$$

(49.21)

for the TE wave and

$$\frac{\partial H_x}{\partial t} = -\frac{1}{\mu}\frac{\partial E_z}{\partial y} + \frac{1}{\mu}\frac{\partial E_y}{\partial z}$$

$$\frac{\partial E_y}{\partial t} = \frac{1}{\varepsilon}\frac{\partial H_x}{\partial z} - \frac{1}{\varepsilon}\frac{\partial P_y}{\partial t}$$

$$\frac{\partial E_z}{\partial t} = -\frac{1}{\varepsilon}\frac{\partial H_x}{\partial y} - \frac{1}{\varepsilon}\frac{\partial P_z}{\partial t}$$

(49.22)

for the TM wave. We can phenomenologically construct the system Hamiltonian corresponding to each of the cases above:

$$\hat{H}_{TE}(t) = \begin{pmatrix} 0 & \Omega_x & 0 \\ \Omega_x & \omega_0 & 0 \\ 0 & 0 & \omega_0 \end{pmatrix}; \hat{H}_{TM}(t) = \begin{pmatrix} 0 & \Omega_y & \Omega_z \\ \Omega_y & \omega_0 & 0 \\ \Omega_z & 0 & \omega_0 \end{pmatrix}. \tag{49.23}$$

The master pseudospin equations (Equation 49.12) for a three-level system with $i, j, k = 1, 2, ..., 8$ are coupled to Maxwell's equations above through the macroscopic polarization, Equation 49.14. It thus becomes possible to derive the following relationships between the polarization components and the pseudospin vector components for the case of a TE wave:

$$P_x = -\wp N_a S_1, \tag{49.24}$$

and for a TM wave:

$$\begin{aligned} P_y &= -\wp N_a S_1 \\ P_z &= -\wp N_a S_3. \end{aligned} \tag{49.25}$$

49.2.5 Numerical Methodology

The Maxwell-pseudospin system is discretized on a Yee grid in 1D (modified Yee grid in 2D, see Figure 49.4) and solved directly in the time domain by the finite-difference time-domain (FDTD) method [46]. At each time step, a solution to all equations is obtained using a predictor-corrector iterative scheme, proven to be very efficient in solving simultaneously a large number of first-order differential equations [47]. The time evolution of a discrete multilevel system under an external perturbation can be viewed as a Goursat-type initial boundary value problem, which is well posed when the time history of the initial electric field is given along, e.g., the left (lower) boundary of the simulation domain in 1D (2D). We impose analytical absorbing boundary conditions at the simulation domain boundaries, based on the Engquist–Majda [48] one-way wave equations discretized by the Mur finite-difference scheme [49] using a two-term Taylor-series approximation [46]. To ensure numerical stability, the time step and the spatial discretization are chosen to satisfy the Courant stability criterion: $\frac{c\Delta t}{\Delta z} \leq 1$ in 1D and $\frac{c\Delta t}{\left(\frac{1}{(\Delta y)^2} + \frac{1}{(\Delta z)^2}\right)^{-1/2}} \leq 1$ in 2D [46]. The FDTD numerical technique has numerous advantages over other methods, such as the beam propagation method (BPM). For instance, modeling structures with rapid longitudinal variations (on the scale of the wavelength, λ) of, e.g., refractive index or energy-level population, is impossible to model with BPM, because it relies on the paraxial approximation which in this case is invalid. Moreover, FDTD can handle strong coupling between forward- and backward-propagating fields and can be used to describe general, nonharmonic electromagnetic fields. It can also be used to simulate ultrashort (sub-picosecond) pulses in situations where the pulse bandwidth is significant compared to material resonances.

49.3 Simulation of Nonlinear Dynamics in Optical Waveguides and Semiconductor Microcavities

In this section, we apply the vector Maxwell-pseudospin formalism to the problem of nonlinear optical pulse propagation in planar waveguides and microcavities with embedded quantum systems. We provide a solution in both 1D and 2D. As a first step, we perform validation studies against well-established theoretical and computational methods.

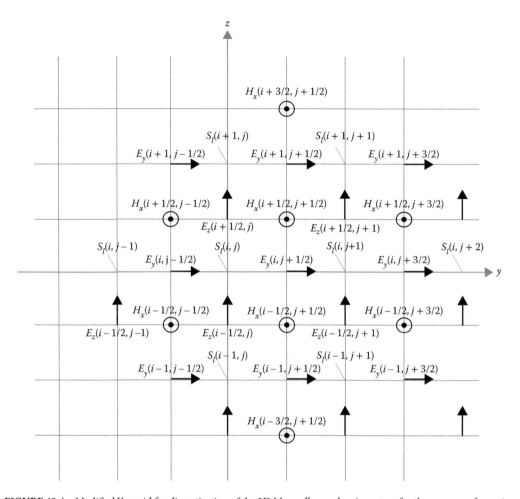

FIGURE 49.4 Modified Yee grid for discretization of the 2D Maxwell-pseudospin system for the purpose of carrying out numerical modeling. The grid is staggered, meaning that the electric and magnetic fields are calculated at every other node. This schematic is for a TM wave: the electric-field vector is in the y–z plane (arrows); the magnetic-field vector is pointing out of the page toward the reader (circle with dot). The electric- and magnetic-field components are spatially separated by $\Delta y/2$ and $\Delta z/2$ in the y–z plane, and temporally by $\Delta t/2$, with the electric-field components situated in the middle of the cells' edges, and magnetic-field components in the center of the cell [46]. The magnetic-field equation is solved at the spatial steps $(i + 1/2)\Delta z$ and $(j + 1/2)\Delta y$ over time step $(k + 1/2)\Delta t$. The electric-field components, E_y and E_z, are solved at the spatial steps $((j + 1/2)\Delta y,\ i\Delta z)$ and $(j\Delta y,\ (i + 1/2)\Delta z)$ over time step $k\Delta t$. Quantum-system variables, S_l, are assigned to the empty nodes in the 2D Yee grid, while the electric-field component values at these empty mesh points are estimated by averaging over the electric fields located at the nearest neighbors in the plane.

49.3.1 One-Dimensional Model of a Two-Level System

Consider a two-level quantum system, such as fundamental heavy-hole transitions in a QD. The coherent Maxwell-pseudospin equations describing the interaction of an optical wave propagating along the z-axis and linearly polarized in a transverse direction along x are given by Equations 49.10, 49.20, and 49.24:

$$\frac{\partial H_y}{\partial t} = -\frac{1}{\mu}\frac{\partial E_x}{\partial z}$$

$$\frac{\partial E_x}{\partial t} = -\frac{1}{\varepsilon}\frac{\partial H_y}{\partial z} - \frac{\wp N_a}{\varepsilon T_2}S_1 + \frac{\wp N_a \omega_0}{\varepsilon}S_2$$

$$\frac{\partial S_1}{\partial t} = -\frac{1}{T_2}S_1 + \omega_0 S_2 \qquad\qquad (49.26)$$

$$\frac{\partial S_2}{\partial t} = -\omega_0 S_1 - \frac{1}{T_2}S_2 + 2\frac{\wp}{\hbar}E_x S_3$$

$$\frac{\partial S_3}{\partial t} = -2\frac{\wp}{\hbar}E_x S_2 - \frac{1}{T_1}\left(S_3 - S_{30}\right).$$

We apply an excitation pulse of duration T_p with either a hyperbolic secant or Gaussian envelope and central frequency, ω_0, in resonance with the two-level transition at the boundary $z = 0$ of the simulation domain. Its time dependence is thus

$$E_x\left(z = 0, t\right) = \begin{cases} E_0 \mathrm{sech}\left(10\frac{t-\frac{T_p}{2}}{\frac{T_p}{2}}\right)\sin\left(\omega_0 t\right) \\[4mm] E_0 \exp\left[\left(\frac{t-t_0}{t_d}\right)^2\right]\sin\left(\omega_0 t\right), \end{cases} \qquad (49.27)$$

where E_0 is the initial pulse amplitude, t_0 is the time moment at which the Gaussian pulse is initially centered, and t_d is the characteristic $1/e$ Gaussian decay.

49.3.2 Passive Cavity Properties

To validate our model, we construct a simple, symmetric test Bragg microcavity containing 5 $\lambda/4$ GaAs/Al$_{0.1}$Ga$_{0.9}$As/AlAs layers, with λ being 1.4 µm (Figure 49.5f) and compute the passive (cold-)cavity modes and the electric-field standing-wave profile within the cavity. In order to compute the resonant cavity modes, we launch a short broad-band pulse of duration ≤ 100 fs from the left boundary of the structure and sample in time the electric-field evolution at the output on the right. This time trace is Fourier-transformed and normalized with respect to the Fourier transform of the input trace, thus yielding the transmission spectrum of the microcavity. The resonant frequency is extracted from the peak in the stop band of the spectrum (Figure 49.5d). Note that reducing the time step allows to resolve higher-order cavity modes. To obtain the electric-field standing-wave profile within the cavity, a continuous sine wave at the resonance frequency (obtained in the previous step) is pumped from the left boundary, and the time evolution of the intracavity field is monitored. The field amplitude increases in time and eventually reaches a steady value. The results shown in Figure 49.5f are compared with the cavity modes and field distributions computed by the transfer matrix method (TMM).

Using the above FDTD-based methodology, we will subsequently investigate passive cavity properties of more complex cavities. For example, we calculate the resonant cavity modes and standing-wave profile of a Coldren-type microcavity structure designed at $\lambda = 1.3$ µm and consisting of 10 GaAs/Al$_{0.9}$Ga$_{0.1}$As pairs for the bottom distributed Bragg reflector (DBR) mirror, followed by an AlAs supply layer, a tuning Al$_{0.7}$Ga$_{0.3}$As layer, a GaAs $\lambda/2$ cavity and 15 GaAs/Al$_{0.9}$Ga$_{0.1}$As pairs for the top DBR mirror. The simulation results for the resonant cavity modes, namely the transmission spectrum as a function of λ and the electric-field standing-wave profile along the structure, are displayed in Figure 49.6.

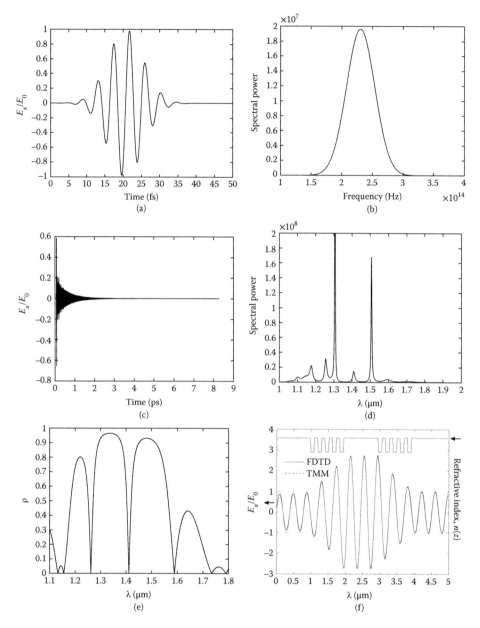

FIGURE 49.5 (a) Initial ultrashort pulse ($T_p \leq 100$ fs). (b) Fourier spectrum of the initial broad-band pulse. (c) Time trace of $E_x(t)$ at the output boundary of the structure. (d) Transmission spectrum as a function of wavelength, λ, computed as the discrete Fast Fourier transform (FFT) of the output time trace in (c), normalized with respect to the input FFT spectrum in (b). The resonant cavity wavelength, $\lambda_0 \approx 1.4\,\mu$m, is extracted from the peak in the DBR cavity stop band. (e) Reflectivity spectrum as a function of wavelength computed by the TMM: the resonant cavity wavelength, $\lambda_0 \approx 1.4\,\mu$m is obtained from the dip in the reflectivity spectrum within the DBR cavity stop band. (f) Comparison of the standing-wave profile within the cavity, as obtained by the TMM (dashed curve) and FDTD (solid black curve) methods. The FDTD-inferred instantaneous field standing-wave profile is computed with continuous sine-wave excitation propagating from the left boundary at the cavity resonance frequency equivalent to $\lambda_{pump} = 1.4116\,\mu$m, obtained from (d) after a sufficiently large number of time steps 2×10^6 (each of duration $\Delta t = 3.33 \times 10^{-3}$ fs) with 31,300 spatial grid points (each spatial step of $\Delta z = 1$ nm), Courant number= 0.75. The scale on the right corresponds to the refractive-index profile of the DBR cavity (top).

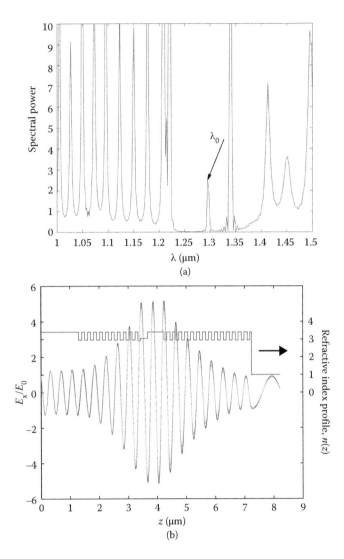

FIGURE 49.6   (a) FFT of the time trace of the signal detected at the output boundary of a Coldren-type microcavity structure, revealing the cavity's (nonnormalized) transmission spectrum. The resonant cavity mode is indicated with λ_0, and corresponds to approximately 1.3 µm. (b) Standing-wave profile after 6 million (dashed curve) and 12 million (solid curve) time steps superimposed on the refractive-index profile of the microcavity (scale on right-hand side). The time step in the simulation was $\Delta t = 3.33 \times 10^{-3}$ fs and the spatial step was $\Delta z = 1$ nm. The amplitude of the standing wave increases as the number of time steps is increased, and eventually reaches a steady maximum value (solid curve). The amplitude is augmented with time due to the build-up of the field caused by multiple reflections at the cavity mirrors.

49.3.3 Active Nonlinear Optical Waveguide/Cavity Properties

We have performed gain validation studies of our 1D semiclassical Maxwell-pseudospin model. This was done by establishing the ability of our model to recover the linear small-signal (unsaturated) gain and nonlinear (saturated) gain of a homogeneously broadened (HB) two-level system, and comparing the result against that of the density-matrix theory, which served as a benchmark (see, e.g., [50]). The test structure used for these validation studies was a 15 µm-long simulation domain consisting of a GaAs slab waveguide with a gain region length of $L_g = 9\,\mu m$, refractive index $n_{GaAs} = 3.59$, and pumping wavelength of $1.5\,\mu m$ corresponding to the resonant transition frequency of a two-level system sandwiched between two free-space regions (Figure 49.7a). The GaAs slab was represented by an initially inverted ($S_{30} = 1$) two-level medium with resonant dipole density $N_a = 1 \times 10^{24}\,m^{-3}$, dipole transition matrix element, $\wp = 1 \times 10^{-29}\,C \cdot m$ and population relaxation and dephasing times, $T_1 = 100$ ps and $T_2 = 50$ fs, respectively. We then applied a switch-on pulse: a continuous, sinusoidal electric-field pulse, with pulse carrier frequency ω_0 in resonance with the two-level system, which is smoothly increased in amplitude from 0 up to unit amplitude ($E_0 = 1\,V \cdot m^{-1}$) within five periods:

$$E_x(t) = E_0 \sin(\omega_0 t) \begin{cases} 0, & \text{for } t < 0 \\ (1 - x^2)^4, & \text{for } 0 \le t \le 5T_p \\ 1, & \text{for } t > 5T_p, \end{cases} \tag{49.28}$$

with $x = (t - 5T_p)/5T_p$, and T_p being the period of one cycle of the sinusoid. The electric-field amplitude is clearly amplified upon the passage of the sinusoidal pulse across the gain region, thus demonstrating linear small-signal gain (Figure 49.7a). In the time domain, the application of an initial sinusoidal pulse with unit amplitude is amplified at the output end of the structure (Figure 49.7b). Let us compute the complex propagation factor between two points in the gain medium, z_1 and z_2, separated by one dielectric wavelength, $l = z_2 - z_1 = \lambda_0/n$:

$$e^{ik_c(z_2 - z_1)} = \frac{E_x(z_2, \omega)}{E_x(z_1, \omega)}, \tag{49.29}$$

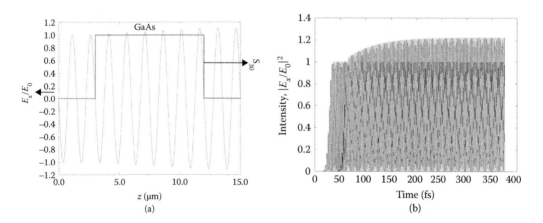

FIGURE 49.7 (a) Slight amplification of a sinusoidal electric-field switch-on pulse upon its passage through an initially inverted two-level GaAs slab medium of refractive index 3.59 after 1×10^6 time steps (each of duration 1×10^{-4} fs) with Courant stability parameter 0.75. The initial population profile across the structure is shown with a thick solid line (top-hat shape). The spatial discretization in this simulation was 1 Å. (b) Intensity evolution in time of the pulse at the input boundary of the simulated structure (fast-oscillating curve on the left) and the amplified time trace of the pulse at the output boundary (time-delayed curve on the right). The fast oscillations are at the resonant pulse carrier frequency, ω_0.

where $E_x(z_i, \omega)$ $i = 1, 2$, are the Fourier transforms of the time traces of the electric field sampled at the two points, and $k_c = \beta + i\gamma$ is the complex wavevector with β-phase shift and γ-gain/absorption coefficient, depending on its sign $(+/-)$. Thus, substituting these back into the left-hand side of Equation 49.29, one can define the amplification and phase factors as $e^{-\gamma \cdot l}$ and $e^{\beta \cdot l}$, respectively. For the amplitude gain coefficient, the steady-state solutions of the density-matrix equations for a two-level system (see [50]) read:

$$\gamma^{ampl}(\omega) = -\frac{\omega_o}{c_o}\frac{N_a \wp^2}{2\varepsilon_o \hbar n_{\mathrm{GaAs}}}g(\omega), \qquad (49.30)$$

where n_{GaAs} is the GaAs refractive index, c_0 is the speed of light, ε_0 is the dielectric permeability in vacuum, and the normalized Lorentzian lineshape function is given by: $g(\omega) = \frac{2T_2}{1+(\omega-\omega_o)^2 T_2^2}$, with normalization condition: $\int\limits_{-\infty}^{\infty} g(\omega)d\omega = 1$. The "atomic" phase shift of the optical wave induced by the two-level system is given by

$$\Delta\varphi_{\mathrm{at}} = \Delta k.l = \frac{\omega}{c_o}\frac{N_a \wp^2 \left(\omega-\omega_o\right)T_2}{4\varepsilon_o \hbar n_{\mathrm{GaAs}}}g(\omega) = \beta l - n_{\mathrm{GaAs}}\frac{\omega}{c_o}l. \qquad (49.31)$$

We can now compare the FDTD simulation results obtained with our Maxwell-pseudospin model (Equation 49.26) with the above analytic expressions (Equations 49.30 and 49.31); the results from the two methods are plotted in Figure 49.8. Our model is in remarkably good agreement with the density matrix-calculated gain/absorption coefficients (a) and phase shift (b) in a HB two-level system.

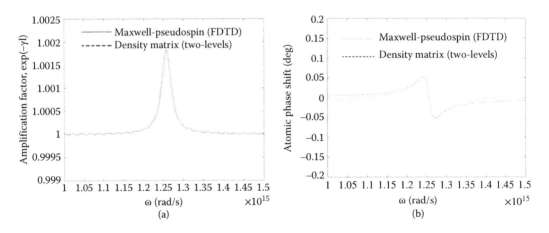

FIGURE 49.8 (a) Amplification spectrum of a switch-on pulse with carrier wave of angular frequency, ω, passing through a GaAs slab waveguide. The spectrum was calculated using two independent methods: by applying the analytical expressions of the density-matrix method to a HB two-level system (Equation 49.30) (dashed curve), and by finding the FDTD solution of the Maxwell-pseudospin set of equations (Equation 49.26) (solid curve). The ω corresponding to the amplification peak is resonant with the optical transition of the two-level quantum system making up the waveguide and is denoted ω_0. Note that the ripple effect observed in the solid curve is an artefact of the FDTD numerical method. (b) ω-dependence of the phase shift (in degrees) of the switch-on pulse emerging from the waveguide. The phase-shift spectrum was inferred independently from the density-matrix method (Equation 49.31) (dashed curve) and the Maxwell-pseudospin method (Equation 49.26) (solid curve). FDTD, finite-difference time-domain.

49.3.4 Resonant Nonlinearities and Gain Saturation

In Section 49.3.3, we employed the analytical steady-state density-matrix equations to perform a successful validation of our Maxwell-pseudospin model against the linear small-signal gain in a HB two-level system. In what follows, we shall show that our model also describes correctly the nonlinear gain dynamics of a HB two-level saturable gain medium. We keep the model structure, source excitation (i.e., input pulse) shape, and material parameters the same as in Section 49.3.3, so that the saturation condition at resonance is satisfied:

$$\frac{\wp^2 E_0^2 T_1 T_2}{\hbar^2} > 1. \tag{49.32}$$

We apply the same source excitation (Equation 49.28) but vary the amplitude, E_0, running separate simulations for different amplitudes, ranging from $E_0 = 5 \times 10^6$ V·m^{-1} to $E_0 = 1 \times 10^8$ V·m^{-1}. The nonlinear gain dynamics is shown in Figure 49.9 [51]. The gain peaks and levels off after some time, due to the depletion of the population residing on the upper level. The higher the initial excitation, the faster the gain reaches the steady-state field amplitude value. We proceed to demonstrate quantitative agreement between density-matrix theory of saturation of a two-level system and our Maxwell-pseudospin model. We choose the initial electric-field amplitude $E_0 = 5 \times 10^7$ V·m^{-1}, $T_1 = 10$ ps, and $T_2 = 10$ fs, again satisfying the resonant saturation condition (Equation 49.32). The saturation value of the population inversion was calculated in [50] to be

$$S_{3\text{sat}} = \frac{1}{1 + \left(\frac{E_0}{E_{s0}}\right)^2}, \quad E_{s0} = \frac{\hbar}{\wp\sqrt{T_1 T_2}}, \tag{49.33}$$

giving a value $S_{3\text{sat}} = 0.3$ for the chosen parameter set. We can calculate the unsaturated intensity gain coefficient from the maximum intensity using $\gamma_{\text{int}}^{\text{unsat}} = \frac{\ln(I\text{max}/I_o)}{2L_g}$. By definition, the saturated intensity is the intensity of the signal passing through the laser medium for which the gain coefficient is reduced down to one-half of its original value [52]:

$$\gamma_{\text{int}}^{\text{sat}} = \frac{\gamma_{\text{int}}^{\text{unsat}}}{1 + I/I_{\text{sat}}}. \tag{49.34}$$

Given that the normalized saturated field intensity is $I_{\text{sat}}/I_0 = exp(\frac{\gamma^{\text{unsat}}}{2}L_g)$ [51], I_{sat}/I_0 is calculated to be 1.01. In order to make quantitative comparison with our model, we run the code with a large number of time steps (4.8×10^7, of 1×10^{-4} fs duration each) to ensure that the field and population dynamics have reached stationary values. In Figure 49.10a, we plot the final instantaneous electric-field distribution and population profile at subsequent time steps, showing the population relaxing to a steady-state value. In the time domain, the time trace at the waveguide output, (Figure 49.10b), and the population inside the gain region (Figure 49.10c), are plotted. It is easy to check that the steady-state values inferred from our model are in excellent agreement with the theoretically computed values. With this we conclude our corroboration of the ability of our model to reproduce correctly the gain saturation dynamics of a HB two-level quantum system.

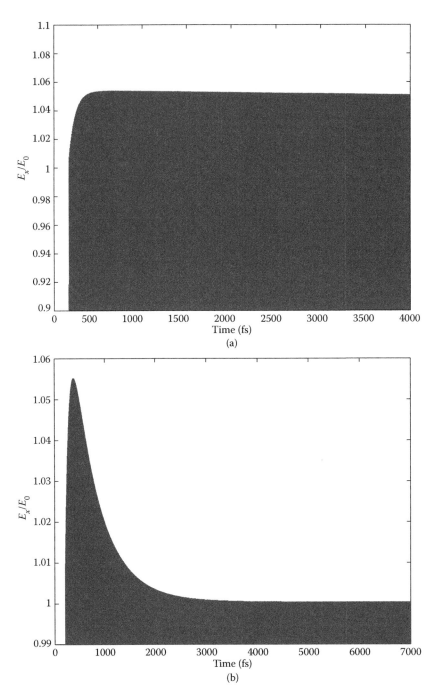

FIGURE 49.9 Gain saturation dynamics of a HB two-level quantum system excited by a source electric field with different amplitude: (a) $E_0 = 5 \times 10^6$ V · m^{-1}; (b) $E_0 = 5 \times 10^7$ V · m^{-1}. The time trace of the amplified electric field is sampled at the output of the simulation domain. In both cases, the amplitude of the oscillations (with frequency ω_0) stabilizes after some time. Note that after peaking, the gain in (a) continues to drop approximately linearly at a relatively slow rate over the entire time span shown on the graph, while the gain in (b) reaches a steady-state value after ~4 ps. (G. M. Slavcheva et al., *IEEE Journal of Selected Topics in Quantum Electronics* © 2004 IEEE.)

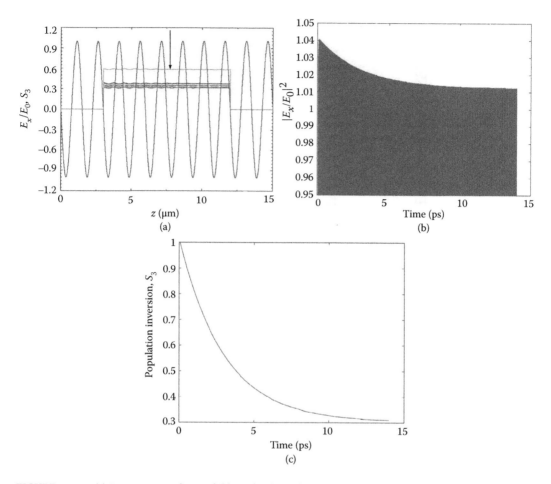

FIGURE 49.10 (a) Instantaneous electric-field amplitude profile across the waveguide (sinusoidal curve), obtained after 4.8×10^7 time steps (corresponding to a total time of ~4.8 ps), and the population inversion profile within the active region at several subsequent time steps (top-hat function with decreasing amplitude). The step in z is $\sim1.8 \times 10^{-10}$ m and the time step is $\sim1 \times 10^{-4}$ fs. The population is initially completely inverted: $S_{30} = 1$, but reduces over time (downward-pointing arrow), eventually converging to the theoretically obtained population steady-state value of 0.3. (b) Time evolution of the normalized field intensity showing convergence toward the theoretical value of 1.01. (c) Population relaxation over time, tending toward the theoretically predicted value of 0.3. (G. M. Slavcheva et al., *IEEE Journal of Selected Topics in Quantum Electronics* © 2004 IEEE.)

49.3.5 Coherent Propagation Effects, Self-Localization and Pattern Formation in 2D Planar Nonlinear Optical Waveguides and Semiconductor Microcavities

49.3.5.1 2D Nonlinear Optical Waveguides

We employ our 2D Maxwell-pseudospin model (see Section 49.2.4), based on self-consistent solution of the time evolution equations of a degenerate three-level quantum system, (Equation 49.4), for $j, k, l = 1, 2, ..., 8$ and Maxwell's curl equations for a TE-guided mode, Equation 49.21 (Equation 49.22 for a TM equivalent), propagating in a planar parallel-plate mirror optical waveguide, or planar semiconductor microcavity [45], of the type shown in Figure 49.11.

It is easy to show that by setting $S_1 = S_1, S_4 = -S_2, S_7 = S_3$, the TE case is equivalent to a two-level system case (Equation 49.7) and the equations for the remaining coherence vector components are decoupled

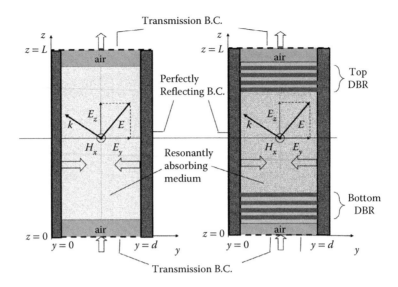

FIGURE 49.11 Planar parallel-mirror waveguide (left panel) and a semiconductor microcavity (right panel) filled with a resonantly absorbing three-level medium. Only TM-guided modes are shown. Perfectly transmitting boundary conditions are imposed on the upper/lower boundaries (upward-pointing arrows) and perfectly reflecting side boundaries are assumed (left/right arrows). The pulse initially propagates through a free-space region (air), then penetrates further, interacting with the degenerate three-level medium, and finally exits through another free-space region. B.C., boundary conditions; DBR, distributed Bragg reflector.

from this system, since there is no coupling to "level 3" in this case. In the FDTD numerical implementation, we use as a source field the TEM (i.e., TM$_0$) guided mode of a parallel-mirror slab waveguide. The electric field is a plane wave linearly polarized in the transverse x-direction perpendicular to the waveguide plane

$$E_y\left(z=0,y,t\right) = \begin{cases} E_0\text{sech}\left(10\Gamma\right)\sin\left(\omega_0 t\right) \\ E_0\exp\left[\left(\frac{t-t_0}{t_d}\right)^2\right]\sin\left(\omega_0 t\right) \end{cases} \quad E_z\left(z=0,y,t\right)=0, \qquad (49.35)$$

where $\Gamma = \frac{t-\frac{T_p}{2}}{\frac{T_p}{2}}$. Here we have allowed for two electric-field envelopes: hyperbolic secant and Gaussian. We should therefore expect that the PAT [18] still holds. The pulse area at a given time is given by

$$\theta\left(z,t\right) = \frac{\wp}{\hbar}\int\limits_{-\infty}^{t} A\left(z,t'\right)dt', \qquad (49.36)$$

where the electric-field envelope is as follows:

$$A\left(z,t\right) = E_0\text{sech}\left(\frac{t-\frac{z}{v}}{\tau}\right) = E_0\text{sech}\left(10\Gamma\right). \qquad (49.37)$$

The area under the entire pulse is given by

$$\theta_{\text{pulse}}\left(z\right) = \frac{\wp}{\hbar} \int_{-\infty}^{\infty} A\left(z, t'\right) dt'. \tag{49.38}$$

According to PAT, pulses with areas less than π should be completely absorbed within several absorption lengths; however, pulses with an area of even multiples of π should continue to propagate undistorted as a solitary wave—known as SIT soliton—through the resonantly absorbing medium. We set all resonant dipoles initially to the ground state, i.e., $S_{7e} = -1, S_{8e} = -\frac{1}{\sqrt{3}}$. We then test the predictions of PAT by calculating the pulse amplitude from the PAT, which is given by the following relation for a pulse with a hyperbolic secant-shaped envelope:

$$E_0 = \frac{2\pi\hbar f_0}{\wp \arctan\left(\sinh\left(u\right)\right)\Big|_{-10}^{+10}} \frac{\theta_{\text{pulse}}\left(z = 0\right)}{2\pi}, \tag{49.39}$$

where $f_0 = 1/\tau_p$ and $\tau_p = T_p/20$, with $T_p = 100$ fs being the pulse duration chosen for this simulation. If we set the initial pulse area $\theta(z = 0)$ to 2π, we get a pulse amplitude corresponding to a soliton pulse of $E_0 = 4.2186 \times 10^9$ V · m^{-1}. The simulated waveguide structure is 150 μm long and 50 μm wide. 3D plots of the soliton field modulus (a) and population relaxation term S_7 (b) distributions in the waveguide y-z plane and a slice through the middle of the waveguide's width are displayed in Figure 49.12.

The stable SIT soliton propagation is shown in Figure 49.13a and b: the initial 2π pulse propagates in the resonantly absorbing medium while maintaining its hyperbolic secant shape without any loss. The atoms are driven into the excited state by absorbing ultrashort pulse energy, and by reradiating this energy to the field they return to the ground state. The corresponding population inversion exhibits complete transition from the ground level to the excited state and back to the initial state within one Rabi period. By contrast, for a 0.9π pulse with pulse amplitude $E_0 = 1.18957 \times 10^9$ V · m^{-1}, the initially symmetric pulse is distorted and reshaped during its propagation (Figure 49.13c). The initial equilibrium between the energy absorbed by the leading pulse edge and the energy reemitted by the trailing edge is violated and the pulse becomes increasingly asymmetric. In addition, the system is only partially inverted (Figure 49.13d). The reemitted energy of the continuously growing trailing edge is transmitted through the lower ($z = 0$) boundary. After a sufficiently long time, this pulse should eventually be completely absorbed, in agreement with the PAT.

Let us consider now the more interesting case of a TM$_1$-guided mode of a parallel-mirror slab waveguide, which cannot be reduced to the two-level system case. We solve self-consistently Maxwell's equations for a TM wave, Equation 49.22, with the pseudospin equations, (Equation 49.4), for a three-level system. As a source field we use the TM$_1$ slab waveguide mode

$$E_y\left(z = 0, y, t\right) = E_0 c_1 \cos\left(\frac{\pi y}{w}\right) \text{sech}\left(10\Gamma\right) \sin\left(\omega_o t\right) = \tilde{E}_y\left(y, t\right) \sin\left(\omega_o t\right)$$

$$E_z\left(z = 0, y, t\right) = -E_0 c_2 \sin\left(\frac{\pi y}{w}\right) \text{sech}\left(10\Gamma\right) \cos\left(\omega_o t\right) = \tilde{E}_z\left(y, t\right) \sin\left(\omega_o t\right) \tag{49.40}$$

$$c_1 = \frac{\sqrt{\frac{\omega_o^2 n^2}{c_0^2} - \left(\frac{\pi}{w}\right)^2}}{\omega_o \varepsilon_o n^2}; \quad c_2 = \frac{\pi/w}{\omega_o \varepsilon_o n^2},$$

where w is the waveguide width. In order to obtain a stable soliton solution, we generalize the PAT to multiple spatial dimensions, defining the pulse area as the area below the modulus of the field envelope,

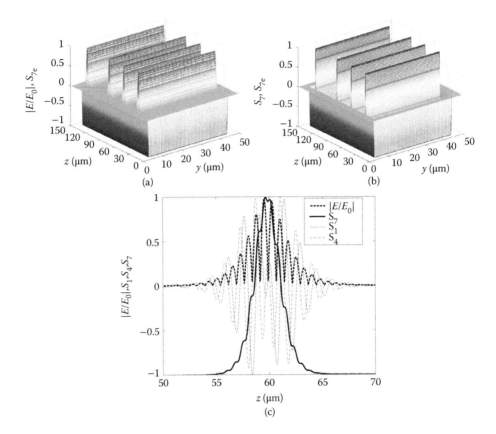

FIGURE 49.12 (a) Superimposed snapshots of the modulus of the electric field for a TEM guided mode propagating in a planar parallel-mirror waveguide at times $t = 150, 250, 350,$ and 500 fs after the arrival of the pulse at the left boundary of the waveguide. The initial population profile, S_{7e}, is also shown as an indication of the 2D boundaries of the active medium; the resonant dipoles are initially in the ground state with $S_{7e} = -1$, absorbing the incoming field radiation. (b) Corresponding population inversion term S_7 distribution across the waveguide plane, showing Rabi flops. (c) Slice straight across the middle of the waveguide along the propagation direction, showing the modulus of the electric field, dispersive, and absorptive polarization components, S_1 and S_4, respectively, and the population, S_7, performing a full Rabi flop along the waveguide axis. $\Delta z = 30$ nm, $\Delta y = 500$ nm, $\Delta t = 9.989 \times 10^{-2}$ fs, the pulse wavelength is $\lambda = 1.5\,\mu$m and the duration $T_p = 100$ fs. (Adapted with permission from G. Slavcheva et al., *Physical Review A*, 66, 063418, 2002. Copyright 2002 by the American Physical Society.)

which depends on the point in the waveguide plane we choose to consider:

$$\theta\left(y, z\right) = \frac{\wp}{\hbar} \int\limits_{-\infty}^{\infty} \left|\tilde{E}\left(y, z, t'\right)\right| dt' = \frac{\wp}{\hbar} \int\limits_{-\infty}^{\infty} \sqrt{\tilde{E}_y^2 + \tilde{E}_z^2}\, dt'. \tag{49.41}$$

Assuming that the initial pulse is injected into the active medium at a point $y = y_0$ at the boundary ($z = 0$), stable pulses in the absorbing medium are obtained when

$$\theta\left(y = y_0, z = 0\right) = \frac{\wp}{\hbar} \int\limits_{-\infty}^{\infty} \sqrt{\tilde{E}_y^2\left(y_0, 0, t'\right) + \tilde{E}_z^2\left(y_0, 0, t'\right)}\, dt' = 2n\pi, \tag{49.42}$$

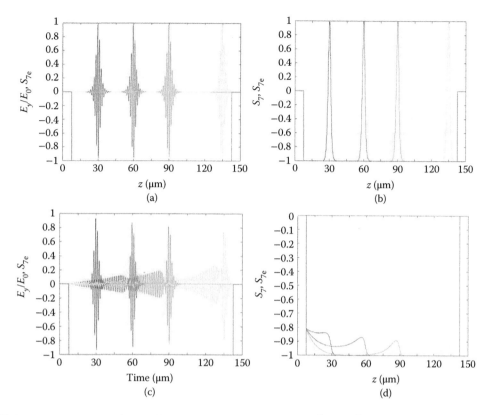

FIGURE 49.13 (a) Time evolution of a 2π initial pulse ($E_0 = 4.2186 \times 10^9$ V · m^{-1}) at $t = 150, 250, 350$, and 500 fs after impinging on the left boundary of a planar parallel-mirror waveguide. The soliton propagates undistorted through the resonant medium, the boundaries of which are shown by the initial population profile, $S_{7e} = -1$, i.e., all dipoles are initially in the ground state. (b) The corresponding population inversion term, S_7, exhibits full Rabi flops, whereby the absorbed energy is fully balanced by the emitted energy over one Rabi period. (c) Time evolution of a pulse with initial pulse area 0.9π ($< \pi$) ($E_0 = 1.18957 \times 10^9$ V · m^{-1}) at the same simulation times. In this case, the pulse becomes distorted due to imbalance of the absorbed and reemitted energies, and will eventually decay to zero in agreement with the PAT. (d) The resonant system is only partially inverted: there is no population swap between the ground and excited levels. (Adapted with permission from G. Slavcheva et al., *Physical Review A*, 66, 063418, 2002. Copyright 2002 by the American Physical Society.)

with n being an integer. A 2π pulse has an initial pulse amplitude $E_0 = 1.123 \times 10^7$ V · m^{-1}. The time evolution of the TM_1 soliton is shown in Figure 49.14.

49.3.5.2 2D Semiconductor Microcavities

Using the 2D model, we numerically investigate conditions of onset of SIT cavity soliton formation in a semiconductor cavity driven by a coherent ultrashort pulse launched from the lower boundary (see Figure 49.11, right panel) [53]. We have designed a number of 2D microcavity structures in order to test the influence of the cavity length on the localization phenomena and pattern formation. A typical 5λ-cavity (where the optical path length between the mirrors of the cavity is equal to an exact multiple of five wavelengths) designed at $\lambda = 1.3$ μm, filled with resonantly absorbing medium sandwiched between a bottom DBR (31.5 pairs of GaAs/AlAs) and a top DBR (24 pairs of $Al_{0.1}Ga_{0.9}As/Al_{0.8}Ga_{0.2}As$) is shown in Figure 49.15a and b. We apply a driving pulse excitation with duration 100 fs at the left boundary

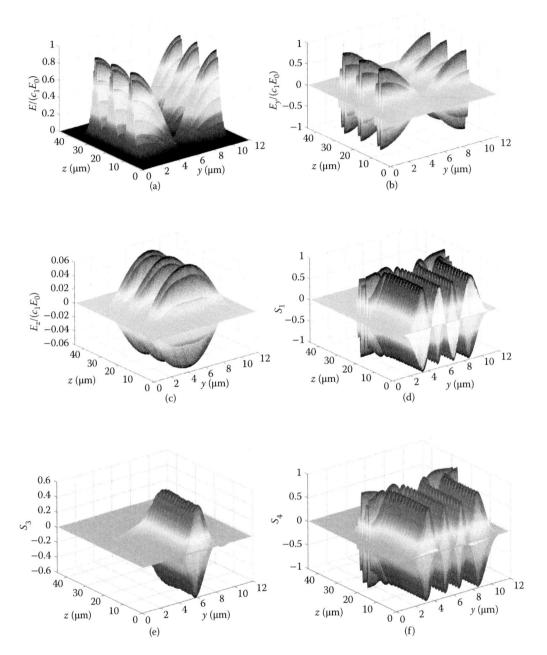

FIGURE 49.14 Time evolution of the (a) modulus, (b) E_y, and (c) E_z electric-field vector component of the TM_1 soliton at times $t = 90, 125$, and 155 fs after its arrival at the left boundary of a planar parallel-mirror waveguide. The 2D pulse travels undistorted in the resonant medium with refractive index, $n = 1$. The waveguide is 50 μm long and 9.9185 μm wide. Discretization grid: $\Delta z = 1.5$ nm; $\Delta y = 119.5$ nm; $\Delta t = 5 \times 10^{-3}$ fs. (d–g) Polarization components of significance (i.e., those corresponding to transitions from one level to another) in a degenerate three-level system. S_1 and S_4: in-phase dispersive components; S_3 and S_6: in-quadrature absorptive components.   (*Continued*)

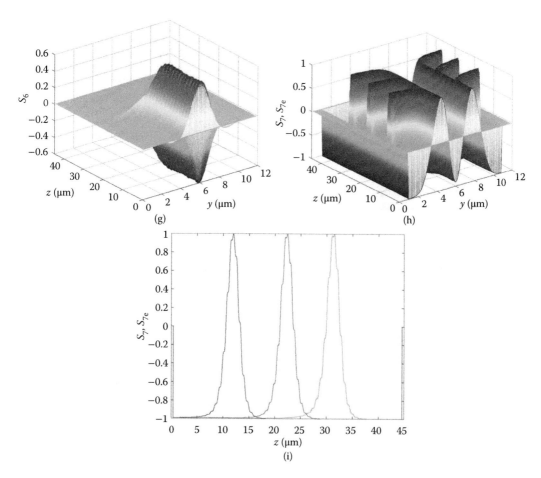

(g)

(h)

(i)

FIGURE 49.14 (*Continued*) (d–g) Polarization components of significance (i.e., those corresponding to transitions from one level to another) in a degenerate three-level system. S_1 and S_4: in-phase dispersive components; S_3 and S_6: in-quadrature absorptive components. (h) Population component S_7 showing population Rabi flops along the waveguide and in a transverse direction (two complete Rabi flops are visible). (i) lengthways cross section of (h) showing complete Rabi flops in the propagation direction, z.

($z = 0$) of the microcavity (Figure 49.15a), which is 12 μm long and $w = 9.92$ μm wide. The dipole number density is $N_a = 10^{24}$ m^{-3} and the dipole matrix element, $\wp = 1 \times 10^{-29}$ C · m. The relaxation times are assumed uniform and equal to 100 ps, satisfying the SIT criterion $T_p \ll T_i$, $i = 1, 2, \ldots, 8$. The initial pulse amplitude $E_0 = 4.8643 \times 10^9$ V · m^{-1} is calculated for a pulse area of 2π, assuming that the PAT still holds. As pointed out in [54–56], this is not a necessary condition for SIT in multilayered media. Even a relatively weak pulse would lead to SIT intracavity patterns after sufficiently long simulation times, due to field enhancement by constructive interference from multiple reflections in the DBR mirrors. The population inversion gradually builds up with time and begins to perform Rabi flopping. In Figure 49.15c through f, the electric-field modulus and the corresponding population inversion profile are plotted for the cavity region. The clearly visible instantaneous roll patterns for the electric field and population inversion, S_7, feature a quasi-stationary standing-wave (i.e., quiescent) SIT soliton. The cavity induces a quasi-stationary population "grating," and the number of the quasi-complete Rabi flops (10) of the population equals the number of half-wavelengths along the cavity length, $L_c = (10\lambda) / (2n)$. At the

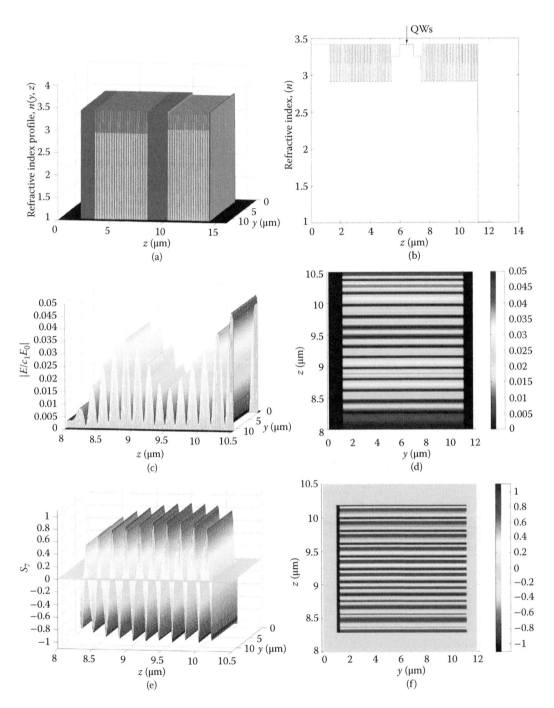

FIGURE 49.15 (a) 2D refractive-index profile across the semiconductor microcavity. (b) Longitudinal slice of (a) at a distance $y_0 = 6\,\mu m$ from the left interface, $y = 0$. (c) A snapshot at time $t = 3.6055 \times 10^{-13}$ s of the electric-field modulus distribution across the cavity. (d) Top view of (c) showing the roll patterns formed. (e) Snapshot of the corresponding population inversion profile across the cavity, showing 10 complete Rabi flops. (f) Top view showing the intracavity population "grating" (roll pattern). (G. Slavcheva et al., *IEEE Journal of Selected Topics in Quantum Electronics* © 2003 IEEE.)   (*Continued*)

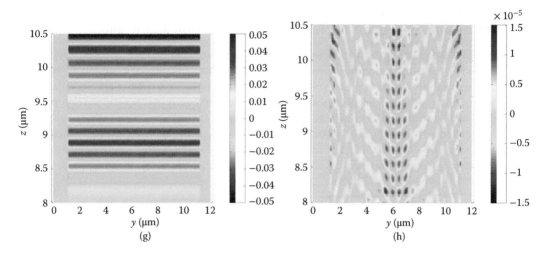

FIGURE 49.15 (*Continued*) (g) Top view of the E_y-component roll pattern. (h) Top view of the 2D pattern of the E_z electric-field component. (G. Slavcheva et al., *IEEE Journal of Selected Topics in Quantum Electronics* © 2003 IEEE.)

same time, a roll pattern for the E_y electric-field component, Figure 49.15g, and a more elaborate 2D one, Figure 49.15h, for the E_z field component are observed. Performing simulations with different driving pulse durations on a number of different microcavity designs, we demonstrate numerically the existence of a new type of 2D spatiotemporally localized standing-wave soliton ("light bullet"). The SIT cavity soliton results from the interplay of the resonant nonlinearity and the Bragg reflection interference and dispersion in the microcavity.

49.4 Optical Orientation in Quantum Wells and QDs

Circularly polarized optical excitations of semiconductor nanostructures excite particular interest due to their ability to transfer angular momentum and thus to drive the quantum system to a specific angular-momentum state. The conversion between the flying qubit encoded in photon polarization and the material qubit encoded, for example, in the carrier spin, is known as optical orientation, and this technique has been successfully applied to spin-polarized states in semiconductors [14]. This, in turn, opens up the possibility for faithful transmission of flying qubits between distant locations. If sufficiently short (and therefore high-intensity) optical pulses are used, so that they can interact with the system before it can be affected by its environment, the spin does not have time to decohere (i.e., lose its phase information) and could be coherently manipulated by such light pulses. In this high-intensity, coherent regime, optical Rabi oscillations take place; one-half of each oscillation consists of the population being driven from the ground to the excited state (or vice versa), hence performing a Rabi flop. A Rabi flop corresponds to a one-qubit rotation, whereby the spin is rotated through π; this entails a transition either from spin-down (state 0) to spin-up (state 1) or vice versa (Figure 49.1). The achievement of control and manipulation of the single charge-carrier spin dynamics and the collective many-body dynamics of multiple entangled spins is the key to the physical implementation of quantum computation.

49.4.1 Circularly Polarized Optical-Pulse Interactions with Two-Level Quantum Systems: Selective Spin Initialization and Readout

In bulk semiconductors and in semiconductor structures with reduced dimensionality, dipole optical transitions (including excitonic transitions) between energy levels are only allowable subject to satisfying the angular-momentum optical selection rules, $\Delta J_z = 0, \pm 1$, for linear and circular photon polarization respectively, where J_z is the total angular-momentum projection on the electromagnetic wave propagation direction coinciding with the QD growth axis. We shall consider optical transitions with $\Delta J_z = \pm 1$ that are excited by circularly (or elliptically) polarized light, and will therefore apply the pseudospin equations derived for circularly polarized light in Equation 49.8, coupled to Maxwell's vector equations for an electromagnetic wave propagating along the z-direction, circularly (elliptically) polarized in a plane perpendicular to z [57]:

$$
\begin{aligned}
\frac{\partial H_x(z,t)}{\partial t} &= \frac{1}{\mu} \frac{\partial E_y(z,t)}{\partial z} \\[4pt]
\frac{\partial H_y(z,t)}{\partial t} &= -\frac{1}{\mu} \frac{\partial E_x(z,t)}{\partial z} \\[4pt]
\frac{\partial E_x(z,t)}{\partial t} &= -\frac{1}{\varepsilon} \frac{\partial H_y(z,t)}{\partial z} - \frac{1}{\varepsilon} \frac{\partial P_x(z,t)}{\partial t} \\[4pt]
\frac{\partial E_y(z,t)}{\partial t} &= \frac{1}{\varepsilon} \frac{\partial H_x(z,t)}{\partial z} - \frac{1}{\varepsilon} \frac{\partial P_y(z,t)}{\partial t}.
\end{aligned}
\tag{49.43}
$$

We phenomenologically construct the Hamiltonian corresponding to initially absorbing energy-level schemes (i.e., $S_{30} = -1$), shown in Figure 49.16:

$$
\hat{H} = \hat{H}_0 + \hat{H}_{\text{int}} = \hbar \begin{pmatrix} 0 & -\frac{1}{2}\left(\Omega_x - i\Omega_y\right) \\[6pt] -\frac{1}{2}\left(\Omega_x + i\Omega_y\right) & \omega_0 \end{pmatrix}.
\tag{49.44}
$$

The Hamiltonian for an initially amplifying two-level system (i.e., $S_{30} = 1$) is the transposed matrix. Here we have introduced a complex Rabi frequency, whose real and imaginary parts correspond to the $E_{x(y)}$ electric-field vector components, $\Omega_x = \frac{\wp}{\hbar} E_x$ and $\Omega_y = \frac{\wp}{\hbar} E_y$.

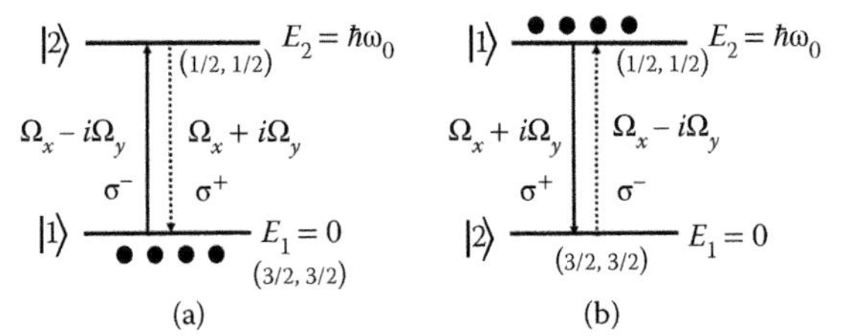

FIGURE 49.16 Energy-level diagram of (a) initially absorbing (i.e., entire charge-carrier population in the ground state) and (b) amplifying (i.e., entire charge-carrier population in the excited state) two-level system, with energy-level separation $\hbar\omega_0$ and allowed dipole optical transitions subject to the condition $\Delta J_z = \pm 1$. States are additionally labeled (J, J_z), where J is the total angular momentum and J_z is its projection along z. (a) A left circularly polarized optical field, σ^-, drives the population into the upper excited level, $|2\rangle$; the population is returned to the ground level by an optical pulse with opposite helicity, σ^+. (b) Having been initially prepared in the excited state, $|1\rangle$, the population in the system is driven by a right circularly polarized pulse, σ^+, to the ground state $|2\rangle$ and returned to the initial excited state by a left circularly polarized pulse, σ^-.

Employing the general methodology in Section 49.2.3, we can easily show that

$$P_x = -\wp N_a S_1,$$
$$P_y = -\wp N_a S_2 .$$

(49.45)

We solve numerically by the FDTD method the system of equations, Equations 49.43 and 49.8, coupled through Equation 49.45. The source optical fields for left and right circularly polarized light are each given by two orthogonal, linearly polarized pulses, one time-delayed relative to the other by $\pi/2$, with carrier frequency ω_0 tuned in resonance with the two-level system:

$$\sigma^- \begin{cases} E_x(z=0, t) = \tilde{E}_x(t)\cos(\omega_0 t) \\ E_y(z=0, t) = -\tilde{E}_y(t)\sin(\omega_0 t) \end{cases} ; \quad \sigma^+ \begin{cases} E_x(z=0, t) = \tilde{E}_x(t)\cos(\omega_0 t) \\ E_y(z=0, t) = \tilde{E}_y(t)\sin(\omega_0 t), \end{cases}$$

(49.46)

where the pulse shape is determined by the envelopes:

$$\tilde{E}_{x,y}(t) = \begin{cases} E_0 \operatorname{sech}(10\Gamma) \\ E_0 \exp\left[-(t-t_0)^2/t_d^2\right], \end{cases}$$

(49.47)

which we choose to be either hyperbolic secant or Gaussian. We consider a 15 μm-long simulation structure filled with either a resonantly absorbing or amplifying ($S_{30} = -1$ or $S_{30} = +1$) two-level medium, assumed to be GaAs (refractive index, $n = 3.3827$). We use a pulse wavelength, $\lambda_0 = 1.5$ μm and a pulse duration, $T_p = 30$ fs; the dipole coupling $\wp = 1 \times 10^{-29}$ C · m, and two-level system ensemble density $N_a = 1 \times 10^{24}$ m^{-3} are taken from [47]. We restrict ourselves to the linear regime (initial electric-field amplitude, $E_0 = 1$ V · m^{-1}). The material relaxation times are $T_1 = 100$ ps and $T_2 = 70$ fs. We launch a left circularly polarized pulse, σ^-, from the left boundary of the simulation domain and sample the electric field at two locations, z_1 and z_2, along the active medium. The complex propagation factor (see Equation 49.29) over one dielectric wavelength $l = z_2 - z_1 = \lambda/n$ is calculated for the E_x and E_y components, and the results for the gain/absorption coefficient spectra are compared with the steady-state density-matrix theory spectra (see Equation 49.30). A snapshot of the electric-field components, E_x and E_y, at the input and output boundaries are shown in the first column of Figure 49.17. The corresponding Fourier spectra and gain/absorption coefficients are shown in the second and third columns, respectively, of Figure 49.17 for an initially absorbing (Figure 49.17a through f) and initially amplifying (gain) medium (Figure 49.17g through l). The optical wave interacts with the two-level system, driving the population into the upper state (or returning the population to the ground state) by absorbing (or emitting) pulse energy. Therefore, we clearly observe gain/absorption in the Fourier spectra. The gain/absorption coefficients coincide with those for linear polarization. In addition, we have shown that when pumping an absorbing system with a right circularly polarized pulse, the pulse fails to excite the system, and similarly, when pumping an initially inverted system (the whole population residing on the upper level) by a left circularly polarized pulse, the system remains unaffected. Therefore our model correctly describes the excitation using circularly polarized light of dipole optical transitions subject to the optical selection rules, $\Delta J = \pm 1$, in a two-level system. We have thus numerically demonstrated selective excitation of specific spin states by circularly polarized pulses with predefined helicity.

49.4.1.1 SIT and Polarized SIT Soliton Formation

The results in Section 49.4.1 were obtained in the linear (small-signal) regime. Let us now consider the nonlinear coherent regime of ultrashort pulse propagation through resonant media, whose atomic lifetimes greatly exceed the pulse duration (i.e., such that $T_1, T_2 \gg T_p$). The simulation domain is 150 μm long and a two-level resonantly absorbing medium with refractive index $n = 1$ is embedded between two symmetric free-space regions, each with length 7.5 μm. We inject a resonant, left circularly polarized pulse,

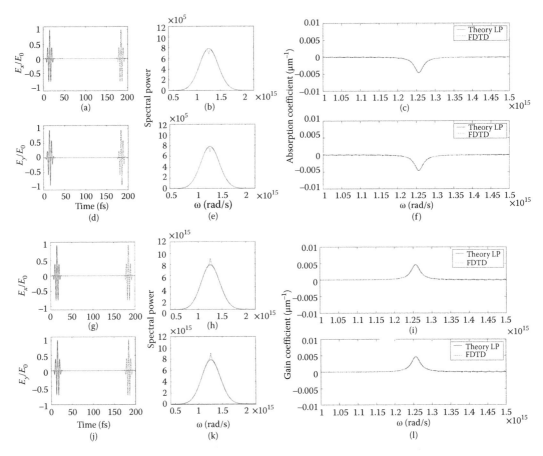

FIGURE 49.17 Time evolution at the input (solid line) and output (dashed line) structure boundaries of the electric-field components, E_x (a) and E_y (d), of a circularly polarized optical pulse with unit amplitude and pulse duration 30 fs for an initially absorbing two-level medium. (b), (e) Fourier spectra corresponding to the E_x and E_y field components, respectively, showing absorption of the pulse (dashed curve dip). (c), (f) Absorption coefficients for E_x and E_y, respectively; in this case, the dashed curve indicates the numerically calculated coefficients for linear polarization, while the solid curve signifies the analytically computed ones (Equation 49.30). (g–l) Respective plots for an initially amplifying (gain) medium; notations are the same as in (a–f). The gain is clearly visible in the Fourier spectra in (h) and (k) (see overshoot of dashed curve). (Adapted with permission from G. Slavcheva and O. Hess. *Physical Review A*, 72, 053804, 2005. Copyright 2005 by the American Physical Society.) (FDTD, finite-difference time-domain.)

σ^-, with hyperbolic secant envelope, since this is the well-known stable soliton solution of the Maxwell–Bloch system (see, e.g., [58]). We choose the electric-field component amplitudes so that the pulse area, according to the PAT, is 2π. This condition implies that $E_0 = 4.2186 \times 10^9$ V·m^{-1}. The time evolution of a 2π pulse is shown in Figure 49.18a. Both circularly polarized pulse components travel undistorted through the resonantly absorbing two-level medium as a solitary wave, and the population inversion is driven through complete Rabi flops by the leading and trailing pulse edges. An expanded view of a snapshot of a circularly polarized SIT soliton at the simulation time $t = 0.2$ ps is shown in Figure 49.18b. By virtue of the PAT, stable soliton solutions in gain (initially inverted, $S_{30} = 1$) medium are obtained for pulse areas which are odd multiples of π. In order to confirm this prediction, we launch a right circularly polarized pulse with amplitude $E_0 = 2.1093 \times 10^9$ V · m^{-1}, which corresponds to a pulse area of π. As expected, a π-pulse completely de-excites the population back to the ground state (Figure 49.18c); the pulse amplitude increases,

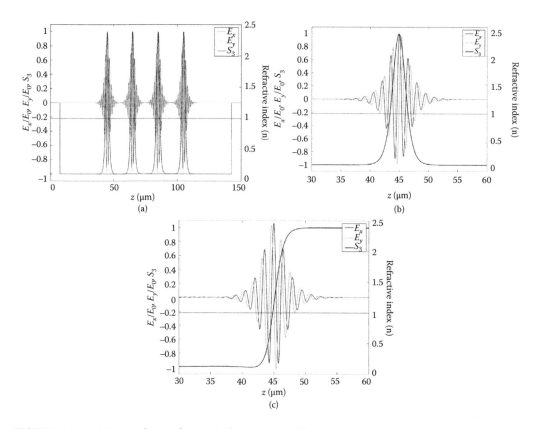

FIGURE 49.18 (a) Time evolution of a 2π, σ^- pulse propagating (from left to right) in a resonantly absorbing medium. The region where the initial charge carrier population profile is $S_{30} = -1$ demarcates the extent of the active medium on the diagram. The normalized electric-field components, E_x and E_y, and the instantaneous population profile, S_3, are plotted at the simulation times $t = 200, 267, 334$ and 400 fs from the moment the pulse is injected at the left boundary of the medium. The pulse travels without loss: it is a SIT soliton. (b) Expanded view of the σ^- SIT soliton in (a) at $t = 267$ fs showing the full Rabi flop of the population inversion. (c) A snapshot of a σ^+ SIT π pulse at $t = 267$ fs that completely inverts the population in the resonantly amplifying medium. (Adapted with permission from G. Slavcheva and O. Hess. *Physical Review A*, 72, 053804, 2005. Copyright 2005 by the American Physical Society.)

while the pulse duration decreases, thereby conserving the pulse area during the pulse propagation in the amplifying medium.

49.4.2 Circularly Polarized Optical Pulse Interactions with Degenerate Four-Level Systems

Let us consider now a more realistic discrete-level quantum system, such as excitonic transitions in bulk or low-dimensional semiconductor systems. For instance, we consider fundamental heavy-hole dipole optical transitions in QDs or quantum wells (QWs), whereby the band structure in the vicinity of the Γ-point ($\mathbf{k} = 0$) in the Brillouin zone (BZ) is isomorphic with a pair of two-level systems corresponding to σ^- and σ^+ heavy-hole excitonic transitions, as shown in Figure 49.19.

Consider a plane electromagnetic wave propagating along z direction, circularly polarized in a plane perpendicular to z, and tuned in resonance with the fundamental heavy-hole transition of the semiconductor system. The resonant nonlinearity is modeled by an ensemble of degenerate four-level systems of the type shown in Figure 49.19, with density N_a. The time-evolution of a four-level system under a dipole-coupling

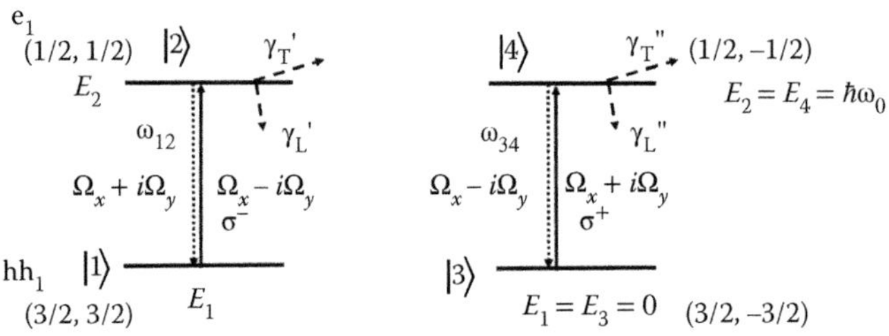

FIGURE 49.19 Energy-level diagram of σ^- and σ^+ heavy-hole transitions in a quantum well near the Γ-point ($\mathbf{k} = 0$) or between s-shell heavy-hole and electron levels in a QD. The energy separation between the upper and lower levels is $\hbar\omega_0$, and the allowed dipole transitions satisfy $\Delta J_z = \pm 1$. Ω_x and Ω_y are, respectively, the real and imaginary part of the Rabi frequencies associated with the coherent transitions. γ_L and γ_T are, respectively, the longitudinal relaxation and transverse dephasing rates of the excited states. e_1 is the bottommost electron energy level in the conduction band, and hh_1 is the topmost heavy-hole energy level in the valence band; together, these energy levels make up the fundamental bandgap.

external perturbation is described in terms of the SU(4) group generators by an equation of motion for a 15-dimensional real state pseudospin vector (see Equation 49.12 for $j, k, l = 1, 2, ..., 15$), where the dissipation in the system is accounted for by introducing nonuniform relaxation times, T_j. Without making any assumptions about the initial population redistribution between the levels, the system Hamiltonian is given by [59]

$$\hat{H} = \hbar \begin{pmatrix} 0 & -\frac{1}{2}\left(\Omega_x - i\Omega_y\right) & 0 & 0 \\ -\frac{1}{2}\left(\Omega_x + i\Omega_y\right) & \omega_0 & 0 & 0 \\ 0 & 0 & 0 & -\frac{1}{2}\left(\Omega_x + i\Omega_y\right) \\ 0 & 0 & -\frac{1}{2}\left(\Omega_x - i\Omega_y\right) & \omega_0. \end{pmatrix}. \tag{49.48}$$

We investigate the nonlinear coherent propagation regime, where the SIT condition $T_p \ll T_j$ is satisfied with $T_p = 100\,\text{fs}$ and $T_j = 100\,\text{ps}$. A circularly polarized pulse, amplitude-modulated by a hyperbolic secant function and whose carrier-wave frequency is resonant with the optical transition (at $\lambda = 1.5\,\mu\text{m}$), Equation 49.46 and 49.47, is applied at the left boundary $z = 0$ of a 150-μm-long simulation domain, with either resonantly absorbing or amplifying media embedded between two free-space regions of thickness 7.5 μm each. We choose the initial pulse to be a π-pulse, giving $E_0 = 1.093 \times 10^9\,\text{V} \cdot \text{m}^{-1}$ according to the PAT. We wish to achieve selective excitation of dipole optical transitions with $\Delta J_z = -1$ or $\Delta J_z = +1$, with a view to manipulate a particular spin state simply by selecting the helicity of the impinging optical pulse. The two leftmost columns of Figure 49.20 show all four possible cases where the initial state population of each two-level system is either entirely in the ground or entirely in the excited state. The two rightmost columns of the same figure also show the outcome of pulse excitation of each respective system. The conclusion that can be drawn is that we can indeed coherently control the spin population of specific spin states in the four-level system by choosing the proper polarization of the injected pulse. We have numerically demonstrated SIT and SIT soliton propagation in a four-level system initially prepared in a state with uniformly distributed populations in either the ground or excited states of the two two-level systems [60]. The ultrashort pulse travels undistorted in the resonant, degenerate four-level medium, Figure 49.21a, and the population is locally driven by the electric field through full Rabi flops simultaneously in both two-level systems, Figure 49.21b (compare with the two-level system soliton, Figure 49.18b).

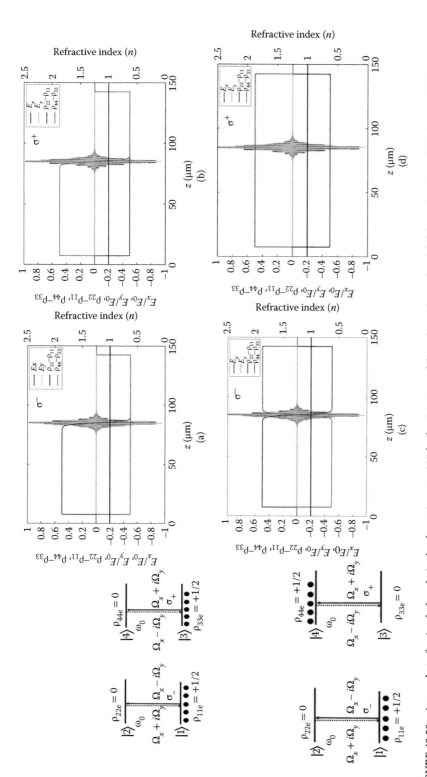

FIGURE 49.20 A snapshot of a circularly polarized pulse at time $t = 333$ fs after injection of the pulse at the medium's left boundary. E_x and E_y electric-field components and the population inversion $\rho_{22} - \rho_{11}$ in the first, and $\rho_{44} - \rho_{33}$ in the second two-level systems are plotted, tracing out the boundaries of the active medium with refractive index, $n = 1$. The leftmost two columns display the initial population distribution between the levels (equal split of the population between the two sets of two-level systems is assumed at the outset). (a) A σ^- π-pulse completely excites the first two-level system, but does not affect the second system. (b) A σ^+ π-pulse completely de-excites the initially inverted second two-level system, but does not affect the first one. (c) A σ^- π-pulse completely excites the first and de-excites the second system. (d) A σ^+ π-pulse does not affect either of the systems: the population in the first remains in the ground state and the population of the second remains in the excited state.

(Continued)

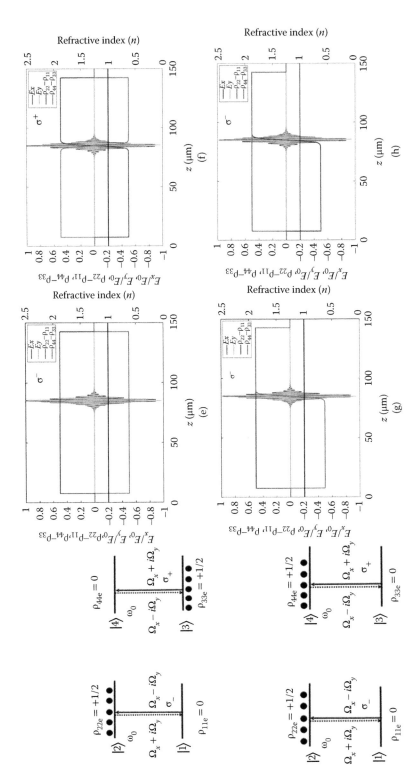

FIGURE 49.20 (*Continued*)   (e) A σ^- π-pulse does not change the population in either of the systems. (f) A σ^+ π-pulse inverts simultaneously the population in both systems. (g) A σ^- π-pulse does not affect the first system, but de-excites the second. (h) A σ^+ π-pulse inverts the population in the first system, but does not affect the second system.

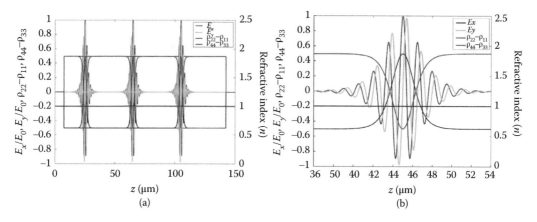

FIGURE 49.21 (a) Polarized SIT soliton propagation through a four-level medium resonant with the carrier wave of the injected pulse. The diagram shows a superimposition of the soliton at three different times: $t = 133, 266$ and 400 fs from the moment the pulse is injected at the left boundary of the medium. (b) A snapshot of the polarized soliton in the four-level medium, consisting of two pairs of two-level systems, at $t = 200$ fs. The medium is initially prepared in a state with the charge-carrier population residing in the lower-lying level, $|1\rangle$, of the first two-level system and in the upper level, $|4\rangle$, of the second two-level system (see Figure 49.19).

49.5 Modeling Coherent Spin Generation, Manipulation, and Readout in Charged QDs

In Section 49.4, we demonstrated coherent, selective spin excitation dependent on the helicity orientation of an impinging ultrashort optical pulse. In this section, we will show how spin states in low-dimensional semiconductor structures can be generated, manipulated, and read out using ultrashort circularly polarized pulses. Our goal will be to compute the time evolution of a discrete-level system and map the allowed dipole optical transitions that occur in a semiconductor nanostructure (e.g., a single QD or an ensemble of QDs) upon ultrashort, circularly polarized optical excitation. The pulse leaves behind a long-lived coherence, and we shall look for characteristic signatures in the dynamics that would allow us to determine the initial spin state and find useful regimes and schemes for high-fidelity optical manipulation of the spin confined in a QD. One promising approach is to address optically individual carrier spins in semiconductor QDs and manipulate them through optically excited states (charged excitons) by employing the techniques of coherent quantum control and optical orientation.

49.5.1 Single-Charged QD Embedded in a Waveguide

We will study n-doped QDs with an overall negative charge of a single electron. Injection of a single electron into a QD can be achieved either by modulation doping of the barrier region, adjusting the impurity doping within a delta-doped layer to transfer on average one electron per dot to the lowest energy states, or by electrical injection. A large class of self-assembled QDs posses quasi-cylindrical geometry about the quantization axis, z, and the electron and hole single-particle states in the conduction and valence band, respectively, can be approximated by harmonic oscillator potentials [61] (Figure 49.22). The resonant, circularly polarized optical excitation of a charged QD leads to the formation of a three-particle complex—a trion—consisting of two electrons occupying the same lowest conduction-band electron level in a spin singlet state, and a hole occupying the lowest valence-band hole level (Figure 49.22c and d). There is strong evidence from both experiment [62] and theory [63,64] that an intense resonant excitation of the trion transition suppresses the electron spin relaxation due to hyperfine interaction of the electron spins with

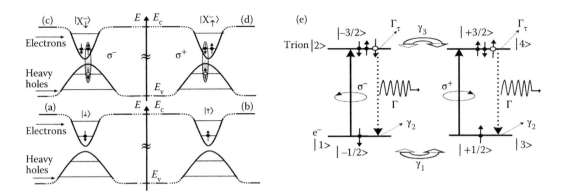

FIGURE 49.22 (a), (b) Schematic of the initial electron states (spin down and spin up) in a singly negatively charged QD and (c), (d) the ground singlet negatively charged exciton, X^-, which forms upon left circularly polarized, σ^-, and right circularly polarized, σ^+, optical excitation of the heavy-hole excitonic transitions. (e) Energy-level diagram of a negatively charged trion (a three-particle complex of two electrons and a hole), equivalent to (a–d); the levels are labeled with the J_z quantum number, and the allowed dipole transitions correspond to $\Delta J_z = \pm 1$ with an energy gap of $\hbar \omega_0$. [Solid upward arrows, coherent optical transitions excited by σ^- or by σ^+; dashed arrows, spontaneous emission rate, Γ; double-headed arrows, transitions due to electron- and hole-spin relaxation, with rates γ_1 and γ_3, respectively; γ_2 and Γ_τ, spin decoherence rates for electrons and holes, respectively.]

the lattice ions' nuclear spins. The latter limits the ability to measure accurately the electron spin orientation at low temperatures. It is obvious that the longer the spin lifetime, the better the chances for optical manipulation and readout of that spin state. We shall investigate both low- and high-intensity nonlinear excitation regimes.

In our treatment of a single quantum object, we shall assume that the ergodic hypothesis holds—i.e., the equivalence of time averages of an observable (in this case, the single-dot polarized, time-resolved photoluminescence TR [PL]) and the ensemble averages at one time over a large number of identical replicas of the dot system. Under such conditions, quantum mechanics allows the prediction of the single-system properties on the basis of macroscopic ensemble averages. Therefore, we shall assume that the time dependence of the optically induced coherent spin generation and subsequent relaxation in a single QD, averaged over a large number of successive measurements, is equivalent to the corresponding spin dynamics of an ensemble of identically prepared QDs. The ensemble of degenerate four-level systems describing the QD resonant nonlinearities is resonantly coupled to an optical wave propagating along the QD growth axis, z, and circularly polarized in a plane perpendicular to z. We use our general methodology to tackle this problem [44] and solve the 1D Maxwell's curl equations (Equation 49.43) coupled to the master pseudospin equations for a four-level system (Equation 49.12) through the induced macroscopic polarization given by

$$
\begin{aligned}
P_x &= -\wp N_a S_1 \\
P_y &= -\wp N_a S_7.
\end{aligned}
\tag{49.49}
$$

The system under investigation is a GaAs/AlGaAs self-assembled, molecular beam epitaxy (MBE)-grown QD with a height of 5 nm and refractive index, $n_{\text{GaAs}} = 3.63$, sandwiched between two $Al_{0.3}Ga_{0.7}As$ barrier regions, each with a width of 50 nm and refractive index, $n_{\text{AlGaAs}} = 3.46$, at the trion fundamental transition resonance wavelength of 757 nm. The circularly polarized pulse, whose carrier wave is resonant with the trion transition, is described by a hyperbolic secant envelope (see Equation 49.46 and first line of Equation 49.47) and has a duration of 1.3 ps. The simulations are run repeatedly with a different initial pulse amplitude, E_0, which is varied from a low value, $E_0 = 550\,\text{V} \cdot \text{m}^{-1}$, through $E_0 = 3 \times 10^6\,\text{V} \cdot \text{m}^{-1}$, representing a π-pulse completely inverting the spin population in the system, to $E_0 = 4 \times 10^7\,\text{V} \cdot \text{m}^{-1}$, corresponding to a pulse area of 12π and inducing six full Rabi flops. The fundamental trion transition dipole moment is estimated to be $\wp = 4.8 \times 10^{-28}\,\text{C} \cdot \text{m}$, and the 3D number density of the resonant four-level dipoles, $N_a = 2.5 \times 10^{24}\,\text{m}^{-3}$, is calculated from the QD surface density to give on average one dot within the microscopic single-dot volume. The trion recombination (spontaneous emission) rate, $\Gamma = 400\,\text{ps}^{-1}$, is taken from [62]; the electron spin-flip rate due to hyperfine interaction with the lattice ions' spins, $\gamma_1 = 0.5\,\text{ns}^{-1}$, has been calculated in [65,66]; the hole spin-flip relaxation is widely assumed to be due to phonon-assisted processes, and an estimate of $\gamma_3 \sim 170\,\text{ps}^{-1}$ [67,68] has been given. The trion-bound electron and hole spin phonon-assisted decoherences are taken as $\gamma_2 = 450\,\text{ps}^{-1}$ and $\Gamma_\tau = 340\,\text{ps}^{-1}$ [69]. We shall assume that the initial spin population resides either in state $|1\rangle$ with spin-down, or in state $|3\rangle$ with spin up. We initially excite the $|1\rangle \rightarrow |2\rangle$ transition by a σ⁻ pulse. The simulation results for the time evolution of the pulse's electric-field vector components and the populations of all four levels for three initial electric-field amplitudes, E_0, are summarized in Figure 49.23. Excitation of the $|3\rangle \rightarrow |4\rangle$ optical transition is, of course, identical to exciting the $|1\rangle \rightarrow |2\rangle$ transition. The simulated detection of the time-resolved signals shows that there is a sufficiently long time interval—on the order of 400 ps—within which it is possible to determine the spin of the initial states. Indeed, there are two reliable ways for initial spin-state identification: through examining the PL trace profile and through the presence or absence of a second echo pulse after a σ⁺/⁻ π-pulse excitation. The observation of a nonmonotonic PL trace, rather than an exponential decay, is always associated with a spin-up state; in addition, the appearance of a second echo pulse post σ⁻ π-pulse excitation alludes to a spin up, whereas a repeated echo after a σ⁺ π-pulse excitation is indicative of a spin down. The simulations show the onset of the high-intensity Rabi oscillations regime, which suppresses the spin-relaxation processes.

49.5.2 Coherent Spin Manipulation through Hot-Charged Exciton States in QDs

Optical manipulation of the spin of a QD-confined single electron or hole through the resonantly driven trion ground singlet transition (described in the previous section) is considered to be one of the most promising schemes for implementation of spin-based quantum computing, due to the extended spin life-times which are limited only by the hyperfine interaction. However, the requirement of a resonantly driven ground trion transition in an inhomogeneously broadened ensemble of charged QDs represents an obstacle for the scalability of architectures based on this quantum system. This problem can be overcome by optical excitation into the excited trion states, taking advantage of a recently discovered effect, herein referred to as "spin filtering". The effect is observed in p-doped QDs under nonresonant, circularly polarized optical excitation, and constitutes an enhanced, photo-induced circular dichroism in the excited trion-state emission, compared to the dichroism exhibited by the ground singlet trion state [70]. The degree of spin polarization is nearly doubled when a high-energy level in the QD is excited resonantly. This is due to an increased spin "injection" efficiency—i.e., the resonant excitation prevents relaxation processes from occurring and the desired spin is maintained. In addition, nonmonotonic dependence of the degree of spin polarization on the optical pulse power is observed, allowing to maximize it by optimizing the pulse characteristics. This spin-filtering effect is promising for realization of high-fidelity schemes for all-optical spin manipulation, since the increased polarization contrast for the two pulse helicities enables highly efficient selective excitation and readout of the spin-up and spin-down populations.

We consider quasi-resonant σ⁻ or σ⁺ circularly polarized excitation of a p-doped QD ensemble into the p-shell in the presence of a resident s-shell hole, and subsequent cascade relaxation to the bright trion ground singlet state, whose decay is detected as PL. The optical excitation creates X^{+*} states, consisting of one electron–hole pair in the s-shell and a resident hole in the p-shell ($1e^1 1h^1 2h^1$), grouped in four degenerate doublets [71]. The exchange interactions can be revealed in the excited exciton spectra, and the energy-level diagram and the spin configurations of the X^{+*} trion are given in Figure 49.24.

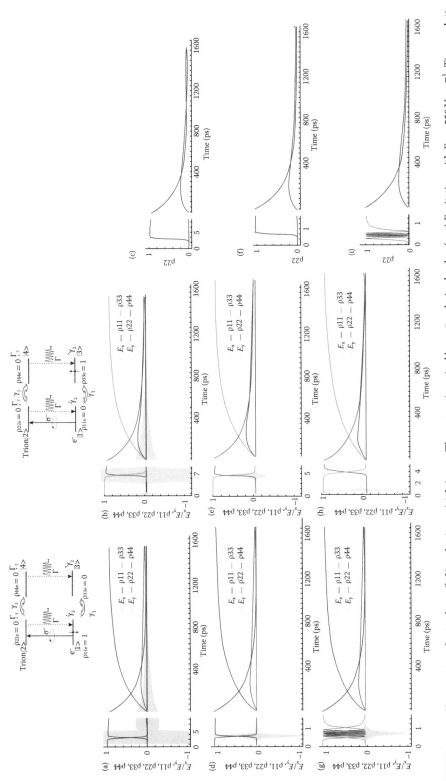

FIGURE 49.23 Top row: initial spin-down (left) and spin-up (right) state. The system is excited by a σ^- polarized pulse. (a–c) Excitation with $E_0 = 550\ V \cdot m^{-1}$. Time evolution of (a) initial spin-down state $|1\rangle$; (b) initial spin-up state in $|1\rangle$; ρ_{22} (blue curve) represents the trion $|-3/2\rangle$ state spin population proportional to σ^- polarized PL signal detected experimentally. (c, f, i) TRPL signal for $\Downarrow$ (blue curve) and $\Uparrow$ (red curve) exhibiting different dynamics in the short term: ρ_{22} populations from the graphs to their respective left are plotted; (d–f) Excitation with a π-pulse with $E_0 = 3 \times 10^6\ V \cdot m^{-1}$. Time-resolved dynamics for (d) spin-down initial state; (e) spin-up initial state. (g–i) Excitation with a 12π-pulse with $E_0 = 4 \times 10^7\ V \cdot m^{-1}$. Time-resolved dynamics for (g) spin-down and (h) spin-up initial states. Excitation of initial spin-down state results in six full spin population Rabi flops. (Adapted with permission from G. Slavcheva, *Physical Review B*, 77, 115347, 2008. Copyright 2008 by the American Physical Society.)

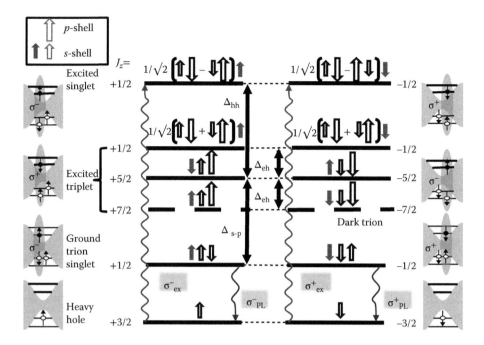

FIGURE 49.24 Energy levels and spin configurations of the hot X^{+*} trion states under σ^- (left six-level system) and σ^+ (right six-level system) optical excitation. The excited singlet is energetically split from the triplet by Δ_{hh}; the excited triplet states are separated from each other by energy Δ_{eh}. The levels are labeled with the total spin projection, J_z, of the electrons and holes. [Small solid arrows, s-shell electron-spin projection; small open arrows, s-shell hole-spin projection; large open arrows, p-shell hole-spin projection; upward wavy arrows, either σ^- or σ^+ excitation; downward wavy arrows, photoluminescence (PL); shaded ellipses, exciton coupling in the radiative states; Δ_{s-p}, energy separation between p-shell and s-shell trion states.]

Spin decoherence mechanisms that need to be considered include the electron–hole anisotropic exchange interaction (AEI), which remains relevant for excited trion states, and the hyperfine interaction between nuclei and either electrons [72] or holes through dipole–dipole interaction [73]. The two degenerate six-level systems for σ^- and σ^+ excitation are coupled through transverse spin decoherence mechanisms, as shown in Figure 49.25.

The system Hamiltonian of a circularly polarized pulse resonantly coupled to an ensemble of six-level resonant absorbers with density N_a is given by

$$
\hat{H}^{\mp} = \hbar \begin{pmatrix} 0 & 0 & 0 & 0 & 0 & -\frac{1}{2}\left(\Omega_x \mp i\Omega_y\right) \\ 0 & \omega_0 - \Delta_{hh} - \Delta_{sp} & 0 & 0 & 0 & 0 \\ 0 & 0 & \omega_0 - \Delta_{hh} - \Delta_{eh} & 0 & 0 & 0 \\ 0 & 0 & 0 & \omega_0 - \Delta_{hh} & 0 & 0 \\ 0 & 0 & 0 & 0 & \omega_0 - \Delta_{hh} + \Delta_{eh} & 0 \\ -\frac{1}{2}\left(\Omega_x \pm i\Omega_y\right) & 0 & 0 & 0 & 0 & \omega_0 \end{pmatrix}, \quad (49.50)
$$

where $-/+$ correspond to σ^-/σ^+ polarization, $\Omega_x = \wp\frac{E_x}{\hbar}$ and $\Omega_y = \wp\frac{E_y}{\hbar}$ are the time-dependent Rabi frequencies associated with the E_x and E_y electric-field components, and ω_0 and $\wp$ are the resonant transition frequency and the optical dipole matrix element of the ground to excited singlet state transition ($|1\rangle \rightarrow |6\rangle$).

In the presence of relaxation processes (Figure 49.25), under time-dependent external perturbation, e.g., a laser pulse, the time evolution of the six-level quantum system is governed by a master equation,

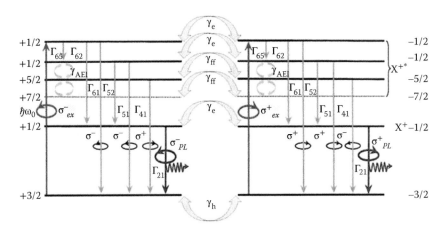

FIGURE 49.25 Discrete-level model of the X^{+*} states in a p-doped QD. The two degenerate six-level systems are coupled via spin decoherence mechanisms represented by double-headed arrows. [γ_h, hole-spin decoherence; γ_e, electron-spin decoherence; γ_{ff}, spin decoherence due to spin flip-flop processes, i.e., anisotropic electron-hole exchange interactions (AEI); $\circlearrowright$ or $\circlearrowleft$, circularly polarized transitions; upward arrows, σ_{ex}^- and σ_{ex}^+ resonant pumping; downward arrows, radiative or nonradiative transitions; downward arrows denoted Γ_{21}, detected polarized PL; vertical small curved arrows, spin flip-flop coupling due to AEI.]

Equation 49.12, for the $N^2 - 1 = 35$-dimensional real state pseudospin vector, S_j, coupled to Maxwell's curl equations, Equation 49.43 [74], through the macroscopic polarization induced by the electromagnetic pulse:

$$P_x = -\wp N_a S_5$$
$$P_y = \mp \wp N_a S_{20}$$

(49.51)

where $-/+$ again correspond to σ^- or σ^+ excitation, respectively.

The QD samples modeled are chosen to be identical to the ones used in TRPL experiments, namely MBE grown on a semi-insulating GaAs (001) substrate. The QD layers are sandwiched between two GaAs barriers and the dot areal density is approximately 2×10^{10} cm^{-2} with an average uncapped height of 4 nm, equivalent to a volume dot density, $N_a = 5 \times 10^{22}$ m^{-3}. The simulation domain consists of an InAs QD layer with nominal thickness given by the height of the typical QD (i.e., 4 nm) embedded between two GaAs barrier regions, each with thickness 50 nm (see Figure 49.26).

The circularly polarized Gaussian pulse at the left boundary of the simulation domain, $z = 0$, is modeled using two x and y linearly polarized waves with resonant carrier frequency, ω_0, phase-shifted by $\pi/2$ according to Equation 49.46 and bottom row of Equation 49.47. The system of equations is discretized in space and time, with spatial and temporal steps, $\Delta z = 1$ Å and $\Delta t = 3.33 \times 10^{-4}$ fs, respectively, and solved numerically directly in the time domain using the FDTD technique.

For comparison with the theory, PL experiments with short (i.e., 50 ps) σ^- or σ^+ optical excitation pulses with pumping wavelength $\lambda_{res} = 1065$ nm inducing a resonant transition to the excited dot states were carried out on QD ensembles nominally doped with one hole. The polarized PL was detected at the ground singlet X^+ transition, exhibiting a peak at $\lambda_{det} = 1148$ nm. In our simulations, an optical pulse of duration 50 ps is injected at $z = 0$. The pulse center frequency, ω_0, is tuned in resonance with the energy splitting between the ground heavy-hole level $|1\rangle$ and the excited singlet trion state $|6\rangle$: $\Delta E_{1 \to 6} = \Delta_{hh} + E_{X^+} + \Delta_{s-p}$, where Δ_{s-p} is the energy separation between the s- and p-shell trion states. The ground trion singlet energy, $E_{X^+} = 1.0815$ eV, is determined from the resonant PL spectra at λ_{det}. $\Delta_{s-p} \approx 73$ meV is inferred from the PL spectra [70], and $\Delta_{hh} = 12$ meV is taken in agreement with [71]. In order to calculate the Hamiltonian in Equation 49.50, $\Delta_{eh} \approx 0.5$ meV is used, as obtained in [71,75].

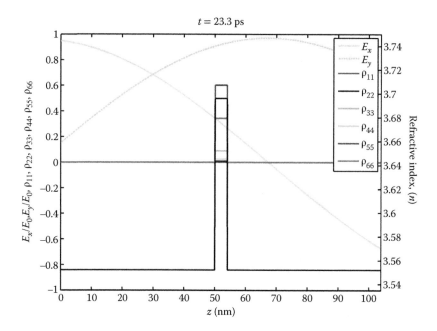

FIGURE 49.26 A snapshot of the spatial distribution of the circularly polarized electric-field components of an excitation pulse interacting with a six-level quantum system. The timing of the snapshot is $t = 23.3$ ps after the source pulse impinges on the left boundary ($z = 0$) of the medium incorporating the quantum system. The charge-carrier spin populations (diagonal density-matrix elements) of all six levels are shown; these indicate the fractional occupancy of each level compared to the total number of charge carriers spread across the different levels. The refractive-index profile of the simulation domain (bottommost curve, right-hand side scale) is produced by an InAs QD layer of thickness 4 nm embedded between two 50-nm-thick GaAs barrier regions.

The pulse area is chosen to be π, so that the pulse completely excites the ground-state spin population to the excited singlet level $|6\rangle$; this corresponds to choosing $E_0 = 2.69 \times 10^5$ V $\cdot$ m^{-1}, assuming a dipole matrix element $\wp = 9.83 \times 10^{-29}$ C $\cdot$ m. The dipole matrix element's value is comparable to the dipole moment of the X^+ ground singlet transition, and is calculated using the Fermi Golden Rule [50] from the energy separation between the levels and the spontaneous emission rate $\Gamma_{61} \sim 1.2$ ns^{-1}, which is close to the experimentally observed value.

The longitudinal relaxation times are taken as follows: the radiative spontaneous decay, $\tau_{21} \approx 1.27$ ns, is experimentally measured from a sample with one hole; spontaneous emission times, $\tau_{41} = 1.35$ ns and $\tau_{51} = \tau_{61} = 1.2$ ns are estimated from the energy-level separation using the Wigner–Weisskopf formula [50]. The nonradiative decay times are taken from theory [76], the transverse spin decoherence rates are obtained from experiment: $\tau_e = 500$ ps [72] and $\tau_h = 14$ ns [73], and a theoretical estimate for $\tau_{ff} = 125$ ps is made on the basis of the Heisenberg uncertainty relations. We should note that there are only two adjustable parameters in our model—namely the largely unknown nonradiative spin relaxation times, τ_{52} and τ_{65}.

The theoretically computed TRPL traces that are displayed in Figure 49.27 clearly exhibit photo-induced circular dichroism in the polarized TRPL emission; this is manifested as a different evolution of PL in time, depending on the pulse helicity used. In contrast to the widely used rate equations model which assumes mono-exponential decays, the dynamics of the quantum system described by the present model is nonlinear and, as a result, the simulated PL decay is not mono-exponential.

The experimentally detected TRPL trace shape strongly depends on detector characteristics: the faster the detector's response and the higher its sensitivity, the better the agreement between theory and

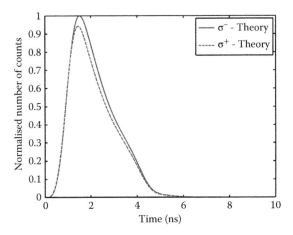

FIGURE 49.27 Theoretically computed polarized TRPL traces for σ^- and σ^+ excitation of a six-level QD. The presence of circular dichroism is clearly visible.

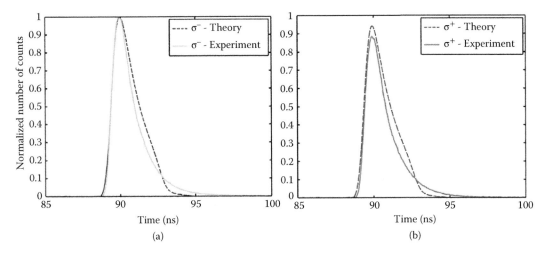

FIGURE 49.28 Comparison between the simulated TRPL trace (after convolution with the detector's response function) and experimentally detected TRPL trace for (a) σ^- excitation and (b) σ^+ excitation of a six-level QD.

experiment. For proper comparison with experiment, we convolve our theoretically calculated polarized traces with the detector's response function and plot them along with the experimental traces in Figure 49.28.

There is excellent agreement between the theoretical and the experimental results. The rising slope and initial decay after the peak in the experimental trace are almost perfectly reproduced by our simulations. The small discrepancy at longer times can be attributed to several factors. Firstly, it is not possible during the course of an experiment to determine precisely which hot trion state is being resonantly excited, so the assumed λ_{res} is only approximate. Secondly, Δ_{s-p} is inferred from the resonant PL spectra and is only a rough estimate, since the excited state does not correspond to a sharp spectral line, but rather a broadened line of finite width; in addition, the inhomogeneous broadening has not been taken into account. The lower peak height for σ^+ excitation can be attributed to the estimated value for the spin flip-flop coupling rate and requires further investigation.

Our theoretical model explains quantitatively the origin of the spin-filtering effect. The experimentally observed enhanced time-resolved circular dichroism in the excited trion state emission is shown to emerge from the dynamical spin flip-flop coupling between the two degenerate sets of excited charged excitonic states. The theory is in very good agreement with experiment, thereby allowing to obtain an estimate for largely unknown intra- and intershell spin-relaxation timescales. The approach also allows to predict optimum pulse parameters, such as power and duration, for efficient control of spin dynamics.

49.6 Quantum Stochastic Formalism for Modeling Cavity-Emitter Systems

Future advances in information systems are expected to exploit the quantum nature of their constituent parts—e.g., in the development of quantum simulators and design of architectures for quantum cryptography. The implementation of these technologies necessitates a fuller understanding of the underlying quantum systems, which in turn will require several fundamental aspects of quantum theory to be tested.

One example of a quantum-optical phenomenon is quantum noise. In contrast to the more familiar classical noise, which can be regarded as spontaneous random fluctuations from a steady state, quantum noise arises from the Heisenberg uncertainty relations, which are fundamental to quantum theory [77]. The concept of quantum noise stems from the statistical interpretation of quantum mechanics and our inability to access any particular one out of the infinite number of degrees of freedom of the electromagnetic field propagating in free space. Quantum noise is closely related to spontaneous emission, irreversible decay and the origin of the spectral line width described by the theory of Weisskopf and Wigner [78], who are considered to have laid the foundations of the quantum noise theory.

At large photon densities, quantum fluctuations are necessary to explain laser linewidth and the threshold properties. By contrast, at low photon densities, light is generated by spontaneous emission described in terms of single-photon emission events within the single-particle Dirac [79] picture. Control over such noise effects will be crucial in the design of devices for quantum information processing.

49.6.1 Langevin Formalism for Modeling Quantum-Optical Effects

Considering a two-level system with a few degrees of freedom immersed in a radiation-field heat bath with an infinite number of degrees of freedom, we exploit the quantum-classical correspondence in the presence of noise and employ the Langevin formalism. The essence of the Langevin approach lies in the realization that the action that the many unknown "bath" (i.e., external) variables have on the system is to modify its deterministic equation of motion, and that this can be accommodated by the inclusion of apparently random terms, known as "Langevin forces."

49.6.2 Spontaneous Emission of a QD Ensemble in a Semiconductor Microcavity and Onset of Lasing

Consider the semiconductor microcavity structures shown in Figure 49.29, designed at $\lambda = 850$ nm and $\lambda = 1.29\,\mu$m. The cavity of the first structure (bottom DBR: 35.5 pairs AlAs/Al$_{0.3}$Ga$_{0.7}$As; top DBR: five pairs AlO/Al$_{0.3}$Ga$_{0.7}$As) is filled with a resonant two-level medium (Al$_{0.5}$Ga$_{0.5}$As), whereas within the second cavity, there are six (Ga$_{0.63}$In$_{0.37}$N$_{0.012}$As) quantum wells filled with the resonant medium. To describe the interaction of an optical wave with these systems, we use the semiclassical 1D Maxwell–Bloch equations in the real vector pseudospin picture, Equation 49.26, for an optical wave linearly polarized along x propagating in the z-direction. In order to model the spontaneous emission of two-level atoms, we add a random electric-field fluctuation, δE_x, at each time step:

$$\left(E_x\left(z,t\right) + \delta E_x\left(z,t\right)\right) = -\frac{1}{\varepsilon}\frac{\partial H_y}{\partial z} - \frac{N_a\wp}{\varepsilon T_2}S_1 + \frac{N_a\wp\omega_0}{\varepsilon}S_2. \tag{49.52}$$

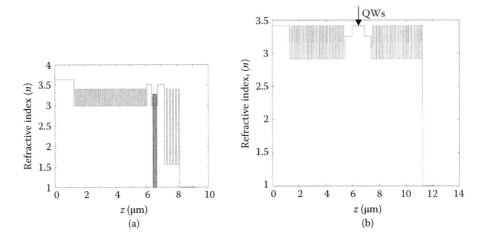

FIGURE 49.29 Refractive-index profile along a semiconductor microcavity (a) filled with gain medium (shaded area) with $n = 3.2736$ at $\lambda = 850$ nm; (b) with 6 GaInNAs-QWs, filled with gain medium of refractive index $n = 3.4$ at $\lambda = 1.29$ μm. (G. Slavcheva et al., *IEEE Journal of Selected Topics in Quantum Electronics* © 2004 IEEE.)

We use a pseudorandom number generator based on the Box–Müller method for generating random deviates with a normal (Gaussian) distribution from uniformly distributed in the interval (0, 1) random numbers a and b. At each time step, j, white Gaussian noise with variance $\xi_E = \sigma^2 = 1 \times 10^{-3}$ V^2m^{-2} is implemented according to

$$E_x(z)_j = E_x(z)_j + \sqrt{-2\xi_E \ln(a)} \cos(2\pi b). \tag{49.53}$$

White Gaussian noise has a mean zero and δ-correlated second moment of the random distribution:

$$\langle \delta E_x(z, t) \rangle = 0$$

$$\left\langle \delta E_x(z, t) \delta E_x\left(z, t'\right) \right\rangle = \tilde{\xi}_E \delta\left(t - t'\right) = \xi_E R_{sp} \delta\left(\frac{t - t'}{\Delta t}\right), \tag{49.54}$$

where an expression for the spontaneous emission rate has been derived:

$$R_{sp} = \frac{\xi_E(\varepsilon_0 \varepsilon)^2}{N_a^2 \wp^2 T_2^2}. \tag{49.55}$$

Consider the first microcavity: we assume that the resonant medium is initially inverted ($S_{30} = 1$), thus providing gain, and apply solely random white Gaussian noise at each point within the cavity according to Equation 49.53. The spatial distribution of the intracavity electric field as a function of time, and the time evolution of the electric field at the microcavity structure's output boundary are calculated. The parameter set used for this simulation is $\wp = 4.8 \times 10^{-28}$ C · m, $n = 3.2736$, $N_a = 1 \times 10^{24}$ m^{-3}, initial field amplitude, $E_0 = 700$ V · m^{-1}, and relaxation and dephasing times, $T_1 = 10$ ps and $T_2 = 70$ fs, respectively. The cavity field enhancement provided by the DBR mirrors leads to build-up of the electric field within the cavity borne out of the quantum noise. A snapshot of the intracavity electric-field build-up is shown in Figure 49.30, along with the corresponding Rabi flopping of the population dynamics. The amplitude that has arisen solely due to noise gradually builds up towards a standing-wave mode in the cavity.

After this initial amplitude build-up process, the lasing threshold is reached, and coherent oscillations appear at the output boundary. This is the onset of the lasing regime, shown for the unsaturated

$$\Omega_R^2 T_1 T_2 \approx 1.59 \tag{49.56}$$

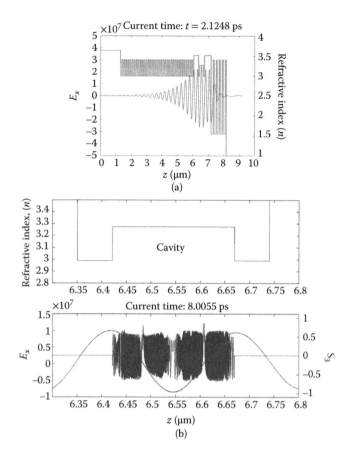

FIGURE 49.30 (a) Stationary electric-field standing-wave profile arisen from quantum noise in the cavity. (b) (Top) Refractive-index profile within the cavity and in its vicinity. (Bottom) A snapshot of the spatial distribution of the intracavity E_x electric field (left axis, slowly-varying sinusoidal curve) and the corresponding population inversion (right axis, fast-oscillating curve) exhibiting fast Rabi oscillations at $t = 8$ ps after the instant when the quantum noise arises. These oscillations result in the fast modulation of the electric-field envelope observed at the output boundary of the structure, which is shown in Figure 49.31a and b.

and saturated

$$\Omega_R^2 T_1 T_2 \approx 16241 \gg 1 \tag{49.57}$$

cases in Figure 49.31 (a,b), respectively. The charge-carrier population performs multiple Rabi flops, oscillating about a steady population inversion value close to zero. The net gain and absorption are therefore similar, resulting in a gain saturation in time.

Let us now consider the structure designed at $\lambda = 1.29\,\mu$m. The population in all six quantum wells is initially inverted into the upper state, $S_{30} = 1$, representing a nonlinear gain medium, while the barrier regions within the cavity are assumed to be in the ground state, $S_{30} = -1$. Random noise is generated within the cavity with a parameter set corresponding to the unsaturated case above (49.56). In Figure 49.32, a snapshot of the spatial distribution of the electric-field amplitude and the steady-state population inversion are shown. The population inversion relaxes to this value after a sufficiently large number of time steps. The time evolution of the electric field at the output boundary is shown in Figure 49.33, showing continuous build-up of coherent self-sustained oscillations and lasing. The lasing threshold is clearly discernible, and the electric-field envelope exhibits rapid damped relaxation oscillations with a decay rate,

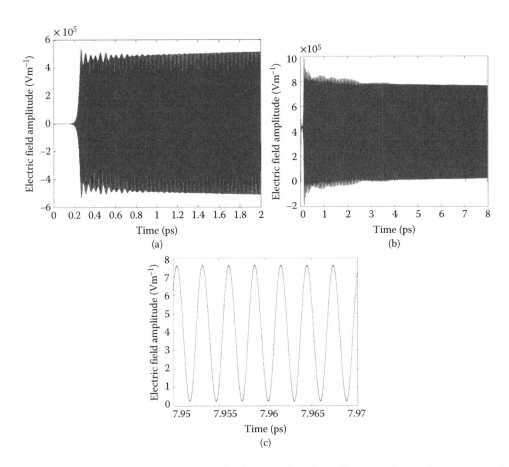

FIGURE 49.31 Time evolution of the electric field at the output boundary of a semiconductor microcavity, showing the noise-induced build-up of coherent oscillations and onset of lasing of the structure displayed in Figure 49.29a. Simulations were run for two different sets of relaxation and dephasing times (T_1 and T_2, respectively). (a) $T_1 = 10$ ps, $T_2 = 70$ fs; (b) $T_1 = 1$ ns, $T_2 = 10$ ps. (c) Expanded view of (b) in the steady-state region, showing single-mode oscillations and lasing at a frequency corresponding to $\lambda = 850$ nm. (G. Slavcheva et al., *IEEE Journal of Selected Topics in Quantum Electronics* © 2004 IEEE.)

$\Gamma_R \approx 1.5 \times 10^{11}$ s^{-1}. These rapidly varying oscillations are a new feature which is not predicted by the usual rotating-wave (RWA) and slowly-varying envelope (SVEA) approximations. The oscillations eventually reach the steady-state gain saturation value, $S_{3sat} = 0.0025$, at simulation times exceeding 16 ps.

We perform a Fourier transform of the output electric-field trace in Figure 49.33 and obtain the laser spectral linewidth, shown in Figure 49.34. The obtained linewidth exhibits superfine structure: the main mode is centered at a frequency ω_0 corresponding to $\lambda = 1.289\,\mu$m, and the wings exhibit satellite peaks at $\omega_0 + n\Omega_{\text{rel}}$. The frequency of the fast relaxation oscillations, $f_{\text{rel}} \approx 2.54$ THz, is determined from the figure.

We showed that an extension within the Langevin framework of our Maxwell-pseudospin model predicts the build-up of coherent, self-sustained oscillations and the onset of lasing, which is a sole consequence of noise being introduced into the cavity. This emphasizes the importance of quantum fluctuations for triggering the lasing regime of cavity operation. The simulations allow to investigate the lasing threshold behavior directly in the time domain and provide an estimate for the coherence time of the laser emission and the laser linewidth. We showed that several important parameters can be extracted from the simulations and that they can be optimized using numerical experiments on a variety of design geometries. Most importantly, the developed quantum stochastic approach is capable of modeling quantum effects, such as

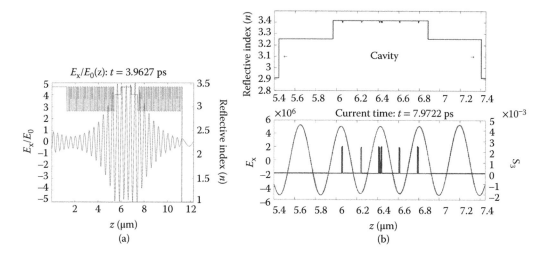

FIGURE 49.32 (a) Instantaneous profile of the electric-field standing wave borne out of quantum noise in the cavity. (b) (Top) Refractive-index profile implying the cavity boundaries and the location of the quantum wells. (Bottom) Spatial distribution of the stationary electric-field standing wave (left axis, sinusoidal curve) and the corresponding saturated population inversion, $S_{3sat} = 0.0025$, in the quantum wells within the cavity (right axis, spiky curve). (G. Slavcheva et al., *IEEE Journal of Selected Topics in Quantum Electronics* © 2004 IEEE.)

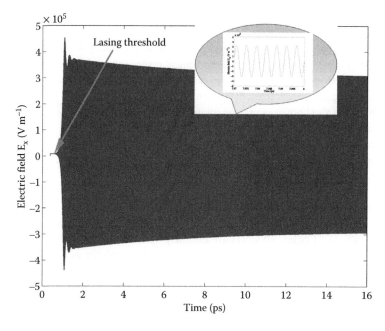

FIGURE 49.33 Build-up in time of coherent, self-sustained oscillations arisen from noise introduced into the cavity of the structure displayed in Figure 49.29b. The lasing threshold is apparent. The decay rate of the electric-field envelope is $\Gamma_R \approx 1.5 \times 10^{11}$ s^{-1}. Inset: expanded view of the fast carrier-frequency oscillations with $\lambda = 1.289$ μm; the steady-state region occurs for $t > 16$ ps. (G. Slavcheva et al., *IEEE Journal of Selected Topics in Quantum Electronics* © 2004 IEEE.)

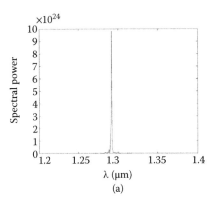

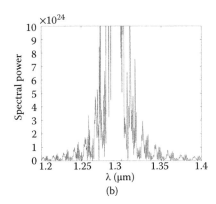

FIGURE 49.34 (a) Laser spectral line obtained by Fourier-transforming the output trace of the electric-field amplitude centered at the lasing frequency, corresponding to $\lambda = 1.289\,\mu$m. (b) Magnified view of the spectrum in (a) showing relaxation oscillation sidebands. (G. Slavcheva et al., *IEEE Journal of Selected Topics in Quantum Electronics* © 2004 IEEE.)

spontaneous emission, and thus can be used to model the system dynamics in the few-photon limit as well as demonstrate and predict quantum-optical effects.

49.7 Modeling Optical Rotation in Chiral Carbon Nanotubes with Master Maxwell-Pseudospin Equations

In this section, we shall employ our vector Maxwell-pseudospin model [44] to investigate how chirality affects the ultrafast nonlinear optical and magneto-optical response of a single chiral CNT [80]. We shall show that a simple discrete-level model of the optically active states near the bandgap edge of a chiral nanotube is sufficient to explain the circular dichroism and birefringence, as well as to quantify the optical rotation per unit length (also known as specific rotatory power), either in the absence or in the presence of an external magnetic field permeating the tube. The model predicts a giant natural gyrotropy for the specific chirality considered in this case, which is comparable with or exceeding that of artificially fabricated helical photonic structures. This is remarkable because the latter are considered to achieve the largest optical rotation of all known materials: orders of magnitude larger than that of crystal birefringent materials or of liquid crystals. Finally, we will give a quantitative estimate of the nonlinear coherent magneto-chiral optical effect in an axial magnetic field.

Chirality is one of the main symmetries of the CNT geometry that determines the optical properties of single-walled CNTs (SWCNTs). SWCNTs are uniquely determined by the chiral vector, $C_h = na_1 + ma_2 \equiv (n, m)$, with $0 < |m| < n$, or equivalently by a pair of integer numbers (n, m) in the planar graphene hexagonal lattice unit-vector basis (Figure 49.35). Two helical forms (enantiomers), e.g., AL $(5, 4)$ and AR $(4, 5)$, can exist. Topologically, the CNT can be viewed as a graphene sheet rolled up along the chiral vector into a cylinder. Chirality serves as a primary classification criterion. Achiral nanotubes, whose mirror image is superimposable, are subdivided into two classes: zigzag $(m = 0)$ and armchair $(m = n)$ nanotubes. The rest of the nanotubes, whose mirror reflection is not superimposable, are subsumed into the most general class of chiral nanotubes.

There is a fundamental relationship between chirality and optical activity. The interaction of polarized light with chiral materials gives rise to the phenomenon of optical rotation (or optical activity), whereby the polarization plane is rotated continuously during the propagation of the light through the nanotube. In the presence of an external magnetic field, magnetically induced optical activity—also known as Faraday effect—takes place. Both effects manifest themselves as a rotation of the transmitted light; however, the origin of the two effects is fundamentally different. While the natural optical activity is a result of

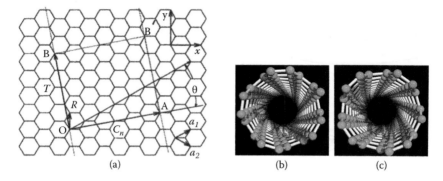

FIGURE 49.35 (a) Chiral vector, C_h (C_n on schematic) and chiral angle, $\theta = \angle(a_1, C_h)$, $0° < |\theta| < 30°$, where a_1 and a_2 are the elementary Bravais cell unit vectors. (b) Left-handed AL (5, 4) (with $m > n$; m, n – integers, see text) and (c) right-handed AR (4, 5) (with $m < n$) single-walled CNT molecular structure calculated by the tight-banding method [81]. Shown is a view along the tube axis looking in the negative z-direction. Chirality is determined by the rotation of the so-called armchair chains of carbon atoms either clockwise or counterclockwise when looking in the negative z-direction.

the nonlocal optical response of a medium lacking mirror symmetry, the magnetic optical activity results from time-reversal symmetry breaking by the magnetic field. The two phenomena are linked through the magneto-chiral optical effect which takes place when both symmetries are broken simultaneously [82].

The electronic band structure of an SWCNT is described by the quantization of the wavevector along the tube circumference perpendicular to the tube axis. This results in a discrete spectrum of allowed k-vector states forming subbands in the valence and conduction bands labeled by the quasi-angular momentum quantum number, μ [83]: $k_\perp = \frac{2\pi\mu}{L}$, $\mu = 0, \pm1, \pm2, ...$, where $L = |C_n|$ is the tube circumference (Figure 49.36a). We should note that due to symmetry in the graphene hexagonal unit cell, there are two equivalent points, K (with $\mu > 0$) and K' (with $\mu < 0$) in the BZ (Figure 49.36c). The energy dispersion of a CNT can be visualized as intersections of the graphene conical dispersion at the quantized transverse wavevector, $k_\perp$ (Figure 49.36b).

The most widely studied excitation geometry in experiments on individual CNT is the one of linearly polarized excitation perpendicular to the tube axis, shown in Figure 49.37a. In this excitation geometry, the longitudinal E_z electric-field component is dominant, because the transverse E_x component is suppressed by depolarization effects, and linearly polarized intersubband optical transitions from the valence μ subband to the conduction μ subband with $\Delta m = 0$ can be excited (here m signifies the eigenvalue of the quasi-angular momentum operator projection, $\hat{J}_z$). We shall be interested, however, in the unconventional circularly polarized excitation geometry shown in Figure 49.37b, whereby a circularly x–y plane-polarized laser pulse propagates along the tube axis, exciting only one of the two allowed transitions, $\mu \to \mu \pm 1$, where "$+/-$" corresponds to right and left circularly polarized light, respectively (Figure 49.38b). The molecular structure and the allowed dipole optical intersubband transitions of a (20, 10) chiral CNT are shown in Figure 49.38.

We shall consider aggregates or bundles of aligned SWCNTs grown by chemical-vapor deposition in an electric field [84], or aligned in a polymer matrix [85]. We model the single carbon-nanotube bandgap-edge structure at the K point of the BZ (where the fundamental bandgap opens) by a large ensemble of identical four-level systems, corresponding to the allowed transitions of the two valence subbands closest to the Fermi level for the AL and AR nanotube enantiomers shown in Figure 49.39. We note that the energy-level diagrams for the left- and right-handed helical forms in Figure 49.39 are nonsuperimposable: absorption of σ^+ light excites the $\mu \to \mu + 1$ transition in AL-handed SWCNT, while absorption of light with the same helicity excites the $\mu \to \mu - 1$ transition in AR-handed SWCNT. The difference between the optical selection rules for left and right circularly polarized light gives rise to the optical activity [86].

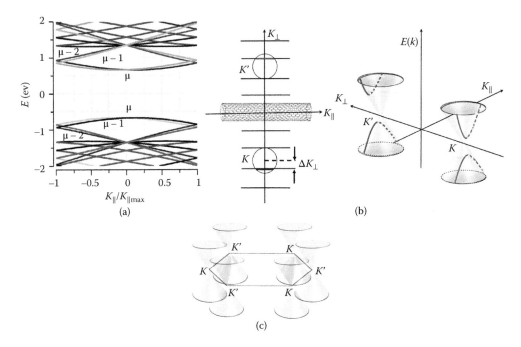

(a)

(b)

(c)

FIGURE 49.36 (a) Energy dispersion as a function of the longitudinal wavevector, $k_{||}$, normalized with respect to the wavevector at the boundary of the first BZ of a (5, 4) SWCNT. $k_{||}$ points along the tube axis. The tube for this chirality is semiconducting—i.e., a bandgap is opened at the center (Γ point) of the BZ. (b) Parabolic cross sections of the graphene linear dispersion, corresponding to points K and K′ in (a) representing the nanotube energy dispersion curves. (c) Unfolded 2D first BZ of graphene, showing the equivalent points K and K′ in the band structure.

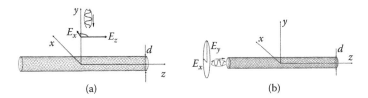

(a)   (b)

FIGURE 49.37 (a) Linearly polarized excitation perpendicular to the tube axis exciting intersubband optical transitions with $\Delta m = 0$. (b) Circularly polarized excitation along the tube axis exciting $\Delta m = \pm 1$ optical transitions.

49.7.1 Natural Optical Activity

In this section, we compute the natural optical activity rotation angle (rotatory power per unit length) in the absence of an external magnetic field and demonstrate the possibility to manipulate it by engineering the nanotube chirality. We consider the specific case of a chiral, left-handed AL (5, 4) SWCNT, which is chosen for illustration of the general method valid for an arbitrary chirality. The nanotube length is 500 nm and all edge effects will be ignored in our 1D model. We performed tight-binding calculations of the electronic band structure (Figure 49.36a) and determined the fundamental bandgap, $E_g = E_{\mu,\mu} = 1.321$ eV, corresponding to a pumping wavelength of $\lambda = 939$ nm. The tube diameter is 0.611 nm, and the chiral angle, 26.33°. The resonant transition energy for circularly polarized excitations is $E_{\mu,\mu\pm1} = 1.982$ eV, corresponding to a resonant wavelength, $\lambda_0 = 626.5$ nm.

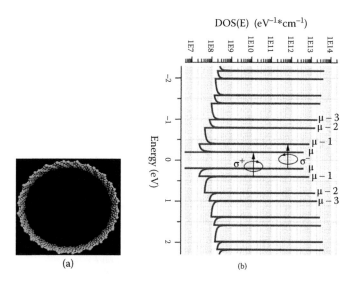

(a)   (b)

FIGURE 49.38 (a) Molecular structure of an AL (20, 10) left-handed chiral CNT. (b) 1D electron density of states (DOS) plotted as a function of energy at the K point of the BZ ($\mu > 0$). The allowed dipole optical transitions for circularly polarized light are denoted by arrows for left- (σ^-) and right-handed (σ^+) pulse helicity.

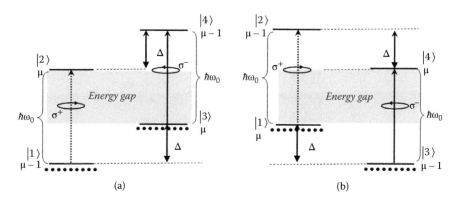

(a)   (b)

FIGURE 49.39 Energy-level structure at the K (or K') point of the lowest subbands labeled by the subband index μ for an (a) AL (le-handed) and (b) AR (right-handed) SWCNT. The fundamental energy gap is shaded. The dipole optical transitions excited by σ^- and σ^+ circularly polarized light are designated by arrows. Only one of the two transitions is allowed for circularly polarized light, denoted by solid arrow (by contrast, the forbidden transition is indicated by a dashed arrow). Valence-band states below the bandgap are populated. (ω_0, resonant transition frequency; Δ, energy separation between the lowest subband and the second lowest subband near the bandgap.)

We consider a circularly polarized ultrashort pulse of duration 60 fs matching experiments [87], and a pulse area of π, giving an initial electric-field amplitude $E_0 = 6.098 \times 10^8$ V $\cdot$ m^{-1}. We inject the pulse from the left boundary of the simulation domain, consisting of the tube embedded between two free-space regions of thickness 50 nm each. The pulse is resonant with the energy $E_{\mu,\mu\pm1}$ of the allowed dipole optical transition and is coupled to an ensemble of identical, HB four-level systems (Figure 49.39a for a left-handed AL SWCNT) used to describe the optical transitions involving the lowest-energy subbands lying nearest to the bandgap edge (Figure 49.40a).

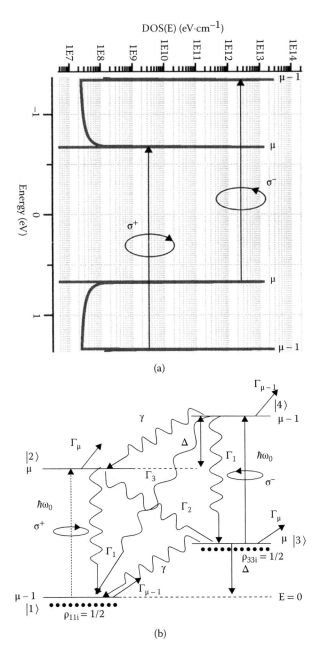

FIGURE 49.40 (a) Plot of the 1D density of states (DOS) showing the lowest-energy subbands near the bandgap edge involved in dipole optical transitions excited by circularly polarized light. Only one transition at a time can be excited by each helicity (σ^- or σ^+). (b) Energy-level diagram in the general case of a resonant optical excitation of $\mu - 1 \rightarrow \mu$ and $\mu \rightarrow \mu - 1$ interband transitions by a σ^+ and σ^- polarized pulse, respectively. The resonant transition energy is $\hbar\omega_0$. The initial population of the lowest valence states below the bandgap, ρ_{11i} and ρ_{33i}, is equally distributed between levels $|1\rangle$ and $|3\rangle$ (here, the subscript "i" denotes "initial"). (Δ, the energy separation between the first and the second lowest conduction (or valence) band states; wavy lines denote longitudinal relaxation processes between the levels, associated with population transfer; Γ_1, spontaneous emission (radiative decay) rate of $|2\rangle \rightarrow |1\rangle$ transitions, assumed to be equal to that of $|4\rangle \rightarrow |3\rangle$; Γ_2, spontaneous emission rate for the linearly polarized transition from the conduction μ subband to the valence μ subband; Γ_3, decay rate of the linearly polarized transition from the conduction $\mu - 1$ subband to the valence $\mu - 1$ subband; γ, intraband relaxation rate; Γ_μ and $\Gamma_{\mu-1}$, transverse relaxation (dephasing) rates.)

The corresponding energy-level diagram is shown in Figure 49.40b, with all coherent and spontaneous transitions indicated. Following our coherent vector Maxwell-pseudospin formalism, the system Hamiltonian for a four-level system that applies for either excitation by σ^- or σ^+ is given by

$$\hat{H} = \hbar \begin{pmatrix} 0 & -\frac{1}{2}\left(\Omega_x + i\Omega_y\right) & 0 & 0 \\ -\frac{1}{2}\left(\Omega_x - i\Omega_y\right) & \omega_0 & 0 & 0 \\ 0 & 0 & \Delta & -\frac{1}{2}\left(\Omega_x - i\Omega_y\right) \\ 0 & 0 & -\frac{1}{2}\left(\Omega_x + i\Omega_y\right) & \Delta + \omega_0 \end{pmatrix}, \quad (49.58)$$

where $\Omega_x = \frac{\wp}{\hbar} E_x$ and $\Omega_y = \frac{\wp}{\hbar} E_y$ are the time-dependent Rabi frequencies associated with the E_x and E_y electric-field components, and $\wp$ is the optical dipole matrix element for $\mu \to \mu \pm 1$ transitions excited by circularly polarized light. This parameter is largely unknown. We have provided an estimate for it,

$$\wp \approx 3 \times 10^{-29}\ \text{C} \cdot \text{m}, \quad (49.59)$$

based on an extension of the effective-mass method applied to chirality effects in carbon nanotubes (CNTs) [88], in agreement with the measured radiative lifetime of typical molecular transition in a single CNT [89], which is on the order of $\tau_{\text{spont}} \approx 10\ \text{ns}$.

We consider a four-level resonant medium consisting of a large number of identical aligned CNTs with an average density $N_a = 6.8 \times 10^{24}\ \text{m}^{-3}$, calculated in such a way that the volume of the simulated nanotube with diameter 0.611 nm and length 500 nm contains just a single nanotube on average. A method called the effective medium approximation—used for describing the macroscopic properties of composite materials, such as the dielectric permittivity of a single CNT—has been successfully applied in FDTD modeling of the thermal radiative properties of vertical arrays of multi-walled CNTs [90]. We use two independent theoretical approaches to obtain an estimate for the effective dielectric constant of an isolated SWCNT: the aforementioned effective medium approximation [80] and tight-binding model calculations of the axial component of the imaginary part of the dielectric function tensor [91,92]. This gives a value for the refractive index, $n \approx 2.3$ along the nanotube axis. With dipole matrix element from (49.59), we use the expression for the spontaneous emission rate: $\tau_{\text{spont}} = \frac{3\pi\varepsilon_0 \hbar c^3}{n\omega_0^3 \wp^2}$, taking into account the energy separation of each transition, to obtain an estimate for the relaxation rates: $\Gamma_1 = 2.907\ \text{ns}^{-1}$, $\Gamma_2 = 9.812\ \text{ns}^{-1}$, and $\Gamma_3 = 1.227\ \text{ns}^{-1}$. We take the experimental value obtained in [87] for the intraband optical transitions, $\gamma = 130\ \text{fs}^{-1}$. Since the dephasing rates for the states involved are largely unknown, we treat them as phenomenological parameters adopting the following values: $\Gamma_\mu = 800\ \text{fs}^{-1}$ and $\Gamma_{\mu-1} = 1.6\ \text{ps}^{-1}$; however, our simulations show that the dynamics is largely insensitive to the choice of these dephasing rates.

We consider separately the cases of an ultrafast resonant optical excitation of the $|1\rangle \to |2\rangle$ transition with σ^+ helicity, and of the $|3\rangle \to |4\rangle$ transition with σ^- helicity (Figure 49.40b). Note that in the former case, level $|4\rangle$ does not participate in the relaxation dynamics and the system is effectively a three-level Λ-system, rather than a four-level system. The initial population is assumed to be equally distributed between the lower-lying levels. The ultrashort circularly polarized source pulse is injected into the medium, and the temporal dynamics of the electric-field vector components and population of all four levels are sampled at four different locations along the nanotube z-axis. The spatially resolved temporal dynamics at a point $z = 300\ \text{nm}$ from the left boundary of the simulated structure for σ^+ and σ^- excitations is shown in Figure 49.41.

It would be interesting from an experimental point of view to test whether nanotube chirality could be determined from the ultrafast nonlinear response using ultrashort pulses with either helicity. We compare the Fourier spectra at the output boundary for each helicity of the optical pulse excitation in Figure 49.42. Both transmission spectra exhibit a sharp peak at the resonant wavelength, indicating

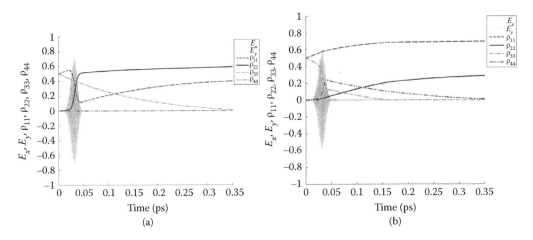

FIGURE 49.41 Time evolution of the electric-field components E_x and E_y, and the populations of all four energy levels of a SWCNT (a) for a circularly polarized σ^+ excitation pulse (note that level $|4\rangle$ is not involved in the dynamics) and (b) for a circularly polarized σ^- excitation pulse. The location is $z = 300$ nm. The left boundary of the SWCNT is at $z = 0$. The SWCNT is 500 nm long and is placed between two free-space regions, each 50 nm long.

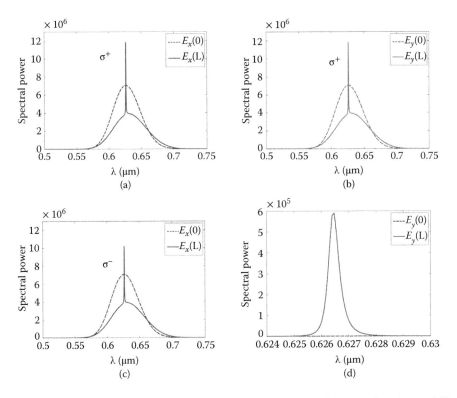

FIGURE 49.42 Transmission Fourier spectra plotted against the wavelength of the E_x and E_y electric-field components at the input, $E_{x,y}(0)$, and output, $E_{x,y}(L)$, of the structure for a (a), (b) σ^+ polarized pulse, and a (c), (d) σ^- polarized pulse. Note that the peak output amplitudes are approximately 20% higher in (a) and (b) compared to in (c) and (d). The simulated structure is as in Figure 49.41.

resonant amplification; however, the peak intensity corresponding to σ^+ excitation is nearly 20% higher. This difference can be exploited in an experiment aiming to determine unambiguously the chirality of a single CNT by first injecting a pulse of one helicity, then the other, and observing the consequences at the output. We show that the difference in the polarization-resolved transmission spectra of a linearly polarized pulse is much more pronounced and allows identification of the precise helical form of the SWCNT.

In order to demonstrate rotation of the electric-field polarization plane during the pulse propagation across the resonant four-level medium, we launch a linearly polarized pulse along x, expressed by:

$$X \begin{cases} E_x\,(z=0,t) = E_0 e^{-(t-t_0)^2/t_d^2} \cos(\omega_0 t) \\ E_y\,(z=0,t) = 0. \end{cases} \tag{49.60}$$

The system's temporal dynamics induced by the ultrashort linearly polarized pulse is shown in Figure 49.43a and b on an expanded scale, where the appearance of a second E_y component, and therefore also the appearance of optical rotation of the electric-field vector polarization, is clearly visible. The maximum amplitude of the E_y component continuously increases as the pulse propagates along the nanotube structure. We should therefore expect the maximum optical rotation angle to occur at the output boundary.

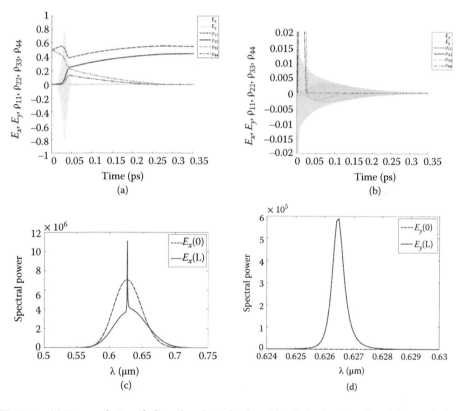

FIGURE 49.43 (a) Time evolution of a linearly polarized pulse with initial value $E_y = 0$, and the populations of all four levels at $z = 550$ nm from the left boundary. The simulated structure is as in Figure 49.41. (b) Magnified view of (a) showing the build-up of the E_y component with time, which is equivalent to a rotation of the electric-field polarization. The transmitted pulse is elliptically polarized and displays different power spectra for (c) the E_x component (sharp resonant peak superimposed on a broader line) and (d) the E_y component (single spectral line).

The polarization-resolved transmission spectra of the E_x and E_y components are quite distinct: while the former is similar to the spectrum obtained for circularly polarized optical excitation, the E_y spectrum exhibits a single spectral line. Conversely, for excitation with light of the opposite helicity (data not shown), the E_x and E_y profiles are interchanged. This difference in the polarization-resolved nonlinear optical response can therefore be used for unambiguous determination of the nanotube chirality.

In order to obtain a quantitative estimate of the natural optical activity in a single chiral CNT, we follow the general methodology developed in Section 49.3.3. We calculate the gain/absorption coefficient from the complex propagation factor, $e^{ik_c(z_2-z_1)}$, for the E_x and E_y electric-field components of a circularly polarized pulse over a distance of one dielectric wavelength, $l = z_2 - z_1 = \lambda_0/n$, where λ_0 is the resonant wavelength. The wavevector is $k_c = \beta + i\gamma$, where β is the phase shift per unit length induced in the optical pulse by the interaction with the resonant medium, and γ is the gain/absorption coefficient. The gain/absorption coefficient allows us to calculate the magnitude of the circular dichroism, whereas the phase shift represents a measure of the rotation angle. A comparison between the spatially resolved gain coefficients for σ^+ and σ^- pulse excitation is shown in Figure 49.44a; the analytically computed gain coefficient of a HB two-level system has also been plotted on the same graph, for reference. The maxima of the gain coefficients for both σ^+ and σ^- occur at the resonant wavelength. The circular dichroism, $\Delta A = G_R - G_L$, is therefore calculated as the average of the difference between the maximal gain coefficients for σ^+ and σ^- excitation (R and L subscripts, respectively) computed at several points along the nanotube. This yields a value for the circular dichroism of $0.083\ \mu m^{-1}$. By comparison, the absolute value of the circular dichroism of an artificial helicoidal bilayered structure varies in the range of 5–9 dB, which is equivalent to a linear amplitude gain/absorption coefficient of 0.58–$1.04\ \mu m^{-1}$. The atomic phase shift ($\beta_{R(L),x(y)}l$) and the average rotation angle are plotted in Figure 49.44b. The specific rotatory power per unit length is calculated using $\rho = \pi(n_L - n_R)/\lambda_0$, where n_L and n_R are the refractive indices for left and right circularly polarized light. Computing this value gives $\approx 2962.24°/mm$.

Although the circular dichroism is low compared to the artificial chiral photonic structures, the optical rotation, as computed from the phase shift, is enormous compared to other birefringent materials. The

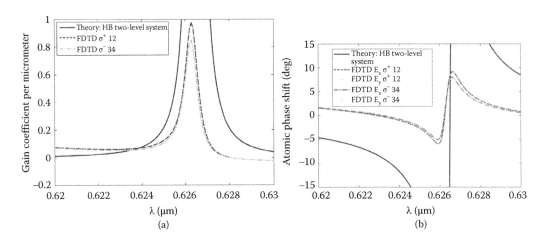

FIGURE 49.44 (a) Plot against wavelength of the spatially resolved gain coefficient per micron for a σ^+ (dashed curve) and σ^- (dash-dotted curve) circularly polarized ultrashort optical pulse, and the theoretical gain coefficient of a homogeneously broadened (HB) two-level system (solid curve) (Equation 49.30). The simulated structure is as in Figure 49.41. (b) Spectrum of the phase shift, averaged over the entire length of the structure, of the E_x (dashed curve) and E_y (dotted curve) electric-field components of σ^+ excitation, E_x (dash-dotted curve) and E_y (dotted curve) of σ^- optical excitation, and theoretical calculations of this phase shift (Equation 49.31), assuming a HB, resonant two-level system.

specific rotatory power of crystals varies widely from 2.24°/mm for $NaBrO_3$ to 522°/mm for $AgGaS_2$. Liquid substances exhibit much lower values of specific rotatory power—e.g., $\rho = -0.37°$/mm for turpentine ($T = 10°$, $\lambda = 589.3$ nm); $\rho = 1.18°$/mm for corn syrup, etc. Cholesteric liquid crystals and sculptured thin films exhibit large rotatory power in the visible spectrum: $\sim 1000°$/mm and $\sim 6000°$/mm, respectively, which is comparable with the calculated rotatory power. We should note that the dependence of the optical rotation angle on chirality opens up the exciting possibility to engineer this angle through manipulating the chirality of the structure.

49.7.2 Faraday Rotation and Magneto-Chiral Effects

In this section, we shall develop a theoretical model of the resonant, coherent nonlinear optical activity when a static magnetic field, $B_\parallel$, threads the nanotube (Figure 49.45).

We shall be interested in the Faraday effect, or the rotation of the polarization of a plane-polarized electromagnetic wave propagating in a medium permeated by a static magnetic field, **B**, oriented along the direction of propagation. In the presence of an axial magnetic field, the electronic band structure of a single CNT—and the electronic states near the bandgap edge in particular—change significantly, owing to the combined action of two effects: the spin-B interaction, resulting in Zeeman splitting of the energy levels [93–95], and the appearance of the Aharonov–Bohm phase in the wavefunction [83,96,97]. The two symmetric subbands at the K (or K') point of the BZ are degenerate at $B_\parallel = 0$ (Figure 49.45b). An applied magnetic field along the nanotube axis lifts this degeneracy by shifting the energy levels. As a result, the bandgap of one of the subbands (K') becomes larger while the bandgap of the other subband (K) becomes smaller [94] (Figure 49.45c).

Without loss of generality, we shall consider the electronic states near the bandgap at the K point in the BZ. Therefore, the overall effect will be a bandgap reduction. We consider and calculate separately the contributions to the bandgap from the Zeeman splitting and the Aharonov–Bohm effect.

The effect of the Aharonov–Bohm flux on the bandgap is to induce oscillations in its energy. The energy bandgap thus varies between zero and a fixed value with a period of the flux quantum, $\Phi_0 = h/e$ [83,93],

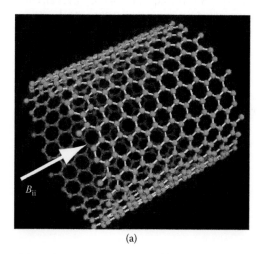
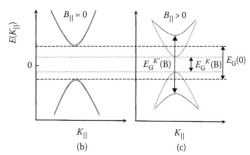

FIGURE 49.45 (a) Single (20, 10) chiral nanotube threaded by an axial magnetic field (a so-called Faraday configuration). (b) Magnetic energy bands in the absence of a magnetic field, $B_\parallel$; the energy subbands at the point K (K') are degenerate. (c) When an axial magnetic field is applied, the degeneracy is lifted and the energy bandgap of the subband at K' becomes larger, while that of the other subband at K diminishes.

resulting in periodic oscillations of the magneto-optical absorption spectra given by

$$
E_G(\Phi) = \begin{cases} 3E_G(0)\left|\dfrac{1}{3} - \dfrac{\Phi}{\Phi_0}\right|, & 0 \le \Phi/\Phi_0 \le 1/2 \\[2mm] 3E_G(0)\left|\dfrac{2}{3} - \dfrac{\Phi}{\Phi_0}\right|, & 1/2 \le \Phi/\Phi_0 \le 1, \end{cases} \tag{49.61}
$$

where $E_G(0) = \hbar\omega_0$ is the energy bandgap at zero magnetic field and ω_0 is the resonant transition frequency.

For a magnetic field $B = 8\,\text{T}$ with a flux Φ threading a $(5,4)$ nanotube with a diameter $0.61145\,\text{nm}$, the ratio $\Phi/\Phi_0 = 0.00057$, and therefore the first of the equations (49.61) above holds. Due to the orbital Aharonov–Bohm effect, this leads to an energy-level shift, or bandgap renormalization of $E_{AB} = 3.37\,\text{meV}$, corresponding to a resonant angular frequency, $\omega_{AB} = 5.12 \times 10^{12}\,\text{rad/s}$. At a fixed value of the static magnetic field, the orbital Aharonov–Bohm effect leads to a uniform shift in the energy levels, so the resonant transition frequency ω_0 is replaced by $\omega_0 - \omega_{AB}$.

In an external magnetic field, the energy levels near the bandgap of Figure 49.40b split and the spin degeneracy is lifted. The resulting energy-level system can be split into two reduced systems of levels, each of which represents a mirror image of the other. The symmetry is broken only by the fact that the allowed optical transitions are different in each case (see Figure 49.46).

The Zeeman splitting, or the spin-B interaction energy, is given by

$$
E_z = \mu_B g_e \sigma B_\parallel, \tag{49.62}
$$

where $\mu_B = \dfrac{e\hbar}{2m_e}$ is the Bohr magneton; the electron g-factor, g_e, is taken to be the same as that of pure graphite (≈ 2), $\sigma = \pm 1/2$ is the z-axis projection of the electron spin (spin-up/spin-down state), and m_e is the free electron mass.

We have performed simulations with magnetic fields in the range 2–100 T and found for the considered $(5,4)$ tube a nonmonotonous dependence of the optical rotation angle on the magnitude of **B**, with maximum rotation reached for magnetic fields in the range 8–10 T. We therefore selected $B_\parallel = 8\,\text{T}$ to demonstrate the maximum angle of Faraday rotation. Another reason for choosing this value is that it was the maximum magnetic field achievable by the static magnet available in the laboratory of our collaborators at the time when we performed these calculations. For $B_\parallel = 8\,\text{T}$, the Equation (49.62) gives an energy shift of $E_z \approx 0.46\,\text{meV}$, which corresponds to a Zeeman resonant frequency, $\omega_z = 7.026 \times 10^{11}\,\text{rad/s}$.

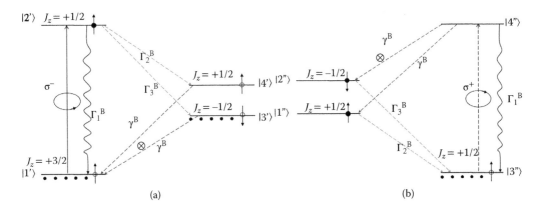

FIGURE 49.46 A four-level system at $B \neq 0$ splits into two nonsuperimposable reduced systems, involving levels energetically closest to the bandgap edge. (a) Left-handed reduced energy-level system excitable by a σ^- optical pulse. (b) Right-handed reduced energy-level system excitable by a σ^+ optical pulse. (Coherent and spontaneous transitions indicated as in Figure 49.40b. Forbidden dipole optical transitions are denoted by $\otimes$. In fact, there is a X-circle over the dashed lines in Figure 49.46, denoting forbidden transitions between levels).

The system Hamiltonians corresponding to the energy-level schemes displayed in Figure 49.46 are given by

$$
\hat{H}' = \hbar
\begin{pmatrix}
\omega_z & -\frac{1}{2}\left(\Omega_x - i\Omega_y\right) & 0 & 0 \\
-\frac{1}{2}\left(\Omega_x + i\Omega_y\right) & \omega_0 - \omega_z & 0 & 0 \\
0 & 0 & \Delta - \omega_z & 0 \\
0 & 0 & 0 & \Delta + \omega_z
\end{pmatrix}
\tag{49.63}
$$

and

$$
\hat{H}'' = \hbar
\begin{pmatrix}
\omega_0 - \omega_z & 0 & 0 & 0 \\
0 & \omega_0 + \omega_z & 0 & 0 \\
0 & 0 & \Delta + \omega_z & -\frac{1}{2}\left(\Omega_x + i\Omega_y\right) \\
0 & 0 & -\frac{1}{2}\left(\Omega_x - i\Omega_y\right) & \Delta + \omega_0 - \omega_z
\end{pmatrix},
\tag{49.64}
$$

respectively.

We employ our vector Maxwell-pseudospin master equations model for the four-level systems above. The simulation structure is the same as the one described for the $B = 0$ case. Note that due to the presence of a magnetic field, the energy levels $|1\rangle$, $|2\rangle$, $|3\rangle$ and $|4\rangle$ are split into two sets, which we choose to distinguish by adding a $'$ and $''$ to the notation; the first set (denoted by $'$) is accessed only using σ^- light, whereas the second (denoted by $''$) is accessible only via σ^+ light, due to the optical selection rules. We then resonantly excite the $|1'\rangle \rightarrow |2'\rangle$ and $|3''\rangle \rightarrow |4''\rangle$ transitions using, respectively, a left (σ^-) and right (σ^+) circularly polarized π-pulse of duration 60 fs. The pulse's central frequency is tuned in resonance with the transition frequency, $\omega_0 - \omega_{AB} - 2\omega_z$, and the pulse envelope is a Gaussian function. Owing to the combined effect of the Aharonov-Bohm and Zeeman energy-level shift, the dipole matrix element is modified. An estimate of the optical dipole matrix element in an axial magnetic field can be obtained from the theory developed in [88], taking into account the bandgap reduction at $B = 8\,T$, giving $\wp = 3.6205 \times 10^{-29}\,C \cdot m$. We recalculate the relaxation times using the above mentioned magnetic-field dipole coupling, thus obtaining the following relaxation rates: $\Gamma_1 = 2.914\,ns^{-1}$, $\Gamma_2 = 9.79\,ns^{-1}$ and $\Gamma_3 = 9.77\,ns^{-1}$. The intraband relaxation rate, $\gamma = 130\,fs^{-1}$, and the dephasing rates, $\Gamma_\mu = 800\,fs^{-1}$ and $\Gamma_{\mu-1} = 1.6\,ps^{-1}$, are taken to be the same as for the zero-field case. Note that the initial charge-carrier populations are different for excitation by the σ^- and σ^+ pulses: while in the former case the initial population is assumed to be equally distributed between the valence-band levels $\rho_{1'1'i} = \rho_{3'3'i} = 1/2$ nearest the bandgap edge, in the latter case the whole population is in the single valence-band state $\rho_{4'4'i} = 1$.

Similar to the zero-field case, the time evolution of the electric-field components and populations of all four levels are sampled at several points along the nanotube axis, and the magnetic circular dichroism spectra of the gain/absorption coefficients and phase shifts are calculated and displayed in Figure 49.47. Note that the resonance is shifted toward longer wavelengths due to the bandgap reduction. We should point out that the magnetic circular dichroism spectra at $B = 8\,T$ are quite distinct from the zero magnetic field ones for natural optical activity (Figure 49.44). While the σ^- polarized pulse is amplified during its propagation along the nanotube, the σ^+ polarized pulse is absorbed, resulting in a much larger net circular dichroism. The average value of the magnetic circular dichroism is $0.706\,\mu m^{-1}$, an order of magnitude larger than the natural circular dichroism. The different behavior of the calculated gain and absorption spectra under σ^- compared to σ^+ polarized optical pulse excitation is a direct consequence of the different energy-level configuration describing the two cases. While the energy-level system for the σ^- excitation is a four-level system, the one corresponding to the σ^+ excitation is a three-level Λ-system, due to level $|2''\rangle$ being completely decoupled from the rest of the levels as a result of the dipole optical selection rules. The calculated spectra in the latter case are, in fact, a signature of EIT and coherent population trapping

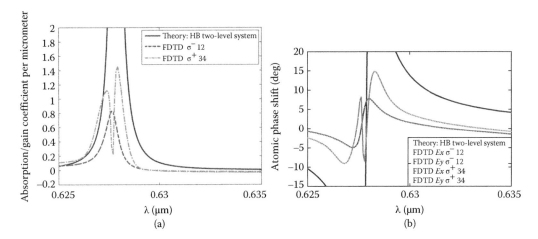

FIGURE 49.47 (a) Spectrum of the gain coefficient per micron, averaged spatially over the entire length of the structure, of the E_x and E_y electric-field vector components of σ^- (dashed curve) and σ^+ (dash-dotted curve) circularly polarized ultrashort optical excitation. The theoretical gain coefficient of a homogeneously broadened (HB), resonant two-level system (solid curve) at $B = 8\,\mathrm{T}$ is also shown. Note that, while the σ^- pulse is amplified during its propagation along the nanotube, the σ^+ pulse is absorbed, resulting in an absorption dip in the spectrum. (b) Spectrum of the phase shift, averaged spatially over the entire length of the structure, at $B = 8\,\mathrm{T}$ of the E_x (dashed curve) and E_y (dots) electric-field components under σ^- excitation, and of E_x (dash-dotted curve) and E_y (dots) under σ^+ circularly polarized ultrashort optical excitation. The phase spectra exhibit double-peaked refraction characteristic of EIT. The theoretical phase shift of a HB resonant two-level system (solid curve), is also given. (FDTD, finite-difference time-domain.)

effects in a three-level system [98], as will be explained subsequently. The absorption at resonance is close to zero, and the spectrum exhibits a dip (Figure 49.47a) similar to the absorption dip characteristic of EIT. The dip is caused by destructive interference between the excitation and emission fields, and leads to the trapping of the charge-carrier population in the ground state (hence, the term "coherent population trapping"). Whereas the phase-shift spectrum for a σ^- pulse excitation is of the type of a two-level atomic phase shift, the spectrum of the phase shift for σ^+ excitation is double peaked (Figure 49.47b). The latter is characteristic of the real part of the dielectric susceptibility in a three-level system exhibiting EIT. The predicted destructive interference in an external axial magnetic field after the passage of the ultrashort pulse is a direct consequence of the timescales of the processes involved in the relaxation dynamics and the phase shift which they introduce in the emission. The calculated average specific rotatory power in a magnetic field $B = 8\,\mathrm{T}$ is $-32580.4°/\mathrm{mm}$, corresponding to an average refractive-index anisotropy $(n_R - n_L)$ of 6.497 (the purpose of the minus sign in the rotatory power is to denote counterclockwise rotation of the field when viewed from a direction opposite to that of its propagation). This is nearly an order of magnitude greater than the natural optical rotation. We should note that the calculated rotation is a combined effect of the chirality of the nanotube and the magnetic field–induced rotation, and thus can be considered as an estimate for the magneto-chiral effect in a single nanotube. Therefore, our model can be used to study the magnetic field and chirality dependence of the optical rotation angle.

49.8 Conclusion

We have developed a general methodology for description of ultrashort optical-pulse interaction with open, discrete multilevel quantum systems, based on a self-consistent solution of the vector Maxwell's equations and equations for the time evolution of the quantum system in the real-vector representation of the density-matrix geometrical picture. An advantage of this theoretical method is that the coupled vector Maxwell-pseudospin equations can be generated on the fly for a quantum system with an arbitrary number of energy levels using the corresponding unique SU(N) Lie algebra generators. Unlike the widely accepted approach of using simplified two-level systems for description of the system dynamics, our approach allows to model multiple optical transitions and time-dependent gain effects that occur in a realistic quantum system. Our theoretical framework goes beyond the usual rate equations approach, which is unable to model and predict coherent quantum effects that could arise due to couplings and interplay between multiple optical transitions in the system. Most other approaches use the SVEA and RWA which consider only events occurring on a relatively slow timescale—such as the pulse envelope—but do not take account of the pulse's fast carrier wave. By contrast, the proposed approach allows to model rigorously ultrashort optical pulses down to a few optical cycles. Contrary to BPM, no approximations are made to the electromagnetic field equations and their solutions.

We have shown that the model can describe coherent propagation effects: self-induced and EIT, and self-localization effects, such as soliton and pattern formation in cavities. The method can be applied to a wide range of quantum-system geometries by imposing macroscopic boundary conditions and, hence, it can be used to simulate the performance of not only individual idealized devices, but whole interacting systems. We have shown that the model can successfully describe the complex dynamics of an open quantum system, and at the same time provides the capability to observe the feedback mechanisms—including non-Markovian memory effects—as they occur in space and time. In addition, within the Langevin model extension, we were able to model quantum-optical effects such as spontaneous emission. This approach can be used to simulate quantum correlations, nonclassical light generation and manipulation, and few-photon optical nonlinearities.

The model can describe interactions of matter with circularly/elliptically polarized optical pulses and was successfully applied to the problem of generation, manipulation, and readout of single spins and their dynamics. This success is critical for the development of quantum information technologies—currently a very hot research topic which could potentially revolutionize technological applications. Possible application areas include—but are not limited to—quantum computing and information technologies, such as spin-qubit manipulation in QDs, cold atoms, and NV centers in nanodiamond. CNT and arrays of aligned nanotubes have the potential to serve as functional elements in polarization-sensitive devices and polarization switches, as well as for metamaterial applications in the visible part of the electromagnetic spectrum. Predicted spatio-temporal coherent effects, such as cavity solitons, can be used as integrated low-loss optical switches and for all-optical information processing. Finally, the proposed methodology can be used as a design tool for future quantum-optical devices and for the advancement of quantum-photonic technologies.

References

1. R. J. Glauber. The quantum theory of optical coherence. *Physical Review*, 130:2529, 1963.
2. R. J. Glauber. *Quantum Optics and Electronics*. New York, NY: Gordon Breach, 1965.
3. R. Hanbury Brown and R. Q. Twiss. *Nature*, 177:27, 1956.
4. L. Mandel and S. Wolf. *Optical Coherence and Quantum Optics*. Cambridge University Press, 1995.
5. B. Lounis and M. Orrit. Single-photon sources. *Report on Progress in Physics*, 68:1129, 2005.
6. A. Einstein, B. Podolsky, and N. Rosen. Can quantum mechanical description of physical reality be considered complete? *Physical Review*, 47:777, 1935.
7. J. L. O'Brien, A. Furusawa, and J. Vuckovic. Photonic quantum technologies. *Nature Photonics*, 3:687, 2009.
8. T. M. Babinec, B. J. M. Hausmann, M. Khan, Y. Zhang, J. R. Maze, Ph. R. Hemmer, and M. LonÄar. A diamond nanowire single-photon source. *Nature Nanotechnology*, 5:195, 2010.
9. J. Claudon, J. Bleuse, N. S. Malik, M. Bazin, P. Jaffrennou, N. Gregersen, C. Sauvan, P. Lalanne, and J.-M. Gerard. A highly efficient single-photon source based on a quantum dot in a photonic nanowire. *Nature Photonics*, 4:174, 2010.

10. D. Dalacu, K. Mnaymneh, V. Sazonova, P. J. Poole, G. C. Aers, J. Lapointe, R. Cheriton, A. J. SpringThorpe, and R. Williams. Deterministic emitter-cavity coupling using a single-site controlled quantum dot. *Physical Review B*, 82:033301, 2010.

11. K. G. Lee, X. W. Chen, H. Eghlidi, P. Kukura, R. Lettow, A. Renn, V. Sandoghdar, and S. Götzinger. A planar dielectric antenna for directional single-photon emission and near-unity collection efficiency. *Nature Photonics*, 5:166, 2011.

12. S. Strauf, N. G. Stoltz, M. T. Rakher, L. A. Coldren, P. M. Petroff, and D. Bouwmeester. High-frequency single-photon source with polarization control. *Nature Photonics*, 1:704, 2007.

13. D. Loss and D. P. diVincenzo. Quantum computation with quantum dots. *Physical Review A*, 57:120, 1998.

14. B. Meier and B. P. Zakharchenya editors. *Optical Orientation*. North-Holland, Amsterdam: Elsevier, 1984.

15. W. S. Warren, H. Rabitz, and M. Dahleh. Coherent control of quantum dynamics: The dream is alive. *Science*, 259:1581, 1993.

16. K. De Greve, D. Press, P. L. McMahon, and Y. Yamamoto. Ultrafast optical control of individual quantum dot spin qubits. *Reports on Progress in Physics*, 76:092501, 2013.

17. M. Sargent III, M. O Scully, and W. E. Lamb Jr. *Laser Physics*. Reading, MA: Addison-Wesley Publishing Company Advanced Book Program, 1974.

18. S. L. McCall and E. L. Hahn. Self-induced transparency by pulsed coherent light. *Physical Review Letters*, 18:908, 1967.

19. S. Schneider, P. Borri, W. Langbein, U. Woggon, J. Förstner, A. Knorr, R. L. Sellin, D. Ouyang, and D. Bimberg. Self-induced transparency in InGaAs quantum-dot waveguides. *Applied Physics Letters*, 83:3668, 2003.

20. S. Hughes. Breakdown of the area theorem: Carrier-wave Rabi flopping of femtosecond optical pulses. *Physical Review Letters*, 81:3363, 1998.

21. D. Pinotsi and A. Imamoglu. Single photon absorption by a single quantum emitter. *Physical Review Letters*, 100:093603, 2008.

22. M. Sondermann, R. Maiwald, H. Konermann, N. Lindlein, U. Peschel, and G. Leuchs. Design of a mode converter for efficient light-atom coupling in free space. *Applied Physics B*, 89:489, 2007.

23. S. J. van Enk. Atoms, dipole waves, and strongly focused light beams. *Physical Review A*, 69:043813, 2004.

24. M. Brune, F. Schmidt-Kaler, A. Maali, J. Dreyer, E. Hagley, J. M. Raimond, and S. Haroche. Quantum Rabi oscillation: A direct test of field quantization in a cavity. *Physical Review Letters*, 76:1800, 1996.

25. K. M. Birnbaum, A. Boca, R. Miller, A. D. Boozer, T. E. Nortup, and H. J. Kimble. Photon blockade in an optical cavity with one trapped atom. *Nature*, 436, 87:87, 2005.

26. R. Young, R. M. Stevenson, P. Atkinson, K. Cooper, D. A. Ritchie, A. J. Shields. *Advances in Solid State Physics*, Vol. 46:55. Berlin: Springer Verlag, 2008.

27. O. Gazzano, S. Michaelis de Vasconcellos, C. Arnold, A. Nowak, E. Galopin, I. Sagnes, L. Lanco, A. Lemaître, and P. Senellart. Bright solid-state sources of indistinguishable single photons. *Nature communications*, 4:1425, 2012.

28. K. Srinivasan and O. Painter. Linear and nonlinear optical spectroscopy of a strongly coupled micro disk-quantum dot system. *Nature*, 450:862, 2007.

29. N. Le Thomas, U. Woggon, O. Schöps, M. V. Artemyev, M. Kazes, and U. Banin. Cavity QED with semiconductor nanocrystals. *Nano Letters*, 6:557, 2006.

30. M. Heiss, Y. Fontana, A. Gustafsson, G. Wust, C. Magen, D. D. O'Regan, J. W. Luo, B. Ketterer, S. Conesa-Boj, A. V. Kuhlmann, J. Houel, E. Russo-Averchi, J. R. Morante, M. Cantoni, N. Mazari, J. Arbiol, A. Zunger, R. J. Warburton, and A. Fontcuberta i Morral. Self-assembled quantum dots in a nanowire system for quantum photonics. *Nature Materials*, 12:439, 2013.

31. H. Kim, R. Bose, T. C. Shen, G. S. Solomon, and E. Waks. A quantum logic gate between a solid-state quantum bit and a photon. *Nature Photonics*, 7:373, 2013.

32. A. Faraon, I. Fushman, D. Englund, N. Stoltz, P. Petroff, and J. Vučković. Coherent generation of non-classical light on a chip via photon-induced tunnelling and blockade. *Nature Physics*, 4:859, 2008.

33. J. Kasprzak, S. Reitzenstein, E. A. Muljarov, C. Kistner, C. Schneider, M. Strauss, S. Hing, A. Forchel, and W. Langbein. Up on the Jaynes-Cummings ladder of a quantum-dot/microcavity system. *Nature Materials*, 9:304, 2010.

34. P. Michler, A. Kiraz, C. Becher, W. V. Schoenfeld, P. M. Petroff, L. Zhang, E. Hu, and A. Imamoglu. A quantum dot single-photon turnstile device. *Science*, 290:2282, 2000.

35. C. Santori, D. Fattal, J. Vuckovic, G. S. Solomon, and Y. Yamamoto. Indistinguishable photons from a single-photon device. *Nature*, 419:594, 2002.

36. E. Knill, R. Laamme, and G. J. Milburn. A scheme for efficient quantum computation with linear optics. *Nature*, 409:46, 2001.

37. L. Allen and J. H. Eberly. *Optical Resonance and Two-Level Atoms*. New York, NY: Wiley, 1975.

38. R. P. Feynman, F. L. Vernon, and R. W. Hellwarth. Geometrical representation of the Schrödinger equation for solving laser problems. *Journal of Applied Physics*, 28:49, 1957.

39. R. J. Cook and B. W. Shore. Coherent dynamics of N-level atoms and molecules. III. An analytically soluble periodic case. *Physical Review A*, 20:539, 1979.

40. J. N. Elgin. Semiclassical formalism for the treatment of three-level systems. *Physics Letters*, 80A:140, 1980.

41. M. Gell-Mann and Y. Neeman. *The Eightfold Way*. New York, NY: Benjamin, 1964.

42. F. T. Hioe and J. H. Eberly. N-level coherence vectors and higher conservation laws in quantum optics and quantum mechanics. *Physical Review Letters*, 47:838, 1981.

43. S. L. McCall and E. L. Hahn. Self-induced transparency. *Physical Review*, 183:457, 1969.

44. G. Slavcheva. Model for the coherent optical manipulation of a single spin state in a charged quantum dot. *Physical Review B*, 77:115347, 2008.

45. G. Slavcheva, J. M. Arnold, I. Wallace, and R. W. Ziolkowski. Coupled Maxwell-pseudospin equations for investigation of self-induced transparency effects in a degenerate three-level quantum system in two dimensions: Finite-difference time-domain study. *Physical Review A*, 66:063418, 2002.

46. A. Taove. *Computational Electrodynamics: The Finite-Difference Time-Domain Method*. Norwood, MA: Artech House, 2000.

47. R. W. Ziolkowski, J. M. Arnold, and D. M. Gogny. Ultrafast pulse interactions with two-level atoms. *Physical Review*, 52:3082, 1995.

48. B. Engquist and A. Majda. Absorbing boundary conditions for the numerical simulation of waves. *Mathematics of Computation*, 31:629, 1977.

49. G. Mur. Absorbing boundary conditions for the finite-difference approximation of the time-domain electromagnetic-field equations. *IEEE Transactions on Electromagnetic Compatibility*, 23:377, 1981.

50. A. Yariv. *Quantum Electronics*. New York, NY: Wiley, 1989.

51. G. M. Slavcheva, J. M. Arnold, and R.W. Ziolkowski. FDTD simulation of the nonlinear gain dynamics in active optical waveguides and semiconductor microcavities. *IEEE Journal of Selected Topics in Quantum Electronics*, 10(5):1052, 2004.

52. A. E. Siegman. *Lasers*. Mill Valley, CA: University of Science Books, 1986.

53. G. Slavcheva, J. M. Arnold, and R. W. Ziolkowski. Ultrashort pulse lossless propagation through a degenerate three-level medium in nonlinear optical waveguides and semiconductor microcavities. *IEEE Journal of Selected Topics in Quantum Electronics*, 9(3):929, 2003.

54. M. Blaaboer, G. Kurizki, and B. A. Malomed. Spatiotemporally localized solitons in resonantly absorbing Bragg reflectors. *Physical Review E*, 62:R57, 2000.

55. A. Kozhekin and G. Kurizki. Self-induced transparency in Bragg reflectors: Gap solitons near absorption resonances. *Physical Review Letters*, 74:5020, 1995.

56. A. Kozhekin, G. Kurizki, and B. A. Malomed. Standing and moving gap solitons in resonantly absorbing Bragg gratings. *Physical Review Letters*, 81:3647, 1998.

57. G. Slavcheva and O. Hess. Dynamical model of coherent circularly polarized optical pulse interactions with two-level quantum systems. *Physical Review A*, 72:053804, 2005.

58. A. I. Maimistov, A. M. Besharov, O. Elyutin, and Yu. M. Sklyarov. Present state of self-induced transparency theory. *Physics Reports*, 191:1–108, 1990.

59. G. Slavcheva and O. Hess. Spin-dependent dynamics of ultrafast polarised optical pulse propagation in coherent semiconductor quantum systems. *Physica Status Solidi C*, 3:2414, 2006.

60. G. Slavcheva and O. Hess. All-optical coherent control of spin dynamics in semiconductor quantum dots. *Optical and Quantum Electronics*, 31(12–14):973, 2007.

61. A. Wojs, P. Hawrylak, S. Fafard, and L. Jacak. Electronic structure and magneto-optics of self-assembled quantum dots. *Physical Review B*, 54:5604, 1996.

62. A. Greilich, R. Oulton, E. A. Zhukov, I. A. Yugova, D. R. Yakovlev, M. Bayer, A. Shabaev, A. L. Efros, I. A. Merkulov, V. Stavarache, D. Reuter, and A. Wieck. Optical control of spin coherence in singly charged (In,Ga)As/GaAs quantum dots. *Physical Review Letters*, 96:227401, 2006.

63. P. Chen, C. Piermarocchi, L. J. Sham, D. Gammon, and D. G. Steel. Theory of quantum optical control of a single spin in a quantum dot. *Physical Review B*, 69:075320, 2004.

64. A. Shabaev, A. L. Efros, D. Gammon, and I. A. Merkulov. Optical readout and initialization of an electron spin in a single quantum dot. *Physical Review B*, 68:201305(R), 2003.

65. A. V. Khaetskii, D. Loss, and L. Glazman. Electron spin decoherence in quantum dots due to interaction with nuclei. *Physical Review Letters*, 88:186802, 2002.

66. I. A. Merkulov, A. L. Efros, and M. Rosen. Electron spin relaxation by nuclei in semiconductor quantum dots. *Physical Review B*, 65:205309, 2002.

67. D. V. Bulaev and V. Loss. Spin relaxation and decoherence of holes in quantum dots. *Physical Review Letters*, 95:076805, 2005.

68. T. Flissikowski, I. A. Akimov, A. Hundt, and F. Henneberger. Single-hole spin relaxation in a quantum dot. *Physical Review B*, 68:161309(R), 2003.

69. S. Economou, R.-B. Liu, L. J. Sham, and D. G. Steel. Unified theory of consequences of spontaneous emission in a λsystem. *Physical Review B*, 71:195327, 2005.

70. M. Taylor, E. Harbord, P. Spencer, E. Clarke, G. Slavcheva, and R. Murray. Optical spin-filtering effect in charged InAs/GaAs quantum dots. *Applied Physics Letters*, 97:171907, 2010.

71. K. Kavokin. Fine structure of the quantum-dot trion. *Physica Status Solidi: A*, 195:592, 2003.

72. P. F. Braun, X. Marie, L. Lombez, B. Urbaszek, T. Amand, P. Renucci, V. K. Kalevich, K. V. Kavokin, O. Krebs, P. Voisin, and Y. Masumoto. Direct observation of the electron spin relaxation induced by nuclei in quantum dots. *Physical Review Letters*, 94:116601, 2005.

73. B. Eble, C. Testelin, P. Desfonds, F. Bernardot, A. Balocchi, T. Amand, A. Miard, A. Lemaître, X. Marie, and M. Chamarro. Hole-nuclear spin interaction in quantum dots. *Physical Review Letters*, 102:146601, 2009.

74. G. Slavcheva. Model of coherent optical spin manipulation through hot trion states in p-doped InAs/GaAs quantum dots. *ArXiv:1301.7018v1*, 2013.

75. T. Warming, E. Siebert, A. Schliwa, E. Stock, R. Zimmermann, and D. D. Bimberg. Hole-hole and electron-hole exchange interactions in single InAs/GaAs quantum dots. *Physical Review B*, 79:125316, 2009.

76. G. A. Narvaez, G. Bester, and A. Zunger. Carrier relaxation mechanisms in self-assembled (In, Ga) As/GaAs quantum dots: Efficient P→S auger relaxation of electrons. *Physical Review B*, 74:075403, 2006.

77. C. W. Gardiner and P. Zoller. *Quantum Noise*, 3rd ed. Berlin: Springer, 2004.

78. V. Weisskopf and E. Wigner. Berechnung der natürlichen linienbreite auf grund der diracschen lichttheorie. *Zeitschrift für Physik*, 63:54, 1930.
79. P. A. M. Dirac. The quantum theory of the emission and absorption of radiation. *Proceedings of the Royal Society London A*, 114:243, 1927.
80. G. Slavcheva and Ph. Roussignol. Nonlinear coherent magneto-optical response of a single chiral carbon nanotube. *New Journal of Physics*, 12:103004, 2010.
81. L. Yang, M. P. Anantram, J. Han, and J. P. Lu. Band-gap change of carbon nanotubes: Effect of small uniaxial and torsional strain. *Physical Review B*, 60:13874, 1999.
82. G. L. A. Rikken and E. Raupach. Observation of magneto-chiral dichroism. *Nature*, 390:493, 390.
83. A. Ajiki and T. Ando. Electronic states of carbon nanotubes. *Journal of the Physical Society of Japan*, 62:1255, 1993.
84. Y. Zhang, A. Chang, J. Cao, Q. Wang, W. Kim, Y. Li, N. Morris, E. Yenilmez, J. Kong, and H. Daia. Electric-field-directed growth of aligned single-walled carbon nanotubes. *Applied Physics Letters*, 79:3155, 2001.
85. L. Jin, C. Bower, and O. Zhou. Alignment of carbon nanotubes in a polymer matrix by mechanical stretching. *Applied Physics Letters*, 73:1197, 1998.
86. Ge. G. Samsonidze, A. Grüneis, R. Saito, A. Jorio, A. G. Souza Filho, G. Dresselhaus, and M. S. Dresselhaus. Interband optical transitions in left- and right-handed single-wall carbon nanotubes. *Physical Review B*, 69:205402, 2004.
87. J.-S. Lauret, C. Voisin, G. Cassabois, C. Delalande, Ph. Roussignol, O. Jost, and L. Capes. Ultrafast carrier dynamics in single-wall carbon nanotubes. *Physical Review Letters*, 90:057404, 2003.
88. E. L. Ivchenko and B. Spivak. Chirality effects in carbon nanotubes. *Physical Review B*, 66:155404, 2002.
89. F. Wang, G. Dukovic, L. E. Brus, and T. F. Heinz. Time-resolved fluorescence of carbon nanotubes and its implication for radiative lifetimes. *Physical Review Letters*, 92:177401, 2000.
90. H. Bao, X. Ruan, and T. S. Fisher. Optical properties of ordered vertical arrays of multi-walled carbon nanotubes from FDTD simulations. *Optics Express*, 18:6347, 2010.
91. V. N. Popov and L. Henrard. Comparative study of the optical properties of single-walled carbon nanotubes within orthogonal and nonorthogonal tight-binding models. *Physical Review B*, 70:115407, 2004.
92. V. N. Popov. Curvature effects on the structural, electronic and optical properties of isolated single-walled carbon nanotubes within a symmetry-adapted non-orthogonal tight-binding model. *New Journal of Physics*, 6:17, 2004.
93. J. Jiang, J. Ding, and D. Y. Xing. Zeeman effect on the electronic spectral properties of carbon nanotubes in an axial magnetic field. *Physical Review B*, 62:13209, 2000.
94. E. D. Minot, Y. Yaish, V. Sazonova, and P. L. McEuen. Determination of electron orbital magnetic moments in carbon nanotubes. *Nature*, 428:536, 2004.
95. F. L. Shuy, C. P. Chang, R. B. Chen, C. W. Chiu, and M. F. Lin. Magnetoelectronic and optical properties of carbon nanotubes. *Physical Review B*, 67:045405, 2003.
96. H. Ajiki. Magnetic-field effects on the optical spectra of a carbon nanotube. *Physical Review B*, 65:233409, 2002.
97. R. Saito, G. Dresselhaus, and M. S. Dresselhaus. Magnetic energy bands of carbon nanotubes. *Physical Review B*, 50:14698, 1994.
98. K. J. Boller, A. Imamoglu, and S. E. Harris. Observation of electromagnetically induced transparency. *Physical Review Letters*, 66:2593, 1991.

Mathematical Methods

50

Drift-Diffusion Models

Patricio Farrell

Nella Rotundo

Duy Hai Doan

Markus Kantner

Jürgen Fuhrmann

and

Thomas Koprucki

50.1 Introduction

The semiconductor technology is undeniably one of the most important branches in modern industry. Apart from its obvious significance to our daily lives, this technology is an excellent example of a broad collaboration among various disciplines. The development of novel technologies and devices has not only been driven by engineers, but also by physicists, mathematicians, and numerical analysts. On the one hand, rigorous mathematical models allow sophisticated predictions, which might be difficult to observe experimentally. On the other hand, numerical simulations have the potential to optimize device designs without the costly and time-consuming development of prototypes.

In 1950, van Roosbroeck introduced the fundamental semiconductor device equations [1] as a system of three nonlinearly coupled partial differential equations (PDEs). They describe the semiclassical transport of free electrons and holes in a self-consistent electric field using a drift-diffusion approximation. Since then, the so-called *van Roosbroeck system* (frequently also called *drift-diffusion system*) became the standard model to describe the current flow in semiconductor devices at a macroscopic scale. Typical devices modeled by these equations range from diodes and transistors to LEDs, solar cells, and lasers [2,3]. In recent years, the emergence of quantum nanostructures [4–7] and organic semiconductors [8–10] has invigorated the research activities in this area.

In this chapter, we focus on solving the van Roosbroeck system numerically. From a mathematical point of view, the challenge lies in the strong nonlinearities, the drift dominated nature of the underlying physics, the formation of internal and boundary layers as well as the need to accurately mirror qualitative physical properties, such as nonnegativity of carrier densities and consistency with thermodynamical principles. The finite difference scheme invented by Scharfetter and Gummel [11] was the first to deal appropriately with most of these difficulties for charge carrier densities described by Boltzmann statistics. Later on, Fichtner, Rose, and Bank [12] extented it to higher dimensions as a finite volume scheme, providing geometric flexibility. Selberherr combined a lot of these results along with his own contributions in his textbook [13] which has become one of the standard reference works in the field.

However, often the Boltzmann approximation is insufficient and one has to impose Fermi–Dirac statistics or other constitutive laws for the charge carrier densities. For this reason, we present recent attempts to generalize the Scharfetter–Gummel approach while still reflecting the physics correctly.

In order to enable readers not familiar with semiconductor simulations to quickly grasp the main points in the discretization process of the van Roosbroeck system, we start by studying a one-dimensional (1D) problem in Section 50.2. Even though we have to postpone some of the solid-state physics to the third section, we can already introduce many key concepts of the standard finite difference ansatz proposed by Scharfetter and Gummel, which we will interpret as a 1D finite volume scheme.

In Section 50.3, we introduce the full van Roosbroeck system in higher spatial dimensions along with its physical concepts. In Section 50.4, we look at its numerical solution. We present the two-point flux finite volume method for the stationary case, which allows to use the Scharfetter–Gummel scheme in higher dimensions and discuss its generalization to more general carrier statistics. The remaining sections are devoted to special numerical aspects: nonlinear solvers, mesh generation, time stepping, the correct computation of terminal currents, and the finite element method. In the final section, we discuss generalizations of the approach and its embedding into more complex physical models.

50.2 How to Get Started: Numerical Solution of the Stationary 1D Semiconductor Equations

This section serves as a quick introduction to convey the main idea behind the van Roosbroeck system and its discretization. The van Roosbroeck system consists of three equations modeling the charge carrier flow as well as the electrostatic potential distribution in a semiconductor device. However, since the physics behind these equations is rather cumbersome and the focus of this chapter are numerical methods, we omit some of these physical details for now and start with a relatively simple 1D problem. Thus, we can highlight the key elements (and difficulties!) in the discretization of the van Roosbroeck system. This approach may appeal to newcomers to the field of optoelectronics. Experienced readers may jump directly to Section 50.3, where we will dive into the physical meaning of this model.

Here, we wish to model the stationary state of a 1D semiconductor device consisting of a homogeneous material on an interval $\Omega = [0, L]$ with Ohmic contacts at each end. This configuration already covers the simulation of p-n and p-i-n junctions.

50.2.1 1D van Roosbroeck Model

The stationary 1D van Roosbroeck system consists of three nonlinear ordinary differential equations for the unknown electrostatic potential $\psi(x)$, the quasi-Fermi potential for electrons $\varphi_n(x)$ and the quasi-Fermi potential for holes $\varphi_p(x)$. It links Poisson's equation for the electric field to the continuity equations for the carrier densities as follows:

$$-\frac{d}{dx}\left(\varepsilon_s \frac{d}{dx}\psi\right) = q\Big(C + p(\psi, \varphi_p) - n(\psi, \varphi_n)\Big), \tag{50.1}$$

$$\frac{d}{dx}j_n = qR(\psi, \varphi_n, \varphi_p), \tag{50.2}$$

$$\frac{d}{dx}j_p = -qR(\psi, \varphi_n, \varphi_p). \tag{50.3}$$

The fluxes or current densities are given by

$$j_n = -q\mu_n n(\psi, \varphi_n)\frac{d}{dx}\varphi_n \quad \text{and} \quad j_p = -q\mu_p p(\psi, \varphi_p)\frac{d}{dx}\varphi_p. \tag{50.4}$$

The electron density n and the hole density p are related to the electric potential ψ as well as the quasi-Fermi potentials of electrons and holes via

$$n(\psi, \varphi_n) = N_c \exp\left(\frac{q(\psi - \varphi_n) - E_c}{k_B T}\right) \quad \text{and} \quad p(\psi, \varphi_p) = N_v \exp\left(\frac{q(\varphi_p - \psi) + E_v}{k_B T}\right). \tag{50.5}$$

We stress here that in (50.5) we have used the so-called Boltzmann approximation. In general, the exponentials will be replaced by some monotonically increasing *statistical distribution function* $\mathcal{F}$. This function will be discussed in greater detail in Section 50.3.

Next, we explain the constants and functions in the van Roosbroeck system. The elementary charge q and the Boltzmann constant k_B are universal constants. The (absolute) dielectric permittivity $\varepsilon_s = \varepsilon_0 \varepsilon_r$ is given as the product of the vacuum dielectric permittivity ε_0 and the relative permittivity of the semiconductor ε_r in static (low frequency) limit. The carrier mobilities μ_n and μ_p, the conduction and valence band densities of states N_c and N_v as well as the conduction and valence band-edge energies E_c and E_v are assumed to be constant even though in general they can vary with the material (e.g., due to abrupt or graded heterojunctions). The temperature T is also assumed to be constant; in general it can be space and even time dependent. The doping profile $C = C(x)$ describes material properties and the function $R(\psi, \varphi_n, \varphi_p)$ models the recombination and generation of electron–hole pairs. The electric field $\mathbf{E}$ is determined by the derivative (in higher dimensions the gradient) of the electrostatic potential

$$\mathbf{E} = -\frac{d}{dx}\psi\, \mathbf{e}_x,$$

where $\mathbf{e}_x$ denotes the unit normal along the x axis. A more detailed discussion on the physics behind these different quantities will be given in Section 50.3.

In the literature, authors frequently use a different set of unknowns for the van Roosbroeck system by replacing the quasi-Fermi potentials with the carrier densities n and p. However, we prefer to use the quasi-Fermi potentials for several reasons: they allow to easily model abrupt or graded heterostructures and appear naturally in the thermodynamic description of the van Roosbroeck system since the negative gradients of the quasi-Fermi potentials are the driving force of the currents [14]. Even mathematically, they are more beautiful as they make it possible to write the whole van Roosbroeck system in a gradient form [15] consistent with the principles of nonequilbrium thermodynamics [16]. Also carrier densities as opposed to quasi-Fermi potentials vary drastically in magnitude, which may lead to numerical difficulties.

The system of Equations 50.1 through 50.3 needs to be supplied with boundary conditions at $x = 0$ and $x = L$. For applied external voltages U_1 and U_2, we require Dirichlet boundary conditions at the *Ohmic contacts*, i.e.,

$$\psi(0) = \psi_0(0) + U_1, \qquad\qquad \psi(L) = \psi_0(L) + U_2, \qquad (50.6)$$

$$\varphi_n(0) = U_1, \qquad\qquad \varphi_n(L) = U_2, \qquad (50.7)$$

$$\varphi_p(0) = U_1, \qquad\qquad \varphi_p(L) = U_2. \qquad (50.8)$$

We discuss the meaning of the potential ψ_0 in Section 50.2.2.

50.2.2 Thermodynamic Equilibrium and Local Charge Neutrality

The goal is to enforce that the van Roosbroeck system of (50.1) through (50.3) is consistent with the *thermodynamic equilibrium*, which is a physical state defined by vanishing currents, namely

$$j_n = j_p = 0 \qquad \text{implying} \qquad \varphi_0 := \varphi_n = \varphi_p = \text{const.} \qquad (50.9)$$

Without loss of generality, we can set $\varphi_0 = 0$. We also discuss the thermodynamic equilibrium in Section 50.3.4. As a consequence, the three differential equations (50.1) through (50.3) reduce to the 1D nonlinear Poisson equation

$$-\frac{d}{dx}\left(\varepsilon_s \frac{d}{dx}\psi\right) = q\left(N_v \exp\left(\frac{E_v - q\psi}{k_B T}\right) - N_c \exp\left(\frac{q\psi - E_c}{k_B T}\right) + C\right). \qquad (50.10)$$

We supply it with Dirichlet boundary conditions (50.6) for zero voltages $U_1 = U_2 = 0$. The solution of Equation 50.10 with these boundary conditions is the *built-in potential* denoted with ψ_{eq}. In general, we cannot expect to find an analytic expression for ψ_{eq}.

Another important physical concept is *local charge neutrality*, characterized by a vanishing left-hand side in the Poisson equation. Combined with the equilibrium condition (50.9), this leads to

$$0 = q\left(N_v \exp\left(\frac{E_v - q\psi_0}{k_B T}\right) - N_c \exp\left(\frac{q\psi_0 - E_c}{k_B T}\right) + C\right), \qquad (50.11)$$

which can be solved for ψ_0 by solving a quadratic equation and omitting the unphysical solution, yielding

$$\psi_0(x) = \frac{E_c + E_v}{2q} - \frac{1}{2}U_T \log\left(\frac{N_c}{N_v}\right) + U_T \operatorname{arcsinh}\left(\frac{C}{2N_{intr}}\right), \qquad (50.12)$$

where the intrinsic carrier density N_{intr} and the *thermal voltage* U_T are given by

$$N_{intr}^2 = N_c N_v \exp\left(-\frac{E_c - E_v}{k_B T}\right) \qquad \text{and} \qquad U_T = \frac{k_B T}{q}. \qquad (50.13)$$

The potential ψ_0 serves two purposes. In Section 50.2.4, it will be used to obtain an initial guess for the nonlinear iteration to solve a discretized version of Equation 50.10. On the other hand, modeling (ideal) Ohmic contacts requires local charge neutrality. Hence, the boundary values in (50.6) are obtained by evaluating the function ψ_0 at the contacts $x = 0$ and $x = L$. The voltage difference between the boundary values $\psi_0(0)$ and $\psi_0(L)$ is called *built-in voltage*.

50.2.3 Discretizing the Poisson Problem

Our goal is to find a discrete solution to Equation 50.1. For this, we will use the *finite volume method* which naturally preserves many important physical properties. To see this, the reader may have a quick look in the table of contents and compare the next two sections devoted to the van Roosbroeck system and its numerical solution. It will become apparent that for many aspects the continuous model and its discretized counterpart are closely related.

We discretize the interval using a (possibly) nonuniform mesh of the form

$$0 =: x_1 < \cdots < x_N := L.$$

We associate each node x_k with a *control volume* ω_k defined by the interior subintervals

$$\omega_k = \left[x_{k-1,k}, x_{k,k+1}\right] \quad \text{for} \quad k = 2, \ldots, N-1$$

and the boundary subintervals

$$\omega_1 = \left[x_1, x_{1,2}\right] \quad \text{and} \quad \omega_N = \left[x_{N-1,N}, x_N\right].$$

The boundaries of the kth control volume are given by

$$x_{k-1,k} := \frac{x_{k-1} + x_k}{2} \quad \text{and} \quad x_{k,k+1} := \frac{x_k + x_{k+1}}{2}$$

for $k = 2, \ldots, N-1$. We point out that we have generated a disjoint partition of the interval, $\Omega = \bigcup_{k=1}^{N} \omega_k$. This discretization along with its notation is illustrated in Figure 50.1.

We introduce a finite volume discretization for Poisson's equation. Using the fundamental theorem of calculus, we integrate Equation 50.1 over the interior control volumes ω_k and obtain Gauss's law of electrodynamics:

$$-\varepsilon_s \left(\left.\frac{d\psi}{dx}\right|_{x_{k,k+1}} - \left.\frac{d\psi}{dx}\right|_{x_{k-1,k}} \right) = \int_{\omega_k} q\Big(C + p(\psi, \varphi_p) - n(\psi, \varphi_n)\Big)\, dx. \tag{50.14}$$

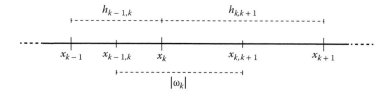

FIGURE 50.1 1D discretization around the kth node along with the corresponding notation.

This equation constitutes a balance law that can be interpreted as follows: the total charge in the kth control volume is given by the difference of the electric displacement $D = -\varepsilon_s \frac{d}{dx}\psi$ at its interfaces.

So far, we have derived an integral form of Poisson's equation for every interior control volume. We need approximations of the electric displacements at the boundary of each control volume. For this we employ central finite differences:

$$\frac{d\psi}{dx}\bigg|_{x_{k,k+1}} \approx \frac{\psi_{k+1} - \psi_k}{h_{k,k+1}}, \tag{50.15}$$

with the distance between two neighboring nodes given by

$$h_{k,k+1} := x_{k+1} - x_k.$$

Moreover, the integral over the net charge density is approximated by the rectangle method, evaluating the integrand at node x_k times the size of the control volume $|\omega_k| = x_{k,k+1} - x_{k-1,k}$. That is, the balance equation (50.14) is approximated by

$$-\varepsilon_s \left(\frac{\psi_{k+1} - \psi_k}{h_{k,k+1}} - \frac{\psi_k - \psi_{k-1}}{h_{k-1,k}} \right) = q\left(C_k + p(\psi_k, \varphi_{p;k}) - n(\psi_k, \varphi_{n;k}) \right) |\omega_k|$$

or, slightly rearranged,

$$0 = -\varepsilon_s \left(\frac{1}{h_{k,k+1}}\psi_{k+1} - \left[\frac{1}{h_{k,k+1}} + \frac{1}{h_{k-1,k}} \right] \psi_k + \frac{1}{h_{k-1,k}}\psi_{k-1} \right)$$
$$- q\left(C_k + p(\psi_k, \varphi_{p;k}) - n(\psi_k, \varphi_{n;k}) \right) |\omega_k|, \tag{50.16}$$

where $k = 2, \ldots, N-1$. The subindices denote evaluation at a given node. So, for example, ψ_k means $\psi(x_k)$ as well as $\varphi_{n;k}$ indicates $\varphi_n(x_k)$. We use this nodal notation frequently within this chapter. To incorporate the Dirichlet boundary conditions (50.6), we simply impose two more equations for the endpoints, namely

$$\psi_1 = \psi_0(0) + U_1 \quad \text{and} \quad \psi_N = \psi_0(L) + U_2. \tag{50.17}$$

50.2.4 Solving the Discrete Nonlinear Poisson Problem

Using the finite volume discretization scheme we have just described, we solve the nonlinear Poisson equation (50.16) with Dirichlet boundary conditions (50.17) in thermodynamic equilibrium ($U_1 = U_2 = 0$) to obtain the built-in potential ψ_{eq}. Recalling that there are $N-2$ interior and two boundary nodes, we end up with a system of N variables which can be summarized by the nonlinear discrete system

$$\mathbf{0} = \mathbf{F}_1(\boldsymbol{\psi}_{eq}),$$

where $\boldsymbol{\psi}_{eq} = (\psi_{eq}(x_k))_{k=1}^N$ is the vector of the nodal values. Solving this system is not straightforward due to its inherent nonlinearity. One approach to obtain a solution is Newton's method, which we discuss in more detail in Section 50.5. Newton's method converges quadratically—if one has a starting guess sufficiently close to the solution [17].

50.2.4.1 Solution Procedure: Nonlinear Poisson Problem

1. The easiest way to obtain a starting guess for the solution of Equation 50.16 with Dirichlet boundary conditions (50.17) is to neglect the left-hand side. That is, we use the local charge neutrality condition (50.11) for each node, which we can solve explicitly for ψ_0 via (50.12). With this function, we determine the components (nodal values) of our starting guess for the nonlinear Poisson problem $\boldsymbol{\psi}_0$ via

$$\boldsymbol{\psi}_0 = (\psi_0(x_k))_{k=1}^{N}.$$

2. Solving the discrete nonlinear Poisson problem via Newton's method with the starting guess $\boldsymbol{\psi}_0$ yields the built-in potential vector solution $\boldsymbol{\psi}_{\mathrm{eq}}$.

This solution procedure for the nonlinear Poisson equation for a p-i-n structure is illustrated in Figure 50.2. In the following, we address how to discretize the continuity equations. A task which turns out to be nontrivial.

50.2.5 Discretizing the Continuity Equations Using the Scharfetter–Gummel Scheme

Now, we turn our attention to the continuity equations (50.2) and (50.3). We proceed just as for Poisson's equation: we integrate over the control volumes $\omega_k = [x_{k-1,k}, x_{k,k+1}]$, apply the fundamental theorem of calculus for the left-hand sides and approximate the integral of the recombination term as before. For the interior control volumes we obtain

$$
\begin{aligned}
0 &= j_{n;k,k+1} - j_{n;k-1,k} - qR(\psi_k, \varphi_{n;k}, \varphi_{p;k})|\omega_k|, \\
0 &= j_{p;k,k+1} - j_{p;k-1,k} + qR(\psi_k, \varphi_{n;k}, \varphi_{p;k})|\omega_k|,
\end{aligned}
\tag{50.18}
$$

where $k = 2, \dots, N-1$. Similar to before, these equations constitute a balance law in integral form which we can interpret as follows: the carrier densities within the control volume ω_k change either due to in- and outflow or by recombination.

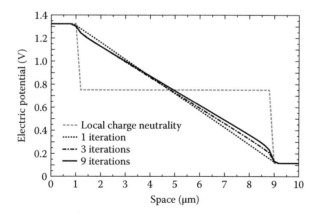

FIGURE 50.2 Convergence of the electric potential toward the thermodynamic equilibrium solution (built-in potential) for a GaAs p-i-n diode on an equidistant mesh with 51 nodes. The locally electroneutral initialization quickly converges to the built-in potential by means of a Newton iteration. The device has total length of $10\,\mu\mathrm{m}$ ($1\,\mu\mathrm{m}$ n-doped with $N_D = 10^{16}\,\mathrm{cm}^{-3}$, $8\,\mu\mathrm{m}$ intrinsic, $1\,\mu\mathrm{m}$ p-doped with $N_A = 10^{17}\,\mathrm{cm}^{-3}$. Standard parameters for GaAs at $T = 300\,\mathrm{K}$: $E_v = 0\,\mathrm{eV}$, $E_c = 1.424\,\mathrm{eV}$, $N_v = 9.0 \times 10^{18}\,\mathrm{cm}^{-3}$, $N_c = 4.7 \times 10^{17}\,\mathrm{cm}^{-3}$, and $\varepsilon_r = 12.9$.

The fluxes (50.18) need to be expressed in terms of our set of unknowns, namely the nodal values of the electrostatic and the quasi-Fermi potentials. Deriving a suitable approximation for the current expressions $j_{n;k,k+1}$ and $j_{p;k,k+1}$ describing the carrier exchange between neighboring control volumes ω_k and ω_{k+1} is crucial for numerical simulations. Since standard discretization techniques such as central finite differences may lead to unphysical behavior (see Section 50.2.7), the derivation of such an approximation requires particular care.

For the Boltzmann approximation (50.5), a numerically stable scheme was invented in 1969 by Scharfetter and Gummel [11]. In fact, it has been discovered independently several times [18–20]. Its derivation is based on integrating the flux along the interval $[x_k, x_\ell]$ between two neighboring discretization nodes x_k and x_ℓ under the assumption of constant current density and constant electric field. That is, either $\ell = k-1$ or $\ell = k + 1$. We illustrate the approach for the electron current density j_n.

In order to obtain a flux approximation, we will make two important assumptions, namely

1. The electric field along each edge is constant. This implies that the derivative of the electric potential $\frac{d\psi}{dx}$ along the edge $[x_k, x_\ell]$ is approximated by $\frac{\psi_\ell - \psi_k}{h_{k,\ell}}$, where again $\psi_\ell = \psi(x_\ell)$ and $\psi_k = \psi(x_k)$. This is consistent with the central difference approximation for the electric field in the Poisson equation, see (50.15).
2. The current density j_n is assumed to be constant along the edge.

The second assumption means that the current fluxes through the boundary from one control volume to another are conserved, which implies that the derivative of the flux along the interval $[x_k, x_\ell]$ vanishes. This means that we assume that there is no recombination along the edge between neighboring nodes. Hence, we obtain for the electron flux (the hole flux follows analogously)

$$\frac{d}{dx}j_n = \frac{d}{dx}\left(-q\mu_n N_c \exp\left(\frac{q(\psi(x) - \varphi_n(x)) - E_c}{k_B T}\right)\frac{d}{dx}\varphi_n(x)\right) = 0. \tag{50.19}$$

We supply the differential equation with the boundary conditions

$$\varphi_n(x_k) = \varphi_{n;k} \quad \text{and} \quad \varphi_n(x_\ell) = \varphi_{n;\ell}. \tag{50.20}$$

The values $\varphi_{n;k}$ and $\varphi_{n;\ell}$ have to be determined. But for now we assume they are given. Equations 50.19 and 50.20 constitute a two-point boundary value problem. Upon integrating once and multiplication by the integrating factor

$$M(x) = \exp\left(-\frac{\psi_\ell - \psi_k}{h_{k,\ell} U_T}x\right),$$

we obtain for the integration constant $j_{n;k,\ell}$ the equation

$$j_{n;k,\ell} M(x) = q\mu_n N_c U_T \frac{d}{dx}\left(M(x)\exp\left(\frac{q(\psi(x) - \varphi_n(x)) - E_c}{k_B T}\right)\right).$$

Integrating once more from x_k to x_ℓ and using the boundary conditons (50.20) yields after some algebraic manipulations the Scharfetter–Gummel expression for the current flux along the edge $[x_k, x_\ell]$ between neighboring control volumes

$$j_{n;k,\ell} = j_{n;k,\ell}(\psi_k, \psi_\ell, \varphi_{n;k}, \varphi_{n;\ell}) = -\frac{q\mu_n U_T}{h_{k,\ell}} \left[B\left(-\frac{\psi_\ell - \psi_k}{U_T}\right) N_c \exp\left(\frac{q(\psi_k - \varphi_{n;k}) - E_c}{k_B T}\right) \right.$$
$$\left. - B\left(\frac{\psi_\ell - \psi_k}{U_T}\right) N_c \exp\left(\frac{q(\psi_\ell - \varphi_{n;\ell}) - E_c}{k_B T}\right) \right],$$

(50.21)

where we have introduced the *Bernoulli function*

$$B(x) = \frac{x}{\exp(x) - 1}.$$

This function smoothly interpolates between zero and $-x$, see Figure 50.3. Remembering the relationships (50.5), the flux can be brought into the simpler form

$$j_{n;k,\ell} = -\frac{q\mu_n U_T}{h_{k,\ell}} \left[B\left(-\frac{\psi_\ell - \psi_k}{U_T}\right) n_k - B\left(\frac{\psi_\ell - \psi_k}{U_T}\right) n_\ell \right].$$

(50.22)

An important property of this scheme is its consistency with the thermodynamic equilibrium. That is, similarly to (50.9), we have

$$j_{n;k,\ell} = 0 \quad \text{for} \quad \varphi_{n;k} = \varphi_{n;\ell} = \text{const.}$$

(50.23)

For vanishing electric fields corresponding to $\delta\psi_{k,\ell} := \psi_\ell - \psi_k = 0$, the flux $j_{n;k,\ell}$ reduces to a purely diffusive flux

$$j_{n;k,\ell} = -q\left(-D_n \frac{n_\ell - n_k}{h_{k,\ell}}\right).$$

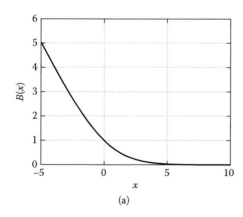
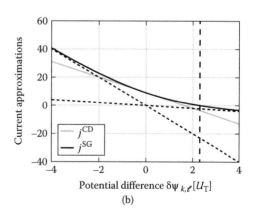

(a)   (b)

FIGURE 50.3 On the left the Bernoulli function $B(x)$ is shown. The picture on the right shows different current approximations for $q\mu_n U_T/h_{k,\ell} = 1$, $n_\ell = 10$, and $n_k = 1$ in terms of the potential difference $\delta\psi_{k,\ell}$. The dashed black line denotes the equilibrium potential difference. The dotted lines show the drift currents, i.e., the asymptotics of the Scharfetter–Gummel flux (50.24) for large positive and for large negative arguments. The central-difference flux is denoted with j^{CD} and the Scharfetter–Gummel flux with j^{SG}.

The diffusion constant D_n is defined via the so-called *Einstein relation* $D_n = \mu_n U_T$, which we discuss in Section 50.3.5 in more detail. For large electric fields, we obtain asymptotically a drift flux

$$j_{n;k,\ell} = \begin{cases} -q\mu_n n_k \frac{\psi_\ell - \psi_k}{h_{k,\ell}}, & \text{for } \delta\psi_{k,\ell} \gg +U_T, \\ -q\mu_n n_\ell \frac{\psi_\ell - \psi_k}{h_{k,\ell}}, & \text{for } \delta\psi_{k,\ell} \ll -U_T. \end{cases} \qquad (50.24)$$

These two extreme cases show that the Scharfetter–Gummel scheme naturally encompasses drift and diffusive currents depending on the strength of the local electric field, see Figure 50.3.

Using analogous considerations, we can derive a current expression for the holes

$$j_{p;k,\ell} = \frac{q\mu_p U_T}{h_{k,\ell}} \left[B\left(\frac{\psi_\ell - \psi_k}{U_T}\right) p_k - B\left(-\frac{\psi_\ell - \psi_k}{U_T}\right) p_\ell \right].$$

Since we have an approximation for the fluxes at the control volume boundary, we can derive from the discrete system (50.18) nonlinear equations for the interior nodes in terms of our set of unknowns $(\psi, \varphi_n, \varphi_p)$.

We stress once more that the previous discussion is only valid for the Boltzmann approximation made in (50.5). A discussion on more general distribution functions follows in Sections 50.3.5 and 50.4.2.

50.2.6 Solving the Full Discretized van Roosbroeck System

So far we have described how to derive a set of discrete and nonlinear equations from the 1D van Roosbroeck system ((50.1)–(50.3)), namely

$$0 = -\varepsilon_s \left(\frac{1}{h_{k,k+1}} \psi_{k+1} - \left[\frac{1}{h_{k,k+1}} + \frac{1}{h_{k-1,k}} \right] \psi_k + \frac{1}{h_{k-1,k}} \psi_{k-1} \right)$$
$$- q\Big(C_k + p(\psi_k, \varphi_{p;k}) - n(\psi_k, \varphi_{n;k}) \Big)|\omega_k|,$$

$$0 = j_{n;k,k+1}(\psi_k, \psi_{k+1}, \varphi_{n;k}, \varphi_{n;k+1}) - j_{n;k-1,k}(\psi_{k-1}, \psi_k, \varphi_{n;k-1}, \varphi_{n;k}) - qR(\psi_k, \varphi_{n;k}, \varphi_{p;k})|\omega_k|,$$

$$0 = j_{p;k,k+1}(\psi_k, \psi_{k+1}, \varphi_{p;k}, \varphi_{p;k+1}) - j_{p;k-1,k}(\psi_{k-1}, \psi_k, \varphi_{p;k-1}, \varphi_{p;k}) + qR(\psi_k, \varphi_{n;k}, \varphi_{p;k})|\omega_k|, \qquad (50.25)$$

with $k = 2, \ldots, N-1$. For the boundary nodes x_1 and x_N, the equations are given by the boundary conditions ((50.6)–(50.8)):

$$\psi_1 = \psi(0) = \psi_0(0) + U_1, \qquad \varphi_{n,1} = \varphi_n(0) = U_1, \qquad \varphi_{p,1} = \varphi_p(0) = U_1,$$

$$\psi_N = \psi(L) = \psi_0(L) + U_2, \qquad \varphi_{n,N} = \varphi_n(L) = U_2, \qquad \varphi_{p,N} = \varphi_p(L) = U_2.$$

Just as for the Poisson problem we assume there are N control volumes in total. We end up with a system of $3N$ variables which we can summarize by the nonlinear discrete system

$$0 = \mathbf{F}(\boldsymbol{\psi}, \boldsymbol{\varphi}_n, \boldsymbol{\varphi}_p) := \begin{pmatrix} \mathbf{F}_1(\boldsymbol{\psi}, \boldsymbol{\varphi}_n, \boldsymbol{\varphi}_p) \\ \mathbf{F}_2(\boldsymbol{\psi}, \boldsymbol{\varphi}_n, \boldsymbol{\varphi}_p) \\ \mathbf{F}_3(\boldsymbol{\psi}, \boldsymbol{\varphi}_n, \boldsymbol{\varphi}_p) \end{pmatrix},$$

where

$$\boldsymbol{\psi} = (\psi(x_k))_{k=1}^N, \qquad \boldsymbol{\varphi}_n = (\varphi_n(x_k))_{k=1}^N, \qquad \boldsymbol{\varphi}_p = (\varphi_p(x_k))_{k=1}^N.$$

Again, we employ Newton's method to solve this system. We aim to guarantee an appropriate starting guess by starting from thermodynamic equilibrium where the three equations reduce to one. The solution of the full van Roosbroeck system with an applied bias is then obtained via an iteration technique that we state here.

50.2.6.1 Solution Procedure: Full van Roosbroeck System

1. We assume thermodynamic equilibrium and set the constant quasi-Fermi potentials to $\varphi_0 = 0$. With the solution procedure introduced in Section 50.2.4, we obtain a solution vector $\boldsymbol{\psi}_{\mathrm{eq}}$ for the nonlinear Poisson problem with Dirichlet boundary conditions.
2. We denote with $\boldsymbol{\varphi}_n^0$ and $\boldsymbol{\varphi}_p^0$ the constant vectors, consisting of the equilibrium value for the quasi-Fermi potentials, namely

$$\boldsymbol{\varphi}_n^0 = (\varphi_0)_{i=1}^N \qquad \text{and} \qquad \boldsymbol{\varphi}_p^0 = (\varphi_0)_{i=1}^N.$$

We choose zero boundary voltages $U_1 = U_2 = 0$ and remember that we have obtained $\psi_0(0)$ and $\psi_0(L)$ via the charge neutrality condition. Thus we have determined the boundary conditions ((50.6)–(50.8)). Using the equilibrium vector $(\boldsymbol{\psi}_{\mathrm{eq}}, \boldsymbol{\varphi}_n^0, \boldsymbol{\varphi}_p^0)$ as a new starting guess, we can now solve the full discrete system (50.25) with these boundary conditions via Newton's method, yielding a solution $(\boldsymbol{\psi}^1, \boldsymbol{\varphi}_n^1, \boldsymbol{\varphi}_p^1)$. That is, we compute the equilibrium state of the full van Roosbroeck system using the equilibrium of the Poisson equation as a starting guess. We point out that solution and starting guess should agree up to machine precision by construction.
3. We slightly increase U_1 and U_2 to generate a "small" bias $\delta U = U_2 - U_1$. What small here means is not very precise and needs to be determined via trial and error. Now we solve the full discrete system (50.25) via Newton's method with the starting guess $(\boldsymbol{\psi}^1, \boldsymbol{\varphi}_n^1, \boldsymbol{\varphi}_p^1)$ to obtain a new solution $(\boldsymbol{\psi}^2, \boldsymbol{\varphi}_n^2, \boldsymbol{\varphi}_p^2)$.
4. Now we iterate the last step: gradually increase the bias and use the old solution as starting guess for the new one.

Figures 50.4 and 50.5 show the numerical results for the p-i-n structure, already introduced in Section 50.2.4. They depict the band edge energies, the quasi-Fermi energy levels $E_{F_n} = -q\varphi_n$ and $E_{F_p} = -q\varphi_p$ as well as the carrier densities for different bias values. Additionally, the current voltage characteristics for different n-dopings are shown.

50.2.7 Why Not Use Central Finite Differences for the Fluxes?

Thus far, we have discussed the Scharfetter–Gummel scheme to approximate the flux at the control volume interface. One might ask why not use a simpler flux approximation, for example, central finite differences. This question we address now.

For simplicity we assume a linear electrostatic potential. In this case

$$\frac{\mathrm{d}}{\mathrm{d}x}\psi = \psi(1) - \psi(0) =: \delta\psi.$$

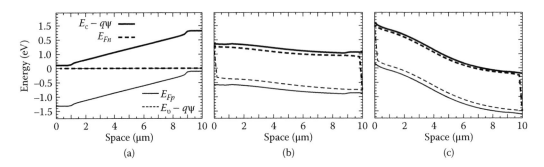

FIGURE 50.4 Computed band edge energy and quasi-Fermi levels for a GaAs p-i-n diode at different values of the applied voltage. (a) Thermodynamic equilibrium (off-state, 0V), (b) $U = 1.5$ V (flat band), (c) $U = 3.0$ V. Device has total length of 10 μm (1 μm n-doped with $N_D = 10^{16}$ cm^{-3}, 8 μm intrinsic, 1 μm p-doped with $N_A = 10^{17}$ cm^{-3}. The parameters used are standard parameters for GaAs as given in the caption of Figure 50.2 and $\mu_n = 8500$ cm^2 V^{-1} s^{-1}, $\mu_p = 400$ cm^2 V^{-1} s^{-1}, $\tau_n = \tau_p = 1$ ns, $C_n = C_p = 1 \times 10^{-29}$ cm^6 s^{-1}, and $r_{\text{spont}} = 1 \times 10^{-10}$ cm^3 s^{-1}.

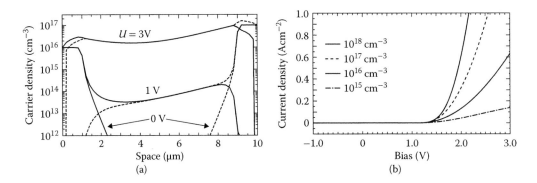

FIGURE 50.5 (a) Carrier densities in the p-i-n-diode at different applied voltages. Electron densities are shown as solid lines, holes are illustrated by dashed lines. One observes how the intrinsic region gets populated by free carriers while increasing the applied bias. (b) Current–voltage curves for the same device with altered values of the donor density N_D. Same parameters as in Figure 50.4.

We focus on the continuity equation for electrons on the domain $\Omega = [0, 1]$ which results in a linear differential equation of second order,

$$\frac{d}{dx} j_n = q D_n \frac{d^2}{dx^2} n - q \mu_n \delta \psi \frac{d}{dx} n = 0. \tag{50.26}$$

In this equation, we have introduced the drift-diffusion form of the (electron) flux

$$j_n = q \left(D_n \frac{d}{dx} n - n \mu_n \frac{d}{dx} \psi \right), \tag{50.27}$$

where we have used again the Einstein relation $D_n = \mu_n U_T$. We omit the physical derivation of the flux here. It is only important to know that this way of characterizing the flux is equivalent to Equation 50.4.

We have already seen how to solve this type of equation numerically by solving *local* two-point boundary value problems for the flux, which yields the Scharfetter–Gummel scheme (50.22). Of course, there are many different ways to discretize the flux (50.27) at a control volume interface $x_{k,\ell}$, needed in (50.18). For example, looking at Equation 50.27 one might consider replacing the derivative of the electron density $\frac{d}{dx} n$ with a central finite difference and the density n with an average. That is, at the control volume interface

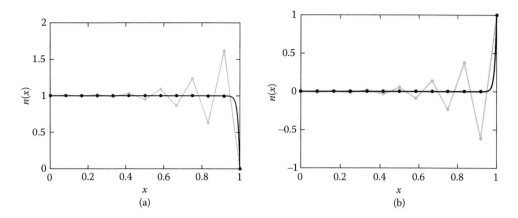

FIGURE 50.6 The exact solution (black) of Equation 50.26 and its central difference approximation (gray) on a uniform mesh with mesh width $h = 1/12$. The boundary conditions (50.28) lead to a numerical solution which violates the maximum principle (left). On the other hand, the boundary conditions (50.29) produce a numerical solution which violates the positivity of the density (right). Here $qD_n = 0.01$ and $q\mu_n\delta\psi = 1$.

we make the approximation

$$j^{CD}_{n;k,\ell} = qD_n \frac{n_\ell - n_k}{h_{k,\ell}} - q\mu_n\delta\psi \frac{n_\ell + n_k}{2} = -\frac{q\mu_n U_T}{h_{k,\ell}} \left(n_k - n_\ell + \frac{\delta\psi_{k,\ell}}{U_T} \frac{n_\ell + n_k}{2} \right).$$

This, however, is a very bad idea since the scheme will become unstable, resulting under certain conditions in large oscillations. To show this, we supply Equation 50.26 with two different types of boundary conditions:

$$n(0) = 1 \quad \text{and} \quad n(1) = 0 \tag{50.28}$$

as well as

$$n(0) = 0 \quad \text{and} \quad n(1) = 1. \tag{50.29}$$

Then, the discretization using central finite differences shows large oscillations for $\mu_n\delta\psi/D_n = 100$, which can be seen in Figure 50.6. For the first set of boundary conditions, we can see that the *maximum principle* is violated, i.e., the solution exceeds both boundary values [21,22]. For the second set of boundary conditions, the numerically computed density becomes negative. Both observations constitute huge violations of basic physical laws and are unacceptable in applications.

The problem here is that the diffusion constant D_n is small compared to the drift constant $\mu_n\delta\psi$. We point out, that in this special 1D case of vanishing recombination the Scharfetter–Gummel scheme, on the other hand, yields *exact* nodal values due to its construction. The Scharfetter–Gummel and the central difference flux are compared in Figure 50.3.

50.3 Van Roosbroeck System

In this section, we make several generalizations to the van Roosbroeck system ((50.1)–(50.3)). In particular, we make the model time dependent and allow higher spatial dimensions. Before stating the equations, we discuss the physical concepts we want to encode in the model.

As in the 1D case, the van Roosbroeck system consists of three equations: the Poisson equation and two continuity equations for the carrier densities. The Poisson equation describes the electric field $\mathbf{E} = -\nabla\psi$ which is generated by a scalar electric potential $\psi(\mathbf{x}, t)$ in the presence of a free charge carrier density. In a (doped) bipolar semiconductor device, this charge density has three ingredients: the density of free (negatively charged) electrons $n(\mathbf{x}, t)$ occupying the conduction band, the density of (positively charged) holes $p(\mathbf{x}, t)$ occupying the valence band and the density of ionized built-in dopants $C(\mathbf{x}) = N_{\mathrm{D}}^{+}(\mathbf{x}) - N_{\mathrm{A}}^{-}(\mathbf{x})$, where N_{D}^{+} denotes the density of singly ionized donor atoms and N_{A}^{-} is the density of singly ionized acceptor atoms.

The continuity equations (in differential form), on the other hand, model the flow of the charge carrier densities due to diffusion and drift governed by the self-consistent electric field, which is generated by the net charge density. Furthermore, the recombination and generation of electron–hole pairs influences the electron and hole densities.

50.3.1 The van Roosbroeck System of Equations

For a bounded spatial domain $\Omega \subset \mathbb{R}^d$ where $d \in \{1, 2, 3\}$, the van Roosbroeck system consists of three coupled nonlinear PDEs of the form

$$-\nabla \cdot \left(\varepsilon_s \nabla\psi\right) = q\left(p - n + C\right), \tag{50.30}$$

$$-q\partial_t n + \nabla \cdot \mathbf{j}_n = qR, \tag{50.31}$$

$$q\partial_t p + \nabla \cdot \mathbf{j}_p = -qR, \tag{50.32}$$

for $\mathbf{x} \in \Omega$ and $t \in [0, T]$. The current densities in Equations 50.31 and 50.32 are given by the usual expressions

$$\mathbf{j}_n = -q\mu_n n\nabla\varphi_n \quad \text{and} \quad \mathbf{j}_p = -q\mu_p p\nabla\varphi_p. \tag{50.33}$$

That is, the negative gradients of the quasi-Fermi potentials are the driving forces of the currents [14]. These relationships correspond directly to Equation 50.4.

Within the framework of effective mass approximation [2], the densities of free carriers in a solid are given by

$$n = N_c \mathcal{F}\left(\frac{q(\psi - \varphi_n) - E_c}{k_{\mathrm{B}}T}\right) \quad \text{and} \quad p = N_v \mathcal{F}\left(\frac{E_v - q(\psi - \varphi_p)}{k_{\mathrm{B}}T}\right). \tag{50.34}$$

Equations 50.34 indicate that the electric potential effectively leads to a bending of the energy band edge levels and thus a nonlinear, self-consistent coupling to the carrier densities is achieved. We assume a globally constant temperature for both carrier species and the crystal lattice.

The function $\mathcal{F}$ describes the occupation of energy states in the semiconductor. Since it plays an important role in the following discussion, we will define and discuss it later in Section 50.3.5. We only point out here that Equations 50.34 is analogous to Equations 50.5 when choosing for $\mathcal{F}$ the exponential function, the so-called *Boltzmann approximation*.

50.3.2 Initial and Boundary Conditions

The system ((50.30)–(50.32)) must be supplemented with initial and boundary conditions. The initial conditions at time $t = 0$ are given by the initial distributions ψ^I, φ_n^I and φ_p^I, i.e.,

$$\psi(\mathbf{x}, 0) = \psi^I(\mathbf{x}), \qquad \varphi_n(\mathbf{x}, 0) = \varphi_n^I(\mathbf{x}), \qquad \varphi_p(\mathbf{x}, 0) = \varphi_p^I(\mathbf{x}) \quad \text{for } \mathbf{x} \in \Omega.$$

Regarding the boundary conditions, we briefly discuss the most important case where the boundary of the domain Ω can be decomposed into Ohmic contacts, a gate contact and *artificial* interfaces, i.e.,

$$\partial\Omega = \left(\bigcup_{\alpha=1}^{N_O} \Gamma_{O,\alpha} \right) \cup \Gamma_G \cup \Gamma_A.$$

Semiconductor–metal interfaces, such as Ohmic contacts, are modeled by Dirichlet boundary conditions. For any Ohmic contact $\Gamma_{O,\alpha}$ with $\alpha = 1, \dots, N_O$, we set for all $\mathbf{x} \in \Gamma_{O,\alpha}$ and $t \in [0, T]$

$$\psi(\mathbf{x}, t) = \psi_0(\mathbf{x}) + U_\alpha(t), \tag{50.35}$$

$$\varphi_n(\mathbf{x}, t) = U_\alpha(t), \tag{50.36}$$

$$\varphi_p(\mathbf{x}, t) = U_\alpha(t), \tag{50.37}$$

where U_α denotes the corresponding externally applied contact voltage. The value ψ_0 at the boundary is defined by local charge neutrality similar to (50.11), where the exponential is now replaced by $\mathcal{F}$:

$$0 = N_v \mathcal{F}\left(\frac{E_v - q\psi_0(\mathbf{x})}{k_B T} \right) - N_c \mathcal{F}\left(\frac{q\psi_0(\mathbf{x}) - E_c}{k_B T} \right) + C(\mathbf{x}).$$

We just remark that, in general, this equation yields no analytical solution, and therefore its solution needs to be obtained by a nonlinear solver. The boundary conditions for the more advanced nonlinear semiconductor–metal interfaces (Schottky contacts) can be found in [13,23].

Gate contacts are modeled by Robin boundary conditions for the electrostatic potential and homogeneous Neumann boundary conditions for the quasi-Fermi potentials

$$\begin{aligned} \varepsilon_s \nabla\psi(\mathbf{x}, t) \cdot \boldsymbol{\nu} + \frac{\varepsilon_{ox}}{d_{ox}}(\psi(\mathbf{x}, t) - U_G(t)) &= 0, \\ \mathbf{j}_n(\mathbf{x}, t) \cdot \boldsymbol{\nu} = \mathbf{j}_p(\mathbf{x}, t) \cdot \boldsymbol{\nu} &= 0, \end{aligned} \qquad \text{for all } \mathbf{x} \in \Gamma_G \text{ and } t \in [0, T], \tag{50.38}$$

where ε_{ox} and d_{ox} are the absolute dielectric permittivity and the thickness of the oxide, respectively. The function $U_G(t)$ is the applied voltage at the outside of the insulating gate oxide at Γ_G. Here $\boldsymbol{\nu}$ denotes the outer normal vector on the interface.

On the remaining (artificial) interfaces, one typically imposes homogeneous Neumann boundary conditions (natural boundary conditions), namely

$$\nabla\psi(\mathbf{x}, t) \cdot \boldsymbol{\nu} = \mathbf{j}_n(\mathbf{x}, t) \cdot \boldsymbol{\nu} = \mathbf{j}_p(\mathbf{x}, t) \cdot \boldsymbol{\nu} = 0 \qquad \text{for all } \mathbf{x} \in \Gamma_A \text{ and } t \in [0, T]. \tag{50.39}$$

50.3.3 Recombination Processes

The net recombination rate R on the right-hand side of Equations 50.31 and 50.32 describes the radiative and nonradiative generation or recombination of carriers due to thermal excitation and various scattering effects. We assume that the recombination rate $R(n, p)$ is given by the sum of the most common processes, namely the Shockley–Read–Hall recombination R_{SRH}, the spontaneous radiative recombination R_{rad} and the Auger recombination R_{Auger}. All of these rates are of the form

$$R(n, p) = r(n, p)np\left(1 - \exp\left(\frac{q\varphi_n - q\varphi_p}{k_B T}\right)\right), \tag{50.40}$$

where $r(n, p)$ is a model-dependent generation-recombination rate [2,13,24]. In Figure 50.7, one finds the definitions of these rates together with a schematic illustration of the corresponding processes. For Boltzmann statistics, Equation 50.40 is equivalent to the widely used $R(n, p) = r(n, p)(np - N_{intr}^2)$, where N_{intr} is the intrinsic carrier density defined in (50.13).

50.3.4 Thermodynamic Equilibrium

The thermodynamic equilibrium is characterized by vanishing current fluxes $\mathbf{j}_n(\mathbf{x}, t) = \mathbf{j}_p(\mathbf{x}, t) = \mathbf{0}$. As a consequence, $R(n, p) = 0$ and the quasi-Fermi potentials assume constant values which due to (50.40) are equal:

$$\varphi_0 := \varphi_n = \varphi_p = \text{const.}$$

Without loss of generality, we set $\varphi_0 = 0$. Therefore, the van Roosbroeck system of (50.30)–(50.32) reduces to the nonlinear Poisson equation

$$-\nabla \cdot \varepsilon_s \nabla \psi = q\left(N_v \mathcal{F}\left(\frac{E_v - q\psi}{k_B T}\right) - N_c \mathcal{F}\left(\frac{q\psi - E_c}{k_B T}\right) + C\right). \tag{50.41}$$

We supply it with Dirichlet boundary conditions (50.35) for zero voltages $U_\alpha = 0$ for $\alpha = 1, \ldots, N_O$ and homogeneus Neumann boundary conditions (50.39) elsewhere. The solution of (50.41) with these boundary conditions defines the *built-in potential* denoted with ψ_{eq}.

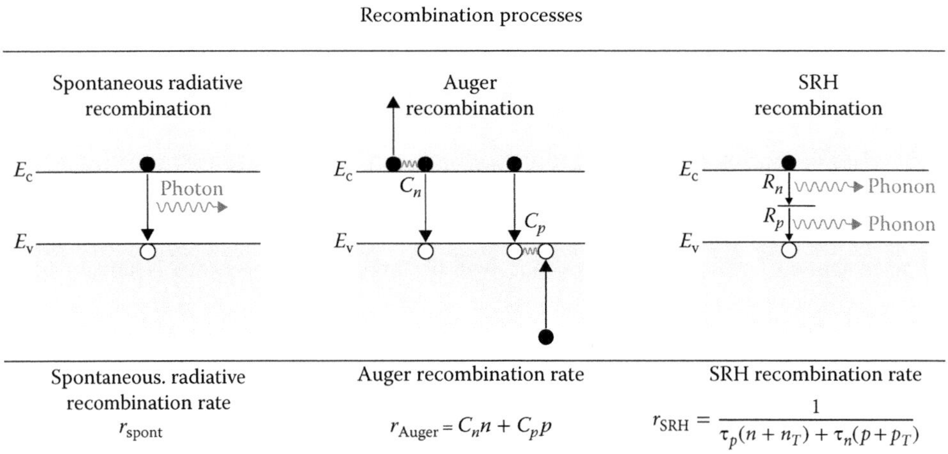

FIGURE 50.7 This figure illustrates three different recombination processes. The carrier life times τ_n, τ_p, the reference carrier densities n_T, p_T, and the coefficients r_{spont}, C_n, C_p are material-dependent parameters.

50.3.5 Non-Boltzmann Statistics and Generalized Einstein Relation

Now, we address the statistical distribution function $\mathcal{F}$ which appears in Equation 50.34. It describes the relationship between the density of the free carriers as well as the electrostatic potential and the quasi-Fermi potentials [2].

Assuming that the electrons in the conduction band are in quasi equilibrium, i.e., they are described by the quasi-Fermi level E_{F_n}, the electron density can be introduced as a convolution integral of the density of states $\mathrm{DOS}(E)$ with the Fermi–Dirac distribution function

$$n = \int_{-\infty}^{\infty} \mathrm{DOS}(E) \frac{1}{\exp\left(\frac{E - E_{F_n}}{k_B T}\right) + 1} dE. \tag{50.42}$$

It is possible to express this convolution as a product of the effective density of states N_c and a nondimensionalized statistical distribution function $\mathcal{F}$, i.e.,

$$n(\eta) = N_c \mathcal{F}(\eta). \tag{50.43}$$

We consider organic and inorganic semiconductors. For inorganic, three-dimensional (3D) bulk semiconductors with parabolic bands the statistical distribution function is given by the Fermi–Dirac integral of order 1/2 [2], which can be approximated by the Blakemore [25] or the Boltzmann distribution in the low density limit. For organic semiconductors, it is given by the Gauss–Fermi integral [10], which reduces to a Boltzmann type distribution function (up to some normalizing factor) in the low-density limit or a Blakemore distribution function for vanishing variance σ. The latter corresponds to a δ-shaped density of states [10,26], describing a single electronic transport level. See Figure 50.8 for definitions and a relationship between these functions and Figure 50.9 for a graphical representation.

For inorganic, 3D bulk semiconductors assuming parabolic bands with effective mass m_e^* the effective density of states is given by

$$N_c = 2 \left(\frac{m_e^* k_B T}{2\pi\hbar^2}\right)^{3/2},$$

where T is the temperature and $\hbar$ the Planck constant. For organic semiconductors, the effective density of states N_c is given by the density of transport states N_t.

Introducing a general relationship between carrier density as well as quasi-Fermi and electrostatic potential via Equation 50.34 has quite a few implications. The nonlinear Poisson equation (50.41) becomes more complicated. In Section 50.2.4, we have studied how to design useful starting guesses for the Boltzmann approximation. For general distribution functions $\mathcal{F}$, it is no longer possible to obtain an explicit expression for ψ_0, like we have achieved in Equation 50.12. One rather needs to solve a nonlinear equation, for example, via Newton's methods at a given point $\mathbf{x}$. For highly doped regions, the Boltzmann approximation strongly overestimates the carrier densities if $\eta > 0$, see Figure 50.9. This implies that the more realistic Fermi–Dirac distribution leads to a higher built-in voltage. Moreover, from a numerical point of view, the Scharfetter–Gummel method needs to be adjusted. We discuss this in Section 50.4.2.

Finally, the distribution function $\mathcal{F}$ influences the ratio between diffusion and drift. To see this, we write the currents in *drift-diffusion* form. Using the relations for the carrier densities (50.34), we obtain

$$\mathbf{j}_n = -q\mu_n n \nabla\psi + qD_n\nabla n, \qquad \mathbf{j}_p = -q\mu_p p \nabla\psi - qD_p\nabla p.$$

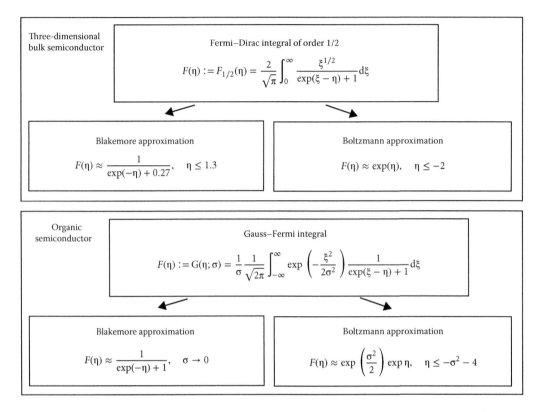

FIGURE 50.8 This figure shows two important choices of the statistical distribution function: for three-dimensional bulk semiconductors and for organic semiconductors [10] described by a Gaussian density of states with variance σ. Both can be approximated using the Blakemore or Boltzmann distribution function.

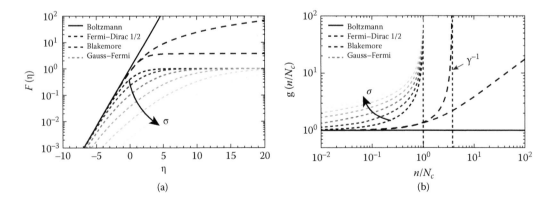

FIGURE 50.9 (a) Some frequently considered distribution functions for the density of free carriers. For the Blakemore and the Gauss–Fermi functions one obtains a saturation of the carrier density at high values of η. (b) Illustration of the diffusion enhancement factors for the distribution functions shown in (a). In the case of the Blakemore and the Gauss–Fermi functions the saturation values of the carrier density directly lead to a divergent behavior of the nonlinear diffusion factor.

The diffusion coefficients D_n and D_p are connected to the carrier mobilities by a *generalized* Einstein relation

$$\frac{D_n}{\mu_n} = \frac{k_B T}{q} g\left(\frac{n}{N_c}\right), \qquad \frac{D_p}{\mu_p} = \frac{k_B T}{q} g\left(\frac{p}{N_v}\right)$$

(50.44)

with a density-dependent nonlinear factor

$$g(\xi) = \xi \left(\mathcal{F}^{-1}\right)'(\xi),$$

leading in general to a nonlinear diffusion coefficient. For the Boltzmann approximation, we immediately see that $g(\xi) = 1$, which gives the classical Einstein relation $D_n = k_B T \mu_n / q$. For non-exponential distribution function and $\xi \geq 1$, however, we note $g(\xi) \geq 1$, see [27]. For this reason, we call this factor *diffusion enhancement* as proposed in [9]. The dependence of the diffusion enhancement g on the density is depicted in Figure 50.9 for various distribution functions.

50.3.6 Free Energy and Dissipation Rate

We consider consistency with fundamental principles of thermodynamics to be a quality measure for models describing natural processes. So it is of significant interest to study the consistency for both continuous and discretized models. Due to our simplifying assumption of constant temperature, we cannot have energy conservation in the proper sense. The second law of thermodynamics for nonequilibrium processes [16,28] requires nonnegativity of the local entropy production which, multiplied by the temperature T, gives the *dissipation rate*.

For the sake of readability, we discuss these concepts for Boltzmann statistics [29] here. See [30] and [31] for more general statistics. For a triple (ψ, n, p) and a thermodynamic equilibrium solution $(\psi_{eq}, n_{eq}, p_{eq}) := (\psi_{eq}, n(\psi_{eq}), p(\psi_{eq}))$ of the full van Roosbroeck system the *free energy* is defined as

$$\mathbb{F}(\psi, n, p) = \int_\Omega \left(n \log \frac{n}{n_{eq}} - n + p \log \frac{p}{p_{eq}} - p + n_{eq} + p_{eq} \right) dx$$

$$+ \int_\Omega \frac{\varepsilon_s}{2} \left| \nabla \left(\psi - \psi_{eq}\right) \right|^2 dx + \frac{1}{2} \int_{\Gamma_G} \frac{\varepsilon_{ox}}{d_{ox}} \left(\psi - \psi_{eq}\right)^2 ds.$$

(50.45)

For a transient solution $(\psi(t), n(t), p(t))$ of $((50.30)-(50.32))$, the function $\mathbb{L}(t) = \mathbb{F}(\psi(t), n(t), p(t))$ decays exponentially as t tends to infinity and one has

$$\mathbb{L}(t) = \mathbb{L}(0) - \int_0^t \mathbb{D}(\tau) d\tau,$$

where $\mathbb{D}(t)$ is the nonnegative dissipation rate [29]

$$\mathbb{D}(t) = \int_\Omega \left(n\mu_n |\nabla \varphi_n|^2 + p\mu_p |\nabla \varphi_p|^2 + r(n, p)(np - 1) \log(np) \right) dx \geq 0$$

(50.46)

depending on t via the time evolution of n and p. This result confirms the consistency with the second law of thermodynamics. Incidentally, the function $\mathbb{L}$ is a Lyapunov function, allowing in certain situations to prove the global stability of the thermodynamical equilibrium. Furthermore, it can be used as a tool to prove uniqueness of solutions to the system $((50.30)-(50.32))$. This leads us naturally to the following subsection.

50.3.7 Existence and Uniqueness Results

When modeling complex physical phenomena, it is often necessary to simplify the underlying physical model. Therefore, a sound mathematical investigation is necessary to assess the implications of such simplifications. Take, for example, the electrostatic potential of a device. Physically, we know that such a potential exists. Hence, a sound model should guarantee its existence—also from a mathematical point of view. If the existence cannot be shown, then the model does not represent the physical world accurately. Apart from existence, uniqueness of the solution is often desirable as well as continuous dependence on the initial data.

The mathematical technique used to prove existence and uniqueness depends on the device geometry and the model. The first existence result on the van Roosbroeck system was shown by Mock [32]. Since then several results have been obtained by Gajewski and Gröger, we refer the interested reader to [33,34]. In [34], the key tool to show the existence and uniqueness of the time-dependent system is based on finding a Lyapunov function. Moreover, Gajewski and Gröger presented the first result which considered Fermi–Dirac statistics instead of Boltzmann statistics. There are other important existence results. We would like to mention explicitly the results studied by Markowich see, for example, [35], Jüngel [36], and Jerome [37].

We would like to show the reader that these results are not of purely analytical interest but can also be used to design numerical methods. Gummel's method [38], for example (see Section 50.5), is based on the same fixed point iteration technique used to obtain the existence result which we discuss next. The connection to numerics can be found in [37]. We would like to give the reader an idea regarding the proof of a standard result without going into the tricky details. We follow ideas from Markowich's textbook [39].

We consider the stationary van Roosbroeck system with Boltzmann statistics. Using a scaling that can be found in [39, Section 2.4] and making a change of variables, we can rewrite the steady-state van Roosbroeck system in the following way:

$$-\lambda^2 \Delta \psi = \delta^2 e^{-\psi} v - \delta^2 e^{\psi} u + C, \tag{50.47}$$

$$\nabla \cdot (\mu_n e^{\psi} \nabla u) = \tilde{R}, \tag{50.48}$$

$$\nabla \cdot (\mu_p e^{-\psi} \nabla v) = \tilde{R}. \tag{50.49}$$

The densities n and p are related to the so-called *Slotboom variables* u and v via $n = \delta^2 e^{\psi} u$ and $p = \delta^2 e^{-\psi} v$. In Equations 50.47 through 50.49, the parameter λ denotes the normalized characteristic Debye length of the device, δ^2 indicates the scaled intrinsic carrier density and $\tilde{R}$ is the scaled recombination term. It is of the form $\tilde{R} = c(\psi, u, v)(uv - 1)$ with $c(\psi, u, v) > 0$. The chosen scaling and change of variables make it possible to rewrite the boundary conditions. We consider only Ohmic contacts and homogeneous Neumann boundaries.

In order to prove the existence of solutions for the system ((50.47)–(50.49)) equipped with suitable boundary conditions, we need to make several technical assumptions. We refer the interested reader to [39] where these assumptions are well explained as well as physically and mathematically justified. Here, we would rather like to focus on the key idea of the proof which is exploited numerically, omitting mathematical detail. The existence proof is based on an iteration scheme which considers the Poisson equation (50.47) decoupled from the continuity equations (50.48) and (50.49). The proof consists of the following steps:

1. We fix some $u_0, v_0 > 0$, and consider the semilinear elliptic problem

$$-\lambda^2 \Delta \psi = \delta^2 e^{-\psi} v_0 - \delta^2 e^{\psi} u_0 + C, \quad \text{in } \Omega, \qquad \text{(plus BC for } \psi \text{ on } \partial\Omega\text{)}. \tag{50.50}$$

Using standard analytical results for semilinear elliptic equations (namely the Leray–Schauder fixed point theorem and the maximum principle [40]) we prove that there exists a unique solution of the problem (50.50) that we denote with ψ_1.

2. We insert the solution ψ_1 of Equation 50.50 into the decoupled linear elliptic equations

$$\nabla \cdot (\mu_n e^{\psi_1} \nabla u) = c(\psi_1, u_0, v_0)(uv_0 - 1), \text{ in } \Omega, \qquad \text{(plus BC for } u \text{ on } \partial\Omega), \qquad (50.51)$$

$$\nabla \cdot (\mu_p e^{-\psi_1} \nabla v) = c(\psi_1, u_0, v_0)(u_0 v - 1), \text{ in } \Omega, \qquad \text{(plus BC for } v \text{ on } \partial\Omega). \qquad (50.52)$$

We point out that in the right-hand side of Equation 50.51 the rate c depends only on ψ_1 determined from the previous step and on the fixed u_0, v_0. The other term, $uv_0 - 1$, depends linearly on the unknown u. Analogous considerations are valid for Equation 50.52. Physically, this means that in the factor describing relaxation to equilibrium, $uv - 1$, we freeze the hole density, more precisely the Slotboom variable v_0, in the continuity equation for the electrons and vice versa. Thanks to the nonnegativity of $c(\psi_1, u, v)$ there exist unique solutions u_1 and v_1 to Equations 50.51 and 50.52. This is ensured by standard results for linear elliptic equations (see, e.g., [40]).
3. Based on the first two steps, we can now define a map H which maps (u_0, v_0) onto (u_1, v_1). This map is known as *Gummel map*. It is possible to prove that this map has a fixed point (u^*, v^*) which determines a (weak) solution (ψ^*, u^*, v^*) of the coupled system ((50.47)–(50.49)). The electrostatic potential ψ^* that solves Poisson's equation (50.47) can be determined using the first step by substituting (u_0, v_0) with (u^*, v^*).

Finally, we discuss the uniqueness of solutions to the van Roosbroeck system. It is well known that for the steady-state system, the uniqueness of the solution cannot be shown without additional assumptions, for example, on the applied voltage. This is no surprise, as in fact, some semiconductor devices (e.g., thyristors [41]) are designed to have multiple steady states. However, for a device in thermal equilibrium (see Section 50.3.4), we have $u = v = 1$. In this case, there is a unique function which satisfies the Poisson equation (50.47). Hence, the solution $(\psi_e, 1, 1)$ is the unique equilibrium solution of the system ((50.47)–(50.49)), see [32]. For sufficiently small bias voltages, the uniqueness of the solution is shown in [39] under some specific assumptions on the recombination rates. All recombination mechanisms appearing in Figure 50.3.3 fulfill these assumptions. However, impact ionization rates, for example, are excluded.

50.3.8 Maximum Principle

An important mathematical tool, often used for proofs similar to the ones in the previous section, is the maximum principle. Intuitively, the maximum principle states that if the domain is bounded and the right-hand side of the elliptic equation is positive, then the maximum of the solution is attained on the boundary of the domain [40].

50.4 Discretizing the van Roosbroeck System

In this section, we introduce a method for discretizing the van Roosbroeck system which is close to the physicist's approach to derive PDEs based on a subdivision of the computational domain into *representative elementary volumes* or *control volumes*. The *two-point flux* finite volume method described here can be seen as a straightforward generalization, preserving the properties of the 1D Scharfetter–Gummel scheme in higher dimensions. Figure 50.10 shows a 2D simulation. The 2D variant of this approach was introduced as *box method* in [12]. Historically, it goes back to [42]. The 3D variant of this method was probably first investigated in [29,43].

The method has two main ingredients: a geometry-based approach to obtain a system describing communicating control volumes and a consistent description of the fluxes between two adjacent control volumes. These are discussed in the following two subsections. We finish this section by describing various properties of the finite volume scheme.

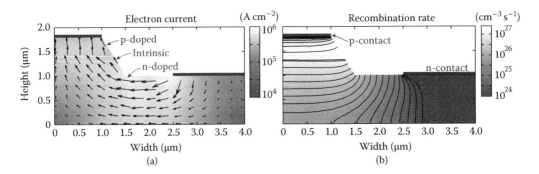

FIGURE 50.10 Exemplary results of a 2D simulation. The example is inspired by a ridge-waveguide laser with Ohmic contacts on the top of a mesa and at the side. For simplicity, we restrict to a homogeneous material which is here GaAs with the parameters as given in Figure 50.2. Within the mesa, an intrinsic domain is enclosed by a p-doped top-layer ($N_A = 2 \times 10^{18}$ cm^{-3}) and the n-doped substrate ($N_D = 2 \times 10^{18}$ cm^{-3}). The boundaries of the doped domains are shown in the pictures as dashed lines. (a) At an applied voltage of 1.5 V (flat band conditions) a significant electron current flow can be observed. The absolute value of the current density is shown by the gray scale, the arrows indicate the direction of particle flow (in the case of electrons the current density vector points in the opposite direction). (b) Plot of the total recombination rate at the same bias. The maximal recombination rate is observed in the vicinity of the intrinsic domain.

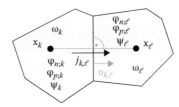

FIGURE 50.11 Two adjacent control volumes ω_k and ω_ℓ with corresponding data.

50.4.1 Finite Volume Method in Higher Dimensions

We partition the domain into N nonintersecting, convex polyhedral control volumes ω_k such that $\Omega = \bigcup_{k=1}^{N} \omega_k$. We associate with each control volume a node $\mathbf{x}_k \in \omega_k$. If the control volume intersects with the boundary of our domain, we demand that the node lies on the boundary, that means $\mathbf{x}_k \in \partial\Omega \cap \omega_k$. We assume that the partition is *admissible* [44], i.e., the *edge* $\overline{\mathbf{x}_k \mathbf{x}_\ell}$ with length $h_{k,\ell}$ is orthogonal to $\partial\omega_k \cap \partial\omega_\ell$. Thus, the normal vectors to $\partial\omega_k$ can be calculated by $\boldsymbol{\nu}_{k,\ell} = (\mathbf{x}_\ell - \mathbf{x}_k)/\|\mathbf{x}_\ell - \mathbf{x}_k\|$. See Figure 50.11 for details.

The set of all nodes and control volumes is called the *mesh* of the domain Ω. For a discussion of constructive ways to obtain such mesh partitions, see Section 50.6. In order to keep the equations simple, we introduce two abbreviations,

$$\eta_n\left(\psi, \varphi_n\right) = \frac{q\left(\psi - \varphi_n\right) - E_c}{k_B T} \quad \text{and} \quad \eta_p\left(\psi, \varphi_p\right) = \frac{E_v - q\left(\psi - \varphi_n\right)}{k_B T},$$

and assume that we have only a single gate contact with gate voltage U_G. We integrate the van Roosbroeck system ((50.30)–(50.32)) over ω_k and apply Gauss's divergence theorem, resulting in the integral equations

$$-\int_{\partial\omega_k} \varepsilon_s \nabla\psi \cdot \boldsymbol{\nu} \mathrm{d}s + \int_{\partial\omega_k \cap \Gamma_G} \frac{\varepsilon_{ox}}{d_{ox}}(\psi - U_G)\mathrm{d}s = q\int_{\omega_k} \left(C - N_c \mathcal{F}\left(\eta_n\left(\psi, \varphi_n\right)\right) + N_v \mathcal{F}\left(\eta_p\left(\psi, \varphi_p\right)\right)\right) \mathrm{d}\mathbf{x},$$

$$-q\frac{\partial}{\partial t}\int_{\omega_k} N_c \mathcal{F}\left(\eta_n\left(\psi,\varphi_n\right)\right)d\mathbf{x} + \int_{\partial\omega_k} \mathbf{j}_n \cdot \boldsymbol{\nu} ds = q\int_{\omega_k} Rd\mathbf{x},$$

$$q\frac{\partial}{\partial t}\int_{\omega_k} N_v \mathcal{F}\left(\eta_p\left(\psi,\varphi_p\right)\right)d\mathbf{x} + \int_{\partial\omega_k} \mathbf{j}_p \cdot \boldsymbol{\nu} ds = -q\int_{\omega_k} Rd\mathbf{x},$$

for $k = 1, \ldots, N$. Here, $\boldsymbol{\nu}$ is outward-pointing unit normal to the control volume ω_k; see Figure 50.11. These equations represent an integral form of the van Roosbroeck system on every control volume. In particular, the first equation is Gauss's law of electrodynamics. The other two equations are balance laws for the carrier densities. The densities in each control volume change only due to in- and outflow through the boundary or recombination.

Next, the surface integrals are split into the sum of integrals over the planar interfaces between the control volume ω_k and its neighbors. Employing one point quadrature rules for the surface and volume integrals, we deduce the finite volume scheme

$$\sum_{\omega_\ell \in \mathcal{N}(\omega_k)} |\partial\omega_k \cap \partial\omega_\ell| D_{k,\ell} = q|\omega_k|\left(C_k - N_c \mathcal{F}\left(\eta_n\left(\psi_k,\varphi_{n;k}\right)\right) + N_v \mathcal{F}\left(\eta_p\left(\psi_k,\varphi_{p;k}\right)\right)\right)$$

$$+|\partial\omega_k \cap \Gamma_G|\frac{\varepsilon_{\text{ox}}}{d_{\text{ox}}}(U_G - \psi_k), \tag{50.53}$$

$$-q|\omega_k|N_c\frac{d}{dt}\mathcal{F}\left(\eta_n\left(\psi_k,\varphi_{n;k}\right)\right) + \sum_{\omega_\ell \in \mathcal{N}(\omega_k)} |\partial\omega_k \cap \partial\omega_\ell|j_{n;k,\ell} = q|\omega_k|R_k, \tag{50.54}$$

$$q|\omega_k|N_v\frac{d}{dt}\mathcal{F}\left(\eta_p\left(\psi_k,\varphi_{p;k}\right)\right) + \sum_{\omega_\ell \in \mathcal{N}(\omega_k)} |\partial\omega_k \cap \partial\omega_\ell|j_{p;k,\ell} = -q|\omega_k|R_k. \tag{50.55}$$

In the above mentioned formulas, $\mathcal{N}\left(\omega_k\right)$ denotes the set of all control volumes neigboring ω_k. In 2D, the measure $|\partial\omega_k \cap \partial\omega_\ell|$ corresponds to the length of the boundary line segment and in 3D to the area of the intersection of the boundary surfaces. The measure $|\omega_k|$ is in 2D given by the area and in 3D by the volume of the control volume ω_k. The unknowns ψ_k, $\varphi_{n;k}$, and $\varphi_{p;k}$ are approximations of the electric potential as well as the quasi-Fermi potentials for electrons and holes evaluated at node $\mathbf{x}_k$ like already introduced for the 1D case in Section 50.2. Accordingly, R_k is defined as

$$R_k = R\left(N_c\mathcal{F}\left(\eta_n\left(\psi_k,\varphi_{n;k}\right)\right), N_v\mathcal{F}\left(\eta_p\left(\psi_k,\varphi_{p;k}\right)\right)\right).$$

The doping is defined by the integral average

$$C_k = \frac{1}{|\omega_k|}\int_{\omega_k} C(\mathbf{x})d\mathbf{x},$$

which can be estimated by its nodal value $C(\mathbf{x}_k)$. The numerical fluxes $D_{k,\ell}$, $j_{n;k,\ell}$ and $j_{p;k,\ell}$ approximate $-\varepsilon\nabla\psi \cdot \boldsymbol{\nu}_{k\ell}$, $\mathbf{j}_n \cdot \boldsymbol{\nu}_{k\ell}$ and $\mathbf{j}_p \cdot \boldsymbol{\nu}_{k\ell}$, respectively, on the interfaces between two adjacent control volumes ω_k and ω_ℓ, see Figure 50.11. As in the 1D case, we assume that these fluxes can be expressed as nonlinear functions depending on the values $\psi_k, \varphi_{n;k}, \varphi_{p;k}$ and $\psi_\ell, \varphi_{n;\ell}, \varphi_{p;\ell}$.

The electric displacement flux is approximated by

$$D_{k,\ell} = -\varepsilon_s\frac{\psi_\ell - \psi_k}{h_{k,\ell}},$$

where

$$h_{k,\ell} = \|\mathbf{x}_\ell - \mathbf{x}_k\|$$

is the edge length, compare with Equation 50.15.

For the Boltzmann approximation, the numerical charge carrier fluxes can be approximated via the Scharfetter–Gummel expression (50.21). In theory, any classical technique such as central differences or upwinding could be used to discretize the numerical fluxes. However, the former suffers from instability issues (as seen in Section 50.2.7) and neither scheme is consistent with the thermodynamic equilibrium. We will explain what this means in Section 50.4.3. In the next section, we discuss thermodynamically consistent schemes for non-Boltzmann statistics.

The finite volume scheme ((50.53)–(50.55)) yields a nonlinear system of $3N$ equations depending on $3N$ variables. We can directly substitute Dirichlet boundary values in the equations. In practice, however, this way of handling Dirichlet values is very technical to implement. Exploiting floating point arithmetic, the Dirichlet penalty method [45] provides a reasonable alternative. Physically, it replaces the Dirichlet boundary conditions by gate boundary conditions with very high oxide permittivity.

50.4.2 Scharfetter–Gummel Fluxes and Their Non-Boltzmann Generalizations

In 1D, we have already discussed the highly effective flux discretization scheme proposed by Scharfetter and Gummel for Boltzmann statistics [11]. They derive the numerical flux by locally solving a linear two-point boundary value problem. The 1D idea immediately carries over to higher dimensions if we insert the Scharfetter–Gummel (50.21) into the discrete system ((50.53)–(50.55)). However, if the Boltzmann approximation is not valid anymore, we can no longer derive the flux like Scharfetter and Gummel suggested.

This motivated the work in [46] where the Scharfetter–Gummel idea was generalized to a large class of nonlinear convection-diffusion problems, allowing to define consistent numerical fluxes from nonlinear two-point boundary value problems. Unfortunately, these *generalized Scharfetter–Gummel schemes* cannot be expressed by closed formulas. Sometimes, however, the local fluxes can be obtained iteratively [27]. It is also possible to approximate the two-point boundary value problems by simpler ones (e.g., by freezing some coefficients). This leads to *modified Scharfetter–Gummel schemes*. We address some of these schemes now.

For the sake of readability, we provide the formulas for electrons only. The formulas for holes follow similarly.

50.4.2.1 Generalized Scharfetter–Gummel Schemes

By the construction of our mesh, it suffices to study the 1D flux j_n along the edge $\overline{\mathbf{x}_k \mathbf{x}_\ell}$. For general distribution functions $\mathcal{F}$, we want to solve the ordinary differential equation

$$\frac{\mathrm{d}}{\mathrm{d}x} j_n = \frac{\mathrm{d}}{\mathrm{d}x}\left(- q\mu_n N_c \mathcal{F}\left(\eta_n\left(\psi, \varphi_n\right)\right) \nabla \varphi_n \right) = 0$$

on the interval $[0, h_{k,\ell}]$ with boundary conditions

$$\varphi_n\left(0\right) = \varphi_{n;k} \quad \text{and} \quad \varphi_n\left(h_{k,\ell}\right) = \varphi_{n;\ell},$$

where $\varphi_{n;k}$ and $\varphi_{n,\ell}$ are the values of the quasi-Fermi potentials at nodes $\mathbf{x}_k$ and $\mathbf{x}_\ell$, respectively, see Figure 50.11. We assume now that the flux is constant between both nodes and denote it with $j_{n;k,\ell}$. Integrating twice leads to an integral equation [27] for the unknown current, namely

$$\int_{\eta_{n;k}}^{\eta_{n;\ell}} \left(\frac{j_{n;k,l}/j_0}{\mathcal{F}(\eta)} + \frac{\psi_\ell - \psi_k}{U_T} \right)^{-1} d\eta = 1, \tag{50.56}$$

where $j_0 = q\mu_n N_c \dfrac{U_T}{h_{k,\ell}}$ and the limits are given by $\eta_{n;k} = \eta_n \left(\psi_k, \varphi_{n;k} \right)$ and $\eta_{n;\ell} = \eta_n \left(\psi_\ell, \varphi_{n;\ell} \right)$. For strictly monotonously increasing $\mathcal{F}(\eta)$ this equation has always a unique solution [47].

For the Boltzmann approximation $\mathcal{F}(\eta) = \exp(\eta)$, this integral equation can be solved analytically and yields the classical Scharfetter–Gummel expression for the flux (50.21). In [27], it was shown that for the Blakemore distribution function $\mathcal{F}(\eta) = \dfrac{1}{\exp(-\eta) + \gamma}$, the integral equation yields a fixed point equation

$$\widehat{j}_{n;k,\ell} = B \left(\gamma \widehat{j}_{n;k,\ell} + \frac{\psi_\ell - \psi_k}{U_T} \right) e^{\eta_{n;\ell}} - B \left(-\left[\gamma \widehat{j}_{n;k,\ell} + \frac{\psi_\ell - \psi_k}{U_T} \right] \right) e^{\eta_{n;k}} \tag{50.57}$$

for the nondimensionalized edge current $\widehat{j}_{n;k,\ell} = j_{n;k,l}/j_0$. The right-hand side is a Scharfetter–Gummel expression where the argument of the Bernoulli function is shifted by $\gamma \widehat{j}_{n;k,\ell}$. Hence, this reduces to the classical Scharfetter–Gummel scheme for $\gamma = 0$.

Since the Bernoulli function is strictly decreasing, this fixed point equation possesses a unique solution $\widehat{j}_{n;k,\ell}$. If we want to use the flux given by (50.57) in the discrete system ((50.53)–(50.55)), we need to solve for the fluxes twice (once for electrons and once for holes) on each discretization edge $\overline{\mathbf{x}_k \mathbf{x}_\ell}$. A few Newton steps are sufficient to solve this equation iteratively.

Even though (50.57) is restricted to the Blakemore approximation, it provides a useful scheme in the context of organic semiconductors. There it arises naturally as a model for materials with δ-shaped density of states [10,26], describing a single electronic transport level, see Figures 50.8 and 50.9. This has been described in Section 50.3.5.

Unfortunately, for a general statistical distribution function, no corresponding equation has been derived so far. Therefore, in [47] it was proposed to use piecewise approximations of $\mathcal{F}$ by Blakemore type or rational approximations of Padé type in order to obtain piecewise integrable expression from the local boundary value problem.

50.4.2.2 Modified Scharfetter–Gummel Schemes

Bessemoulin-Chatard [48] derived a finite volume scheme for convection-diffusion problems by averaging the nonlinear diffusion term appropriately. This idea was generalized to more general distribution functions in [49], introducing a logarithmic average of the nonlinear diffusion enhancement

$$g_{n;k,\ell} = \frac{\eta_{n;k} - \eta_{n;\ell}}{\log \mathcal{F}\left(\eta_{n;k} \right) - \log \mathcal{F}\left(\eta_{n;\ell} \right)}$$

along the discretization edge. Using the generalized Einstein relation (50.44), one immediately observes that the diffusion enhancement g can be seen as a modification factor of the thermal voltage U_T. Replacing U_T in the Scharfetter–Gummel expression (50.21) by $U_T^* = U_T g_{n;k,\ell}$, we deduce the following modified Scharfetter–Gummel scheme

$$j_{n;k,\ell} = -\frac{q\mu_n U_T}{h_{k,\ell}} g_{n;k,\ell} \left(N_c \mathcal{F}\left(\eta_{n;k} \right) B \left(-\frac{\psi_\ell - \psi_k}{U_T g_{n;k,\ell}} \right) - N_c \mathcal{F}\left(\eta_{n;\ell} \right) B \left(\frac{\psi_\ell - \psi_k}{U_T g_{n;k,\ell}} \right) \right), \tag{50.58}$$

approximating the current along the edge.

Previously, we have replaced the thermal voltage by a suitable average along the edge. Now, we want to approximate $\mathcal{F}(\eta)$ along the edge by an exponential (Boltzmann approximation) and adjusting the effective density of states N_c accordingly, i.e.,

$$N_c \mathcal{F}(\eta) \approx N^*_{c;k,\ell} \exp(\eta).$$

This choice makes it possible to use the original Scharfetter–Gummel flux (50.21) by replacing N_c with $N^*_{c;k,\ell}$. One choice for the modified density of states is

$$N^*_{c;k,\ell}(\eta^*) = N_c \frac{\mathcal{F}(\eta^*)}{\exp(\eta^*)},$$

where $\eta^* \in [\eta_k, \eta_\ell]$, assuming $\eta_k \leq \eta_\ell$. In practice, we might consider taking the geometric average between $N_c(\eta_k)$ and $N_c(\eta_\ell)$, which leads to another modified Scharfetter–Gummel scheme:

$$j_{n;k,\ell} = -\frac{q\mu_n U_T}{h_{k,\ell}} N_c \sqrt{\frac{\mathcal{F}(\eta_{n;k}) \, \mathcal{F}(\eta_{n;\ell})}{\exp(\eta_{n;k}) \exp(\eta_{n;\ell})}} \left(B\left(-\frac{\psi_\ell - \psi_k}{U_T}\right) e^{\eta_{n;k}} - B\left(\frac{\psi_\ell - \psi_k}{U_T}\right) e^{\eta_{n;\ell}} \right). \quad (50.59)$$

The idea behind this scheme was introduced in [50] for the numerical solution of the generalized Nernst–Planck system which is similar to the van Roosbroeck system ((50.30)–(50.33)). A variant of this scheme for Fermi–Dirac statistics is described in [14,51] and numerically implemented in the semiconductor device simulation package WIAS-TeSCA [52].

50.4.3 Flux Expressions Consistent with the Thermodynamic Equilibrium

As introduced in Section 50.3.4, the currents of holes and electrons vanish in the thermodynamic equilibrium when no bias is applied. If a numerical scheme for zero bias boundary conditions results into vanishing numerical fluxes, we call it *consistent with the thermodynamical equilibrium*. In practice, one can examine this consistency by checking whether $j_{n;k,\ell} = j_{p;k,\ell} = 0$ if $\varphi_{n;k} = \varphi_{n;\ell}$ and $\varphi_{p;k} = \varphi_{p;\ell}$. Violating this property causes unphysical dissipation (spurious Joule heating) in the steady state attained for zero bias boundary conditions, which by definition is supposed to be the thermodynamic equilibrium [48].

All schemes introduced in Section 50.4.2 are consistent with the thermodynamical equilibrium. Indeed, let us assume that the quasi-Fermi potentials φ_n and φ_p between two adjacent control volumes ω_k and ω_ℓ are equal. The consistency of the classical Scharfetter–Gummel (50.21) and the scheme using a modified density of states (50.59) are obvious since the Bernoulli function satisfies $B(-x) = \exp(x) B(x)$.

Due to the definition from the solution of the two-point boundary value problem, the generalized Scharfetter–Gummel scheme defined in Equation 50.56 and its specialization for the Blakemore approximation (50.57) are consistent with the thermodynamic equilibrium. And finally, the logarithmic average of the diffusion enhancement is the only possible average which guarantees consistency with thermodynamics in the scheme (50.58). For details see [27,47,49].

50.4.4 Free Energy and Dissipation Rate

For Boltzmann statistics, it is rather straightforward to define discrete analoga of the free energy (50.45) and the dissipation rate (50.46). The positivity of the discrete dissipation rate was shown in [29], and the exponential decay of the free energy to its equilibrium value was proven in [53]. An overview of the entropy method for finite volume schemes has been given in [54]. First results on more general statistics functions in this respect have been obtained in [55].

These pioneering works strongly indicate that the chosen discretization approach results in discrete models, which are consistent with the structural assumptions of nonequilibrium thermodynamics. A full account of these issues in the context of general statistics functions remains an open research topic.

50.4.5 Existence, Uniqueness, and Convergence

There are very few existence proofs for the solutions of the discretized system ((50.53)–(50.55)). For the Boltzmann approximation, Gärtner [56] proved that the discretized steady-state system has a solution, which becomes unique if a small bias is applied. A similar result for Fermi statistics has been obtained in [47].

A convergence theory for the finite volume scheme for the full discrete system ((50.53)–(50.55)) and general flux functions is still missing. However, practical experience and a number of results make its convergence plausible.

For example, in one space dimension, second-order convergence in the discrete maximum norm for the Scharfetter–Gummel scheme has been shown in [19]. Under the assumption that second derivatives of the continuous solution exist, in [57] for moderately sized drift terms and two-dimensional (2D), square grids, first-order convergence for the simple upwind scheme (see e.g., [21]), and second-order convergence for the exponential fitting scheme in the L_2-norm has been shown. Reinterpretations of the finite volume Scharfetter–Gummel scheme as a nonstandard finite element method allowed to obtain convergence estimates for Scharfetter–Gummel schemes on Delaunay grids (see Section 50.6) [58,59]. For a general approach to the convergence theory of finite volume schemes, see [44]. In [46], weak convergence (no order estimate) for a generalization of the Scharfetter–Gummel scheme to nonlinear convection-diffusion problems has been shown. A convergence proof for a variant of the van Roosbroeck system discretized with the simple upwind scheme was given in [60].

50.4.6 Maximum Principle

When applied to the drift-diffusion formulation, compared to various variants of the stabilized finite element method, the two-point flux finite volume scheme is outstanding in the sense that it guarantees positivity of densities and absense of unphysical oscillations [61,62]. It allows to carry over the discussions of Section 50.2.7 to the higher dimensional case.

50.5 Nonlinear Solvers

In Section 50.2, we have already briefly mentioned that we need to solve a nonlinear discrete system of the form

$$0 = \mathbf{F}(\boldsymbol{\psi}, \boldsymbol{\varphi}_n, \boldsymbol{\varphi}_p) := \begin{pmatrix} \mathbf{F}_1(\boldsymbol{\psi}, \boldsymbol{\varphi}_n, \boldsymbol{\varphi}_p) \\ \mathbf{F}_2(\boldsymbol{\psi}, \boldsymbol{\varphi}_n, \boldsymbol{\varphi}_p) \\ \mathbf{F}_3(\boldsymbol{\psi}, \boldsymbol{\varphi}_n, \boldsymbol{\varphi}_p) \end{pmatrix}.$$

Here, we discuss two ways to do this: Newton's method and *Gummel's iteration method* [38], which is based on a blockwise decoupling of the discrete system.

50.5.1 Newton's Method

Assuming that a good starting guess has been provided, for example, by following the ideas in Section 50.2.6, Newton's method constructs a new $(k + 1)$th iterate from the kth one by solving the linear system

$$
\begin{pmatrix}
\frac{\partial \mathbf{F}_1}{\partial \psi} & \frac{\partial \mathbf{F}_1}{\partial \varphi_n} & \frac{\partial \mathbf{F}_1}{\partial \varphi_p} \\
\frac{\partial \mathbf{F}_2}{\partial \psi} & \frac{\partial \mathbf{F}_2}{\partial \varphi_n} & \frac{\partial \mathbf{F}_2}{\partial \varphi_p} \\
\frac{\partial \mathbf{F}_3}{\partial \psi} & \frac{\partial \mathbf{F}_3}{\partial \varphi_n} & \frac{\partial \mathbf{F}_3}{\partial \varphi_p}
\end{pmatrix}^k
\begin{pmatrix}
\delta\psi^k \\
\delta\varphi_n^k \\
\delta\varphi_p^k
\end{pmatrix}
= -
\begin{pmatrix}
\mathbf{F}_1(\psi^k, \varphi_n^k, \varphi_p^k) \\
\mathbf{F}_2(\psi^k, \varphi_n^k, \varphi_p^k) \\
\mathbf{F}_3(\psi^k, \varphi_n^k, \varphi_p^k)
\end{pmatrix}
\tag{50.60}
$$

for the update vector

$$
\begin{pmatrix}
\delta\psi^k \\
\delta\varphi_n^k \\
\delta\varphi_p^k
\end{pmatrix}
:=
\begin{pmatrix}
\psi^{k+1} - \psi^k \\
\varphi_n^{k+1} - \varphi_n^k \\
\varphi_p^{k+1} - \varphi_p^k
\end{pmatrix}.
$$

Then the new iterate is obtained by adding the previous vector to the update. The advantage of Newton's method is that it converges quadratically if the starting guess is sufficiently close to the solution [17]. This allows to obtain highly accurate discrete solutions at low additional cost. The major drawback is that it might converge very slowly or even fail to converge if the starting guess is too far from the actual solution. Damping—multiplying the update with a factor less than 1—is known to increase the convergence region. Another remedy is parameter embedding where one slowly changes a parameter, always using the old solution as a new starting guess. We have already employed this embedding technique in Section 50.2.6 when gradually increasing the bias voltage in the solution procedure.

An important advantage of finite element and finite volume discretizations is the fact that they create only next-neighbor couplings in the discretized systems. The resulting linearized systems are therefore sparse, i.e., the maximum number of nonzero elements in a matrix row is bounded by a constant independent of the number of discretization cells, making it possible to use highly economic storage schemes. For 2D applications, one can usually solve the resulting sparse linear system (50.60) via sparse direct solvers, such as PARDISO [63–66] or UMFPACK [67,68]. Direct solvers calculate a representation of the system matrix as a product of easily solvable lower and upper triangular matrices, which unfortunately are not anymore sparse, increasing memory consumption and computational time especially in large 3D applications. This fill-in phenomenon is avoided by preconditioned Krylov subspace methods [69], which compared to the direct solvers need significant effort in tuning and adaptation to the problem at hand.

50.5.2 Gummel's Method

Gummel [13,38] suggested decoupling the three equations in the van Roosbroeck system. He devised an iterative method at the continuous level, which for Boltzmann statistics and drift-diffusion form leads to alternating between solving linear differential equations for the electric potential as well as the charge carrier densities. Suppose, one already knows the iterate (ψ^k, n^k, p^k). In order to obtain the electric potential at the new level, ψ^{k+1}, one solves

$$
0 = -\frac{1}{q} \nabla \cdot \left(\varepsilon_s \nabla \psi^{k+1} \right) + \left(n^k + p^k \right) \left(\frac{\psi^{k+1} - \psi^k}{U_{\mathrm{T}}} \right) + n^k - p^k - C.
$$

This formulation is motivated by linearizing a fixed point problem that one obtains from the Poisson equation [13]. Once one has solved for the electrostatic potential at the new level, one can successively solve the continuity equations from

$$0 = \nabla \cdot \mathbf{j}_n(\psi^{k+1}, n^{k+1}) - qR(\psi^{k+1}, n^{k+1}, p^k)$$

$$0 = \nabla \cdot \mathbf{j}_p(\psi^{k+1}, p^{k+1}) + qR(\psi^{k+1}, n^{k+1}, p^{k+1}).$$

Gummel's method is known to have a larger convergence region and a slower asymptotic convergence rate compared to Newtons's method. Depending on the software environment chosen, it also may be easier to implement. Note that standard existence and uniqueness proofs for the van Roosbroeck system rely on a similar decoupling strategy, see Section 50.3.7.

50.6 Mesh Generation for the Finite Volume Method

While appearing to be just a technical footnote compared to the questions of modeling, analysis and discretization, mesh generation in reality is a hard problem which deserves special attention. We focus on methods to construct admissible subdivisions of the computational domain Ω as used in Section 50.4.

50.6.1 Boundary Conforming Delaunay Meshes on Polyhedral Domains

Assume a partition (triangulation in 2D, tetrahedralization in 3D) of the polyhedral domain $\Omega = \bigcup_{k=1}^{N_\Sigma} \Sigma_k$ into nonoverlapping simplices Σ_k as it is commonly used for finite element methods [70]. We require that this simplicial partition of Ω has the *boundary conforming Delaunay property* [71]. For a triangulation of a 2D domain this property is equivalent to:

(1) For any two triangles with a common edge, the sum of their respective angles opposite to that edge is less or equal to 180°.
(2) For any triangle sharing an edge with $\partial\Omega$, its angle opposite to that edge is less or equal to 90°.

For a given vertex $x_i \in X$, where X is the set of vertices, the *Voronoi cell* around X is the set

$$V_i = \left\{ \mathbf{x} \in \mathbb{R}^d \,:\, \|\mathbf{x} - \mathbf{x}_i\| < \|\mathbf{x} - \mathbf{x}_j\| \text{ for all } \mathbf{x}_j \in X \text{ with } \mathbf{x}_j \neq \mathbf{x}_i \right\}.$$

The *Voronoi diagram*—the set of Voronoi cells for all vertices in X—is dual to the *Delaunay triangulation* of the point set X in the sense that for each edge $\overline{\mathbf{x}_i\mathbf{x}_j}$ in the Delaunay triangulation, $\partial V_i \cap \partial V_j \neq \emptyset$. If the simplicial partition is boundary conforming Delaunay, the restricted Voronoi cells $\omega_i = \Omega \cap V_i$ are well defined and can be obtained by joining the circumcenters of the simplices adjacent to the vertex $\mathbf{x}_i$. These restricted Voronoi cells provide an admissible control volume partition as required in Section 50.4. We note that in order to implement the finite volume method as described in Section 50.4, there is no need for an explicit construction of the control volumina ω_i as geometrical objects. Given the simplical partition, it is sufficient to base the calculations on the simplicial contributions $s_{ij}^k = |\partial\omega_i \cap \partial\omega_j \cap \Sigma_k|$ and $|\omega_i \cap \Sigma_K|$, and to use a simplex-based assembly loop as common for finite elements [29]. Figure 50.12 shows (boundary conforming) Delaunay triangulations.

There are several efficient algorithms to construct Delaunay triangulations for a given point set X [73,74]. These are good starting points of devising meshing algorithms in general. Many different problems complicate the generation of a mesh. For example, the boundary conforming Delaunay property is rather difficult to achieve, in particular it requires the careful insertion of additional points on the boundary. In 3D, slivers (very flat tetrahedra) must be avoided. And finally, it may be very complicated to fulfill additional requirements like constraints on the minimum angle or the local element size. Though there are still unsolved problems, the triangle ([72], 2D, free for noncommercial use) and TetGen ([75], 3D, open source) mesh generators help to create boundary conforming Delaunay meshes based on algorithms which are proven to

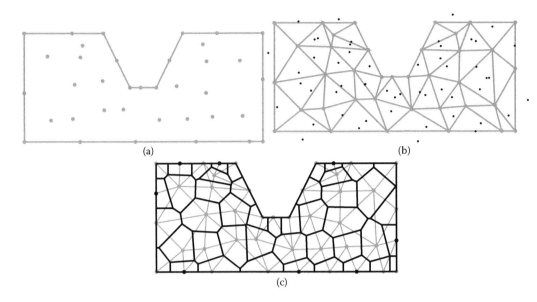

FIGURE 50.12   (a) Piecewise linear description of computational domain with given point cloud. (b) Delaunay triangulation of domain and triangle circumcenters. As some boundary triangles have angles larger than 90° angles opposite to the boundary, their circumcenters (small dots) are outside of the domain. (c) Boundary conforming Delaunay triangulation with automatically inserted additional points at the boundary (large dots), and Voronoi and restricted Voronoi cells (black). Created using triangle [72].

deliver meshes with the desired properties in finite time for a broad class of geometries given by a piecewise linear description of the boundary.

Other mesh generation approaches, in particular the advancing front [76] and the octree method [77], are similarly widespread. However, while popular in finite element community, their design makes it much harder to use them as a starting point for generating boundary conforming Delaunay meshes.

50.6.2 Tensor Product Approaches

In simplified geometrical situations, rectangular and cuboid mesh structures are a straightforward way to generate admissible finite volume partitions of the computational domain [12]. Extrusion of a 2D boundary conforming Delaunay base mesh into a 3D mesh consisting of prisms (and optional subsequent subdivision of these prisms into tetrahedra) provides another method to create admissible finite volume partitions [78].

50.6.3 Open Problems

Charge densities and potentials in semiconductors exhibit interior and boundary layers, i.e., thin regions charaterized by rapid gradients in one direction and slow gradients in the orthogonal direction. This may happen in space charge regions in the vicinity of gate contacts and at p-n junctions. These layers and other possible "hotspots" due to recombination, for example, call for local mesh adaptation. Element size control via the mesh generators using *a priori* knowledge is often used in this situation [72,75]. The development of reliable a posteriori mesh refinement criteria is an open problem in the context of the strongly coupled van Roosbroeck system. A physically motivated heuristic approach based on the equidistribution of the dissipation rate was suggested in [29] and investigated in [79].

Mesh generation tools mostly fail to create boundary conforming Delaunay meshes which use thin, anisotropic simplices to resolve interior and boundary layers with an optimal number of points. A partial remedy in some situations can be provided by node offsetting [80], a technique still under development. In certain cases, tensor product approaches [78] give reasonable results.

50.7 Time Discretization

In order to discretize the time derivative in the van Roosbroeck system ((50.30)–(50.32)), we define a series of discrete time values $0 = t^0, t^1, \ldots, t^M = T$ with step lengths $\tau^m = t^{m+1} - t^m$. We describe the temporal semi-discretization of the (spatially) continuous problem. For practical purposes it needs to be combined with the space discretization approach described in Section 50.4. The implict Euler time discretization scheme assumes given values $\psi^m, \varphi_n^m, \varphi_p^m, n^m, p^m$ of the variables at time t^m. Values at the subsequent time t^{m+1} are calculated by solving the nonlinear system of equations

$$-\nabla \cdot (\varepsilon_s \nabla \psi^{m+1}) = q(C - n^{m+1} + p^{m+1}), \tag{50.61}$$

$$-q\frac{n^{m+1} - n^m}{\tau^m} - q\nabla \cdot \mu_n n^{m+1} \nabla \varphi_n^{m+1} = qR(n^{m+1}, p^{m+1}), \tag{50.62}$$

$$q\frac{p^{m+1} - p^m}{\tau^m} - q\nabla \cdot \mu_p p^{m+1} \nabla \varphi_p^{m+1} = -qR(n^{m+1}, p^{m+1}), \tag{50.63}$$

where

$$n^{m+1} = N_c \mathcal{F}(\eta_n(\psi^{m+1}, \varphi_n^{m+1})),$$

$$p^{m+1} = N_v \mathcal{F}(\eta_p(\psi^{m+1}, \varphi_p^{m+1})).$$

As a consequence, the nonlinear system ((50.61)–(50.63)), which is structurally similar to Equations 50.53 through 50.55, has to be solved for each time step by iterative methods described in Section 50.5. An initial guess for these methods can be obtained from the old time step t^m or a linear extrapolation involving data from several previous time steps [81].

It is advisable to choose the size of the time steps τ^m according to multiple criteria, including the convergence radius of Newton's method, the free energy, and local changes of the potentials.

At first glance, the computational cost of having to solve a nonlinear system for each time step appears to be rather high. However, this scheme is stable independent of the time step size and allows to carry over all advantages of the finite volume scheme from the stationary to the transient case. The transient discrete solution converges to the stationary one, in particular, for the corresponding boundary condition to the thermodynamic equilibrium solution. The free energy of the system decays during this approach to the equilibrium. In several cases, it has been proven that this decay is exponential [55,82]. By preserving the physics, the implicit Euler scheme fits well to the thermodynamically motivated discretization schemes. The use of other schemes implies giving up some of these physical properties. If the resulting deviations from thermodynamic properties can be kept under control, higher order schemes (e.g., backward differentiation formula (BDF) methods [83]) might help to reduce computational costs due to a higher temporal convergence order.

Due to the perceived complexity and high computational costs of solving coupled nonlinear PDE systems, linear implicit methods, mixing linearization, and time discretizations are alternatives. Most prominent is the scheme devised in [84].

50.8 Contact Terminal Currents

In simulations, one is usually interested in IV-curves, i.e., the dependency of terminal currents on applied voltages. Therefore, calculating terminal currents accurately is crucial to a successful postprocessing of the simulated field data. The *total current density* is given by

$$\mathbf{j} = \mathbf{j}_d + \mathbf{j}_n + \mathbf{j}_p,$$

where $\mathbf{j}_d = -\varepsilon_s \nabla(\partial_t \psi) = \partial_t \mathbf{D}$ is the *displacement current density*. Taking the time derivative of the Poisson equation (50.30) yields

$$-\nabla \cdot \varepsilon_s \nabla(\partial_t \psi) = q(\partial_t n - \partial_t p) = -\nabla \cdot \mathbf{j}_n - qR - \nabla \cdot \mathbf{j}_p + qR$$

$$\nabla \cdot \partial_t \mathbf{D} = -\nabla \cdot \mathbf{j}_n - \nabla \cdot \mathbf{j}_p,$$

resulting in $\nabla \cdot \mathbf{j} = 0$, which physically implies charge conservation.

Given a set of contacts $\Gamma_\alpha \subset \partial\Omega$ for $\alpha = 1, \ldots, N_C$, the terminal current through contact α is defined as

$$I_\alpha = \int_{\Gamma_\alpha} \mathbf{j} \cdot \boldsymbol{\nu} \, ds.$$

The correct calculation of this integral presupposes a regularity of the solution and its derivative which is not supported by analytical theory. Therefore, in [51,52] a test function–based technique was proposed, which uses the weak formulation of the problem. For a physical motivation of this approach and an interpretation as a generalized Shockley–Ramo theorem we refer to [85]. For any α, let the function $T_\alpha : \Omega \to \mathbb{R}$ solve the boundary value problem

$$-\nabla^2 T_\alpha = 0 \qquad\qquad \text{in } \Omega, \tag{50.64}$$

$$\nabla T_\alpha \cdot \boldsymbol{\nu} = 0 \qquad\qquad \text{on } \partial\Omega \setminus \bigcup_{\beta=1}^{N_C} \Gamma_\beta, \tag{50.65}$$

$$T_\alpha = \delta_{\alpha\beta} \qquad\qquad \text{on } \Gamma_\beta \text{ (for } \beta = 1, \ldots, N_C), \tag{50.66}$$

where $\delta_{\alpha\beta}$ denotes the Kronecker delta. As a consequence,

$$I_\alpha = \int_{\Gamma_\alpha} \mathbf{j} \cdot \boldsymbol{\nu} \, ds = \int_{\partial\Omega} T_\alpha \mathbf{j} \cdot \boldsymbol{\nu} \, ds = \int_\Omega \nabla \cdot (T_\alpha \mathbf{j}) \, d\mathbf{x}$$

$$= \int_\Omega \nabla T_\alpha \cdot \mathbf{j} \, d\mathbf{x} + \int_\Omega T_\alpha \nabla \cdot \mathbf{j} \, d\mathbf{x}$$

$$= \int_\Omega \nabla T_\alpha \cdot (\mathbf{j}_n + \mathbf{j}_p) \, d\mathbf{x} + \partial_t \int_\Omega \nabla T_\alpha \cdot \mathbf{D} \, d\mathbf{x}.$$

Given the semi-discretization in time from Section 50.7, one arrives at

$$I_\alpha^{m+1} = \int_\Omega \nabla T_\alpha \cdot (\mathbf{j}_n^{m+1} + \mathbf{j}_p^{m+1}) \, d\mathbf{x} + \frac{1}{\tau^m} \int_\Omega \nabla T_\alpha \cdot \left(\mathbf{D}^{m+1} - \mathbf{D}^m\right) \, d\mathbf{x}.$$

Like in Section 50.4, we approximate the system (50.64)–(50.66) by a finite volume discretization. Using partial integration and the notation introduced in Section 50.4 we come up with an expression for the total current through the α-th contact of the form

$$I_\alpha^{m+1} \approx \sum_{\substack{\omega_\ell \in \mathcal{N}(\omega_k) \\ k < \ell}} |\partial\omega_k \cap \partial\omega_\ell| \left(j_{n;k,\ell}^{m+1} + j_{p;k,\ell}^{m+1} + \frac{D_{k,\ell}^{m+1} - D_{k,\ell}^m}{\tau^m} \right) (T_{\alpha;k} - T_{\alpha;\ell}). \qquad (50.67)$$

50.9 An Alternative: The Finite Element Method

We have put special emphasis on the finite volume method. However, it is worth pointing out that there are other approaches. The finite difference method that has been used to solve the van Roosbroeck system [13] on tensor product meshes is equivalent to the previously introduced finite volume method.

The most popular ansatz to deal with unstructured meshes is the finite element method [70]. It starts with a *weak formulation* of the steady-state van Roosbroeck problem ((50.30)–(50.32)). Each equation of the system is multiplied with a so-called test function and integrated in space. Integrating by parts, each equation of the steady-state system can be restated as the following problem:

find $u \in U$ such that $a(u, v) = f(v)$ for all $v \in V$,

where U is a function space containing the ansatz functions u and V is a function space containing the test functions v. The map $a: U \times V \to \mathbb{R}$ has the structure

$$a(u, v) = \int_\Omega \alpha(u)\nabla u \cdot \nabla v \, d\mathbf{x} - \int_{\partial\Omega} \nabla u \cdot \mathbf{n} \, v \, ds \qquad (50.68)$$

and the functional $f: V \to \mathbb{R}$ is continuous. The idea of a finite element method is to approximate this continuous problem with a discrete version. We approximate the infinite dimensional spaces U and V with finite dimensional subspaces U_h and V_h spanned by basis functions with localized support. Then we obtain a finite dimensional version of Equation 50.68, namely

find $u_h \in U_h$ such that $a_h(u_h, v_h) = f(v_h)$ for all $v_h \in V_h$.

As it is sufficient to fulfill this condition for finitely many localized basis functions in V_h, we obtain a sparse (containing many zeros) nonlinear system of equations for the basis coefficients of u_h, assuming we are able to calculate the integrals. One usually approximates these integrals by quadrature rules.

Auf der Maur [86] suggested to write the stationary van Roosbroeck system in this framework, using the quasi-Fermi potential formulation of the flux. In this case one has

$$a^\psi(\psi, v) = \int_\Omega \varepsilon_s \nabla\psi \cdot \nabla v \, d\mathbf{x} + \int_{\Gamma_G} \frac{\varepsilon_{ox}}{d_{ox}}(\psi - U_G)v \, ds, \qquad f^\psi(v) = q \int_\Omega (p - n + C)v \, d\mathbf{x}, \qquad (50.69)$$

$$a^{\varphi_n}(\varphi_n, v) = \int_\Omega \mu_n n \nabla\varphi_n \cdot \nabla v \, d\mathbf{x}, \qquad f^{\varphi_n}(v) = \int_\Omega R v \, d\mathbf{x}, \qquad (50.70)$$

$$a^{\varphi_p}(\varphi_p, v) = \int_\Omega \mu_p p \nabla\varphi_p \cdot \nabla v \, d\mathbf{x}, \qquad f^{\varphi_p}(v) = -\int_\Omega R v \, d\mathbf{x}. \qquad (50.71)$$

At the Dirichlet parts of the boundary, namely at the Ohmic contacts $\Gamma_{O,\alpha}$ introduced in Section 50.3, we assume that test functions v vanish and the ansatz functions denoted by ψ, φ_n, and φ_p fulfill the Dirichlet boundary conditions ((50.35)–(50.37)). Note, that the gate boundary conditions (50.38) have been incorporated into the boundary part of the quadratic form (50.69).

We point out that for Poisson's equation $\alpha^\Psi(\psi) = \varepsilon_s$ and for the continuity equations we have $\alpha^{\varphi_n}(\varphi_n) = \mu_n n$ and $\alpha^{\varphi_p}(\varphi_p) = \mu_p p$, respectively. Due to the formulation of the problem with quasi-Fermi potentials as unknowns, this means that $\alpha^{\varphi_n}(\varphi_n)$ and $\alpha^{\varphi_p}(\varphi_p)$ introduce nonlinearities in a^{φ_n} and a^{φ_p}. The integrals on the left-hand side of Equations 50.69 through 50.71 have no analytical expressions and are approximated using quadrature rules [70].

Essential for this finite element scheme is the underlying potential-based formulation of the flux in terms of φ_n and φ_p. If one attempts to apply the finite element method using the density-based, drift-diffusion formulation of the flux, one encounters a lot of technical difficulties and one has to stabilize the finite element method [61,87]. A convergence proof of the finite element method for a similar system in a somewhat different context (and without recombination) is available in [88].

50.10 Extensions and Outlook

In this chapter, we discussed the most important numerical solution techniques of the van Roosbroeck system, describing the transport of electrons and holes in semiconductors in a self-consistent electric field. This approach provides the core functionalities for a solver, which can be enhanced by models describing additional physical phenomena and more complex device structures including, for example, heterostructures. We mention a number of additional extensions and point to the corresponding literature.

50.10.1 Additional Physical Models

Many devices require a more accurate description of the carrier mobilities. For example, models of ionized impurity scattering, high-field drift velocity saturation and similar effects [13,89] introduce dependencies of the mobilities on carrier densities and the electric field strength, resulting in even more nonlinear couplings in the system.

The carrier transport in organic semiconductors is governed by hopping transport between the energy states of neighboring molecules. The van Roosbroeck system can be used to describe this effectively [8,9]. This may require the use of statistical distribution functions that reflect the distribution of the energy transport levels such as Blakemore for δ-shaped densities of states or the Gauss–Fermi integral [10]. In this context, the correct treatment of the diffusion enhancement is vital. One also has to account for the nonlinear mobilities [8,90] related to the energetic disorder characteristic for organic semiconductors.

Spin-polarized drift-diffusion models have been proposed in [91] for the description of spintronic devices. They generalize the van Roosbroeck system, introducing spin-resolved densities for electron and holes and additional mechanisms describing the spin relaxation. They have been recently studied also from a numerical point of view in [92–94].

50.10.2 Coupling the van Roosbroeck System to Other Models

When modeling lasers one has to couple the van Roosbroeck system to equations for the optical modes. This introduces additional recombination processes, describing stimulated emission. Additionally, one has to consider balance equations for the photon number. A comprehensive model (also including heating effects) along with a numerical solution strategy is given in [14].

In semiconductor devices with embedded nanostructures (such as quantum dots and quantum wells), one has to couple the drift-diffusion equations to equations describing the dynamics of the carriers localized in the nanostructures [4,5,7]. A multiscale approach for coupling atomistic with continuum drift-diffusion models is presented in [95].

If heating effects become important, the van Roosbroeck system needs to be extended by an energy transport model for the heat flow in the device. For a thermodynamically consistent extension of the van Roosbroeck system to account for this effect we refer to [14,31].

50.10.3 Methods for Doping Optimization

An important task when designing semiconductor devices is finding a suitable doping profile. This is an analytical and numerical challenge. Electric properties of the device can be improved by optimizing the doping profiles using suitable objective functionals [96–98]. Recently, the approach introduced in [96] has been extented for the optimization of doping profiles in lasers [99].

50.10.4 Alternative Modeling Approaches for Carrier Transport

The van Roosbroeck system uses implicitly the assumption that the carrier ensemble is locally in quasi equilibrium. When this assumption is not met, for example, for hot electrons, one has to consider alternative approaches. The most common ones are hydrodynamic models [100] and approximations of the Boltzmann transport equation which are derived from a spherical harmonics expansion [101]. The book [102] gives a mathematically oriented overview of these topics.

Acknowledgments

The authors are grateful for innumerable discussions with K. Gärtner and H. Gajewski, who helped to shape their still incomplete knowledge on the topic. This work has been supported by ERC-2010-AdG no. 267802 "Analysis of Multiscale Systems Driven by Functionals" (N.R.), by the Deutsche Forschungsgemeinschaft DFG within CRC 787 "Semiconductor Nanophotonics" (T.K., N.R., M.K) and partially funded in the framework of the project "Macroscopic Modeling of Transport and Reaction Processes in Magnesium-Air-Batteries" (grant 03EK3027D) under the research initiative "Energy storage" of the German Federal government.

References

1. W. van Roosbroeck. Theory of the flow of electrons and holes in germanium and other semiconductors. *Bell Syst. Tech. J.*, 29(4):560–607, 1950.
2. S. M. Sze and K. K. Ng. *Physics of Semiconductor Devices*, 3rd edition. Wiley, 2006. ISBN: 978-0-471-14323-9.
3. J. Piprek. *Optoelectronic Devices: Advanced Simulation and Analysis*. Berlin: Springer, 2005.
4. M. Grupen and K. Hess. Simulation of carrier transport and nonlinearities in quantum-well laser diodes. *IEEE J. Quant. Electron.*, 34(1):120–140, 1998.
5. S. Steiger, R. G. Veprek, and B. Witzigmann. Unified simulation of transport and luminescence in optoelectronic nanostructure. *J. Comput. Electron.*, 7(4):509–520, 2008.
6. T. Koprucki, H.-C. Kaiser, and J. Fuhrmann. Electronic states in semiconductor nanostructures and upscaling to semi-classical models. In A. Mielke, editor, *Analysis, Modeling and Simulation of Multiscale Problems*, pp. 365–394. Heidelberg, Berlin: Springer, 2006.
7. T. Koprucki, A. Wilms, A. Knorr, and U. Bandelow. Modeling of quantum dot lasers with microscopic treatment of Coulomb effects. *Opt. Quant. Electron.*, 42:777–783, 2011.
8. R. Coehoorn, W. F. Pasveer, P. A. Bobbert, and M. A. J. Michels. Charge-carrier concentration dependence of the hopping mobility in organic materials with Gaussian disorder. *Phys. Rev. B.*, 72(15):155206, 2005.
9. S. L. M. van Mensfoort and R. Coehoorn. Effect of Gaussian disorder on the voltage dependence of the current density in sandwich-type devices based on organic semiconductors. *Phys. Rev. B.*, 78(8):085207, 2008.
10. G. Paasch and S. Scheinert. Charge carrier density of organics with Gaussian density of states: Analytical approximation for the Gauss-Fermi integral. *J. Appl. Phys.*, 107(10):104501, 2010.

11. D. L. Scharfetter and H. K. Gummel. Large-signal analysis of a silicon Read diode oscillator. *IEEE Trans. Electron. Dev.*, 16(1):64–77, 1969.

12. W. Fichtner, D. J. Rose, and R. E. Bank. Semiconductor device simulation. *SIAM J. Sci. Stat. Comput.*, 4(3):391–415, 1983.

13. S. Selberherr. *Analysis and Simulation of Semiconductor Devices*. Berlin: Springer, 1984.

14. U. Bandelow, H. Gajewski, and R. Hünlich. Fabry–Perot lasers: Thermodynamics-based modeling. In J. Piprek, editor, *Optoelectronic Devices*. Berlin: Springer, 2005.

15. A. Mielke. A gradient structure for reaction-diffusion systems and for energy-drift diffusion systems. *Nonlinearity*, 24(4):1329, 2011.

16. S. R. de Groot and P. Mazur. *Non-Equilibrium Thermodynamics*. North Holland: Dover, 1962.

17. P. Deuflhard *Newton Methods for Nonlinear Problems: Affine Invariance and Adaptive Algorithms*, volume 35. Berlin: Springer, 2011.

18. D. N. Allen and R. V. Southwell. Relaxation methods applied to determine the motion, in two dimensions, of a viscous fluid past a fixed cylinder. *Quart. J. Mech. Appl. Math.*, 8:129–145, 1955.

19. A. M. Il'in. A difference scheme for a differential equation with a small parameter multiplying the second derivative. *Mat. Zametki.*, 6:237–248, 1969.

20. J. S. Chang and G. Cooper. A practical difference scheme for Fokker-Planck equations. *J. Comput. Phys.*, 6(1):1–16, 1970.

21. H.-G. Roos, M. Stynes, and L. Tobiska. *Robust Numerical Methods for Singularly Perturbed Differential Equations*, volume 24. Berlin: Springer, 2008.

22. K. W. Morton. *Numerical Solution of Convection-Diffusion Problems: Applied Mathematics*. London: Taylor & Francis, 1996.

23. D. Schröder. *Modelling of Interface Carrier Transport for Device Simulation*. Wien: Springer, 1994.

24. S. Y. Karpov. *Visible Light-Emitting Diodes*, Chapter 14. In J. Piprek, editor, *Nitride Semiconductor Devices: Principles and Simulation*, pp. 303–325. Weinheim, Germany: Wiley-VCH Verlag GmbH & Co. KGaA, 2007.

25. J. S. Blakemore. The parameters of partially degenerate semiconductors. *Proc. Phys. Soc. London A.*, 65:460–461, 1952.

26. M. Gruber, E. Zojer, F. Schürrer, and K. Zojer. Impact of materials versus geometric parameters on the contact resistance in organic thin-film transistors. *Adv. Funct. Mater.*, 23(23):2941–2952, 2013.

27. T. Koprucki and K. Gärtner. Discretization scheme for drift-diffusion equations with strong diffusion enhancement. *Opt. Quant. Electron.*, 45(7):791–796, 2013.

28. R. Haase. *Thermodynamics of Irreversible Processes*. Berlin: Addison-Wesley, 1968.

29. H. Gajewski and K. Gärtner. On the discretization of van Roosbroeck's equations with magnetic field. *Z. Angew. Math. Mech.*, 76(5):247–264, 1996.

30. H. Gajewski and K. Gröger. Semiconductor equations for variable mobilities based on Boltzmann statistics or Fermi-Dirac statistics. *Math. Nachr.*, 140(1):7–36, 1989.

31. G. Albinus, H. Gajewski, and R. Hünlich. Thermodynamic design of energy models of semiconductor devices. *Nonlinearity*, 15(2):367, 2002.

32. M. S. Mock. On equations describing steady-state carrier distributions in a semiconductor device. *Comm. Pure Appl. Math.*, 25(25):781–792, 1972.

33. H. Gajewski and K. Gröger. On the basic equations for carrier transport in semiconductors. *J. Math. Anal. Appl.*, 113:12–35, 1986.

34. H. Gajewski. On existence, uniqueness and asymptotic behavior of solutions of the basic equations for carrier transport in semiconductors. *Z. Angew. Math. Mech.*, 65(2):101–108, 1985.

35. P. A. Markowich. A nonlinear eigenvalue problem modelling the avalanche effect in semiconductor diodes. *SIAM J. Math. Anal.*, 6:1268–1283, 1985.

36. A. Jüngel. On the existence and uniqueness of transient solutions of a degenerate nonlinear drift-diffusion model for semiconductors. *Math. Models Methods Appl. Sci.*, 04:677, 1994.

37. J. W. Jerome. *Analysis of Charge Transport. A Mathematical Study of Semiconductor Devices.* Berlin Heidelberg: Springer, 1996.

38. H. K. Gummel. A self-consistent iterative scheme for one-dimensional steady state transistor calculations. *IEEE Trans. Electron Dev.*, 11(10):455–465, 1964.

39. P. A. Markowich. *The Stationary Semiconductor Device Equations.* Berlin: Springer-Verlag, 1986.

40. D. Gilbarg and N. S. Trudinger. *Elliptic Partial Differential Equations of Second Order.* Berlin: Springer, 2001. Reprint of the 1998 edition.

41. I. Rubinstein. *Electro-Diffusion of Ions*, volume 11. Philadelphia, PA: SIAM, 1990.

42. R. H. Macneal. An asymmetrical finite difference network. *Quart. Math. Appl.*, 11:295–310, 1953.

43. P. Fleischmann and S. Selberherr. Three-dimensional Delaunay mesh generation using a modified advancing front approach. In *Proceedings of IMR97*, pp. 267–278, 1997.

44. R. Eymard, T. Gallouët, and R. Herbin. Finite volume methods. In *Handbook of Numerical Analysis*, P.G. Ciarlet and J.L. Lions (Eds.), 7:713–1018, 2000.

45. J. W. Barrett and C. M. Elliott. Finite element approximation of the Dirichlet problem using the boundary penalty method. *Numer. Math.*, 49(4):343–366, 1986.

46. R. Eymard, J. Fuhrmann, and K. Gärtner. A finite volume scheme for nonlinear parabolic equations derived from one-dimensional local Dirichlet problems. *Numer. Math.*, 102(3):463–495, 2006.

47. K. Gärtner. Existence of bounded discrete steady state solutions of the van Roosbroeck system with monotone Fermi-Dirac statistic functions. *J. Comput. Electron.*, 14(3):773–787, 2015.

48. M. Bessemoulin-Chatard. A finite volume scheme for convection–diffusion equations with nonlinear diffusion derived from the Scharfetter–Gummel scheme. *Numer. Math.*, 121(4):637–670, 2012.

49. T. Koprucki, N. Rotundo, P. Farrell, D. H. Doan, and J. Fuhrmann. On thermodynamic consistency of a Scharfetter–Gummel scheme based on a modified thermal voltage for drift-diffusion equations with diffusion enhancement. *Opt. Quant. Electron.*, 47(6):1327–1332, 2015.

50. J. Fuhrmann. Comparison and numerical treatment of generalised Nernst-Planck models. *Comput. Phys. Commun.*, 196:166–178, 2015.

51. H. Gajewski. *Analysis und Numerik von Ladungstransport in Halbleitern.* WIAS Report No. 6. Berlin: WIAS, 1993. ISSN 0942-9077.

52. H. Gajewski, M. Liero, R. Nürnberg, and H. Stephan. *WIAS-TeSCA User Manual 1.2.* WIAS Technical Report No. 14. Berlin: WIAS, 2016. ISSN: WIAS 1618-7776.

53. A. Glitzky and K. Gärtner. Energy estimates for continuous and discretized electro-reaction-diffusion systems. *Nonlinear Anal.*, 70(2):788–805, 2009.

54. C. Chainais-Hillairet. Entropy method and asymptotic behaviours of finite volume schemes. In J. Fuhrmann, M. Ohlberger, and C. Rohde, editors, *Finite Volumes for Complex Applications VII – Methods and Theoretical Aspects*, pp. 17–35. Berlin: Springer, 2014.

55. M. Bessemoulin-Chatard and C. Chainais-Hillairet. Exponential decay of a finite volume scheme to the thermal equilibrium for drift–diffusion systems. *J. Numer. Math.*, 2016. (Online first, doi: 10.1515/jnma-2016-0007)
M. Bessemoulin-Chatard and C. Chainais-Hillairet. Exponential decay of a finite volume scheme to the thermal equilibrium for drift–diffusion systems. *arXiv preprintarXiv:1601.00813*, 2016.

56. K. Gärtner. Existence of bounded discrete steady-state solutions of the van Roosbroeck system on boundary conforming delaunay grids. *SIAM J. Sci. Comput.*, 31(2):1347–1362, 2009.

57. R. D. Lazarov, I. D. Mishev, and P. S. Vassilevski. Finite volume methods for convection diffusion problems. *SIAM J. Numer. Anal.*, 33(1):31–55, 1996.

58. J. J. H. Miller and S. Wang. An analysis of the Scharfetter-Gummel box method for the stationary semiconductor device equations. *RAIRO-Math. Num.*, 28(2):123–140, 1994.

59. J. Xu and L. Zikatanov. A monotone finite element scheme for convection-diffusion equations. *Math. Comput. Am. Math. Soc.*, 68(228):1429–1446, 1999.

60. C. Chainais-Hillairet, J.-G. Liu, and Y.-J. Peng. Finite volume scheme for multidimensional drift-diffusion equations and convergence analysis. *ESAIM: Math. Model. Numer. Anal. Math. Num.*, 37(2):319–338, 2003.

61. R. E. Bank, W. M. Coughran Jr, and L. C. Cowsar. The finite volume Scharfetter-Gummel method for steady convection diffusion equations. *Comput. Vis. Sci.*, 1(3):123–136, 1998.

62. M. Augustin, A. Caiazzo, A. Fiebach, J. Fuhrmann, V. John, A. Linke, and R. Umla. An assessment of discretizations for convection-dominated convection-diffusion equations. *Comp. Meth. Appl. Mech. Eng.*, 200:3395–3409, 2011.

63. A. Kuzmin, M. Luisier, and O. Schenk. Fast methods for computing selected elements of the greens function in massively parallel nanoelectronic device simulations. In F. Wolf, B. Mohr, and D. Mey, editors, *Euro-Par 2013 Parallel Processing*, volume 8097, *Lecture Notes in Computer Science*, pp. 533–544. Berlin: Springer, 2013.

64. O. Schenk, M. Bollhöfer, and R. A. Römer. On large-scale diagonalization techniques for the Anderson model of localization. *SIAM Rev.*, 50(1):91–112, 2008.

65. O. Schenk, A. Wächter, and M. Hagemann. Matching-based preprocessing algorithms to the solution of saddle-point problems in large-scale nonconvex interior-point optimization. *Comput. Optim. Appl.*, 36(2–3):321–341, 2007.

66. O. Schenk. PARDISO version 5.0.0. http://www.pardiso-project.org. Accessed on February 22, 2016.

67. T. A. Davis. *Direct Methods for Sparse Linear Systems (Fundamentals of Algorithms 2)*. Philadelphia, PA: Society for Industrial and Applied Mathematics, 2006.

68. T. Davis. SuiteSparse version 4.5.1. http://faculty.cse.tamu.edu/davis/suitesparse.html. Accessed on February 22, 2016.

69. Y. Saad. *Iterative Methods for Sparse Linear Systems*. Philadelphia, PA: SIAM, 2003.

70. P. Ciarlet. *The Finite Element Method for Elliptic Problems*. Philadelphia, PA: SIAM, 2002.

71. H. Si, K. Gärtner, and J. Fuhrmann. Boundary conforming Delaunay mesh generation. *Comput. Math. Math. Phys.*, 50(1):38–53, 2010.

72. J. Shewchuk. Triangle: A two-dimensional quality mesh generator and Delaunay triangulator. http://www.cs.cmu.edu/~quake/triangle.html. Accessed on December 1, 2015.

73. H. Edelsbrunner. *Algorithms in Combinatorial Geometry*. Heidelberg: Springer-Verlag, 1987.

74. S.-W. Cheng, T. K. Dey, and J. R. Shewchuk. *Delaunay Mesh Generation*. Boca Raton, FL: Chapman & Hall/CRC, 1st edition, 2012.

75. H. Si. TetGen version 1.5. http://tetgen.org/. Accessed on 2015-12-01.

76. J. Schöberl. Netgen an advancing front 2d/3d-mesh generator based on abstract rules. *Comput. Vis. Sci.*, 1(1):41–52, 1997.

77. M. S. Shephard and M. K. Georges. Automatic three-dimensional mesh generation by the finite octree technique. *Int. J. Numer. Meth. Eng.*, 32(4):709–749, 1991.

78. K. Gärtner and R. Richter. DEPFET sensor design using an experimental 3d device simulator. *Nucl. Instrum. Meth. A.*, 568(1):12–17, 2006.

79. B. Schmithüsen, K. Gärtner, and W. Fichtner. Grid adaptation for device simulation according to the dissipation rate. In *Simulation of Semiconductor Processes and Devices 1998*, pp. 197–200. Berlin: Springer, 1998.

80. J. Krause, N. Strecker, and W. Fichtner. Boundary-sensitive mesh generation using an offsetting technique. *Int. J. Numer. Meth. Eng.*, 49(1–2):51–59, 2000.

81. A. Fiebach, A. Glitzky, and A. Linke. Uniform global bounds for solutions of an implicit voronoi finite volume method for reaction–diffusion problems. *Numer. Math.*, 128(1):31–72, 2014.

82. A. Glitzky. Exponential decay of the free energy for discretized electro-reaction–diffusion systems. *Nonlinearity*, 21(9):1989, 2008.

83. E. Hairer, S. P. Norsett, and G. Wanner. *Solving Ordinary Differential Equations: Nonstiff Problems. v. 2: Stiff and Differential-Algebraic Problems*. Berlin: Springer Verlag, 2010.

84. M. S. Mock. A time-dependent numerical model of the insulated-gate field-effect transistor. *Solid-State Electron.*, 24(10):959–966, 1981.

85. P. D. Yoder, K. Gärtner, U. Krumbein, and W. Fichtner. Optimized terminal current calculation for Monte Carlo device simulation. *IEEE Trans. Comput-Aided Design Integr. Circuits Syst.*, 16(10):1082–1087, 1997.

86. M. Auf Der Maur. *A Multiscale Simulation Environment for Electronic and Optoelectronic Devices.* PhD thesis, University of Tor Vergata, 2008.

87. V. John and P. Knobloch. On spurious oscillations at layers diminishing (SOLD) methods for convection–diffusion equations: Part I–A review. *Comput. Methods Appl. Mech. Eng.*, 196(17):2197–2215, 2007.

88. A. Prohl and M. Schmuck. Convergent discretizations for the Nernst-Planck-Poisson system. *Numer. Math.*, 111(4):591–630, 2009.

89. V. Palankovski and R. Quay. *Analysis and Simulation of Heterostructure Devices.* Computational Microelectronics. Vienna: Springer Science & Business Media, 2004.

90. H. Bässler. Charge transport in disordered organic photoconductors a Monte Carlo simulation study. *Phys. Stat. Sol. (B).*, 175(1):15–56, 1993.

91. J. Fabian, S. C. Erwin, and I. Zutic. Bipolar spintronics: Fundamentals and applications. *IBM J. Res. Dev.*, 50(1):121–139, 2006.

92. A. Glitzky and K. Gärtner. Existence of bounded steady state solutions to spin-polarized drift-diffusion systems. *SIAM J. Math. Anal.*, 41(6):2489–2513, 2010.

93. A. Jüngel, C. Negulescu, and P. Shpartko. Bounded weak solutions to a matrix drift diffusion model for spin-coherent electron transport in semiconductors. *Math. Models Methods Appl. Sci.*, 25(5):929–958, 2015.

94. C. Chainais-Hillairet, A. Jüngel, and P. Shpartko. A finite-volume scheme for a spinorial matrix drift-diffusion model for semiconductors. *Numer. Methods for Partial Differ. Equ.*, 32(3):819–846, 2016.

95. M. Auf der Maur, A. Pecchia, G. Penazzi, F. Sacconi, and A. Carlo. Coupling atomistic and continuous media models for electronic device simulation. *J. Comput. Electron.*, 12(4):553–562, 2013.

96. M. Hinze and R. Pinnau. An optimal control approach to semiconductor design. *Math. Models Methods Appl. Sci.*, 12(1):89–107, 2002.

97. M. Hinze and R. Pinnau. Second-order approach to optimal semiconductor design. *J. Optim. Theory. Appl.*, 133(2):179–199, 2007.

98. M. Burger and R. Pinnau. Fast optimal design of semiconductor devices. *SIAM J. Appl. Math.*, 64(1):108–126, 2003.

99. D. Peschka, N. Rotundo, and M. Thomas. Towards doping optimization of semiconductor lasers. To appear in. *J. Comput. Theor. Trans.*, 45(5):410–423, 2016.

100. T. Grasser, T.-W. Tang, H. Kosina, and S. Selberherr. A review of hydrodynamic and energy-transport models for semiconductor device simulation. *Proc. IEEE*, 91(2):251–274, 2003.

101. S.-M. Hong and C. Jungemann. A fully coupled scheme for a Boltzmann-Poisson equation solver based on a spherical harmonics expansion. *J. Comput. Electron.*, 8(3):225–241, 2009.

102. A. Jüngel. *Transport Equations for Semiconductors*, volume 773. Berlin: Springer, 2009.

51

Monte Carlo Device Simulations

Katerina Raleva

Abdul R. Shaik

Raghuraj Hathwar

Akash Laturia

Suleman S. Qazi

Robin Daugherty

Dragica Vasileska

and

Stephen M. Goodnick

51.1 Introduction

The semiconductor device–based electronics industry has been the largest industry in the world with global sales of over a trillion dollars since 1998. The revolution in the semiconductor industry, a large portion of the electronics industry, began in 1947 with the fabrication of bipolar devices on slabs of polycrystalline germanium (Ge). Single-crystalline materials were later proposed and introduced that made possible the fabrication of grown junction transistors. Migration to silicon (Si)–based devices was initially hindered by the stability of the Si/SiO_2 materials system, necessitating a new generation of crystal pullers with improved environmental controls to prevent SiO_2 formation. Later, the stability and low interface-state density of the Si/SiO_2 materials system provided passivation of junctions and eventually, the migration from bipolar devices to field-effect devices in 1960. By 1968, both complementary metal–oxide–semiconductor (CMOS) devices and poly-Si gate technology that allowed self-alignment of the gate to the source/drain of the device had been developed. These innovations permitted a significant reduction in power dissipation and a reduction of the overlap capacitance, improving frequency performance and resulting in the essential components of the modern CMOS device.

Since the invention of the bipolar transistor in 1947, the number and variety of semiconductor devices have increased tremendously as advanced technology, new materials, and broadened comprehension have been applied to the creation and innovation of new devices. Dr. William Shockley suggested in the fifties the use of semiconductors of different bandgap for the fabrication of heterostructure devices. Professor Herbert Kroemer's contributions to heterostructures—from heterostructures bipolar transistors to lasers—culminated in a Nobel Prize in Physics in 2000 and have paved the way for novel heterostructure devices including those in silicon. The unique properties of semiconductor materials have enabled the development of a wide variety of ingenious devices that have literally changed our world. To date, there are about

TABLE 51.1   Major Semiconductor Device Discoveries

- 1874: Metal-semiconductor contact
- 1907: Light-emitting diode
- 1947: Bipolar junction transistor (BJT)
- 1954: Solar cell
- 1957: Heterojunction bipolar transistor (HBT)
- 1958: Tunnel diode
- 1959: Integrated circuits
- 1960: Field-effect transistors (FETs)
- 1962: Semiconductor lasers
- 1966: Metal-semiconductor FET (MESFET)
- 1967: Nonvolatile semiconductor memory
- 1974: Resonant tunneling diode (RTD)
- 1980: Modulation (MOD) FET
- 1994: Room-temperature single-electron memory cell (SEMC)
- 2001: 15 nm MOSFET
- 2006: 3 nm FinFET
- 2012: Single atom transistor in Si technology

---

60 major devices, with over 100 device variations related to them. A list of most of the basic semiconductor devices discovered and used over the past century with the date of their introduction is shown in Table 51.1.

The metal–oxide–semiconductor field-effect transistor (MOSFET) and related integrated circuits (ICs) now constitute about 90% of the semiconductor device market. Combining Si with the elegance of the field-effect transistor (FET) structure has allowed simultaneously making devices smaller, faster, and cheaper—the mantra that has driven the modern semiconductor microelectronics industry. Indeed, the single factor driving continuous device improvement has been the semiconductor industry's relentless effort to reduce the cost per function on a chip. The way this is done is to put more devices on a chip while either reducing manufacturing costs or holding them constant. This leads to three methods of reducing the cost per function. The first is *transistor scaling*, which involves reducing the transistor size in accordance with some goal, i.e., keeping the electric field constant from one generation to the next. With smaller transistors, more can fit into a given area than in previous generations. The second method is *circuit cleverness*, which is associated with the physical layout of the transistors with respect to each other. If the transistors can be packed into a tighter space, then more devices can fit into a given area than before. The third method is to make the die larger. More devices can be fabricated on a *larger die*. All the while, the semiconductor industry is constantly looking for technological breakthroughs to decrease the manufacturing cost. All of this effort serves to reduce the cost per function on a chip.

51.1.1 Transistor Scaling

Device engineers are most concerned with the method of scaling introduced in Section 51.1. The semiconductor industry has been so successful in providing continued system performance improvement year after year that the Semiconductor Industry Association (SIA) has been publishing roadmaps for semiconductor technology since 1992. These roadmaps represent a consensus outlook of industry trends, taking history as a guide. SIA roadmaps [1] incorporate participation from the global semiconductor industry, including the United States, Europe, Japan, Korea, and Taiwan. They basically affirm the desire of the industry to continue with *Moore's law* [2], which is often stated as the doubling of transistor performance and quadrupling of the number of devices on a chip every 3 years. The phenomenal progress signified by Moore's law has been achieved through scaling of the MOSFET from larger to smaller physical dimensions [3].

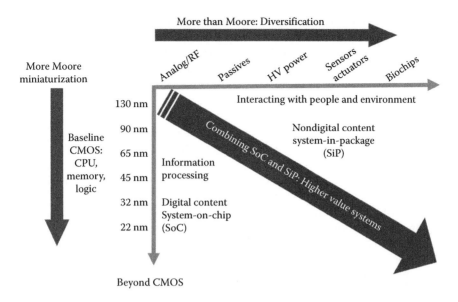

FIGURE 51.1 Schematic description of More Moore, More than Moore, and beyond complementary metal–oxide–semiconductor (CMOS). (RF, radio frequency; HV, high-voltage. [Reprinted from ITRS Roadmap for Semiconductors. Available at: www.itrs.org. With permission.])

The *More Moore* domain is internationally defined as an attempt to further develop advanced CMOS technologies and reduce the associated cost per function by introducing strain and silicon-on-insulator (SOI) technology (see Figure 51.1). Almost 70% of the total semiconductor components market is directly impacted by advanced CMOS miniaturization achieved in the More Moore domain. This 70% comprises three component groups of similar size, namely microprocessors, mass memories, and digital logic. As an example, in recent years, technology has advanced to fabricate ICs at 14-nm gate length commercially [4]. Fab industry giants such as Intel, TSMC, Samsung, and Global Foundries have plans to fabricate ICs at 10 nm by 2017. Samsung has already fabricated and tested 128 Mb SRAM in 10 nm [5] and is hoping to commercialize the process by the end of 2016. This aggressive scaling of technology is possible because of the advent of FinFETs and fully-depleted silicon-on-insulator (FD-SOI) device technology [6].

More than Moore (MtM) refers to a set of technologies that enable nondigital micro-/ nanoelectronic functions. MtM radio frequency (RF), high-voltage power, solid-state lighting (SSL), medical ultrasound, biochips and microfluidics, energy scavenging, electronic imaging, sensors, and actuators on CMOS platforms are some of the applications. They are based on, or derived from, Si technology but do not necessarily scale with Moore's Law. MtM devices typically provide conversion of nondigital as well as nonelectronic information, such as mechanical, thermal, acoustic, chemical, optical, and biomedical functions to digital data, and vice versa.

51.1.2 More Moore Technology

For digital circuits, a figure of merit for MOSFETs for unloaded circuits is CV/I, where C is the gate capacitance, V the voltage swing, and I the current drive of the MOSFET. For loaded circuits, the current drive of the MOSFET is of paramount importance. Keeping in mind both the CV/I metric and the benefits of a large current drive, one notes that device performance may be improved [7] by (1) inducing a larger charge density for a given gate voltage drive; (2) enhancing the carrier transport by improving the mobility, saturation velocity, or ballistic transport; (3) ensuring device scalability to achieve a shorter channel length; and (4) reducing parasitic capacitances and resistances. For capitalizing these opportunities, the proposed technology options generally fall into two categories: new materials and new device structures. In many cases, the introduction of a new material requires the use of a new device structure, or vice versa. To fabricate devices beyond current scaling limits, IC companies have simultaneously pushed the planar, bulk Si CMOS design while incorporating alternative gate stack materials (high-*k* dielectric [8] and metal gates), band engineering methods (using strained Si [9] or SiGe [10]), and alternative transistor structures. The concept of a band-engineered transistor is to enhance the mobility of electrons and/or holes in the channel by modifying the band structure of Si in the channel in a way such that the physical structure of the transistor remains substantially unchanged. This enhanced mobility increases the transistor transconductance (g_{m}) and on-drive current (I_{on}). A SiGe layer or a strained Si on a relaxed SiGe layer is used as an enhanced-mobility channel layer. It has already been demonstrated experimentally that at $T = 300$ K (room temperature), effective hole enhancement of about 50% can be achieved using the SiGe technology [10]. Intel has adopted strained Si technology for the first time for its 90-nm process [11]. This resulted in nearly 20% performance improvement per technology generation, with only a few additional process steps. The challenge in identifying suitable high-*k* dielectrics and metal gates for both conventional p-channel MOS (PMOS) and n-channel MOS (NMOS) transistors has led to early adoption of alternative transistor designs. These include primarily partially-depleted (PD) SOI and FD-SOI devices. Today, there is extensive research in double-gate (DG) structures and FinFET transistors [12], which have better electrostatic integrity and, theoretically, have better transport properties than single-gated FETs. A FinFET is a form of a DG transistor having surface conduction channels on two opposite vertical surfaces and having current flow in the horizontal direction. The channel length is given by the horizontal separation between source and drain and is usually determined by a lithographic step combined with a side-wall spacer etch process. Many innovative structures, involving structural challenges such as fabrication on nanometer-scale fins and nanometer-scale planarization over an entire wafer are currently being implemented. Some new and revolutionary technology such as carbon nanotubes or molecular transistors might be on the horizon, but it is not clear, in view of the predicted future capabilities of CMOS, whether they will be competitive. In conclusion, the semiconductor industry is approaching the end of an era of scaling gains by rote shrinkage of device dimensions and entering a postscaling era, a new phase of CMOS evolution in which innovation is demanded simply to compete. The trends in benefits to density, performance, and power will be continued through such innovations. Rather than coming to a close, a new era of CMOS technology is just beginning.

51.1.3 Importance of Modeling and Simulation

As semiconductor devices are scaled into nanoscale regimes, first, velocity saturation starts to limit the carrier mobility due to pronounced intervalley scattering; when the device dimensions are scaled below 200 nm, velocity overshoot (velocity of the carriers in the channel is larger than the steady-state saturation velocity) starts to dominate the device behavior leading to larger on-state currents. Alongside with the developments in the semiconductor nanotechnology, in recent years, there has been significant progress in physically based modeling of semiconductor devices. There are two issues that make simulation important. Product cycles are getting shorter with each generation, and the demand for production wafers shadows development efforts in the factory. Consider the product cycle issue first. In order for companies to maintain their competitive edge, products have to be taken from design to production in less than 18 months. As a result, the development phase of the cycle is getting shorter. Contrast this requirement with the fact that it takes 2–3 months to run a wafer lot through a factory, depending on its complexity. The specifications for experiments run through the factory must be near the final solution. While simulations may not be completely predictive, they provide a good initial guess. This can ultimately reduce the number of iterations during the device development phase. The second issue that reinforces the need for simulation is the production pressures that factories face. To meet customer demand, development factories are making way for production space. It is also expensive to run experiments through a production facility. Such resources could have otherwise been used to produce sellable products. Again, device simulation can be used to decrease the number of experiments run through a factory. Device simulation can be used as a tool to guide manufacturing down the right path, thereby decreasing the development time and costs. Besides offering the possibility to test hypothetical devices that have not (or could not) yet been manufactured, device simulation offers unique insight into device behavior by allowing the observation of phenomena that cannot be measured on real devices. It is related to, but usually separate from, process simulation, which deals with various physical processes such as material growth, oxidation, impurity diffusion, etching, and metal deposition inherent in device fabrication leading to ICs. Device simulation is distinct from another important aspect of computer-aided design (CAD), device modeling, which deals with compact behavioral models for devices and subcircuits relevant for circuit simulation in commercial packages such as simulation program with integrated circuit emphasis (SPICE). Device simulation can provide the parameters that are used to generate the compact behavioral models, and when coupled with process simulation and circuit simulation, it provides a hierarchical approach to technology computer-aided design (TCAD).

The main components of semiconductor device simulation at any level of approximation are illustrated in Figure 51.2 [13]. There are two main kernels, which must be solved self-consistently with one another, the *transport equations* governing charge flow, and the *fields* driving charge flow. Both are coupled strongly to one another, and hence must be solved simultaneously. The fields arise from external sources, as well as the charge and current densities which act as sources for the time-varying electric and magnetic fields obtained from the solution of Maxwell's equations. Under appropriate conditions, only the quasi-static electric fields arising from the solution of Poisson's equation are necessary. The fields, in turn, are driving forces for charge transport as illustrated in Figure 51.3 for the various levels of approximation within a hierarchical structure ranging from compact modeling at the top to an exact quantum-mechanical description at the bottom. For devices for which gradual channel approximation cannot be used due to the two-dimensional (2D) nature of the electrostatic potential and the electric fields driving the carriers from source to drain, drift-diffusion models have been exploited. These models are valid, in general, for large devices in which the fields are not that high so that there is no degradation of the mobility due to the electric field. The validity of the drift-diffusion models can be extended to take into account the velocity saturation effect with the introduction of field-dependent mobility and diffusion coefficients. When velocity overshoot becomes important (for length scales less than 200 nm—see Figure 51.4), the drift-diffusion model is no longer valid and the hydrodynamic model must be used.

The hydrodynamic model has been the workhorse for technology development and several high-end commercial device simulators have appeared including Silvaco, Synopsys, Crosslight, etc. The advantages of the hydrodynamic model are that it allows quick simulation runs, but the problem is that the amount of the velocity overshoot depends upon the choice of the energy relaxation time. The smaller the device,

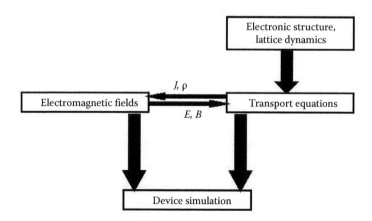

FIGURE 51.2 A schematic description of the building blocks of device simulator.

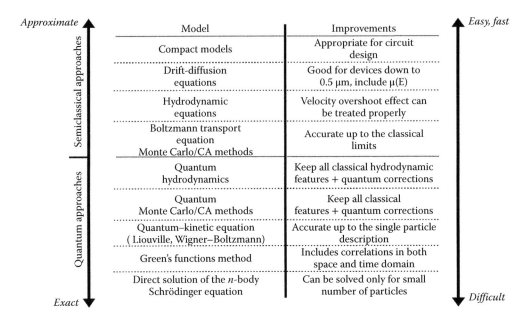

	Model	Improvements	
Approximate ↑ / Easy, fast ↑			
Semiclassical approaches	Compact models	Appropriate for circuit design	
	Drift-diffusion equations	Good for devices down to 0.5 μm, include μ(E)	
	Hydrodynamic equations	Velocity overshoot effect can be treated properly	
	Boltzmann transport equation Monte Carlo/CA methods	Accurate up to the classical limits	
Quantum approaches	Quantum hydrodynamics	Keep all classical hydrodynamic features + quantum corrections	
	Quantum Monte Carlo/CA methods	Keep all classical features + quantum corrections	
	Quantum–kinetic equation (Liouville, Wigner–Boltzmann)	Accurate up to the single particle description	
	Green's functions method	Includes correlations in both space and time domain	
Exact ↓ / Difficult ↓	Direct solution of the n-body Schrödinger equation	Can be solved only for small number of particles	

FIGURE 51.3 Illustration of the hierarchy of transport models.

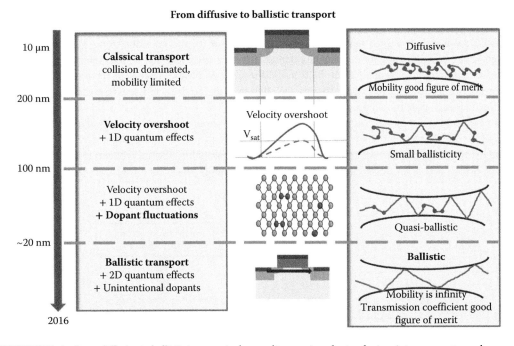

FIGURE 51.4 From diffusive to ballistic transport when scaling semiconductor devices into nanometer scale.

the larger the deviation when using the same set of energy relaxation times. A standard way in calculating the energy relaxation times is to use bulk Monte Carlo simulations. However, the energy relaxation times are material-, device geometry-, and doping-dependent parameters, so their determination ahead of time is not possible.

To avoid the problem of the proper choice of the energy relaxation times in hydrodynamic models, a direct solution of the Boltzmann transport equation (BTE)

$$\frac{\partial f}{\partial t} + \frac{1}{\hbar} \nabla_k E(k) \cdot \nabla_r f + \left(-\frac{e\mathbf{E}}{\hbar} \right) \cdot \nabla_k f = \left(\frac{\partial f}{\partial t} \right)_{\text{coll}} \qquad (51.1)$$

using the Monte Carlo method is the best method of choice for nanoscale devices in which position and momentum can still be treated as independent variables. In Equation 51.1, $f(r, k, t)$ is the semiclassical distribution function, $E(k)$ the dispersion relation for the electrons, and $\mathbf{E}$ the electric field. When, for example, collisional broadening of the states is important, quantum transport approaches must be pursued.

This chapter focuses on Monte Carlo methods for the solution of the BTE and is organized as follows. Monte Carlo solution of the BTE for homogeneous systems is given in Section 51.2. Particle-based device simulator is described in Section 51.3. Representative simulation results for bulk material and for semiconductor devices are given in Section 51.4. Conclusions and future directions of research are presented in Section 51.5.

51.2 Bulk Monte Carlo Method Description

The ensemble Monte Carlo (EMC) techniques have been used for well over 40 years as a numerical method to simulate nonequilibrium transport in semiconductor materials and devices and have been the subject of numerous books and reviews [14–16]. In the applications to transport problem, a random walk is generated using the random number–generating algorithms common to modern computers to simulate the stochastic motion of particles subject to collision processes. This process of random walk generation is part of a very general technique used to evaluate integral equations and is connected to the general random sampling technique used in the evaluation of multidimensional integrals [17].

The basic technique as applied to transport problems is to simulate the free particle motion (referred to as the *free flight*) terminated by instantaneous random *scattering events*. The Monte Carlo algorithm consists of generating random free-flight times for each particle, choosing the type of scattering occurring at the end of the free flight, changing the final energy and momentum of the particle after scattering, and then repeating the procedure for the next free flight. Sampling the particle motion at various times throughout the simulation allows for the statistical estimation of physically interesting quantities such as the single particle distribution function, the average drift velocity in the presence of an applied electric field, the average energy of the particles, etc. By simulating an *ensemble* of particles, representative of the physical system of interest, the nonstationary time-dependent evolution of the electron and hole distributions under the influence of a time-dependent driving force may be simulated. This particle-based picture, in which the particle motion is decomposed into free flights terminated by instantaneous collisions, is basically the same approximate picture underlying the derivation of the semiclassical BTE. In fact, it may be shown that the one-particle distribution function obtained from the random walk Monte Carlo technique satisfies the BTE for a homogeneous system in the long-time limit [18]. This semiclassical picture breaks down when quantum-mechanical effects become pronounced, and one cannot unambiguously describe the instantaneous position and momentum of a particle. In the following, we describe first the standard Monte Carlo algorithm used to simulate charge transport in semiconductors. We then discuss how this basic model for charge transport within the BTE is self-consistently solved with the appropriate field equations to perform particle-based device simulation (Section 51.3).

In the bulk Monte Carlo method [19], particle motion is assumed to consist of free flights terminated by instantaneous scattering events, which change the momentum and energy of the particle after scattering. So, the first task is to generate free flights of random time duration for each particle. To simulate this process, the probability density, $P(t)$, is required, in which $P(t)\text{d}t$ is the joint probability that a particle will arrive at time t without scattering after a previous collision occurring at time $t = 0$, and then suffer a collision in a time interval dt around time t. The probability of scattering in the time interval dt around t may be written as $\Gamma[\mathbf{k}(t)]$dt, where $\Gamma[\mathbf{k}(t)]$ is the scattering rate of an electron or hole of wavevector $\mathbf{k}$. The scattering rate, $\Gamma[\mathbf{k}(t)]$, represents the sum of the contributions from each individual scattering mechanism, which are usually calculated quantum mechanically using perturbation theory [20]. The implicit dependence of $\Gamma[\mathbf{k}(t)]$ on time reflects the change in $\mathbf{k}$ due to acceleration by internal and external fields. For electrons subject to time-independent electric and magnetic fields, the time evolution of $\mathbf{k}$ between collisions is represented as

$$\mathbf{k}(t) = \mathbf{k}(0) - \frac{e(\mathbf{E} + \mathbf{v} \times \mathbf{B})t}{\hbar}, \tag{51.2}$$

where $\mathbf{E}$ is the electric field, $\mathbf{v}$ the electron velocity, and $\mathbf{B}$ the magnetic flux density. In terms of the scattering rate, $\Gamma[\mathbf{k}(t)]$, the probability that a particle has not suffered a collision after a time t is given by $\exp\left(-\int_0^t \Gamma\left[k(t')\right] dt'\right)$. Thus, the probability of scattering in the time interval dt after a free flight of time t may be written as the joint probability:

$$P(t)\mathrm{d}t = \Gamma[k(t)] \exp\left[-\int_0^t \Gamma\left[k\left(t'\right)\right] \mathrm{d}t'\right]\mathrm{d}t. \tag{51.3}$$

Random flight times may be generated according to the probability density $P(t)$ above using, for example, the pseudo-random number generator implicit on most modern computers, which generate uniformly distributed random numbers in the range [0, 1]. Using a direct method, random flight times sampled from $P(t)$ may be generated according to

$$r = \int_0^{t_r} P(t)\mathrm{d}t, \tag{51.4}$$

where r is a uniformly distributed random number and t_r is the desired free-flight time. Integrating Equation 51.4 with $P(t)$ given by Equation 51.3 above yields

$$r = 1 - \exp\left[-\int_0^{t_r} \Gamma\left[k(t')\right]dt'\right]. \tag{51.5}$$

Since $1 - r$ is statistically the same as r, Equation 51.5 may be simplified to

$$-\ln r = \int_0^{t_r} \Gamma\left[\mathbf{k}\left(t'\right)\right] \mathrm{d}t'. \tag{51.6}$$

Equation 51.6 is the fundamental equation used to generate the random free-flight time after each scattering event, resulting in a random walk process related to the underlying particle distribution function. If there is no external driving field leading to a change of $\mathbf{k}$ between scattering events (e.g., in ultrafast photoexcitation experiments with no applied bias), the time dependence vanishes, and the integral is trivially evaluated. In the general case, where this simplification is not possible, it is expedient to introduce the so-called self-scattering method [21] in which one introduces a fictitious scattering mechanism whose scattering rate always adjusts itself in such a way that the total (self-scattering plus real scattering) rate is energy-independent:

$$\Gamma = \Gamma\left[\mathbf{k}(t')\right] + \Gamma_{\text{self}}\left[\mathbf{k}(t')\right], \tag{51.7}$$

where $\Gamma_{\text{self}}[\mathbf{k}(t')]$ is the self-scattering rate. The self-scattering mechanism itself is defined such that the final state before and after scattering is identical. Hence, it has no effect on the free-flight trajectory of a particle when selected as the terminating scattering mechanism, yet results in the simplification of Equation 51.6 such that the free flight is given by

$$t_r = -\frac{1}{\Gamma}\ln r. \tag{51.8}$$

The constant total rate (including self-scattering) Γ must be chosen at the start of the simulation interval (there may be multiple such intervals throughout an entire simulation) so that it is larger than the maximum scattering encountered during the same time interval. In the simplest case, a single value is chosen at the beginning of the entire simulation (constant gamma method), checking to ensure that the real rate never exceeds this value during the simulation. Other schemes may be chosen that are more computationally efficient, and which modify the choice of Γ at fixed time increments [22].

The algorithm described above determines the random free-flight times during which the particle dynamics is treated semiclassically. For the scattering process itself, we need the type of scattering (i.e., impurity, acoustic phonon, photon emission, etc.) which terminates the free flight, and the final energy and momentum of the particle(s) after scattering. The type of scattering that terminates the free flight is chosen using a uniform random number between 0 and Γ, and using this pointer to select among the relative total scattering rates of all processes including self-scattering at the final energy and momentum of the particle:

$$\Gamma = \Gamma_{\text{self}}[n, \mathbf{k}] + \Gamma_1[n, \mathbf{k}] + \Gamma_2[n, \mathbf{k}] + \cdots \Gamma_N[n, \mathbf{k}], \tag{51.9}$$

with n the band index of the particle (or sub-band in the case of reduced-dimensionality systems), and $\mathbf{k}$ the wavevector at the end of the free flight. This process is illustrated schematically in Figure 51.5.

Once the type of scattering terminating the free flight is selected, the final energy and momentum (as well as band or subband) of the particle due to this type of scattering must be selected. For elastic scattering processes such as ionized impurity scattering, the energy before and after scattering is the same. For the interaction between electrons and the vibrational modes of the lattice described as quasi-particles known as phonons, electrons exchange finite amounts of energy with the lattice in terms of emission and absorption of phonons. For determining the final momentum after scattering, the scattering rate, $\Gamma_j[n,\mathbf{k};m,\mathbf{k}']$ of the jth scattering mechanism is needed, where n and m are the initial and final band indices, and $\mathbf{k}$ and $\mathbf{k}'$ are the particle wavevectors before and after scattering. Defining a spherical coordinate system

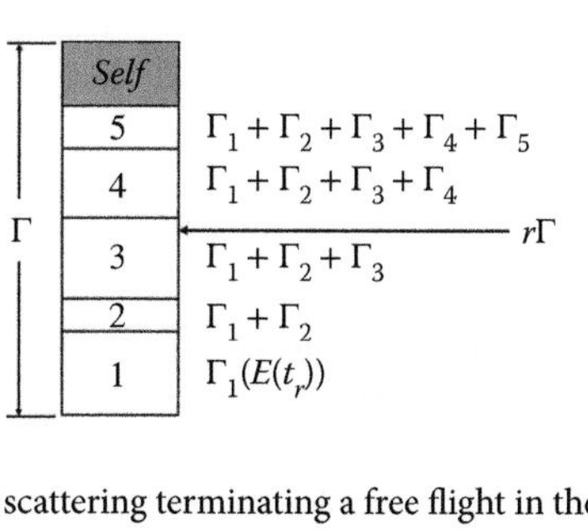

FIGURE 51.5 Selection of the type of scattering terminating a free flight in the Monte Carlo algorithm.

around the initial wavevector $\mathbf{k}$, the final wavevector $\mathbf{k}'$ is specified by $|\mathbf{k}'|$ (which depends on conservation of energy) as well as the azimuthal and polar angles, ϕ and θ around $\mathbf{k}$. Typically, the scattering rate, $\Gamma_j[n,\mathbf{k};m,\mathbf{k}']$, only depends on the angle θ between $\mathbf{k}$ and $\mathbf{k}'$. Therefore, ϕ may be chosen using a uniform random number between 0 and 2π (i.e., $2\pi r$), while θ is chosen according to the angular dependence for scattering arising from $\Gamma_j[n,\mathbf{k};m,\mathbf{k}']$. If the probability for scattering into a certain angle $P(\theta)d\theta$ is integrable, then random angles satisfying this probability density may be generated from a uniform distribution between 0 and 1 through a prescribed recipe. Otherwise, a rejection technique may be used to select random angles according to $P(\theta)$. Common scattering mechanisms that contribute to transport are summarized in Figure 51.6. The corresponding scattering rate expressions for general nonparabolic bands are summarized in Table 51.2 [23–25].

A general Monte Carlo code is then developed as follows: First, a subroutine is typically called that contains all material and scattering rates parameters for the scattering mechanisms included in the theoretical model. After the material and run parameters are read, in the first step of the Monte Carlo simulation procedure, it is necessary to construct scattering tables for the Γ, L, and X valleys (for GaAs as a prototypical example) that initialize a series of events that are summarized in Figure 51.7. Then, for each energy, the cumulative scattering rates for each valley are stored in separate look-up tables and renormalized according to the maximum scattering rate (including self-scattering) that occurs over the range of energies stored.

Having constructed the scattering table and after renormalizing the table, examples of which are given for GaAs in Figures 51.8 and 51.9 for the Γ, L, and X valleys, the next step is to initialize the carrier's wavevector and energy and the initial free-flight time. This is accomplished by calling the initialization subroutine. Energy and wavevector histograms of the initial carrier energy and the components of the wavevector along the x-, y-, and z-axes are shown in Figure 51.10. For good statistics, the number of particles simulated is typically taken to be 10,000, and one can see the statistical fluctuation of these average quantities associated with the finite number of particles. Notice that the variance is inversely proportional to the square-root of the number of particles. Also note that the initial y-component for the wavevector is symmetric around the y-axis, which means that the average wavevector along the y-axis is zero, which should be expected since the electric field along the y-component is zero at $t = 0$. Identical distributions have been obtained for the x- and for the z-components of the wavevector. Also note that the energy distribution has the Maxwell–Boltzmann form, as it should be expected. One can also estimate from this graph that the average energy of the carriers is on the order of $(3/2)k_B T$ (k_B is the Boltzmann constant and T the temperature).

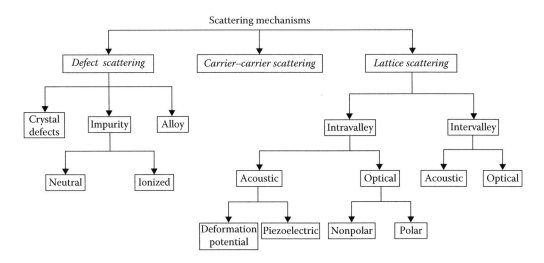

FIGURE 51.6 Scattering mechanisms in a typical semiconductor.

TABLE 51.2   Scattering Rate Expressions $\Gamma(k) = W(k) = W(E)$ for Electrons for General Nonparabolic Energy Bands

1. *Ionized impurity scattering:* Scattering from ionized Coulomb charges in the Brooks–Herring approach.

$$W(E) = \left(\frac{\sqrt{2}e^4 N_I m_d^{\frac{3}{2}}}{\pi \varepsilon_s^2 \gamma^4} \right) \times \left(\sqrt{E(1+\alpha E)} \times (1+2\alpha E) \right) \times \left(\frac{1}{q_D^2 \left(q_D^2 + \frac{8m_d E(1+\alpha E)}{\hbar^2} \right)} \right) \quad q_D = \sqrt{\frac{e^2(n+p)}{\varepsilon_s K_B T_L}},$$ where n and p are the electron and

hole concentrations, respectively, k_B is the Boltzmann constant, m_d is the density of states mass, T_L is the lattice temperature, ε_s is the permittivity of the material, E is the energy, and α is the nonparabolicity factor.

2. *Alloy disorder scattering* (e.g., $Al_x Ga_{1-x} As$): Scattering due to the presence of a random alloy.

$$W(E) = \left(\frac{x(1-x)a^3}{\pi} \right) \times \left(\frac{D_{alloy}^2 d}{\hbar^4} \right) \times m_d \sqrt{2m_d E(1+\alpha E)} \times (1+2\alpha E),$$

where x is the mole fraction, d the lattice disorder ($0 \le d \le 1$), a the lattice constant, and D_{alloy} the alloy disorder scattering potential.

3. *Dislocation scattering* (e.g., GaN): Scattering due to crystal defects/dislocations (Coulomb-type interaction).

$$W(E) = \left(\frac{N_{dis} m_d e^4}{4\hbar^3 \varepsilon^2 c^2} \right) \times \left(\frac{\lambda^4}{\left(1 + \frac{8\lambda^2 m_d E(1+\alpha E)}{\hbar^2} \right)^{\frac{3}{2}}} \right) \times \left(1 + \frac{4\lambda^2 m_d E(1+\alpha E)}{\hbar^2} \right) \times (1+2\alpha E) \quad \lambda = \sqrt{\frac{\varepsilon K_B T_L}{e^2 n'}},$$ where n' is the effective screening

concentration and N_{dis} the line dislocation density.

4. *Acoustic deformation potential scattering:* Describes fluctuations in the band structure due to deformation of the unit cell because of the strain due to the different amounts of displacement of the atoms (in the same direction).

$$W(E) = \left(\frac{2\pi D_{ac}^2 K_B T_L}{\hbar C_1} \right) \times \left(\frac{(2m_d)^{\frac{3}{2}} \sqrt{E(1+\alpha E)}}{4\pi^2 \hbar^3} \right) \times (1+2\alpha E),$$

where D_{ac} is the acoustic deformation potential constant and C_1 is the sound velocity.

5. *Piezoelectric scattering:* Dipole-type interaction in zinc-blende materials in which there is charge transfer between the two atoms in the unit cell. The optical modes of vibration (vibration of the atoms in opposite direction) directly affect the magnitude of the dipole moment, which in turn leads to scattering potential.

$$W(E) = \left(\frac{m_d^{\frac{1}{2}} K_B T_L}{4\sqrt{2}\pi \rho v_s^2 \hbar^2} \right) \times \left(\frac{1+2\alpha E}{\sqrt{E(1+\alpha E)}} \right) \times \left(\frac{e e_{pz}}{\varepsilon_\infty} \right)^2 \times \ln\left(1 + \frac{8m_d E(1+\alpha E)}{\hbar^2 q_D^2} \right) \quad q_D = \sqrt{\frac{e^2(n+p)}{\varepsilon_s K_B T_L}},$$ where ρ is the crystal density, v_s sound

velocity, e_{pz} the piezoelectric constant, and ε_∞ the high-frequency dielectric constant.

6. *Polar optical phonon (POP) scattering:* In polar materials (where there is transfer of charge from one atom to the other, which in turn, leads to a dipole moment), optical modes of vibration cause direct modification of the dipole moment, which then leads to polar optical phonon scattering which is a Coulomb-type interaction.

$$W(E) = \left(\frac{\sqrt{m_d} e^2 W_{LO}}{4\sqrt{2}\pi \hbar \varepsilon_p} \right) \times \left(N_0 + \frac{1}{2} \mp \frac{1}{2} \right) \times \left(\frac{1+2\alpha E_k'}{\gamma_k} \right) \times F\left(E_k, E_k' \right)$$

$$N_0 = \frac{1}{e^{\frac{\hbar W_{LO}}{K_B T_{L-1}}}} \quad \varepsilon_p = \frac{1}{\frac{1}{\varepsilon_{high}} - \frac{1}{\varepsilon_{low}}} \quad F(E_k, E_k') = \ln\left(\mathrm{mod}\left(\frac{\sqrt{\gamma_k} + \sqrt{\gamma_{k'}}}{\sqrt{\gamma_k} - \sqrt{\gamma_{k'}}} \right) \right)$$

$$\gamma_K = E_k(1+\alpha)E_k \quad E_k = E_k \pm \hbar W_{LO},$$

where W_{LO} is the optical phonon energy and the indices high/low stand for high frequency/static dielectric constants.

7. *Intervalley phonon scattering:* Scattering between equivalent or nonequivalent valleys.

$$W(E) = \left(\frac{\pi D_{ij}^2 Z_j}{\rho W_{ij}} \right) \times \left(n(W_{ij}) \frac{1}{2} \mp \frac{1}{2} \right) \times \left(\frac{(2m_d)^{\frac{3}{2}} \sqrt{E_f(1+\alpha E_f)}}{4\pi^2 \hbar^3} \right) \times (1+2\gamma E_f) \quad E_f = F \pm \hbar w_{ij} - \Delta E_{ij},$$

where D_{ij} is the coupling constant, Z_j the number of final valleys, ρ the crystal density, and w_{ij} the energy of the phonon involved in the process.

Sources: M. Lundstrom, Fundamentals of Carrier Transport, Cambridge University Press, London, 2000; B. K. Ridley, *Quantum Processes in Semiconductors, Oxford University Press, 1982;* D. Vasileska nanoHUB.org page: nanohub.org/groups/dragica_vasileska/semiclassical For more details on the derivation of these formulas, please refer to Refs. [26–28].

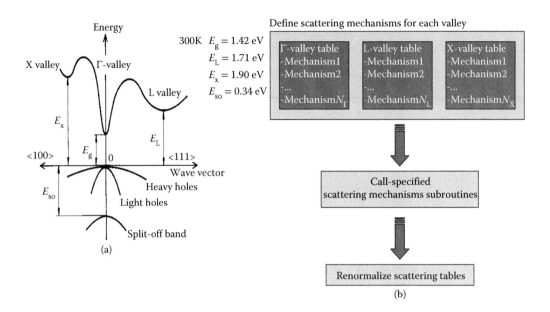

FIGURE 51.7   (a) Schematics of a bandstructure for GaAs. (b) Procedure for the creation of the scattering tables for electron transport.

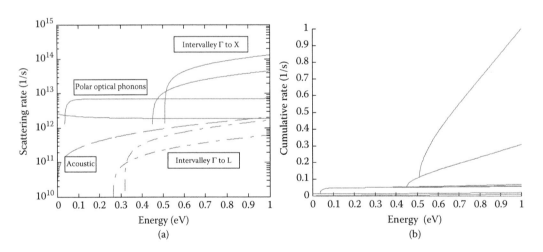

FIGURE 51.8   (a) Scattering rates for the Γ valley. For simplicity, we have omitted Coulomb scattering in these calculations. In (a) the dashed line corresponds to the acoustic phonon scattering rate, solid lines correspond to polar optical phonon scattering (absorption and emission), and the dashed-dotted line corresponds to intervalley scattering from Γ to L valleys. Since the L valley is along the [111] direction, there are eight equivalent directions, and since these valleys are shared, there are a total of four equivalent L valleys. The dotted line corresponds to scattering from Γ to X valleys. The X valleys are in the [100] direction and since there are six equivalent [100] directions and the valleys are shared between Brillouin zones, there are three equivalent X valleys. (b) Normalized cumulative scattering table for the Γ valley. Everything above the top line up to $\Gamma = 1$ is self-scattering so it is advisable when checking the scattering mechanisms to first check whether the scattering mechanism chosen is self-scattering or not. This is in particular important for energies below 0.5 eV for this particular scattering table when the Γ to X intervalley scattering (absorption and emission) takes over.

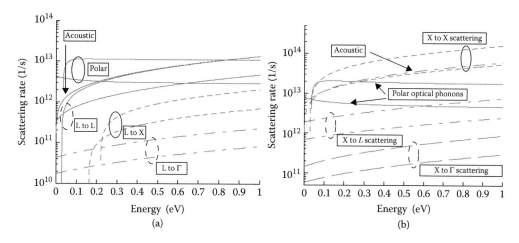

FIGURE 51.9 Scattering rates for the L (a) and X (b) valleys used to create the corresponding normalized scattering tables (not shown here).

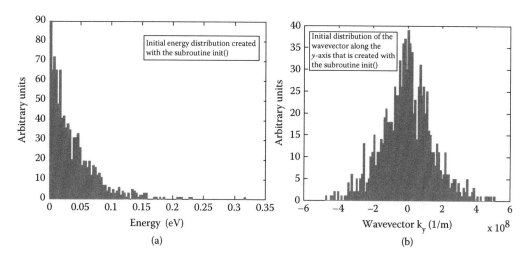

FIGURE 51.10 Initial carrier distribution for an ensemble of 10,000 particles. (a) Energy distribution. (b) Distribution of wavevector k_y.

When the initialization process is finished, the main free-flight-scatter procedure takes place until the completion of the simulation time. There are two components in this process: first, the carriers accelerate freely due to the electric field, accomplished by calling the **drift()** subroutine (which implements Equation 51.2), and then their free-flights are interrupted by random scattering events that are managed by the **scatter_carrier()** subroutine. The flowchart for performing the free-flight-scatter process within one time-step Δt is shown diagrammatically in Figure 51.11.

In the **scatter_carrier()** subroutine from Figure 51.11, first the scattering mechanism terminating the free flight is chosen, to which certain attributes are associated such as the change in energy after scattering. For inelastic scattering processes, we have the change in energy due to emission or absorption of phonons, for example. Also, the nature of the scattering process is identified as isotropic or anisotropic. Note that when performing acoustic phonon and intervalley scattering for GaAs, both of which are isotropic scattering processes, no coordinate system transformation is needed to determine the final wavevector after scattering. Because polar optical phonon and Coulomb scattering mechanisms are anisotropic, it is necessary to do a rotation of the coordinate system, scatter the carrier in the rotated system, and then perform

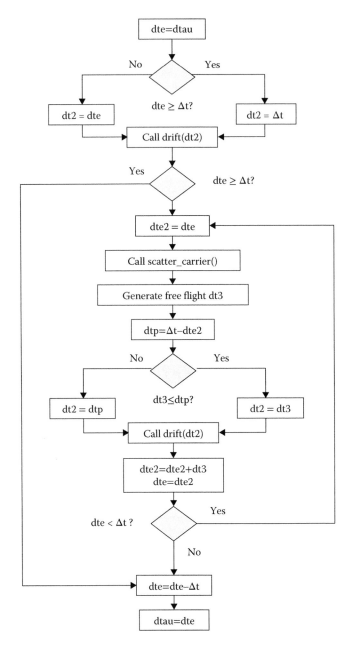

FIGURE 51.11 Free-flight-scatter procedure within one time-step. Time variable dtau is the remaining free-flight time from previous time-step Δt.

inverse coordinate transformation. This procedure is needed because it is much easier to determine final carrier momentum in the rotated coordinate system in which the initial wavevector k is aligned with the z-axis. For this case, one can calculate that the final polar angle of scattering with polar optical phonons for parabolic bands in the rotated coordinate system as

$$\cos\theta = \frac{(1+\xi)-(1+2\xi)^r}{\xi}, \quad \xi = \frac{2\sqrt{E_k\left(E_k \pm \hbar\omega_0\right)}}{\left(\sqrt{E_k} - \sqrt{E_k \pm \hbar\omega_0}\right)^2}, \quad (51.10)$$

where E_k is the carrier energy, $\hbar\omega_0$ the polar optical phonon energy, and r a random number uniformly distributed between 0 and 1. The final angle for scattering with ionized impurities (Coulomb scattering) and for parabolic bands is

$$\cos\theta = 1 - \frac{2r}{1 + 4k^2 L_D^2(1-r)}, \tag{51.11}$$

where k is the carrier wavevector and L_D is the Debye screening length. The azimuthal angle for both scattering processes is simply calculated using $\varphi = 2\pi r$. The importance of properly calculating the polar angle θ after scattering to describe small angles of deflection in the case of Coulomb or polar optical phonon scattering is illustrated in Figure 51.13. The histogram of the polar angle (from 0 to $\pi = 3.141592654$) after scattering for electron–polar optical phonon scattering is presented in Figure 51.12, from where we can clearly see the preference for small angle deflections that are characteristic for any Coulomb-type interaction (polar optical phonon is in fact electron–dipole interaction). Graphical representation of the determination of the final angle after scattering for both isotropic and anisotropic scattering processes is shown in Figure 51.13.

The direct technique described above can be applied when the integrals describing $\cos\theta$ can be analytically calculated. For most cases of interest, the integral cannot be easily inverted. In these cases, a rejection technique may be employed. The procedure for the rejection technique goes as follows:

- Choose a maximum value C, such that $C > f(x)$ for all x in the interval (a, b).
- Choose pairs of random numbers, one between a and b $\left(x_1 = a + r_1(b - a)\right)$ and another one $f_1 = r_1' C$ between 0 and C, where r_1 and r_1' are random numbers uniformly distributed between 0 and 1.
- If $f_1 \leq f(x_1)$, then the number x_1 is accepted as a suitable value, otherwise it is rejected.

The three steps described above are schematically shown in Figure 51.14. For $x = x_1$, $r_1 C$ is larger than $f(x_1)$ and in this case, if this represents the final polar angle for scattering, this angle is rejected and a new sequence of two random numbers is generated to determine x_2 and $r_2 C$. In this second case, $f(x_2) > r_2 C$ and the polar angle $\theta = x_2$ is selected (for polar angle selection $a = 0$ and $b = \pi$).

After the simulation is completed, typical results to check are the velocity–time, the energy–time, and the valley occupation versus time characteristics, such as those shown in Figure 51.15, where the velocity–time characteristics for applied electric fields ranging from 0.5 to 7 kV/cm, with an electric field increment

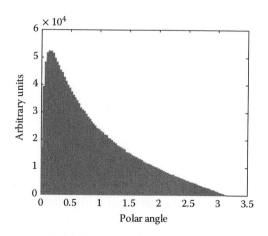

FIGURE 51.12 Histogram of the polar angle for electron–polar optical phonon scattering.

1. Isotropic scattering processes

$$\cos\theta = 1 - 2r, \varphi = 2\pi r$$

2. Anisotropic scattering processes (Coulomb, POP)

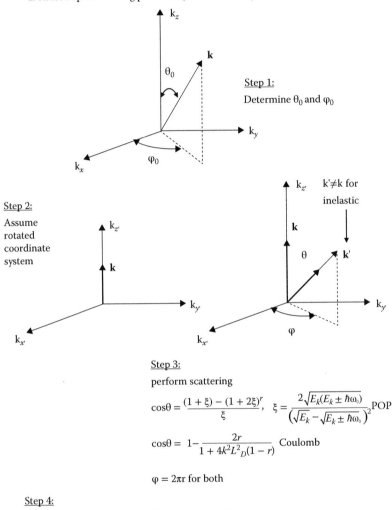

Step 1:

Determine θ_0 and φ_0

Step 2:

Assume rotated coordinate system

$k' \neq k$ for inelastic

Step 3:

perform scattering

$$\cos\theta = \frac{(1 + \xi) - (1 + 2\xi)^r}{\xi}, \quad \xi = \frac{2\sqrt{E_k(E_k \pm \hbar\omega_0)}}{\left(\sqrt{E_k} - \sqrt{E_k \pm \hbar\omega_0}\right)^2} \text{POP}$$

$$\cos\theta = 1 - \frac{2r}{1 + 4k^2 L^2{}_D(1 - r)} \quad \text{Coulomb}$$

$$\varphi = 2\pi r \text{ for both}$$

Step 4:

$k_{xp} = k'\sin\theta\cos\varphi, k_{yp} = k'\sin\theta^*\sin\varphi, k_{zp} = k'\cos\theta$

Return back to the original coordinate system:
$k_x = k_{xp}\cos\varphi_0\cos\theta_0 - k_{yp}\sin\varphi_0 + k_{zp}\cos\varphi_0\sin\theta_0$
$k_y = k_{xp}\sin\varphi_0\cos\theta_0 + k_{yp}\cos\varphi_0 + k_{zp}\sin\varphi_0\sin\theta_0$
$k_z = -k_{xp}\sin\theta_0 + k_{zp}\cos\theta_0$

FIGURE 51.13 Description of final angle selection for isotropic and anisotropic scattering processes using the direct technique.

of 0.5 kV/cm, are shown. These clearly demonstrate that after a transient phase, the system reaches a stationary steady state, after which time we can start taking averages for calculating steady-state quantities.

From the results shown in Figure 51.15, one can see that steady state is achieved for larger time intervals when the electric field value is increased and the carriers are still sitting in the Γ valley. Afterward, time needed to get to steady state decreases. This trend is related to the valley repopulation and movement of the

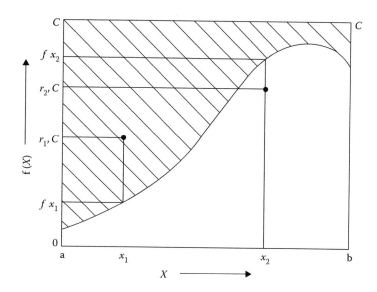

FIGURE 51.14 Schematic description of the rejection technique.

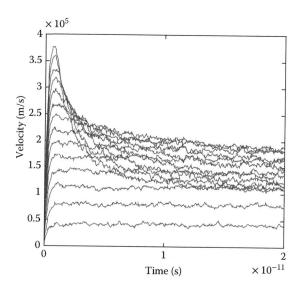

FIGURE 51.15 Time evolution of the drift velocity for electric field strengths ranging between 0.5 and 7 kV/cm, in 0.5 kV/cm increments.

carriers from Γ, into the X and finally into the L valley. The steady-state velocity-field and valley population versus electric field characteristics are shown in Figures 51.16 and 51.17, respectively. One can clearly see on the velocity–field characteristics that a low field mobility of about 8000 cm^2/V-s is correctly reproduced for GaAs without the use of any adjustable parameters.

At this point, it is advisable to check the energy and wavevector histograms (Figure 51.18) to ensure that the energy range chosen in the scattering tables is correct or not for the particular maximum electric field strength being considered, which gives the worst case scenario. Since, as already noted, we apply the electric field in the y-direction, for comparative purposes, we plot the histograms of the x-component of the wavevector, y-component of the wavevector, and the histogram of the final carrier energy distribution

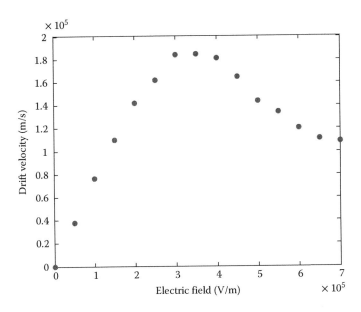

FIGURE 51.16 Steady-state drift velocity versus electric field.

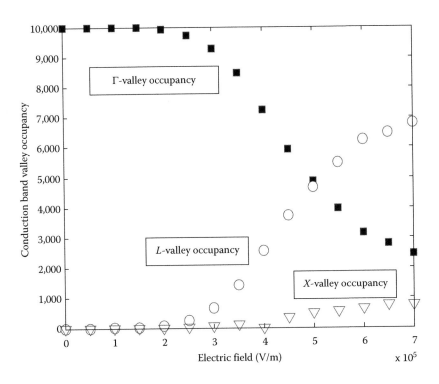

FIGURE 51.17 Different valley occupancy versus electric field.

for which a drifted Maxwellian form is evident. Since there is no field applied in the x-direction, we see that the average wavevector in the x-direction is 0. Due to the application of the field in the y-direction, there is a finite positive shift in the y-component of the velocity, which is yet another signature for the displaced Maxwellian form of the energy distribution in the bottom histogram.

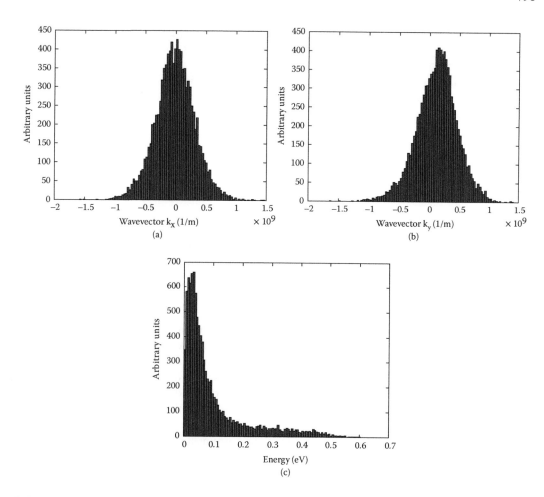

FIGURE 51.18 (a) Histogram of the *x*-component of the wavevector. (b) Histogram of the *y*-component of the wavevector. (c) Histogram of the carrier energy. Applied electric field is 7 kV/cm.

51.3 Particle-Based Device Simulation

In Section 51.2, we introduced the numerical solution of the BTE using Monte Carlo method. Within a device, both the transport kernel and the field solver are coupled to each other. In Poisson's equation, the field associated with the potential is the driving force accelerating particles in the Monte Carlo phase, for example, while the distribution of mobile (both electrons and holes) and fixed charges (e.g., donors and acceptors) provides the source of the electric field. Below, we give a brief description of the Monte Carlo particle-based device simulator development.

Within the particle-based EMC method, with its time-marching algorithm, Poisson's equation may be decoupled from the BTE over a suitably small time step (typically less than the inverse plasma frequency corresponding to the highest carrier density in the device). Over this time interval, carriers accelerate according to the frozen field profile from the previous time-step solution of Poisson's equation, and then Poisson's equation is solved at the end of the time interval with the frozen configuration of charges arising from the Monte Carlo phase [6,26]. Note that Poisson's equation is solved on a mesh, whereas the solution of charge motion using EMC occurs over a continuous range of coordinate space in terms of the particle position. Therefore, a particle-mesh (PM) coupling is needed for both the charge assignment and the force interpolation. The PM coupling is broken into four steps: (1) assign particle charge to the mesh, (2) solve the Poisson equation on the mesh, (3) calculate the mesh-defined forces, and (4) interpolate to find forces on the particle.

The motion in real space of particles under the influence of electric fields is somewhat more complicated due to the band structure. The velocity of a particle in real space is related to the E-**k** dispersion relation defining the bandstructure as

$$
\begin{aligned}
\mathbf{v}(t) &= \frac{d\mathbf{r}}{dt} = \frac{1}{\hbar}\nabla_k E(\mathbf{k}(t)) \\
\frac{d\mathbf{k}}{dt} &= \frac{q\mathbf{E}(\mathbf{r})}{\hbar}
\end{aligned}
\tag{51.12}
$$

where the rate of change of the crystal momentum is related to the local electric field acting on the particle through the acceleration theorem expressed by the second equation. In turn, the change in crystal momentum, $\mathbf{k}(t)$, is related to the velocity through the gradient of E with respect to $\mathbf{k}$. If one has to use the full bandstructure of the semiconductor, then integration of these equations to find $\mathbf{r}(t)$ is only possible numerically, using, for example, a Runge–Kutta algorithm [27].

To simulate the steady-state behavior of a device, the system must be initialized in some initial condition, with the desired potentials applied to the contacts, and then the simulation proceeds in a time stepping manner until steady state is reached. This process may take several picoseconds of simulation time, and consequently several thousand time-steps based on the usual time increments required for stability. Clearly, the closer the initial state of the system is to the steady-state solution, the quicker the convergence. If one is, for example, simulating the first bias point for a transistor simulation, and has no *a priori* knowledge of the solution, a common starting point for the initial guess is to start out with charge neutrality, i.e., to assign particles randomly according to the doping profile in the device and based on the super-particle charge assignment of the particles, so that initially the system is charge neutral on the average. (For 2D device simulation, one should keep in mind that each particle actually represents a rod of charge into the third dimension.) Subsequent simulations at the same device at different bias conditions can use the steady-state solution at the previous bias point as a good initial guess. After assigning charges randomly in the device structure, charge is then assigned to each mesh point using the nearest-grid-point (NGP), cloud-in-cell (CIC), or nearest-element-cell (NEC) PM coupling methods, and Poisson's equation is solved. The forces are then interpolated on the grid, and particles are accelerated over the next time-step. A flowchart of a typical Monte Carlo device simulation is shown in Figure 51.19.

As the simulation evolves, charge will flow in and out of the contacts, and depletion regions internal to the device will form until steady state is reached. The charge passing through the contacts at each time step can be tabulated, and a plot of the cumulative charge as a function of time gives the steady-state current. Figure 51.20 shows the particle distribution in three dimension of a metal-semiconductor field-effect transistor (MESFET), where the dots indicate the individual simulated particles for two different gate biases. Here, the heavily doped MESFET region (shown by the inner box) is surrounded by semi-insulating GaAs forming the rest of the simulation domain. Figure 51.20a corresponds to no net gate bias (i.e., the gate is positively biased to overcome the built-in potential of the Schottky contact), while Figure 51.20b corresponds to a net negative bias applied to the gate, such that the channel is close to pinch-off. One can see the evident depletion of carriers under the gate under the latter conditions.

51.3.1 Calculation of the Current

The device output current can be determined using two different yet consistent methods. First, by keeping track of the charges entering and exiting each terminal/contact, the net number of charges over a period of the simulation can be used to calculate the terminal current. The net charge crossing a terminal boundary

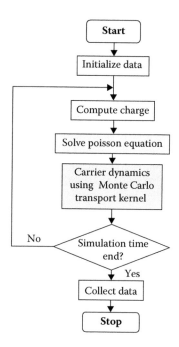

FIGURE 51.19 Flowchart of a typical particle-based device simulation.

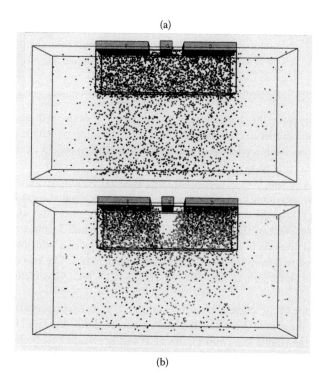

FIGURE 51.20 Example of the particle distribution in an MESFET structure simulated in three dimension using an EMC approach. (a) The device with zero gate voltage applied and (b) the device with a negative gate voltage applied, close to pinch-off.

is determined by

$$Q(t) = e(n_{abs}(t) - n_{injec}(t)) + \varepsilon \int E_y(x, t) dy, \qquad (51.13)$$

where n_{abs} is the number of particles that are absorbed by the contact (exit), n_{injec} the number of particles that have been injected at the contact, and E_y the vertical field at the contact. The second term in Equation 51.13 on the right-hand side is used to account for the displacement current due to the changing field at the contact. Equation 51.13 assumes the contact is at the top of the device and that the fields in the x- and z- directions are negligible. The charge e in Equation 51.13 should be multiplied by the particle charge if it is not unity. The slope of $Q(t)$ versus time gives a measure of the terminal current. In steady state, the current can be found by

$$I = \frac{dQ(t)}{dt} = \frac{e(n_{net})}{\Delta t}, \qquad (51.14)$$

where $n_{net} = n_{abs} - n_{injec}$ is the net number of particles exiting the contact over a fixed period of time Δt.

In a second method, the sum of the electron velocities in a portion of the channel region of the device is used to calculate the current. The electron current density through a cross-section of the device is given by

$$J = env_d, \qquad (51.15)$$

where v_d is the average electron drift velocity and n the carrier concentration. If there are a total of N particles in a differential volume, $dV = dL \cdot dA$, the current found by integrating Equation 51.15 over the cross-sectional area, dA, is

$$I = \frac{eNv_d}{dL}, \text{ or } I = \frac{e}{dL} \sum_{i=1}^{N} v_x(i), \qquad (51.16)$$

where $v_x(i)$ is the velocity along the channel of the ith electron. The device is divided into several sections along the x-axis, and the number of electrons and their corresponding velocity is added for each section after each free flight. The total x-velocity in each section is then averaged over several time-steps to determine the current for that section. Total device current can be determined from the average of several sections, which gives a much smoother result compared to counting terminal charges. By breaking the device into sections, individual currents can be compared to verify that there is conservation of particles (constant current) throughout the device. In addition, sections near the source and drain regions may have a high y-component in their velocity and should be excluded from the current calculations. Finally, by using several sections in the channel, the average energy and velocity of electrons along the channel can be observed to ensure the proper physical characteristics. The two methods for the calculation of the current are performed, as illustrated in Figure 51.17, on a 50-nm channel length MOSFET device.

Extrapolating the slope of the curve shown in Figure 51.21a, that represents the cumulative electron charge that enters/exits the source/drain contact, leads to source/drain current of 0.5205/0.5193 mA/µm. When compared with the results shown in Figure 51.21b, it is evident that both current measurement techniques discussed in this section give current values with relative error less than 1%.

51.3.2 Ohmic Contacts

Another issue that has to be addressed in particle-based simulations is the real space boundary conditions for the particle part of the simulation. Reflecting boundary conditions are usually imposed at the artificial

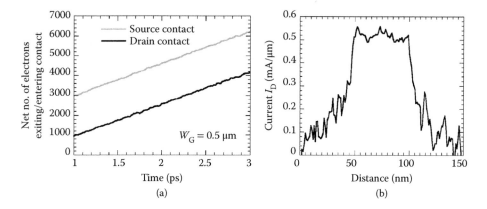

FIGURE 51.21 (a) Net charge entering/exiting the source/drain contact. (b) Average current along the channel. The gate length of the device being modeled equals 50 nm. $V_G = 1.4$ V and $V_D = 1$ V in these simulations. Width of the gate is $W = 0.5$ μm.

boundaries. As far as the ohmic contacts are concerned, they require more careful consideration because electrons crossing the source and drain contact regions contribute to the corresponding terminal current. Commonly employed models for the contacts include the following [28]:

- Electrons are injected at the opposite contact with the same energy and wavevector **k**. If the source and drain contacts are in the same plane, as in the case of MOSFET simulations, the sign of **k**, normal to the contact will change. This is an unphysical model, however [29].
- Electrons are injected at the opposite contact with a wavevector randomly selected based upon a thermal distribution. This is also an unphysical model.
- Contact regions are considered to be in thermal equilibrium. The total number of electrons in a small region near the contact are kept constant, with the number of electrons equal to the number of dopant ions in the region. This is a very good model most commonly employed in actual device simulations.
- Another method uses "reservoirs" of electrons adjacent to the contacts. Electrons naturally diffuse into the contacts from the reservoirs, which are not treated as part of the device during the solution of Poisson's equation. This approach gives results similar to the velocity weighted Maxwellian but at the expense of increased computational time due to the extra electrons simulated. It is an excellent model employed in few most sophisticated particle-based simulators.

There are also several possibilities for the choice of the distribution function—Maxwellian, displaced Maxwellian, and velocity-weighted Maxwellian [10].

51.3.3 Time-Step

As in the case of solving the drift-diffusion, hydrodynamic, or full Maxwell's equations, for a stable Monte Carlo device simulation, one has to choose the appropriate time step, Δt, and the spatial mesh size (Δx, Δy, and/or Δz). The time-step and the mesh size may correlate to each other in connection with the numerical stability. For example, as discussed in the context of solving drift-diffusion simulations, the time-step Δt must be related to the plasma frequency:

$$\omega_p = \frac{e^2 n}{\varepsilon_s m^*}, \tag{51.17}$$

where n is the carrier density. From the viewpoint of the stability criterion, Δt must be much smaller than the inverse plasma frequency. The highest carrier density specified in the device model is used to estimate

Δt. If the material is a multivalley semiconductor, the smallest effective mass to be experienced by the carriers must be used in Equation 51.17 as well. In the case of GaAs, with the doping of 5×10^{17} cm^{-3}, $\omega_p \cong 5 \times 10^{13}$; hence, Δt must be smaller than 0.02 ps.

The mesh size for the spatial resolution of the potential is dictated by the charge variations. Hence, one has to choose the mesh size to be smaller than the smallest wavelength of the charge variations. The smallest distance is approximately equal to the Debye length, given as

$$\lambda_D = \sqrt{\frac{\varepsilon_s k_B T}{e^2 n}}. \tag{51.18}$$

The highest carrier density specified in the model should be used to estimate λ_D from the stability criterion. The mesh size must be chosen to be smaller than the value given by Equation 51.18. In the case of GaAs, with the doping density of 5×10^{17} cm^{-3}, $\lambda_D \cong 6$ nm.

Based on the discussion above, the time step (Δt) and the mesh size (Δx, Δy, and/or Δz) can be specified separately. However, the Δt chosen must be checked again by calculating the distance l_{max}, defined as

$$l_{max} = \mathbf{v}_{max} \times \Delta t, \tag{51.19}$$

where $\mathbf{v}_{max}$ is the maximum carrier velocity that can be approximated by the maximum group velocity of the electrons in the semiconductor (on the order of 10^8 cm/s). Therefore, the distance l_{max} is regarded as the maximum distance the carriers can propagate during Δt. The time step chosen must be small enough so that l_{max} is smaller than the spatial mesh size chosen using Equation 51.19. This is because large Δt chosen may cause substantial change in the charge distribution, while the field distribution in the simulation is only updated every Δt.

51.3.4 PM Coupling

As mentioned earlier, the position of charge as described by the EMC algorithm is continuous, whereas Poisson's equation is solved on a mesh, hence the charge associated with the individual particles must be mapped onto the field mesh in some manner. The charge assignment and force interpolation schemes usually employed in self-consistent Monte Carlo device simulations are the NGP, CIC, and NEC schemes [30]. In the NGP scheme, the particle position is mapped into the charge density at the closest grid point to a given particle. This has the advantage of simplicity, but leads to a noisy charge distribution, which may exacerbate numerical instability. Alternately, within the CIC scheme, a finite volume is associated with each particle spanning several cells in the mesh, and a fractional portion of the charge per particle is assigned to grid points according to the relative volume of the "cloud" occupying the cell corresponding to the grid point. This method has the advantage of smoothing the charge distribution due to the discrete charges of the particle-based method, but may result in an artificial "self-force" acting on the particle, particularly if an inhomogeneous mesh is used. For the case of inhomogeneous mesh or spatially varying dielectric permittivity, the NEC scheme is the best choice. Within the NEC scheme, the charge is moved at the center of the cell and then CIC-like charge assignment to the node points is performed. The PM coupling sequence is presented in Figure 51.22.

51.4 Representative Simulation Results

51.4.1 Bulk Monte Carlo Simulations of Different Materials

In most semiconductors, in order to properly simulate high field transport, it is necessary to consider more than one conduction band valley. To calculate the drift velocity along any direction, the effective mass along

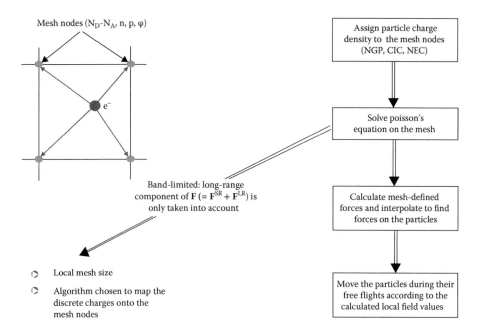

FIGURE 51.22 Particle-mesh coupling sequence. (SR, short range; LR, long range.)

that particular direction is required. In GaAs, for example, a subvalley of the L valley lies along the [111] direction. We know the effective masses along the transverse and longitudinal directions of the subvalley but we do not know the effective mass along the [100] direction. This makes it hard to calculate the drift velocity along the [100] direction.

Assume the Monte Carlo simulation is run on the x, y, z- coordinate system where the x-direction is [100], y-direction is [010], and z-direction is [001]. Let the three perpendicular directions that describe the subvalley be $[a_1, b_1, c_1]$, $[a_2, b_2, c_2]$, and $[a_3, b_3, c_3]$. The electrons in the Monte Carlo simulation will be drifted according to the x, y, z coordinate system so there will be k_x the wavevector along [100], k_y the wavevector along [010], and k_z the wavevector along [001]. The drift velocity of the electron is first calculated along the three directions that describe the subvalley and along which we know the effective masses.

$$v_{\mathrm{d}}, [[a]_1 b_1 c_1] = \frac{\hbar k_{[[a]_1 b_1 c_1]}}{m_1 (1 + 2\alpha E)} v_{\mathrm{d}}, [[a]_1 b_1 c_1] = \frac{\hbar k_{[[a]_1 b_1 c_1]}}{m_1 (1 + 2\alpha E)} \tag{51.20}$$

Here m_1 is the effective mass of the electron along $[a_1, b_1, c_1]$. The final expression of drift velocity along that direction is then calculated by calculating $k_{[[a]_1 b_1 c_1]} k_{[[a]_1 b_1 c_1]}$ using a simple transformation of coordinates to give [31]

$$k_{[[a]_1 b_1 c_1]} = \frac{k_x a_1}{\sqrt{(a_1^2 + b_1^2 + c_1^2)}} + \frac{k_y b_1}{\sqrt{(a_1^2 + b_1^2 + c_1^2)}} + \frac{k_z a_1}{\sqrt{(a_1^2 + b_1^2 + c_1^2)}}. \tag{51.21}$$

Using similar methods we get $v_{\mathrm{d}}, [[a]_2 b_2 c_2]$ and $v_{\mathrm{d}}, [[a]_3 b_3 c_3]$. The coordinates system is then transformed once again back to the x, y, z coordinate system to get the drift velocities along $x, y,$ and z.

$$v_x = \frac{a_1 v_{\mathrm{d},[[a]_1 b_1 c_1]}}{\sqrt{\left(a_1^2 + b_1^2 + c_1^2\right)}} + \frac{a_2 v_{\mathrm{d},[[a]_2 b_2 c_2]}}{\sqrt{\left(a_2^2 + b_2^2 + c_2^2\right)}} + \frac{a_3 v_{\mathrm{d},[[a]_3 b_3 c_3]}}{\sqrt{\left(a_3^2 + b_3^2 + c_3^2\right)}}$$

$$v_y = \frac{b_1 v_{\mathrm{d},[[a]_1 b_1 c_1]}}{\sqrt{\left(a_1^2 + b_1^2 + c_1^2\right)}} + \frac{b_2 v_{\mathrm{d},[[a]_2 b_2 c_2]}}{\sqrt{\left(a_2^2 + b_2^2 + c_2^2\right)}} + \frac{b_3 v_{\mathrm{d},[[a]_3 b_3 c_3]}}{\sqrt{\left(a_3^2 + b_3^2 + c_3^2\right)}} \qquad (51.22)$$

$$v_x = \frac{c_1 v_{\mathrm{d},[[a]_1 b_1 c_1]}}{\sqrt{\left(a_1^2 + b_1^2 + c_1^2\right)}} + \frac{c_2 v_{\mathrm{d},[[a]_2 b_2 c_2]}}{\sqrt{\left(a_2^2 + b_2^2 + c_2^2\right)}} + \frac{c_3 v_{\mathrm{d},[[a]_3 b_3 c_3]}}{\sqrt{\left(a_3^2 + b_3^2 + c_3^2\right)}}$$

Also, as the three directions are mutually perpendicular we have

$$a_1 a_2 + b_1 b_2 + c_1 c_2 = 0$$
$$a_1 a_3 + b_1 b_3 + c_1 c_3 = 0 \qquad (51.23)$$
$$a_3 a_2 + b_3 b_2 + c_3 c_2 = 0.$$

For N electrons in the simulation, an average drift velocity is then calculated as

$$\langle v_x \rangle = \frac{1}{N} \sum_{i=1}^{N} v_{x,i}, \quad \langle v_y \rangle = \frac{1}{N} \sum_{i=1}^{N} v_{y,i}, \quad \langle v_z \rangle = \frac{1}{N} \sum_{i=1}^{N} v_{z,i}, \qquad (51.24)$$

where each drift velocity depends on the subvalley the electron is currently in. The mobility is calculated as the slope of the velocity versus field curve for low fields. The range of "low fields" varies from material to material, but in all cases, it is the range of fields for which the velocity linearly varies with the field.

The results presented in Figure 51.23 shows the difference in velocity of electrons for different materials. For GaN, the peak velocity of 3×10^7 cm/s occurs at around 150 kV/cm so that in the electric field range shown in Figure 51.19 it is clear that for electric fields below 30 kV/cm, GaN carriers in the bulk GaN materials are still in the low-field regime.

As the drift velocities saturate at much smaller electric fields in Si and Ge, they are plotted on a separate graph (see Figure 51.24) [32].

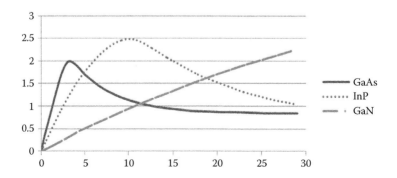

FIGURE 51.23 Drift velocity (10^7 cm/s) versus electric field (kV/cm).

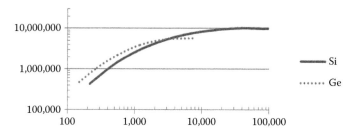

FIGURE 51.24 Electron drift velocity (cm/s) versus electric field (V/cm) in Si and Ge along the principal crystallographic directions.

51.4.2 Modeling Thermal Effects in SOI Devices Using Particle-Based Device Simulator

There are many applications of the Monte Carlo approach to device simulations that can be found in the literature. Here, we show the example of Monte Carlo device simulation to self-heating which requires coupling of the Monte Carlo device solver self-consistently with a solver for the phonon bath. Self-heating (despite the fact that supply voltages are reduced when reducing the size of the channel of nanoscale transistors) is still a problem, since there are sufficiently large electric fields in the smallest devices that can accelerate the carriers to average energies of 0.5–0.8 eV. These hot carriers then interact with the lattice and transfer their energy to the phonon bath (both acoustic and optical), thus heating the lattice and increasing the lattice vibrations (scattering). This leads to a negative feedback on the carrier drift velocity or mobility (if transport is scattering dominated). Thus, we may conclude that when transistors are in the ON-state, there is a strong interaction of the electrons with the lattice on a very short time scale (in particular with the optical phonons). The optical phonons in a longer time scale decay into acoustic phonons through a harmonic multiple phonon processes, thus removing the heat from the hot spot. Therefore, to properly treat the operation of nanoscale devices it is necessary to, at least, consider the electrons within the BTE picture and, in the lowest approximation, via energy balance equations including the acoustic and the optical phonon baths separately. This is what the Arizona State University (ASU) group has implemented couple of years ago [33]. Our ultimate goal is to have coupled the developed phonon BTE solver [34] with the electron BTE solver in a self-consistent manner.

51.4.2.1 Theoretical Model and Computation Details

A schematic description of our 2D/3D electrothermal simulator is shown in Figure 51.25. It is important to note that the EMC transport kernel for the electrons is in the loop for the solution of the energy balance equations for the acoustic and optical phonon temperatures. The criterion for convergence of the whole self-consistent loop (measured in terms of Gummel cycles) is convergence in the current to the third significant digit.

Between iterations, there are variables that are being exchanged between the electron and phonon solvers. Namely, after running the EMC for the electrons in the device for about 5–10 ps for 2D device analysis and 10 ps for the 3D nanowire simulations, the electron transport solver passes to the phonon energy balance solver the information about the spatial variation of the electron density, spatial variation of the average drift velocity, and spatial variation of the average electron temperature, where an assumption is made that the drift energy is much smaller than the thermal energy. These variables are then used in the solution of the energy balance equations for the acoustic and optical phonons to get updated values for the spatial variation of the optical phonon and acoustic phonon temperatures, which are then used as inputs, in the choice of the scattering table for the electrons (proper scattering table is chosen with the acoustic and optical phonon temperatures at the NGP). Then, when the proper table is selected, based on the energy of the electron and a selection of a random number, the scattering mechanism is selected.

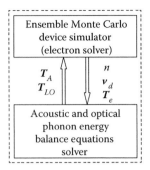

FIGURE 51.25 Exchange of variables between the two kernels. The electron density (n), electron drift velocity (v_d), and electron temperature (T_e), obtained from the electron solver, are input variables for the phonon kernel. The output variables of the phonon solver are acoustic and optical phonon temperature profiles (T_A and T_{LO}, respectively). They enter in the beginning of the free-flight-scattering phase through phonon temperature-dependent scattering tables.

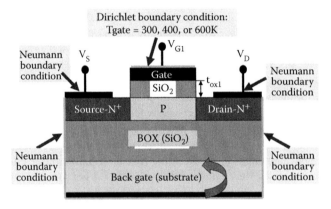

FIGURE 51.26 Schematic of the simulated device structure which illustrates the boundary conditions used in some of the simulations presented. Note also that the boundary conditions at the bottom of the substrate can be mapped as a boundary condition at the bottom of the BOX.

The question of the proper boundary conditions for the electronic part of the problem is rather clear and has been discussed in many papers and texts in the literature [35]. The problem in properly specifying the phonon boundary conditions is the selection of acoustic and optical phonon temperatures either at the artificial boundaries or at the contacts. Typically, acoustic phonon temperature is equated with the lattice temperature. To better understand the choice of boundary conditions we have considered in various works [15], we want to point out that lattice temperature is analogous to electrostatic potential and heat flux is analogous to current. As it is well known, when considering electrical behavior of the device, at least one node within the structure has to have Dirichlet boundary conditions specified (specify the potential for Poisson). Analogously, for the lattice temperature, we need at least one node that is a thermal contact and whose temperature is set to 300 K. As illustrated in Figure 51.26 for the case of FD-SOI devices, the bottom of the BOX is assumed to be at thermal equilibrium. Also, the gate contact is assumed to be at thermal equilibrium, but not necessarily at 300 K. For example, in one set of simulations, we vary the temperature on the gate to be 300, 400, and 600 K. Now comes the question: What happens to the source and drain contacts and the side artificial boundaries? Specifying Dirichlet boundary conditions at the ohmic contacts is not accurate from a standpoint that there is current flowing through the contacts, and since the contacts have finite resistance, there will be Joule heating (so the problem becomes unconstrained). The best solution to this, as we have done in several studies, is to extend the metal contact to become part of the simulation

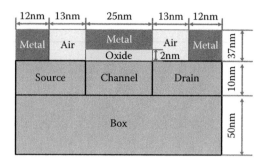

FIGURE 51.27 Device structure with extended domain. The geometrical dimensions are for the simulated 25-nm channel length FD-SOI nMOSFET.

domain and to apply at the very ends of the contact isothermal boundary conditions (see Figure 51.27). With the top and bottom specified, we have vertical transport of heat through the structure.

The next question is, is there lateral transfer of heat across the artificial boundaries? Here we can consider two cases: case when the neighboring device is ON, in which case Neumann boundary conditions are appropriate (derivative of the temperature), and case when the neighboring device is OFF, in which case Dirichlet boundary conditions are appropriate. In the case when Neumann boundary conditions are applied across the artificial boundaries, the heat transport remains vertical, but for the case when Dirichlet boundary conditions are applied across the boundary, the heat transport has both vertical and horizontal component. The case of Neumann boundary conditions can happen in analog circuits such as current mirror in which both transistors are simultaneously ON, whereas the case of Dirichlet boundary conditions occurs in digital circuits in which most of the time the transistors are OFF.

Yet another issue that deserves attention in getting physically correct results is the proper choice of the thermal conductivity for thin Si slabs and for nanowires. Asheghi and coworkers [36] and Li Shi and coworkers [37] have demonstrated via experimental measurements that thermal conductivities of thin Si films and Si nanowires, respectively, strongly depend of the geometry as for smaller geometries phonon boundary scattering can reduce the thermal conductivity of the Si film or nanowire by a factor of 10 or more from its bulk value. Moreover, thermal conductivity is temperature-dependent quantity. We made extensive efforts, using the theory of Sondhaimer for conductivity of metals [38], to arrive at an empirical formula that simultaneously described the thickness and temperature dependence of the thermal conductivity [39]. Our empirical expression perfectly matched the experimental data of Asheghi and coworkers [16]. In the model, it was assumed that phonon boundary scattering was perfectly diffusive.

51.4.2.2 Simulation Results

To validate our premise that self-heating effects become smaller as we scale devices into the nanometer regime (due to more pronounced velocity overshoot effects and thinner BOX layer that leads to smaller thermal resistance), in Table 51.3 we present the parameters of the devices being simulated. Simulation results for the lattice temperature profile in the Si film for devices with channel lengths from 25 to 180 nm are shown in Figure 51.28. We clearly see that the hot spot is in the channel for 180-nm channel length device and moves into the drain contact for 25-nm channel length device. The current degradation for isothermal temperatures on the gate of 300, 400, and 600 K are shown in Figure 51.29. These are the worst case scenario results as we use Neumann boundary conditions at the side boundaries.

In obtaining the results presented in Figures 51.28 and 51.29, constant thermal conductivity value of 13 W/m-K was used in the simulations [40]. The use of constant thermal conductivity model with value of 13 W/m-K leads to underestimation (25 nm channel length) or overestimation (180 nm channel length) of the average maximum temperature of the lattice in the channel region of the device.

TABLE 51.3   Parameters for Various Simulated Device Technology Nodes (Constant Field Scaling)

L(nm)	t_{ox} (nm)	t_{si} (nm)	t_{box} (nm)	N_{ch} (cm^{-3})	$V_{GS} = V_{DS}$ (V)	I_D (mA/ μm)
25	2	10	50	1×10^{18}	1.2	1.82
45	2	18	60	1×10^{18}	1.2	1.41
60	2	24	80	1×10^{18}	1.2	1.14
80	2	32	100	1×10^{17}	1.5	1.78
90	2	36	120	1×10^{17}	1.5	1.67
100	2	40	140	1×10^{17}	1.5	1.57
120	3	48	160	1×10^{17}	1.8	1.37
140	3	56	180	1×10^{17}	1.8	1.23
180	3	72	200	1×10^{17}	1.8	1.03

Note: *Parameters of the simulated structure given in Figure 51.26 are L, gate length; t_{ox}, gate oxide thickness; t_{Si}, active Si-layer thickness; t_{box}, BOX thickness; N_{ch}, channel doping concentration; I_D, isothermal current value (300 K).*

51.4.2.3 Summary of Results

In this section, we have presented modeling of self-heating effects in FD-SOI devices. Larger velocity over-shoot and smaller BOX thickness lead to smaller degradation due to self-heating effects in 25-nm channel length FD-SOI devices. The amount of self-heating significantly depends upon the magnitude of the thermal conductivity used. Bulk values are inadequate and proper thickness and temperature dependence of

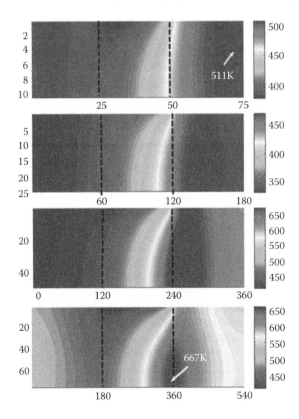

FIGURE 51.28   Lattice temperature profile in the active Si-layer for the FD-SOI devices from Table 51.2 ranging from 25 nm (top) to 180 nm (bottom) channel length. Shown here is only the thin Si film of the SOI device, namely the active area of the device.

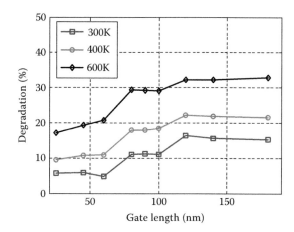

FIGURE 51.29 Current degradation (abs($I_{th} - I_o$)/I_o, where I_{th} is the on current when the self-heating model is incorporated and I_o is current in the isothermal case) versus technology generation for the FD-SOI devices from Table 51.2 ranging from 25- to 180-nm channel length. Different curves correspond to different temperatures on the gate electrode (bottom set of results: T_{gate} = 300 K, middle set of results: T_{gate} = 400 K and top set of results: T_{gate} = 600 K).

the thermal conductivity must be taken into consideration. The choice of the BOX material makes significant impact on the amount of observed self-heating. Both diamond and Si are good materials to be used as BOX. Diamond, when compared to aluminum nitride (AlN) is a better heat spreader.

51.5 Outlook

The drift-diffusion solvers are applicable in situations in which the bias conditions and the device geometry are such that electric fields relatively low and velocity saturation model is applicable [41]. Situations in which drift-diffusion models are applicable are the Si-based power MOSFET devices, bipolar junction transistors, light-emitting diodes (that are used more and more in solid-state lightning), and crystalline solar cells, just to name a few. In some of these devices, such as power transistors and light-emitting diodes (LEDs) used for solid-state lighting, it is of paramount importance to incorporate self-heating models within the drift-diffusion framework [42]. The accuracy of simple heating models in conjunction with drift-diffusion models is to some degree questionable so that it is in many circumstances justifiable to use particle-based device simulators with more exact self-heating models [43].

On the other hand, the hydrodynamic models do not suffer from the limitations of the drift-diffusion approaches and the incorporation of the additional energy balance equation allows one to include velocity overshoot in the model. Velocity overshoot and nonstationary transport are key features of conventional MOSFET devices with gate lengths of 200 nm and below [44]. However, the magnitude of the velocity overshoot observed via simulations depends strongly upon the choice of the energy relaxation time, mostly in sub-100-nm channel length devices. This, in turn, affects the magnitude of the drain current. The reason for such drastic differences in the results when different energy relaxation times are used is due to the fact that the energy relaxation time is a material-dependent as well as device geometry-dependent parameter. So, to calculate better estimates for the velocity overshoot, higher moments of the BTE are needed [45]. These, in turn involve parameters that are more and more ambiguous on the expense of increased computational cost and when the computational cost of hydrodynamic models exceeds one of particle-based device simulators, there is no point in using moment methods. In these circumstances, the direct solution of the BTE via the Monte Carlo method becomes a method of choice. Thus, we might conclude that it is advisable to use particle-based device simulators when nanoscale devices are concerned [33].

But how far down in the scaling can we go? Particle-based device simulators capture on an equal footing both ballistic and diffusive transport, so if the ballisticity factor in the device increases [46], there is no problem that ballistic transport is effectively captured with particle-based device simulators [47]. Quantum-mechanical size quantization effects can also be captured by solving in slices the corresponding one-dimensional (1D) or 2D Schrödinger equation if one is concerned with conventional or FD-SOI MOSFETs, or nanowire transistors, respectively. What cannot be captured with particle-based device simulators is if there are local strains and stresses in the ultra-nanoscale devices, but that can also be cured via coupling of Monte Carlo device simulators with atomistic models [48,49] for bandstructure calculation.

In summary, Monte Carlo device simulators are a powerful tool for modeling devices ranging from the nanoscale regime to the microscale regime. What cannot be modeled with particle-based device simulators are resonant tunneling diodes in which quantum interference effects dominate the device behavior [50]. Efforts have been made along this direction as well [51], but the inclusion of the quantum-mechanical phase alongside with well-defined particle trajectory still remains open field of research.

References

1. ITRS International Technology Roadmap for Semiconductors. Available at: http://www.itrs.org.
2. G. Moore. Progress in digital integrated electronics. *IEDM Tech. Digest*, 11–13, 1975.
3. R. Dennard, F. H. Gaensslen, H. N. Yu, L. Rideout, E. Bassous, and A. R. LeBlanc. Design of Ion-Implanted MOSFET's with very small physical dimensions. *IEEE J. Solid State Circ.*, Vol. 9 (5), 256–268, 1974.
4. E. Fayneh, M. Yuffe, E. Knoll, M. Zelikson, M. Abozaed, Y. Talker, Z. Shmuely, and S. A. Rahme. 14 nm 6th-generation core processor SoC with low power consumption and improved performance. *IEEE International Solid-State Circuits Conference*, San Francisco, pp. 72–74, 2016.
5. T. Song, W. Rim, S. Park, Y. Kim, J. Jung, G. Yang, S. Baek, J. Choi, B. Kwon, Y. Lee, S. Kim, G. Kim, H. S. Won, J. H. Ku, S. S. Paak, E. Jung, S. S. Park, and K. Kim. A 10 nm FinFET 128 Mb SRAM with assist adjustment system for power, performance, and area optimization. *IEEE International Solid-State Circuits Conference*, San Francisco, pp. 306–309, 2016.
6. H.-Y. Chen, F.-C. Chen, Y.-L. Chan, K.-N. Yang, F.-L. Yang, and C. Hu. Method of fabricating a necked FINFET device. US 7122412 B2, 2006.
7. L. Geppert. A quantum leap for photonics. *IEEE Spectr.*, 28–33, April 9, 2004.
8. W. Zhu, J. P. Han, and T. P. Ma. Mobility measurement and degradation mechanisms of MOSFETs made with ultrathin high-k dielectrics. *IEEE Trans. Electron Dev.*, Vol. 51, 98–105, 2004.
9. J. Welser, J. L. Hoyt, and J. F. Gibbons. Temperature and scaling behavior of strained-Si N-MOSFET's. *IEDM Tech. Dig.*, 1000–1002, 1992.
10. R. Oberhuber, G. Zandler, and P. Vogl. Subband structure and mobility of two-dimensional holes in strained Si/SiGe MOSFET's. *Phys. Rev. B.*, Vol. 58, 9941–9948, 1998.
11. R. Chau, J. Kavalieros, B. Doyle, A. Murthy, N. Paulsen, D. Lionberger, D. Barlage, R. Arghavani, B. Roberts, and M. Doczy. A 50 nm depleted-substrate CMOS transistor (DST). *IEDM Techn. Dig.*, 29.1.1–29.1.4, 2001.
12. Y.-K. Choi, K. Asano, N. Lindert, V. Subramanian, T.-J King., J. Bokor, and C. Hu. Ultrathin-body SOI MOSFET for deep-sub-tenth micron era. *IEEE Electron Dev. Lett.*, Vol. 21 (5), 254–255, 2000.
13. D. Vasileska and S. M. Goodnick. Computational electronics. *Mater. Sci. Eng. R Rev. J.*, Vol. R38, 181–236, 2002.
14. C. Jacoboni and L. Reggiani. The Monte Carlo method for the solution of charge transport in semiconductors with applications to covalent materials. *Rev. Mod. Phys.*, Vol. 55, 645–705, 1983.
15. C. Jacoboni and P. Lugli. *The Monte Carlo Method for Semiconductor Device Simulation*. Vienna, VA: Springer-Verlag, 1989.
16. K. Hess. *Monte Carlo Device Simulation: Full Band and Beyond*. Boston, MA: Kluwer Academic Publishing, 1991.
17. M. H. Kalos and P. A. Whitlock. *Monte Carlo Methods*. New York, NY: Wiley, 1986.

18. D. K. Ferry. *Semiconductors*. New York, NY: Macmillan, 1991.
19. D. Vasileska, S. M. Goodnick, and G. Klimeck. *Computational Electronics: Semiclassical and Quantum Transport Modeling*. New York, NY: Taylor & Francis, 2010.
20. D. K. Ferry. *Quantum Mechanics: An Introduction for Device Physicists and Electrical Engineers*. New York, NY: Taylor & Francis, 2001.
21. H. D. Rees. Calculation of distribution functions by exploiting the stability of the steady state. *J. Phys. Chem. Solids*, Vol. 30, 643, 1969.
22. R. M. Yorston. Free-flight time generation in the Monte Carlo simulation of carrier transport in semiconductors. *J. Comp. Phys.*, Vol. 64, 177, 1986.
23. M. Lundstrom. *Fundamentals of Carrier Transport*. London: Cambridge University Press, 2000.
24. B. K. Ridley. *Quantum Processes in Semiconductors*. Oxford: Oxford University Press, 1982.
25. D. Vasileska. https://nanohub.org/groups/dragica_vasileska/semiclassical.
26. K. Tomizawa. *Numerical Simulation of Submicron Semiconductor Devices*. Boston, MA: Artech House, 1993.
27. M. V. Fischetti and S. E. Laux. Band structure, deformation potentials and carrier mobility in strained Si, Ge and SiGe alloys. *J. Appl. Phys.*, Vol. 80, 2234–2252, 1996.
28. T. Gonzalez and D. Pardo. Physical models of ohmic contact for Monte Carlo device simulation. *Solid State Electron.*, Vol. 39, 555–562, 1996.
29. P. A. Blakey, S. S. Cherensky, and P. Sumer. *Physics of Submicron Structures*. New York, NY: Plenum Press, 1984.
30. S. E. Laux. On particle-mesh coupling in Monte Carlo device simulation. *IEEE Trans. Comp. Aided Des. Int. Circ. Sys.*, Vol. 15, 1266, 1996.
31. R. Hathwar. Generalized Monte Carlo tool for investigating low-field and high field properties of materials using non-parabolic band structure model. MS Thesis, Arizona State University, June 2011.
32. S. S. Qazi. Electrical and thermal transport in alternative device technologies. MS Thesis, Arizona State University, November 2013.
33. K. Raleva, D. Vasileska, S. M. Goodnick, and M. Nedjalkov. Modeling thermal effects in nanodevices. *IEEE Trans. Electron Dev.*, Vol. 55 (6), 1306, 2008.
34. A. R. Shaik. Multi-scale study of heat transfer using Monte Carlo technique for phonon transport. MS Thesis, Arizona State University, April 2016.
35. D. Vasileska, K. Raleva, and S. M. Goodnick. Self-heating effects in nanoscale FD SOI devices: The role of the substrate, boundary conditions at various interfaces, and the dielectric material type for the box. *IEEE Trans. Electron Dev.*, Vol. 56, 3064, 2009.
36. W. Liu and M. Asheghi. Phonon-boundary scattering in ultrathin single-crystal silicon layers. *Appl. Phys. Lett.*, Vol. 84, 3819, 2004.
37. D. Li, Y. Wu, P. Kim, L. Shi, P. Yang, and A. Majumdar. Thermal conductivity of individual silicon nanowires. *Appl. Phys. Lett.*, Vol. 83, 2934, 2003.
38. E. H. Sondheimer. The mean free path of electrons in metals. *Adv. Phys.*, Vol. 1, 1, 1952, reprinted in *Adv. Phys.*, 50, 499, 2001.
39. D. Vasileska, K. Raleva, and S. M. Goodnick. Electrothermal studies of FD SOI devices that utilize a new theoretical model for the temperature and thickness dependence of the thermal conductivity. *IEEE Transactions on Electron Devices*, Vol. 57, 726–728, 2010.
40. K. Raleva and D. Vasileska. The importance of thermal conductivity modeling for simulations of self-heating effects in FD SOI devices. *J. Comput. Elec.*, Vol. 12 (4), 601–610, 2013.
41. D. Vasileska and S. M. Goodnick, Editors. *Nano-Electronic Devices: Semiclassical and Quantum Transport Modeling*. Vienna, VA: Springer Verlag, June 2011.
42. *www.silvaco.com*.
43. D. Vasileska. Modeling thermal effects in nano-devices. *Microelectron. Eng.*, Vol. 109 (9), 163–167 2013.

44. T. Grasser, T. W. Tang, H. Kosina, and S. Selberherr. A review of hydrodynamic and energy-transport models for semiconductor device simulation. *Proc. IEEE*, Vol. 91 (2), 251–274, 2003.

45. T. Grasser, H. Kosina, C. Heitzinger, and S. Selberherr. Characterization of the hot electron distribution function using six moments. *J. Appl. Phys*, Vol. 91, 3869, 2002.

46. M. Lundstrom. Elementary scattering theory of the Si MOSFET. *IEEE Electron Dev. Lett.,* Vol. 18, 361–363, 1997.

47. M. V. Fischetti and S. E. Laux. Monte Carlo analysis of electron transport in small semiconductor devices including band-structure and space-charge effects. *Physical Review B*, Vol. 38(14), 9721, 1988.

48. T. Frauenheim, G. Seifert, M. Elstner, T. Niehaus, C. Köhler, M. Amkreutz et al. Atomistic simulations of complex materials: Ground-state and excited-state properties. *J. Phys. Condens. Matter*, Vol. 14 (11), 3015, 2002.

49. G. Klimeck, S. S. Ahmed, H. Bae, N. Kharche, S. Clark, B. Haley, S. Lee et al. Atomistic simulation of realistically sized nanodevices using NEMO 3-D—Part 1: models and benchmarks. *IEEE Trans. Electron Dev.*, Vol. 54 (9), 2079–2089, 2007.

50. R. Lake, G. Klimeck, R. C. Bowen, and D. Jovanovic. Single and multiband modeling of quantum electron transport through layered semiconductor devices. *J. Appl. Phys.*, Vol. 81 (12), 7845–7869, 1997.

51. L. Shifren and D. K. Ferry. Wigner function quantum Monte Carlo. *Physica B: Condens. Matter*, Vol. 314 (1), 72–75, 2002.

52

Photonics

Frank Schmidt

This chapter deals with numerical simulation in photonics. It provides an overview of the most widely used methods, explains their concepts, and gives some details, as long as it is necessary for a qualitative understanding. It is far from being complete but tries to describe the landscape of simulation tools. Parts of this review are closely related to the book *Numerical Methods in Photonics* [1], which itself is a comprehensive representation of the field covered by many very detailed monographs. The methods we describe in the following are the *finite-difference time-domain FDTD method*, closely related to the *finite integration technique* (FIT), the *Fourier modal method* (FMM), the *finite element method* (FEM), and the *discontinuous galerkin* (DG) method. We omit other very useful methods often specialized to certain applications, in particular:

- Beam propagation method (BPM), for simulation of paraxial beam propagation [2]
- Integral methods such as boundary integral methods, for simulation problems in grating design [3]
- Plane wave expansion (PWE) for simulations of band structures [4]

and refer the reader to the given monographs.

52.1 Landscape of Numerical Methods

There are many aspects to consider if one has to choose a numerical method to solve a particular problem in photonics. We now discuss a few.

52.1.1 Time-Harmonic versus Time-Dependent Modeling

In the most general case, one wants to solve time-dependent Maxwell's equations with general, i.e., non-local, nonlinear, dispersive media. Both FDTD and DG methods in time provide such general solutions. On the other hand, especially in the case of high accuracy requirements, strict conditions on the maximal stepsizes in time may cause a high amount of central processing unit (CPU)-time consumption. Therefore, in case of linear and nondispersive problems, it is often useful to consider time-harmonic problems, and, if needed, to combine time-harmonic solutions via Fourier transform with different frequencies to obtain a time-dependent behavior.

52.1.2 Exact versus Approximate Models

It is common to approximate Maxwell's equations by structurally simpler equations. Especially, the simplification in terms of polarizations have a long tradition, where the full vectorial equations are replaced by semivectorial or scalar approximations. The scalar Helmholtz equation, e.g., is well investigated and a number of efficient numerical approaches exists, hence at first glance it seems reasonable to try Helmholtz solvers instead of Maxwell solvers. But the full vectorial Maxwell solvers available today do not have significant disadvantages with respect to speed or memory consumption, thus it is obsolete in many cases to use such semivectorial or scalar approximations. The case is different for special situations such as paraxial field propagations along distinguished propagation directions. As long as reflections can be neglected, in the case of waveguide tapers for example, so-called BPM are useful. They map a time-harmonic three-dimensional (3D) scattering problem approximately to a time-like evolution problem with 2D cross sections, which results in a substantial reduction of the computational effort.

52.1.3 Order of Approximation and Flexibility of Meshes

All methods described in the following offer, at least in principle, high-order approximations in space and time, thus resulting in high-order convergence. Nevertheless, this is not always used in practice. FDTD methods, e. g., are typically linked to a second-order discretization in time and space on staggered grids. Therefore, once geometrical flexibility is needed, one would typically, instead of enhancing FDTD, directly apply DG. Implementation of DG is more involved than of FDTD, but offers the flexibility of FEM meshes by design and yields the time evolution with costs comparable to FDTD.

52.1.4 Computing Aspects: CPU-Time versus Memory

The time-dependent solvers FDTD and DG can be treated by explicit time integrators. Thus, the evolution procedure results in much less memory consumption than that is typical for FMM or FEM, where either full or large sparse matrices have to be inverted. Hence, on memory-restricted systems, FDTD or DG are advantageous. The price to pay is longer computing times.

52.1.5 Special Adaptation to Photonics

FDTD, DG, and FEM are general methods for solving electromagnetic problems; they are not specially tailored to photonics. Since mode propagation is often the central physical effect, numerical methods based on the interaction of waveguide or cavity modes are often very effective. The dominant class of methods here are the FMMs. Subdividing the structure into waveguide sections and relating waveguide modes to each other result in algorithms which often offer surprisingly fast convergence in the range of lower, but in practice sufficient accuracy.

52.2 Maxwell's Equations in Nondispersive, Chargeless Media

In the most general case, the numerical methods of photonics solve time-dependent Maxwell's equations with general media, i.e., with nonlocal, nonlinear, dispersive media. For a concise and clear presentation of the different numerical methods, we use only a subset of the general equations. Our discussions are based on

$$
\begin{aligned}
\nabla \times \mathbf{H} &= \mathbf{J} + \frac{\partial \mathbf{D}}{\partial t} \\
\nabla \times \mathbf{E} &= -\frac{\partial \mathbf{B}}{\partial t} \\
\nabla \cdot \mathbf{D} &= 0 \\
\nabla \cdot \mathbf{B} &= 0
\end{aligned}
\tag{52.1}
$$

completed by the corresponding constitutive equations

$$
\mathbf{D} = \varepsilon \mathbf{E} \qquad \mathbf{B} = \mu \mathbf{H} \qquad \mathbf{J} = \sigma \mathbf{E}.
\tag{52.2}
$$

These equations describe the fields in linear, nondispersive, and chargeless media, where the magnetic field $\mathbf{H}$, the electric field $\mathbf{E}$, the electric displacement $\mathbf{D}$, and the magnetic flux density $\mathbf{B}$ are vectorial quantities defined for the entire space and positive times, $\mathbf{H}, \mathbf{E}, \mathbf{D}, \mathbf{B} : \mathbb{R}^3 \times \mathbb{R}^+ \longrightarrow \mathbb{R}^3$, and the permittivity ε, the permeability μ, and the conductivity σ are scalar, piecewise continuous functions in space, $\varepsilon, \mu, \sigma : \mathbb{R}^3 \longrightarrow \mathbb{R}$.

The field equations combined with the constitutive equations yield the system, on which our representation of the FDTD method and the DG method are based:

$$
\begin{pmatrix} -\sigma & \nabla \times \\ \nabla \times & \end{pmatrix} \begin{pmatrix} \mathbf{E} \\ \mathbf{H} \end{pmatrix} = \frac{\partial}{\partial t} \left[\begin{pmatrix} \varepsilon & \\ & -\mu \end{pmatrix} \begin{pmatrix} \mathbf{E} \\ \mathbf{H} \end{pmatrix} \right].
\tag{52.3}
$$

The other methods, FEM, FMM, etc., are based on the time-harmonic form of Equation 52.3. The concept is to use a Fourier transform of the fields in Equation 52.3 for a single frequency ω. Assuming the single frequency dependence to be $\exp(-i\omega t)$, the Fourier transformed fields become complex but live in space only, $\mathbf{H}, \mathbf{E} : \mathbb{R}^3 \longrightarrow \mathbb{C}^3$. We do not introduce a special notation for the time-harmonic fields, since the meaning becomes clear from the context. In the frequency domain, Equation 52.3 reads

$$\begin{pmatrix} & \nabla \times \\ \nabla \times & \end{pmatrix} \begin{pmatrix} \mathbf{E} \\ \mathbf{H} \end{pmatrix} = -i\omega \left[\begin{pmatrix} \epsilon - \frac{\sigma}{i\omega} & \\ & -\mu \end{pmatrix} \begin{pmatrix} \mathbf{E} \\ \mathbf{H} \end{pmatrix} \right]. \tag{52.4}$$

52.2.1 Scattering, Mode Propagation, and Resonance Problems

Maxwell's equations take different forms depending on the different modeling situations. In the case of time-dependent problems, these are the different types of linear and nonlinear scattering problems. In the case of time-harmonic problems, where mostly but not exclusively, linear problems are considered, we have scattering, mode propagation, and resonance problems. Time-dependent solutions to Equation 52.3 including also nonlocal, dispersive, or nonlinear materials are the most general classical solutions to Maxwell's equations—the corresponding computational effort is usually relatively large. Solutions to the time-harmonic system (52.4) are often much cheaper.

52.2.1.1 Scattering on Unbounded Domain

Figure 52.1 shows the situation of a time-harmonic scattering problem, the situation for time-dependent scattering is very similar. There is a source field $\mathbf{E}_{\mathrm{source}}$ traveling through the unbounded space, being itself a solution to Maxwell's equations (Equation 52.4). If no scatterer would be present, this would be the only field we have. But if it hits an object, scattered fields $\mathbf{E}_{\mathrm{scattered}}$ (or in short $\mathbf{E}_s$) are generated. We need to compute the field inside a computational domain Ω, which we also call the interior domain, in contrast to the exterior domain surrounding Ω. Further, we denote the boundary of the computational domain with $\partial\Omega$.

Let the source field $\mathbf{E}_{\mathrm{source}}(\mathbf{r})$, with $\mathbf{r}$ in the exterior domain, $\mathbf{r} \in \mathbb{R}^3 \setminus \bar{\Omega}$ be given. Then, the general scattering problem in 3D is defined by the following conditions:

1. Interior problem. The field on the computational domain obeys Equation 52.4.
2. Exterior problem. The fields in the exterior domain are a superposition of the source and the scattered fields $\mathbf{E}_{\mathrm{source}}$ and $\mathbf{E}_s$. The source field and the scattered field must also obey Maxwell's equations.
3. Continuity of the tangential data and their normal derivative along the boundary of the computational domain.
4. In the case of time-harmonic fields, the Silver–Müller radiation condition for the scattered field

$$\lim_{r \to \infty} r \left((\nabla \times \mathbf{E}_s) \times \mathbf{r}^0 - i |\mathbf{k}| \mathbf{E}_s \right) = 0 \tag{52.5}$$

holds true uniformly in all directions $\mathbf{r}$. Here, $\mathbf{r}^0$ denotes the radial vector of unit length.

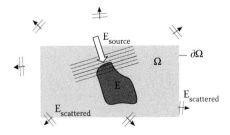

FIGURE 52.1 Scattering problem.

Typically, the radiation condition is not directly applied. But the methods that treat the exterior part of the problem ensure that the scattered fields satisfy the condition. The most prominent and successful method here is the perfectly matched layer (PML) method, see the Section 52.2.4.

52.2.1.2 Resonance Problems

In resonance problems, we consider Equation 52.4 as an eigenvalue problem with the unknown eigenvalue ω and the unknown resonance mode, the eigenvector, $(\mathbf{E}, \mathbf{H})^T$. The special challenge lies in the treatment of the exterior domain. Either the boundary conditions on the boundary of the computational domain are explicitly known, e. g., as a magnetic or electric wall, or methods from scattering problems are applied.

52.2.1.3 Mode Propagation Problems

Again, we consider Equation 52.4 as an eigenvalue problem. This time, we specialize the problem further to z-invariant problems: $\varepsilon = \varepsilon(x, y)$. We take a fixed frequency ω and look for solutions of the type $\mathbf{E}(x, y, z) = \mathbf{E}(x, y)e^{ik_z z}$ with a (unknown) phase velocity k_z. Two common forms of the eigenvalue problem for k_z are the following: Since any z-dependence is given by factors $\exp(ik_z z)$, one can introduce a specialized nabla-operator $\nabla_{k_z} = \mathbf{e}_x \partial_x + \mathbf{e}_y \partial_y + ik_z \mathbf{e}_z$ and Equation 52.4 reads

$$
\begin{pmatrix} & \nabla_{k_z} \times \\ \nabla_{k_z} \times & \end{pmatrix} \begin{pmatrix} \mathbf{E} \\ \mathbf{H} \end{pmatrix} = -i\omega \left[\begin{pmatrix} \epsilon - \frac{\sigma}{i\omega} & \\ & -\mu \end{pmatrix} \begin{pmatrix} \mathbf{E} \\ \mathbf{H} \end{pmatrix} \right].
$$

Here, the eigenvector consists of both $\mathbf{E}$ and $\mathbf{H}$, the eigenvalue k_z appears in linear form. Alternately, eliminating, e. g., the magnetic field, and using the decompositions $\mathbf{E} = \mathbf{E}_\perp + \mathbf{e}_z E_z$ and $\nabla = \nabla_\perp + i\mathbf{e}_z k_z$, one derives from Equation 52.4 the modal eigenvalue problem

$$
\begin{bmatrix} \frac{1}{\epsilon}\nabla_\perp \times \frac{1}{\mu}\nabla_\perp \times \ +\frac{k_z^2}{\epsilon\mu} & \frac{ik_z}{\epsilon\mu}\nabla_\perp \\ i\frac{k_z}{\epsilon}\nabla_\perp \cdot \frac{1}{\mu} & -\frac{1}{\epsilon}\nabla_\perp \cdot \frac{1}{\mu}\nabla_\perp \end{bmatrix} \begin{bmatrix} \mathbf{E}_\perp \\ E_z \end{bmatrix} = \omega^2 \begin{bmatrix} \mathbf{E}_\perp \\ E_z \end{bmatrix} \tag{52.6}
$$

The eigenvector consists just of the E-field, but this is a quadratic eigenvalue problem in k_z. The same procedure for the magnetic field yields

$$
\begin{bmatrix} \frac{1}{\mu}\nabla_\perp \times \frac{1}{\epsilon}\nabla_\perp \times \ +\frac{k_z^2}{\epsilon\mu} & \frac{ik_z}{\epsilon\mu}\nabla_\perp \\ i\frac{k_z}{\mu}\nabla_\perp \frac{1}{\epsilon} & -\frac{1}{\mu}\nabla_\perp \cdot \frac{1}{\epsilon}\nabla_\perp \end{bmatrix} \begin{bmatrix} \mathbf{H}_\perp \\ H_z \end{bmatrix} = \omega^2 \begin{bmatrix} \mathbf{H}_\perp \\ H_z \end{bmatrix}. \tag{52.7}
$$

52.2.2 Simplification: 2D Geometries and 2D Fields

In a 2D geometry, the permittivity ε depends only on two coordinates, say the x- and y-coordinates: $\varepsilon = \varepsilon(x, y)$, just as in the mode propagation case in Section 52.2.1. In the mode propagation case, all fields depend on a common phase factor $\exp(ik_z z)$. If we set $k_z = 0$, i. e., we consider only fields with transversal position dependence $\mathbf{E} = \mathbf{E}(x, y)$, we obtain from (Equations 52.6 and 52.7) the decoupled 2D equations

$$
\nabla_\perp \times \nabla_\perp \times \mathbf{E}_\perp(x, y) = \varepsilon(x, y)\mu\omega^2 \mathbf{E}_\perp(x, y)
$$

$$
-\nabla_\perp \cdot \nabla_\perp E_z(x, y) = \varepsilon(x, y)\mu\omega^2 E_z(x, y). \tag{52.8}
$$

If we additionally assume uniformity in y-direction, we find the eigenvalue problem for waveguides with a 1D cross section

$$\partial_x^2 E_z + \varepsilon(x)\mu\omega^2 E_z = k_y^2 E_z. \tag{52.9}$$

The same applies to the magnetic field:

$$\nabla_\perp \times \frac{1}{\varepsilon(x,y)} \nabla_\perp \times \mathbf{H}_\perp = \mu\omega^2 \mathbf{H}_\perp(x,y)$$

$$\varepsilon(x)\partial_x \frac{1}{\varepsilon(x)} \partial_x H_z + \varepsilon(x)\mu\omega^2 H_z = k_y^2 H_z. \tag{52.10}$$

These equations are simpler than the original ones and they build the basis of many (approximate) solvers. They are simpler in the operator structure and they are linear in the eigenvalue k_y^2.

Note that, conventionally, the propagation direction of waveguide modes along the axis of waveguides is chosen to be the z-axis. To meet this convention, one has to interchange the y- and z-indices in Equations 52.9) and 52.10.

As a consequence of decoupling, we can split the entire field globally into a transversal electric (TE or s-polarization) and a transversal magnetic (TM or p-polarization) field:

$$\begin{pmatrix} \mathbf{E}(x,y) \\ \mathbf{H}(x,y) \end{pmatrix} = \underbrace{\begin{pmatrix} E_z(x,y)\mathbf{e}_z \\ H_x(x,y)\mathbf{e}_x + H_y(x,y)\mathbf{e}_y \end{pmatrix}}_{\text{s-polarization (TE)}} + \underbrace{\begin{pmatrix} E_x(x,y)\mathbf{e}_x + E_y(x,y)\mathbf{e}_y \\ H_z(x,y)\mathbf{e}_z \end{pmatrix}}_{\text{p-polarization (TM)}}. \tag{52.11}$$

52.2.3 Transformation Rules

Transformation techniques often yield considerable simplifications, e.g.,

- Exploiting symmetries by adapting coordinate systems (cylindrical, spherical coordinate systems, etc.)
- Using precomputations on reference elements as in FEM, cf. Section 52.5
- Handling of PML, cf. Section 52.2.4, by a complex coordinate stretching

The transformation rules describe the change of the equations when we change a local coordinate system. The PML, seen as a complex scaling of coordinates, can be treated as change of a coordinate system. The computations in FEM simplify if they are done in the reference coordinate systems and then mapped to the physical systems.

Geometric quantities and field quantities transform differently. For an easy *algebraic* notation, it is useful to write geometric quantities in row vectors and field quantities in column vectors of their coefficients. Expressions like $\mathbf{dl} \cdot \mathbf{E}$ with $\mathbf{dl} = (dx, dy, dz)$ (a row vector of path components) and $\mathbf{E} = (E_x, E_y, E_z)^T$ (a column vector of field components) have then the right algebraic form and can be programed directly as they appear in the formula.

52.2.3.1 Mapping of Geometric Quantities

Let an arbitrary domain mapping $Q : \widehat{K} \to K$ be given by

$$\begin{pmatrix} x \\ y \\ z \end{pmatrix} = Q\left(\widehat{x}, \widehat{y}, \widehat{z}\right), \tag{52.12}$$

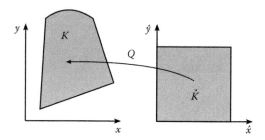

FIGURE 52.2 Example of a domain mapping. A square is mapped to a curvilinear quadrilateral. Quantities belonging to the reference domain are marked with a hat.

see Figure 52.2. This way we map *points* from the reference domain to *points* in the original domain, where the points are defined by the coefficients $\widehat{x}, \widehat{y}, \widehat{z}$ in the reference system and by the coefficients x, y, z in the original system.

We study how infinitesimally small path, face, and volume elements dl, dA, dV, transform. In 3D, we describe dl and dA by three-coefficient row vectors. A linearization of the mapping at point $(\widehat{x}, \widehat{y}, \widehat{z})$ yields

$$\begin{pmatrix} dx \\ dy \\ dz \end{pmatrix} = J(\widehat{x}, \widehat{y}, \widehat{z}) \begin{pmatrix} d\widehat{x} \\ d\widehat{y} \\ d\widehat{z} \end{pmatrix}, \qquad J(\widehat{x}, \widehat{y}, \widehat{z}) := \begin{pmatrix} \partial_{\widehat{x}} Q_1 & \partial_{\widehat{y}} Q_1 & \partial_{\widehat{z}} Q_1 \\ \partial_{\widehat{x}} Q_2 & \partial_{\widehat{y}} Q_2 & \partial_{\widehat{z}} Q_2 \\ \partial_{\widehat{x}} Q_3 & \partial_{\widehat{y}} Q_3 & \partial_{\widehat{z}} Q_3 \end{pmatrix}. \tag{52.13}$$

Defining the path element as the row vector $dl = (dx, dy, dz)$, we obtain the path element transformation

$$dl = \widehat{dl} J^T. \tag{52.14}$$

A similar computation for a face element $dA = \left(dA_x, dA_y, dA_z \right)$ spanned by two path elements gives, with $|J|$ the determinant of J, the transformation rule

$$dA = \widehat{dA} J^{-1} |J| \tag{52.15}$$

and for a volume element dV spanned by three path elements we obtain

$$dV = d\widehat{V} |J|. \tag{52.16}$$

52.2.3.2 Transformation of Grad, Curl, Div

We use the nabla notation describing a column vector

$$\mathbf{grad}\, v(x, y, z) = \nabla_{xyz} v(x, y, z) = \begin{pmatrix} \partial_x v(x, y, z) \\ \partial_y v(x, y, z) \\ \partial_z v(x, y, z) \end{pmatrix}.$$

Let an arbitrary function scalar $v(x, y, z)$ be given which lives on K and allows for the computation of the gradient. We compute the scalar quantity $dl \cdot \nabla_{xyz} v$. This is a physical quantity with the meaning of a voltage (e.g., $\mathbf{E} \cdot \mathbf{ds}$) or a current ($\mathbf{H} \cdot \mathbf{ds}$). This quantity must not change if we change the coordinate system, hence, we require

$$dl \cdot \nabla_{xyz} v(x, y, z) = \widehat{dl} \cdot \nabla_{\widehat{xyz}} \widehat{v} \left(\widehat{x}, \widehat{y}, \widehat{z} \right),$$

TABLE 52.1 Transformation of Geometric and Field Quantities

	Geometric Quantity	Field Quantity	Nabla Operation
Segment	$d\mathbf{l} = \hat{d\mathbf{l}} J^T$	$\mathbf{E} = J^{-T}\hat{\mathbf{E}}$ $\mathbf{H} = J^{-T}\hat{\mathbf{H}}$	$\nabla_{xyz}v = J^{-T}\nabla_{\hat{x}\hat{y}\hat{z}}\hat{v}$
Face	$d\mathbf{A} = \hat{d\mathbf{A}}\,\lvert J\rvert\, J^{-1}$	$\mathbf{D} = \frac{J}{\lvert J\rvert}\hat{\mathbf{D}}$ $\mathbf{B} = \frac{J}{\lvert J\rvert}\hat{\mathbf{B}}$	$\nabla_{xyz}\times\mathbf{v} = \frac{J}{\lvert J\rvert}\nabla_{\hat{x}\hat{y}\hat{z}}\times\hat{\mathbf{v}}$
Volume	$dV = d\hat{V}\,\lvert J\rvert$	$\rho = \frac{1}{\lvert J\rvert}\hat{\rho}$	$\nabla_{xyz}\cdot\mathbf{v} = \frac{1}{\lvert J\rvert}\nabla_{\hat{x}\hat{y}\hat{z}}\cdot\hat{\mathbf{v}}$

where $\hat{v}\left(\hat{x},\hat{y},\hat{z}\right)$ in the reference domain follows from $v\left(\hat{x},\hat{y},\hat{z}\right)$ in the original domain by $v(x,y,z) = v\left(Q\left(\hat{x},\hat{y},\hat{z}\right)\right) = (v\circ Q)\left(\hat{x},\hat{y},\hat{z}\right)$, hence $\hat{v} = v\circ Q$, based on the coordinate transform Equation 52.12. It follows from Equation 52.14 that

$$\hat{d\mathbf{l}} J^T \cdot \nabla_{xyz}v(x,y,z) = \hat{d\mathbf{l}} \cdot \nabla_{\hat{x}\hat{y}\hat{z}}\hat{v}\left(\hat{x},\hat{y},\hat{z}\right).$$

Since this should hold true for all functions v, we find the mapping of the gradient

$$\nabla_{xyz} = J^{-T}\nabla_{\hat{x}\hat{y}\hat{z}}. \tag{52.17}$$

The coefficient of the nabla operator transform like the coefficients of the field strength. The same way one computes the transformations rules for **curl** and **div**, as collected in Table 52.1.

52.2.3.3 Transformation of μ and ε

Rewriting Maxwell's **curl** equations from the physical to the reference domain we get

$$\frac{J}{\lvert J\rvert}\nabla_{\hat{x}\hat{y}\hat{z}}\times\hat{\mathbf{H}} = i\omega\varepsilon J^{-T}\hat{\mathbf{E}}$$

$$\frac{J}{\lvert J\rvert}\nabla_{\hat{x}\hat{y}\hat{z}}\times\hat{\mathbf{E}} = -i\omega\mu J^{-T}\hat{\mathbf{H}},$$

which shows that the entire transformation can be traced back to a mapping of the material quantities

$$\hat{\varepsilon}: = \lvert J\rvert J^{-1}\varepsilon J^{-T} \tag{52.18}$$

$$\hat{\mu}: = \lvert J\rvert J^{-1}\mu J^{-T}. \tag{52.19}$$

52.2.4 Perfectly Matched Layers

A standard problem in the simulation of scattering problems is the treatment of scattered fields outside the domain of interest Ω_{int}, cf. Figure 52.3. The predominantly used method is the PML technique introduced by Berenger [5]. The PML treatment of Berenger relies on a split field formulation of Maxwell's equations in the PML region. This has been shown in [6] to be only weakly stable. In contrast, the uniaxial PML (UPML) [7] does not require such a splitting of Maxwell's fields and maintains the strong well-posedness of the original Maxwell system. Both formulations are equivalent to an elegant and general approach: the stretched-coordinate PML [8,9]. We describe briefly only the latter one. It has among others two remarkable properties

1. It can be used in a similar way in higher space dimensions.
2. It can be plugged into existing codes provided that the codes allow for sufficient general materials and field representations.

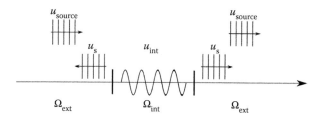

FIGURE 52.3 Sketch of a 1D time-harmonic scattering problem. The source fields are given, the interior field u_{int} and the scattered field u_S has to be computed.

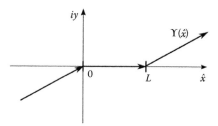

FIGURE 52.4 Complex PML path. Within the interior domain $(0, L)$ the x-coordinate is real but becomes complex outside.

52.2.4.1 PML in 1D

Let us consider the right exterior domain $x > L$, where L marks the right boundary point of the situation depicted in Figure 52.3. The general solution of the scattered field in the scalar 1D version of Equation 52.8 is

$$u_s = c_- e^{-ikx} + c_+ e^{ikx}, \quad x > L.$$

We have to discriminate between the incoming version of the scattered field $c_- \exp(-ikx)$, which we need to drop and the outgoing scattered field $c_+ \exp(ikx)$. Here, x is the real space coordinate. Now, we allow x to be a path in the complex plane, not just on the real axis. We parameterize path x with the real parameter $\hat{x}$ to derive a path in the complex plane; now $\hat{x}$ takes the role of the reference coordinate as in Section 52.2.3:

$$x(\hat{x}) = L + (1 + i\sigma)\hat{x}$$

$$\sigma > 0, \text{ real, fixed} \tag{52.20}$$

$$\hat{x} > 0, \text{ real reference coordinate;}$$

compare Figure 52.4, which yields the following mapping:

$$c_- \exp(-ikx) \;\mapsto\; c_- \exp(-ikL)\exp(-ik\hat{x})\exp(\sigma k\hat{x})$$
$$c_+ \exp(ikx) \;\mapsto\; c_+ \exp(ikL)\exp(ik\hat{x})\exp(-\sigma k\hat{x}).$$

The first expression, the incoming field, blows up for $\hat{x} \to \infty$, due to the factor $\exp(\sigma k\hat{x})$, whereas the second one decreases exponentially with $\exp(-\sigma k\hat{x})$. So for $\hat{x}$ large enough, we have a clear discrimination:

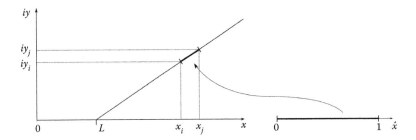

FIGURE 52.5 Transformation from the unit interval to the complex-valued PML interval.

nearly the entire field value comes from the left traveling scattered field. Hence, a proper formulation of Sommerfeld's radiation condition in terms of the complexified path $x(\widehat{x})$ is

$$u\left(x\left(\widehat{x}\right)\right) \longrightarrow 0 \text{ as } \widehat{x} \longrightarrow \infty.$$

Let us stay in the right exterior domain and introduce the complex distance variable x:

$$x(\widehat{x}) = \widehat{x} + i\sigma\widehat{x}.$$

We consider the complexification of the spatial variable within the Helmholtz equation:

$$\partial_x^2 u + k^2 u = 0 \qquad \Rightarrow \qquad \frac{1}{(1 + i\sigma)^2} \frac{\partial^2 u}{\partial \widehat{x}^2} + k^2 u = 0.$$

This step is decisive. If we are allowed to generalize the real derivative to a complex derivative, i. e., if the derivative is independent of the direction in the complex plane, we can use the linear path transformation, see Figure 52.5, (or others) to obtain the last equation.

This transformation is what we in fact exploit within the algorithmic realization. We define the complexification of the real x-coordinate to a complex coordinate x by means of the real reference coordinate $\widehat{x}$

$$x = \begin{cases} \widehat{x}(1 + i\sigma), & \widehat{x} < 0 \\ \widehat{x}, & 0 \le \widehat{x} < L \\ L + (\widehat{x} - L)(1 + i\sigma), & \widehat{x} \ge L. \end{cases} \tag{52.21}$$

52.2.4.2 PML in 2D and 3D

The (tensor product type) generalization of Equation 52.20 to 2D is obviously the following (σ_x, σ_y are real, positive constants):

$$x = \begin{cases} \widehat{x}\left(1 + i\sigma_x\right), & \widehat{x} < 0 \\ \widehat{x}, & 0 \le \widehat{x} < L \\ L + (\widehat{x} - L)\left(1 + i\sigma_x\right), & \widehat{x} \ge L \end{cases}, \quad y = \begin{cases} \widehat{y}\left(1 + i\sigma_y\right), & \widehat{y} < 0 \\ \widehat{y}, & 0 \le \widehat{y} < L \\ L + (\widehat{y} - L)\left(1 + i\sigma_y\right), & \widehat{y} \ge L. \end{cases} \tag{52.22}$$

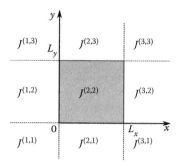

FIGURE 52.6 PML-regions in 2D lexicographically enumerated.

The Jacobian of the mapping $(\hat{x}, \hat{y}) \mapsto (x, y)$ depends on the position, see Figure 52.6, and evaluates to, e.g.,

$$J^{(2,2)} = \begin{pmatrix} 1 & \\ & 1 \end{pmatrix} \qquad J^{(3,2)} = \begin{pmatrix} 1 + i\sigma_x & \\ & 1 \end{pmatrix} \qquad J^{(3,3)} = \begin{pmatrix} 1 + i\sigma_x & \\ & 1 + i\sigma_y \end{pmatrix}.$$

The realization of the PML now consists in using the transformed permittivity (Equation 52.18) and permeability (Equation 52.19), where the Jacobians are computed from the complex stretching of the space variables (Equation 52.22). We proceed now as follows. We consider a subdomain of the exterior, say $\Omega^{(2,1)} \subset \mathbb{R}^2$ from Figure 52.6 and the complex solution $u^{(2,1)} : \Omega^{(2,1)} \to \mathbb{C}$ living there. Then, we use the mapping from the real to the complex coordinates to define a complex-valued spatial domain: $\Omega^{(2,1)} \to \Omega^{c,(2,1)} \subset \mathbb{C}^2$. On the complexified domain $\Omega^{c,(2,1)}$, we have a complex extension of $u^{(2,1)}$ which we denote by $u^{c,(2,1)}$. The solution $u^{c,(2,1)}$ is different to the one on the real space $\Omega^{(2,1)}$, it is the complex continuation of the original function. The key point is that our original function $u^{(2,1)}(x, y)$ has a complex extension by its nature, and we point to its values by introducing complex instead of only real spatial coordinates.

52.3 FDTD Method

The FDTD method is the most widely used method for the simulation of electromagnetic field propagation. It combines algorithmic simplicity with a deep universality. Nonlinear, nonlocal, and dispersive material responses can be treated. The reason for this generality lies in the fact that the structure of the FDTD scheme is very close to the structure of Maxwell's equations and has several algorithmic features:

- It is simple to implement and fast.
- High-order discretizations are feasible.
- It is explicit in time.
- There is a comprehensive theory.

Besides these pros, there are also severe drawbacks:

- The simple local approximation based on Cartesian staggered grids restricts geometric flexibility unless the grid is prohibitively fine (staircasing problem).
- The inherent second-order accuracy in time and space of the original method limits long-time computations on large domains.

52.3.1 Historical Notes

The initial and most important step was done by Yee in 1966 [10]. He used a so-called staggered mesh discretization in space and time, see the following subsections, and replaced the differential operators **grad**, **curl**, and **div** by finite-difference approximations to obtain a discrete approximation of second-order accuracy. Since then, a huge number of contributions improved the method and extended the fields of applications. An interesting view on the derivation of the FDTD equations has been given by Weiland in 1977 [11] starting from an integral formulation of Maxwell's equations. This approach is called FIT and yields the same set of equations. We use this approach to motivate the finite difference equations.

A number of efforts have been aimed at addressing the above-mentioned problems. The staircasing problem has been treated in many papers, e.g., [12]. High-order methods have been proposed, e.g., in [13, 14].The *generalized finite difference method* introduced by Bossavit [15] in 2001 takes ideas from differential geometry and topology and stimulates new interpretations of classical finite element schemes in terms of finite differences. Clemens and Weiland [16] follow a similar spirit in the framework of FIT to extend classical schemes.

A comprehensive overview of FDTD methods and further improvements along with detailed applications is given in [17] which is the standard reference book on FDTD methods. A thoroughly different approach to time-dependent problems is given by the class of DG methods which by its inherent construction offer high-order approximations in space and time as well as flexible meshes, cf. Section 52.6.

52.3.2 Concept: Discrete Differential Operators on Staggered Grids

Figure 52.7 explains the basic construction principle. We consider two entangled loops. One, say the light gray loop, is associated with the electric field, the other one with the magnetic field (left part of Figure 52.7). These loops will be used later for line integrals. The straight edges of the loops themselves are the edges of a cube, see the middle part of Figure 52.7. Then, by adding cubes related to the electric and magnetic fields, the complete spatial Yee mesh is generated. It can be considered as composed from two independent Cartesian meshes where the "electric" mesh and the "magnetic" mesh are shifted compared to each other by half of an elementary edge length of a cube.

52.3.2.1 Degrees of Freedom

A central question in any numerical scheme is the proper definition of the degrees of freedom (DoF). Typically, the stability of a method essentially relies on this. In FDTD methods, it is common to project the fields to the edges and to take the values at the midpoints of the edges as DoF. For example, we project **E** to the ith edge $\mathbf{e}_i$, $\mathbf{E} \cdot \mathbf{e}_i$, take the value at the midpoint, and denote it as E_i. We motivate the DoFs in the following by approximations of integral expressions (e.g., by the midpoint rule) as it is done in the FIT. This offers additional insight into the structure of the equations. The same results are obtained if the differential operators are approximated by finite difference stencils.

52.3.2.2 Discrete Gradient

For a direct discretization of Maxwell's equations, it is not necessary to consider the discrete approximation of the gradient but it completes the insight into the structure of the subsequent discretization process. Consider an edge **e** of length h connecting two neighboring vertices v_1, v_2 of the Yee mesh. The fundamental theorem of calculus

$$\int_{v_1}^{v_2} \mathbf{grad}\,\phi\,\mathrm{d}\mathbf{e} = \phi(v_2) - \phi(v_1),$$

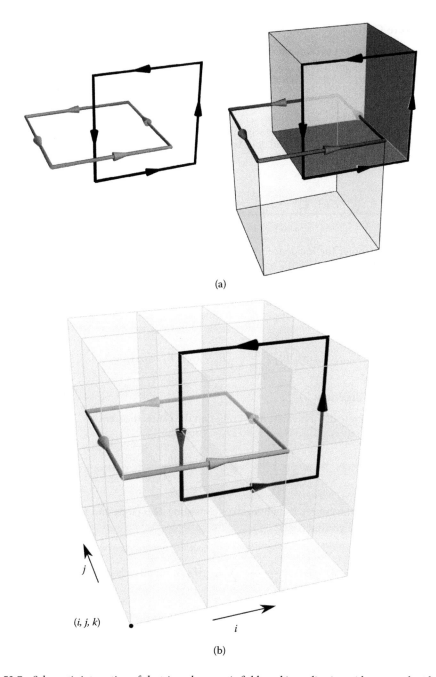

(a)

(b)

FIGURE 52.7 Schematic interaction of electric and magnetic fields and its realization with staggered grids.

motivates the definition of the discrete gradient operator $\mathbf{grad}_h$:

$$\mathbf{grad}_h \, \phi\big|_\mathbf{e} \; : \; = \frac{\mathbf{e}^0}{h} \left(\phi(v_2) - \phi(v_1) \right) \; \text{forall edges e,}$$

where $\mathbf{e}^0$ is the normalization of $\mathbf{e}$ to unit length. The $\mathbf{grad}_h$ -operator maps quantities defined on *vertices* to quantities defined on *edges*.

52.3.2.3 Discrete Curl

The definition of the discrete **curl** -operator, $\mathbf{curl}_h$, relies on the Stokes theorem

$$\int_{\mathbf{f}} \mathbf{curl\,E}\ d\mathbf{f} = \int_{\partial \mathbf{f}} \mathbf{E}\ d\mathbf{e}$$

with **f** a plane, rectangular face of a cube and $\partial \mathbf{f}$ its boundary consisting of four edges $\mathbf{e}_i$. The discrete version reads

$$\mathbf{curl}_h\ \mathbf{E}\big|_{\mathbf{f}}\ :\ = \frac{\mathbf{f}^0}{h_1 h_2} \sum_{i=1}^{4} \iota_{e_i,f}\ |\mathbf{e}_i|\ E_i\ \text{forall faces } \mathbf{f}.$$

The iota-symbol $\iota_{e_i,f}$ denotes the incidence number of edge $\mathbf{e}_i$ with respect to the orientation of the face **f**. It holds $\iota_{e_i,f} = 1$, if edge $\mathbf{e}_i$ shares the same direction as that induced by the face vector **f**, otherwise $\iota_{e_i,f} = -1$. The $\mathbf{curl}_h$ -operator maps quantities defined on *edges* to quantities defined on *faces*. For the light gray loop of Figure 52.7, e.g., starting with the edge in front of the figure and assuming the edge lengths of the rectangle are all the same $h_1 = h_2 = h$, this is

$$\mathbf{curl}_h\ \mathbf{E}\big|_{\mathbf{f}}\ :\ = \frac{\mathbf{f}^0}{h}\left(E_1 + E_2 - E_3 - E_4\right).$$

52.3.2.4 Discrete Divergence

The definition of the discrete **div** -operator, $\mathbf{div}_h$, relies on the Gauss theorem

$$\int_{V} \mathbf{div\,D}\ dV = \int_{\partial V} \mathbf{D}\ d\mathbf{f}.$$

Here, V is the volume of the cube, ∂V its boundary consisting of 6 faces, and **f** is the outward pointing surface vector. The discrete version reads

$$\mathbf{div}_h\ \mathbf{D}\big|_{c} = \frac{1}{h_1 h_2 h_3} \sum_{i=1}^{6} \iota_{e_f,c}\ |\mathbf{f}_i|\ D_i\ \text{for all cells } c.$$

The iota-symbol $\iota_{f_i,c}$ denotes the incidence number of face $\mathbf{f}_i$ with respect to the orientation of cell c, defined positive by the outward orientation of all surface vectors. Hence, it holds $\iota_{f_i,c} = 1$, if face $\mathbf{f}_i$ points outward c, otherwise $\iota_{f_i,c} = -1$. The $\mathbf{div}_h$ -operator maps quantities defined on *faces* to quantities defined on *cells*.

52.3.2.5 Exact Sequences

As a direct consequence of the fact that the Yee mesh is orientable, it follows

$$\mathbf{curl}_h \cdot \mathbf{grad}_h = 0$$
$$\mathbf{div}_h \cdot \mathbf{curl}_h = 0. \tag{52.23}$$

Hence, the discrete operators share the same properties as the continuous ones. For the Yee mesh, the property follows easily from a direct computation.

52.3.3 Spatial Discretization of Maxwell's Equations

Let us denote by $\mathbf{E_h}$ the column vector of all DoFs associated with the primal mesh and by $\mathbf{H_h}$ the column vector of all DoFs associated with the dual mesh. Then, the spatial discrete version of Equation 52.3, setting $\sigma = 0$ for simplicity, reads

$$\varepsilon \partial_t \mathbf{E_h}\big|_{e_p} = \mathbf{curl}_h\, \mathbf{H_h}\big|_{f_d} \qquad \text{for all faces } \mathbf{f_d} \text{ of the dual mesh,} \tag{52.24}$$

$$\mu \partial_t \mathbf{H_h}\big|_{e_d} = -\,\mathbf{curl}_h\, \mathbf{E_h}\big|_{f_p} \qquad \text{for all inner faces } \mathbf{f_p} \text{ of the primal mesh.} \tag{52.25}$$

Here, $\mathbf{e_p}$ denotes the edge of the primal mesh that penetrates face $\mathbf{f_d}$ of the dual mesh and $\mathbf{e_d}$ denotes the edge of the dual mesh going through $\mathbf{f_p}$. For implementations, it is convenient to consider both the primal and the dual meshes as submeshes of a common Cartesian mesh.

52.3.4 Time Evolution

The time evolution of the semidiscretized equations (Equations 52.24 and 52.25) can be computed by a number of different time integrators. Frequently used is the so-called leapfrog scheme. It fits perfectly to the spatial discretization and offers the advantage of being an explicit method.

52.3.4.1 Leapfrog Scheme in Time

We introduce a uniform mesh in time

$$0 = t_0 < t_{\frac{1}{2}} < t_1 < \ldots < t_i < t_{i+\frac{1}{2}} < t_{i+1} \ldots < t^N = T,$$

with $t_{i+1} - t_i = \Delta t$ for all time steps and indicate the discrete time by a superscript at the quantities to use, e. g., the discrete electric field vector at time $t_{n+\frac{1}{2}}$ is $\mathbf{E}_h^{n+\frac{1}{2}}$. Using a central finite difference approximation of the time derivative in Equation 52.24, we obtain its fully discretized version

$$\mathbf{E_h}\big|_{e_p}^{n+\frac{1}{2}} = \mathbf{E_h}\big|_{e_p}^{n-\frac{1}{2}} + \frac{\Delta t}{\varepsilon}\,\mathbf{curl}_h\, \mathbf{H_h}\big|_{f_d}^{n} \qquad \text{for all faces } \mathbf{f_d} \text{ of the dual mesh.} \tag{52.26}$$

The time derivative is centered around n, hence it is of second-order accuracy. Using the updated electric field of Equation 52.26, Equation 52.25 is updated accordingly:

$$\mathbf{H_h}\big|_{e_d}^{n+1} = \mathbf{H_h}\big|_{e_d}^{n} - \frac{\Delta t}{\mu}\,\mathbf{curl}_h\, \mathbf{E_h}\big|_{f_p}^{n+\frac{1}{2}} \qquad \text{for all inner faces } \mathbf{f_p} \text{ of the primal mesh.} \tag{52.27}$$

Equations 52.26 and 52.27 are typical FDTD update equations.

52.3.5 Dispersion, Stability, and Courant–Friedrichs–Lewy Condition

Let us consider a monochromatic plane wave of frequency ω in a homogeneous medium, with speed of light $c = 1/\sqrt{\varepsilon\mu}$, ε and μ independent of position, and a wavevector $\mathbf{k} = (k_x, k_y, k_z)$. Then the plane wave

$$\mathbf{E}e^{-i\omega t + i\mathbf{kr}}$$

with a constant vector **E** is a solution of Maxwell's equations without sources, if the dispersion relation

$$\omega^2 = \mathbf{k}^2 c^2 \tag{52.28}$$

and orthogonality condition

$$\mathbf{k}\mathbf{E} = 0$$

hold true. The first one follows from the **curl-curl** equations, the second one from the divergence condition. For the Yee mesh, there is a discrete counterpart. The discrete plane wave satisfies Equations 52.26 and 52.27 takes the natural form

$$\mathbf{E}_\mathrm{h} e^{-i\omega t_j + i\mathbf{k}\mathbf{r}_\mathrm{h}}.$$

The corresponding discrete dispersion relation, however, contains the stepsizes in time, Δt, and space, h_x, h_y, h_z,

$$\frac{1}{\Delta t^2} \sin^2 \frac{\omega \Delta t}{2} = c^2 \left(\left(\frac{1}{h_x^2} \sin^2 \frac{k_x h_x}{2} \right) + \left(\frac{1}{h_y^2} \sin^2 \frac{k_y h_y}{2} \right) + \left(\frac{1}{h_z^2} \sin^2 \frac{k_z h_z}{2} \right) \right). \tag{52.29}$$

For a derivation see, e. g.,[17] or [1]. Clearly, this discrete dispersion relation approaches Equation 52.28 as $\Delta t, h_x, h_y, h_z \to 0$ as it must be. But the sin-function of the left-hand side of Equation 52.29 has a further consequence. Let k_x, k_y, k_z be real, hence the plane wave is spatially bounded everywhere as in the case of the the continuous plane wave. Multiplying with Δt^2 from the left-hand side of Equation 52.29 might result in a real number larger than 1, if Δt is large enough. Then ω onthe left-hand side becomes imaginary, and since the expression involves $\sin^2$, both positive and negative imaginary frequencies occur. The positive one will cause an exponential increase of the discrete plane wave in time, hence the solution becomes unstable. Consequently, for stability we must require

$$c^2 \Delta t^2 \left(\frac{1}{h_x^2} + \frac{1}{h_x^2} + \frac{1}{h_z^2} \right) \leq 1$$

or, for uniform meshes, $h = h_x = h_y = h_z$,

$$\Delta t \leq \frac{h}{c\sqrt{3}}.$$

These inequalities are the Courant–Friedrichs–Lewy (CFL) conditions for the wave equation.

52.3.6 Discrete Div-Conditions

It is a remarkable property that the Yee scheme preserves the **div**-conditions of Equation 52.1 for homogeneous media in discrete form [18]. Computing the discrete divergence of the updated Equation 52.26

$$\mathbf{div}_\mathrm{h}\, \varepsilon \mathbf{E}_\mathrm{h}\big|_{\mathbf{e}_\mathrm{p}}^{n+\frac{1}{2}} = \mathbf{div}_\mathrm{h}\, \varepsilon \mathbf{E}_\mathrm{h}\big|_{\mathbf{e}_\mathrm{p}}^{n-\frac{1}{2}} + \Delta t \mathbf{div}_\mathrm{h}\, \mathbf{curl}_\mathrm{h}\, \mathbf{H}_\mathrm{h}\big|_{\mathbf{f}_\mathrm{d}}^{n} \quad \text{for all cells of the dual mesh.}$$

The last term vanishes due to property (Equation 52.23), hence we have the convergence

$$\mathbf{div}_h \, \varepsilon \mathbf{E}_h \big|_{e_p}^{n+\frac{1}{2}} = \mathbf{div}_h \, \varepsilon \mathbf{E}_h \big|_{e_p}^{n-\frac{1}{2}} = \ldots = \mathbf{div}_h \, \varepsilon \mathbf{E}_h \big|_{e_p}^{-\frac{1}{2}} \quad \text{for all cells of the dual mesh.}$$

The divergence attributed to a cell is preserved over all time steps. For the magnetic field, we obtain the corresponding result the same way:

$$\mathbf{div}_h \, \mu \mathbf{H}_h \big|_{e_d}^{n+1} = \mathbf{div}_h \, \mu \mathbf{H}_h \big|_{e_d}^{n} = \ldots = \mathbf{div}_h \, \mu \mathbf{H}_h \big|_{e_d}^{0} \quad \text{for all cells of the primal mesh.}$$

If the initial data at time step $n = -1/2$ and $n = 0$ are divergence-free,

$$\mathbf{div}_h \, \varepsilon \mathbf{E}_h^{-\frac{1}{2}} = 0 \qquad \mathbf{div}_h \, \mu \mathbf{H}_h^0 = 0,$$

this property is preserved for all following time steps.

52.3.7 Convergence

Let, as before, $\mathbf{E}_h$ be the vector of field values attributed to the edges of the primal mesh with values computed at the midpoint of edges. Let us denote the vector of the exact field values, projected to the edges, by $\mathbf{E}$; hence $\mathbf{E}_h$ approximates $\mathbf{E}$ pointwise. We define the following, mesh-dependent discrete norm

$$\left\| \mathbf{E} - \mathbf{E}_h \right\|^2 \; : \, = h^3 \left(\mathbf{E}_h - \mathbf{E} \right)^2,$$

and the analog for $\mathbf{H}$ with values attributed to the centroids of the cell faces. The following holds true. Let $\mathbf{E}$ and $\mathbf{H}$ be three times continuously differentiable in Ω. Then, there exists a constant C independent of h such that

$$\left\| \mathbf{E} - \mathbf{E}_h \right\| + \left\| \mathbf{H} - \mathbf{H}_h \right\| \leq C h^2.$$

FDTD converges with second order. This classical result extends also to nonuniform meshes as shown by Monk and Süli [19].

52.4 Fourier Modal Method

This summary of FMMs follows Chapter 6 of [1]; for more details see that textbook and the references given therein.

52.4.1 Historical Notes

The FMM is a family of methods that belong to the class of spectral methods for the frequency domain and has been well established for more than three decades now. The term spectral method means that the field approximation is realized with functions which have a global support. In contrast, in FDTD methods, we consider only the pointwise-defined functions, and in FEM methods, we have basis functions with a very small, local support.

The origins of the FMM dates back to the 1960s and were devoted to grating analyses [20,21]. In 1981, Moharam and Gaylord established in their seminal paper [22] a computational framework they called rigorous coupled wave analysis (RCWA). Other synonyms or variations of the method include the PWE, the eigenmode expansion (EME), and eigenmode expansion technique (EET). A typical observation in RCWA methods has been that the convergence of TM fields is notably worse than that of TE fields. In 1996,

Lalanne and Granet and coworkers [23,24] found a solution to overcome this TM convergence problem and in the same year Li published his factorization rules [25] formalizing the results. Currently, different proposals are being investigated to establish a fast convergent 3D FMM, e. g., [26].

52.4.2 Concept: Segmentwise Waveguide Treatment

The family of FMMs are characterized by the following building blocks:

1. The geometry is subdivided into a sequence of waveguide-type sections, see Figure 52.8b and c.
2. The field within each waveguide section is represented by up and down propagating modes as well as evanescent modes.
3. The transversal fields of the waveguide-type eigenmodes are computed in Fourier representation.
4. The interaction of the individual sections is realized by posing proper continuity conditions at the interfaces and a scattering-matrix formalism.

52.4.3 Transfer-Matrix Approach

We consider the fields in two subsequent layers i and $i+1$, cf. Figure 52.9. At the moment, we do not focus on the special structure of the electromagnetic field except that we can decompose it into upward ($\mathbf{E}^+$) and downward ($\mathbf{E}^-$) propagating parts with respect to the z-direction. Let $\mathbf{E}_i^\pm(x,z)$ be arbitrary normalized solutions of Maxwell's equations split into propagation directions and $a_i^\pm \in \mathbb{C}$ their weights. It holds that

$$
\begin{aligned}
\text{Field in layer } i \quad &: \quad a_i^+ \mathbf{E}_i^+(x,z) \quad + \quad a_i^- \mathbf{E}_i^-(x,z) \\
&\Downarrow \text{ implies} \\
\text{Field in layer } i+1 \quad &: \quad a_{i+1}^+ \mathbf{E}_{i+1}^+(x,z) \quad + \quad a_{i+1}^- \mathbf{E}_{i+1}^-(x,z).
\end{aligned}
\tag{52.30}
$$

Given the complete field in one layer, Maxwell's equations including continuity conditions enforce the complete field in the other layer. This can be expressed by a transfer-matrix relation of the coefficients

$$
\mathbf{A}_{i+1} \begin{pmatrix} a_{i+1}^+ \\ a_{i+1}^- \end{pmatrix} = \mathbf{A}_i \begin{pmatrix} a_i^+ \\ a_i^- \end{pmatrix},
\tag{52.31}
$$

where $\mathbf{A}_i$, $\mathbf{A}_{i+1}$ are complex 2×2 matrices. We introduced this symmetric form since the count sequence of layers (top-down or bottom-up) is arbitrary. The typical application of the 1D transfer-matrix method is, given n layers of finite thickness with indices 0 and $n+1$ referring to substrate and superstrate, the

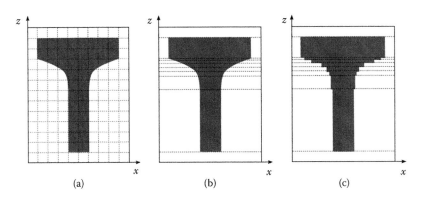

FIGURE 52.8 (a) A schematic geometry and a Cartesian mesh as used in FDTD methods. (b) Slicing of the geometry into segments along z. (c) Waveguide-like approximation of the geometry used in Fourier modal methods.

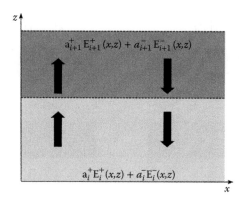

FIGURE 52.9 Superposition of up and down propagating fields in two subsequent layers.

update procedure

$$\begin{pmatrix} a_{n+1}^{+} \\ a_{n+1}^{-} \end{pmatrix} = \mathbf{T}_{n,n+1} \cdot \ldots \cdot \mathbf{T}_{i,i+1} \cdot \ldots \cdot \mathbf{T}_{0,1} \begin{pmatrix} a_{0}^{+} \\ a_{0}^{-} \end{pmatrix}.$$

This procedure may become unstable due to the exponentials involved in $T_{i,i+1}$.

52.4.4 Scattering-Matrix Approach

The scattering-matrix approach uses merely the same representation as the transfer-matrix approach, except that the computation of the coefficients $a_{i}^{\pm}$ is reordered (Figure 52.10). The scattering-matrix notation corresponding to Equation 52.31 is

$$\begin{pmatrix} a_{i}^{-} \\ a_{i+1}^{+} \end{pmatrix} = \mathbf{S}_{i,i+1} \begin{pmatrix} a_{i}^{+} \\ a_{i+1}^{-} \end{pmatrix} \tag{52.32}$$

with the obvious computation of the coefficients of $\mathbf{S}_{i,i+1}$ from $\mathbf{A}_{i}$ and $\mathbf{A}_{i+1}$. The notion scattering follows from the fact that we can consider the coefficients $\left(a_{i}^{+}, a_{i+1}^{-}\right)$ as fields *incident* to the layer separating boundary and the coefficients $\left(a_{i}^{-}, a_{i+1}^{+}\right)$ as fields *scattered* away from the boundary, respectively. The scattering matrix for an n-layer system is derived as follows. We rewrite Equation 52.31 as

$$\begin{pmatrix} 0 \\ 0 \end{pmatrix} = \begin{pmatrix} \mathbf{A}_{i}\left(d_{i}\right) & -\mathbf{A}_{i+1}\left(0\right) \end{pmatrix} \begin{pmatrix} a_{i}^{+} \\ a_{i}^{-} \\ a_{i+1}^{+} \\ a_{i+1}^{-} \end{pmatrix}.$$

Here, the top coefficient a_{i}^{+} of the column vector and the bottom coefficient a_{i+1}^{-} are the incident coefficients. We repeat this for the next layer, increasing the index i by 1

$$\begin{pmatrix} 0 \\ 0 \\ 0 \\ 0 \end{pmatrix} = \begin{pmatrix} \mathbf{A}_{i}\left(d_{i}\right) & -\mathbf{A}_{i+1}\left(0\right) & \\ & \mathbf{A}_{i+1}\left(d_{i+1}\right) & -\mathbf{A}_{i+2}\left(0\right) \end{pmatrix} \begin{pmatrix} a_{i}^{+} \\ a_{i}^{-} \\ a_{i+1}^{+} \\ a_{i+1}^{-} \\ a_{i+2}^{+} \\ a_{i+2}^{-} \end{pmatrix}.$$

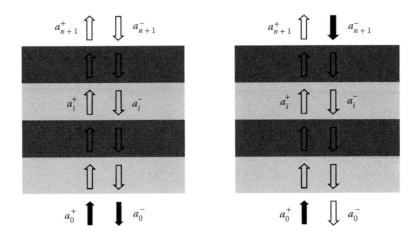

FIGURE 52.10 (a) Transfer-matrix approach. The data a_0^+, a_0^- are given (filled arrows), the others are computed. (b) Scattering-matrix approach. The data a_0^+, a_{n+1}^- are given, the others are computed.

Again, the top and bottom coefficients of the column vector are the incident coefficients of the layer system. Moving these coefficients to the left-hand side, deleting the first and last column of the system matrix, and inverting it yields the scattering-matrix formulation with scattering matrix **S**:

$$
\mathbf{S}
\begin{pmatrix}
a_0^+ \\
0 \\
\vdots \\
0 \\
a_{n+1}^-
\end{pmatrix}
=
\begin{pmatrix}
a_0^- \\
a_1^+ \\
a_1^- \\
\vdots \\
a_{n+1}^+
\end{pmatrix}.
\tag{52.33}
$$

52.4.5 Layered Media Stack as Model Case

A layered media stack is a 1D problem, if we choose the z-axis normal to the faces of the layers. To discuss the structure of FMM algorithms, we construct the standard transfer-matrix algorithm in scattering-matrix formulation from the viewpoint of the FMM. To find the field representation in layer i, we have to solve Maxwell's equations in layer i, i. e., either the equation for TE or TM. Let us consider the TE case. The modal ansatz

$$
\text{layer } i: \quad E_y(x, z) = E(x)e^{ik_z z} = E_0 e^{ik_x x + ik_z z}
$$

satisfies Equation 52.9 if $n_i^2 k_0^2 - k_x^2 = k_{z,i}^2$. Hence, given k_x, we have

$$
\text{layer } i: \quad E_{y,i}(x, z) = a_i^+ e^{ik_x x + ik_{z,i} z} + a_i^- e^{ik_x x - ik_{z,i} z},
\tag{52.34}
$$

which is the first line in Equation 52.30. Equally, we have for layer $i + 1$

$$
\text{layer } i + 1: \quad E_{y,i+1}(x, z) = a_{i+1}^+ e^{ik_x x + ik_{z,i+1} z} + a_{i+1}^- e^{ik_x x - ik_{z,i+1} z}.
$$

We compute matrices A_i and A_{i+1}. Let the boundary between the two layers be placed at $z = 0$. The tangential electric and magnetic fields must be continuous. It follows for the electric field

$$a_i^+ e^{ik_x x} + a_i^- e^{ik_x x} = a_{i+1}^+ e^{ik_x x} + a_{i+1}^- e^{ik_x x} \quad \text{for all } x \in \mathbb{R},$$

hence

$$a_i^+ + a_i^- = a_{i+1}^+ + a_{i+1}^-. \tag{52.35}$$

Given the electric field in each layer, we compute the tangential component of the magnetic field

$$\begin{aligned}
\text{layer } i: \quad & \omega\mu H_{z,i}(x,z) = a_i^+ k_{z,i} e^{ik_x x + ik_{z,i} z} &- a_i^- k_{z,i} e^{ik_x x - ik_{z,i} z} \\
\text{layer } i+1: \quad & \omega\mu H_{z,i+1}(x,z) = a_{i+1}^+ k_{z,i+1} e^{ik_x x + ik_{z,i+1} z} &- a_{i+1}^- k_{z,i+1} e^{ik_x x - ik_{z,i+1} z}
\end{aligned}$$

and obtain the continuity condition

$$k_{z,i} a_i^+ - k_{z,i} a_i^- = k_{z,i+1} a_{i+1}^+ - k_{z,i+1} a_{i+1}^-. \tag{52.36}$$

Consequently, taking Equations 52.35 and 52.36,

$$\begin{pmatrix} 1 & 1 \\ k_{z,i+1} & -k_{z,i+1} \end{pmatrix} \begin{pmatrix} a_{i+1}^+ \\ a_{i+1}^- \end{pmatrix} = \begin{pmatrix} 1 & 1 \\ k_{z,i} & -k_{z,i} \end{pmatrix} \begin{pmatrix} a_i^+ \\ a_i^- \end{pmatrix}. \tag{52.37}$$

Repeating this calculation with a finite layer thickness d_i between the two boundaries, we obtain

$$A_i(d_i) = \begin{pmatrix} 1 & 1 \\ k_{z,i} & -k_{z,i} \end{pmatrix} \begin{pmatrix} e^{ik_{z,i} d_i} & 0 \\ 0 & -e^{ik_{z,i} d_i} \end{pmatrix}. \tag{52.38}$$

Collecting the steps, we obtain the classical transfer-matrix algorithm in scattering formulation Algorithm 52.1. In 1D it is not necessary to save the eigenvectors since they are the same for all layers. But in 2D and 3D, the eigenvectors are different in each layer, so we introduced the step here for full parallelism with Algorithm 52.1.

Algorithm 52.1: Transfer-Matrix Algorithm in Scattering Formulation

---

Require: a_0^+ and a_{n+1}^- to define source fields.
  for all layers **do**
    Solve each layer's eigenvalue problem (Equation 52.9).
    Save: eigenvectors $\exp(ik_x x)$ and eigenvalues $k_{z,i}$
    for later field representation (Equation 52.34).
    Compute the matrices A_i by (Equation 52.38).
  end for
  Assemble the system (Equation 52.33).
  Solve.
  return coefficients $a_0^-, \ldots, a_{n+1}^+$.

52.4.6 FMM in 2D

We generalize the foregoing, where ε has been constant in each layer. Now, let ε vary with x, each layer may have a different dependence $\varepsilon_i(x)$. The eigenvalue problems are Equations 52.9 and 52.10 with layerwise z-independent permittivities. In RCWA and FMM methods, these eigenvalue equations are solved based on a Fourier representation of the fields and the permittivities. This implies essentially two properties:

- Advantage: There is no lateral mesh needed. The nonlocal (or spectral) basis functions of fields and permittivities are simply constructed which results in a straightforward implementation.
- Disadvantage: The convergence of RCWA and FMM depend on the convergence properties of the Fourier representation which are not advantageous for nonsmooth functions.

52.4.6.1 Fourier Series Representation of Functions and Products of Functions

We consider a function $f(x)$ which is periodic with L: $f(x + L) = f(x)$. Hence, $f(x)$ has a Fourier series representation

$$f(x) = \sum_{m=-\infty}^{\infty} f_m e^{im\frac{2\pi}{L}x}. \tag{52.39}$$

We denote its Fourier transform by $\widehat{f}$

$$\widehat{f}(k) = \frac{1}{L}\int_0^L f(x)e^{-ikx}\,dx \text{ and compute } f_m = \widehat{f}\left(\frac{2\pi}{L}m\right), \quad m \in \mathbb{Z}.$$

We consider a second function $g(x)$, also periodic with interval L, and study the product $h(x) = f(x)g(x)$, which is also L-periodic and consequently has a Fourier representation

$$h(x) = \sum_{m=-\infty}^{\infty} h_m e^{im\frac{2\pi}{L}x}.$$

In terms of the Fourier coefficients of f and g, this is

$$h(x) = \left(\sum_{m=-\infty}^{\infty} f_m e^{im\frac{2\pi}{L}x}\right)\left(\sum_{n=-\infty}^{\infty} g_n e^{in\frac{2\pi}{L}x}\right) = \sum_{m=-\infty}^{\infty}\left(\sum_{n=-\infty}^{\infty} f_{m-n}g_n\right)e^{im\frac{2\pi}{L}x}$$

$$h_m = \sum_{n=-\infty}^{\infty} f_{m-n}g_n. \tag{52.40}$$

We write the last equation in matrix-vector form

$$\mathbf{h} = \mathbf{Fg}$$

$$\mathbf{F} = \begin{pmatrix} \cdots & \cdots & \cdots & \cdots & \cdots & \cdots \\ \cdots & f_0 & f_{-1} & f_{-2} & f_{-3} & \cdots \\ \cdots & f_1 & f_0 & f_{-1} & f_{-2} & \cdots \\ \cdots & f_2 & f_1 & f_0 & f_{-1} & \cdots \\ \cdots & f_3 & f_2 & f_1 & f_0 & \cdots \\ \cdots & \cdots & \cdots & \cdots & \cdots & \cdots \end{pmatrix},$$

where the convolution matrix $\mathbf{F}$ is called the Toeplitz matrix.

52.4.6.2 Fourier Series Convergence in Dependence of the Smoothness of the Functions

In electromagnetics, we have to deal with piecewise smooth functions which have different global smoothness properties. Roughly, we have three classes:

Discontinuous: Functions that are piecewise smooth, but may have jumps or singularities. Example: Normal field components in a 3D setting with material discontinuities.

Continuous: Functions that are continuous but not differentiable at positions, where different materials meet. Example: H_y in TM fields as solution of, e.g., Equation 52.10.

Smooth: Functions that are continuous *and* the first derivative exists everywhere, even at positions, where different materials meet. Example: E_y in s-polarization, e.g., Equation 52.9.

52.4.6.3 Truncation

For an algorithmic application, the infinite series has to be truncated. The truncated Fourier series (Equation 52.39) then reads

$$f^M(x) = \sum_{m=-M}^{M} f_m e^{im\frac{2\pi}{L}x}, \quad f_m = \frac{1}{L}\int_0^L f(x)e^{-ik_m x}\, dx, \quad k_m = \frac{2\pi}{L}m \tag{52.41}$$

and the symmetric truncation of (Equation 52.40) yields

$$h^M(x) = \sum_{m=-M}^{M} \left(\sum_{n=-M}^{M} f_{m-n}g_n \right) e^{im\frac{2\pi}{L}x}, \quad h_m^M = \sum_{n=-M}^{M} f_{m-n}g_n.$$

Not only does the number of terms of the convolution sum change, but the individual coefficients h_n^M change too. The truncation changes the problem from an infinite to a finite one. The finite dimensional problem will differ from the exact, infinite dimensional one, as always in numerics, and the quality of this difference depends on the nature of the applied series truncation. The truncated matrix-vector form reads

$$\mathbf{h}^M = \mathbf{F}^M \mathbf{g}^M \quad \text{"direct rule"} \tag{52.42}$$

$$\mathbf{F}^M = \left(f_{m-n}\right)_{-M\leq m,n\leq M} \tag{52.43}$$

with $\mathbf{F}^M$ the finite dimensional Toeplitz matrix and $f_{m-n} = 0$ for $|m - n| > M$. As shorthand for $\mathbf{F}^M$ generated by coefficients f_m we also write $[\![f]\!] := \mathbf{F}^M$.

52.4.6.4 Li's Factorization Rules

It has turned out that the originally observed slow convergence rate of TM problems is related to the smoothness properties of the products of field and permittivity and a special, polarization-dependent treatment can be used to remove the difficulty [23–25]. To this end, we introduce the discontinuity-types shown in Figure 52.11:

1. Type: I The product $f(x)g(x)$ results in $h(x)$ which exhibits also jump discontinuities.
2. Type: II The product $f(x)g(x)$ yields a continuous function $h(x)$. The discontinuities are removed in $h(x)$.

For a formulation of the FMM algorithm, taking into account the accelerated TM technique, we need additionally, the finite Fourier series representation of the inverse function $f^\dagger$:

$$f^\dagger(x) := \frac{1}{f(x)}$$

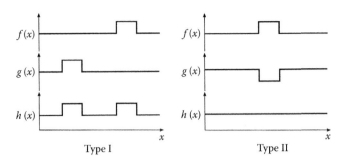

FIGURE 52.11 Two different types of jump discontinuities. Type I: The product $f(x)g(x)$ results in $h(x)$ which exhibits also jump discontinuities. Type II: The product $f(x)g(x)$ yields a continuous function $h(x)$.

together with its matrix-vector representation as counterpart to Equation 52.42

$$\mathbf{h}^M = \left(\mathbf{F}^{M\dagger}\right)^{-1}\mathbf{g}^M \qquad \text{"inverse rule."} \tag{52.44}$$

The factorization rules are as follows:

1. Type: I If the discontinuities are not removable (Type I), apply the direct rule (Equation 52.42) for fastest convergence.
2. Type: II If the discontinuities are removable (Type II), apply the inverse rule (Equation 52.44) for fastest convergence.

52.4.6.5 Eigenvalue Problems in Fourier Space and Application of the Factorization Rules

The eigenvalue problems Equations 52.9 and 52.10 with ε independent of z hold true in each layer. We consider the TE equation Equation 52.9. Its continuous Fourier transform reads

$$\partial_x^2 E_y + \varepsilon(x)\mu\omega^2 E_y = k_z^2 E_y \quad \overset{\text{Fourier transform}}{\Longrightarrow} \quad (ik_x)^2\widehat{E}_y + \mu\omega^2\widehat{\varepsilon} * \widehat{E}_y = k_z^2\widehat{E}_y. \tag{52.45}$$

Restricting $E_y(x)$ to functions periodic over L (discretization in k-space), truncating the number of Fourier basis functions, and using the Toeplitz matrix to compute the convolution we obtain the eigenvalue equation in Fourier space

$$(i\mathbf{K}_x)^2\widehat{\mathbf{E}}_y + \mu\omega^2 \left[\!\left[\widehat{\varepsilon}\right]\!\right]\widehat{\mathbf{E}}_y = k_z^2\widehat{\mathbf{E}}_y. \tag{52.46}$$

Equation 52.46 yields the generalization of the plane waves (TE) used in Equation 52.34 for the derivation of the transfer-matrix algorithm. The same procedure applies to the TM case, where strict consideration of the factorization rules is important. We compute

$$\varepsilon(x)\partial_x\frac{1}{\varepsilon(x)}\partial_x H_y + \varepsilon(x)\mu\omega^2 H_y = k_z^2 H_y.$$

$$\overset{\text{Fourier transform}}{\Longrightarrow} \quad \widehat{\varepsilon} * (ik_x)\widehat{\left(\frac{1}{\varepsilon}\right)} * (ik_x)\widehat{H}_y + \mu\omega^2\widehat{\varepsilon} * \widehat{H}_y = k_z^2\widehat{H}_y. \tag{52.47}$$

For proper discretization based on the factorization rules, we rewrite Equation 52.47, factoring out $\hat{\varepsilon}$,

$$\hat{\varepsilon} * \left((ik_x)\widehat{\left(\frac{1}{\varepsilon}\right)} * (ik_x) + \mu\omega^2 \right) \hat{H}_y = k_z^2 \hat{H}_y. \tag{52.48}$$

First, $\partial_x H_y$ and $1/\varepsilon$ are discontinuous, their product is continuous, hence factorization rule II applies. Second, the large parenthesis on the left-hand side of Equation 52.48 must be discontinuous since ε is discontinuous. The product, however, is continuous, since the right-hand side is continuous. Hence, the factorization rule type II applies again:

$$\left[\left[\widehat{\left(\frac{1}{\varepsilon}\right)} \right] \right]^{-1} \left(i\mathbf{K}_x \left[\left[\widehat{(\varepsilon)} \right] \right]^{-1} (i\mathbf{K}_x) + \mu\omega^2 \right) \hat{\mathbf{H}}_y = k_z^2 \hat{\mathbf{H}}_y. \tag{52.49}$$

52.4.6.6 Continuity Condition

We derive the continuity condition Equation 52.31 for the 2D case. In two subsequent waveguide sections, the multimode generalization of Equation 52.30 holds:

$$\begin{array}{rl}
\text{field in section } i & : \quad \sum_m a_{i,m}^+ \mathbf{E}_{i,m}^+(x, z) \quad + \quad \sum_m a_{i,m}^- \mathbf{E}_{i,m}^-(x, z) \\
\Downarrow \text{ implies} & \\
\text{field in section } i+1 & : \quad \sum_m a_{i+1,m}^+ \mathbf{E}_{i+1,m}^+(x, z) \quad + \quad \sum_m a_{i+1,m}^- \mathbf{E}_{i+1,m}^-(x, z).
\end{array} \tag{52.50}$$

The generalization from plane waves (Equation 52.34) to waveguide modes reads

$$\text{section } i : \quad E_{y,i}(x, z) = \sum_m a_{i,m}^+ \mathbf{E}_{i,m}(x) e^{ik_{z,i,m} z} + \sum_m a_{i,m}^- \mathbf{E}_{i,m}(x) e^{-ik_{z,i,m} z}. \tag{52.51}$$

Each mth normalized mode in section i has a Fouriermode representation $\mathbf{E}_{i,m} = \sum_l b_{i,m,l} \exp ik_{x,l} x$. Here, the weights $b_{i,m,l}$ (ith section, mth waveguide mode, lth Fouriermode of this waveguide mode) are, in case of TE fields, the coefficients of the eigenvectors (Equation 52.46), i.e., $b_{i,m,l} = \left(\hat{\mathbf{E}}_y \right)_l$, if $\hat{\mathbf{E}}_y$ is the mth waveguide mode in the ith section. Hence,

$$E_{y,i}(x, z) = \sum_m a_{i,m}^+ \sum_l b_{i,m,l} e^{ik_{x,i,m,l} x} e^{ik_{z,i,m} z} + \sum_m a_{i,m}^- \sum_l b_{i,m,l} e^{ik_{x,i,m,l} x} e^{-ik_{z,i,m} z}.$$

Setting the interface position between sections i and $i+1$ to $z = 0$ and requiring continuity of E_y yields

$$\sum_{m=-M}^{M} b_{i,m,l} \left(a_{i,m}^+ + a_{i,m}^- \right) = \sum_{m=-M}^{M} b_{i+1,m,l} \left(a_{i+1,m}^+ + a_{i+1,m}^- \right), \qquad -M \leq l \leq M.$$

In the same way, one computes the 2D counterpart to Equation 52.36

$$\sum_{m=-M}^{M} b_{i,m,l} \left(k_{z,i,m} a_{i,m}^+ - k_{z,i,m} a_{i,m}^- \right) = \sum_{m=-M}^{M} b_{i+1,m,l} \left(k_{z,i+1,m} a_{i+1,m}^+ - k_{z,i+1,m} a_{i+1,m}^- \right)$$

$$-M \leq l \leq M$$

Introducing the matrix

$$\mathbf{B}_i = \left(b_{i,l,m} \right)^T = \left(\begin{array}{ccccc} \widehat{\mathbf{E}}_{y,-M}, & \cdots & , \widehat{\mathbf{E}}_{y,m}, & \cdots & , \widehat{\mathbf{E}}_{y,M} \end{array} \right)^T$$

allows writing the two continuity conditions in compact block matrix form:

$$\left(\begin{array}{cc} \mathbf{B}_i & \\ & \mathbf{B}_i \end{array} \right) \left(\begin{array}{cc} \mathbf{I}_d & \mathbf{I}_d \\ \mathrm{diag}\,\mathbf{k}_{z,i} & -\mathrm{diag}\,\mathbf{k}_{z,i} \end{array} \right) \left(\begin{array}{c} \mathbf{a}_i^+ \\ \mathbf{a}_i^- \end{array} \right)$$

$$= \left(\begin{array}{cc} \mathbf{B}_{i+1} & \\ & \mathbf{B}_{i+1} \end{array} \right) \left(\begin{array}{cc} \mathbf{I}_d & \mathbf{I}_d \\ \mathrm{diag}\,\mathbf{k}_{z,i+1} & -\mathrm{diag}\,\mathbf{k}_{z,i+1} \end{array} \right) \left(\begin{array}{c} \mathbf{a}_{i+1}^+ \\ \mathbf{a}_{i+1}^- \end{array} \right)$$

In this form, the similarity to the continuity condition Equation 52.37 of the transfer-matrix algorithm becomes obvious. For compact algorithmic notation, we define

$$\mathbf{A}_i := \left(\begin{array}{cc} \mathbf{B}_i & \\ & \mathbf{B}_i \end{array} \right) \left(\begin{array}{cc} \mathbf{I}_d & \mathbf{I}_d \\ \mathrm{diag}\,\mathbf{k}_{z,i} & -\mathrm{diag}\,\mathbf{k}_{z,i} \end{array} \right) \tag{52.52}$$

and have exactly as in the 1D case

$$\mathbf{A}_i \left(\begin{array}{c} \mathbf{a}_i^+ \\ \mathbf{a}_i^- \end{array} \right) = \mathbf{A}_{i+1} \left(\begin{array}{c} \mathbf{a}_{i+1}^+ \\ \mathbf{a}_{i+1}^- \end{array} \right).$$

Collecting the steps, we obtain the FMM algorithm in 2D, Algorithm 52.2.

Algorithm 52.2: Fourier Modal Method in 2D

> **Require:** $\mathbf{a}_0^+$ and $\mathbf{a}_{n+1}^-$ to define source fields.
>   **for all** sections i **do**
>   Solve each sections's eigenvalue problem, either (Equation 52.46), TE, or (Equation 52.49), TM.
>   Save: eigenvectors $\widehat{\mathbf{E}}_{y,i,m}$, eigenvalues $k_{z,i,m}$
>   for later field representation (Equation 52.51).
>   Compute the matrices A_i by (Equation 52.52).
>   **end for**
>   Assemble the system (Equation 52.33).
>   Solve.
>   **return** vectors $\mathbf{a}_0^-, \dots, \mathbf{a}_{n+1}^+$.

52.4.7 Convergence

The two significant approximation errors in FMM are the geometrical approximation by staircasing and the trunction in the number of Fourier modes. These problems have been addressed in numerous papers (see also the introduction). Especially, the representation of permittivities, which are not continuous in lateral direction, by global Fourier modes poses a convergence problem. Since in this case, the corresponding

Fourier coefficients decay only as $1/n$, where n is the number of the Fouriermode, the spatial approximation is worse than the $1/n^2$ we know, e. g., from the FDTD method. On the other hand, physics very often behaves in a way that only a few modes are sufficient to describe reasonably wave propagation and scattering. In these cases, FMM has been proven to be a very efficient method.

52.5 Finite Element Method

Several excellent books treating the finite element analysis of Maxwell's equations are available; the books of Peter Monk [27] and Leszek Demkowicz [28] are very recommendable. The summary of FEM here follows Chapter 8 of [1].

52.5.1 Historical Notes

The deep understanding of the relation between the scalar Helmholtz case, which enforces the use of classical, nodal based Lagrange shape functions, and the vectorial Maxwell case relies greatly on the work of Whitney and Nédélec [29,30], Bosssavit [31], and Webb [32]. Many important contributions of other authors have added to the field, which makes the FEM today a very flexible, effective, and well-studied method for electrodynamical simulations. An early survey of finite element studies with focus on nano-optical applications has been given in [33].

52.5.2 Concept: Patchwise Tangential Continuous, Polynomial Approximation on Unstructured Grids

The FEM is based on a *variational formulation* of Maxwell's equations that involves integral expressions on the computational domain. The FEM does not use any approximation to Maxwell's equations itself, except the discretization of the geometry, but approximates the solution space in which one tries to find a reasonable approximation to the exact solution. This is in contrast to the finite difference method which approximates Maxwell's equations directly via an approximation of the differential operators by finite difference stencils. In some respects, the FEM is comparable to FMMs: one takes a number of trial functions (the Fourier modes in case of FMM) and asks in which way a superposition of them gives a reasonable approximation to the exact solutions.

The solution space of FEM, however, is obtained by subdividing the computational domain into small patches and by providing a number of polynomials on each patch for the approximation of the solution. The patches together with the local polynomials defined on them are called *finite elements*. The most common examples of finite elements are triangles and rectangles in 2D and tetrahedrons and cubes in 3D together with constant, linear, quadratic, and cubic polynomials. These locally defined polynomial spaces have to be pieced together to ensure tangential continuity of the electric and magnetic field across the boundaries of neighboring patches. Once these local approximations with proper continuity conditions have been defined, these are inserted into the variational equation. This results in a linear system whose solution is a piecewise polynomial approximation of the exact solution. Hence, the two basic ingredients in each FEM for Maxwell's equations are

1. A variational formulation of Maxwell's equations.
2. A suitable construction of finite elements based on polynomials defined on local geometric patches to transform the variational formulation into a discrete, algebraic problem.

This concept has a number of remarkable properties:

- Complex geometrical shapes can be treated without geometrical approximations, e.g., rounding can be well and easily approximated, cf. Figure 52.12.

FIGURE 52.12 Sketch of a 2D triangulation for FEM, cf. also Figure 52.8 with staircase approximation. The triangles are shaped and placed such that the geometry is well covered by relatively few triangles.

- The finite element mesh can easily be adapted to the behavior of the solution, e.g., to singularities at corners.
- High-order approximations are available and ensure fast convergence.

52.5.3 1D: Weak Form and Variational Formulation

For the introduction of the finite element concept, we start with the most simple approximation of the time-harmonic scattering problem—the scalar Helmholtz equation in 1D, defined on the whole real axis, see Figure 52.3. The exterior domain should have a spatially invariant k. We let our computational domain Ω be the interval $(0, L)$. The source field is a plane wave coming from the left. A possible problem formulation reads: Find $u(x), x \in \Omega = (0, L)$ such that

$$\partial_x^2 u(x) + k^2 u(x) = 0, \qquad\qquad x \in \mathbb{R} \qquad\qquad (52.53)$$

$$u_{\text{source}}(x) = e^{ikx}, \qquad\qquad x \in \mathbb{R} \setminus \Omega. \qquad\qquad (52.54)$$

Let us choose the interval $(0, L)$. We multiply Equation 52.53 with the complex conjugate of a so-called test function $v(x)$ and integrate over the domain Ω to arrive at

$$\int_0^L v^*(x) \left(\partial_x^2 u(x) + k^2 u(x) \right) dx = 0 \qquad\qquad (52.55)$$

where nothing has been specified so far about $v(x)$, except that we must be allowed to compute the integral. This formulation cannot be equivalent to Equation 52.53, since very special choices of v would equate the integral to zero even for functions, which do not satisfy Equation 52.53. Hence, in order to have the chance to get something equivalent to Equation 52.53, Equation 52.55 must hold true for many functions $v(x)$. We call the class of functions from which we take $v(x)$, the *test space*. In contrast, the space of functions containing the desired solution $u(x)$ is called the *trial* or *ansatz space*. In general, both spaces can be different. It is subject to the convergence theory of finite elements to define precisely the test and trial spaces and to conclude convergence rates. For our purpose here, it is sufficient to define these spaces in a way that makes the construction of the numerical algorithms reasonable. In the first attempt, we require $\int_0^L v^* v \, dx < \infty$, which is the class of square-integrable functions, the $L^2(0, L)$-functions. We will make the choice of the space more precise below, but only in a framework that introduces usual FEM notions and gives a first look at the topic. A comprehensive treatment can be found, e. g., in [27]. The functions $u(x)$ should be twice differentiable since the second derivative is contained in Equation 52.53. Hence, we seek them in the class $C^2(0, L)$. So in a first attempt to construct a variational equation, we formulate find a function $u(x) \in C^2(0, L)$ such that Equation 52.55 holds true for all $v \in L^2(0, L)$.

52.5.3.1 Weak Form

We perform an integration by parts on the left-hand side of Equation 52.55:

$$\int_0^L v^*(x) \left(\partial_x^2 u(x) + k^2 u(x) \right) dx = \int_0^L \left(-\partial_x v^*(x) \partial_x u(x) + k^2 v^*(x) u(x) \right) dx + v^*(x) \partial_x u(x) \big|_0^L . \quad (52.56)$$

It is convenient to abbreviate the negative of the integral without boundary terms by

$$a_{\text{int}}(v, u) := \int_0^L \left(\partial_x v^*(x) \partial_x u(x) - k^2 v^*(x) u(x) \right) dx \quad (52.57)$$

to rewrite (Equation 52.56) to

$$\int_{\Omega_{\text{int}}} v^*(x) \left(\partial_x^2 u(x) + k^2 u(x) \right) dx =: -a_{\text{int}}(v, u) + \sum_{x \in \partial\Omega_{\text{int}}} v^*(x) \partial_n u(x). \quad (52.58)$$

The term Equation 52.57 is called a *sesquilinear form* because it is antilinear in its first argument (the v, due to complex conjugation) and linear in its second argument (the u). Hence, $a_{\text{int}}(v, u)$ maps two complex functions via integration to a complex number. We may repeat this for the scattered part of the exterior field

$$a_{\text{int}}(v, u) - \sum_{x \in \partial\Omega_{\text{int}}} v^*(x) \partial_n u(x) = 0 \quad \text{in } \Omega_{\text{int}} \quad (52.59)$$

$$a_{\text{ext}}(v, u_{\text{s}}) - \sum_{x \in \partial\Omega_{\text{int}}} v^*(x) \partial_n u_{\text{s}}(x) = 0 \quad \text{in } \mathbb{R} \setminus \Omega_{\text{int}}, \quad (52.60)$$

where we ensure (see discussion below) that the boundary term at infinity vanishes and only the boundary $\partial\Omega_{\text{int}}$ contributes to the equation. Using on this boundary, $u = u_{\text{source}} + u_{\text{s}}$ and introducing an auxiliary function g with

$$u_s = \underbrace{u_s + g}_{w} - g$$

$$g(\partial\Omega_{\text{int}}) = u_{\text{source}}(\partial\Omega_{\text{int}})$$

$$g(\infty) = 0$$

$$\quad (52.61)$$

we obtain

$$a_{\text{int}}(v, u) - \sum_{x \in \partial\Omega_{\text{int}}} v^*(x)\partial_n \left(u_{\text{source}} + u_s\right) = 0 \quad \text{in } \Omega_{\text{int}} \tag{52.62}$$

$$a_{\text{ext}}(v, w - g) - \sum_{x \in \partial\Omega_{\text{int}}} v^*(x)\partial_n u_s(x) = 0 \quad \text{in } \mathbb{R} \setminus \Omega_{\text{int}}. \tag{52.63}$$

Adding these two equations, taking into account the different normal directions, we end up with

$$a_{\text{int}}(v, u) + a_{\text{ext}}(v, w) = \sum_{x \in \partial\Omega_{\text{int}}} v^*(x)\partial_n u_{\text{source}} + a_{\text{ext}}(v, g). \tag{52.64}$$

Next, using the function u, which lives only on the interior domain, and the function w, which lives only on the exterior domain, we define a joint, continuous function, again called u. This gives rise to the variational formulation of the scattering problem.

52.5.3.2 Variational Formulation

Let the source field data u_{source} and $\partial_n u_{\text{source}}$ at the boundary $\partial\Omega$ of the computational domain Ω be given. The variational form of the scattering problem reads

$$\text{Find } u \in V : \quad a(v, u) = \sum_{x \in \partial\Omega_{\text{int}}} v^*(x)\partial_n u_{\text{source}} + a_{\text{ext}}(v, g) \quad \text{for all } v \in V, \tag{52.65}$$

with $a(\cdot, \cdot) = a_{\text{int}}(\cdot, \cdot) + a_{\text{ext}}(\cdot, \cdot)$, with a_{int} and a_{ext} given by Equation 52.59, Equation 52.60 and g defined by Equation 52.61. Note that in Equation 52.60 we neglected the contribution from the outer boundary $\partial\Omega_{\text{ext}}$. The reason is that the expression is established for the scattered field u_s and we perform the integration over the complex extended exterior domain as described in Section 52.2.4. The complex extension leads to an exponential decay of u_s, hence the contribution of $\partial\Omega_{\text{ext}}$ vanishes.

52.5.4 1D: Linear Finite Elements, Local Matrices, and Assembly Process

The discretization of a given variational form to get the FEM equations is a straight forward procedure: just restrict the infinite dimensional test and trial space V to a finite dimensional space V_h:

$$\text{Find } u \in V : \quad a(v, u) \quad = \quad b(v) \quad \text{for all } v \in V.$$

$$\downarrow$$

$$\text{Find } u_h \in V_h : \quad a(v_h, u_h) \quad = \quad b(v_h) \quad \text{for all } v_h \in V_h. \tag{52.66}$$

The space V_h is not uniquely predetermined. The different types of Galerkin-type methods differ in the construction of the elements of V_h. Let V_h be spanned by N linearly independent functions $v_j(x)$, $j = 1, \dots, N$. We represent the discrete approximation of the exact solution u by $u_h(x) = \sum_{j=1}^{N} u_j v_j(x)$, insert this into Equation 52.66:

$$a\left(v_h, \sum_{j=1}^{N} u_j v_j(x)\right) = b(v_h) \tag{52.67}$$

and get the linear system

$$i = 1, \ldots, N : \quad \sum_{j=1}^{N} a_{ij} u_j = b_i \quad \text{with } a_{ij} = a\left(v_i, v_j\right) \quad \text{and } b_i = b\left(v_i\right). \tag{52.68}$$

52.5.4.1 Linear Finite Elements

We want to construct the matrix $\left(a_{ij}\right)$ as a sparse matrix, i.e., a matrix with predominantly zero entries, which is a very useful numerical property. To this end we construct test functions v_j that exist only on small patches. The most common finite elements are the *piecewise linear "hat"*-functions shown in Figure 52.13a:

$$v_i(x) = \begin{cases} \frac{x-x_{i-1}}{h_{i-1}} & \text{for} \quad x \in \left[x_{i-1}, x_i\right] \\ -\frac{x-x_{i+1}}{h_i} & \text{for} \quad x \in \left[x_i, x_{i+1}\right] \\ 0 & \text{elsewhere,} \end{cases} \tag{52.69}$$

where we introduced the lengths of the subdomains with $h_{i-1} = x_i - x_{i-1}$ and $h_i = x_{i+1} - x_i$.

52.5.4.2 Assembly Process: Patchwise Computation of $A = \left(a_{ij}\right)$

We start from Equation 52.67 with the summation moved to front of $a(\cdot, \cdot)$:

$$i = 1, \ldots, N \quad \sum_{j=1}^{N} a\left(v_i(x), v_j(x)\right) u_j = b(v_i).$$

Next we denote $a(\cdot, \cdot)$ restricted to patch K by $a(\cdot, \cdot)|_K = a_K(\cdot, \cdot)$ and rewrite the above in terms of a sum over patches:

$$i = 1, \ldots, N \quad \sum_{j=1}^{N} \sum_{k} a_k\left(v_i(x), v_j(x)\right) u_j = b(v_i).$$

Interchanging the order of summation we have

$$i = 1, \ldots, N \quad \sum_{k} \sum_{j=1}^{N} a_k\left(v_i(x), v_j(x)\right) u_j = b(v_i).$$

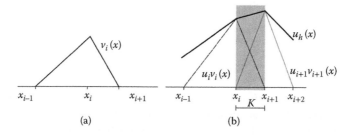

(a)   (b)

FIGURE 52.13 (a) Piecewise linear test function $v_i(x)$. (b) Piecewise linear function u_h composed from trial functions $v_i(x)$ and $v_{i+1}(x)$. Only trial functions v_i and v_{i+1} have support on the interval K, but no other trial function.

In principle, this is a summation over k different matrices $A_k = a_k\left(v_i(x), v_j(x)\right)$, each of dimension $N \times N$ multiplied with a vector $u = (u)_i$ of dimension N, hence

$$\left(\sum_k A_k\right) u = b(v_i).$$

The matrices A_k have full dimension. However, since they are computed only over a single patch K they have entries different from zero only if both v_i and v_j have support over K, hence most entries but a few are zero. Suppose we have a set of (few) trial functions $v_{i_1}, v_{i_2}, \ldots, v_{i_n}$ with index set $\{i_1, i_2, \ldots, i_n\}$ which contribute to A_k assigned to patch K. We count these indices as follows:

$$\text{local DOFs} \quad \rightarrow \quad \text{global DOFs}$$

$$(1, 2, \ldots, n) \quad \rightarrow \quad \left(i_1, i_2, \ldots, i_n\right)$$

and store the connectivity data, using MATLAB ® notation, in an element connectivity vector

$$c = [i_1, i_2, \ldots, i_n]. \tag{52.70}$$

We have such a mapping for each patch connecting the *local* DOFs to the *global* ones. With the help of the local DOFs we define dense, so-called *local* matrices $A_k^{\text{loc}} \in \mathbb{C}^{n \times n}$ of small dimension n and corresponding *local* vectors $b_k^{\text{loc}} \in \mathbb{C}^n$:

$$A_k^{\text{loc}}(l, m) \quad : = \quad A_k\left(i_l, i_m\right) \quad \text{for all indices with non-vanishing values of } A_k, \text{ and}$$

$$b_k^{\text{loc}}(l) \quad : = \quad b_k(i_l).$$

The process of distributing the entries of the local matrices to the global matrices is called index scattering. The MATLAB ® statement

$$A(c, c) = A^{\text{loc}}$$

scatters the entries of the local matrix A^{loc} to their global counterparts.

52.5.4.3 Computation of Local Matrices

For a convenient computation of the local vectors and matrices, it is useful to decompose the form $a(v, u)$, e.g., Equation 52.57, into the so-called *stiffness* and *mass* parts $s(v, u)$, $m(v, u)$, respectively:

$$a(v, u) = s(v, u) - m(v, u) \quad \text{with}$$

$$s(v, u) : = \int_0^L \partial_x v^* \partial_x u \quad \text{and} \quad m(v, u) : = \int_0^L k^2 v^* u.$$

The terms stiffness and mass come from mechanical systems and have no meaning in electrodynamics. But, as we will see, both parts have very different mathematical properties.

Let us compute the stiffness matrix based on the hat-functions. We consider the patch K from Figure 52.13 which gives us the local$\longrightarrow$global assignment of indices

$$(1, 2) \longrightarrow (i, i + 1).$$

Using Equation 52.69, we get

$$
s^{\text{loc}} = \begin{pmatrix} s\left(v_i, v_i\right)\big|_K & s\left(v_i, v_{i+1}\right)\big|_K \\ s\left(v_{i+1}, v_i\right)\big|_K & s\left(v_{i+1}, v_{i+1}\right)\big|_K \end{pmatrix}.
$$

We compute directly

$$
s^{\text{loc}} = \frac{1}{h_i} \begin{pmatrix} 1 & -1 \\ -1 & 1 \end{pmatrix} \quad \text{and the in same way} \quad m^{\text{loc}} = h_i \begin{pmatrix} \frac{1}{3} & \frac{1}{6} \\ \frac{1}{6} & \frac{1}{3} \end{pmatrix}.
$$

The structure of a finite element algorithm is given in Algorithm 52.3.

52.5.5 2D and 3D: Weak Form and Variational Formulation

The Maxwell scattering problem with PML is derived in complete analogy to Section 52.5.3. Let the source field data $\mathbf{E}_{\text{source}}$ and $\nabla \times \mathbf{E}_{\text{source}}$ at the boundary $\partial\Omega$ of the computational domain Ω be given. The variational form of the scattering problem reads:

$$
\text{Find } \mathbf{E} \in V : \quad a(\mathbf{v}, \mathbf{E}) = \int_{\partial\Omega} d\mathbf{s} \cdot \left(\mathbf{v}^* \times \left[\left(\mu^c\right)^{-1} \nabla \times \mathbf{E}_{\text{source}}\right]\right) + a_{\text{ext}}\left(\mathbf{v}, \mathbf{g}\right) \quad \text{for all } \mathbf{v} \in V, \quad (52.71)
$$

with

$$
a(\cdot, \cdot) = a_{\text{int}}(\cdot, \cdot) + a_{\text{ext}}(\cdot, \cdot)
$$

$$
a_{\text{int}}(\mathbf{v}, \mathbf{E}) = \int_{\Omega} dx\, dy\, \left(\nabla \times \mathbf{v}^*\right) \cdot \mu^{-1} \left(\nabla \times \mathbf{E}\right) - \omega^2 \int_{\Omega} dx\, dy\, \mathbf{v}^* \varepsilon \mathbf{E}
$$

$$
a_{\text{ext}}(\mathbf{v}, \mathbf{g}) = \int_{\Omega_{\text{ext}}} dx\, dy\, \left(\nabla \times \mathbf{v}^*\right) \cdot \left(\mu^c\right)^{-1} \left(\nabla \times \mathbf{g}\right) - \omega^2 \int_{\Omega_{\text{ext}}} dx\, dy\, \mathbf{v}^* \varepsilon^c \mathbf{E}
$$

and $\mathbf{g}$ is an arbitrary tangentially continuous function with

$$
\mathbf{g}(\partial\Omega) = \mathbf{E}_{\text{source}}(\partial\Omega)
$$

$$
\mathbf{g}(\infty) = 0.
$$

The transformed permeabilities and permittivities are

$$
\left(\mu^c\right)^{-1} = \frac{1}{|J|} J^T \mu^{-1} J, \qquad \varepsilon^c = |J| J^{-1} \varepsilon J^{-T},
$$

with J the Jacobian of the complex extension. Here, we let Ω_{ext} be the original, real exterior domain and the PML-complexification is hidden in the transformed materials μ^c and ε^c, see Equations 52.18 and 52.19.

52.5.6 2D and 3D: Linear Finite Elements, Local Matrices, and Assembly Process

The basic scheme of deriving the numerical algorithm is exactly the same as in 1D, see Section 52.5.4, and also Algorithm 52.3:

1. Construct a mesh that describes the geometry plus surrounding domain. The meshes in 2D are triangular or quadrilateral meshes, in 3D these are meshes based on tetrahedrons, bricks, prisms, and pyramids. Typically, special mesh generators are employed. The construction of high-quality meshes is often a nontrivial task and a research field in its own right.
2. On each patch of the mesh, say on a triangle, a polynomial space is defined which is the basis of the construction of so-called shape functions. The local space of shape functions must be capable of approximating the solutions on the patch. Given a polynomial space, different families of shape functions can be constructed which all span the same space. A local matrix and a local right-hand side vector is attributed to each patch. The computation of these involves mapping from reference elements.
3. The shape functions of neighboring patches are glued together to build the discrete test and trial functions v_i. These functions must be tangentially continuous across the patch boundaries to be a global basis function of the solution space for Maxwell's equations. One DoF is attributed to each v_i.
4. Having obtained all test and trial functions v_i, the assembly process is carried out as in 1D, see Equations 52.67 and 52.68. The scattering of local to global matrices depends on the neighborhood relation of patches.

Algorithm 52.3: Basic Finite Element Algorithm

> construct a tessellation $\bar{\Omega} = \cup_i \bar{K}_i$
> **for all** patches K_i **do**
>   map $\hat{K} \to K_i$
>   compute local matrices $\mathbf{A}_i^{\mathrm{loc}}$
>   compute local vectors $\mathbf{b}_i^{\mathrm{loc}}$
>   scatter the local matrices and vectors to global ones
>   update global matrices $\mathbf{A}$ and right-hand side $\mathbf{b}$
> **end for**
> solve $\mathbf{A}\mathbf{u}_h = \mathbf{b}$
> **return** $\mathbf{u}_h$

52.5.6.1 Finite Elements

Figure 52.14 recalls the construction of the global test and trial functions from the local shape functions. The process is the same in higher dimensions, but there are many more possibilities for how to do it.

We illustrate these possibilities by means of some examples in 2D. First, we consider a mesh built from quadrilaterals, see Figure 52.15a. For a technical simplification of finite element algorithms, each quadrilateral is considered to be a map from a unit square, see Figure 52.16. The test and trial functions are constructed over the unit square and then mapped to their original positions. The lowest order trial functions are linear functions, in the case of a unit square, we use simply tensor products of 1D linear functions, see Section 52.5.4.

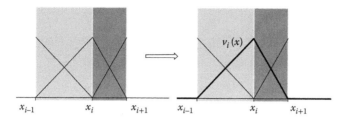

FIGURE 52.14 Construction of a global trial function v_i. Two neighboring intervals with two linear shape functions, each have the common boundary point x_i. One of the shape functions of each interval is taken and glued to the other and continued to a global, continuous trial function with local support.

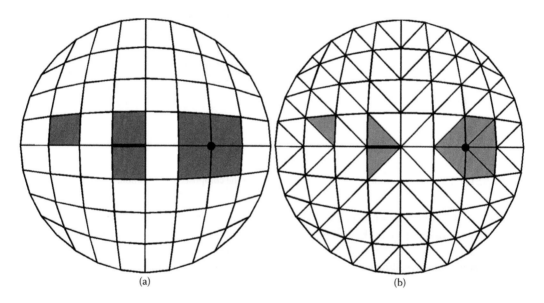

FIGURE 52.15 Structured mesh based on quadrilaterals (a) and a triangular mesh with the same vertices (b). The gray shaded areas show the geometric possibilities of constructing trial functions based on the interior of a single patch, neighboring patches with a common edge and neighboring patches with a common vertex.

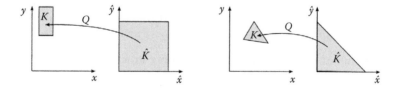

FIGURE 52.16 Mappings from reference elements to desired shapes.

52.5.6.2 Scalar Helmholtz Equation

Scalar trial functions v_i are used to approximate the solution of Helmholtz equations (Equation 52.8). In the case of a quadrilateral mesh with a unit square as reference element, the tensor product of the 1D shape functions gives four bilinear shape functions:

$$\left\{ N_i : i = 1, \ldots, 4 \right\} = \left\{ \left(1 - \hat{x}\right)\left(1 - \hat{y}\right), \hat{x}\left(1 - \hat{y}\right), \hat{x}\hat{y}, \left(1 - \hat{x}\right)\hat{y} \right\}.$$

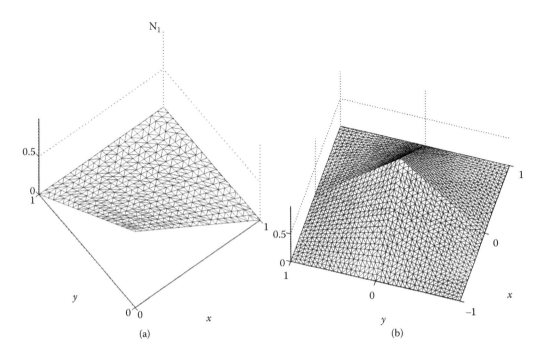

FIGURE 52.17   Bilinear shape function N_1 on a unit square and the composed trialfunction.

Figure 52.17a shows the shape function N_1, which is associated to the vertex $(0, 0)$ of the reference coordinate system $\hat{K}$. In the spirit of the 1D construction (Figure 52.14), neighboring shape functions will be connected to continuous, global trial functions v_i as shown in Figure 52.17b. The common vertex is the defining feature of the resulting global trial function. The DoF attributed to this function is usually taken as its nodal value at the vertex position.

52.5.6.3   Maxwell's Equations

In the case of Maxwell's equations, the fields are generally no longer continuous. They are tangentially continuous but may have jumps or even singularities in normal field components at material discontinuities. Hence, nodal values, at least along material discontinuities, do not represent the physical situation properly. Therefore, the lowest order vectorial shape functions for Maxwell'equations are so-called edge functions. They are constructed in a way that they allow for an easy extension to tangential continuous, global trial functions. The four lowest order edge functions attributed to the unit square are shown in Figure 52.18.

52.5.6.4   Computation of Local Matrices and Assembly Process

This is very similar to the procedure discussed for the 1D Helmholtz problem Section 52.5.4. According to the Maxwell scattering formulation (Equation 52.71), we have to evaluate the variational formulation (Equation 52.71) on a finite dimensional test and trial space V_h. The computation of the local matrices and vectors goes the same way as in Section 52.5.4 with

$$a_{i,j} = a\left(\mathbf{v}_i, \mathbf{v}_j\right) = \underbrace{\int_{\Omega_{cd}} dx\, dy\, (\nabla \times \mathbf{v}_i^*) \cdot \mu^{-1} \left(\nabla \times \mathbf{v}_j\right)}_{\text{stiffness part } s_{ij}} - \underbrace{\omega^2 \int_{\Omega_{cd}} dx\, dy\, \mathbf{v}_i^* \varepsilon \mathbf{v}_j}_{\text{mass part } m_{ij}}$$

e_x (1−y)

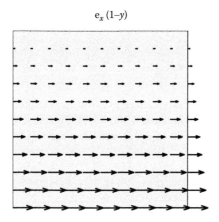

e_x y

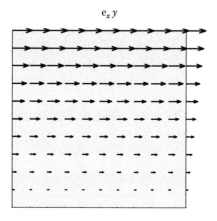

e_y (1−x)

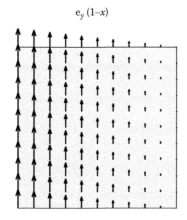

e_y x

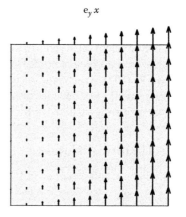

FIGURE 52.18 Vectorial, first-order edge elements of the unit square.

and

$$(b)_i = \int_{\partial\Omega} d\mathbf{f} \cdot \left(\mathbf{v}_i^* \times \left[\mu^{-1}\nabla \times \mathbf{E}_{\text{source}}\right]\right) + a_{\text{ext}}\left(\mathbf{v}_i, \mathbf{g}\right). \tag{52.72}$$

52.5.7 Convergence

The convergence analysis for edge element discretizations of time-harmonic scattering problems yields very useful insights into the behavior of numerical codes. The analysis itself is a challenging task since even in the lossless case the variational equations are neither elliptic nor definite. As a typical example we refer to Monk [34], who investigated a lossless scattering problem with sources on a bounded domain with electric boundary conditions. The error estimation uses the **curl** -norm

$$\left\|\mathbf{E} - \mathbf{E}_h\right\|_{\mathbf{curl}}^2 := \left\|\mathbf{E} - \mathbf{E}_h\right\|^2 + \left\|\nabla \times (\mathbf{E} - \mathbf{E}_h)\right\|^2,$$

where $\|\cdot\|$ is the standard L^2-norm, h denotes again a typical meshsize and as subscript it indicates discrete approximation. Depending on the properties of the computational domain and the smoothness of the exact solution there are constants C and δ independent of h and a constant $h_0 > 0$ independent of $\mathbf{E}$ and $\mathbf{E}_h$ such that for all $0 < h < h_0$

$$\left\| \mathbf{E} - \mathbf{E}_h \right\|_{\mathbf{curl}} \leq \frac{1}{1 - Ch^{1/2+\delta}} \inf_{v_h \in V_h} \left\| \mathbf{E} - \mathbf{v}_h \right\|_{\mathbf{curl}}.$$

This result shows that for h less than a threshold value h_0, the FEM scheme converges to the exact solution, but in the worst case without a polynomial rate. However, if the solution is smooth quantified as $\mathbf{E} \in H^s(\mathbf{curl}; \Omega)$ and s is a smoothness parameter, then, we obtain in fact the desired polynomial convergence if we use high-order finite elements

$$\left\| \mathbf{E} - \mathbf{E}_h \right\|_{\mathbf{curl}} \leq Ch^s.$$

52.6 DG Method

The FDTD method has become the method of choice in physics and engineering when time-dependent Maxwell's equations have to be solved. The reasons are its simplicity, robustness and sufficient accuracy for most practical problems. On the other hand, it has severe limitations. These are mainly the difficultie incorporating complicated geometry due to its link to staggered grids unless the grid is prohibitively fine, the inherent second-order accuracy in time and space. Both topics have been addressed successfully in the frame work of FDTM methods, see the section about FDTD. However, DG methods present a successful alternative with the two main features geometrical flexibility and high-order approximations both in time and space inherently anchored in the method from the beginning.

52.6.1 Historical Notes

As any other method described in this chapter, the DG method has a long history with many contributions from many sides. Reed and Hill [35] applied a first form of the DG method to the neutron transport equation in 1973.

Since then, it has undergone a fast development and finds numerous applications to e.g., the Euler equations of gas dynamics, the shallow water equations, the equations of magneto-hydrodynamics, the compressible Navier–Stokes equations, and Maxwell's equations only to mention a few; see e.g., [36] for an overview and analysis of many of these applications.

The combination of the DG method for spatial discretization and of an explicit Runge–Kutta scheme for time-integration has been introduced by Cockburn and Shu in several papers, [37] maybe being one of the latest. [38] gives an introduction to the DG method and contains a more complete list of references on the (RKDG) method; it can be downloaded online. We also mention [39–41], which is a review paper about RKDG methods, as references on the DG method. [42] gives a review of the DG method applied to nanophotonics.

52.6.2 Concept: Combine Adaptive Spatial FEM Discretization with the Time Treatment in Finite Volume Methods

The FEM treats successfully complicated geometries in a natural way. There is no basic obstacle in combining local high-order approximations on unstructured meshes composed from different elements like tetrahedrons, prisms, bricks, and pyramids. The difficulties here lie on the side of implementations and is linked to the complexity of the problem: nontrivial geometries have to be described, decomposed into geometric patches, a detailed and efficient bookkeeping has to be realized for fast traversing. A direct application of FEM to time-dependent problems are possible but has a severe drawback: due to the occurrence of the mass matrix in conjunction with the time-derivative, an inversion of the mass matrix is needed. This can be done in several ways; nevertheless, it causes a problem not present in FDTD methods. The reason for the necessity to invert a mass matrix lies in the construction of finite elements. Typically, neighboring elements couple to each other in a way that the resulting system matrices are sparse but yield a global coupling of DoF which does not break down to structures which are cheaply invertible.

Finite volume methods, on the other hand, offer explicit time-integration schemes like those in FDTD methods. The key idea is to localize most of all spatial discretization into cells which are treated completely independent from each other. The cells are connected to each other by an explicit kind of continuity statement. This connection has to satisfy natural requirements like continuity or convergence properties but is not derived from first principles. There are many choices which can be adapted to given situations. The most intriguing feature is that by construction, each cell "sees" its own time evolution which results in an explicit scheme. The key idea of DG methods now is to combine this kind of time-treatment with a high-order spatial discretization localized to individual cell which are connected only by explicitly forced interface conditions on the cell boundaries.

52.6.3 Scalar Advection Equation as Model Equation: The Finite Volume Method in 1D

Consider the scalar advection equation

$$\partial_t u + \partial_x f(u) = 0$$

$$f(u) = au, \quad a \text{ is a real constant,} \tag{52.73}$$

$$u(x, 0) = u_0(x), \quad x \in \mathbb{R},$$

where $\partial_x f(u)$ is the 1D model for divergence in higher dimensions $\nabla \cdot \mathbf{F}(\mathbf{u})$. The solution to Equation 52.73 is $u(x, t) = u_0(x - at)$. If $a > 0$, this is a right traveling wave, if $a < 0$, this is a left traveling wave. In order to analyze the stability of the FVM described below, we consider energy conservation. We multiply the wave equation (Equation 52.73) with u and integrate over a subdomain $\Omega_i = (x_i, x_{i+1})$:

$$\frac{d}{dt} \frac{1}{2} u^2 + \frac{d}{dx} \frac{1}{2} au^2 = 0$$

$$\frac{d}{dt} \int_{x_i}^{x_{i+1}} \frac{1}{2} u^2 \, dx + \int_{x_i}^{x_{i+1}} \frac{d}{dx} \frac{1}{2} au^2 \, dx = 0$$

$$\frac{d}{dt} \|u\|_{\Omega_i}^2 = -a \left(u^2 \left(x_{i+1} \right) - u^2 \left(x_i \right) \right). \tag{52.74}$$

52.6.3.1 Direction of Information Flow

From Equation 52.74 we deduce the direction of information flow and stability under perturbation. Let $a > 0$, the wave is right propagating. At a fixed time, we consider the function $u(x)$ on the interval (x_i, x_{i+1}). Since (Equation 52.73) is linear, we can subtract a constant such that we obtain a new $u(x)$ with $u(x_i) = 0$. When we consider this value as a given zero boundary condition at the left side, we read from (Equation 52.74) $\frac{d}{dt} \|u\|_{\Omega_i}^2 \leq 0$ independently from the right value. This right value might be the result of a computation, hence it might be contaminated by an error. Independent of contamination, there is no increase in the norm of u, hence a perturbation does not lead to instability. If we had chosen the right boundary to be the place for the exact boundary condition, the opposite would be the case. There would always be an increase in $\|u\|_{\Omega_i}^2$ in time.

52.6.3.2 Link to Finite Volume Methods

The whole computational domain Ω is decomposed into small control volumes $\bar{\Omega} = \cup \bar{\Omega}_i$, $\Omega_i \cap \Omega_j = 0$ for $i \neq j$. We consider one subdomain Ω_i and use the conservation form of Equation 52.73 applying Gauss's theorem:

$$\partial_t \int_{\Omega_i} u \, dV + \int_{\partial \Omega_i} f \, dS = 0, \tag{52.75}$$

where in higher dimensions the surface integral takes the form $\int_S \mathbf{f} \cdot \mathbf{n} \, dS$ and $\mathbf{n}$ is the outwardly directed unit normal vector. Next, let the surface $\partial \Omega_i$ of cell Ω_i be subdivided into surfaces S_k which connect cell i to neighboring cells. Let $|\Omega_i|$ be the volume of Ω_i. We introduce the cell average u_i as an unknown target quantity

$$\partial_t u_i + \frac{1}{|\Omega_i|} \sum_k \int_{S_k} f \, dS = 0 \tag{52.76}$$

$$u_i = \frac{1}{|\Omega_i|} \int_{\Omega_i} u \, dV. \tag{52.77}$$

On the 1D mesh this results in the numerical scheme:

$$\partial_t u_i + \frac{1}{|\Omega_i|} \left(f_{i+1/2} - f_{i-1/2} \right) = 0; \tag{52.78}$$

see Figure 52.19. As in the FDTD method, the solution is based on a combination of quantities defined in primal meshes (the function u_h at integer positions) and on dual meshes (the discrete flux f_h at half counting positions). The main question is how do we define the values of $f = au$ at positions $x_{i-1/2}$ and $x_{i+1/2}$?

The answer depends, as discussed above, on the direction of wave propagation. It is natural to select the points in the solution at current time that are "upwind" of the solution at the position i for increasing time to maintain the causal connection. Depending then on the direction in which the solution is translated, we have different numerical schemes.

52.6.3.3 Finite Volume Method with First-Order Upwind Scheme

If $a > 0$, there is propagation in the positive x-direction. It holds $u_{i-1/2} \approx u_{i-1}$ and $u_{i+1/2} \approx u_i$, hence $f_{i+1/2} = au_i$, $f_{i-1/2} = au_{i-1}$, and the scheme (Equation 52.78) would be

$$\partial_t u_i + \frac{a}{\Delta x} \left(u_i - u_{i-1} \right) = 0.$$

In contrast, if $a < 0$, there is propagation in the negative x-direction. It holds $u_{i-1/2} \approx u_i$ and $u_{i+1/2} \approx u_{i+1}$, hence $f_{i+1/2} = au_{i+1}$, $f_{i-1/2} = au_i$, and the scheme (Equation 52.78) would be

$$\partial_t u_i + \frac{a}{\Delta x} \left(u_{i+1} - u_i \right) = 0.$$

Let us fix position $x_{i+1/2}$. The approximation of the flux just used is called the numerical flux, f^{num},

$$f^{\text{num}}_{i+1/2} \left(u_i, u_{i+1} \right) = \begin{cases} au_i & \text{if } a > 0 \\ au_{i+1} & \text{if } a < 0. \end{cases} \tag{52.79}$$

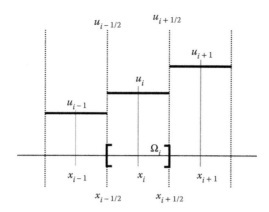

FIGURE 52.19 Vertex-centered finite volume method in 1D.

This can be compactly denoted as

$$f_{i+1/2}^{num}\left(u_i, u_{i+1}\right) = a \underbrace{\frac{1}{2}\left(u_i + u_{i+1}\right)}_{:=\{u\}_{i+1/2}} + \underbrace{\frac{|a|}{2}\left(u_i - u_{i+1}\right)}_{:=[\![u]\!]_{i+1/2}}$$

$$= a\{u\}_{i+1/2} + \frac{|a|}{2}\,[\![u]\!]_{i+1/2}.$$

We describe the typical finite volume discretization and give explicitly the 1D scheme:

1. Subdivide the domain into finite control volumes Ω_i. In one space dimension this is $\Omega_i = \left(x_{i-1/2}, x_{i+1/2}\right)$.
2. The conservation law is computed independently on each individual control volume Ω_i (52.76)

$$\partial_t u_i = -\frac{1}{|\Omega_i|}\sum_k \int_{S_k} f\,dS$$

with u_i the approximation to the average (Equation 52.77)

$$u_i = \frac{1}{|\Omega_i|}\int_{\Omega_i} u\,dV.$$

The resulting upwind scheme is, with flux (Equation 52.79),

$$\partial_t u_i = -\frac{1}{|\Omega_i|}\left(f_{i+1/2} - f_{i-1/2}\right)$$

$$= -\frac{1}{|\Omega_i|}\left(f_{i+1/2}^{num}\left(u_i, u_{i+1}\right) - f_{i+1/2}^{num}\left(u_{i-1}, u_i\right)\right). \tag{52.80}$$

3. Solve the semidiscrete system (Equation 52.80) with a suitable time-integrator. There are many possibilities, the most often used scheme in conjunction with time-dependent Maxwell's equations are low-storage Runge–Kutta methods. The resulting scheme after discretization in time must be a so-called monotonic scheme. Monotonic numerical schemes are ones which, given an initial distribution which is monotonic, produce a monotonic distribution after advection. A consequence of this property is that monotonic schemes neither create new extrema in the solution nor amplify existing extrema.

52.6.3.4 DG Method in 1D

The goal is to replace the low-order (piecewise constant) solution by a high-order approximation. We consider cell Ω_i. Exactly as in the FEM, we multiply (Equation 52.73) with a test function $v \in V$:

$$\int_{\Omega_i} v \partial_t u \, dx + \int_{\Omega_i} v \partial_x f(u) \, dx = 0.$$

An integration by parts gives

$$\int_{\Omega_i} v \partial_t u \, dx - \int_{\Omega_i} \partial_x v f(u) \, dx = - \int_{\partial\Omega_i} v f(u) \, ds.$$

Again as in FEM, we obtain the numerical method in *weak form* if we restrict the general space V to a finite dimensional space (of possibly high order) $V_h \subset V$:

$$\int_{\Omega_i} v_h \partial_t u_h \, dx - \int_{\Omega_i} \partial_x v_h f(u_h) \, dx = - \int_{\partial\Omega_i} v_h f^{\text{num}}(u_h) \, ds \quad \text{for all } v_h \in V_h. \tag{52.81}$$

The corresponding strong version results, if we undo the partial integration:

$$\int_{\Omega_i} v_h \partial_t u_h \, dx + \int_{\Omega_i} v_h \, \partial_x f(u_h) \, dx = \int_{\partial\Omega_i} v_h \left(f\left(u_h\right) - f^{\text{num}}(u_h) \right) \, ds \quad \text{for all } v_h \in V_h. \tag{52.82}$$

52.6.4 Maxwell's Equations in Conservation Form

We generalize the above step by step and present the framework of DG for Maxwell's equation's in 3D. The matrix-valued flux function $\mathbf{F} := \left(\mathbf{F}_1, \mathbf{F}_2, \mathbf{F}_3\right)$ is composed of three column vectors of length n, $\mathbf{F}_i$: $\Omega \subset \mathbb{R}^3 \to \mathbb{R}^n$, with $i = 1, 2, 3$, and $\mathbf{F}_i = \left(F_{1i}, \ldots, F_{ni}\right)^T$. The divergence $\nabla \cdot \mathbf{F}$ of this matrix-valued field $\mathbf{F} : \Omega \subset \mathbb{R}^3 \to \mathbb{R}^{n\times3}$ is defined as

$$\nabla \cdot \mathbf{F} := \begin{pmatrix} \partial_{x_1} F_{11} & +\partial_{x_2} F_{12} + & \partial_{x_3} F_{13} \\ & \vdots & \\ \partial_{x_1} F_{n1} & +\partial_{x_2} F_{n2} + & \partial_{x_3} F_{n3} \end{pmatrix}.$$

The generalization of the 1D advection equation (Equation 52.73) reads

$$\mathbf{Q}\partial_t u + \nabla \cdot \mathbf{F}(\mathbf{u}) = 0$$
$$u(x, 0) = u_0(x), \quad x \in \mathbb{R}^3. \tag{52.83}$$

Next, we bring Maxwell's equations into this conservation form. From

$$\begin{pmatrix} \varepsilon & \\ & \mu \end{pmatrix} \begin{pmatrix} \mathbf{E} \\ \mathbf{H} \end{pmatrix} = \begin{pmatrix} & \nabla\times \\ -\nabla\times & \end{pmatrix} \begin{pmatrix} \mathbf{E} \\ \mathbf{H} \end{pmatrix}$$

we read

$$\mathbf{Q} = \begin{pmatrix} \varepsilon & \\ & \mu \end{pmatrix} \qquad \mathbf{F}_i = \begin{pmatrix} -\mathbf{e}_i \times \mathbf{H} \\ \mathbf{e}_i \times \mathbf{E} \end{pmatrix}$$

with $\mathbf{e}_i$ being one of the three Cartesian unit vectors. Hence, there is in fact a conservation form (Equation 52.83) of Maxwell's equations.

52.6.5 Concept of DG-Algorithms for 3D Maxwell's Equations

The 3D algorithm follows the 1D concept. The only missing link is the upwind flux for the situation of Maxwell's equations. Following the same strategy as in 1D, Hesthaven and Warburg [43] derive the following expression:

$$\mathbf{n} \cdot \left(\mathbf{F} - \mathbf{F}^{\text{num}}\right) = \left(\begin{array}{c} \bar{Z}^{-1}\mathbf{n} \times \left(-Z^+ [\![\mathbf{H}]\!] + \mathbf{n} \times [\![\mathbf{E}]\!]\right) \\ \bar{Y}^{-1}\mathbf{n} \times \left(-Y^+ [\![\mathbf{E}]\!] + \mathbf{n} \times [\![\mathbf{H}]\!]\right) \end{array} \right). \tag{52.84}$$

Here Z and Y are the wave impedance and admittance, respectively,

$$Z^\pm = \sqrt{\frac{\mu^\pm}{\varepsilon^\pm}}, \quad Y^\pm = \left(Z^\pm\right)^{-1}, \quad \bar{Z} = \frac{Z^+ + Z^-}{2}, \quad \bar{Y} = \frac{Y^+ + Y^-}{2}.$$

The superscript "$-$" denotes the limit from the interior to the cell boundary, and the superscript "$+$" the limit from the neighboring cell to the boundary.

Algorithm 52.4: Basic Discontinuous Galerkin Algorithm

---

    1. Discretization in space
construct a tessellation $\bar{\Omega} = \cup_i \bar{\Omega}_i$
for all patches Ω_i **do**
    map $\hat{K} \rightarrow K_i$
    compute local matrices on local volumes
    assemble them (independently) to a global matrix
    compute local matrices on interior interfaces
    connect them to the global matrix
    compute local matrices on boundary interfaces
    connect them to build up the right-hand side
end for
2. Time evolution
use a time integrator (e. g., Runge–Kutta, leap-frog) to solve the semidiscrete, block-diagonal system in time
return vector of field values and fluxes in time

52.6.6 Convergence

We consider Maxwell's equations with homogeneous, isotropic materials and a convex domain $\Omega \subset \mathbb{R}^3$ and arbitrary smooth initial data. This is a simplified model case which has no singularities. In this case optimal L^2-convergence for a fixed time interval is obtained [44]:

$$\|\mathbf{u} - \mathbf{u}_h\| \leq Ch^{p+1}.$$

This is the same as we would have, under the same conditions, for finite elements.

Symbol Definition

$\mathbf{r}^0$ Vector of unit length.

Ω Computational domain.

$\partial\Omega$ Boundary of Ω.

$\mathbf{E}_\perp$ $\mathbf{e}_x E_x + \mathbf{e}_y E_y$

$\nabla_\perp$ $\mathbf{e}_x \partial_x + \mathbf{e}_y \partial_y$

∇_{k_z} $\mathbf{e}_x \partial_x + \mathbf{e}_y \partial_y + i k_z \mathbf{e}_z$

$\iota_{e,f}$ incidence number edge-face

c speed of light

$\mathbf{k}$ wavevector

$\mathbf{r}$ position

$[\![f]\!]$ Töplitz matrix generated by f

$[\![\cdot]\!]$ jump across a face

$\{\!\{\cdot\}\!\}$ average across a face

References

1. Andrei V Lavrinenko, Jesper Lagsgaard, Niels Gregersen, Frank Schmidt, and Thomas Søndergaard. *Numerical Methods in Photonics*. Boca Raton, FL: CRC Press, 2014.
2. Junji Yamauchi. *Propagating Beam Analysis of Optical Waveguides*. Electronic and Electrical Engineering Research Studies. Optoelectronics and Microwaves Series. Exeter, UK: Research Studies Press, 2003.
3. Jean-Claude Nédélec. *Acoustic and Electromagnetic Equations: Integral Representations for Harmonic Problems*, Vol. 144. Springer Science & Business Media, 2013.
4. Richard A Norton and Robert Scheichl. Planewave expansion methods for photonic crystal fibres. *Applied Numerical Mathematics*, 63:88–104, 2013.
5. Jean-Pierre Berenger. A perfectly matched layer for the absorption of electromagnetic waves. *Journal of Computational Physics*, 114(2):185–200, 1994.
6. Saul Abarbanel and David Gottlieb. On the construction and analysis of absorbing layers in CEM. *Applied Numerical Mathematics*, 27(4):331–340, 1998.
7. Stephen D Gedney. An anisotropic perfectly matched layer-absorbing medium for the truncation of FDTD lattices. *IEEE Transactions on Antennas and Propagation*, 44(12):1630–1639, 1996.
8. Weng Cho Chew and William H Weedon. A 3D perfectly matched medium from modified Maxwell's equations with stretched coordinates. *Microwave and Optical Technology Letters*, 7(13):599–604, 1994.
9. Fernando L Teixeira and Weng C Chew. General closed-form PML constitutive tensors to match arbitrary bianisotropic and dispersive linear media. *IEEE Microwave and Guided Wave Letters*, 8(6):223–225, 1998.

10. Kane S Yee. Numerical solution of initial boundary value problems involving Maxwell's equations in isotropic media. *IEEE Transactions on Antennas Propagation*, 14(3):302–307, 1966.

11. Thomas Weiland. A discretization model for the solution of Maxwell's equations for six component fields. *Archiv Elektronik und Uebertragungstechnik*, 31:116–120, 1977.

12. Adi Ditkowski, K Dridi, and Jan S Hesthaven. Convergent Cartesian grid methods for Maxwell's equations in complex geometries. *Journal of Computational Physics*, 170(1):39–80, 2001.

13. Jan S Hesthaven. High-order accurate methods in time-domain computational electromagnetics: A review. *Advances in Imaging and Electron Physics*, 127:59–125, 2003.

14. Allen Taflove. *Advances in Computational Electrodynamics, the Finite-Difference time Domain.* Boston: Artech House, 1998.

15. Alain Bossavit. 'Generalized finite differences' in computational electromagnetics. *Progress in Electromagnetics Research*, 32:45–64, 2001.

16. Markus Clemens and Thomas Weiland. Discrete electromagnetism with the finite integration technique. *Progress in Electromagnetics Research*, 32:65–87, 2001.

17. Allen Taflove and Susan C Hagness. *Computational Electrodynamics: The Finite Difference Time-Domain Method*, 3rd edition. Boston, MA: Artech House Publishers, 2005.

18. Anders Bondeson, Thomas Rylander, and Par Ingelstrm. *Computational Electromagnetics*, Vol. 51. New York, NY: Springer Verlag, 2005.

19. Peter Monk and Endre Süli. A convergence analysis of Yee's scheme on nonuniform grids. *SIAM Journal on Numerical Analysis*, 31(2):393–412, 1994.

20. Theodor Tamir, HC Wang, and AA Oliner. Wave propagation in sinusoidally stratified dielectric media. *IEEE Transactions on Microwave Theory and Techniques*, 12(3):323–335, 1964.

21. C Yeh, KF Casey, and ZA Kaprielian. Transverse magnetic wave propagation in sinusoidally stratified dielectric media. *IEEE Transactions on Microwave Theory and Techniques*, 13(3):297–302, 1965.

22. MG Moharam and TK Gaylord. Rigorous coupled-wave analysis of planar-grating diffraction. *Journal of the Optical Society of America*, 71(7):811–818, 1981.

23. Philippe Lalanne and G Michael Morris. Highly improved convergence of the coupled wave method for tm polarization. *JOSA A*, 13(4):779–784, 1996.

24. G Granet and B Guizal. Efficient implementation of the coupled-wave method for metallic lamellar gratings in tm polarization. *Journal of the Optical Society of America A*, 13(5):1019–1023, 1996.

25. Lifeng Li. Use of Fourier series in the analysis of discontinuous periodic structures. *Journal of the Optical Society of America A*, 13(9):1870–1876, 1996.

26. Jens Küchenmeister. Three-dimensional adaptive coordinate transformations for the Fourier modal method. *Optics Express*, 22(2):1342–1349, 2014.

27. Peter Monk. *Finite element Methods for Maxwell's Equations*. Oxford University Press, 2003.

28. Leszek Demkowicz. *Computing with hp-Adaptive Finite Elements: One and Two Dimensional Elliptic and Maxwell Problems*, Vol. 1. CRC Press, 2006.

29. Jean-Claude Nédélec. Mixed finite elements in R^3. *Numerische Mathematik*, 35(3):315–341, 1980.

30. Jean-Claude Nédélec. A new family of mixed finite elements in R^3. *Numerische Mathematik*, 50:57–81, 1986.

31. Alain Bossavit. Whitney forms: A class of finite elements for three-dimensional computations in electromagnetism. *IEE Proceedings A (Physical Science, Measurement and Instrumentation, Management and Education, Reviews)*, 135(8):493–500, 1988.

32. Jon P Webb. Hierarchal vector basis functions of arbitrary order for triangular and tetrahedral finite elements. *IEEE Transactions on Antennas and Propagation*, 47(8):1244–1253, 1999.

33. Jan Pomplun, Sven Burger, Lin Zschiedrich, and Frank Schmidt. Adaptive finite element method for simulation of optical nano structures. *Physica Status Solidi (B)*, 244(10):3419–3434, 2007.

34. Peter Monk. A simple proof of convergence for an edge element discretization of Maxwell's equations. In *Computational Electromagnetics*, pp. 127–141. Springer, 2003.

35. WH Reed and TR Hill. Triangular mesh methods for the neutron transport equation. Report LA-UR-73-479. Los Alamos, NM: Los Alamos Scientific Laboratory, 1973.
36. Bernardo Cockburn, George E Karniadakis, and Chi-Wang Shu. *The Development of Discontinuous Galerkin Methods*. Lecture notes in computational science and engineering. Berlin, Heidelberg: Springer-Verlag, 2000.
37. Bernardo Cockburn and Chi-Wang Shu. The Runge–Kutta discontinuous Galerkin method for conservation laws V: Multidimensional systems. *Journal of Computational Physics*, 141(2):199–224, 1998.
38. Bernardo Cockburn. An introduction to the discontinuous Galerkin method for convection-dominated problems. In *Advanced Numerical Approximation of Nonlinear Hyperbolic Equations*, pp. 151–268. Berlin Heidelberg: Springer, 1998.
39. Bernardo Cockburn, George E Karniadakis, and Chi-Wang Shu. *Discontinuous Galerkin Methods: Theory, Computation and Applications*. Berlin Heidelberg: Springer, 2000.
40. Bernardo Cockburn and Chi-Wang Shu. Runge–Kutta discontinuous Galerkin methods for convection-dominated problems. *Journal of Scientific Computing*, 16(3):173–261, 2001.
41. Jan S Hesthaven and Tim Warburton. *Nodal Discontinuous Galerkin Methods: Algorithms, Analysis, and Applications*. Springer Science & Business Media, 2007.
42. Kurt Busch, Michael Koenig, and Jens Niegemann. Discontinuous Galerkin methods in nanophotonics. *Laser & Photonics Reviews*, 5(6):773–809, 2011.
43. Jan S Hesthaven and Timothy Warburton. Nodal high-order methods on unstructured grids: I. Time-domain solution of Maxwell's equations. *Journal of Computational Physics*, 181(1):186–221, 2002.
44. Marcus J Grote, Anna Schneebeli, and Dominik Schtzau. Interior penalty discontinuous Galerkin method for Maxwell's equations: Optimal l2-norm error estimates. *IMA journal of Numerical Analysis*, 28(3):440–468, 2008.

Index

Printed and bound by CPI Group (UK) Ltd, Croydon, CR0 4YY

24/10/2024

01778292-0017